PEARSON
Physics

James S. Walker

PEARSON

Boston   Columbus   Indianapolis   New York   San Francisco   Upper Saddle River
Amsterdam   Cape Town   Dubai   London   Madrid   Milan   Munich   Paris   Montreal   Toronto
Delhi   Mexico City   Sao Paolo   Sydney   Hong Kong   Seoul   Singapore   Taipei   Tokyo

ISBN-13: 978-0-13-137115-6
ISBN-10: 0-13-137115-0

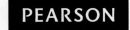

PearsonSchool.com

1 2 3 4 5 6 7 8 9 10 CRK 18 17 16 15 14 13

About the Author

James Walker obtained his Ph.D. in theoretical physics from the University of Washington in 1978. He subsequently served as a post-doc at the University of Pennsylvania, the Massachusetts Institute of Technology, and the University of California at San Diego before joining the physics faculty at Washington State University in 1983. Professor Walker's research interests include statistical mechanics, critical phenomena, and chaos. His many publications on the application of renormalization group theory to systems ranging from absorbed monolayers to binary-fluid mixtures have appeared in *Physical Review, Physical Review Letters, Physica,* and a host of other publications. He has also participated in observations on the summit of Mauna Kea, looking for evidence of extrasolar planets.

Jim Walker likes to work with students at all levels, from judging elementary school science fairs to writing research papers with graduate students, and has taught introductory physics for many years. His enjoyment of this course and his empathy for students have earned him a reputation as an innovative, enthusiastic, and effective teacher. Jim's educational publications include "Reappearing Phases" (*Scientific American,* May 1987) as well as articles in the *American Journal of Physics* and *The Physics Teacher.* In recognition of his contributions to the teaching of physics at Washington State University, Jim was named Boeing Distinguished Professor of Science and Mathematics Education for 2001–2003.

When he is not writing, conducting research, teaching, or developing new classroom demonstrations and pedagogical materials, Jim enjoys amateur astronomy, eclipse chasing, bird and dragonfly watching, photography, juggling, unicyling, boogie boarding, and kayaking. Jim is also an avid jazz pianist and organist. He has served as ballpark organist for a number of Class A minor league baseball teams, including the Bellingham Mariners, an affiliate of the Seattle Mariners, and the Salem-Keizer Volcanoes, an affiliate of the San Francisco Giants. He can play "Take Me Out to the Ball Game" in his sleep.

Reviewers

Hakan Armagan
Burke High School
Omaha, Nebraska

Michael Blair
Theodore Roosevelt High
  School
Des Moines, Indiana

Michael Brickell
Somerset High School
Galloway, Ohio

Mark Buesing
Libertyville High School
Libertyville, Illinois

Beverly Cannon
Highland Park High School
Dallas, Texas

Chris Chiaverina
New Trier High School
Winnetka, Illinois

Anthony Cutaia
White Plains High School
White Plains, New York

John Dell
Thomas Jefferson High School
  for Science and Technology
Alexandria, Virginia

Jim Dillon
Madison High School
Mansfield, Ohio

Eleanor Dorso
Brentwood High School
Brentwood, New York

Stan Eisenstein
Centennial High School
Ellicott City, Maryland

Paul Gathright
Willis High School
Willis, Texas

Oommen George
San Jacinto College Central
Pasadena, Texas

Jack Giannattasio
A. L. Johnson High School
Clark, New Jersey

Bernard Gilroy
The Hun School of Princeton
Princeton, New Jersey

Marla Glover
Rossville High School
Rossville, Indiana

David Hees
Leon M. Goldstein High School
Brooklyn, New York

Thomas Henderson
Glenbrook South High School
Glenview, Illinois

Charles Hibbard
Lowell High School
San Francisco, California

Lana Hood
Robert E. Lee High
Tyler, Texas

Janie Horn
Cleveland High School
Cleveland, Texas

Robert Juranitch
University School of Milwaukee
Milwaukee, Wisconsin

Jackie Kelly
El Toro High School
Lake Forest, California

Boris Korsunsky
Weston High School
Weston, Massachusetts

James Maloy
Bethlehem Center High School
Fredericktown, Pennsylvania

David Martin
Masuk High School
Monroe, Connecticut

John McCann
Waynesboro Area Senior High
Waynesboro, Pennsylvania

Theodore Neill
Senior High School
Harmony, Pennsylvania

Mary Norris
Stephenville High School
Stephenville, Texas

Matthew Ohlson
Green Local Schools
Green, Ohio

Steve Oppman
West High School
Oshkosh, Wisconsin

Chris Peoples
Sunny Hills High School
Fullerton, California

Pamela Perry
Lewiston High School
Lewiston, Maine

Susan Poland
Dysart High School
El Mirage, Arizona

Gloria Reche
Success Academy
Houston, Texas

Diane Riendeau
Deerfield High School
Deerfield, Illinois

Brian Shock
Powhatan High School
Powhatan, Virginia

Linda Singley
Greencastle-Antrim High
  School
Greencastle, Pennsylvania

Larry Stookey
Antigo High School
Antigo, Wisconsin

Martin Teachworth
La Jolla High School
La Jolla, California

Richard Thompson
Somerset High School
Somerset, Wisconsin

Blythe Tipping
Sylvania Southview High School
Sylvania, Ohio

Connie Wells
Pembroke Hill School
Kansas City, Kansas

Jeff Wetherhold
Parkland High School
Allentown, Pennsylvania

Matt Wilson
Holly High School
Holly, Michigan

A New Force in Physics

Pearson Physics offers a new path to mastery— a "concepts first" approach that supports a superior, step-by-step problem solving process.

In your new program, you'll find:

- **Example problems** that build reasoning and problem-solving skills.
- **Relevant connections** that tie abstract concepts to everyday experiences and modern technologies.
- **Rich lab explorations** and **study support** that allow students to practice and reinforce essential skills.
- **Cutting-edge technology** that offers multiple options for interacting with—and mastering—the content.

The following pages showcase several key elements of **Pearson Physics** *that will lead students to success.*

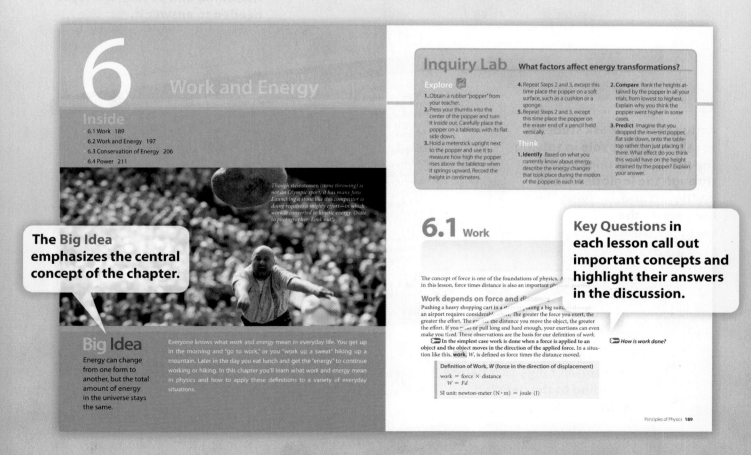

> The **Big Idea** emphasizes the central concept of the chapter.

> **Key Questions** in each lesson call out important concepts and highlight their answers in the discussion.

Leading by Example

Every class contains a unique and diverse group of students. Pearson Physics supports each student's unique learning style, offering all students a path to success. A key element of this approach is the program's use of four distinct Example types, each with a particular purpose.

QUICK Example 5.2 What's the Force?

An 1800-kg car has an acceleration of 3.8 m/s². What is the force acting on the car?

Solution
Substitute $m = 1800$ kg and $a = 3.8$ m/s² in $F = ma$:

$$F = ma$$
$$= (1800 \text{ kg})(3.8 \text{ m/s}^2)$$
$$= 6800 \text{ N}$$

Quick Examples offer simple and concise solutions that model how newly introduced equations and units are used.

CONCEPTUAL Example 5.1 Which String Breaks?

A heavy anvil hangs from a string attached to a ceiling, as shown on the right. An identical string hangs from the bottom of the anvil. Which string breaks if you jerk the lower string downward rapidly?

Reasoning and Discussion
If the lower string is pulled downward rapidly, the inertia of the massive anvil keeps it from responding quickly. Since the anvil barely moves, the force in the lower string quickly becomes large. As a result, the lower string breaks before the anvil has a chance to move. (Pulling slowly on the lower string causes the upper string to break, instead.)

Answer
The lower string breaks if you jerk the lower string downward rapidly.

Conceptual Examples pose a thought-provoking question and then explain the logical reasoning and physics concepts needed to answer it.

ACTIVE Example 6.8 Determine the Final Speed

A boy does 19 J of work as he pulls a 6.4-kg sled through a distance of 2.0 m. No other work is done on the sled. If the initial speed of the sled is 0.50 m/s, what is its final speed?

Solution (Perform the calculations indicated in each step.)

1. Rearrange the work-energy theorem to solve for the final kinetic energy:
$$\tfrac{1}{2}mv_f^2 = W_{total}$$

2. Now solve the equation for the final
$$v_f = \sqrt{2 \cdot \frac{?}{?}}$$

$$v_f = 2.5$$

Active Examples ask students to take an active role in solving the problem by thinking through the logic described on the left and verifying their answers on the right.

GUIDED Example 17.6 | Prismatics

Dispersion

A flint-glass prism has a cross section in the shape of a 30°-60°-90° triangle, as shown in the diagram. Red and violet light are incident on the prism at right angles to its vertical side. Given that the index of refraction of flint glass is 1.66 for red light and 1.70 for violet light, find the difference in the refraction angles as the rays emerge from the prism.

Picture the Problem
The prism and the red and violet rays are shown in our sketch. Notice that the angle of incidence on the vertical side of the prism is 0°. Therefore, the angle of refraction is also 0° for both rays. On the slanted side of the prism, the rays have an angle of incidence equal to 30.0°. Their angles of refraction are different, however.

Strategy
To find the final angle of refraction for each ray, we apply Snell's law with the appropriate index of refraction. We then subtract the angles to find the difference.

Solution

1. Solve Snell's law ($n_1 \sin \theta_1 = n_2 \sin \theta_2$) for the angle of refraction, θ_2. Next, substitute the known values of $n_1 = 1.66$, $\theta_1 = 30.0°$, and $n_2 = 1.00$ to calculate θ_2 for red light:

2. Repeat Step 1 for violet light, with $n_1 = 1.70$, $\theta_1 = 30.0°$, and $n_2 = 1.00$:

3. Subtract 56.1° from 58.2° to find the difference in the refraction angles:

Insight
This kind of difference in refraction angles is the reason light is seen with a prism.

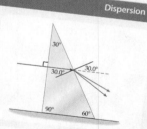

Known
angles for the triangle: 30°, 60°, 90°
$n = 1.66$ (red light)
$n = 1.70$ (violet light)

Unknown
difference in refraction angles = ?

$$n_1 \sin \theta_1 = n_2 \sin \theta_2$$

Math HELP
Trigonometric Functions
See Math Review, Section VI

$$\theta_2 = \sin^{-1}\left(\frac{n_1}{n_2}\sin\theta_1\right)$$
$$= \sin^{-1}\left(\frac{1.66}{1.00}\sin 30.0°\right)$$
$$= 56.1°$$

$$\theta_2 = \sin^{-1}\left(\frac{n_1}{n_2}\sin\theta_1\right)$$
$$= \sin^{-1}\left(\frac{1.70}{1.00}\sin 30.0°\right)$$
$$= 58.2°$$

$$58.2° - 56.1° = \boxed{2.1°}$$

Guided Examples present a visual model of the physical situation and outline the key concepts that apply to it before proceeding to the detailed step-by-step solution.

Relevant Connections

Pearson Physics emphasizes the fact that physics applies to everything in your world, connecting ideas and concepts to everyday experience.

Physics & You features throughout the book explain the physics behind interesting technologies, the impact of technology on society, and the role of physics in various careers.

Physics & You: Technology passages in the discussion explain how various modern technologies make use of the physics concepts just learned.

Physics & You: Technology The wheels on older cars often lock during panic braking, causing the car to skid uncontrollably. In general, sliding or skidding tires are subject to kinetic friction, whereas tires that roll experience static friction, as discussed in Conceptual Example 5.13. Since static friction is usually greater than kinetic friction, a car will stop in a shorter distance if its wheels are *rolling* (static friction) than if its wheels are locked up and skidding (kinetic friction)!

This is the idea behind antilock braking systems (ABS). When the brakes are applied in a car with ABS, an electronic rotation sensor in each wheel detects when the wheel is about to skid. To prevent the skid, a small computer automatically begins to pump the brakes. This pumping allows the wheels to continue rotating, even in an emergency stop, and thus static friction determines the stopping distance. **Figure 5.17** shows a comparison of braking distances for cars with and without ABS.

In-text Labs and Study Tools

Pearson Physics provides hands-on lab explorations in the text itself and through a separate Lab Manual. Extra study support features appear throughout the chapters when students need them most.

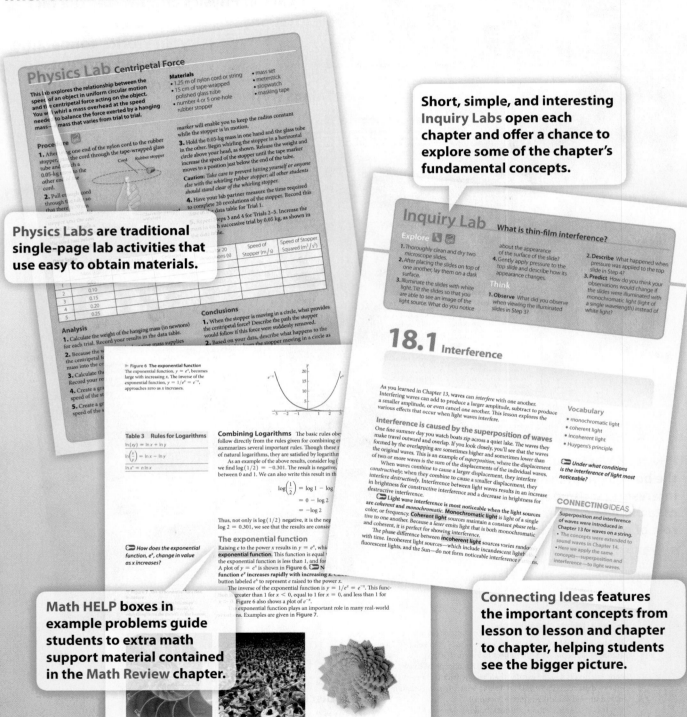

Physics Labs are traditional single-page lab activities that use easy to obtain materials.

Short, simple, and interesting Inquiry Labs open each chapter and offer a chance to explore some of the chapter's fundamental concepts.

Math HELP boxes in example problems guide students to extra math support material contained in the Math Review chapter.

Connecting Ideas features the important concepts from lesson to lesson and chapter to chapter, helping students see the bigger picture.

MasteringPhysics®

The Mastering platform is the most effective and widely used online homework, tutorial, and assessment system for physics.

- **Students interact with self-paced tutorials that focus on course objectives, provide individualized coaching, and respond to their progress.**

- **Instructors use the Mastering system to maximize class time with easy-to-assign, customizable, and automatically graded assessments that motivate students to learn outside of class and arrive prepared for lecture and lab.**

Prelecture Questions

Assignable Prelecture Concept Questions encourage students to read the textbook so they're more engaged in class.

Gradebook Diagnostics

The Gradebook Diagnostics screen provides instructors with weekly diagnostics. With a single click, charts identify the most difficult problems, vulnerable students, and grade distribution.

Tutorials with Hints and Feedback

Mastering's easy-to-assign tutorials provide students with individualized coaching.

- Hints and Feedback offer "scaffolded" instruction similar to what students would experience in an after-school study session.

- Hints often provide problem-solving strategies or break the main problem into simpler exercises.

- Wrong-answer-specific feedback gives students exactly the help they need by addressing their particular mistake without giving away the answer.

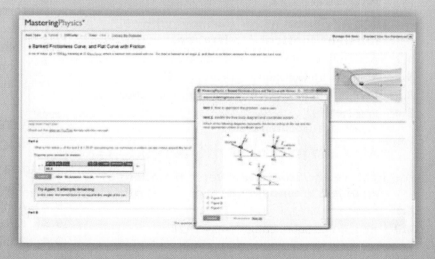

Contents

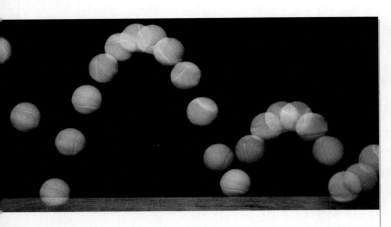

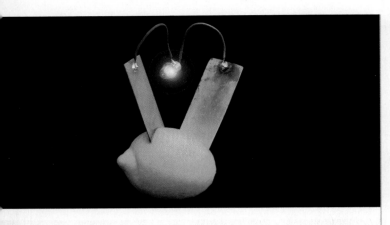

Program Components

MasteringPhysics®

MasteringPhysics® is the most effective and widely used online homework, tutorial, and assessment system for science courses. It delivers self-paced tutorials that focus on your course objectives, provides individualized coaching, and responds to each student's progress. The Mastering system helps teachers maximize class time with easy-to-assign, customizable, and automatically graded assessments that motivate students to learn.

Upon textbook purchase, students and teachers are granted access to MasteringPhysics with Pearson eText. Teachers can obtain preview or adoption access for MasteringPhysics in one of the following ways:

Preview Access

• Teachers can request preview access online by visiting **PearsonSchool.com/ Access_Request** (choose option 2). Preview Access information will be sent to the teacher via email.

Adoption Access

• A Pearson Adoption Access Card, with codes and complete instructions, will be delivered with your textbook purchase (ISBN: 0-13-034391-9).
• Ask your sales representative for an Adoption Access Code Card (ISBN: 0-13-034391-9).
OR
• Visit **PearsonSchool.com/Access_Request** (choose option 3). Adoption access information will be sent to the teacher via email.

Students, ask your teacher for access.

For the Student

Laboratory Manual, available for purchase.

For the Teacher

Annotated Teacher's Edition

Laboratory Manual, Teacher's Edition

ExamView® CD-ROM

Classroom Resource DVD-ROM

Teacher's Solutions Manual (electronic format only)

Some of the teacher supplements and resources for this text are available electronically to qualified adopters on the Instructor Resource Center (IRC). Upon adoption or to preview, please go to **www.pearsonschool.com/access_request** and select Instructor Resource Center. You will be required to complete a brief one-time registration subject to verification of educator status. Upon verification, access information and instructions will be sent to you via email. Once logged into the IRC, enter ISBN 0-13-137115-0 in the "Search our Catalog" box to locate resources.

Electronic teacher supplements are also available within the Instructor's tab of MasteringPhysics.

INQUIRY LABS

Use readily available materials and easy procedures to produce reliable lab results.

PHYSICS LABS

Apply physics concepts and skills with these quick, effective hands-on opportunities.

PHYSICS & YOU FEATURE PAGES

Learn more about how physics applies to real-world situations. You'll read about the impact physics has on society and technology, and survey some interesting careers that apply physics.

PHYSICS & YOU TECHNOLOGY

Learn how chapter content applies to a wide range of devices and technologies.

Guide to Examples

Guide to Examples

Guide to Examples

Guide to Examples

Guide to Examples

Guide to Examples

1 Introduction to Physics

Inside

Physics is a quantitative science, based on careful measurements of quantities such as mass, length, and time. This baby elephant has a mass of about 480 kilograms.

Big Idea

Physics applies
to everything.

The goal of physics is to gain a deeper understanding of the world in which we live. In fact, everything in nature—from atoms and subatomic particles to solar systems and galaxies—obeys the laws of physics. Everything!

We begin our study of physics with a few fundamental topics. These topics provide a basic language of physics that describes its units, measurements, equations, and logical thinking. This language is used throughout the book and can be applied to any science you study. With this in mind, let's start on a wonderful journey of discovery into physics.

1.1 Physics and the Scientific Method

What Is Physics?

Physicists want to know how things work. They observe nature and figure out the rules—or laws—that govern its operation. This basic curiosity is at the heart of all the advances made in physics over the centuries.

Physics studies the laws of nature

Physics is the study of the fundamental laws of nature. Physicists have found that these laws can be expressed in terms of mathematical equations. As a result, it is possible to compare the predictions of theories with the observations of experiments. Physics, then, is rooted equally in theory and experiment, as indicated in **Figure 1.1**. As physicists make new observations, they constantly test and—if necessary—refine the present theories.

Creativity plays an important role in interpreting nature, and in finding ways to solve problems. Physics helps develop creative, logical, and consistent ways of thinking. Because of these attributes, people who study physics go on to careers in many interesting fields.

Vocabulary

- physics
- science
- scientific method
- observation
- inference
- hypothesis
- independent variable
- dependent variable
- theory

▶ **Figure 1.1 Physics combines theory and experiment**
(a) A physics theory is expressed in terms of mathematical equations. The equations give predictions that can be tested with experiments. **(b)** Careful experiments are required to verify a physics theory.

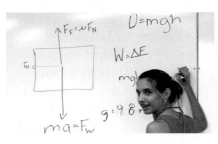

The Major Principles of Physics

What makes physics particularly fascinating is that it applies to all of nature. Physics shows that the complexity and variety in the world around us, and in the universe as a whole, are manifestations of just a few fundamental laws and principles. The fact that we can discover and apply these basic laws of nature is both astounding and exhilarating.

The snowboarder featured here illustrates several important principles, including Newton's laws of motion, energy, and momentum. Several important physics principles are described in this Visual Summary.

These men accelerate the bobsled before jumping on board.

The pole-vaulter is propelled upward by energy stored in the bent pole.

The momentum of the cue ball is used to "break" the other balls apart.

Force, Acceleration, and Motion

Motion does not require a force. A force is needed to cause a change in motion, however. The connection between a force and the resulting change in motion is given by Newton's laws of motion.
Chapter 5

Energy and Energy Conservation

The concept of energy is a surprisingly recent addition to physics. In fact, Galileo and Newton knew nothing about energy. Energy was difficult to discover because it can't be seen or touched, and because it takes so many different forms. Nevertheless, energy is of central importance to modern society. In fact, the total energy of the universe always stays the same — it is conserved.
Chapter 6

Momentum and Momentum Conservation

When Newton wrote his laws of motion, he expressed them in terms of an object's mass times its velocity — a quantity known as momentum. Momentum has been of central importance in physics ever since. The total momentum of the universe is conserved, just like the total energy.
Chapter 7

Hot exhaust gases increase the disorder, or entropy, of the universe.

Entropy and Thermodynamics

Physicists studying heat engines discovered a new physical quantity — entropy. The future of the universe is shaped by the fact that the total entropy can only increase.
Chapter 11

Lightning strikes transfer large amounts of electric charge.

This image of a fly was made using a beam of electrons, not with light.

Electricity and Magnetism

Electricity and magnetism at first seem quite unrelated. Physicists have discovered, however, that they are actually different aspects of the same physical force. The discovery of electromagnetism paved the way for much of our modern technology, including electronics and telecommunications.
Chapters 19–23

Waves and Particles

We usually think of waves (like a water wave or a sound wave) as being completely different from particles (like a baseball or a billiard ball). Modern physics has shown that they are not so different after all. We now know that waves have particle-like properties, and particles have wave-like properties. This insight forms the basis of quantum physics.
Chapter 24

▲ Positron emission tomography (PET) scans of the inside of the human body are used by doctors when making diagnoses.

What Is Science?

Physics is one of many natural sciences. Others include chemistry, biology, and geology. These disciplines study different aspects of nature, but they all share certain key characteristics that define a science.

Science is a way of understanding nature

Science is an organized way of thinking about nature and understanding how it works. Thus, science is a process—it is not a thing or an object. Science provides a method for gaining knowledge and increasing our understanding of the natural world. Science is never applied to supernatural phenomena of any kind.

Science is constantly evolving and developing. In fact, some subjects of research, like positron-emission tomography, didn't even exist a short time ago. Even well-established disciplines like physics continue to grow and change over time. Scientific breakthroughs and revolutions—like the development of quantum physics and the theory of relativity—expand our knowledge of nature. When a scientific revolution occurs, it revises and deepens our previous understanding of nature. It also produces a whole new set of questions that may lead to even more breakthroughs. The progress of science is a fascinating story, and where it will take us next can never be predicted.

Science seeks explanations

It's important to note that science isn't just a collection of facts. Scientists attempt to find explanations for the knowledge that has been gained about natural processes. These explanations provide the basis for a better understanding of nature, as well as a means of predicting the outcomes of future natural events. For example, when a powerful earthquake occurs, scientific knowledge about the behavior of water waves gives scientists the ability to predict areas that are vulnerable to a tsunami—and even when the tsunami will arrive. Predictions like these can save lives and protect property.

The Scientific Method

Have you ever tried to learn a new video game when no one was around and you didn't have a user's manual? You might say to yourself, "I wonder what happens if I push this button?" or "What happens if I move this joystick?" You try the button and the joystick, and you observe what happens on the screen. After a while you begin to learn the "rules" that govern the game—the rules of its make-believe world.

Science is a lot like that, only with science you're trying to learn the rules of the *real world*. There's no user's manual, and no one to tell you all the answers. You have to figure out the rules—the laws of nature—on your own. Of course, the real world is a bit more complicated than a video-game world, but it's a lot more interesting, too.

The **systematic** approach scientists use to learn about the laws of nature is referred to as the **scientific method**. Though each situation is handled a bit differently, the scientific method has certain steps that are always taken when conducting a scientific study. These steps are as follows:

- Observe
- Infer and hypothesize
- Test
- Conclude

Reading Support ✓
Vocabulary Builder
systematic
[sis tuh MAT ik]
(adjective) acting according to a system, set plan, or method; methodical
The gardener used a systematic approach to rid his yard of weeds.

Science begins with careful observations

The starting point of any scientific investigation is careful **observation**, in which you describe events in a logical and orderly way. For example, you might observe how an object moves. Does it speed up or slow down? Does it move in a straight line or on a curved path? Does it start and stop or move constantly? All of these properties are relevant to a physical description of the motion.

It's also important to be creative in your observations. You may be looking at something that people have looked at a thousand times, but perhaps you see it in a way that no one has thought of before. For example, people had seen apples falling from trees for millennia. They had also seen the Moon in the night sky. What Isaac Newton realized as he observed a falling apple was that the Moon moves in a way that is similar to the apple—both objects fall toward the center of the Earth (see Chapter 9 for details). The simple observation of a falling apple led Newton to a completely new way of thinking about the Moon and the force of gravity.

Observations lead to inferences and a hypothesis

Thinking about your observations often leads to inferences about what is going on. In general, an **inference** is a logical interpretation of your observations. Your observations combined with your inferences might allow you to develop a hypothesis. A **hypothesis** is a detailed scientific explanation for a set of observations that can be verified or rejected by careful experiments.

Hypotheses are tested with experiments

A useful hypothesis makes predictions that can be tested with experiments. If an experiment verifies a prediction, the hypothesis gains support—though no one experiment can prove a hypothesis to be correct. If an experiment disagrees with a prediction, the hypothesis must be rejected or modified.

This is an important aspect of the scientific method. A hypothesis must be rejected if it disagrees with experiment, even if the hypothesis has agreed with other experiments, and even if the hypothesis is very popular. Scientists must be open-minded, willing to let the results of experiments guide their thinking, even if the results are not what they expected. Significant breakthroughs in science often start off as hypotheses that seem to go against intuition (are counterintuitive). For example, most people thought Galileo was wrong when he said that heavy objects fall at the same rate as light objects. He was right, however, as he knew from his own careful experiments.

▲ No matter how simple or complex the laboratory equipment, careful observation is the key to accurate and reproducible experimental results.

Applying the scientific method: a case study

Figure 1.2 shows the testing of a physics hypothesis about a pendulum, a system that Galileo also studied. (A pendulum is basically a weight that swings back and forth on a string.) Suppose you want a pendulum that takes a specific amount of time to complete one back-and-forth swing. This amount of time is called the *period* of the pendulum, as illustrated in Figure 1.2 (a). Let's see how the scientific method might apply to this case.

Observe the Pendulum Let a pendulum swing back and forth and measure the period. Change the length of the pendulum and repeat the time measurement. What effect does changing the length have on the pendulum?

Form a Hypothesis Observations indicate that a longer pendulum takes more time to swing back and forth, as indicated in Figure 1.2 (b). Thus, we hypothesize that the period of a pendulum increases with increasing length.

Test the Hypothesis Conduct an experiment in which you measure the period of a pendulum for a variety of lengths. The variable you change in the experiment—the length of the pendulum in this case—is called the **independent variable**. The variable you measure to see how it depends on the independent variable is called the **dependent variable**. The period is the dependent variable in this case. Changing only one variable at a time lets you isolate the effect of that change on the system. Record the results of your experiments and then create a graph of period versus length, as in Figure 1.2 (c).

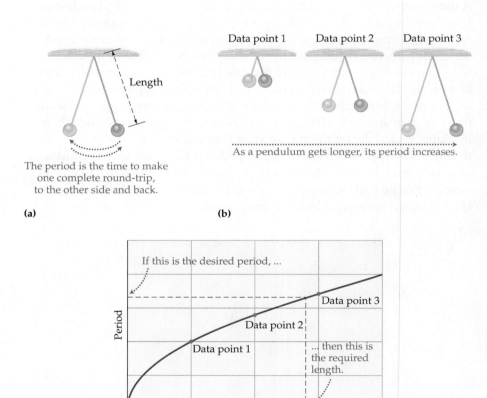

► **Figure 1.2 Studying a pendulum**
(a) A simple pendulum is a weight (bob) that swings back and forth at the end of a string. The length of the pendulum is the length of the string (for a small bob), and its period is the time to complete one round-trip, from one side of its swing to the other side and back to its starting point. **(b)** Observations indicate that the period of a pendulum increases with increasing length. Experiments measure the period for different lengths, yielding data points that can be plotted. **(c)** A graph of period versus length for a simple pendulum. The data points from part (b) are shown, as is a smooth curve that connects the points. (The equation for this curve is given in Chapter 13.) The curve can be used to determine the length required to give a desired period.

Draw a Conclusion The experimental results in Figure 1.2 (c) show that the period of a pendulum does indeed increase with increasing length. (See Chapter 13 for details.) You could use that graph to select the length of the pendulum that has the desired period.

Again, although the precise way to implement the scientific method varies from case to case, the basic elements remain the same:

- Make observations.
- Form inferences and hypotheses.
- Conduct detailed tests of the hypotheses, using experiments.

The validity of a hypothesis is based solely on its ability to account for known observations and to correctly predict new observations.

Well-tested hypotheses lead to theories

A scientific **theory** is a detailed explanation of some aspect of nature that accounts for a set of well-tested hypotheses. For example, Galileo made the hypothesis that falling objects move with constant acceleration. He verified his hypothesis with a variety of experiments. Later, Newton proposed a mathematical theory of gravity that explained why the acceleration of a falling object is constant. His theory also predicted other effects of gravity, like the orbits of planets, moons, and comets. These predictions were verified by later observations. Building up from tested hypotheses to verified theories is the hallmark of the scientific method.

In general, theories are well-supported *explanations* of nature. An example is the theory of gravity, which explains why objects fall with constant acceleration on Earth's surface. On the other hand, laws of nature—like the law of conservation of energy—are well-supported *descriptions* of nature. Laws do not provide explanations, but instead describe specific relationships under given conditions. As you study physics, you will encounter a number of theories and laws.

If new observations disagree with a theory, the theory has to be discarded or revised. It's exciting when this happens, because it usually signals a profound breakthrough in science. It also means that an entirely new area of research, full of things to be discovered, will be opened for exploration.

1.1 LessonCheck (MP)

Checking Concepts

1. Explain How do you verify a scientific hypothesis?

2. Describe How are the fundamental laws and principles of physics related to the complexity that we see in nature?

3. Big Idea How do the laws of physics apply to other sciences such as biology, chemistry, and earth science? Give a specific example to show the connection.

Solving Problems

4. Solve Einstein's most famous equation is $E = mc^2$. In this equation, E stands for energy, m stands for mass, and c stands for the speed of light. Use algebra to solve this equation for the mass. That is, complete this equation:

$$m = ?$$

1.2 Physics and Society

Vocabulary

- bias
- peer review

Science is a human endeavor. As such, it is a part of the fabric of human society. Let's take a look at some of the ways science, technology, and physics impact our everyday life.

Science in Modern Society

Modern society runs on developments in science and technology, and more breakthroughs are coming all the time. How should we use these advances for the greatest benefit of society as a whole? This is a question that goes beyond science. It involves economics, practicality, morals, and laws.

Ethics is an important part of science

A scientific discovery gives added insight into nature. It shows how a certain part of nature works. But knowing how something works isn't the same as knowing how best to use that knowledge. That is something scientists, politicians, and an informed public must decide together.

For example, the development of electrical power systems has greatly improved the quality of life for millions of people. This great benefit also comes with a downside, however, because electricity can be deadly if it is not used with proper care. Society makes decisions about how to reduce the dangers and what level of danger is acceptable. It must also consider the advantages and disadvantages of the various ways of producing electricity, such as coal-fired power plants, hydroelectric dams, and nuclear power plants, illustrated in **Figure 1.3**.

In making the decisions that affect society, people need to avoid bias. A **bias** is a preference for a particular point of view for personal rather than logical or scientific reasons. Because scientists are human too, they can be affected by bias as much as politicians, business leaders, and others. If enough of us in the general public are educated about the various aspects of modern science, however, we can see through biases and make informed decisions that benefit all of humanity.

▼ **Figure 1.3 Electric power plants**
Electricity can be produced in a number of different ways. The power plants shown here use (from left to right) nuclear energy, the Sun, and water to generate electricity. Each of these methods has benefits and drawbacks.

Scientists share knowledge, insights, and ideas

A scientific research project can last months or even years. Once it is complete, the next step is to publish the results to make them known to the scientific community. A great deal of effort goes into writing the report, making sure that the new ideas are expressed clearly and that proper credit is given to previous work on the subject.

Respected scientific journals don't publish every report or paper that is submitted to them, however. Before a report can be published, it must undergo peer review. **Peer review** means that a report is sent to several experts in the field so that they can look for errors, biases, and oversights. It's a bit like our judicial system, where a person is tried by a jury of peers. Only after being recommended for publication by these experts can the report be published. Peer review doesn't guarantee that a report is correct, but it does guarantee that it meets the minimum standards of quality and scholarship demanded by the scientific journal.

Scientific research and technology affect society

Science and technology are an important part of modern society. New technologies depend on scientific advances, some of which occur years or even decades before the practical applications.

For example, Einstein's theory of relativity (published in 1905) predicts that moving clocks run slower than clocks at rest. That's interesting, but because the time differences are only a few billionths of a second, it was thought at the time that there would be no practical application of the result. Decades later the Global Positioning System (GPS) was developed. For this system to function properly, it must keep track of time to within billionths of a second. Each satellite in the GPS system carries an atomic clock, which is moving at high speed. Therefore, to make the GPS system work, it is necessary to take into account the relativistic slowing down of moving clocks.

Now, Einstein didn't develop the theory of relativity as the basis for a GPS system—no one had even thought of that idea in 1905. Even so, his advancement in basic science became indispensable decades later when the GPS was developed. This is one of the reasons society supports basic scientific research, even when the researcher has no definite practical application in mind. It has been proven time and again that advances in basic knowledge inevitably lead to technological advances down the road. In short, investment in basic science is an investment in future technologies. And the best technologies are precisely the ones we can't anticipate in advance.

The Visual Summary on the next page presents several areas of physics in which significant research is ongoing.

Areas of Physics Research

There are many specialties within physics that concentrate on particular aspects of nature. High-energy physicists, for example, use particle accelerators like the Large Hadron Collider (LHC), shown in this aerial view, to study the elementary particles of matter. The LHC, located in Switzerland and built by the European Organization for Nuclear Research (CERN), is the world's largest and highest-energy particle accelerator. This Visual Summary describes other important areas of physics research.

Lasers are useful in research and for making precise measurements.

The Giant Magellan Telescope in Chile will use seven separate mirrors.

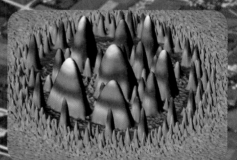

An atomic force microscope image of yttrium oxide (Y_2O_3).

Laser Physics

Lasers have revolutionized physics. They produce light that is intense and highly focused. Current research centers on using lasers to provide ultra-short pulses of light and as a means of cooling atoms to extremely low temperatures.

Optical Physics

Optics have been an important part of physics since Galileo made his first telescope and Newton split light into its component colors with a prism. Today, optics is important in astronomy, camera design, and communications.

Atomic Physics

Learning the structure of atoms was at the heart of modern physics. The ability to manipulate atoms has led to the development of miniaturized electronic components, as well as new types of microscopes that can produce pictures of individual atoms.

A nuclear reactor glows blue due to Cherenkov radiation.

Nuclear Physics

The nucleus of the atom continues to be an area of intense research. The forces that act within a nucleus provide clues as to the forces that acted in the early universe, shortly after the Big Bang.

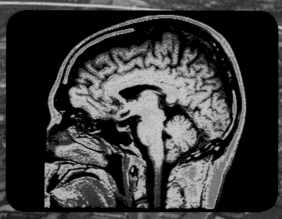

This magnetic resonance image (MRI) shows normal brain anatomy.

Medical Physics

The first application of physics to medicine was the invention of X-ray imaging. Today, physics supplies radioactive elements to treat cancers and to image the living, thinking brain. Physics has also made possible computerized tomography, a technological breakthrough that gave physicians and surgeons a noninvasive way to look inside the human body.

A theoretical physicist must be adept at working with abstract mathematical models.

Theoretical Physics

Theoretical physics is the field that applies mathematics to the search for a fundamental understanding of the laws of nature. In physics theory the goal is a better understanding of nature, rather than a specific real-world application. Theoretical research has led to the breakthroughs of quantum physics, relativity, string theory, and many more.

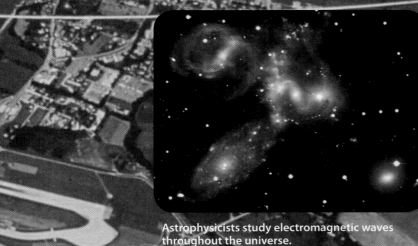

Astrophysicists study electromagnetic waves throughout the universe.

Astrophysics and Cosmology

It's ironic, but scientists who study the smallest objects in nature — the fundamental particles—have provided deep insight into the history and fate of the largest of objects—the universe. Physicists who study string theory and general relativity are pushing the limits of our knowledge to the earliest times of the universe, and to the possible future that awaits us all.

Performing Physics Investigations

As you study physics, you will conduct a number of experiments. In these experiments you will make measurements and employ safety procedures. Let's take a look at the common system of measurement used in all sciences and the basic procedures that ensure experimental safety.

Scientists use the metric system

Why is the metric system important in science?

An important part of science is the ability of scientists in different locations to reproduce an experiment and verify that the reported results are consistent. **To aid in collaboration and verification, scientists have adopted a system of measurement referred to as the *metric system*. Scientists throughout the world use this system when collecting data and performing experiments.** The metric system uses a decimal system of units that are related to one another by multiples of 10. For example, the scientific unit of length is the meter, and a kilometer is 1000 meters. Using multiples of 10 simplifies converting from one unit to another. The basic units for the metric system are presented in detail in the next lesson.

All safety procedures must be followed

The number-one priority in any physics experiment is safety. Some experiments involve moving objects; others apply forces or use electric currents. In all cases certain safety procedures must be followed to ensure not only your own safety, but also the safety of your classmates and teacher. Appendix D presents the basic safety rules for any experiment you will do in your class.

Before you start a laboratory exercise, be sure to read all the steps and make sure that you understand the entire procedure. If you have any doubt or uncertainty about any part of the exercise, be sure to ask your teacher for an explanation. Following your teacher's instructions and the directions in this book is the most important safety rule of all.

1.2 LessonCheck (MP)

Checking Concepts

5. **Explain** What makes the metric system so convenient for science?

6. **Explain** What is a bias? Why should biases be avoided?

7. **Describe** What is meant by peer review?

Solving Problems

8. **Calculate** How many meters are in 15 kilometers?

9. **Calculate** How many kilometers are in 12,000 meters?

1.3 Units and Dimensions

Suppose you want to measure the length of your pet cat. You might use a tape measure stretched from the cat's nose to the tip of its tail. Remember to straighten out the tail to obtain its full length! You can now read the length on the tape measure as your cat goes back to cleaning its fur. Measurements like this are an important part of conducting experiments in physics.

Base Units

Meaningful experimental results require careful measurements of quantities such as length, mass, and time. Physicists have worked hard to development accurate and repeatable ways to make these measurements.

Physical quantities are measured using base units

Measurements involve units. 🔑 **The standard "measuring stick" for a physical quantity is referred to as its *base unit*.** The base units are defined precisely, and measurements are expressed as multiples of these base units.

For example, the base unit of length is the **meter**, which we abbreviate as m. It follows that a person who is 1.94 m tall has a height 1.94 times this base unit of length. An example of someone making this type of measurement is shown in **Figure 1.4**. Similar comments apply to the **second** (abbreviated as s), the base unit of time, and the **kilogram** (abbreviated as kg), the base unit of mass.

The detailed system of units used in this book was established in 1960 at the Eleventh General Conference of Weights and Measures in Paris, France. This system goes by the name Système International d'Unités, or SI for short. Thus, when we refer to **SI units**, we mean base units of meters (m), kilograms (kg), and seconds (s). Taking the first letter of each of these units gives a common alternate name—the mks system.

The base unit of length is the meter

The units of length have become more precise over the centuries. Early units were usually associated with the human body. For example, the Egyptians defined the *cubit* to be the distance from the elbow to the tip of the middle finger. Similarly, the *foot* was once defined to be the length of the royal foot of King Louis XIV of France. As colorful as such units may be, they are not particularly reproducible—at least not to great precision.

In 1793 the French Academy of Sciences, seeking a more objective and reproducible standard, defined a unit of length equal to one ten-millionth of the distance from the North Pole to the equator. This new unit was named the *metre* (from the Greek *metron* for "measure"). The preferred spelling in the United States is *meter*. This definition was widely accepted, and in 1799 a standard meter was produced: It consisted of a platinum-iridium rod with two marks on it 1 meter apart.

Vocabulary

- meter
- second
- kilogram
- SI unit
- mass
- weight
- dimensional analysis

🔑 *What are base units?*

▲ **Figure 1.4 Using base units**
The height of this person is being measured as a multiple of the base unit of length, the meter. Similarly, her mass is measured as a multiple of the base unit for mass, the kilogram.

Table 1.1 Typical Distances

Distance from Earth to the nearest large galaxy (the Andromeda galaxy, M31)	2×10^{22} m
Diameter of our galaxy (the Milky Way)	8×10^{20} m
Distance from Earth to the nearest star (other than the Sun)	4×10^{16} m
One light-year	9.46×10^{15} m
Average radius of Pluto's orbit	6×10^{12} m
Distance from Earth to the Sun	1.5×10^{11} m
Radius of Earth	6.37×10^{6} m
Length of a football field	10^{2} m
Height of a person	2 m
Diameter of a CD	0.12 m
Diameter of human aorta	0.018 m
Diameter of a period at the end of a sentence in this book	5×10^{-4} m
Diameter of a red blood cell	8×10^{-6} m
Diameter of a hydrogen atom	10^{-10} m
Diameter of a proton	2×10^{-15} m

Since 1983 we have used an even more precise definition of the meter, based on the speed of light in a vacuum (empty space):

> One meter is defined to be the distance traveled by light in a vacuum in $1/299{,}792{,}458$ of a second.

For comparison purposes, a meter is about 3.28 feet. This is roughly 10% longer than a yard. A list of typical lengths is given in **Table 1.1**. Lengths from the microscopic to the galactic scale are shown in **Figure 1.5**.

The base unit of mass is the kilogram

Mass is a measure of the amount of matter in an object. It is measured in units defined by a metal cylinder in France. This cylinder, known as the *standard mass,* is shown in **Figure 1.6**. Notice that the cylinder is kept under two glass bell jars to prevent even a speck of dust from landing on it and changing its mass.

In SI units, mass is measured in kilograms.

> One kilogram is defined to be the mass of a platinum-iridium cylinder at the International Bureau of Weights and Standards in Sèvres, France.

To put the kilogram in everyday terms, a quart of milk has a mass slightly less than 1 kilogram. Other masses are given in **Table 1.2**.

Note that we *do not* define the kilogram to be the weight of the platinum-iridium cylinder. In fact, weight and mass are quite different quantities, even though they are often confused in everyday language. Mass is an **intrinsic**, unchanging property of an object. **Weight**, in contrast, is a measure of the gravitational force acting on an object and varies depending on the object's location. If you could travel to Mars someday, you would find your weight to be less than it is on Earth. Your mass, however, would be unchanged. The force of gravity is discussed in detail in Chapter 9.

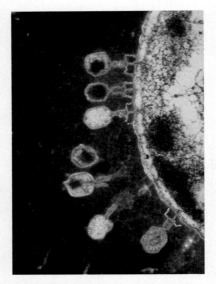

▲ **Figure 1.5 Length scales in the universe**
(a) The size of these viruses, which are attacking a bacterial cell, is about one ten-millionth of a meter (10^{-7} m). **(b)** The diameter of this typical galaxy is about 10^{21} m.

Table 1.2 Typical Masses

Galaxy (Milky Way)	4×10^{41} kg
Sun	2×10^{30} kg
Earth	5.97×10^{24} kg
Space shuttle	2×10^{6} kg
Elephant	5400 kg
Automobile	1200 kg
Human	70 kg
Baseball	0.15 kg
Honeybee	1.5×10^{-4} kg
Red blood cell	10^{-13} kg
Bacterium	10^{-15} kg
Hydrogen atom	1.67×10^{-27} kg
Electron	9.11×10^{-31} kg

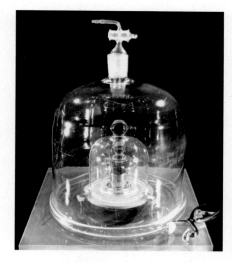

▲ **Figure 1.6 The standard kilogram**
The standard kilogram, a platinum-iridium cylinder 0.039 m in height and diameter, is kept under carefully controlled conditions in Sèvres, France. Exact replicas are maintained in other laboratories around the world.

The base unit of time is the second

Nature has provided us with a fairly accurate timepiece in the revolving Earth. In fact, prior to 1956 the *mean solar day* was defined to consist of 24 hours, with 60 minutes per hour, and 60 seconds per minute, for a total of $(24)(60)(60) = 84{,}400$ seconds. Even the rotation of the Earth is not completely regular, however.

Today, time is measured in units based on the operation of atomic clocks, the world's most accurate timepieces. An example of an atomic clock is shown in **Figure 1.7**. These clocks have typical accuracies of about 1 second in 300,000 years. The atomic clock used for defining the second is designed to operate with cesium-133 atoms. Thus, the second is defined as follows:

> One second is defined to be the time it takes for radiation from a cesium-133 atom to oscillate 9,192,631,770 times.

A range of typical time intervals appears in **Table 1.3**.

Table 1.3 Typical Time Spans

Age of the universe	5×10^{17} s
Age of the Earth	1.3×10^{17} s
Existence of the human species	6×10^{13} s
Human lifetime	2×10^{9} s
One year	3×10^{7} s
One day	8.6×10^{4} s
Time between heartbeats	0.8 s
Human reaction time	0.1 s
One cycle of a high-pitched sound wave	5×10^{-5} s
One cycle of an AM radio wave	10^{-6} s
One cycle of a visible light wave	2×10^{-15} s

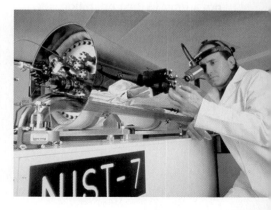

▲ **Figure 1.7 An atomic clock**
This atomic clock, which keeps time on the basis of radiation from cesium atoms, is accurate to about three millionths of a second per year.

In the United States, the time standard is determined by a *cesium fountain atomic clock* developed at the National Institute of Standards and Technology (NIST) in Boulder, Colorado. The clock, designated NIST-F1, shoots cesium atoms upward in an evacuated chamber, producing a "fountain" about a meter high. It takes roughly a second for the atoms to rise and fall through this height, as you will see in the next chapter. The NIST-F1 clock gains or loses no more than 1 second in every 20 million years of operation. Now that's accurate!

Prefixes modify the SI base units

Although SI units are used throughout this book, we will occasionally refer to other systems. For example, a system of units similar to the mks system, but with smaller units, is the cgs system, whose name stands for centimeter (cm), gram (g), and second (s). In addition, multiples of the basic units are common no matter which system is used. **Standard prefixes are used to designate common multiples in powers of 10.** For example, the prefix *kilo* (abbreviated k) means one thousand, or equivalently 10^3. Thus, 1 kilogram is the same as 1000 grams, or 10^3 grams.

$$1 \text{ kg} = 10^3 \text{ g}$$

Likewise, 1 kilometer equals 1000 meters, or 10^3 meters.

$$1 \text{ km} = 10^3 \text{ m}$$

Similarly, the prefix *milli* (abbreviated m) means one thousandth, or 10^{-3}. Thus, 1 millimeter (mm) is the same as 0.001 m, or 10^{-3} meter.

$$1 \text{ mm} = 10^{-3} \text{ m}$$

The most common unit prefixes are listed in **Table 1.4**.

How are prefixes used to modify base units?

C⊙L *PHYSICS*
Atomic Clocks

Atomic clocks are almost common-place these days. For example, did you know that each orbiting satellite used in the Global Positioning System (GPS) carries its own atomic clock? This allows the satellites to make the precision time measurements that are needed for equally precise determinations of position. Also, the "atomic clocks" that are advertised for use in the home, while not atomic in their operation, set themselves according to radio signals sent out from the atomic clocks at NIST. You can access the official U.S. time on the Internet at www.time.gov. Now that's timely physics.

Table 1.4 Common Unit Prefixes

Power	Prefix	Abbreviation
10^{15}	peta	P
10^{12}	tera	T
10^{9}	giga	G
10^{6}	mega	M
10^{3}	kilo	k
10^{2}	hecto	h
10^{1}	deka	da
10^{-1}	deci	d
10^{-2}	centi	c
10^{-3}	milli	m
10^{-6}	micro	μ
10^{-9}	nano	n
10^{-12}	pico	p
10^{-15}	femto	f

A typical *E. coli* bacterium is about 0.005 mm in length. Convert this length to meters (m).

Solution

First, notice that 0.005 mm is 5 thousandths of a millimeter, or 5×10^{-3} mm. Next, Table 1.4 shows that 1 mm $= 10^{-3}$ m, which means that 1000 mm $= 1$ m. With these observations, we can convert from millimeters to meters as follows:

$$(5 \times 10^{-3} \text{ mm})\left(\frac{1 \text{ m}}{1000 \text{ mm}}\right) = \boxed{5 \times 10^{-6} \text{ m}}$$

Insight

The fraction in parentheses is a conversion factor that changes mm to m. The desired unit (m) is in the numerator of the conversion factor, and the unit to be canceled (mm) is in the denominator. This is a general rule for conversion factors, as we shall see later in this lesson.

CONNECTINGIDEAS

The three dimensions introduced in this chapter—mass, length, and time—are the only ones you'll use until Chapter 10.
• Other quantities you will encounter in upcoming chapters, like force, momentum, and energy, are combinations of these basic dimensions.

Practice Problems

10. Follow-up What is the length of an *E. coli* bacterium in kilometers? (*Hint:* Follow the same procedure. Use Table 1.4 to write a conversion factor that will convert 5×10^{-6} m to kilometers.)

11. A minivan sells for 33,200 dollars. Give the price of the minivan in (**a**) kilodollars and (**b**) megadollars.

12. A honeybee flaps its wings 200 times per second. How much time is required for one wingbeat? Give your answer in milliseconds.

Dimensional Analysis and Unit Conversion

The *dimension* of a physical quantity refers to the *type* of quantity it is, regardless of the units used in the measurement. For example, consider the two distance measurements of 25 km and 1.25 cm. Though each has a different unit, they both have the same dimension—length. The same is true of a quantity like velocity. The velocities 12 m/s and 3 mm/century both have dimensions of length divided by time.

As explained below, the dimensions found in an equation lead to useful techniques for verifying a solution and for converting from one type of unit to another. The dimensions of some common physical quantities are summarized in **Table 1.5**.

Physics equations are dimensionally consistent

Any valid equation in physics must be *dimensionally consistent*. **What dimensional consistency means is that each term in a physics equation must have the same dimensions. If a physics equation isn't dimensionally consistent, it isn't correct.** After all, it doesn't make sense to add a distance to a time, any more than it makes sense to add apples and oranges. They are different things.

Table 1.5 Dimensions of Some Common Physical Quantities

Quantity	Dimension
Distance	L
Area	L^2
Volume	L^3
Velocity	L/T
Acceleration	L/T^2

What is dimensional consistency, and how does it apply to physics equations?

As an example, let's check the dimensions of a physics equation that you'll learn about in the next chapter:

$$x_f = x_i + vt$$

In this equation, x_f and x_i represent lengths, v is velocity (length/time), and t is time. Writing out the dimensions of each term, we have

$$\text{length} = \text{length} + \left(\frac{\text{length}}{\text{time}}\right)(\text{time})$$

It might seem that the last term has different dimensions than the other two, but dimensions obey the same rules of algebra as other quantities. This means that the dimensions of time cancel in the last term:

$$\left(\frac{\text{length}}{\text{time}}\right)(\text{time}) = \text{length}$$

It follows that each term in the equation has the same dimension, as required:

$$\text{length} = \text{length} + \text{length}$$

This type of calculation—one written in terms of dimensions—is referred to as **dimensional analysis**. You can use this technique to verify that a physics equation doesn't contain any obvious typographical errors.

Converting units is a straightforward process

In Quick Example 1.1 you learned how to convert from one SI unit to another. Occasionally, you might need to convert from a non-SI unit to an SI unit. For example, suppose you would like to convert 316 ft to its equivalent in meters. The process is the same as before. Looking at the conversion factors on the inside front cover of the text, you find that

$$1 \text{ m} = 3.281 \text{ ft}$$

Equivalently,

$$\frac{1 \text{ m}}{3.281 \text{ ft}} = 1$$

Now, to make the conversion, simply multiply 316 ft by this expression, which is equivalent to multiplying by 1:

$$(316 \text{ ft})\left(\frac{1 \text{ m}}{3.281 \text{ ft}}\right) = 96.3 \text{ m}$$

Note that the conversion factor is written in this particular way, as 1 m divided by 3.281 ft, so that the units of feet cancel out, leaving the final result in the desired units of meters.

Of course, you can just as easily convert from meters to feet. To do so, use the reciprocal of the above conversion factor—which is also equal to 1. For example, you convert a distance of 26.4 m to feet by canceling out the units of meters, as follows:

$$(26.4 \text{ m})\left(\frac{3.281 \text{ ft}}{1 \text{ m}}\right) = 86.6 \text{ ft}$$

Thus, you can see that converting units is as easy as multiplying by 1— because that's really what you're doing.

An everyday example of length conversions, this time from miles to kilometers, is shown in **Figure 1.8**.

▲ **Figure 1.8 Converting distances**
Using this sign, you can create factors for converting miles to kilometers, and vice versa.

A warehouse is 1980 cm long, 1050 cm wide, and 495 cm high. What is its volume in cubic meters (m^3)?

Picture the Problem

Our sketch of the warehouse labels each of its dimensions—its length L, width W, and height H.

Strategy

We begin by converting each dimension to meters using a conversion factor derived from the relationship $100 \text{ cm} = 1 \text{ m}$. Once this is done, the volume V in cubic meters is simply the product of the three converted dimensions.

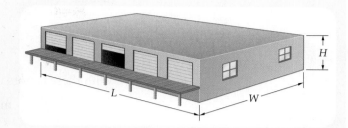

Known	Unknown
$L = 1980$ cm	$V = ?$
$W = 1050$ cm	
$H = 495$ cm	

Solution

1 Convert the length of the warehouse to meters:

$$L = (1980 \text{ cm})\left(\frac{1 \text{ m}}{100 \text{ cm}}\right)$$
$$= 19.8 \text{ m}$$

2 Convert the width to meters:

$$W = (1050 \text{ cm})\left(\frac{1 \text{ m}}{100 \text{ cm}}\right)$$
$$= 10.5 \text{ m}$$

3 Convert the height to meters:

$$H = (495 \text{ cm})\left(\frac{1 \text{ m}}{100 \text{ cm}}\right)$$
$$= 4.95 \text{ m}$$

> **Math HELP**
> Calculating Volume
> See Math Review, Section V

4 Calculate the volume of the warehouse:

$$V = L \times W \times H$$
$$= (19.8 \text{ m})(10.5 \text{ m})(4.95 \text{ m})$$
$$= \boxed{1030 \text{ m}^3}$$

Insight

Thus, the warehouse has a volume of 1030 cubic meters—the same as 1030 cube-shaped boxes that are 1 meter on a side.

Practice Problems

13. **Follow-up** What is the volume of the warehouse in cubic meters if its length is 0.012 km and the other dimensions are unchanged?

14. How many milliliters are in 1.2 L? ($1 \text{ mL} = 0.001 \text{ L}$)

15. **Challenge** The Star of Africa, a diamond in the royal scepter of the British crown jewels, has a mass of 530.2 carats, where 1 carat $= 0.20$ g. Given that 1 kg has a weight of 2.21 lb, what is the weight of the Star of Africa in pounds?

Multiple conversions can be done at once

The procedure described above can be applied to conversions involving any number of units. For instance, if you jog at 3.00 meters per second, how fast is that in kilometers per hour? In this case we need the following conversion factors:

$$1 \text{ km} = 1000 \text{ m}$$
$$60 \text{ s} = 1 \text{ min}$$
$$60 \text{ min} = 1 \text{ h}$$

With these factors in hand, we carry out the conversion as follows:

$$\left(3.00 \ \frac{\cancel{m}}{\cancel{s}}\right)\left(\frac{1 \text{ km}}{1000 \ \cancel{m}}\right)\left(\frac{60 \ \cancel{s}}{1 \ \cancel{min}}\right)\left(\frac{60 \ \cancel{min}}{1 \text{ h}}\right) = 10.8 \ \frac{\text{km}}{\text{h}}$$

Notice that in each conversion factor the numerator is equal to the denominator. In addition, each conversion factor is written in such a way that the unwanted units cancel, leaving just kilometers per hour in our final result.

ACTIVE Example 1.3 Convert the Units of Speed

Blood in the human aorta can attain a speed of 35.0 cm/s. How fast is this in kilometers per hour?

Solution *(Perform the calculations indicated in each step.)*

1. First, convert centimeters to kilometers: $\quad 3.50 \times 10^{-4} \text{ km/s}$

2. Next, convert seconds to hours: $\quad 9.72 \times 10^{-8} \text{ km/h}$

Insight
Of course, the two conversions can be carried out in a single calculation if desired.

1.3 LessonCheck (MP)

Checking Concepts

16. **Recall** What are the SI base units for mass, length, and time?

17. **Describe** What does the prefix *kilo* (k) mean? How can this prefix be used to modify the description of an obstacle course that is 1450 m long?

18. **Explain** Why must all terms in a physics equation have the same dimensions?

19. **Contrast** What is the difference between a unit and a dimension? Give an example of each to illustrate your point.

Solving Problems

20. **Convert** The speed of light in a vacuum is approximately 0.3 Gm/s. What is the speed of light in meters per second?

21. **Think & Calculate** Many highways in the United States have a speed limit of 65 mi/h.
(a) Is this speed greater than, less than, or equal to 65 km/h? Explain.
(b) Find the speed limit in kilometers per hour that corresponds to 65 mi/h.

22. **Verify** Show that the equation $v_f = v_i + at$ is dimensionally consistent. In this equation, v_f and v_i are velocities, a is an acceleration, and t is time.

1.4 Basic Math for Physics

Although physics concepts can be expressed in words, the real power of physics comes when these concepts are expressed mathematically. This lesson introduces you to the basic math skills used throughout this book.

Equations

Mathematics is the language of physics. The math doesn't have to be complicated, and usually it isn't. In fact, much of physics is expressed elegantly in simple *equations*. Applying basic math to physics equations allows you to derive new and interesting results.

Equations are used to relate physical quantities

Physics is a science built on concepts, many of which can be described by an equation. In general, an **equation** is a mathematical expression that relates physical quantities. A *physical quantity* is a property of a physical system that can be measured, like length, speed, acceleration, or time duration. Sometimes the equations of physics can be the trees that block the view of the forest. For a physicist, these equations are simply different ways of expressing a few fundamental ideas. It is the forest—the basic laws and principles of physics—that is the focus of this text.

As an example, consider the equation for calculating the speed of an object. As you may know, the speed of an object is the distance covered divided by the time required to cover it. For example, if you travel 60 kilometers in 1 hour, your speed is 60 kilometers per hour.

To express this with mathematical symbols, we say that speed v is equal to distance d divided by time t. We can write this as follows:

$$v = \frac{d}{t}$$

Notice that $v = d/t$ is simply a shorthand abbreviation that saves us from writing out the full statement "speed is equal to distance divided by time." It would be tedious to repeat this sentence over and over when we can just as well abbreviate it as $v = d/t$. It's similar to what you do when you send a text message and use abbreviations like fyi, lol, imho, and idk in place of writing out the full phrases. Equations like $v = d/t$ and $F = ma$ are just like that, a convenient shorthand.

Equations can be solved for different quantities

Any equation can be rearranged to give information about any of the physical quantities in it. For example, suppose you want to find the distance traveled by an object moving with a certain speed for a certain period of time. To do this, you simply solve the basic equation $v = d/t$ for the distance (d).

Vocabulary

- equation
- accuracy
- precision
- significant figure
- scientific notation
- linear relationship
- parabola
- inverse relationship
- scalar
- speed
- vector
- velocity

What is a physical quantity?

The steps used to solve $v = d/t$ for d are shown below.

Start with the basic equation:

$$v = \frac{d}{t}$$

Multiply each side of the equation by t:

$$v \times t = \frac{d}{t} \times t$$

Cancel the t's on the right side of the equation:

$$v \times t = \frac{d}{\not{t}} \times \not{t}$$

$$v \times t = d$$

Rewrite the result so that d is on the left side: $\quad d = vt$

Now we have a "new" equation, $d = vt$. It's not really new, of course—it's just a rearrangement of our starting equation, $v = d/t$.

What if you want to know how much time it takes an object to travel a certain distance at a certain speed? Again, you simply solve the basic equation $v = d/t$ for the appropriate quantity—in this case the time (t).

Start with the basic equation:

$$v = \frac{d}{t}$$

Multiply each side of the equation by t:

$$v \times t = \frac{d}{t} \times t$$

Cancel the t's on the right side of the equation:

$$v \times t = \frac{d}{\not{t}} \times \not{t}$$

$$v \times t = d$$

Divide each side of the equation by v and cancel the v's on the left side to isolate t:

$$\frac{\not{v} \times t}{\not{v}} = \frac{d}{v}$$

$$t = \frac{d}{v}$$

We now have an equation for the time, $t = d/v$.

Thus, it looks like we have several different equations:

$$v = \frac{d}{t} \qquad d = vt \qquad t = \frac{d}{v}$$

However, these equations are really just the one basic idea, $v = d/t$, expressed in different ways. Thus, there is no reason to memorize all the equations. Each of them expresses the same idea—speed is distance divided by time.

Measured and Calculated Values

You will make quite a few measurements in the physics laboratory and use the data to draw conclusions and calculate answers. Needless to say, making accurate measurements and accurately expressing calculated results is an important skill.

Measurements require accuracy and precision

When a mass, a length, or a time is measured in a scientific experiment, the result is always subject to some uncertainty. The inaccuracy, or uncertainty, can be caused by a number of factors:

- limitations in the measuring device,
- limitations in human perception, and
- limitations in the skill of the experimenter.

An example of an everyday measurement with a simple measuring device is shown in **Figure 1.9.**

▲ **Figure 1.9 Measuring a length**
Every measurement has some degree of uncertainty associated with it. How precise would you expect this measurement to be? How accurate?

▲ **Figure 1.10 Accuracy and precision**
The results of three tosses in a game of lawn darts. The object is to get your darts within the yellow plastic circle. Darts within the circle represent good accuracy; darts closely packed together represent good precision.

The quality of a scientific measurement is determined by two different characteristics. The first is **accuracy**, which is a measure of how close the measured value of a quantity is to its actual value. The second is **precision**, which is a measure of how close together the values of a series of measurements are to one another. **Figure 1.10** illustrates the differences between situations where accuracy and precision are either good or poor.

Significant figures indicate accuracy

Let's suppose you want to determine the walking speed of your pet tortoise. To do so, you measure the time it takes for the tortoise to walk a given distance. The speed of the tortoise is simply the distance divided by the time.

To measure the distance you use a ruler that has one tick mark per centimeter. Looking carefully at the ruler you find that the distance is just over 21 cm. You estimate that the amount over 21 cm is about 0.2 cm, so your best estimate of the distance is 21.2 cm. We say that the measured distance of 21.2 cm contains three significant figures. The number of **significant figures** in a measurement is defined to be the number of digits that are actually measured plus one estimated digit.

The finish of a 100-meter hurdles race. If the timing had been accurate to only tenths of a second—as would probably have been the case before electronic devices came into use—these two runners might "clock" the same time.

To measure the walking time of your tortoise, you use an old pocket watch. The time is over 8 s, and to your best estimate it is halfway between 8 s and 9 s. Therefore, your time measurement is 8.5 s, accurate to two significant figures. An example of a time measurement that requires more than three significant figures is shown in **Figure 1.11**.

Notice that if you were to measure your tortoise's walking time with a digital watch that had a readout giving the time to $1/100$ of a second, the accuracy of the result would still be limited by the reaction time of the experimenter (you). The reaction time would have to be predetermined in a separate experiment. (See Problem 114 in Chapter 3 for a simple way to determine your reaction time.)

Mathematical operations affect significant figures

🔑 *How many significant figures result from addition and subtraction or from multiplication and division?*

Now that we've determined the distance and time for your tortoise, let's calculate its speed. Using the equation for speed, the distance and time found above, and a calculator with an eight-digit display, we find:

$$v = \frac{d}{t} = \frac{21.2 \text{ cm}}{8.5 \text{ s}} = 2.4941176 \text{ cm/s}$$

Considering the limitations of our measurements, the accuracy of this result is simply not justified. After all, you can't expect measurements with two and three significant figures to give an answer with eight significant figures.

Rule for Multiplication and Division So the question is, "How many significant figures *should* the answer have?" The guiding principle is that the answer cannot be more accurate than the *least* accurately known measurement. 🔑 **For multiplication and division, the number of significant figures in the answer is the same as the number of significant figures in the least accurately known input value.**

Let's apply this principle to our speed calculation. We know the distance to three significant figures and the time to two significant figures. As a result, the speed should be given with just two significant figures,

$$v = \frac{21.2 \text{ cm}}{8.5 \text{ s}} = 2.5 \text{ cm/s}$$

Rounding Off Notice that we didn't just keep the first two digits in 2.4941176 cm/s and drop the rest. Instead, we rounded up. We did this because the first digit to be dropped (9 in this case) is greater than or equal to 5. As a result, we increased the previous digit (4 in this case) by 1. Thus, 2.5 cm/s is our best estimate for the tortoise's speed.

A tortoise races a rabbit by walking with a constant speed of 2.51 cm/s for 52.23 s. How much distance does the tortoise cover?

Picture the Problem

The race between the rabbit and the tortoise is shown in our sketch. The rabbit pauses to eat a carrot while the tortoise walks with a constant speed.

Strategy

The distance covered by the tortoise is the speed of the tortoise multiplied by the time during which it walks. That is, $d = vt$.

Known

$v = 2.51$ cm/s
$t = 52.23$ s

Unknown

$d = ?$

Solution

1 Multiply the speed by the time to find the distance d:

$$d = vt$$
$$= (2.51 \text{ cm/s})(52.23 \text{ s})$$
$$= 131.0973 \text{ cm}$$
$$= \boxed{131 \text{ cm}}$$

Insight

If we simply multiply 2.51 cm/s by 52.23 s, we obtain 131.0973 cm. We don't give all of these digits in our answer, however. Because the quantity that is known with the least accuracy (the speed) has only three significant figures, we give a result with three significant figures.

Practice Problems

23. Follow-up How long does it take for the tortoise to walk 17 cm? (*Hint*: Solve $d = vt$ for the time t.)

24. What is the area of a circle with a radius of 12.77 m? Recall that the area of a circle is given by area $= \pi(\text{radius})^2$.

25. The triangular sail on a boat has a height of 4.1 m and a base of 6.15 m. What is the area of the sail? Recall that the area of a triangle is given by area $= \frac{1}{2}(\text{base})(\text{height})$.

Rule for Addition and Subtraction When we add or subtract numerical values, we use a slightly different rule to determine the number of significant figures in the answer. In this case the rule involves the number of decimal places in each of the original values. ⬤ **The number of decimal places in a result from addition or subtraction is equal to the *smallest* number of decimal places in any of the input values.** Thus, if you make a time measurement of 16.74 s, and then a second time measurement of 5.1 s, the total time of the two measurements must have just one decimal place. Therefore, the correct answer is 21.8 s, rather than 21.84 s.

You and Brittany pick some raspberries. Your flat has a mass of 5.7 kg, and Brittany's has a mass of 3.25 kg. What is the combined mass of the raspberries?

Solution
Adding the two numbers gives 8.95 kg. According to the above rule, however, the final result must have only a single decimal place (corresponding to the value with the smallest number of decimal places). Rounding off to one decimal place, then, gives 9.0 kg as the correct result.

Practice Problems

26. On a fishing trip you catch a bass, a rock cod, and a salmon with masses of 1.07 kg, 6.0 kg, and 6.05 kg, respectively. What is the total mass of your catch?

27. What is the perimeter of a sheet of paper that is 25.2 cm tall and 18.1 cm wide?

Scientific notation specifies the number of significant figures

What is the main reason for using scientific notation?

Sometimes the number of significant figures in a numerical value is unclear because of zeros at the end of the number. For example, if a distance is stated to be 2500 m, the two zeros may be significant or they may simply indicate the location of the decimal point. If the zeros are significant, the measurement has four significant figures, and the uncertainty in the distance is roughly a meter. If the zeros simply locate the decimal point, the distance has only two significant figures. In this case the uncertainty is about 100 m. Writing numbers in scientific notation avoids this confusion.

Scientific notation expresses a value as a number between 1 and 10 times an appropriate power of 10. In our example we express the distance 2500 m as 2.5×10^3 m if there are two significant figures, or as 2.500×10^3 m if there are four significant figures. Likewise, a time given as 0.000036 s has only two significant figures—the four zeros preceding the digits 36 serve only to fix the decimal point. If the time were known to three significant figures, we would write it as 3.60×10^{-5} s to prevent any possible confusion.

Scientific notation is used mostly to save time and effort when writing out very large and very small numbers. For example, instead of writing the speed of light as 300,000,000 m/s, we write 3.0×10^8 m/s. Similarly, the mass of an electron is written as 9.1×10^{-31} kg. Writing this mass out with all of its zeroes would be quite an effort, and would likely lead to error. See Section I of Math Review for a more detailed discussion of scientific notation.

How many significant figures are there in **(a)** 21.00, **(b)** 21, **(c)** 2.1×10^{-2}, and **(d)** 2.10×10^{-3}?

Solution

(a) Four. The zeros to the right of the decimal point are significant, as are the 2 and the 1.
(b) Two. Nonzero numbers are always significant.
(c) Two. The number that precedes the power of 10 determines the number of significant figures. In this case that number has two significant figures.
(d) Three. The number preceding the power of 10 has three significant figures.

Practice Problems

28. How many significant figures are there in **(a)** 0.000054 and **(b)** 3.001×10^5?

29. The first six digits of the square root of 2 are 1.41421. What is the square root of 2 to four significant figures?

Rounding off can produce small discrepancies

Let's say you work on a numerical calculation. You perform all your calculations with great care and round to the appropriate number of significant figures, as described above. Even so, you find that your answer differs in its last digit from the one given in the book. What's going on? In most cases it's not a cause for concern at all—usually the discrepancy is due to unavoidable *round-off error*.

Round-off errors occur when numerical results are rounded off at different times during a calculation. To see how this works, suppose you buy a knickknack for $2.21, plus 8% sales tax. The total price is $2.3868, or, rounded off to the nearest penny, $2.39. Later, you buy another item for $1.35. With tax this becomes $1.458, or, again to the nearest penny, $1.46. The total expenditure for these two items is $2.39 + $1.46 = $3.85.

Now, let's do the rounding off in a different way. Suppose you buy both items at the same time for a total before-tax price of $2.21 + $1.35 = $3.56. Adding the 8% tax gives $3.8448, which rounds off to $3.84. This is one penny different from the previous amount!

This type of discrepancy can occur in physics problems. In general, it's a good idea to keep one extra digit throughout your calculations whenever possible, rounding off only the final result. But while this practice can help to reduce the likelihood of round-off error, there is no way to avoid it in every situation.

Order-of-magnitude estimates offer insight

An *order-of-magnitude calculation* is a rough estimate designed to be accurate to within the nearest power of 10. Such a ballpark estimate gives you a quick idea of the order of magnitude to expect from a complete, detailed calculation. If an order-of-magnitude calculation indicates an answer on the order of 10^4 m and your detailed solution gives an answer on the order of 10^7 m, then an error has occurred that needs to be resolved.

For example, suppose you want to estimate the speed of a cliff diver on entering the water. First, the cliff may be about 10 m high—certainly not 1 m or 10^2 m. Next, the diver hits the water something like a second later—certainly not 0.1 s later or 10 s later. Thus, a reasonable order-of-magnitude estimate of the diver's speed is

$$v = \frac{d}{t} = \frac{10 \text{ m}}{1 \text{ s}} = 10 \text{ m/s}$$

This is roughly 20 mi/h. If you do a detailed calculation and your answer is on the order of 10^4 m/s, you probably entered one of your numbers incorrectly.

ACTIVE Example 1.7   Estimate the Speed of Hair Growth

Give an order-of-magnitude estimate (in m/s) for the speed at which your hair grows after a haircut.

Solution (*Perform the calculations indicated in each step.*)

1. Identify the equation to use for an order-of-magnitude estimate of speed:	$v = \dfrac{d}{t}$
2. Estimate the rate of hair growth:	about 1 cm of growth in 1 month (The order of magnitude is 10^{-2} m.)
3. Determine the number of seconds in 1 month:	$t = 2.6 \times 10^6$ s (The order of magnitude is 10^6 s.)
4. Divide the order-of-magnitude distance (10^{-2} m) by the order-of-magnitude time (10^6 s) to obtain an estimate for the speed:	$v = 10^{-8}$ m/s

Insight
A more detailed calculation, using the correct number of seconds in a month, yields a speed of 3.9×10^{-9} m/s. This level of accuracy is not justified, however, because we are only estimating when we say "about 1 cm of growth in 1 month." The best we can say is that the speed of hair growth is on the order of 10^{-9} to 10^{-8} m/s.

Appendix C provides a number of typical values for length, mass, speed, acceleration, and many other quantities. You may find these to be useful in making your own order-of-magnitude estimates.

Graphs are useful for visualizing data

Suppose you measure the speed of a falling soccer ball at various times. You might find the speeds after 1, 2, 3, and 4 seconds to be 9.8 m/s, 19.6 m/s, 29.4 m/s, and 39.2 m/s, respectively. This is all well and good. Your results are presented in **Table 1.6**.

Linear Graphs Let's see what a graph of the data can tell us. The four data points in Table 1.6 are plotted in **Figure 1.12**. Notice that the points lie on a straight line. This means that the speed of the soccer ball increases *linearly* with time. That is, speed = constant × time:

$$v = (\text{constant})t$$

With this information we can predict the speed of the soccer ball at 7 seconds, 2.5 seconds, or any other time. In general, a **linear relationship** exists between two quantities that form a straight line on a graph. Notice the visual impact of a graph, how it conveys so much information at just a glance—more so than a simple list or table of numbers.

Table 1.6

Time (s)	Speed (m/s)
1	9.8
2	19.6
3	29.4
4	39.2

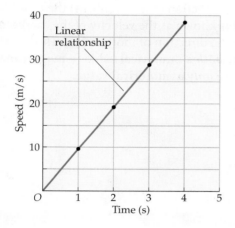

◀ **Figure 1.12 A linear relationship**
The data points in this plot indicate a linear relationship between the speed of a falling soccer ball and the time.

Parabolic Graphs Next, suppose we measure the distance covered by the soccer ball. We might find that it falls 4.9 m in 1 second, 19.6 m in 2 seconds, 44.1 m in 3 seconds, and so on. Plotting these points gives the graph shown in **Figure 1.13**. In this case the graph isn't linear. The upward-curving shape shown on the graph is a **parabola**. This shape indicates that the position depends on the time *squared*. That is, position = constant × time × time:

$$x = (\text{constant})t^2$$

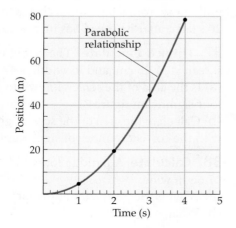

◀ **Figure 1.13 A parabolic relationship**
The data points in this plot indicate a parabolic relationship between the position of a falling soccer ball and the time.

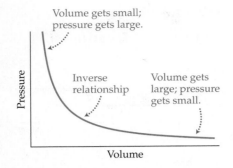

Volume gets small;
pressure gets large.

Inverse
relationship

Volume gets
large; pressure
gets small.

Pressure

Volume

▲ **Figure 1.14 An inverse relationship**
The data points in this plot indicate an
inverse relationship between the pressure
of a gas and its volume.

Inverse Graphs Finally, measurement of the volume of a container
and the pressure of the gas within it gives the graph shown in **Figure 1.14**.
This relationship is *inverse*. In an **inverse relationship**, as one quantity gets
larger, the other gets smaller. In this case, pressure = constant/volume:

$$P = \frac{\text{constant}}{V}$$

In general, a graph helps you see the relationship between physical quantities.
We shall make frequent use of graphs as we study physics.

Physical quantities can be scalars or vectors

Quantities known as **scalars** are represented by a number with a unit.
Examples include the temperature of an ice cube ($-10\ °C$) and the speed of
a car (25 m/s). Notice that **speed** is a scalar quantity that describes the rate
of motion.

Other quantities, known as **vectors**, consist of both a numerical value
with its unit and a direction. For example, when the speed of a car (25 m/s)
is combined with its direction of travel (north) the result is a vector (25 m/s
north). This vector is known as the velocity of the car. **Velocity** describes the
rate of motion *and* the direction of motion. ⏍ **The key thing to remember
is that vectors have *both* a numerical value (a magnitude) and a direction,
whereas scalars have *only* a numerical value.**

⏍ *How are vector quantities
different from scalar quantities?*

1.4 LessonCheck ⓂⓅ

Checking Concepts

30. ⏍ **Give Examples** Give three examples of a
physical quantity.

31. ⏍ **Identify** The height of a picture frame is
known to three significant figures, and the width is
known to two significant figures. How many significant
figures are there in the area of the picture frame?

32. ⏍ **Apply** How can a speed of 100 m/s be written so that it has three significant figures?

33. ⏍ **Compare and Contrast** How are speed and
velocity similar? How are they different?

34. Apply A poster is 0.95 m high and 1.0 m wide.
How many digits follow the decimal point when the
perimeter of the poster is expressed with the correct
number of significant figures?

Solving Problems

35. Apply The speed of light to five significant figures
is 2.9979×10^8 m/s. What is the speed of light to three
significant figures?

36. Calculate A school bus moves at 2.2 m/s as it
pulls away from a bus stop. What is the speed of the bus
20 s later, after its speed has increased by 5.225 m/s?

37. [**Think & Calculate**] The height of a computer
screen is 31.25 cm, and its width is 47 cm.
(**a**) Is the area of the screen known to one, two, three, or
four significant figures?
(**b**) Calculate the area of the screen, giving your answer
with the correct number of significant figures.

38. Calculate A parking lot is 144.3 m long and
47.66 m wide.
(**a**) What is the perimeter of the lot?
(**b**) What is its area?

1.5 Problem Solving in Physics

Physics is a lot like riding a bicycle—you have to learn by doing. You could read a book on riding bicycles and memorize every word in it, but when you get on a bicycle the first time, you are going to have problems. You have to practice with a bicycle to get the feel of it.

Similarly, you could read this book carefully and memorize every equation in it, but when you finished, you still wouldn't have learned physics. To learn physics, you have to go beyond passive reading—you have to interact with physics and experience it by doing problems.

Problem solving requires creativity and logic

Here we present a general overview of problem solving in physics. The suggestions given below, which apply to problems in all areas of physics, should help you develop a systematic approach.

We should emphasize at the outset that there is no recipe for solving problems in physics. Solving physics problems is a creative activity. *In fact, the opportunity to be creative is one of the attractions of physics.* The following suggestions, then, are not intended as a rigid set of steps that must be followed like the steps in a computer program. Rather, they provide general guidelines that experienced problem solvers find to be effective.

- **Read the problem carefully.** Before you can solve a problem, you need to know exactly what information it gives and what it asks you to determine. Some information is given directly, as when a problem states that a person has a mass of 70 kg. Other information is implied. For example, saying that a ball is dropped from rest means that its initial speed is zero. Clearly, a *careful* reading is the essential first step in problem solving.

- **Sketch the system.** This may seem like a step you can skip—but don't. A sketch helps you acquire a physical feeling for the system. It also provides an opportunity to label those quantities that are known and those that are to be determined. All the Guided Examples in this text begin with a sketch of the system, accompanied by a brief description, in a section labeled "Picture the Problem."

- **Visualize the physical process.** Try to visualize what is happening in the system as if you were watching it in a movie. Your sketch should help. This step ties in closely with the next step.

- **Strategize.** This is the most creative part of the problem-solving process. From your sketch and visualization, try to identify the physical processes at work in the system. Ask yourself what concepts or principles are involved in this situation. Then, develop a strategy—a game plan—for solving the problem. All the Guided Examples in this book have a "Strategy" section before the "Solution" section.

▲ The best way to learn physics is by doing physics. Doing physics involves conducting experiments, like the one shown here, as well as solving problems.

- **Identify appropriate equations.** Once you have developed a strategy, find the specific equations that are needed to carry it out.
- **Solve the equations.** Use basic algebra to solve the equations identified in the previous step. In general, you should work with symbols (such as x or v) as much as possible, substituting numerical values near the end of the calculations. Working with symbols will make it easier to go back over a problem to locate and identify mistakes, and to explore other cases of interest.
- **Check your answer.** Once you have an answer, check to see if it makes sense. Does your answer have the correct dimensions? Is its numerical value reasonable?
- **Explore limits and special cases.** Getting the correct answer is nice, but it's not all there is to physics. You can learn a great deal about physics and about the connection between physics and mathematics by checking various limits of your answer. For example, if you have two masses in your system, m_1 and m_2, what happens in the special case where $m_1 = 0$ or $m_1 = m_2$? Check to see whether your answer and your physical intuition agree.

This text has four distinct types of examples

This text includes four different example types, each with a distinct role in helping to increase your understanding of physics. They are described below.

Guided Examples The Guided Examples in this text are designed to deepen your understanding of physics and at the same time develop your problem-solving skills. They all have the same elements: Problem Statement, Picture the Problem, Strategy, Solution, and Insight. As you work through these examples in the chapters to come, notice how the basic problem-solving guidelines outlined above are implemented in a consistent way.

Active Examples The purpose of an Active Example is to act as a bridge between examples and homework problems. An analogy would be to think of examples as a tricycle, with no balancing required. Homework problems are like a bicycle, where balancing is initially difficult to master. Active Examples are like a bicycle with training wheels that give just enough help to prevent a fall.

Quick Examples Quick Examples present a short, simple calculation to aid in understanding a new equation that has been introduced. They are followed by Practice Problems that give further experience with the new equation.

Conceptual Examples When you're first learning physics, it's important to make connections between numerical calculations and basic concepts. Thinking through a physical situation, without doing a detailed calculation, helps you understand what the physics really means. This is the role of the Conceptual Examples. You can think of Conceptual Examples as helping with a conceptual understanding of physics in the same way that the other examples help with a numerical understanding.

Physics & You

Atmospheric Modeling and Weather Prediction

Physics in the Atmosphere The laws of physics apply everywhere and are constantly at work in Earth's atmosphere. At times these laws produce astonishing results, like the raw power of a tornado. However, understanding the laws of physics and using them to predict future weather patterns is no easy task.

Meteorologists and climatologists must gather and process huge amounts of information before they can make reliable weather predictions. Coming from satellites and thousands of other instruments around the world, the data include velocities (in meters per second), temperatures (in degrees Celsius), and pressures (in newtons per square meter). Scientists use this information to run computer simulations, or models, of weather patterns. The logical interpretation of these models allows scientists to make predictions about how the weather or climate might change.

Supercomputers and Society
Advances in computer technology have vastly improved the ability of meteorologists to make accurate predictions. These predictions help to protect the lives of millions of people each year. For example, prediction of a hurricane's path or the likelihood of a tornado allows communities to take steps to protect lives and property.

The National Oceanic and Atmospheric Administration (NOAA) uses supercomputers to analyze enormous amounts of data. These computers can make trillions of calculations per second and have allowed scientists to make ever more accurate and timely weather predictions. NOAA scientists are also using supercomputers to help them understand climate change. Accurate climate models could play a role in raising environmental awareness, shaping environmental laws, and guiding energy technology.

When a powerful tornado touches down, wind speeds over 100 m/s (~250 mph) are possible.

Take it Further

1. Hypothesize *Even with supercomputers processing a huge amount of real-time data, weather predictions are not 100% accurate. Suggest several reasons why this is the case.*

2. Infer *Is it necessary for the various pieces of pressure data processed by a computer to have the same units?*

Physics Lab Measuring Devices and Units

This lab compares the effectiveness of using body dimensions and standardized measuring sticks to measure the lengths of objects. Even though the lab does not use standard SI units, your results will allow you to conclude which approach produces more consistent results.

Materials
- yardstick
- foot-long ruler with inch markings
- various objects to measure

Procedure

According to history, units of measure were often derived from various parts of the body. For example, an inch was taken to be the width of a thumb, a foot was the length of a foot, and the span from fingertips to fingertips, with arms spread wide, was a yard. Today, we use *standardized* units of measure that are always the same. This lab will make you appreciate modern standardized units.

Part I: Establishing the Standards
The Inch
1. Place your thumb on a sheet of paper and record its width with two pencil marks. This represents *your* inch.

2. Use a modern foot-long ruler with inch markings to draw a 1-inch-long line next to the marks you made in Step 1. Make sure the line is exactly 1 inch long.

The Foot
3. Place your foot on a piece of paper and record its length with two pencil marks. This represents *your* foot.

4. Using a modern foot-long ruler, draw a 1-foot-long line next to the marks made in Step 3. Make sure the line is exactly 1 foot long.

The Yard
5. Stretch your arms out along a chalkboard or whiteboard and record the span using chalk or a marker. This is *your* yard.

6. Using a standardized ruler, draw a line that is exactly 1 yard long next to the marks in Step 5.

Part II: Using Your Standards
7. Using your body as your measuring tool, measure the items listed in the data table. You and your partner should decide which standard (inch, foot, or yard) to use to measure each object. You should each measure the same object using *your own* individual standards of measure. Record your data in the data table.

8. Repeat Step 7, this time using the modern standardized measures.

9. Share your data with your partner.

Data Table

Item	Length measured *by you* using *your* standard unit of measure	Length measured *by your partner* using *your partner's* standard unit of measure	Length measured *by you* using a modern standardized ruler	Length measured *by your* partner using a modern standardized ruler
Pencil				
The room				
Desk				
Paperclip				
Hallway				
Height of person				

Analysis

1. If you paid for carpet by the foot, would you get a better deal if you used your foot or the standard foot? Explain.

2. Compare your "inch" to that of your lab partner. How close were your "inches"?

3. What unit of measurement did you use to measure the length of the room? Why did you choose this unit?

4. Which of the items were difficult to measure in your units? Explain why.

5. How close did you and your partner come to getting the same measurements for the objects when you each used your personal units?

6. How close did you and your partner come to getting the same measurements for the objects when you each used standardized units?

Conclusion

1. Explain why it is important for scientists to use standard units of measure.

1 Study Guide

Big Idea

Physics applies to everything.

Physics is the study of the laws of nature. Everything in the natural world obeys these laws.

1.1 Physics and the Scientific Method

🔑 The validity of a hypothesis is based solely on its ability to account for known observations and to correctly predict new observations.

• Physics is based on a small number of fundamental laws and principles.

• The scientific method involves observations, a hypothesis, and experiments designed to verify the hypothesis. If the experimental results do not support the hypothesis, then the hypothesis must be revised.

1.2 Physics and Society

🔑 Scientists have adopted a common system of measurement, referred to as the *metric system*, that they use when collecting data and performing experiments.

• A bias is a belief based on personal preference, rather than objective facts.

• Peer review is a process in which a scientific report is recommended for publication only after a panel of experts in the field have ensured that the report meets standards of quality and scholarship.

1.3 Units and Dimensions

🔑 The standard "measuring stick" for a physical quantity is referred to as its base unit.

🔑 Standard prefixes are used to designate common multiples in powers of ten.

🔑 Dimensional consistency means that each term in a physics equation has the same dimensions. If a physics equation isn't dimensionally consistent, it isn't correct.

• One meter is the distance traveled by light in a vacuum in 1/299,792,458 second.

• One kilogram is the mass of a platinum-iridium cylinder kept at the International Bureau of Weights and Standards.

• One second is the time required for radiation from a cesium-133 atom to complete 9,192,631,770 oscillations.

• To convert from one unit to another, multiply by a conversion factor that relates the two units.

1.4 Basic Math for Physics

🔑 A physical quantity is a property of a physical system that can be measured, like length, speed, acceleration, or time duration.

🔑 For multiplication and division, the number of significant figures in the answer is the same as the number of significant figures in the least accurately known input value.

🔑 The number of decimal places in a result from addition or subtraction is equal to the *smallest* number of decimal places in any of the input values.

🔑 Scientific notation is used to save time and effort when writing out very large and very small numbers.

🔑 Vectors have *both* a numerical value (a magnitude) and a direction, whereas scalars have *only* a numerical value.

• Round-off errors are discrepancies caused by rounding off numbers at intermediate steps of a calculation.

• An order-of-magnitude calculation is a ballpark estimate designed to be accurate to within the nearest power of 10.

• Physical relationships can be displayed visually in a graph.

1.5 Problem Solving in Physics

• The following steps are general guidelines for solving problems.

> Read the problem carefully.
>
> Sketch the system.
>
> Visualize the physical process.
>
> Strategize.
>
> Identify appropriate equations.
>
> Solve the equations.
>
> Check your answer.
>
> Explore limits and special cases.

1 Assessment

ANSWERS TO SELECTED ODD-NUMBERED PROBLEMS APPEAR IN APPENDIX A.

Lesson by Lesson

1.1 Physics and the Scientific Method

Conceptual Questions

39. What is an inference? A hypothesis?

40. What does it mean to say that a quantity is conserved?

41. Does a force cause motion or a change in motion?

1.2 Physics and Society

Conceptual Questions

42. The metric system is based on powers of ten. What is the advantage to such a system?

43. If a scientific report is published after peer review, does that guarantee that its conclusions are correct? Explain.

1.3 Units and Dimensions

Conceptual Questions

44. Use dimensional analysis to determine which of the following expressions gives the area of a circle: Is it πr^2 or $2\pi r$? Explain.

45. If a distance d has units of meters and a time T has units of seconds, does the quantity $T + d$ make sense physically? What about the quantity d/T? Explain in both cases.

46. Is it possible for two quantities to **(a)** have the same units but different dimensions or **(b)** have the same dimensions but different units? Explain.

47. Which of the following quantities has the same dimension as a distance?

A. vt B. $\frac{1}{2}at^2$ C. $2at$ D. v^2/a

Note that v is speed, t is time, and a is acceleration. Refer to Table 1.5 for the corresponding dimensions.

48. Which of the following quantities has the same dimensions as a speed?

A. $\frac{1}{2}at^2$ B. at C. $(2x/a)^{1/2}$ D. $(2ax)^{1/2}$

Note that v is speed, t is time, and a is acceleration. Refer to Table 1.5 for the corresponding dimensions.

Problem Solving

49. A new movie earns $114,000,000 in its opening weekend. Express this amount in **(a)** gigadollars and **(b)** teradollars.

50. Peacock mantis shrimps (*Odontodactylus scyllarus*) feed largely on snails. They shatter the shells of their prey by delivering a sharp blow with their front legs, which have been observed to reach a peak speed of 23 m/s. What is this speed in kilometers per hour?

51. The largest building in the world by volume is the Boeing 747 plant in Everett, Washington. It measures approximately 631 m long, 646 m wide, and 34 m high. What is its volume in cubic centimeters?

52. Radiation from a cesium-133 atom completes 9,192,631,770 cycles each second. How long does it take for this radiation to complete 1.5 million cycles?

53. The blue whale (*Balaenoptera musculus*) is thought to be the largest animal ever to inhabit Earth. The longest blue whale ever observed had a length of 33 m. What is this length in millimeters?

54. A human hair has a thickness of about 70 μm. What is this thickness in **(a)** meters and **(b)** kilometers?

55. A supercomputer can do 136.8 teracalculations per second. How many calculations can it do in a microsecond?

56. The American physical chemist Gilbert Newton Lewis (1875–1946) proposed a unit of time called the *jiffy*. According to Lewis, 1 jiffy is the time it takes light to travel 1 centimeter. **(a)** If you perform a task in a jiffy, how long does it take in seconds? **(b)** How many jiffys are in 1 minute? Use the fact that the speed of light is approximately 3.00×10^8 m/s.

57. Suppose 1.0 m³ of oil is spilled into the ocean. Find the area of the resulting slick, assuming that it is one molecule thick and that each molecule occupies a cube 0.50 μm on a side.

58. The acceleration due to gravity is approximately 9.81 m/s² (depending on your location). What is the acceleration due to gravity in centimeters per second squared?

59. Velocity can be related to acceleration and distance by the following equation: $v^2 = 2ax^p$. Find the power p that makes this equation dimensionally consistent.

60. Acceleration is related to distance and time by the following equation: $a = 2xt^p$. Find the power p that makes this equation dimensionally consistent.

1.4 Basic Math for Physics

Conceptual Questions

61. Give an order-of-magnitude estimate for the time in seconds of the following: **(a)** a year, **(b)** a baseball game, **(c)** a heartbeat, **(d)** the age of Earth, **(e)** your age.

62. Give an order-of-magnitude estimate for the length in meters of the following: **(a)** your height, **(b)** a fly, **(c)** a car, **(d)** a jetliner, **(e)** an interstate highway stretching from coast to coast.

Problem Solving

63. The first several digits of π are known to be $\pi = 3.14159265358979\ldots$. What is π to **(a)** three significant figures and **(b)** five significant figures?

64. What is the area of a circle of radius 24.87 m?

65. Give a ballpark estimate of the number of seats in a typical Major League ballpark (see **Figure 1.15**). Show your reasoning.

Figure 1.15 Shea Stadium, in New York City.

66. Milk is often sold by the gallon in plastic containers. **(a)** Estimate the number of gallon containers of milk that are purchased in the United States each year. **(b)** What approximate weight of plastic does this represent?

67. New York City is roughly 4800 km from Seattle. When it is 10:00 A.M. in Seattle, it is 1:00 P.M. in New York. Using this information, estimate **(a)** the rotational speed of the surface of Earth, **(b)** the circumference of Earth, and **(c)** the radius of Earth.

Mixed Review

68. Which of the following equations are dimensionally consistent?

A. $v = at$ B. $v = \frac{1}{2}at^2$ C. $t = a/v$ D. $v^2 = 2ax$

Note that v is speed, t is time, and a is acceleration. Refer to Table 1.5 for the corresponding dimensions.

69. Which of the following quantities have the dimensions of an acceleration?

A. xt^2 B. v^2/x C. x/t^2 D. v/t

Note that v is speed, t is time, and a is acceleration. Refer to Table 1.5 for the corresponding dimensions.

70. The light that plants absorb to perform photosynthesis has a wavelength that peaks near 675 nm. Express this distance in **(a)** millimeters and **(b)** meters.

71. On June 9, 1983, the lower part of the Variegated Glacier in Alaska (**Figure 1.16**) was observed to be moving at a rate of 64 m per day. What is this speed in kilometers per hour?

Figure 1.16 Alaska's Variegated Glacier.

72. Male mosquitoes find female mosquitoes by listening for the characteristic "buzzing" frequency of the females' wingbeats. This frequency is about 605 wingbeats per second. **(a)** How many wingbeats occur in 1 minute? **(b)** How many cycles of oscillation does the radiation from a cesium-133 atom complete during one mosquito wingbeat? (See Problem 52.)

73. When Coast Guard pararescue jumpers leap from a helicopter to save a person in the water, they like to jump when the helicopter is flying "ten and ten," which means it is 10 feet above the water and moving forward with a speed of 10 knots. What is "ten and ten" in SI units? (A knot is 1 nautical mile per hour, and a nautical mile is 1.852 km.)

74. Type A nerve fibers in humans can conduct nerve impulses at speeds up to 140 m/s (see **Figure 1.17**). **(a)** How fast are the nerve impulses moving in kilometers per hour? **(b)** How far (in meters) can the impulses travel in 5.0 ms?

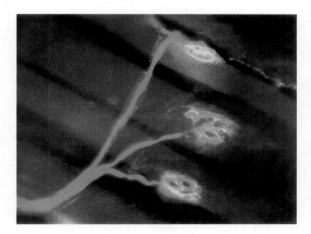

Figure 1.17 The impulses in these nerve axons, which carry commands to the skeletal muscle fibers in the background, travel at speeds of up to 140 m/s.

75. The mass of a newborn baby's brain has been found to increase by about 1.6 mg per minute. **(a)** How much does the brain's mass increase in 1 day? **(b)** How long does it take for the brain's mass to increase by 0.0075 kg?

76. On December 25, 2004, during a NASA mission to Saturn, the spacecraft *Cassini* released a probe named Huygens, which landed on the Saturnian moon Titan on January 14, 2005. Huygens was released from the main spacecraft at a gentle relative speed of 31 cm/s. As Huygens moved away, it rotated at a rate of seven revolutions per minute. **(a)** How many revolutions had Huygens completed when it was 150 m from *Cassini*? **(b)** How far did Huygens move away from *Cassini* during each revolution? Give your answer in meters.

77. Acceleration is related to velocity and time by the following expression: $a = vt^p$. Find the power p that makes this equation dimensionally consistent.

78. The period T of a simple pendulum is the amount of time required for it to undergo one complete oscillation. If the length of the pendulum is L and the acceleration due to gravity is g, then the period is given by

$$T = 2\pi L^p g^q$$

Find the powers p and q required for dimensional consistency.

Writing about Science

79. Write a short report on the metric system and SI units. What are some of the advantages of the metric system? What are some of the disadvantages of changing from a system like British units to the metric system? What are some of the countries that do and do not use the metric system?

80. **Connect to the Big Idea** Give several examples of how physics applies to chemistry, meteorology, and biology.

Read, Reason, and Respond

A Cricket Thermometer All chemical reactions, whether organic or inorganic, proceed at a rate that depends on temperature—the higher the temperature, the higher the rate of reaction. This can be explained in terms of molecules moving with increased energy as the temperature is increased and colliding with other molecules more frequently. In the case of organic reactions, the result is that metabolic processes speed up with increasing temperature.

An increased or decreased metabolic rate can manifest itself in a number of ways. For example, a cricket trying to attract a mate chirps at a rate that depends on the overall rate of its metabolism. As a result, the chirping rate of crickets depends directly on temperature. In fact, some people even use a pet cricket as a thermometer.

The cricket that is most accurate as a thermometer is the snowy tree cricket (*Oecanthus fultoni* Walker). Its rate of chirping is described by the following equation:

$$N = \text{number of chirps in 7.0 seconds}$$
$$= T - 5.0$$

In this expression, T is the temperature in degrees Celsius.

81. How many chirps will a snowy tree cricket give in 21 s at a temperature of 22 °C?

   A. 17   B. 22   C. 27   D. 51

82. Which plot in **Figure 1.18**—A, B, C, D, or E—represents the chirping rate of the snowy tree cricket?

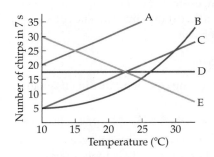

Figure 1.18 Chirping rate versus temperature.

83. If the temperature is 15 °C, how many seconds does it take for the cricket to chirp 80 times?

   A. 8   C. 56

   B. 10   D. 80

84. Your pet cricket chirps 113 times in 1 min. What is the temperature in degrees Celsius?

   A. 8.2   C. 18

   B. 13   D. 113

Standardized Test Prep

Multiple Choice

1. Which equation shows the kinetic-energy equation, $KE = \frac{1}{2}mv^2$, correctly solved for v?

   (A) $v = \dfrac{2(KE)}{m}$

   (B) $v = \sqrt{\dfrac{2(KE)}{m}}$

   (C) $v = \sqrt{2(KE)(m)}$

   (D) $v = \sqrt{(KE)(m)}$

2. What is the SI base unit of time?
   (A) the mean solar day
   (B) the second
   (C) the time it takes light to travel 10^8 m
   (D) the rate at which an atom oscillates

3. Which is *not* an SI base unit of measurement?
   (A) kilogram   **(C)** kilometer
   (B) second   **(D)** meter

4. Use dimensional analysis (see Table 1.5) to determine which of the following is *not* a valid equation. In these equations, d is distance, v is velocity, t is time, and a is acceleration.
   (A) $d = vt + \frac{1}{2}at^2$
   (B) $v = at$
   (C) $v^2 = \frac{1}{2}ad^2$
   (D) $v^2 = 2ad$

5. Which statement uses SI unit prefixes correctly?
   (A) 100 cm = 10 km
   (B) 2.5 kg = 2500 cg
   (C) 1000 mL = 1 dL
   (D) 150 cm = 1.5 m

6. Which quantity is *not* equivalent to 100 m?
   (A) 1.00×10^4 cm
   (B) 1.00×10^{-1} km
   (C) 1.00×10^2 m
   (D) 1.00×10^6 mm

Test-Taking Hint

When solving problems, use dimensional analysis to check your work and to detect mathematical or unit conversion errors.

7. Which is a reasonable order-of-magnitude estimate of the number of golf balls that would fill a home swimming pool?
   (A) 10^3
   (B) 10^5
   (C) 10^8
   (D) 10^{10}

Use the graph below to answer Questions 8 and 9.

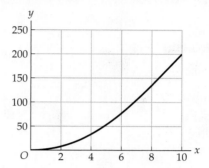

8. Which equation describes the line on the graph?
   (A) $y = 2x^2$
   (B) $y = 20x$
   (C) $x = 0.2y$
   (D) $y = \frac{1}{2}x^2$

9. Use the graph to estimate the value of y when $x = 3$.
   (A) $y = 0$
   (B) $y = 10$
   (C) $y = 20$
   (D) $y = 30$

Extended Response

10. In a laboratory experiment, students recorded the following length data: 12.2 cm, 12.1 cm, 12 cm, 11.9 cm, and 12.20 cm. **(a)** Determine the average length and express the answer using the correct number of significant figures. **(b)** Based on your average length calculation in part (a), discuss the importance of recording measurements to the appropriate number of significant figures.

If You Had Difficulty With . . .

Question	1	2	3	4	5	6	7	8	9	10
See Lesson	1.3	1.3	1.3	1.3	1.3	1.3	1.4	1.4	1.4	1.4

2

Introduction to Motion

Inside

These sprinters, running in a straight line to the finish, are engaging in one-dimensional motion. They are moving with constant velocity at a speed of more than 10 m/s.

Big Idea

Motion can be represented by a position-time graph.

Galileo started his studies of physics with the swinging motion of a pendulum, Newton started with the motion of falling objects, and Einstein considered objects moving at high speed. In each case motion played the central role. In this chapter you'll learn how to describe motion and how to represent it with graphs and equations.

Read the activity and obtain the required materials.

Explore

1. Identify the three cars as a push car, a pull-back car, and a battery-operated car.
2. Push the push car and observe its motion.

3. Gently push down on the pull-back car and pull it back 10–15 cm. Let the car go and observe its motion.
4. Turn on the battery-operated car and observe its motion.

Think

1. **Summarize** Write out a description of each car's motion using everyday language.
2. **Compare** How are the motions similar? How are they different?
3. **Cite** Give examples of everyday objects that have the same motion as each car.

2.1 Describing Motion

Motion is more than a common everyday experience—it's essential to life. You can't live without the food that grows on farms and is transported to your house. You can't live without the water that flows through the public water system into your drinking glass. We begin this chapter with a careful study of motion, the foundation of all of physics.

A coordinate system defines an object's position

How do you describe the motion of an object? For a soccer ball during a game, you might say, "The soccer ball rolled 25 meters in 2 seconds after it was kicked." For a car on the highway, you might say, "We drove 35 kilometers in half an hour." In either case, describing motion requires a frame of reference to specify how the object's position changes over time. In physics we refer to a frame of reference as a **coordinate system**.

An example of a coordinate system in one dimension is shown in **Figure 2.1** on the next page. This is simply an x axis, with an *origin* at the point where x is equal to zero ($x = 0$) and an arrow indicating the positive direction. When we say "positive direction," we mean the direction in which x increases. In setting up a coordinate system, you can choose the origin and the positive direction as you like. Once you've made your choice, however, you must use it consistently in any calculations that follow.

Vocabulary

- coordinate system
- distance
- displacement

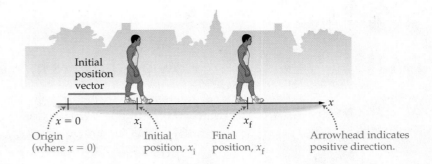

Symbols help to organize information

In general, an object's location is defined by a vector drawn from the origin to the object, as indicated in Figure 2.1. This is referred to as the object's *position vector*. As a label for the position vector—or position, for short—we use the letter x.

Any time an object is in motion, its position is changing. For example, the initial position of the "object" (person) in Figure 2.1 is x_i, where the subscript i stands for the *i*nitial value of a quantity. The person is moving to the right, and his final position is indicated with x_f, where the subscript f stands for the *f*inal value. Because the positive direction is to the right, the final position in Figure 2.1 is greater than the initial position. This means that x_f has a larger value than x_i.

Another coordinate system is shown in **Figure 2.2**. This one shows the positions of your house, your friend's house, and the local grocery store. The origin is at your friend's house, and the positive direction is from your friend's house toward the grocery store. As before, a position on this coordinate system is denoted by the symbol x.

Distance is the length traveled

When you walk from one location to another, it's interesting to know the length of your trip. Let's say you take a shopping trip to the local mall, and walk from store to store. Though you eventually return to your starting point at the car, you will have taken many steps and covered a path of considerable length in the meantime. In general, the total length of the path that is taken on a trip is called the **distance**.

> **Definition of Distance**
>
> distance = total length traveled
>
> SI unit: meter (m)

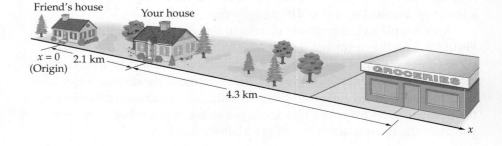

► **Figure 2.2 One-dimensional coordinates**
The locations of your house, your friend's house, and the grocery store in terms of a one-dimensional coordinate system.

As an example, suppose you leave your house in Figure 2.2, walk to the grocery store, and then return home. The distance you've covered is

$$4.3 \text{ km} + 4.3 \text{ km} = 8.6 \text{ km}$$

The distance you cover when walking is given by the reading on a **pedometer**. (In a car the distance is given by the odometer.) If you walk a mile, your pedometer reads 1 mile regardless of the direction in which you have walked. Therefore, distance has no direction associated with it—distance is a number (a *scalar*). In addition, your pedometer always gives a positive reading; that is, distance is always positive.

QUICK Example 2.1 What's the Distance?

Referring to Figure 2.2, what is the distance from the grocery store to your friend's house?

Solution
The total distance is 4.3 km to your house plus 2.1 km to your friend's house:

$$\text{distance} = 4.3 \text{ km} + 2.1 \text{ km}$$
$$= \boxed{6.4 \text{ km}}$$

The distance is positive (as it always is) even though your trip was in the negative direction.

Practice Problem

1. What is the distance of a trip from your house in Figure 2.2 to the grocery store, then to your friend's house, and finally back to your house?

Displacement is the change in an object's position

If you walk from your house to the store and back in Figure 2.2, the total distance is 8.6 km. Even so, you end up right where you started, with the same initial and final positions. On the other hand, if you walk 8.6 km in a straight line, your starting and ending positions are 8.6 km apart. How can we indicate not only the distance of your trip, but also the change in your position? To do so, we introduce a new quantity.

In physics, we define the **displacement** of an object to be the change in its position. We indicate the displacement with the symbol Δx.

Definition of Displacement, Δx

displacement = change in position
$$\Delta x = x_\text{f} - x_\text{i}$$

SI unit: meter (m)

The notation for displacement, Δx, does not mean Δ times x. It is simply a convenient shorthand notation that stands for $x_\text{f} - x_\text{i}$. It is read "delta x." For the two trips mentioned above, we say that $\Delta x = 0$ (delta x equals zero) for the round-trip to the store and that $\Delta x = 8.6 \text{ km}$ (delta x equals 8.6 kilometers) for the trip in a straight line.

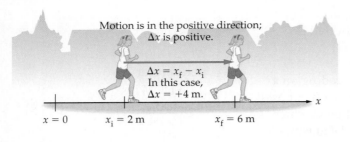

Motion is in the positive direction;
Δx is positive.

$\Delta x = x_f - x_i$
In this case,
$\Delta x = +4$ m.

$x = 0$ $x_i = 2$ m $x_f = 6$ m

(a)

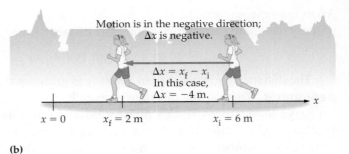

Motion is in the negative direction;
Δx is negative.

$\Delta x = x_f - x_i$
In this case,
$\Delta x = -4$ m.

$x = 0$ $x_f = 2$ m $x_i = 6$ m

(b)

▲ **Figure 2.3 Positive and negative displacements**
A person who moves in the positive direction has a positive displacement. Moving in the negative direction produces a negative displacement.

🔑 *How do distance and displacement differ?*

Notice that Δx can be positive or negative depending on the direction in which an object has moved:

- Δx is positive when the change in position is in the positive direction.
- Δx is negative when the change in position is in the negative direction.

Positive and negative displacements are illustrated in **Figure 2.3**.

Because displacement has both a magnitude and a direction, it is a *vector*. Its direction is given by its sign, as shown in Figure 2.3.

Distance and displacement are not the same

The SI unit of displacement is the meter—the same as for distance—but displacement and distance are really very different. Whenever you return to your starting point, your displacement is zero, even though you may have traveled a considerable distance. Similarly, you may move in the negative direction, in which case your displacement is negative even though the distance you covered is positive.

As an example, consider the school hallway shown in **Figure 2.4**. Suppose you go from the math classroom to the library, and then to the physics classroom. For this trip the distance is

$$\text{distance} = 8.0 \text{ m} + 5.0 \text{ m}$$
$$= 13.0 \text{ m}$$

On the other hand, the displacement is *negative*:

$$\Delta x = x_f - x_i$$
$$= 5.0 \text{ m} - 8.0 \text{ m}$$
$$= -3.0 \text{ m}$$

The minus sign means that your displacement is in the negative direction (to the left in this case). 🔑 **In general, distance is the total length traveled, and displacement is the net change in position.** These distinctions between distance and displacement should be kept in mind as we continue our study of motion.

▶ **Figure 2.4 The difference between distance and displacement**
The distance traveled in going from the math classroom to the library and then to the physics classroom is 13.0 m. In contrast, the displacement is −3.0 m.

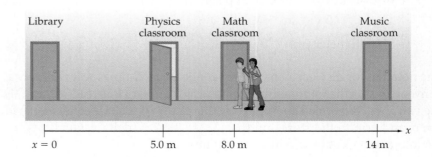

QUICK Example 2.2 What's the Displacement?

Referring to Figure 2.4, what is the displacement for a trip from the physics classroom to the music classroom and then to the math classroom?

Solution

To determine the displacement, we first identify the initial and final positions. The initial position is $x_i = 5.0$ m, and the final position is $x_f = 8.0$ m. Therefore, the displacement is

$$\Delta x = x_f - x_i$$
$$= 8.0 \text{ m} - 5.0 \text{ m}$$
$$= \boxed{3.0 \text{ m}}$$

The displacement is only 3.0 m, but the distance is 15 m.

Math HELP
Delta, Δ
See Math Review, Section I

Practice Problem

2. What is the displacement for a trip from the math classroom in Figure 2.4 to the music classroom and then to the library? What is the distance for this trip?

2.1 LessonCheck ⓜⓟ

Checking Concepts

3. 🔑 **Explain** (a) Is the distance on a round-trip positive, negative, or zero? (b) Is the displacement on a round-trip positive, negative, or zero?

4. Identify Does an odometer in a car measure distance or displacement? Explain.

5. Analyze Is it possible for you to take a hike and have the distance you cover be equal to the magnitude of your displacement? If yes, give an example to justify your answer.

6. Analyze You and your dog go for a walk to a nearby park. On the way your dog takes many short side trips to chase squirrels, examine fire hydrants, and so on.
(a) When you arrive at the park, do you and your dog have the same displacement? Explain.
(b) Have you and your dog traveled the same distance? Explain.

Solving Problems

7. Calculate A golfer putts on the eighteenth green at a distance of 5.0 m from the hole. The ball rolls straight, in the positive direction, but overshoots the hole by 1.2 m. The golfer then putts back to the hole and sinks the putt for par.
(a) What is the distance traveled by the ball?
(b) What is the displacement of the ball?

8. Calculate A billiard ball travels 22 cm in the positive direction, hits the cushion and rebounds in the negative direction, and finally comes to rest 7.5 cm behind its original position.
(a) What is the distance covered by the ball?
(b) What is the displacement of the ball?

9. Think & Calculate A train on a straight track goes in the positive direction for 5.9 km, and then backs up for 3.8 km.
(a) Is the distance covered by the train greater than, less than, or equal to its displacement? Explain.
(b) What is the distance covered by the train?
(c) What is the train's displacement?

2.2 Speed and Velocity

Vocabulary

- average speed
- average velocity

Life would be pretty dull if the objects around us stayed at the same locations all the time. Fortunately, this is not the case. Given that many objects do move, we would like to know more about the details of motion. For example, how fast does an object move? In what direction does it move? You'll learn how to answer these questions in this lesson.

Average speed is distance per time

The kangaroo in **Figure 2.5** is traveling rapidly across the Australian outback. How fast do you think the kangaroo is moving? How far can it travel in 1 minute? How much time does it take for it to cover 1 kilometer? All of these questions involve the rate of motion—that is, the distance covered divided by the time. In physics, the rate of motion is referred to as the *speed*. Speed simply describes how fast or slow something moves.

Knowing an object's speed is the next step in describing its motion. The simplest way to describe the rate of motion is with average speed. The **average speed** of an object is defined as its speed averaged over a given period of time.

> **Definition of Average Speed**
>
> $$\text{average speed} = \frac{\text{distance}}{\text{elapsed time}}$$
>
> SI units: meters per second (m/s)

The dimensions of average speed are distance per unit of time—or in SI units, meters per second (m/s). Like distance, average speed is *always* positive.

▶ **Figure 2.5 A speedy kangaroo**
This kangaroo can hop as fast as 65 km/h.

A kingfisher is a bird that catches fish by plunging into water from a height of several meters. If a kingfisher dives from a height of 7.0 m with an average speed of 4.00 m/s, how long does it take for it to reach the water?

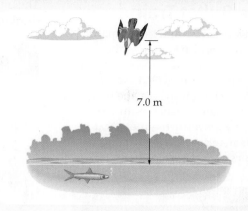

7.0 m

Picture the Problem

As shown in our sketch, the kingfisher moves in a straight line through a vertical distance of 7.0 m. The average speed of the bird is 4.00 m/s.

Strategy

We rearrange the average-speed equation to solve for the elapsed time:

$$\text{average speed} = \frac{\text{distance}}{\text{elapsed time}}$$

Known

distance = 7.0 m
average speed = 4.00 m/s

Unknown

elapsed time = ?

Solution

1 Rearrange the average-speed equation to solve for the elapsed time:

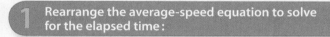

$$\text{average speed} = \frac{\text{distance}}{\text{elapsed time}}$$

$$\text{elapsed time} = \frac{\text{distance}}{\text{average speed}}$$

2 Substitute the numerical values for the distance and the average speed:

$$\text{elapsed time} = \frac{7.0 \ \text{m}}{4.00 \ \text{m/s}} = \boxed{1.8 \ \text{s}}$$

Insight

Notice that the equation average speed = distance/elapsed time is not just a formula for calculating the average speed. It relates the three quantities speed, time, and distance. Given any two of these quantities, we can find the third with this equation.

Practice Problems

10. Follow-up Suppose the kingfisher dives with an average speed of 4.6 m/s for 1.4 s before hitting the water. What was the height from which the bird dove?

11. It was a dark and stormy night, when suddenly you saw a flash of lightning. Three-and-a-half seconds later you heard the thunder. Given that the speed of sound in air is 340 m/s, how far away did the lightning bolt strike? Give your answer in both meters and kilometers.

12. The red kangaroo (*Macropus rufus*, shown in Figure 2.5) is the largest marsupial in the world. It has been clocked hopping at a speed of 65 km/h.

(a) How far (in kilometers) can a red kangaroo hop in 3.2 minutes at this speed?
(b) How much time will it take the kangaroo to hop 0.25 km at this speed?

13. Challenge A finch rides on the back of a Galapagos tortoise, which walks at the stately pace of 0.060 m/s. After 1.2 minutes the finch tires of the tortoise's slow pace, and it takes flight, traveling in the same direction for another 1.2 minutes at 13 m/s. What was the average speed of the finch over the entire 2.4-minute interval?

COOLPHYSICS
The Bullet Train

The world's fastest train is Japan's *Shinkansen*, also known as the *bullet train*. This train has reached a speed of 581 km/h (361 mph). To attain such high speeds, the train floats above its tracks, levitated there by powerful magnets. You will learn about the interesting physics of magnets in Chapter 22.

The following Conceptual Example considers the average speed of a trip consisting of two parts of equal distance, each traveled at a different speed. The question can be answered without a calculation by applying logical thinking and using the basic concept of average speed.

CONCEPTUAL Example 2.4 What Is the Average Speed?

You ride your bicycle 1 km at 10 km/h and then another 1 km at 30 km/h. Is your average speed for the total 2-km trip greater than, less than, or equal to 20 km/h?

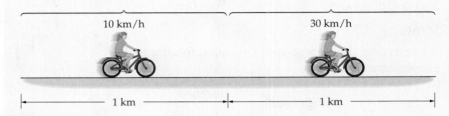

Reasoning and Discussion

At first glance it seems that your average speed must be 20 km/h (the average of 10 km/h and 30 km/h). Notice, however, that it takes more time to travel 1 km at 10 km/h than it does to travel 1 km at 30 km/h. Therefore, you are traveling at the lower speed for most of your trip. As a result, your average speed is closer to 10 km/h than to 30 km/h.

Answer

Your average speed is less than 20 km/h.

Average velocity is displacement per time

There is a quantity that is even more useful in describing motion than the average speed. An object's **average velocity** is defined as its displacement per unit of time. In other words, average velocity is displacement divided by elapsed time.

> **Definition of Average Velocity, v_{av}**
>
> $$\text{average velocity} = \frac{\text{displacement}}{\text{elapsed time}}$$
>
> $$v_{av} = \frac{\Delta x}{\Delta t} = \frac{x_f - x_i}{t_f - t_i}$$
>
> SI units: meters per second (m/s)

The average velocity tells us, on average, how fast something is moving. But it also tells us the *average direction* in which the object is moving. Like displacement, the average velocity is a *vector*, and for straight-line motion its direction is given by its sign:

- If an object moves in the positive direction, the average velocity is positive.
- If an object moves in the negative direction, the average velocity is negative.

For example, suppose object 1 has a velocity of 5 m/s and object 2 has a velocity of −5 m/s. Both objects are moving with a speed of 5 m/s. Their velocities have opposite signs, however, and therefore the objects are traveling in opposite directions.

In the next Guided Example, pay close attention to the positive and negative signs of the average velocity. They are the key to solving the problem.

An athlete sprints 50.0 m in 6.00 s, stops, and then walks slowly back to the starting line in 40.0 s. If the direction of the sprint is taken to be positive, what are (**a**) the average sprint velocity, (**b**) the average walking velocity, and (**c**) the average velocity for the complete round-trip?

Picture the Problem

In our sketch we set up a coordinate system with the sprint in the positive x direction, as described in the problem statement. For convenience we choose the origin to be at the starting line. The finish line, then, is at $x = 50.0$ m.

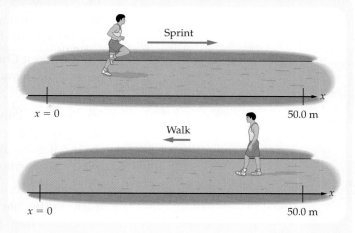

Strategy

In each part of the problem we are asked for the average velocity and we are given information on the distances and times. All that is needed, then, is to determine the displacement, $\Delta x = x_f - x_i$, and identify the elapsed time, $\Delta t = t_f - t_i$, in each case. With these results, we can use the defining equation, $v_{av} = \Delta x / \Delta t$, to calculate the average velocity.

Known

(**a**) $x_i = 0$ $x_f = 50.0$ m $\Delta t = 6.00$ s
(**b**) $x_i = 50.0$ m $x_f = 0$ $\Delta t = 40.0$ s
(**c**) $x_i = 0$ $x_f = 0$ $\Delta t = 46.0$ s

Unknown

$v_{av} = ?$

Solution

1 (**a**) Apply $v_{av} = \Delta x / \Delta t$ to the sprint, with $x_f = 50.0$ m, $x_i = 0$, and $\Delta t = 6.00$ s:

$$v_{av} = \frac{\Delta x}{\Delta t} = \frac{x_f - x_i}{\Delta t}$$

$$= \frac{50.0 \text{ m} - 0 \text{ m}}{6.00 \text{ s}}$$

$$= \boxed{8.33 \text{ m/s}}$$

> **Math HELP**
> Delta, Δ
> See Math Review, Section I

2 (**b**) Apply $v_{av} = \Delta x / \Delta t$ to the walk. In this case, $x_f = 0$, $x_i = 50.0$ m, and $\Delta t = 40.0$ s:

$$v_{av} = \frac{\Delta x}{\Delta t} = \frac{x_f - x_i}{\Delta t}$$

$$= \frac{0 \text{ m} - 50.0 \text{ m}}{40 \text{ s}}$$

$$= \boxed{-1.25 \text{ m/s}}$$

3 (**c**) For the round-trip, $x_f = x_i = 0$; thus, $\Delta x = 0$:

$$v_{av} = \frac{\Delta x}{\Delta t}$$

$$= \frac{0 \text{ m}}{46.0 \text{ s}}$$

$$= \boxed{0 \text{ m/s}}$$

Insight

Notice that the signs of the velocities in parts (a) and (b) indicate the directions of motion: positive for motion to the right, and negative for motion to the left. Also, notice that the average *speed* for the entire 100.0-m trip is 100.0 m/46.0 s = 2.17 m/s. Thus, the average speed is nonzero, even though the average velocity is zero.

14. Follow-up If the average velocity during the athlete's walk back to the starting line in Guided Example 2.5 is -1.50 m/s, how much time does it take for the athlete to walk back to the starting line?

15. Concept Check After a tennis match the two players dash to the net to shake hands. If they both run with a speed of 3 m/s, are their velocities equal? Explain.

16. In a well-known novel a person travels around the world in 80 days. (**a**) What is the person's approximate average speed during the adventure? (**b**) What is the approximate average velocity for the entire trip? (Note that Earth's circumference at the equator is 40,075 km.)

17. Rank Four trains travel on different sections of a long straight track. Taking north to be the positive direction, rank the trains in order by velocity, from most negative to most positive. The trains move as follows:

Train A moves north with a speed of 10 m/s.

Train B heads south and covers 900 m in 1 min.

Train C also heads south and has twice the speed of train A.

Train D travels north and covers 24 m in 2 s.

🗝 *What distinguishes velocity from speed?*

Average speed and average velocity are not the same

As we have seen, speed and velocity are different. In fact, they differ in the same way as distance and displacement. To make this clear, note that speed has the following characteristics:

Characteristics of Speed

- Speed is the rate of motion.
- Speed is always positive and gives no information about the direction of motion.
- The greater the speed of an object, the faster it moves.

The speedometer in a car, like the one shown in **Figure 2.6**, gives the speed of the car but not the direction of motion. It truly is a "speed meter" and not a velocity meter. For comparison, the characteristics of velocity are as follows:

Characteristics of Velocity

- Velocity gives both the rate of motion *and* its direction.
- The sign of the velocity gives the direction of motion.
- The magnitude of the velocity is the speed of motion.

🗝 **Thus, velocity gives the rate of motion *and* its direction, whereas speed gives only the rate of motion.**

When we say "the magnitude of the velocity," as in the last bulleted item above, we mean its absolute value. For example, we saw that the average velocity for the walking portion of the trip in Guided Example 2.5 was

▲ Figure 2.6 **Speedometer**
The speed of a car is measured with its speedometer. The speed does not depend on the direction of travel, and therefore the meter measures only speed and not velocity.

-1.25 m/s. The absolute value of this velocity is 1.25 m/s. Thus, the average speed of the athlete when walking was 1.25 m/s—the magnitude of the average velocity. This relationship is just like that between distance and displacement. A displacement of -3 m corresponds to a straight-line distance of 3 m, the magnitude of the displacement.

2.2 LessonCheck ⓂⓅ

Checking Concepts

18. 🔒 **Describe** What is the main difference between velocity and speed?

19. State What are the SI units of speed?

20. Analyze Friends tell you that on a recent trip their average velocity was $+20$ m/s. Is it possible that at any time during the trip their velocity was -20 m/s? Explain, and give an example to justify your answer.

21. Triple Choice Suppose you ride a bicycle around the block, returning to your starting point. At the end of your trip, is your average speed greater than, less than, or equal to the magnitude of your average velocity? Explain.

Solving Problems

22. Calculate In 2009, Usain Bolt of Jamaica set a world record of 9.58 s in the 100-m dash. What was his average speed? Give your answer in meters per second and kilometers per hour.

23. Calculate Radio waves travel at the speed of light, approximately 300,000,000 m/s. How much time does it take for a radio message to travel from the Earth to the Moon and back? (See the inside back cover for the necessary astronomical data.)

24. Think & Calculate A train travels in a straight line at 20.0 m/s for 2 km, then at 30.0 m/s for another 2 km.
(**a**) Is the average speed of the train greater than, less than, or equal to 25 m/s? Explain.
(**b**) Verify your answer to part (a) by calculating the average speed.

25. Verify Determine the average speed of the bicycle in Conceptual Example 2.4 for the entire 2-km trip. Verify that it is less than 20 km/h, as expected.

2.3 Position-Time Graphs

Vocabulary

• slope

How is a position-time graph drawn?

Table 2.1 Position and Time Data

Time (s)	Position (m)
0.0	0.0
1.0	0.5
2.0	1.0
3.0	1.5
4.0	2.0
5.0	2.5

Visualization is often helpful in sports. If you visualize yourself hitting a home run or kicking a soccer ball into the net, it may help you accomplish these feats in real life. It should come as no surprise, then, that visualization is also helpful in science. To see the quantities position and time plotted on a graph is useful in understanding the motion the graph represents. You'll learn how this is done in this lesson.

Graphs help visualize motion

Let's start with a simple example of motion. Consider a person in a wagon being pulled from one side of your classroom to the other. One way to describe the motion of the wagon is to record its position over a period of time. **Table 2.1** shows times and positions for a typical experiment with the wagon.

A table like Table 2.1 contains a lot of information, but it's not much use in visualizing the motion. A graph of the same data, however, helps us visualize the motion. **Plotting the position data on the *y* axis and the time data on the *x* axis creates a *position-time graph* of the motion.** The position-time graph of the wagon's motion is shown in **Figure 2.7**. Notice how each pair of data points (such as 3.0 s and 1.5 m) corresponds to a single plotted point, (x, y), on the graph.

The visual impact of the position-time graph in Figure 2.7 is obvious. The straight line that slants upward on the graph corresponds exactly to the increasing distance covered by the wagon. The steady climb of the straight line represents the constant speed of the wagon.

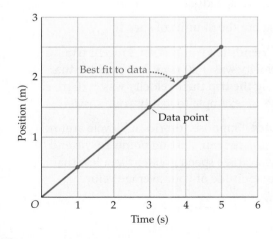

▲ Figure 2.7 **A position-time graph**
Plotting the position and time data from Table 2.1 produced this position-time graph. The best-fit line drawn through the set of plotted data points is a straight line.

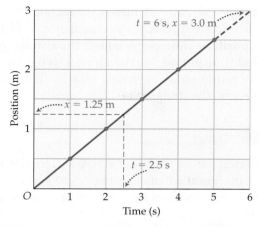

▲ **Figure 2.8 Gathering additional information from a graph**
We can use the best-fit line for our data points to gain additional information. To find the position at $t = 2.5$ s (not an original data point), for example, we trace vertically from 2.5 s on the time axis to the straight line, then trace horizontally to the position axis. This gives $x = 1.25$ m. Similarly, extending the best-fit line gives approximate positions for later times.

Additional data can be obtained from a graph

We plotted only six data points in Figure 2.7, but much more information than this can be obtained from the graph. For example, suppose you would like to estimate the wagon's position at the time 2.5 s, which is not one of the original data points used to make the graph. To do this you simply find 2.5 s on the x axis (time axis) and trace from that point upward until you reach the straight line. Then, trace sideways from there until you reach the y axis (position axis). As **Figure 2.8** shows, the wagon's position at 2.5 s is 1.25 m.

Similarly, you can extend the best-fit line to times greater than 5.0 s. In this way you can approximate the position of the wagon at later times—assuming you don't run out of space in the classroom. For example, if you extend the straight line in Figure 2.8 to 6.0 s (as shown by the dashed part of the line), you will find that the corresponding position is 3.0 m. Give it a try!

Average velocity can be determined graphically

You can use a position-time graph to determine the average velocity of a moving object. To see how, examine the version of the wagon's position-time graph shown in **Figure 2.9**. The wagon's motion produces a straight line with a positive slope. The **slope** of a line is equal to its rise over its run:

$$\text{slope} = \frac{\text{rise}}{\text{run}}$$

From the graph in Figure 2.9, we see that the rise corresponds to the vertical axis, which is the wagon's position, x, in meters. Likewise, the run corresponds to the horizontal axis, which is the elapsed time, t, in seconds. Therefore,

$$\text{slope} = \frac{\text{rise}}{\text{run}} = \frac{\text{change in vertical axis value}}{\text{change in horizontal axis value}} = \frac{\Delta x}{\Delta t}$$

You can use any two points you like on a straight line to calculate its slope. As an example, let's try the points (1.0 s, 0.5 m) and (4.0 s, 2.0 m). This gives

$$\text{slope} = \frac{\Delta x}{\Delta t} = \frac{(2.0 \text{ m} - 0.5 \text{ m})}{(4.0 \text{ s} - 1.0 \text{ s})} = \frac{1.5 \text{ m}}{3.0 \text{ s}} = 0.50 \text{ m/s}$$

Try another set of points to see that you get the same value.

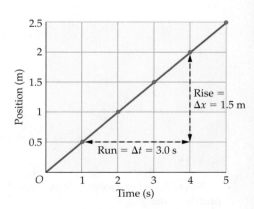

▲ **Figure 2.9 Constant velocity corresponds to constant slope on a position-time graph**
The slope of a position-time graph represents the average velocity. Because the slope is the same anywhere along a straight line, the average velocity for motion corresponding to a straight line is constant.

How can average velocity be determined from a position-time graph?

Notice that the units of slope are meters per second (m/s), the same as the units of average velocity. This is not a coincidence. **The slope of a straight line on a position-time graph equals the average velocity of the motion during that time period.** Furthermore, a straight line on a position-time graph represents motion with *constant* velocity. The velocity (slope) is the same everywhere on the straight line.

Practice Problem

26. A rose-covered parade float is at $x = 0$ at time $t = 0$. The float moves in a straight line at 2.0 m/s for the next 5 s before coming to a stop. After a 5-s stop, the float moves again at 1.0 m/s in the same direction as before.
(**a**) Sketch the position-time graph for the float from the time $t = 0$ until the time $t = 15$ s.
(**b**) From your graph, determine the positions of the float at $t = 2$ s and $t = 11$ s.

Different slopes represent different types of motion

A straight line on a position-time graph can have positive, negative, or zero slope. The positive slope in Figure 2.9 means that the wagon moves with a positive velocity. Not surprisingly, a negative slope on a position-time graph means that the object moves with a negative velocity. What about a horizontal line? A horizontal line results when the position does not change over time. Thus, a horizontal line has zero slope, which corresponds to zero average velocity.

The steepness of the slope on a position-time graph tells us even more about the object's motion. The steeper the slope, the greater the average speed. Clearly, you can tell a lot about an object's motion just by looking at its position-time graph. **Figure 2.10** summarizes the meanings of the different slopes you will see on position-time graphs.

Recall that the average velocity is the velocity averaged over a period of time, Δt. If the interval of time is made smaller and smaller, the average velocity approaches the velocity at a given instant of time. This is referred to as the *instantaneous velocity* of an object. The instantaneous velocity can be obtained from a position-time graph by drawing a *tangent* line to the graph at a given instant of time. The slope of the tangent line equals the instantaneous velocity at that time.

▶ **Figure 2.10 Interpreting slopes on a position-time graph**
Velocity corresponds to the slope on a position-time graph. A positive slope means a positive velocity, and the greater the slope, the greater the velocity. Zero slope (a horizontal line) means zero velocity, and negative slope means negative velocity.

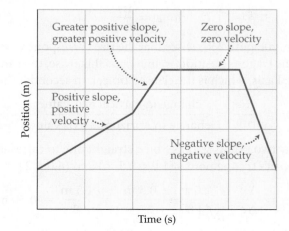

2.3 LessonCheck ⒨ℙ

Checking Concepts

27. 🔘 **Identify** What physical quantities are measured along the axes of a position-time graph?

28. 🔘 **Explain** What does the slope of a position-time graph tell you about the motion of an object?

29. **Big Idea** What does the position-time graph look like for an object moving with constant velocity?

30. **Rank** A tennis player moves back and forth along the baseline while waiting for her opponent to serve, producing the position-time graph shown in **Figure 2.11**.

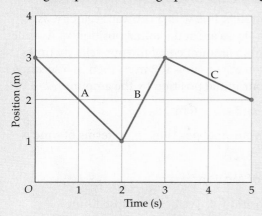

▲ **Figure 2.11**

(a) Without performing a calculation, rank the segments of the graph—A, B, and C—in order of increasing average speed.
(b) Rank the three segments of the graph in order of increasing average velocity, from most negative to most positive.

31. **Rank** The position-time graphs in **Figure 2.12** show four different motions, labeled A, B, C, and D.

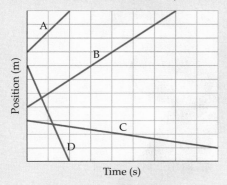

▲ **Figure 2.12**

(a) Rank these motions in order of increasing speed.
(b) Rank these motions in order of increasing velocity, from most negative to most positive.

Solving Problems

32. **Interpret Graphs** Referring to Problem 30 and Figure 2.11, what is the average velocity of the tennis player between $t = 0$ and $t = 5$ s?

33. **Interpret Graphs** A small-gauge train moves slowly back and forth along a straight segment of track. The position-time graph for the train is shown in **Figure 2.13**.

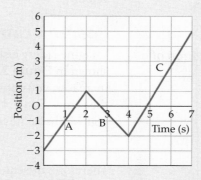

▲ **Figure 2.13**

(a) During which portion of its motion—A, B, or C—does the train have the smallest speed?
(b) What is this smallest speed?

34. **Interpret Graphs** Referring to Problem 33, what is the average velocity of the train from $t = 0$ to $t = 7.0$ s?

2.4 Equation of Motion

This lesson introduces the important idea that events happening in the real world can be described by mathematical equations. Once you understand the underlying concepts, the equations are easy to apply. We begin with a simple equation describing the position of an object moving with constant velocity.

Position can be described as a function of time

🔑 **What does the equation of motion tell you?**

Imagine riding a skateboard in a straight line with a constant velocity, as shown in **Figure 2.14**. At time $t = 0$ you are at the initial position x_i. As time increases, your position changes by an amount equal to your velocity times the elapsed time. For example, if you are moving at 2 m/s, then in 3 seconds you will have moved away from your initial position by the amount

$$(2 \text{ m/s})(3 \text{ s}) = 6 \text{ m}$$

Therefore, the equation that gives your final position, x_f, in terms of time is the following:

> **Position-Time Equation of Motion**
>
> final position = initial position + (velocity)(elapsed time)
> $$x_f = x_i + vt$$
>
> SI unit: m

🔑 **As you can see, if the initial position and the constant velocity of an object are known, the position-time equation of motion gives its position x at any time t.** A graph of the equation of motion is shown in **Figure 2.15**.

▶ **Figure 2.14 A skateboarder's position in terms of elapsed time**
The position of a skateboarder as a function of time can be found using the position-time equation of motion.

$x = 0$

$x = 1.5 \text{ m}$

x

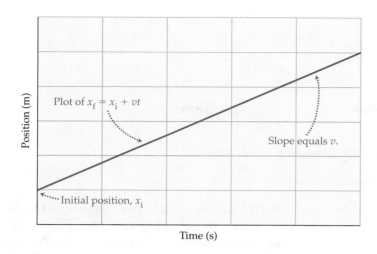

Figure 2.15 The position-time equation of motion
The position-time equation of motion gives a straight line for the case of constant velocity. The intercept of the straight line is the initial position, and its slope is the velocity.

QUICK Example 2.6 What's the Equation of Motion?

A skateboarder with an initial position of 1.5 m moves with a constant velocity of 3.0 m/s. **(a)** Write the position-time equation of motion for the skateboarder. **(b)** What is the position of the skateboarder at $t = 2.5$ s?

Solution
(a) Substitute 1.5 m for the initial position and 3.0 m/s for the constant velocity in the equation of motion. This gives

$$x_f = 1.5 \text{ m} + (3.0 \text{ m/s})t$$

(b) Substitute $t = 2.5$ s into the equation in part (a).

The corresponding position is $x_f = 9.0$ m.

Practice Problems

35. The position-time equation of motion for a bunny hopping across a yard is

$$x_f = 8.3 \text{ m} + (2.2 \text{ m/s})t$$

(a) What is the initial position of the bunny?
(b) What is the bunny's velocity?

36. A bowling ball moves with constant velocity from an initial position of 1.6 m to a final position of 7.8 m in 3.1 s.
(a) What is the position-time equation for the bowling ball?
(b) At what time is the ball at the position 8.6 m?

The equation for constant velocity is a straight line

You may recall from math class that the equation for a straight line is often written as $y = mx + b$. This is known as the slope-intercept form of the line equation. In this equation, m is the slope of the line and b is the y intercept. An equivalent way to write this equation is with a slight rearrangement, as $y = b + mx$. With some minor changes of labels, this is the same equation we have written for motion with constant velocity.

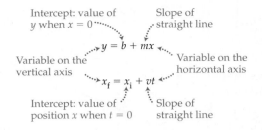

Intercept: value of y when $x = 0$ ⟶ Slope of straight line

$$y = b + mx$$

Variable on the vertical axis ⟶ Variable on the horizontal axis

$$x_f = x_i + vt$$

Intercept: value of position x when $t = 0$ ⟶ Slope of straight line

▲ **Figure 2.16 Comparing straight lines**
The position-time equation of motion for constant velocity corresponds exactly to the straight-line equation studied in math class.

🔑 **How is motion with constant velocity related to the equation for a straight line?**

A detailed comparison of the slope-intercept equation and the equation of motion for constant velocity is shown in **Figure 2.16**. 🔑 **Everything you know about the slope-intercept line equation from math class applies equally well to the equation of motion for constant velocity.** All we've changed are the letters in the equations.

Practice Problem

37. Triple Choice The position-time graph in **Figure 2.17** shows the motion of two cars, A and B.

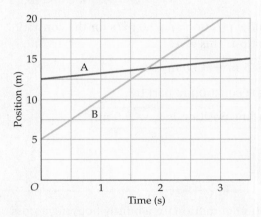

▲ **Figure 2.17**

(**a**) Is the velocity of car A greater than, less than, or the same as the velocity of car B?
(**b**) Is the initial position of car A greater than, less than, or equal to the initial position of car B?
(**c**) In the time period from $t = 0$ to $t = 1$ s, is car A ahead of car B, behind car B, or at the same position as car B?

Reading Support ✓
Origin of
inter-
a prefix used in many Latin-based words, meaning *between, among,* or *during*
(examples) intersect, interactive, international
An international airport has flights to destinations in foreign countries.

Position-time graphs intersect when objects have the same location

You may have solved problems in math class where you find the point at which two lines **intersect**. The point of intersection is found by setting the equations for the two lines equal to one another. The same process is often used to solve motion problems in physics. In physics, though, the equations represent real-life objects in motion, such as the travelers in the following Guided Example.

You and a friend are at the airport, hurrying to the gate for your vacation flight to Hawaii. Your friend is well ahead and seems sure to get to the gate first. Just then you spot a moving walkway going your way. At the moment you step onto the walkway, your friend is 7.8 m ahead and walking with a speed of 2.30 m/s. Your speed on the walkway is 4.20 m/s. Show that you catch up with your friend at $t = 4.10$ s, and determine the location the two of you share at that time.

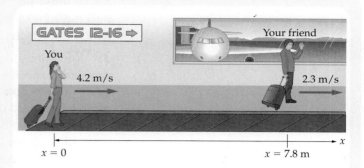

Picture the Problem

Our sketch shows you (red bag) and your friend (blue bag) moving in the same direction, which we take to be the positive direction. We choose $x = 0$ to be the point where you step onto the walkway. The corresponding time is $t = 0$. At that time your friend is at the position $x = 7.8$ m. The velocities are also indicated: 4.20 m/s for you and 2.30 m/s for your friend.

We also show a position-time graph with two straight lines, one for you (red) and one for your friend (blue).

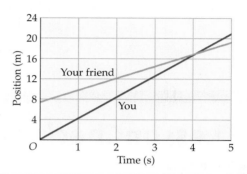

Strategy

Start by writing the position-time equation of motion for yourself. Substitute $t = 4.10$ s and evaluate the position. Repeat the process for your friend, and check to see that the positions are indeed the same.

Known

you: $x_i = 0$ $v = 4.20$ m/s
your friend: $x_i = 7.8$ m $v = 2.30$ m/s

Unknown

location where you catch up $= ?$

Solution

1 Write the equation of motion for your final position, noting that $x_i = 0$ and $v = 4.2$ m/s:

$$x_f = x_i + vt$$
$$= 0 + (4.20 \text{ m/s})t$$
$$= (4.20 \text{ m/s})t$$

> **Math HELP**
> Significant Figures
> **See Lesson 1.4**

2 Evaluate your final position at $t = 4.1$ s:

$$x_f = (4.20 \text{ m/s})(4.10 \text{ s})$$
$$= 17.2 \text{ m}$$

3 Write the equation of motion for your friend's final position, noting that $x_i = 7.8$ m and $v = 2.3$ m/s:

$$x_f = x_i + vt$$
$$= (7.8 \text{ m}) + (2.30 \text{ m/s})t$$

4 Evaluate your friend's final position at $t = 4.1$ s:

$$x_f = 7.8 \text{ m} + (2.30 \text{ m/s})(4.10 \text{ s})$$
$$= 7.8 \text{ m} + 9.43 \text{ m} = \boxed{17.2 \text{ m}}$$

Insight

You catch up with your friend in 4.10 s at the position $x = 17.2$ m, that is, 17.2 meters from the point where you stepped onto the moving walkway. These results agree with the intersection point shown in the above position-time graph.

38. **Follow-up** Suppose in Guided Example 2.7 that your friend stops walking 2.2 s after you step onto the moving walkway. How long does it take you to catch up with your friend in this case?

39. You are riding in a car on a straight stretch of a two-lane highway with a speed of 26 m/s. At a certain time, which we will choose to be $t = 0$, you notice a truck moving toward you in the other lane. The truck has a speed of 31 m/s and is 420 m away at $t = 0$.
(**a**) Write the position-time equations of motion for your car and for the truck in the other lane.
(**b**) Plot the two equations of motion on a position-time graph.
(**c**) At what time do you and the truck pass one another, going in opposite directions?

2.4 LessonCheck _{MP}

Checking Concepts

40. **Describe** What type of information can be obtained from the position-time equation of motion?

41. **Explain** A straight line on a position-time graph represents motion with constant velocity. What does the slope of the line represent? What does the y intercept of the line represent?

42. **Analyze** Two bicycles have the following equations of motion:

$$x_1 = -4.0 \text{ m} + (2.5 \text{ m/s})t$$
$$x_2 = 6.7 \text{ m} + (-3.5 \text{ m/s})t$$

Which bicycle has the greater speed?

Solving Problems

43. **Calculate** Referring to Problem 42, what is the distance between the bicycles at $t = 0$? At $t = 1$ s?

44. **Calculate** The position of a ball as a function of time is given by

$$x = 3.0 \text{ m} + (-5.0 \text{ m/s})t$$

What is the position of the ball at 1.5 s?

45. **Think & Calculate** Two bumper cars move in a straight line with the following equations of motion:

$$x_1 = -4.0 \text{ m} + (1.5 \text{ m/s})t$$
$$x_2 = 8.8 \text{ m} + (-2.5 \text{ m/s})t$$

(**a**) Which bumper car is has the greater speed?
(**b**) At what time do the bumper cars collide?

Physics & You

Climate Modelers

Climate describes the weather conditions in a region over a long period of time. Interactions between ocean, land, biosphere, and atmosphere produce Earth's climates. Climate modelers use complex computer models to simulate past and future climatic conditions. Leaders in governments, communities, and a range of industries study the predictions made by such climate simulations to help them plan how to deal with possible future climate fluctuations.

Climate modelers work closely with other scientists, including oceanographers, atmospheric chemists, and meteorologists, to obtain the most accurate environmental data. The data are used in the complex mathematical calculations of the climate model, which are carried out by supercomputers. The results are resolved over a grid that covers Earth's surface. The points that lie between the grid lines are filled in through *interpolation*, a mathematical technique for estimating values that lie between known data points.

Climate modeling is a rapidly changing field. Every new generation of climate model incorporates even more accurate environmental data, and the simulations are refined to finer and finer grids. A climate modeler must stay informed about the latest modeling techniques and technologies.

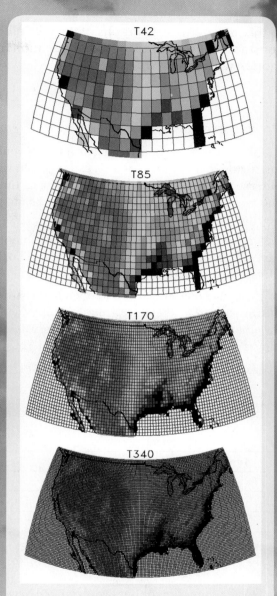

▲ The amount of detail in climate models has increased in recent years, largely because of the calculating power of newer supercomputers. These images show how the precision of the model is improved as the grid size is reduced.

Take It Further

1. Infer *What types of data do you think climate modelers use as input for their simulation programs?*

2. Assess *How might the predictions of a climate model help a country and its citizens plan for possible changes in climate?*

3. Summarize *Research the basic physics principles that affect the weather, and then use what you learn to infer how this knowledge is used by climate modelers.*

Physics Lab

Position versus Time for a Constant-Velocity Car

In this lab you will use graphs of the motion of a constant-velocity car to derive an equation for motion with constant velocity.

Materials
- stopwatch
- metric measuring tape
- masking tape
- three constant-velocity cars

Procedure

1. Obtain three constant-velocity cars from your instructor. Identify the cars as 1, 2, and 3. You will observe the motion of each car, recording its time and position, and then use the recorded data to plot a position-time graph.

2. Stretch out the measuring tape along a smooth flat surface. Use masking tape to secure each end of the measuring tape to the surface.

3. Designate one lab partner to operate the stopwatch during each trial. The other group members will observe the position of the car at 2-s intervals as it moves along the measuring tape.

4. Turn on the car. Position the car on the measuring tape as specified by your teacher and start the timer. Observe the position of the car every 2 s for a total of 10 s. Record your data in the data table.

5. Repeat Step 4 for the each of the other two cars. Make sure to start each car at the location specified by your instructor.

Analysis

1. Write a description of each car's motion during its trial. Include both the direction of motion and the speed.

2. Use your recorded data to plot position-time graphs for all three cars on the same axes. Plot the data points for each car in a different color. Draw a best-fit straight line for each set of points.

Data Table

Car 1 Starting Position: ___ m		Car 2 Starting Position: ___ m		Car 3 Starting Position: ___ m	
Time (s)	Position (m)	Time (s)	Position (m)	Time (s)	Position (m)

Conclusions

1. Do all of the straight lines you graphed have the same y intercept? What is the significance of the y intercept?

2. By looking at the graphs, how can you tell which car was the fastest?

3. By looking at the graphs, how can you tell which car went backward?

4. Determine the equation of each straight line. Write each equation in slope-intercept form:

$$y = mx + b$$

5. Predict where car 2 would be after 16 s if you had allowed it to continue to run.

6. Write the equation of motion for a car that is placed at the 4.7-m position on the measuring tape and travels backward at a speed of 0.4 m/s.

7. Generalize your equations from Question 4 by writing a single general equation in terms of the following variables: v = velocity, t = time, x_i = initial position, and x_f = final position.

2 Study Guide

Big Idea

Motion can be represented by a position-time graph.

A graph of position versus time is useful in analyzing the motion of an object. The position-time graph for an object moving with constant velocity is a straight line, with a slope equal to the velocity and a y intercept equal to the initial position.

2.1 Describing Motion

🔑 In general, distance is the total length traveled, and displacement is the net change in position.

• The total length of the path that is taken on a trip is called the *distance*.

• The displacement of an object is the change in its position.

• Displacement is calculated by subtracting the *initial* position from the *final* position.

Key Equation

Displacement, Δx, is the change in position:

$$\Delta x = x_f - x_i$$

2.2 Speed and Velocity

🔑 Velocity gives the rate of motion and its direction, whereas speed gives only the rate of motion.

• The rate of motion is referred to as the *speed*.

• Average velocity is positive if the displacement is in the positive direction or negative if the displacement is in the negative direction.

Key Equation

Average velocity, v_{av}, is displacement divided by time:

$$v_{av} = \frac{\Delta x}{\Delta t} = \frac{x_f - x_i}{t_f - t_i}$$

2.3 Position-Time Graphs

🔑 Plotting the position data on the y axis and the time data on the x axis creates a position-time graph of the motion.

🔑 The slope of a straight line on a position-time graph equals the average velocity of the motion during that time period.

• A position-time graph that is a straight line corresponds to motion with constant velocity.

2.4 Equation of Motion

🔑 If the initial position and the constant velocity of an object are known, the position-time equation of motion gives its position x at any time t.

🔑 Everything you know about the slope-intercept line equation from math class applies equally well to the equation of motion for constant velocity.

• The intersection of the graphs of two equations of motion for two different objects represents the time at which the objects are at the same location.

Key Equation

The position of an object at any time is described by the equation of motion:

$$x_f = x_i + vt$$

ANSWERS TO SELECTED ODD-NUMBERED PROBLEMS APPEAR IN APPENDIX A.

Lesson by Lesson

2.1 Describing Motion

Conceptual Questions

46. Can you take a walk in such a way that the distance you cover is greater than the magnitude of your displacement? Give an example if your answer is yes; explain why not if your answer is no.

47. Can you take a bicycle ride in such a way that the distance you cover is less than the magnitude of your displacement? Give an example if your answer is yes; explain why not if your answer is no.

48. Who has the greater displacement, an astronaut who has just completed an orbit of the Earth or you when you have just traveled from home to school? Explain.

49. Triple Choice You travel along the x axis from the location $x_i = 10$ m to the location $x_f = 25$ m. Your friend travels from $x_i = 35$ m to $x_f = 40$ m. Is the distance you cover greater than, less than, or equal to the distance covered by your friend? Explain.

50. You travel from the location $x_i = 20$ m to the location $x_f = 25$ m. Your friend travels from $x_i = 35$ m to $x_f = 30$ m. Which of you has a positive displacement? Which of you has a negative displacement? Explain.

Problem Solving

51. Suppose you walk from your home in **Figure 2.18** to the library, then to the park. (a) What is the distance traveled? (b) What is your displacement?

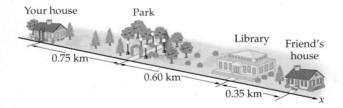

Figure 2.18

52. You walk from the park in Figure 2.18 to your friend's house, then back to your house. What is your (a) distance traveled and (b) displacement?

53. Suppose you start at the position $x_i = 4.5$ m. If you undergo a displacement of 6.2 m, what is your final position?

54. Suppose you start at the position $x_i = 7.5$ m. If you undergo a displacement of -8.3 m, what is your final position?

55. After a displacement of -26 m, a train on a straight track is at the position $x_f = 4.3$ m. What was the train's initial position?

56. After a displacement of 17 m, a train on a straight track is at the position $x_f = -2.2$ m. What was the train's initial position?

57. The two tennis players shown in **Figure 2.19** walk to the net to shake hands. (a) Find the distance traveled and the displacement of player A. (b) Repeat for player B.

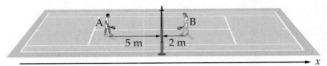

Figure 2.19

58. The golfer in **Figure 2.20** sinks the ball in two putts, as shown. What is (a) the distance traveled by the ball and (b) the displacement of the ball?

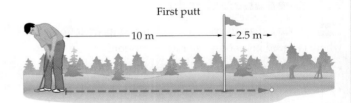

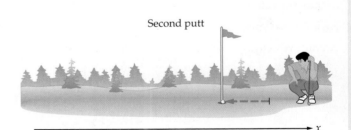

Figure 2.20

59. A jogger runs on the track shown in **Figure 2.21**. Neglecting the curvature of the corners, (a) what is the distance traveled and the displacement in running from point A to point B? (b) Find the distance and displacement for a complete circuit of the track.

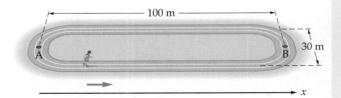

Figure 2.21

2.2 Speed and Velocity
Conceptual Questions

60. If the average velocity of your dog on its daily walk is zero, is its displacement positive, negative, or zero? Explain.

61. Who has the greater average velocity, an astronaut who has just completed an orbit of the Earth or you when you have just traveled from home to school? Explain.

62. Is it possible for two different objects to have the same speed but different velocities?

63. Is it possible for two different objects to have the same velocity but different speeds?

64. **Predict & Explain** You drive your car in a straight line at 15 m/s for 10 km, then at 25 m/s for another 10 km. **(a)** Is your average speed for the entire trip more than, less than, or equal to 20 m/s? **(b)** Choose the *best* explanation from the following:

A. More time is spent driving at 15 m/s than at 25 m/s.

B. The average of 15 m/s and 25 m/s is 20 m/s.

C. Less time is spent driving at 15 m/s than at 25 m/s.

65. **Predict & Explain** You drive your car in a straight line at 15 m/s for 10 min, then at 25 m/s for another 10 min. **(a)** Is your average speed for the entire trip more than, less than, or equal to 20 m/s? **(b)** Choose the *best* explanation from the following:

A. More time is required to drive the same distance at 15 m/s than at 25 m/s.

B. Less distance is covered driving at 25 m/s than at 15 m/s.

C. Equal time is spent driving at 15 m/s and 25 m/s.

Problem Solving

66. Britta Steffen of Germany set the women's Olympic record for the 100-m freestyle swim with a time of 53.12 s. What was her average speed? Give your answer in meters per second and miles per hour.

67. **Rubber Ducks** A severe storm on January 10, 1992, near the Aleutian Islands, caused a cargo ship to spill 29,000 rubber ducks and other bath toys into the ocean. Ten months later hundreds of rubber ducks began to appear along the shoreline near Sitka, Alaska, roughly 2600 km away. What was the approximate average speed, in meters per second, of the ocean current that carried the ducks to shore? (Rubber ducks from the same spill began to appear on the coast of Maine in July 2003.)

68. A roller coaster moves on a certain section of its track with an average speed of 12 m/s. How much distance does it cover in 5.5 s?

69. In a game of billiards you give the ball a speed of 0.76 m/s. How much time does it take for the ball to cover a distance of 0.23 m?

70. You jog at 9.50 km/h for 8.00 km; then you jump into a car and ride an additional 16.0 km. What average speed must the car have for the average speed for the entire 24.0-km trip to be 22.0 km/h?

71. A dog runs back and forth between its two owners, who are walking toward one another as shown in **Figure 2.22**. The dog starts running when the owners are 10.0 m apart. If the dog runs with a speed of 3.0 m/s, and each owner walks with a speed of 1.3 m/s, how far has the dog traveled when the owners meet?

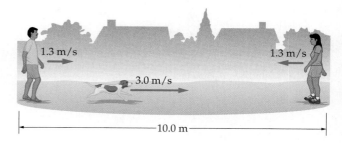

Figure 2.22

72. **Think & Calculate** You drive in a straight line at 20.0 m/s for 10.0 min, then at 30.0 m/s for another 10.0 min. **(a)** Is your average speed 25.0 m/s, more than 25.0 m/s, or less than 25.0 m/s? Explain. **(b)** Verify your answer to part (a) by calculating the average speed.

73. **Think & Calculate** You drive in a straight line at 20.0 m/s for 10.0 mi, then at 30.0 m/s for another 10.0 mi. **(a)** Is your average speed 25.0 m/s, more than 25.0 m/s, or less than 25.0 m/s? Explain. **(b)** Verify your answer to part (a) by calculating the average speed.

2.3 Position-Time Graphs
Conceptual Questions

74. **(a)** Can the position-time graph for the motion of an object be a horizontal line? **(b)** Can the position-time graph be a vertical line? Explain your answer in each case.

75. **Triple Choice** An object's position-time graph is a straight line with a positive slope. Is the velocity of this object positive, negative, or zero? Explain.

76. **Triple Choice** An object's position-time graph is a straight line with a negative slope. Is the speed of this object positive, negative, or zero? Explain.

77. **Triple Choice** The position-time graph for the motion of a certain particle is a smooth curve, like a parabola. At a given instant of time, the tangent line to the position-time graph has a negative slope. Is the instantaneous velocity of the particle at this time positive, negative, or zero? Explain.

78. Two trains run on a straight level track. The motions of the trains for three cases—A, B, and C—are described by the position-time graphs shown in **Figure 2.23**. In which case(s) do the trains **(a)** travel in opposite directions, **(b)** get steadily farther apart as they travel, or **(c)** collide?

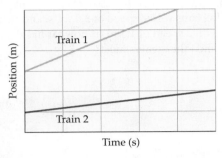

Case A

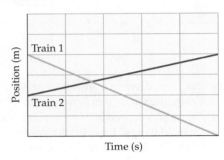

Case B

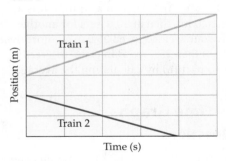

Case C

Figure 2.23

79. Referring to the position-time graphs in Figure 2.23, state whether each train has **(a)** a positive velocity or **(b)** a negative velocity in each of the three cases, A, B, and C.

80. An expectant father paces back and forth, producing the position-time graph shown in **Figure 2.24**. Without performing a calculation, indicate whether the father's velocity is positive, negative, or zero on each of the following segments of the graph: **(a)** A, **(b)** B, **(c)** C, and **(d)** D.

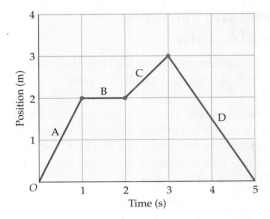

Figure 2.24

Problem Solving

81. Referring to Problem 80 and Figure 2.24, calculate the father's velocity for each segment: **(a)** A, **(b)** B, **(c)** C, and **(d)** D.

82. Make a position-time graph for a particle that is at $x = 5.0$ m at $t = 0$ and moves with a constant velocity of 3.5 m/s. Plot the motion for the range $t = 0$ to $t = 6.0$ s.

83. Make a position-time graph for a particle that is at $x = 3.1$ m at $t = 0$ and moves with a constant velocity of -2.7 m/s. Plot the motion for the range $t = 0$ to $t = 6.0$ s.

84. Sketch a position-time graph for an object that starts at $x = 1.5$ m, moves with a velocity of 2.2 m/s from $t = 0$ to $t = 1$ s, has a velocity of 0 m/s from $t = 1$ s to $t = 2$ s, and has a velocity of -3.7 m/s from $t = 2$ s to $t = 5$ s.

85. **Figure 2.25** shows the position-time graphs for two different objects, A and B. What is the initial position of **(a)** object A and **(b)** object B?

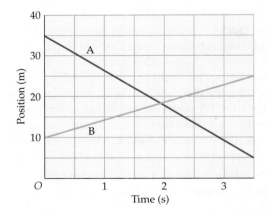

Figure 2.25

86. Referring to Figure 2.25, what is the velocity of (a) object A and (b) object B?

87. In Figure 2.24, what is the average velocity of the expectant father between the times $t = 1$ s and $t = 3$ s?

2.4 Equation of Motion
Conceptual Questions

88. Is it possible for two different objects to have the same initial positions but different velocities?

89. Is it possible for two different objects to have the same velocity but different initial positions?

90. The initial position of an object that moves with constant velocity is increased. Does this change the intercept or the slope of the position-time graph of the object's motion? Explain.

91. The velocity of an object that moves with constant velocity is increased. Does this change the intercept or the slope of the position-time graph of the object's motion? Explain.

92. Two fish swimming in a river have the following equations of motion:

$$x_1 = -6.4 \text{ m} + (-1.2 \text{ m/s})t$$
$$x_2 = 1.3 \text{ m} + (-2.7 \text{ m/s})t$$

Which fish is moving faster?

93. Two people walking on a sidewalk have the following equations of motion:

$$x_1 = 8.2 \text{ m} + (-1.1 \text{ m/s})t$$
$$x_2 = 5.9 \text{ m} + (1.7 \text{ m/s})t$$

(a) Which person is moving faster? (b) Which person will be at $x = 0$ at some time in the future?

Problem Solving

94. Referring to Figure 2.25, write the equation of motion for object A.

95. Referring to Figure 2.25, write the equation of motion for object B.

96. Consider a rabbit that is at $x = 8.1$ m at $t = 0$ and moves with a constant velocity of -1.6 m/s. What is the equation of motion for the rabbit?

97. The equation of motion for a person riding a bicycle is $x = 6.0$ m $+$ $(4.5 \text{ m/s})t$. (a) Where is the bike at $t = 2.0$ s? (b) At what time is the bike at the location $x = 24$ m?

98. The equation of motion for a float in a parade is $x = -9.2$ m $+ (1.5 \text{ m/s})t$. (a) Where is the float at $t = 3.5$ s? (b) At what time is the float at $x = 0$?

99. Cleo the black lab runs to pick up a stick on the ground at the location $x = 3.0$ m. The equation of motion for Cleo is $x = -12.1$ m $+ (5.2 \text{ m/s})t$. (a) Where is Cleo at $t = 1.6$ s? (b) At what time does Cleo reach the stick?

100. Two football players move in a straight line directly toward one another. Their equations of motion are as follows:

$$x_1 = 0.1 \text{ m} + (-3.1 \text{ m/s})t$$
$$x_2 = -6.3 \text{ m} + (2.8 \text{ m/s})t$$

(a) Which player is moving faster? (b) At what time do the players collide?

101. A soccer ball rests on the field at the location $x = 5.0$ m. Two players run along the same straight line toward the ball. Their equations of motion are as follows:

$$x_1 = -8.2 \text{ m} + (4.2 \text{ m/s})t$$
$$x_2 = -7.3 \text{ m} + (3.9 \text{ m/s})t$$

(a) Which player is closer to the ball at $t = 0$? (b) At what time does one player pass the other player? (c) What is the location of the players when one passes the other?

Mixed Review

102. A golf cart moves with a velocity of 8 m/s. Is the displacement of the golf cart from $t = 0$ to $t = 5$ s greater than, less than, or equal to its displacement from $t = 5$ s to $t = 10$ s? Explain.

103. Your average velocity over a 10-min period is 2.2 m/s. Is it possible that you were at rest at some point during the 10 min?

104. Two dragonflies have the following equations of motion:

$$x_1 = 2.2 \text{ m} + (0.75 \text{ m/s})t$$
$$x_2 = -3.1 \text{ m} + (-1.1 \text{ m/s})t$$

(a) Which dragonfly is moving faster? (b) Which dragonfly starts out closer to $x = 0$ at $t = 0$?

105. Estimate the speed of a garden snail.

106. **Nerve Impulses** The human nervous system can propagate nerve impulses at about 10^2 m/s. Estimate the time it takes for a nerve impulse generated when your finger touches a hot object to travel to your brain.

107. Which has the greater displacement, object 1, which moves from 5.0 m to 7.0 m in 2.0 s, or object 2, which moves from 15 m to 16 m in 25 s? Explain.

108. What is the final position of an object that starts at 7.3 m and moves with a velocity of -1.1 m/s for 3.5 s?

109. (a) What is the velocity of an object that moves from 73 m to 62 m in 12 s? (b) What is its speed?

110. A horse is at $x = 4.3$ m at $t = 0$ and moves with a constant velocity of 6.7 m/s. (a) Plot the motion of the horse for the range of times from $t = 0$ to $t = 5.0$ s. (b) Write the corresponding equation of motion.

111. The equation of motion for a train on a straight track is $x = 11$ m $+ (6.5 \text{ m/s})t$. (a) Plot the position-time graph for the train from $t = 0$ to $t = 5.0$ s. (b) At what time is the train at $x = 32$ m?

112. **Think & Calculate** A child rides a pony on a circular track whose radius is 4.5 m. **(a)** Find the distance traveled and the displacement after the child has gone halfway around the track. **(b)** Has the distance traveled increased, decreased, or stayed the same when the child has completed one circuit of the track? Explain. **(c)** Has the displacement increased, decreased, or stayed the same when the child has completed one circuit of the track? Explain. **(d)** Find the distance and displacement after a complete circuit of the track.

113. In heavy rush-hour traffic you drive in a straight line at 12 m/s for 1.5 min, then you have to stop for 3.5 min, and finally you drive at 15 m/s for another 2.5 min. **(a)** Plot a position-time graph for this motion. Your graph should extend from $t = 0$ to $t = 7.5$ min. **(b)** Use your graph from part (a) to calculate the average velocity between $t = 0$ and $t = 7.5$ min.

114. Object 1 starts at 5.4 m and moves with a velocity of 1.3 m/s. Object 2 starts at 8.1 m and moves with a velocity of -2.2 m/s. The two objects are moving directly toward one another. **(a)** At what time do the objects collide? **(b)** What is the position of the objects when they collide?

115. Object 1 starts at 25 m and moves with a velocity of -5.6 m/s. Object 2 starts at 13 m and moves directly toward object 1. The two objects collide 0.61 s after starting. **(a)** What is the velocity of object 2? **(b)** What is the position of the objects when they collide?

116. On your cousin's wedding day you leave for the church 30.0 min before the ceremony is to begin, which should leave plenty of time since the church is only 17.0 km away. On the way, however, you have to make an unanticipated stop because of road construction work. As a result, your average speed for the first 15 min is only 12.0 km/h. What average speed do you need to have on the remainder of the trip to get you to the church on time?

Writing about Science

117. Research the motion of land on either side of the San Andreas fault. What is the average speed of slipping between the sides of the fault? How long will it take for a location near Los Angeles to reach San Francisco?

118. **Connect to the Big Idea** Write position-time equations of motion for some common objects at your school. Determine the speeds of the objects, and set up a coordinate system to define the initial positions of the objects.

Read, Reason, and Respond

Robot Walking Walking is difficult, especially on just two legs. Think of how long it takes for a human baby to "graduate" from crawling on all fours to standing upright and walking. Engineers working to develop robots capable of walking on two legs, like humans do, face many challenges. Still, much progress has been made. Some two-legged robots can run, and some can even walk up or down a flight of stairs.

The position-time graph in **Figure 2.26** shows the progress of a typical two-legged robot that walked forward, stopped for a few seconds, and then walked backward.

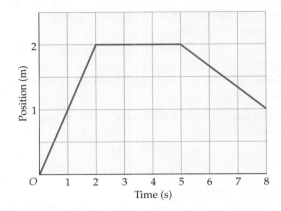

Figure 2.26

119. What was the velocity of the robot as it walked forward?

A. 0.5 m/s   C. 2.0 m/s

B. 1.0 m/s   D. 4.0 m/s

120. What was the velocity of the robot as it walked backward?

A. -0.33 m/s   C. -1.0 m/s

B. -0.67 m/s   D. -1.5 m/s

121. What was the displacement of the robot between the times $t = 2.0$ s and $t = 8.0$ s?

A. 0 m   C. -2.0 m

B. -1.0 m   D. -6.0 m

122. What was the average velocity of the robot between the times $t = 2.0$ s and $t = 8.0$ s?

A. -0.17 m/s   C. -2.0 m/s

B. -0.33 m/s   D. -3.0 m/s

Standardized Test Prep

Multiple Choice

Use the graph below to answer Questions 1–5. The graph plots the displacement of an object versus time.

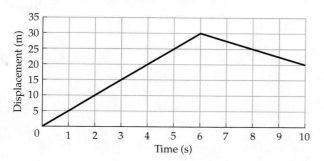

1. What is the displacement of the object whose motion is represented in the graph over the time period from $t = 0$ to $t = 3$ s?
   (A) 5 m
   (B) 10 m
   (C) 15 m
   (D) 20 m

2. What does the constant slope of the graph from $t = 0$ to $t = 6$ s indicate?
   (A) no change in position
   (B) constant velocity
   (C) constant acceleration
   (D) constantly changing velocity

3. What is the displacement of the object over the time period from $t = 0$ to $t = 10$ s?
   (A) 5 m   (C) 15 m
   (B) 10 m   (D) 20 m

4. What is the average velocity of the object from $t = 8$ s to $t = 10$ s?
   (A) −5.0 m/s   (C) 0
   (B) −2.5 m/s   (D) 5.0 m/s

5. Which statement *best* describes the change in motion that occurs at $t = 6$ s?
   (A) The object changes direction and speed.
   (B) The object changes direction, but its speed remains the same.
   (C) The object changes speed but not direction.
   (D) The object speeds up.

6. Runner A starts at $x = 0$ and runs with a velocity of +3.0 m/s toward runner B. Runner B starts at the same time at $x = 20$ m and runs with a velocity of −2.0 m/s toward runner A. Where do they meet, and how much time has elapsed since they started running?
   (A) $t = 3.0$ s and $x = 9.0$ m
   (B) $t = 3.0$ s and $x = 12$ m
   (C) $t = 4.0$ s and $x = 9.0$ m
   (D) $t = 4.0$ s and $x = 12$ m

7. Which of the following is an example of motion with constant velocity?
   (A) a car that comes to a stop at a stop sign
   (B) an object moving in a circle with a constant speed
   (C) a car that drives with a constant speed up and over a hill and then down the other side
   (D) a box sliding with a constant speed down a ramp

8. An object's equation of motion is given by $x = 12 + 5t$, where x is the final position. Which statement is true?
   (A) The object's initial velocity could be 12 m/s.
   (B) The object moves with a constant velocity of 12 m/s.
   (C) The object moves with a constant velocity of 5 m/s.
   (D) The object accelerates at 5 m/s^2.

Extended Response

9. A hiker starts at his campsite and walks at 2 m/s northward for 30 min, then turns and walks due west at 1 m/s for 1 h. **(a)** Calculate the distance the hiker walks in 1.5 h. **(b)** Calculate the hiker's displacement from the campsite after walking for 1.5 h. **(c)** Explain the difference between distance and displacement.

Test-Taking Hint

When the position-time graph is linear, velocity is constant.

If You Had Difficulty With . . .

Question	1	2	3	4	5	6	7	8	9
See Lesson(s)	2.1, 2.3	2.2, 2.3	2.1, 2.3	2.2, 2.3	2.3	2.4	2.2	2.4	2.1

3

Acceleration and Accelerated Motion

Inside

This bungee jumper is experiencing the exhilaration of free fall—at least until the cord begins to stretch and slow him down. During free fall, the jumper's speed increases by 9.81 m/s every second.

Big Idea

All objects in free fall move with the same constant acceleration.

In this chapter we continue our study of motion. This time, however, we consider situations in which the velocity of an object changes with time. Examples of changing velocity are all around us, from a car that slows down when a traffic light turns red to a ball that speeds up when it is dropped. As we will see, motion with changing velocity is of central importance in physics.

3.1 Acceleration

Have you changed your velocity today? Most likely you have. Every time you start, stop, turn around, speed up, or slow down, your velocity changes. In physics, an object with a changing velocity is said to be *accelerating*. Thus, you accelerate many times during a typical day.

To be more specific, **acceleration** is the rate at which velocity changes with time. Notice that both velocity and acceleration are rates of change. Velocity is the rate at which *displacement* changes with time. Acceleration is the rate at which *velocity* changes with time. Both velocity and acceleration have a direction associated with them, and thus both are vectors.

Changes in speed and/or direction cause acceleration

Let's review the ways in which velocity can change:

- Velocity changes when the speed of an object changes.
- Velocity changes when the direction of motion changes.

Both of these changes cause an acceleration. 🔑 **To summarize, acceleration occurs when there is a change in speed, a change in direction, or a change in both speed and direction.**

Let's look at a simple example of acceleration. **Figure 3.1** on the next page shows a cyclist who accelerates from rest by increasing his speed by 2 m/s every second. After 1 second his speed is 2 m/s, after 2 seconds his speed is 4 m/s, and so on. Because his speed is steadily increasing, the amount of distance he covers each second is increasing as well. We apply this idea of changing velocity in the following Conceptual Example.

Vocabulary

- acceleration
- average acceleration
- instantaneous acceleration
- constant acceleration

🔑 *What kind of motion results in an acceleration?*

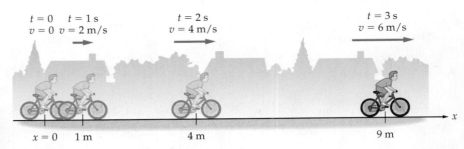

▶ Figure 3.1 An accelerating cyclist
A cyclist starts at rest and increases his speed by 2 m/s every second. As a result, the distance covered during each second is greater than the distance covered in the previous second.

(a)

(b)

(c)

▲ Figure 3.2 The sensation of acceleration
(a) A person rides comfortably at high speed in a jetliner. (b) The space shuttle *Discovery* accelerates upward on the initial phase of its journey into orbit. During this time the astronauts experience an acceleration of 29 m/s², almost three times the acceleration due to gravity. (c) A test dummy in a crash test experiences an extreme acceleration.

CONCEPTUAL Example 3.1 Accelerating or Not?

Shown below is a toy car moving along a track. The motion of the car is shown with a series of "snapshots" taken at 1-s intervals. The green arrows indicate the velocity of the car at each instant—the longer the arrow, the greater the speed. On which sections of the track does the car accelerate?

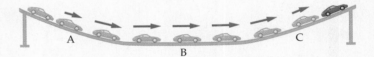

Reasoning and Discussion
On section A the car speeds up as it goes downhill. This is a change in velocity, and hence the car accelerates. On section B the car moves with constant speed in the same direction. Therefore, the car does not accelerate here. Finally, on section C the car slows down as it goes uphill. This is also a change in velocity, and hence the car accelerates.

Answer
The car accelerates on sections A and C.

Detecting Velocity and Acceleration The human body cannot detect velocity. Not at all. However, the body is great at sensing acceleration. For example, imagine that you are a passenger on a jetliner flying at 900 km/h (~560 mph), as shown in **Figure 3.2 (a)**. You do not sense the plane's speed. In fact, you feel the same as if you were at rest on the ground. Of course, you know the airplane is moving when it hits some turbulence and bounces around. Each of the bounces is an acceleration. It's the acceleration that you sense, and not the velocity itself.

Think about riding in a car. When the car speeds away from a stoplight, the seat *pushes forward* on you. When the car brakes suddenly, the seat belt *pushes backward* on you. When the car rounds a corner quickly, you *lean* to one side. In each of these cases you are accelerating, and the greater the acceleration, the greater the sensation. For example, the large acceleration that occurs during the launch of a space shuttle (see **Figure 3.2 (b)**) presses the astronauts so firmly into their seats that it is hard for them to move. Similarly, the dummy in the crash test in **Figure 3.2 (c)** definitely "senses" a large acceleration. Though you may not have thought about it before, your body is an excellent "acceleration meter."

Average acceleration is velocity change over time

Let's look at a precise definition of acceleration over a period of time. In physics, we say that the **average acceleration** of an object is the change in its velocity divided by the change in time. For a given time interval, Δt, we have the following definition:

Definition of Average Acceleration, a_{av}

$$\text{average acceleration} = \frac{\text{change in velocity}}{\text{change in time}}$$

$$a_{av} = \frac{\Delta v}{\Delta t} = \frac{v_f - v_i}{\Delta t}$$

SI units: meters per second per second (m/s^2)

Average acceleration is a vector, just like displacement and average velocity. Therefore, the average acceleration can be positive, negative, or zero, depending on the values of the initial and final velocities, v_i and v_f. For example, when the final velocity is greater than the initial velocity, the average acceleration is positive.

Units of Acceleration

The dimensions of average acceleration are the dimensions of velocity per time, or (meters per second) per second:

$$\frac{\text{meters per second}}{\text{second}} = \frac{\text{m/s}}{\text{s}} = \frac{\text{m}}{\text{s}^2}$$

As a shorthand, this is generally expressed in words as "meters per second squared" and is written symbolically as m/s^2. For example, if we say that an object has an acceleration of 5 m/s^2, we mean that its velocity increases by 5 m/s every second. Typical magnitudes of acceleration are given in **Table 3.1**.

Table 3.1 Typical Accelerations (m/s^2)

Ultracentrifuge	3×10^6
Bullet fired from a rifle	4.4×10^5
Batted baseball	3×10^4
Click beetle righting itself	400
Acceleration required to deploy air bags	60
Bungee jump	30
High jump	15
Acceleration due to gravity on Earth	9.81
Emergency stopping of a car	8
Airplane during takeoff	5
An elevator	3
Acceleration due to gravity on the Moon	1.62

QUICK Example 3.2 What's the Acceleration?

(a) An advertisement says a car can go from 0 to 26.8 m/s (the SI equivalent of 60 mph) in 6.2 s. What is the average acceleration of the car? **(b)** A drag racer deploys a parachute and slows down from 130 m/s to 45 m/s in 3.0 s. What is the acceleration of the racer?

Solution

(a) Substitute $v_i = 0$ m/s, $v_f = 26.8$ m/s, and $\Delta t = 6.2$ s into the average acceleration equation. This yields the following:

$$a_{av} = \frac{v_f - v_i}{\Delta t} = \frac{26.8 \text{ m/s} - 0}{6.2 \text{ s}} = \boxed{4.3 \text{ m/s}^2}$$

(b) In this case, substitute $v_i = 130$ m/s, $v_f = 45$ m/s, and $\Delta t = 3.0$ s into the average acceleration equation. This gives

$$a_{av} = \frac{v_f - v_i}{\Delta t} = \frac{45 \text{ m/s} - 130 \text{ m/s}}{3.0 \text{ s}} = \boxed{-28 \text{ m/s}^2}$$

Notice that the velocities of the drag racer are positive but the acceleration is negative. As we shall see later in this chapter, opposite signs for velocity and acceleration means that the object is slowing down.

1. Rank Three cars accelerate while moving in the positive direction. Rank the cars in order of increasing acceleration, from most negative to most positive.

Car 1 speeds up from 25 m/s to 35 m/s in 10 s.

Car 2 speeds up from 0 to 30 m/s in 15 s.

Car 3 slows down from 32 m/s to 12 m/s in 5 s.

2. The winner of the drag race shown in **Figure 3.3** was traveling with a speed of 140.3 m/s (313.9 mph) at the end of the quarter-mile course. The car started from rest and had an average acceleration of 30.4 m/s^2. What was the winning time for this race?

▲ Figure 3.3 **A drag race**

3. A chameleon extends its tongue to capture a tasty insect. The chameleon's tongue accelerates at 33 m/s^2 for 0.12 s to make the capture. What is the speed of the chameleon's tongue when it grabs the insect?

Acceleration can change with time or be constant

When a car pulls away from a stoplight, the seat pushes on you as the car accelerates. It can't maintain this acceleration for long, though, or it would exceed the speed limit. Thus, the acceleration of the car must change with time. We refer to the acceleration of an object at a given instant of time as the **instantaneous acceleration**. In contrast, the average acceleration is the acceleration averaged over a period of time of duration Δt.

Most of the situations in this text involve **constant acceleration**, which means the acceleration is the same at every instant of time. When an object's acceleration is constant, its instantaneous and average accelerations are the same.

Acceleration can be determined graphically

In Chapter 2 you learned how to describe motion with a position-time graph, with position plotted versus time. Here we consider a *velocity-time graph*, where velocity is plotted on the vertical axis and time on the horizontal axis. Just as you can find an object's velocity from its position-time graph, you can find an object's acceleration from its velocity-time graph.

To see how this works, imagine sliding down a snow-covered hill on a plastic sled, as shown in **Figure 3.4**. Measurements of your velocity at different times yield the data shown in **Table 3.2**. To better visualize your acceleration, we plot the data points in **Figure 3.5**. This is a velocity-time graph for your sled ride.

🔑 *What does a velocity-time graph of constant acceleration look like?*

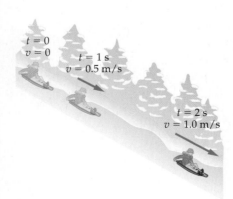

◀ Figure 3.4 **A sled ride**
Sliding down a snow-covered hill gives you a steadily increasing velocity.

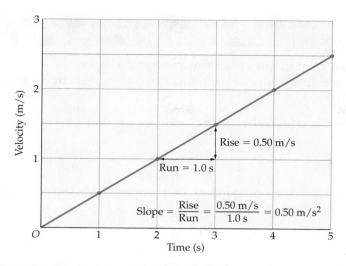

Figure 3.5 A velocity-time graph for the sled ride
A velocity-time graph for constant acceleration is a straight line. The slope of the line is equal to the acceleration.

Table 3.2 Velocity and Time Data	
Time (s)	Velocity (m/s)
0.0	0.0
1.0	0.50
2.0	1.00
3.0	1.50
4.0	2.00
5.0	2.50

The first thing to notice about the velocity-time graph in Figure 3.5 is that it is a straight line. Straight lines always have a constant slope—that is, they always rise by a constant amount for a given distance on the horizontal axis. The fact that the velocity-time graph is a straight line means that the velocity increases by the same amount during every second on the horizontal axis. In other words, the *straight line* means a *constant* acceleration.

To be specific, the acceleration in the case of the sled ride is 0.50 m/s per second, or 0.50 m/s^2. In fact, the acceleration of the sled is equal to the slope of the straight line in the velocity-time graph, as we see in Figure 3.5. **In general, the velocity-time graph of an object with constant acceleration is a straight line with a slope equal to the acceleration.** The steeper the straight line, the greater its slope.

Notice that velocity-time graphs use green lines, while position-time graphs use blue lines. These color conventions are used consistently throughout this book.

CONCEPTUAL Example 3.3 Comparing Accelerations

Shown at the right are velocity-time graphs for two different cars (A and B) that are accelerating. Is the acceleration of car A greater than, less than, or equal to the acceleration of car B?

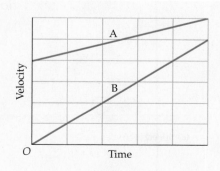

Reasoning and Discussion
The acceleration of each car is equal to the slope of its velocity-time graph. Thus, the acceleration of car B is greater than the acceleration of car A, even though car A has the greater velocity.

Answer
Car B has the greater acceleration.

A subway train speeds up with an average acceleration of 1.2 m/s^2. How much time does it take for the train's speed to increase by 4.1 m/s?

Solution

Solve the average acceleration equation for the time interval, Δt:

Math HELP
Solving Simple
Equations
**See Math
Review,
Section III**

$$a_{av} = \frac{\Delta v}{\Delta t} \quad \text{becomes} \quad \Delta t = \frac{\Delta v}{a_{av}}$$

Substitute $\Delta v = 4.1 \text{ m/s}$ and $a_{av} = 1.2 \text{ m/s}^2$ and calculate the time:

$$\Delta t = \frac{\Delta v}{a_{av}}$$

$$= \frac{4.1 \text{ m/s}}{1.2 \text{ m/s}^2}$$

$$= \boxed{3.4 \text{ s}}$$

Practice Problems

4. How much time is required for an airplane to reach its takeoff speed of 77 m/s if it starts from rest and its average acceleration is 8.2 m/s^2?

5. A horse runs with an initial velocity of 11 m/s and slows to 5.2 m/s over a time interval of 3.1 s. What is the horse's average acceleration?

6. Challenge As a train accelerates away from a station, it reaches a speed of 4.7 m/s in 5.0 s. If the train's acceleration is constant, what is its speed after an additional 6.0 s have elapsed?

Objects can accelerate or decelerate

🔑 *What determines whether an acceleration causes an increase or a decrease in speed?*

Suppose you ride a mountain bike along a tricky trail. At some points on the trail you pedal harder and increase your speed. At other points you brake a little and decrease your speed. Both pedaling and braking produce accelerations, but notice that an acceleration doesn't always mean an increase in speed. The acceleration due to braking reduces your speed.

To better understand the connection between acceleration and changing speed, consider **Figure 3.6**. Here you see a bike moving in the positive direction. This means that its velocity is positive, since velocity always points in

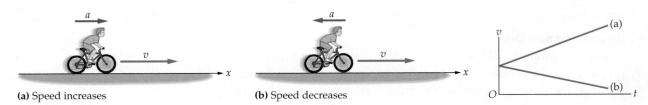

(a) Speed increases   **(b)** Speed decreases

▲ **Figure 3.6 A bicycle moving in the positive direction**
The speed of the bicycle increases when its velocity and acceleration are in the same direction. Its speed decreases when its velocity and acceleration are in opposite directions.

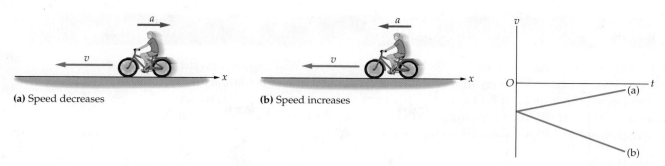

(a) Speed decreases   **(b)** Speed increases

▲ **Figure 3.7 A bicycle moving in the negative direction**
The speed of the bicycle increases when its velocity and acceleration are in the same direction.
Its speed decreases when its velocity and acceleration are in opposite directions.

the direction of motion. The bike is also accelerating, with either a positive
or a negative acceleration. When the acceleration is positive (like the velocity), the speed increases. When the acceleration is negative (opposite to the
velocity), the speed decreases.

Similarly, consider **Figure 3.7**. In this case the bike moves in the negative direction. When its acceleration is also negative, the speed increases,
but when the acceleration is positive, the speed decreases. What common
pattern can we see in these results? It is the following: 🔑 **Speed increases
when velocity and acceleration have the same sign. Speed decreases when
velocity and acceleration have opposite signs.**

Notice that the sign of the acceleration by itself doesn't tell us if an object
speeds up or slows down. In particular, a negative acceleration doesn't automatically mean a decrease in speed. What is important is whether velocity
and acceleration have the same or opposite signs. In general, when the speed
of an object decreases (velocity and acceleration have opposite signs), we say
that it is *decelerating*.

Practice Problem

7. Concept Check The velocity-time graphs of four cars are shown in
Figure 3.8.

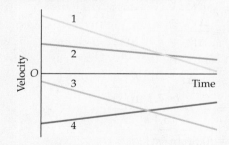

▲ **Figure 3.8**

(a) Which of the four lines indicate(s) a car that is slowing down?
Explain.
(b) Which of the four lines indicate(s) a car that is speeding up? Explain.

A ferry makes a short run between two docks: one in Anacortes, Washington, and the other on Guemes Island. The ferry moves in the positive x direction as it approaches Guemes Island with a speed of 7.4 m/s. (**a**) If the ferry slows to a stop in 12.3 s, what is its average acceleration? (**b**) The ferry returns to the Anacortes dock with a speed of 7.3 m/s. If it comes to rest in 13.1 s, what is its average acceleration?

Picture the Problem

Our sketch shows the positive direction and the ferry heading for Guemes Island. The distance between the docks is not given, and it is not needed.

Strategy

Because the ferry comes to a stop in each case, its final velocity is zero. Therefore, the initial velocity, the final velocity, and the amount of time required to come to rest are known. We can find the average acceleration in each case using the equation $a_{av} = \Delta v / \Delta t$. Pay careful attention to the sign of each velocity.

Known

(**a**) $v_i = 7.4$ m/s
     $v_f = 0$
     $\Delta t = 12.3$ s

(**b**) $v_i = -7.3$ m/s
     $v_f = 0$
     $\Delta t = 13.1$ s

Unknown

(**a**) $a_{av} = ?$
(**b**) $a_{av} = ?$

Solution

1 (a) Calculate the average acceleration, noting that $v_i = 7.4$ m/s and $v_f = 0$:

$$a_{av} = \frac{\Delta v}{\Delta t}$$

$$= \frac{v_f - v_i}{\Delta t}$$

$$= \frac{0 - 7.4 \text{ m/s}}{12.3 \text{ s}}$$

$$= \boxed{-0.60 \text{ m/s}^2}$$

Math HELP
Negative Numbers
See Math Review, Section I

2 (b) Repeat the calculation, this time with $v_i = -7.3$ m/s and $v_f = 0$:

$$a_{av} = \frac{\Delta v}{\Delta t}$$

$$= \frac{v_f - v_i}{\Delta t}$$

$$= \frac{0 - (-7.3 \text{ m/s})}{13.1 \text{ s}}$$

$$= \boxed{0.56 \text{ m/s}^2}$$

Insight

In each case the acceleration of the ferry is opposite in sign to its velocity. As a result, the ferry decelerates.

8. Follow-up When the ferry leaves Guemes Island and heads back toward Anacortes, its speed increases from 0 to 5.8 m/s in 9.25 s. What is its average acceleration?

9. How much time does it take for a car to come to rest if it has an initial speed of 22 m/s and slows with a deceleration of 6.1 m/s²?

10. An astronaut on the Moon releases a rock from rest and allows it to drop straight downward. If the acceleration due to gravity on the Moon is 1.62 m/s² and the rock falls for 2.4 s before hitting the ground, what is its speed just before it lands?

3.1 LessonCheck (MP)

Checking Concepts

11. 🔑 Explain Can an object accelerate even if its speed is constant?

12. 🔑 Describe What is the shape of a velocity-time graph for an object with constant acceleration?

13. 🔑 Give Examples Does a negative acceleration always mean a decrease in speed? Does a positive acceleration always mean an increase in speed? Give specific examples to support your answers.

14. Triple Choice If the same change in velocity occurs in less time, does the magnitude of the corresponding average acceleration increase, decrease, or stay the same? Explain.

15. Rank The table below summarizes four situations (A, B, C, and D) in which a motorcycle has a given change in velocity in a given period of time. Rank these four situations in order of acceleration, from most negative to most positive.

	A	B	C	D
v_f	5 m/s	25 m/s	25 m/s	−15 m/s
v_i	15 m/s	30 m/s	20 m/s	−10 m/s
Δt	10 s	15 s	2 s	3 s

16. Explain The velocity of a ball is 1 m/s at a given instant of time. Is it possible for the ball's acceleration to be −1 m/s² at this instant?

Solving Problems

17. Calculate A car moves with an initial velocity of 18 m/s due north. Find the velocity of the car after 7.0 s if (**a**) its acceleration is 1.5 m/s² due north and (**b**) its acceleration is 1.5 m/s² due south.

18. Interpreting Graphs A truck moves as shown by the velocity-time graph in **Figure 3.9**. Find the average acceleration of the truck during each period, A, B, and C.

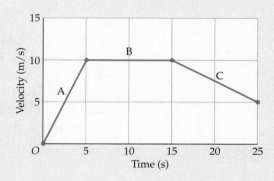

▲ Figure 3.9

19. Challenge An object has an average acceleration of +6.24 m/s² for 0.300 s. At the end of this time the object's velocity is +9.31 m/s. What was the object's initial velocity?

3.2 Motion with Constant Acceleration

This lesson builds on your knowledge of position, velocity, and acceleration. As you will see, simple equations describe the connections between these physical quantities.

Multiplying acceleration and time gives the change in velocity

The longer the accelerator pedal of a car is held down, the faster the car goes. For example, suppose a car has an initial velocity of 10 m/s. If the car accelerates at 5 m/s², its speed increases by 5 m/s each second. Thus, after 1 second its speed is 15 m/s, after 2 seconds its speed is 20 m/s, and so on. Physicists express this observation with the following equation:

> **Velocity-Time Equation**
>
> final velocity = initial velocity + (acceleration)(time)
> $$v_f = v_i + at$$

You can verify this equation by applying it to the car mentioned above. With an initial velocity of $v_i = 10$ m/s and an acceleration of $a = 5$ m/s², the velocity of the car after 1 second is

$$v_f = v_i + at$$
$$= 10 \text{ m/s} + (5 \text{ m/s}^2)(1 \text{ s})$$
$$= 15 \text{ m/s}$$

A similar calculation shows that the velocity after 2 seconds is 20 m/s.

In Lesson 3.1 you learned that constant acceleration corresponds to a velocity-time graph that is a straight line. The velocity-time equation given above describes that line. Specifically, the line crosses the velocity axis at a value equal to the initial velocity, v_i, and has a slope equal to the acceleration, a. This is illustrated in **Figure 3.10**.

▶ **Figure 3.10 Features of a velocity-time graph**
The intercept of the straight line indicates the initial velocity. The slope of the line indicates the acceleration.

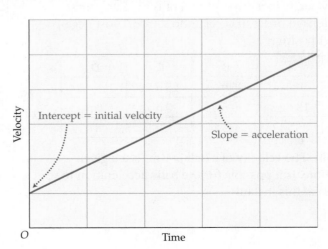

Intercept = initial velocity

Slope = acceleration

Velocity

Time

O

Whenever the acceleration is constant, the average acceleration is equal to the instantaneous acceleration. This is the case shown in Figure 3.10. Therefore, we use the symbol a for acceleration rather than the symbol for average acceleration, a_{av}.

After being tossed straight up in the air, a baseball experiences a constant downward acceleration of 9.81 m/s².

QUICK Example 3.6 What's the Velocity?

You throw a baseball upward with an initial velocity of 8.2 m/s. Gravity gives the ball a downward acceleration of −9.81 m/s². (Notice that we have assumed upward to be the positive direction.) What is the velocity of the ball 1.5 s after it was thrown?

Solution

The problem statement gives us the initial velocity (v_i = 8.2 m/s), the acceleration (a = −9.81 m/s²), and the time (t = 1.5 s). Substituting these values into the velocity-time equation, $v_f = v_i + at$, yields the final velocity:

$$v_f = v_i + at$$
$$= 8.2 \text{ m/s} + (-9.81 \text{ m/s}^2)(1.5 \text{ s})$$
$$= \boxed{-6.5 \text{ m/s}}$$

The negative sign for the final velocity means that the ball is moving downward (the negative direction) at 1.5 s.

Practice Problems

20. Concept Check The velocity-time equation for a golf cart is

$$v_f = (1 \text{ m/s}) + (-0.5 \text{ m/s}^2)t$$

(a) What is the cart's initial velocity?
(b) What is the cart's acceleration?

21. An eagle flies with an initial velocity of 5.0 m/s and has a constant acceleration of 1.3 m/s². What is the velocity of the eagle at t = 2.0 s?

22. Rank Three objects have velocities that vary with time.

Object	Velocity
1	$v = 2 \text{ m/s} + (3 \text{ m/s}^2)t$
2	$v = -8 \text{ m/s} - (4 \text{ m/s}^2)t$
3	$v = 1 \text{ m/s} - (5 \text{ m/s}^2)t$

(a) Rank the velocities of the three objects at t = 3 s, from most negative to most positive. Indicate ties where appropriate.
(b) Rank the speeds of the three objects at t = 3 s, from smallest to largest. Indicate ties where appropriate.

► **Figure 3.11 A train's average velocity**
This graph shows that the average velocity
for constant acceleration is equal to
one half the sum of the initial and final
velocities.

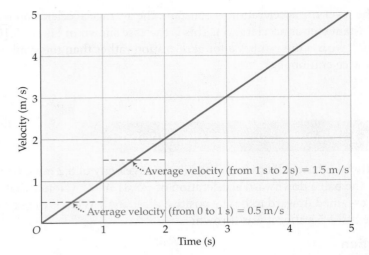

*How is average velocity
calculated when the acceleration
is constant?*

Average velocity is easy to calculate

Suppose a train pulls out of the station and increases its velocity with constant acceleration. The train's velocity is zero initially, after 1 second it is 1 m/s, after 2 seconds it is 2 m/s, and so on. The velocity is increasing steadily, as shown in **Figure 3.11**.

Notice that the velocity changes constantly during the first second, increasing from 0 to 1 m/s. What is the average velocity during this time? As you can see from Figure 3.11, the average velocity is 0.5 m/s, as you might expect. After all, the velocity is less than 0.5 m/s for half a second and greater than 0.5 m/s for half a second. Notice that the average velocity is the average of the initial and final velocities:

$$v_{av} = \tfrac{1}{2}(0 + 1 \text{ m/s}) = 0.5 \text{ m/s}$$

When the acceleration is constant, the average velocity is equal to the sum of the initial and final velocities divided by 2.

> **Average Velocity**
> average velocity = $\tfrac{1}{2}$(initial velocity + final velocity)
> $$v_{av} = \tfrac{1}{2}(v_i + v_f)$$

Figure 3.11 also shows that the average velocity of the train between 1 s and 2 s is 1.5 m/s. This agrees with the average velocity equation, given an initial velocity of 1 m/s and a final velocity of 2 m/s:

$$v_{av} = \tfrac{1}{2}(1 \text{ m/s} + 2 \text{ m/s}) = 1.5 \text{ m/s}$$

The average velocity is useful in calculating the position of an accelerating object, as we now show.

Average velocity is used in equations of motion

In Chapter 2 you learned that the position-time equation for constant velocity is

$$x_f = x_i + vt$$

We can apply this equation to situations in which velocity is changing by replacing the constant velocity, v, with the average velocity, v_{av}.

> **Position-Time Equation with Average Velocity**
>
> final position = initial position + (average velocity)(time)
> $$x_f = x_i + v_{av}t$$

Expressing average velocity in terms of the initial and final velocities gives

> **Position-Time Equation with Initial and Final Velocity**
>
> $$\text{final position} = \text{initial position} + \frac{1}{2}\left(\text{initial velocity} + \text{final velocity}\right)(\text{time})$$
> $$x_f = x_i + \tfrac{1}{2}(v_i + v_f)t$$

This equation is used to find the position of an accelerating object.

QUICK Example 3.7 What's the Velocity?

A boat moves slowly inside a marina with a constant speed of 1.50 m/s. As soon as it leaves the marina, it throttles up and moves with a constant acceleration of 2.40 m/s². **(a)** What is the velocity of the boat after accelerating for 5.00 s? **(b)** How far does the boat travel in this time?

Solution
(a) The velocity-time equation for constant acceleration is $v_f = v_i + at$. To find the final velocity, substitute the given values for initial velocity ($v_i = 1.50$ m/s), acceleration ($a = 2.40$ m/s²), and time ($t = 5.00$ s):

$$v_f = v_i + at$$
$$= 1.50 \text{ m/s} + (2.40 \text{ m/s}^2)(5.00 \text{ s})$$
$$= \boxed{13.5 \text{ m/s}}$$

Math HELP
Dimensional Analysis
See Lesson 1.3

(b) The position-time equation for changing velocity is $x_f = x_i + \tfrac{1}{2}(v_i + v_f)t$. To find the final position, substitute the given values for initial velocity ($v_i = 1.50$ m/s), final velocity ($v_f = 13.5$ m/s), and time ($t = 5.00$ s). For convenience, take the initial position to be $x_i = 0$.

$$x_f = x_i + \tfrac{1}{2}(v_i + v_f)t$$
$$= 0 + \tfrac{1}{2}(1.50 \text{ m/s} + 13.5 \text{ m/s})(5.00 \text{ s})$$
$$= \boxed{37.5 \text{ m}}$$

Practice Problems

23. A motor scooter is stopped at a traffic light. When the light turns green, the scooter accelerates at 4.2 m/s².
(a) How fast is the scooter moving after 3.0 s?
(b) How far does the scooter travel in this time?

24. A soccer ball accelerates from rest and rolls 6.5 m down a hill in 3.1 s. It then bumps into a tree. What is the speed of the ball just before it hits the tree?

25. Challenge As you ride a bicycle on the sidewalk with a speed of 6.4 m/s, a ball suddenly rolls out in front of you. You hit the brakes and come to rest in 3.8 m. How much time does it take you to stop?

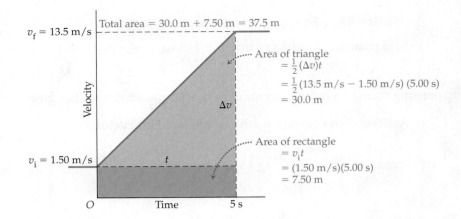

▶ **Figure 3.12 Velocity-time graph for the boat in Quick Example 3.7**
The distance traveled by the boat between $t = 0$ and $t = 5.00$ s is equal to the corresponding area under the velocity curve.

A velocity-time graph indicates distance traveled

🔑 *How is distance related to the area under a velocity-time curve?*

Velocity-time graphs provide a convenient way to find the distance traveled by an object. The connection is very simple: 🔑 **The distance traveled by an object is equal to the area under its velocity-time curve.** This is true for any velocity-time curve and any time interval.

As a specific example, **Figure 3.12** shows the velocity-time curve for the boat in Quick Example 3.7. The acceleration of the boat starts at $t = 0$ and ends at $t = 5.00$ s. The distance traveled by the boat in this time is equal to the area under the velocity-time curve. Looking at Figure 3.12, you can see that the area is equal to the area of a rectangle plus the area of a triangle. In particular,

- The area of the rectangle is the base times the height. The base is 5.00 s and the height is 1.50 m/s; thus, the area is 7.50 m.
- The area of the triangle is one-half the base times the height. The base is 5.00 s and the height is 12.0 m/s; thus, the area is 30.0 m.

The sum of these areas is 37.5 m, in agreement with the result in Quick Example 3.7.

Position is related to acceleration and time

You've learned in this lesson that position can be calculated in terms of the initial and final velocities as follows:

$$x_f = x_i + \tfrac{1}{2}(v_i + v_f)t$$

You've also learned that velocity changes with time according to the following equation:

$$v_f = v_i + at$$

Let's combine these equations to get an expression for position in terms of acceleration. Specifically, we substitute the expression for final velocity into the equation for position:

$$
\begin{aligned}
x_f &= x_i + \tfrac{1}{2}(v_i + v_f)t \\
&= x_i + \tfrac{1}{2}(v_i + (v_i + at))t \\
&= x_i + v_i t + \tfrac{1}{2}at^2
\end{aligned}
$$

This gives a very useful result for relating position to acceleration and time:

> **Position-Time Equation for Constant Acceleration**
>
> $$\begin{matrix} \text{final} \\ \text{position} \end{matrix} = \begin{matrix} \text{initial} \\ \text{position} \end{matrix} + \left(\begin{matrix} \text{initial} \\ \text{velocity} \end{matrix} \right)(\text{time}) + \frac{1}{2}(\text{acceleration})(\text{time})^2$$
>
> $$x_f = x_i + v_i t + \tfrac{1}{2}at^2$$

GUIDED Example 3.8 | Put the Pedal to the Metal   **Constant Acceleration**

A drag racer starts from rest and accelerates at 7.40 m/s². How far will it travel in (**a**) 1.00 s, (**b**) 2.00 s, and (**c**) 3.00 s?

Picture the Problem

The drag racer starts from rest at the origin and accelerates in the positive x direction. It follows that $x_i = 0$, $v_i = 0$, and $a = +7.40$ m/s². The positions of the drag racer in our sketch are drawn to scale.

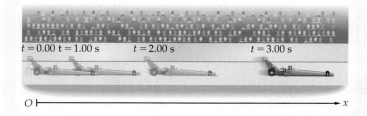

Strategy

This problem gives the acceleration and asks for the position at various times. Therefore, we use the equation $x_f = x_i + v_i t + \frac{1}{2}at^2$, which contains both acceleration and time.

Known

$x_i = 0$
$v_i = 0$
$a = +7.40$ m/s²

Unknown

(**a**) x_f at $t = 1.00$ s
(**b**) x_f at $t = 2.00$ s
(**c**) x_f at $t = 3.00$ s

Solution

1 (**a**) To find the position after 1.00 s, evaluate $x_f = x_i + v_i t + \frac{1}{2}at^2$ with $a = 7.40$ m/s² and $t = 1.00$ s:

$$x_f = x_i + v_i t + \tfrac{1}{2}at^2$$
$$= 0 + (0)t + \tfrac{1}{2}at^2 = \tfrac{1}{2}at^2$$
$$= \tfrac{1}{2}(7.40 \text{ m/s}^2)(1.00 \text{ s})^2 = \boxed{3.70 \text{ m}}$$

2 (**b**) From the calculation in part (a), we see that $x_f = x_i + v_i t + \frac{1}{2}at^2$ reduces to $x_f = \frac{1}{2}at^2$ in this situation. Thus, to find the position after 2.00 s, evaluate $x_f = \frac{1}{2}at^2$ at $t = 2.00$ s:

$$x_f = \tfrac{1}{2}at^2$$
$$= \tfrac{1}{2}(7.40 \text{ m/s}^2)(2.00 \text{ s})^2 = \boxed{14.8 \text{ m}}$$

3 (**c**) Repeat part (b) with $t = 3.00$ s:

$$x_f = \tfrac{1}{2}at^2$$
$$= \tfrac{1}{2}(7.40 \text{ m/s}^2)(3.00 \text{ s})^2 = \boxed{33.3 \text{ m}}$$

Insight

This example illustrates one of the key features of accelerated motion—position does not change uniformly with time when an object accelerates. In fact, the distance traveled in the first 2 seconds is 4 times the distance traveled in the first second, and the distance traveled in the first 3 seconds is 9 times the distance traveled in the first second. This behavior is a direct result of the fact that x depends on t^2 when an object accelerates.

26. **Follow-up** How much time does it take for the drag racer in Guided Example 3.8 to travel 5.00 m from its starting point?

27. A child slides down a hill on a toboggan with an acceleration of 1.8 m/s². If she starts with an initial push of 1.2 m/s, how far does she travel in 4.0 s?

28. **Triple Choice** A horse accelerates from rest for 1 s and covers a distance *D*. If, instead, the horse accelerates from rest with the same acceleration for 2 s, will the distance it covers be equal to 2*D*, 4*D*, or 9*D*? Explain.

29. **Challenge** A cheetah can accelerate from rest to 25.0 m/s in 6.00 s. Assuming that the cheetah moves with constant acceleration, what distance does it cover in the first 3.00 s?

CONNECTINGIDEAS

In this chapter we develop equations for motion with constant acceleration.

• These equations will be used again when we study rotational motion in Chapter 8. The only difference will be a slight change in notation—the concepts are exactly the same.

Acceleration causes velocity to change with position

It takes time for an object to move from one position to another. Therefore, if an object's velocity changes with time, it also changes with position. Our final constant acceleration equation shows the connection between velocity and position for an accelerating object:

Velocity-Position Equation

$$\left(\begin{array}{c}\text{final}\\\text{velocity}\end{array}\right)^2 = \left(\begin{array}{c}\text{initial}\\\text{velocity}\end{array}\right)^2 + 2(\text{acceleration})(\text{change in position})$$

$$v_f^2 = v_i^2 + 2a(x_f - x_i)$$
$$= v_i^2 + 2a\Delta x$$

This equation allows us to relate the velocity at one position to the velocity at another position, without knowing how much time is involved. Guided Example 3.9 shows how to apply the equation to a practical situation.

All of our constant acceleration equations are collected for easy reference in **Table 3.3**.

Table 3.3 Constant Acceleration Equations of Motion

Variables Related	Equation
velocity, time, acceleration	$v_f = v_i + at$
initial, final, and average velocity	$v_{av} = \frac{1}{2}(v_i + v_f)$
position, time, velocity	$x_f = x_i + \frac{1}{2}(v_i + v_f)t$
position, time, acceleration	$x_f = x_i + v_i t + \frac{1}{2}at^2$
velocity, position, acceleration	$v_f^2 = v_i^2 + 2a\Delta x$

A park ranger driving on a back country road suddenly sees a deer "frozen" in the headlights 20.0 m ahead. The ranger, who is driving at 11.4 m/s, immediately applies the brakes and slows with an acceleration of magnitude 3.80 m/s². How much distance is required for the ranger's vehicle to come to rest?

Picture the Problem

We choose the positive x direction to be the direction of motion. With this choice it follows that the initial velocity is $v_i = +11.4$ m/s. In addition, the fact that the ranger's vehicle is slowing down means that its acceleration points in the *opposite* direction to the velocity. Therefore, the vehicle's acceleration is $a = -3.80$ m/s². Finally, when the vehicle comes to rest, its final velocity is zero, $v_f = 0$.

Strategy

We can find the stopping distance by rearranging $v_f^2 = v_i^2 + 2a\Delta x$ to solve for Δx.

Known

$v_i = +11.4$ m/s
$a = -3.80$ m/s²
$v_f = 0$

Unknown

$\Delta x = ?$

Solution

1 Solve $v_f^2 = v_i^2 + 2a\Delta x$ for the distance, Δx:

$$\Delta x = \frac{v_f^2 - v_i^2}{2a}$$

2 Substitute the numerical values for v_i, v_f, and a and evaluate Δx:

$$\Delta x = \frac{(0)^2 - (11.4 \text{ m/s})^2}{2(-3.80 \text{ m/s}^2)}$$

$$= \frac{-129.96 \text{ m}^2/\text{s}^2}{-7.60 \text{ m/s}^2}$$

$$= \boxed{17.1 \text{ m}}$$

Math HELP
Negative
Numbers
See Math
Review,
Section I

Insight

Whew! The vehicle stops in front of the deer, with 20.0 m −17.1 m = 2.9 m to spare. Notice that if the initial speed of the vehicle is doubled, the distance needed to stop increases by a factor of 4. This is because Δx depends on the square of the initial velocity. This result indicates why speed on the highway has such a great influence on safety.

Practice Problems

30. **Follow-up** How much time is required for the ranger's vehicle in Guided Example 3.9 to stop?

31. On October 9, 1992, a 12.25-kg meteorite struck a car in Peekskill, New York, leaving a dent 22 cm deep in the trunk. If the meteorite struck the car with a speed of 130 m/s, what was the magnitude of its deceleration, assuming it to be constant?

32. A model rocket rises from rest with a constant acceleration of 106 m/s². What is the rocket's speed at a height of 3.2 m?

33. **Challenge** Coasting due west on your bicycle at 8.4 m/s, you encounter a sandy patch of road 7.2 m across. When you leave the sandy patch, your speed has been reduced to 6.4 m/s. What is the bicycle's acceleration in the sandy patch? Assume that the acceleration is constant and that the direction of travel is the positive direction.

A velocity-position graph for the ranger's vehicle in Guided Example 3.9 is shown in **Figure 3.13**. The graph shows the vehicle's velocity versus the distance traveled after the brakes are applied. Notice that although the acceleration is constant, the vehicle's velocity changes more in the second half of the stopping distance than in the first half. This is because more time is required to cover the second half of the stopping distance—and the greater the time, the greater the change in velocity.

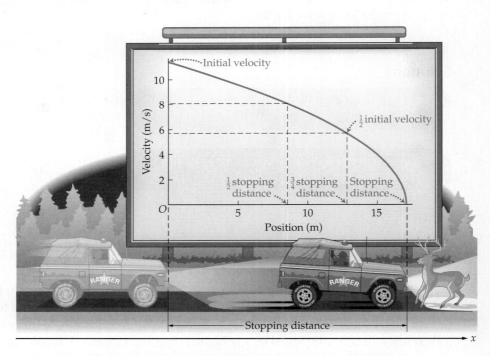

▲ Figure 3.13 **Velocity-position graph for the vehicle in Guided Example 3.9**
The ranger's vehicle in Guided Example 3.9 comes to rest with constant acceleration. This means that its velocity decreases uniformly with time. The velocity does not decrease uniformly with *distance*, however. In particular, notice how rapidly the velocity decreases in the final one-quarter of the stopping distance.

3.2 LessonCheck (MP)

Checking Concepts

34. 🔵 **Explain** An object moves with constant acceleration. How is the average velocity of this object related to its initial and final velocities?

35. 🔵 **Describe** An object moves 5 m between the times $t = 1$ s and $t = 2$ s. What is the area under the object's velocity-time curve between these times?

36. | Triple Choice | The acceleration of a train increases. As a result, does the rate at which its velocity changes increase, decrease, or stay the same? Explain.

37. Apply Assume that the brakes in your car create a constant deceleration. If you double your driving speed, how does this affect **(a)** the time required to come to a stop and **(b)** the distance needed to stop? Explain.

Solving Problems

38. Calculate When you see a traffic light turn red, you apply the brakes and stop with constant acceleration. If your initial speed was 12 m/s, what is your average speed during braking?

39. Calculate A baseball player is dashing toward home plate with a speed of 5.8 m/s when she decides to hit the dirt. She slides for 1.1 s, just reaching the plate as she stops (safe, of course).
(a) What is her acceleration? (Assume that her running direction is the positive direction.)
(b) How far does she slide?

40. Calculate A firefighter slides down a pole in a fire station with an acceleration of 6.2 m/s². If the firefighter starts at rest and slides for a distance of 3.7 m, what is his final speed?

41. Interpret Graphs The velocity-time graph for an ice skater is shown in **Figure 3.14**. What is the distance covered by the skater **(a)** between 0 and 2 s and **(b)** between 2 s and 5 s?

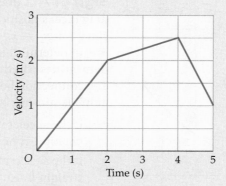

▲ Figure 3.14

3.3 Position-Time Graphs for Constant Acceleration

The shape of a position-time graph tells a lot about the type of motion it depicts. This is true whether the motion has constant velocity or constant acceleration.

Position-time graphs of accelerated motion are not linear

🗝 *What are the characteristics of a position-time graph for constant acceleration?*

In the previous lesson we calculated the position of a drag racer at three different times in Guided Example 3.8. These results are collected in **Table 3.4**. While a table like this is useful, a far better way to visualize the motion of the drag racer is with a position-time graph, as shown in **Figure 3.15**.

The first thing you notice about Figure 3.15 is that the motion does not produce a straight line. This makes sense. After all, a straight line indicates constant velocity, not the constant acceleration of the drag racer. Because the accelerating dragster covers an increasing distance with each passing second, the position-time graph is an upward-curving *parabola*. 🗝 **Constant acceleration produces a parabolic position-time graph.**

Table 3.4 Positions of the Drag Racer in Guided Example 3.8

Time (s)	Position (m)
1.00	3.70
2.00	14.8
3.00	33.3

Sign of Acceleration and Curvature The curved shape of the drag racer's position-time graph is due to the t^2 term in the position-time equation, $x_f = x_i + v_i t + \frac{1}{2}at^2$. In addition, the acceleration plays a key role in determining the type of curvature. 🗝 **The sign of the acceleration determines whether the parabola has an upward or downward curvature.** The connection between acceleration and the shape of the position-time graph is summarized below and illustrated in **Figure 3.16**.

- If the acceleration is positive ($a > 0$), then the parabola of the position-time graph curves upward.
- If the acceleration is zero ($a = 0$), then the position-time graph is a straight line.
- If the acceleration is negative ($a < 0$), then the parabola of the position-time graph curves downward.

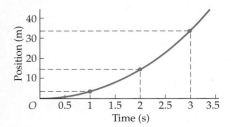

▲ **Figure 3.15 Position-time graph for drag racer in Guided Example 3.8**
The upward-curving, parabolic shape of this position-time graph indicates a positive, constant acceleration. The red dots are the data points from Table 3.4.

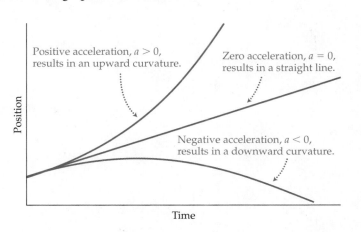

▶ **Figure 3.16 Positive, negative, and zero acceleration and the shape of the position-time graph**

Positive acceleration, $a > 0$, results in an upward curvature.

Zero acceleration, $a = 0$, results in a straight line.

Negative acceleration, $a < 0$, results in a downward curvature.

Magnitude of Acceleration and Curvature
How sharply a position-time graph curves tells you the magnitude of the acceleration. The sharper the curvature, the greater the magnitude of the acceleration. A graph that curves sharply upward indicates a large positive acceleration; a graph that curves sharply downward indicates a large negative acceleration.

To visualize the connection between acceleration and curvature, take a close look at the six parabolic curves shown in **Figure 3.17**. Notice that as the acceleration becomes increasingly positive, the parabolas become more sharply curved upward. Similarly, as the acceleration becomes more negative (which increases the magnitude), the parabolas become more sharply curved downward. ⌐⊙⊐ **In general, the greater the curvature of the parabola, the greater the magnitude of the acceleration.**

Term-by-Term Graph Characteristics
You can get a good understanding of each term in the equation $x_f = x_i + v_i t + \frac{1}{2}at^2$ by examining the graph in **Figure 3.18**. Notice the following important points:

- The vertical-axis intercept of the curve is equal to the initial position, x_i.
- The initial slope of the curve is equal to the initial velocity, v_i.
- The sharpness of the curvature indicates the magnitude of the acceleration, a.

Clearly, a wealth of information can be obtained from a position-time graph.

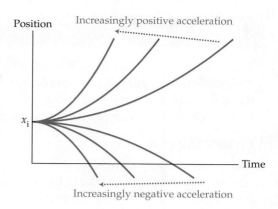

▲ **Figure 3.17 Position-time curves and magnitude of acceleration**
The greater the magnitude of an acceleration, the more sharply curved the position-time parabola.

Practice Problem

42. Rank The motions of four different objects are shown by the position-time graphs in **Figure 3.19**. The curves are labeled A, B, C, and D.

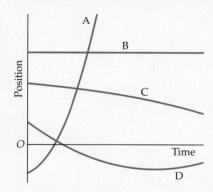

▲ **Figure 3.19**

(a) Rank the curves in order of initial position, from most negative to most positive.
(b) Rank the curves in order of initial velocity, from most negative to most positive.
(c) Rank the curves in order of acceleration, from most negative to most positive.

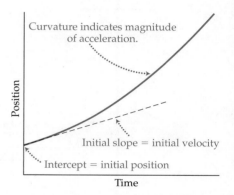

▲ **Figure 3.18 Position-time graph for constant acceleration**
Each term in the position-time equation has a specific meaning. The intercept is the initial position, the initial slope is the initial velocity, and the curvature indicates the acceleration.

Position-time graphs can track multiple objects
You can compare the motions of several objects by plotting their position-time graphs on the same set of axes. The next Guided Example considers one such case, a police car catching up with a speeder. Notice that the intersection of the two position-time curves indicates the catch-up time and location.

A speeder travels at 18.2 m/s. The instant the speeder passes a parked police car, the police begin their pursuit. The speeder maintains a constant velocity, and the police car has a constant acceleration of 4.50 m/s². Show that the police car catches up with the speeder after 8.09 s, and determine the location where this occurs.

Picture the Problem

Our sketch shows the two cars at the moment the speeder passes the resting police car. At this instant, which we take to be $t = 0$, both the speeder and the police car are at the origin, $x_i = 0$. In addition, we choose the positive x direction to be the direction of motion. Therefore, the speeder's initial velocity is $v_{speeder} = 18.2$ m/s, and the police car's initial velocity is zero. The speeder's acceleration is zero, but the police car has an acceleration of $a_{police} = 4.50$ m/s².

Our plot shows a straight-line position-time graph for the speeder (constant velocity) and a parabolic position-time graph for the police car (constant acceleration).

Math HELP
The Quadratic Equation, Graphical Solution
See Math Review, Section III

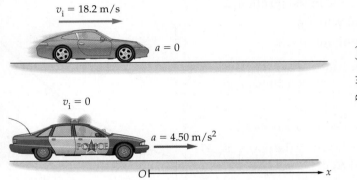

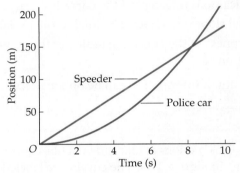

Strategy

To solve this problem we apply the position-time equation to both the speeder and the police car. We substitute the given time and show that each car has the same position at that time.

Known

$v_{speeder} = +18.2$ m/s
$a_{police} = +4.50$ m/s²

Unknown

$x_f = ?$

Solution

1 Write the position-time equation for the speeder. In this case, $x_i = 0$, $v_i = v_{speeder} = 18.2$ m/s, and $a = 0$:

$$x_f = x_i + v_i t + \tfrac{1}{2}at^2$$
$$= 0 + (18.2 \text{ m/s})t + \tfrac{1}{2}(0)t^2$$
$$= (18.2 \text{ m/s})t$$

2 Evaluate the position of the speeder at the time $t = 8.09$ s:

$$x_f = (18.2 \text{ m/s})(8.09 \text{ s})$$
$$= \boxed{147 \text{ m}}$$

⎫ Speeder

3 Write the position-time equation for the police car. In this case, $x_i = 0$, $v_i = 0$, and $a = a_{police} = 4.50$ m/s²:

$$x_f = x_i + v_i t + \tfrac{1}{2}at^2$$
$$= 0 + (0)t + \tfrac{1}{2}(4.50 \text{ m/s}^2)t^2$$
$$= \tfrac{1}{2}(4.50 \text{ m/s}^2)t^2$$

4 Evaluate the position of the police car at the time $t = 8.09$ s:

$$x_f = \tfrac{1}{2}(4.50 \text{ m/s}^2)(8.09 \text{ s})^2$$
$$= \boxed{147 \text{ m}}$$

⎫ Police car

Insight

Notice that the time and location at which the police car catches up are in agreement with the intersection point in the position-time graph.

43. **Follow-up** Show that the velocity of the police car at the moment it catches up with the speeder is twice the speeder's velocity. Use the average velocity equation to show that this should not be a surprise.

44. **Think & Calculate** Suppose the police car accelerates at 5.00 m/s^2. **(a)** Is the time until it catches up with the speeder greater than, less than, or equal to $t = 8.09$ s? Explain. **(b)** Find the catch-up time if the acceleration of the police car is 5.00 m/s^2.

45. **Challenge** Two cars drive on a straight highway. At time $t = 0$, car 1 passes road marker 0 traveling due east with a speed of 20.0 m/s. At the same time, car 2 is 1.0 km east of road marker 0 traveling at 30.0 m/s due west. Car 1 is speeding up, with an acceleration of 2.5 m/s^2, and car 2 is slowing down, with an acceleration of -3.2 m/s^2. **(a)** Write position-time equations for both cars. Let east be the positive direction. **(b)** At what time do the two cars meet?

Graphs can show acceleration and deceleration

What happens to the velocity of a ball when you throw it straight up into the air? The ball slows down on the way up, stops at its peak height (called the **apogee**), and speeds up on the way down. The downward acceleration of the ball is the same at all times, however.

The situation is illustrated with the position-time graph in **Figure 3.20**. On the upward part of the flight, the ball's velocity is upward but its acceleration is downward. As you learned in Lesson 3.1, objects slow down (or decelerate) when their velocity and acceleration are in opposite directions. On the way down, the ball's velocity is downward—as is the acceleration. With the velocity and acceleration in the same direction, the ball speeds up. You can tell that the acceleration doesn't change during the ball's flight because the curvature of the parabola in Figure 3.20 does not change. Thus, a single parabola in a position-time graph can show both deceleration and acceleration. Situations like this are the main focus of the next lesson.

Reading Support ✓

Vocabulary Builder

apogee [AP uh jee]

(noun) a point in the orbit of the Moon or a satellite at which it is farthest from Earth; the farthest or highest point; the culmination

At its apogee the rocket was 300 km from Earth.

The tremendous success of her third novel was the apogee of her career.

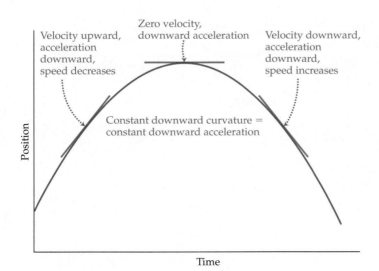

Velocity upward, acceleration downward, speed decreases

Zero velocity, downward acceleration

Velocity downward, acceleration downward, speed increases

Constant downward curvature = constant downward acceleration

Position

Time

◄ **Figure 3.20 A ball moves upward but accelerates downward**

A ball tossed up in the air moves upward while gravity accelerates it downward. This causes the speed of the ball to decrease. Eventually the ball begins to move downward, and then the velocity and acceleration are in the same direction. The ball's speed now begins to increase.

3.3 LessonCheck ⏀

Checking Concepts

46. 🖼 **Apply** Use your knowledge of position-time graphs of accelerated motion to answer the following:
(a) A plot shows two position-time graphs, one a straight line and one a parabola. Which graph corresponds to an object with constant acceleration?
(b) The acceleration of a car changes from positive to negative. How does this affect the shape of the car's position-time graph?
(c) For each position-time graph in **Figure 3.21**, explain how the motions of object A and object B differ.

Graph 1

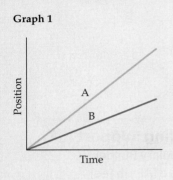

Graph 2

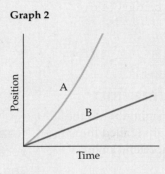

Graph 3

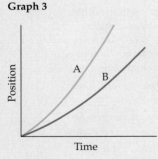

Graph 4

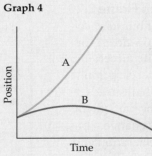

▲ **Figure 3.21**

47. |Triple Choice| The acceleration of an object becomes more negative. Is the sharpness of the downward curvature of its position-time graph greater than, less than, or the same as it was before?

Solving Problems

48. Apply You throw a ball upward with an initial speed of 3.0 m/s from an initial height of 1.5 m. After you throw the ball, its acceleration is 9.81 m/s downward. Taking upward to be the positive direction, write the position-time equation for the ball's motion.

49. Calculate A bicyclist is finishing her repair of a flat tire when a friend rides by with a constant speed of 3.5 m/s. Two seconds later the bicyclist hops on her bike and accelerates at 2.4 m/s^2 until she catches up with her friend.
(a) How much time does it take until she catches up?
(b) How far does she travel in this time?
(c) What is her speed when she catches up?

50. Calculate Two motorcycles travel along a straight road heading due north. At $t = 0$ motorcycle 1 is at $x = 50.0$ m and moves with a constant speed of 6.50 m/s; motorcycle 2 starts from rest at $x = 0$ and moves with constant acceleration. Motorcycle 2 passes motorcycle 1 at the time $t = 10.0$ s.
(a) Figure 3.22 shows a position-time graph for the two motorcycles. Which motorcycle does the red line represent?
(b) What is the position of the two motorcycles when motorcycle 2 passes motorcycle 1?
(c) What is the acceleration of motorcycle 2?
(d) What is the speed of motorcycle 2 when it passes motorcycle 1?

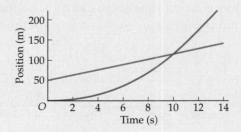

▲ **Figure 3.22**

3.4 Free Fall

Suppose you step from a diving board and drop straight down into a swimming pool, executing a perfect cannonball. Air resistance has little effect on your motion as you fall. For the most part your motion is determined solely by the force of gravity. Cases like this, where an object's motion is determined by gravity alone, are referred to as **free fall**. The word *free* indicates that the object is free from influences other than gravity. We explore a number of examples of free fall in this lesson.

Freely falling objects have constant acceleration

Whether the Italian scientist Galileo Galilei (1564–1642) dropped objects from the Leaning Tower of Pisa to study free fall will probably never be known for certain. What we do know, however, is that he performed extensive experiments on motion. **From his experiments Galileo concluded that if the effects of air resistance can be neglected, then all objects in free fall have the same constant acceleration.** His conclusion applies to all objects, regardless of size or weight—provided only that air resistance is small enough to ignore.

Today it is easy to verify Galileo's assertion by dropping objects in a vacuum chamber, where the effects of air resistance are removed. A novel version of this experiment was carried out in 1971 on the Moon. In the near-perfect vacuum on the Moon's surface, an astronaut dropped a feather and a hammer and showed a worldwide television audience that they fell to the ground in the same time.

Many situations approximate free fall

The effects of air resistance are easy to observe. For example, try dropping a sheet of paper and a rubber ball at the same time, as shown in **Figure 3.23**. The paper drifts slowly to the ground, taking much longer to fall than the ball. Now, wad the sheet of paper into a tight ball and repeat the experiment. This time the paper and the ball reach the ground in nearly the same time. What was different in the two experiments? When the sheet of paper is wadded into a ball, the effect of air resistance is greatly reduced. As a result, both objects fall almost as they would in a vacuum. In general, the motion of an object can be treated as free fall whenever the effects of air resistance are small enough to ignore. In situations where air resistance or other forms of friction are significant, the motion of a falling object deviates from the ideal behavior of free fall.

Vocabulary

- free fall

What did Galileo conclude about objects in free fall?

(a) Dropping a sheet of paper and a rubber ball

(b) Dropping a wadded-up sheet of paper and a rubber ball

▲ Figure 3.23 **Free fall and air resistance**

▲ **Figure 3.24 Blanket toss**
Whether she is on the way up, at the peak of her flight, or on the way down, this girl is in free fall, accelerating downward with the acceleration due to gravity. Only when she is in contact with the blanket does her acceleration change.

Freely falling objects are always accelerating

Here's something most people don't realize: The word *fall* in *free fall* does not mean that the object is necessarily moving downward. The girl tossed into the air in **Figure 3.24** is a perfect example. She is in free fall even when she is moving upward. In fact, her acceleration is the same at all times while she is in the air. Only when she is in contact with the blanket does her acceleration change.

What about a thrown ball? Does the direction in which you throw it affect its acceleration? Not at all. It doesn't matter if you throw the ball upward or downward, or simply let it drop—the ball is in free fall as soon as it leaves your hand. The ball's acceleration is caused by gravity, and gravity produces the same downward acceleration regardless of how the ball is moving. The ball is even accelerating downward at the top of its flight, when it is momentarily at rest. After all, gravity doesn't turn off just because the ball is at rest for an instant.

CONCEPTUAL Example 3.11 Does the Separation Change?

Imagine dropping two rocks from rest, one after the other. When the first rock has fallen 4 m, you drop a second rock. As the two rocks continue their free fall, does their separation increase, decrease, or stay the same?

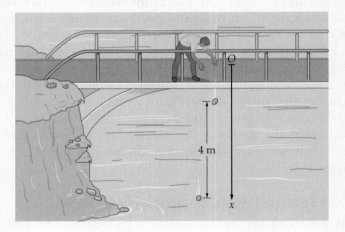

Reasoning and Discussion
It might seem that since both rocks are in free fall, their separation remains the same. This is not so. The first rock always has a greater velocity than the second rock. As a result, the first rock covers more distance in any interval of time than the second rock. It follows that the separation between the rocks increases with time.

Answer
The separation between the rocks increases.

The acceleration due to gravity is denoted as *g*

The acceleration produced by gravity at the Earth's surface is denoted with the symbol *g*. As shorthand, we refer to *g* as "the acceleration due to gravity." The acceleration due to gravity varies slightly from location to location on the Earth, and also with altitude above the surface. **Table 3.5** gives the value of *g* at a variety of locations.

Table 3.5 Values of *g* at Different Locations on Earth (m / s²)

Location	Latitude	*g*
North Pole	90° N	9.832
Oslo, Norway	60° N	9.819
Hong Kong	30° N	9.793
Quito, Ecuador	0°	9.780

In all our calculations we will use $g = 9.81 \text{ m/s}^2$ for the acceleration due to gravity. Note that g always stands for $+9.81 \text{ m/s}^2$, and never -9.81 m/s^2. For example, if we choose a coordinate system with the positive direction upward, the acceleration in free fall is $a = -g$. If the positive direction is downward, then free-fall acceleration is $a = g$.

GUIDED Example 3.12 | Do the Cannonball!   Free Fall

A boy steps off the end of a 3.00-m-high diving board and drops to the water below. (**a**) How much time does it take for him to reach the water? (**b**) What is the boy's speed on entering the water?

Picture the Problem

In our sketch we choose the origin to be at the height of the diving board, and we let the positive direction be downward. With these choices, $x_i = 0$, $a = g$, and the water is at $x_f = 3.00$ m. Of course, $v_i = 0$ since the boy simply steps off the board.

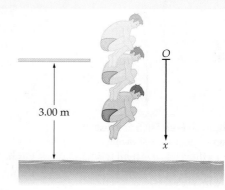

3.00 m

Strategy

Because the boy is in free fall, with a constant acceleration equal to g, the equations in Table 3.3 apply. In part (a) we substitute the known distance and acceleration into $x_f = x_i + v_i t + \frac{1}{2}at^2$ and find the time of the fall. In part (b) we calculate the final velocity, v_f, using $v_f = v_i + at$.

Known

$x_i = 0 \quad x = 3.00 \text{ m} \quad v_i = 0$
$a = g = 9.81 \text{ m/s}^2$

Unknown

(**a**) $t = ?$ (**b**) $v = ?$

Solution

Part (a)

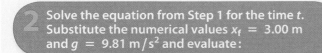

1 Substitute the known values $x_i = 0$, $v_i = 0$, and $a = g$ into $x_f = x_i + v_i t + \frac{1}{2}at^2$:

$$x_f = x_i + v_i t + \tfrac{1}{2}at^2$$
$$= 0 + (0)t + \tfrac{1}{2}gt^2$$
$$= \tfrac{1}{2}gt^2$$

> **Math HELP**
> Solving Simple Equations
> See Math Review, Section III

2 Solve the equation from Step 1 for the time t. Substitute the numerical values $x_f = 3.00$ m and $g = 9.81 \text{ m/s}^2$ and evaluate:

$$t = \sqrt{\frac{2x_f}{g}}$$
$$= \sqrt{\frac{2(3.00 \text{ m})}{9.81 \text{ m/s}^2}}$$
$$= \boxed{0.782 \text{ s}}$$

Part (b)

3 Use the time found in part (a) in $v_f = v_i + at$:

$$v_f = v_i + gt$$
$$= 0 + (9.81 \text{ m/s}^2)(0.782 \text{ s})$$
$$= \boxed{7.67 \text{ m/s}}$$

Insight

Let's express these results in more common, everyday units. If you step off a diving board 9.84 ft (3.00 m) above the water, you enter the water with a speed of 17.2 mi/h (7.67 m/s).

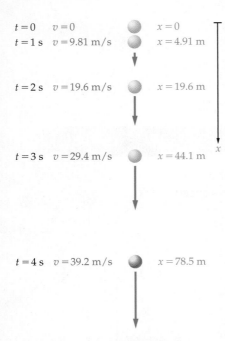

$t = 0$ $v = 0$ $x = 0$
$t = 1$ s $v = 9.81$ m/s $x = 4.91$ m

$t = 2$ s $v = 19.6$ m/s $x = 19.6$ m

$t = 3$ s $v = 29.4$ m/s $x = 44.1$ m x

$t = 4$ s $v = 39.2$ m/s $x = 78.5$ m

▲ **Figure 3.25 Free fall from rest**
Position and velocity are shown as functions of time. It is apparent that velocity increases with t, whereas position increases with t^2.

Practice Problems

51. Follow-up Repeat part (b) of Guided Example 3.12 using the equation $v_f^2 = v_i^2 + 2a\Delta x$. Verify that this method yields the same final speed of 7.67 m/s.

52. Follow-up What is the speed of the boy in Guided Example 3.12 on entering the water if he steps off a 10.0-m-high diving tower?

53. Triple Choice A survival package is dropped from a hovering helicopter to stranded hikers. If the package is dropped from a height H, it lands with a speed V. If the package is dropped from a height $2H$ instead, is its landing speed $\sqrt{2}V$, $2V$, or $4V$?

54. One fine summer day a group of students were jumping from a railroad bridge into the Snohomish River. They stepped off the bridge when they "jumped," and they hit the water 1.5 s later. How high was the bridge?

Position and velocity change differently in free fall

The velocity of an object in free fall increases by the same amount with each passing second, but its position changes more with each second. This behavior is illustrated in **Figure 3.25**.

- Velocity increases linearly with time (t); this is referred to as a *linear relationship*. Doubling the time (say from $t = 1$ s to $t = 2$ s) doubles the velocity (from 9.81 m/s to 19.6 m/s).

- Distance increases with time squared (t^2); this is referred to as a *parabolic relationship*. Doubling the time (from 1 s to 2 s) increases the distance by a factor of $2^2 = 4$ (from 4.91 m to 19.6 m).

Symmetry of Motion The motion of objects in free fall has a wonderful symmetry to it. This is illustrated beautifully in **Figure 3.26**, which shows the impressive sight of fountains of lava shooting out from a volcano. Notice the smooth parabolic paths taken by the individual lava bombs.

▶ **Figure 3.26 Parabolic paths**
In the absence of air resistance, these lava bombs from the Kilauea volcano on the big island of Hawaii would strike the water with the same speed they had when they were blasted into the air.

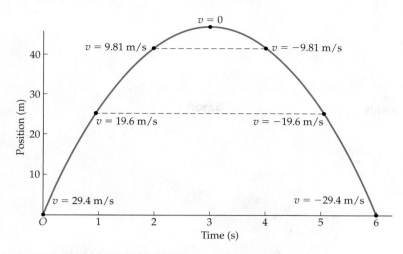

▲ Figure 3.27 **Position and velocity of a lava bomb**
This lava bomb is in the air for 6 s. Notice the symmetry about the midpoint of the bomb's flight.

The symmetry of free fall can also be seen in a position-time graph. **Figure 3.27** shows the position-time graph for a lava bomb whose flight time is 6 seconds. At $t = 3$ s the lava is at its highest point and its velocity is zero. At times equally spaced before and after $t = 3$ s, the lava is at the same height and has the same speed, but is moving in opposite directions. Thus, a lava bomb lands with the same speed it had when it was shot out of the volcano. Because of this symmetry, a movie of the lava bomb's flight looks exactly the same whether viewed forward or in reverse. Cool.

3.4 LessonCheck MP

Checking Concepts

55. ▭ **Compare** Two rocks with different weights are dropped on the surface of the Moon (where there is a near vacuum). How does the acceleration of the heavier rock compare to that of the lighter rock?

56. Apply Why is a person doing a cannonball dive an example of free fall while a person descending to Earth on a parachute is not?

57. Big **Idea** What property do all free-falling objects have in common?

58. ☐ Triple Choice ☐ At the end of a baseball game Zoe drops her glove to the ground. Mickey tosses his glove into the air. While the gloves are still in the air, is the acceleration of Zoe's glove greater than, less than, or equal to the acceleration of Mickey's glove? Explain.

59. Determine A girl on a backyard trampoline bounces straight upward with an initial velocity of 4.5 m/s. What is the girl's velocity when she returns to the trampoline?

Solving Problems

60. Calculate Seagulls are often observed dropping clams and other shellfish from a height onto rocks below, as a means of opening the shells. If a seagull drops a shell from rest from a height of 14 m, how fast is the shell moving when it hits the rocks?

61. Calculate A volcano launches a lava bomb straight upward with an initial speed of 28 m/s. Taking upward to be the positive direction, find the speed and direction of motion of the lava bomb (**a**) 2.0 s and (**b**) 3.0 s after it is launched.

62. Analyze Bernardo steps off a 3.0-m-high diving board and drops to the water below. At the same time Michi jumps upward with a speed of 4.2 m/s from a 1.0-m-high diving board. Taking the origin to be at the water's surface and upward to be the positive x direction, write position-time equations for Bernardo and Michi.

Physics & You

Microbursts

What Are They? A microburst is a powerful stream of cool air accelerating downward from a thunderstorm. When the air hits the ground, it "splashes" and accelerates outward and upward.

Why Are They Important? If a microburst occurred over an airport runway, it could dangerously accelerate an airplane that was landing or taking off. Though accidents caused by microbursts are rare, they have happened, and they can pose extreme risks to aircraft.

A Close Call In 1983 *Air Force One* landed safely on a dry runway at Andrews Air Force Base near Washington DC with President Ronald Reagan on board. It was a smooth safe landing, but just 7 minutes later a microburst hit the same runway. Winds in that microburst had speeds of more than 240 kmh (about 150 mph)! If *Air Force One* had been caught in the microburst as it was landing, the microburst could have accelerated the plane toward the ground, resulting in an almost unavoidable crash. After this close call the Federal Aviation Agency (FAA) invested millions of dollars in improving the technology for detecting microbursts, focusing on the use of Doppler radar.

How Can They Be Avoided? In the 1990s airports began using Doppler radar systems to scan for weather conditions that might lead to microbursts. As the use of Doppler radar spread, meteorologists became more skilled at analyzing the data it provided, which has allowed more accurate predictions of high-risk areas. Thus, better weather-monitoring technology and increased forecaster expertise have greatly improved the ability of pilots to avoid microbursts. And, as an added safety measure, United States commercial pilots are required to take special training concerning what to do in case they are caught in a microburst.

▲The downward thrust of air from a microburst usually lasts only a few minutes.

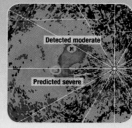

▲Predicting areas at high risk for microbursts requires Doppler radar data and expert analysis.

Take It Further

1. Research *Conduct research to find out how the cold air column in a microburst forms and why it accelerates downward.*

2. Research *Use the Internet to search for recent news stories related to microbursts. Summarize your findings for your classmates in the form of a poster or written report.*

3. Infer *Explain how the effect of a microburst on an airplane will vary depending on which portion of the microburst the airplane passes through.*

Physics Lab Investigating Acceleration

In this lab you will use a motion detector to discover the nature of positive and negative acceleration.

Materials
• track   • cart   • motion detector

Procedure

1. Construct a ramp on a tabletop by placing one end of the track on top of one or two books.

2. You need to know how to operate the motion detector, especially how to zero it. You also need to be familiar with the graphing capabilities of your software.

3. For each trial, position the motion detector and zero it as indicated in the chart below.

(a) Sketch a predicted position-time graph.

(b) Sketch a predicted velocity-time graph.

(c) Predict whether the acceleration will be positive or negative.

(d) Run the trial with the motion detector.

(e) Sketch the actual position-time graph.

(f) Sketch the actual velocity-time graph.

(g) Check the acceleration-time graph and note the actual acceleration.

Trial	Location of Motion Detector	Motion of the Cart	Zero Position
1	Top of ramp	Cart rolls down from the top of ramp.	Top of ramp
2	Bottom of ramp	Cart rolls up ramp. Push the cart before you start the motion detector and catch the cart at the top	Bottom of ramp
3	Bottom of ramp	Cart rolls down from the top of ramp. Catch the cart before it hits the motion detector.	Bottom of ramp
4	Top of ramp	Cart rolls up ramp. Push the cart before you start the motion detector and catch the cart at the top.	Top of ramp
5	Bottom of ramp	Cart rolls up ramp. Push the cart before you start the motion detector and catch the cart at the top.	Top of ramp

Data Table

Trial	x_0	x_f	v_0	v_f	a
1					
2					

Analysis

1. What is the shape of the position-time graph when the cart is accelerating?

2. Based on your observations, describe the motion of the cart when the velocity is positive.

3. For each trial, identify the quantities in the data table as positive (+), negative (−), or zero (0).

Conclusions

1. A position-time graph can be concave (like a dip in the road) or convex (like the top of a hill). Which shape corresponds to positive acceleration and which to negative acceleration?

2. Describe two different situations in which the cart experienced a positive acceleration.

3. Describe two different situations in which the cart experienced a negative acceleration.

4. Evaluate the following statements. If they are false, indicate how you would correct them.

(a) A positive acceleration means that the cart is speeding up.

(b) If the acceleration is negative, the magnitude of the velocity must be decreasing.

(c) If the cart travels in the negative direction and slows down, the acceleration is positive.

3 Study Guide

Big Idea

All objects in free fall move with the same constant acceleration.

This is true regardless of whether the object is moving upward or downward or is thrown, kicked, batted, or simply dropped. It also applies no matter what the mass of the object. To be in free fall, an object must move under the influence of gravity alone, with negligible effects from air resistance or other forms of friction.

3.1 Acceleration

🔑 Acceleration occurs when there is a change in speed, a change in direction, or a change in both speed and direction.

🔑 The velocity-time graph of an object with constant acceleration is a straight line with a slope equal to the acceleration.

🔑 Speed increases when velocity and acceleration have the same sign. Speed decreases when velocity and acceleration have opposite signs.

• An object with decreasing speed is said to be *decelerating*.

• Average acceleration is positive when $v_f > v_i$, negative when $v_f < v_i$, and zero when $v_f = v_i$.

Key Equation

Average acceleration is the change in velocity divided by the change in time:

$$a_{av} = \frac{\Delta v}{\Delta t} = \frac{v_f - v_i}{\Delta t}$$

3.2 Motion with Constant Acceleration

🔑 When the acceleration is constant, the average velocity is equal to the sum of the initial and final velocities divided by 2.

🔑 The distance traveled by an object is equal to the area under its velocity-time curve.

Key Equations

Velocity with constant acceleration is described by the following equations:

$$v_f = v_i + at$$
$$v_{av} = \tfrac{1}{2}(v_i + v_f)$$
$$v_f^2 = v_i^2 + 2a\Delta x$$

Position with constant acceleration is described by the following equations:

$$x_f = x_i + \tfrac{1}{2}(v_i + v_f)t$$
$$x_f = x_i + v_i t + \tfrac{1}{2}at^2$$

3.3 Position-Time Graphs for Constant Acceleration

🔑 Constant acceleration produces a parabolic position-time graph.

🔑 The sign of the acceleration determines whether the parabola has an upward or downward curvature.

🔑 The greater the curvature of the parabola, the greater the magnitude of the acceleration.

3.4 Free Fall

🔑 From his experiments Galileo concluded that if the effects of air resistance can be neglected, then all objects in free fall have the same constant acceleration.

• In general, the motion of an object can be treated as free fall whenever the effects of air resistance are small enough to ignore.

• The acceleration due to gravity at the Earth's surface is $g = 9.81 \text{ m/s}^2$.

3 Assessment

For instructor-assigned homework, go to www.masteringphysics.com MP™

ANSWERS TO SELECTED ODD-NUMBERED PROBLEMS APPEAR IN APPENDIX A.

Lesson by Lesson

3.1 Acceleration

Conceptual Questions

63. Why are the units of acceleration meters per second squared?

64. Is it possible to round a corner with constant speed? With constant velocity? Explain in each case.

65. The brakes on a train create a constant deceleration, regardless of how fast it's moving. If the speed of the train is doubled, how does this affect the time required for it to come to a stop?

66. If the velocity of a ball is zero at a given instant of time, can its acceleration at that instant be positive? Give an example if your answer is yes; explain why not if your answer is no.

67. If the velocity of a ball is positive, can its acceleration be negative? Give an example if your answer is yes; explain why not if your answer is no.

68. Triple Choice An object has a position-time graph that is a straight line with a negative slope. Is the acceleration of this object positive, negative, or zero? Explain.

69. Sketch a velocity-time graph for an object whose initial velocity is negative, but whose acceleration is positive.

Problem Solving

70. A jet that is making a landing is traveling due east with a speed of 115 m/s. If the jet comes to rest in 13.0 s, what are the magnitude and the direction of its average acceleration?

71. At the starting gun a runner accelerates from rest at 1.9 m/s² for 2.2 s. What is the runner's speed 2.0 s after she starts running?

72. A skier starts from rest and accelerates down a slope at 1.2 m/s². How much time is required for the skier to reach a speed of 7.3 m/s?

73. An object is moving initially with a velocity of 5.6 m/s. After 3.1 s the object's velocity is −2.7 m/s. What is the object's average acceleration during this time?

74. A car with an initial velocity of 12 m/s comes to rest in 3.5 s. What is the car's average acceleration during braking? Give both magnitude and sign.

75. A motorcycle moves according to the velocity-time graph shown in **Figure 3.28**. Find the average acceleration of the motorcycle during each of the following segments of the motion: **(a)** A, **(b)** B, and **(c)** C.

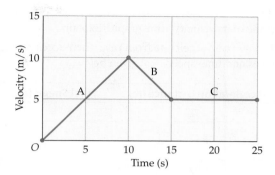

Figure 3.28

76. A person on horseback moves according to the velocity-time graph shown in **Figure 3.29**. Find the average acceleration of the horse and rider for each of the following segments of the motion: **(a)** A, **(b)** B, and **(c)** C.

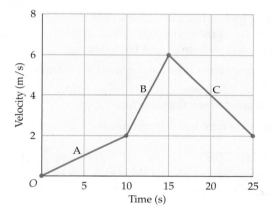

Figure 3.29

77. Think & Calculate Assume that the brakes in your car create a constant deceleration of 4.2 m/s² regardless of how fast you are driving. **(a)** If you double your driving speed from 16 m/s to 32 m/s, does the time required to come to a stop increase by a factor of 2 or a factor of 4? Explain. Verify your answer to part (a) by calculating the stopping times for initial speeds of **(b)** 16 m/s and **(c)** 32 m/s.

3.2 Motion with Constant Acceleration

Conceptual Questions

78. What is the difference between average acceleration and instantaneous acceleration?

79. Truck 1 accelerates from 5 m/s to 10 m/s in 10 s. Truck 2 accelerates from 10 m/s to 25 m/s in 15 s. Which truck has the greater acceleration?

80. Triple Choice Truck 1 accelerates from 5 m/s to 10 m/s in 10 m. Truck 2 accelerates from 15 m/s to 20 m/s in 10 m. Is the acceleration of truck 1 greater than, less than, or equal to the acceleration of truck 2? Explain.

81. An object moves with a constant acceleration. What is the shape of its velocity-time graph? Explain.

82. Rank Four cars accelerate from rest. Their acceleration distances and final speeds are given below.

Car	Acceleration Distance (m)	Final Speed (m)
1	10	5
2	20	10
3	1	3
4	100	20

Rank the cars in order of increasing acceleration. Indicate ties where appropriate.

83. Predict & Explain Two bows shoot identical arrows with the same initial speed. To accomplish this, the string in bow 1 must be pulled back farther than the string in bow 2. **(a)** Is the acceleration of the arrow shot by bow 1 greater than, less than, or equal to the acceleration of the arrow shot by bow 2? **(b)** Choose the *best* explanation from among the following:

A. The arrow from bow 2 accelerates over a greater distance.

B. Both arrows have the same final speed.

C. The arrow from bow 1 accelerates over a greater distance.

Problem Solving

84. An airplane lands with a speed of 81.9 m/s traveling due south. It comes to rest in 949 m. Assuming that the airplane slows with constant acceleration, find the magnitude and the direction of its acceleration.

85. On an amusement park ride passengers accelerate straight downward from rest to 20.1 m/s in 2.2 s (see **Figure 3.30**). What is the average acceleration of the passengers on this ride?

Figure 3.30

86. In a novel from 1866 the author describes a spaceship that is blasted out of a cannon with a speed of about 11,000 m/s. The spaceship is approximately 274 m long, but part of it is packed with gunpowder, so it accelerates over a distance of only 213 m. What was the acceleration (in m/s^2) experienced by the occupants of the spaceship during launch? (Though the novel's author realized that the "travelers would . . . encounter a violent recoil," he probably didn't know that people generally lose consciousness if they experience accelerations greater than about $7g$, or ~70 m/s^2.)

87. When a chameleon captures an insect, its tongue can extend 16 cm in 0.10 s (see **Figure 3.31**). Find the magnitude of the tongue's acceleration, assuming it to be constant.

Figure 3.31

88. Air Bags Air bags are designed to deploy in 10 ms. Given that the air bags expand 20 cm as they deploy, estimate the acceleration of the front surface of an expanding bag. Express your answer in terms of the acceleration due to gravity, $g = 9.81$ m/s^2.

89. A person on horseback moves according to the velocity-time graph shown in Figure 3.29. Find the displacement of the horse and rider for each of the following segments of the motion: **(a)** A, **(b)** B, and **(c)** C.

90. Think & Calculate Assume that the brakes in your car create a constant deceleration of 3.7 m/s^2 regardless of how fast you are driving. **(a)** If you double your driving speed from 11 m/s to 22 m/s, does the distance required to come to a stop increase by a factor of 2 or a factor of 4? Explain. Verify your answer to part (a) by calculating the stopping distances for initial speeds of **(b)** 11 m/s and **(c)** 22 m/s.

91. Think & Calculate You are driving through town at 12.0 m/s when suddenly a ball rolls out in front of you. You apply the brakes and begin decelerating at 3.5 m/s^2. **(a)** How far do you travel before stopping? **(b)** When you have traveled only half the distance in part (a), is your speed 6.0 m/s, greater than 6.0 m/s, or less than 6.0 m/s? **(c)** Support your answer with a calculation.

92. Surviving a Large Deceleration On July 13, 1977, while on a test drive at Britain's Silverstone racetrack, the throttle on David Purley's car stuck wide open. The resulting crash subjected Purley to the greatest *g*-force ever survived by a human—he decelerated from 173 km/h to zero in a distance of only about 0.66 m. Calculate the magnitude of the acceleration experienced by Purley (assuming it to be constant), and express your answer as a multiple of the acceleration due to gravity, $g = 9.81 \text{ m/s}^2$.

3.3 Position-Time Graphs for Constant Acceleration

Conceptual Questions

93. An object moves with a constant, negative acceleration. What is the shape of the position-time graph? Explain.

94. What does the intercept of a position-time graph represent? What does the initial slope represent? Assume motion with constant acceleration.

95. What is the acceleration of an object whose position-time graph is a straight line? Explain.

96. Triple Choice The position-time graph for an object moving with constant acceleration starts off with a positive slope. At later times the slope becomes negative. Is the initial velocity of this object positive, negative, or zero? Is the acceleration positive, negative, or zero? Explain.

97. Triple Choice The position-time graph for an object moving with constant acceleration starts off with zero slope. At later times the slope is positive. Is the initial velocity of this object positive, negative, or zero? Is the acceleration of the object positive, negative, or zero? Explain.

Problem Solving

98. A car has an initial position of 5.5 m, an initial velocity of 2.1 m/s, and a constant acceleration of 0.75 m/s^2. What is the position of the car at the time $t = 2.5$ s?

99. A car has an initial position of 3.2 m, an initial velocity of -8.4 m/s, and a constant acceleration of 1.1 m/s^2. What is the position of the car at the time $t = 1.5$ s?

100. The position-time equation for a certain train is

$$x_f = 2.1 \text{ m} + (8.3 \text{ m/s})t + (2.6 \text{ m/s}^2)t^2$$

(a) What is the initial velocity of this train? (b) What is its acceleration?

101. What is the position of the train in the previous problem at the time $t = 4.1$ s?

102. The position-time equation for a cheetah chasing an antelope is

$$x_f = 1.6 \text{ m} + (1.7 \text{ m/s}^2)t^2$$

(a) What is the initial position of the cheetah? (b) What is the initial velocity of the cheetah? (c) What is the cheetah's acceleration? (d) What is the position of the cheetah at $t = 4.4$ s?

103. A ball starts from rest at the initial position $x_i = 0$. The ball has a constant acceleration of 2.4 m/s^2. (a) Write the position-time equation for the ball. (b) What is the position of the ball at $t = 1.0$ s? (c) What is its position at $t = 2.0$ s?

104. A fishing boat leaves a marina with a constant speed of 3.4 m/s. A speedboat leaves the marina 12 s later with an initial speed of 2.8 m/s and an acceleration of 1.7 m/s^2. Let the time the speedboat leaves the marina be $t = 0$ and let the direction of travel of both boats be the positive *x* direction. (a) Write position-time equations for both boats. (b) When does the speedboat catch up with the fishing boat?

3.4 Free Fall

Conceptual Questions

105. What does it mean to say that an object is in free fall?

106. Can an object that is moving upward be in free fall?

107. Triple Choice An object is dropped at $t = 0$ and allowed to fall freely toward the ground. Is the distance covered by the object between $t = 0$ and $t = 1$ s greater than, less than, or equal to the distance covered by the object between $t = 1$ s and $t = 2$ s? Explain.

108. A batter hits a pop fly straight up. (a) Is the acceleration of the ball on the way up different from its acceleration on the way down? (b) Is the acceleration of the ball at the top of its flight different from its acceleration just before it lands?

109. Triple Choice Standing at the edge of a cliff, you drop ball 1 from rest. Later you throw ball 2 downward with a certain initial speed. When the balls reach the base of the cliff, is the *increase* in speed of ball 1 greater than, less than, or equal to the increase in speed of ball 2? Explain.

Problem Solving

110. How far does a freely falling apple drop in 5.0 s after being released from rest?

111. What is the speed of a freely falling baseball 6.0 s after it is dropped from rest?

112. A cartoon shows two friends watching an unoccupied car in free fall after it has rolled off a cliff. One friend says to the other, "It goes from zero to sixty [mph] in about three seconds." Is this statement correct? (Note that 60 mph converts to 26.8 m/s.)

113. A grapefruit falls from a tree and hits the ground 0.75 s later. **(a)** How far did the grapefruit drop? **(b)** What was its speed when it hit the ground?

114. An astronaut drops a rock on the surface of an asteroid. The rock is released from rest at a height of 0.95 m above the ground, and hits the ground 1.39 s later. What is the acceleration due to gravity on this asteroid?

115. Measure Your Reaction Time Here's something you *can* try at home—an experiment to measure your reaction time. Have a friend hold a ruler by one end, letting the other end hang down vertically. Hold your thumb and index finger on either side the lower end of the ruler, ready to grip it (see **Figure 3.32**). Have your

Figure 3.32

friend release the ruler without warning. Catch it as quickly as you can. If you catch the ruler 5.2 cm (~2 in) from the lower end, what is your reaction time?

116. Highest Water Fountain The world's highest-shooting water fountain is located, appropriately enough, in Fountain Hills, Arizona. The water rises to a height of 171 m (about 560 ft, or 5 ft higher than the Washington Monument). **(a)** What is the initial speed of the water? **(b)** How much time does it take for water to reach the highest point?

117. Think & Calculate Standing side by side, you and a friend step off a bridge at different times and fall for 1.6 s to the water below. Your friend goes first, and you follow after she has dropped a distance of 2.0 m. **(a)** When your friend hits the water, is the separation between the two of you 2.0 m, less than 2.0 m, or more than 2.0 m? **(b)** Verify your answer to part (a) with a calculation.

118. A hot-air balloon is ascending at a rate of 7.5 m/s when a passenger drops a camera. If the camera is 25 m above the ground when it is dropped, **(a)** how much time does it take for the camera to reach the ground, and **(b)** what is its velocity just before it lands? Let upward be the positive direction for this problem.

119. A hot-air balloon is descending at a rate of 2.0 m/s when a passenger drops a camera. If the camera is 45 m above the ground when it is dropped, **(a)** how much time does it take for the camera to reach the ground, and **(b)** what is its velocity just before it lands? Let upward be the positive direction for this problem.

Mixed Review

120. **Analyze & Extend** A car starts from rest and moves with an acceleration a_0 for a time t_0. In that time it covers a distance of 5 m. How much distance does the car cover if it starts from rest and has the following acceleration and time: **(a)** $2a_0$ and $2t_0$ and **(b)** $4a_0$ and $0.5t_0$?

121. Lava Bombs A volcano shoots out blobs of molten lava, called *lava bombs*, from its summit, as shown in **Figure 3.33**. A geologist observing the eruption uses a stopwatch to time the flight of a lava bomb that is projected straight upward. If the time for it to rise to its maximum height is 2.38 s and its acceleration is 9.81 m/s² downward, what was its initial speed?

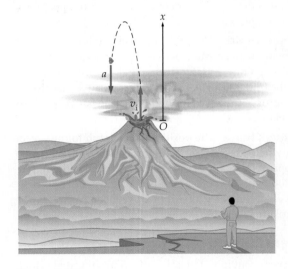

Figure 3.33

122. Ball A is dropped from rest. At the same time ball 2 is thrown upward with an initial velocity v_i. Which of the five graphs in **Figure 3.34** shows velocity versus time for **(a)** ball A and **(b)** ball B? (In this problem we have chosen downward to be the positive direction.)

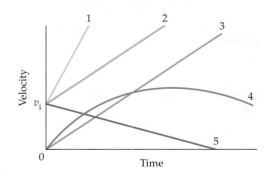

Figure 3.34

123. To celebrate a victory, a pitcher throws her glove straight upward with an initial speed of 6.0 m/s. **(a)** How much time does it take for the glove to reach its maximum height? **(b)** How much time does it take for the glove to return to the pitcher?

124. **Predict & Explain** A carpenter on the roof of a building accidentally drops her hammer. As the hammer falls, it passes two windows of equal height, as shown in **Figure 3.35**. **(a)** Is the *increase* in speed of the hammer as it drops past window 1 greater than, less than, or equal to the *increase* in speed as it drops past window 2? **(b)** Choose the *best* explanation from among the following:

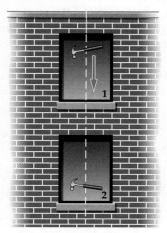

Figure 3.35

A. The greater speed at window 2 results in a greater increase in speed.

B. Constant acceleration means that the hammer speeds up the same amount for each window.

C. The hammer spends more time dropping past window 1.

125. **Predict & Explain** **Figure 3.36** shows a velocity-time graph for the hammer dropped by the carpenter in Problem 123. Notice that the times when the hammer passes the two windows are indicated by shaded areas. **(a)** Is the area of the shaded region corresponding to window 1 greater than, less than, or equal to the area of the shaded region corresponding to window 2? **(b)** Choose the *best* explanation from among the following:

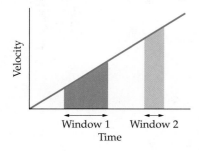

Figure 3.36

A. The shaded area for window 2 is higher than the area for window 2.

B. The windows are equally tall.

C. The shaded area for window 1 is wider than the area for window 2.

126. While riding on an elevator descending with a constant speed of 3.0 m/s, you accidentally drop a book from under your arm. **(a)** How much time does it take for the book to reach the elevator floor, 1.2 m below your arm? **(b)** What is the book's speed relative to you when it hits the elevator floor?

127. **Think & Calculate** You are driving through town at 16 m/s when suddenly a car backs out of a driveway in front of you. You apply the brakes and begin decelerating at 3.2 m/s². **(a)** How much time does it take you to stop? **(b)** After braking for half the time found in part (a), is your speed 8.0 m/s, greater than 8.0 m/s, or less than 8.0 m/s? **(c)** Support your answer with a calculation.

128. **Think & Calculate** A boat is cruising in a straight line at a constant speed of 2.6 m/s when it is shifted into neutral. After coasting 12 m, the boat is put back into gear and resumes cruising at the reduced constant speed of 1.6 m/s. **(a)** Assuming that the acceleration was constant during coasting, how much time did it take for the boat to coast the 12 m? **(b)** What was the boat's acceleration while it was coasting? **(c)** When the boat had coasted for 6.0 m, was its speed 2.1 m/s, more than 2.1 m/s, or less than 2.1 m/s? Explain.

129. A car in stop-and-go traffic starts at rest, moves forward 13 m in 8.0 s, then comes to rest again. The velocity-time graph for this car is given in **Figure 3.37**. What is the constant speed, *V*, that occurs in the middle portion of its motion?

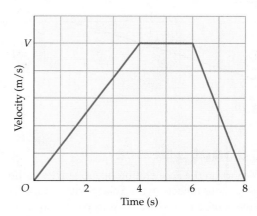

Figure 3.37

130. Suppose the motion of a second car corresponds to Figure 3.37, with a velocity in the middle portion of $V = 1.5$ m/s. What is the distance covered by this car **(a)** from 0 to 4 s, **(b)** from 4 s to 6 s, and **(c)** from 6 s to 8 s?

131. Astronauts on a distant planet throw a rock straight upward and record its motion with a video camera. After digitizing their video, they produce the graph of height, *y*, versus time, *t*, shown in **Figure 3.38**. **(a)** What is the acceleration of gravity on this planet? **(b)** What was the initial speed of the rock?

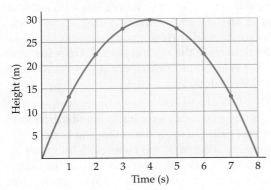

Figure 3.38

132. A car and a truck are heading directly toward one another on a straight and narrow street, but they avoid a head-on collision by simultaneously applying their brakes at $t = 0$. The resulting velocity-time graphs for the two vehicles are shown in **Figure 3.39**. What is the separation between the car and the truck when they have come to rest, given that at $t = 0$ the car is at $x = 15$ m and the truck is at $x = -35$ m. (Notice that this information determines which line in the graph corresponds to which vehicle.)

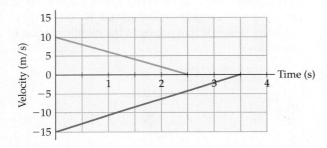

Figure 3.39

Writing about Science

133. Write a report on the acceleration due to gravity at the surface of various objects in the Solar System. Include some planets, moons, and asteroids in your report, and compare their values with the acceleration due to gravity on the Earth.

134. Connect to the Big Idea All objects in free fall move with the same constant acceleration. Given that astronauts in orbit are in free fall, explain why they feel as if they are "weightless." Compare with the situation inside an elevator that drops after the cable breaks.

Read, Reason, and Respond

Bam!—*Apollo 15* lands on the Moon The first word spoken on the surface of the Moon after *Apollo 15* landed on July 30, 1971, was "Bam!" This was James Irwin's involuntary reaction to their rather bone-jarring touchdown. "We did hit harder than any of the other flights!" says Irwin. "And I was startled, obviously, when I said, 'Bam!'"

The reason for the "firm" touchdown of *Apollo 15* was that the rocket engine was shut off earlier than planned. In fact, the engine was shut off when the lander was still 1.31 m above the lunar surface and moving downward with a speed of 0.152 m/s. From that point on the lander descended in lunar free fall, with an acceleration of 1.62 m/s². As a result, the landing speed of *Apollo 15* was by far the largest of any of the *Apollo* missions.

To visualize the descent of *Apollo 15*, consider its position-time curve during the final stages of landing, shown in **Figure 3.40**.

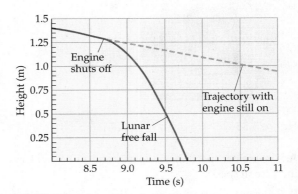

Figure 3.40

135. How much time did it take for the lander to drop the final 1.31 m to the Moon's surface?

A. 1.18 s   C. 1.78 s

B. 1.37 s   D. 2.36 s

136. What was the impact speed of the lander when it touched down?

A. 0.735 m/s   C. 3.03 m/s

B. 2.07 m/s   D. 3.23 m/s

Standardized Test Prep

Multiple Choice

Use the graph below to answer Questions 1–3. The graph shows the position of a moving object over a period of time.

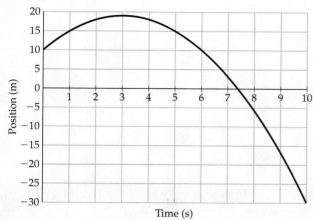

1. The velocity of the object at 6 s is approximately
   (A) 2 m/s.
   (B) 5 m/s.
   (C) −2 m/s.
   (D) −5 m/s.

2. At what time does the object change direction?
   (A) 3 s
   (B) 5 s
   (C) 7.5 s
   (D) It does not change direction during the 10-s period shown on the graph.

3. Which equation best describes the motion plotted on the graph?
   (A) $x = 10 + 6t - t^2$
   (B) $x = 10 + 5t + 2t^2$
   (C) $x = 6t - t^2$
   (D) $x = 5t + 10t^2$

4. A baseball is thrown straight upward with a velocity of 4.0 m/s. At the peak of its flight, the ball's
   (A) velocity and acceleration are both zero.
   (B) velocity is −4.0 m/s and acceleration is zero.
   (C) velocity is zero and acceleration is 9.8 m/s².
   (D) velocity is zero and acceleration is −9.8 m/s².

5. When a rock is dropped from a cliff, as it falls, its
   (A) velocity and acceleration both increase in magnitude.
   (B) velocity and acceleration both decrease in magnitude.
   (C) velocity increases in magnitude and acceleration remains constant.
   (D) velocity and acceleration both remain constant.

6. The graph of velocity versus time for a moving object is a straight line. Which of the following could be true?
   (A) The velocity is constant.
   (B) The acceleration is constant.
   (C) The acceleration and velocity are both zero.
   (D) All of the above could be true.

7. Which of the following could be true of an object moving with constant acceleration?
   (A) The object moves with constant velocity.
   (B) The object is increasing its velocity.
   (C) The object is decreasing its velocity.
   (D) All of the above could be true.

Extended Response

8. A student throws a ball straight upward. Describe the speed, velocity, and acceleration of the ball as it passes a tree limb above the student's head—both on the way up and on the way back to the student's hand.

> **Test-Taking Hint**
>
> When the position-time graph is curved, the motion is accelerated.

If You Had Difficulty With . . .

Question	1	2	3	4	5	6	7	8				
See Lesson	3.1	3.3	3.2	3.4	3.4	3.3	3.1	3.4				

4

Motion in Two Dimensions

Inside

This time-lapse photo shows a tennis ball bouncing up and down as it moves with uniform speed toward the right. The combination of vertical and horizontal motions produces a parabolic path between bounces, as you'll learn in this chapter.

Big Idea

The horizontal and vertical motions of an object are independent of one another.

The main idea of this chapter is quite simple: Horizontal and vertical motions are independent. That's it. For example, a ball thrown horizontally continues to move with the same speed in the horizontal direction, even as it falls with an increasing speed in the vertical direction. Simply put, each motion continues as if the other motion were not present.

To develop this idea we begin the chapter with a discussion of vectors in physics. As we will see, the components of a vector are the key to understanding the independence of horizontal and vertical motions.

Inquiry Lab   What does independence of motion mean?

Explore

1. Place a "target" coin so that it partially hangs over the edge of a table.
2. Place an identical "launch" coin just behind and off to one side of the target coin.
3. Use your finger to flick the launch coin off the table. The launch coin should graze the edge of the target coin, causing it to fall vertically to the floor.

The launch coin should travel horizontally and arc to the floor. Observe the motions of the coins, noting the times required for them to reach the floor.

Think

1. **Analyze** Describe the motions of the coins after they left the table. Which coin traveled the greater distance? How did the times of fall to the floor compare for the two coins?

2. **Identify** Name the force or forces acting on each of the two coins after they left the table.

3. **Predict** What do you think would happen if you were to apply a larger force to the launch coin, giving it a greater speed? How would using coins of different masses affect the motions of the coins? Explain your reasoning.

4.1 Vectors in Physics

Of all the mathematical tools used in this book, none is more important than the vector. In fact, vector s are *indispensable* in describing many of the key quantities in physics, like velocity, acceleration, and force.

Vector Characteristics

There's nothing particularly difficult about vectors. They just have to be handled a bit differently than ordinary numbers. This lesson provides you with the basic tools necessary to use vectors with confidence.

Vectors have a length and a direction

Suppose you wanted to determine the volume of a container, the temperature of the air, or the time of an event. Each of these quantities would be represented by a number. For example, the volume might be 2 m^3, the temperature 23 °C, and the time 37 s. Recall that numbers in physics are referred to as *scalars*.

Sometimes a scalar isn't enough to adequately describe a physical quantity. There are many cases where a direction is needed as well. For example, suppose you're in an unfamiliar city and you ask a person on the street, "Do you know where the library is?" If the person replies, "Yes, it is 1 kilometer from here," you still don't know where it is. The library could be anywhere on a circle

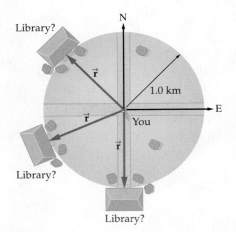

What are the key characteristics of a vector and its magnitude?

Figure 4.1 Distance and direction
If you know only that the library is 1.0 km from you, it could lie anywhere on a circle of radius 1.0 km. If, instead, you are told that the library is 1.0 km northwest, you know its precise location.

of radius 1 kilometer, as shown in **Figure 4.1**. To pin down the location, you need a reply such as "Yes, the library is 1 kilometer northwest of here." With both a distance *and* a direction, you know the location of the library.

Now, if you walk northwest for 1 kilometer, you arrive at the library, as indicated by the upper left arrow in Figure 4.1. The arrow points in the direction traveled and has a length of 1.0 kilometer. The arrow is a vector. **A vector is a quantity that is specified by both a length and a direction. The length of a vector is referred to as its magnitude.**

Representing Vectors   When we indicate a vector on a diagram or a sketch, we draw an arrow, as in Figure 4.1. To represent a vector with a written symbol, we use a **boldface** letter with a small arrow above it to remind us of its vector nature. Thus, for example, the upper-left vector in Figure 4.1 is designated by the symbol $\vec{r}$. The magnitude of a vector is represented by the *italic* version of the same letter. For example, the magnitude of $\vec{r}$ in Figure 4.1 is $r = 1.0$ km. It is common in handwritten material to draw a small arrow over the vector's symbol, which is very similar to the way vectors are represented in this text.

QUICK Example 4.1 What's the Velocity?

As you ride in a van on a straight highway, the speedometer reads 45 km/h. What additional information is required to specify the van's velocity?

Solution
The direction of travel is also needed. For example, if the van is headed due north, its velocity is specified as follows:

$$\vec{v} = 45 \text{ km/h, north}$$

The speed of the van is the magnitude of the velocity. In this case,

$$\text{speed} = v = 45 \text{ km/h}$$

Practice Problems

1. Concept Check   For each of the following quantities, indicate whether it is a scalar or a vector: (**a**) the time it takes you to run the 100-m dash, (**b**) your displacement after running the 100-m dash, (**c**) your average velocity while running, (**d**) your average speed while running.

2. Concept Check   Rank the vectors in **Figure 4.2** in order of increasing magnitude. (The magnitude of a vector is its length.)

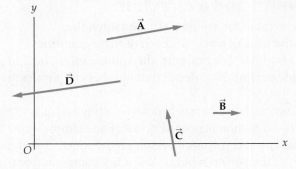

Figure 4.2

Reading Support ✓
Multiple Meanings of **magnitude**

scientific meanings: (astronomy) a scale used to describe the relative brightness of a star (geology) a value on the Richter scale that describes the energy released by an earthquake (math) the length of a vector; relative size expressed in terms of an order of magnitude

Stars of the sixth magnitude are just visible to the unaided human eye.

An earthquake of magnitude 8.0 struck the coastal area last night.

The magnitude of the vector is 2.5 meters.

common meanings: of great importance or significance; relative size or extent

The magnitude of the discovery was hard to comprehend.

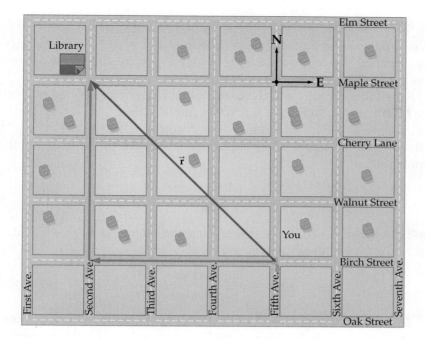

Vectors are described by their components

When we talked about finding the library in an unfamiliar city, we left out one key factor in actually *getting* to the library. In most cities it would be impossible to simply walk in a straight line directly to the library. To do so would take you through buildings and other obstructions.

However, if the city streets are laid out along north–south and east–west directions, you might instead walk west for a certain distance, then proceed north an equal distance, as illustrated in **Figure 4.3**. By walking this way you *resolve* the displacement vector $\vec{r}$ into its east–west and north–south *components*. These are important concepts. 🔑 **To resolve a vector means to find its components; a vector's components are the lengths of the vector along specified directions.**

As an example, suppose an ant leaves its nest at the origin and, after foraging for some time, is at the location given by the vector $\vec{r}$ in **Figure 4.4 (a)**. This vector has the magnitude $r = 1.50$ m and points in the direction $\theta = 25.0°$ above the x axis. Equivalently, $\vec{r}$ can be defined by saying that it extends a distance $r_x = 1.36$ m in the x direction and a distance $r_y = 0.634$ m in the y direction, as shown in **Figure 4.4 (b)**. The quantities r_x and r_y are the x and y components of the vector $\vec{r}$.

🔑 *What is meant by resolving a vector into its vector components?*

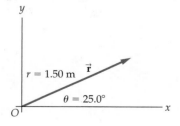

(a) A vector defined in terms of its length and direction angle

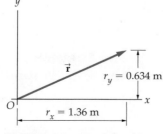

(b) The same vector defined in terms of its x and y components

◀ **Figure 4.4 A vector and its components**
(a) The vector $\vec{r}$ is defined by its length ($r = 1.50$ m) and its direction angle ($\theta = 25.0°$). **(b)** Alternatively, the vector $\vec{r}$ can be defined by its x component ($r_x = 1.36$ m) and its y component ($r_y = 0.634$ m).

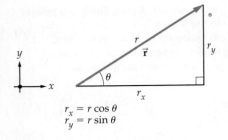

$r_x = r \cos \theta$
$r_y = r \sin \theta$

▲ **Figure 4.5 A vector and its components**
Given the magnitude and the direction of a vector, its components can be found with basic trigonometry.

Vector components depend on length and direction

So, how do you find the values $r_x = 1.36$ m and $r_y = 0.634$ m for the ant's displacement? It's really quite simple. You find them by using some of the most basic relations from trigonometry.

Recall that the sine and the cosine of an angle θ are defined in terms of a right triangle. In **Figure 4.5**, for example, we see a right triangle with angle θ and sides r_x, r_y, and r. The cosine (cos) of the angle θ is defined to be the length of the adjacent side divided by the length of the hypotenuse:

$$\cos \theta = \frac{\text{adjacent side}}{\text{hypotenuse}} = \frac{r_x}{r}$$

Similarly, the sine (sin) of θ is defined to be the length of the opposite side divided by the length of the hypotenuse:

$$\sin \theta = \frac{\text{opposite side}}{\text{hypotenuse}} = \frac{r_y}{r}$$

This is the way you usually see sin θ and cos θ defined in math books. For physics it is generally more useful to rearrange these definitions, and solve for the sides r_x and r_y in terms of the hypotenuse r and the angle θ. Thus, the side r_x is equal to

$$r_x = r \cos \theta$$

Similarly, the side r_y is equal to

$$r_y = r \sin \theta$$

Applying these results to the ant's displacement in Figure 4.4, we find the expected components:

$$r_x = r \cos 25.0°$$
$$= (1.50 \text{ m})(0.906)$$
$$= 1.36 \text{ m}$$
$$r_y = r \sin 25.0°$$
$$= (1.50 \text{ m})(0.423)$$
$$= 0.634 \text{ m}$$

These relations for finding the sides of a right triangle in terms of its hypotenuse and angle are used so often that we have summarized them in **Table 4.1** for easy reference. They also appear on the inside front cover of the book.

Thus, we can say that the ant's final displacement is equivalent to what it would be if the ant had simply walked 1.36 m in the x direction and then 0.634 m in the y direction. This illustrates the true importance of vector components.

Table 4.1 Vector Conversions: Length and Angle → Components

Given the length (r) and direction angle (θ) of a vector, its components are

$$r_x = r \cos \theta$$
$$r_y = r \sin \theta$$

At a local ski resort you take a rope tow to the top of the bunny slope. Your displacement vector for this trip has a length of 190 m and points at an angle of 26° above the horizontal. During the ride to the top, how far have you gone in (**a**) the horizontal direction and (**b**) the vertical direction?

Picture the Problem

Our sketch shows your displacement vector, which we designate with the symbol $\vec{d}$. It has a length of $d = 190$ m and points at an angle of $\theta = 26°$ above the horizontal. The horizontal distance you cover is d_x, and the vertical distance is d_y.

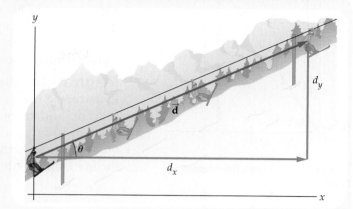

Strategy

Notice that d_x, d_y, and $\vec{d}$ form a right triangle. Therefore, the relations given in Table 4.1 apply. The hypotenuse of this triangle is d, and thus $d_x = d \cos \theta$ and $d_y = d \sin \theta$.

Known	Unknown
$d = 190$ m	(**a**) $d_x = ?$
$\theta = 26°$	(**b**) $d_y = ?$

Solution

1 (a) Calculate the horizontal distance using $d_x = d \cos \theta$:

$$d_x = d \cos \theta$$
$$= (190 \text{ m})(\cos 26°)$$
$$= \boxed{170 \text{ m}}$$

Math HELP
Trigonometric Functions
See Math Review, Section VI

2 (b) Calculate the vertical distance using $d_y = d \sin \theta$:

$$d_y = d \sin \theta$$
$$= (190 \text{ m})(\sin 26°)$$
$$= \boxed{83 \text{ m}}$$

Insight

Your displacement is exactly the same as if you had walked 170 m in the horizontal direction and then climbed a ladder 83 m in the vertical direction. Of course, it's a lot more fun to be pulled directly to the top—and it's a shorter trip that way too!

Practice Problems

3. **Follow-up** (**a**) If the angle of the chair lift is decreased, will the horizontal distance d_x increase, decrease, or stay the same? Assume that the length of the lift remains the same, $d = 190$ m. (**b**) Find d_x for the angle $\theta = 15°$.

4. **Concept Check** The vectors $\vec{A}$ and $\vec{B}$ in **Figure 4.6** are equal in magnitude, but point in different directions.

▶ **Figure 4.6**

(**a**) Which vector has the larger x component?
(**b**) Which vector has the larger y component?

5. Find the x and y components of a position vector that has a magnitude of $r = 75.0$ m and an angle relative to the x axis of (**a**) 35.0° and (**b**) 65.0°.

6. **Challenge** The press box at a baseball park is 9.75 m above the ground. A reporter in the press box looks at an angle of 15.0° below the horizontal to see second base. What is the horizontal distance from the press box to second base?

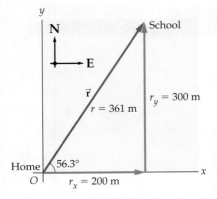

▲ Figure 4.7 A walk from home to school

Your walk from home to school covers a distance of 500 m, but the straight-line distance from home to school is only 361 m. Thus, the magnitude of your displacement vector is 361 m. The angle of your displacement vector is 56.3° north of east.

Vector components can be converted to vectors

Suppose on your walk to school you go 2 blocks east (200 m) and then 3 blocks north (300 m). You can think of these distances as the components of your displacement vector from home to school. How do you convert the components of a vector to the magnitude and direction of the vector? As we will see, simple trigonometry is all that's needed.

Magnitude The magnitude of the vector is found with the familiar Pythagorean theorem. In **Figure 4.7** we show the vector $\vec{r}$ from your home to school and its components, r_x and r_y. The magnitude of the vector is its length, which is the hypotenuse of the triangle. The components are the other two sides of the right triangle. Therefore, according to the Pythagorean theorem, the length (magnitude) of the vector from home to school is

$$r = \sqrt{r_x^2 + r_y^2}$$

Let's apply this result. Using $r_x = 200$ m and $r_y = 300$ m, we find

$$r = \sqrt{r_x^2 + r_y^2}$$
$$= \sqrt{(200\text{m})^2 + (300 \text{ m})^2} = \sqrt{130{,}000 \text{ m}^2} = 361 \text{ m}$$

We've just converted from components to magnitude. This result tells us that the straight-line distance from your home to school is 361 m, though the distance you walk is 500 m.

Direction Next, we find the angle θ by applying the definition of the tangent (tan) of an angle. Specifically, the tangent is the length of the opposite side divided by the length of the adjacent side:

$$\tan \theta = \frac{\text{opposite side}}{\text{adjacent side}} = \frac{r_y}{r_x}$$

Solving for the angle gives

$$\theta = \tan^{-1}\left(\frac{r_y}{r_x}\right)$$

Notice that $\tan^{-1}$ in this expression is the inverse tangent function, as represented on a button on your calculator. It is *not* 1 divided by the tangent. Applying this result to your displacement, we find

$$\theta = \tan^{-1}\left(\frac{300 \text{ m}}{200 \text{ m}}\right) = \tan^{-1}(1.50) = 56.3°$$

Thus, the displacement vector for your walk from home to school is at an angle of 56.3° north of east.

The conversion from components to magnitude and direction are summarized in **Table 4.2** for easy reference. These relations are also given on the inside front cover of the book. Together, the relations given in Tables 4.1 and 4.2 are the basic tools you will need to deal with vectors throughout this text.

Table 4.2 Vector Conversions: Components → Magnitude and Direction

Given the x and y components of a vector, its magnitude and direction angle are

$$r = \sqrt{r_x^2 + r_y^2}$$
$$\theta = \tan^{-1}\left(\frac{r_y}{r_x}\right)$$

You plan to cut a piece of plywood to make a skateboard ramp. The ramp will cover a horizontal distance of 1.33 m and will rise a vertical distance of 0.380 m. What are (**a**) the required length of the ramp and (**b**) the angle the ramp makes with the horizontal?

Picture the Problem

Our sketch shows the ramp, which covers a horizontal distance of $d_x = 1.33$ m and has a vertical rise of $d_y = 0.380$ m. It makes an angle θ with the horizontal and has a length d.

Strategy

Notice that d_x, d_y, and d form a right triangle. Therefore, the relations given in Table 4.2 apply. The hypotenuse of this triangle is given by $d = \sqrt{d_x^2 + d_y^2}$, and the angle is given by $\theta = \tan^{-1}(d_y/d_x)$.

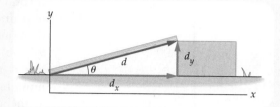

Known	Unknown
$d_x = 1.33$ m	(**a**) $d = ?$
$d_y = 0.380$ m	(**b**) $\theta = ?$

Solution

1 (**a**) Calculate the length of the ramp using the Pythagorean theorem:

$$d = \sqrt{d_x^2 + d_y^2}$$
$$= \sqrt{(1.33 \text{ m})^2 + (0.380 \text{ m})^2}$$
$$= \boxed{1.38 \text{ m}}$$

2 (**b**) Use the inverse tangent to calculate the angle of the ramp:

$$\theta = \tan^{-1}\left(\frac{d_y}{d_x}\right)$$
$$= \tan^{-1}\left(\frac{0.380 \text{ m}}{1.33 \text{ m}}\right)$$
$$= \boxed{15.9°}$$

> **Math HELP**
> Finding the Hypotenuse and Angle
> **See Math Review, Section VI**

Insight

Notice that the length of the ramp is only slightly greater than the horizontal distance it covers. This is because the angle θ has a rather small value. In the special case where $\theta = 0$, the ramp lies flat on the ground, and its length is then equal to the horizontal distance.

Practice Problems

7. Follow-up (**a**) If the horizontal distance is doubled but the vertical rise remains the same, will the angle θ increase, decrease, or stay the same? (**b**) Calculate θ for $d_x = 2(1.33 \text{ m}) = 2.66$ m and $d_y = 0.380$ m.

8. A road that rises 1 ft for every 100 ft traveled horizontally is said to have a 1% grade. Portions of the Lewiston grade, near Lewiston, Idaho, have a 6% grade. At what angle is this road inclined above the horizontal?

9. Challenge You slide a box up a loading ramp that is 3.7 m long. At the top of the ramp the box has risen a height of 1.1 m. What is the angle of the ramp above the horizontal?

Suppose that each component of a vector $\vec{A}$ is tripled. Does the direction angle of the vector increase, decrease, or stay the same?

Reasoning and Discussion

If you walk three blocks east and then three blocks north, the *direction* of your displacement is the same as if you walk one block east and then one block north. In terms of the direction angle, note that $\theta = \tan^{-1}(A_y/A_x)$. If each component of $\vec{A}$ is tripled, the direction angle is $\theta = \tan^{-1}(3A_y/3A_x) = \tan^{-1}(A_y/A_x)$. Notice that the factors 3 in the numerator and the denominator cancel. Therefore, the direction angle is unchanged.

Answer

The direction angle stays the same.

One note of caution regarding calculators: Take a close look at the angle your calculator gives when you use the $\tan^{-1}$ button. Make sure it seems reasonable based on a drawing of the vector in question. If the angle seems off, just add 180° to the calculator's result to obtain the correct angle. Why? The reason is that adding 180° to an angle doesn't change the tangent, and so the calculator doesn't know which angle is appropriate for you—both angles of such a pair give the same tangent. It's just like calculating the square root of 4. Both 2 and −2 are correct answers, but the calculator just gives 2. You have to decide if −2 is actually more appropriate in a particular case.

4.1 LessonCheck (MP)

Checking Concepts

10. ⚬ **Contrast** What distinguishes a vector from a scalar?

11. ⚬ **Identify** Does the magnitude of a vector refer to its length or its direction?

12. ⚬ **Relate** How are the components of a vector related to its magnitude?

13. ⚬ **Explain** What is the result of resolving a vector?

14. Triple Choice Suppose that each component of a vector is doubled.
(a) Does the magnitude of the vector increase, decrease, or stay the same? Explain.
(b) Does the direction angle of the vector increase, decrease, or stay the same?

Solving Problems

15. Calculate As an airplane descends toward an airport, it drops a vertical distance of 24 m and moves forward a horizontal distance of 320 m. What is the distance covered by the plane during this time?

16. Calculate You are driving up an inclined road. After 2.4 km you notice a roadside sign that indicates that your elevation has increased by 160 m. What is the angle of the road above the horizontal?

17. Calculate The displacement vector from your house to the library is 760 m long, pointing 35° north of east. What are the components of this displacement vector?

18. Calculate In gym class you run 22 m horizontally, then climb a rope vertically for 4.8 m. What is the direction angle of your total displacement, as measured from the horizontal?

4.2 Adding and Subtracting Vectors

Just as numbers can be added and subtracted, so too can vectors. In this lesson you'll learn how to add and subtract vectors in two different ways, graphically and in terms of components.

Vectors are added by placing them head to tail

One day you open an old chest in the attic and find a treasure map inside. To locate the treasure, the map says that you must "go to the sycamore tree in the backyard, march 5 paces north, then 3 paces east." These two displacements are represented by the vectors $\vec{A}$ and $\vec{B}$ in **Figure 4.8**. The total displacement from the tree to the treasure is given by the vector $\vec{C}$. We say that $\vec{C}$ is the *resultant*, or vector sum, of $\vec{A}$ and $\vec{B}$; that is, $\vec{C} = \vec{A} + \vec{B}$. A **resultant** vector is the result of adding two or more vectors.

In general, vectors are added graphically according to the following rule:

- 🔑 To add two vectors $\vec{A}$ and $\vec{B}$, place the tail end of $\vec{B}$ at the head of $\vec{A}$. The sum $\vec{C} = \vec{A} + \vec{B}$ is the vector that extends from the tail of $\vec{A}$ to the head of $\vec{B}$.

What if instructions to find the treasure are a bit more involved? For instance, you might be told to go 5 paces north, 3 paces east, and then 4 paces southeast, as shown in **Figure 4.9** on the next page.

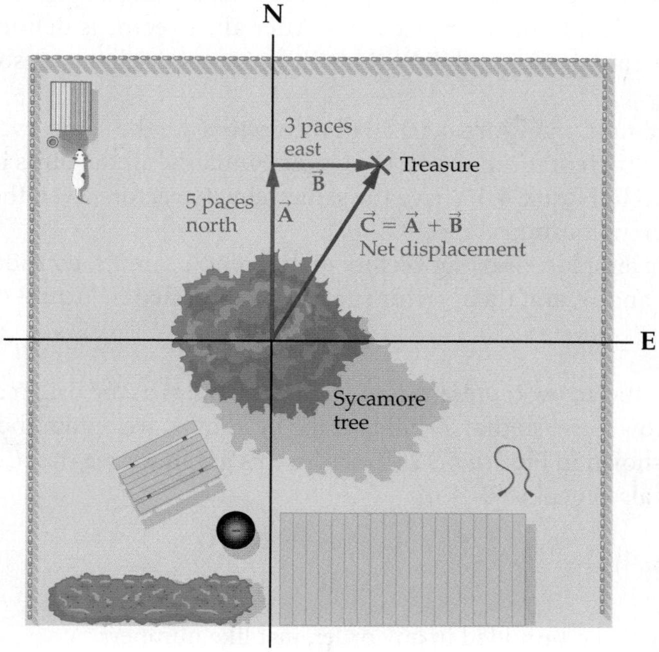

Vocabulary

- resultant

🔑 **How are vectors added graphically?**

◀ **Figure 4.8 The sum of two vectors**
To go from the sycamore tree to the treasure, you must first go 5 paces north ($\vec{A}$) and then 3 paces east ($\vec{B}$). The net displacement from the tree to the treasure is given by the vector sum $\vec{C} = \vec{A} + \vec{B}$.

▶ Figure 4.9 Adding several vectors
Searching for a treasure that is 5 paces north ($\vec{A}$), 3 paces east ($\vec{B}$), and 4 paces southeast ($\vec{C}$) of the sycamore tree. The net displacement from the tree to the treasure is $\vec{D} = \vec{A} + \vec{B} + \vec{C}$.

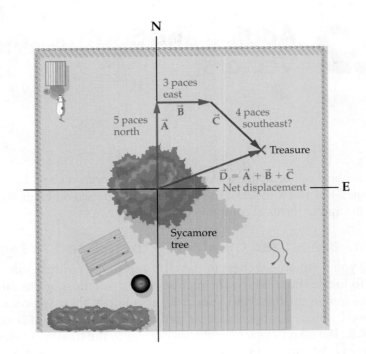

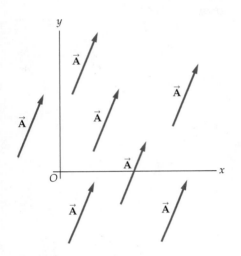

▲ Figure 4.10 Identical vectors at different locations
A vector is defined by its direction and length; its location doesn't matter.

▲ Figure 4.11 Geese with the same velocity
These geese are all moving in the same direction with the same speed. As a result, their velocity vectors are equal, even though their positions are different.

Figure 4.9 shows that the total displacement $\vec{D}$ is the sum of the three vectors $\vec{A}, \vec{B}$, and $\vec{C}$; that is, $\vec{D} = \vec{A} + \vec{B} + \vec{C}$. In general, the following rule applies for adding more than two vectors:

- 🔑 **To add more than two vectors, first place all of the vectors head to tail, head to tail, and so on. Then draw the resultant vector from the tail of the first vector to the head of the last vector.**

Vectors are unchanged if moved to other locations

In order to place a given pair of vectors head to tail, it may be necessary to move the corresponding arrows. This is fine, as long as you don't change their length or their direction. After all, a vector is defined by its length (magnitude) and direction—if these are unchanged, so is the vector.

For example, in **Figure 4.10** all of the vectors are the same, even though they are at different locations on the graph. Similarly, all the birds in the flock shown in **Figure 4.11** have the same velocity vectors, even though they are in different locations.

As an example of moving vectors to different locations, consider two vectors, $\vec{A}$ and $\vec{B}$, and their vector sum, $\vec{C}$, as indicated in **Figure 4.12 (a)**:

$$\vec{C} = \vec{A} + \vec{B}$$

By moving the arrow representing $\vec{B}$ so that its tail is at the origin and moving the arrow for $\vec{A}$ so that its tail is at the head of $\vec{B}$, we obtain the construction shown in **Figure 4.12 (b)**. From this graph we see that $\vec{C}$, which is $\vec{A} + \vec{B}$, is also equal to $\vec{B} + \vec{A}$:

$$\vec{C} = \vec{A} + \vec{B}$$
$$= \vec{B} + \vec{A}$$

Thus, vectors can be added in any order, just like numbers.

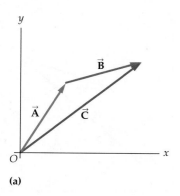

(a)

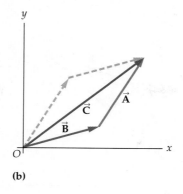

(b)

◀ **Figure 4.12 Summing two vectors**
The vector $\vec{C}$ is equal to **(a)** $\vec{A} + \vec{B}$ and **(b)** $\vec{B} + \vec{A}$. The order of addition doesn't matter.

CONCEPTUAL Example 4.5 Which Is the Vector Sum?

Consider the vectors $\vec{A}$ and $\vec{B}$ shown to the right. Which of the other four vectors in the figure ($\vec{C}$, $\vec{D}$, $\vec{E}$, or $\vec{F}$) is the sum $\vec{A} + \vec{B}$?

Reasoning and Discussion
The vector $\vec{A}$ points upward, and the vector $\vec{B}$ points to the right. Therefore, the sum of $\vec{A}$ and $\vec{B}$ will point upward and to the right. The only vector among $\vec{C}$, $\vec{D}$, $\vec{E}$, and $\vec{F}$ that satisfies this condition is $\vec{C}$.

Answer
The vector $\vec{C}$ is the sum $\vec{A} + \vec{B}$.

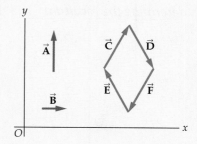

Vectors can be added graphically

Adding vectors on a graph yields approximate results that are limited by how accurately the vectors can be drawn and measured. The results are still useful as a rough guide, and as a check on the more detailed calculations presented later in this lesson.

As a specific example of graphical addition, suppose a vector $\vec{A}$ has a magnitude of 5.00 m and a direction angle of 60.0° above the x axis, and a vector $\vec{B}$ has a magnitude of 4.00 m and a direction angle of 20.0° above the x axis. These two vectors and their sum $\vec{C}$ are shown in **Figure 4.13**. The question is: What are the length and the direction angle of $\vec{C}$?

The graphical way to answer this question is to simply measure the length and direction of $\vec{C}$ in Figure 4.13. With a ruler we find that the length of $\vec{C}$ is approximately 1.75 times the length of $\vec{A}$. This means that $\vec{C}$ has a magnitude of roughly 1.75(5.00 m) = 8.75 m. Similarly, with a protractor we find that the angle θ is about 45.0° above the x axis. These are approximate results that can be improved by using components, as we show next.

Vectors can be added with components

The simplest way to add vectors—with components—is also the most precise. In this method you first find the components of the vectors to be added. You then add the x components of the original vectors to find the x component of the sum. You do the same for the y components.

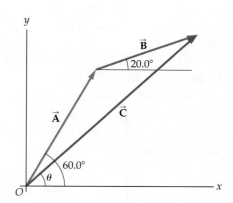

▲ **Figure 4.13 Graphical addition of vectors**
The vector $\vec{A}$ has a magnitude of 5.00 m and a direction angle of 60.0°; the vector $\vec{B}$ has a magnitude of 4.00 m and a direction angle of 20.0°. The magnitude and the direction of $\vec{C} = \vec{A} + \vec{B}$ can be measured on the graph with a ruler and a protractor.

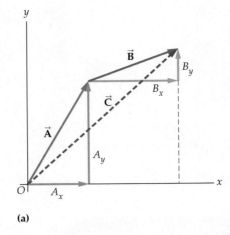

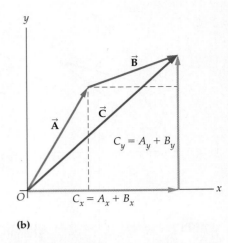

(a)

(b)

▶ **Figure 4.14 Component addition of vectors**
(a) The x and y components of $\vec{A}$ and $\vec{B}$.
(b) The x and y components of $\vec{C}$. Notice that $C_x = A_x + B_x$ and $C_y = A_y + B_y$. Thus, vectors are added by adding their components.

🔑 How are vector components used to determine the resultant vector?

The component method is illustrated in **Figure 4.14**. First, **Figure 4.14 (a)** shows the components of $\vec{A}$ and $\vec{B}$. Next, **Figure 4.14 (b)** shows the components of the sum $\vec{C} = \vec{A} + \vec{B}$. The figure shows that the components of $\vec{C}$ are indeed the sums of the components of $\vec{A}$ and $\vec{B}$. That is,

$$C_x = A_x + B_x$$

and

$$C_y = A_y + B_y$$

🔑 In general, the sum of the x components gives the x component of the resultant, and the sum of the y components gives the y component of the resultant. Let's apply the component method to the example in Figure 4.13. In this case the components of $\vec{A}$ and $\vec{B}$ are

$$A_x = (5.00 \text{ m})(\cos 60.0°) = 2.50 \text{ m}$$
$$A_y = (5.00 \text{ m})(\sin 60.0°) = 4.33 \text{ m}$$

and

$$B_x = (4.00 \text{ m})(\cos 20.0°) = 3.76 \text{ m}$$
$$B_y = (4.00 \text{ m})(\sin 20.0°) = 1.37 \text{ m}$$

Adding component by component yields the components of $\vec{C} = \vec{A} + \vec{B}$:

$$C_x = A_x + B_x$$
$$= 2.50 \text{ m} + 3.76 \text{ m} = 6.26 \text{ m}$$

and

$$C_y = A_y + B_y$$
$$= 4.33 \text{ m} + 1.37 \text{ m} = 5.70 \text{ m}$$

With these results we can now find *precise* values for the magnitude and the direction angle of vector $\vec{C}$. In particular,

$$C = \sqrt{C_x^2 + C_y^2} = \sqrt{(6.26 \text{ m})^2 + (5.70 \text{ m})^2} = \sqrt{71.7 \text{ m}^2} = 8.47 \text{ m}$$

and

$$\theta = \tan^{-1}\left(\frac{C_y}{C_x}\right) = \tan^{-1}\left(\frac{5.70 \text{ m}}{6.26 \text{ m}}\right) = \tan^{-1}(0.911) = 42.3°$$

Notice that these exact values are in agreement with the approximate results that were found by graphical addition.

PROBLEM-SOLVING NOTE

From here on we will always add vectors using components; graphical addition is useful primarily as a rough check on the results obtained with components.

What are the magnitude and direction of the total displacement for the treasure hunt shown below, where you need to go 6 paces east ($\vec{A}$) and 3 paces north ($\vec{B}$)? Assume that each pace is 0.500 m in length. Let east (E) be the positive x direction and north (N) be the positive y direction.

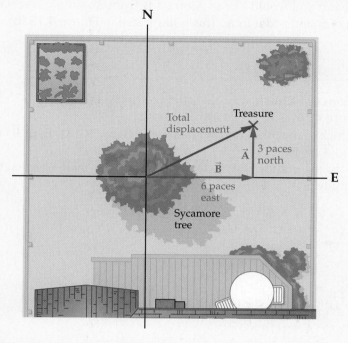

Solution *(Perform the calculations indicated in each step.)*

1. Find the x and y components of $\vec{A}$:

$A_x = 3.00$ m
$A_y = 0$

2. Find the x and y components of $\vec{B}$:

$B_x = 0$
$B_y = 1.50$ m

3. Add the components of $\vec{A}$ and $\vec{B}$ to find the components of $\vec{C}$:

$C_x = A_x + B_x = 3.00$ m
$C_y = A_y + A_B = 1.50$ m

4. Determine C and θ:

$C = \sqrt{C_x^2 + C_y^2} = 3.35$ m

$\theta = \tan^{-1}(C_y/C_x) = 26.6°$

> **Math HELP**
> Finding the Hypotenuse and Angle
> **See Math Review, Section VI**

Vectors are subtracted by subtracting components

As you might expect, you subtract vectors by subtracting their components. For example, suppose you want to find $\vec{A}$ minus $\vec{B}$. Let's call the result $\vec{D}$:

$$\vec{D} = \vec{A} - \vec{B}$$

The components of $\vec{D}$ are simply the components of $\vec{A}$ minus the components of $\vec{B}$:

$$D_x = A_x - B_x$$
$$D_y = A_y - B_y$$

Once the components of $\vec{D}$ are found, its magnitude and direction angle can be calculated as usual (see Table 4.2).

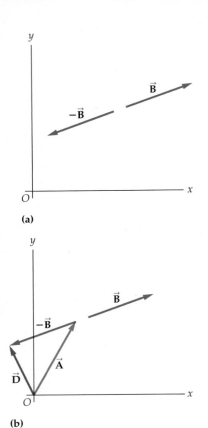

(a)

(b)

▲ **Figure 4.15 Vector subtraction**
(a) The vector $\vec{B}$ and its negative $-\vec{B}$.
(b) A vector construction for $\vec{D} = \vec{A} - \vec{B}$.

Vectors can also be subtracted graphically, of course. First, we note that the negative of a vector has a very simple graphical interpretation:

- The *negative* of a vector is represented by an arrow of the same length as the original vector, but pointing in the *opposite* direction. That is, multiplying a vector by -1 *reverses its direction*.

For example, the vectors $\vec{B}$ and $-\vec{B}$ are indicated in **Figure 4.15 (a)**. It follows that if you would like to subtract $\vec{B}$ from $\vec{A}$, simply reverse the direction of $\vec{B}$ and add it to $\vec{A}$. This is illustrated in **Figure 4.15 (b)**.

Practice Problem

19. Concept Check Consider the vectors $\vec{A}$ and $\vec{B}$ shown in **Figure 4.16**.
(a) Which of the other four vectors in the figure ($\vec{C}, \vec{D}, \vec{E}$, or $\vec{F}$) is equal to $\vec{A} - \vec{B}$?
(b) Is the magnitude of $\vec{A} - \vec{D}$ greater than, less than, or equal to the magnitude of $\vec{A} - \vec{E}$?

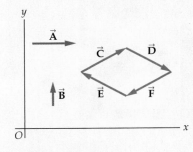

▲ **Figure 4.16**

4.2 LessonCheck (MP)

Checking Concepts

20. Describe How do you place the heads and tails of two vectors that you want to add?

21. Explain How do you add vectors using their components?

22. Triple Choice Referring to Figure 4.16, is the magnitude of $\vec{A} + \vec{D}$ greater than, less than, or equal to the magnitude of $\vec{A} + \vec{E}$?

23. Triple Choice Referring to Figure 4.16, is the magnitude of $\vec{B} - \vec{C}$ greater than, less than, or equal to the magnitude of $\vec{B} - \vec{F}$?

24. Diagram Use a sketch to show that two vectors of unequal magnitude cannot add to zero, but that three vectors of unequal magnitude can.

Solving Problems

25. Graph Vector $\vec{A}$ has a magnitude of 27 m and points in a direction 32° above the positive x axis. Vector $\vec{B}$ has a magnitude of 35 m and points in a direction 55° below the positive x axis. Sketch the vectors $\vec{A}, \vec{B}$, and $\vec{C} = \vec{A} + \vec{B}$. Estimate the magnitude and the direction angle of $\vec{C}$ from your sketch.

26. Calculate Use the component method of vector addition to find the magnitude and the direction angle of the vector $\vec{C}$ in Problem 25.

27. Calculate Vector $\vec{A}$ points in the positive y direction and has a magnitude of 12 m. Vector $\vec{B}$ has a magnitude of 33 m and points in the negative x direction. Find the direction and the magnitude of **(a)** $\vec{A} + \vec{B}$, **(b)** $\vec{A} - \vec{B}$, and **(c)** $\vec{B} - \vec{A}$.

4.3 Relative Motion

Suppose an airline pilot wants to fly from Denver to Dallas. If the air is still, the pilot can simply head the plane toward the destination. If there is a wind blowing from west to east, however, the pilot must use vectors to determine the correct heading. Otherwise, the plane and its passengers may end up in Little Rock rather than Dallas. This is a classic problem of relative motion, the topic of this lesson.

Vectors can describe the relative motion of objects

In general, **relative motion** simply means the motion of one object *relative* to another object. For example, suppose you are standing on the ground as a train goes by at 15.0 m/s, as shown in **Figure 4.17**. Inside a car of the train a railroad worker is walking in the forward direction at 1.2 m/s relative to the train. How fast is the worker moving relative to the ground? Clearly, the answer is the sum of the velocities:

$$1.2 \text{ m/s} + 15.0 \text{ m/s} = 16.2 \text{ m/s}$$

What if the worker is walking with the same speed, but toward the back of the train instead? In this case the velocity of the worker relative to the ground is equal to the difference of the original velocities:

$$-1.2 \text{ m/s} + 15.0 \text{ m/s} = 13.8 \text{ m/s}$$

We can apply this same idea to any relative velocities. For example, let the velocity of the train relative to the ground be $\vec{v}_{tg}$, the velocity of the worker relative to the train be $\vec{v}_{wt}$, and the velocity of the worker relative to the ground be $\vec{v}_{wg}$. According to the previous discussion, the velocity of the worker relative to the ground is

$$\vec{v}_{wg} = \vec{v}_{wt} + \vec{v}_{tg}$$

This vector addition is illustrated in **Figure 4.18** for the two cases discussed above.

Vocabulary

Vocabulary

- relative motion

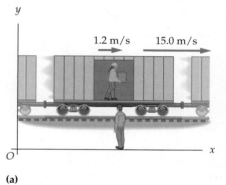

(a)

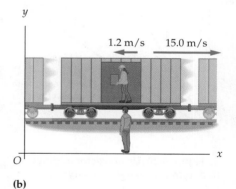

(b)

▲ **Figure 4.17 Relative velocity**
(a) The worker walks toward the front of the train. The velocity of the worker relative to the ground is 16.2 m/s. (b) The worker walks toward the rear of the train. In this case the velocity of the worker relative to the ground is 13.8 m/s.

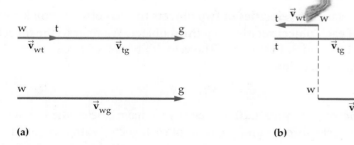

(a)   (b)

▲ **Figure 4.18 Adding velocity vectors**
Vector addition to find the velocity of the worker with respect to the ground for
(a) Figure 4.17 (a) and (b) Figure 4.17 (b).

A worker climbs up a ladder on a moving train with velocity $\vec{v}_{wt}$ relative to the train. If the train moves relative to the ground with a velocity $\vec{v}_{tg}$, the velocity of the worker on the train relative to the ground is $\vec{v}_{wg} = \vec{v}_{wt} + \vec{v}_{tg}$.

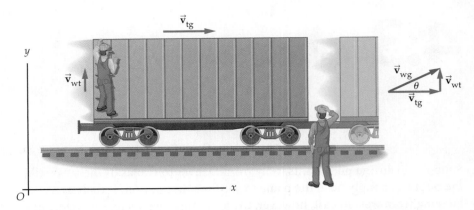

If a worker climbs a ladder to the roof of the railroad car, as shown in **Figure 4.19**, the same idea applies. You just add the two velocity vectors. In general, the velocity of one object relative to another is obtained by adding the velocity vectors.

ACTIVE Example 4.7 Determining the Speed and Direction

The worker in Figure 4.19 climbs the ladder with a speed of $v_{wt} = 0.20$ m/s, as the train coasts forward at $v_{tg} = 0.70$ m/s. Find the speed and the direction of the worker relative to the ground.

Solution *(Perform the calculations indicated in each step.)*

1. Find the x and y components of $\vec{v}_{wt}$:

$$v_{wt,x} = 0$$
$$v_{wt,y} = 0.20 \text{ m/s}$$

2. Find the x and y components of $\vec{v}_{tg}$:

$$v_{tg,x} = 0.70 \text{ m/s}$$
$$v_{tg,y} = 0$$

3. Add the components of $\vec{v}_{wt}$ and $\vec{v}_{tg}$ to find the components of $\vec{v}_{wg}$:

$$v_{wg,x} = v_{wt,x} + v_{tg,x} = 0.70 \text{ m/s}$$
$$v_{wg,y} = v_{wt,y} + v_{tg,y} = 0.20 \text{ m/s}$$

4. Determine v_{wg} and θ:

$$v_{wg} = \sqrt{(v_{wg,x})^2 + (v_{wg,y})^2} = 0.73 \text{ m/s}$$
$$\theta = \tan^{-1}(v_{wg,y}/v_{wg,x}) = 16°$$

Insight
The worker has a speed of 0.73 m/s relative to the ground and a direction that is 16° above the horizontal.

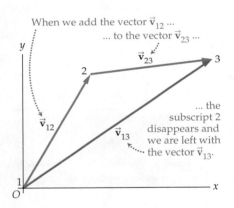

▲ **Figure 4.20 Using vector addition to determine relative velocity**

When we add the vector $\vec{v}_{12}$ ... ... to the vector $\vec{v}_{23}$ ... ... the subscript 2 disappears and we are left with the vector $\vec{v}_{13}$.

🔑 *How are relative velocity vectors related to one another?*

Subscripts help to clarify relative motion

🔑 You can relate the velocities of two objects to each other if you know the velocity of each object relative to a third object. We refer to the objects of interest as 1, 2, and 3. In this case the velocity of object 1 relative to object 3 is written as follows:

$$\vec{v}_{13} = \vec{v}_{12} + \vec{v}_{23}$$

This sum is shown in **Figure 4.20**. For clarity we have labeled the tail of each vector with its first subscript and the head of each vector with its second subscript. In the train example we used 1 as the *worker*, 2 as the *train*, and 3 as the *ground*. In the following example we again apply the relative velocity equation, this time with different identifications for 1, 2, and 3. The idea is exactly the same, however.

You are riding in a boat whose speed relative to the water is 6.1 m/s. The boat points at an angle of 25° upstream on a river flowing at 1.4 m/s. (**a**) What is your speed relative to the ground? (**b**) What is your direction of travel relative to the ground? Give your answer in terms of degrees upstream.

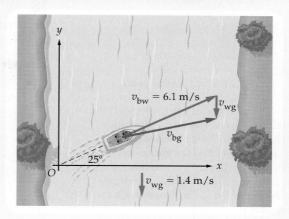

Picture the Problem

We choose the x axis to be perpendicular to the river, and the y axis to point upstream. With these choices the velocity of the boat relative to the water is 25° above the x axis. In addition, the velocity of the water relative to the ground has a magnitude of 1.4 m/s and points in the negative y direction.

Strategy

To find the velocity of the boat we use $\vec{v}_{13} = \vec{v}_{12} + \vec{v}_{23}$ but replace 1 with b (for *boat*), 2 with w (for *water*), and 3 with g (for *ground*). This gives $\vec{v}_{bg} = \vec{v}_{bw} + \vec{v}_{wg}$. (**a**) The speed of the boat relative to the ground is $v_{bg} = \sqrt{v_{bg,x}^2 + v_{bg,y}^2}$. (**b**) The direction angle is $\theta = \tan^{-1}(v_{bg,y}/v_{bg,x})$.

Known	Unknown
$v_{bw} = 6.1$ m/s	(**a**) $v_{bg} = ?$
$v_{wg} = 1.4$ m/s	(**b**) direction of
direction of boat	boat relative to
relative to water $= 25°$	ground: $\theta = ?$

Solution

Part (a)

1 Rewrite $\vec{v}_{13} = \vec{v}_{12} + \vec{v}_{23}$ with $1 \rightarrow b$, $2 \rightarrow w$, and $3 \rightarrow g$:

$$\vec{v}_{bg} = \vec{v}_{bw} + \vec{v}_{wg}$$

2 Find the x and y components of the velocity of the boat relative to the water:

$$v_{bw,x} = (6.1 \text{ m/s})(\cos 25°) = 5.5 \text{ m/s}$$
$$v_{bw,y} = (6.1 \text{ m/s})(\sin 25°) = 2.6 \text{ m/s}$$

3 Find the x and y components of the velocity of the water relative to the ground:

$$v_{wg,x} = 0$$
$$v_{wg,y} = -1.4 \text{ m/s}$$

4 Add the components of $\vec{v}_{bw}$ and $\vec{v}_{wg}$ to find the components of $\vec{v}_{bg}$:

$$v_{bg,x} = 5.5 \text{ m/s} + 0 = 5.5 \text{ m/s}$$
$$v_{bg,y} = 2.6 \text{ m/s} - 1.4 \text{ m/s} = 1.2 \text{ m/s}$$

5 Calculate the speed of the boat relative to the ground:

$$v_{bg} = \sqrt{(v_{bg,x})^2 + (v_{bg,y})^2}$$
$$= \sqrt{(5.5 \text{ m/s})^2 + (1.2 \text{ m/s})^2}$$
$$= \boxed{5.6 \text{ m/s}}$$

Part (b)

6 Calculate the direction angle for the velocity of the boat relative to the ground:

$$\theta = \tan^{-1}\left(\frac{v_{bg,y}}{v_{bg,x}}\right)$$
$$= \tan^{-1}\left(\frac{1.2 \text{ m/s}}{5.5 \text{ m/s}}\right)$$
$$= \boxed{12° \text{ upstream}}$$

> **Math HELP**
> Finding the Hypotenuse and Angle
> **See Math Review, Section VI**

Insight

The flow of water has slowed the speed of the boat relative to the ground to only 5.6 m/s. It has also changed the direction of travel of the boat to make it closer to the x axis.

28. **Follow-up** Find the speed and the direction of the boat in Guided Example 4.8 relative to the ground if the river flows at 4.5 m/s and everything else remains the same.

29. The pilot of an airplane wants to fly due north, but there is a 65 km/h wind blowing from east to west. In what direction should the pilot head her plane if its speed relative to the air is 340 km/h? Give your answer in degrees north of east.

30. **Concept Check** Should the captain of the boat in Guided Example 4.8 increase or decrease its speed in order to travel straight across the river (along the x axis)? Explain.

▼ **Figure 4.21 Wind speeds in a hurricane**
A hurricane combines two motions: (1) the motion of the eye, and (2) the circulation around the eye. The velocities of these two motions add to give a greater total velocity in the upper-right quadrant of the hurricane. This is where the hurricane poses the greatest risk to people on the ground.

Physics & You: Technology Meteorologists use computer models and atmospheric data to track hurricanes and predict risk levels for cities in their paths. They know that relative velocity plays a key role in determining where the damage from a hurricane will be most severe. This is illustrated in **Figure 4.21**, where we show a hurricane heading for the coast of Florida. The center of the hurricane (the eye) moves toward the northwest with a speed v_1. In addition, the hurricane circulates in a counterclockwise direction with a speed v_2 all around the circumference. The wind speeds relative to the ground are strongest in the upper-right quadrant of the hurricane, where v_1 and v_2 point in the same direction. The trailing edge of the hurricane has weaker winds, because v_1 and v_2 point in opposite directions there. Thus, damage is usually greatest in the upper-right section of a hurricane—a direct consequence of relative velocity.

4.3 LessonCheck (MP)

Checking Concepts

31. **Determine** If the velocity of object 1 relative to object 2 is v_{12} and the velocity of object 2 relative to object 3 is v_{23}, what is the velocity of object 1 relative to object 3?

32. **Apply** When running around a track on a windy day, you feel the wind is stronger when you are running into it than when you run with it. Explain in terms of relative velocity.

33. **Explain** Why does the boat in Guided Example 4.8, which is pointed 25° upstream, move in a direction that is only 12° upstream?

Solving Problems

34. **Calculate** A ferry approaches shore, moving north with a speed of 6.2 m/s relative to the dock. A person on the ferry walks from one side of the ferry to the other, moving east with a speed of 1.1 m/s relative to the ferry. What is the speed of the person relative to the dock?

35. **Calculate** A person on a cruise ship is doing laps on the promenade deck. On one portion of the track the person is moving north with a speed of 3.8 m/s relative to the ship. The ship moves east with a speed of 12 m/s relative to the water. What is the direction of motion of the person relative to the water?

4.4 Projectile Motion

At the beginning of this chapter we saw that motions in the horizontal and vertical directions are independent of one another. In this lesson we explore this concept in greater detail and apply it to *projectiles*—objects that are launched or projected into the air.

Horizontal and vertical motions are independent

Here's a simple demonstration that illustrates the independence of horizontal and vertical motions. While standing still, drop a rubber ball to the floor and catch it on the rebound. Notice that the ball goes straight down, lands near your feet, and returns almost to your hand in about a second.

Next, walk or roller skate with constant speed as you drop the ball, as in **Figure 4.22**. From your viewpoint the ball again goes straight down below your hand and lands near your feet, as shown in **Figure 4.23**. The ball then bounces straight back up and returns in about a second. The fact that you are moving in the horizontal direction the whole time has no effect on the ball's vertical motion—the motions are completely independent.

To an observer who sees you walking or skating by, the ball follows a curved path, as is also shown in Figure 4.23. This path is a parabola.

Projectiles illustrate independence of motion

A perfect application of the independence of horizontal and vertical motions is projectile motion. To be specific, a **projectile** is an object that is thrown, kicked, batted, or otherwise launched into motion and then allowed to follow a path determined solely by the influence of gravity. In studying projectile motion we make the following assumptions:

• Air resistance can be ignored.
• The acceleration due to gravity is constant and directed downward with a magnitude equal to $g = 9.81 \text{ m/s}^2$.

▲ **Figure 4.22 A dropped object keeps up with you**
This rollerblader may not be thinking about independence of motion, but the ball she releases illustrates the concept perfectly. Notice that the ball continues to move horizontally with constant speed—even though the rollerblader is no longer touching it. The ball keeps up with her, directly below her hand, at the same time that it accelerates downward.

The moving person sees the ball fall straight down below her hand ...

... but a stationary observer sees the ball follow a curved path.

◄ **Figure 4.23 Independence of vertical and horizontal motions**
When you drop a ball while walking, running, or skating with constant velocity, you see the ball dropping straight down from the point where you released it. To a person at rest, the ball follows a curved path that combines horizontal and vertical motions.

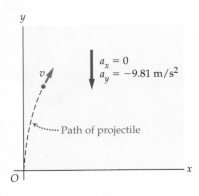

$a_x = 0$
$a_y = -9.81 \text{ m/s}^2$

······ Path of projectile

▲ **Figure 4.24 Coordinate system for projectiles**
When studying projectile motion in this text, we always use a coordinate system with the positive y direction pointing upward and the positive x direction pointing to the right. In this case, the acceleration due to gravity is in the negative y direction. There is no acceleration in the x direction.

CONNECTINGIDEAS

The basic ideas of projectile motion are used again in Chapter 9, where we explore orbital motion. In particular, see Figure 9.18.

These assumptions define an *ideal projectile* and are quite accurate for most everyday projectiles. In cases where air resistance is important, projectiles deviate from ideal behavior. The frictional force of air resistance reduces the speed of a projectile, causing it to rise to a lesser height and travel a shorter distance in the horizontal direction. We will look at these effects in more detail later in this lesson.

Because the horizontal and vertical motions are independent for an ideal projectile, its time of flight depends *only* on its vertical speed. This is because the vertical speed determines the time required to rise to maximum height and fall back to the initial level. The horizontal distance of travel is simply the horizontal speed multiplied by the time of flight.

Motion Equations for Projectiles To see how independence of motion is applied, let's start with the position-time equation you learned about in Chapter 3:

$$x_f = x_i + v_i t + \tfrac{1}{2} a t^2$$

Independence of motion means that this equation applies equally well to both the x and y motions. To apply it to the y motion, we first change the x's to y's. In addition, we note that the initial velocity and acceleration can be different in the x and y directions. Therefore, we replace v_i with $v_{x,i}$ for the x equation and with $v_{y,i}$ for the y equation. Similarly, we replace a with a_x for the x equation and with a_y for the y equation. This gives us the following position-time equations:

$$x_f = x_i + v_{x,i} t + \tfrac{1}{2} a_x t^2$$
$$y_f = y_i + v_{y,i} t + \tfrac{1}{2} a_y t^2$$

These equations specify the x and y positions at any given time.

For projectile motion we'll use the coordinate system shown in **Figure 4.24**, with x increasing to the right and y increasing upward. A projectile experiences a downward acceleration, which means the acceleration is in the negative y direction. There is no acceleration in the x direction. Therefore, it follows that $a_x = 0$ and $a_y = -g$. Our x and y equations simplify to the following:

> **Position-Time Equations for Projectiles**
>
> $$x_f = x_i + v_{x,i} t$$
> $$y_f = y_i + v_{y,i} t - \tfrac{1}{2} g t^2$$

We apply these equations to a specific projectile by identifying the initial position (x_i, y_i) and initial velocity ($v_{x,i}$, $v_{y,i}$) of the projectile.

As an example, consider the situation shown in **Figure 4.25**. Here a person skateboards in the positive x direction with a speed of 1.30 m/s and drops a basketball from a height of 1.25 m. The initial position of the ball (where it is released) is $x_i = 0$ and $y_i = 1.25$ m. The initial velocity of the ball is in the positive x direction; thus, $v_{x,i} = 1.30$ m/s and $v_{y,i} = 0$. Substituting these values into our equations gives

$$x_f = (1.30 \text{ m/s}) t$$
$$y_f = 1.25 \text{ m} - \tfrac{1}{2} g t^2$$

These equations are specific to this particular dropped basketball. We use these equations in the following Active Example.

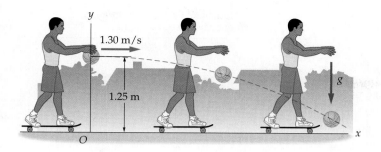

► **Figure 4.25 A dropped basketball**
The skateboarder moves with a constant speed of 1.30 m/s in the positive x direction. At time zero ($t = 0$) the ball is released at $x_i = 0$ and $y_i = 1.25$ m. The ball then falls with an acceleration of magnitude g in the negative y direction.

ACTIVE Example 4.9 Determining the Position of a Dropped Ball

A person skateboarding with a constant speed of 1.30 m/s releases a ball from a height of 1.25 m above the ground, as in Figure 4.25. Given that $x_i = 0$ and $y_i = 1.25$ m, find x_f and y_f for the times **(a)** $t = 0.250$ s and **(b)** $t = 0.500$ s. Recall that $g = 9.81$ m/s^2.

Solution *(Perform the calculations indicated in each step.)*

Part (a)

1. Begin with the position-time equations for the projectile given above:

$$x_f = (1.30 \text{ m/s})t$$
$$y_f = 1.25 \text{ m} - \tfrac{1}{2}gt^2$$

2. Substitute $t = 0.250$ s and $g = 9.81$ m/s^2 into the equations and solve:

$$x_f = 0.325 \text{ m}$$
$$y_f = 0.943 \text{ m}$$

Part (b)

3. Substitute $t = 0.500$ s and $g = 9.81$ m/s^2 into the equations and solve:

$$x_f = 0.650 \text{ m}$$
$$y_f = 0.0238 \text{ m}$$

Insight

Notice that the ball is only about an inch above the ground at $t = 0.500$ s. It will hit the ground shortly after this time.

Practice Problems

36. Follow-up How much time does it take for the ball to land? (*Hint:* Set $y = 0$ and solve for the time t.)

37. An archer shoots an arrow horizontally at a target 15 m away. The arrow is aimed directly at the center of the target, but it hits 52 cm lower. How long did it take for the arrow to reach the target?

Notice that the x position of the ball in Active Example 4.9 does not depend on the acceleration due to gravity. Similarly, the y position does not depend on the initial horizontal speed of the ball. If the skateboarder is moving faster when he drops the ball, for example, it still falls to the ground in exactly the same time as before. It will go farther in the x direction, though.

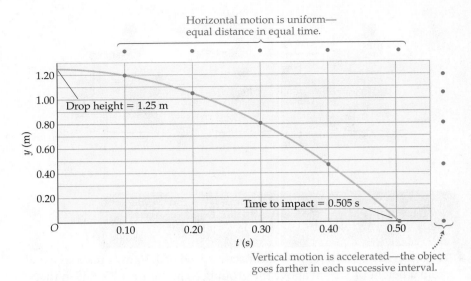

▶ **Figure 4.26 Trajectory of a dropped basketball**
This plot shows the x and y positions of the basketball that was dropped in Figure 4.25. The ball was released at the height 1.25 m and moved with constant speed in the x direction. At the same time it accelerated downward, just as if it had been dropped from rest.

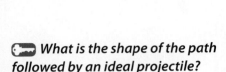 *What is the shape of the path followed by an ideal projectile?*

Projectiles follow parabolic paths

You've seen that the dropped basketball in Figure 4.25 follows a curved path. Let's examine this motion more carefully.

Figure 4.26 shows the x and y positions of the basketball at a variety of times during its fall to the ground. Notice that the horizontal position of the ball increases steadily with time, as expected for constant speed. The vertical position changes by an increasing amount with each time interval. In fact, the vertical motion is the same as the free fall of an object dropped from rest. Combining these two motions yields a curved path with the shape of a parabola. **In general, an ideal projectile follows a parabolic path regardless of how it is launched.** The parabolic path is produced by the combined effects of the horizontal and vertical motions acting independently.

Independence of motion also applies to velocity

We've seen that the position-time equation from Chapter 3 can be applied to x and y motions separately. This is also true for the velocity-time equation:

$$v_f = v_i + at$$

To apply this equation to x and y components of velocity, we replace v_f with $v_{x,f}$ in the x equation and with $v_{y,f}$ in the y equation. We make similar replacements for v_i. In addition, we replace a with a_x in the x equation and with a_y in the y equation. This yields

$$v_{x,f} = v_{x,i} + a_x t$$
$$v_{y,f} = v_{y,i} + a_y t$$

For the specific case of a projectile, we know that the acceleration is $a_x = 0$ and $a_y = -g = -9.81 \text{ m/s}^2$. Therefore, the velocity-time equations for projectile motion are as follows:

> **Velocity-Time Equations for Projectiles**
>
> $$v_{x,f} = v_{x,i}$$
> $$v_{y,f} = v_{y,i} - gt$$

All we need to apply these equations to a particular projectile are the initial velocities in the x and y directions. These equations, along with the corresponding position-time equations, are collected in **Table 4.3** for easy reference.

Table 4.3 Equations of Motion for an Ideal Projectile

Position-time equations for an ideal projectile:

$$x_f = x_i + v_{x,i}t$$
$$y_f = y_i + v_{y,i}t - \tfrac{1}{2}gt^2$$

Velocity-time equations for an ideal projectile:

$$v_{x,f} = v_{x,i}$$
$$v_{y,f} = v_{y,i} - gt$$

QUICK Example 4.10 What's the Speed in Each Direction?

You throw a baseball from the roof of a house to a friend on the ground. The ball has an initial velocity of 12.0 m/s in the horizontal direction. After 1.00 s, how fast is the ball moving in **(a)** the x direction and **(b)** the y direction?

Solution

(a) The velocity in the x direction doesn't change. Therefore, after 1.00 s we have

$$v_{x,f} = v_{x,i} = \boxed{12.0 \text{ m/s}}$$

(b) For the y direction we have the following at $t = 1.00$ s:

$$
\begin{aligned}
v_{y,f} &= v_{y,i} - gt \\
&= 0 - (9.81 \text{ m/s}^2)(1.00 \text{ s}) \\
&= \boxed{-9.81 \text{ m/s}}
\end{aligned}
$$

Practice Problems

38. Follow-up What is the speed of the ball at the time $t = 1.00$ s? (*Hint:* Speed is the magnitude of the velocity vector, $v = \sqrt{v_x^2 + v_y^2}$.)

39. Follow-up What is the direction of motion of the ball at the time $t = 1.00$ s? (*Hint:* The direction of motion is given by $\theta = \tan^{-1}(v_y/v_x)$.)

CONCEPTUAL Example 4.11 Comparing Splashdown Speeds

Two youngsters dive off an overhang into a lake. Diver 1 drops straight down; diver 2 runs off the cliff with an initial horizontal speed v. Is the splashdown speed of diver 2 greater than, less than, or equal to the splashdown speed of diver 1?

Reasoning and Discussion

The two divers fall for the same amount of time, and their y components of velocity are the same at splashdown. Since diver 2 also has an x component of velocity, the speed of diver 2 is greater.

Answer

The splashdown speed of diver 2 is greater than that of diver 1.

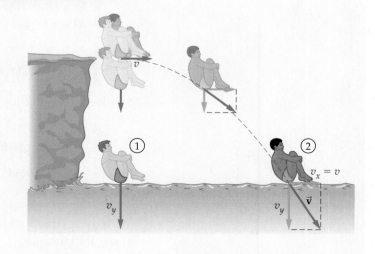

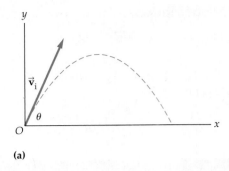

(a)

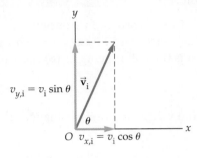

(b)

CONNECTINGIDEAS

You will work with motion in two dimensions again in Chapter 20 when we study the motion of charged particles (like electrons) in electric fields. The same basic principles apply.

A projectile can be launched at any angle

Projectiles don't always start off in the horizontal direction, of course. Examples of projectiles launched at different angles include balls thrown from one person to another, water spraying from a hose, salmon leaping over rapids, and divers jumping from a springboard. As we will see, it is straightforward to take different launch angles into account.

Figure 4.27 (a) shows a projectile launched with an initial speed v_i at an angle θ. The components of the initial velocity are determined using basic trigonometry, as indicated in **Figure 4.27 (b)**:

$$v_{x,i} = v_i \cos \theta$$
$$v_{y,i} = v_i \sin \theta$$

Substituting these results into the position-time equations yields the following:

Position-Time Equations for Projectiles Launched at an Angle θ

$$x_f = x_i + (v_i \cos \theta)t$$
$$y_f = y_i + (v_i \sin \theta)t - \tfrac{1}{2}gt^2$$

Similarly, substituting into the velocity-time equations gives

Velocity-Time Equations for Projectiles Launched at an Angle θ

$$v_{x,f} = v_i \cos \theta$$
$$v_{y,f} = v_i \sin \theta - gt$$

With these equations we can solve problems involving projectiles launched at any initial angle.

Chipping from the rough, a golfer sends the ball over a 3.00-m-high tree that is 14.0 m away. The ball lands—on the green, of course—at the same level at which it was struck, after traveling a horizontal distance of 17.8 m. (**a**) If the ball left the club at an angle of 54.0° above the horizontal and landed on the green 2.24 s later, what was its initial speed? (**b**) The ball passes directly over the tree at $t = 1.76$ s. How high was the ball at that time?

Picture the Problem

Our sketch shows the ball taking flight from the origin, $x_i = y_i = 0$, with a launch angle of 54.0°, and arcing over the tree. The ball lands at $x_f = 17.8$ m and $y_f = 0$. The individual points along the parabolic trajectory correspond to equal time intervals.

Strategy

(**a**) We can find initial speed by applying the x position-time equation. In particular, we can solve $x_f = x_i + (v_i \cos \theta)t$ for the initial speed, v_i, since all the other quantities in the equation are known. (We could also find v_i with the y position-time equation, but the algebra is a little messier in that case.) (**b**) We substitute $t = 1.76$ s into the y position-time equation, $y_f = y_i + (v_i \sin \theta)t - \frac{1}{2}gt^2$. This gives the height of the ball as it passes over the tree.

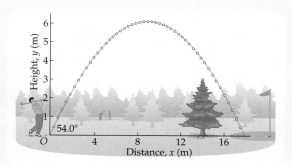

Known	Unknown
(**a**) $d = 17.8$ m	(**a**) $v_i = ?$
$\theta = 54.0°$	(**b**) $y_f = ?$
$t = 2.24$ s	
(**b**) $t = 1.76$ s	

Solution

Part (a)

1 Apply $x_f = x_i + (v_i \cos \theta)t$ to the golf ball, noting that $x_i = 0$:

$$x_f = x_i + (v_i \cos \theta)t$$
$$= 0 + (v_i \cos \theta)t$$
$$= (v_i \cos \theta)t$$

2 Solve the equation from part (a) for the initial speed, v_i, and substitute the numerical values:

$$v_i = \frac{x_f}{(\cos \theta)t}$$

$$= \frac{17.8 \text{ m}}{(\cos 54°)(2.24 \text{ s})}$$

Math HELP
Solving Simple Equations
See Math Review, Section III

$$= \boxed{13.5 \text{ m/s}}$$

Part (b)

3 Evaluate $y_f = y_i + (v_i \sin \theta)t - \frac{1}{2}gt^2$ at the time $t = 1.76$ s. Recall that $y_i = 0$ and that $v_i = 13.5$ m/s from part (a):

$$y_f = 0 + (v_i \sin \theta)t - \frac{1}{2}gt^2$$
$$= (13.5 \text{ m/s})(\sin 54.0°)(1.76 \text{ s})$$
$$\quad - \frac{1}{2}(9.81 \text{ m/s}^2)(1.76 \text{s})^2$$
$$= \boxed{4.03 \text{ m}}$$

Insight

The ball clears the top of the tree by 1.03 m. When it lands on the green, its speed (in the absence of air resistance) is again 13.5 m/s—the same as when it was launched.

40. Follow-up What is the speed of the golf ball in Guided Example 4.12 when it passes over the tree?

41. A soccer ball is kicked from the ground with an initial speed of 12 m/s at an angle of 32° above the horizontal. What are the x and y positions of the ball 0.50 s after it is kicked?

42. Referring to Problem 41, what are the x and y components of the soccer ball's velocity 0.50 s after it is kicked?

CONCEPTUAL **Example 4.13 How Does the Speed of the Ball Change?**

You throw a ball to a friend with an initial speed of 10 m/s. Your friend catches the ball at time T at the same level from which you threw it. Referring to the figure below, which of the plots (1, 2, or 3) best represents the speed of the ball as a function of time?

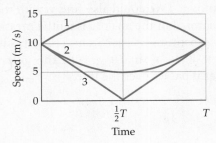

Reasoning and Discussion
The ball is thrown in a direction above the horizontal, and therefore its speed decreases during its flight. This eliminates plot 1, which shows an increase in speed. Of the two remaining plots we can eliminate 3, because it shows the speed going to zero at the halfway point of the flight. The ball wasn't thrown straight upward, however, and thus its speed is nonzero at its highest point. It follows that plot 2 best represents the speed as a function of time.

Answer
Plot 2 shows the ball's speed as a function of time.

Range is affected by the launch angle

What launch angle gives a projectile maximum range?

How far can you throw a baseball or kick a football? The answer depends on both the initial speed that you give the ball and the angle at which it is launched. The **range** of a projectile is the horizontal distance it travels before landing. For a projectile launched with an initial speed v_i at an angle θ above the horizontal, straightforward mathematics shows that the range, R, is given by the following equation:

Range Equation

$$\text{range} = \left(\frac{(\text{initial velocity})^2}{\text{acceleration due to gravity}} \right) \sin(2 \times \text{launch angle})$$

$$R = \left(\frac{v_i^2}{g} \right) \sin 2\theta$$

This expression is valid for projectiles that land at the level from which they were launched. 🔑 **Maximum range occurs when the launch angle is $\theta = 45°$.** In this case the range is

$$R_{\max} = \frac{v_i^2}{g}$$

For example, a projectile with an initial speed of 13 m/s has a maximum range of 17 m. The range is smaller if the launch angle is greater than or less than 45°.

Air resistance can affect the motion of projectiles

As mentioned earlier, air resistance or other forms of friction do not act on ideal projectiles. Most everyday projectiles are good approximations to the ideal case. An example of three ideal projectiles, all launched with the same initial speed of 20 m/s but at different initial angles, is shown in **Figure 4.28**. As expected, the projectile launched at 45° has the greatest range. Notice also that projectiles launched at angles greater than and less than 45° by the same amount (like 60° and 30°) have the same range.

The same three projectiles, this time subject to a significant amount of air resistance, are shown in **Figure 4.29** on the next page. Notice that the range and the maximum height of each projectile are reduced, as one would expect.

C⊙L*PHYSICS*
Moonshot
A projectile launched on the Moon travels about six times farther than it does on Earth. Why? Because the range, *R*, depends inversely on the acceleration of gravity. The smaller *g* is, the larger the range. On the Moon *g* is only about a sixth of what it is on Earth. It was for this reason that astronaut Alan Shepard couldn't resist bringing a golf club and ball with him on the third lunar landing mission. On February 6, 1971, he ambled out onto the Fra Mauro Highlands and became the first person to hit a tee shot on the Moon. His distance was respectable—unfortunately, his ball landed in a sand trap.

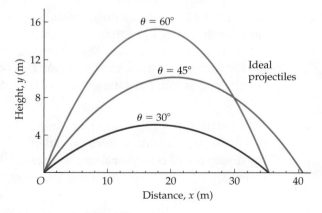

▲ **Figure 4.28 Projectiles without air resistance**
Three ideal projectiles (no air resistance), each with a launch speed of 20 m/s. The greatest range occurs with a launch angle of 45°. Launch angles greater than and less than 45° by the same amount have equal ranges.

Three projectiles, as in Figure 4.28, but with significant air resistance. Air resistance reduces the ranges and changes the shapes of the trajectories.

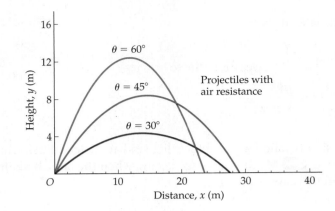

Projectiles with air resistance

Notice also in Figure 4.29 that air resistance has altered the nice symmetry of the ranges at 60° and 30°. The shape of the trajectories has been affected as well; they are no longer perfect parabolas. These are the kind of effects one can expect when air resistance becomes important.

4.4 LessonCheck (MP)

Checking Concepts

43. ▣ **Identify** Which of the following shapes is seen in the path of a projectile: straight line, parabola, circle, ellipse, hyperbola?

44. ▣ **Describe** How must a projectile be launched in order to maximize its range?

45. Big Idea Jennifer's bowling ball is dropped from rest. Janet's bowling ball is thrown horizontally at the same time. After the balls are released, is the acceleration of Jennifer's bowling ball greater than, less than, or equal to the acceleration of Janet's bowling ball? Explain.

46. Concept Check Three projectiles (A, B, and C) are launched with the same initial speed but with different launch angles, as shown in **Figure 4.30**. Rank the projectiles in order of increasing (a) horizontal component of initial velocity and (b) time of flight. Indicate ties where appropriate.

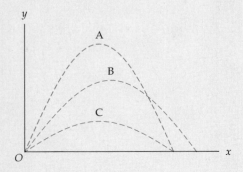

▲ Figure 4.30

Solving Problems

47. Calculate What is the maximum range of the golf ball in Guided Example 4.12?

48. Calculate A snowball is thrown with an initial x velocity of 7.5 m/s and an initial y velocity of 8.2 m/s. How much time is required for the snowball to reach its highest point? (*Hint:* The highest point of a projectile corresponds to the time when $v_{y,f} = 0$.)

49. Sketch A softball is thrown from the origin of an x-y coordinate system with an initial speed of 18 m/s at an angle of 35° above the horizontal.
(**a**) Find the x and y positions of the softball at the times $t = 0.50$ s, 1.0 s, 1.5 s, and 2.0 s.
(**b**) Plot the results from part (a) on an x-y coordinate system, and sketch the parabolic curve that passes through them.

50. Calculate A diver runs horizontally off the end of a diving board with an initial speed of 1.85 m/s. If the diving board is 3.00 m above the water, what is the diver's speed just before she enters the water?

Global Positioning Systems

▲ An assortment of geocaching prizes.

Asking for Directions Today you are just as likely to ask a Global Positioning System (GPS) receiver for directions as a stranger. Signals from Earth-orbiting satellites allow the GPS receiver to determine its current location. Once you tell the receiver where you want to go, it uses satellite signals and on-board mapping and routing software to determine your route. In terms of physics the GPS receiver calculates vectors and converts them into directions that it conveys to you as you proceed.

GPS at Work on Taxiways The cockpit, the workplace of aircraft pilots, has been changed greatly by GPS. Many pilots say that the most stressful and complex part of a flight is moving on the taxiway. Research has produced new GPS technology that helps pilots guide their aircraft safely on the ground.

The technology makes use of a heads-up display, or HUD. The HUD is a sheet of glass in the pilot's line of vision on which images and data are projected. The projection of navigation cones onto the HUD guides the pilot to cleared (safe) routes while taxiing. The system works because GPS can determine the exact position of the plane as it moves. HUD also provides a real-time map showing the position of the plane, the route, and the positions of other aircraft on the ground.

Fun with GPS Geocaching is an outdoor game in which one player hides a cache (a container with a prize in it), then uses GPS to determine the cache's coordinates, and posts those coordinates online. Anyone who wants to hunt for the cache uses GPS to get to the area where it is hidden. Geocaching is becoming very popular globally. Over a million caches reside in over a hundred countries and on all seven continents, including Antarctica!

Take It Further

1. Propose *How could GPS and HUD technologies improve safety on highways? Make a list of possible applications.*

2. Debate *GPS technology makes it possible to track individual people. Is this a good idea? Is such tracking ethical? Write a short argument on each side of the issue.*

Physics Lab Projectile Motion

This lab explores projectile motion. You will analyze the motion of a ball as it rolls and then becomes a projectile and use the experimental data to determine the height of a table.

Materials
- stopwatch
- meterstick
- track
- steel ball or marble
- washer
- string
- several books

Procedure

1. Construct a ramp on a tabletop by placing one end of the track on top of one or two books. Position the track so that its lower end is between 1 and 2 m from the edge of the table.

2. Place a piece of masking tape near the top of the ramp. This will serve as the point of release of the ball.

3. Use the string and washer to form a plumb line that extends from the edge of the table to a point just above the floor. Mark the position directly below the washer on the floor. This point should be directly below the edge of the table.

4. Measure the length of the table from the end of the ramp to the edge of the table. Record the distance in the data table.

5. Release the ball from the starting point on the ramp. Using a stopwatch, measure the time required for the ball to move from the *end* of the ramp to the *edge* of the table. Record the time in the data table. Repeat this procedure until you complete five trials. Have someone catch the ball just after it drops off the edge of the table. If the ball is moving too quickly to accurately time, reduce the incline of the track.

6. Release the ball from the starting point and let it hit the floor. Carefully watch where the ball strikes the floor and mark this spot. This is the point of impact. Measure the distance between the mark on the floor (from Step 3) and the point of impact. Record the distance in the data table. Repeat this procedure until you complete five trials.

Data Table

Length of Table: _____ m					
	Trial 1	Trial 2	Trial 3	Trial 4	Trial 5
Time to roll across table (s)					
Horizontal distance traveled by projectile (m)					

Analysis

Show your work and include units.

1. Calculate the average horizontal speed of the ball while it rolls across the table.

2. Calculate the time of flight of the ball from the tabletop to the floor. (*Hint:* The flight time is calculated using the range of the projectile and its average speed.)

3. Calculate the height of the table. (*Hint:* The height is calculated using the flight time and the position-time equation for a projectile in the vertical direction. The table height is equal to Δy.)

4. Measure the actual height of the table. Determine the percent error for your experimental value of the height:

$$\frac{\text{percent}}{\text{error}} = \left(\frac{\text{actual value} - \text{experimental value}}{\text{actual value}} \right) \times 100$$

Conclusions

1. Explain why your experimental value differs from the actual value for the height of the table.

2. Discuss whether changing the slope of the ramp would affect your data or the value for the height of the table.

3. Explain the importance of marking the starting point on the ramp.

4 Study Guide

Big Idea

The horizontal and vertical motions of an object are independent of one another.

For example, a projectile accelerates downward with the acceleration due to gravity regardless of its horizontal speed. Similarly, the horizontal speed of a projectile remains constant, even though its vertical speed is changing. Each motion proceeds as if the other motion didn't exist.

4.1 Vectors in Physics

🔑 A vector is a quantity that is specified by both a length and a direction. The length of a vector is referred to as its *magnitude*.

🔑 To resolve a vector means to find its components; a vector's components are the lengths of the vector along specified directions.

• Vectors can be specified in terms of their magnitude and direction or in terms of their x and y components.

Key Equations

The x and y components of a vector $\vec{A}$ are

$$A_x = A \cos \theta$$
$$A_y = A \sin \theta$$

The magnitude of a vector $\vec{A}$ is

$$A = \sqrt{A_x^2 + A_y^2}$$

The direction angle of a vector $\vec{A}$ is

$$\theta = \tan^{-1}(A_y/A_x)$$

4.2 Adding and Subtracting Vectors

🔑 To add two vectors $\vec{A}$ and $\vec{B}$, place the tail end of $\vec{B}$ at the head of $\vec{A}$. The sum $\vec{C} = \vec{A} + \vec{B}$ is the vector that extends from the tail of $\vec{A}$ to the head of $\vec{B}$.

🔑 To add more than two vectors, first place all of the vectors head to tail, head to tail, and so on. Then draw the resultant vector from the tail of the first vector to the head of the last vector.

🔑 In general, the sum of the x components gives the x component of the resultant, and the sum of the y components gives the y component of the resultant.

Key Equations

If $\vec{C} = \vec{A} + \vec{B}$, then the components of $\vec{C}$ are determined as follows:

$$C_x = A_x + B_x$$
$$C_y = A_y + B_y$$

4.3 Relative Motion

🔑 You can relate the velocities of two objects to each other if you know the velocity of each object relative to a third object.

• Relative velocities are determined by the simple addition of vectors.

Key Equation

If the velocity of object 1 relative to object 2 is $\vec{v}_{12}$ and the velocity of object 2 relative to object 3 is $\vec{v}_{23}$, then the velocity of object 1 relative to object 3 is

$$\vec{v}_{13} = \vec{v}_{12} + \vec{v}_{23}$$

4.4 Projectile Motion

🔑 In general, an ideal projectile follows a parabolic path regardless of how it is launched.

🔑 Maximum range occurs when the launch angle is $\theta = 45°$.

• Components of motion in the x and y directions can be treated independently of one another.

• The acceleration due to gravity on the Earth's surface is $g = 9.81 \text{ m/s}^2$.

Key Equation

The range of a projectile launched with an initial speed v_i at an angle θ is

$$R = \left(\frac{v_i^2}{g}\right)\sin 2\theta$$

4 Assessment

ANSWERS TO SELECTED ODD-NUMBERED PROBLEMS APPEAR IN APPENDIX A.

Lesson by Lesson

4.1 Vectors in Physics

Conceptual Questions

51. Which, if any, of the vectors shown in **Figure 4.31** are equal?

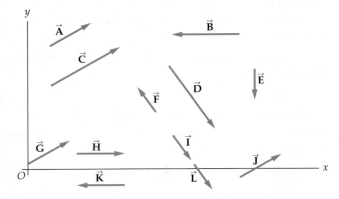

Figure 4.31

52. Given that $\vec{A} + \vec{B} = 0$, **(a)** how does the magnitude of $\vec{B}$ compare with the magnitude of $\vec{A}$? **(b)** How does the direction of $\vec{B}$ compare with the direction of $\vec{A}$?

53. Can a component of a vector be greater than the vector's magnitude?

54. Can a vector with zero magnitude have one or more components that are nonzero? Explain.

Problem Solving

55. The baseball diamond in **Figure 4.32** is a square with sides 27.4 m (90 ft) long. If the positive x axis points from home plate to first base and the positive y axis points from home plate to third base, find the magnitude and the direction angle of the displacement vector of a base runner who has just hit **(a)** a double, **(b)** a triple, and **(c)** a home run.

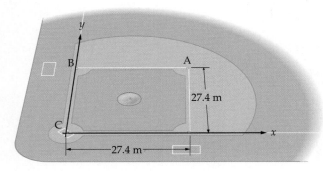

Figure 4.32

56. **Think & Calculate** The x and y components of a vector $\vec{r}$ are $r_x = 14$ m and $r_y = -9.5$ m, respectively. Find **(a)** the direction and **(b)** the magnitude of the vector $\vec{r}$. **(c)** If both r_x and r_y are doubled, how do your answers to parts (a) and (b) change?

57. You drive a car 680 m to the east, then 340 m to the north. **(a)** What is the magnitude of your displacement? **(b)** Using a sketch, estimate the direction of your displacement. **(c)** Verify your estimate in part (b) with a numerical calculation of the direction.

58. Vector $\vec{A}$ has a magnitude of 50 km and points in the positive x direction. A second vector, $\vec{B}$, has a magnitude of 120 km and points at an angle of 70° below the x axis. Which vector has **(a)** the greater x component and **(b)** the greater y component?

59. A treasure map directs you to start at a palm tree and walk due north for 15.0 m. You are then to turn 90° and walk 22.0 m; then turn 90° again and walk 5.00 m. Give the distance from the palm tree and the direction relative to north for each of the four possible locations of the treasure.

60. A whale comes to the surface to breathe and then dives at an angle of 20.0° below the horizontal as shown in **Figure 4.33**. If the whale continues in a straight line for 150 m, **(a)** how deep is it, and **(b)** how far has it traveled horizontally?

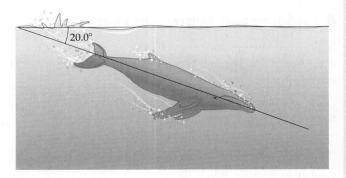

Figure 4.33

4.2 Adding and Subtracting Vectors

Conceptual Questions

61. Suppose that $\vec{A}$ and $\vec{B}$ have nonzero magnitudes. Is it possible for $\vec{A} + \vec{B}$ to be zero?

62. Given that $\vec{A} + \vec{B} = \vec{C}$ and that $A^2 + B^2 = C^2$, how are $\vec{A}$ and $\vec{B}$ oriented relative to one another?

63. Given that $\vec{A} + \vec{B} = \vec{C}$ and that $A + B = C$, how are $\vec{A}$ and $\vec{B}$ oriented relative to one another?

Problem Solving

64. A vector $\vec{A}$ has a magnitude of 40.0 m and points in a direction 20.0° below the positive x axis. A second vector, $\vec{B}$, has a magnitude of 75.0 m and points in a direction 50.0° above the positive x axis. Sketch the vectors $\vec{A}$, $\vec{B}$, and $\vec{C} = \vec{A} + \vec{B}$.

65. Vector $\vec{A}$ points in the positive x direction and has a magnitude of 75 m. The vector $\vec{C} = \vec{A} + \vec{B}$ points in the positive y direction and has a magnitude of 95 m. **(a)** Sketch $\vec{A}$, $\vec{B}$, and $\vec{C}$. **(b)** Estimate the magnitude and the direction of the vector $\vec{B}$ from your sketch.

66. Vector $\vec{A}$ points in the negative y direction and has a magnitude of 5 km. Vector $\vec{B}$ has a magnitude of 15 km and points in the positive x direction. Use components to find the magnitude of **(a)** $\vec{A} + \vec{B}$, **(b)** $\vec{A} - \vec{B}$, and **(c)** $\vec{B} - \vec{A}$.

67. Control Tower An air traffic controller observes two airplanes approaching the airport. The displacement from the control tower to plane 1 is given by the vector $\vec{A}$, which has a magnitude of 220 km and points in a direction 32° north of west. The displacement from the control tower to plane 2 is given by the vector $\vec{B}$, which has a magnitude of 140 km and points 65° east of north. **(a)** Sketch the vectors $\vec{A}$, $-\vec{B}$, and $\vec{D} = \vec{A} - \vec{B}$. Notice that $\vec{D}$ is the displacement from plane 2 to plane 1. **(b)** Use components to find the magnitude and the direction of the vector $\vec{D}$.

68. A basketball player runs down the court, following the path indicated by the vectors $\vec{A}$, $\vec{B}$, and $\vec{C}$ in **Figure 4.34**. The magnitudes of these three vectors are $A = 10.0$ m, $B = 20.0$ m, and $C = 7.0$ m. Find the magnitude and direction of the net displacement of the player using **(a)** the graphical method and **(b)** the component method of vector addition. Compare your results.

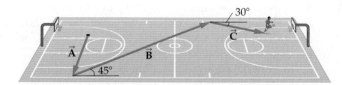

Figure 4.34

4.3 Relative Motion
Conceptual Questions

69. Rain is falling vertically downward, and you are running for shelter. To keep driest, should you hold your umbrella vertically, tilted forward, or tilted backward? Explain.

70. The wind feels stronger when you are sailing a boat upwind (beating) than when you are sailing downwind (running). Explain.

71. Figure 4.35 shows a KC-10A Extender using a boom to refuel an aircraft in flight. If the velocity of the KC-10A is 125 m/s due east relative to the ground, what is the velocity of the aircraft being refueled relative to **(a)** the ground and **(b)** the KC-10A?

Figure 4.35 Air-to-air refueling.

Problem Solving

72. As an airplane taxis on the runway with a speed of 16.5 m/s, a flight attendant walks toward the tail of the plane with a speed of 1.22 m/s. What is the flight attendant's speed relative to the ground?

73. As you hurry to catch your flight at the local airport, you encounter a moving walkway that is 85 m long and has a speed of 2.2 m/s relative to the ground. If it takes you 68 s to cover 85 m when walking on the ground, how long will it take you to cover the same distance on the walkway? Assume that you walk with the same speed on the walkway as you do on the ground.

74. A migrating robin flies due north with a speed of 12 m/s relative to the air. The air moves due east with a speed of 6.1 m/s relative to the ground. What is the robin's speed relative to the ground?

75. A passenger walks from one side of a ferry to the other as it approaches a dock. If the passenger's velocity is 1.50 m/s due north relative to the ferry and 4.50 m/s at an angle of 30.0° west of north relative to the water, what are the direction and magnitude of the ferry's velocity relative to the water?

76. ⎡ **Think & Calculate** ⎤ The pilot of an airplane wishes to fly due north, but there is a 36 km/h wind blowing toward the west. **(a)** In what direction should the pilot head her plane if its speed relative to the air is 350 km/h? **(b)** Draw a vector diagram that illustrates your result in part (a). **(c)** If the pilot decreases the airspeed of the plane, but still wants to head due north, should she increase or decrease the angle found in part (a)?

77. You are riding on a jet ski at an angle of 35° upstream on a river flowing with a speed of 2.8 m/s. If your velocity relative to the ground is 9.5 m/s at an angle of 20.0° upstream, what is the speed of the jet ski relative to the water? (*Note:* Angles are measured relative to the x axis as was done in Guided Example 4.8.)

4.4 Projectile Motion

Conceptual Questions

78. The initial velocity of a projectile has a horizontal component equal to 5 m/s and a vertical component equal to 6 m/s. At the highest point of the projectile's flight, what is **(a)** the horizontal component of its velocity and **(b)** the vertical component of its velocity? Explain.

79. A boy rides on a pony that is walking with constant velocity. The boy leans over to one side, and a scoop of ice cream falls from his ice cream cone. Describe the path of the scoop of ice cream as seen by **(a)** the child and **(b)** his parents standing on the ground nearby.

80. What is the acceleration of a projectile when it reaches its highest point? What is its acceleration just before and just after reaching this point?

81. ⎡Rank⎤ Three projectiles (A, B, and C) are launched with different initial speeds so that they reach the same maximum height, as shown in **Figure 4.36**. Rank the projectiles in order of increasing **(a)** initial speed and **(b)** time of flight. Indicate ties where appropriate.

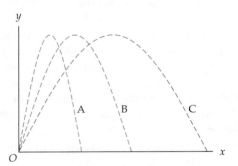

Figure 4.36

Problem Solving

82. A projectile is launched with an initial speed of 12 m/s. At its highest point its speed is 6 m/s. What was the launch angle of the projectile?

83. A racquet ball is struck in such a way that it leaves the racquet with a speed of 4.87 m/s in the horizontal direction. When the ball hits the court, it is a horizontal distance of 1.95 m from the racquet. Find the height of the racquet ball when it left the racquet.

84. **Victoria Falls** The great, gray-green, greasy Zambezi River flows over Victoria Falls in south central Africa. The falls are approximately 108 m high. If the river is flowing horizontally at 3.60 m/s just before going over the falls, what is the speed of the water when it hits the bottom? Assume that the water is in free fall as it drops.

85. Playing shortstop, you pick up a ground ball and throw it to second base. The ball is thrown horizontally, with a speed of 22 m/s, directly toward point A as shown in **Figure 4.37**. When the ball reaches the second baseman

0.45 s later, it is caught at point B. **(a)** How far were you from the second baseman? **(b)** What is the distance of vertical drop, from A to B?

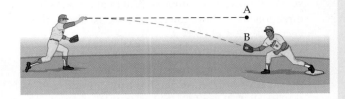

Figure 4.37

86. A crow is flying horizontally with a constant speed of 2.70 m/s when it releases a clam from its beak as shown in **Figure 4.38**. The clam lands on the rocky beach 2.10 s later. Just before the clam lands, what is **(a)** its horizontal component of velocity and **(b)** its vertical component of velocity?

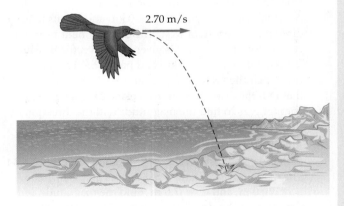

2.70 m/s

Figure 4.38

87. **Pumpkin Toss** In Denver, after Halloween, children bring their jack-o'-lanterns to the top of a tower and compete for accuracy in hitting a target on the ground, as shown in **Figure 4.39**. Suppose that the tower is 9.0 m high and that the bull's-eye is a horizontal distance of 3.5 m from the launch point. If the pumpkin is thrown horizontally, what is the launch speed needed to hit the bull's-eye?

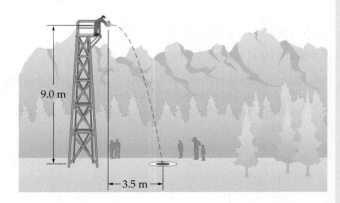

9.0 m

3.5 m

Figure 4.39

88. **Punkin Chunkin** In Sussex County, Delaware, a post-Halloween tradition is Punkin Chunkin, a competition in which contestants build cannons, catapults, trebuchets, and other devices to launch pumpkins to the greatest distance they can. Though hard to believe, pumpkins have been projected a distance of 1245 m (over $\frac{3}{4}$ mi) in this contest. What is the minimum initial speed needed for such a shot?

89. A diver runs horizontally off the end of a 3.0-m-high diving board with an initial speed of 1.8 m/s. **(a)** Given that the diver's initial position is $x_i = 0$ and $y_i = 3.0$ m, find her x and y positions at the times $t = 0.25$ s, $t = 0.50$ s, and $t = 0.75$ s. **(b)** Plot the results from part (a), and sketch the corresponding parabolic path.

90. Fairgoers ride a Ferris wheel with a radius of 5.00 m, as shown in **Figure 4.40**. The wheel completes one revolution every 32.0 s. **(a)** What is the average speed of a rider on this Ferris wheel? **(b)** If a rider accidentally drops a stuffed animal at the top of the wheel, where does it land relative to the base of the ride? (*Note:* The bottom of the wheel is 1.75 m above the ground.)

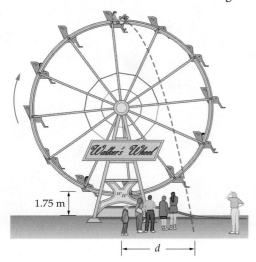

Figure 4.40

91. A golfer gives a ball a maximum initial speed of 34.4 m/s. **(a)** What is the longest possible hole-in-one shot for this golfer? Neglect any distance the ball might roll on the green and assume that the tee and the green are at the same level. **(b)** What is the minimum speed of the ball during this hole-in-one shot?

92. The hang time of a punt is measured to be 4.50 s. If the ball was kicked at an angle of 63.0° above the horizontal and was caught at the same level from which it was kicked, what was its initial speed?

Mixed Review

93. A baseball player hits a high fly ball to the outfield. **(a)** Is there a point during the flight of the ball where its velocity is parallel to its acceleration? **(b)** Is there a point where the ball's velocity is perpendicular to its acceleration? Explain in each case.

94. **Triple Choice** Child 1 throws a snowball horizontally from a rooftop; child 2 throws a snowball straight down from the same rooftop. Once in flight, is the acceleration of snowball 2 greater than, less than, or equal to the acceleration of snowball 1?

95. **Triple Choice** The penguin on the left in **Figure 4.41** is about to land on an ice floe. Just before it lands, is its speed greater than, less than, or equal to its speed when it left the water?

Figure 4.41 A penguin behaves much like a projectile from the time it leaves the water until it touches down on the ice.

96. **Predict & Explain** As you walk briskly down the street, you toss a small ball into the air. **(a)** If you want the ball to land in your hand when it comes back down, should you toss the ball straight upward, in a forward direction, or in a backward direction? **(b)** Choose the *best* explanation from among the following:

A. If you throw the ball straight up, you will leave it behind.

B. You have to throw the ball in the direction in which you are walking.

C. The ball moves in the forward direction at your walking speed at all times.

97. **Predict & Explain** Two divers run horizontally off the edge of a low cliff. Diver 2 runs with twice the speed of diver 1. **(a)** When the divers hit the water, is the horizontal distance covered by diver 2 twice as much as, four times as much as, or equal to the horizontal distance covered by diver 1? **(b)** Choose the *best* explanation from among the following:

A. The drop time is the same for both divers.

B. Drop distance depends on t^2.

C. All divers in free fall cover the same distance.

98. **Think & Calculate** Two of the allowed chess moves for a knight are shown in **Figure 4.42** on the next page. **(a)** Is the magnitude of displacement 1 greater than, less than, or equal to the magnitude of displacement 2? Explain. **(b)** Find the magnitude and direction of the knight's displacement for each of the two moves. Assume that the chessboard squares are 3.5 cm on a side.

Figure 4.42

99. A Lob versus a Bullet A quarterback can throw a receiver a high, lazy lob pass or a low, quick bullet pass. These passes are indicated by curves 1 and 2, respectively, in **Figure 4.43**. **(a)** The lob pass is thrown with an initial speed of 21.5 m/s, and its time of flight is 3.97 s. What is its launch angle? **(b)** The bullet pass is thrown with a launch angle of 25.0°. What is the initial speed of this pass? **(c)** What is the time of flight of the bullet pass?

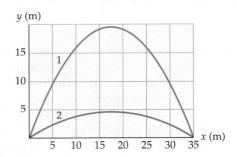

Figure 4.43

Writing about Science

100. Write a report on the maximum range of a projectile. Address the following questions: How does the launch angle for maximum range depend on whether the projectile lands at a level that is higher or lower than its launch level? How is the angle for maximum range affected by air resistance?

101. Connect to the Big Idea You drop a set of keys as you walk down the street. Do the keys land behind you or right at your feet? Is the drop time of the keys affected by your forward motion? Is the acceleration of the keys affected by your forward motion? Explain.

Read, Reason, and Respond

Landing Rovers on Mars When the twin Mars exploration rovers, *Spirit* and *Opportunity*, set down on the surface of the red planet in January 2004, their method of landing was both unique and elaborate. After initial braking with retro rockets, the rovers began their long descent through the thin Martian atmosphere on parachutes until they reached an altitude of about 16.7 m. At that point a set of air bags were inflated, additional retro rocket blasts slowed each craft nearly to a standstill, and the rovers detached from their parachutes. After a period of free fall to the surface, with an acceleration of 3.72 m/s², the rovers bounced about a dozen times before coming to rest. They then deflated their air bags, righted themselves, and began to explore the surface.

Figure 4.44 shows a rover with its surrounding cushion of air bags making its first bounce off the Martian surface. The rover bounces upward with a speed of 9.92 m/s at an angle of 75.0° above the horizontal.

Figure 4.44

102. How much time elapses between the first and second bounces?

   A. 1.38 s   C. 5.15 s

   B. 2.58 s   D. 5.33 s

103. How far does the rover travel in the horizontal direction between its first and second bounces?

   A. 13.2 m   C. 51.1 m

   B. 49.4 m   D. 98.7 m

104. What is the maximum height of the rover between its first and second bounces?

   A. 2.58 m   C. 12.3 m

   B. 4.68 m   D. 148 m

Standardized Test Prep

Multiple Choice

Use the graphs below to answer Questions 1–4. The graphs show a projectile's position and vertical velocity versus time.

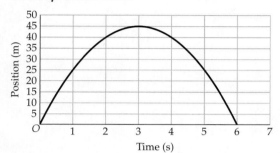

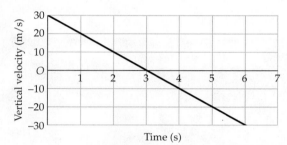

1. What is the maximum altitude, or vertical displacement, of the projectile?
   (A) 5 m
   (B) 15 m
   (C) 30 m
   (D) 45 m

2. Which expression gives a projectile's vertical velocity during flight if the projectile is fired with a speed v at an angle of 30° above the horizontal?
   (A) $v \sin 30° - gt$
   (B) $v \cos 30° - gt$
   (C) $v \tan 30° - gt$
   (D) v

3. Which expression gives a projectile's horizontal velocity during flight if the projectile has an initial launch velocity v at an angle of 30° above the horizontal?
   (A) $v \sin 30°$
   (B) $v \cos 30°$
   (C) $v \tan 30°$
   (D) $v - gt$

4. Which conditions result in a parabolic path for a projectile?
   (A) constant velocity in two dimensions
   (B) constant acceleration in two dimensions
   (C) constant acceleration in one dimension and constant velocity in the other dimension
   (D) increasing acceleration in one dimension

5. An airplane is flying north with an airspeed of 200 km/h when it encounters a crosswind of 70 km/h toward the east. Which expression gives the resultant speed of the airplane?
   (A) $200 \text{ km/h} + 70 \text{ km/h}$
   (B) $200 \text{ km/h} - 70 \text{ km/h}$
   (C) $\sqrt{(200 \text{ km/h})^2 + (70 \text{ km/h})^2}$
   (D) $\sqrt{(200 \text{ km/h})^2 - (70 \text{ km/h})^2}$

6. You are riding your bike northward at 5.0 m/s when you see a friend standing on a corner. What is your friend's velocity relative to you?
   (A) zero
   (B) 2.5 m/s toward the south
   (C) 2.5 m/s toward the north
   (D) 5.0 m/s toward the south

7. Which vector is the result of subtracting $\vec{B}$ from $\vec{A}$?

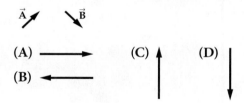

Extended Response

8. Describe how the position-time graph and the velocity-time graph for Questions 1–4 can each be used to determine the vertical position of the projectile at $t = 4$ s.

> **Test-Taking Hint**
>
> The horizontal and vertical motions of a projectile are independent of each other—each has its own equations of motion.

If You Had Difficulty With . . .

Question	1	2	3	4	5	6	7	8			
See Lesson(s)	4.5	4.1, 4.5	4.1	4.4, 4.5	4.2, 4.3	4.2, 4.3	4.4, 4.5	4.2			

5

Newton's Laws
of Motion

Inside

These athletes are exerting a force to accelerate a mass—a direct application of Newton's laws of motion.

Big Idea

All motion is governed by Newton's laws.

We are all subject to Newton's laws of motion, whether we know it or not. You can't do a somersault, ride a bicycle, or throw a baseball in a way that violates these rules. In short, our everyday existence is constrained and regulated by these three basic laws of nature. In this chapter you'll learn about Newton's laws and see how to apply them to practical situations.

5.1   Newton's Laws of Motion

Two of the most important quantities in physics are force and acceleration. These quantities are linked in a very specific way, discovered by the English scientist Isaac Newton (1642–1726). You've learned about acceleration in previous chapters. We now take a look at force.

Newton's First Law

Newton's first law describes how objects are affected by forces. But what exactly is a force? Well, when you *push* on a box to slide it across the floor, you exert a force; when you *pull* on the handle of a drawer to open it, you exert a force. In short, a **force** is a push or a pull. Forces are truly all around us. For example, when you hold this book in your hand, you exert an upward force to oppose the downward pull of gravity. Another example is shown in **Figure 5.1**.

When you push or pull on something, there are two quantities that characterize the force. The first is the strength, or *magnitude,* of your force. The second is the *direction* in which you are pushing or pulling. Because a force has both a magnitude and a direction, it is a vector.

Vocabulary

- force
- net force
- inertia
- newton

▶ Figure 5.1 **Forces**
Forces are a part of everyday life. For example, a tugboat exerts a large force to maneuver a ship into port. Look for additional examples of forces around you. How many different situations can you identify where one object exerts a force on another object?

Objects don't start or stop moving on their own

Have you ever seen a chair start moving across the room by itself? Has one of your textbooks ever moved itself to a new location? Of course not. A force is required to make an object start to move. Similarly, a force is required to stop an object, or to change its direction of motion. These simple observations are the essence of Newton's first law of motion:

> **Newton's First Law**
>
> An object at rest remains at rest as long as no net force acts on it.
>
> An object moving with constant velocity continues to move with the same speed and in the same direction as long as no net force acts on it.

Notice that we say "no net force" in the statement of Newton's first law. This is because most objects have more than one force acting on them. For example, a plate at rest on a table experiences a downward force due to gravity and an upward force due to the table. If you push the plate across the table, it also experiences a horizontal force due to your push. The **net force** exerted on an object is the vector sum of all the individual forces that act on it. The phrase "no net force" means that the net force is equal to zero. (Just to be clear, *net* and *total* mean exactly the same thing and can be used interchangeably.)

How many forces are acting on you right now? Gravity is certainly one. Another is the force of the chair holding you up. Perhaps your arms are resting on the table in front of you—if so, the table is exerting an additional force on you. Similarly, the floor is exerting a force on your feet. At this moment the net force on you is zero, which is why you are sitting at rest. Take a look at some of the objects around you today and think about the forces that act on them. Is the net force acting on each of them zero or nonzero?

An object's motion does not change on its own

The second part of Newton's first law is even more interesting. It says that an object doesn't change its speed or its direction of motion unless a net force causes it to. Now, you might object and say that if you stop pushing a box across the floor, it stops moving. It doesn't stop moving by itself, though. The box stops moving because the force of *friction* between it and the floor causes it to stop. If you could eliminate the friction, the box would continue moving forever. Even if you just reduce the friction, by pushing the box on a sheet of ice, the distance it moves after you stop pushing will increase.

You can see this behavior most clearly with an *air track*, a device that allows objects to move with practically no friction. An example of such a device is shown in **Figure 5.2**. Notice that air is blown through small holes in the track, creating a cushion of air for a small cart to ride on. A cart placed at rest on a level track remains at rest—unless you push it to get it started.

Once set in motion, the cart glides with constant velocity—constant speed in a straight line—until it hits a bumper at the end of the track. The bumper exerts a force on the cart, causing it to change its direction of motion. After bouncing off the bumper, the cart again moves with constant velocity. If the track could be extended to infinite length and made perfectly frictionless, the cart would simply keep moving with constant velocity forever.

CONNECTINGIDEAS

You just learned that a force is a push or a pull. As you might expect, there are many different types of forces.
• Here friction is introduced as the force that often brings a moving object to a stop.
• Lesson 5.2 will describe several common types of forces that act on objects.
• Finally, Lesson 5.3 focuses exclusively on friction and its effects.

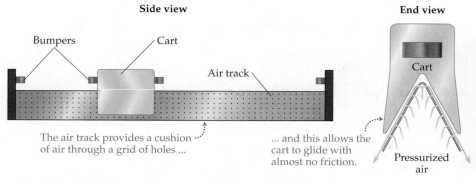

Side view

Bumpers · Cart · Air track

The air track provides a cushion of air through a grid of holes ...

End view

Cart

... and this allows the cart to glide with almost no friction.

Pressurized air

▲ **Figure 5.2 The air track**
An air track provides a cushion of air on which a cart can ride with virtually no friction.

Newton's first law is known as the law of inertia

Newton's first law is sometimes referred to as the *law of inertia*. This is appropriate because the meaning of the word *inertia* is "laziness." Loosely speaking, we can say that matter is "lazy," in that it doesn't change its motion unless forced to do so. For example, an object at rest doesn't start moving on its own. An object moving with constant velocity doesn't alter its speed or direction, unless a force causes the change. The tendency of an object to resist any change in its motion is referred to as its **inertia**.

Thus, any time you see an object changing its motion, you know that a force is at work. A planet changing direction as it orbits a star, a car accelerating away from a stoplight, an apple falling from a tree—these are all examples of an object changing its motion. In each case a force causes the change.

CONCEPTUAL Example 5.1 **Which String Breaks?**

A heavy anvil hangs from a string attached to a ceiling, as shown on the right. An identical string hangs from the bottom of the anvil. Which string breaks if you jerk the lower string downward rapidly?

Reasoning and Discussion
If the lower string is pulled downward rapidly, the inertia of the massive anvil keeps it from responding quickly. Since the anvil barely moves, the force in the lower string quickly becomes large. As a result, the lower string breaks before the anvil has a chance to move. (Pulling slowly on the lower string causes the upper string to break, instead.)

Answer
The lower string breaks if jerked downward rapidly.

Newton's Second Law

The second law of motion is more specific than the first law. It tells exactly how a force changes an object's motion.

Acceleration depends on force and mass

If you throw a baseball, the force required is not too great. On the other hand, if you push a car and give it the same speed as the baseball, the force required is quite large. Why the difference? As you can imagine, the difference is that the car contains a lot more matter than the baseball—the car has more *mass*. Recall that an object's mass is a measure of the

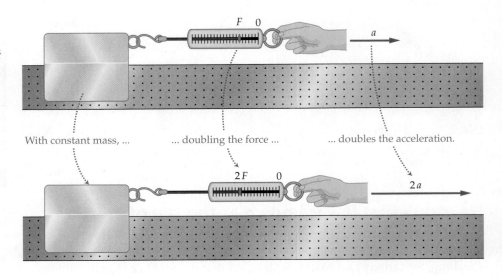

Figure 5.3 Acceleration is proportional to force
A spring scale is used to accelerate a mass on a "frictionless" air track. If the force is doubled, the acceleration is also doubled.

With constant mass, ... doubling the force ...   ... doubles the acceleration.

Table 5.1 Typical Masses in Kilograms (kg)

Earth	5.97×10^{24}
Space shuttle	2,000,000
Blue whale (largest animal on Earth)	178,000
Whale shark (largest fish)	18,000
Elephant (largest land animal)	5400
Automobile	1200
Human (adult)	70
Gallon of milk	3.6
Quart of milk	0.9
Baseball	0.145
Honeybee	0.00015
Bacterium	10^{-15}

quantity of matter it contains. As mentioned in Chapter 1, we measure mass in kilograms (kg), where 1 kilogram is defined as the mass of a standard cylinder of platinum-iridium. A list of typical masses is given in **Table 5.1**.

To see exactly how an object's acceleration is affected by force and mass, consider the experiment illustrated in **Figure 5.3**. In this experiment an air-track cart is pulled with a scale that measures the amount of force, and the resulting acceleration is measured. If the pulling force is *F*, the acceleration of the cart is *a*. If the force is doubled to 2*F*, the acceleration also doubles, to 2*a*. Thus, the acceleration is proportional to the force—the greater the force, the greater the acceleration.

Next, let's return to the original pulling force, *F*, and double the mass instead. We do this by connecting two carts together, as in **Figure 5.4**. In this case the acceleration is halved, to $\frac{1}{2}a$. Thus, we find that acceleration is inversely proportional to mass—the greater the mass, the smaller the acceleration. These are the key observations behind Newton's second law of motion.

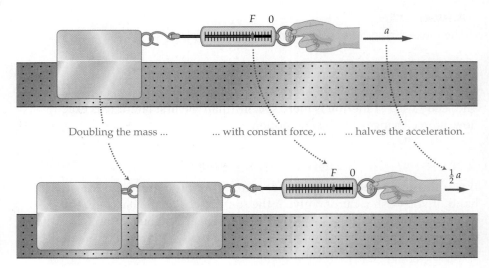

Doubling the mass ...   ... with constant force, ... halves the acceleration.

Figure 5.4 Acceleration is inversely proportional to mass
If the mass of an object is doubled but the force remains the same, the acceleration is halved.

The second law relates force, mass, and acceleration

Combining the preceding results, we can express the acceleration caused by a single force in the form of an equation as follows:

$$a = \frac{F}{m}$$

Notice that the acceleration is proportional to the force and inversely proportional to the mass in this equation, as desired. Rearranging the equation yields the form of Newton's second law that is perhaps best known, F equals m times a:

> **Newton's Second Law (for a single force)**
>
> force = mass × acceleration
>
> $F = ma$
>
> ⚙ **Force is equal to mass times acceleration.**

We use this form when only one force causes an acceleration. This equation relates the magnitude of the force to the magnitude of the acceleration. As for the direction, the acceleration is always in the *same* direction as the force.

Force is measured with a unit called the newton

We measure forces using a unit called, appropriately enough, the **newton** (N). In particular, 1 newton is defined as the force required to give 1 kilogram of mass an acceleration of 1 m/s². Thus,

$$1\,N = (1\,kg)(1\,m/s^2) = 1\,kg \cdot m/s^2$$

In everyday terms a newton is roughly a quarter of a pound. Typical forces and their magnitudes in newtons are listed in **Table 5.2**.

⚙ *How are force, mass, and acceleration related?*

Table 5.2 Typical Forces in Newtons (N)

Main engines of a space shuttle	31,000,000
Pulling force of a locomotive	250,000
Thrust of a jet engine	75,000
Force to accelerate a car	7000
Weight of an adult human	700
Weight of an apple	1
Weight of a rose	0.1
Weight of an ant	0.001

QUICK Example 5.2 What's the Force?

An 1800-kg car has an acceleration of 3.8 m/s². What is the force acting on the car?

Solution

Substitute $m = 1800\,kg$ and $a = 3.8\,m/s^2$ in $F = ma$:

$$F = ma$$
$$= (1800\,kg)(3.8\,m/s^2)$$
$$= \boxed{6800\,N}$$

Practice Problems

1. **Concept Check** Is it possible for an object at rest to have one force acting on it? If your answer is yes, provide an example. If your answer is no, explain why not.

2. You push a 12.3-kg shopping cart with a force of 10.1 N.
(a) What is the acceleration of the cart?
(b) If the cart starts from rest, how far does it move in 2.50 s?

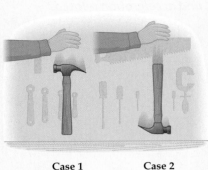

Case 1   Case 2

The metal head of a hammer is loose. To tighten it, you drop the hammer down onto a table. Should you drop the hammer with the handle end down (case 1) or with the head end down (case 2), or will you get the same result either way?

Reasoning and Discussion
It might seem that since the same hammer hits the same table in either case, there shouldn't be a difference. Actually, there is.

In case 1 the handle of the hammer comes to rest when it hits the table, but the head continues downward until a force acts on it to bring it to rest. The force that acts on it is supplied by the handle, which results in the head being wedged more tightly onto the handle. Since the metal head is heavy, the force wedging it onto the handle is great.

In case 2 the head of the hammer comes to rest on the table, but the handle continues to move until a force brings it to rest. The handle is lighter than the head, however; thus, the force acting on it is smaller, resulting in less tightening.

Answer
Drop the hammer with the handle end down to tighten the head (case 1).

🔑 *How is net force related to mass and acceleration?*

The second law also applies to net force

Imagine a scrum in a rugby match, where players are pushing from all directions trying to get the ball. Think how many forces are acting on the ball. In general, there are several forces acting on any given object. To take this into account, we replace F in the equation $F = ma$ with the sum of the force vectors:

$$\text{sum of force vectors} = \vec{\mathbf{F}}_{net} = \Sigma\vec{\mathbf{F}}$$

The notation $\Sigma\vec{\mathbf{F}}$, which uses the Greek letter sigma (Σ), is read "sum of forces." We can now write the complete statement of Newton's second law of motion, which is valid for any number of forces:

> **Newton's Second Law (for multiple forces)**
>
> sum of forces = mass × acceleration
> $$\Sigma\vec{\mathbf{F}} = m\vec{\mathbf{a}}$$
>
> 🔑 **The sum of the forces (the net force) acting on an object is equal to its mass times its acceleration.**

Notice that this is a vector equation. It states that an object's acceleration is always in the same direction as the net force acting on the object. This is a very important aspect of Newton's second law.

The best way to apply this equation is to write it out component by component:

$$\Sigma F_x = ma_x$$
$$\Sigma F_y = ma_y$$

Sometimes only a single component is needed, as in the examples in this lesson. In later lessons you'll see cases where both components are used.

C👁L*PHYSICS*
Walking and Shrinking

Walking is similar to dropping a hammer on a table, as in case 1 of Conceptual Example 5.3. Specifically, each step you take tamps your head down onto your spine. In effect, your head is like the head of the hammer, and your spine is like the handle. This tamping down causes you to become shorter during the day! Test it out. Measure your height first thing in the morning, then again before going to bed. If you're like many people, you'll find that you have shrunk a few centimeters (about an inch) during the day.

Moe, Larry, and Curly push on a 752-kg boat that floats next to a dock. They each exert an 80.5-N force parallel to the dock. (**a**) What is the boat's acceleration if they all push in the same direction? (**b**) What is the boat's acceleration if Larry and Curly push in the opposite direction from Moe?

Picture the Problem

In our sketch we indicate the three relevant forces acting on the boat: $\vec{F}_M$, $\vec{F}_L$, and $\vec{F}_C$. Notice that we have chosen the positive x direction to be toward the right, the direction in which all three push for part (a). Therefore, all three forces have a positive x component in part (a). In part (b), however, the forces exerted by Larry and Curly have negative x components.

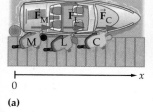
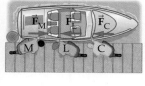

(a)   (b)

Strategy

Since we know the mass of the boat and the forces acting on it, we can find the acceleration using $\sum F_x = ma_x$. Even though this problem is one-dimensional, it is important to think in terms of vector components. For example, when we sum the x components of the forces, we are careful to use the appropriate signs—as always when dealing with vectors.

Known   (**a**) $F_M = F_L = F_C = 80.5$ N
$m = 752$ kg

(**b**) $F_M = 80.5$ N
$F_L = F_C = -80.5$ N
$m = 752$ kg

Unknown   (**a**) and (**b**) $a_x = ?$

Solution

Part (a)

1 Apply Newton's second law in the x direction, and solve for the acceleration, a_x:

$$\sum F_x = ma_x$$

$$a_x = \frac{\sum F_x}{m}$$

Math HELP
Sigma (Σ)
See Math Review, Section I

2 Substitute the known forces and the mass to calculate the acceleration, a_x. Since a_x turns out to be positive, the acceleration is to the right, as expected:

$$a_x = \frac{\sum F_x}{m} = \frac{F_M + F_L + F_C}{m}$$
$$= \frac{80.5 \text{ N} + 80.5 \text{ N} + 80.5 \text{ N}}{752 \text{ kg}}$$
$$= \boxed{0.321 \text{ m/s}^2 \text{ (to the right)}}$$

Part (b)

3 Again, apply Newton's second law in the x direction and solve for the acceleration, a_x:

$$\sum F_x = ma_x \qquad a_x = \frac{\sum F_x}{m}$$

4 Substitute the known forces and the mass to calculate the acceleration, a_x. In this case a_x turns out to be negative, indicating that the acceleration is to the left:

$$a_x = \frac{\sum F_x}{m} = \frac{F_M + F_L + F_C}{m}$$
$$= \frac{80.5 \text{ N} - 80.5 \text{ N} - 80.5 \text{ N}}{752 \text{ kg}}$$
$$= \boxed{-0.107 \text{ m/s}^2 \text{ (to the left)}}$$

Insight

Our results are in agreement with everyday experience: Three forces in the same direction cause more acceleration than three forces in opposing directions. The method of using vector components and being careful about their signs gives the expected results in a simple situation like this. Being careful and consistent like this also works in more complicated situations where everyday experience may be of little help.

3. Follow-up In Guided Example 5.4, Moe, Larry, and Curly all push to the right with 85.0-N forces. If the boat has an acceleration of 0.530 m/s², then what is the mass of the boat?

4. What force must be exerted on a 1500-kg car to make it move with an acceleration of 3.8 m/s²?

5. You pull your little sister on a sled across an icy (nearly frictionless) surface. When you exert a constant horizontal force of 120 N, the sled has an acceleration of 2.5 m/s². If the sled has a mass of 7.4 kg, what is the mass of your little sister?

Newton's Third Law

Newton's third law describes the forces that objects exert on one another.

Forces always come in pairs

How are action and reaction forces related?

Nature never produces just one force at a time. It just doesn't happen. *Forces always come in pairs.* For example, you can't touch someone without being touched in return. In addition, the forces in a pair are equal in magnitude, opposite in direction, and act on *different objects.* These concepts are combined in Newton's third law of motion:

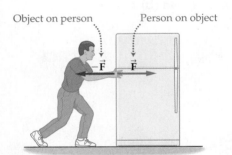

> **Newton's Third Law**
>
> 🔑 For every *action force* acting on an object, there is a *reaction force* acting on a different object.
>
> 🔑 The action and reaction forces are equal in magnitude and opposite in direction.

This law is commonly known in this abbreviated form: For every action there is an equal and opposite reaction.

Figure 5.5 illustrates some action-reaction pairs. Notice that there is always a reaction force, whether the action force pushes on something hard to move, like a refrigerator, or on something that moves with virtually no friction, like an air-track cart. In some cases the reaction force tends to be overlooked, as when the Earth exerts a *downward* gravitational force on a space shuttle and the shuttle exerts an equal and opposite *upward* gravitational force on the Earth. Still, the reaction force always exists.

Action-reaction forces act on different objects

An important part of the third law is that paired action-reaction forces always act on *different* objects. This is illustrated in Figure 5.5. As a result, *the two forces do not cancel.*

Consider a car accelerating from rest. As the car's engine turns the wheels, the tires exert a force on the road. By the third law the road exerts an equal and opposite force on the car's tires. It is this second force—which acts on the car through its tires—that propels the car forward. The force exerted by the tires on the road does not accelerate the car.

Since paired action-reaction forces act on different objects, they generally produce different accelerations, as we see in the next Guided Example.

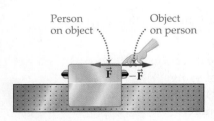

▲ **Figure 5.5** Examples of action-reaction force pairs

Two groups of canoeists meet in the middle of a lake. After a brief visit a person in canoe 1 pushes on canoe 2 with a force of 46 N to separate the canoes. The mass of canoe 1 and its occupants is $m_1 = 150$ kg, and the mass of canoe 2 and its occupants is $m_2 = 250$ kg. What is the acceleration of each canoe?

Picture the Problem

We have chosen the positive x direction to point from canoe 1 to canoe 2. With this choice the force exerted on canoe 2 is $F_2 = 46$ N (to the right). By Newton's third law the force exerted on the person in canoe 1 is $F_1 = -46$ N (to the left). Notice that each force has only an x component.

Strategy

From Newton's third law the force on canoe 1 is equal in magnitude to the force on canoe 2. The masses of the canoes are different, however, and therefore their accelerations are different as well. Since all the forces are in the x direction, we can find the acceleration of each canoe by solving $\sum F_x = ma_x$ for the acceleration.

Known

$F_1 = -46$ N (to the left)
$F_2 = 46$ N (to the right)
$m_1 = 150$ kg
$m_2 = 250$ kg

Unknown

$a_1 = ?$
$a_2 = ?$

Solution

1 Use Newton's second law to find the acceleration of canoe 1. Only one force acts in the x direction on canoe 1, and thus $\sum F_x = F_1$. Notice that the value we obtain for this acceleration is negative, indicating that canoe 1 accelerates to the left:

$$a_1 = \frac{F_1}{m_1}$$
$$= \frac{-46 \text{ N}}{150 \text{ kg}}$$
$$= \boxed{-0.31 \text{ m/s}^2 \text{ (to the left)}}$$

> **Math HELP**
> Solving Simple Equations
> **See Math Review, Section III**

2 Do the same calculation for canoe 2, using the appropriate force and mass. Only one force acts in the x direction on canoe 2, and thus $\sum F_x = F_2$. The acceleration is positive in this case, indicating that canoe 2 accelerates to the right:

$$a_2 = \frac{F_2}{m_2}$$
$$= \frac{46 \text{ N}}{250 \text{ kg}}$$
$$= \boxed{0.18 \text{ m/s}^2 \text{ (to the right)}}$$

Insight

The forces acting on the canoes have the same magnitude, so the lighter canoe has the greater acceleration.

Practice Problems

6. **Follow-up** If the mass of canoe 2 is increased, will its acceleration increase, decrease, or stay the same? Calculate the acceleration for the case where canoe 2 is replaced by a 25,000-kg ship.

7. **Concept Check** As you hold a book at rest in your hand, two forces are being exerted on the book.

(a) Identify these forces.

(b) Are these forces equal in magnitude and opposite in direction?

(c) Are these forces an action-reaction pair? Explain.

8. A 71-kg parent and a 19-kg child meet at the center of an ice rink. They place their hands together and push. If the acceleration of the child has a magnitude of 2.6 m/s^2, what is the magnitude of the parent's acceleration?

Newton's laws have limits

Newton's laws apply to everything you encounter in everyday life. They describe the mechanics of how objects move and how they affect one another. Thus, Newton's laws are often referred to as the *laws of mechanics*, and the behavior they predict determines the properties of the "mechanical universe" in which we all live.

There are limits to Newton's laws, however. Objects that are very small, like atoms and molecules, are described by a different kind of physics known as *quantum mechanics*, which we discuss in Chapter 24. Similarly, objects that have speeds close to the speed of light are described by Einstein's theory of relativity, presented in Chapter 27. We will point out situations where Newton's laws must be modified when necessary—but for objects of everyday size moving at everyday speeds, Newton's laws apply with great accuracy and can be used with complete confidence.

5.1 LessonCheck (MP)

Checking Concepts

9. Identify What is required to make a resting object move or a moving object change its speed or direction?

10. Apply What happens to the acceleration of an object if the net force that acts on it triples in magnitude?

11. Give Examples Describe a situation in which you exert a force on something but it does not move. Identify the action force and the reaction force, the relative size of these forces, and the objects they act on.

12. Big Idea Use Newton's laws to explain (**a**) why a shopping cart remains at rest if you do not push it, (**b**) why the cart accelerates when you push it, and (**c**) how the cart's acceleration is related to its mass and to how hard you push.

13. Find a Way An astronaut on a space walk discovers that his jet pack no longer works, leaving him stranded 50 m from the spacecraft. If the jet pack is removable, explain how the astronaut can still use it to return to the ship.

14. Analyze Is it possible for an object to move in a direction that is different from the direction of the net force acting on the object? Give an example to support your answer.

15. Interpret Diagrams Figure 5.6 shows a front-wheel-drive car accelerating from rest.
(**a**) Which of the forces, $\vec{F}_1$ or $\vec{F}_2$, is the force exerted by the car on the ground and which is the force exerted by the ground on the car?
(**b**) Which of the following correctly expresses the relationship between the two forces: (**1**) $\vec{F}_2 = \vec{F}_1$, (**2**) $\vec{F}_2 = -\vec{F}_1$, or (**3**) $\vec{F}_2 > \vec{F}_1$?

▲ Figure 5.6

Solving Problems

16. Calculate How much force is required to give a 0.15-kg baseball an acceleration of 12 m/s²?

17. Calculate An airplane lands and begins to slow down as it moves along the runway. If its mass is 3.50×10^5 kg and the net braking force is 4.30×10^5 N, what is the airplane's acceleration?

18. Calculate Your 1400-kg car pulls a 560-kg trailer away from a stoplight with an acceleration of 1.85 m/s².
(**a**) What is the net force exerted by the car on the trailer?
(**b**) What force does the trailer exert on the car?
(**c**) What is the net force acting on the car?

As you go about your activities today, look for Newton's laws in action. An accelerating object is a clear case where Newton's second law applies, but even an object at rest is stationary only because the forces acting on it add to zero. Take a moment to analyze situations like these, and you'll gain a greater appreciation of the importance of Newton's laws.

Diagramming Forces

It's not uncommon for an object to be acted on by several forces, each with a different magnitude and direction. In such a case the net force acting on the object is the vector sum of the individual forces. Scientists and engineers have developed a graphical way of keeping track of these forces. The method also makes it easier to apply Newton's laws in a systematic way.

Free-body diagrams are useful in applying Newton's laws

When solving a problem involving Newton's laws, it's helpful to make a sketch showing all the relevant forces. A sketch that shows all the forces acting on an object is referred to as a **free-body diagram**. To simplify the drawing we often represent the object with a point. ⬤ **Once all of the forces are drawn on a free-body diagram, a coordinate system is chosen and each force is resolved into components. At this point Newton's second law can be applied to each coordinate direction separately.**

To see how this works, consider a situation where two astronauts are using jet packs to push a 940-kg satellite toward a space station. The physical system is shown in **Figure 5.7 (a)**, and the free-body diagram is shown in **Figure 5.7 (b)**. With the coordinate system used in the figure, astronaut 1 pushes in the positive x direction, and astronaut 2 pushes in the positive y direction. If each astronaut pushes with a force of 46 N, what are the magnitude and the direction of the satellite's acceleration?

The easiest way to solve a problem like this is to treat each coordinate direction independently of the other. Thus, we first resolve each force into its x and y components. Referring to Figure 5.7, we write the force components for astronauts 1 and 2 as shown on the next page.

Vocabulary

- free-body diagram
- normal force
- apparent weight
- Hooke's law
- spring constant
- tension
- equilibrium

⬤ *How are free-body diagrams used to solve problems?*

CONNECTINGIDEAS

When describing motion in Chapter 2, we treated an object as a single point.

- Here we use the same concept, this time simplifying the real-life situation of an object with forces acting on it by representing the object as a point and the forces as arrows.

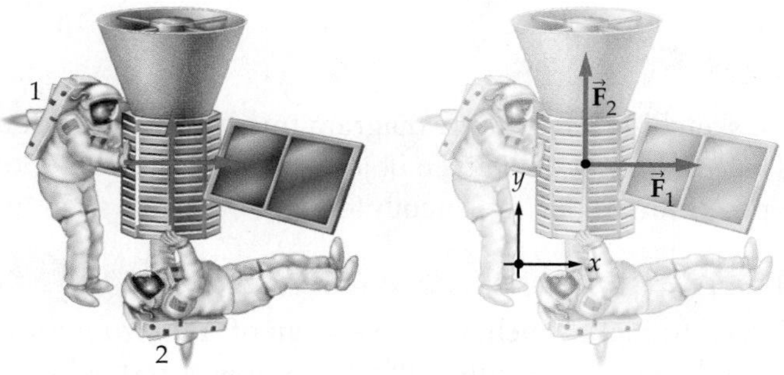

(a) Physical picture   (b) Free-body diagram

◀ **Figure 5.7 Two astronauts push a satellite**
The acceleration of the satellite can be found by calculating its x and y components separately, then combining the components to find a and θ.

Free-body diagram

$$F_{1,x} = 46 \text{ N} \qquad\qquad F_{2,x} = 0$$
$$F_{1,y} = 0 \qquad\qquad F_{2,y} = 46 \text{ N}$$

Next, we find the acceleration in the x and y directions by using the x- and y-component forms of Newton's second law:

$$\sum F_x = ma_x \qquad\qquad \sum F_y = ma_y$$
$$F_{1,x} + F_{2,x} = ma_x \qquad\qquad F_{1,y} + F_{2,y} = ma_y$$
$$46 \text{ N} + 0 = ma_x \qquad\qquad 0 + 46 \text{ N} = ma_y$$
$$46 \text{ N} = ma_x \qquad\qquad 46 \text{ N} = ma_y$$

Solving for the x- and y-components of acceleration yields

$$a_x = \frac{\sum F_x}{m} = \frac{46 \text{ N}}{940 \text{ kg}} = 0.049 \text{ m/s}^2$$

$$a_y = \frac{\sum F_y}{m} = \frac{46 \text{ N}}{940 \text{ kg}} = 0.049 \text{ m/s}^2$$

Thus, the satellite accelerates in both the x and the y directions. Using the Pythagorean theorem gives us the total acceleration:

$$a = \sqrt{a_x^2 + a_y^2} = \sqrt{(0.049 \text{ m/s}^2)^2 + (0.049 \text{ m/s}^2)^2} = \boxed{0.069 \text{ m/s}^2}$$

Simple trigonometry gives us the direction of the acceleration:

$$\theta = \tan^{-1}\left(\frac{a_y}{a_x}\right) = \tan^{-1}\left(\frac{0.049 \text{ m/s}^2}{0.049 \text{ m/s}^2}\right) = \tan^{-1}(1) = \boxed{45°}$$

PROBLEM-SOLVING NOTE

This component-by-component method can be applied whenever a system has forces that point in a variety of directions.

ACTIVE Example 5.6 Determine the Acceleration

Suppose the force exerted by astronaut 1 in Figure 5.7 is increased to 58 N. If everything else in the system remains the same, what are the magnitude and the direction of the satellite's acceleration?

Solution *(Perform the calculations indicated in each step.)*

1. Find the x component of acceleration: $\qquad a_x = 0.062 \text{ m/s}^2$

2. Find the y component of acceleration: $\qquad a_y = 0.049 \text{ m/s}^2$

3. Solve for the magnitude of the acceleration: $\quad a = 0.079 \text{ m/s}^2$

4. Find the angle of the acceleration: $\qquad\qquad \theta = 38°$

Types of Forces

A key part of sketching a free-body diagram involves correctly identifying the various forces that act on a given object. Let's take a closer look at several common types of forces that are encountered in everyday situations.

Normal forces act perpendicular to a surface

As you get ready to make lunch, you take a can of soup from the cupboard and place it on the kitchen counter. The can is now at rest, which means that the net force acting on it is zero. Thus, the downward force of gravity is

CONNECTING IDEAS

Forces are a central theme throughout physics.

• In Chapter 6 you'll learn how a force acting on an object over a distance changes the object's energy.

opposed by an upward force exerted by the counter, as shown in **Figure 5.8**. This force, which is perpendicular to the surface, is referred to as the *normal force*, $\vec{\mathbf{N}}$. In general, the force exerted perpendicular to the surface of contact between any two objects is called the **normal force**. The reason for this is that *normal* means "perpendicular" in mathematics.

The origin of the normal force is the interaction between atoms in a solid that act to maintain its shape. When the can of soup is placed on the countertop, for example, it causes a very small compression of that surface. The greater the mass placed on the countertop, the greater the normal force the countertop exerts to oppose being compressed.

Weight is the force exerted by gravity

When you weigh yourself, the scale gives a measurement of the pull of Earth's gravity. Recall that in general, the weight W of an object is the gravitational force exerted on it by a planet or other large mass. As we know from everyday experience, the greater the mass of an object, the greater its weight.

To see the connection between weight and mass, consider dropping a brick of mass m. As indicated in **Figure 5.9**, the only force acting on the brick is its weight, $F = W$. We know that the acceleration of the falling brick is the acceleration due to gravity, $a = g$, the same as for any object in free fall. Applying Newton's second law for a single force, we have

$$\text{(Newton's second law)} \quad a = \frac{F}{m} \quad \text{becomes} \quad g = \frac{W}{m} \quad \text{or} \quad W = mg$$

🔑 **The weight of an object is equal to its mass times the acceleration due to gravity.**

> **Definition of Weight, W**
>
> weight = mass × acceleration due to gravity
> $$W = mg$$
>
> SI unit: newton (N)

Weight and mass are not the same Though many people are not aware of it, there is a clear distinction between weight and mass. Weight is a gravitational force, measured in newtons; mass is a measure of the inertia of an object, and it is measured in kilograms. For example, if you were to travel to the Moon, your mass would not change—you would have the same amount of matter in you, regardless of your location. On the other hand, the gravitational force on the Moon's surface is less than the gravitational force on the Earth's surface. As a result, you would weigh less on the Moon than on the Earth, even though your mass would be the same.

To be specific, on Earth an 81.0-kg person has a weight given by

$$W_{\text{Earth}} = mg_{\text{Earth}} = (81.0 \text{ kg})(9.81 \text{ m/s}^2) = 795 \text{ N}$$

In contrast, the same person on the Moon, where the acceleration of gravity is 1.62 m/s², would weigh only

$$W_{\text{Moon}} = mg_{\text{Moon}} = (81.0 \text{ kg})(1.62 \text{ m/s}^2) = 131 \text{ N}$$

This is roughly one-sixth the person's weight on Earth. If there is a Lunar Olympics sometime in the future, the Moon's low gravity will be a boon for pole-vaulters, gymnasts, and other athletes.

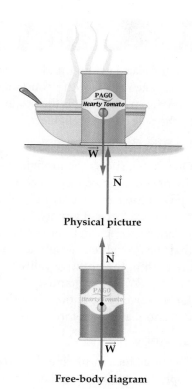

Physical picture

Free-body diagram

▲ **Figure 5.8 The normal force is perpendicular to a surface**
A can of soup rests on a kitchen counter, which exerts a normal (perpendicular) force, $\vec{\mathbf{N}}$, to support the can. In the special case shown here the normal force is equal in magnitude to the weight of the can, $W = mg$, and opposite in direction.

🔑 *How is the weight of an object related to its mass?*

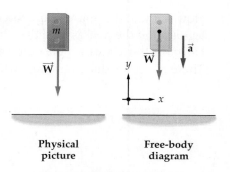

Physical picture   Free-body diagram

▲ **Figure 5.9 Weight and mass**
A brick of mass m has only one force acting on it in free fall—its weight, $\vec{\mathbf{W}}$. The resulting acceleration has the magnitude $a = g$; therefore, $W = mg$.

The fire alarm goes off, and a 97-kg firefighter slides down a pole with a constant acceleration of $a = 4.2$ m/s^2. What is the upward force $\vec{F}$ exerted by the pole on the firefighter?

Picture the Problem

Our sketch shows the firefighter sliding down the pole. Only two forces act on him: (1) the upward force exerted by the pole, $\vec{F}$, and (2) the downward force of gravity, $\vec{W}$. We choose the positive y direction to be upward. This means that the firefighter's weight and acceleration are both in the negative y direction. Therefore, $W_y = -W = -mg$ and $a_y = -4.2$ m/s^2. The force exerted by the pole is in the positive y direction, and therefore $F_y = F$.

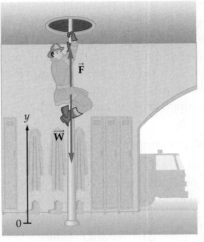

Physical picture   **Free-body diagram**

Strategy

The basic idea in solving this problem is to apply Newton's second law in the y direction: $\sum F_y = ma_y$. Substituting $W_y = -mg$ and $a_y = -4.2$ m/s^2 into Newton's second law allows us to solve for the unknown force, F.

Known	Unknown
$m = 97$ kg	$F = ?$
$W_y = -mg$	
$F_y = F$	
$a_y = -4.2$ m/s^2	

Solution

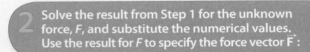

1 Apply Newton's second law in the y direction:

$$\sum F_y = ma_y$$
$$F - mg = ma_y$$

Math HELP
Sigma (Σ)
See Math Review, Section I

2 Solve the result from Step 1 for the unknown force, F, and substitute the numerical values. Use the result for F to specify the force vector $\vec{F}$:

$$F = mg + ma_y$$
$$= m(g + a_y)$$
$$= (97 \text{ kg})(9.81 \text{ m/s}^2 - 4.2 \text{ m/s}^2)$$
$$= 540 \text{ N}$$
$$\vec{F} = \boxed{540 \text{ N, upward}}$$

Insight

How is it that the pole exerts a force on the firefighter? Well, by wrapping his arms and legs around the pole as he slides, the firefighter exerts a downward force on the pole. By Newton's third law the pole exerts an upward force of equal magnitude on the firefighter. These forces are due to friction, which you will learn about in the next lesson.

Practice Problems

19. **Follow-up** What is the firefighter's acceleration if the force exerted on him by the pole is 650 N?

20. You lift a stuffed suitcase with a force of 105 N, giving it an acceleration of 0.705 m/s^2.
(a) What is the mass of the suitcase?
(b) What is its weight?

Your weight seems different when you accelerate

Have you ever ridden in an elevator? If so, you've probably noticed that you feel heavier than normal when the elevator starts moving upward and lighter than normal when it slows down to stop. Why is that? As we'll see, the answer comes directly from Newton's second law.

First, what causes the sensation of weight anyway? If you think about it, the way you experience your weight is from the force exerted upward on your feet by the floor. If the floor exerts a force greater than your weight, you feel heavy; if it exerts a force less than your weight, you feel light. This sensation of having a different weight due to the force exerted on you by the floor is referred to as **apparent weight**.

To relate apparent weight to Newton's second law, imagine you are in an elevator that is moving with an acceleration a, as indicated in **Figure 5.10**. Two forces act on you:

- your weight, $W = mg$, acting downward
- the upward normal force exerted on your feet by the floor of the elevator.

We'll call the second force W_a, since it represents your *apparent* weight.

Now, let's apply Newton's second law in the vertical direction:

$$\sum F_y = ma_y$$
$$W_a - W = ma_y$$

The acceleration of the elevator is $a_y = a$, and therefore

$$W_a - W = ma$$

Solving for the apparent weight, W_a, and substituting mg in place of the weight, W, gives

$$W_a = W + ma$$
$$= mg + ma$$

If the elevator is accelerating upward, which means $a > 0$, you feel heavier because $mg + ma$ is greater than mg. Similarly, if the elevator accelerates downward, which means $a < 0$, you feel lighter. If the elevator doesn't accelerate at all, or $a = 0$, your weight is unchanged. Notice that all of these conclusions are true regardless of the speed of the elevator—as mentioned in Chapter 3, you can only sense acceleration, not velocity.

One special case of particular interest occurs when the elevator has a downward acceleration of g, that is, if the elevator is in free fall. In this case, $a = -g$ and

$$W_a = mg + m(-g) = 0$$

Your apparent weight is zero in free fall! You feel "weightless" in a freely falling elevator, or in a freely falling spaceship in orbit.

A stretched or compressed spring exerts a force

Springs are useful because they exert force when stretched or compressed. The force acts to restore the spring to its natural state (not stretched or compressed). Thus, a compressed spring pushes outward and a stretched spring pulls inward. For example, if you've ever played a keyboard, you know that it takes a force to push a key down. As soon as you release the key, it springs back to its initial position. It is the force exerted by the compressed spring that pushes the key back up.

Physical picture

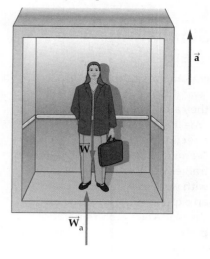

Free-body diagram

▲ **Figure 5.10 Apparent weight** When a person rides in an elevator that is accelerating upward, the net force must also be upward. This means that the force exerted on the person by the floor of the elevator, $\vec{W}_a$, must be greater than the person's weight, $\vec{W}$. As a result, the person feels heavier than normal.

C☉OLPHYSICS
Simulating Weightlessness

NASA trains astronauts in a "weightless" environment before they are sent into orbit. Trainees are taken aloft in a KC-135 airplane affectionately known as the "vomit comet" (since many trainees experience nausea along with weightlessness). To generate an experience of weightlessness, the plane flies along a parabolic path—the same path followed by a projectile in free fall. Each round of weightlessness lasts about half a minute; then the plane pulls up to regain altitude and start the cycle again. On a typical flight trainees experience about 40 cycles of weightlessness.

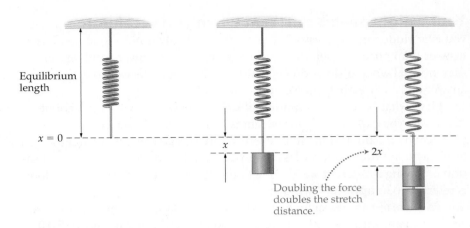

▲ **Figure 5.11 A spring scale**
A spring scale is used to measure the weight of an object. The heavier the object, the more the spring stretches.

Similarly, if you've ever pulled on a spring, you know that the amount of stretch varies with the force you apply. The greater the force, the greater the stretch. In fact, this is the principle behind a spring scale, illustrated in **Figure 5.11**. The greater the weight you place on the scale, the greater the stretch of the spring and the higher the reading.

Let's represent the change in length of a spring with the symbol x, as shown in **Figure 5.12**. When $x = 0$, the spring is relaxed and there is no change in length. When the spring is stretched or compressed, x represents the distance from equilibrium. According to **Hooke's law**, the force exerted by an ideal spring is proportional to the distance of its stretch or compression. This can be written in equation form as follows:

> **Hooke's Law for a Spring**
> force = spring constant × change in length
> $$F = kx$$

The constant k is called the **spring constant**, and its units are newtons per meter, N/m. The larger the spring constant, the greater the force exerted by the spring. Thus, a large spring constant corresponds to a stiff or tough spring.

The stretch or compression distance x is always positive, and thus $F = kx$ is the magnitude of the spring force. Finally, it should be noted that Hooke's law is not a law of nature (like $F = ma$), but it is a very good approximation for real springs that aren't stretched or compressed too far.

QUICK Example 5.8 What's the Force?

A spring with a spring constant of 21 N/m is stretched 3.4 cm. What force is required to cause this amount of stretch?

Solution
Using $k = 21$ N/m and $x = 0.034$ m, we find
$$F = kx = (21\text{ N/m})(0.034\text{ m}) = \boxed{0.71\text{ N}}$$

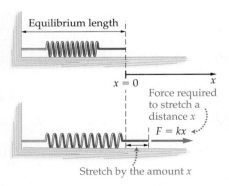

▲ **Figure 5.12 Stretching a spring**
The equilibrium position of the spring is indicated by $x = 0$. The magnitude of the force required to stretch or compress the spring by a distance x is $F = kx$.

21. Pulling on a spring with a force of 1.2 N causes a stretch of 6.4 cm. What is the spring constant for this spring?

22. What is the stretch when you pull with a force of 2.3 N on a spring with a spring constant of 19 N/m?

Strings transmit forces from one place to another

Imagine picking up a string and holding one end in each hand. Now, pull on both ends so that the string becomes taut. Whenever a string is taut, like a tightened guitar string, we say that there is a *tension* in the string. In general, **tension**, T, is the force exerted by a string, rope, or wire that is pulled tight.

Suppose you put tension in a string and your friend cuts the string in the middle. The tension force, T, is the force pulling the ends apart, as illustrated in **Figure 5.13**. Your friend would have to exert the force T on *each* end of the string to hold it together. Notice that the string pulls equally to the right and to the left.

In an ideal string (one with little mass) the tension is the same everywhere in the string. In this sense the string *transmits* a force from one location to another. Unless stated otherwise, all strings, ropes, and wires in this text are assumed to be ideal.

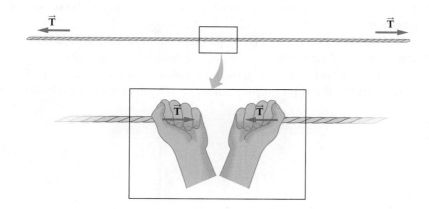

◀ **Figure 5.13 Tension in a string**
A string, pulled from either end, has a tension, T. If the string is cut at any point, the force required to hold the ends together is T.

Equilibrium

When you sit at rest in a classroom, you have zero acceleration. In general, objects with zero acceleration are said to be in **equilibrium**. From Newton's second law we know that an object with zero acceleration is acted on by a net force of zero. This is the required condition for equilibrium. **An object in equilibrium is subject to zero net force.**

$$\sum \vec{\mathbf{F}} = 0$$

Notice that being in equilibrium doesn't necessarily mean that the object is at rest—only that it isn't accelerating. An object moving with constant velocity is also in equilibrium.

As an example of equilibrium, consider the bucket of water suspended by a rope in the next Conceptual Example.

What is the condition for an object to be in equilibrium?

A person hoists a bucket of water from a well and holds the rope, keeping the bucket at rest, as in case 1 below. A short time later the person ties the rope to the bucket's handle so that the rope holds the bucket in place, as in case 2. Is the tension in the rope in case 2 greater than, less than, or equal to the tension in the rope in case 1?

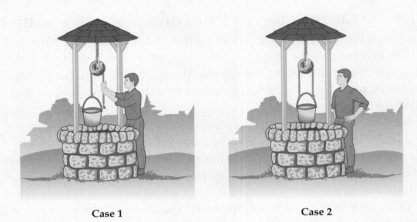

Case 1　　　　　　**Case 2**

Reasoning and Discussion

In case 1 the only upward force exerted on the bucket is the tension in the rope. Since the bucket is at rest, the tension must equal the weight of the bucket.

In case 2 the two ends of the rope exert equal upward forces on the bucket, and thus the tension in the rope is only half the weight of the bucket. To see this more clearly, imagine cutting the bucket in half so that each end of the rope supports half the weight, as illustrated below.

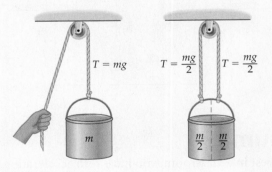

Answer

The tension in the rope is less in case 2 than in case 1. In fact, it's less by a factor of 2.

Systems in equilibrium often involve forces much greater than the weight of the object that is being supported. The next Active Example considers such a case.

A 1.84-kg bag of clothespins hangs in the middle of a clothesline, causing it to sag by an angle $\theta = 3.50°$. Find the tension force, T, in each part of the clothesline.

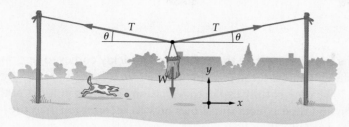

Solution *(Perform the calculations indicated in each step.)*

1. Find the y component of each of the two tension forces:

$$T_y = T \sin \theta$$

2. Find the y component of the weight:

$$W_y = -mg$$

3. Set the net force in the y direction equal to zero, $\sum F_y = 0$:

$$T \sin \theta + T \sin \theta - mg = 0$$

4. Solve for the tension T:

$$T = mg/(2 \sin \theta) = 148 \text{ N}$$

Insight
The tension in the clothesline is fairly large. After all, the tension is 148 N, and the weight of the bag is only 18.1 N. If you and a friend were to pull on the ends of the clothesline in an attempt to straighten it out, you would find that no matter how hard you pulled, the clothesline would still sag in the middle. You might be able to reduce the angle of the sag, but as you did so, the tension would increase rapidly. In principle, it takes an infinite force to completely straighten the clothesline!

PROBLEM-SOLVING NOTE
We considered only the y components of force because the forces in the x direction cancel as a result of the symmetry of the system.

5.2 LessonCheck (MP)

Checking Concepts

23. 🔵 **List** List the steps that are followed when using a free-body diagram to help solve a problem.

24. 🔵 **Identify** What equation is used to determine the weight of an object with mass m?

25. 🔵 **Assess** If an object is at rest, what can you say about the forces acting on it?

26. State Hooke's law describes the connection between the stretch of a spring and the force it exerts. What is this connection?

27. Apply To decrease the angle of the sag in Active Example 5.10, is it necessary to increase or decrease the tension in the clothesline? Explain.

Solving Problems

28. Diagram A 9.3-kg child sits in a 3.7-kg chair. Her feet do not touch the ground.
(a) Draw a free-body diagram for the child, and find the normal force exerted by the chair on the child.
(b) Draw a free-body diagram for the chair, and find the normal force exerted by the floor on the chair.

29. Calculate A spring with a spring constant of 62 N/m is stretched 19 cm by an applied force. What is the magnitude of the force?

30. Challenge Two astronauts push on a satellite. One pushes in the positive x direction with a force of 42 N, and the other pushes in the positive y direction. If the net force is at an angle of 15° above the positive x axis, what is the magnitude of the second astronaut's force?

5.3 Friction

Vocabulary

- friction
- kinetic friction
- static friction

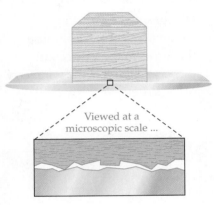

▲ Figure 5.14 The origin of friction
Even "smooth" surfaces have irregularities when viewed at the microscopic level. This type of roughness contributes to friction.

🔑 *How does the magnitude of the normal force affect the amount of kinetic friction that acts?*

CONNECTINGIDEAS

This lesson focuses on the friction that acts when one surface slides over another.
- Recall from Lesson 3.4 that another type of friction—air resistance—acts on objects moving through the air. Also recall that air resistance does not act on objects in free fall.
- Lesson 4.4 explored the effect of air resistance on the path of projectiles.

You know from experience that a food tray quickly comes to rest once you stop pushing it across a cafeteria table. The tray is brought to rest by a type of force we have not yet fully explored—friction. This lesson looks at the types of friction and their effects and shows how to account for their presence in free-body diagrams.

Frictional forces resist motion

To simplify problem solving we often assume that surfaces are smooth and that objects slide without resistance to that motion. No surface is perfectly smooth, however. When viewed on the atomic level, even the "smoothest" surface is actually rough and jagged, as shown in **Figure 5.14**. Sliding one surface past another requires enough force to overcome the resistance caused by microscopic hills and valleys bumping against one another. The force that opposes the motion of one surface over another is referred to as **friction**.

Friction Pros and Cons We often think of friction as something that should be reduced, or even eliminated if possible. For example, roughly 20% of the gasoline you buy does nothing but overcome friction within your car's engine. Clearly, reducing that friction would be most desirable.

On the other hand, friction can be helpful—even indispensable—in some situations. Suppose, for example, that you are standing still and then decide to begin walking forward. The force that accelerates you is the force of friction between your shoes and the ground. We simply couldn't walk or run without friction—it's hard enough when friction is merely reduced, as on an icy sidewalk. Similarly, starting or stopping a car, or even turning a corner, all require friction. Friction is an important and useful aspect of everyday life.

Kinetic friction acts between sliding surfaces

Imagine you're playing a game of baseball. You hit a ball to the outfield and decide to stretch your single into a double. You race for second base and slide in, safe of course. During your slide across the ground you experience a force that slows you down, the force of *kinetic friction*. **Kinetic friction** is the friction encountered when surfaces slide against one another.

The force generated by kinetic friction is designated with the symbol f_k. The f stands for "force," and the subscript k stands for "kinetic." This force acts to oppose the sliding motion at the point of contact between surfaces. To oppose the motion, the force of kinetic friction must act *opposite* to the direction of the velocity of a moving object.

Kinetic Friction and the Normal Force A simple experiment illustrates one of the main characteristics of kinetic friction. Imagine attaching a spring scale to a rough object, like a brick, and pulling the object with constant velocity across a table, as shown in **Figure 5.15**. The reading on the scale tells you the magnitude of the force of kinetic friction, f_k. Now, if you repeat the experiment with a second brick on top of the first one, you'll find that the force needed to pull the brick is doubled, to $2f_k$.

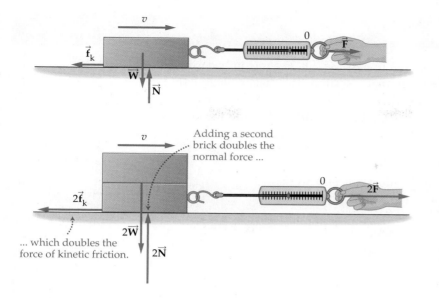

From this experiment we see that when we double the normal force—by stacking up two bricks—the force of kinetic friction is also doubled. 🔋 **In general, the force of kinetic friction is proportional to the normal force, N.**

Kinetic Friction

$$\frac{\text{force of}}{\text{kinetic friction}} = \frac{\text{coefficient of}}{\text{kinetic friction}} \times \text{normal force}$$

$$f_k = \mu_k N$$

The constant μ_k in this equation is referred to as the *coefficient of kinetic friction*. It is pronounced "mew sub kay." The larger the coefficient of kinetic friction, the greater the friction. Typical values for μ_k range between 0 and 1, as indicated in **Table 5.3**. Note that Table 5.3 also gives value for the coefficient of static friction, which is discussed later in this lesson.

Properties of Kinetic Friction

As you know from everyday experience, the force of kinetic friction tends to oppose motion, as shown in Figure 5.15. For example, when you slide into second base, friction opposes your motion and slows you down. Friction is unaffected by the speed of the sliding surfaces, however, or by the area of their contact. To summarize, the kinetic friction between two sliding surfaces has the following properties:

- It opposes the motion of the surfaces.
- It is proportional to the normal force between the surfaces.
- It is the same regardless of the speed of the surfaces.
- It is the same regardless of the area of contact between the surfaces.

The fact that kinetic friction doesn't depend on the area of contact may seem surprising at first. Why doesn't a larger area of contact produce a larger force? One way to think about this is to consider that when the area of contact is large, the normal force is spread out over a large area, giving a small force per area. As a result, the microscopic hills and valleys are not pressed too deeply against one another. On the other hand, if the area is small, the normal force is concentrated in a small region. This presses the surfaces together more firmly. The net effect is roughly the same frictional force in either case.

Table 5.3 Typical Coefficients of Friction

Materials	Kinetic, μ_k	Static, μ_s
Rubber on concrete (dry)	0.80	1–4
Steel on glass	0.57	0.74
Glass on glass	0.40	0.94
Wood on leather	0.40	0.50
Copper on steel	0.36	0.53
Rubber on concrete (wet)	0.25	0.30
Steel on ice	0.06	0.10
Waxed ski on snow	0.05	0.10
Teflon on Teflon	0.04	0.04
Human joints	0.003	0.01

Someone at the other end of the table asks you to pass the salt. Feeling quite dashing, you slide a salt shaker with a mass of 50.0 g in that direction. The shaker slides with an acceleration of magnitude $a = 0.787 \ \text{m/s}^2$ and comes to rest at the desired location. What is the coefficient of kinetic friction between the shaker and the table?

Picture the Problem

There is no acceleration in the vertical direction. Therefore, the weight $W = mg$ and the normal force N must cancel. This means that the magnitude of the normal force is equal to the magnitude of the weight, $N = mg$.

In the horizontal direction the force of kinetic friction causes the shaker to slow down with an acceleration of magnitude $a = 0.787 \ \text{m/s}^2$.

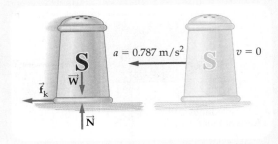

Strategy

From Newton's second law we know that the force of kinetic friction has a magnitude of $f_k = ma$, since it causes a mass m to have an acceleration a. We also know that the force of kinetic friction is given by $f_k = \mu_k N$. Combining these two equations and using the fact that $N = mg$, we can solve for the coefficient of kinetic friction, μ_k.

Known

$m = 50.0 \ \text{g}$

$a = 0.787 \ \text{m/s}^2$

Unknown

$\mu_k = ?$

Solution

1 Solve the equation $f_k = \mu_k N$ for the coefficient of kinetic friction, μ_k:

$$f_k = \mu_k N \quad \text{or} \quad \mu_k = \frac{f_k}{N}$$

2 Substitute ma for the force of kinetic friction, f_k, and mg for the normal force, N. Cancel out mass, m, to simplify:

$$\mu_k = \frac{f_k}{N} = \frac{\cancel{m}a}{\cancel{m}g} = \frac{a}{g}$$

Math HELP
Dimensional Analysis
See Lesson 1.3

3 Substitute the known values for the acceleration of the salt shaker, a, and the acceleration due to gravity, g:

$$\mu_k = \frac{a}{g} = \frac{0.787 \ \cancel{\text{m/s}^2}}{9.81 \ \cancel{\text{m/s}^2}} = \boxed{0.0802}$$

Insight

Notice that mass canceled out in Step 2, so our result for the coefficient of friction is independent of the shaker's mass. For example, if we were to slide a shaker with twice the mass, but with the same initial speed, it would slide exactly the same distance. It's unlikely that this independence of mass would have been apparent if we had worked the problem numerically rather than symbolically.

Practice Problems

31. Follow-up Suppose the coefficient of kinetic friction is equal to 0.120. What is the acceleration of the salt shaker in this case?

32. A baseball player slides into third base with an initial speed of 4.0 m/s. If the coefficient of kinetic friction between the player and the ground is 0.46, how far does she slide before coming to rest?

Static friction acts to prevent motion from starting

If you sit on a playground slide and don't slide to the bottom, static friction is holding you back. As its name suggests, **static friction** is the force that opposes the sliding of one nonmoving surface past another.

Like kinetic friction, static friction is due to the microscopic irregularities of surfaces that are in contact. In fact, static friction is typically stronger than kinetic friction because when surfaces are in static contact, their microscopic hills and valleys can nestle deeply into one another.

Range of Static Friction The force of static friction can have any value from zero to a well-defined maximum, as shown in **Figure 5.16**. For example, suppose you put a brick on a board. When the board is level, the brick has no tendency to move, and the force of static friction is zero. As you slowly tilt the board upward, gravity pulls downhill on the brick. If the tilt isn't too great, the brick stays put—static friction holds it still. As you increase the tilt, however, there comes a point when the brick breaks loose and begins to slide. This is where the force of static friction "maxes out"—where it is as large as it can be.

Maximum Static Friction A stationary object begins to move when the applied force equals the maximum force of static friction. Once the object is moving, kinetic friction takes over. The maximum force that static friction can exert, $f_{s,max}$, is given by the following expression:

> **Maximum Force of Static Friction**
>
> $$\frac{\text{maximum force of}}{\text{static friction}} = \frac{\text{coefficient of}}{\text{static friction}} \times \text{normal force}$$
>
> $$f_{s,max} = \mu_s N$$

In this equation, μ_s (pronounced "mew sub ess") is the *coefficient of static friction*. Typical values of μ_s are given in Table 5.3. 🔑 **In most cases, μ_s is greater than μ_k. This means that the force of static friction is usually greater than the force of kinetic friction.**

The force of static friction can have a magnitude of zero ...

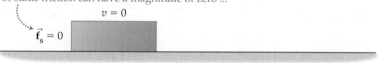

... or greater than zero ...

... up to a maximum value.

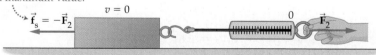

Once sliding begins, however, the friction is kinetic and has a magnitude less than the maximum value for static friction.

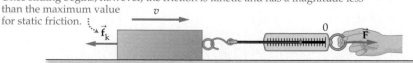

🔑 *How do the coefficients of static friction and kinetic friction compare?*

◀ **Figure 5.16 The maximum limit of static friction**
As the force applied to an object increases, so does the force of static friction—up to a certain limit. Beyond this maximum value static friction can no longer hold the object, and it begins to slide. As this occurs, kinetic friction takes over.

Students in a physics lab design an experiment to determine the coefficient of static friction between a 1.8-kg jack-o'-lantern and the tabletop on which it rests. Their setup is shown in the figure below. The basic procedure is to pour sand slowly into the pail until the pumpkin begins to slide. The pail with the sand in it is then placed on a scale and found to weigh 11 N. What is the coefficient of static friction between the jack-o'-lantern and the table?

Picture the Problem

The pumpkin exerts a downward force mg on the table, with $m = 1.8$ kg. The table exerts an upward force that cancels the weight; thus, $N = mg$. As sand is poured into the pail, an increasing horizontal force is applied to the pumpkin by the string. At the same time the force of static friction, f_s, increases to oppose the force from the string. When the static friction force reaches its maximum value, $f_{s,max} = \mu_s N$, the pumpkin begins to slide. The pail (with sand in it) is then weighed, giving the result $W_{pail} = 11$ N.

Strategy

Solve the equation $f_{s,max} = \mu_s N$ for the coefficient of static friction. Notice that the maximum force of static friction, $f_{s,max}$, is equal to the force exerted by the pail, W_{pail}, at the time the pumpkin begins to slide.

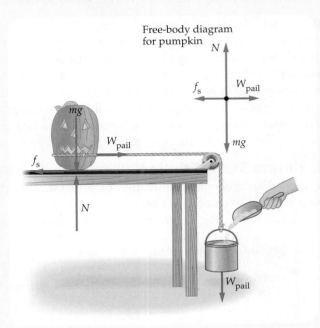

Free-body diagram for pumpkin

Known	$m = 1.8$ kg	$W_{pail} = 11$ N
Unknown	$\mu_s = ?$	

Solution

1 Solve $f_{s,max} = \mu_s N$ for the coefficient of static friction:

$$f_{s,max} = \mu_s N \quad \text{or} \quad \mu_s = \frac{f_{s,max}}{N}$$

2 Substitute W_{pail} for $f_{s,max}$ and mg for the normal force, N:

$$\mu_s = \frac{f_{s,max}}{N} = \frac{W_{pail}}{mg}$$

3 Substitute the known values for W_{pail}, m, and g. Evaluate to obtain the coefficient of static friction:

$$\mu_s = \frac{11 \text{ N}}{(1.8 \text{ kg})(9.81 \text{ m/s}^2)}$$

$$= \frac{11 \cancel{\text{N}}}{18 \cancel{\text{N}}}$$

$$= \boxed{0.61}$$

Insight

Once the pumpkin begins to slide, it experiences kinetic friction. The coefficient of kinetic friction is usually less than the coefficient of static friction, and thus the pumpkin will accelerate as it slides.

33. Follow-up Suppose the coefficient of static friction is 0.55 rather than 0.61.
(**a**) Is the weight of the pail and sand necessary to start the pumpkin moving in this case greater than, less than, or equal to 11 N?
(**b**) Calculate the required weight.

34. When you push a 1.80-kg book that is resting on a tabletop, a force of 2.25 N is required to start the book sliding. Once it is sliding, however, a force of only 1.50 N keeps the book moving with constant speed. What are the coefficients of static and kinetic friction between the book and the tabletop?

Friction plays a key role in driving safety

Since most driving is done with the tires rolling, rather than skidding, it's important to know the characteristics of friction for rolling tires. In Conceptual Example 5.13 we consider whether static or kinetic friction is key in such cases.

CONCEPTUAL Example 5.13 Static or Kinetic?

A car drives with its tires rolling freely. Is the friction between the tires and the road kinetic or static?

Reasoning and Discussion
A reasonable sounding answer is that because the car is moving, the friction between its tires and the road must be kinetic friction—but this is not the case. Instead, the friction is static because the bottom of each tire is in static contact with the road. To understand this, watch your feet as you walk. Even though you are moving forward, each foot is in static contact with the ground once you step down on it. Your foot doesn't move again until you lift it up and move it forward for the next step. A tire can be thought of as a circular arrangement of feet, each of which is momentarily in static contact with the ground.

Answer
The friction between the tires and the road is static friction.

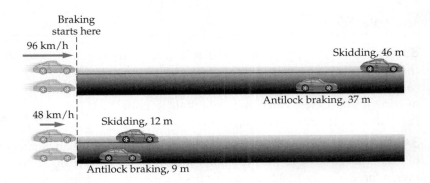

► **Figure 5.17 Stopping distance with and without ABS**
An antilock braking system (ABS) allows a car to stop using static friction rather than kinetic friction—even when the driver slams on the brakes. As a result, the braking distance is reduced, since μ_s is typically greater than μ_k.

Physics & You: Technology The wheels on older cars often lock during panic braking, causing the car to skid uncontrollably. In general, sliding or skidding tires are subject to kinetic friction, whereas tires that roll experience static friction, as discussed in Conceptual Example 5.13. Since static friction is usually greater than kinetic friction, a car will stop in a shorter distance if its wheels are *rolling* (static friction) than if its wheels are locked up and skidding (kinetic friction)!

This is the idea behind antilock braking systems (ABS). When the brakes are applied in a car with ABS, an electronic rotation sensor at each wheel detects when the wheel is about to skid. To prevent the skidding, a small computer automatically begins to pump the brakes. This pumping allows the wheels to continue rotating, even in an emergency stop, and thus static friction determines the stopping distance. **Figure 5.17** shows a comparison of braking distances for cars with and without ABS.

5.3 LessonCheck ⓂⓅ

Checking Concepts

35. 🔑 Apply A book slides along a table and comes to a stop. How are the normal force and the force of kinetic friction affected if a second book is placed on top of the first and the stack of two books is slid across the table?

36. 🔑 Critique A classmate tells you that making the shortest possible stop on your bike requires you to lock the brakes and skid to a stop. Is he correct? Explain.

37. Identify Which type of friction tries to hold objects in place?

38. Give Examples Describe (**a**) two everyday situations in which friction is undesirable and (**b**) two everyday situations in which friction is helpful.

Solving Problems

39. Calculate You want to slide a 0.27-kg book across a table. If the coefficient of kinetic friction is 0.11, what force is required to move the book with constant speed?

40. Calculate You want to slide a 0.39-kg book across a table. If the coefficient of kinetic friction is 0.21, what force is required to move the book with an acceleration of 0.18 m/s^2?

41. Calculate How much force is required to start a 14-kg flowerpot sliding if the coefficient of static friction between the pot and the surface on which it rests is 0.64?

42. Challenge An 11-g coin with an initial speed of 1.8 m/s slides 0.86 m across a table and comes to rest. (**a**) What is the coefficient of kinetic friction between the coin and the table? (**b**) Will two 11-g coins stacked one on top of the other and having the same initial speed as the single coin slide a smaller distance, the same distance, or a greater distance before coming to rest? Explain.

Physics & You

Earthquake Scientists and Engineers

Literally able to move mountains, earthquakes are among nature's greatest hazards. Earthquake scientists and engineers investigate ways to protect lives and property during these events.

Seismologists study the build-up of stresses in the Earth's crust and the earthquakes that result when these stresses are released. They identify areas most at risk for earthquakes and hope in the near future to be able to make short-term predictions of impending quakes. Records of tremors, called *seismograms*, are made constantly at hundreds of stations around the world. Data from three or more stations can be combined to calculate the location and depth of a quake.

Earthquake engineers are usually civil or mechanical engineers who specialize in designing structures that can survive earthquakes. Such design work typically focuses on making these structures strong enough to handle the inertial forces that result when the underlying ground moves from side to side, dragging the building with it.

Sometimes, however, buildings are designed to sit on *base isolators*, which are usually synthetic rubber pads. During a tremor the pads can stretch horizontally and compress vertically, allowing the building to remain relatively still while the ground shakes beneath it. The use of base isolators greatly reduces the forces on the building.

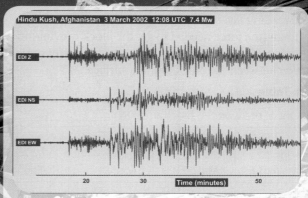

This seismogram of the 2010 Haitian earthquake was made by a seismograph 2800 km away at the James Madison High School in Rochester, New York.

▲ Base isolators, like this one made of rubber, offer modern buildings some degree of protection from earthquake damage.

Take It Further

1. Infer *What are some of the reasons earthquakes generally cause more damage to older buildings?*

2. Hypothesize *Why are buildings much more vulnerable to horizontal shaking than to vertical shaking?*

Physics Lab Static and Kinetic Friction

This lab explores the frictional forces acting between a wooden block and the surface it rests on. You will take force measurements and use them to determine the coefficients of static friction and kinetic friction.

Materials
• wooden block
• 5-N spring scale
• set of standard masses

Procedure

1. Hang the wooden block from the spring scale. Observe the weight of the block and record it in the data table.

2. Place the wooden block on a tabletop or other smooth surface with its largest side down. Attach the spring scale to the block and hold it horizontally. Gradually increase the horizontal pull on the block and note the maximum scale reading *just before* the block begins to move. Record the force (the force needed to get the block moving) in the data table. This force is the *maximum force of static friction* between the block and the surface.

3. Repeat Step 2, except this time pull horizontally on the block so that it moves at a slow, constant speed along the table. Observe the force reading on the spring scale *as the block moves*. Record the force (the force needed to move the block at constant speed) in the data table. This force is the *force of kinetic friction* between the block and the surface.

4. Place a 0.200-kg mass on top of the block and repeat Steps 2 and 3. Continue adding 0.200-kg masses and repeating Steps 2 and 3 until the total mass on the block is 1.000 kg.

Data Table

Weight of the wooden block alone (N): _____

Added Mass (kg)	Weight of Added Mass (N)	Normal Force (N)	Maximum Force of Static Friction (N)	Force of Kinetic Friction (N)	Coefficient of Static Friction, μ_s	Coefficient of Kinetic Friction, μ_k
0	—					
0.200						
0.400						
0.600						
0.800						
1.000						

Analysis

1. Calculate the weight of the added mass used in each trial and record it in the data table. Then, calculate and record the normal force in each trial.

2. Draw a free-body diagram of the block in Step 2 just before it begins to move. Draw another free-body diagram of the block in Step 3 as it is moving.

3. Calculate the coefficients of static friction and kinetic friction for each trial. Enter your values in the data table.

4. Graph the force of static friction and the force of kinetic friction versus the normal force. Plot each curve in a different color and draw a best-fit line for each set of data.

5. Determine the slope of each best-fit line on your graph.

Conclusions

1. Make a generalization regarding the relative magnitudes of the force of kinetic friction and the maximum force of static friction.

2. Describe what happened to the force of kinetic friction when the normal force acting on the block was increased.

3. Describe how an increase in normal force affected the coefficient of kinetic friction.

4. Identify what the slope of each line on your graph represents.

5. Suggest at least one way in which the coefficients of friction could be reduced in this experiment.

5 Study Guide

Big Idea
All motion is governed by Newton's laws.

If you throw a ball, do a somersault, or just walk down the street, you are experiencing Newton's laws in action. These fundamental laws, which we applied in order to send astronauts to the Moon and back, will always be an integral part of everyday life.

5.1 Newton's Laws of Motion

🔑 Newton's first law: An object at rest remains at rest as long as no net force acts on it. An object moving with constant velocity continues to move with the same speed and in the same direction as long as no net force acts on it.

🔑 Newton's second law (for a single force): Force is equal to mass times acceleration, $F = ma$.

🔑 Newton's second law (for multiple forces): The sum of the forces (the net force) acting on an object is equal to its mass times its acceleration.

🔑 Newton's third law: For every *action force* acting on an object, there is a *reaction force* acting on a different object. The action and reaction forces are equal in magnitude and opposite in direction.

• If the net force on an object is zero, its velocity is constant.

• For every action there is an equal and opposite reaction.

Key Equation

The sum of the forces (the net force) acting on an object is equal to its mass times its acceleration:

$$\sum \vec{F} = m\vec{a}$$

5.2 Applying Newton's Laws

🔑 Once all of the forces are drawn in a free-body diagram, a coordinate system is chosen and each force is resolved into components. At this point Newton's second law can be applied to each coordinate direction separately.

🔑 The weight of an object is equal to its mass times the acceleration due to gravity.

🔑 An object in equilibrium is subject to zero net force.

• A normal force is exerted *perpendicular* to the surface of contact between two objects.

• Apparent weight is due to the force exerted through contact with the floor. For example, the sensation of feeling heavier or lighter in an accelerating elevator is an apparent weight.

• Tension is the force exerted by a string, rope, or wire that is pulled tight.

Key Equations

The weight W of an object of mass m is

$$W = mg$$

The force to stretch or compress a spring is given by Hooke's law:

$$F = kx$$

5.3 Friction

🔑 The force of kinetic friction is proportional to the normal force, N.

🔑 In most cases, μ_s is greater than μ_k. This means that the force of static friction is usually greater than the force of kinetic friction.

• Kinetic friction does not depend on the speed or surface area of objects that are sliding past one another.

• Static friction exerts a force whose magnitude can vary from zero to a well-defined maximum.

Key Equations

The force of kinetic friction is given by

$$f_k = \mu_k N$$

The maximum force of static friction is given by

$$f_{s,max} = \mu_s N$$

5 Assessment

For instructor-assigned homework, go to www.masteringphysics.com

ANSWERS TO SELECTED ODD-NUMBERED PROBLEMS APPEAR IN APPENDIX A.

Lesson by Lesson

5.1 Newton's Laws of Motion

Conceptual Questions

43. What is the difference between a force and a net force?

44. What is the inertia of an object?

45. Driving down the road, you hit the brakes suddenly. As a result, your body moves toward the front of the car. Explain, using Newton's laws.

46. (a) If the force acting on an object doubles, what happens to the object's acceleration? (b) If the mass of an object doubles, what happens to the object's acceleration?

47. You've probably seen videos of someone pulling a tablecloth out from under glasses, plates, and silverware set out for a formal dinner. Perhaps you've even tried it yourself. Using Newton's laws of motion, explain how this stunt works.

48. When a dog gets wet, it shakes its body from head to tail to shed the water, as shown in **Figure 5.18**. Explain, in terms of Newton's first law, why this works.

Figure 5.18 A dog takes advantage of inertia to shake water from its coat.

49. ⬛ Predict & Explain ⬛ A small car collides with a large truck. (a) Is the magnitude of the force experienced by the car greater than, less than, or equal to the magnitude of the force experienced by the truck? (b) Choose the *best* explanation from among the following:

A. Action-reaction forces always have equal magnitudes.

B. The truck has more mass, and hence the force exerted on it is greater.

C. The massive truck exerts a greater force on the lightweight car.

50. ⬛ Predict & Explain ⬛ A small car collides with a large truck. (a) Is the acceleration experienced by the car greater than, less than, or equal to the acceleration experienced by the truck? (b) Choose the *best* explanation from among the following:

A. The truck exerts a larger force on the car, giving it the greater acceleration.

B. Both vehicles experience the same force; therefore, the lightweight car experiences the greater acceleration.

C. The greater force exerted on the truck gives it the greater acceleration.

51. Suppose you jump from the cliffs of Acapulco and perform a perfect swan dive. As you fall, you exert an upward force on the Earth equal in magnitude to the downward force the Earth exerts on you. Why, then, does it seem that you are the one doing all the accelerating? Since the forces are the same, why aren't the accelerations?

52. ⬛ Rank ⬛ Each of the three identical hockey pucks shown in **Figure 5.19** is acted on by a 3-N force. Puck A moves with a speed of 7 m/s in a direction opposite to the force; puck B is instantaneously at rest; puck C moves with a speed of 7 m/s at right angles to the force. Rank the three pucks in order of the magnitude of their acceleration, starting with the smallest. Indicate ties where appropriate.

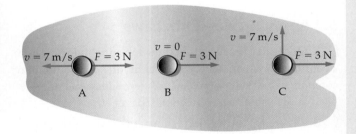

Figure 5.19

Problem Solving

53. What net force must act on a 540-kg golf cart to give it an acceleration of 2.5 m/s^2?

54. The net force acting on a backpack is 23 N. If the acceleration of the backpack is 3.8 m/s^2, what is its mass?

55. A 1.3-kg rainbow trout experiences a net force of 3.4 N. What is its acceleration?

56. An 1800-kg truck pulls a 620-kg trailer away from a stoplight with an acceleration of 1.16 m/s^2. What is the force exerted by the truck on the trailer?

180 Chapter 5 · Assessment

57. An 1800-kg truck pulls a 620-kg trailer away from a stoplight with an acceleration of 1.16 m/s². What is the net force acting on the truck?

58. Triple Choice A 91-kg parent and a 21-kg child meet at the center of an ice rink. They place their hands together and push. **(a)** Is the force experienced by the child more than, less than, or the same as the force experienced by the parent? **(b)** Is the acceleration of the child more than, less than, or the same as the acceleration of the parent? Explain.

59. If the acceleration of the child in Problem 58 is 0.31 m/s² in magnitude, what is the magnitude of the parent's acceleration?

60. **Aircraft Carrier Takeoff** On an aircraft carrier, a jet can be catapulted from 0 to 250 km/h in 2.00 s (see **Figure 5.20**). If the average force exerted by the catapult is 9.35×10^5 N, what is the mass of the jet?

Figure 5.20 A jet takes off from the flight deck of an aircraft carrier.

61. Driving home from school one day, you spot a ball rolling out into the street as shown in **Figure 5.21**. You brake for 1.20 s, slowing your 950-kg car from 16.0 m/s to 9.50 m/s. **(a)** What was the average force exerted on your car during braking? **(b)** How far did you travel while braking?

$v = 9.50$ m/s
$v = 16.0$ m/s

Figure 5.21

62. Responding to an alarm, a 782-N firefighter slides down a pole to the ground floor, 3.3 m below. The firefighter starts at rest and lands with a speed of 4.2 m/s. Find the average force exerted on the firefighter by the pole.

63. When two people push in the same direction on an object of mass m, they cause an acceleration of magnitude a_1. When the same people push on the object in opposite directions, the acceleration of the object has the magnitude a_2. Determine the magnitude of the force exerted by each of the two people in terms of m, a_1, and a_2.

5.2 Applying Newton's Laws
Conceptual Questions

64. Explain why a person of mass m has a weight equal to mg.

65. What is a free-body diagram, and how is it used?

66. A car is parked on a horizontal road. What is the direction of the normal force exerted on the car by the road?

67. Is your apparent weight in an elevator related to your velocity or to your acceleration? Explain.

68. If the force exerted on a spring is doubled, by what factor does the spring's stretch increase?

69. Give an everyday example of a tension force.

70. Can an object moving with a speed of 1000 m/s be in equilibrium? Explain.

71. A clothesline always sags a little, even if nothing hangs from it. Explain.

72. A young girl slides down a rope. As she slides faster and faster, she tightens her grip, increasing the force exerted on her by the rope. What happens when this force is equal in magnitude to her weight? Explain.

73. Triple Choice A brick is at rest on a tabletop. A spring is then attached to the top of the brick and stretched by someone pulling up on it. The brick remains at rest. Is the normal force acting on the brick greater than, less than, or equal to the brick's weight? Explain.

74. Predict & Explain You jump out of an airplane and open your parachute after an extended period of free fall. **(a)** To decelerate your fall, must the force exerted on you by the parachute be greater than, less than, or equal to your weight? **(b)** Choose the *best* explanation from among the following:

A. A parachute can only exert a force that is less than the weight of the skydiver.

B. The parachute exerts a force exactly equal to the skydiver's weight.

C. To decelerate a skydiver in free fall, the net force acting on the skydiver must be upward.

75. **Rank** A hockey puck is acted on by one or more forces, as shown in **Figure 5.22**. Rank the four cases, A, B, C, and D, in order of the magnitude of the puck's acceleration, starting with the smallest. Indicate ties where appropriate.

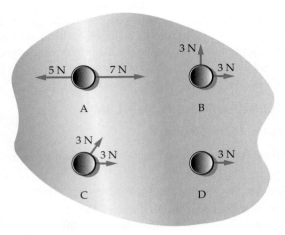

Figure 5.22

Problem Solving

76. On a planet far, far away, an astronaut picks up a rock. The rock has a mass of 5.00 kg, and on this particular planet its weight is 40.0 N. If the astronaut exerts an upward force of 46.2 N on the rock, what is its acceleration?

77. A 5.0-kg bag of potatoes sits in a stationary shopping cart. **(a)** Sketch a free-body diagram for the bag of potatoes. **(b)** Now suppose the cart moves with a constant velocity. How does this affect your free-body diagram? Explain.

78. **Think & Calculate** As part of a physics experiment, you stand on a bathroom scale in an elevator. Though your normal weight is 610 N, the scale reads 730 N. **(a)** Is the acceleration of the elevator at this moment upward, downward, or zero? Explain. **(b)** Calculate the magnitude of the elevator's acceleration. **(c)** What, if anything, can you say about the velocity of the elevator? Explain.

79. How much force is required to stretch a spring 12 cm if the spring constant is 55 N/m?

80. A force of 4.3 N compresses a spring that has a spring constant of 71 N/m. What is the compression distance?

81. What spring constant must a spring have if it is to stretch 9.5 cm when a force of 4.8 N is applied?

82. **Spiderweb** An orb weaver spider with a mass of 0.26 g hangs vertically by one of its threads. The thread has a spring constant of 7.1 N/m. What is the increase in the thread's length caused by the spider?

83. A steel wire 4.7 m long stretches 0.11 cm when it is subjected to a tension of 360 N. What is the spring constant of the wire?

84. A 4.60-kg sled is pulled across a smooth ice surface, as shown in **Figure 5.23**. The force acting on the sled has a magnitude of 6.20 N and points in a direction 35.0° above the horizontal. If the sled starts from rest, how fast is it going after being pulled for 1.15 s?

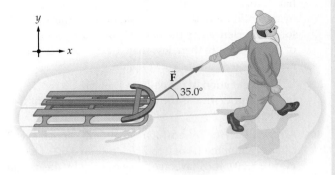

Figure 5.23

85. To give a 19-kg child a ride, two teenagers pull on a 3.7-kg sled with ropes, as indicated in **Figure 5.24**. Both teenagers pull with a force of 55 N at an angle of 35° relative to the forward direction, which is the direction of motion. In addition, the snow exerts a retarding force on the sled that points opposite to the direction of motion and has a magnitude of 57 N. Find the acceleration of the sled and child.

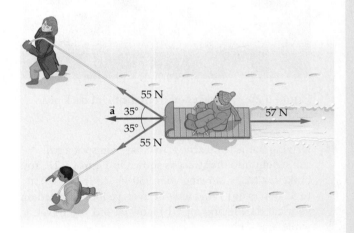

Figure 5.24

86. At the bow of a ship on a stormy sea, a crew member conducts an experiment by standing on a bathroom scale. In calm waters the scale reads 750 N. During the storm the crew member observes a maximum reading of 910 N and a minimum reading of 560 N. Find **(a)** the maximum upward acceleration and **(b)** the maximum downward acceleration experienced by the crew member.

87. Two astronauts use their jet packs to push on a satellite in deep space. One astronaut exerts a force of 45 N in the positive x direction. The second astronaut exerts a force of 68 N at an angle that is 55° above the positive x axis. **(a)** Use a ruler and protractor to determine the approximate direction and magnitude of the net force acting on the satellite. **(b)** If the satellite has an acceleration of magnitude 0.12 m/s², what is its mass?

88. Three astronauts use their jet packs to push on a satellite in deep space. Astronaut 1 exerts a force of 52 N in the positive x direction, astronaut 2 exerts a force of 35 N in the negative y direction, and astronaut 3 exerts a force of 74 N at an angle of 63° above the positive x axis. **(a)** Use a ruler and protractor to determine the approximate direction and magnitude of the net force acting on the satellite. **(b)** If the satellite has a mass of 680 kg, what is the magnitude of its acceleration?

89. A suitcase is at rest on the ground, though a force of magnitude F pulls upward on it at an angle θ to the horizontal. Determine F for each of the three curves shown in **Figure 5.25**. Give your answer in terms of the weight of the suitcase, mg.

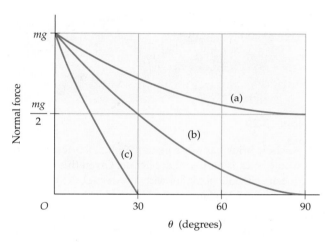

Figure 5.25

5.3 Friction
Conceptual Questions

90. Describe some of the **(a)** similarities and **(b)** differences between kinetic and static friction.

91. A car is stuck on an icy side street. Some students on their way to class see the predicament and help out by sitting on the trunk of the car to increase its traction. Why does this help?

92. The parking brake on a car causes the rear wheels to lock up. What would be the likely consequence of applying the parking brake in a car that is in rapid motion? (*Note:* Do *not* try this at home.)

93. **Rank** The three identical boxes shown in **Figure 5.26** remain at rest on a rough, horizontal surface, even though they are acted on by two different forces, $\vec{F}_1$ and $\vec{F}_2$. All of the forces labeled $\vec{F}_1$ have the same

magnitude; all of the forces labeled $\vec{F}_2$ are identical to one another. Rank the boxes in order of increasing magnitude of the force of static friction between the box and the surface. Indicate ties where appropriate.

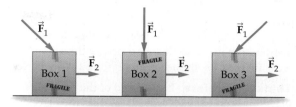

Figure 5.26

94. **Predict & Explain** You push two identical bricks across a tabletop with constant speed, v, as shown in **Figure 5.27**. In case 1 you place the bricks end to end; in case 2 you stack the bricks one on top of the other. **(a)** Is the force of kinetic friction in case 1 greater than, less than, or equal to the force of kinetic friction in case 2? **(b)** Choose the *best* explanation from among the following:

A. The normal force in case 2 is larger, and thus the bricks press down more firmly against the tabletop.

B. The normal force is the same in the two cases, and friction is independent of surface area.

C. The bricks in case 1 have more surface area in contact with the tabletop, and this results in greater friction.

Figure 5.27

95. **Predict & Explain** Two drivers traveling side by side at the same speed suddenly see a deer in the road ahead of them and begin braking. Driver 1 stops by locking his brakes and screeching to a halt; driver 2 stops by applying her brakes just to the verge of locking, so that the wheels continue to turn until her car comes to a complete stop. **(a)** All other factors being equal, is the stopping distance of driver 1 greater than, less than, or equal to the stopping distance of driver 2? **(b)** Choose the *best* explanation from among the following:

A. Locking the brakes gives the greatest possible braking force.

B. The same tires on the same road result in the same force of friction.

C. Locked brakes result in sliding (kinetic) friction, which is less than rolling (static) friction.

Problem Solving

96. A 55-kg baseball player slides into third base with an initial speed of 4.6 m/s. If the coefficient of kinetic friction between the player and the ground is 0.46, what is the player's acceleration?

97. **Think & Calculate** If the mass of the baseball player in Problem 96 is doubled, **(a)** does the player's acceleration increase, decrease, or stay the same? Explain. **(b)** Give the acceleration for a 110-kg player.

98. **Think & Calculate** If the speed of the baseball player in Problem 96 is halved, **(a)** does the player's acceleration increase, decrease, or stay the same? Explain. **(b)** Give the acceleration for a player with an initial speed of 2.3 m/s.

99. A car accelerates at 12 m/s^2 without spinning the tires. Determine the minimum coefficient of static friction between the tires and the road needed to make this possible.

100. A 97-kg sprinter wishes to accelerate from rest to a speed of 13 m/s in a distance of 22 m. What coefficient of static friction is required between the sprinter's shoes and the track?

101. **Coffee to Go** A person places a cup of coffee on the roof of her car while she dashes back into the house for a forgotten item. When she returns to the car, she hops in and takes off with the coffee cup still on the roof. **(a)** If the coefficient of static friction between the coffee cup and the roof of the car is 0.24, what is the maximum acceleration the car can have without causing the cup to slide? Ignore the effects of air resistance. **(b)** What is the smallest amount of time in which the person can accelerate the car from rest to 15 m/s and still keep the coffee cup on the roof?

102. **The da Vinci Code** Leonardo da Vinci (1452–1519) is credited with being the first to perform quantitative experiments on friction. His results weren't known until centuries later, however, in part because of the secret code (mirror writing) he used in his notebooks. Some of his sketches of friction experiments are shown in **Figure 5.28**. In other experiments Leonardo placed a

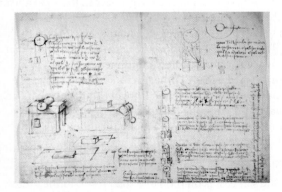

Figure 5.28 Sketches from the notebooks of Leonardo da Vinci showing experiments he performed to study friction.

block of wood on an inclined plane and measured the angle at which the block began to slide. He reported that the coefficient of static friction was 0.25 in his experiments. At what angle did Leonardo's block begin to slide?

103. When you push a 1.92-kg book resting on a tabletop, you have to exert a force of 2.05 N to start the book sliding. Once it is sliding, however, you can use a force of only 1.03 N to keep the book moving with constant speed. What are the coefficients of static and kinetic friction between the book and the tabletop?

Mixed Review

104. In **Figure 5.29** Wilbur is asking Mr. Ed the talking horse to pull a cart. Mr. Ed replies that he would like to, but the laws of nature just won't allow it. According to Newton's third law, he says, if he pulls on the cart, it pulls back on him with an equal force. Clearly, then, the net force is zero, and the cart will stay put. How should Wilbur answer the clever horse?

Figure 5.29

105. A whole brick has more mass than half a brick; thus, the whole brick is harder to accelerate. Given this fact, why doesn't a whole brick fall with a lower acceleration than half a brick? Explain.

106. The force exerted by gravity on a whole brick is greater than the force exerted by gravity on half a brick. Given this fact, why doesn't a whole brick fall faster than half a brick? Explain.

107. Since all objects are "weightless" in orbit, how is it possible for an orbiting astronaut to tell if one object has more mass than another object? Explain.

108. When you weigh yourself on good old *terra firma* (solid ground), your weight is 540 N. In an elevator your apparent weight is 480 N. What are the direction and the magnitude of the elevator's acceleration?

109. Give the direction of the net force acting on each of the following objects. If the net force is zero, state "zero."
(a) a car accelerating northward from a stoplight
(b) a car traveling southward and slowing down
(c) a car traveling westward with constant speed
(d) a skydiver parachuting downward with constant speed
(e) a baseball during its flight from pitcher to catcher (ignoring air resistance)

110. The spring in a cushion of a chair has a force constant of 580 N/m. How much does the spring compress when a force of 51 N pushes on it?

111. A 22-kg chimpanzee hangs from the end of a horizontal, broken branch, as shown in **Figure 5.30**. The branch sags downward through a vertical distance of 13 cm. Treating the branch as a spring satisfying Hooke's law, what is its spring constant?

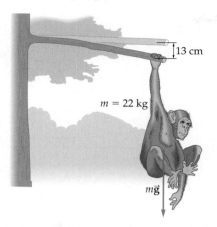

Figure 5.30

112. Two astronauts on a space walk decide to take a break and play catch with a baseball. Describe what happens as the game of catch progresses.

113. A 42.0-kg parachutist is moving straight downward with a speed of 3.85 m/s. **(a)** If the parachutist comes to rest with constant acceleration over a distance of 0.750 m, what force does the ground exert on her? **(b)** If the parachutist comes to rest over a shorter distance, is the force exerted by the ground greater than, less than, or the same as in part (a)? Explain.

114. A drag-racing car accelerates forward because of the force exerted on it by the road. Why, then, does it need an engine? Explain.

115. Flight of the Samara A 1.21-g samara—the winged fruit of a maple tree—falls toward the ground with a constant speed of 1.1 m/s, as shown in **Figure 5.31**. **(a)** What is the force of air resistance exerted on the

Figure 5.31

samara? **(b)** If the constant speed of descent is greater than 1.1 m/s, is the force of air resistance greater than, less than, or the same as in part (a)? Explain.

116. In a daring rescue by helicopter, two men with a combined mass of 172 kg are lifted to safety. **(a)** If the helicopter lifts the men straight up with constant acceleration, is the tension in the rescue cable greater than, less than, or equal to the combined weight of the men? Explain. **(b)** Determine the tension in the cable if the men are lifted with a constant acceleration of 1.10 m/s².

117. Gecko Feet Researchers have found that a gecko's foot, like the one shown in **Figure 5.32**, is covered with hundreds of thousands of small hairs (called *setae*) that allow it to walk up walls and even across ceilings. A single foot, whose toe pads have a total surface area of 1.0 cm², can attach to a wall or ceiling with a force of 11 N.

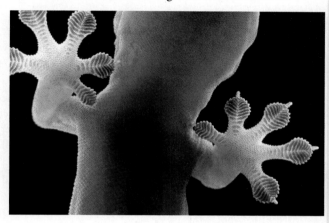

Figure 5.32 A Tokay gecko (*Gekko gecko*) shows off its famous feet.

(a) How many 250-g geckos could be suspended from the ceiling by a single foot? **(b)** Estimate the force per square centimeter that your body exerts on the soles of your shoes, and compare with the 11 N/cm² of the sticky gecko foot.

118. Two workers pull a raft through a lock, as shown in **Figure 5.33**. One worker pulls with a force of 130 N at an angle of 34° relative to the direction of the raft's forward motion. The second worker, on the opposite side of the lock, pulls at an angle of 45°. With what force should the second worker pull so that the net force of the two workers is in the forward direction?

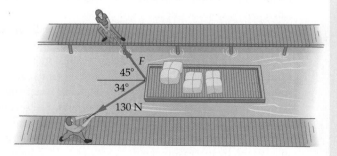

Figure 5.33

119. For a birthday gift you and some friends take a hot-air balloon ride. One friend is late, so the balloon floats a couple of feet off the ground as you wait. Before this person arrives, the combined weight of the basket and passengers is 1220 kg, and the balloon is neutrally buoyant. When the late arrival climbs up into the basket, the balloon begins to accelerate downward at 0.56 m/s². What was the mass of the last person to climb aboard?

120. You push a box along the floor against a constant force of friction. When you push with a horizontal force of 75 N, the acceleration of the box is 0.50 m/s²; when you increase the force to 81 N, the acceleration is 0.75 m/s². Find **(a)** the mass of the box and **(b)** the coefficient of kinetic friction between the box and the floor.

Writing about Science

121. Write a report on some of the advantages and disadvantages of friction in everyday life. Give at least two examples of friction that is useful or helpful and two examples of friction that you would like to reduce.

122. Write a report comparing the coefficients of static and kinetic friction for different substances. What substances have the greatest friction? Which have the least friction? Are there substances for which the coefficient of static friction is less than the coefficient of kinetic friction? If so, which ones?

123. Connect to the Big Idea Does the net force exerted on an object determine its speed or its acceleration? Does an object's acceleration depend on its speed? If you push a heavy box across the floor, does the box push back on you? If so, how does the force exerted by the box on you compare with the force exerted by you on the box?

Read, Reason, and Respond

Increasing Safety in a Collision Safety experts say that an automobile accident is really a succession of three separate collisions: (1) The automobile collides with an obstacle and comes to rest; (2) people within the car continue to move forward until they collide with the interior of the car or are brought to rest by a restraint system like a seat belt or an air bag; (3) organs within the occupant's bodies continue to move forward until they collide with the body wall and are brought to rest. Not much can be done about the third collision, but the effects of the first two can be lessened by increasing the distance over which the car and its occupants are brought to rest.

For example, the severity of the first collision is reduced by building collapsible *crumple zones* into car bodies and by placing compressible collision barriers near dangerous road obstacles like bridge supports. The second collision is addressed primarily through the use of seat belts and air bags. These devices reduce the force that acts on a person by increasing the distance over which he or

she comes to rest. **Figure 5.34** shows the force exerted on a 65.0-kg driver who slows from an initial speed of 18.0 m/s (lower curve) or 36.0 m/s (upper curve) to rest in a distance ranging from 5.00 cm to 1.00 m.

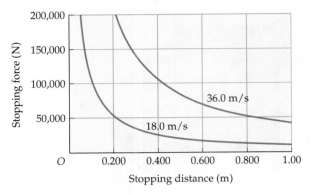

Figure 5.34

124. The combination of crumple zones and air bags and seat belts might increase the distance over which a person stops in a collision to as much as 1.00 m. What is the force exerted on a 65.0-kg driver who decelerates from 18.0 m/s to rest over a distance of 1.00 m?

A. 162 N   C. 1.05×10^4 N

B. 585 N   D. 2.11×10^4 N

125. A driver who does not wear a seat belt continues to move forward with a speed of 18.0 m/s (due to inertia) until something solid like the steering wheel is encountered. The driver then comes to rest in a much shorter distance—perhaps only a few centimeters. Find the net force acting on a 65.0-kg driver who is decelerated from 18.0 m/s to rest in 5.00 cm.

A. 3240 N   C. 2.11×10^5 N

B. 1.17×10^4 N   D. 4.21×10^5 N

126. Suppose the initial speed of the driver is doubled, to 36.0 m/s? If the driver still has a mass of 65.0 kg and comes to rest in 1.00 m, what is the force exerted on the driver during this collision?

A. 648 N   C. 1170 N

B. 2.11×10^4 N   D. 4.21×10^4 N

Standardized Test Prep

Multiple Choice

1. When a 100-N force acts horizontally on an 8.0-kg chair, the chair moves at constant speed across a level floor. Which statement has to be true?
   (A) The chair accelerates at 12.5 m/s².
   (B) The applied force is greater than the friction force.
   (C) The friction force is 100 N.
   (D) There is no friction between the floor and chair.

2. Consider a puck that slides without friction across the ice during a hockey game. Which statement is correct?
   (A) The puck has constant acceleration.
   (B) The puck moves at constant speed.
   (C) The puck will come to a stop.
   (D) A constant horizontal force acts on the puck.

3. A book rests on a level desktop. Which force is equal in magnitude to the normal force exerted on the book?
   (A) the weight of the book
   (B) the force of static friction
   (C) the force of kinetic friction
   (D) the gravitational force exerted by the desktop on the book

Use the diagram below to answer Questions 4 and 5. The diagram shows a box held stationary on a ramp by a string connected to a wall.

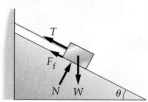

4. Which relationship is true?
   (A) $T = -F_f$ (C) $W \sin \theta = T + F_f$
   (B) $T = W \cos \theta + F_f$ (D) $T + F_f = W$

5. How do the normal force, friction force, and tension change when the angle θ is increased?

	Force		
	Normal, N	Friction, F_f	Tension, T
(A)	increases	increases	increases
(B)	decreases	decreases	decreases
(C)	increases	increases	decreases
(D)	decreases	decreases	increases

Use the graph below to answer Questions 6 and 7. The data describe a cart pulled across a level surface.

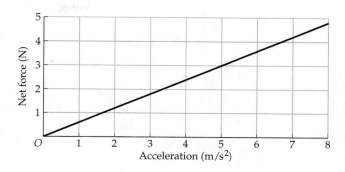

6. What is the mass of the cart?
   (A) 0.4 kg   (C) 1.0 kg
   (B) 0.6 kg   (D) 1.7 kg

7. When a 4-N force pulls on the cart, the cart accelerates at 5 m/s². What is the friction force resisting the pull?
   (A) 0   (C) 0.5 N
   (B) 1.0 N   (D) 2.0 N

8. A 1.6-kg block is pulled across a level surface by a horizontal force of 12 N. Determine the approximate coefficient of friction between the block and the surface if the block moves at constant velocity.
   (A) 0.25   (C) 0.75
   (B) 0.50   (D) 1.30

Extended Response

9. When a box is kicked, it slides 3 m across the floor before coming to a stop. After a second kick the box comes to a stop in 6.0 m. Compare the friction forces and initial velocities for these two cases. Explain your reasoning.

> **Test-Taking Hint**
>
> Draw free-body diagrams to help you analyze problems involving forces.

If You Had Difficulty With . . .

Question	1	2	3	4	5	6	7	8	9				
See Lesson(s)	5.1, 5.2	5.1	5.1, 5.2	5.2, 5.3	5.2, 5.3	5.1	5.1	5.3	5.2, 5.3				

6 Work and Energy

Though steinstossen *(stone throwing) is not an Olympic sport, it has many fans. Launching a stone like this competitor is doing requires a mighty effort—in which work is converted to kinetic energy. (Note to photographer: Look out!)*

Big Idea

Energy can change from one form to another, but the total amount of energy in the universe stays the same.

Everyone knows what *work* and *energy* mean in everyday life. You get up in the morning and "go to work," or you "work up a sweat" hiking up a mountain. Later in the day you eat lunch and get the "energy" to continue working or hiking. In this chapter you'll learn what *work* and *energy* mean in physics and how to apply these definitions to a variety of everyday situations.

Inquiry Lab   What factors affect energy transformations?

Explore 🦴

1. Obtain a rubber "popper" from your teacher.
2. Press your thumbs into the center of the popper and turn it inside out. Carefully place the popper on a tabletop, with its flat side down.
3. Hold a meterstick upright next to the popper and use it to measure how high the popper rises above the tabletop when it springs upward. Record the height in centimeters.

4. Repeat Steps 2 and 3, except this time place the popper on a soft surface, such as a cushion or a sponge.
5. Repeat Steps 2 and 3, except this time place the popper on the eraser end of a pencil held vertically.

Think

1. **Identify** Based on what you currently know about energy, describe the energy changes that took place during the motion of the popper in each trial.

2. **Compare** Rank the heights attained by the popper in all your trials, from lowest to highest. Explain why you think the popper went higher in some cases.
3. **Predict** Imagine that you dropped the inverted popper, flat side down, onto the tabletop rather than just placing it there. What effect do you think this would have on the height attained by the popper? Explain your answer.

6.1 Work

The concept of force is one of the foundations of physics. As you will see in this lesson, force times distance is also an important physical quantity.

Work depends on force and distance

Pushing a heavy shopping cart in a store or pulling a big suitcase through an airport requires considerable effort. The greater the force you exert, the greater the effort. The greater the distance you move the object, the greater the effort. If you push or pull long and hard enough, your exertions can even make you tired. These observations are the basis for our definition of *work*.

🦴 **In the simplest case work is done when a force is applied to an object and the object moves in the direction of the applied force.** In a situation like this, **work**, W, is defined as force times the distance moved.

Vocabulary
- work
- joule

🦴 *How is work done?*

> **Definition of Work, *W* (force in the direction of displacement)**
>
> work = force × distance
> $$W = Fd$$
>
> SI unit: newton-meter $(N \cdot m)$ = joule (J)

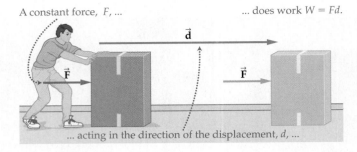

A constant force, F, ...
... does work $W = Fd$.
... acting in the direction of the displacement, d, ...

▲ **Figure 6.1 Work is force times distance**
A constant force, $\vec{F}$, pushes a box through a displacement, $\vec{d}$. In this special case the force and displacement are in the *same* direction, and the work done on the box by the force is $W = Fd$.

For example, if you push a box with a force F through a distance d, then the work you do is $W = Fd$. This is shown in **Figure 6.1**. In this simple case the force is in the same direction as the displacement. We will consider other cases later in this lesson.

Work is measured with a unit called the joule

The dimensions of work are force (newtons) times distance (meters), or $N \cdot m$. The combination $N \cdot m$ is called the **joule** (rhymes with *cool*) in honor of James Prescott Joule (1818–1889). Joule was such a dedicated physicist that he even conducted physics experiments while on his honeymoon. We define a joule as follows:

Definition of the Joule (J)

$$1\,J = 1\,N \cdot m = 1\,kg \cdot m^2/s^2$$

Just how much work is a joule, anyway? Not a lot, actually. You do 1 joule of work when you lift a 1-gallon container of milk 2.54 centimeters (1 inch). A joule of work will lift an apple 1 meter, or light a 100-watt bulb for $1/100$ of a second. Clearly, a joule is a modest amount of work. Additional examples of amounts of work are given in **Table 6.1**.

Table 6.1 Typical Amounts of Work

Activity	Equivalent Work (J)
Annual U.S. energy use	8×10^{19}
Mount St. Helens eruption	10^{18}
Burning 1 gal of gas	10^8
Daily human food intake	10^7
Melting an ice cube	10^4
Lighting a 100-W bulb for 1 min	6000
A human heartbeat	0.5
Turning a page of a book	10^{-3}
One hop of a flea	10^{-7}
Breaking a bond in DNA	10^{-20}

An intern pushes a 72-kg patient on a 15-kg gurney, producing an acceleration of 0.60 m/s^2. How much work does the intern do by pushing the patient and gurney through a distance of 2.5 m? Assume that the gurney moves without friction.

Picture the Problem

Our sketch shows the physical situation for this problem. Notice that the force exerted by the intern is in the same direction as the displacement of the gurney. Therefore, we know that the work is $W = Fd$.

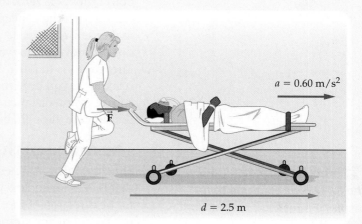

Strategy

We are not given the magnitude of the force F. We can calculate it with $F = ma$, however, since both m and a are given. The work done by the intern is then $W = Fd$.

Known	Unknown
$m = 72 \text{ kg} + 15 \text{ kg}$	$W = ?$
$d = 2.5 \text{ m}$	
$a = 0.60 \text{ m/s}^2$	

Solution

1 First, find the force, F, exerted by the intern using $F = ma$:

$$F = ma$$
$$= (72 \text{ kg} + 15 \text{ kg})(0.60 \text{ m/s}^2)$$
$$= 52 \text{ N}$$

2 The work done by the intern, W, is the force F times the distance d:

$$W = Fd$$
$$= (52 \text{ N})(2.5 \text{ m})$$
$$= \boxed{130 \text{ J}}$$

Insight

You might wonder whether the work done by the intern depends on the speed of the gurney. For example, does it take more work to push the gurney rapidly than to push it slowly? The answer is no, as long as the force and distance are the same. The work done on an object, $W = Fd$, doesn't depend at all on its speed.

Practice Problems

1. Follow-up If the force exerted by the intern is doubled and the distance is halved, does the work done by the intern increase, decrease, or remain the same?

2. There is a species of Darwin's finch on the Galapagos Islands that can exert a force of 205 N with its beak as it cracks open a seed case. If its beak moves through a distance of 0.40 cm during this operation, how much work does the finch do to get at the seed inside the case?

3. Early one October you go to a pumpkin patch to select your Halloween pumpkin. You lift a 3.2-kg pumpkin to a height of 0.80 m to check it out. How much work do you do on the pumpkin when you lift it from the ground?

Work is zero if the distance moved is zero

Suppose you push on a car but it doesn't move. How much work have you done on the car? As long as the car doesn't move, the work done on it is zero. That's right, zero. This is true no matter how hard you push. Why is that? Well, recall that work is force times distance, $W = Fd$. If the distance the object moves is zero, then so is the work.

Though the concept of zero work may seem a bit odd at first, it actually makes sense. Suppose you hold a 5-kg suitcase in your hand. The suitcase doesn't move, so no work is done on it. Even so, you get tired pretty quickly. The reason is that the individual cells in your muscles are contracting and relaxing as you hold the suitcase. The muscle cells are doing work, exerting a force over a distance, and that is what makes you tired. No work is done on the suitcase, though, because it's at rest the whole time.

The angle of the force affects the work done

How is work calculated when the force is at an angle to the displacement?

Figure 6.2 shows a person pulling a suitcase on a level surface with a strap that makes an angle θ with the horizontal. Notice that the force is at an angle to the direction of motion. How do we calculate the work in this case? We use the following simple rule: **Only the *component* of the force in the *direction* of the displacement does work.**

To see how this works, notice in Figure 6.2 that the component of force in the direction of displacement is $F \cos \theta$. In addition, the magnitude of the displacement is d. Therefore, the work is $F \cos \theta$ times d, or $Fd \cos \theta$:

> **Definition of Work, *W* (with force and displacement at an angle θ)**
>
> work = force $\times$ distance $\times$ $\left(\begin{array}{c}\text{cosine of angle between} \\ \text{force and displacement}\end{array}\right)$
>
> $W = Fd \cos \theta$
>
> SI unit: joule (J)

In the case where the force is in the direction of motion, the angle θ is zero. Since $\cos 0° = 1$, it follows that

$$W = Fd \cos 0° = (Fd)(1) = Fd$$

This is in agreement with our earlier definition of work. The new definition is more general, however, and applies for any angle between the force and the displacement.

The component of force in the direction of displacement is $F \cos \theta$. This is the only component of the force that does work.

▶ **Figure 6.2 Work when force is at an angle to displacement**
A person pulls a suitcase with a strap at an angle θ to the direction of motion. The component of force in the direction of motion is $F \cos \theta$, and the work done by the person is $W = (F \cos \theta)d$.

The work in this case is $W = (F \cos \theta)d$.

In a gravity escape system, an enclosed lifeboat on a large ship is deployed by letting it slide down a ramp and then continue in free fall to the water below. Suppose a 4900-kg lifeboat slides a distance of 5.0 m on a ramp that makes an angle of 60° with the vertical. How much work does gravity do on the boat?

Picture the Problem

Our sketch shows that the force of gravity, $m\vec{g}$, and the displacement, $\vec{d}$, are at an angle $\theta = 60°$ relative to one another. In addition, the magnitude of the displacement is $d = 5.0$ m.

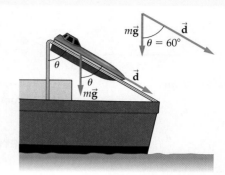

Strategy

By definition, the work done on the lifeboat by gravity is $W = Fd \cos \theta$. The force exerted on the lifeboat by gravity has a magnitude given by $F = mg$. We first calculate F and then substitute that value into our equation for work.

Known

$m = 4900$ kg $\theta = 60°$ $d = 5.0$ m

Unknown:

$W = ?$

Solution

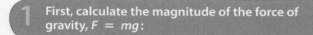

1 First, calculate the magnitude of the force of gravity, $F = mg$:

$F = mg$
$= (4900 \text{ kg})(9.81 \text{ m/s}^2)$
$= 48{,}000 \text{ N}$

Math HELP
Significant Figures
See Lesson 1.4

2 Substitute the numerical values for F, d, and θ into the work equation and calculate the work done by gravity on the boat:

$W = Fd \cos \theta$
$= (48{,}000 \text{ N})(5.0 \text{ m})\cos 60°$
$= \boxed{120{,}000 \text{ J}}$

Insight

Notice that only a fraction ($\cos 60° = 0.5$) of the force of gravity does work on the boat as it slides down the ramp. If the ramp were vertical ($\cos 0° = 1$), then all of the force of gravity would do work. If the ramp were horizontal ($\cos 90° = 0$), then gravity would do no work on the boat at all.

Practice Problems

4. **Follow-up** Suppose the lifeboat slides halfway to the water, gets stuck for a moment, and then starts up again and continues to the end of the ramp. What is the work done by gravity in this case?

5. A parent pulls a child in a little red wagon with constant speed. If the parent pulls with a force of 16 N for 12 m and the handle of the wagon is inclined at an angle of 25° above the horizontal, how much work does the parent do on the wagon?

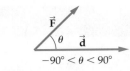

(a) Positive work

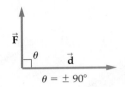

(b) Zero work

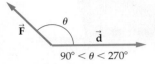

(c) Negative work

▲ **Figure 6.3 Positive, negative, and zero work**
Work is positive when the force is in the same general direction as the displacement and is negative if the force is generally opposite to the displacement. Zero work is done if the force is at right angles to the displacement.

Work can be positive, negative, or zero

As we have seen, work depends on the cosine of the angle between the force $\vec{F}$ and the displacement $\vec{d}$. As a result, work can be positive, negative, or zero. When you calculate work, be sure to think about its sign.

- Work is positive if the force has a component in the direction of motion, as shown in **Figure 6.3 (a)**.
- Work is zero if the force has no component in the direction of motion, as shown in **Figure 6.3 (b)**.
- Work is negative if the force has a component opposite to the direction of motion, as shown in **Figure 6.3 (c)**.

Two special cases deserve particular attention, since they arise in many practical situations.

Zero Work The work done is zero when the force is perpendicular to the displacement ($\theta = 90°$). This is the case because $\cos 90° = 0$.

$$W = 0 \qquad \text{(when } \theta = 90°\text{)}$$

Negative Work The work done is *negative* force times distance when the force acts opposite to the displacement ($\theta = 180°$). This is the case because $\cos 180° = -1$.

$$W = -Fd \qquad \text{(when } \theta = 180°\text{)}$$

CONCEPTUAL Example 6.3 Rank the Work Done

Each of the boxes shown below moves through the same horizontal distance, d. The force applied to each of the boxes has the same magnitude, F. Notice, however, that the direction of the force is different in each of the cases A, B, C, and D. Rank these cases in order of increasing work done by the force F, from most negative to most positive.

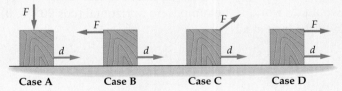

Case A   Case B   Case C   Case D

Reasoning and Discussion
The work done by the force is zero in case A, because F is perpendicular to the displacement. The work is negative in case B, because F points opposite to the displacement. The work is positive in both cases C and D. It is larger in case D, however, because the entire force is in the direction of displacement, rather than just a component of the force, as in case C. The ranking, then, is B, A, C, D.

Answer
The ranking of the four cases in order of increasing work done by the force F is B, A, C, D.

The work done by separate forces can be summed

When more than one force acts on an object, the total work is the sum of the work done by each force separately. Thus, if force $\vec{F}_1$ does work W_1, force $\vec{F}_2$ does work W_2, force $\vec{F}_3$ does work W_3, and so on, the total work is

$$W_{total} = W_1 + W_2 + W_3 + \cdots$$

Since individual amounts of work can be positive or negative, it's possible that the total work may be zero even though the individual amounts are nonzero.

GUIDED Example 6.4 | Jamming with Rock Hero   **Work**

As you and your friends prepare to play Rock Hero, you push on the drum set to slide it into position. You exert a horizontal force of $F = 24$ N, kinetic friction opposes the motion with a force of $f = 22$ N, and you slide the drums a distance of $d = 1.5$ m. The weight of the drum set is $mg = 54$ N. What is the total work done on the drums?

Picture the Problem

Our sketch shows the drum set and the forces that act on it. In addition to the applied force, F, the friction force, f, and the weight, mg, we have added the normal force, N, which is exerted by the floor to cancel the weight. We have also indicated the displacement through the distance d. Notice that the displacement is in the same direction as the applied force.

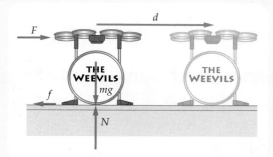

Strategy

To find the total work we calculate the work done by each force separately and then sum the results. First, we note that mg and N are at right angles to the displacement. As a result these forces do zero work: $W_{weight} = 0$, $W_{normal} = 0$. Next, we see that F is in the same direction as the displacement. Therefore, the work it does is $W_{applied} = Fd$. Similarly, f is in the *opposite* direction from the displacement. Thus, the work it does is $W_{friction} = -fd$.

Known	Unknown
$F = 24$ N	$W_{total} = ?$
$f = 22$ N	
$mg = 54$ N	
$d = 1.5$ m	

Solution

1 The vertical forces, mg and N, are at right angles to the displacement. They do zero work:

$$W_{weight} = 0 \quad \text{and} \quad W_{normal} = 0$$

2 The applied force does positive work, Fd, and the friction force does negative work, $-fd$:

$$W_{applied} = Fd \quad \text{and} \quad W_{friction} = -fd$$

3 Sum the amounts of work to obtain the total work:

$$W_{total} = W_{weight} + W_{normal}$$
$$+ W_{applied} + W_{friction}$$
$$= 0 + 0 + Fd - fd$$
$$= Fd - fd$$

4 Substitute the numerical values:

$$W_{total} = Fd - fd$$
$$= (24\text{ N})(1.5\text{ m}) - (22\text{ N})(1.5\text{ m})$$
$$= 36\text{ J} - 33\text{ J}$$
$$= \boxed{3\text{ J}}$$

Insight

Though four forces act on the drum set, only the applied force and the friction force do work. The work from these forces almost cancels, leaving a total work of only 3 J.

Practice Problems

6. **Follow-up** The frictional force acting on the drum set does negative work. Does friction always do negative work? If yes, explain why. If no, give an example of friction doing positive work.

7. You slide a 0.12-kg coffee mug 0.15 m across a table. The force you exert is horizontal and of magnitude 0.10 N. The coefficient of kinetic friction between the mug and the table is 0.05. How much work is done on the mug?

6.1 LessonCheck (MP)

Checking Concepts

8. **State** How is work calculated when force and displacement are in the same direction?

9. **Identify** Which component of force is used to calculate work when the force and the displacement are at an angle to each other?

10. **Assess** Is it possible to do work on an object that remains at rest? Explain.

11. **Give Examples** Give two examples of (a) positive work done by a frictional force (if possible) and (b) negative work done by a frictional force.

Solving Problems

12. **Calculate** A child in a tree house uses a rope attached to a basket to lift a 22-N dog upward through a distance of 4.7 m into the house. How much work does the child do in lifting the dog?

13. **Calculate** To move a suitcase up to the check-in stand at an airport, a student pushes with a horizontal force through a distance of 0.95 m. If the work done by the student is 32 J, what is the magnitude of the force he exerts?

14. **Calculate** A farmhand pushes a bale of hay 3.9 m across the floor of a barn. She exerts a force of 88 N at an angle of 25° below the horizontal. How much work has she done?

15. **Calculate** The coefficient of kinetic friction between a large box and the floor is 0.21. A person pushes horizontally on the box with a force of 160 N for a distance of 2.3 m. If the mass of the box is 72 kg, what is the total work done on the box?

6.2 Work and Energy

When you do work on an object, you change its energy. For example, when you push a shopping cart, your work goes into increasing its *kinetic energy*. When you climb a mountain, your work goes into increasing your *potential energy*. Similarly, energy has the capacity to do work on an object. In this lesson, you'll learn the precise connection between work and energy.

Kinetic Energy

Imagine colliding with another soccer player during a game, as shown in **Figure 6.4**. Ouch. As you know, the intensity of the impact depends on both the mass and the speed of the player. A physicist would say that the intensity of the impact depends on the kinetic energy of the player. Let's use physics to make this observation more precise.

Kinetic energy is the energy of motion

We can use our knowledge of Newton's laws and the equations of motion to write a relationship between work and energy. Suppose you push a box of mass m across an ice-skating rink with a force F, as in **Figure 6.5**. The acceleration of the box is given by Newton's second law:

$$a = \frac{F}{m}$$

As you learned in Chapter 3, the speed and displacement of an accelerating object are related by the equation of motion:

$$v_f^2 = v_i^2 + 2ad$$

A slight rearrangement gives

$$2ad = v_f^2 - v_i^2$$

Vocabulary

- kinetic energy
- potential energy

▲ Figure 6.4 **The energy of motion depends on mass and speed**
The impact of this collision depends on both the mass and the speed of the colliding players.

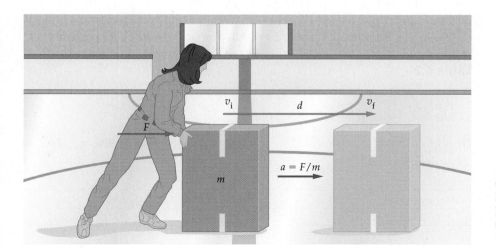

◀ Figure 6.5 **Doing work on an object changes its speed**
Doing work on an object changes its speed. In this case the work, $W = Fd$, changes the speed according to the equation $W = \frac{1}{2}mv_f^2 - \frac{1}{2}mv_i^2$.

Replacing a with F/m and multiplying both sides of the equation by $m/2$ gives

$$Fd = \tfrac{1}{2}mv_f^2 - \tfrac{1}{2}mv_i^2$$

Recalling from the previous lesson that work is $W = Fd$, we have

$$W = \tfrac{1}{2}mv_f^2 - \tfrac{1}{2}mv_i^2$$

This equation is valid whether we use the simpler form of the work equation, $W = Fd$, or the more general form, $W = Fd \cos \theta$.

Notice that the work done on the box (or on any other object) is related to the change in the quantity $\tfrac{1}{2}mv^2$. This quantity has special significance in physics. We refer to $\tfrac{1}{2}mv^2$ as the *kinetic energy*, or *KE*, of an object of mass m and speed v:

> **Definition of Kinetic Energy, *KE***
>
> kinetic energy $= \tfrac{1}{2}(\text{mass}) \times (\text{velocity})^2$
>
> $$KE = \tfrac{1}{2}mv^2$$
>
> SI unit: $\text{kg} \cdot \text{m}^2/\text{s}^2 = \text{J}$

In general, the **kinetic energy** of an object is the energy due to its motion. Kinetic energy is measured with the joule, the same unit used to measure work.

To get a feeling for typical values of kinetic energy, consider a soccer player who runs into you. Assuming a mass of 62 kg and a speed of 2.5 m/s, the player's kinetic energy is

$$
\begin{aligned}
KE &= \tfrac{1}{2}mv^2 \\
&= \tfrac{1}{2}(62 \text{ kg})(2.5 \text{ m/s})^2 \\
&= 190 \text{ J}
\end{aligned}
$$

Notice that the kinetic energy is increased if the mass or the speed is increased. **More specifically, kinetic energy increases linearly with mass and with the square of the velocity.** For example, a slightly larger 76-kg player running slightly faster at 3.5 m/s would have a kinetic energy of 466 J. That's a lot more energy, and the impact would be considerably more unpleasant. Additional examples of kinetic energy are given in **Table 6.2**.

Table 6.2 Typical Kinetic Energies

Object	Approximate Kinetic Energy (J)
Jet aircraft at 500 mph	10^9
Car at 60 mph	10^6
Home-run baseball	10^3
Person at walking speed	50
Housefly in flight	10^{-3}

A 3900-kg truck is moving at 6.0 m/s. **(a)** What is its kinetic energy? **(b)** What is the truck's kinetic energy if its speed is doubled to 12 m/s?

Solution

(a) Substituting $m = 3900$ kg and $v = 6.0$ m/s into $KE = \frac{1}{2}mv^2$ gives

$$KE = \frac{1}{2}mv^2$$
$$= \frac{1}{2}(3900 \text{ kg})(6.0 \text{ m/s})^2$$
$$= \boxed{70{,}000 \text{ J}}$$

(b) Kinetic energy depends on the speed squared. Therefore, doubling the speed quadruples the kinetic energy.

$$KE = 4(70{,}000 \text{ J}) = \boxed{280{,}000 \text{ J}}$$

Practice Problems

16. A 13-g goldfinch has a speed of 8.5 m/s. What is its kinetic energy?

17. The kinetic energy of a small boat is 15,000 J. If the boat's speed is 5.0 m/s, what is its mass?

18. What is the speed of a 0.15-kg baseball whose kinetic energy is 77 J?

Work and kinetic energy are related

There is a very simple connection between work and the change in kinetic energy. Specifically, **the total work done on an object equals the change in its kinetic energy.** This connection is known as the *work-energy theorem*:

How does the work done on an object affect its kinetic energy?

> **Work-Energy Theorem**
>
> total work = change in kinetic energy
> $$W_{\text{total}} = \Delta KE$$
> $$= \frac{1}{2}mv_f^2 - \frac{1}{2}mv_i^2$$

The work-energy theorem is one of the most important and fundamental results in physics. It is also a very handy tool for problem solving.

QUICK Example 6.6 **What's the Work?**

How much work is required for a 74-kg sprinter to accelerate from rest to a speed of 2.2 m/s?

Solution

Since the initial speed is zero, $v_i = 0$, we have

$$W_{\text{total}} = \Delta KE = \frac{1}{2}mv_f^2 - \frac{1}{2}mv_i^2$$
$$= \frac{1}{2}(74 \text{ kg})(2.2 \text{ m/s})^2 - \frac{1}{2}(74 \text{ kg})(0 \text{ m/s})^2$$
$$= \boxed{180 \text{ J}}$$

Math HELP
Delta (Δ)
See Math Review, Section I

Notice how the sign of the work is related to the change in kinetic energy. If the total work is positive, the kinetic energy increases. Similarly, if the total work is negative, the kinetic energy decreases. Finally, if the total work is zero, then there is no change in kinetic energy.

A student lifts a 4.10-kg box of books vertically from rest with an upward force of 52.7 N. The distance of the lift is 1.60 m. Find (**a**) the work done by the student, (**b**) the work done by gravity, and (**c**) the final speed of the box.

Picture the Problem

Our sketch shows that the direction of motion of the box is upward. In addition, we see that the force the student applies, $\vec{F}_{student}$, is upward and the force of gravity, $m\vec{g}$, is downward. Finally, the box is lifted from rest ($v_i = 0$) through a distance $\Delta y = 1.60$ m.

Strategy

The applied force is in the direction of motion, so the work it does is force times distance: $W_{student} = F_{student}\Delta y$. Gravity is opposite in direction to the motion; thus, its work is *negative* force times distance: $W_{gravity} = -mg\Delta y$. The total work is the sum of $W_{student}$ and $W_{gravity}$, and the final speed of the box is found by applying the work-energy theorem, $W_{total} = \Delta KE$.

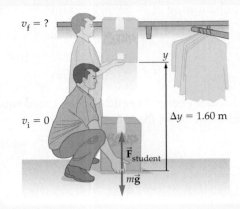

$v_f = ?$

y

$v_i = 0$

$\Delta y = 1.60$ m

$\vec{F}_{student}$

$m\vec{g}$

Known

$m = 4.10$ kg
$\Delta y = 1.60$ m
$F_{student} = 52.7$ N

Unknown

(**a**) $W_{student} = ?$
(**b**) $W_{gravity} = ?$
(**c**) $v_f = ?$

Solution

Part (a)

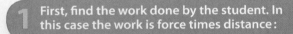

1 First, find the work done by the student. In this case the work is force times distance:

$$W_{student} = F_{student}\Delta y$$
$$= (52.7 \text{ N})(1.60 \text{ m})$$
$$= \boxed{84.3 \text{ J}}$$

Part (b)

2 Next, calculate the work done by gravity. The distance is $\Delta y = 1.60$ m, as before, but now the work is negative force times distance:

$$W_{gravity} = -mg\Delta y$$
$$= -(4.10 \text{ kg})(9.81 \text{ m/s}^2)(1.60 \text{ m})$$
$$= \boxed{-64.4 \text{ J}}$$

Part (c)

3 The total work done on the box is the sum of $W_{student}$ and $W_{gravity}$:

$$W_{total} = W_{student} + W_{gravity}$$
$$= 84.3 \text{ J} - 64.4 \text{ J}$$
$$= 19.9 \text{ J}$$

4 Apply the work-energy theorem to solve for the final speed, v_f. Recall that the box started at rest; thus, $v_i = 0$. Solve for v_f and substitute the known values:

$$W_{total} = \tfrac{1}{2}mv_f^2 - \tfrac{1}{2}mv_i^2$$
$$= \tfrac{1}{2}mv_f^2$$
$$v_f = \sqrt{\frac{2W_{total}}{m}}$$
$$= \sqrt{\frac{2(19.9 \text{ J})}{4.10 \text{ kg}}}$$
$$= \boxed{3.12 \text{ m/s}}$$

Math HELP
Solving Simple Equations
See Math Review, Section III

Insight

The total work in this case is positive. As a result, the final kinetic energy is greater than the initial kinetic energy. Of course, this also means that the speed has increased.

19. **Follow-up** Do you expect the change in kinetic energy of the box to be greater than, less than, or equal to 19.9 J? Verify your answer by calculating the change in its kinetic energy.

20. At $t = 1.0$ s, a 0.40-kg object is falling with a speed of 6.0 m/s. At $t = 2.0$ s, it has a kinetic energy of 25 J.
(**a**) What is the kinetic energy of the object at $t = 1.0$ s?
(**b**) What is the speed of the object at $t = 2.0$ s?
(**c**) How much work was done on the object between $t = 1.0$ s and $t = 2.0$ s?

Notice that the work done by the student in Guided Example 6.7 is partially offset by the work done by gravity. If the student adjusts his force to be equal to the weight of the box, the two amounts of work cancel completely. In this case the total work acting on the box is zero, and the work-energy theorem tells us that the kinetic energy of the box doesn't change—the speed of the box remains constant. In general, an object moving with constant velocity either has zero force acting on it or has more than one force acting on it but the work done by these forces adds up to zero. The net effect is the same in either case.

Objects with Initial Speed In the two previous examples the initial speed was zero. This is not always the case, of course. The following Active Example illustrates how the work-energy theorem applies to an object with an initial speed.

> ### _ACTIVE_ Example 6.8 Determine the Final Speed
>
> A boy does 19 J of work as he pulls a 6.4-kg sled through a distance of 2.0 m. No other work is done on the sled. If the initial speed of the sled is 0.50 m/s, what is its final speed?

Solution (*Perform the calculations indicated in each step.*)

1. Rearrange the work-energy theorem to solve for the final kinetic energy:

$$\tfrac{1}{2}mv_f^2 = W_{total} + \tfrac{1}{2}mv_i^2$$

2. Now, solve the equation for the final speed:

$$v_f = \sqrt{\frac{2W_{total}}{m} + v_i^2}$$

> **Math HELP**
> Solving Simple Equations
> **See Math Review, Section III**

3. Substitute the numerical values and evaluate v_f:

$$v_f = 2.5 \text{ m/s}$$

To accelerate a certain car from rest to the speed v requires the work W. Is the work required to accelerate the car from v to $2v$ equal to W, $2W$, $3W$, or $4W$?

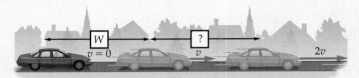

Reasoning and Discussion

A common mistake is to reason that since the speed is increased by the same amount in each case, the work required is the same. It is not. The reason is that work depends on the speed squared rather than on the speed itself.

Because of this dependence on speed squared, the work required to accelerate the car from rest to $2v$ is $4W$. Therefore, the work required to increase the speed from v to $2v$ is the difference between $4W$ and W; that is, $4W - W = 3W$.

Answer

The required work is $3W$.

Potential Energy

You must do work if you want to lift the bowling ball in **Figure 6.6** from the floor onto the shelf. Once on the shelf, the bowling ball has zero kinetic energy, just as it did on the floor. Even so, the work you did in lifting the ball has not been lost—it is stored as *potential energy*. If the ball later falls from the shelf, gravity will do the same amount of work on the way down as you did in lifting the ball up. As a result, the work you did is "recovered" in the form of kinetic energy.

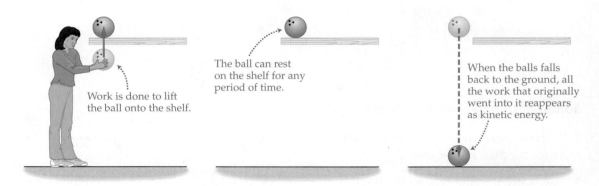

Work is done to lift the ball onto the shelf.

The ball can rest on the shelf for any period of time.

When the balls falls back to the ground, all the work that originally went into it reappears as kinetic energy.

▲ **Figure 6.6 Storing energy for later use**

Lifting a bowling ball onto a shelf requires a certain amount of work. The bowling ball can remain on the shelf for any length of time, but if the ball falls back to the floor, gravity does the same work on the ball that you did to lift it in the first place. Thus, the work done in lifting the ball is stored as potential energy, ready for conversion to kinetic energy when the ball falls.

Potential energy is stored energy

Energy that is stored for later use is referred to as **potential energy**, or *PE*. You can think of potential energy as a storage system for energy. When we increase the separation between the ball and the floor, the work we do is stored in the form of increased potential energy. Not only that, but the storage system is perfect, in the sense that the energy is never lost. The ball could rest on the shelf for a million years, but when it fell, it would gain the same amount of kinetic energy.

One type of potential energy is due to gravity

Potential energy has several forms. One of the most common is *gravitational potential energy*. It's easy to calculate the potential energy for gravity: All that's needed is to calculate the work required to lift an object. All of this work is stored as potential energy.

What determines how much potential energy an object has?

For example, imagine lifting an object of mass m from the ground to a height h. The force you must exert is $F = mg$ in the upward direction. The work you do is force (mg) times distance (h), or

$$W = mgh$$

Thus, the potential energy of an object is determined by the amount of work required to move it from one location to another. In the case of gravity the work is mgh, and this work is stored as gravitational potential energy, $PE_{gravity}$:

Definition of Gravitational Potential Energy, $PE_{gravity}$

$$\frac{\text{potential energy}}{\text{of gravity}} = \text{mass} \times \left(\begin{array}{c} \text{acceleration due} \\ \text{to gravity} \end{array} \right) \times \text{height}$$

$$PE_{gravity} = mgh$$

SI unit: $kg \cdot m^2/s^2 = J$

Notice that potential energy, like all energy, is measured in joules (J), the same unit used for work.

QUICK Example 6.10 What's the Potential Energy?

Find the gravitational potential energy of a 65-kg person sitting on a diving board that is 3.0 m high.

Solution
Substituting $m = 65$ kg and $h = 3.0$ m in $PE_{gravity} = mgh$ yields

$$PE_{gravity} = mgh$$
$$= (65 \text{ kg})(9.81 \text{ m/s}^2)(3.0 \text{ m})$$
$$= \boxed{1900 \text{ J}}$$

Practice Problem

21. As an Acapulco cliff diver drops to the water from a height of 46 m, his gravitational potential energy decreases by 25,000 J. What is the diver's weight in newtons?

The way you obtain energy to convert into kinetic or potential energy is by eating food. A single item of food can be converted into a surprisingly large amount of potential energy, as shown in the next Guided Example.

A candy bar called the Mountain Bar has a nutritional energy content of 8.87×10^5 J. If an 81.0-kg mountain climber eats a Mountain Bar and converts all of its energy to potential energy, what altitude gain can the climber achieve?

Picture the Problem

We show the mountain climber pausing to eat the candy bar at a given level on the mountain. The altitude gain corresponding to the energy of the bar is indicated as h.

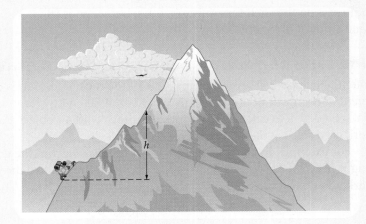

Strategy

To find the altitude gain we set $PE_{gravity} = mgh$ equal to the energy provided by the candy bar, 8.87×10^5 J, and solve for h.

Known	Unknown
$m = 81.0$ kg	$h = ?$
$PE_{gravity} = 8.87 \times 10^5$ J	

Solution

1 Rearrange $PE_{gravity} = mgh$ to solve for the height h:

$$h = \frac{PE_{gravity}}{mg}$$

2 Substitute the numerical values, with $PE_{gravity} = 8.87 \times 10^5$ J:

$$h = \frac{PE_{gravity}}{mg}$$
$$= \frac{8.87 \times 10^5 \text{ J}}{(81.0 \text{ kg})(9.81 \text{ m/s}^2)}$$
$$= \boxed{1120 \text{ m}}$$

Math HELP
Solving Simple Equations
See Math Review, Section III

Insight

This is more than two-thirds of a mile in altitude. Even if we take into account the fact that metabolic efficiency is only about 25%, the gain would still be 280 m—nearly two-tenths of a mile. It's remarkable just how much our bodies can do with so little.

Practice Problems

22. Follow-up If the mass of the mountain climber is increased—by adding more items to his backpack, for example—does the possible altitude gain increase, decrease, or stay the same? Calculate the altitude gain for a climber with a mass of 91.0 kg.

23. What is the gravitational potential energy of a 0.25-kg ball when it is 1.3 m above the floor?

24. The gravitational potential energy of a person on a 3.0-m-high diving board is 1800 J. What is the person's mass?

Elastic materials can also have potential energy

Objects like springs and rubber bands are referred to as *elastic* because they return to their original size and shape after being distorted. Suppose, for example, that you stretch or compress a spring. It takes work to do that. Where does the work go? Well, the work is stored in the spring as potential energy. You can "see" the stored energy by releasing the spring and watching it "pop" back to its original length. Potential energy is also stored in other elastic items, like rubber bands, bungee cords, paper clips, and so on.

We refer to the potential energy stored in a distorted elastic material as *elastic potential energy*. The specific case of a spring is particularly important in physics. When a spring is stretched or compressed by a distance x, the force exerted on the spring increases uniformly from 0 to kx, where k is the *spring constant*, in newtons per meter (N/m). The average force is $\frac{1}{2}kx$. It follows that the work done in changing the length of the spring is the average force times the distance of the stretch or compression:

$$W = \left(\tfrac{1}{2}kx\right)(x)$$
$$= \tfrac{1}{2}kx^2$$

This work is stored as *spring potential energy*.

Definition of Spring Potential Energy, PE_{spring}

$$\text{potential energy of a spring} = \frac{1}{2} \times \left(\text{spring constant}\right) \times \left(\text{distance of stretch or compression}\right)^2$$

$$PE_{spring} = \tfrac{1}{2}kx^2$$

SI unit: $\text{kg} \cdot \text{m}^2/\text{s}^2 = \text{J}$

QUICK Example 6.12 What's the Potential Energy?

A spring with a spring constant of 120 N/m is compressed a distance of 2.3 cm. How much potential energy is stored in the spring?

Solution

Substituting $k = 120$ N/m and $x = 2.3$ cm $= 0.023$ m in $PE_{spring} = \frac{1}{2}kx^2$ yields

$$PE_{spring} = \tfrac{1}{2}kx^2$$
$$= \tfrac{1}{2}(120 \text{ N/m})(0.023 \text{ m})^2$$
$$= \boxed{0.032 \text{ J}}$$

Practice Problems

25. How far must a spring with a spring constant of 85 N/m be stretched to store 0.22 J of potential energy?

26. A hockey stick stores 4.2 J of potential energy when it is bent 3.1 cm. Treating the hockey stick as a spring, what is its spring constant?

6.2 LessonCheck (MP)

Checking Concepts

27. ⚒ **Apply** How does the kinetic energy of an object change if its speed doubles? Triples?

28. ⚒ **Explain** If positive work is done on an object, does its kinetic energy increase, decrease, or stay the same? Explain.

29. [Triple Choice] A pitcher throws a baseball at 40 m/s (~90 mph), and the catcher stops it in her glove. Is the work done on the ball by the catcher positive, negative, or zero? Explain.

30. ⚒ **Analyze** Is the change in potential energy when a box is lifted 1 meter off the ground the same on Earth and the Moon? Explain.

31. Apply How does the potential energy of a spring change if its amount of stretch is doubled?

Solving Problems

32. Calculate A 9.50-g bullet has a speed of 1.30 km/s. What is the kinetic energy of the bullet?

33. Calculate A 0.27-kg volleyball has a kinetic energy of 7.8 J. What is the speed of the volleyball?

34. Calculate How much work is required for a 73-kg runner to accelerate from rest to a speed of 7.5 m/s?

35. Calculate A 0.14-kg pinecone is 16 m above the ground. What is its gravitational potential energy?

36. Calculate It takes 13 N to stretch a certain spring 9.5 cm. How much potential energy is stored in this spring? (*Hint:* Calculate the spring constant first, then the potential energy.)

6.3 Conservation of Energy

Vocabulary

- mechanical energy

Suppose you have a total of $500. Some of that money is in the bank, some is in your wallet as paper money, and some is in your pocket in the form of coins. You might withdraw a $10 bill from the bank and add it to your wallet. You might take the coins in your pocket and convert them to a $1 bill. The total amount of money you have remains the same, of course; you just have different amounts in the different locations. A physicist would say that the money was *conserved* as you changed it from form to another.

A similar situation occurs with energy, which also takes many different forms. A few examples of forms of energy are mechanical, electrical, thermal, and nuclear. One process might transform some kinetic energy to electrical potential energy; another might transform some spring potential energy to kinetic energy. In general, any time work is done, energy is transformed from one form to another. No matter what the process, though, the total amount of energy in the universe remains the same. This is what is meant by the conservation of energy.

Mechanical energy is conserved in some situations

Energy can never be created or destroyed—it can only be transformed from one form to another. For example, when a baseball player slides into second base, her initial kinetic energy is transformed to thermal energy (increased temperature) by friction. In situations where all forms of friction can be ignored, however, no potential or kinetic energy is transformed to thermal energy. When this is the case, we say that the system is ideal, and the sum of the kinetic and potential energies is always the same.

The sum of the potential and kinetic energies of an object is referred to as its **mechanical energy**, E. Thus,

$$\text{mechanical energy} = \text{potential energy} + \text{kinetic energy}$$
$$E = PE + KE$$

➤ In an ideal system (with no form of friction), energy is transformed from potential energy to kinetic energy and vice versa, but the sum of the two is constant. This means that the mechanical energy is conserved. Thus, mechanical energy is conserved when friction is absent, but total energy (including thermal energy) is always conserved, no matter what.

Applying Energy Conservation

Energy conservation simplifies many physics problems. To see how, consider a set of keys of mass m that is dropped to the floor from a height h, as shown in **Figure 6.7**. How fast are the keys moving just before they land?

Ignoring any form of friction—like air resistance—we know that $E = PE + KE$ is constant as the keys fall. Thus, the initial mechanical energy (at the drop point) must be equal to the final mechanical energy (just before the keys land).

$$E_i = E_f$$

Writing this in terms of potential and kinetic energies yields

$$PE_{\text{gravity,i}} + KE_i = PE_{\text{gravity,f}} + KE_f$$

This one equation—which is just bookkeeping—can be used to solve our problem. Notice that the potential energy in this example is the gravitational potential energy, $PE_{\text{gravity}} = mgh$.

Calculating Final Speed

Let's apply the basic physics of energy conservation to find the final speed of the set of keys. To be specific, we break the calculation into the following steps:

- The initial potential energy is determined by the release height, h. $PE_{\text{gravity,i}} = mgh$
- The initial kinetic energy is zero because the keys are dropped from rest. $KE_i = 0$
- The final potential energy is zero because the final point is at ground level. $PE_{\text{gravity,f}} = 0$
- The final kinetic energy is one-half the mass times the final velocity squared. $KE_f = \frac{1}{2}mv_f^2$

We now apply conservation of energy and solve for the final speed v_f:

$$PE_{\text{gravity,i}} + KE_i = PE_{\text{gravity,f}} + KE_f$$
$$mgh + 0 = 0 + \frac{1}{2}mv_f^2$$
$$v_f = \sqrt{2gh}$$

➤ When is mechanical energy conserved?

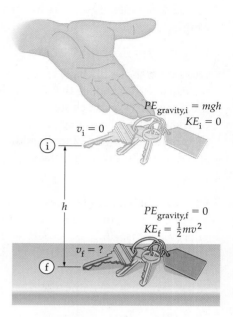

▲ Figure 6.7 **Solving a kinematics problem using conservation of energy** A set of keys falls to the floor. Ignoring frictional forces, we know that the mechanical energy at the initial (i) and final (f) points must be the same: $E_i = E_f$. Using this condition, we can find the speed of the keys just before they land.

PROBLEM-SOLVING NOTE

Multiple Solution Options

Some problems can be solved using several different methods. Notice, for example, that the result we obtained here for the final velocity, v_f, using energy conservation is the same as that obtained using kinematics in Chapter 3.

In the bottom of the ninth inning a player hits a 0.15-kg baseball over the outfield fence. The ball leaves the bat with a speed of 36 m/s, and a fan in the bleachers catches it 7.2 m above the point where it was hit. Assuming that frictional forces can be ignored, find (**a**) the kinetic energy of the ball just before it is caught and (**b**) its speed just before it is caught.

Picture the Problem

Our sketch shows the ball's trajectory. We label the hit point i and the catch point f. Notice that we are measuring heights relative to the hit point. (We can choose any point we like, but the hit point is convenient.) It follows that point i has the height $h = 0$ and point f has the height $h = 7.2$ m. In addition, we are given the initial speed, $v_i = 36$ m/s. The final speed, v_f, is to be determined.

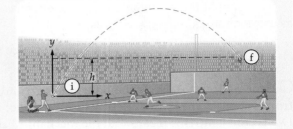

Strategy

(**a**) Because frictional forces can be ignored, it follows that the initial mechanical energy is equal to the final mechanical energy; that is, $PE_{gravity,i} + KE_i = PE_{gravity,f} + KE_f$. Use this relation to find KE_f. (**b**) Once KE_f is determined, use $KE_f = \frac{1}{2}mv_f^2$ to find v_f.

Known	Unknown
$m = 0.15$ kg	(**a**) $K_f = ?$
$v_i = 36$ m/s	(**b**) $v_f = ?$
$h = 7.2$ m	

Solution

Part (a)

1 Begin by writing *PE* and *KE* for point i:

$$PE_{gravity,i} = 0$$
$$KE_i = \frac{1}{2}mv_i^2$$
$$= \frac{1}{2}(0.15 \text{ kg})(36 \text{ m/s})^2$$
$$= 97 \text{ J}$$

2 Next, write *PE* and *KE* for point f:

$$PE_{gravity,f} = mgh$$
$$= (0.15 \text{ kg})(9.81 \text{ m/s}^2)(7.2 \text{ m})$$
$$= 11 \text{ J}$$
$$KE_f = \frac{1}{2}mv_f^2$$

3 Set the total mechanical energy at point i, $E_i = PE_{gravity,i} + KE_i$, equal to the total mechanical energy at point f, $E_f = PE_{gravity,f} + KE_f$, and solve for KE_f:

$$PE_{gravity,i} + KE_i = PE_{gravity,f} + KE_f$$
$$0 + 97 \text{ J} = 11 \text{ J} + KE_f$$
$$KE_f = 97 \text{ J} - 11 \text{ J}$$
$$= \boxed{86 \text{ J}}$$

Part (b)

4 Use $KE_f = \frac{1}{2}mv_f^2$ to find v_f:

$$KE_f = \frac{1}{2}mv_f^2$$
$$v_f = \sqrt{\frac{2KE_f}{m}} = \sqrt{\frac{2(86 \text{ J})}{0.15 \text{ kg}}} = \boxed{34 \text{ m/s}}$$

Insight

To find the ball's speed just before it is caught, all we need to know is the height of point f—we don't need to know any details about the ball's trajectory. For example, it is not necessary to know the angle at which the ball leaves the bat or its maximum height. The energy histograms to the right show the values of *PE* and *KE* at the points i and f. Notice that the energy of the system is still mostly kinetic just before the ball is caught.

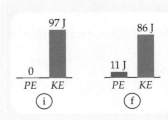

37. Follow-up If the mass of the ball were increased, would its speed just before it is caught be greater than, less than, or the same as the value we just found? Assume that everything else remains the same.

38. In a tennis match a player wins a point by hitting the 0.059-kg ball sharply to the ground on the opponent's side of the net. If the ball bounces upward from the ground with a speed of 16 m/s and is caught by a fan in the stands when it has a speed of 12 m/s, how high above the court is the fan? Ignore air resistance.

39. A crow drops a 0.11-kg clam onto a rocky beach from a height of 9.8 m. What is the kinetic energy of the clam when it is 5.0 m above the ground? What is its speed at that point?

Same Height, Different Paths   Conservation of mechanical energy allows you to analyze a variety of situations. For example, the following Conceptual Example explores how a child's final speed at the bottom of a water slide is related to her initial potential energy.

CONCEPTUAL Example 6.14 Comparing Final Speeds

Swimmers at a water park can enter a pool using one of two frictionless slides of equal height. Slide 1 approaches the water with a uniform slope. Slide 2 dips rapidly at first, then levels out. Is the swimmer's speed at the bottom of slide 2 greater than, less than, or equal to the swimmer's speed at the bottom of slide 1?

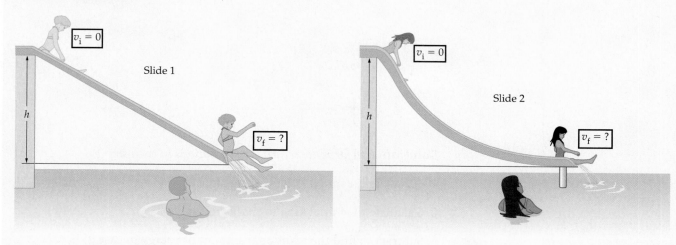

Reasoning and Discussion
The slides have equal heights, and hence the same amount of potential energy, *mgh*, is converted to kinetic energy in either case. Since the conversion of gravitational potential energy to kinetic energy is the *only* energy transaction taking place, it follows that the speed is the same for each slide.

Answer
The speeds are the same.

Interestingly, although the final speeds at the bottom of the slides are the same in Conceptual Example 6.14, the time required to get there is different. The reason is that swimmer 2 reaches a high speed early and maintains it. In contrast, the speed of swimmer 1 increases slowly and steadily. Thus, the average speed is greater on slide 2, and the time to reach the bottom is less.

Same Path, Different Speeds What happens if we change the initial speed of an object by a small amount? Active Example 6.15 explores the effect for a snowboarder.

ACTIVE Example 6.15 Determine the Final Speed

A 75-kg snowboarder coasts on a smooth track that rises from one level to another. If the snowboarder's initial speed is 4.0 m/s, as in case 1 below, the snowboarder just makes it to the upper level and comes to rest. With a slightly greater initial speed of 5.0 m/s, as in case 2, the snowboarder is still moving to the right on the upper level. What is the snowboarder's final speed in case 2?

Case 1
$v_f = 0$
$v_i = 4.0$ m/s
h

Case 2
$v_f = ?$
$v_i = 5.0$ m/s
h

Solution *(Perform the indicated calculations in each step.)*

1. Calculate the initial kinetic energy in case 1:   600 J

2. Calculate the initial kinetic energy in case 2:   940 J

3. Subtract to find the additional amount of energy in case 2: 340 J

4. Find the speed corresponding to a kinetic energy of 340 J: 3.0 m/s

You might be surprised to see that increasing the initial speed by just 1.0 m/s (from 4.0 m/s to 5.0 m/s) increases the final speed by 3.0 m/s (from zero to 3.0 m/s). This occurs because kinetic energy depends on the speed squared. Thus, even a small increase in speed causes a relatively large increase in kinetic energy. This is why even a modest increase in highway speed can have a significant impact on the severity of a car accident.

6.3 LessonCheck (MP)

Checking Concepts

40. ▭ Apply The potential energy of an object decreases by 10 J. What is the change in the object's kinetic energy, assuming there is no friction in the system?

41. Big Idea What is the necessary condition for the mechanical energy of a system to be conserved?

42. Analyze Discuss the various energy conversions that occur when a person performs a pole vault. Include as many conversions as you can. Be sure to consider times before, during, and after the vault itself.

43. Analyze You throw a ball straight up into the air. It reaches a maximum height and returns to your hand. At what location(s) is the kinetic energy of the ball (**a**) a maximum and (**b**) a minimum? At what location(s) is the potential energy of the ball (**c**) a maximum and (**d**) a minimum?

Solving Problems

44. Calculate At an amusement park, a swimmer uses a water slide to enter the main pool. If the swimmer starts at rest, slides without friction, and descends through a vertical height of 2.31 m, what is her speed at the bottom of the slide?

45. Calculate A 0.21-kg apple falls from a tree to the ground, 4.0 m below. Ignoring air resistance, determine the apple's kinetic energy, KE, the gravitational potential energy of the system, $PE_{gravity}$, and the total mechanical energy of the system, E, when the apple's height above the ground is 3.0 m.

46. Analyze You throw a baseball glove straight upward to celebrate a victory. Its initial kinetic energy is KE, and it reaches a maximum height h. What is the kinetic energy of the glove when it is at the height $h/2$?

6.4 Power

You often hear people talk about how "powerful" a certain car is. But just what is *power* anyway? How do you calculate the power of a car? In this lesson you'll learn how to answer these questions and more.

Power is a measure of how quickly work is done

Walking up a flight of stairs is pretty easy. Running up the same stairs can be exhausting. Why the difference? After all, the work is the same in the two cases. It all has to do with how quickly the work is done. ▭ **The faster work is done, the greater the power.**

To be precise, **power**, P, is the amount of work done in a given amount of time. For example, suppose the work W is performed in the time t. The power delivered during this time is defined as follows:

> **Definition of Power, P**
>
> $$\text{power} = \frac{\text{work}}{\text{time}}$$
>
> $$P = \frac{W}{t}$$
>
> SI unit: J/s = watt (W)

Vocabulary

• power

▭ *How is power related to the rate at which work is done?*

Walking up the stairs is easy.

Running up the stairs can leave you winded at the top.

Thus, for an engine to be "powerful," it must produce a lot of work (large W) in a small amount of time (small t). Similarly, as **Figure 6.8** shows, the faster you run up a flight of stairs, the less time it takes you to do the work. As a result, you produce a lot more power when running up the stairs than when walking up.

Power is measured with a unit called the watt

The dimensions of power are work (joules) per time (seconds). In fact, we define 1 joule per second to be a *watt* (W), after James Watt (1736–1819), the Scottish engineer and inventor who played a key role in the development of steam engines. Thus,

$$1 \text{ watt} = 1 \text{ W} = 1 \text{ J/s}$$

A typical compact fluorescent lightbulb (CFL) has a power of 23 W.

Another common unit of power is the horsepower (hp), which is used to rate the output of car engines. It is defined as follows:

$$1 \text{ horsepower} = 1 \text{ hp} = 746 \text{ W}$$

Though it sounds like a horse should be able to produce 1 horsepower, in fact, a horse can generate only about $\frac{2}{3}$ horsepower for a sustained period. The reason for the discrepancy is that when James Watt defined the horsepower—as a way to characterize the output of his steam engines—he purposely chose a unit that was overly generous to the horse. That way, potential investors wouldn't complain that he was overstating the capability of his engines.

Human power output is limited to about 1 hp

Let's get a better feeling for the magnitudes of the watt and the horsepower. Suppose, for example, that an 80-kg person walks up a flight of stairs in 22 s and that the altitude gain is 3.7 m. The work done by the person is

$$W = mgh = (80 \text{ kg})(9.81 \text{ m/s}^2)(3.7 \text{ m}) = 2900 \text{ J}$$

The corresponding power is simply the work divided by the time:

$$P = \frac{W}{t}$$
$$= \frac{2900 \text{ J}}{22 \text{ s}}$$
$$= 130 \text{ W}$$
$$= 0.17 \text{ hp}$$

Thus, a leisurely stroll up the stairs requires about 130 W (about $\frac{1}{6}$ hp). For comparison, the power produced by a sprinter bolting out of the starting blocks is about 746 W (1 hp), and the greatest power most people can produce for a sustained period of time is roughly 370 W ($\frac{1}{2}$ hp). Further examples of power are given in **Table 6.3**.

Table 6.3 Typical Values of Power

Source	Approximate Power (W)
Hoover Dam	1.34×10^9
Car moving at 40 mph	7×10^4
Home stove	1.2×10^4
Sunlight falling on 1 square meter	1380
Refrigerator	615
Television	200
Person walking up a flight of stairs	150
Human brain	20

Practice Problems

47. Concept Check Engine 1 does twice the work of engine 2. Is it correct to conclude that engine 1 produces twice as much power as engine 2? Explain.

48. A record for running up the stairs of the Empire State Building was set on February 3, 2003. The runner completed the 86 flights, with a total of 1576 steps, in 9 min 33 s. If the altitude gain for each step was 0.20 m and the mass of the runner was 70.0 kg, what was his average power output during the climb? Give your answer in both watts and horsepower.

Power determines a car's ability to accelerate

Power output is an important factor in the performance of a car. For example, suppose it takes a certain amount of work, W, to accelerate a car from rest to 26.8 m/s (60 mi/h). If the average power provided by the engine is P, then the amount of time required to reach 26.8 m/s is $t = W/P$. Clearly, the greater the power, the less the time required to accelerate. Thus, in a loose way of speaking, we can say that the power of a car is a measure of "how fast it can go fast."

To pass a slow-moving truck a driver has to accelerate his fancy 1.30×10^3-kg car from 13.4 m/s (30.0 mph) to 17.9 m/s (40.0 mph) in 3.00 s. What is the minimum power required to do this?

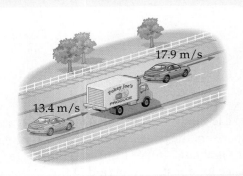

Picture the Problem

Our sketch shows the car accelerating from an initial speed of $v_i = 13.4$ m/s to a final speed of $v_f = 17.9$ m/s. We assume that the road is level, so no work is done against gravity, and that friction and air resistance may be ignored.

Strategy

Power is work divided by time, and work is equal to the change in kinetic energy as the car accelerates. We can determine the change in kinetic energy from the given mass of the car and its initial and final speeds. With this information we can determine the power using $P = W/t = \Delta KE/t$.

Known	Unknown
$m = 1.30 \times 10^3$ kg	$P = ?$
$v_i = 13.4$ m/s	
$v_f = 17.9$ m/s	
$t = 3.00$ s	

Solution

1 First, calculate the change in kinetic energy:

$$\Delta KE = \tfrac{1}{2}mv_f^2 - \tfrac{1}{2}mv_i^2$$
$$= \tfrac{1}{2}(1.30 \times 10^3 \text{ kg})(17.9 \text{ m/s})^2$$
$$- \tfrac{1}{2}(1.30 \times 10^3 \text{ kg})(13.4 \text{ m/s})^2$$
$$= 9.16 \times 10^4 \text{ J}$$

2 Next, use the power equation. First, substitute the change in *KE* for the work, and then, insert the numerical values for (Δ)*KE* and *t*:

$$P = \frac{W}{t}$$
$$= \frac{\Delta KE}{t}$$
$$= \frac{9.16 \times 10^4 \text{ J}}{3.00 \text{ s}}$$
$$= \boxed{3.05 \times 10^4 \text{ W} \quad (40.9 \text{ hp})}$$

> **Math HELP**
> Scientific Notation
> See Math Review, Section I

Insight

Suppose the fancy car continues to produce the same 3.05×10^4 W of power as it accelerates from $v = 17.9$ m/s (40.0 mph) to $v = 22.4$ m/s (50.0 mph). Is the time required for this acceleration more than, less than, or equal to 3.00 s? It will take more than 3.00 s. The reason is that ΔKE is greater for a change in speed from 40.0 mph to 50.0 mph than for a change in speed from 30.0 mph to 40.0 mph, because *KE* depends on speed squared. Since ΔKE is greater, the time, $t = \Delta KE/P$, is also greater.

Practice Problems

49. **Follow-up** Find the time required to accelerate from 17.9 m/s to 22.4 m/s with 3.05×10^4 W of power.

50. A pitcher accelerates a 0.14-kg hardball from rest to 42.5 m/s in 0.060 s.
(a) How much work does the pitcher do on the ball?
(b) What is the pitcher's power output during the pitch?

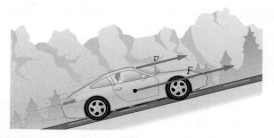

◀ **Figure 6.9 Driving uphill with constant speed**
To move a car uphill with a constant speed v requires a force F. The power output of the engine is force times speed, $P = Fv$.

Power depends on force and speed

Let's see how power is related to force and speed. Consider a car moving uphill with a constant speed v, as shown in **Figure 6.9**. To maintain the constant speed the engine exerts a constant force F equal to the combined effects of friction, gravity, and air resistance. As the car travels a distance d, the work done by the engine is $W = Fd$, and the power it delivers is

How is power related to force and speed?

$$P = \frac{W}{t} = \frac{Fd}{t}$$

Since the car has a constant speed, $v = d/t$, it follows that

$$P = \frac{Fd}{t} = F\left(\frac{d}{t}\right) = Fv$$

Power is equal to force times speed.

For example, suppose you push a heavy wagon with a given force. You produce twice as much power when you push at 2 m/s than when you push at 1 m/s, even though you are pushing no harder. It's just that the amount of work you do in a given time period is doubled. Similarly, if you push a wagon at a given speed, you produce twice as much power if your force is 40 N rather than 20 N. The connection between power, force, and speed is very straightforward.

ACTIVE **Example 6.17 Find the Maximum Speed**

It takes a force of 1280 N to keep a 1500-kg car moving with constant speed up a slope of 5.00°. If the engine delivers 50.0 hp to the drive wheels, what is the maximum speed of the car?

Solution *(Perform the calculations indicated in each step.)*

1. Convert the given power, 50.0 hp, to watts: $P = 3.73 \times 10^4$ W

2. Solve $P = Fv$ for the speed v: $v = P/F$

3. Substitute the numerical values for the power and the force: $v = 29.1$ m/s

Math HELP
Unit Conversion
See
Lesson 1.3

Insight
Thus, the maximum speed of the car on this slope is 29.1 m/s (about 65 mi/h). Notice that you must convert from horsepower to watts in the first step. As a general rule, horsepower must be converted to watts in order to solve problems involving meters and kilograms.

Practice Problem

51. A small motor runs a lift that raises a load of bricks weighing 836 N to a height of 10.7 m in 23.2 s. Assuming that the bricks are lifted with constant speed, what is the minimum power the motor must produce?

6.4 LessonCheck ⓂⓅ

Checking Concepts

52. 🖘 **Explain** If the rate at which work is done on an object is increased, does the power supplied to that object increase, decrease, or stay the same?

53. 🖘 **Assess** System 1 has a force of 10 N and a speed of 5 m/s. System 2 has a force of 20 N and a speed of 2 m/s. Which system has the greater power? Explain.

54. Analyze Engine 1 produces twice the power of engine 2. Is it correct to conclude that engine 1 does twice as much work as engine 2? Explain.

55. Analyze Football player 1 exerts twice the force of football player 2. Is it correct to conclude that football player 1 produces twice as much power as football player 2? Explain.

Solving Problems

56. Calculate You raise a bucket of water from the bottom of a well that is 12 m deep. The mass of the bucket and the water is 5.00 kg, and it takes 15 s to raise the bucket to the top of the well. How much power is required?

57. Calculate What is the power output of a 1.4-g fly as it walks straight up a windowpane at 2.3 cm/s?

58. Calculate A kayaker paddles with a power output of 50.0 W to maintain a steady speed of 1.50 m/s. Find the force exerted by the kayaker.

59. Calculate An ice cube is placed in a microwave oven. Suppose the oven delivers 105 W of power to the ice cube and it takes 32,200 J to melt it. How long does it take for the ice cube to melt?

Physics & You

Hybrid Vehicles

Why Are They Important?

Hybrid vehicles, with their increased fuel efficiency, address the economic problem of increasing fuel costs and the environmental need to decrease carbon dioxide emissions.

How Do They Work?

A typical hybrid vehicle has a secondary energy storage device in addition to its fuel tank. This secondary device, which is usually a battery, is used for temporary storage. Energy normally lost to friction and heat while braking is instead used to turn an electrical generator. The generated electrical energy is then stored as chemical potential energy in the battery. This process is known as *regenerative braking*.

A regenerative system recovers kinetic energy $\left(\frac{1}{2}mv^2\right)$ when the brakes are used to slow the car or to maintain a constant speed when descending a hill. The stored energy is available for use when the vehicle is accelerated or driven uphill. At these times the battery powers an electric motor that converts the stored chemical energy back to mechanical energy.

The power from the electric motor adds to the power produced by the internal combustion engine and thus works to decrease fuel consumption. Fuel economy is further increased by an electromechanical transmission that keeps the engine speed within a range of maximum efficiency.

Flywheel Hybrids

An alternative to storing chemical energy in a battery is storing kinetic energy in a flywheel. A flywheel is simply a spinning disk that stores energy (you will learn about the kinetic energy of rotation in Chapter 8). If the flywheel is built with an internal motor-generator, it becomes an electromechanical battery that can be used in place of a standard chemical battery. Currently, pound for pound, the most efficient flywheels cannot store as much energy as the best batteries. Flywheels, however, can be charged and discharged very quickly, and their lifetimes far exceed those of chemical batteries. A flywheel hybrid power system is well suited for a bus that makes frequent stops and starts along a relatively flat route.

Take It Further

1. Assess *Would the adoption of purely electric (completely battery-powered) vehicles eliminate automobile-related carbon dioxide emissions?*

2. Infer *Gasoline-powered cars are often most efficient during highway driving. Hybrid vehicles are often most efficient in city driving. Explain why you think this is the case.*

3. Debate *Some people argue that there are ethical reasons for buying a hybrid vehicle instead of a gasoline-only vehicle. Do you agree? Use logic, skepticism, and your own values to write a short essay about this argument.*

The city of Nashville, Tennessee, runs clean diesel hybrid buses on short downtown loops.

Physics Lab Investigating Work on Inclined Planes

This lab explores the work done when an inclined plane is used to raise an object against the force of gravity. You will collect data while using ramps of varying steepness to raise a weight to a specified height. Analysis of the collected force and distance data will illustrate the value of using an inclined plane.

Materials
- meterstick
- spring scale
- masking tape
- smooth board
- stack of books
- block of wood with hook

Procedure

1. Trial 1: Suspend the block from the spring scale. Slowly lift the block at a constant speed in the vertical direction, from the floor, to a height of 0.50 m. Observe the pulling force required to lift the block and record this value for Trial 1 in the data table.

2. Trial 2: Place a stack of books against a wall and use the board to make an incline that is at least 1.0 m high. Use the meterstick to measure straight up from the floor to where the height

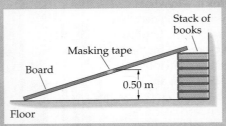

of the ramp is 0.50 m. Place a small piece of masking tape at this point on the side of the board—then draw a line on it to indicate the exact point where the height is 0.50 m.

3. Attach the spring scale to the block and place the block at the bottom of the ramp. Pull the block up the incline at a constant speed. Note the reading on the spring scale and measure the distance you had to pull the block *up the incline* to reach the 0.50-m height. Record your data for Trial 2 in the data table.

4. Trials 3 through 6: Repeat Steps 2 and 3 using a smaller stack of books so that the angle of the incline changes. Make sure that the block can be pulled up to a height of 0.50 m. Continue this process until you have five trials using different inclines.

Data Table

Trial	Pulling Force (N)	Pulling Distance (m)	Work Done (J)
1		0.50	
2			
3			
4			
5			
6			

Analysis

1. Complete the data table by calculating the amount of work done for each trial.

2. Use the data from Trials 2–6 to plot pulling force versus pulling distance.

Conclusions

1. From your graph, what type of relationship exists between the force required to pull the block up the ramp and the distance the block is pulled?

2. Ideally, the work done should be the same for each trial. What factors prevented you from obtaining ideal results? List as many factors as you can and explain how each would affect your calculations for the amount of work done.

3. Imagine you pulled the block up to a height of 0.50 m along a 1° incline. How much work would you do in the process? How would the force required in this case compare to the values you obtained in this lab?

4. Where do you experience inclines in your everyday life? What is their purpose? Why are they helpful?

6 Study Guide

Big Idea

Energy can change from one form to another, but the total amount of energy in the universe stays the same.

We express this fact by saying that energy is conserved. In systems with no friction the mechanical energy is conserved. If friction is present, some mechanical energy will be converted to thermal energy.

6.1 Work

🔑 In the simplest case work is done when a force is applied to an object and the object moves in the direction of the applied force.

🔑 Only the *component* of the force in the *direction* of the displacement does work.

• If the displacement is zero, then so is the work.

• If more than one force does work, the total work is the sum of the amounts of work done by the forces separately.

• The SI unit of work and energy is the joule (J), where $1\,J = 1\,N \cdot m$.

Key Equations

In the simplest case work is force times distance:

$$W = Fd$$

When the force and displacement are at an angle θ, the work done is

$$W = Fd \cos \theta$$

6.2 Work and Energy

🔑 Kinetic energy increases linearly with mass and with the square of the velocity.

🔑 The total work done on an object equals the change in its kinetic energy.

🔑 The potential energy of an object is determined by the amount of work required to move it from one location to another.

• Kinetic energy is the energy due to motion.

• Energy stored for later use is potential energy.

Key Equations

Kinetic energy, KE, is one-half mass times speed squared:

$$KE = \tfrac{1}{2}mv^2$$

Total work, W, is equal to the change in kinetic energy:

$$W = \Delta KE = \tfrac{1}{2}mv_f^2 - \tfrac{1}{2}mv_i^2$$

The gravitational potential energy of a mass m at a height h is

$$PE_{\text{gravity}} = mgh$$

The potential energy of a spring stretched or compressed by a distance x is

$$PE_{\text{spring}} = \tfrac{1}{2}kx^2$$

6.3 Conservation of Energy

🔑 In an ideal system (with no form of friction), energy is transformed from potential energy to kinetic energy and vice versa, but the sum of the two is constant.

• The sum of the potential and kinetic energies of an object is its mechanical energy, E.

Key Equation

Mechanical energy, E, is the sum of the kinetic and potential energies:

$$E = PE + KE$$

6.4 Power

🔑 The faster work is done, the greater the power.

🔑 Power is equal to force times speed.

• The SI unit of power is the watt (W), where $1\,W = 1\,J/s$.

• Horsepower (hp) is defined as follows: $1\,hp = 746\,W$.

Key Equations

Power is work divided by the time required to do the work:

$$P = \frac{W}{t}$$

Equivalently, power is force times speed:

$$P = Fv$$

ANSWERS TO SELECTED ODD-NUMBERED PROBLEMS APPEAR IN APPENDIX A.

Lesson by Lesson

6.1 Work

Conceptual Questions

60. A friend makes this statement: "Only the component of force perpendicular to the direction of motion does work." Is this statement true or false? Explain.

61. In the equation $W = Fd \cos \theta$, what is the angle θ?

62. The general equation for work is $W = Fd \cos \theta$. For what angle is the work $W = Fd$? For what angle is the work $W = -Fd$?

63. Your favorite uncle makes this statement: "A force that is always perpendicular to the velocity of a particle does no work on the particle." Is this statement true or false? If it is true, state why. If it is false, give a counterexample.

64. ‖Triple Choice‖ The International Space Station orbits the Earth in an approximately circular orbit at a height of $h = 375$ km above the Earth's surface. In one complete orbit, is the work done by Earth's gravity on the space station positive, negative, or zero? Explain.

65. ‖Triple Choice‖ A pendulum bob swings from point A to point B along the circular arc indicated in **Figure 6.10**. (a) Is the work done on the bob by gravity positive, negative, or zero? Explain. (b) Is the work done on the bob by the string positive, negative, or zero? Explain.

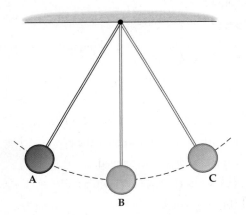

Figure 6.10

66. ‖Triple Choice‖ A pendulum bob swings from point B to point C along the circular arc indicated in Figure 6.10. (a) Is the work done on the bob by gravity positive, negative, or zero? Explain. (b) Is the work done on the bob by the string positive, negative, or zero? Explain.

Problem Solving

67. A dog lifts a 0.75-kg bone straight up through a distance of 0.11 m. How much work was done by the dog?

68. A weightlifter does 9.8 J of work while lifting a weight straight upward through a distance of 0.12 m. What was the force exerted by the weightlifter?

69. You do 13 J of work as you lift a 35-N pail of water. Through what height did you lift the pail?

70. You push a book 0.45 m across a desk with a 5.2-N force that is at an angle of 21° below the horizontal. How much work did you do on the book?

71. You pick up a 3.4-kg can of paint from the ground and lift it to a height of 1.8 m. (a) How much work do you do on the can of paint? (b) You hold the can stationary for half a minute, waiting for a friend on a ladder to take it. How much work do you do during this time? (c) Your friend decides not to use the paint, so you lower it back to the ground. How much work do you do on the can as you lower it?

72. ‖Think & Calculate‖ A tow rope, parallel to the water, pulls a water skier directly behind a boat with constant velocity for a distance of 65 m before the skier falls. The tension in the rope is 120 N. (a) Is the work done on the skier by the rope positive, negative, or zero? Explain. (b) Calculate the work done by the rope on the skier.

73. ‖Think & Calculate‖ In Problem 72, (a) is the work done on the boat by the rope positive, negative, or zero? Explain. (b) Calculate the work done by the rope on the boat.

74. A child pulls a friend in a little red wagon. If the child pulls with a force of 16 N for 12 m and the handle of the wagon is inclined at an angle of 25° above the horizontal, how much work does the child do on the wagon?

75. A 51-kg packing crate is pulled across a rough floor with a rope that is at an angle of 43° above the horizontal. If the tension in the rope is 120 N, how much work is done on the crate to move it 18 m?

76. Water skiers often ride to one side of the center line of a boat, as shown in **Figure 6.11**. In this case the boat is traveling at 15 m/s, and the tension in the rope is 75 N. If the boat does 2500 J of work on the skier in 42 m, what is the angle θ between the tow rope and the center line of the boat?

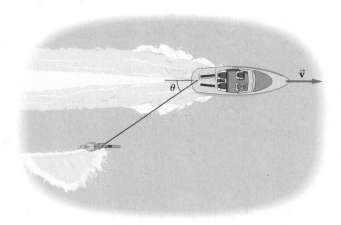

Figure 6.11

6.2 Work and Energy
Conceptual Questions

77. A package rests on the floor of an elevator that is rising with constant speed. The elevator exerts an upward force on the package and thus does positive work on it. Why doesn't the kinetic energy of the package increase?

78. An object moves with constant velocity. Is it safe to conclude that no force acts on the object? Why, or why not?

79. Rank Four joggers have the following masses and speeds:

Jogger	Mass	Speed
A	m	v
B	$m/2$	$3v$
C	$3m$	$v/2$
D	$4m$	$v/2$

Rank the joggers in order of increasing kinetic energy. Indicate ties where appropriate.

80. Is it possible for the kinetic energy of an object to be negative? Explain.

81. Predict & Explain The work required to accelerate a car from 0 to 50 km/h is W. (a) Is the work required to accelerate the car from 50 km/h to 150 km/h equal to $2W$, $3W$, $8W$, or $9W$? (b) Choose the *best* explanation from among the following:

A. The work to accelerate the car depends on the speed squared.

B. The final speed is three times the speed that was produced by the work W.

C. The increase in speed from 50 km/h to 150 km/h is twice the increase in speed from 0 to 50 km/h.

82. Predict & Explain Ball 1 is dropped to the ground from rest. Ball 2 is thrown to the ground with an initial downward speed. Assuming that the balls have the same mass and are released from the same height, is the change in gravitational potential energy of ball 1 greater than, less than, or equal to the change in gravitational potential energy of ball 2? (b) Choose the *best* explanation from among the following:

A. Ball 2 has the greater total energy, and therefore more of its energy can go into gravitational potential energy.

B. The gravitational potential energy depends only on the mass of the ball and its initial height above the ground.

C. All of the initial energy of ball 1 is gravitational potential energy.

83. Triple Choice Referring to Figure 6.10, is the gravitational potential energy of the bob at point C greater than, less than, or equal to the gravitational potential energy at (a) point A and (b) point B? Explain.

84. Triple Choice The potential energy of a stretched spring is positive. Is the potential energy of a compressed spring positive, negative, or zero? Explain.

Problem Solving

85. *Skylab's* **Reentry** When NASA's *Skylab* reentered the Earth's atmosphere on July 11, 1979, it broke into a myriad of pieces. One of the largest fragments was a 1770-kg lead-lined film vault, and it landed with an estimated speed of 120 m/s. What was the kinetic energy of the film vault when it landed?

86. A 7.3-kg bowling ball is placed on a shelf 1.7 m above the floor. What is its gravitational potential energy?

87. A 0.15-kg baseball has a kinetic energy of 18 J. What is its speed?

88. The gravitational potential energy of a 0.12-kg bird in a tree is 6.6 J. What is the height of the bird above the ground?

89. A spring with a spring constant of 92 N/m is compressed by 2.8 cm. How much potential energy is stored in the spring?

90. A force of 27 N stretches a given spring by 4.4 cm. How much potential energy is stored in the spring when it is compressed 3.5 cm?

91. A spring that is stretched 2.6 cm stores a potential energy of 0.053 J. What is the spring constant of this spring?

92. Think & Calculate A 1100-kg car is coasting on a horizontal road with a speed of 19 m/s. After passing over an unpaved, sandy stretch 32 m long, the car's speed has decreased to 12 m/s. (a) Was the net work done on the car positive, negative, or zero? Explain. (b) Find the magnitude of the average net force on the car in the sandy section of the road.

93. A 65-kg bicyclist rides his 8.8-kg bicycle with a speed of 14 m/s. (a) How much work must be done by the brakes to bring the bike and rider to a stop? (b) What is the magnitude of the braking force if the bicycle comes to rest in 3.5 m?

94. After hitting a long fly ball that goes over the right fielder's head and lands in the outfield, a batter decides to keep going past second base and try for third base. The 62-kg player begins sliding 3.4 m from the base with a speed of 4.5 m/s. **(a)** If the player comes to rest at third base, how much work was done on the player by friction with the ground? **(b)** What was the coefficient of kinetic friction between the player and the ground?

95. An object has a speed of 3.5 m/s and a kinetic energy of 14 J at $t = 0$. At $t = 5.0$ s the object has a speed of 4.7 m/s. **(a)** What is the mass of the object? **(b)** What is the kinetic energy of the object at $t = 5.0$ s? **(c)** How much work was done on the object between $t = 0$ and $t = 5.0$ s?

96. A 0.33-kg pendulum bob is attached to a string 1.2 m long. What is the change in the gravitational potential energy of the system as the bob swings from point A to point B in **Figure 6.12**?

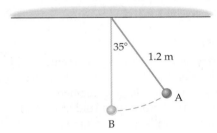

Figure 6.12

6.3 Conservation of Energy

Conceptual Questions

97. An object moves with no friction or air resistance. Initially, its kinetic energy is 10 J, and its gravitational potential energy is 20 J. What is its kinetic energy when its potential energy has decreased to 15 J? What is its potential energy when its kinetic energy has decreased to 5 J?

98. An object moves with no friction or air resistance. Initially, its kinetic energy is 10 J, and its gravitational potential energy is 30 J. What is the greatest potential energy possible for this object? What is the greatest kinetic energy possible for this object?

99. Predict & Explain You throw a ball upward and let it fall to the ground. Your friend drops an identical ball straight down to the ground from the same height. **(a)** Is the change in kinetic energy (from just after the ball is released until just before it hits the ground) of your ball greater than, less than, or equal to the change in kinetic energy of your friend's ball? **(b)** Choose the *best* explanation from among the following:

A. Your friend's ball converts all of its initial energy into kinetic energy.

B. Your ball is in the air longer, which results in a greater change in kinetic energy.

C. The change in gravitational potential energy is the same for each ball, which means that the change in kinetic energy must also be the same.

100. Three balls are thrown upward with the same initial speed, v_i, but at different angles relative to the horizontal, as shown in **Figure 6.13**. Ignoring air resistance, indicate which of the following statements (A, B, C, or D) is correct when the balls are at the dashed level?

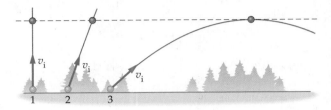

Figure 6.13

A. Ball 3 has the lowest speed.

B. Ball 1 has the lowest speed.

C. All three balls have the same speed.

D. The speed of the balls depends on their mass.

Problem Solving

101. A player passes a 0.600-kg basketball down court for a fast break. The ball leaves the player's hands with a speed of 8.30 m/s and slows down to 7.10 m/s at its highest point. Ignoring air resistance, how high above the release point is the ball when it is at its maximum height?

102. A 5.76-kg rock is dropped and allowed to fall freely. Find the initial kinetic energy, the final kinetic energy, and the change in kinetic energy for **(a)** the first 2.00 m of fall and **(b)** the second 2.00 m of fall.

103. A 0.26-kg rock is thrown vertically upward from the top of a cliff that is 32 m high. When it hits the ground at the base of the cliff, the rock has a speed of 29 m/s. Assuming that air resistance can be ignored, find **(a)** the initial speed of the rock and **(b)** the greatest height of the rock as measured from the base of the cliff.

104. A block with a mass of 3.7 kg slides with a speed of 2.2 m/s on a frictionless surface. The block runs into a stationary spring and compresses it a certain distance before coming to rest. What is the compression distance, given that the spring has a spring constant of 3200 N/m?

105. A 1.3-kg block is pushed up against a stationary spring, compressing it a distance of 4.2 cm. When the block is released, the spring pushes it away across a frictionless, horizontal surface. What is the speed of the block, given that the spring constant of the spring is 1400 N/m?

106. Suppose the pendulum bob in Figure 6.12 has a mass of 0.33 kg and is moving to the right at point B with a speed of 2.4 m/s. Air resistance is negligible. **(a)** What is the change in the system's gravitational potential energy when the bob reaches point A? **(b)** What is the speed of the bob at point A?

107. (a) In Problem 106, what is the bob's kinetic energy at point B? **(b)** At some point the bob will come to rest momentarily. Without doing an additional calculation, determine the change in the system's gravitational potential energy between point B and the point where the bob comes to rest. **(c)** Find the maximum angle the string makes with the vertical as the bob swings back and forth. Ignore air resistance.

108. The two masses in the device shown in **Figure 6.14** are initially at rest at the same height. After they are released, the large mass, m_2, falls through a height h and hits the floor, and the small mass, m_1, rises through a height h. **(a)** Find the speed of the masses just before m_2 lands, giving your answer in terms of m_1, m_2, g, and h. Assume that the ropes and pulley have negligible mass and that friction can be ignored. **(b)** Evaluate your answer to part (a) for the case where $h = 1.2$ m, $m_1 = 3.7$ kg, and $m_2 = 4.1$ kg.

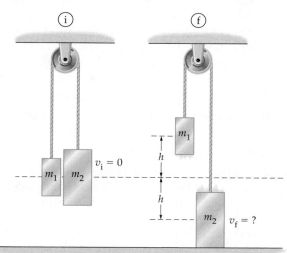

Figure 6.14

6.4 Power

Conceptual Questions

109. Engine 1 produces twice the power of engine 2. If it takes engine 1 the time T to do the work W, how long does it take engine 2 to do the same work? Explain.

110. Engine 1 produces twice the power of engine 2. If it takes engine 1 the time T to do the work W, how long does it take engine 2 to do the work $3W$? Explain.

111. Rank Four forces do the following amounts of work in the indicated times:

Force	Work	Time
A	5 J	10 s
B	3 J	5 s
C	6 J	18 s
D	25 J	125 s

Rank these forces in order of increasing power produced. Indicate ties where appropriate.

112. Rank Four forces do the following amounts of work and produce the indicated powers:

Force	Work	Power
A	40 J	80 W
B	35 J	5 W
C	75 J	25 W
D	60 J	30 W

Rank these forces in order of increasing time required to do the work. Indicate ties where appropriate.

Problem Solving

113. How many joules of energy are in a kilowatt-hour?

114. The Power You Produce Estimate the power you produce as you walk leisurely up a flight of stairs. Give your answer in both watts and horsepower (1 hp = 746 W).

115. As you lift an 88-N box straight upward, you produce a power of 72 W. What is the speed of the box?

116. In order to keep a leaking ship from sinking, it is necessary to pump 12 kg of water each second from below deck 2.1 m upward and over the side. What is the minimum horsepower motor that can be used to save the ship (1 hp = 746 W)?

117. Human-Powered Flight Human-powered aircraft require a pilot to pedal, as on a bicycle, and to produce a sustained power output of about 0.30 hp (1 hp = 746 W). The *Gossamer Albatross* flew across the English Channel on June 12, 1979, in 2 h 49 min. **(a)** How much energy did the pilot expend during the flight? **(b)** How many candy bars (280 Cal per bar) would the pilot have to consume to be "fueled up" for the flight? Note that a nutritional calorie (1 Cal) is equivalent to 1000 calories (1000 cal) as defined in physics. In addition, the conversion factor between calories and joules is as follows: 1 Cal = 1000 cal = 1 kcal = 4186 J.

118. Think & Calculate A grandfather clock is powered by the descent of a 4.35-kg weight. **(a)** If the weight descends through a distance of 0.760 m in 3.25 days, how much power does it deliver to the clock? **(b)** To increase the power delivered to the clock, should the time it takes for the mass to descend be increased or decreased? Explain.

Mixed Review

119. To get out of bed in the morning, do you have to do work? Explain.

120. A leaf falls to the ground with constant speed. Is $PE_i + KE_i$ for this system greater than, less than, or equal to $PE_f + KE_f$? Explain.

121. A ball is dropped from rest. Which of the three graphs (A, B, or C) in **Figure 6.15** corresponds to **(a)** the potential energy, **(b)** the kinetic energy, and **(c)** the total mechanical energy for the ball as it falls to the ground? Assume that the system is ideal, with no form of friction.

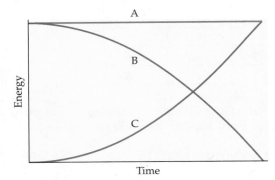

Figure 6.15

122. A spring has a spring constant of 310 N/m. Plot the potential energy for this spring when it is stretched by 1.0 cm, 2.0 cm, 3.0 cm, and 4.0 cm. Draw a curve that goes through your plotted points.

123. A small motor runs a lift that raises a load of bricks weighing 836 N to a height of 10.7 m in 23.2 s. Assuming that the bricks are lifted with constant speed, what is the minimum power the motor must produce?

124. Brain Power The human brain uses about 22 W of power under normal conditions (though more power may be required during exams!). How long can one Snickers bar (see Problem 117) power the normally functioning brain?

125. The Atmos Clock The Atmos clock (the so-called perpetual motion clock) gets its name from the fact that it runs off pressure variations in the atmosphere, which drive a bellows containing a mixture of gas and liquid ethyl chloride. Because the power to drive these clocks is so limited, they have to be very efficient. In fact, a single 60.0-W lightbulb could power 240 million Atmos clocks simultaneously. Find the amount of energy, in joules, required to run an Atmos clock for 1 day.

126. You push a 67-kg box across a floor, where the coefficient of kinetic friction is $\mu_k = 0.55$. The force you exert is horizontal. How much power is needed to push the box at a speed of 0.50 m/s?

127. Think & Calculate To clean a floor, a janitor pushes on a mop handle with a force of 43 N. **(a)** If the mop handle is at an angle of 55° above the horizontal, how much work is required to push the mop 0.50 m? **(b)** If the angle the mop handle makes with the horizontal is increased to 65°, does the work done by the janitor increase, decrease, or stay the same? Explain.

128. A small airplane tows a glider at constant speed and altitude. If the plane does 2.00×10^5 J of work to tow the glider 145 m and the tension in the tow rope is 2560 N, what is the angle between the tow rope and the horizontal?

129. Cookie Power To make a batch of cookies, you mix half a bag of chocolate chips into a bowl of cookie dough, exerting a 21-N force on the stirring spoon. Assume that your force is always in the direction of motion of the spoon. **(a)** What power is needed to move the spoon at a speed of 0.23 m/s? **(b)** How much work do you do if you stir the mixture for 1.5 min?

130. A particle moves without friction. At point A the particle has a kinetic energy of 12 J; at point B the particle is momentarily at rest, and the potential energy of the system is 25 J; at point C the potential energy of the system is 5 J. **(a)** What is the potential energy of the system when the particle is at point A? **(b)** What is the kinetic energy of the particle at point C?

131. Predict & Explain When a ball of mass m is dropped from rest from a height h, its kinetic energy just before landing is KE. Now, suppose a second ball of mass $4m$ is dropped from rest from a height $h/4$. **(a)** Just before ball 2 lands, is its kinetic energy $4KE$, $2KE$, KE, $KE/2$, or $KE/4$? **(b)** Choose the *best* explanation from among the following:

A. The two balls have the same initial energy.

B. The more massive ball will have the greater kinetic energy.

C. The lower drop height results in a reduced kinetic energy.

132. Meteorite On October 9, 1992, a 12-kg meteorite struck a car in Peekskill, New York, creating a dent about 22 cm deep, as shown in **Figure 6.16**. If the initial speed of the meteorite was 550 m/s, what was the average force exerted on the meteorite by the car?

Figure 6.16 An interplanetary fender-bender.

133. (a) At what rate must you lift a 3.6-kg container of milk (1 gallon) if the power output of your arm is to be 22 W? **(b)** How long does it take to lift the milk container through a distance of 1.0 m at this rate?

134. Catapult Launcher A catapult launcher on an aircraft carrier accelerates a jet from rest to 72 m/s. The work done by the catapult during the launch is 7.6×10^7 J. **(a)** What is the mass of the jet? **(b)** If the jet is in contact with the catapult for 2.0 s, what is the power output of the catapult?

135. The water skier in Figure 6.11 is at an angle of 35° with respect to the center line of the boat and is being pulled at a constant speed of 14 m/s. **(a)** If the tension in the tow rope is 90.0 N, how much work does the rope do on the skier in 10.0 s? **(b)** How much work does the resistive force of water do on the skier in the same time?

136. Calculate the power output of a 1.8-g spider as it walks up a windowpane at 2.3 cm/s. The spider walks on a path that is at 25° to the vertical, as illustrated in **Figure 6.17**.

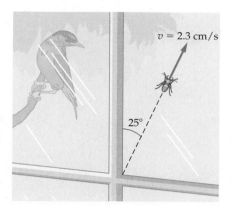

$v = 2.3$ cm/s

25°

Figure 6.17

137. ⟨Think & Calculate⟩ A pitcher accelerates a 0.14-kg hardball from rest to 25.5 m/s in 0.075 s. **(a)** How much work does the pitcher do on the ball? **(b)** What is the pitcher's power output during the pitch? **(c)** Suppose the ball reaches 25.5 m/s in less than 0.075 s. Is the power produced by the pitcher in this case more than, less than, or the same as the power found in part (b)? Explain.

138. The Beating Heart The average power output of the human heart is 1.33 W. **(a)** How much energy does the heart produce in a day? **(b)** Compare the energy found in part (a) with the energy required to walk up a flight of stairs. Estimate the height a person could attain on a set of stairs using nothing more than the daily energy produced by the heart.

139. ⟨Think & Calculate⟩ A sled slides without friction down a small, ice-covered hill. If the sled starts from rest at the top of the hill, its speed at the bottom is 7.50 m/s. **(a)** On a second run, the sled starts with a speed of 1.50 m/s at the top. When it reaches the bottom of the hill, is its speed 9.00 m/s, more than 9.00 m/s, or less than 9.00 m/s? Explain. **(b)** Find the speed of the sled at the bottom of the hill after the second run.

140. An 1865-kg airplane starts at rest on an airport runway at sea level. What is the change in mechanical energy of the airplane if it climbs to a cruising altitude of 2420 m and maintains a constant speed of 96.5 m/s?

141. The water slide shown in **Figure 6.18** ends at a height of 1.50 m above the pool. If the person starts from rest at point A and lands in the water at point B, what is the height h of the water slide? (Assume that the water slide is frictionless.)

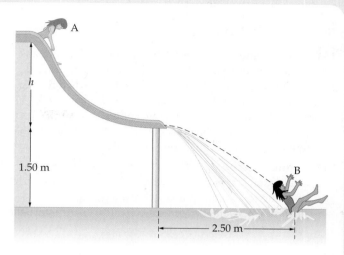

A

h

1.50 m

B

2.50 m

Figure 6.18

142. A skateboarder starts at point A in **Figure 6.19** and rises to a height of 2.64 m above the top of the ramp at point B. What was the skateboarder's initial speed at point A?

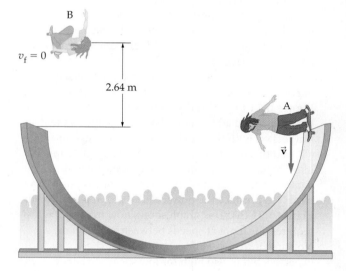

B

$v_f = 0$

2.64 m

A

$\vec{v}$

Figure 6.19

143. A 1.9-kg block slides down a frictionless ramp, as shown in **Figure 6.20**. The top of the ramp is 1.5 m above the ground; the bottom of the ramp is 0.25 m above the ground. The block leaves the ramp moving horizontally, and lands a horizontal distance d away. Find the distance d.

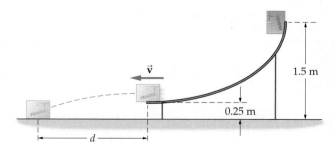

$\vec{v}$

1.5 m

0.25 m

d

Figure 6.20

Writing about Science

144. Research and write a report on the power that humans produce in everyday life. Include the power produced by the brain when thinking and the heart when resting. Also select several strenuous activities such as track-and-field events, bicycle racing, and swimming. In each case, explain how the power is determined. Make a table to compare the various power outputs.

145. Connect to the Big Idea The mechanical energy of a falling ball stays the same, even though it is constantly speeding up. Similarly, if you throw a ball upward, it slows down, even though its mechanical energy stays the same. Explain.

Read, Reason, and Respond

***Microraptor gui*: The Biplane Dinosaur** The evolution of flight is a subject of intense interest in paleontology. Some subscribe to the *cursorial (or ground-up) hypothesis*, which says that flight began with ground-dwelling animals running and jumping after prey. Others favor the *arboreal (or tree-down) hypothesis*, which says that tree-dwelling animals, like modern-day flying squirrels, developed flight as an extension of gliding from tree to tree.

A fossil of a small dinosaur from the Cretaceous period was recently discovered in China (see **Figure 6.21 (a)**). This discovery supports the arboreal hypothesis and adds a new element—it appears that feathers on both the wings and the lower legs and feet allowed this dinosaur, *Microraptor gui*, to glide much like a biplane. Researchers have created a detailed computer simulation of the flight of *Microraptor gui* and used it to produce the power-speed graph presented in **Figure 6.21 (b)**. This graph shows how much power is required for flight at speeds between 0 and 30 m/s. Notice that the power increases at high speeds, as expected, but is also high for low speeds, where the dinosaur is almost hovering. A minimum of 8.1 W is needed for flight at 10 m/s. The lower horizontal line shows the estimated 9.8-W power output of the dinosaur, indicating the small range of speeds for which flight would be possible. The upper horizontal line shows the wider range of flight speeds that would be available if the dinosaur were able to produce 20 W of power.

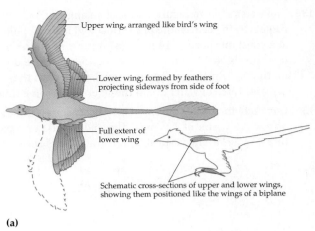

(a)

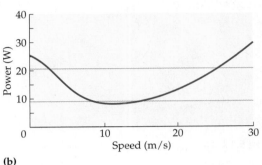

(b)

Figure 6.21 (a) Reconstruction showing what *Microraptor gui* may have looked like in life (top view on left and side view on right). The wings, plus the lower legs and feet, are thought to have acted like a pair of airfoils, much like the airfoils in a biplane. (b) A plot of the power output required for the dinosaur to fly at various speeds.

146. Estimate the range of flight speeds for *Microraptor gui* if its power output is 9.8 W.

A. 0−7.7 m/s   C. 15−30 m/s

B. 7.7−15 m/s   D. 0−15 m/s

147. What approximate range of flight speeds would be possible if *Microraptor gui* could produce 20 W of power?

A. 0−25 m/s   C. 2.5−25 m/s

B. 25−30 m/s   D. 0−2.5 m/s

148. How much energy would *Microraptor gui* have to expend to fly with a speed of 10 m/s for 1.0 min?

A. 8.1 J   C. 490 J

B. 81 J   D. 600 J

Standardized Test Prep

Multiple Choice

1. In which case is the most work done by a force F that pushes a box a distance d across a level surface at constant speed?
   (A) F is horizontal.
   (B) F is at angle of $30°$ above the floor.
   (C) F is at angle of $60°$ above the floor.
   (D) F is at angle of $90°$ to the floor.

2. In which case is the net work positive?
   (A) work done by the tension in a string as a ball attached to it whirls in a horizontal circle
   (B) work done to raise and lower a set of 25-kg barbells to and from the floor ten times
   (C) work done to hold a set of 25-kg barbells at a constant height for 3 min
   (D) work done to kick a set of 25-kg barbells and cause them to roll across the floor

Use the graph below to answer Questions 3–5. The graph shows the energy of a brick dropped from a height of 8 m.

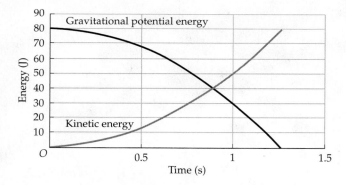

3. Based on the energy data in the graph, what is the approximate mass of the brick?
   (A) 1 kg
   (B) 2 kg
   (C) 3 kg
   (D) 4 kg

4. What are the kinetic energy and the total energy of the brick 1 s after it is dropped?
   (A) $KE = 10$ J, $E = 10$ J
   (B) $KE = 20$ J, $E = 20$ J
   (C) $KE = 50$ J, $E = 80$ J
   (D) $KE = 30$ J, $E = 50$ J

5. What is the approximate speed of the brick 1 s after it is dropped?
   (A) 3 m/s
   (B) 10 m/s
   (C) 30 m/s
   (D) 50 m/s

6. Which situation requires the greatest average power?
   (A) lifting a 5-kg block to a height of 2 m in 2 s
   (B) pushing a block across a level surface with a net force of 10 N at a velocity of 3 m/s
   (C) changing the kinetic energy of a rolling wheel from 15 J to 55 J in 20 s
   (D) burning a 10-watt lightbulb for 20 h

Extended Response

7. A 2.0-kg ball dropped from a height of 4.0 m loses 10% of its mechanical energy to thermal energy when it hits the floor. Determine (a) the kinetic energy of the ball just before it hits the floor (assuming no air resistance), (b) the kinetic energy of the ball just after it hits the floor, and (c) the height to which the ball returns after its first bounce. Show all your work.

> **Test-Taking Hint**
>
> Energy that is removed from a system is expressed with a negative sign.

Question	1	2	3	4	5	6	7					
See Lesson(s)	6.1	6.1	6.2	6.2, 6.3	6.2	6.4	6.3					

If You Had Difficulty With . . .

7

Linear Momentum and Collisions

Inside

Collisions are common in nature, whether they involve bighorn sheep, atoms, or even galaxies. The key to understanding collisions is momentum.

Big Idea

Momentum is conserved in all collisions, as long as external forces do not act.

Conservation laws play an important role in physics. In the last chapter you learned that energy is conserved. In this chapter you'll learn about *momentum* and see that it too is a conserved quantity. Nothing you can do—in fact, nothing that occurs in nature—can change the total energy or the total momentum of the universe.

Inquiry Lab Which shoots the fastest and the farthest?

Read the activity, obtain the required materials and create a data table.

Explore

1. Select a tube and place a marshmallow inside one end. Take the tube to the designated marshmallow shooting range, hold it level, and blow into the *other* end of the tube. Observe the speed of the marshmallow as it exits the tube and the distance it travels.
2. Repeat Step 1, this time blowing into the end of the tube that holds the marshmallow.

3. Decide which method is best for shooting marshmallows. Use this method in Step 4.
4. Shoot marshmallows using tubes of each length provided. Record the length of each tube you use and the distance the marshmallow travels in your data table.

Think

1. **Identify** Which method of shooting marshmallows used in Steps 1 and 2 worked better? Explain why you think this method worked better.

2. **Analyze** What force pushes the marshmallow out of the tube? Is the force exerted on the marshmallow the same in each trial? Does the force act on the marshmallow for the same period of time in each trial?
3. **Analyze** Describe how the length of the tube is related to the distance the marshmallow travels in Step 4. Explain why this is so. List all the factors that you think affect the exit speed of the marshmallow and the distance it travels.

7.1 Momentum

A moving object has two main characteristics—its mass and its velocity. In this lesson you'll learn how mass and velocity combine to yield a new physical quantity of great importance.

Moving objects have momentum

Imagine standing on a skateboard while a friend tosses a heavy, slow-moving ball to you. Catching the ball causes you and the skateboard to move with a certain speed. If, instead, your friend throws a lightweight but fast-moving ball, the net result can be the same, as indicated in **Figure 7.1**. How is it that a fast, lightweight ball can have the same effect as a slow-moving, heavy ball?

Vocabulary

● momentum

◀ **Figure 7.1 Equal momentum**
The effect of catching a slow, heavy object can be the same as catching a fast, lightweight object.

How do changes in mass and velocity affect an object's momentum?

The answer is that the balls have the same effect if they have the same momentum. Specifically, **momentum** is defined to be mass times velocity. The symbol for momentum is $\vec{p}$:

> **Definition of Momentum, $\vec{p}$**
>
> momentum = mass × velocity
> $$\vec{p} = m\vec{v}$$
>
> SI units: kg · m/s

Because momentum is the product of mass and velocity, an object's momentum changes whenever its mass or velocity changes. These changes can sometimes offset one another and leave the momentum the same. In our skateboard example, if the heavy ball has twice the mass of the light ball, but the light ball has twice the speed of the heavy ball, the momenta of the two balls are equal. That is, $(2m)\vec{v} = m(2\vec{v})$.

The magnitude of the momentum, p, is the mass times the speed:

> **Magnitude of the Momentum, p**
>
> magnitude of momentum = mass × speed
> $$p = mv$$
>
> SI units: kg · m/s

The units of momentum, kg · m/s, are simply the unit of mass (kg) times the units of velocity or speed (m/s). This combination of units has no special shorthand name.

We sometimes refer to $\vec{p}$ (and p) as *linear momentum*. This is to distinguish the momentum of an object in straight-line motion from the *angular momentum* that is associated with a rotating object. You'll learn more about angular momentum in Chapter 8.

COOLPHYSICS
Thrust Reversers

A real-life application that illustrates the importance of momentum vectors is the thrust reverser used to slow an airplane when it lands. The thrust reversers shown below change the direction of the air coming out of the engine. This process is basically the same as a ball bouncing off the floor and changing its direction of motion. (See the rubber ball experiment described on the next page.) The force required to reverse the direction of the air is *twice* the original thrust, and this reversing force results in a net backward force on the plane.

> ### QUICK Example 7.1 What's the Momentum?
>
> A 1200-kg car is driven through town with a speed of 15 m/s. What is the magnitude of the car's momentum?
>
> **Solution**
> Applying the definition $p = mv$, we find
> $$p = mv$$
> $$= (1200 \text{ kg})(15 \text{ m/s})$$
> $$= \boxed{18{,}000 \text{ kg} \cdot \text{m/s}}$$

Practice Problems

1. A Major League pitcher can give a 0.142-kg baseball a speed of 45.1 m/s. Find the magnitude of the baseball's momentum.

2. The momentum of a 12-kg dog racing to greet its owner has a magnitude of 37 kg · m/s. What is the dog's speed?

3. Rank The masses and speeds of four objects are given in the following table. Rank the objects in order of increasing magnitude of their momentum. Indicate ties where appropriate.

	Object A	Object B	Object C	Object D
Mass	10 kg	15 kg	5 kg	60 kg
Speed	10 m/s	4 m/s	20 m/s	3 m/s

Vectors for momentum and velocity are related

In many situations only the magnitude of the momentum is of interest—the direction doesn't really matter. In other cases the direction of the momentum plays a key role. But what is the direction of the momentum? From the definition, $\vec{p} = m\vec{v}$, **it follows that the momentum vector points in the same direction as the velocity vector.** If the velocity changes direction, the momentum changes direction as well.

To see how the momentum's direction can be important, consider the experiments shown in **Figure 7.2**. In Figure 7.2 (a) a 1-kg beanbag bear is dropped to the floor. It hits with a speed of 4 m/s and sticks. In Figure 7.2 (b) a 1-kg rubber ball hits the floor with the same speed of 4 m/s. The ball bounces off the floor and moves upward with a speed of 4 m/s (assuming an ideal bounce). Now the question is "What is the change in momentum in each case?"

Analyzing the Beanbag Experiment

To determine the momentum change, we first need to introduce a coordinate system. Let's use the one shown in Figure 7.2. With this choice the beanbag is moving in the negative y direction (downward) with a speed of 4 m/s just before hitting the floor. Thus, its initial momentum is

$$\vec{p}_i = (1 \text{ kg})(-4 \text{ m/s}) = -4 \text{ kg} \cdot \text{m/s} \quad \text{(downward)}$$

After hitting the floor the beanbag is at rest ($\vec{v} = 0$), so its final momentum is zero, $\vec{p}_f = 0$. Therefore, the beanbag's change in momentum is

$$\text{Beanbag} \quad \Delta\vec{p} = \vec{p}_f - \vec{p}_i$$
$$= 0 - (-4 \text{ kg} \cdot \text{m/s})$$
$$= 4 \text{ kg} \cdot \text{m/s} \quad \text{(upward)}$$

Notice that the change in momentum is positive—that is, in the upward direction. This makes sense because, before the beanbag landed, it had a negative (downward) momentum in the y direction. In order to increase the momentum from a negative value to zero, it is necessary to add a positive (upward) momentum.

Analyzing the Rubber Ball Experiment

Next, consider the rubber ball in Figure 7.2 (b). Before bouncing, its momentum is the same as it was for the beanbag: $\vec{p}_i = -4 \text{ kg} \cdot \text{m/s}$ (downward). After bouncing, its momentum is upward with the same magnitude: $\vec{p}_f = 4 \text{ kg} \cdot \text{m/s}$ (upward). Therefore, the change in momentum for the ball is

$$\text{Rubber ball} \quad \Delta\vec{p} = \vec{p}_f - \vec{p}_i$$
$$= (4 \text{ kg} \cdot \text{m/s}) - (-4 \text{ kg} \cdot \text{m/s})$$
$$= 8 \text{ kg} \cdot \text{m/s} \quad \text{(upward)}$$

This is *twice* the change in momentum of the beanbag! The reason is that the momentum of the ball must first be increased from $-4 \text{ kg} \cdot \text{m/s}$ to 0, and then increased again from 0 to $4 \text{ kg} \cdot \text{m/s}$. For the beanbag, the change was only from $-4 \text{ kg} \cdot \text{m/s}$ to 0.

Clearly, recognizing the vector nature of the momentum was the key to solving this problem. Otherwise, you might conclude—erroneously—that the ball had zero change in momentum, since the *magnitude* of its momentum was unchanged by the bounce. The ball's momentum did change, however, due to fact that its *direction* of motion changed.

What is the direction of momentum?

CONNECTINGIDEAS

Here you learn about the vector nature of momentum.
- Later in this chapter you'll use components of the momentum vector to verify momentum conservation.

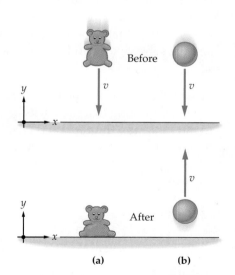

▲ **Figure 7.2 Change in momentum**
A beanbag bear and a rubber ball, with the same mass m and the same downward speed v, hit the floor. **(a)** The beanbag bear comes to rest after hitting the floor. Its change in momentum is mv upward. **(b)** The rubber ball bounces upward with a speed v. Its change in momentum is $2mv$ upward.

Total momentum is a vector sum

In physics the systems you analyze will often contain more than one moving object. The *total momentum* of a system of objects is the *vector sum* of the momentums of all the individual objects:

Total Momentum, $\vec{p}_{total}$

total momentum = momentum 1 + momentum 2 + $\cdots$

$$\vec{p}_{total} = \vec{p}_1 + \vec{p}_2 + \cdots$$

SI units: $kg \cdot m/s$

Because total momentum is a vector sum, the momentums of objects headed in opposite directions can partially or totally cancel one another. Thus, it is possible for a system of several moving objects to have a total momentum that is positive, negative, or even zero.

GUIDED Example 7.2 | Duck, Duck, Goose   **Total Momentum**

At a city park a person throws some bread into a duck pond. Two 4.0-kg ducks and a 7.6-kg goose paddle rapidly toward the bread from opposite directions. If the ducks swim at 1.1 m/s and the goose swims with a speed of 1.3 m/s, find the magnitude and the direction of the total momentum of the three birds.

Picture the Problem

In our sketch we place the origin where the bread floats on the water. The two ducks swim in the negative x direction, with speed $v_d = 1.1$ m/s and velocity $\vec{v}_d = -1.1$ m/s (to the left). The goose swims in the positive x direction, with speed $v_g = 1.3$ m/s and velocity $\vec{v}_g = 1.3$ m/s (to the right). Notice that the subscript d stands for "duck" and the subscript g stands for "goose."

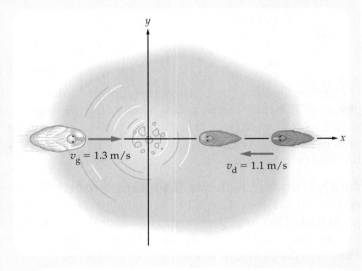

Strategy

First, calculate the momentum of each bird separately, noting that the ducks have negative momentum and the goose has positive momentum. Next, add the three momentums, keeping their appropriate signs, to obtain the total momentum.

Known		Unknown
$m_d = 4.0$ kg	$m_g = 7.6$ kg	$\vec{p}_{total} = ?$
$\vec{v}_d = -1.1$ m/s (to the left)		
$\vec{v}_g = 1.3$ m/s (to the right)		

Solution

1   **Find the momentum of each duck:**

$$\vec{p}_d = m_d\vec{v}_d$$
$$= (4.0 \text{ kg})(-1.1 \text{ m/s})$$
$$= -4.4 \text{ kg} \cdot \text{m/s} \quad \text{(to the left)}$$

2 Find the momentum of the goose:

$$\vec{p}_g = m_g \vec{v}_g$$
$$= (7.6 \text{ kg})(1.3 \text{ m/s})$$
$$= 9.9 \text{ kg} \cdot \text{m/s}$$
$$\text{(to the right)}$$

Math HELP
Negative Numbers
See Math Review, Section I

3 Sum the individual momentums to obtain the total momentum:

$$\vec{p}_{total} = \vec{p}_d + \vec{p}_d + \vec{p}_g$$
$$= (-4.4 \text{ kg} \cdot \text{m/s}) + (-4.4 \text{ kg} \cdot \text{m/s})$$
$$+ (9.9 \text{ kg} \cdot \text{m/s})$$
$$= \boxed{1.1 \text{ kg} \cdot \text{m/s} \qquad \text{(to the right)}}$$

Insight

The total momentum of the system of three birds has a much smaller magnitude than the momentum of any of the individual birds. This is because the momentums are in opposite directions and partially cancel one another.

Practice Problems

4. **Follow-up** Should the speed of the goose be increased or decreased to make the total momentum of the three birds add up to zero? Verify your answer by calculating the required speed.

5. A car and a bicycle move toward each other, traveling in opposite directions. Each has a speed of 12.0 m/s. The mass of the car and its passengers is 1250 kg, and the mass of the bicycle and its rider is 81.6 kg. What is the total momentum of the system, assuming that the car moves in the positive direction?

6. A 1200-kg car moves due north with a speed of 15 m/s. An identical car moves due east with the same speed. What are the direction and the magnitude of the system's total momentum?

7.1 LessonCheck (MP)

Checking Concepts

7. **Apply** How does momentum change if the mass of an object is doubled?

8. **Explain** How is the direction of momentum related to the direction of velocity?

9. **Compare** The speed of an object doubles. How does the change in the magnitude of the object's momentum compare to the change in its kinetic energy?

10. **Analyze** A system consisting of two particles is known to have zero total momentum. Does it follow that the kinetic energy of the system is zero as well? Explain.

Solving Problems

11. **Calculate** What is the momentum of a 0.16-kg apple that moves with a speed of 2.7 m/s?

12. **Calculate** A school bus has a mass of 18,200 kg. The bus moves at 12.5 m/s. How fast must a 0.142-kg baseball move in order to have the same momentum as the bus?

13. **Calculate** A dog running with a speed of 1.9 m/s has a momentum of 17 kg · m/s. What is the mass of the dog?

14. **Calculate** Two carts move directly toward one another on an air track. Cart 1 has a mass of 0.35 kg and a speed of 1.2 m/s. Cart 2 has a mass of 0.61 kg and a speed of 0.85 m/s. What is the total momentum of the system, assuming that cart 1 moves in the positive direction?

7.2 Impulse

Vocabulary

- impulse

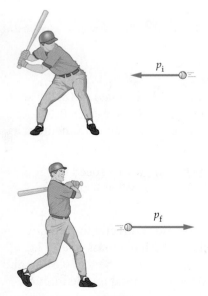

🔑 **What determines the direction of an impulse?**

The next time you push a shopping cart at the grocery store, notice how long it takes to accelerate it from rest to walking speed. If you push lightly, it might take 5 seconds of pushing. If you push harder, you might speed the cart up in only 1 second. The net result is the same—that is, a small force acting for a long time has the same effect as a large force acting for a short time. In this lesson you will learn about the effects of a force acting over a period of time.

A force acting for a time produces an impulse

The pitcher delivers a fastball, the batter takes a swing, and with a crack of the bat the ball that was approaching home plate at 42 m/s is heading toward the pitcher at 54 m/s. In the language of physics we say that the bat has delivered an *impulse* to the ball.

The result of this impulse is a change in the ball's momentum, as shown in **Figure 7.3**. The actual change in momentum occurs when the bat is exerting a force on the ball. This happens during the brief period of time, Δt, when the two are in contact. In general, the product of a force and the time over which it acts is defined as the **impulse, $\vec{\textbf{I}}$**.

> **Definition of Impulse, $\vec{\textbf{I}}$**
>
> impulse = force × time interval
>
> $$\vec{\textbf{I}} = \vec{\textbf{F}}\Delta t$$
>
> SI units: N · s = kg · m/s

Notice that impulse takes into account not just the force that acts on an object, but also the period of time over which the force acts. Both quantities are important. Also notice that the units of impulse (N · s) are the same as the units of momentum (kg · m/s); that is, N · s = 1 kg · m/s. This makes sense when you recall that the impulse changed the momentum of the baseball.

What about the direction of impulse? 🔑 **As defined, impulse is a vector that points in the same direction as the force.** In cases where the direction is not important, we can use the magnitudes of the force and the impulse:

> **Magnitude of the Impulse, I**
>
> magnitude of impulse = magnitude of force × time
>
> $$I = F\Delta t$$
>
> SI units: N · s = kg · m/s

▲ **Figure 7.3 Hitting a baseball**
A batter hits a ball, sending it back toward the pitcher's mound. The impulse delivered to the ball by the bat changes the ball's momentum.

You push on a 22-kg shopping cart with a force of 6.5 N. If you push for 1.9 s, what is the magnitude of the impulse you deliver to the cart?

Solution

Since direction doesn't play a role in this problem, we use magnitudes only. Thus, we find that the impulse is

$$I = F\Delta t$$
$$= (6.5\,\text{N})(1.9\,\text{s})$$
$$= \boxed{12\,\text{kg}\cdot\text{m/s}}$$

Notice that the mass of the cart is not needed.

Practice Problems

15. You kick a soccer ball, delivering an impulse of $23\,\text{kg}\cdot\text{m/s}$. If the force you exert on the ball lasts for 0.25 s, what is the magnitude of the force?

16. Triple Choice Apple 1 falls from a tree and drops 2 m to the ground. Apple 2, which has the same mass as apple 1, falls from the tree and drops 3 m to the ground. Is the impulse delivered by gravity to apple 1 greater than, less than, or equal to the impulse delivered by gravity to apple 2? Explain.

An impulse changes an object's momentum

When a force acts on an object, like a baseball or a shopping cart, it changes the object's momentum. Thus, there must be a connection between impulse and momentum change. In fact, the general form of Newton's second law is expressed in terms of momentum as follows:

$$\vec{\mathbf{F}} = \frac{\Delta\vec{\mathbf{p}}}{\Delta t}$$

This is how Newton originally stated his second law—force is equal to the rate of change of momentum. (This is the same as $\vec{\mathbf{F}} = m\vec{\mathbf{a}}$ when the mass is constant.)

Rearranging this equation, we get

$$\vec{\mathbf{F}}\Delta t = \Delta\vec{\mathbf{p}}$$

This shows that the product of force and elapsed time is equal to the change in momentum. But recall that impulse, $\vec{\mathbf{I}} = \vec{\mathbf{F}}\Delta t$, is also the product of force and elapsed time. Therefore, the connection between impulse and momentum change is as follows:

> **Momentum-Impulse Theorem**
>
> impulse = force × time = change in momentum
> $$\vec{\mathbf{I}} = \vec{\mathbf{F}}\Delta t = \Delta\vec{\mathbf{p}}$$

Simply put, the impulse, $\vec{\mathbf{I}}$, that acts on an object is equal to the change in the object's momentum, $\Delta\vec{\mathbf{p}}$:

$$\vec{\mathbf{I}} = \Delta\vec{\mathbf{p}} = \vec{\mathbf{p}}_f - \vec{\mathbf{p}}_i$$

CONNECTING IDEAS

In Chapter 5 you learned how Newton's laws relate force, mass, and acceleration.
• Here you learn that force is related to a new quantity called *impulse* and that impulses are related to changes in momentum.

How is impulse related to momentum?

Reading Support ✓

Multiple Meanings of impulse

scientific meanings: a force acting for a finite period of time; a short pulse of electric current

The impulse exerted by the baseball dented the car's fender.

Nerve impulses are responsible for your sense of pain.

common meanings: a strong and sudden desire to act (sometimes instinctual); an impelling force; an impetus

Candy bars are displayed near the cashier to stimulate impulse buying.

For example, an impulse caused by a 5-N force acting on a shopping cart for 1 s produces a change in momentum equal to 5 kg · m/s. Thus, if the cart starts at rest, its final momentum is 5 kg · m/s. If the cart starts with an initial momentum of 2 kg · m/s in the same direction as the impulse, its final momentum is 7 kg · m/s.

QUICK Example 7.4 What's the Force?

A force acts on a bicycle for 12 s and changes its momentum from 15 kg · m/s in the positive x direction to 38 kg · m/s in the positive x direction. What are the magnitude and the direction of the force?

Solution
First, calculate the change in momentum:

$$\Delta\vec{p} = \vec{p}_f - \vec{p}_i$$
$$= (38\ \text{kg} \cdot \text{m/s}) - (15\ \text{kg} \cdot \text{m/s})$$
$$= 23\ \text{kg} \cdot \text{m/s} \quad (+x\ \text{direction})$$

Math HELP
Delta (Δ)
See Math Review, Section I

Next, divide by the time to find the force:

$$\vec{F} = \frac{\Delta\vec{p}}{\Delta t} = \frac{23\ \text{kg} \cdot \text{m/s}}{12\ \text{s}} = \boxed{1.9\ \text{N} \quad (+x\ \text{direction})}$$

Practice Problem

17. An impulse of 12.2 kg · m/s is delivered to an object whose initial momentum is 4.5 kg · m/s. The impulse has the same direction as the initial momentum. What is the object's final momentum?

The force creating an impulse can be large

The forces associated with impulses are often large and complex. **Figure 7.4** shows the force exerted on a baseball when it is struck by a bat. While the ball and bat are in contact—for as little as a thousandth of a second—the

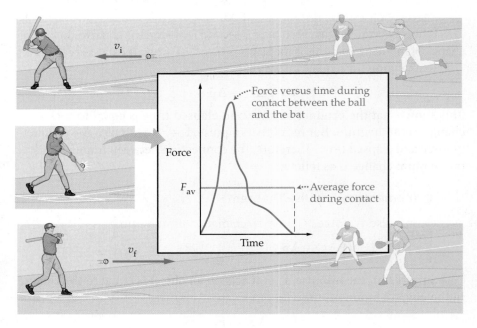

▶ **Figure 7.4 The average force during a collision**
The force between two objects that collide, such as a bat and a baseball, rises rapidly to a very large value, then drops again to zero in a matter of milliseconds. Rather than trying to describe the complex behavior of the force, we focus on its average value, F_{av}. Notice that the area under the F_{av} rectangle is the same as the area under the actual force curve.

force rises to a peak value and then falls to zero. Fortunately, we do not need to worry about the complexities of the force. Instead, we can use the *average* force, $\vec{F}_{av}$, and the time over which the force acts, Δt, to analyze problems.

As an example, let's calculate the impulse and the average force for the baseball considered at the beginning of this lesson. First, we set up a coordinate system with the positive x axis pointing from home plate toward the pitcher's mound, as shown in **Figure 7.5**. If the ball's mass is 0.145 kg, its initial momentum—which is in the negative x direction—is

$$\vec{p}_i = -mv_i$$
$$= -(0.145 \text{ kg})(42 \text{ m/s})$$
$$= -6.1 \text{ kg} \cdot \text{m/s} \quad (-x \text{ direction})$$

After being hit, the ball's final momentum is in the positive x direction:

$$\vec{p}_f = mv_f$$
$$= (0.145 \text{ kg})(54 \text{ m/s})$$
$$= 7.8 \text{ kg} \cdot \text{m/s} \quad (+x \text{ direction})$$

The impulse equals the change in momentum:

$$\vec{I} = \Delta\vec{p}$$
$$= \vec{p}_f - \vec{p}_i$$
$$= (7.8 \text{ kg} \cdot \text{m/s}) - (-6.1 \text{ kg} \cdot \text{m/s})$$
$$= 14 \text{ kg} \cdot \text{m/s} \quad (+x \text{ direction})$$

If ball and bat are in contact for $1.20 \text{ ms} = 1.20 \times 10^{-3}$ s, a typical time, the average force is

$$\vec{F}_{av} = \frac{\Delta\vec{p}}{\Delta t}$$
$$= \frac{14 \text{ kg} \cdot \text{m/s}}{1.20 \times 10^{-3} \text{ s}}$$
$$= 12{,}000 \text{ N} \quad (+x \text{ direction})$$

Notice that the average force is in the positive x direction, toward the pitcher, as expected. The magnitude of the average force is remarkably large—about 2700 pounds in everyday units! As shown in **Figure 7.6**, a golf club exerts a comparably large force when it hits a golf ball, giving the ball a squashed shape.

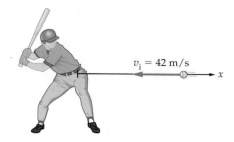
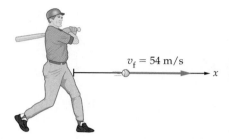

▲ **Figure 7.5 Impulse and force when a baseball is hit**
The change in momentum of the baseball determines the impulse delivered to it. Knowing the time of contact allows calculation of the average force.

◀ **Figure 7.6 A squashed golf ball**
The force exerted by a golf club is sufficient to cause significant deformation of the ball.

We saw in Lesson 7.1 that the change in momentum is different for an object that hits something and sticks (the beanbag bear) than it is for an object that hits something and bounces off (the rubber ball). This means that the impulse, and hence the force, is different in the two cases as well. We explore this difference in the following Conceptual Example.

CONCEPTUAL Example 7.5 Rain versus Hail

A person stands under an umbrella during a rain shower. A few minutes later the raindrops turn to hail—though the number of "drops" hitting the umbrella per unit of time and their speed remains the same. Is the force required to hold the umbrella upright in the hail greater than, less than, or equal to the force required to hold it in the rain?

Rain   Hail

Reasoning and Discussion
When raindrops strike the umbrella, they tend to splatter and run off; when hailstones hit the umbrella, they bounce back upward. As a result, the change in momentum is greater for the hail—just as the change in momentum is greater for a rubber ball bouncing off the floor than it is for a beanbag landing on the floor. Thus, the impulse and the force are greater in the hail.

Answer
The force required to hold the umbrella is greater in the hail.

Three friends push a stalled car and its driver to a nearby service station. Assume that the car rolls without friction on a smooth and level road. (**a**) If the car is at rest initially and the three students give a combined push of 305 N, what is the momentum of car and driver after 12.0 s? (**b**) If the mass of the car and driver is 1360 kg, what is the car's velocity after 12.0 s of pushing?

Picture the Problem

Our sketch shows the three students combining to exert a force $\vec{F}$ on the car. The car moves in the direction of the force, which we choose to be the positive x direction.

Strategy

(**a**) The momentum of the car after 12.0 s can be found using the momentum-impulse theorem, $\vec{I} = \vec{F}\Delta t = \Delta\vec{p} = \vec{p}_f - \vec{p}_i$. The change in momentum equals the force times the time. The car is initially at rest, and so its initial momentum is zero, $\vec{p}_i = 0$. (**b**) The final momentum, $\vec{p}_f$, is determined in part (a). The corresponding velocity can be found by rearranging the momentum equation, $\vec{p} = m\vec{v}$.

Known	Unknown
$\vec{F} = 305$ N $(+x$ direction$)$	(**a**) $\vec{p}_f = ?$
$\Delta t = 12.0$ s	(**b**) $\vec{v}_f = ?$
$m = 1360$ kg	

Solution

Part (a)

1 Apply the momentum-impulse equation to the car and set $\vec{p}_i = 0$:

$$\vec{F}\Delta t = \Delta\vec{p}$$
$$= (\vec{p}_f - \vec{p}_i)$$
$$= (\vec{p}_f - 0)$$
$$= \vec{p}_f$$

> **Math HELP**
> Solving Simple Equations
> See Math Review, Section III

2 Substitute the numerical values for $\vec{F}$ and Δt and calculate the final momentum, $\vec{p}_f$:

$$\vec{p}_f = \vec{F}\Delta t$$
$$= (305 \text{ N})(12.0 \text{ s})$$
$$= \boxed{3660 \text{ kg} \cdot \text{m/s}} \quad (+x \text{ direction})$$

Part (b)

3 Rearrange the momentum equation, $\vec{p} = m\vec{v}$, to solve for the velocity. Substitute the numerical values and evaluate:

$$\vec{p}_f = m\vec{v}_f$$
$$\vec{v}_f = \frac{\vec{p}_f}{m}$$
$$= \frac{3660 \text{ kg} \cdot \text{m/s}}{1360 \text{ kg}}$$
$$= \boxed{2.69 \text{ m/s}} \quad (+x \text{ direction})$$

Insight

The impulse, momentum, and velocity are all in the same direction in this system. Therefore, we could have solved this problem with magnitudes only. It's good to keep vectors in mind, however, because they are needed when different directions are involved, as they were for the batted baseball considered earlier in this lesson.

18. **Follow-up** Suppose one of the three students in Guided Example 7.6 stops pushing the car after 6.00 s. In this case the force exerted on the car is 305 N for the first 6.00 s and 215 N for the second 6.00 s.
(a) What is the final momentum of the car?
(b) What is the final speed of the car?

19. A baseball player bunts (hits softly) a 0.144-kg baseball thrown at 43.0 m/s. If the bat exerts an average force of 6.50×10^3 N on the ball for 0.00122 s, what is the final speed of the ball? Assume that the ball is bunted directly back toward the pitcher.

20. When spiking a volleyball, a player changes the velocity of the ball from 4.2 m/s to −24 m/s along a certain direction. If the impulse delivered to the ball by the player is −9.3 kg · m/s, what is the mass of the volleyball?

21. As an orange falls vertically downward, its momentum changes by 2.4 kg · m/s per second.
(a) What is the weight of the orange?
(b) How much time does it take for the orange's momentum to change by 6.1 kg · m/s?

⌕ How does a bike safety helmet work?

Increasing the time of impact decreases the force

Physics & You: Technology Have you ever looked inside a bike safety helmet to see how it is made? Beneath the hard, outer shell is a firm, yet deformable, material. This material protects your head during an impact. We can see how this protection is accomplished by applying the momentum-impulse theorem.

The theorem shows that increasing the time over which a given impulse occurs (the time of impact) decreases the average force exerted.

$$\vec{\mathbf{I}} = \vec{\mathbf{F}}\Delta t$$
$$= \vec{\mathbf{F}}\Delta t = \vec{\mathbf{F}}\Delta t$$

Thus, a large force acting for a short time creates the same impulse as a small force acting for a large time.

During a biking accident the momentum of your head produces impact forces when it strikes something. Your head undergoes a large change in momentum as it is brought to rest. Though the change in momentum is the same whether you wear a helmet or not, the average force exerted on your head is reduced by the energy-absorbing materials in the helmet. ⌕ **The materials inside a bike safety helmet *increase* the time of impact, thereby *reducing* the force—and the extent of injury—to your head.**

An example of a material protecting an object during a collision is shown in **Figure 7.7**. Here we see two eggs that are dropped from equal heights. The egg that strikes a ceramic plate breaks, whereas the egg that lands on a foam pillow does not. Though the momentum change of each egg is the same, as is the impulse received, the average force exerted on each egg is very different. Simply put, the foam pillow reduces the force by extending the stopping time. The smaller force brings the egg to rest with no damage.

▲ Figure 7.7 **Protecting an egg**
Extending the stopping time of a falling egg reduces the force required to bring it to rest. The result is a safe landing rather than a "splat."

7.2 LessonCheck (MP)

Checking Concepts

22. ▭ **Identify** Which (one or more) of the following quantities has the same direction as impulse: momentum, change in momentum, velocity, force, kinetic energy?

23. ▭ **Explain** Does impulse determine an object's momentum or the change in an object's momentum?

24. ▭ **Assess** Why would a thin steel helmet offer little protection during an accident?

25. | Rank | Impulses are delivered to systems A through D as described below. Rank the systems in order of increasing impulse. Indicate ties where appropriate.

	System A	System B	System C	System D
Magnitude of Force	F	$2F$	$5F$	$10F$
Duration	Δt	$\Delta t/3$	$\Delta t/10$	$\Delta t/100$

26. | Triple Choice | An unattended car rolls slowly across an empty parking lot. Consider the following two cases: case 1, the car hits a light pole and comes to rest; case 2, the car hits a pile of plastic garbage bags and comes to rest.
(a) Is the impulse in case 1 greater than, less than, or equal to the impulse in case 2?
(b) Is the average force in case 1 greater than, less than, or equal to the average force in case 2?

Solving Problems

27. Calculate A safety helmet extends the time of impact from 0.005 s to 0.020 s. By what factor is the average force that causes the impact reduced?

28. Calculate In a typical golf swing the club is in contact with the ball for about 0.0010 s. If the 45-g ball acquires a speed of 67 m/s, estimate the magnitude of the force exerted by the club on the ball.

29. Calculate A 0.50-kg croquet ball is initially at rest on the grass. When the ball is struck with a croquet mallet, the average force exerted on the ball is 230 N. If the ball's speed after being struck is 3.2 m/s, how long was the mallet in contact with the ball?

7.3 Conservation of Momentum

Like energy, momentum is a *conserved* quantity—the total momentum in the universe always remains the same. In this lesson you will learn how to apply conservation of momentum to common, everyday situations.

Momentum conservation requires zero total force

🔑 *How is momentum related to the total external force?*

The momentum of an object can't change unless an external force acts on the object. As an example, imagine putting a skateboard on the sidewalk. If you don't touch it, or exert any other kind of force on it, it stays at rest. It certainly doesn't start moving by itself. Its momentum remains equal to zero until a nonzero total external force causes it to change. This everyday observation is in agreement with the definition of impulse given in Lesson 7.2. Recall that

$$\vec{I} = \vec{F}\Delta t = \vec{p}_f - \vec{p}_i$$

🔑 If the total force is zero, $\vec{F}_{total} = 0$, then the initial and final momentums must be the same, $\vec{p}_f = \vec{p}_i$. This is momentum conservation.

> ### Conservation of Momentum
>
> If the total force acting on an object is zero, its momentum cannot change. In other words, its momentum is conserved.
>
> $$\vec{p}_f = \vec{p}_i$$

Internal and external forces can act on a system

🔑 *How do internal and external forces affect a system's momentum?*

One day you take your bicycle out for a ride. Let's say that you and the bicycle are a system and that everything else is outside, or external to, this system. Many forces act on you and the bicycle, and it's useful to distinguish between those that are internal to the system and those that are external.

Internal Forces *Internal forces* act between objects *within* a system. If you push forward on the handlebars as you ride, you exert an internal force between you and the bicycle. The handlebars push back on you with an equal and opposite force, as we know from Newton's third law. This is shown in **Figure 7.8**. These forces are internal, because you and the handlebars are both part of the system.

External Forces *External forces*, on the other hand, are exerted on a system by something *outside* the system. When you pedal the bicycle, the rear tire exerts a force against the road, and the road exerts an equal and opposite force against the tire, as indicated in Figure 7.8. Since the road is outside the system, the force it exerts on the tire is an external force. Similarly, the force of gravity pulling down on you and the bicycle is an external force.

◀ **Figure 7.8 Internal forces can't change the momentum of a system, but external forces can**
In a system consisting of you and a bicycle, pushing or pulling on the handlebars (internal forces) can't change the system's momentum. It's a bit like trying to lift yourself by your bootstraps. On the other hand, external forces—like the force exerted on the tires by the road—can change the momentum of the system.

Changes in Momentum
The reason we are careful to distinguish internal forces from external forces is that they play very different roles when it comes to the momentum of a system. 🔑 **It may seem odd at first, but *only* external forces can change a system's momentum. Internal forces have no effect on a system's momentum.** For example, pushing forward on the handlebars of your bicycle (an internal force) doesn't make it go faster. Pedaling harder, however, which results in external forces between the road and the tires, does change the speed and momentum.

To see the difference between internal and external forces, let's write an expression for the total force acting on a system. This includes forces acting within the system (internal forces, $\vec{\mathbf{F}}_{\text{int}}$) and forces coming from outside the system (external forces, $\vec{\mathbf{F}}_{\text{ext}}$). Using the Greek letter sigma (Σ) to represent a sum, we can write the total force as follows:

$$\vec{\mathbf{F}}_{\text{total}} = \sum \vec{\mathbf{F}}_{\text{int}} + \sum \vec{\mathbf{F}}_{\text{ext}}$$

Let's apply these summations to the case of two canoes floating at rest next to one another on a lake, as shown in **Figure 7.9**. We'll take the system to be the two canoes and the people inside them. When a person in canoe 1 pushes on canoe 2, a force $\vec{\mathbf{F}}_2$ is exerted on canoe 2. An equal and opposite force, $\vec{\mathbf{F}}_1 = -\vec{\mathbf{F}}_2$, is exerted on the person in canoe 1. These forces are internal, since they act between objects in the system, and they add up to zero:

$$\vec{\mathbf{F}}_1 + \vec{\mathbf{F}}_2 = (-\vec{\mathbf{F}}_2) + \vec{\mathbf{F}}_2 = 0$$

CONNECTING IDEAS

Newton's laws, covered in Chapter 5, also apply to momentum.
• Newton's third law is key to the conservation of momentum.

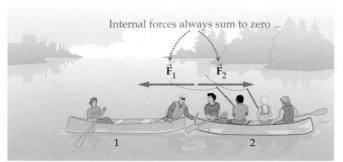

Internal forces always sum to zero ...

$\vec{\mathbf{F}}_1$ $\vec{\mathbf{F}}_2$

1   2

... and hence they have no effect on the total momentum of the system.

◀ **Figure 7.9 Separating two canoes**
A system composed of two canoes and their occupants. The forces $\vec{\mathbf{F}}_1$ and $\vec{\mathbf{F}}_2$ are internal to the system. They sum to zero.

This special case demonstrates the following general principles:

- Internal forces, like all forces, always occur in action-reaction pairs.
- Because the forces in action-reaction pairs are equal and opposite—as required by Newton's third law—internal forces *always* sum to zero:

$$\sum \vec{\mathbf{F}}_{\text{int}} = 0$$

- Because internal forces always cancel, the total force acting on a system is equal to the sum of the external forces acting on it:

$$\vec{\mathbf{F}}_{\text{total}} = \sum \vec{\mathbf{F}}_{\text{ext}}$$

This is a great simplification—it means that we never have to consider internal forces when calculating the total force on a system.

Momentums can vary as the total is conserved

Let's summarize our conclusions regarding forces (both internal and external) and momentum.

- *Internal* forces have no effect on the total momentum of a system.
- If the *total external* force acting on a system is zero, the system's total momentum is conserved. That is,

$$\vec{\mathbf{p}}_{1,f} + \vec{\mathbf{p}}_{2,f} + \vec{\mathbf{p}}_{3,f} + \cdots = \vec{\mathbf{p}}_{1,i} + \vec{\mathbf{p}}_{2,i} + \vec{\mathbf{p}}_{3,i} + \cdots$$

It's important to note that the above statements apply only to the *total* momentum of a system, not to the momentum of each individual object. For example, suppose a system consists of two objects, 1 and 2, and the total external force acting on the system is zero. As a result, the total momentum must remain constant:

$$\vec{\mathbf{p}}_{\text{total}} = \vec{\mathbf{p}}_1 + \vec{\mathbf{p}}_2 = \text{constant}$$

This does not mean, however, that $\vec{\mathbf{p}}_1$ is constant or that $\vec{\mathbf{p}}_2$ is constant. All we can say is that the *sum* of $\vec{\mathbf{p}}_1$ and $\vec{\mathbf{p}}_2$ does not change. Thus, if the momentum of one of the objects increases, the momentum of the other object must decrease by the same amount. In Figure 7.9 both canoes experience a change in momentum, but the two changes cancel.

PROBLEM-SOLVING NOTE

In this equation for momentum conservation, the f subscripts stand for "final" and the i subscripts stand for "initial." The subscripts 1, 2, 3, . . . refer to different objects within the system, like the two canoes in Figure 7.9.

GUIDED Example 7.7 | Tippy Canoe   **Conservation of Momentum**

Two groups in canoes float motionless in the middle of a lake. After a brief visit someone in canoe 1 pushes on canoe 2 with a force of 46 N. The force lasts for 1.20 s and moves the canoes in opposite directions. What is the momentum of each canoe after the push?

Picture the Problem

The two canoes and their occupants are the system. The positive x direction is chosen to point from canoe 1 to canoe 2. Thus, the force exerted on canoe 2 is $\vec{\mathbf{F}}_2 = +46$ N ($+x$ direction), and the force exerted on canoe 1 is $\vec{\mathbf{F}}_1 = -46$ N ($-x$ direction).

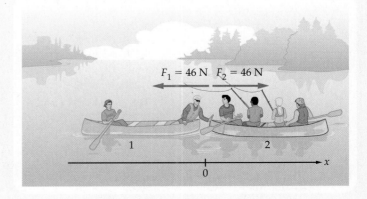

Strategy

Each canoe starts at rest, with zero initial momentum. Determine the change in momentum of each canoe using the momentum-impulse theorem, $\vec{F}\Delta t = \vec{p}_f - \vec{p}_i$.

Known

$\vec{F}_1 = -46 \text{ N } (-x \text{ direction})$
$\vec{p}_{1,i} = 0$
$\vec{F}_2 = 46 \text{ N } (+x \text{ direction})$
$\vec{p}_{2,i} = 0$
$\Delta t = 1.20 \text{ s}$

Unknown

$\vec{p}_{1,f} = ?$
$\vec{p}_{2,f} = ?$

Solution

1 Solve $\vec{F}\Delta t = \vec{p}_f - \vec{p}_i$ for $\vec{p}_f$. Substitute the numerical values for canoe 1, including $\vec{p}_{1,i} = 0$ and $\vec{F}_1 = -46 \text{ N } (-x \text{ direction})$:

$\vec{p}_{1,f} = \vec{F}_1 \Delta t + \vec{p}_{1,i}$
$= (-46 \text{ N})(1.20 \text{ s}) + 0$
$= \boxed{-55 \text{ kg} \cdot \text{m/s}} \quad (-x \text{ direction})$

2 Solve $\vec{F}\Delta t = \vec{p}_f - \vec{p}_i$ for $\vec{p}_f$. Substitute the numerical values for canoe 2, including $\vec{p}_{2,i} = 0$ and $\vec{F}_2 = 46 \text{ N } (+x \text{ direction})$:

$\vec{p}_{2,f} = \vec{F}_2 \Delta t + \vec{p}_{2,i}$
$= (46 \text{ N})(1.20 \text{ s}) + 0$
$= \boxed{55 \text{ kg} \cdot \text{m/s}} \quad (+x \text{ direction})$

Insight

The sum of the momentums of the two canoes is zero $(-55 \text{ kg} \cdot \text{m/s} + 55 \text{ kg} \cdot \text{m/s} = 0)$. This is as expected. After all, the system (the two canoes and their occupants) starts at rest, with zero momentum, and the net external force acting on it is zero. As a result, the momentum of the system can't change—it must stay equal to zero at all times. Notice that each canoe does experience a change in momentum (since each canoe has an external force acting on it), but those changes cancel one another.

Practice Problems

30. Follow-up Calculate the final velocity of each canoe given that the total mass of canoe 1 is 130 kg and the total mass of canoe 2 is 250 kg. Use your results to determine whether velocity is a conserved quantity.

31. Follow-up What is the final momentum of each canoe if they are pushed apart with a force of 56 N for 1.3 s?

32. Two ice skaters stand at rest in the center of an ice rink. When they push off against each other, the 45-kg skater acquires a speed of 0.62 m/s. If the speed of the other skater is 0.89 m/s, what is that skater's mass?

33. A 92-kg astronaut holds onto a 1200-kg satellite; both are at rest relative to a nearby space shuttle. The astronaut pushes on the satellite, giving it a speed of 0.14 m/s directly away from the shuttle. The astronaut comes into contact with the shuttle 7.5 s after pushing away from the satellite. What was the initial distance from the shuttle to the astronaut?

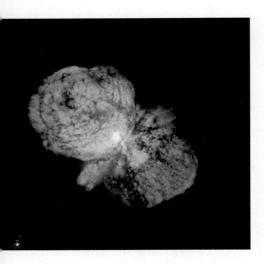

Figure 7.10 An exploding star
As this star blows up, it sends material out in opposite directions. This ensures that its total momentum is unchanged.

Momentum conservation applies to all systems

Is momentum conserved when you jump upward into the air? Although it might not seem so at first, the answer is yes. Jumping into the air is an example of one object pushing away from another, just like the canoes in Guided Example 7.7. In the case of you jumping, the system is you and the Earth. When you jump upward, your momentum is canceled by a corresponding downward momentum of the Earth. Because the Earth is so massive, however, its movement is imperceptible.

Momentum conservation also applies to the largest possible system—the universe. By definition, nothing is external to the universe. Thus, the universe has no external forces acting on it. This means that the total momentum of the universe is conserved. No matter what happens—a comet collides with the Earth, a star explodes and becomes a supernova, a black hole swallows part of a galaxy—the total momentum of the universe simply cannot change. **Figure 7.10** provides a vivid illustration of momentum conservation as a star explodes.

Momentum conservation results in recoil

Have you ever gone ice-skating or rollerblading with a friend? If so, you know that pushing your friend away will send you moving in the opposite direction. This backward motion is known as *recoil*. Recoil also occurs when a gun is fired or, as shown in **Figure 7.11**, when a firefighter directs the stream of water from a fire hose. Recoil is a result of momentum conservation.

Though you may not realize it, your body is recoiling as you read this. Each time your heart pumps blood in one direction, your body recoils in the opposite direction. The recoil is small, but measurable. You may have noticed the recoil yourself at home. When you rest quietly in a rocking chair, the chair will wobble back and forth slightly about once a second, one wobble for each beat of your heart. Doctors use this effect in a medical instrument called a *ballistocardiograph*, which assesses the health of a heart by analyzing the recoil it produces.

Physics & You: Technology Another application of heartbeat recoil is in security, for example, at the Riverbend Maximum Security Institution in Tennessee. After four inmates escaped by hiding in a delivery truck, the facility began using heartbeat recoil detectors. All vehicles leaving the prison stop at a checkpoint where a small motion detector is attached with a suction cup. Anyone hiding in the vehicle can be detected because of the recoil produced by his or her beating heart. Similar systems are used at international border crossings.

Figure 7.11 Recoil
These firefighters are exerting a large force to hold the hose still against the strong recoil delivered by the high-speed stream of water.

In Guided Example 7.7 the final momentum of the system (consisting of the two canoes and their occupants) is equal to the initial momentum of the system. What about the kinetic energy? Is the final kinetic energy of the system greater than, less than, or equal to the initial kinetic energy of the system?

Reasoning and Discussion

The final momentum of the two canoes is zero because one canoe has a positive momentum and the other has a negative momentum of the same magnitude. The two momentums sum to zero.

Kinetic energy, which is $\frac{1}{2}mv^2$, cannot be negative, and hence no such cancellation is possible. Both canoes have positive kinetic energies after the push. Therefore, the final kinetic energy is greater than the initial kinetic energy (which was zero).

Where does the increase in kinetic energy come from? It comes from the muscular work done by the person who pushes the canoes apart.

Answer

The final kinetic energy is greater than the initial kinetic energy.

7.3 LessonCheck (MP)

Checking Concepts

34. Explain If the total external force acting on a system is zero, what can you say about its total momentum?

35. Describe Internal forces can change the momentum of individual objects within a system. How do they affect the total momentum of the system?

36. Identify Two ice skaters at rest in the center of an ice rink push off each other in opposite directions. Identify the system in which momentum is conserved, and list the internal and external forces acting on the system.

37. Assess An object resting on a frictionless surface is struck by a second object.
(a) Is it possible for both objects to be at rest after the collision? Explain.
(b) Is it possible for one of the objects to be at rest after the collision? Explain.

38. Explain If you drop your keys, their momentum increases as they fall. Is the momentum of the keys conserved, or does the momentum of the universe increase as the keys fall? Explain.

Solving Problems

39. Calculate Two canoes are touching and at rest on a lake. The occupants push away from each other in opposite directions, giving canoe 1 a speed of 0.58 m/s and canoe 2 a speed of 0.42 m/s. If the mass of canoe 1 is 320 kg, what is the mass of canoe 2?

40. Calculate A 94-kg lumberjack stands on a 1200-kg log in a pool of water. Initially the lumberjack and the log are at rest. The lumberjack then starts jogging toward shore along the length of the log. If the lumberjack's speed relative to the shore is 2.2 m/s, what is the log's speed relative to the shore?

41. Calculate A young hockey player stands at rest on the ice holding a 1.3-kg helmet. The player tosses the helmet directly in front of him with a speed of 6.5 m/s and recoils with a speed of 0.25 m/s. What is the mass of the hockey player?

7.4 Collisions

Vocabulary

- elastic collision
- inelastic collision
- completely inelastic collision

Momentum is conserved when objects collide. Even so, some of their kinetic energy may be converted to other kinds of energy. In this lesson you will learn about different types of collisions and see how to analyze them using momentum conservation.

Types of Collisions

In physics a *collision* is a situation where two objects free from external forces strike one another. Common examples include one billiard ball hitting another, a baseball bat hitting a ball, and one car smashing into another at an intersection. In cases like these momentum is conserved because external forces are either zero or much smaller than the forces involved in the collision.

It may seem surprising at first, but just because momentum is conserved during a collision doesn't necessarily mean that kinetic energy is conserved as well. In fact, momentum can be conserved even though most—or even all—of the system's kinetic energy is converted to other forms of energy, like sound and heat.

Collisions are either elastic or inelastic

Collisions are categorized according to what happens to the kinetic energy of the system. There are just two possibilities—the kinetic energy is conserved, or it is not.

Elastic Collisions A collision in which the kinetic energy is conserved (unchanged) is referred to as an **elastic collision**. Specifically, the final kinetic energy of the system is equal to its initial kinetic energy. We say that the colliding objects rebound *elastically* after the collision, with no loss of energy to heat or deformation. Heading a soccer ball, as shown in **Figure 7.12**, is an example of a collision that is essentially elastic.

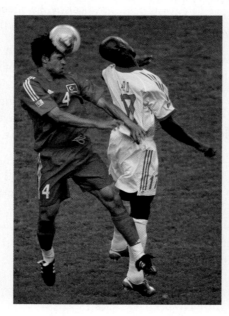

▶ **Figure 7.12 Collisions**
Collisions can conserve kinetic energy or convert it to other forms of energy. The soccer ball bouncing off a player's head is an example of a collision that is approximately elastic. The collision of the football players is clearly inelastic, with a lot of energy going into deforming pads, making sounds, and so on.

Inelastic Collisions Kinetic energy is not conserved in some collisions, however. In particular, an **inelastic collision** is one in which the kinetic energy changes as a result of the collision. This means that the final kinetic energy is not the same as the initial kinetic energy. In a typical inelastic collision part of the initial kinetic energy is converted to sound, heat, and deformation. The collision during a football game shown in Figure 7.12 is clearly inelastic. Of course, the total energy of the universe is always the same—the "lost" kinetic energy is simply converted to other forms of energy.

Completely Inelastic Collisions In some collisions objects stick together after colliding. Examples include a lump of putty colliding with a beanbag, two railroad cars colliding and latching onto one another, and one football player tackling and hanging onto another player. An inelastic collision where the colliding objects stick together is referred to as a **completely inelastic collision**.

As you might expect, a lot of the initial kinetic energy is converted to other forms when two objects hit so hard that they stick together. In fact, completely inelastic collisions convert the *maximum* amount of kinetic energy to other forms. In some cases all of the initial kinetic energy goes to other forms.

Momentum conservation applies to all collisions

The basic idea in solving any collision problem is to conserve momentum. For example, suppose two objects have a collision. The total momentum of the two objects is the sum of mass times velocity for each object. Using subscripts 1 and 2 for the objects, we can write the total momentum as follows:

$$\vec{\mathbf{p}}_{\text{total}} = m_1\vec{\mathbf{v}}_1 + m_2\vec{\mathbf{v}}_2$$

This is the basic equation we use to analyze collision problems.

Analyzing Completely Inelastic Collisions

Let's explore a completely inelastic collision in detail. Consider a system of two identical train cars of mass m on a smooth, level track. One car is at rest initially, while the other moves toward it with a speed v, as shown in **Figure 7.13**. When the cars collide, the coupling mechanism latches them together. Because the cars stick together and move as a unit, the collision is completely inelastic. What is the speed of the cars after the collision?

What problem-solving approach is used to solve collision problems?

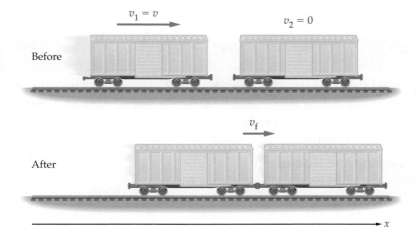

$v_1 = v$

$v_2 = 0$

Before

v_f

After

x

◄ **Figure 7.13 Railroad cars collide and stick together**
A moving train car collides with an identical car that is stationary. After the collision the cars stick together and move with the same speed.

Momentum To determine the speed of the train cars, we apply momentum conservation. The initial momentum is just the momentum of the one moving car. Thus, the initial conditions are as follows:

Train car 1 (moving initially): $m_1 = m, \vec{v}_1 = v$ (to the right)

Train car 2 (at rest initially): $m_2 = m, \vec{v}_2 = 0$

Substituting these values into our momentum equation yields the following:

$$\vec{p}_{total} = m_1\vec{v}_1 + m_2\vec{v}_2$$

$$\vec{p}_{total,i} = mv + 0 = mv \quad \text{(to the right)}$$

Let's do the same for the final conditions. After the collision both train cars have the same speed, which we call v_f. Therefore, the final conditions are as follows:

Train car 1: $m_1 = m, \vec{v}_1 = v_f$ (to the right)

Train car 2: $m_2 = m, \vec{v}_2 = v_f$ (to the right)

As a result, the final momentum is

$$\vec{p}_{total} = m_1\vec{v}_1 + m_2\vec{v}_2$$

$$\vec{p}_{total,f} = mv_f + mv_f = 2mv_f \quad \text{(to the right)}$$

Setting the initial and final momentums equal to one another, we have

$$\vec{p}_{total,f} = \vec{p}_{total,i} \quad \text{or} \quad 2mv_f = mv$$

Canceling the mass (m) and solving for the final speed, we find

$$v_f = \tfrac{1}{2}v$$

Thus, the two train cars stick together and move with half the speed of the incoming car. Not a surprising result, but now you can see how it comes directly from momentum conservation.

Kinetic Energy During the collision of the railroad cars in Figure 7.13, some of the initial kinetic energy is converted to other forms. Some of the energy is given off as sound, some is converted to heat, and some creates permanent deformations in the metal of the latching mechanism. The precise amount of kinetic energy that is lost is addressed in the following Conceptual Example.

CONCEPTUAL Example 7.9 **How Much Kinetic Energy Is Lost?**

A railroad car of mass m moving with speed v collides and sticks to an identical railroad car that is initially at rest. After the collision, is the kinetic energy of the system one-half, one-third. or one-quarter of its initial kinetic energy?

Reasoning and Discussion
Before the collision, the kinetic energy of the system is

$$KE_i = \tfrac{1}{2}mv^2$$

After the collision the mass doubles and the speed is halved. Thus, the final kinetic energy is

$$KE_f = \frac{1}{2}(2m)\left(\frac{v}{2}\right)^2 = \tfrac{1}{4}mv^2 = \tfrac{1}{2}(KE_i)$$

Therefore, one-half of the initial kinetic energy is converted to other forms of energy.

Answer
The final kinetic energy is one-half of the initial kinetic energy.

It's interesting that we know the precise amount of kinetic energy that was lost, even though we don't know just how much went into sound, how much went into heat, and so on. It's not necessary to know all the intricate details to determine how much kinetic energy is lost—we get the answer from momentum conservation.

Analyzing a More Complex Collision
So far we've only looked at collisions between objects of equal mass moving in the same direction. When the colliding objects are moving in opposite directions, however, we must pay close attention to the signs of the velocities and momentums. We must also include different masses for the objects where appropriate. The next Guided Example shows how to address both of these issues.

GUIDED Example 7.10 | Goal-Line Stand   **Completely Inelastic Collision**

On a touchdown attempt a 95.0-kg running back runs toward the end zone at 3.75 m/s. A 111-kg linebacker moving at 4.10 m/s meets the runner in a head-on collision. If the two players stick together, what is their velocity immediately after the collision?

Picture the Problem

In our sketch we let subscript 1 refer to the running back (in red and gray), who carries the ball in the positive x direction. Subscript 2 refers to the linebacker (in blue and gold), who will make the tackle and is moving in the negative x direction.

Strategy

We can find the final velocity by applying momentum conservation to the system consisting of the two players. Initially, the players have momentums in opposite directions. After the collision the players move together with a combined mass $m_1 + m_2$ and a velocity $\vec{v}_f$.

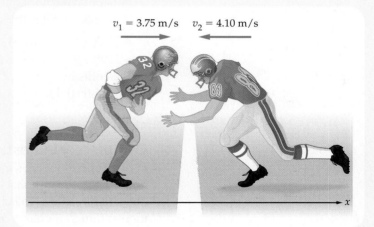

$v_1 = 3.75\,\text{m/s} \qquad v_2 = 4.10\,\text{m/s}$

Known

$m_1 = 95.0$ kg $\vec{v}_1 = 3.75$ m/s (to the right)
$m_2 = 111$ kg $\vec{v}_2 = -4.10$ m/s (to the left)

Unknown

$\vec{v}_f = ?$

Solution

1 Set the initial momentum equal to the final momentum:

$$\vec{p}_i = \vec{p}_f$$
$$m_1\vec{v}_1 + m_2\vec{v}_2 = (m_1 + m_2)\vec{v}_f$$

Math HELP
Solving Simple Equations
See Math Review, Section III

2 Rearrange to solve for the final velocity:

$$\vec{v}_f = \frac{m_1\vec{v}_1 + m_2\vec{v}_2}{m_1 + m_2}$$

3 Substitute the numerical values for the masses and velocities. Be careful to use the appropriate signs for the velocities:

$$\vec{v}_f = \frac{(95.0 \text{ kg})(3.75 \text{ m/s}) + (111 \text{ kg})(-4.10 \text{ m/s})}{95.0 \text{ kg} + 111 \text{ kg}}$$
$$= \boxed{-0.480 \text{ m/s}} \quad (\text{to the left})$$

Insight

After the collision the two players are moving in the negative x direction—away from the end zone. This is because the linebacker had more negative momentum than the running back had positive momentum. As for the kinetic energy, the initial kinetic energy of the system was 1600 J. After the collision the system has only 23.7 J of kinetic energy. This means that more than 98% of the original kinetic energy has been converted to other forms. Even so, *none* of the momentum is lost.

Practice Problems

42. Follow-up If the final speed of the two players is to be zero, should the running back run faster or slower? Explain.

43. Follow-up Check your answer to Problem 42 by calculating the required speed for the running back.

44. A 1.35-kg block of wood is at rest on a smooth table. A 0.0105-kg bullet moving horizontally with a speed of 715 m/s embeds itself within the block. What is the speed of the bullet-block system after the collision?

45. A 1200-kg car moving at 2.5 m/s is struck directly from behind by a 2600-kg truck moving at 6.2 m/s. If the vehicles stick together after the collision, what is their speed immediately after colliding? (Assume that external forces may be ignored.)

Completely Inelastic Collisions in Two Dimensions Collisions in two dimensions are similar to those in one dimension. The main difference is that a two-dimensional collision requires momentum conservation in two different directions. To handle this we set up a coordinate system and resolve the momentum into x and y components. Momentum conservation is then applied in the x and y directions separately. Specifically:

initial momentum (x direction) = final momentum (x direction)

$$p_{x,i} = p_{x,f}$$

initial momentum (y direction) = final momentum (y direction)

$$p_{y,i} = p_{y,f}$$

The following Guided Example shows how to carry out such a calculation for an all too common type of two-dimensional collision.

A 950-kg red car with a speed of 16 m/s approaches an intersection, traveling in the positive x direction. A 1300-kg blue minivan traveling at 21 m/s is heading for the same intersection in the positive y direction. The car and minivan collide and stick together, producing wreckage that moves at an angle of $\theta = 61°$ relative to the positive x axis. Find the speed of the wrecked vehicles just after the collision, assuming that external forces can be ignored.

Picture the Problem

In our sketch we align the x and y axes with the crossing streets. With this choice, $\vec{v}_1$ (the red car's velocity) is in the positive x direction, and $\vec{v}_2$ (the blue minivan's velocity) is in the positive y direction. After the collision the two vehicles move together with a speed v_f in a direction at an angle θ with respect to the positive x axis. This means that the x component of the final velocity is $v_f \cos \theta$ and the y component is $v_f \sin \theta$.

Strategy

Because external forces can be ignored, the total momentum of the system must be conserved during the collision. To determine the final speed we can apply momentum conservation in either the x or the y direction. We will use the x direction in the calculations that follow.

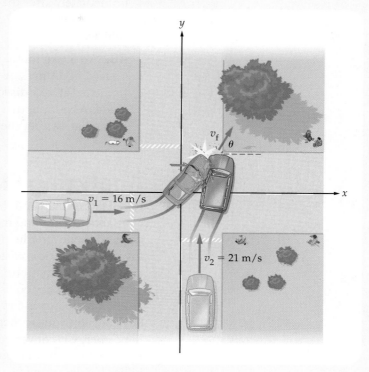

Known

$m_1 = 950$ kg $\vec{v}_{1,i} = 16$ m/s ($+x$ direction)
$m_2 = 1300$ kg $\vec{v}_{2,i} = 21$ m/s ($+y$ direction)

Unknown

$v_f = ?$

Solution

1 Apply momentum conservation in the x direction:

$$p_{x,i} = p_{x,f}$$
$$m_1 v_{1,i} = (m_1 + m_2)(v_f \cos \theta)$$

2 Rearrange to solve for the final speed, v_f, and insert the numerical values:

$$v_f = \frac{m_1 v_{1,i}}{(m_1 + m_2)(\cos \theta)}$$
$$= \frac{(950 \text{ kg})(16 \text{ m/s})}{(950 \text{ kg} + 1300 \text{ kg})(\cos 61°)}$$
$$= \boxed{14 \text{ m/s}}$$

Math HELP
Solving Simple
Equations
**See Math
Review,
Section III**

Insight

After real-world collisions, skid marks at the scene are often measured. Such measurements, along with some basic physics, allow the initial vehicle speeds and directions to be estimated. The data might be used in a court of law to determine which driver is at fault.

As a check, you should verify that the momentum equation for the y direction gives precisely the same value for v_f.

46. Follow-up Suppose the initial speed of the red car in Guided Example 7.11 was greater than 16 m/s. Would the angle in which the vehicles move after the collision be greater than, less than, or equal to 61°? Explain.

47. Follow-up Suppose the speed and the direction of the vehicles immediately after the collision in Guided Example 7.11 are known to be $v_f = 12.5$ m/s and $\theta = 42°$, respectively. Find the initial speed of each vehicle.

48. An 82-kg football player is running in the positive y direction with a speed of 3.1 m/s. A 96-kg player from the opposing team is running in the negative x direction with a speed of 2.4 m/s when the tackle is made. Assuming that the players remain together, what is their speed immediately after the tackle?

Analyzing Elastic Collisions

Recall from Chapter 6 that the word *elastic* refers to a property of objects like rubber bands and metal springs. Such objects return to their original size and shape after being stretched by a small amount. Similarly, objects involved in elastic collisions return to their original shape and lose no energy to deformation.

Most everyday collisions are far from elastic. Usually a significant amount of energy is converted to other forms during the collision. In fact, any collision you can hear is inelastic. Why? Because some of the initial kinetic energy is converted to sound energy that moves away from the system. However, objects that bounce off each other with little deformation—like billiard balls—provide a good approximation to an elastic collision.

Elastic collisions conserve energy and momentum

In an elastic collision objects collide and bounce off each other. The system's momentum is conserved ($\vec{p}_f = \vec{p}_i$), and its kinetic energy is also conserved ($KE_f = KE_i$). The nature of elastic collisions is illustrated in **Figure 7.14**. The device shown has five identical metal balls suspended by strings. When a ball is pulled out and released, it falls back and strikes the remaining balls. This causes a series of elastic collisions among the balls. Because the collisions are elastic, momentum and kinetic energy are conserved. This produces interesting results.

▶ **Figure 7.14 Elastic collisions**
The collisions between these metal balls are approximately elastic. Therefore, if one ball is pulled away to one side and released, the result will be one ball bouncing out on the other side.

For example, when one ball is pulled back and released, one ball swings out from the opposite end. If two balls are pulled back and released, two balls swing out from the opposite end. These results are due to energy and momentum conservation. The only way to conserve *both* energy and momentum is for an equal number of balls on each end to swing out.

Elastic collisions are analyzed using conservation laws

The analysis of an elastic collision is similar to that of an inelastic collision— you apply conservation of momentum. 🔗 **For an elastic collision, however, conservation of kinetic energy must also be applied to the situation.** Let's see how this works in a simple case.

🔗 *What additional information is used to solve elastic collision problems?*

Momentum Conservation Consider a head-on collision of two carts on an air track, as pictured in **Figure 7.15**. The carts are provided with bumpers that give an elastic bounce when they collide. Let's suppose that cart 1 is initially moving to the right with a speed v toward cart 2, which is at rest. If the masses of the carts are m_1 and m_2, respectively, then momentum conservation can be expressed as follows:

$$m_1 v = m_1 v_{1,f} + m_2 v_{2,f}$$

In this expression, $v_{1,f}$ and $v_{2,f}$ are the final velocities of the two carts. Note that we say velocities, not speeds, since it is possible for cart 1 to reverse direction, in which case $v_{1,f}$ would be negative.

Kinetic Energy Conservation Next, the fact that this is an elastic collision means that the final velocities must also satisfy energy conservation:

$$\tfrac{1}{2} m_1 v^2 = \tfrac{1}{2} m_1 v_{1,f}^2 + \tfrac{1}{2} m_2 v_{2,f}^2$$

Thus, we now have two equations for the two unknowns, $v_{1,f}$ and $v_{2,f}$. Straightforward algebra yields the following results:

$$v_{1,f} = \left(\frac{m_1 - m_2}{m_1 + m_2}\right)v \qquad v_{2,f} = \left(\frac{2m_1}{m_1 + m_2}\right)v$$

Notice that the final velocity of cart 1 can be positive, negative, or zero, depending on whether m_1 is greater than, less than, or equal to m_2, respectively. The final velocity of cart 2, however, is always positive.

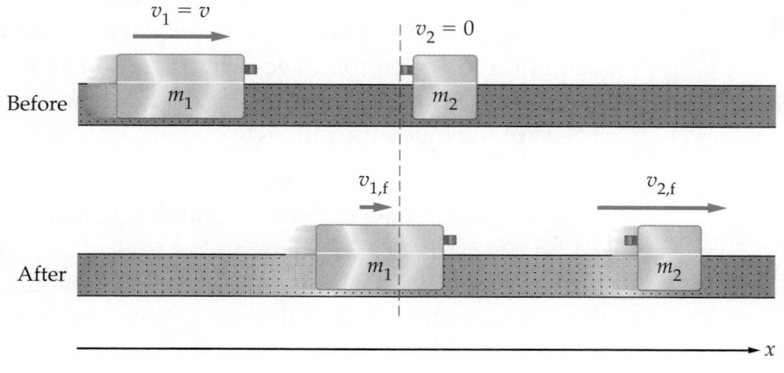

◀ **Figure 7.15 An elastic collision between two air-track carts**
Two air-track carts collide elastically. Their final velocities must satisfy both momentum conservation and energy conservation.

At an amusement park a 96.0-kg bumper car moving with a speed of 1.24 m/s bounces elastically off a 135-kg bumper car that is at rest initially. Find the final velocities of the two cars.

Solution *(Perform the calculations indicated in each step.)*

1. Substitute $m_1 = 96.0$ kg, $m_2 = 135$ kg, and $v = 1.24$ m/s into the equation for the final speed of car 1: $v_{1,f} = -0.209$ m/s

2. Substitute the same values into the equation for the final velocity of car 2: $v_{2,f} = 1.03$ m/s

Insight
Notice that the direction of travel of car 1 has been reversed—it has bounced backward.

7.4 LessonCheck (MP)

Checking Concepts

49. Relate When a collision occurs, how are the initial and final momentums of the system related?

50. Describe How is the initial kinetic energy related to the final kinetic energy in an elastic collision?

51. Explain What happens to the kinetic energy that is lost in an inelastic collision?

52. Identify What do we call a collision in which the colliding objects stick together?

53. Big Idea What condition is required for momentum to be conserved in a collision?

54. Assess Two cars collide at an intersection. If the cars do not stick together, can you conclude that the collision was elastic? Explain.

Solving Problems

55. Calculate A cart of mass $m = 0.12$ kg moves with a speed $v = 0.45$ m/s on a frictionless air track and collides with an identical cart that is stationary. The carts stick together after the collision. What are (**a**) the initial kinetic energy and (**b**) the final kinetic energy of the system?

56. Calculate The collision between a geological hammer and a rock lying loose on the ground can be considered to be approximately elastic. Calculate the final speed of a 0.19-kg rock when it is struck by a 0.55-kg hammer moving with an initial speed of 4.5 m/s. The rock is initially at rest.

57. Calculate Suppose the red car in Guided Example 7.11 has an initial speed of 20.0 m/s and the direction in which the wrecked vehicles move after the collision is 40.0° above the x axis. Find the initial speed of the blue minivan.

58. Calculate Brittany stands at rest on a skateboard. The combined mass of Brittany and the skateboard is 61 kg. Dave tosses a 3.7-kg pumpkin to Brittany, and when she catches it, she and the skateboard begin to roll backward with a speed of 0.16 m/s. What was the speed of the pumpkin?

Ballistic Pendulum

What is It? A ballistic pendulum is a simple device used to measure the momentum of a projectile. The momentum data are then used to determine the projectile's velocity and kinetic energy. All of this is accomplished without the need to collect any time data whatsoever.

Who Invented It? Benjamin Robins, an English mathematician and engineer, invented the device in 1742. Though it is now obsolete, the ballistic pendulum was an important invention that led to great advances in ballistics.

How Does It Work? The ballistic pendulum is based on the law of conservation of momentum. The pendulum consists of a wooden block of mass m suspended by strings. A projectile of mass m is fired at the wooden block. In the resulting collision the projectile embeds itself in the block and causes it to rise through a height h.

Because much of the projectile's energy is converted into heat (warming the block and projectile) and work (deforming the block and projectile), kinetic energy is not conserved. Total horizontal momentum is conserved, however, because there are no *external* horizontal forces acting on the system. Applying conservation of momentum before and after the collision and solving for the final speed of the system (projectile and block) yields

$$v = mv/(m + M)$$

The kinetic energy of the system immediately after the collision is $KE = \frac{1}{2}(m + M)v^2$. All of this kinetic energy is converted to potential energy, $PE = (m + M)gh$, as the block rises to its maximum height h. Equating KE and PE, substituting the previous expression for v, and solving for v yields

$$v = \frac{M + m}{m}\sqrt{2gh}$$

As you can see, the speed of the projectile is calculated without using time data.

False-color, high-speed image showing the air flow around a bullet traveling at 500 m/s.

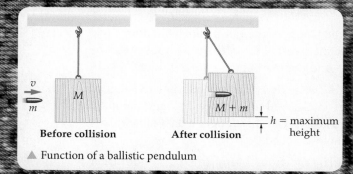

Before collision   After collision h = maximum height

▲ Function of a ballistic pendulum

Take It Further

1. Calculate *The block of a ballistic pendulum has a mass M, and it rises 1.00 cm when struck by a 0.00260-kg projectile traveling at 330.0 m/s* **(a)** *What is M?* **(b)** *What percentage of the projectile's kinetic energy is converted to heat and deformation of the block?*

2. Design *An experiment requires using a ballistic pendulum to measure the momentum of an object falling vertically downward. Sketch a design suitable for this purpose and explain how it works.*

Physics Lab Momentum Conservation during a Collision

This lab explores momentum conservation during a one-dimensional collision. You will measure and record projectile data and then analyze the data in order to determine if momentum is conserved.

Materials
- ramp
- glass marble
- steel ball bearing
- masking tape
- balance
- meterstick
- two sheets of carbon paper
- stack of books
- three sheets of $8\frac{1}{2}$-by-11-inch paper

Procedure 🖐

1. Read the entire lab. Create a data table to record your experimental data.

2. Set up the ramp as shown and tape the flat portion of the ramp to the table.

3. Tape the $8\frac{1}{2}$-by-11-inch sheets of paper together end-to-end. This is your target. Place the target on the floor so that one end is *directly* below the end of the ramp. Tape the target to the floor.

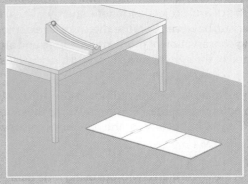

4. Measure the masses of the steel ball bearing and the glass marble to the nearest gram. Measure the height of the ramp above the floor to the nearest millimeter. Record your data.

Analysis

1. Use the meterstick to measure the horizontal distance traveled (the range) for each trial in Parts I and II. Record this data in a data table titled "Data Table 1: Range."

2. Use the height of the ramp above the floor to calculate the free-fall time (the time it takes for the steel ball and the marble to strike the floor).

3. Use the range data and free-fall time to calculate the horizontal speed for the steel ball and the marble in each trial. Record this data in a data table titled "Data Table 2: Speed."

4. Calculate the momentum (in kg · m/s) for each trial of the steel ball in Part I and for each trial of the steel ball and the marble in Part II. In Part II,

Part I

5. Release the ball bearing from the top of the ramp and note where it strikes the target on the floor. Place the carbon paper (carbon side down) at the point of impact.

6. Complete five trials of releasing the ball bearing from the top of the ramp. Catch the ball after impact—it should not bounce twice. When done, remove the carbon paper and number each impact mark from 1 to 5 and label them "Steel ball alone."

Part II

7. Place the glass marble so that it rests at the end of the ramp. Release the ball bearing from the top of the ramp and note the impact locations of the marble and the ball bearing. Place a sheet of carbon paper at each impact location.

8. Complete five trials using the ball bearing and the marble. Again, be sure each bounces only once. When done, label each impact mark "Steel ball after collision" or "Marble after collision."

determine the total momentum for each trial by summing the momentum of the steel ball and the momentum of the marble. Record this data in a data table titled "Data Table 3: Momentum."

5. Calculate the average total momentum for the trials in Part I and Part II based on the values in Data Table 3. If the momentum in Part I is the accepted value, what is the percent error?

Conclusions

1. Was momentum conserved in the collision? How do you know? Use data to support your answer.

2. Identify possible sources of experimental error and describe how they might have affected your results.

7 Study Guide

Big Idea

Momentum is conserved in all collisions, as long as external forces do not act.

Kinetic energy may or may not be conserved in a collision. If kinetic energy is lost during a collision, it is converted to other forms of energy, like heat, sound, and deformation.

7.1 Momentum

🔑 Because momentum is the product of mass and velocity, an object's momentum changes whenever its mass or velocity changes.

🔑 From the definition, $\vec{p} = m\vec{v}$, it follows that the momentum vector points in the same direction as the velocity vector.

• In physics momentum is defined as mass times velocity.

Key Equations

The linear momentum of an object of mass m moving with velocity $\vec{v}$ is

$$\vec{p} = m\vec{v}$$

For a system of several objects the total linear momentum is the vector sum of the individual momenta:

$$\vec{P}_{\text{total}} = \vec{p}_1 + \vec{p}_2 + \cdots$$

7.2 Impulse

🔑 Impulse is a vector that points in the same direction as the force.

🔑 The impulse, $\vec{I}$, that acts on an object is equal to the change in the object's momentum, $\Delta\vec{p}$.

🔑 The materials inside a bike safety helmet *increase* the time of impact, thereby *reducing* the force—and the extent of injury—to your head.

• The product of a force and the time over which it acts is defined as the impulse, $\vec{I}$.

• Since an impulse is often delivered in a very short time interval, the average force over that time can be large.

Key Equations

The impulse delivered to an object by a force $\vec{F}$ acting for a time Δt is

$$\vec{I} = \vec{F}\Delta t$$

The impulse delivered to an object is equal to the change in its momentum:

$$\vec{I} = \Delta\vec{p}$$

7.3 Conservation of Momentum

🔑 If the total force is zero, $\vec{F} = 0$, then the initial and final momentums must be the same, $\vec{p}_f = \vec{p}_i$.

🔑 *Only* external forces can change a system's momentum. Internal forces have no effect on a system's momentum.

7.4 Collisions

🔑 The basic idea in solving any collision problem is to conserve momentum.

🔑 For an elastic collision conservation of kinetic energy must also be applied to the situation.

• A collision in which the kinetic energy is conserved (unchanged) is an elastic collision.

• An inelastic collision is one in which the kinetic energy changes as a result of the collision.

• An inelastic collision where the colliding objects stick together is a completely inelastic collision.

• In a two-dimensional collision both the x and the y components of momentum are conserved separately.

Key Equations

When two objects collide, their momentum is

$$\vec{p} = m_1\vec{v}_1 + m_2\vec{v}_2$$

This equation can be applied to the initial and final conditions to analyze a collision.

For an elastic collision in one dimension, where mass m_1 is moving with an initial velocity v and mass m_2 is initially at rest, the velocities of the masses after the collision are

$$v_{1,f} = \left(\frac{m_1 - m_2}{m_1 + m_2}\right)v \quad \text{and} \quad v_{2,f} = \left(\frac{2m_1}{m_1 + m_2}\right)v$$

7 Assessment

For instructor-assigned homework, go to www.masteringphysics.com

ANSWERS TO SELECTED ODD-NUMBERED PROBLEMS APPEAR IN APPENDIX A.

Lesson by Lesson

7.1 Momentum

Conceptual Questions

59. Is it possible for a baseball to have more momentum than a truck? Explain.

60. A system of particles is known to have zero kinetic energy. What can you say about the momentum of the system?

61. A system of particles has a kinetic energy of 10,000 J but a total momentum of zero. Explain why this is possible.

62. What is the difference between velocity and momentum? How are velocity and momentum similar?

63. Rank Mass and speed data for four objects are given below. Rank the objects in order of increasing magnitude of momentum. Indicate ties where appropriate.

	A	B	C	D
Mass (kg)	1	2	0.1	5
Speed (m/s)	1	3	20	0.2

64. Rank Momentum and speed data for four objects are given below. Rank the objects in order of increasing mass. Indicate ties where appropriate.

	A	B	C	D
Momentum (kg · m/s)	4	5	1	2
Speed (m/s)	1	2	4	0.5

Problem Solving

65. A 0.15-kg baseball has a momentum of 0.78 kg · m/s just before it lands on the ground. What was the ball's speed just before landing?

66. A car moving with a speed of 25 m/s has a momentum of 32,000 kg · m/s. What is the mass of the car?

67. What is the momentum of a 5.2-kg salmon swimming upstream with a speed of 2.3 m/s?

68. Soccer player 1 has a mass of 45 kg and moves to the right with a speed of 1.4 m/s. Soccer player 2 has a mass of 32 kg and moves to the left with a speed of 2.1 m/s. What are the direction and the magnitude of the total momentum of the two players?

69. Soccer player 1 has a mass of 47 kg and moves to the right with a speed of 1.1 m/s. Soccer player 2 has a mass of 38 kg. If the total momentum of the two players is 2.2 kg · m/s to the right, what are the speed and the direction of motion of player 2?

70. Two carts move directly toward one another on an air track. Cart 1 has a mass of 0.35 kg and a speed of 1.2 m/s. Cart 2 has a mass of 0.61 kg. What speed must cart 2 have if the total momentum of the system is to be zero?

71. A 280-g ball falls vertically downward, hitting the floor with a speed of 2.5 m/s and then rebounding upward with a speed of 2.0 m/s. **(a)** What was the ball's momentum just before it hit the floor? **(b)** What was the ball's momentum just after it rebounded from the floor? **(c)** What was the ball's change in momentum during its impact with the floor?

72. A 26-kg dog is running northward at 2.7 m/s, while a 5.3-kg cat is running eastward at 3.0 m/s. Find the magnitude and direction of the total momentum for this system.

7.2 Impulse

Conceptual Questions

73. How does a force differ from an impulse?

74. Is it possible for a small force to deliver a greater impulse than a large force? If so, explain how.

75. Triple Choice As a school bus approaches a stop sign, the driver applies the brakes and brings the bus to a slow, gradual stop. If the driver had instead stomped on the brakes and brought the bus to a sudden stop, would the magnitude of the impulse be greater than, less than, or equal to the magnitude of the impulse with the gradual stop? Explain.

76. Automobile air bags, which deploy during a collision, are designed to protect the vehicle's occupants. Using the concept of impulse, explain how air bags protect a car's passengers.

77. Predict & Explain A net force of 200 N acts on a 100-kg boulder, and a force of the same magnitude acts on a 100-g pebble. **(a)** Is the change in the boulder's momentum in 1 s greater than, less than, or equal to the change in the pebble's momentum in the same time period? **(b)** Choose the *best* explanation from among the following:

A. The larger mass of the boulder gives it the greater change in momentum.

B. The force causes a much greater speed in the 100-g pebble, resulting in a larger change in momentum.

C. Equal force means equal change in momentum for a given time period.

260 Chapter 7 · Assessment

78. **Predict & Explain** Referring to the boulder and the pebble in Problem 77, (a) is the change in the boulder's speed in 1 s greater than, less than, or equal to the change in speed of the pebble in the same time period? (b) Choose the *best* explanation from among the following:

A. The larger mass of the boulder results in a smaller acceleration.

B. The same force results in the same change in speed for a given time.

C. Once the boulder gets moving, it is harder to stop than the pebble.

79. **Rank** Force and time data for four different impulses are given below. Rank the impulses in order of increasing magnitude. Indicate ties where appropriate.

	A	B	C	D
Force (N)	1 N	2 N	5 N	10 N
Time (s)	1 s	0.3 s	0.1 s	0.07 s

Problem Solving

80. A 0.14 kg baseball is dropped from rest. It has a momentum of 0.78 kg·m/s just before it lands on the ground. For what amount of time was the ball in the air?

81. Find the magnitude of the impulse delivered to a soccer ball when a player kicks it with a force of 1250 N. Assume that the player's foot is in contact with the ball for 5.95×10^{-3} s.

82. In a typical golf swing the club is in contact with the ball for about 0.0010 s. If the 45-g ball acquires a speed of 67 m/s, estimate the magnitude of the force exerted by the club on the ball.

83. During an intense game of croquet, a 0.52-kg ball at rest on the grass is struck by a mallet with an average force of 190 N. If the mallet is in contact with the ball for 7.2×10^{-3} s, what is the ball's speed just after it is hit?

84. During a ballistics test a bullet is fired into a thick gel to bring it to a stop. A 5.5-g bullet traveling at 325 m/s is brought to a stop in the gel by an average force of 550 N. What amount of time is needed for the bullet to come to rest?

85. When spiking a volleyball, a player changes the velocity of the ball from 4.2 m/s to −24 m/s in a certain direction. If the impulse delivered to the ball by the player is −9.3 kg·m/s, what is the mass of the volleyball?

86. A 15-g marble is dropped from rest onto the floor 1.4 m below. (a) If the marble bounces straight upward to a height of 0.64 m, what are the magnitude and the direction of the impulse delivered to the marble by the floor? (b) If the marble bounces to a greater height, is the impulse delivered to it greater than or less than the impulse found in part (a)? Explain.

7.3 Conservation of Momentum

Conceptual Questions

87. Which types of forces (internal or external) can change the total momentum of a system?

88. A system consists of two objects. If the total momentum of the system is conserved, does this mean that the momentum of each object must remain the same? Explain.

89. Give an example of a situation where momentum is conserved but velocity is not.

90. Explain how a rocket functions in space even though there is no matter for its exhaust gases to push against.

91. An astronaut carrying a large wrench is tethered to a space shuttle by a wire. Explain how the astronaut can use the wrench to return to the space shuttle after the tether wire breaks.

92. At the instant a bullet is fired from a gun, the bullet and the gun have equal and opposite momentums. Which object, the bullet or the gun, has the greater kinetic energy? Use your answer to explain why the recoil of the gun is not harmful but the speeding bullet is dangerous.

Problem Solving

93. Two ice skaters stand at rest in the center of an ice rink. When they push off against one another, the 45-kg skater acquires a speed of 0.62 m/s. If the speed of the other skater is 0.89 m/s, what is that skater's mass?

94. Two rollerbladers face each other and stand at rest on a flat parking lot. Tracey has a mass of 32 kg, and Jonas has a mass of 45 kg. When they push off against one another, Jonas acquires a speed of 0.55 m/s. What is Tracey's speed?

95. **Think & Calculate** A 63-kg canoeist stands in the middle of her 32-kg canoe. Initially, both the canoeist and the canoe are at rest, and the canoe is pointing directly toward the shore. (a) If the canoeist begins to walk toward one end of the canoe with a speed of 0.95 m/s relative to the shore, what is the speed of the canoe relative to the shore? (b) If the mass of the canoe were increased, would the speed found in part (a) be greater than, less than, or the same as before? Explain. (c) Calculate the speed in part (a) for a 39-kg canoe.

96. A 92-kg astronaut and a 1200-kg satellite are at rest relative to a space station. The astronaut pushes on the satellite, giving it a speed of 0.14 m/s directly away from the station. Seven and a half seconds later the astronaut comes into contact with the station. What was the initial distance from the station to the astronaut?

97. **Think & Calculate** An 85-kg lumberjack stands at one end of a 380-kg floating log, as shown in **Figure 7.16**. Both the log and the lumberjack are at rest initially. **(a)** The lumberjack then trots toward the shore with a speed of 2.7 m/s relative to the shore. What is the speed of the log relative to the shore? Ignore friction between the log and the water. **(b)** If the mass of the log were greater, would its speed relative to the shore be greater than, less than, or the same as the speed found in part (a)? Explain. **(c)** Check your answer to part (b) by calculating the speed relative to the shore for a 450-kg log.

$v = 2.7$ m/s

Figure 7.16

7.4 Collisions

Conceptual Questions

98. What distinguishes a completely inelastic collision from other inelastic collisions?

99. Momentum is conserved during the collision of two billiard balls. Is the momentum of each ball conserved during the collision? Explain.

100. What two physical quantities are conserved in an elastic collision?

101. What physical quantity is not conserved in an inelastic collision?

102. In a two-dimensional elastic collision momentum must be conserved in both the x and the y directions. What about the kinetic energy? Must it be conserved in both directions too? Explain.

103. Can two objects on a horizontal frictionless surface have a collision in which all the initial kinetic energy of the system is lost? Explain why not if your answer is no, and give a specific example if your answer is yes.

Problem Solving

104. A 0.10-kg cart moves with a speed of 0.66 m/s on a frictionless air track and collides with a stationary cart whose mass is 0.20 kg. If the two carts stick together after the collision, what is the final kinetic energy of the system?

105. A 25,000-kg train car moving at 2.50 m/s collides with and connects to a train car of equal mass moving in the same direction at 1.00 m/s. **(a)** What is the speed of the

connected cars? **(b)** How much does the kinetic energy of the system decrease during the collision?

106. A 732-kg car stopped at an intersection is rear-ended by a 1720-kg truck moving with a speed of 15.5 m/s. If the car was in neutral and its brakes were off, so the collision is approximately elastic, find the final speed of both vehicles after the collision.

107. A 92-kg rugby player running at 7.5 m/s collides in midair with a 112-kg player moving in the opposite direction. After the collision each player has zero velocity. **(a)** Diagram the situation before and after the collision, identifying masses and speeds. **(b)** What is the initial momentum of the 92-kg player? **(c)** What is the change in momentum of the 92-kg player due to the collision? **(d)** How is the change in momentum of the 92-kg player related to the change in momentum of the 112-kg player? **(e)** What was the initial speed of the 112-kg player before the collision?

108. Two 72.0-kg hockey players skating at 5.45 m/s collide and stick together. If the angle between their initial directions was 90°, what is their speed after the collision?

109. Two curling stones collide on an ice rink. Stone 1 has a mass of 21 kg and an initial velocity of 1.7 m/s to the north. Stone 2 has a mass of 16 kg, and was at rest initially. The stones collide dead center. **(a)** What was the final velocity of stone 2? **(b)** What was the final velocity of stone 1?

110. Two curling stones collide on an ice rink. Stone 1 has a mass of 16 kg and an initial velocity of 1.5 m/s to the north. Stone 2 was at rest initially. The stones collide dead center, giving stone 2 a final velocity of 0.69 m/s to the north. **(a)** What was the mass of stone 2? **(b)** What was the final velocity of stone 1?

111. **Think & Calculate** A bullet with a mass of 4.0 g and a speed of 650 m/s is fired at a block of wood with a mass of 0.095 kg. The block rests on a frictionless surface, and it is thin enough that the bullet passes completely through it. Immediately after the bullet exits the block, the speed of the block is 23 m/s. **(a)** What is the speed of the bullet when it exits the block? **(b)** Is the final kinetic energy of this system greater than, less than, or equal to the initial kinetic energy? Explain. **(c)** Verify your answer to part (b) by calculating the initial and final kinetic energies of the system.

112. A 0.430-kg block is attached to a horizontal spring that is at its equilibrium length and has a spring constant of 20.0 N/m. The block rests on a frictionless surface. A 0.0500-kg wad of putty is thrown horizontally at the block with a speed of 2.30 m/s. The putty sticks to the block. How far does the putty-block system compress the spring?

113. A velocity amplifier The three air-track carts shown in **Figure 7.17** have masses, from left to right, of $4m$, $2m$, and m, respectively. The most massive cart has an initial speed of v; the other two carts are at rest initially. All three carts are equipped with spring bumpers that give elastic collisions. **(a)** Find the final speed of each cart. Notice that the smallest cart has a final speed greater than v. We refer to this effect as an *amplification* of the initial velocity. **(b)** Verify that the final kinetic energy of the system is equal to the initial kinetic energy. Assume that the air track is long enough to accommodate all collisions. (An example of velocity amplification is shown in **Figure 7.18**. Notice that the speed of the lightweight green ball is increased by its collision with the heavier orange ball, which sends it much higher than the original drop height.)

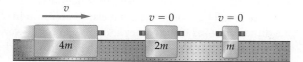

Figure 7.17

Figure 7.18

Mixed Review

114. Rank Object A has a mass m, object B has a mass $2m$, and object C has a mass $m/2$. Rank these objects in order of increasing kinetic energy, given that they all have the same momentum. Indicate ties where appropriate.

115. Rank Object A has a mass m, object B has a mass $4m$, and object C has a mass $m/4$. Rank these objects in order of increasing momentum, given that they all have the same kinetic energy. Indicate ties where appropriate.

116. Predict & Explain A block of wood is struck by a bullet. **(a)** Is the block more likely to be knocked over if the bullet is metal and embeds itself in the wood or if the bullet is rubber and bounces off the wood? **(b)** Choose the *best* explanation from among the following:

A. The change in momentum when a bullet rebounds is twice as much as when it is brought to rest.

B. The metal bullet does more damage to the block.

C. Since the rubber bullet bounces off, it has little effect on the block.

117. Small navigational rockets are used to change the speed and direction of the space probe *Voyager II*. If a rocket exerts a force of 25 N, how long must the rocket fire to change the speed of the 722-kg probe by 1.0 m/s?

118. A car moving with an initial speed v collides with a stationary car that is one-half as massive. After the collision the first car moves in the same direction as before with a speed $v/3$. **(a)** Find the final speed of the second car. **(b)** Is this collision elastic or inelastic?

119. An apple that weighs 2.7 N falls vertically downward from rest for 1.4 s. **(a)** What is the change in the apple's momentum per second? **(b)** What is the total change in its momentum during the 1.4-s fall?

120. A young hockey player stands at rest on the ice holding a 1.3-kg helmet. The player tosses the helmet with a speed of 6.5 m/s in a direction 11° above the horizontal, and then recoils with a speed of 0.25 m/s. Find the mass of the hockey player.

121. A friend climbs an apple tree and drops a 0.22-kg apple from rest to you, standing 3.5 m below. When you catch the apple, you bring it to rest in 0.28 s. **(a)** What was the speed of the apple just before you caught it? **(b)** What average force did you exert on the apple to bring it to rest? (*Hint*: Be sure to include both the weight of the apple and the force needed to bring it to rest.)

122. The Force of a Storm During a severe storm in Palm Beach, Florida, on January 2, 1999, 79 cm (31 in) of rain fell in a period of 9 hours. Assuming that the raindrops hit the ground with a speed of 10 m/s, estimate the average upward force exerted by 1 square meter of ground to stop the falling raindrops during the storm. (One cubic meter of water has a mass of 1000 kg.)

123. At a busy intersection a 1540-kg car traveling west with a speed of 12 m/s collides head-on with a minivan traveling east with a speed of 9.4 m/s. The cars stick together and move with an initial velocity of 1.5 m/s to the east after the collision. What is the mass of the minivan?

124. A 1.7-kg block of wood rests on a rough surface. A 0.011-kg bullet strikes the block with a speed of 670 m/s and embeds itself. The bullet-block system slides forward 2.4 m before coming to rest. What is the coefficient of kinetic friction between the block and the surface?

125. A 1.35-kg block of wood sits at the edge of a table, 0.782 m above the floor. A 0.0105-kg bullet moving horizontally with a speed of 715 m/s embeds itself within the block. What horizontal distance does the block cover before hitting the ground?

126. The three air-track carts shown in **Figure 7.19** have masses, from left to right, of m, $2m$, and $4m$, respectively. Initially, the cart on the right is at rest, whereas the other two carts are moving to the right with a speed v. All three carts are equipped with putty bumpers that make them have completely inelastic collisions. **(a)** Find the final speed of the carts. **(b)** Calculate the ratio of the final kinetic energy of the system to the initial kinetic energy.

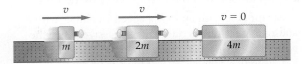

Figure 7.19

Writing about Science

127. Write a paragraph describing in detail the forces involved when two objects collide. Choose a specific collision to write about, such as a tennis ball being hit with a tennis racket. Be sure to discuss action and reaction forces, time of contact, and impulse.

128. Connect to the Big Idea Collisions conserve momentum, but not velocity. What is the difference between momentum and velocity? To illustrate your answer, give an example of objects that have the same momentum but different velocities. Similarly, give an example of objects that have different momentums but the same velocity.

Read, Reason, and Respond

The Gravitational Slingshot Spacecraft often use a navigational maneuver known as the *gravitational slingshot effect*. In this maneuver a close encounter with a planet produces a significant increase in the speed of a spacecraft, helping to send it to its destination.

The first use of this effect occurred on February 5, 1974, when *Mariner 10* made a close flyby of Venus on its way to Mercury. More recently, the probe *Cassini*, which was launched on October 15, 1997 and arrived at Saturn on July 1, 2004, made two close passes by Venus, followed by a flyby of Earth and a flyby of Jupiter.

A simplified version of the slingshot maneuver is shown in **Figure 7.20**, where we see a spacecraft moving to the left with an initial speed v_i, a planet moving to the right with a speed u, and the same spacecraft moving to the right with a final speed v_f after the encounter. This interaction can be thought of as an elastic collision in one dimension—as if the planet and the spacecraft were two carts on an air track. As a result, the relative speed of approach is equal to the relative speed of departure. This condition, along with the fact that the speed of the massive planet is essentially unchanged, can be used to determine the final speed of the spacecraft.

Figure 7.20

129. From the perspective of an observer on the planet, what is the spacecraft's speed of approach?

    A. $v_i + u$ C. $u - v_i$

    B. $v_i - u$ D. $v_f - u$

130. From the perspective of an observer on the planet, what is the spacecraft's speed of departure?

    A. $v_f + u$ C. $u - v_f$

    B. $v_f - u$ D. $v_i - u$

131. Set the speed of departure from Problem 130 equal to the speed of approach from Problem 129. Solving this relation for the final speed, v_f, yields

    A. $v_f = v_i + u$ C. $v_f = v_i + 2u$

    B. $v_f = v_i - u$ D. $v_f = v_i - 2u$

Standardized Test Prep

Multiple Choice

Use the graph below to answer Questions 1 and 2. The graph describes the motion of a 0.1-kg air-track cart as it collides with an elastic band.

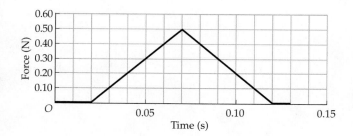

1. What is the total change in the cart's momentum during the time it is in contact with the elastic band?
   (A) 0
   (B) 0.025 kg·m/s
   (C) 0.05 kg·m/s
   (D) 0.5 kg·m/s

2. Calculate the change in velocity of the cart from 0.02 s to 0.07 s.
   (A) 0
   (B) 0.050 m/s
   (C) 0.125 m/s
   (D) 0.250 m/s²

3. Object A, which is moving to the right with a speed v, collides head-on and elastically with an identical object, B, moving to the left with a speed $2v$. After the collision object A moves to the left with a speed $2v$. What are the speed and the direction of motion of object B after the collision?
   (A) $2v$ to the left
   (B) v to the left
   (C) $2v$ to the right
   (D) v to the right

4. During an inelastic collision the system's kinetic energy decreases because
   (A) kinetic energy is changed to other forms of energy.
   (B) kinetic energy has to decrease as momentum increases.
   (C) kinetic energy cannot be conserved during any collision.
   (D) kinetic energy decreases any time a force is applied to a system.

5. Which expression gives the change in velocity?
   (A) $\Delta p/t$
   (B) $Fa/\Delta t$
   (C) $ma/\Delta t$
   (D) $\Delta p/m$

6. On an air track a 1.0-kg cart moving at 2.0 m/s toward the right collides with a 2.0-kg cart moving at 3.0 m/s toward the left. After the collision the 1.0-kg cart is moving at 2.0 m/s toward the left. The velocity of the 2.0-kg cart after the collision is
   (A) 1.0 m/s toward the right.
   (B) 1.0 m/s toward the left.
   (C) 2.0 m/s toward the right.
   (D) 2.0 m/s toward the left.

Extended Response

7. In a controlled demonstration a bowling ball is rolled along the floor of a bus moving at constant velocity. The ball hits the back of the bus. What is the effect on the velocity of the bus? Explain your answer.

> **Test-Taking Hint**
>
> Force, impulse, and momentum are vectors that must be resolved into components when solving problems involving two-dimensional collisions.

If You Had Difficulty With . . .

Question	1	2	3	4	5	6	7						
See Lesson(s)	7.1	7.2	7.3, 7.4	7.4	7.2	7.3	7.1						

8 Rotational Motion and Equilibrium

Inside

The riders on this swing carousel are enjoying the experience of rotational motion. In this chapter you will learn how to calculate both their angular speed and their linear speed.

Big Idea

Forces can produce torques, and torques can produce rotation.

It's no exaggeration to say that *rotation* is a part of everyday life. After all, we live on a planet that rotates once a day and revolves around the Sun once a year. Automobile engines have moving parts that rotate rapidly, as do devices that play CDs and DVDs. In this chapter we'll explore the basic principles that apply to rotating objects of all types.

Read the activity and obtain the required materials.

Explore

1. Suspend a small mass (about 100 g) about 10 cm from one end of a half-meterstick. Hold that end of the stick with one hand.
2. With your arm extended and your forearm parallel to the floor, use your wrist to rotate the half-meterstick up and down. Note the effort required. This is Trial 1.

3. Repeat Steps 1 and 2, this time placing the mass about 40 cm away from an end. This is Trial 2.
4. Repeat Steps 1 and 2 and then Step 3 using a larger mass (about 200 g). These are Trials 3 and 4.

Think

1. **Rank** List the four trials, from easiest to hardest, according to how difficult it was to rotate the stick. Was the heaviest mass always the hardest to rotate? Why or why not?

2. **Identify** What two factors did you change in the trials? Based on your observations, explain how each of these factors affected how difficult it was to rotate the stick.
3. **Apply** Explain how the factors you identified in Question 2 apply to a child and an adult riding up and down on opposite ends of a seesaw. Where must each person sit?

8.1 Describing Angular Motion

Every time you ride a bicycle or drive a car, you use the circular motion of wheels to get from one place to another. To gain a better understanding of circular motion—also referred to as *rotational motion*—we begin by considering the position, speed, and acceleration of a rotating object.

Angular position is related to an object's orientation

If you want to describe the motion of an object moving in a straight line, you set up a coordinate system with an origin (where the position is zero) and a positive direction. We do the same kind of thing to describe the motion of a rotating object.

For example, consider a bicycle wheel that is free to rotate about its axle, as shown in **Figure 8.1**. The axle is the *axis of rotation* for the wheel. As the wheel rotates, every point on it moves in a circular path centered on the axis of rotation.

Now, suppose there is a small spot of red paint on the bicycle tire, and we would like to describe its rotational motion. The first step is to define its *angular position*. The **angular position** of an object is the angle θ that it makes with respect to a given reference line. This is illustrated in Figure 8.1 for the spot of red paint. Note that the reference line defines where the angular position is zero, $\theta = 0$. This is just like the origin in a linear coordinate system, which defines where the linear position is zero, $x = 0$.

Vocabulary

- angular position
- radian
- average angular velocity
- angular speed
- average angular acceleration

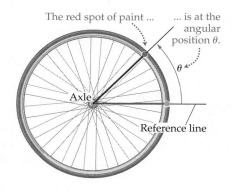

▶ **Figure 8.1 Angular position**
The angular position, θ, of a spot of paint on a bicycle wheel. The reference line is where $\theta = 0$. Positive angles correspond to counterclockwise rotation; negative angles correspond to clockwise rotation.

Definition of Angular Position, θ

$\theta =$ angle measured from reference line

SI unit: radian (rad), which is dimensionless

CONNECTINGIDEAS

In Chapters 2 and 3 we defined position, velocity, and acceleration.

• Here we revise these definitions to apply to rotational motion. The basic equations of motion are the same; only the names have been changed.

Sign of Angular Position

The sign of the angular position depends on its orientation relative to the reference line.

- **Counterclockwise rotation from the reference line corresponds to** *positive* angles, $\theta > 0$.
- **Clockwise rotation from the reference line corresponds to** *negative* angles, $\theta < 0$.

For example, the angle shown in Figure 8.1 is positive.

The radian is a useful unit for measuring angles

Now that we have established a reference line (for $\theta = 0$) and a positive direction of rotation (counterclockwise), we must choose units with which to measure angles. Common everyday units are degrees (°) and revolutions (rev). One revolution—that is, going completely around a circle—corresponds to 360°:

$$1 \text{ rev} = 360°$$

You may have noticed that some news programs use 360 in their titles—this implies that the program considers an issue from all possible directions.

The most convenient unit for angle measurements in scientific calculations, however, is the radian. A **radian** (rad) is the angle for which the length of a circular arc is equal to the radius of the circle. An example of a radian is shown in **Figure 8.2**. Here we show a piece of pie cut with an angle of 1 radian. All three sides of the piece of pie have the same length. In particular, the arc length of the outer crust edge is equal in length to the radius of the pie. Thus, the next time a server asks how big a piece of pie you want, reply "One radian, please" if you're feeling hungry.

A comparison between angles measured in degrees and radians is shown in **Figure 8.3**.

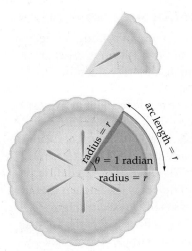

▲ **Figure 8.2 The definition of a radian**
The angle for this piece of pie is equal to 1 radian (about 57.3°). Thus, all three of its sides are of equal length.

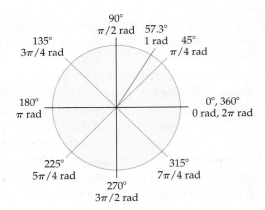

▲ **Figure 8.3 Comparing degrees and radians**
Angles are indicated around the circumference of this circle in both degrees and radians.

Radians are useful in scientific calculations

The radian is useful because it provides a simple relationship between angles and arc lengths, as illustrated in **Figure 8.4**. Multiplying the angle θ (measured in radians) by the radius gives the arc length, s. In equation form:

$$s = r\theta$$

For example, the arc length, s, for an angle of 1 radian ($\theta = 1$) is equal to the radius, $s = r$. This simple and straightforward relation does not hold for degrees or revolutions.

In one revolution an object rotates through a complete circle, 360°. The corresponding arc length is the circumference of a circle, $2\pi r$. Comparing this relationship with $s = r\theta$, we see that a complete revolution corresponds to $\theta = 2\pi$ radians. Therefore,

$$1 \text{ rev} = 360° = 2\pi \text{ rad}$$

One final note on the units for angles: Radians, as well as degrees and revolutions, are dimensionless. In the relation $s = r\theta$, for example, the arc length (s) and the radius (r) both have the SI unit of meter, and the angle has no dimensions. For clarity we will write an angle of 3 radians as $\theta = 3$ rad, just to remind us of which angular unit is being used.

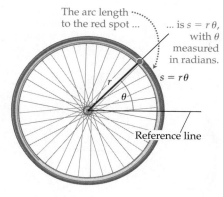

▲ **Figure 8.4 Arc length**
The arc length, s, from the reference line to the spot of paint is given by $s = r\theta$ if the angular position, θ, is measured in radians.

QUICK Example 8.1 How Many Degrees?

How many degrees correspond to 1 radian?

Solution
We have seen that 360° equals 2π rad. That is, $360° = 2\pi$ rad. Dividing both sides of this equation by 2π yields

$$1 \text{ rad} = \frac{360°}{2\pi} = \boxed{57.3°}$$

Practice Problems

1. The following angles are given in degrees. Convert them to radians: 30°, 45°, 90°, 180°.

2. The following angles are given in radians. Convert them to degrees: $\pi/6$ rad, 0.70 rad, 1.5π rad, 5π rad.

Angular velocity: how angular position changes

In linear motion velocity is the change in position divided by the change in time. The same idea applies to angular motion. For example, as the bicycle wheel in Figure 8.1 rotates, the angular position of the spot of red paint changes. This is illustrated in **Figure 8.5**. The *angular displacement* of the spot, $\Delta\theta$, is the difference between its final angle and its initial angle:

$$\Delta\theta = \theta_f - \theta_i$$

The angular displacement divided by the time during which the displacement occurs is the **average angular velocity**, ω_{av}. (Recall that the subscript av stands for "average.")

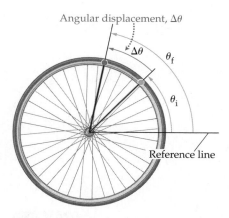

▲ **Figure 8.5 Angular displacement**
As the wheel rotates, the spot of paint undergoes an angular displacement, $\Delta\theta = \theta_f - \theta_i$.

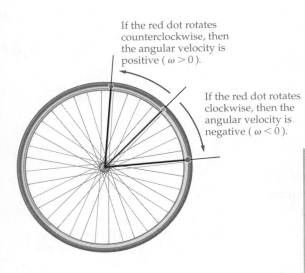

If the red dot rotates counterclockwise, then the angular velocity is positive ($\omega > 0$).

If the red dot rotates clockwise, then the angular velocity is negative ($\omega < 0$).

◀ **Figure 8.6 Angular speed and velocity**
Counterclockwise rotation corresponds to a positive angular velocity ($\omega > 0$). Similarly, clockwise rotation corresponds to a negative angular velocity ($\omega < 0$). The magnitude of the angular velocity is the angular speed.

If the angular displacement, $\Delta\theta$, occurs in the time Δt, the average angular velocity is as follows:

> **Definition of Average Angular Velocity, ω_{av}**
>
> $$\text{average angular velocity} = \frac{\text{change in angular position}}{\text{change in time}}$$
>
> $$\omega_{av} = \frac{\Delta\theta}{\Delta t}$$
>
> SI units: radian per second $(\text{rad}/\text{s}) = \text{s}^{-1}$

This is just like the definition of the average linear velocity: $v_{av} = \Delta x/\Delta t$. The units for angular velocity are rad/s, which is simply 1/s since the radian has no dimensions.

How is the sign of angular velocity determined?

Sign of Angular Velocity Notice that we call ω the angular velocity, not the angular speed. The reason is that ω can be positive or negative, depending on the direction of rotation. For example, if the red paint spot rotates in the counterclockwise direction, the angular position, θ, increases. As a result, $\Delta\theta$ is positive and, therefore, so is ω. Similarly, clockwise rotation corresponds to a negative $\Delta\theta$ and hence a negative ω.

- **Counterclockwise rotation corresponds to positive angular velocity, $\omega > 0$.**
- **Clockwise rotation corresponds to negative angular velocity, $\omega < 0$.**

The *sign* of ω indicates the *direction* of the angular velocity, as shown in **Figure 8.6**. Similarly, the magnitude of the angular velocity is called the **angular speed**.

> ### QUICK Example 8.2 What's the Angular Velocity?
>
> An antique long-playing (LP) phonograph record rotates clockwise at $33\frac{1}{3}$ rpm (revolutions per minute). What is its angular velocity in radians per second?
>
> **Solution**
> Convert from revolutions per minute to radians per second. This can be done by using one conversion factor to convert from revolutions to radians and another factor to convert from minutes to seconds. Notice that clockwise rotation corresponds to a negative angular velocity:
>
> $$\omega = -33\frac{1}{3}\frac{\text{rev}}{\text{min}}\left(\frac{2\pi\ \text{rad}}{1\ \text{rev}}\right)\left(\frac{1\ \text{min}}{60\ \text{s}}\right)$$
>
> $$= \boxed{-3.49\ \text{rad}/\text{s}}$$

Math HELP
Unit Conversion
See Lesson 1.3

3. A CD rotates at 22.0 rad/s. What is its angular speed in revolutions per minute (rpm)?

4. A ceiling fan rotates at the rate of 45° every 0.75 s. What is the angular speed of the fan in radians per second?

5. An airplane propeller rotates with an angular speed of 260 rad/s. Through what angle does the propeller rotate in 5.0 s? Give your answer in both radians and degrees.

6. How much time does it take for a spinning baseball with an angular speed of 38 rad/s to rotate through 15°?

Angular and linear velocities are related

Let's say you go to a county fair and ride on the merry-go-round. Using your digital watch, you find that it takes 7.5 s to complete one revolution (2π radians). It follows that your angular velocity on the merry-go-round is

$$\omega = \frac{\Delta\theta}{\Delta t} = \frac{2\pi \text{ rad}}{7.5 \text{ s}} = 0.84 \text{ rad/s}$$

Your path is circular, with the center of the circle at the axis of rotation of the merry-go-round. In addition, notice that at any instant of time you are moving in a direction that is **tangential** to the circular path, as **Figure 8.7** shows. What is your tangential speed, v_t? In other words, what is the speed of the wind in your face?

Tangential Speed To find your tangential speed, divide the distance corresponding to the circumference of the circular path, $d = 2\pi r$, by the time required to complete one circuit, $t = T$. This gives the following:

$$v_t = \frac{d}{t} = \frac{2\pi r}{T} = r\left(\frac{2\pi}{T}\right)$$

The quantity $2\pi/T$ is your angular velocity, ω. After all, 2π is your angular displacement for one revolution, and T is the corresponding amount of time. Therefore, we can express your tangential speed as follows:

> **Tangential Speed of a Rotating Object**
>
> tangential speed = radius × angular speed
> $$v_t = r\omega$$
>
> SI units: m/s

Important note: The angular speed, ω, must be given in radians per second (rad/s) for this relation to be valid.

In the case of the merry-go-round, if the radius of your circular path is $r = 4.3$ m, your tangential speed is

$$v_t = r\omega = (4.3 \text{ m})(0.84 \text{ rad/s}) = 3.6 \text{ m/s}$$

When it is clear that we are referring to the tangential (or linear) speed, we will drop the subscript t and simply write $v = r\omega$.

An interesting application of the relation between linear and angular speeds is provided by the playing of a compact disk (CD), as we see in the next Guided Example.

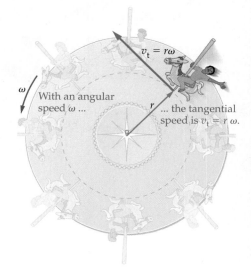

▲ **Figure 8.7 Angular and linear speed** Overhead view of a child riding on a merry-go-round. The child's path is a circle centered on the merry-go-round's axis of rotation. At any given time the child is moving tangentially to the circular path with a speed $v_t = r\omega$.

A CD is played by shining a laser beam onto the disk and converting the pattern of reflected light into a pattern of sound waves. For proper operation the linear (tangential) speed of the disk where the laser beam shines on it must be maintained at the constant value of 1.25 m/s. As the CD is played, the laser beam scans the disk in a spiral track beginning near the center and proceeding outward to the rim. What angular speed must a CD have when the laser beam is shining on the disk (**a**) 2.50 cm and (**b**) 6.00 cm from the axis of rotation?

Picture the Problem

A spinning CD is shown in our sketch. In addition, we show two different locations for the laser beam, corresponding to parts (a) and (b) of the problem.

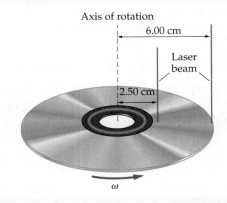

Axis of rotation
6.00 cm
Laser beam
2.50 cm
ω

Strategy

The *angular* speed of the CD varies as the laser beam moves outward from the center to the rim. The *linear* (tangential) speed where the laser beam shines on the CD is always the same, however. Therefore, we solve the equation $v = r\omega$ for the angular speed, ω, and substitute the appropriate values for the radius, r. The value of the linear speed is $v = 1.25$ m/s.

Known	Unknown
$v = 1.25$ m/s	$\omega = ?$
(a) $r = 2.50$ cm	
(b) $r = 6.00$ cm	

Solution

1 (a) Solve $v = r\omega$ for ω and substitute $r = 2.50$ cm $= 0.0250$ m:

$$\omega = \frac{v}{r} = \frac{1.25 \text{ m/s}}{0.0250 \text{ m}} = \boxed{50.0 \text{ rad/s}}$$

2 (b) Substitute $r = 6.00$ cm $= 0.0600$ m in $\omega = v/r$:

$$\omega = \frac{v}{r} = \frac{1.25 \text{ m/s}}{0.0600 \text{ m}} = \boxed{20.8 \text{ rad/s}}$$

Math HELP
Significant Figures
See
Lesson 1.4

Insight

Thus, a CD slows from 50.0 rad/s (about 500 rpm) to roughly 20.8 rad/s (about 200 rpm) as it plays.

Practice Problems

7. **Follow-up** At what distance from the axis of rotation does the laser beam shine when the CD's angular speed is 44.0 rad/s?

8. The hour hand on a certain clock is 8.2 cm long. Find the tangential speed of the tip of this hand during normal operation.

How do the angular and tangential speeds of an object vary from one point to another? We explore this question in the following Conceptual Example.

Two children ride on a merry-go-round, with child 1 at a greater distance from the axis of rotation than child 2. Is the angular speed of child 1 greater than, less than, or the same as the angular speed of child 2?

Reasoning and Discussion

At any given time the angle θ for child 1 is the same as the angle θ for child 2, as shown. Therefore, when the angle for child 1 has gone through 2π, for example, so has the angle for child 2. As a result, both children have the same angular speed. In fact, *each and every point on the merry-go-round has exactly the same angular speed.*

Answer

The children's angular speeds are the same.

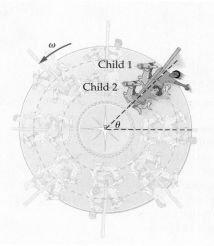

Though the angular speeds of the children are the same in Conceptual Example 8.4, their tangential speeds are different. Child 1 has the greater tangential speed since he travels around a larger circle in the same time that child 2 travels around a smaller circle. This is in agreement with the equation $v = r\omega$, since the radius to child 1 is greater than the radius to child 2. The dependence of linear speed on distance from the axis of rotation is illustrated in **Figure 8.8**.

Angular acceleration: changing angular velocity

A rotating object accelerates if its rate of spinning changes. For example, if you attach a bucket to a pulley and let it drop, the pulley will spin at an increasing rate. The acceleration of the pulley is similar to the acceleration of an object moving in a straight line.

For example, if the angular velocity of a rotating bicycle wheel increases or decreases with time, we say that the wheel has an *angular acceleration*. In general, the **average angular acceleration**, α_{av}, is the change in angular velocity, $\Delta\omega$, in a given interval of time, Δt.

▲ **Figure 8.8 Dependence of linear speed on distance**
In the photo on the left, the blurring of the cars at the outer rim of the Ferris wheel clearly shows that they are moving faster than the parts of the wheel closer to its central axis of rotation. Similarly, in the photo on the right, the boy near the rim of the playground merry-go-round is moving faster than the girl near the hub.

$\alpha > 0$ $\omega > 0$

$\omega < 0$ $\alpha < 0$

$\alpha > 0$ $\omega < 0$

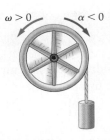
$\omega > 0$ $\alpha < 0$

(a) Angular speed increases   **(b)** Angular speed increases   **(c)** Angular speed decreases   **(d)** Angular speed decreases

▲ **Figure 8.9 Angular acceleration and angular speed**
When angular velocity and acceleration have the same sign, as in **(a)** and **(b)**, the angular speed increases. When angular velocity and angular acceleration have opposite signs, as in **(c)** and **(d)**, the angular speed decreases.

🔑 *What determines whether the speed of rotation increases or decreases?*

Definition of Average Angular Acceleration, α_{av}

$$\text{average angular acceleration} = \frac{\text{change in angular velocity}}{\text{change in time}}$$

$$\alpha_{av} = \frac{\Delta\omega}{\Delta t}$$

SI units: radians per second per second $(\text{rad}/\text{s}^2) = \text{s}^{-2}$

Notice that the SI units for angular acceleration are rad/s^2. Since the radian is dimensionless, these units simplify to s^{-2}. If the angular acceleration is constant, which is often the case, we will drop the subscript av, since the average angular acceleration is then the same as the angular acceleration.

The sign of the angular acceleration is determined by whether the change in angular velocity is positive or negative. For example, if ω is becoming more positive with time, it follows that α is positive. Similarly, if ω is becoming more negative with time, it follows that α is negative. We can summarize these relationships as follows:

- 🔑 If ω and α have the *same sign*, then the speed of rotation is *increasing*.
- 🔑 If ω and α have *opposite signs*, then the speed of rotation is *decreasing*.

These relations are illustrated in **Figure 8.9**.

QUICK Example 8.5 What's the Stopping Time?

As the wind dies, a windmill that was rotating at 2.1 rad/s begins to slow down with a constant angular acceleration of -0.45 rad/s². How much time does it take for the windmill to come to a complete stop?

Solution
The initial angular velocity is $\omega_i = 2.1$ rad/s, and the final angular velocity is $\omega_f = 0$ rad/s. Solving $\alpha = \Delta\omega/\Delta t$ for the time yields

$$\Delta t = \frac{\Delta\omega}{\alpha}$$

$$= \frac{\omega_f - \omega_i}{\alpha}$$

$$= \frac{0 - 2.1 \text{ rad/s}}{-0.45 \text{ rad/s}^2}$$

$$= \boxed{4.7 \text{ s}}$$

Math HELP
Negative Numbers
See Math Review, Section I

9. The wheels of a car speed up from 5.2 rad/s to 7.9 rad/s in 1.3 s. What is the angular acceleration of the wheels?

10. As you start riding a bicycle, the wheels begin at rest and have an angular acceleration of 2.3 rad/s². What is the angular speed of the wheels after 3.8 s?

Angular and linear accelerations are related

Just as linear speed is equal to the radius times the angular speed, the linear acceleration is equal to the radius times the angular acceleration. Specifically, if a_t is the tangential (linear) acceleration and α is the angular acceleration, the relation between the two is

> **Tangential Acceleration of a Rotating Object**
>
> tangential acceleration = radius × angular acceleration
> $$a_t = r\alpha$$
>
> SI units: m/s

Important note: The angular acceleration, α, must be given in radians per second squared (rad/s²) for this relation to be valid.

When it is clear that we are referring to the tangential (linear) acceleration, we will drop the subscript t and simply write $a = r\alpha$.

8.1 LessonCheck (MP)

Checking Concepts

11. ⊂▭ **Identify** A bicycle wheel has rotated 32° counterclockwise from the reference line. Is this angular position positive or negative?

12. ⊂▭ **Analyze** You observe the wheels of a car as it moves past you from right to left. Do the wheels have a positive or a negative angular velocity?

13. ⊂▭ **Determine** An object has an angular velocity of 1.0 rad/s and an angular acceleration of −0.5 rad/s². Is the speed of its rotation increasing or decreasing?

14. Identify What is the arc length for an angle that is equal to 2 rad?

15. Triple Choice Is the angular speed of the hour hand of a clock greater than, less than, or equal to the angular speed of the minute hand?

Solving Problems

16. Apply Through how many radians does the minute hand of a clock rotate in 15 min?

17. Apply Find the angular speed of the Earth as it orbits about the Sun. Give your answer in radians per second (rad/s).

18. Calculate A bicycle wheel with a radius of 0.62 m rotates with an angular speed of 21 rad/s about its axle, which is at rest. What is the linear speed of a point on the rim of the wheel?

19. Calculate A propeller on a ship has an initial angular velocity of 5.1 rad/s and an angular acceleration of 1.6 rad/s². What is the angular velocity of the propeller after 3.0 s?

8.2 Rolling Motion and the Moment of Inertia

Vocabulary

- moment of inertia
- rotational kinetic energy
- angular momentum

We began this chapter with a bicycle wheel rotating about its axle. In that case the axle was at rest, and every point on the wheel moved in a circular path about the axle. In this lesson you'll learn what happens when the wheel rolls along the ground.

Rolling Motion

The invention of the wheel was a huge step forward for humanity. Today, we take for granted the rolling wheels on everything from cars to trains to bicycles. Still, the behavior of a rolling wheel is more interesting than you might imagine.

Rolling combines two kinds of motion

What two kinds of motion are combined in a rolling wheel?

Suppose a bicycle wheel rolls freely, with no slipping between the tire and the ground. The wheel rotates about its axle, and at the same time the axle moves in a straight line. **As a result, the *rolling motion* of a wheel is a combination of both rotational motion and linear motion.** This motion is illustrated in **Figure 8.10**.

Notice that the bottom of the rolling wheel is in static contact with the ground. The forward motion of the axle exactly cancels the backward motion of the bottom of the wheel, resulting in an instantaneous speed of zero at the bottom. On the other hand, the top of the wheel has twice the speed of the axle. Thus, if a car's speed is v (as read on the speedometer), the tops of its wheels have the speed $2v$.

The fact that the bottom of the wheel is instantaneously at rest, and therefore in static contact with the ground, is precisely what is meant by "rolling without slipping." In fact, a wheel that rolls without slipping is just like a person who is walking without slipping—even though the body moves forward, the soles of the shoes are momentarily at rest every time they are placed on the ground. Watch your feet carefully the next time you walk down the hallway to see that this is what happens. (Be careful not to bump into anyone, though! Also, see Conceptual Example 5.13 in Chapter 5.)

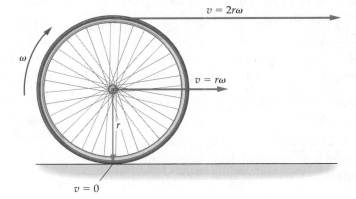

▶ Figure 8.10 Velocities associated with rolling motion
In a wheel that rolls without slipping, the point in contact with the ground is instantaneously at rest. The center of the wheel moves forward with the speed $v = r\omega$, and the top of the wheel moves forward with twice that speed, $v = 2r\omega$.

A car whose tires have a radius of 32 cm travels down the highway at 25 m/s (about 55 mph). What is the angular speed of the tires?

Solution

First, convert 32 cm to 0.32 m. Then, rearrange $v = r\omega$ to solve for the angular speed, ω:

$$\omega = \frac{v}{r} = \frac{25 \ \cancel{m}/s}{0.32 \ \cancel{m}} = \boxed{78 \ \text{rad/s}}$$

This is about 12 revolutions per second.

Practice Problems

20. Follow-up What is the linear speed of the top of each tire?

21. A wooden plank rests on two soup cans laid on their sides, as shown in the upper part of **Figure 8.11**. The plank is then pulled 1.0 m to the right, as shown in the lower part of the figure. The cans roll without slipping. How far does the center of each can move?

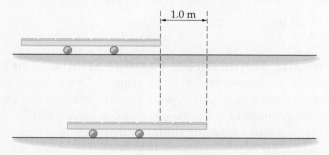

▶ **Figure 8.11**

Moment of Inertia

An object in motion resists changes to that motion. With linear motion we say that the object has inertia. The greater the mass of the object, the more difficult it is to change its motion and the greater its inertia. A similar relationship occurs with rotational motion.

Rotating objects resist changes in rotation

Some objects are hard to start or stop rotating, like a merry-go-round at a park. The rotation of other objects, like baseballs or bicycle wheels, can be started or stopped rather easily. Each object has a **moment of inertia**, I, which determines how easy or hard it is to change its rotation. 🗝 **An object with a large moment of inertia is difficult to start or stop rotating.** The moment of inertia plays the same role for rotational motion that mass does for linear motion.

As an example, hold onto a meterstick with your hand at its center. Rotate the stick back and forth rapidly. Not too hard, is it? Next, grip the stick at one end and try rotating it back and forth just as rapidly as before. Considerably harder now, isn't it? It's harder because more of the mass is farther from your hand than it was before. Thus, the farther an object's mass is from the axis of rotation, the greater the object's moment of inertia.

🗝 *How is an object's moment of inertia related to changes in its rotation?*

Moment of inertia depends on mass and distance

Experiments show that if the distance from the axis of rotation to a mass is doubled, the moment of inertia becomes four times greater. Similarly, if the mass of an object is doubled, the moment of inertia doubles. **It follows that the moment of inertia depends linearly on the mass and on the distance squared.** In fact, for a mass m at a distance r from the axis of rotation, the moment of inertia is $I = mr^2$.

> **Moment of Inertia, I
> (for a single object of mass m at a distance r)**
>
> moment of inertia = mass $\times$ (radius)^2
> $$I = mr^2$$
>
> SI units: $kg \cdot m^2$

In a system with several particles at different distances from the axis of rotation, the moment of inertia is the sum of the mr^2 for each object separately:

> **Moment of Inertia, I
> (for a collection of objects)**
> $$I = m_1 r_1^2 + m_2 r_2^2 + m_3 r_3^2 + \cdots$$
> SI units: $kg \cdot m^2$

Moments of Inertia for Various Shapes

The above results are valid for individual particles or groups of particles. What about a solid object, like a disk or a sphere, with a continuous distribution of mass? The moments of inertia for objects like this can be determined as well. **Table 8.1** collects the results for some common shapes.

Notice, for example, that the moment of inertia of a hoop of mass m and radius r is mr^2. This is because all the mass of a hoop is at the radius r. In contrast, the mass of a disk is spread uniformly from the axis of rotation out to the rim. This means that a lot of mass is near the axis of rotation, which reduces the moment of inertia. In fact, the moment of inertia for a disk is $\frac{1}{2}mr^2$, exactly half the result for the hoop.

> **Moments of Inertia for a Hoop and a Disk**
> $$I_{\text{hoop}} = mr^2$$
> $$I_{\text{disk}} = \tfrac{1}{2}mr^2$$
> SI units: $kg \cdot m^2$

This difference in moment of inertia means that it's twice as hard to start or stop a hoop compared to a disk, given equal mass and radius.

Table 8.1 also shows why it is easier to rotate a meterstick about its center than about one end. A meterstick is basically a long, thin rod. From Table 8.1 we see that the moment of inertia of a thin rod about its end is four times greater than the moment of inertia about its center. This increase in the moment of inertia caused the effect you felt as you tried to rotate the meterstick about its end.

Table 8.1 Moments of Inertia for Uniform, Rigid Objects of Various Shapes and Total Mass *m*

Hoop or
cylindrical shell
$I = mr^2$

Disk or
solid cylinder
$I = \frac{1}{2}mr^2$

Disk or
solid cylinder
(axis at rim)
$I = \frac{3}{2}mr^2$

Long thin rod
(axis through midpoint)
$I = \frac{1}{12}mL^2$

Long thin rod
(axis at one end)
$I = \frac{1}{3}mL^2$

Hollow sphere
$I = \frac{2}{3}mr^2$

Solid sphere
$I = \frac{2}{5}mr^2$

Solid sphere
(axis at rim)
$I = \frac{7}{5}mr^2$

Solid plate
(axis through center,
in plane of plate)
$I = \frac{1}{12}mL^2$

Solid plate
(axis perpendicular
to plane of plate)
$I = \frac{1}{12}m(L^2 + W^2)$

CONCEPTUAL Example 8.7 Which Object Wins the Race?

A disk and a hoop of the same mass and radius are released at the same time at the top of an inclined plane. Does the disk reach the bottom of the plane before, after, or at the same time as the hoop?

Reasoning and Discussion
The hoop has the larger moment of inertia. As a result, the hoop starts rotating more slowly than the easier-to-rotate disk. Thus, the disk reaches the bottom first and wins the race.

Answer
The disk reaches the bottom before the hoop.

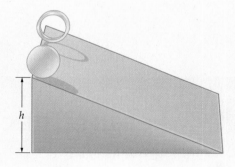

Rotational *KE* depends on the moment of inertia

Recall that an object in motion has kinetic energy. This is true whether the motion is linear or rotational. For linear motion the kinetic energy of a mass *m* moving with a speed *v* is $KE = \frac{1}{2}mv^2$. To find the kinetic energy for rotational motion, we first replace the linear speed, *v*, with the angular speed, ω. Next, we replace the mass, *m*, with the moment of inertia, *I*, which we can think of as "rotational mass." Thus, the **rotational kinetic energy**, KE_{rot}, of an object is one-half the product of its moment of inertia and the square of its angular speed.

> **Rotational Kinetic Energy**
>
> $\text{rotational kinetic energy} = \frac{1}{2}\left(\text{moment of inertia}\right) \times \left(\text{angular speed}\right)^2$
>
> $KE_{\text{rot}} = \frac{1}{2}I\omega^2$
>
> SI unit: J

CONNECTINGIDEAS

Kinetic energy and conservation of energy were introduced in Chapter 6, and conservation of momentum was discussed in Chapter 7.
• Here kinetic energy plays a key role in defining the moment of inertia.
• Later in this chapter we apply conservation of energy and angular momentum to rotational motion.

Table 8.2 Connections between Linear and Angular Quantities

Linear Quantity	Angular Quantity
m	I
v	ω
$KE = \frac{1}{2}mv^2$	$KE = \frac{1}{2}I\omega^2$
$p = mv$	$L = I\omega$
$F = ma$	$\tau = I\alpha$

Connections like this between linear quantities (m and v) and angular quantities (I and ω) are summarized in **Table 8.2**. These connections make it easier to remember the angular quantities, since they are all directly related to the corresponding linear quantities.

Angular momentum is similar to linear momentum

We've seen that the kinetic energy of a rotating object is very similar to the kinetic energy of linear motion. In fact, all we need to do is replace mass with moment of inertia (m with I) and linear speed with angular speed (v with ω). We can do exactly the same thing with momentum. Recall that linear momentum is $p = mv$. If we replace m with I and v with ω, we obtain the *angular momentum*. **Angular momentum**, L, is the product of an object's moment of inertia, I, and its angular speed, ω.

> **Angular Momentum, L**
>
> angular momentum = moment of inertia × angular speed
> $$L = I\omega$$
>
> SI units: $kg \cdot m^2/s$

Notice, for example, that a hoop has twice the angular momentum of a disk, given the same mass, radius, and angular speed.

8.2 LessonCheck (MP)

Checking Concepts

22. ⟵ **Explain** How does rolling motion differ from pure rotational motion? How does it differ from pure linear motion?

23. ⟵ **Explain** How does increasing the moment of inertia affect how easily an object rotates?

24. ⟵ **Generalize** How will doubling the mass of an object affect its moment of inertia?

25. Predict A disk and a solid sphere are raced down the inclined plane in Conceptual Example 8.7. Use Table 8.1 to predict the winner of this race.

26. Identify All wheels that roll without slipping have the same speed at the bottom of the wheel. What is this speed?

27. Triple Choice A hoop and a disk have the same mass and radius. In addition, they both spin about their centers with the same angular speed. Is the kinetic energy of the hoop greater than, less than, or equal to the kinetic energy of the disk? Explain. (Refer to Table 8.1.)

Solving Problems

28. Calculate As a car travels along a road, the speed of the tops of its wheels is 46 m/s. What is the speed of the car and its occupants?

29. Calculate A basketball has a radius of 0.12 m and a mass of 0.57 kg. Assuming the ball to be a hollow sphere, what is its moment of inertia?

30. Calculate A chef spins a disk of pizza dough over her head, giving it an angular speed of 7.2 rad/s. If the moment of inertia of the pizza dough is 6.3×10^{-6} kg · m², what is its rotational kinetic energy? (Assume that the disk of dough is uniform.)

31. Calculate The moment of inertia of a ball is 1.6×10^{-8} kg · m². If the ball spins with an angular speed of 8.2 rad/s, what is its angular momentum?

8.3 Torque

A force can cause a rotation. How effective the force is depends on where the force is applied and in what direction. The effectiveness of forces in causing rotation is the subject of this lesson.

A force can cause rotation

Suppose you want to loosen a nut by rotating it counterclockwise with a wrench, as shown in **Figure 8.12**. You probably know that the nut is more likely to turn if you apply your force as far from it as possible. This is indicated in **Figure 8.13 (a)**, where we see that the required force far from the nut (F_1) is much less than the required force near the nut (F_2). Similarly, it is much easier to open a revolving door if you push far from the axis of rotation, as **Figure 8.13 (b)** indicates.

These observations show that the effectiveness of a force in causing a rotation depends on both the magnitude of the force and the distance from the axis of rotation to the force. Taking this into account, we define a new physical quantity, called *torque*. **Torque**, τ, is the product of force and distance:

Vocabulary

- torque
- moment arm

▲ **Figure 8.12 Applying a torque**
The long handle of this wrench enables the user to produce a large torque without having to exert a very great force.

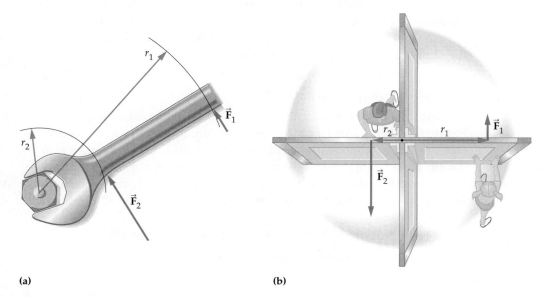

(a)   **(b)**

▲ **Figure 8.13 Torque depends on distance**
(a) When a wrench is used to loosen a nut, less force is required if it is applied far from the nut. **(b)** Similarly, less force is required to open a revolving door if it is applied far from the door's axis of rotation.

What does a torque do?

You can think of torque as the physical quantity that causes rotation. It increases with both force and distance.

The equation $\tau = rF$ is valid only when the applied force is *tangential* to a circle of radius r centered on the axis of rotation. This is true for the cases shown in Figure 8.13. The more general case is considered later in this lesson. First, let's use $\tau = rF$ to determine how much force is needed to open a revolving door, depending on where we apply the force.

ACTIVE Example 8.8 Find the Required Force

To open the door in Figure 8.13 (b) a tangential force F is applied at a distance r from the axis of rotation. If the minimum torque required to open the door is 3.1 N · m, what force must be applied if r is **(a)** 0.94 m or **(b)** 0.35 m?

Solution *(Perform the calculations indicated in each step.)*

(a) Solve $\tau = r_1 F_1$ for the force F_1; then substitute the numerical values: $\qquad F_1 = \dfrac{\tau}{r_1} = 3.3$ N

(b) Repeat the calculation, this time with $r_2 = 0.35$ m: $\qquad F_2 = \dfrac{\tau}{r_2} = 8.9$ N

As expected, the required force is less when it is applied farther from the axis of rotation.

Math HELP
Solving Simple Equations
See Math Review, Section III

Tangential components of force produce a torque

To this point we have considered tangential forces only. Let's look now at the case where a force is exerted in a direction that is not tangential.

For example, suppose you want to rotate a playground merry-go-round to give some friends a ride. If you pull in a direction that is directly toward or away from the axis of rotation, your force causes no rotation, as indicated in **Figure 8.14 (a)**. You can pull as hard as you want, but the merry-go-round just won't rotate. In this case we say that your force is *radial*, because it is in the same direction as the radius, which extends from the axis of rotation to the outer edge of the circular merry-go-round. We conclude the following:

- Radial forces produce zero torque.

But what if your force is at an angle θ relative to a radial line, as in **Figure 8.14 (b)**? To analyze this case, we resolve the force vector $\vec{\mathbf{F}}$ into radial and tangential components. The radial component ($F \cos \theta$) produces zero torque, as we have just seen. The tangential component of $\vec{\mathbf{F}}$ is $F \sin \theta$, and we know that tangential forces do produce torques. Therefore, the torque in the general case is distance times the tangential component of the force, $r(F \sin \theta)$:

> **Definition of Torque, τ (for a nontangential force)**
>
> $$\text{torque} = \text{radius} \times \text{force} \times \sin \theta$$
> $$\tau = rF \sin \theta$$
>
> SI units: N · m

The angle θ is the angle between the radius vector and the force vector, as indicated in Figure 8.14 (b).

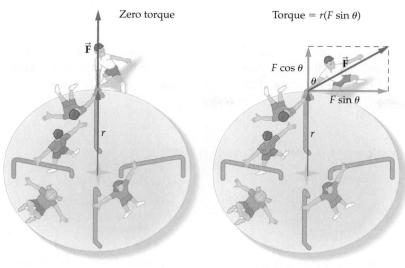

Zero torque

Torque = $r(F \sin \theta)$

$F \cos \theta$

$\vec{F}$

$F \sin \theta$

r

r

(a) A radial force produces zero torque.

(b) Only the tangential component of a force causes a torque.

◀ **Figure 8.14 Only the tangential component of a force causes a torque**
(a) A radial force causes no rotation. In this case the force $\vec{F}$ is opposed by an equal and opposite force exerted by the axle of the merry-go-round. The merry-go-round does not rotate. **(b)** A force applied at an angle θ with respect to the radial direction. The radial component of this force, $F \cos \theta$, causes no rotation; the tangential component, $F \sin \theta$, does produce rotation.

As a quick check, notice that a radial force corresponds to $\theta = 0$. In this case,

$$\tau = r(F \sin 0) = 0$$

This is as expected for a radial force. If the force is tangential, however, it follows that $\theta = \pi/2$. This gives

$$\tau = r(F \sin \pi/2) = rF$$

This agrees with our previous definition of torque for a tangential force. The general result applies for all possible angles.

Torque can be defined in terms of a moment arm

Figure 8.15 shows a geometric way of thinking about torque. The idea here is to first extend a line through the force vector. Next, draw a second line from the axis of rotation perpendicular to the line of the force. This is the shortest distance between the axis of rotation and the line of force. The perpendicular distance from the axis of rotation to the line of the force is defined as the **moment arm**, $r_\perp$. From Figure 8.15 you can see that the length of the moment arm is

$$r_\perp = r \sin \theta$$

Referring to the general equation for torque, $\tau = rF \sin \theta$, we can rearrange as follows:

$$\tau = rF \sin \theta$$
$$= (r \sin \theta)F$$
$$= r_\perp F$$

🔑 **Thus, the torque is equal to the moment arm times the force.** In equation form we write

$$\tau = r_\perp F$$

This gives us two equivalent ways to calculate the torque. One is to multiply the distance by the tangential component of the force: $\tau = r(F \sin \theta)$. The second is to multiply the force by the moment arm: $\tau = (r \sin \theta)F$. You can choose the method that is easier for a given situation.

🔑 *How is torque defined in terms of a moment arm?*

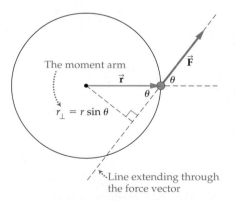

The moment arm

$\vec{r}$

$\vec{F}$

θ

θ

$r_\perp = r \sin \theta$

Line extending through the force vector

▲ **Figure 8.15 The moment arm**
To find the moment arm, $r_\perp$, for a given force, first extend a line through the force vector. Next, draw a perpendicular line from the axis of rotation to the line of the force. The perpendicular distance is the moment arm, $r_\perp = r \sin \theta$.

32. Concept Check A mass m is attached to a meterstick that you hold by the left end. Rank the cases shown in **Figure 8.16** in order of increasing difficulty of holding the meterstick still. Indicate ties where appropriate.

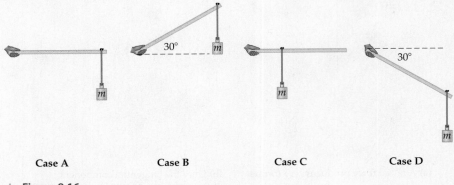

Case A   Case B   Case C   Case D

▲ **Figure 8.16**

33. A force of 8.8 N pushes on the rim of a wheel of radius 0.41 m. The direction of the force is at an angle of 22° relative to the radial direction. What is the torque produced by this force?

34. In Problem 33, what is the maximum torque the force can produce? What is its direction relative to the radial direction in this case?

Sign of Torque
If you pull on a merry-go-round in the direction shown in Figure 8.14 (b), it will rotate in the clockwise direction. Pulling in the opposite direction causes a counterclockwise rotation. We take these different directions of rotation into account by giving the torque a sign.

- A torque that causes a counterclockwise rotation is defined to be positive.
- A torque that causes a clockwise rotation is defined to be negative.

In a system with more than one torque, as in **Figure 8.17**, the sign of each torque is determined by the type of angular rotation *it alone* would produce. The net torque acting on the system, then, is the sum of the individual torques, taking into account the appropriate sign for each. This is illustrated in the following Guided Example.

▶ **Figure 8.17 Multiple torques**
The net torque on the wheel of this ship is the sum of the torques exerted by the two helmsmen. At the moment pictured, they are both exerting negative torques on the wheel, causing it to rotate in the clockwise direction. This will turn the boat to its left—or, in nautical terms, to port.

Two helmsmen, in disagreement about which way to turn a ship, exert the forces shown below on the ship's wheel. The wheel has a radius of 0.74 m, and the two forces have the magnitudes $F_1 = 72$ N and $F_2 = 58$ N. Find (**a**) the torque caused by $\vec{F}_1$ and (**b**) the torque caused by $\vec{F}_2$. (**c**) In which direction does the wheel turn as a result of these two forces?

Picture the Problem

Our sketch shows that both forces are applied at the distance $r = 0.74$ m from the axis of rotation. However, $F_1 = 72$ N is at an angle of 50.0° relative to the radial direction, whereas $F_2 = 58$ N is tangential, which means that its angle relative to the radial direction is 90.0°.

Strategy

For each force we find the magnitude of the corresponding torque, using $\tau = rF \sin \theta$. As for the signs of the torques, we must consider the angular rotation each force alone would cause. $\vec{F}_1$ acting by itself would cause the wheel to rotate counterclockwise, and hence its torque is positive. $\vec{F}_2$ alone would rotate the wheel clockwise, and hence its torque is negative. If the sum of the two torques is positive, the wheel rotates counterclockwise; if the sum of the two torques is negative, the wheel rotates clockwise.

Known	Unknown
$r = 0.74$ m	(**a**) $\tau_1 = ?$
$F_1 = 72$ N	(**b**) $\tau_2 = ?$
$F_2 = 58$ N	(**c**) direction $= ?$
$\theta_1 = 50.0°$	
$\theta_2 = 90.0°$	

Solution

1 (**a**) Use the general equation $\tau = r(F \sin \theta)$ to calculate the torque due to $\vec{F}_1$. This torque is positive:

$$\tau_1 = rF_1 \sin 50.0°$$
$$= (0.74 \text{ m})(72 \text{ N})(\sin 50.0°)$$
$$= \boxed{41 \text{ N} \cdot \text{m}}$$

2 (**b**) Similarly, calculate the torque due to $\vec{F}_2$. This torque is negative:

$$\tau_2 = -rF_2 \sin 90.0°$$
$$= -(0.74 \text{ m})(58 \text{ N})$$
$$= \boxed{-43 \text{ N} \cdot \text{m}}$$

3 (**c**) Sum the torques from parts (a) and (b) to find the total torque:

$$\tau_{\text{total}} = \tau_1 + \tau_2$$
$$= 41 \text{ N} \cdot \text{m} - 43 \text{ N} \cdot \text{m}$$
$$= \boxed{-2 \text{ N} \cdot \text{m}}$$

Insight

Because the total torque is negative, the wheel rotates clockwise. Thus, even though $\vec{F}_2$ is the smaller force, it has the greater effect in determining the wheel's direction of rotation. This is because $\vec{F}_2$ is applied tangentially, whereas $\vec{F}_1$ is applied in a direction that is partially radial.

35. Follow-up What magnitude must $\vec{F}_2$ have to result in zero net torque on the wheel in Guided Example 8.9?

36. To tighten a spark plug, it is recommended that a torque of 15 N · m be applied. If a mechanic tightens the spark plug with a wrench that is 25 cm long, what is the force necessary to create the desired torque?

Torque is related to angular acceleration

What effect does a torque have on an object?

If you pull on a rope wrapped around a pulley, the pulley starts spinning with an increasing angular speed. This effect is similar to a force causing a mass to speed up with a linear acceleration. **In general, a torque, τ, acting on an object produces an angular acceleration, α.**

For linear motion we know that Newton's second law is $F = ma$. By substituting moment of inertia for mass (I for m) and angular acceleration for linear acceleration (α for a), we obtain the rotational version of Newton's second law:

> **Newton's Second Law for Rotational Motion**
>
> sum of all torques = moment of inertia × angular acceleration
> $$\sum \tau = I\alpha$$
> SI units: N · m

If only a single torque, τ, acts on a system, we will simply write

$$\tau = I\alpha$$

This result is given along with the other linear-angular connections in Table 8.2.

GUIDED Example 8.10 | A Fish Takes the Line   **Angular Acceleration**

A fisherman is dozing when a fish takes the line and pulls tangentially on the spool of the fishing reel with a tension $T = 8.2$ N. The disk-shaped spool is at rest initially and rotates without friction. The radius of the spool is $r = 6.6$ cm, and its mass is $m = 1.8$ kg. What is the angular acceleration of the spool?

Picture the Problem

Our sketch shows the fishing line being pulled tangentially from the spool with the tension $T = 8.2$ N. Because the radius of the spool is r and the tangential force is $F = T$, the torque produced by the line is $\tau = rF = rT$.

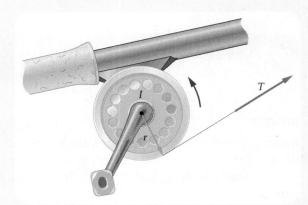

Strategy

To find the angular acceleration we apply Newton's second law for rotational motion, $\tau = I\alpha$. Rearranging to obtain the angular acceleration, we find $\alpha = \tau/I$. As mentioned above, the torque is $\tau = rT$. The moment of inertia is calculated with the equation for a disk, $I = \frac{1}{2}mr^2$, as given in Table 8.1.

Solution

1 Calculate the torque applied to the disk using $\tau = rF = rT$:

$$\tau = rT$$
$$= (0.066 \text{ m})(8.2 \text{ N})$$
$$= \boxed{0.54 \text{ N} \cdot \text{m}}$$

> **Math HELP**
> Scientific Notation
> See Math Review, Section I

2 Since the spool is a disk, its moment of inertia is given by $I = \frac{1}{2}mr^2$ (see Table 8.1). Substitute the numerical values to determine I:

$$I = \tfrac{1}{2}mr^2$$
$$= \tfrac{1}{2}(1.8 \text{ kg})(0.066 \text{ m})^2$$
$$= \boxed{3.9 \times 10^{-3} \text{ kg} \cdot \text{m}^2}$$

3 The angular acceleration of the spool is $\alpha = \tau/I$:

$$\alpha = \frac{\tau}{I}$$
$$= \frac{0.54 \text{ N} \cdot \text{m}}{3.9 \times 10^{-3} \text{ kg} \cdot \text{m}^2}$$
$$= \boxed{140 \text{ rad/s}^2}$$

Insight

The corresponding linear acceleration of the fishing line is $a = r\alpha = 9.06 \text{ m/s}^2$, slightly less than the acceleration due to gravity. To understand the final units in Step 3, recall that the newton has the dimensions of mass times acceleration; that is, $\text{N} = \text{kg} \cdot \text{m/s}^2$. Substituting this result and canceling where possible yields s^{-2} for the units of angular acceleration. We insert the radian (which is dimensionless) as a reminder that this quantity involves angles. Thus, the angular acceleration has units of rad/s^{-2}.

Practice Problems

37. **Follow-up** What tension is required to give the spool of the fishing reel an angular acceleration of 120 rad/s²?

38. A torque of 0.97 N · m is applied to a bicycle wheel of radius 35 cm and mass 0.75 kg. Treating the wheel as a hoop, find its angular acceleration.

COOLPHYSICS

Eye of the Storm

Winds in tornados and hurricanes also speed up near their centers of rotation. As air moves toward the center of the storm, its speed of rotation increases, just as a skater's rotational speed increases when the skater's arms are pulled inward toward the body. This is the main cause of the violent winds that accompany these storms.

The rotating systems shown below differ only in that the two adjustable masses are positioned either far from the axis of rotation (left) or near the axis of rotation (right). The hanging blocks are released simultaneously from rest at the same height. Will the block on the left land first, the block on the right land first, or both blocks land at the same time?

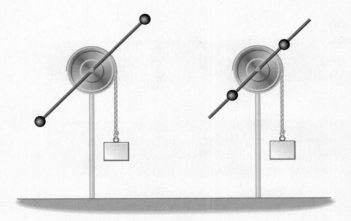

Reasoning and Discussion

The external torque supplied by the hanging blocks is the same for each of these systems. However, the moment of inertia of the system on the right is less than that of the system on the left because the adjustable masses are closer to the axis of rotation. Since angular acceleration is inversely proportional to the moment of inertia ($\alpha = \tau/I$), the system on the right has the greater angular acceleration. As a result, the block on the right lands first.

Answer

The block on the right lands first.

Angular momentum can be conserved

Physical quantities that are conserved play an important role in physics. Recall from Chapter 7, for example, that linear momentum is conserved when the total force is zero. This greatly simplified the analysis of collisions.

Conservation of angular momentum is another one of the fundamental principles of physics. It plays a role similar to that of conservation of linear momentum and conservation of energy. As we shall see, angular momentum is conserved when the net torque is zero.

As an example of angular momentum conservation, consider the ice skater in **Figure 8.18**. The torque exerted on her by the ice is approximately zero. As a result, her angular momentum remains constant. But if her angular momentum is constant, why does she spin so much faster when she pulls her arms in? Doesn't she have more angular momentum when she spins faster?

▲ Figure 8.18 **Using momentum conservation to spin faster**
After jumping into the air, this skater pulls her arms in as she spins. Why? Pulling her arms in reduces her moment of inertia. Her angular momentum must stay the same, however, and the only way this can happen is for her angular speed to increase. A rapidly spinning skater can complete two or even three full revolutions before landing.

Not at all. In fact, the skater spins faster with her arms pulled in precisely *because* her angular momentum is constant. When she pulls her arms in, she has more mass close to her axis of rotation. This makes her moment of inertia smaller. To maintain the same angular momentum, her angular speed must increase. In general, a large I and a small ω give the same angular momentum as a small I and a large ω:

$$I\omega = I\boldsymbol{\omega}$$

The same principle applies to divers and gymnasts who go into a tucked position to increase their rate of spin.

8.3 LessonCheck ⓜⓟ

Checking Concepts

39. Describe How do force and torque differ?

40. Explain How is torque calculated using a moment arm?

41. Explain How is torque related to angular acceleration?

42. Big Idea What causes an object to rotate?

43. Triple Choice Suppose both systems in Conceptual Example 8.11 are rotating with the same angular speed. Is the angular momentum of the system on the left greater than, less than, or equal to the angular momentum of the system on the right? Explain.

44. Critique A classmate states that two forces of equal magnitude must produce equal torques. Is the classmate correct? Explain why or why not.

Solving Problems

45. Calculate A force of 5.5 N is applied to an object. The moment arm for the force is 0.84 m. What is the torque produced by the force?

46. Calculate A torque of 7.4 N · m is applied to a wheel with a moment of inertia of 0.092 kg · cm². What is the resulting angular acceleration?

47. Calculate A ceiling fan has an angular acceleration of 62 rad/s² when acted on by a torque of 8.3 N · m. What is the moment of inertia of the fan?

48. Calculate What torque is required to give a disk of mass 6.1 kg and radius 0.58 m an angular acceleration of 17 rad/s²?

8.4 Static Equilibrium

Vocabulary

● center of mass

⚷ What conditions ensure that a system has no linear or angular acceleration?

Though you may not have thought about it, you are probably in *static equilibrium* at this very moment. What does that mean? Well, basically it means that you are at rest and are staying at rest. You'll learn about the basic physics of static equilibrium in this lesson.

Objects in static equilibrium do not accelerate

The parents of a young boy support him on a long, lightweight plank, as illustrated in **Figure 8.19**. The mass of the child is m, and therefore the upward forces exerted by the parents must sum up to the child's weight, mg. That is,

$$F_1 + F_2 = mg$$

This condition ensures that the total force acting on the plank is zero. It *does not*, however, guarantee that the plank remains at rest.

To see why, imagine for a moment that the parent on the right lets go of the plank and that the parent on the left increases her force until it is equal to the weight of the child. In this case, $F_1 = mg$ and $F_2 = 0$, which clearly satisfy the force equation we have just written. The right end of the plank is no longer supported, however, and hence it drops toward the ground while the left end rises. In other words, the plank rotates in a clockwise sense.

For an object like this plank to remain completely at rest, with no up or down motion and no rotation, the following *two* conditions must be satisfied:

● **⚷ The total force acting on the object must be zero. This ensures that there is no linear acceleration.**

● **⚷ The total torque acting on the object must also be zero. This ensures that there is no angular acceleration.**

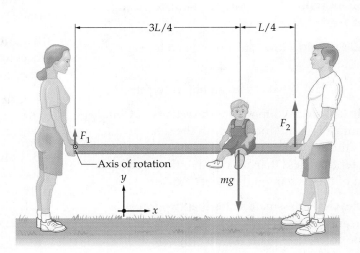

▶ **Figure 8.19 Forces required for static equilibrium**
Two parents support a child on a lightweight plank of length L. For the text discussion we chose the axis of rotation to be at the left end of the plank.

If both of these conditions are met, the plank remains completely at rest—it is in static equilibrium. These conditions also apply to the objects shown in **Figure 8.20**.

Forces can be calculated for objects that are in static equilibrium

Let's apply these two conditions to the plank that supports the child. First, consider the forces acting on the plank, with upward chosen as the positive direction, as in Figure 8.19. Setting the total force equal to zero yields

$$F_1 + F_2 - mg = 0$$

This agrees with the force equation we wrote earlier.

Next, we apply the torque condition. To do so, we must first choose an axis of rotation. For example, we might take the left end of the plank to be the axis, as in Figure 8.19. With this choice we see that the force F_1 exerts zero torque, since it acts directly through the axis of rotation ($r = 0$). On the other hand, F_2 acts at the far end of the plank, a distance L from the axis. Notice that F_2 would cause a counterclockwise (positive) rotation if it acted alone, as we can see in Figure 8.19. Therefore, the torque due to F_2 is

$$\tau_2 = F_2 L$$

The weight of the child, mg, acts at a distance $\frac{3}{4}L$ from the axis. This force would cause a clockwise (negative) rotation if it acted alone. Thus, its torque is negative:

$$\tau_{mg} = -mg\left(\tfrac{3}{4}L\right)$$

We can now set the total torque equal to zero. Summing the torques we've just calculated, and setting the sum equal to zero, we find

$$0 + F_2 L - mg\left(\tfrac{3}{4}L\right) = 0 \qquad \text{or} \qquad F_2 L = mg\left(\tfrac{3}{4}L\right)$$

We next cancel the length, L, in the above equation to obtain F_2:

$$F_2 = \tfrac{3}{4}mg$$

Substituting this result into the force equation gives F_1:

$$F_1 + \tfrac{3}{4}mg - mg = 0$$
$$F_1 = \tfrac{1}{4}mg$$

These two forces support the plank *and* keep it from rotating. As you might expect, the force nearer the child (F_2) is larger.

Any point can be chosen for the axis of rotation

In general, you are free to choose an axis of rotation that is most convenient for a given problem. For example, it's a good idea to have the axis be at the location of one of the unknown forces. This eliminates that force from the torque condition and simplifies the remaining algebra. We consider an alternative choice for the axis of rotation in the following Active Example.

▲ **Figure 8.20 Objects in static equilibrium**
The objects shown here are in static equilibrium because both the total force and the total torque acting on them are zero.

A child of mass m is supported on a light plank by his parents, who exert the forces F_1 and F_2 as in Figure 8.19. Find the forces required to keep the plank in static equilibrium. (*Note:* Use the *right end* of the plank as the axis of rotation for this calculation.)

Solution (*Perform the calculations indicated in each step.*)

1. Set the net force acting on the plank equal to zero: $\qquad F_1 + F_2 - mg = 0$

2. Set the net torque acting on the plank equal to zero: $\qquad -F_1(L) + mg\left(\tfrac{1}{4}L\right) = 0$

3. Cancel the length, L, and solve for F_1 in the torque equation: $\qquad F_1 = \tfrac{1}{4}mg$

4. Substitute F_1 into the force equation to obtain F_2: $\qquad F_2 = \tfrac{3}{4}mg$

Insight

As expected, the results are identical to those obtained previously. Notice that in this case the torque produced by the child would cause a counterclockwise rotation; hence, it is positive. Thus, the magnitude *and* the sign of the torque produced by a given force depend on the location chosen for the axis of rotation.

Objects balance at their center of mass

Acrobats add excitement to a circus. Suppose an acrobat lifts a stack of chairs over his head and supports it with one hand. Where should his hand be placed to balance the chairs? In general, the point where an object can be balanced is referred to as the **center of mass** (CM). 🔑 **In many ways an object behaves as if all of its mass were concentrated at its center of mass.**

For example, suppose you are making a mobile. At one stage in its construction, you want to balance a light rod with objects of mass m_1 and m_2 on either end, as in **Figure 8.21**. To make the rod balance, you should attach a string at the center of mass of the two masses—just as if all of the mass were concentrated there.

🔑 *How is an object's mass distribution related to its center of mass?*

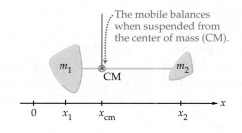

The mobile balances when suspended from the center of mass (CM).

▶ **Figure 8.21 Balancing a mobile**
Consider a portion of a mobile, with masses m_1 and m_2 at the locations x_1 and x_2, respectively. The object balances when a string is attached at the center of mass, x_{cm}.

We use the simple equation shown below to determine the location of the center of mass in Figure 8.21. In this equation, x_1 and x_2 represent the locations of the two masses, and x_{cm} is the location of their center of mass.

Center of Mass for Two Objects

$$x_{cm} = \frac{m_1 x_1 + m_2 x_2}{m_1 + m_2}$$

If the masses are equal, the center of mass is halfway between them. If one mass is greater than the other, the center of mass is closer to the larger mass. This behavior is illustrated in **Figure 8.22**.

The balance point corresponds to zero net torque

To see how balance is related to torque, consider the small mobile in **Figure 8.23**. The rod is in static equilibrium—it is balanced. As a result, the net torque acting on it is zero. Taking the point where the thread is tied to the rod as the axis of rotation, we can express the net torque as follows:

$$m_1 g(r_1) - m_2 g(r_2) = 0$$

In this equation, r_1 is the distance of mass m_1 from the axis of rotation, and r_2 is the distance of mass m_2 from the axis of rotation. Notice that the force $m_1 g$ acting alone would cause a counterclockwise rotation; therefore, its torque is positive. In contrast, the torque exerted by the force $m_2 g$ is negative, since it would cause a clockwise rotation if it acted alone.

Rearranging the net torque equation, we find

$$m_1 g r_1 = m_2 g r_2$$

This equation says that the torques exerted by the two masses have the same magnitude. The two torques act in opposite directions, however, and hence they cancel. We can say that the torques are *balanced*. Thus, *balancing* the rod means *balancing* the torques.

To make things a little simpler, we can cancel the g's on opposite sides of the balancing equation:

$$m_1 r_1 = m_2 r_2$$

We apply this condition for balance to a mobile in the next Guided Example.

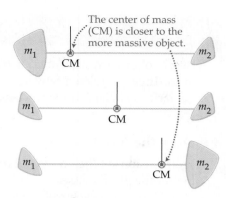

▲ **Figure 8.22 The center of mass of two objects**
The center of mass is closer to the larger mass, or equidistant between the masses if they are equal.

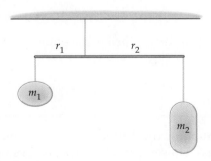

▲ **Figure 8.23 Zero torque and balance**
One section of a mobile. The rod is balanced when the net torque acting on it is zero. This is equivalent to having the center of mass directly under the suspension point.

As a school project, students construct a mobile representing some of the major food groups. Their completed artwork is shown below. Find the mass m_1 that is required for a perfectly balanced mobile. Assume that the strings and the horizontal rods have negligible mass.

Picture the Problem

The dimensions of the horizontal rods and the values of the given masses are indicated in our sketch. Notice that each rod is balanced at its suspension point.

Strategy

The unknown mass can be found by applying the condition for balance, $m_1r_1 = m_2r_2$. In this case, m_1 is unknown, and r_1 is 31 cm. Similarly, m_2 is equal to the sum of all the masses on the left side of the mobile, and r_2 is 18 cm.

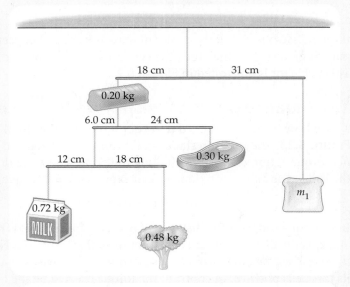

Known	Unknown
masses and distances on mobile	$m_1 = ?$

Solution

1 Sum all of the masses on the left side of the mobile to find m_2:

$$m_2 = (0.20 \text{ kg} + 0.30 \text{ kg} \\ + 0.72 \text{ kg} + 0.48 \text{ kg})$$
$$= \boxed{1.70 \text{ kg}}$$

2 Solve the condition for balance for the unknown mass, m_1:

$$m_1r_1 = m_2r_2$$
$$m_1 = \frac{m_2r_2}{r_1}$$

3 Substitute the known masses and distances to find m_1:

$$m_1 = \frac{(1.70 \text{ kg})(18 \text{ cm})}{31 \text{ cm}}$$
$$= \boxed{0.99 \text{ kg}}$$

Math HELP
Significant Figures
See
Lesson 1.4

Insight

With this value for m_1, the center of mass of the *entire* mobile is directly below the point where the uppermost string attaches to the ceiling.

Practice Problems

49. Follow-up Suppose the 0.30-kg mass on the mobile is replaced with a 0.40-kg mass. Do you expect the required value of m_1 to increase, decrease, or stay the same? Find the new value of m_1 that results in a balanced mobile.

50. A lightweight plastic rod has a mass of 1.0 kg attached to one end and a mass of 1.5 kg attached to the other end. The rod has a length of 0.80 m. How far from the 1.0-kg mass should a string be attached to balance the rod?

51. A 0.34-kg meterstick balances at its center. If a necklace is suspended from one end of the stick, the balance point moves 9.5 cm toward that end. What is the mass of the necklace?

52. A lightweight wooden stick balances when it is suspended from a string attached 0.25 cm from its left end. The stick has a length of 0.90 m. A 0.75-kg weight is attached to the left end of the stick, and an unknown mass m is attached to the right end of the stick.
(**a**) Is m greater than, less than, or equal to 0.75 kg? Explain.
(**b**) Determine the mass m.

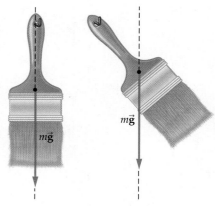

(**a**) Zero torque (**b**) Nonzero torque

▲ **Figure 8.24 Equilibrium of a suspended object**
(**a**) If an object's center of mass is directly below the suspension point, its weight creates zero torque and the object is in equilibrium. (**b**) When the object is rotated, so that the center of mass is no longer directly below the suspension point, the object's weight creates a torque. The torque tends to rotate the object to bring the center of mass under the suspension point.

The suspension point locates the center of mass

In general, if you allow an object with any shape at all to hang freely, its center of mass is directly below the suspension point. To see why, note that when the center of mass is directly below the suspension point, the torque due to gravity is zero. This is because the force of gravity extends right through the axis of rotation, as shown in **Figure 8.24 (a)**. If the object is rotated slightly, as in **Figure 8.24 (b)**, the force of gravity is not in line with the axis of rotation—thus, gravity produces a torque. This torque tends to rotate the object, bringing the center of mass back under the suspension point.

For example, suppose you cut a piece of wood into the shape of the continental United States, as shown in **Figure 8.25**. Next, you drill a small hole in it and hang it from the point A. The result is that the center of mass lies somewhere on the line aa'. Similarly, if you drill a second hole at point B, you will find that the center of mass lies somewhere on the line bb'. The *only* point that is on both the line aa' *and* the line bb' is the center of mass (CM), near Smith Center, Kansas. This point not only marks the center-of-mass location for a model of the continental United States, but also gave the town its name.

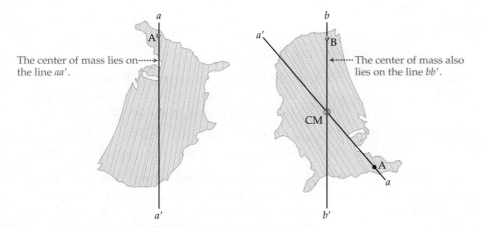

The center of mass lies on
the line aa'.

The center of mass also
lies on the line bb'.

▲ **Figure 8.25 The geometric center of the United States**
To find the center of mass of an irregularly shaped object, such as a wooden model of the continental United States, suspend it from two or more points. The center of mass lies on a vertical line extending downward from the suspension point. The intersection of these vertical lines gives the precise location of the center of mass.

A croquet mallet balances when suspended from its center of mass, as indicated in the upper part of the figure below. If you cut the mallet in two at its center of mass, as in the lower part of the figure, is the mass of the piece with the head of the mallet greater than, less than, or equal to the mass of the other piece?

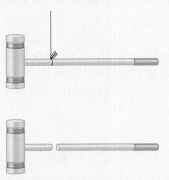

Reasoning and Discussion

The mallet balances because the torques due to the two pieces are of equal magnitude—the torques balance. The piece of the mallet with the head extends a smaller distance from the point of suspension than the other piece does. Since a large mass at a small distance creates the same torque as a small mass at a large distance, it follows that the piece with the head of the mallet has the greater mass.

Answer

The mass of the piece with the head of the mallet is greater.

8.4 LessonCheck (MP)

Checking Concepts

53. 🔑 **Summarize** What can you say about a system that (**a**) has no linear acceleration and (**b**) has no angular acceleration?

54. 🔑 **Explain** Why does an object balance at its center of mass?

55. Triple Choice Suppose in Conceptual Example 8.14 you attach the string at the point on the croquet mallet where there is equal mass on either side of the suspension point. When you allow the mallet to hang freely from the string, does it remain horizontal, rotate clockwise, or rotate counterclockwise? Explain.

Solving Problems

56. Calculate If the two parents in Figure 8.19 exert upward forces of 18 N on the left end of the plank and 71 N on the right end, how much does the child weigh?

57. Calculate In Problem 56, how far is the child from the left end of the plank? The plank has a length of 2.2 m.

58. Challenge Two students sit on either side of a teeter-totter that is 2.8 m in length. The teeter-totter balances when the student on the left side is 1.1 m from the center and the student on the right is 1.4 m from the center. The total mass of the two students is 84 kg. What is the mass of the student on the left side of the teeter-totter? (Assume that the teeter-totter itself pivots at the center and produces zero torque.)

Physics & You

Flying an airliner requires understanding and controlling rotational motion. The pilot has rotated this airplane so that it is pointing upward during takeoff. A pilot actually says "rotate" just before lifting the nose off the runway.

▲ This simulator is just like an actual cockpit and provides training under realistic conditions.

Commercial Pilot

The cockpit is a pilot's "office." It has the controls, instruments, displays, and communications equipment a pilot needs in order to do the job. As you can see, the cockpit of a commercial airliner is a complex place. Commercial airliners have two experienced pilots in the cockpit, a captain and a first officer. They work closely together as a team, especially while landing and taking off, the most complex and busy parts of a typical flight.

A pilot's workplace extends far beyond the cockpit. Pilots are in constant contact with air-traffic controllers in control towers on the ground. The air-traffic controllers inform pilots about weather conditions and direct them to follow specific flight paths.

In terms of education, the major domestic airlines require that pilots have a four-year college degree. Beyond that, they require extensive training and thousands of hours of flying time. Some of this training includes learning about the physics of aerodynamics. Prospective pilots can pursue either civilian training or military training. Each has its advantages and disadvantages.

Training offered by the U.S. military is simply part of your job (if you qualify), and it offers the opportunity to fly some of the world's most sophisticated aircraft. This training, however, requires a commitment of up to ten years of service—some of which might be very dangerous. Civilian training can be pursued on your own schedule and does not require an extended service commitment. It is, however, very expensive.

Take It Further

1. Research *Write a detailed report about one of the two ways to train to become an airline pilot. Research the coursework required and the specific number of flight hours required.*

2. Research *The job of an air-traffic controller is extremely important. Research what is required in order to become an air-traffic controller. Be sure to describe the pros and cons of the job.*

Physics Lab Investigating Torque and Equilibrium

In this lab you will use a meterstick and hanging weights to experiment with torques and rotational equilibrium.

Materials
- masking tape
- pen or pencil
- meterstick
- meterstick clamp
- meterstick stand
- hooked masses
- mass hangers
- balance or spring scale

Procedure

Before you begin, read the entire lab and create the required data tables.

1. Identify each of the three mass hangers using a piece of masking tape and a pen or pencil. Label the hangers 1, 2, and 3. Measure the mass of each mass hanger and record the masses in Data Table 1.

2. Slide the clamp that *does not* have a wire hanger onto the meterstick and then place it on the support stand. Adjust the position of the clamp until the meterstick balances and then tighten the screw. Note the location of the center of the clamp. This point is the fulcrum.

Analysis

1. For each mass in Trials 1–3, calculate the force that the mass and its hanger exerted on the meterstick. Then, calculate the torque that the force produced. Indicate whether the direction of the torque was clockwise or counterclockwise. Finally, calculate the net torque acting on the meterstick. Record all of your results in Data Table 3.

2. Determine the unknown mass. Show all your work!

3. Trial 1: Use a mass hanger to suspend a 100-g mass at the 20-cm mark on the meterstick. Position a mass hanger with a 100-g mass on the other end of the meterstick; then, using trial and error, find the location where the system is in rotational equilibrium. Record the masses m_1 and m_2 (don't forget about the mass hangers) and the moment arms r_1 and r_2 (the distances from the fulcrum) in Data Table 2. Remove the masses and make sure your meterstick still balances.

4. Trial 2: Repeat Step 3, this time using a mass hanger with a 200-g mass at the 35-cm mark and then establishing equilibrium using a mass hanger with a 100-g mass.

5. Trial 3: Repeat Step 3, this time using a mass hanger with a 200-g mass at the 7-cm mark and a mass hanger with a 100-g mass at the 30-cm mark and then establishing equilibrium using a mass hanger with a 50-g mass.

6. Trial 4: Use a mass hanger to suspend the unknown mass at the 15-cm mark. Use a mass hanger with a 200-g mass to establish equilibrium. Record the masses and moment arms in Data Table 2.

Conclusions

1. Based on your results, describe the general relationship between rotational equilibrium and net torque.

2. Does the largest force always produce the largest torque? Explain your answer.

3. Draw a free-body diagram showing *all* of the forces acting on the meterstick in Trial 1. Your diagram should show two forces that were ignored in this lab. Why were you able to ignore these forces?

4. Suppose a uniform meterstick is at rotational equilibrium when a 200-g mass is suspended at the 5.0-cm mark, a 130-g mass is suspended at the 85-cm mark, and the support stand is placed at the 40-cm mark. What is the mass of the meterstick?

8 Study Guide

Big Idea

Forces can produce torques, and torques can produce rotation.

A force can cause a torque if it acts at a distance from an object's axis of rotation and has a tangential component. An object that experiences a torque rotates with an angular acceleration.

8.1 Describing Angular Motion

🔑 Counterclockwise rotation from the reference line corresponds to *positive* angles, $\theta > 0$. Clockwise rotation from the reference line corresponds to *negative* angles, $\theta < 0$.

🔑 Counterclockwise rotation corresponds to *positive* angular velocity, $\omega > 0$. Clockwise rotation corresponds to *negative* angular velocity, $\omega < 0$.

🔑 If ω and α have the *same sign*, then the speed of rotation is *increasing*. If ω and α have *opposite signs*, then the speed of rotation is *decreasing*.

Key Equations

The arc length, s, is equal to the radius, r, times the angle θ measured in radians:

$$s = r\theta$$

Angular velocity, ω, is the rate of change of angular position:

$$\omega = \frac{\Delta\theta}{\Delta t}$$

Angular acceleration, α, is the rate of change of angular velocity:

$$\alpha = \frac{\Delta\omega}{\Delta t}$$

The tangential (linear) speed, v_t, of a rotating object is

$$v_t = r\omega$$

8.2 Rolling Motion and the Moment of Inertia

🔑 The rolling motion of a wheel is a combination of both rotational motion and linear motion.

🔑 An object with a large moment of inertia is difficult to start or stop rotating.

🔑 The moment of inertia depends linearly on the mass and on the distance squared.

Key Equations

The moment of inertia for a single object of mass m located a distance r from the axis of rotation is

$$I = mr^2$$

The moment of inertia for a collection of objects is

$$I = m_1 r_1^2 + m_2 r_2^2 + m_3 r_3^2 + \ldots$$

The kinetic energy of a rotating object is

$$KE = \tfrac{1}{2}I\omega^2$$

The angular momentum of a rotating object is

$$L = I\omega$$

8.3 Torque

🔑 Torque is the physical quantity that causes rotation.

🔑 The torque is equal to the moment arm times the force.

🔑 A torque, τ, acting on an object produces an angular acceleration, α.

• The perpendicular distance from the axis of rotation to the line of the force is the moment arm, $r_\perp$.

Key Equations

A tangential force F applied at a distance r from the axis of rotation produces a torque:

$$\tau = rF$$

A force F exerted at an angle θ with respect to the radial direction and applied at a distance r from the axis of rotation produces a torque given by

$$\tau = r(F \sin \theta)$$

Torque is equal to the moment arm times the applied force:

$$\tau = r_\perp F$$

The relationship between torque and angular acceleration is

$$\tau = I\alpha$$

8.4 Static Equilibrium

The conditions for static equilibrium are as follows:

🔑 The total force acting on an object must be zero. This ensures that there is no linear acceleration. The total torque acting on an object must also be zero. This ensures that there is no angular acceleration.

🔑 An object behaves as if all of its mass were concentrated at its center of mass.

Key Equation

The balance point between two masses is determined using

$$m_1 r_1 = m_2 r_2$$

8 Assessment

For instructor-assigned homework, go to www.masteringphysics.com

ANSWERS TO SELECTED ODD-NUMBERED PROBLEMS APPEAR IN APPENDIX A.

Lesson by Lesson

8.1 Describing Angular Motion

Conceptual Questions

59. A rigid object rotates about a fixed axis. Do all points on the object have the same angular speed? Do all points on the object have the same linear speed? Explain.

60. The fact that the Earth rotates gives people in New York a linear speed of about 335 m/s (750 mi/h). Where should you stand on the Earth to have the smallest possible linear speed? The greatest possible linear speed?

61. **Predict & Explain** Two children, Jason and Betsy, ride on the same merry-go-round. Jason is a distance R from the axis of rotation; Betsy is a distance $2R$ from the axis. **(a)** Is the angular speed of Jason greater than, less than, or equal to the angular speed of Betsy? **(b)** Choose the *best* explanation from among the following:

A. The angular speed is greater for Jason because he moves more slowly than Betsy.

B. The angular speed is greater for Betsy since she must go around a circle with a larger circumference.

C. The angular speeds are the same because it takes the same amount of time for Jason and Betsy to complete a revolution.

Problem Solving

62. The following angles are given in degrees. Convert them to radians: 20°, 35°, 80°, 270°.

63. The following angles are given in radians. Convert them to degrees: $\pi/3$, 0.40π, 1.7π, 6π.

64. Express the angular velocity of the second hand on a clock in the following units: **(a)** rev/hr, **(b)** deg/min, and **(c)** rad/s.

65. A carousel at the local carnival rotates once every 45 s. **(a)** What is the linear speed of an outer horse on the carousel, which is 2.75 m from the axis of rotation? **(b)** What is the linear speed of an inner horse that is 1.75 m from the axis of rotation?

66. **The Crab Nebula** One of the most studied objects in the night sky is the Crab Nebula. It is the remains of a supernova explosion observed by the Chinese in 1054. In 1968 it was discovered that a pulsar—a rapidly rotating neutron star that emits a pulse of radio waves with each revolution—lies near the center of the Crab Nebula. The amount of time required for each rotation of this pulsar is 33 ms. What is the angular speed (in rad/s) of the pulsar?

67. **Think & Calculate** As Tony the fisherman reels in a "big one," he turns the spool on his fishing reel at the rate of 3.0 complete revolutions every second. See **Figure 8.26.** **(a)** If the radius of the reel is 3.7 cm, what is the linear speed of the fishing line as it is reeled in? **(b)** How would your answer to part (a) change if the radius of the reel were doubled?

Figure 8.26

8.2 Rolling Motion and the Moment of Inertia

Conceptual Questions

68. Give a common, everyday example for each of the following: **(a)** an object that has zero rotational kinetic energy but nonzero translational (linear) kinetic energy, **(b)** an object that has zero translational (linear) kinetic energy but nonzero rotational kinetic energy, **(c)** an object that has nonzero rotational and translational kinetic energies.

69. At the grocery store you pick up a can of beef broth and a can of chunky beef stew. The cans are identical in diameter and weight. Rolling both of them down the aisle with the same initial speed, you notice that the can of chunky stew rolls much farther than the can of broth. Why?

70. **Predict & Explain** The minute and hour hands of a clock have a common axis of rotation and equal masses. The minute hand is long, thin, and uniform; the hour hand is short, thick, and uniform. **(a)** Is the moment of inertia of the minute hand greater than, less than, or equal to the moment of inertia of the hour hand? **(b)** Choose the *best* explanation from among the following:

A. The hands have equal masses, and hence equal moments of inertia.

B. Having mass farther from the axis of rotation results in a greater moment of inertia.

C. The more compact hour hand has its mass more concentrated and thus has the greater moment of inertia.

Problem Solving

71. A child pedals a tricycle, giving the large wheel an angular speed of 0.373 rev/s, as shown in **Figure 8.27**. If the radius of the wheel is 0.260 m, what is the child's linear speed?

Figure 8.27

72. An electric fan spinning with an angular speed of 13 rad/s has a kinetic energy of 4.6 J. What is the moment of inertia of the fan?

73. A rotating disk has a mass of 0.51 kg, a radius of 0.22 m, and an angular speed of 0.40 rad/s. What is the angular momentum of the disk?

74. As a car travels down a road, the linear speed at the tops of its wheels is 43 m/s. What is the linear speed of **(a)** the axles and **(b)** the bottoms of the wheels?

75. A soccer ball, which has a circumference of 70.0 cm, rolls 14.0 m in 3.35 s. What was the average angular speed of the ball during this time?

76. After you pick up a spare, your bowling ball rolls without slipping back toward the ball rack with a linear speed of 2.8 m/s as shown in **Figure 8.28**. **(a)** If the diameter of the bowling ball is 0.22 m, what is its angular speed? **(b)** To reach the rack, the ball rolls up a ramp. If the angular speed of the ball when it reaches the top of the ramp is 1.2 rad/s, what is the linear speed of the ball?

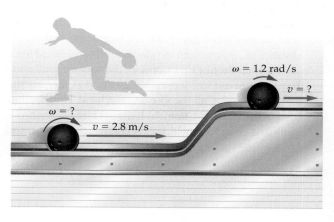

Figure 8.28

8.3 Torque

Conceptual Questions

77. Two forces produce the same torque. Does it follow that they have the same magnitude? Explain.

78. Does a larger force always produce more torque than a smaller force? Explain why not if your answer is no; give an example if your answer is yes.

79. [Rank] The tangential force acting on four different objects is given below. Also given is the distance from the force to the axis of rotation. Rank the objects in order of increasing torque. Indicate ties where appropriate.

System	A	B	C	D
Tangential force	5 N	10 N	1 N	20 N
Distance to axis	10 m	1.5 m	0.5 m	2.5 m

Problem Solving

80. A mechanic uses a wrench that is 22 cm long to tighten a spark plug. If the mechanic exerts a force of 58 N to the end of the wrench, what is the maximum torque she can apply to the spark plug?

81. A person slowly lowers a 3.6-kg crab trap over the side of a dock, as shown in **Figure 8.29**. What torque does the trap exert about the person's shoulder?

82. A torque of 0.97 N · m is applied to a bicycle wheel of radius 35 cm and mass 0.75 kg. Treating the wheel as a hoop, find its angular acceleration.

Figure 8.29

83. Force to Hold a Baseball A person holds a 1.42-N baseball in his hand, a distance of 34.0 cm from the elbow joint, as shown in **Figure 8.30**. The biceps, attached at a distance of 2.75 cm from the elbow, exerts an upward force of 12.6 N on the forearm. Consider the forearm and hand to be a uniform rod with a mass of 1.20 kg. **(a)** Calculate the net torque acting on the forearm and hand. Use the elbow joint as the axis of rotation. **(b)** If the net torque obtained in part (a) is nonzero, in which direction will the forearm and hand rotate?

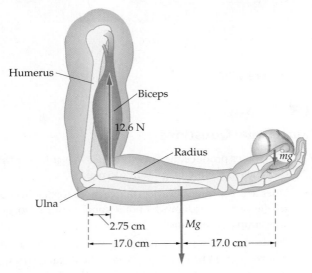

Figure 8.30

84. At the local playground, a 16-kg child sits on the end of a horizontal teeter-totter, 1.5 m from the pivot point. On the other side of the pivot an adult pushes straight down on the teeter-totter with a force of 95 N. In which direction does the teeter-totter rotate if the adult applies the force at a distance of **(a)** 3.0 m, **(b)** 2.5 m, or **(c)** 2.0 m from the pivot? (Assume that the teeter-totter itself pivots at the center and produces zero torque.)

8.4 Static Equilibrium
Conceptual Questions

85. A 25-kg child sits on one side of a teeter-totter, at a distance of 2 m from the pivot point. A mass m is placed at a distance d on the other side of the pivot, in an effort to balance the teeter-totter. Which of the following combinations of mass and distance (A, B, C, or D) balances the teeter-totter? (Assume that the teeter-totter itself pivots at the center and produces zero torque.)

	A	B	C	D
Mass, m	10 kg	50 kg	40 kg	20 kg
Distance, d	2 m	1 m	1.5 m	2.5 m

86. Two trophies are made half of plastic and half of metal, as shown in **Figure 8.31**. The first-place trophy has the metal half on the top, and the second-place trophy has the metal half on the bottom. A force is exerted on each trophy until it begins to tip. Is the force required to tip the first-place trophy (F_1) greater than, less than, or equal to the force required to tip the second-place trophy (F_2)? Explain.

Figure 8.31

Problem Solving

87. A Person's Center of Mass To determine the location of her center of mass, a physics student lies on a lightweight plank supported by two scales 2.50 m apart, as indicated in **Figure 8.32**. If the left scale reads 290 N and the right scale reads 122 N, find **(a)** the student's mass and **(b)** the distance from the student's head to her center of mass.

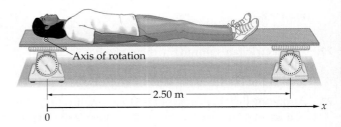

Figure 8.32

88. Triceratops A set of fossilized triceratops footprints discovered in Texas indicated that the front and rear feet were 3.2 m apart. The rear footprints were observed to be twice as deep as the front footprints. Assuming that the rear feet pressed down on the ground with twice the force exerted by the front feet, find the horizontal distance from the rear feet to the triceratops's center of mass.

89. A schoolyard teeter-totter with a total length of 5.2 m and a mass of 38 kg is pivoted at its center. A 19-kg child sits on one end of the teeter-totter. Where should a parent push vertically downward with a force of 210 N in order to hold the teeter-totter level?

90. A 0.122-kg remote control 23.0 cm long rests on a table, as shown in **Figure 8.33**, with a length L overhanging its edge. To operate the power button on this remote requires a force of 0.365 N. How far can the remote extend beyond the edge of the table and still not tip over when you press the power button? Assume that the mass of the remote is distributed uniformly and that the power button is 1.41 cm from the overhanging end of the remote.

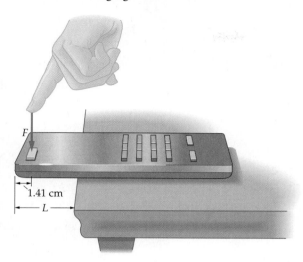

Figure 8.33

91. | **Think & Calculate** | A meterstick balances at its center. If an 86-g necklace is suspended from one end of the meterstick, the balance point moves 8.2 cm toward that end. **(a)** Is the mass of the meterstick greater than, less than, or equal to the mass of the necklace? Explain. **(b)** Find the mass of the meterstick.

Mixed Review

92. Consider the two rotating systems shown in **Figure 8.34**, each consisting of a mass m attached to a rod of negligible mass pivoted at one end. On the left, the mass is attached at the midpoint of the rod. On the right, it is attached to the free end of the rod. The rods are released from rest in the horizontal position at the same time. When the rod on the left reaches the vertical position, is the rod on the right not yet vertical (A), vertical (B), or past vertical (C)? Explain.

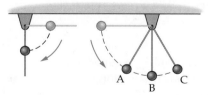

Figure 8.34

93. **Spin-Dry Dragonflies** Dragonflies often dip into the water on the surface of a lake or pond, presumably to drink or to clean themselves. Recently, it has been discovered that after dipping the dragonflies fly straight upward, then pitch forward and tumble head over tail several times until the water has been shed from their bodies. Observations show that the dragonflies complete one revolution every 0.017 s during this "spin dry"

maneuver. What is their angular speed in radians per second and revolutions per minute? (For comparison, a typical car engine runs at about 1500 rpm.)

94. Place two quarters on a table with their rims touching, as shown in **Figure 8.35**. While holding one quarter fixed, roll the other one—without slipping—around the circumference of the fixed quarter until it has completed one round trip. How many revolutions has the rolling quarter made about its center?

Figure 8.35

95. **Rotating Tray** To provide uniform cooking microwave ovens have a glass tray that sits on top of a circular ring with three small wheels, as shown in **Figure 8.36**. When the tray rests on top of the wheels, it is rotated easily by a small motor in the base of the microwave. The rotation of the tray ensures even heating. When the tray completes one full revolution, how many revolutions has the circular ring underneath it completed? Explain.

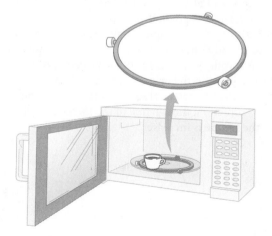

Figure 8.36

Writing about Science

96. Write a report on conservation of momentum in figure skaters and high divers. What angular speeds do these athletes have before and after making their bodies more compact? Find other similar examples of angular momentum conservation.

97. Connect to the Big Idea How would you explain the difference between force and torque to a family member? Address the following questions: Can a force produce zero torque? Can a torque be produced without a force? Why is a force more effective in producing a torque when it is tangential than when it is radial?

Read, Reason, and Respond

Correcting Torsiversion *Torsiversion* is a medical condition in which a tooth is rotated away from its normal position about the long axis of the root. Studies show that about 2% of the population suffer from this condition to some degree. For those who do, the improper alignment of the tooth can lead to tooth-to-tooth collisions during eating, as well as other problems. Typical patients display a rotation ranging from 20° to 60°, with an average around 30°.

An example is shown in **Figure 8.37 (a)**, where the first premolar is not only displaced slightly from its proper location in the negative y direction, but is also rotated clockwise from its normal orientation. To correct this condition, an orthodontist might use an archwire and a bracket to apply both a force and a torque to the tooth. In the simplest case, two forces are applied to the tooth in different locations, as indicated by F_1 and F_2 in Figure 8.37 (a). These two forces, if chosen properly, can reposition the tooth by exerting a net force in the positive y direction, and also reorient it by applying a torque in the counterclockwise direction.

In a typical case it may be desired to have a net force in the positive y direction of 1.8 N. In addition, the distances in Figure 8.37 (a) can be taken to be $d = 3.2$ mm and $D = 4.5$ mm. Given these conditions, a range of torques is possible for various values of the forces F_1 and F_2. For example, **Figure 8.37 (b)** shows the values of F_1 and F_2 necessary to produce a given torque, where the torque is measured about the center of the tooth (which is also the origin of the coordinate system). Notice that the two forces always add to 1.8 N in the positive y direction, though one of the forces changes sign as the torque is increased.

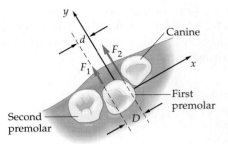

(a)

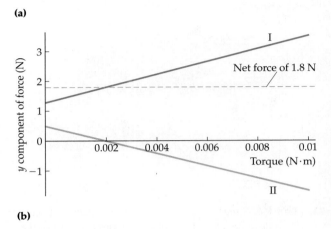

(b)

Figure 8.37

98. The two solid, straight lines in Figure 8.37 (b) represent the two forces applied to the tooth. Which line corresponds to which force?

A. I is F_1, and II is F_2.

B. I is F_2, and II is F_1.

99. What value of the torque corresponds to one of the forces being equal to zero?

A. 0.0023 N · m   C. 0.0081 N · m

B. 0.0058 N · m   D. 0.017 N · m

100. Find the values of F_1 and F_2 required to give zero net torque.

A. $F_1 = -1.2$ N, $F_2 = 3.0$ N

B. $F_1 = 1.1$ N, $F_2 = 0.75$ N

C. $F_1 = -0.73$ N, $F_2 = 2.5$ N

D. $F_1 = 0.52$ N, $F_2 = 1.3$ N

101. Find the values of F_{1y} and F_{2y} required to give a net torque of 0.0099 N · m. This is a torque that would be effective in rotating the tooth.

A. $F_1 = -1.7$ N, $F_2 = 3.5$ N

B. $F_1 = -3.8$ N, $F_2 = 5.6$ N

C. $F_1 = -0.23$ N, $F_2 = 2.0$ N

D. $F_1 = 4.0$ N, $F_2 = -2.2$ N

Standardized Test Prep

Multiple Choice

Use the diagram below to answer Questions 1–3.
The diagram shows a wheel (with a radius of 10 cm)
that rotates around an axle (with a radius of 2 cm).
Tangential forces *A*, *B*, *C*, and *D* act on the wheel and
axle as shown.

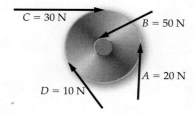

C = 30 N

B = 50 N

A = 20 N

D = 10 N

1. Which force produces the torque with the largest
magnitude?
(A) force *A*
(B) force *B*
(C) force *C*
(D) force *D*

2. What is the net torque on the system?
(A) 100 N · m, clockwise
(B) 100 N · m, counterclockwise
(C) 200 N · m, clockwise
(D) 300 N · m, counterclockwise

3. What magnitude must force *C* have in order for
the system to be in equilibrium?
(A) 10 N
(B) 20 N
(C) 40 N
(D) 50 N

4. A ball with a diameter of 20 cm is rolling across
the floor with an angular velocity of 4 rad/s. How
fast is the center of mass of the ball moving?
(A) The center of mass is not moving.
(B) 0.04 m/s
(C) 0.4 m/s
(D) 4.0 m/s

5. A thin 1-m long rod has a 200-g object attached to
one end and a 300-g object attached to the other
end. At what point should the rod be supported so
that it remains in horizontal equilibrium?
(A) 20 cm from the 300-g object
(B) 40 cm from the 300-g object
(C) 60 cm from the 300-g object
(D) 40 cm from the 200-g object

6. A ball with mass *m* and radius *r* has a moment of
inertia equal to $\frac{2}{5}mr^2$. The ball rolls without slip-
ping, and its center has a linear speed of *v*. What is
the total kinetic energy of the ball?
(A) $\frac{1}{2}mv^2$
(B) mv^2
(C) $I\omega^2$
(D) $0.7mv^2$

7. A turntable with a radius of 8.0 cm rotates at 45 rpm.
What is the linear speed of a point on the outside
rim of the turntable?
(A) 22.6 m/s
(B) 5.6 m/s
(C) 3.6 m/s
(D) 0.38 m/s

8. Which of the following is *not* a vector quantity?
(A) angular acceleration
(B) moment of inertia
(C) angular velocity
(D) torque

9. Which of the following *is* a vector quantity?
(A) angular position
(B) angular speed
(C) torque
(D) moment of inertia

Test-Taking Hint

Remember that net torque and net force are zero for
an object in static equilibrium.

If You Had Difficulty With . . .

Question	1	2	3	4	5	6	7	8	9				
See Lesson(s)	8.3	8.3	8.3	8.1	8.4	8.2	8.1	8.1–8.3	8.1, 8.2				

9 Gravity and Circular Motion

The rings of Saturn are one of the wonders of the Solar System. Early observers were mystified by them, but we now know that the rings are composed of innumerable pieces of ice orbiting the planet.

Big Idea

Gravity acts on everything in the universe.

The study of gravity has always been of central importance in physics. From Galileo's early experiments on free fall, to Einstein's general theory of relativity, to Stephen Hawking's work on black holes, scientists have striven for a better understanding of gravity. Perhaps the greatest milestone in this endeavor was the discovery by Newton of the *law of universal gravitation*. With just one simple equation to describe the force of gravity, Newton was able to determine the orbits of planets, moons, and comets. He was also able to explain such earthly phenomena as the tides and the fall of an apple.

Inquiry Lab What keeps an object moving along a circular path?

Read the activity and obtain the required materials.

Explore

1. Place a stiff paper plate on a flat table. Roll a marble around the inner rim of the plate and observe the motion.
2. Use scissors to cut out a wedge equal to one-quarter of the plate.
3. Once again roll the marble around the inner rim of the plate and observe what happens.
4. Repeat Step 3, rolling the marble with a greater speed; then repeat that step again, with a reduced speed. Observe the results.

Think

1. **Compare** How does the motion of the marble before the plate was cut compare to the motion after the plate was cut?
2. **Apply** Is the marble accelerating in Step 1 as it rolls at a constant speed? Explain why or why not.
3. **Identify** Describe each of the forces acting on the marble (**a**) in Step 1 (while the marble rolls around the rim of the plate) and (**b**) in Step 2 (after the marble leaves the plate and rolls on the table).
4. **Summarize** What effect did changing the speed of the marble have on its motion?
5. **Predict** Where on the table should you place a small target—such as a pen cap—so that the marble will hit it after it leaves the plate in Step 3?

9.1 Newton's Law of Universal Gravity

We're all familiar with the old adage that "what goes up must come down." While true up to a point, this statement only scratches the surface when it comes to explaining gravity. Newton's law of gravity gives us a complete understanding, and it can be expressed with just one equation. That equation, and what it means physically, is the subject of this lesson.

Gravity affects everything in the universe

If you let go of your pencil, it drops to the ground. What causes the pencil to fall? The answer is gravity. **Gravity** is the force of nature that attracts one mass to another mass. This is the attraction that pulls your pencil downward toward Earth.

Gravity is the force most apparent to us in our everyday lives. Gravity holds you on Earth, causes apples to fall, and accelerates sledders down snowy slopes, as shown in **Figure 9.1** on the next page. On a larger scale, gravity is responsible for the motion of the Moon, Earth, and other planets. In fact, the very fate of the universe itself is determined by the pull of gravity.

Vocabulary

- gravity
- superposition

Law of Gravity To describe the force of gravity, Newton proposed the following simple law:

> ### Newton's Law of Universal Gravitation
>
> The force of gravity between any two objects with masses m_1 and m_2 separated by a distance r is attractive and has a magnitude F given by
>
> $$\text{magnitude of gravitational force} = G \times \frac{(\text{mass 1}) \times (\text{mass 2})}{(\text{distance between the masses})^2}$$
>
> $$F = G\frac{m_1 m_2}{r^2}$$

The constant G in this equation is referred to as the *universal gravitation constant*. The numerical value of G is

$$G = 6.67 \times 10^{-11}\ \text{N} \cdot \text{m}^2/\text{kg}^2$$

This is an incredibly tiny number.

According to Newton's law of gravity, all objects in the universe attract all other objects in the universe. It is in this sense that the force law is "universal." Thus, the total gravitational force acting on you right now is due not only to the planet on which you live, but also to people nearby, other planets, and even stars in far-off galaxies. In short, everything in the universe "feels" everything else, thanks to gravity.

Line of Action Newton's law of gravity also states that the force between two masses is directed along the line connecting the masses. This is illustrated in **Figure 9.2**. Both masses in this figure experience an attractive force of the same magnitude, $F = Gm_1m_2/r^2$, but the two attractive forces act in opposite directions. Thus, the force of gravity between two objects forms an action-reaction pair.

Gravity is the weakest force of nature

Though the effects of gravity are all around us, it is still the weakest of all the known forces of nature. Its weakness is due to the numerical value of G being only 0.0000000000667 N · m²/kg². As a result, gravity is imperceptible between objects of everyday size. It only becomes important for large objects like planets and stars.

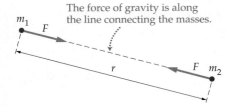

The force of gravity is along the line connecting the masses.

▲ **Figure 9.2 Gravitational force between point masses**
Two point masses, m_1 and m_2, separated by a distance r exert equal and opposite attractive forces on one another. The forces are along the line connecting the masses, and the magnitude of each force is $F = Gm_1m_2/r^2$.

A girl takes her dog for a walk on a sunny beach, as shown below. Find the force of gravity between the 45-kg girl and her 11-kg dog when they are separated by a distance of 1.0 m. (Treat the girl and her dog as point masses.)

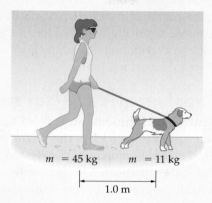

$m = 45$ kg $m = 11$ kg

1.0 m

Solution

Substitute $m_1 = 45$ kg, $m_2 = 11$ kg, and $r = 1.0$ m into $F = Gm_1m_2/r^2$:

$$F = G\frac{m_1m_2}{r^2}$$

$$= (6.67 \times 10^{-11}\,\text{N} \cdot \text{m}^2/\text{kg}^2)\frac{(45\,\text{kg})(11\,\text{kg})}{(1.0\,\text{m})^2}$$

$$= \boxed{3.3 \times 10^{-8}\,\text{N}}$$

This is roughly 3 ten-millionths of a newton—much too small to be noticed. (Though we chose $m_1 = 45$ kg and $m_2 = 11$ kg, the same answer is obtained with $m_1 = 11$ kg and $m_2 = 45$ kg.)

> **Math HELP**
> Dimensional Analysis
> See Lesson 1.3

Practice Problems

1. A 6.1-kg bowling ball and a 7.2-kg bowling ball rest on a rack. If the force of gravity pulling each bowling ball toward the other is 3.1×10^{-9} N, what is the separation between the balls?

2. Two identical cars are parked 4.7 m apart on a car dealer's showroom floor. The force of gravity between the cars is 4.5×10^{-6} N. What is the mass of each car?

The force in Quick Example 9.1 is imperceptibly small. To put it in perspective, the force of gravity between the girl and her dog is about 10 billion times smaller than the girl's weight. Gravity is so weak, in fact, that it takes the mass of the *entire Earth* just to exert a force of about 450 N (~100 lb) on the girl!

Gravity decreases rapidly with distance

Not only is gravity a weak force, but it gets weaker with distance. In fact, its strength decreases quite rapidly with distance. You can see this yourself by examining the gravitational force equation:

$$F = G\frac{m_1m_2}{r^2}$$

CONNECTINGIDEAS

You just learned that the force of gravity between two masses equals a constant (G) times a product ($m_1 \times m_2$) divided by the square of the distance between the masses (r^2).

• Later, in Chapter 19, you will learn that the force between two electric charges has the same general relationship—a constant times a product divided by the square of the distance.

🔑 **What relationship describes how gravity decreases with distance?**

🔑 **The force of gravity decreases as the inverse of the square of the distance, $1/r^2$. Because of this, we say that gravity obeys an *inverse square* force law.** For example, increasing the distance r by a factor of 10 decreases the gravitational force by a factor of 100. Why? Because $1/r^2 = 1/10^2 = 1/100$.

A plot of the force of gravity versus distance is given in **Figure 9.3**. Notice that the force diminishes rapidly with distance. Still, it never completely vanishes. For this reason we say that gravity is a force of infinite range. There is always some force due to gravity no matter how great the distance. There is no cut-off point where gravity stops acting.

Total gravitational force is a vector sum

🔑 **How do gravitational forces add together?**

🔑 **If a mass experiences gravitational forces from a number of other masses, the total force acting on it is the vector sum of all those individual forces.** The fact that the forces of gravity add together like vectors is referred to as **superposition**. As an example, superposition means that the net gravitational force exerted on you at this moment is the vector sum of the force exerted on you by Earth, plus the force exerted by the Moon, plus the force exerted by the Sun, and so on.

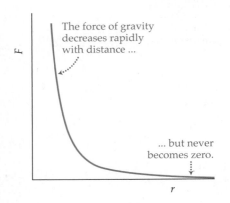

The force of gravity decreases rapidly with distance ...

... but never becomes zero.

▲ **Figure 9.3 Dependence of gravity on the separation distance**
The gravitational force decreases rapidly with distance, but it never completely vanishes. We say that gravity is a force of infinite range.

CONCEPTUAL Example 9.2 Rank the Forces

Each of the three equally spaced and equally massive objects shown below experiences gravitational forces due to the other two objects. Rank the objects in order of increasing magnitude of the total gravitational force they experience. Indicate ties where appropriate.

$$m \quad m \quad m$$
$$\text{A} \quad \text{B} \quad \text{C}$$

Reasoning and Discussion
The total force acting on object A is toward the right, and the total force acting on object C is toward the left. Because of the symmetry of the system, the magnitudes of these two total forces are the same. The total force acting on object B is different, however. Object B is attracted equally to the left and to the right by objects A and C, respectively. These two attractive forces cancel, leaving a total force of zero acting on object B.

Answer
The ranking is B < A = C.

GUIDED Example 9.3 | How Much Force Is with You? **Universal Gravitation**

As part of a daring rescue attempt, the *Millennium Eagle* passes between twin asteroids. The sketch shows the situation and relevant distances. The mass of the spaceship is $m = 2.50 \times 10^7$ kg, and the mass of each asteroid is $M = 3.50 \times 10^{11}$ kg. (**a**) Find the magnitudes of the gravitational forces F_1 and F_2 acting on the *Millennium Eagle* when it is at location A. (**b**) What is the total force acting on the *Millennium Eagle* at location B?

Picture the Problem

Our sketch shows the spaceship as it follows a path between the twin asteroids. The relevant distances and masses are indicated, as are the two points of interest, A and B.

Strategy

(a) We find the magnitude of the gravitational forces at location A by using $F = Gm_1m_2/r^2$. To reduce the number of subscripts in the problem, we set $m_1 = m$ and $m_2 = M$. With these substitutions we have $F = GmM/r^2$. The masses are given in the problem statement, and the distance r can be found from the distances in the sketch. **(b)** At location B we note that the forces act in opposite directions.

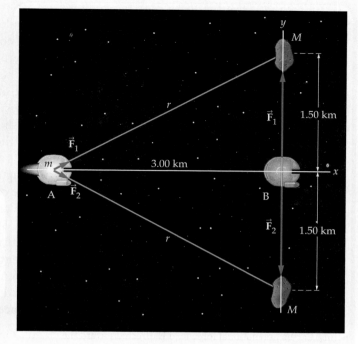

Known

$m = 2.50 \times 10^7$ kg

$M = 3.50 \times 10^{11}$ kg

distances given in the sketch

Unknown

(a) magnitudes of F_1 and F_2 (at point A)

(b) total force (at point B)

Solution

Part (a)

1 Use the Pythagorean theorem to find the distance r from point A to each asteroid:

$$r = \sqrt{(3.00 \times 10^3 \text{ m})^2 + (1.50 \times 10^3 \text{ m})^2}$$
$$= 3350 \text{ m}$$

2 Because the distance to each asteroid is the same and the asteroids have equal masses, the gravitational forces have equal magnitudes, F_1 and F_2. Substitute the known values for G, r, m, and M to calculate the magnitudes of $\vec{F}_1$ and $\vec{F}_2$ at point A:

Math HELP
Significant Figures
See Lesson 1.4

$$F_1 = F_2 = G\frac{mM}{r^2}$$

$$= (6.67 \times 10^{-11} \text{ N} \cdot \text{m}^2/\text{kg}^2)$$
$$\times \left(\frac{(2.50 \times 10^7 \text{ kg})(3.50 \times 10^{11} \text{ kg})}{(3350 \text{ m})^2}\right)$$
$$= \boxed{52.0 \text{ N}}$$

Part (b)

3 There is no need to calculate the magnitudes of the forces at point B. Instead, use the fact that $\vec{F}_1$ and $\vec{F}_2$ have equal magnitudes and point in opposite directions to determine the total force, $\vec{F}_{total}$, acting on the spaceship there:

$$\vec{F}_{total} = \vec{F}_1 + \vec{F}_2$$
$$= \boxed{0}$$

Insight

The total force at location A is directed in the positive x direction, as one would expect from the symmetry of the situation. At location B, where the force exerted by each asteroid is about five times greater than it is at location A, the *total* force is zero. This is because the attractive forces exerted by the two asteroids are equal and opposite, and thus cancel one another.

Practice Problems

3. Follow-up Find the magnitude of the gravitational forces F_1 and F_2 acting on the spaceship when it is on the x axis, a distance of 2.00 km from point B.

4. Ceres, the largest asteroid known, has a mass of roughly 8.7×10^{20} kg. If Ceres passes within 14,000 km of a spaceship in which you are traveling, what force does it exert on a 95-kg astronaut in your ship?

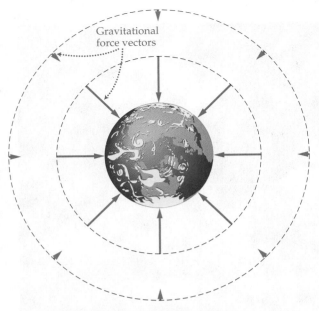

Gravitational force vectors

▲ **Figure 9.4 Earth's gravitational force field**
Any massive object sets up a gravitational force field in its vicinity. Earth's gravitational force field is directed toward the center of Earth and decreases in strength with distance.

Gravity is a force field

You often hear about "force fields" in science fiction movies. Of course, these movies rarely get the science right. Still, there *are* force fields in nature, and gravity is an example.

Gravity exerts its force from one end of the universe to the other, even across the vacuum of space. **Figure 9.4** gives a visual representation of Earth's gravitational force field. At each location in space the force exerted by Earth is indicated by the appropriate force vector. Notice that the force vectors all point toward the center of Earth and become shorter in length (magnitude) as their distance from Earth increases. We can think of Earth as setting up a force field in its vicinity, and any object entering this field feels its effect.

As you will learn in later chapters, electricity and magnetism also create fields. In fact, the analysis of electric and magnetic fields is an important part of our study of those phenomena. Furthermore, it turns out that light is a wave made up of electric and magnetic fields that move though space. Thus, force fields may sound like science fiction, but they are, truly, science fact.

9.1 LessonCheck (MP)™

Checking Concepts

5. 🔧 **Apply** If the separation between two masses is doubled, does the gravitational force between them increase or decrease? By what factor?

6. 🔧 **Explain** Two gravitational forces act on a given object. How do you determine the total gravitational force acting on the object?

7. Describe In what direction does the force of gravity act on a pair of masses?

8. Big Idea Describe how factors such as mass and distance affect the force of gravity throughout the universe.

9. Critique Suppose someone makes the following comment: "Astronauts are weightless in orbit because they are beyond Earth's gravitational pull." How would you respond to this assertion?

10. Triple Choice The three equally spaced objects in **Figure 9.5** have the indicated masses. Is the total gravitational force acting on object B toward the left, toward the right, or zero? Explain.

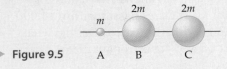

▶ **Figure 9.5**   A   B   C

11. Rank Referring to Figure 9.5, rank the objects in order of increasing magnitude of the total gravitational force. Indicate ties where appropriate.

Solving Problems

12. Calculate You place a 0.18-kg can of soup and a 0.34-kg jar of pickles on the kitchen counter, separated by a distance of 0.42 m. What is the magnitude of the force of gravity that (**a**) the can of soup exerts on the jar of pickles and (**b**) the jar exerts on the can?

13. Compare What is the magnitude of the force of gravity between two 1.6-kg flower pots if their separation is (**a**) 1.0 m or (**b**) 3.0 m?

14. Calculate The gravitational force between two volleyball players is 3.3×10^{-7} N. If the masses of the players are 66 kg and 72 kg, what is their separation?

15. Calculate A batter checks her swing as a 0.15-kg baseball crosses home plate outside the strike zone. If the separation between the batter and the ball is 0.77 m and the gravitational force exerted on the batter by the ball is 1.1×10^{-9} N, what is the mass of the batter?

9.2 Applications of Gravity

When Newton introduced his law of gravitation, he applied it to a number of interesting situations. We consider some of these applications in this lesson.

A spherical mass can be treated like a point mass

One of the first applications considered by Newton was the gravitational force exerted by a spherical mass on other masses outside the sphere. Using the methods of calculus—which he invented for this purpose—Newton was able to prove the following:

- 🔑 **A spherical mass exerts the same gravitational force on masses outside it as it would if all the mass of the sphere were concentrated at its center.**

For example, the net force between the sphere of mass M and the point mass m in **Figure 9.6** is

$$F = G\frac{mM}{r^2}$$

In the above equation r is the distance from the mass m to the center of the sphere.

Earth's Gravity A perfect application of the above result is the gravitational force produced by Earth. Earth is approximately spherical, with a mass M_E and a radius R_E. The gravitational force exerted on an object of mass m on the surface of Earth is the same as if all the mass of Earth were concentrated at its center, a distance R_E from the surface. Therefore, the force exerted by Earth on the object is

$$F = G\frac{mM_E}{R_E^2}$$

🔑 *How is the force of gravity on a spherical mass calculated?*

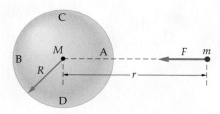

▲ **Figure 9.6 Gravitational force between a point mass and a sphere**
The force between the point mass and the sphere is the same as it would be if all the mass of the sphere were concentrated at its center.

We also know that the gravitational force experienced by a mass m on Earth's surface is equal to its weight, $W = mg$. Substituting mg for the force, F, and canceling m on each side of the equation, gives the following:

$$\cancel{m}g = \frac{G\cancel{m}M_E}{R_E^2}$$

$$g = \frac{GM_E}{R_E^2}$$

$$= \frac{(6.67 \times 10^{-11}\ \text{N} \cdot \cancel{\text{m}^2}/\cancel{\text{kg}^2})(5.97 \times 10^{24}\ \cancel{\text{kg}})}{(6.37 \times 10^6\ \cancel{\text{m}})^2}$$

$$= 9.81\ \text{m/s}^2$$

(Recall that the units of the newton are $N = \text{kg} \cdot \text{m/s}^2$.)

This result for g assumes a perfectly spherical Earth, with a uniform distribution of mass. In reality, small deviations in the shape and structure of Earth cause small variations in the acceleration due to gravity from place to place. **Figure 9.7** shows a global map of the acceleration due to gravity. Areas in yellow and red have stronger local gravity than areas in blue and green.

Gravity maps like the one in Figure 9.7 are important in many fields of scientific research. For example, maps of the gravitational force over the oceans shed light on ocean currents, which play the vital role of transporting thermal energy around the planet. Gravity maps also give valuable information on seismic activity and plate tectonics, by indicating areas of uplift and subsidence in Earth's crust. Finally, petroleum deposits are generally found in low-density rocks, which means that exploration for new petroleum reserves can be guided by looking for areas with weaker gravitational force.

The calculation of g given above applies to a mass on Earth's surface—a distance R_E from the center of Earth. But what if the mass is at a height h above Earth's surface? In this case we calculate g by replacing the radius of Earth with $R_E + h$. Let's call this result g_h, the acceleration due to gravity at the height h. The next Guided Example explores this case.

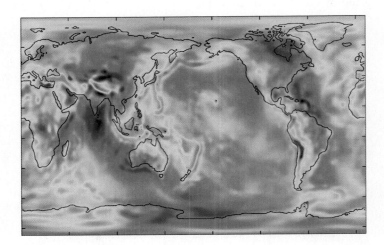

▲ **Figure 9.7 A gravity map**
This global model of Earth's gravitational strength was constructed from a combination of surface gravity measurements and satellite tracking data. It shows how the acceleration due to gravity varies from one location to another. Gravity is strongest in the red areas and weakest in the blue areas.

If you climb to the top of Mt. Everest, you will be 8850 m (about 5.50 mi) above sea level. What is the acceleration due to gravity at this altitude?

Picture the Problem

At the top of the mountain your distance from the center of Earth is $r = R_E + h$, where $h = 8850$ m is the altitude.

Strategy

First, use $F = GmM_E/r^2$ to find the force due to gravity on the mountaintop. Then, set $F = mg_h$ to find the acceleration due to gravity at that height.

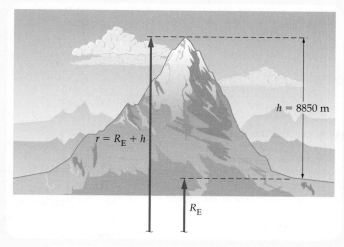

Solution

Known	Unknown
$h = 8850$ m	$g_h = ?$

1 Begin with $F = GmM_E/r^2$ and substitute $R_E + h$ for r. This gives the force F due to gravity at a height h above Earth's surface:

$$F = G\frac{mM_E}{r^2}$$

$$F = G\frac{mM_E}{(R_E + h)^2}$$

> **Math HELP**
> Solving Simple Equations
> **See Math Review, Section III**

2 Substitute mg_h for the force F at the height h. Next, cancel the mass m on each side and solve for g_h:

$$\cancel{m}g_h = G\frac{\cancel{m}M_E}{(R_E + h)^2}$$

$$g_h = G\frac{M_E}{(R_E + h)^2}$$

3 Substitute the numerical values and solve for g_h:

$$g_h = (6.67 \times 10^{-11}\,\text{N}\cdot\cancel{\text{m}}^2/\cancel{\text{kg}}^2)$$
$$\times \left(\frac{(5.97 \times 10^{24}\,\cancel{\text{kg}})}{(6.37 \times 10^6\,\cancel{\text{m}} + 8850\,\cancel{\text{m}})^2}\right)$$
$$= \boxed{9.79\,\text{m/s}^2}$$

Insight

As expected, the acceleration due to gravity decreases as one moves farther from the center of Earth. Thus, if you were to climb to the top of Mt. Everest, you would lose weight—not only because of the physical exertion required for the climb, but also because of the reduced gravity. In fact, a person with a mass of 80 kg (about 180 lb) would weigh almost half a pound less standing on the summit of the mountain.

Practice Problems

16. **Follow-up** Find the acceleration due to gravity at the altitude of the International Space Station's orbit, 370 km above Earth's surface.

17. At what altitude above Earth's surface is the acceleration due to gravity equal to $g/2$?

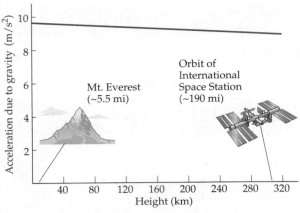

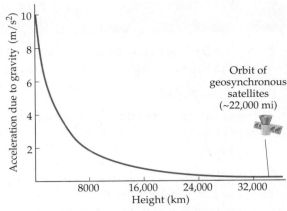

(a) Acceleration due to gravity near the Earth's surface

(b) Acceleration due to gravity far from the Earth

▲ **Figure 9.8 The acceleration due to gravity at a height *h* above Earth's surface**
(a) In this plot the peak of Mt. Everest is at about $h = 8850$ m (5.5 mi), and the International Space Station's orbit is at roughly $h = 310$ km (190 mi). **(b)** This plot shows the decrease in the acceleration due to gravity from the surface of Earth to an altitude of about 40,000 km (25,000 mi). The orbit of geosynchronous satellites—ones that remain constantly above a fixed point on Earth—is at roughly $h = 36{,}000$ km (22,000 mi).

A plot of the acceleration due to gravity as a function of *h* is shown in **Figure 9.8 (a)**. The plot indicates the altitude of Mt. Everest and the orbit of the International Space Station as points of comparison. The acceleration due to gravity changes only slightly over that altitude range. **Figure 9.8 (b)** shows the acceleration out to the altitude of communications and weather satellites, which orbit at roughly 36,000 km (22,000 mi) above Earth's surface. The acceleration due to gravity is very small at that large altitude.

The Moon's Gravity Our equation for *g* on Earth's surface ($g = GM_E/R_E^2$) can be rewritten to apply to any mass and radius as follows:

$$g = \frac{GM}{R^2}$$

We can use this result to calculate the acceleration due to gravity on other objects in the Solar System. For example, to calculate the acceleration due to gravity on the Moon, g_m, we use the mass of the Moon, M_m, and the radius of the Moon, R_m. This gives

$$g_m = \frac{GM_m}{R_m^2}$$

Once g_m is known, the weight of an object of mass *m* on the Moon is found by using $W_m = mg_m$.

The low gravity of the Moon is apparent in **Figure 9.9 (a)**, where we see an astronaut doing a standing high jump. Impressive jump, considering the heavy suit. **Figure 9.9 (b)** shows a Mars lander, whose design must take into account the reduced gravity on that planet.

▶ **Figure 9.9 Gravity on other astronomical bodies**
(a) The weak lunar gravity permits astronauts, even encumbered by their massive space suits, to bound over the Moon's surface. The low gravitational pull, only about one-sixth that of Earth, is due to the Moon's smaller size and its lower average density. **(b)** The rover *Sojourner*, brought to Mars on the *Pathfinder* mission of 1997, exploring the Martian surface. The gravity of Mars is only about 38% that of Earth.

(a)

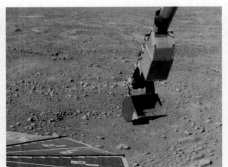

(b)

Find the acceleration due to gravity on the surface of the Moon. The mass of the Moon is $M_m = 7.35 \times 10^{22}$ kg, and its radius is $R_m = 1.74 \times 10^6$ m.

Solution

On the surface of the Moon, the acceleration due to gravity is

$$g_m = \frac{GM_m}{R_m^2} = \frac{(6.67 \times 10^{-11} \text{ N} \cdot \text{m}^2/\text{kg}^2)(7.35 \times 10^{22} \text{ kg})}{(1.74 \times 10^6 \text{ m})^2} = \boxed{1.62 \text{ m/s}^2}$$

This is about one-sixth of the acceleration due to gravity on Earth.

Math HELP
Scientific
Notation
See Math
Review,
Section I

Practice Problems

18. The lunar rover on the *Apollo* missions had a mass of 225 kg. What was its weight on Earth and on the Moon?

19. At a certain distance from the center of Earth, a 4.6-kg object has a weight of 2.2 N.
(a) Find this distance.
(b) If the object is released at this location and allowed to fall toward Earth, what is its initial acceleration?

The Cavendish experiment approximated *G*

The British physicist Henry Cavendish performed a delicate experiment in 1798 that is often referred to as "weighing the Earth." What he did, in fact, was measure the value of the universal gravitation constant, *G*, that appears in Newton's law of gravity. He was the first to obtain a numerical value for this quantity. As we saw earlier, *G* is a very small number. This is why it took more than 100 years after Newton published the law of gravitation before *G* could be measured.

In the Cavendish experiment, illustrated in **Figure 9.10**, two masses *m* are suspended from a thin thread. Near each suspended mass is a large stationary mass *M*, as shown. Each suspended mass is attracted by the force of gravity toward the large mass near it. As a result, the rod holding the suspended masses tends to rotate and twist the thread. The angle through which the thread twists can be measured by bouncing a beam of light from a mirror attached to the thread. This, in turn, gives the amount of force acting on the suspended masses. Using this experiment, Cavendish found $G = 6.754 \times 10^{-11}$ N $\cdot$ m^2/kg^2, in good agreement with the currently accepted value.

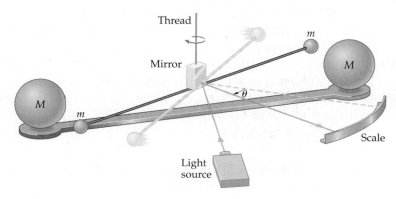

◀ **Figure 9.10 The Cavendish experiment**
The gravitational attraction between the masses *m* and *M* causes the rod and the suspending thread to twist. Measurement of the twist angle allows for a direct measurement of the gravitational force.

So why do people say that Cavendish "weighed the Earth." Well, recall that the acceleration due to gravity at the surface of Earth can be written as follows:

$$g = G\frac{M_E}{R_E^2}$$

Rearranging this equation to solve for the mass of Earth, M_E, yields

$$M_E = \frac{gR_E^2}{G}$$

Before the Cavendish experiment the quantities g and R_E were known from direct measurements, but G had yet to be determined. When Cavendish measured G, he didn't actually "weigh" Earth, of course. Instead, he made it possible to calculate its mass, M_E.

QUICK Example 9.6 What's the Mass of Earth?

Use $M_E = gR_E^2/G$ to calculate the mass of Earth.

Solution
Substituting the known values, we find

$$M_E = \frac{gR_E^2}{G} = \frac{(9.81 \text{ m/s}^2)(6.37 \times 10^6 \text{ m})^2}{6.67 \times 10^{-11} \text{ N} \cdot \text{m}^2/\text{kg}^2}$$

$$= \boxed{5.97 \times 10^{24} \text{ kg}}$$

Math HELP
Scientific
Notation
**See Math
Review,
Section I**

Practice Problem

20. The acceleration due to gravity at the Moon's surface is known to be about one-sixth of that on Earth. Given that the radius of the Moon is roughly one-quarter of Earth's radius, find the mass of the Moon in terms of the mass of Earth.

Earth's internal structure can be inferred from *G*

As soon as Cavendish determined the mass of Earth, geologists were able to use his result to calculate the average mass of Earth per volume—that is, Earth's average density. They found the following:

average density of Earth $= 5.53 \text{ g/cm}^3$

This is an interesting result because typical rocks found near the surface of Earth, such as granite, have a density of only about 3.00 g/cm^3. Therefore, the interior of Earth must have a greater density than its surface. In fact, from analysis of the propagation of seismic waves around the world, we now know that Earth has a rather complex interior structure, including a solid inner core with a density of about 15.0 g/cm^3.

A similar calculation for the Moon yields an average density of about 3.33 g/cm^3. This is essentially the same as the density of the lunar rocks brought back by *Apollo* missions. Thus, it is likely that the Moon does not have an internal structure similar to that of Earth.

Massive stars can become black holes

What is the defining characteristic of a black hole?

You've probably heard about black holes from television shows or science fiction movies. They really do exist. In fact, black holes play an important role in the history of the universe.

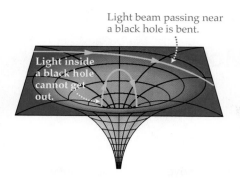

Light beam passing near a black hole is bent.

Light inside a black hole cannot get out.

◀ **Figure 9.11 A black hole**
The gravitational field of a black hole is so strong that not even light can escape. Light passing nearby is bent along a curved trajectory.

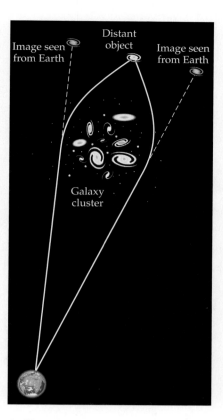

Distant object

Image seen from Earth

Image seen from Earth

Galaxy cluster

According to Einstein's theory of general relativity, the gravity of a massive star can become so strong that nothing—not even light—can escape. This is shown in **Figure 9.11.** 🔒 **An object with gravity so strong that light can't escape from it is known as a *black hole*.** Anything entering a black hole is making a one-way trip to an unknown destiny.

Evidence for Black Holes Since black holes cannot be seen directly, our evidence for their existence is indirect. However, we can predict that as matter is drawn toward a black hole, it should become heated to the point where it would emit strong beams of X-rays before disappearing from view. X-ray beams matching this prediction have in fact been observed. These observations, coupled with a variety of others, give astronomers confidence that massive black holes reside at the core of most galaxies—including our own!

Gravitational Lensing Einstein's theory of general relativity predicts that any amount of mass can bend light—at least a little. For example, light from distant stars is deflected by the mass of the Sun. The deflection is rather small, however—only the size of a quarter at a distance of 2.9 km (1.8 mi). Light passing an entire galaxy of stars or a black hole can be bent by much larger amounts, as **Figure 9.12** indicates. This effect is referred to as *gravitational lensing*, because the galaxy or black hole changes the direction of light—just like a lens does. Because of gravitational lensing, photographic images of distant galaxies often show them in duplicate, quadruplicate, or even spread out into circular arcs.

▲ **Figure 9.12 Gravitational lensing**
Astronomers often find that very distant objects seem to produce multiple images in photographs. The cause is the gravitational attraction of intervening galaxies or clusters of galaxies, which are so massive that they can bend the light from remote objects as it travels toward Earth.

9.2 LessonCheck 🅼🅿

Checking Concepts

21. 🔒 **Describe** What assumption can be used to model a large spherical mass like the Sun?

22. 🔒 **Explain** Why is a black hole called "black"?

23. [Triple Choice] If both the mass and the radius of a planet were doubled, would the acceleration due to gravity at its surface increase, decrease, or stay the same? Explain.

24. Apply A spaceship orbits a distant star. The star suddenly collapses to half its original size, with no change in its mass. What effect does this have on the spaceship? Explain.

Solving Problems

25. Calculate The radius of Mercury is 2400 km, and the acceleration due to gravity at its surface is 3.7 m/s^2. What is the mass of Mercury?

26. Calculate What is the acceleration due to gravity at a distance of two Earth radii from Earth's center?

27. Calculate What is the acceleration due to gravity at the surface of Mars? The mass of Mars is 6.4×10^{23} kg, and its radius is 3400 km.

28. Calculate At what distance from the center of the Moon is the acceleration due to the Moon's gravity equal to 0.50 m/s^2?

9.3 Circular Motion

Vocabulary

- centripetal acceleration
- centripetal force
- spring tides
- neap tides

🔑 *What is required to move an object along a circular path?*

According to Newton's second law, an object moves with constant speed in a straight line unless acted on by a force. Simply put, a force is required to change an object's speed, direction, or both. In this lesson we consider the consequences of changing the direction of motion—as occurs when an object moves in a circle.

Circular motion requires constant acceleration

If you drive a car on a circular track, the direction of the car's motion changes continuously—even if its speed remains the same. A force must act on the car to cause this change in direction. Similarly, a force must act on a satellite to make it move in a circular orbit, or on a roller coaster to make it do a loop-the-loop. There are two things we would like to know about a force that causes circular motion. First, what is its direction? Second, what is its magnitude?

Let's start with the direction of the force. Imagine swinging a ball tied to a string in a circle over your head, as shown in **Figure 9.13**. As you swing the ball, you feel a tension in the string pulling outward. Of course, on the other end of the string, where it attaches to the ball, the tension pulls inward, toward the center of the circle. Thus, the force the ball experiences is a force that is always directed toward the center of the circle.

- 🔑 **To make an object move in a circle with constant speed, a force that is directed toward the center of the circle must act on the object.**

Since the ball is acted on by a *force* directed toward the center of the circle, it follows that it must be *accelerating* toward the center of the circle. This might seem odd at first: How can a ball that moves with constant speed have an acceleration? The answer is that acceleration is produced whenever the speed or direction of the velocity changes. In circular motion the direction changes continuously. The center-directed acceleration of an object in circular motion is referred to as centripetal acceleration. We represent it with a_{cp}, where the subscript "cp" stands for **centripetal**.

Reading Support ✓
Word origin of
centripetal

from the Latin words *centrum* meaning "center" and *-petus* meaning "seeking"

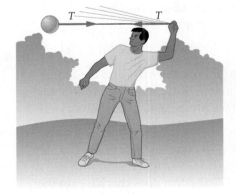

▶ **Figure 9.13 Swinging a ball in a circle overhead**
The tension in the string pulls inward on the ball, giving it a centripetal acceleration.

CONCEPTUAL Example 9.7 Choose the Path

A puck attached to a string undergoes circular motion on the horizontal surface of a frictionless air table. If the string breaks at the point indicated in the figure, is the subsequent motion of the puck best described by path A, B, C, or D?

Reasoning and Discussion

Once the string breaks, there is zero total force acting on the puck, which is moving on the frictionless table. An object with zero total force acting on it moves in a straight line with constant speed. When the string breaks, the puck is moving upward and to the right. It will continue in this direction with constant speed after the string breaks, and hence it will follow path B.

Answer

The motion of the puck will be along path B.

Speed and radius determine centripetal acceleration

Perhaps you've noticed that when a car rounds a curve you feel more acceleration the faster the car is moving. That is, the faster the car goes around the curve, the more you feel pushed or pulled to one side or the other. Similarly, the tighter the curve—that is, the smaller the radius of the circle—the greater the acceleration. Combining these observations, we see that the magnitude of an object's centripetal acceleration depends on both the speed of the object and the radius of the circle in which it moves.

To be precise, experiments and mathematical calculations show that the magnitude of the centripetal acceleration is equal to the speed squared divided by the radius. Thus, when an object moves with a speed v in a circle of radius r its centripetal acceleration is

---

Centripetal Acceleration

$$\text{centripetal acceleration} = \frac{(\text{speed})^2}{\text{radius}}$$

$$a_{cp} = \frac{v^2}{r}$$

Since the centripetal acceleration depends on the speed squared, it follows that doubling the speed increases the acceleration by a factor of 4. The acceleration is also inversely dependent on the radius. Therefore, halving the radius doubles the acceleration.

Centripetal force depends on mass, speed, and radius

As we learned in Chapter 5, Newton's second law states that force equals mass times acceleration. For circular motion the acceleration is the centripetal acceleration. Therefore, the force that causes circular motion, referred to as the **centripetal force**, is mass times the centripetal acceleration.

C**OO**L PHYSICS
Banked Curves

Have you noticed that the curves in many roads are tilted, or banked? Pay attention during your next car trip and you'll notice that the banking always tilts toward the center of the circular path. This is by design. On a banked curve the normal force exerted by the road contributes to the required centripetal force. In fact, if the tilt angle is just right, the normal force provides *all* of the centripetal force. With the correct banking a car can negotiate the turn even if there is *no friction at all* between its tires and the road.

What are the defining characteristics of the centripetal force?

The magnitude of the centripetal force, f_{cp}, is given by the following equation.

> **Newton's Second Law for Circular Motion**
>
> centripetal force = mass × centripetal acceleration
> $$f_{cp} = ma_{cp}$$

Substituting v^2/r for the centripetal acceleration, a_{cp}, allows us to write Newton's second law as follows:

$$f_{cp} = \frac{mv^2}{r}$$

During circular motion the centripetal force has a constant magnitude (mv^2/r) and is always directed toward the center of the circle. In addition, the centripetal force is always perpendicular to the tangential velocity of the object in circular motion.

An important aspect of centripetal force is that it can be produced in a number of different ways. For example, the centripetal force might be the tension in a string, as in the example of the ball on a string in Figure 9.13. It could also be the tension in the chains suspending the passengers on the carnival ride in **Figure 9.14**. Or the centripetal force could be due to friction between tires and the road, as when a car goes around a corner. It could even be the force of gravity causing a satellite to orbit Earth. Thus, the centripetal force is a force that must be present to cause circular motion, but its specific source varies from system to system.

▲ Figure 9.14 **Carnival ride**
The people enjoying this carnival ride are experiencing a centripetal acceleration of roughly 10 m/s² directed inward, toward the axis of rotation.

GUIDED Example 9.8 | Rounding a Corner   **Circular Motion**

A 1200-kg car rounds a corner of radius $r = 45$ m with a speed $v = 12$ m/s. Find (**a**) the centripetal acceleration of the car and (**b**) the centripetal force required to keep it on the circular path.

Picture the Problem

The first sketch shows a bird's-eye view of the car as it moves around the circular path. The next sketch shows the car moving directly toward the observer. The positive x direction points toward the center of the circular path and the positive y axis points vertically upward. We also indicate the three forces acting on the car: gravity, $\vec{W}$; the normal force, $\vec{N}$; and the force of static friction, $\vec{f}_s$, between the tires and the road.

Strategy

Static friction provides the centripetal force required for the car to move in a circular path. This friction force, directed toward the center of the circle, keeps the car from skidding off the road. (**a**) The centripetal acceleration of the car is given by $a_{cp} = v^2/r$. (**b**) The magnitude of the centripetal force is given by $f_{cp} = ma_{cp}$, or $f_{cp} = mv^2/r$.

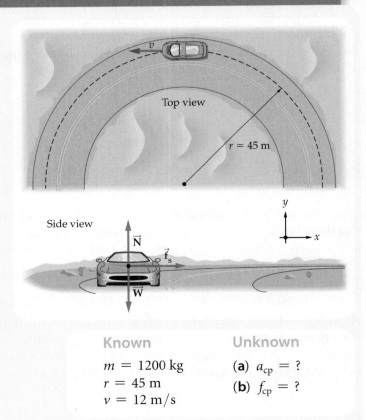

Known	Unknown
$m = 1200$ kg	(**a**) $a_{cp} = ?$
$r = 45$ m	(**b**) $f_{cp} = ?$
$v = 12$ m/s	

Solution

1 (a) Substitute $v = 12$ m/s and $r = 45$ m into the equation for centripetal acceleration, $a_{cp} = v^2/r$:

$$a_{cp} = \frac{v^2}{r}$$
$$= \frac{(12 \text{ m/s})^2}{45 \text{ m}}$$
$$= \boxed{3.2 \text{ m/s}^2}$$

2 (b) The centripetal force exerted on the car is $f_{cp} = ma_{cp}$, with $m = 1200$ kg and $a_{cp} = 3.2$ m/s²:

$$f_{cp} = ma_{cp}$$
$$= (1200 \text{ kg})(3.2 \text{ m/s}^2)$$
$$= \boxed{3800 \text{ N}}$$

Insight

Static friction between the tires and the road holds the car on the circular path. If the car goes too fast, friction won't be strong enough to hold it on the road. Similarly, if rain or ice makes the road slick, the force of static friction will be reduced. This, in turn, lowers the maximum safe speed for the corner. (Again, recall that $N = \text{kg} \cdot \text{m/s}^2$.)

Practice Problems

29. Follow-up What is the minimum coefficient of static friction needed to hold the car on the road?

30. When riding in a 1300-kg car, you go around a curve of radius 59 m with a speed of 16 m/s. The coefficient of static friction between the car and the road is 0.88. Assuming that the car doesn't skid, what is the force exerted on it by static friction?

Turning "into the skid" can help regain control

If a driver rounds a corner too rapidly, the car might start to skid—that is, the car might begin to slide sideways across the road. A common bit of road wisdom is to turn the car in the direction of the skid to regain control. To most people this sounds completely counterintuitive. The advice is sound, however.

Suppose a car is turning to the left and begins to skid to the right. If the driver turns more sharply to the left, in an effort to correct for the skid, it will simply reduce the turning radius of the car. The result is that the centripetal acceleration becomes larger and the tendency to skid is increased.

On the other hand, turning the car slightly to the right at the start of the skid *increases* the turning radius. This decreases the centripetal acceleration. In this case the car stops skidding, and control is regained.

A dip in the road makes you feel heavier

Another situation where you can experience the effects of circular motion is riding in a car as it encounters a dip in the road. As you probably know, you feel momentarily heavier near the bottom of the dip. This change in apparent weight is due to the approximately circular motion of the car, as shown in the next Active Example.

While traveling along a country lane with a constant speed of 18 m/s, you encounter a dip in the road. The dip can be approximated as a circular arc, with a radius of 65 m. What is the normal force exerted by a car seat on an 81-kg passenger when the car is at the bottom of the dip?

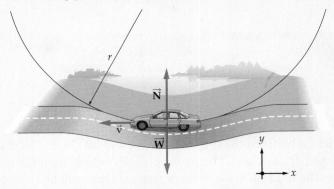

Solution *(Perform the calculations indicated in each step.)*

1. Apply Newton's second law ($\sum F = ma$) in the *y* direction—for the passenger:

$$N - mg = ma_y$$

2. Replace a_y with the expression for centripetal acceleration:

$$N - mg = mv^2/r$$

3. Solve for the normal force, *N*:

$$N = mg + mv^2/r$$

4. Substitute the numerical values and calculate the normal force:

$$N = 1200 \text{ N}$$

Notice that the passenger in Active Example 9.9 feels about 50% heavier than usual at the bottom of the dip. This effect occurs because the normal force must supply the centripetal force. The same situation occurs when a jet pilot pulls a plane out of a high-speed dive. In that case the magnitude of the effect can be much larger, resulting in a decrease of blood flow to the brain and eventually to loss of consciousness.

Suppose you stand on a bathroom scale and get a reading of 700 N. If Earth didn't rotate, would the reading on the scale be greater than, less than, or equal to 700 N?

Reasoning and Discussion
On the rotating Earth you are moving along a circular path. The centripetal force required for your circular motion is supplied by the gravitational attraction of Earth. So, a portion of the gravitational attraction is needed to supply the centripetal force, and the rest pulls down on the scale to give you the reading of 700 N. Without the rotation of Earth the entire gravitational attraction would pull down on the scale, giving a reading greater than 700 N.

Answer
The reading on the scale would be greater than 700 N.

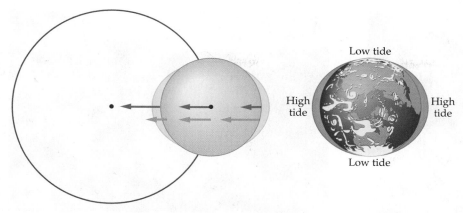

(a) The mechanism responsible for tides   **(b)** Tidal deformations on Earth

▲ **Figure 9.15 The reason why there are two tides a day**
(a) Tides are caused by a mismatch between the gravitational force exerted at various points on a finite-sized object (dark red arrows) and the centripetal force needed for circular motion (light red arrows). **(b)** On Earth the water in the oceans responds more to the deforming effects of tides than the solid land does. The result is two high tides and two low tides daily on opposite sides of Earth.

Gravity and circular motion create the tides

For thousands of years the cause of ocean tides was a complete mystery. That changed once Newton introduced his law of universal gravitation. With the understanding that a force is required to cause an object to move in a circular path and that the force of gravity becomes weaker with distance, Newton was able to explain the tides in detail. Let's investigate the basic ideas behind tides.

Forces of Varying Magnitude   To begin, consider the idealized situation shown in **Figure 9.15 (a)**. Here we see an object of finite size (a moon or a planet, for example) orbiting a point mass. If all the mass of the object were concentrated at its center, the gravitational force exerted on it by the point mass would be precisely the amount needed to cause it to move in its circular path. Since the object is of finite size, however, the force exerted on various parts of it has different magnitudes.

Deformation   To see the effect of this variation in force, we use a dark red vector in Figure 9.15 (a) to indicate the force exerted by the central point mass at three different points on the object. In addition, we use a light red vector to show the force that is required at each of these three points to cause a mass at that distance to orbit the central mass. Comparing these vectors, we see that the forces are identical at the center of the object—as expected. On the near side of the object, however, the force exerted by the central mass is larger than the force needed to hold the object in orbit. On the far side, the force due to the central mass is less than the force needed to hold the object in orbit. The result is that the near side of the object is pulled closer to the central mass and the far side tends to move farther away. This causes a deformation of the object into an egglike shape, as indicated in Figure 9.15 (a).

Any two objects orbiting one another cause deformations of this type. For example, Earth causes a deformation in the Moon, and the Moon causes a similar deformation in Earth. In **Figure 9.15 (b)** we show Earth and the water of its oceans deformed into an egg shape. Since the water in the oceans can flow, the oceans deform much more than the underlying rocky surface of Earth. As a result, the water level relative to the surface of Earth is greater at the *tidal bulges*, as shown in the figure.

▲ **Figure 9.16 Extreme tides**
Tides on Earth are caused chiefly by the Moon's gravitational pull, though at full and new moons, when the Moon and Sun are aligned, the Sun's gravity can enhance the effect. In some places on Earth, such as the Bay of Fundy between Maine and Nova Scotia, local topographic conditions produce abnormally large tides.

Spring Tides and Neap Tides As Earth rotates about its axis, a person at a given seaside location will observe two high tides and two low tides each day. This is the basic mechanism of the tides on Earth. If the Sun aligns with the Moon, the tidal effect is enhanced, and we experience **spring tides**. When the Sun and the Moon are at right angles relative to Earth, the tides are smaller and are referred to as **neap tides**. Certain locations on Earth experience particularly large tides due to their local topography. **Figure 9.16** shows an example of extreme tides in the Bay of Fundy.

9.3 LessonCheck (MP)

Checking Concepts

31. 🔗 **Explain** What must be true of a force that produces circular motion?

32. 🔗 **Describe** What is the magnitude of a centripetal force? What is its direction?

33. Compare and Contrast How does centripetal acceleration differ from other accelerations? How is it similar to other accelerations?

34. Assess The gas pedal and the brake pedal are capable of causing a car to accelerate. Can the steering wheel also produce an acceleration? Explain.

35. Triple Choice An object moves with a speed v on a circular path of radius r. If both the speed and the radius are doubled, does the centripetal acceleration of the object increase, decrease, or stay the same? Explain.

Solving Problems

36. Rank The masses of four objects and their speeds on circular paths of the indicated radii are shown below. Rank these objects in order of increasing centripetal force. Indicate ties where appropriate.

	Object A	Object B	Object C	Object D
Mass (kg)	60	5	24	2
Speed (m/s)	1	2	2	4
Radius (m)	5	2	8	1

37. Rank Use the information in Problem 36 to rank the objects in order of increasing centripetal acceleration. Indicate ties where appropriate.

38. Calculate You swing a bucket of water in a vertical circle of radius 1.3 m. What speed must the bucket have if it is to complete the circle without any water spilling?

39. Calculate A car experiences a centripetal acceleration of 4.4 m/s² as it rounds a corner with a speed of 15 m/s. What is the radius of the corner?

9.4 Planetary Motion and Orbits

Before Newton it was generally accepted that the heavens were completely separate from Earth. Objects in the night sky were thought to obey their own "heavenly" laws. Newton showed that, on the contrary, the same law of gravity that operates at the surface of Earth applies to the Moon and to other astronomical objects. This lesson explores the basic laws that govern the orbits of astronomical objects, including comets and planets.

Gravitation and Advances in Physics

The law of gravitation extended the realm of physics

So successful was Newton's law of gravitation that Edmond Halley (1656–1742) was able to use it to predict the return of the comet that today bears his name. Though he did not live to see its return in 1758, the fact that the comet did reappear as predicted was an event unprecedented in history.

Roughly a hundred years later, Newton's theory of gravity scored an even more impressive success. Astronomers observing the planet Uranus noticed small deviations in its orbit, which they thought might be due to the gravitational tug of a previously unknown planet. Using Newton's law to calculate the predicted position of the new planet—now called Neptune—they found it on the very first night of observations. The fact that Neptune was precisely where the law of gravitation said it should be still stands as one of the most astounding triumphs in the history of science.

The Moon is constantly falling toward Earth

A brilliant flash of insight came to Isaac Newton in 1666. After seeing an apple fall to the ground, Newton came to the following realization:

- The force that causes a falling apple to accelerate downward is the same force that causes the Moon to move in a circular path around Earth.

To put it another way, Newton was the first to realize that the Moon is *constantly falling* toward Earth, though it never gets any closer. This phenomenon is illustrated in **Figure 9.17**. In addition, the Moon falls for exactly the same reason that an apple falls—the force of gravity exerted by Earth.

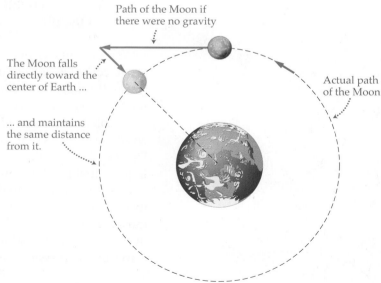

Path of the Moon if there were no gravity

The Moon falls directly toward the center of Earth ...

... and maintains the same distance from it.

Actual path of the Moon

▲ Figure 9.17 **The falling Moon**
The Moon constantly falls toward the center of Earth, but it never gets any closer. (Drawing is not to scale.)

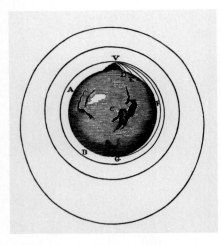

▲ **Figure 9.18 Understanding orbits**
Imagine throwing a projectile horizontally from the top of a mountain. The greater the initial speed of the projectile, the farther it travels in free fall before striking the ground. In the absence of air resistance, a great enough initial speed results in the projectile circling Earth and returning to its starting point.

🔑 *What is the shape of planetary orbits?*

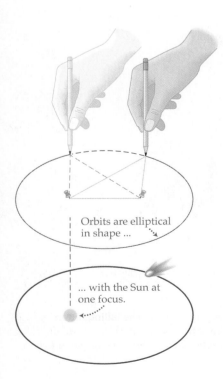

Orbits are elliptical in shape ...

... with the Sun at one focus.

Figure 9.18 presents Newton's classic illustration of this point. It shows how an object that is projected horizontally with an increasingly large speed lands farther and farther from the launch point. Eventually, with a high enough launch speed, the object goes into orbit. What this really means is that as the object falls toward Earth, the surface of Earth curves away under it, leaving the object always the same distance above the surface.

Kepler's Laws

A problem of long-standing interest and contention throughout history was the description of planetary motion. Ptolemy—the ancient Greek astronomer—claimed that the other planets orbit Earth. Later, Johannes Kepler (1571–1630) showed with experimental results that all of the planets—including Earth—orbit the Sun. This was in agreement with the ideas proposed by Copernicus (1473–1543).

Kepler was also able to deduce three laws that planets obey in their orbits. Why the planets obeyed Kepler's laws no one knew—not even Kepler. Decades after Kepler's death Newton was able to show that each of Kepler's laws follows as a direct consequence of the law of universal gravitation. We'll look at each of Kepler's laws.

Kepler's first law describes the shape of orbits

Kepler searched to find circular orbits that would match the observations of the planets, and of Mars in particular. Though the orbit of Mars was exasperatingly close to circular, the small differences between a circular path and the experimental observations just could not be ignored. Eventually, after a great deal of hard work, Kepler discovered that Mars followed an orbit that was elliptical rather than circular. The same was true of the orbits of the other planets. This observation became Kepler's first law:

> **Kepler's First Law**
> 🔑 **Planets follow elliptical orbits, with the Sun at one focus of the ellipse.**

This is a fine example of the scientific method in action. Though Kepler expected and wanted to find circular orbits, he would not allow himself to ignore the data. He had to discard a treasured—but incorrect—theory and move on to an unexpected, but ultimately correct, view of nature.

Kepler's first law is illustrated in **Figure 9.19**, along with a definition of an *ellipse* in terms of its two *foci*. In the case where the two foci merge, as in **Figure 9.20**, the ellipse becomes a circle. Thus, a circular orbit *is* allowed by Kepler's first law, but only as a special case.

Newton was able to show mathematically that, because the force of gravity decreases with the inverse square of the distance, or $1/r^2$, closed orbits must have the form of ellipses or circles. He also showed that orbits that are not closed—say, the orbit of a comet that passes by the Sun once and then leaves the solar system—are either parabolas or hyperbolas.

◀ **Figure 9.19 Drawing an ellipse**
To draw an ellipse, put two tacks in a piece of cardboard. The tacks define the foci of the ellipse. Now connect a length of string to the two tacks, and use a pencil and the string to sketch out a smooth closed curve, as shown. This closed curve is an ellipse. In its orbit a planet follows an elliptical path, with the Sun at one focus. Nothing is at the other focus.

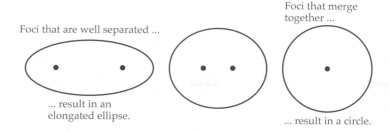

Foci that are well separated ...

... result in an elongated ellipse.

Foci that merge together ...

... result in a circle.

As the two foci of an ellipse approach one another, the ellipse becomes more circular. In the limit where the foci merge, the ellipse becomes a circle.

Kepler's second law relates speed and distance

When Kepler plotted the positions of a planet on its elliptical orbit, indicating at each position the time the planet was there, he made an interesting observation. First, he drew a line from the Sun to a planet at a given time. Then a certain time later—perhaps a month—he drew a line again from the Sun to the new position of the planet. The result was that the planet "swept out" a wedge-shaped area, as indicated in **Figure 9.21 (a)**. When this procedure was repeated when the planet was on a different part of its orbit, another wedge-shaped area was generated. Kepler's observation was that the areas of these two wedges are equal:

> **Kepler's Second Law**
>
> 🔑 **As a planet moves in its orbit, it sweeps out an equal amount of area in an equal amount of time.**

Kepler's second law follows from the fact that the force of gravity on a planet pulls directly toward the Sun. As a result, gravity exerts zero torque about the Sun, and hence the angular momentum of an orbiting planet must be conserved. Newton was able to show that conservation of angular momentum is equivalent to the equal-area law stated by Kepler.

Though we have stated the first two laws in terms of planets, they apply equally well to any object orbiting the Sun. For example, a comet might follow a highly elliptical orbit, as shown in **Figure 9.21 (b)**. When it is near the Sun, it moves very quickly, for the reason discussed in Conceptual Example 9.11, sweeping out a broad wedge-shaped area in a month's time. Later in its orbit, the comet is far from the Sun and moving slowly. In this case the area it sweeps out in a month is a long, thin wedge. Still, the two wedges have equal areas.

CONCEPTUAL Example 9.11 Comparing Orbital Speeds

Earth's orbit is slightly elliptical. In fact, Earth is closer to the Sun during winter (in the northern hemisphere) than it is during summer. Is the speed of Earth during winter greater than, less than, or the same as its speed during summer?

Reasoning and Discussion
According to Kepler's second law, the area swept out by Earth per month is the same in winter as it is in summer. In winter, however, the radius from the Sun to Earth is less than it is in summer. Therefore, if this smaller radius is to sweep out the same area, Earth must move more rapidly.

Answer
The speed of Earth is greater during the winter.

CONNECTING IDEAS

In Chapter 8 you learned that angular momentum is a conserved quantity.
• Here, conservation of angular momentum plays a key role in planetary motion, leading to Kepler's second law.

🔑 *How is a planet's motion related to the area it sweeps out?*

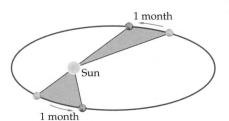

1 month

Sun

1 month

(a) Equal areas in equal times

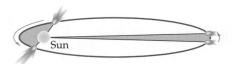

Sun

(b) Equal areas in equal times for a highly elliptical orbit

▲ **Figure 9.21 Kepler's second law**
(a) The second law states that a planet sweeps out equal areas in equal times. **(b)** In this highly elliptical orbit the long, thin area is equal to the broad, fan-shaped area.

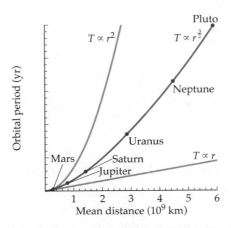

Figure 9.22 Kepler's third law compared with other possibilities These plots represent three possible mathematical relationships between period of revolution, T (in years), and mean distance from the Sun, r (in kilometers). The lower curve shows $T = $ (constant)r. The upper curve is $T = $ (constant)r^2. The middle curve, which fits the data, is $T = $ (constant)$r^{3/2}$. This is Kepler's third law.

🔑 *How is the period of a planet related to its mass and its distance from the Sun?*

Kepler's third law relates orbital radius and period

Kepler also studied the relation between the distance of a planet from the Sun and its *orbital period*—that is, the time it takes for the planet to complete one orbit. **Figure 9.22** shows a plot of period versus distance for the planets of the Solar System. Kepler tried to get these results to fit a simple dependence between period and distance. When he tried a linear fit (the bottom curve in Figure 9.22), he found that the period did not increase rapidly enough with distance. When he tried a quadratic fit (the top curve in Figure 9.22), the period increased too rapidly. Splitting the difference, Kepler found an excellent fit (the middle curve in Figure 9.22) when the period depends on the $\frac{3}{2}$ power of distance.

> **Kepler's Third Law**
>
> 🔑 **The period, T, of a planet increases as its distance from the Sun, r, raised to the 3 / 2 power. That is, $T = $ (constant)$r^{3/2}$.**

Newton was able to show that Kepler's third law also follows from the law of gravitation. In fact, he was able to derive a specific mathematical expression for the constant in Kepler's third law. Newton's result is as follows:

$$T = \left(\frac{2\pi}{\sqrt{GM_s}}\right)r^{3/2}$$

Notice that the period of a planet depends on the mass of the Sun, M_s, but not on the mass of the planet. 🔑 **In general, the period of an orbiting planet or moon depends on the mass being orbited. The period does not depend on the mass of the planet or the moon itself.**

Geosynchronous satellites obey Kepler's third law

Kepler's third law can be applied to a moon or a satellite orbiting a planet. To do so, we simply note that it is the planet that is being orbited, and not the Sun. To find the period of a satellite orbiting Earth, we replace the mass of the Sun in Kepler's third law with the mass of Earth, M_E:

$$T = \left(\frac{2\pi}{\sqrt{GM_E}}\right)r^{3/2}$$

Similarly, the speed of a satellite orbiting Earth can be shown to be

$$v = \sqrt{\frac{GM_E}{r}}$$

Notice that the orbital speed decreases with increasing distance from Earth.

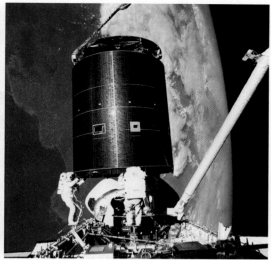

A particularly interesting application of this result is a geosynchronous satellite. A *geosynchronous satellite* is one that orbits above Earth's equator with a period equal to 1 day. From Earth such a satellite appears to be in the same location in the sky at all times. This makes it particularly useful for applications such as communications and weather forecasting. **Figure 9.23** shows a communications satellite that is being fitted with a rocket engine to take it up to a geosynchronous orbit.

From Kepler's third law we know that a satellite has a period of 1 day only if its orbital radius has a particular value. A satellite in a lower orbit will orbit too quickly; one in a higher orbit will orbit too slowly. We determine the correct orbital height for a geosynchronous satellite in the following Active Example.

▲ **Figure 9.23 Geosynchronous satellites**
Many weather and communications satellites are placed in geosynchronous orbits that allow them to remain stationary in the sky—that is, fixed over one point on Earth's equator. Because Earth rotates, the period of such a satellite must exactly match that of Earth. The photo at left shows the communications satellite *Intelsat VI* just prior to its capture in 1992 by astronauts of the space shuttle *Endeavour*. A launch failure had left the satellite stranded in low orbit. The astronauts snared the satellite (right) and fitted it with a new engine that boosted it to its geosynchronous orbit, where it is still in operation today.

ACTIVE Example 9.12 Find the Altitude of the Orbit

Find the altitude above Earth's surface where a satellite orbits with a period of 1 day. Note that $R_E = 6.37 \times 10^6$ m, $M_E = 5.97 \times 10^{24}$ kg, and $T = 1$ day $= 8.64 \times 10^4$ s.

Solution *(Perform the calculations indicated in each step.)*

1. Write Kepler's third law for the case of a satellite orbiting Earth:

$$T = \left(\frac{2\pi}{\sqrt{GM_E}}\right)r^{3/2}$$

2. Rearrange and solve for the orbital radius, r:

$$r = \left(\frac{T\sqrt{GM_E}}{2\pi}\right)^{2/3}$$

3. Substitute the numerical values:

$$r = 4.22 \times 10^7 \text{ m}$$

4. Subtract the radius of Earth to find the altitude above Earth's surface:

$$r - R_E = 3.58 \times 10^7 \text{ m}$$

Math HELP
Exponents
See Math Review, Section I
Solving Simple Equations
See Math Review, Section III

Insight
Thus, *all* geosynchronous satellites orbit at an altitude of 3.58×10^7 m $\approx 22,000$ mi.

Not all spacecraft are placed in geosynchronous orbits, however. The International Space Station orbits at an altitude of only about 320 km. At that altitude, it takes less than an hour and a half to complete one orbit.

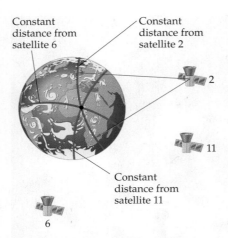

Constant
distance from
satellite 6

Constant
distance from
satellite 2

2

11

Constant
distance from
satellite 11

6

▲ **Figure 9.24 How the Global
Positioning System works**
A system of 24 satellites in orbit about
Earth makes it possible to determine a
person's location with great accuracy.
Measuring the distance of a person from
satellite 2 places the person somewhere
on the red circle. Similar measurements
using satellite 11 place the person's position
somewhere on the green circle, and further
measurements can pinpoint the person's
location.

GPS satellites have relatively low orbits

The 24 satellites of the Global Positioning System (GPS) are in relatively low
orbits. These satellites have an average altitude of 20,200 km (12,550 mi) and
orbit Earth every 12 hours. They are used to provide a precise determination
of an observer's position anywhere on Earth.

The operating principle of the GPS is illustrated in **Figure 9.24**. Imagine,
for example, that satellite 2 in this figure emits a radio signal at a particular
time (all GPS satellites carry atomic clocks on board). This signal travels
away from the satellite with the speed of light and is detected a short time
later by an observer's GPS receiver. Multiplying the time delay by the speed
of light gives the distance of the receiver from satellite 2. Thus, in our example, the observer must lie somewhere on the red circle in Figure 9.24. Additional time-delay measurements for signals from satellite 11 show that the
observer is also somewhere on the green circle. It follows that the observer
is either at the point shown in Figure 9.24 or at the second intersection of
the red and green circles on the other side of the planet. Measurements
from satellite 6 can resolve the ambiguity and place the observer at the point
shown in the figure. Measurements from additional satellites can even supply the observer's altitude.

GPS receivers, which were originally used by hikers, boaters, and others
who need to know their precise location, are now common in cars. Coupled
with navigation software, they have become an extremely popular application of satellite technology. A GPS receiver typically uses signals from as
many as 12 satellites. Currently, GPS gives positions with an accuracy of
about 1 meter.

9.4 LessonCheck ⓜⓟ™

Checking Concepts

40. 💬 **Describe** What is the shape of a planet's
orbit?

41. 💬 **Explain** How does the area a planet sweeps
out in a given amount of time change throughout its
orbit?

42. 💬 **Describe** How does the orbital period of a
planet change if the radius of its orbit is increased?

43. Choose Does the orbital period of a planet depend
on the mass of the planet or on the mass of the star that
it orbits?

44. State How long does it take for a geosynchronous
satellite to complete one orbit?

45. Decide When a communications satellite is placed
in a geosynchronous orbit above the equator, it remains
fixed over a given point on the ground. Is it possible to
put a satellite into an orbit such that it will remain fixed
above the North Pole? Explain.

Solving Problems

46. Think & Calculate Suppose the orbital radius of a
satellite is quadrupled.
(**a**) Does the period of the satellite increase, decrease,
or stay the same?
(**b**) By what factor does the period of the satellite
change?
(**c**) By what factor does the orbital speed change?

47. Calculate Venus orbits the Sun with a period of
1.94×10^7 s. What is its average distance from the Sun?

48. Calculate On the *Apollo* missions to the Moon,
the command module orbited constantly at an altitude
of 110 km above the lunar surface. How much time did
it take for the command module to complete one orbit?

49. Calculate How much time does it take for Jupiter
to complete one orbit around the Sun? Does your result
depend on the mass of Jupiter?

50. Calculate A GPS satellite orbits at an altitude of
20,200 km above the surface of Earth. What is the speed
of the satellite? (Recall that $R_E = 6.37 \times 10^6$ m.)

Physics & You

In the tidal stream system illustrated here, energy from flowing tides powers turbines that generate electricity.

Tidal Energy

What Is It? Tides are the periodic rises and falls of sea level caused by the gravitational tug-of-war between the Sun, the Moon, and Earth. Tides provide a source of natural, clean, renewable energy. Tidal energy is harvested by converting the kinetic energy of the moving water into electricity.

When Was It Invented? Tidal power plants known as *barrage plants* began harnessing the power of tides in the 1960s. Tidal stream systems, which use a different technology, are planned or under development in several countries.

How Does It Work? Tidal stream systems, like the one illustrated here, are one way to produce electrical power from tides. They use a shrouded turbine to harvest the kinetic energy of water flowing in the tides. The tidal stream system is placed along a coastline or in a river that is free of features that could obstruct or deflect the tidal flow. The latest tidal stream systems are designed to pivot, allowing them to follow the direction of peak tidal flow.

Tidal energy is generated when the force of tidal water turns the blades on a turbine. The turbine converts tidal energy first into mechanical energy and then into electrical energy. The amount of energy contained in a flowing tide is related to the cube of the tide's velocity. Thus, a slight increase in tidal velocity corresponds to a very large increase in available energy. For example, compare the power generated by tidal velocities of 1.5 m/s and 3.0 m/s. Though the water velocities differ by a factor of 2, the faster velocity yields 8 times more tidal energy.

Why Is It Important? Tidal energy is a consistent form of energy that can be harnessed to provide a clean energy alternative to coastal communities. Once in place, tidal power plants require little maintenance. This renewable energy resource does not produce any waste or greenhouse gases.

Take It Further

1. Compare *Use information from the Internet to evaluate the pros and cons of tidal, solar, and wind energy. Summarize your findings in a one-page written report.*

2. Critical Thinking *Research the environmental and economic implications of using a barrage power plant versus a tidal stream system. Which type of tidal energy system would you recommend to the city council of a coastal community?*

Physics Lab Centripetal Force

This lab explores the relationship between the speed of an object in uniform circular motion and the centripetal force acting on the object. You will whirl a mass overhead at the speed needed to balance the force exerted by a hanging mass—a mass that varies from trial to trial.

Materials
- 1.25 m of nylon cord or string
- 15 cm of tape-wrapped polished glass tube
- number 4 or 5 one-hole rubber stopper
- mass set
- meterstick
- stopwatch
- masking tape

Procedure 🐾

1. After tying one end of the nylon cord to the rubber stopper, pass the cord through the tape-wrapped glass tube and attach a 0.05-kg mass to the other end of the cord.

2. Pull enough cord through the tube so that there is 70 cm of cord between the stopper and the end of the tube. Attach a small piece of tape to the cord just below the tube. This *tape*

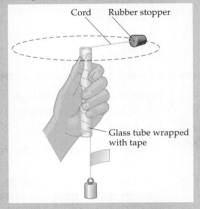

Cord   Rubber stopper

Glass tube wrapped with tape

marker will enable you to keep the radius constant while the stopper is in motion.

3. Hold the 0.05-kg mass in one hand and the glass tube in the other. Begin whirling the stopper in a horizontal circle above your head, as shown. Release the weight and increase the speed of the stopper until the tape marker moves to a position just below the end of the tube.

Caution: *Take care to prevent hitting yourself or anyone else with the whirling rubber stopper; all other students should stand clear of the whirling stopper.*

4. Have your lab partner measure the time required to complete 20 revolutions of the stopper. Record this time in the data table for Trial 1.

5. Repeat Steps 3 and 4 for Trials 2–5. Increase the mass in each successive trial by 0.05 kg, as shown in the data table.

Data Table

Trial	Hanging Mass (kg)	Weight of Hanging Mass (N)	Centripetal Force (N)	Time for 20 Revolutions (s)	Speed of Stopper (m/s)	Speed of Stopper Squared (m²/s²)
1	0.05					
2	0.10					
3	0.15					
4	0.20					
5	0.25					

Analysis

1. Calculate the weight of the hanging mass (in newtons) for each trial. Record your results in the data table.

2. Because the weight of the hanging mass supplies the centripetal force, copy the weights of the hanging mass into the centripetal force column.

3. Calculate the speed of the stopper for each trial. Record your results in the data table.

4. Create a graph of the centripetal force versus the speed of the stopper.

5. Create a graph of the centripetal force versus the speed of the stopper squared.

Conclusions

1. When the stopper is moving in a circle, what provides the centripetal force? Describe the path the stopper would follow if this force were suddenly removed.

2. Based on your data, describe what happens to the force needed to keep the stopper moving in a circle as the speed of the stopper is increased.

3. What is the shape of the curve on the graph of centripetal force versus speed squared? Based on the graph, what can you conclude about the relationship between centripetal force and speed (when the revolving mass and the radius are held constant)?

4. Explain how you think centripetal force and mass are related. Based on this relationship, predict how much more centripetal force would be required if two identical stoppers were used instead of one.

9 Study Guide

Big Idea
Gravity acts on everything in the universe.

Gravity is a force with an infinite range. The strength of the gravitational force decreases rapidly with distance, but it never goes to zero. Thus, every object in the universe exerts some small force on us.

9.1 Newton's Law of Universal Gravity

🔑 The force of gravity decreases as the inverse of the square of the distance, $1/r^2$. Because of this, we say that gravity is an *inverse square* force law.

🔑 If a mass experiences gravitational forces from a number of other masses, the total force acting on it is the vector sum of all those individual forces.

• Gravity is the force of nature that attracts one mass to another mass.

• The fact that the forces of gravity add together like vectors is referred to as *superposition*.

• The gravitational forces exerted by two objects on one another form an action-reaction pair. The forces are equal in magnitude but opposite in direction.

Key Equations
The force of gravity between two masses m_1 and m_2 separated by a distance r is attractive and has a magnitude given by the following:

$$F = G\frac{m_1 m_2}{r^2}$$

The constant G in the gravitational force equation is the universal gravitation constant. It has the following value:

$$G = 6.67 \times 10^{-11}\ \text{N} \cdot \text{m}^2/\text{kg}^2$$

9.2 Applications of Gravity

🔑 A spherical mass exerts the same gravitational force on masses outside it as it would if all the mass of the sphere were concentrated at its center.

🔑 An object with gravity so strong that light can't escape it is known as a *black hole*.

• Tides result from the variation of the gravitational force from one side of an astronomical object to the other side.

Key Equation
The acceleration due to gravity at the surface of Earth is determined by the following equation:

$$g = \frac{GM_E}{R_E^2}$$

9.3 Circular Motion

🔑 To make an object move in a circle with constant speed, a force that is directed toward the center of the circle must act on the object.

🔑 During circular motion the centripetal force has a constant magnitude (mv^2/r) and is always directed toward the center of the circle. In addition, the centripetal force is always perpendicular to the tangential velocity of the object in circular motion.

• The center-directed acceleration of an object in circular motion is the centripetal acceleration.

Key Equations
The magnitude of the centripetal acceleration, a_{cp}, is given by the following:

$$a_{cp} = v^2/r$$

• The magnitude of the centripetal force, f_{cp}, is mass times the centripetal acceleration:

$$f_{cp} = ma_{cp} \quad \text{or} \quad f_{cp} = mv^2/r$$

9.4 Planetary Motion and Orbits

🔑 Planets follow elliptical orbits, with the Sun at one focus of the ellipse.

🔑 As a planet moves in its orbit, it sweeps out an equal amount of area in an equal amount of time.

🔑 The period, T, of a planet increases as its distance from the Sun, r, raised to the 3/2 power. That is, $T = (\text{constant})r^{3/2}$.

🔑 The period of an orbiting planet or moon depends on the mass being orbited. The period does not depend on the mass of the planet or the moon itself.

• A geosynchronous satellite is one that orbits above the equator of Earth with a period equal to 1 day.

Key Equations
The period of a satellite's orbit around Earth depends on its average distance from Earth, r, and the mass of Earth, M_E:

$$T = \left(\frac{2\pi}{\sqrt{GM_E}}\right)r^{3/2}$$

The speed of a satellite orbiting Earth is

$$v = \sqrt{\frac{GM_E}{r}}$$

These results also apply to planets orbiting the Sun if we replace the mass of Earth, M_E, with the mass of the Sun, M_S.

9 Assessment

ANSWERS TO SELECTED ODD-NUMBERED PROBLEMS APPEAR IN APPENDIX A.

Note: Many problems in this chapter require astronomical data from Appendix C.

Lesson by Lesson

9.1 Newton's Law of Universal Gravity
Conceptual Questions

51. When a person passes you on the street, you do not feel a gravitational tug. Explain.

52. What happens to the gravitational force between two masses if their separation is halved?

53. Two objects experience a gravitational attraction. Give a reason why the gravitational force between them depends on the product of their masses and not on the sum of their masses. (*Hint:* What happens if one of the masses goes to zero? What happens if one of the masses is doubled?)

54. Rank Four two-mass systems are described below. Rank the systems in order of increasing gravitational force. Indicate ties where appropriate.

System	A	B	C	D
Mass 1	m	m	$2m$	$4m$
Mass 2	m	$2m$	$3m$	$5m$
Separation	r	$2r$	$2r$	$3r$

Problem Solving

55. In each hand you hold a 0.16-kg peach. What is the gravitational force exerted by one peach on the other when their separation is **(a)** 0.25 m and **(b)** 0.50 m?

56. A 6.8-kg bowling ball and a 7.1-kg bowling ball rest on a rack 0.75 m apart. What is the force of gravity pulling each ball toward the other?

57. In one hand you hold a 0.11-kg apple, in the other hand a 0.24-kg orange. The apple and orange are separated by 0.85 m. What is the magnitude of the force of gravity that **(a)** the orange exerts on the apple and **(b)** the apple exerts on the orange?

58. Think & Calculate A spaceship of mass m travels from Earth to the Moon along a line that passes through the center of Earth and the center of the Moon. **(a)** At what distance from the center of Earth does the gravitational force due to Earth have twice the magnitude of that due to the Moon? **(b)** How does your answer to part (a) depend on the mass of the spaceship? Explain.

59. At new moon, Earth, Moon, and Sun are in a line, as shown in **Figure 9.25**. Find the direction and magnitude of the net gravitational force exerted on the Moon.

Figure 9.25

60. Three 6.75-kg masses are at the corners of an equilateral triangle and located in space far from any other masses. If the sides of the triangle are 1.25 m long, find the magnitude of the total force exerted on each mass.

61. Four masses are positioned at the corners of a rectangle, as indicated in **Figure 9.26**. Find the magnitude and the direction of the net force acting on the 2.0-kg mass.

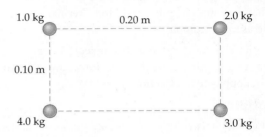

Figure 9.26

9.2 Applications of Gravity
Conceptual Questions

62. Rank Examine objects A, B, and C shown in **Figure 9.27**. Rank the objects in order of increasing net gravitational force experienced. Indicate ties where appropriate.

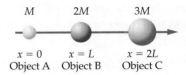

Figure 9.27

63. How does the acceleration due to gravity at the surface of a planet change if the planet's mass is doubled?

64. How does the acceleration due to gravity at the surface of a planet change if the planet's radius is doubled?

65. Rank Examine the objects in Figure 9.27. Rank them in order of increasing initial acceleration if each object alone was allowed to move. Indicate ties where appropriate.

Problem Solving

66. **Gravity on Titan** Titan is the largest moon of Saturn and the only moon in the solar system known to have a substantial atmosphere. Find the acceleration due to gravity at Titan's surface, given that its mass is 1.35×10^{23} kg and its radius is 2570 km.

67. Find the acceleration due to gravity at the surface of (a) Mercury, and (b) Venus.

68. At what altitude above Earth's surface is the acceleration due to gravity equal to $g/4$?

69. What is the acceleration due to Earth's gravity at a distance from the center of Earth equal to the orbital radius of the Moon?

70. A communications satellite with a mass of 480 kg is in a circular orbit about Earth. The radius of the orbit is 35,000 km, measured from the center of Earth. Calculate the gravitational force exerted on the satellite by Earth.

71. At a certain distance from the center of Earth, a 12-kg object has a weight of 15 N. Find this distance.

72. **Think & Calculate** In one of his novels author Jules Verne imagined that astronauts inside a spaceship walked on the floor of the cabin when the force exerted on the ship by Earth was greater than the force exerted by the Moon. When the force exerted by the Moon was greater, he thought the astronauts walked on the ceiling of the cabin. (a) At what distance from the center of Earth would the forces exerted on the spaceship by Earth and the Moon be equal? (b) Explain why Verne's description of gravitational effects is incorrect.

73. The acceleration due to gravity at the surface of Mars is 38% of the acceleration due to gravity on Earth. Given that the radius of Mars is 0.53 of that of Earth, find the mass of Mars in terms of the mass of Earth.

9.3 Circular Motion
Conceptual Questions

74. Discuss the physics involved in the spin cycle of a washing machine. In particular, how is circular motion related to the removal of water from the clothes?

75. In many science fiction stories a rotating space station, like that shown in **Figure 9.28**, provides "artificial gravity" for its inhabitants. How does this work?

Figure 9.28

76. **Rank** A car drives with *constant speed* on an elliptical track, as shown in **Figure 9.29**. Rank the points A, B, and C in order of increasing likelihood that the car might skid. Indicate ties where appropriate.

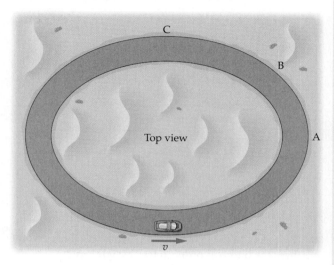

Figure 9.29

77. A car is driven with constant speed around a circular track. Answer the following questions with "Yes" or "No". (a) Is the car's velocity constant? (b) Is its speed constant? (c) Is the magnitude of its acceleration constant? (d) Is the direction of its acceleration constant?

Problem Solving

78. **A Human Centrifuge** To test the effects of high acceleration on the human body, the National Aeronautics and Space Administration (NASA) constructed a large centrifuge at the Manned Spacecraft Center in Houston. Astronauts are placed in a capsule that moves in a circular path with a radius of 15 m. If the astronauts in this centrifuge experience a centripetal acceleration 9.0 times that due to gravity, what is the linear speed of the capsule?

79. The rotating drum in a clothes-dryer has a radius of 0.31 m. If the acceleration at the rim of the drum is 27 m/s^2, what is the tangential speed of the rim?

80. Find the linear speed of the bottom of a test tube in a centrifuge if the centripetal acceleration there is 52,000 times the acceleration due to gravity. The distance from the axis of rotation to the bottom of the test tube is 7.5 cm.

81. A driver takes a 1400-kg car out for a spin, going around a corner with a radius of 63 m at a speed of 18 m/s. The coefficient of static friction between the car and the road is 0.85. Assuming the car doesn't skid, what is the force exerted on it by static friction?

82. Driving in a car with a constant speed of 12 m/s, you encounter a bump in the road that has a circular cross-section, as shown in **Figure 9.30**. If the radius of curvature of the bump is 35 m, find the apparent weight of a 67-kg person in your car as you pass over the top of the bump.

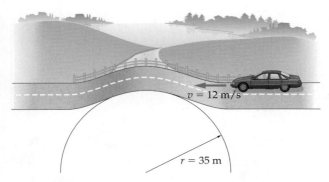

Figure 9.30

83. Referring to the Problem 82, at what speed must you go over the bump in order to feel "weightless"?

84. Jill of the Jungle swings on a vine 6.9 m long. What is the tension in the vine if Jill, whose mass is 63 kg, is moving at 2.4 m/s when the vine is vertical?

85. **Think & Calculate** (a) As you ride on a Ferris wheel like the one in **Figure 9.31**, your apparent weight is different at the top than at the bottom. Explain. (b) Calculate your apparent weight at the top and bottom of a Ferris wheel, given that the radius of the wheel is 7.2 m, it completes one revolution every 28 s, and your mass is 55 kg.

Figure 9.31 A Ferris wheel.

9.4 Planetary Motion and Orbits
Conceptual Questions

86. Does the radius of Mars's orbit sweep out the same amount of area per time as that of Earth? Explain.

87. In 2 months a planet sweeps out the area A. How much area does it sweep out in 1 month? In 3 months?

88. Friend 1 says that an orbiting satellite is in free fall. Friend 2 disagrees, reasoning that a satellite in free fall would crash to the ground. Who is right? How would you explain this phenomenon to your friends?

89. **Mass of Pluto** On June 22, 1978, James Christy made the first observation of a moon orbiting Pluto. Until that time the mass of Pluto was not known, but the discovery of its moon, Charon, allowed its mass to be calculated with some accuracy. Explain.

90. One day in the future you may take a pleasure cruise to the Moon. While there you might climb a lunar mountain and throw a rock horizontally from its summit. If, in principle, you could throw the rock fast enough, it might end up hitting you in the back. Explain.

91. The force exerted by the Sun on the Moon is more than twice the force exerted by Earth on the Moon. Should the Moon be thought of as orbiting Earth or the Sun? Explain.

92. **Predict & Explain** Laser reflectors left on the surface of the Moon by the *Apollo* astronauts show that the average distance from Earth to the Moon is increasing at the rate of 3.8 cm per year. (a) As a result, will the length of the month increase, decrease, or remain the same? (b) Choose the *best explanation* from among the following:

A. The greater the radius of an orbit, the greater the period, which implies a longer month.

B. The length of the month will remain the same because of conservation of angular momentum.

C. The speed of the Moon increases with increasing orbital radius; therefore, the length of the month will be less.

Problem Solving

93. **GPS** Satellites that make up the Global Positioning System, or GPS, orbit at an altitude of 2.02×10^7 m. Find the orbital period of a GPS satellite.

94. In July of 1999 a planet was reported to be orbiting the Sun-like star Iota Horologii with a period of 320 days. Find the radius of the planet's orbit, assuming that Iota Horologii has the same mass as the Sun.

95. Phobos, one of the moons of Mars, orbits at a distance of 9378 km from the center of the red planet. What is the orbital period of Phobos?

96. The largest moon in the solar system is Ganymede, a moon of Jupiter. Ganymede orbits at a distance of 1.07×10^9 m from the center of Jupiter with an orbital period of about 6.18×10^5 s. Using this information, find the mass of Jupiter.

97. **Think & Calculate** (a) Calculate the orbital period of a satellite that orbits two Earth radii above the surface of Earth. (b) How does your answer to part (a) depend on the mass of the satellite? Explain.

98. **Think & Calculate** The Martian moon Deimos has an orbital period that is greater than the other Martian moon, Phobos. Both moons have approximately circular orbits. (a) Is Deimos closer to or farther from Mars than

Phobos? Explain. **(b)** Calculate the distance from the center of Mars to Deimos, given that its orbital period is 1.10×10^5 s.

99. Find the orbital speed of a satellite in a geosynchronous circular orbit 3.58×10^7 m above the surface of Earth.

100. What is the orbital speed of Earth around the Sun?

101. Think & Calculate The asteroid 243 Ida has its own small moon, Dactyl. Ida and Dactyl are shown in **Figure 9.32**. **(a)** Outline a strategy to find the mass of 243 Ida, given that the orbital radius of Dactyl is 89 km and its period is 19 hr. **(b)** Use your strategy to calculate the mass of 243 Ida.

Figure 9.32 Ida and its moon, Dactyl.

102. Given that the orbital speed of a satellite depends only on G, M_E, and r, use dimensional analysis to find a formula for the orbital speed. (The simplest dimensionally consistent formula is the correct result.)

103. Think & Calculate Two satellites orbit Earth, with satellite 1 at a greater altitude than satellite 2. **(a)** Which satellite has the greater orbital speed? Explain. **(b)** Calculate the orbital speed of a satellite that orbits at an altitude of one Earth radius above the surface of Earth.

Mixed Review

104. You weigh yourself on a scale inside an airplane that is flying due east above the equator. If the airplane turns around and heads due west with the same speed, will the reading on the scale increase, decrease, or stay the same?

105. Triple Choice A small satellite orbits at the same altitude as the International Space Station. Is the orbital speed of the satellite greater than, less than, or equal to the orbital speed of the space station? Explain.

106. The period of a planet or satellite increases with distance to the 3/2 power. How does the orbital speed depend on distance? Does increasing the distance increase or decrease the orbital speed?

107. **Clearview Screen** Large ships often have circular structures in their windshields, as shown in **Figure 9.33**. What is their purpose? Called *clearview screens*, they consist of a glass disk that is rotated at high speed by a motor (in the center of the circle) to disperse rain and spray. If the screen has a diameter of 0.39 m and rotates at 1700 rpm, the speed at the rim of the screen is 35 m/s.

What is the centripetal acceleration at the rim of the screen? (A large centripetal acceleration will keep liquid from remaining on the screen. For comparison, recall that the acceleration due to gravity is 9.81 m/s^2.)

Figure 9.33 A rotating disk of glass remains clear of rain and spray, providing a clear view.

108. **Spin-Dry Dragonflies** Some dragonflies splash down on the surface of a lake and then fly upward, spinning rapidly to spray the water off their bodies. When the dragonflies spin, they tuck themselves into a "ball." They spin with a linear speed of 2.3 m/s and produce a centripetal acceleration of 250 m/s^2. What is the radius of the ball they form?

109. **The Crash of Skylab** *Skylab*, the largest spacecraft ever to fall back to Earth, met its fiery end on July 11, 1979, after flying directly over Everett, Washington, on its last orbit. On the news that night before the crash, broadcaster Walter Cronkite, in his rich baritone voice, said, "NASA says there is *a* little chance it will land in a populated area." (Italics added.) After a commercial he immediately corrected himself, saying, "I meant to say '*there is little chance*' *Skylab* will hit a populated area." In fact, it landed primarily in the Indian Ocean off the west coast of Australia, though several pieces were recovered near the town of Esperance, Australia, which sent the U.S. State Department a $400 bill for littering. The cause of *Skylab*'s crash was the friction it experienced in the upper reaches of Earth's atmosphere. As the radius of *Skylab*'s orbit decreased, did its speed increase, decrease, or stay the same?

110. An astronaut exploring a distant solar system lands on an unnamed planet with a radius of 3860 km. When the astronaut jumps upward with an initial speed of 3.10 m/s, she rises to a height of 0.580 m. What is the mass of the planet?

111. A child sits on a rotating merry-go-round, 2.3 m from its center. If the speed of the child is 2.2 m/s, what is the minimum coefficient of static friction between the child and the merry-go-round that will prevent the child from slipping?

112. **Exploring Mars** In the future astronauts may travel to Mars to carry out scientific explorations. As part of their mission, it is likely that a "geosynchronous" satellite will be placed above a given point on the Martian equator to facilitate communications. At what altitude above the surface of Mars should such a satellite orbit? (*Note*: The Martian "day" is 24.6229 hr, and the mass of Mars is 6.4×10^{23} kg.)

113. A hockey puck of mass m is attached to a string that passes through a hole in the center of a table, as shown in **Figure 9.34**. The hockey puck moves in a circle of radius r. Tied to the other end of the string, and hanging vertically beneath the table, is a mass M. Assuming that the tabletop is perfectly smooth, what speed must the hockey puck have if the mass M is to remain at rest?

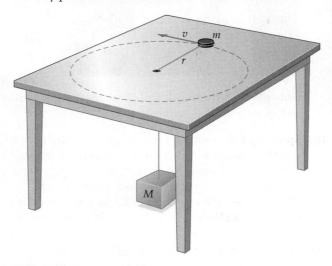

Figure 9.34

Writing about Science

114. Write a report describing the operation of the global positioning system (GPS). Your report should address the following questions: How many satellites are used? What are their orbital altitudes? What are their periods? How fast do the GPS satellites move? How accurate are they?

115. **Connect to the Big Idea** If gravity has an infinite range, how is it that astronauts feel weightless in space? Is the weightlessness of an astronaut any different from the "weightlessness" you feel as you go over the top of a hill on a high-speed roller coaster? Explain.

Read, Reason, and Respond

Exploring Comets On February 7, 1999, NASA launched a spacecraft with the ambitious mission of making a close encounter with a comet, collecting samples from its tail, and returning the samples to Earth for analysis. This spacecraft, appropriately named *Stardust*, took almost 5 years to rendezvous with its objective—Comet Wild 2 (pronounced "Vilt Two"), shown in **Figure 9.35**—and another 2 years to return with its samples. The reason for the long round trip is that the spacecraft had to make three orbits around the Sun and also an Earth Gravity Assist (EGA) flyby, to increase its speed enough to put it in an orbit appropriate for the encounter.

When *Stardust* finally reached Comet Wild 2 on January 2, 2004, it flew within 237 km of the comet's nucleus, snapping pictures and collecting tiny specks of dust in the glistening coma. The approach speed between the spacecraft and the comet at the encounter was a relatively "slow" 6200 m/s, so that dust particles could be collected safely without destroying the vehicle. Notice that "slow" is in quotation marks; after all, 6200 m/s is still about 6 times the speed of a rifle bullet!

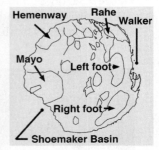

Figure 9.35 Comet Wild 2 and some of its surface features, including the Walker basin.

The roughly spherical Comet Wild 2 has a radius of 2.7 km, and the acceleration due to gravity at its surface is 0.00010 g. The two curves in **Figure 9.36** show the surface acceleration as a function of radius for a spherical comet with two different masses, one of which corresponds to Comet Wild 2. Also indicated are radii at which these two hypothetical comets have densities equal to that of ice and granite.

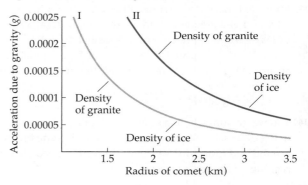

Figure 9.36

116. Which curve in Figure 9.36 corresponds to Comet Wild 2?
    A. Curve I   B. Curve II

117. What is the mass of Comet Wild 2?
    A. 1.1×10^8 kg   C. 1.1×10^{14} kg
    B. 1.1×10^{12} kg   D. 1.1×10^{18} kg

118. Suppose Comet Wild 2 had a small satellite, like the asteroid Ida in Figure 9.32. If this satellite were to orbit at twice the radius of the comet, what would be its period of revolution?
    A. 0.93 h   C. 5.8 h
    B. 2.9 h   D. 8.2 h

Standardized Test Prep

Multiple Choice

1. A satellite is in orbit at a distance r above Earth's surface. If Earth's radius is R and the period of the orbit is T, what is the satellite's orbital speed?

   (A) $\dfrac{2\pi R}{T}$

   (B) $\dfrac{2\pi r}{T}$

   (C) $\dfrac{2\pi(R + r)}{T}$

   (D) $\dfrac{T}{2\pi r}$

2. When short-track speed skaters round a circular curve on the track, they lean inward toward the center of the curve. Which is the *best* explanation for their use of this technique?

   (A) It places the skater's weight more directly over the skates.

   (B) It creates a force between the ice and skates that acts toward the center of the track.

   (C) It places more weight on the skate blades and melts the ice under the skates.

   (D) It reduces the normal force of the ice on the skates, increasing the skater's speed.

3. The free-body diagram below represents a car driving around a level circular curve. Which set of identities correctly specifies the forces acting in the diagram?

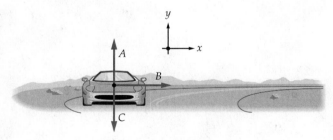

   (A) $A = N,\ B = mg,\ C = f_{cp}$

   (B) $A = mg,\ B = F_{cp},\ C = N$

   (C) $A = f_{cp},\ B = mg,\ C = N$

   (D) $A = N,\ B = f_{cp},\ C = mg$

4. The circular orbit of a space station is changed in such a way that it experiences a lower value of g. Which of the following actions led to this result?

   (A) orbit closer to Earth, increase orbital speed

   (B) orbit farther from Earth, decrease orbital speed

   (C) orbit closer to Earth, decrease orbital speed

   (D) orbit farther from Earth, increase orbital speed

5. Which is *best* explained by Kepler's second law of planetary motion?

   (A) the increase in the orbital speed of Earth as it approaches a position closer to the Sun

   (B) the fact that all objects orbiting the Sun have orbital periods proportional to their orbital radii

   (C) the observation that all planets have elliptical orbits

   (D) the decrease in the Moon's orbital speed as it approaches a position closer to Earth

6. The gravitational force exerted on you by Earth is

   (A) inversely proportional to the ratio of your mass to the mass of Earth.

   (B) much larger than the gravitational force you exert on Earth.

   (C) exactly equal in magnitude to the gravitational force you exert on Earth.

   (D) independent of your position relative to Earth.

Extended Response

7. A distant planetary system contains two planets (A and B) that orbit a star. The orbital radius of planet B is four times that of planet A; that is, $R_B = 4R_A$. **(a)** What is the orbital speed of planet B relative to that of planet A? Show your solution. **(b)** Explain how the mass of each planet affects its orbital speed.

> **Test-Taking Hint**
>
> A centripetal force is always directed toward the center of the circular path it causes.

Question	1	2	3	4	5	6	7						
If You Had Difficulty With . . .													
See Lesson(s)	9.2, 9.4	9.2, 9.3	9.3	9.4	9.4	9.1	9.1, 9.4						

10 Temperature and Heat

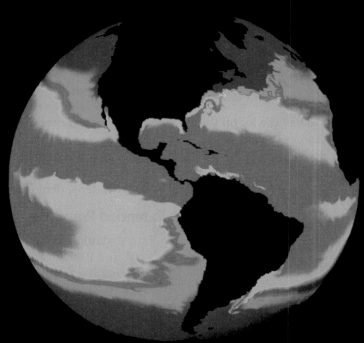

This map shows the global variation in ocean temperatures, with red indicating the warmest areas and blue the coolest areas. Differences in temperature result in the flow of thermal energy, which in turn affects worldwide climate and weather patterns.

Big Idea

Heat is a form of energy that is transferred because of temperature differences.

We are all familiar with the everyday meanings of *temperature* and *heat*. Even so, the physics behind these concepts may hold a few surprises. In this chapter you'll learn about the physics of temperature and heat and explore some of the interesting behavior associated with these quantities.

Inquiry Lab   How can you make a cloud?

Read the activity and obtain the required materials.

Explore

1. Add a small amount of very warm tap water to a 2-L plastic soda bottle and replace the cap.
2. Shake the bottle vigorously until small water droplets stick to the inside of the bottle. Remove the cap and pour the excess water into a sink. Recap the bottle.
3. Squeeze and release the sides of the bottle and observe what happens.

4. Remove the cap from the bottle. Light a match and then quickly blow it out. Immediately insert the head of the match into the neck of the bottle and allow the smoke to collect inside. Replace the cap.
5. Once again, squeeze and release the walls of the bottle and observe what happens.

Think

1. **Observe** Describe your observations for Steps 3 and 5.

2. **Compare** How does the atmosphere inside the bottle differ in Steps 3 and 5? Attempt to explain how this change in atmosphere affects the observed result in Step 5.
3. **Predict** How would the outcomes of Steps 3 and 5 change if you used cold water instead of warm water? If time permits, try this and find out.

10.1   Temperature, Energy, and Heat

We begin this lesson with a few key terms related to temperature, energy, and heat. We take care to define these terms, because they are often used differently in everyday conversation. Fortunately, a brief discussion is all that is needed.

The Relationships between Temperature, Energy, and Heat

As a small child, you learned to avoid objects that are "hot." You also discovered that if you forget to wear your coat outside, you can become "cold." These basic notions about hot and cold carry over into physics.

Temperature depends on average kinetic energy

If you touch a hot object, you feel an immediate sensation of pain. What causes the pain? It turns out that the pain is related to the speed of the particles (typically molecules) in the hot object. Experiments show that the particles in a hot object move *faster* than those in a cold object. This observation leads to the physics definition of temperature:

- **Temperature** is proportional to the average kinetic energy of particles in a substance.

Vocabulary

- temperature
- thermal energy
- thermal equilibrium
- heat
- absolute zero

CONNECTING IDEAS

The concept of temperature is introduced in this lesson.

- Temperature is a key concept in Chapter 11 on thermodynamics and in the discussion of states of matter in Chapter 12.

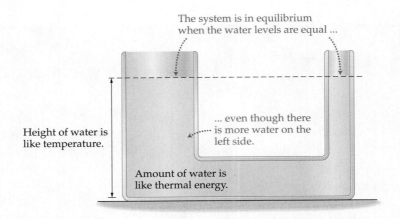

Figure 10.1 A U-tube analogy
The height of water on the two sides of the U-tube is like temperature—the system is in equilibrium when the temperatures (water levels) are the same. The total amount of water is like the thermal energy of a system.

The system is in equilibrium when the water levels are equal ...

Height of water is like temperature.

... even though there is more water on the left side.

Amount of water is like thermal energy.

The kinetic energy of a particle changes with time as it is jostled and bumped randomly by its neighbors—only the average value of its kinetic energy is related to the temperature.

The total amount of energy in a substance—the sum of all of its kinetic and potential energy—is referred to as its *internal energy*, or **thermal energy**. Thus, an object's thermal energy refers to both the random motion of its particles (kinetic energy) and the separation and orientation of its particles relative to one another (potential energy). Adding thermal energy to a system is known as *heating* and removing thermal energy is known as *cooling*.

A temperature difference causes energy to flow

The water-filled U-tube in **Figure 10.1** provides a useful analogy for temperature and thermal energy. Think of the height of the water as the temperature and the amount of water as the thermal energy. The U-tube is in equilibrium when the height of water is the same on the two sides.

Suppose you start with the water at a greater height on the right side, as in **Figure 10.2**. As you know, water will flow to the left side until the water levels are the same—even though there is more water on the left side to begin with. Similar behavior occurs when a hot object is placed in contact with a cold object. Thermal energy flows from the hot object to the cool object until the temperatures are the same—even if the cool object starts out with more energy. **Temperature doesn't depend on the total amount of energy in an object; it depends on the average kinetic energy of the particles in the object.**

How is temperature related to total energy and average kinetic energy?

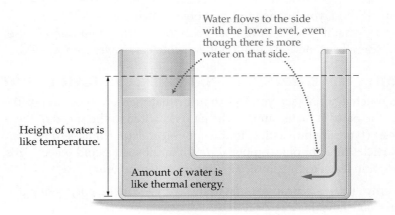

Figure 10.2 Restoring equilibrium in a U-tube
When the water level (temperature) is greater on one side of the U-tube, water flows to the other side of the tube—even if that side has more water. Similarly, thermal equilibrium requires that the temperature be the same in two objects, regardless of the amount of energy in each object.

Water flows to the side with the lower level, even though there is more water on that side.

Height of water is like temperature.

Amount of water is like thermal energy.

Energy Flow Imagine putting a hot brick in contact with a cold brick, as in **Figure 10.3**. When a rapidly vibrating molecule in the hot brick collides with a slowly vibrating molecule in the cold brick, the slow molecule picks up speed and the fast molecule slows down. As a result, energy is transferred from the hot brick to the cold brick. Occasionally, a molecule in the cold brick gives some energy to a molecule in the hot brick, but overall it is much more likely that molecules in the cold brick will gain energy.

Thermal Equilibrium Energy transfer continues between the bricks until they reach the same temperature. When this happens, the amount of energy transferred between the bricks is the same in either direction, and we say that the system is in *thermal equilibrium*. To be specific, a system is in **thermal equilibrium** when its temperature is constant and there is no net transfer of energy. So, although the bricks exchange energy back and forth as long as they are in contact, the total energy in each brick remains the same when they are in equilibrium.

Now imagine that you place a small block of metal in a large bucket of water. Will thermal energy flow between the block and the water? To answer this question, you only need to measure the temperature of each. If the temperatures are the same, then there is no net flow of thermal energy. If the temperatures are different, thermal energy flows from the warmer object to the cooler one. Nothing else matters—not the type of metal, its mass, its shape, the amount of water, whether the water is fresh or salt, and so on. All that matters is the temperature of each. This conclusion is referred to as *the zeroth law of thermodynamics*:

> **The Zeroth Law of Thermodynamics**
>
> Objects in contact with one another are in thermal equilibrium if they have the same temperature. Nothing else matters.

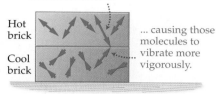

▲ **Figure 10.3 Establishing thermal equilibrium**
The two bricks are not in thermal equilibrium initially, since they start off with different temperatures. The high-speed molecules in the hot brick collide with the slow-moving molecules in the cold brick and transfer energy to them. This energy transfer increases the temperature of the cold brick and lowers the temperature of the hot brick. When the temperatures of the two bricks are equal, the system is in thermal equilibrium.

Heat is the flow of thermal energy

When you put a cool pan of water on a hot stove burner, we say that you are "heating" the water. The fast particles in the burner collide with the slow particles in the pan of water and cause them to speed up. This kind of transfer of energy is referred to as *heat*. In general, **heat** is the energy that is transferred between objects because of a temperature difference. Because heat is energy, it is measured in joules.

When we say that there is a "transfer of thermal energy" or a "flow of thermal energy" from a hot brick to a cold brick, we simply mean that the total energy of the hot brick decreases and the total energy of the cold brick increases. Thus, an object does not "contain" heat—what it does contain is thermal energy. **An object has a certain amount of thermal energy, and the thermal energy it *exchanges* with other objects because of a temperature difference is called *heat*.**

How is heat related to thermal energy?

Practice Problems

1. A cup of hot coffee is placed on a table. Is it in thermal equilibrium? What condition determines when the coffee is in equilibrium?

2. The particles in a hot object move more rapidly on average than those in a cold object. Is it valid to say that a hot object contains more heat than a cold object? Is it valid to say that a hot object contains more energy than a cold object?

Measuring Temperature

A variety of temperature scales are used in everyday situations and in physics. Let's look at the most common temperature scales and see how to convert from one to another.

The Celsius scale is used in most of the world

Perhaps the easiest temperature scale to remember is the Celsius scale, named in honor of the Swedish astronomer Anders Celsius (1701–1744). By definition, water freezes at zero degrees Celsius, which we abbreviate as 0 °C. In addition, water boils at one hundred degrees Celsius, or 100 °C.

The choice of the zero level for a temperature scale is completely arbitrary, as is the number of degrees between any two reference points. In the Celsius scale, as in others, there is no upper limit to the value a temperature may have. There is a lower limit, however. In the Celsius scale the lowest possible temperature is −273.15 °C, as we shall see later in this lesson.

The Fahrenheit scale is used in the United States

The Fahrenheit scale was developed by Gabriel Fahrenheit (1686–1736), who chose zero to be the lowest temperature he was able to achieve in his laboratory. He also chose 96 degrees to be body temperature, though why he made this choice is not known. In the modern version of the Fahrenheit scale, body temperature is 98.6 °F. In addition, water freezes at 32 °F and boils at 212 °F.

The Fahrenheit scale not only has a different zero point than the Celsius scale, it also has a different size for its degree. For example, 180 Fahrenheit degrees make up the span from the freezing to the boiling point of water. Only 100 degrees are needed for this span on the Celsius scale, however. Therefore, the Fahrenheit degrees are almost one-half the size of the Celsius degrees. The precise ratio is

$$\frac{100}{180} = \frac{5}{9}$$

To convert to a Fahrenheit temperature, T_F, from a Celsius temperature, T_C, we can use the following relationship:

> **Conversion between Degrees Celsius and Degrees Fahrenheit**
>
> Fahrenheit temperature $= \frac{9}{5}($Celsius temperature$) + 32$
>
> $$T_F = \frac{9}{5}T_C + 32$$

Similarly, a conversion in the opposite direction is given by the following:

> **Conversion between Degrees Fahrenheit and Degrees Celsius**
>
> Celsius temperature $= \frac{5}{9}($Fahrenheit temperature $- 32)$
>
> $$T_C = \frac{5}{9}(T_F - 32)$$

The following Guided Example shows how the conversion equations are used.

(**a**) On a fine spring day you notice that the temperature is 68 °F. What is the corresponding temperature on the Celsius scale? (**b**) If the temperature on a brisk winter morning is −2.0 °C, what is the corresponding Fahrenheit temperature?

Picture the Problem

A thermometer with both Fahrenheit and Celsius scales allows for quick approximate conversions. For this problem, we will calculate the exact conversions. (Notice that the scales on the circular thermometer shown here extend well beyond the temperatures 68 °F and −2.0 °C that this problem deals with.)

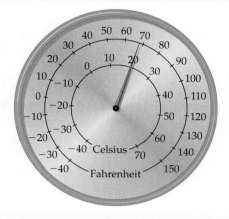

Strategy

The conversions asked for in this problem are straightforward applications of the relations between T_F and T_C. In particular, for (**a**) we use $T_C = \frac{5}{9}(T_F - 32)$, and for (**b**) we use $T_F = \frac{9}{5}T_C + 32$.

Known

(**a**) $T_F = 75\ °\text{F}$

(**b**) $T_C = -2.0\ °\text{C}$

Unknown

(**a**) $T_C = ?$

(**b**) $T_F = ?$

Solution

1 (**a**) Substitute $T_F = 68\ °\text{F}$ into $T_C = \frac{5}{9}(T_F - 32)$:

$$T_C = \frac{5}{9}(T_F - 32)$$
$$= \frac{5}{9}(68 - 32)$$
$$= \boxed{20\ °\text{C}}$$

Math HELP
Significant Figures
See Lesson 1.4

2 (**b**) Substitute $T_C = -2.0\ °\text{C}$ into $T_F = \frac{9}{5}T_C + 32$:

$$T_F = \frac{9}{5}T_C + 32$$
$$= \frac{9}{5}(-2.0) + 32$$
$$= \boxed{28\ °\text{F}}$$

Insight

Notice that the precise results from our calculations agree with the approximate results that can be obtained directly from the scales on the circular thermometer.

Practice Problems

3. **Follow-up** Find the Celsius temperature that corresponds to 110 °F.

4. The lowest temperature ever recorded on Earth was set at Vostok, Antarctica, on July 21, 1983. The temperature on that day fell to −89.2 °C, well below the temperature of dry ice. What is this temperature in degrees Fahrenheit?

5. One day you notice that the outside temperature increases by 27 °F between your early morning jog and your lunch at noon. What is the corresponding change in temperature on the Celsius scale?

6. **Challenge** What temperature is the same on both the Celsius and Fahrenheit scales? (*Hint:* Set the Fahrenheit and Celsius temperatures equal to the same value, T, in $T_F = \frac{9}{5}T_C + 32$.)

Absolute zero is the lowest temperature

Experiments show that there is a lowest temperature, just as there is a lowest possible kinetic energy. The lowest possible temperature, called **absolute zero**, is the temperature below which it is impossible to cool an object. Though absolute zero can be approached, it can never be attained.

A simple experiment allows us to estimate the location of absolute zero on the Celsius scale. Suppose you have a volume of air in a balloon at 100 °C. If you cool the balloon to −100 °C, the volume will shrink to one-half its original value. Cooling the balloon another 200 °C (to −300 °C) will reduce the volume to zero. Since a volume less than zero doesn't make sense, we conclude that absolute zero must be close to −300 °C. Careful measurements show, in fact, that absolute zero is −273.15 °C, in good agreement with our simple observations.

> **Absolute Zero**
>
> Absolute zero is −273.15 °C. It is impossible to have a temperature lower than this.

The Kelvin scale is based on absolute zero

The Kelvin temperature scale, named for the Scottish physicist William Thomson, Lord Kelvin (1824–1907), is based on the existence of absolute zero. In fact, the zero point of the Kelvin scale, abbreviated 0 K, is set exactly at absolute zero. Thus, in this scale there are no negative temperatures. A degree on the Kelvin scale (called a *kelvin*) has the same size as a Celsius degree.

As mentioned, absolute zero is −273.15 °C. Thus, the conversion between a Kelvin temperature, T, and a Celsius temperature, T_C, is as follows:

> **Conversion between a Celsius Temperature and a Kelvin Temperature**
>
> Kelvin temperature = Celsius temperature + 273.15
>
> $$T = T_C + 273.15$$

The difference between the Celsius and Kelvin scales is simply a difference in the zero level.

> *QUICK* **Example 10.2 What's the Kelvin Temperature?**
>
> Convert 55 °F to the Kelvin temperature scale.
>
> **Solution**
> First, use $T_C = \frac{5}{9}(T_F - 32)$ to convert from °F to °C:
>
> $$T_C = \frac{5}{9}(55 - 32)$$
> $$= 13\ °C$$
>
> Next, use $T = T_C + 273.15$ to convert °C to K:
>
> $$T = 13 + 273.15$$
> $$= \boxed{286\ K}$$

7. The temperature at the surface of the Sun is about 6000 K. Convert this temperature to (**a**) the Celsius scale and (**b**) the Fahrenheit scale.

8. What is the boiling point of water on the Kelvin scale?

The three temperature scales presented in this section are shown side by side in **Figure 10.4**, with temperatures of particular interest indicated. This permits a useful visual comparison between the scales. Though the Celsius and Fahrenheit scales are most common in everyday usage, the Kelvin scale is the one that is used in the SI system.

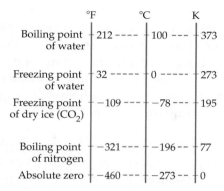

▲ **Figure 10.4 Temperature scales**
A comparison of the Fahrenheit, Celsius, and Kelvin temperature scales. Physically significant temperatures, such as the freezing and boiling points of water, are indicated for each scale.

10.1 LessonCheck (MP)

Checking Concepts

9. Explain Does the average kinetic energy of particles in a substance increase, decrease, or stay the same when the temperature of the substance is increased?

10. Critique A classmate tells you that his thermos of coffee contains a lot of heat. Is this statement correct? Explain.

11. Big Idea What do we call energy that is transferred because of a temperature difference?

12. Describe How is the thermal energy of a substance defined?

13. State What are the key characteristics of thermal equilibrium?

14. Identify What is the name used for zero on the Kelvin temperature scale?

Solving Problems

15. Convert The filament of a glowing incandescent lightbulb has a temperature of about 4500 °F, which is close to the surface temperature of the Sun. What is this temperature in degrees Celsius?

16. Convert As a cold front moves through your area, the outside temperature drops by 35 °F. What is the corresponding temperature change in (**a**) degrees Celsius and (**b**) kelvins?

17. Convert A high-pressure weather system moves into your area and increases the temperature by 29 °C. What is the corresponding temperature change on (**a**) the Fahrenheit scale and (**b**) the Kelvin scale?

10.2 Thermal Expansion and Energy Transfer

Vocabulary

- coefficient of thermal expansion
- conduction
- conductor
- insulator
- convection
- radiation

🔑 How is temperature change related to length change?

Things tend to get larger, or *expand*, when their temperature increases. Likewise, objects tend to get smaller, or *contract*, when their temperature decreases. This lesson explores the physics of thermal expansion and contraction, and the ways thermal energy is transferred.

Thermal Expansion

Most substances expand when heated. You may have noticed, for example, that power lines sag much lower on a hot summer day than they do on a cold winter day. Some high-speed airplanes get so hot during flight that they are 15 cm (about half a foot) longer when they land than they were when they took off. Determining how much an object expands when its temperature is increased turns out to be rather easy.

An increase in temperature usually causes expansion

Have you ever used a liquid-filled thermometer to take someone's temperature? If so, you've noticed that the red liquid in the thermometer rises as the temperature rises. The liquid expands as it gets warmer. Most substances do exactly the same thing.

To be more specific, consider a rod whose initial length is L_i at a given temperature. Experiments show that when the rod is heated or cooled, its length changes in proportion to the change in temperature. Thus, if the temperature changes by the amount ΔT, the change in length of the rod is

$$\Delta L = (\text{constant})\Delta T$$

The constant of proportionality in this equation depends, among other things, on the substance from which the rod is made. Iron has one constant, aluminum another constant, and so on.

CONCEPTUAL Example 10.3 Comparing Expansions

When rod 1 is heated by an amount ΔT, its length increases by ΔL. Rod 2 is twice as long as rod 1 and made of the same material. If rod 2 is heated by the same amount as rod 1, does its length increase by ΔL, $2\Delta L$, or $\Delta L/2$?

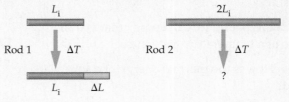

Reasoning and Discussion

We can imagine rod 2 to be composed of two copies of rod 1 placed end to end, as shown in the following figure.

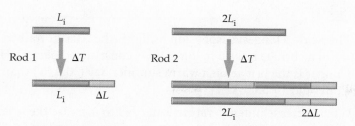

When the temperature is increased by ΔT, each copy of rod 1 expands by ΔL. Hence, the total expansion of the two copies is $2\Delta L$, as is the total expansion of rod 2.

Answer

The rod that is twice as long expands twice as much, or $2\Delta L$.

Conceptual Example 10.3 shows that the change in length with heating depends on the initial length of an object. 🔵 The *thermal expansion* of **an object (its change in length) is proportional to *both* the initial length, L_i, and the temperature change,** ΔT. The constant of proportionality that relates temperature change and length change is known as the **coefficient of thermal expansion,** α. The symbol α is the Greek letter alpha.

> **Definition of Coefficient of Thermal Expansion, α**
>
> $$\begin{matrix}\text{change} \\ \text{in length}\end{matrix} = \begin{matrix}\text{coefficient of} \\ \text{thermal expansion}\end{matrix} \times \begin{matrix}\text{initial} \\ \text{length}\end{matrix} \times \begin{matrix}\text{temperature} \\ \text{change}\end{matrix}$$
>
> $$\Delta L = \alpha L_i \Delta T$$
>
> SI unit for α: $\dfrac{1}{K}$ (or K^{-1})

Table 10.1 gives some values of the coefficient of thermal expansion. Notice that the change in temperature, ΔT, can be given in kelvins or degrees Celsius, since the size of a degree is the same in both of these temperature scales.

Table 10.1 Coefficients of Thermal Expansion

Substance	Coefficient of Thermal Expansion, α (K^{-1})
Lead	29×10^{-6}
Aluminum	24×10^{-6}
Brass	19×10^{-6}
Copper	17×10^{-6}
Iron (steel)	12×10^{-6}
Concrete	12×10^{-6}
Window glass	11×10^{-6}
Pyrex glass	3.3×10^{-6}
Quartz	0.50×10^{-6}

QUICK Example 10.4 What's the Increase in Height?

The Eiffel Tower, shown in **Figure 10.5**, is made of iron. If the tower is 301 m high on a day when the temperature is 22 °C, how much does its height decrease when the temperature cools to 0.0 °C?

Solution

We can calculate the change in height with $\Delta L = \alpha L_i \Delta T$. The coefficient of thermal expansion for iron is $12 \times 10^{-6} \ K^{-1}$, and the change in temperature is $\Delta T = -22 \ °C = -22 \ K$. Therefore, the change in length is

$$\Delta L = \alpha L_i \Delta T$$
$$= (12 \times 10^{-6} \ K^{-1})(301 \ m)(-22 \ K)$$
$$= -0.079 \ m = \boxed{-7.9 \ cm}$$

Notice that we converted the change in temperature from degrees Celsius to kelvins so that the units of temperature would cancel out.

▲ **Figure 10.5 The Eiffel Tower**
The Eiffel Tower in Paris was constructed in 1887–1889 and designed by Alexandre Eiffel. It gains or loses about four-tenths of a centimeter in height for each Celsius degree that the temperature rises or falls.

18. The world's longest suspension bridge is the Akashi Kaikyo Bridge in Japan. The bridge is 3910 m long and is constructed of steel. How much longer is the bridge on a warm summer day (30.0 °C) than on a cold winter day (−5.00 °C)?

19. What increase in temperature is needed to increase the length of an aluminum meterstick by 1.0 mm?

Thermal expansion has many practical uses

An interesting application of thermal expansion is the behavior of a bimetallic strip. As the name suggests, a bimetallic strip consists of two metals bonded together to form a single strip of metal. Such a strip is illustrated in **Figure 10.6**. Since two different metals usually have different coefficients of thermal expansion, the two sides of the strip will change lengths by different amounts when heated or cooled. This causes the strip to bend.

For example, suppose metal B in **Figure 10.6 (a)** has the larger coefficient of thermal expansion. This means that its length changes by a greater amount than metal A's length for the same temperature change. Hence, if this bimetallic strip is cooled, the B side shrinks more than the A side, resulting in the strip bending toward the B side, as in **Figure 10.6 (b)**. On the other hand, if the strip is heated, the B side expands by a greater amount than the A side, and the strip curves toward the A side, as in **Figure 10.6 (c)**. Thus, the deflection of the bimetallic strip depends sensitively on temperature.

Physics & You: Technology Because of their behavior, bimetallic strips are used in a variety of thermal applications. For example, a bimetallic strip can be used as a thermometer—as the strip bends, it can move a needle to indicate the temperature. Similarly, many household thermostats use a bimetallic strip to turn on or shut off a heater. This is shown in **Figure 10.6 (d)**.

▶ **Figure 10.6 A bimetallic strip**
(a) A bimetallic strip composed of metals A and B. If metal B has a larger coefficient of thermal expansion than metal A, it will **(b)** shrink more when cooled and **(c)** expand more when heated.
(d) A bimetallic strip can be used to construct a thermostat. If the temperature falls, the strip bends downward and closes the electrical circuit, which then operates a heater. When the temperature rises, the strip deflects in the opposite direction, breaking the circuit and turning off the heater.

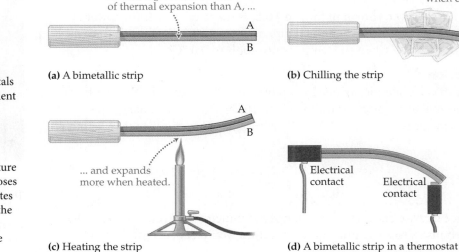

If B has a greater coefficient of thermal expansion than A, ...

(a) A bimetallic strip

... then it shrinks more when cooled ...

(b) Chilling the strip

... and expands more when heated.

(c) Heating the strip

Electrical contact

Electrical contact

(d) A bimetallic strip in a thermostat

As the temperature of the room changes, the bimetallic strip deflects in one direction or the other, which either closes or breaks the electrical circuit connected to the heater.

Thermal expansion can also cause problems. Have you ever noticed that many bridges have interlocking gaps between different sections of the roadway, as shown in **Figure 10.7 (a)**. When the air temperature rises in the summer, the sections of the bridge can expand freely into these gaps. If the gaps were not present, the expansion of the sections could cause the bridge to buckle and warp. Thus, these gaps, called *expansion joints*, are a way to avoid this type of heat-related damage. Expansion joints are also used in railroad tracks and oil pipelines, to name just two other examples. **Figure 10.7 (b)** shows how expansion is handled in a pipeline.

(a)

CONCEPTUAL Example 10.5 Expand or Contract?

A washer has a hole in the middle. As the washer is heated, does the hole expand, shrink, or stay the same?

Reasoning and Discussion

We can make a washer out of a disk of metal by removing a smaller inner disk, as shown below. If we then heat these pieces, both the washer and the inner disk expand. Notice that if we had left the inner disk in place and heated the original disk, it would still have expanded. Thus, removing the *heated* inner disk creates an expanded washer with an expanded hole in the middle. We obtain the same result whether we remove the inner disk and then heat, or heat first and then remove the inner disk.

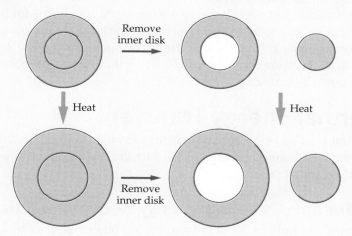

Answer

The hole expands along with the washer.

(b)

▲ **Figure 10.7 Compensating for thermal expansion**
Thermal expansion, though small, is far from negligible for many everyday objects, especially long structures such as railroad tracks, bridges, and pipelines. **(a)** Bridges and elevated highways must include expansion joints to prevent the roadway from buckling when it expands in hot weather. **(b)** Pipelines typically include loops that allow for expansion and contraction when the temperature changes.

Water has special properties

You might wonder what bizarre substance could possibly be an exception to the general rule of expanding with heating and shrinking with cooling. The most important exception is a substance you drink every day—water. This exceptional behavior is just one of water's many special properties that set it apart from most other substances.

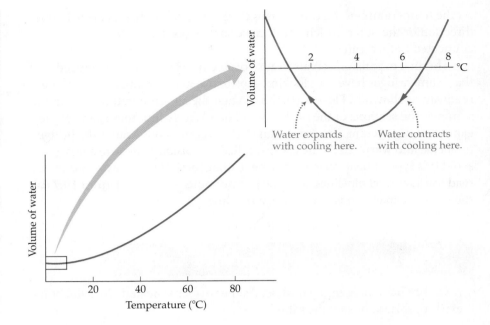

▶ **Figure 10.8 The unusual behavior of water near 4 °C**
Water expands with cooling below 4 °C. This is what causes water pipes to burst in freezing weather.

Figure 10.8 shows the volume of a certain amount of water over a wide range of temperatures. As the water is cooled, it shrinks in volume, as one would expect. At least at first. Notice, however, that the volume reaches a minimum at about 4 °C. If we cool the water below 4 °C, it actually expands with cooling! This is because the water molecules are starting to form microscopic clumps of crystalline ice. Since the crystal structure of ice is very open, the molecules must move farther apart to form the clumps. As a result, the volume of the water increases.

The expansion of water between 4 °C and 0 °C is to blame for bursting water pipes in the winter. Even a steel water pipe is not strong enough to keep from rupturing when the ice forming within it expands outward. Later, when the temperature rises above freezing again, a burst pipe makes its condition known by springing a leak.

Thermal Energy Transfer

Energy must be added to or removed from an object in order to change its temperature and cause it to expand or contract. There are several ways in which this can occur.

Conduction transfers energy between particles

How is energy exchanged by conduction?

If you hold one end of a metal rod and put the other end in a fire, it doesn't take long before your end of the rod starts to feel warm. The warmth you feel on your end of the rod is *conducted* along the length of the rod. **Conduction** is a form of thermal energy transfer that occurs through collisions between particles of matter.

Let's look at conduction from a microscopic point of view. When you place one end of the rod in the fire, the high temperature at that location causes the particles there to vibrate with increased kinetic energy. **Conduction occurs as higher energy particles collide with and jostle neighboring lower energy particles, transferring kinetic energy from one particle to the next.** Eventually, the conduction process transfers energy from particle to particle along the entire length of the rod.

If you repeat the experiment, this time with a wooden rod, the hot end of the rod will heat up and catch on fire, but your end will still be comfortably cool. Thus, conduction depends on the type of material involved. Materials that are good at conducting thermal energy are called **conductors**; those that conduct thermal energy poorly are called **insulators**.

Nature provides many examples of animals that have to deal with the conduction of thermal energy. For example, the lizard in **Figure 10.9 (a)** is trying to minimize the flow of thermal energy from the hot sand to its body by standing on only two legs at a time. The polar bear in **Figure 10.9 (b)** is comfortable on the frozen iceberg because its fur is a good insulator, preventing the thermal energy in its body from being conducted into the ice.

CONCEPTUAL Example 10.6 Warmer or Cooler?

You get up in the morning and walk barefoot from the bedroom to the bathroom. In the bedroom you walk on carpet, but in the bathroom the floor is tile. Does the tile feel warmer, cooler, or the same temperature as the carpet?

Reasoning and Discussion
Everything in the house is at the same temperature, so it might seem that the carpet and the tile would feel the same. As you probably know from experience, however, the tile feels cooler. The reason is that tile has a much larger thermal conductivity than carpet. In fact, carpet is a good insulator. As a result, more thermal energy flows from your skin to the tile than from your skin to the carpet. To your feet, then, it feels as if the tile is much cooler than the carpet.

Answer
The tile feels cooler than the carpet.

(a)

(b)

▲ **Figure 10.9 Natural ways to reduce the conduction of thermal energy**
Maintaining proper body temperature in an environment that is often too hot or too cold is a problem for many animals. (**a**) When the sand is blazing hot, this lizard keeps its contact with the ground to a minimum. By standing on two legs instead of four, it reduces conduction of thermal energy from the ground to its body. (**b**) Polar bears have the opposite problem. The loss of body warmth to their surroundings is retarded by their thick fur, which is actually made up of hollow fibers. Air trapped within these fibers provides enhanced insulation, just as it does in our thermal blankets and double-paned windows.

How does convection differ from conduction?

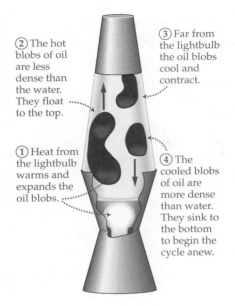

② The hot blobs of oil are less dense than the water. They float to the top.

③ Far from the lightbulb the oil blobs cool and contract.

① Heat from the lightbulb warms and expands the oil blobs.

④ The cooled blobs of oil are more dense than water. They sink to the bottom to begin the cycle anew.

▲ **Figure 10.10 The physics of lava lamps**
Lava lamps illustrate convection in motion.

Why does sunlight feel warm on your skin?

Convection transfers energy by physical movement

Suppose you want to heat a small room. To do so, you bring a portable electric heater into the room and turn it on. The heating coils get red hot and warm the air in their vicinity. As this air warms, it expands and becomes less dense. Because of its lower density, the warm air rises, to be replaced by cold dense air descending from overhead. This sets up a circulating flow of air that transfers thermal energy from the heating coils to the air throughout the room. The transfer of thermal energy due to the physical movement of material is referred to as **convection.** **Convection differs from conduction in that it transfers thermal energy through the physical movement of particles from one place to another, rather than between particles that vibrate in place.**

A nice way to visualize convection is with a lava lamp, which consists of a water-filled container in which there is also some dense, taffy-like oil. Turning on the light in a lava lamp heats the oil at the bottom of the container, as shown in **Figure 10.10**. As the oil warms, it expands and becomes less dense than the water in the lamp. It then floats to the top in blobs. Once an oil blob reaches the top of the lamp, it begins to cool. This makes it more dense than water again, and it sinks to the bottom to begin the cycle anew. The mesmerizing movement of the oil blobs is a perfect illustration of convection in action.

Convection is also responsible for keeping a candle burning by bringing in a continuous supply of fresh air. When convection is cut off by allowing a candle to drop in free fall, as shown in **Figure 10.11**, the supply of fresh air is also cut off, and the candle goes out in a fraction of a second.

Radiation transfers energy as electromagnetic waves

When you go outside on a nice sunny day, you immediately feel the warmth of the Sun. **The warmth you feel from the Sun is due to light and infrared waves that it emits. These waves (referred to as *electromagnetic waves*) carry energy from the Sun across the vacuum of space to Earth.** The transfer of thermal energy in the form of electromagnetic waves (like light and infrared rays) is referred to as **radiation.**

Since radiation can include visible light, it is often possible to "see" the temperature of an object. This is the physical basis of the *optical pyrometer*, invented by Josiah Wedgwood (1730–1795), the renowned English potter. When objects reach a temperature of about 800 °C, they appear to be a dull red, which we call "red hot." Examples include the heating coils of a range or oven and the molten lava in **Figure 10.12 (a)**. The filament in an incandescent lightbulb glows "yellow hot" at about 3000 °C, about the same temperature as the surface of the red supergiant Betelgeuse, a star in the constellation Orion. The surface of the Sun, in comparison, is about 6000 °C. Very hot stars, with surface temperatures in the range 10,000–30,000 °C, are "blue hot" and actually appear bluish in the night sky. Rigel, also in the constellation Orion, is an example of such a star. Look above and to the left of Orion's belt in **Figure 10.12 (b)** to see the red star Betelgeuse, and below and to the right to see the blue star Rigel.

◀ **Figure 10.11 A candle in free fall**
A burning candle heats the air near the flame. Because hot air is less dense than cool air, a circulation pattern is established (like that in the lava lamp), with hot air rising and being replaced from below with cool, oxygenated air. When a burning candle in a jar is dropped, it suddenly finds itself in a "weightless" environment where hot air no longer rises. As a result, convection ceases, and the flame quickly goes out as it consumes all the oxygen in its vicinity.

(a)

(b)

◀ **Figure 10.12 "Seeing" the temperature**
(a) Red-hot volcanic lava is just hot enough (about 1000 °C) to radiate in the visible range. Even when it cools enough to stop glowing, it still emits energy, but most of it is in the form of invisible infrared radiation. (b) Stars have colors that indicate their surface temperatures. For example, the red supergiant Betelgeuse (upper left) in the constellation Orion is much cooler than the blue supergiant Rigel (lower right).

10.2 LessonCheck (MP)

Checking Concepts

20. 🔒 **Apply** Two identical metal rods are at room temperature. One rod is heated by 10 °C, and the other is cooled by 10 °C. How do their new lengths compare?

21. 🔒 **Explain** Describe the process by which conduction occurs.

22. 🔒 **Identify** Which exchange process carries thermal energy through boiling water as it heats on top of a stove?

23. 🔒 **Infer** Does thermal energy exchange by radiation require a medium (particles of matter) through which the energy is transferred?

24. Identify If you want to exchange a lot of thermal energy by conduction, should you use a conductor or an insulator?

25. Apply A bimetallic strip is made from copper and aluminum. When the strip is heated, does it bend toward the copper side of the strip or the aluminum side?

26. Rank Each of the four systems described below consists of a metal rod (either aluminum or steel) with a given initial length. The rods are subjected to the indicated increases in temperature. Rank the systems in order of increasing change in length. Indicate ties where appropriate. (Refer to Table 10.1 for the coefficients of thermal expansion.)

	System A	System B	System C	System D
Metal	aluminum	steel	steel	aluminum
Initial length	2 m	2 m	1 m	1 m
Temperature increase	40 °C	20 °C	30 °C	10 °C

Solving Problems

For Problems 27–30, refer to Table 10.1 for the coefficients of thermal expansion.

27. Calculate A hole in a copper plate has a diameter of 1.325 cm at 21.00 °C. What is the diameter of the hole at 224.0 °C?

28. Calculate A hole in a steel plate has a diameter of 1.166 cm at 23.00 °C. At what temperature is the diameter of the hole equal to 1.164 cm?

29. Calculate An aluminum rod changes its length by 0.0032 cm when the temperature changes by 120 °C. What was the initial length of the aluminum rod?

30. Calculate A metal rod has an initial length of 2.5 m. When its temperature is increased by 85 °C, its length increases by 0.36 cm. What is the metal's coefficient of thermal expansion? Which metal is it most likely to be?

10.3 Heat Capacity

Vocabulary

- kilocalorie
- specific heat capacity
- calorimeter

🔑 *How are heat and work related?*

CONNECTING IDEAS

In Chapter 6 you learned about the concept of work—a force acting over a distance.
- Work plays a key role in this lesson, when we determine the mechanical equivalent of heat.

You've learned a lot so far about heat and how it affects the temperature of an object. In this lesson you'll find out how heat is related to the work done by your muscles or a machine.

Thermal energy is related to work

As we have seen, when objects of different temperature are placed in contact, thermal energy flows from the hotter one to the cooler one. At one time it was thought—erroneously—that an object contained a certain amount of "heat fluid," or "caloric," that could flow from one place to another. This idea was overturned by Benjamin Thompson (1753–1814), also known as Count Rumford. At one point in his career Count Rumford supervised the boring of cannon barrels by large drills. He observed that as long as mechanical work was done to turn the drill bits, the drill bits continued to produce thermal energy.

With this observation it became clear that heat (the transfer of thermal energy) was another form of energy. As such, heat must be taken into account when applying conservation of energy. For example, if you rub sandpaper back and forth over a piece of wood, you do work against friction. The energy associated with that work is not lost. Instead, it produces an increase in temperature in the wood and in the sandpaper. Taking into account the energy associated with this temperature change, we find that energy is indeed conserved. In fact, no observation has ever indicated a situation in which energy is not conserved.

Mechanical Equivalent of Heat The equivalence between heat and work was first explored by James Prescott Joule (1818–1889). In one of his experiments Joule observed the increase in temperature in a device similar to the one shown in **Figure 10.13**. Here, two masses fall through a certain distance. The work done by gravity on the falling masses turns the paddles in the water, resulting in a slight warming of the water. By measuring the mechanical work and the increase in the water's temperature, Joule was able to show that energy was indeed conserved.

Joule's experiments established the *mechanical equivalent of heat.* 🔑 **The mechanical equivalent of heat is the precise amount of mechanical work that has the same effect as the transfer of a given amount of thermal energy.** The relationship is as follows:

> **The Mechanical Equivalent of Heat**
>
> One calorie of heat is the equivalent of 4.186 J of mechanical work.
>
> $1 \text{ cal} = 4.186 \text{ J}$
> $1 \text{ kcal} = 4186 \text{ J}$

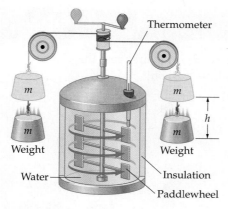

▲ Figure 10.13 **The mechanical equivalent of heat**
A device of this type was used by James Joule to measure the mechanical equivalent of heat.

Units of Heat

The customary unit for measuring heat is the calorie (cal). In fact, one **kilocalorie** (kcal) is defined as the amount of heat needed to raise the temperature of 1 kilogram of water from 14.5 °C to 15.5 °C. This amount of heat is equal to 4186 joules of energy.

In nutrition a different unit of energy is used. It is the Calorie (C), spelled with a capital C. By definition, one nutritional calorie (1 C) is the same as one kilocalorie. That is,

$$1 \text{ C} = 1 \text{ kcal}$$

Perhaps using nutritional calories helps people feel a little better about their calorie intake. After all, a 250-C candy bar sounds a lot less fattening than a 250,000-cal candy bar. They are equivalent, however. (Notice that one nutritional calorie, 1 C, is not the same as one degree Celsius, 1 °C. The degree, °, is missing in the symbol for the nutritional calorie.)

Heat Is Represented by Q

The symbol Q is used to denote heat—the thermal energy transferred because of a temperature difference.

Heat, Q

Q = heat
SI unit: J

Using the mechanical equivalent of heat as the conversion factor, we will sometimes express heat in calories and sometimes in joules, whichever is more convenient for a particular problem. The sign of heat is determined as follows:

- Heat is *positive* when *thermal energy is added* to a system.
- Heat is *negative* when *thermal energy is removed* from a system.

A good analogy for the relationship between thermal energy and heat is money in a bank account. The total amount of money in the account is like the thermal energy in an object. Adding money to the account is like positive heat, which adds thermal energy to the object. Similarly, withdrawing money from the account is like negative heat, which removes thermal energy from the object.

A 74.0-kg person drinks a thick, rich, 305-C milkshake. How many stairs must this person climb to work off the shake? Let the height of a stair be 20.0 cm.

Picture the Problem

Our sketch shows the person climbing the stairs after drinking the milkshake. Notice that each stair step has a height $h = 20.0$ cm. In addition, the total height to which the person climbs to work off the milkshake is designated by H. (The horizontal distance is irrelevant here, as it does not affect the work done against gravity.)

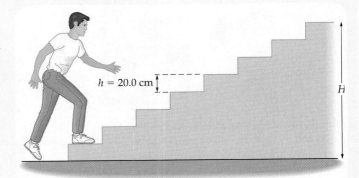

Strategy

We know that the energy gained from drinking the milkshake is equivalent to a heat $Q = 305,000$ cal. This energy can be converted to joules using the relationship 1 cal $= 4.186$ J. Finally, we set the energy of the shake equal to the work done against gravity, mgH, in climbing to a height H.

Known

$m = 74.0$ kg

$h = 20.0$ cm

$Q = 305$ C

Unknown

$H = ?$ (number of stairs)

Solution

1 Convert the energy of the milkshake to joules:

$Q = 305,000$ cal

$= 305,000 \text{ cal} \left(\dfrac{4.186 \text{ J}}{1 \text{ cal}} \right)$

$= 1.28 \times 10^6$ J

2 Equate the energy of the milkshake to the work done against gravity in climbing to a height H:

$Q = mgH$

3 Solve for the height, H, substitute the numerical values, and calculate the answer:

$H = \dfrac{Q}{mg}$

$= \dfrac{1.28 \times 10^6 \text{ J}}{(74.0 \text{ kg})(9.81 \text{ m/s}^2)}$

$= 1760$ m

> **Math HELP**
> Solving Simple Equations
> **See Math Review, Section III**

4 Multiply the height in meters by the number of stairs per meter to find the number of stairs:

$1760 \text{ m} \left(\dfrac{1 \text{ stair}}{0.200 \text{ m}} \right) = \boxed{8800 \text{ stairs}}$

Insight

This is clearly a lot of stairs, and a significant height to climb. In fact, 1760 m is more than a mile. Even assuming a metabolic efficiency of only 25%, a height of about a quarter of a mile must be climbed to work off the shake. Put another way, the shake would be enough "fuel" for the person to walk to the top of the Empire State Building—with a little left over for celebrating.

31. Follow-up How many Calories are burned when the person climbs 100 stairs? Assume a metabolic efficiency of 100%.

32. During a workout a person repeatedly lifts a 6.2-kg barbell through a distance of 0.58 m. How many reps of this lift are required to burn off 150 C?

Substances change temperature at different rates

Have you ever eaten a slice of apple pie right out of the oven? If you have, you know that the crust cools down much faster than the apples inside. The apples stay hot for a long time and can be painful to bite into, even when the crust is cool and easy to eat. The apples stay hot longer because they contain more water than the crust. As a result, more thermal energy must be removed from apples than from crust in order to decrease the temperature to a safe level.

In general, different substances require different amounts of thermal energy for the same change in temperature. The **specific heat capacity** of a substance is the thermal energy required to change the temperature of 1 kilogram of the substance by 1 °C. **A substance with a high specific heat capacity requires a lot of thermal energy to show a given change in temperature.** Water is such a substance.

A useful analogy for the specific heat capacity is provided by the vases shown in **Figure 10.14**. In this analogy the height of the water corresponds to temperature, and the amount of water corresponds to thermal energy. To raise the water level (temperature) of vase A by a certain amount requires more water (thermal energy) than is required for vase B. This is because vase A is much wider—it has a greater capacity for water than vase B. Similarly, a substance with a large specific heat capacity requires more heat for a given temperature change than does a substance with a low specific heat capacity.

How does specific heat capacity describe a substance's response to heat?

A lot of water is needed to fill this vase to a given level.

This vase is like a substance with a large specific heat capacity.

Less water is needed to fill this vase to a given level.

This vase is like a substance with a small specific heat capacity.

Water height is like temperature.

Amount of water is like the thermal energy.

Vase A   Vase B

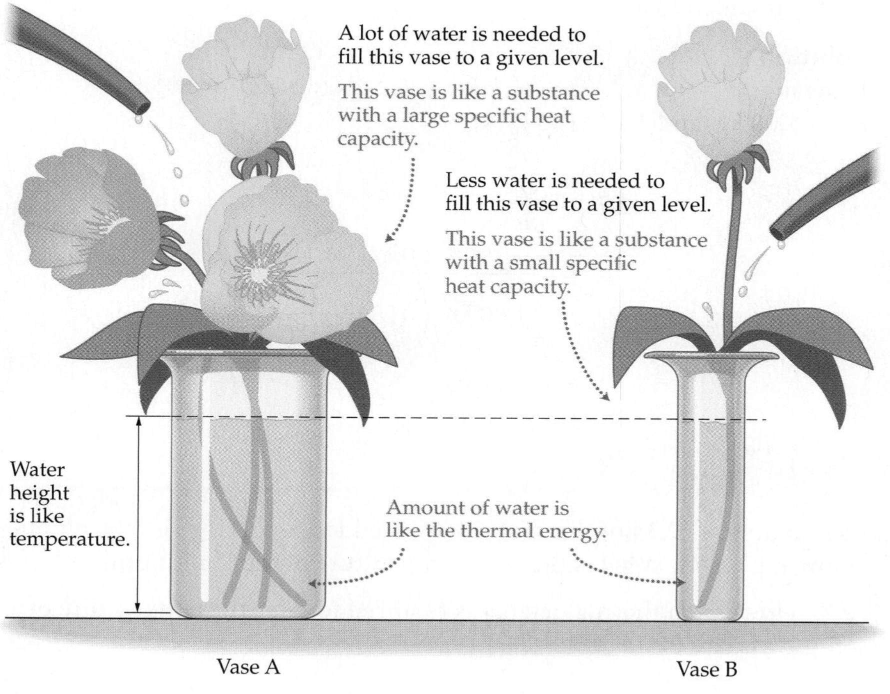

◀ **Figure 10.14 An analogy for specific heat capacity**
Specific heat capacity is a measure of how much thermal energy is needed to change an object's temperature by a given amount. This is similar to how much water is needed to raise the water level of a vase by a given amount. A vase with a large capacity for water is like a substance with a large capacity for thermal energy.

Specific heat relates heat, mass, and temperature

We designate the specific heat capacity with the symbol c. If a certain amount of heat, Q, is required to change the temperature of a substance of mass m by the amount ΔT, the specific heat capacity of the substance is defined as follows:

Table 10.2 Specific Heat Capacities

Substance	Specific Heat Capacity, c $\left(\dfrac{J}{kg \cdot {}^\circ C}\right)$
Water	4186
Ice	2090
Steam	2010
Beryllium	1820
Air	1004
Aluminum	900
Glass	837
Silicon	703
Iron (steel)	448
Copper	387
Silver	234
Gold	129
Lead	128

Definition of Specific Heat Capacity, c

$$\text{specific heat capacity} = \frac{\text{heat}}{\text{mass} \times \text{temperature change}}$$

$$c = \frac{Q}{m\Delta T}$$

SI units: $J/(kg \cdot K) = J/(kg \cdot {}^\circ C)$

For example, the specific heat capacity of water is $c_{water} = 4186\ J/(kg \cdot {}^\circ C)$. Therefore, the heat (thermal energy) that must be transferred to change the temperature of 1 kg of water by 1 °C is

$$Q = mc\Delta T$$
$$= (1\ kg)\left(4186\frac{J}{kg \cdot {}^\circ C}\right)(1\ {}^\circ C)$$
$$= 4186\ J$$

Specific heat capacities for some common substances are listed in **Table 10.2**. Notice that the specific heat capacity of water is by far the largest of any common material. This is just another of the many unusual properties of water. Having such a large specific heat capacity means that water can give off or take in large quantities of thermal energy with little change in temperature.

QUICK Example 10.8 What Is the Temperature Change?

What is the temperature change of 1.00 kg of water if 505 J of thermal energy is added to it?

Solution
Rearrange $c = Q/m\Delta T$ to solve for ΔT. Substitute $Q = 505$ J, $m = 1.00$ kg, and $c = 4186\ J/(kg \cdot {}^\circ C)$:

$$\Delta T = \frac{Q}{mc}$$
$$= \frac{505\ J}{(1.00\ kg)\left(4186\dfrac{J}{kg \cdot {}^\circ C}\right)}$$
$$= \boxed{0.121\ {}^\circ C}$$

Math HELP
Solving Simple Equations
See Math Review, Section III

Practice Problems

33. Suppose 79.3 J of thermal energy is added to a 111-g piece of aluminum at 22.5 °C. What is the final temperature of the aluminum?

34. How much thermal energy is required to raise the temperature of a 55-g glass ball by 15 °C?

The difference in specific heat capacity for two substances is illustrated in **Figure 10.15**. Here we see two blocks of metal, one aluminum and one lead. These blocks have equal volumes and equal initial temperatures of 100 °C. Notice that the aluminum block melted more paraffin wax than the lead block, indicating it gave off more thermal energy—even though the lead has four times the mass. This occurred because the specific heat capacity of aluminum is considerably greater than that of lead, as you can see in Table 10.2.

Climate Implications
Water's unusually large specific heat capacity accounts for the moderate climates experienced in regions near large bodies of water. In particular, the enormous volume and large heat capacity of an ocean serve to maintain a nearly constant temperature in its water, which in turn tends to even out the temperature of adjacent coastal areas. For example, the West Coast of the United States benefits from the moderating effect of the Pacific Ocean, aided by the prevailing breezes that come from the ocean onto the coastal regions. Changes in water temperature, such as the El Niño effect shown in **Figure 10.16**, can have strong influences on coastal climates. In the Midwest, on the other hand, temperature variations can be considerably greater as the land (with a relatively small specific heat capacity) quickly heats up in the summer and cools off in the winter.

Calorimetry is used to measure specific heat

Suppose you would like to measure the specific heat capacity of a substance. To do so, you might use a **calorimeter**, which is basically a lightweight, insulated flask. The mass of the calorimeter is assumed to be small enough to be ignored. It is also assumed that no thermal energy is transferred between the calorimeter and its surroundings. Calorimeters are standard laboratory devices that are easy to use.

Here is how you can use a calorimeter to find the specific heat capacity of a block of an unknown metal. First, place a known mass of water in the calorimeter and measure its temperature. Next, measure the mass of the metal block, and then heat it to a convenient temperature greater than that of the water. Finally, drop the warm block into the cool water and wait for the system to reach thermal equilibrium. Measure the final temperature of the water—which is the same as that of the block. These measured quantities can be used to calculate the specific heat capacity of the block, as we shall see.

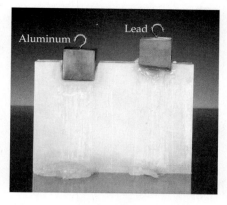

▲ Figure 10.15 **Visualizing specific heat capacity**
These two metal blocks have equal volumes and were heated to the same temperature before being placed on the block of paraffin wax. Notice, however, that the aluminum block melted more wax—and hence gave off more thermal energy—even though the lead block is about four times heavier. The reason is that lead has a very small specific heat capacity (about one-seventh of that of aluminum), and thus it gives off considerably less thermal energy for every degree of temperature change.

CONNECTINGIDEAS

In Chapter 6 you learned that energy is always conserved.
• Here, calorimetry is shown to be a practical application of energy conservation.

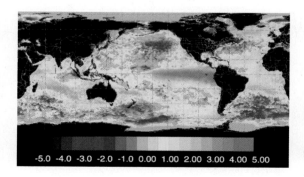

◀ Figure 10.16 **The El Niño effect**
This photo shows an El Niño event—the appearance of a mass of unusually warm water in the equatorial region of the eastern Pacific Ocean. An El Niño causes worldwide changes in climate and weather patterns. In this satellite image the warmest water temperatures are represented by red, and the coolest by violet. The El Niño is the red spike extending westward from South America.

First, we note that there are two basic ideas involved in this experiment. They are as follows:

- 🔲 **The final temperatures of the block and the water are equal, since the system is in thermal equilibrium.**
- 🔲 **The total energy of the system is conserved.**

In particular, the second condition means that the amount of thermal energy *lost* by the block is equal to the amount of thermal energy *gained* by the water. We can write this condition as follows:

$$Q_b + Q_w = 0$$

In this case the quantity Q_b is negative because the block loses thermal energy. Similarly, the quantity Q_w is positive because the water gains thermal energy. The sum of these positive and negative energies is zero—meaning that there has been no net gain or loss of thermal energy. We use this result in the next Guided Example to calculate the specific heat capacity of a block made from an unknown metal.

GUIDED Example 10.9 | Cooling Off

Calorimetry

A 0.50-kg block of metal with an initial temperature of 54.5 °C is dropped into a calorimeter holding 1.1 kg of water at 20.0 °C. If the final temperature of the block-water system is 21.4 °C, what is the specific heat capacity of the metal? Assume that the calorimeter can be ignored and that no thermal energy is exchanged with the surroundings.

Picture the Problem

Initially, when the block is first dropped into the water, the temperatures of the block and water are $T_b = 54.5\,°C$ and $T_w = 20.0\,°C$, respectively. When thermal equilibrium is established, both the block and the water have the same temperature, $T = 21.4\,°C$.

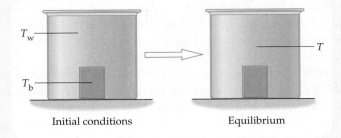

Initial conditions   Equilibrium

Strategy

Thermal energy flows from the block to the water. Setting the energy flow *out of the block* plus the energy flow *into the water* equal to zero (conservation of energy) yields the block's specific heat capacity.

The final temperature for both the block and the water is T. The initial temperature of the block is T_b, and the initial temperature of the water is T_w. Therefore, the change in temperature, $\Delta T = T_f - T_i$, for the block is $\Delta T = T - T_b$, and for the water it is $\Delta T = T - T_w$.

Known	Unknown
$m_b = 0.50\ \text{kg}$	$c_b = ?$
$m_w = 1.1\ \text{kg}$	
$T_b = 54.5\,°C$	
$T_w = 20.0\,°C$	
$T = 21.4\,°C$	

Solution

1 Use the definition of specific heat capacity to write an expression for the thermal energy flow out of the block. Note that Q_{block} is negative, since T is less than T_b:

$$Q_{block} = m_b c_b (T - T_b)$$

2 Use the definition of specific heat capacity to write an expression for the thermal energy flow into the water. Note that Q_{water} is positive, since T is greater than T_w:

$$Q_{\text{water}} = m_w c_w (T - T_w)$$

3 Apply conservation of energy by setting the sum of the energies equal to zero:

$$Q_{\text{block}} + Q_{\text{water}} = 0$$
$$m_b c_b (T - T_b) + m_w c_w (T - T_w) = 0$$

4 Rearrange and solve for the specific heat capacity of the block, c_b:

$$c_b = \frac{m_w c_w (T - T_w)}{m_b (T_b - T)}$$

5 Substitute the numerical values and calculate c_b:

$$c_b = \frac{(1.1\ \text{kg})\left(4186\,\dfrac{\text{J}}{\text{kg}\cdot{}^\circ\text{C}}\right)(21.4\,{}^\circ\text{C} - 20.0\,{}^\circ\text{C})}{(0.50\ \text{kg})(54.5\,{}^\circ\text{C} - 21.4\,{}^\circ\text{C})}$$

$$= \boxed{390\ \text{J}/(\text{kg}\cdot{}^\circ\text{C})}$$

Insight

We note from Table 10.2 that the block is probably made of copper. A value of the specific heat capacity can often be used to identify an unknown substance.

Practice Problems

35. **Follow-up** What is the final equilibrium temperature if the mass of the water in the calorimeter is decreased to 0.50 kg (the same as the mass of the block)?

36. A 235-g lead ball at a temperature of 84.2 °C is placed in a light calorimeter containing 177 g of water at 21.5 °C. Find the equilibrium temperature of the system.

Notice that the final temperature in Guided Example 10.9 is much closer to the initial temperature of the water than to the initial temperature of the block. This is due primarily to the fact that the water's specific heat capacity is more than 10 times greater than that of the copper block.

ACTIVE Example 10.10 Find the Final Temperature

Suppose 550 g of water at 32 °C is poured into a 210-g aluminum can with an initial temperature of 15 °C. Find the final temperature of the system, assuming that no thermal energy is exchanged with the surroundings.

Solution *(Perform the calculations indicated in each step.)*

1. Write an expression for the energy flow out of the water:

$$Q_w = m_w c_w (T - T_w)$$

2. Write an expression for the energy flow into the aluminum:

$$Q_a = m_a c_a (T - T_a)$$

3. Apply energy conservation:

$$Q_w + Q_a = 0$$

4. Solve for the final temperature:

$$T = 31\,{}^\circ\text{C}$$

10.3 LessonCheck (MP)

Checking Concepts

37. ⬭ **Assess** Your friend claims that stirring a glass of water with a spoon increases the temperature of the water. Is he right? Explain.

38. ⬭ **Explain** What does the specific heat capacity tell you about a substance?

39. ⬭ **Explain** Why is it possible to assume that the thermal energy gained by the water in a calorimeter is equal to the thermal energy lost by the sample being tested?

40. ⬚ Triple Choice ⬚ Objects A and B have the same mass. When they receive the same amount of thermal energy, the temperature of object A increases more than the temperature of object B. Is the specific heat capacity of object A greater than, less than, or equal to the specific heat capacity of object B? Explain.

41. ⬚ Triple Choice ⬚ Is the specific heat capacity of a large block of gold greater than, less than, or equal to the specific heat capacity of a small gold coin? Explain.

Solving Problems

42. Calculate How much thermal energy is required to raise the temperature of a 0.75-kg piece of copper pipe by 15 °C?

43. Calculate An orange is mostly water. Estimate the thermal energy required to warm a 0.20-kg orange from 15 °C to 22 °C.

44. Calculate A 1.4-kg block of ice is at a temperature of −10 °C. If 6200 J of thermal energy is added to the ice, what is its final temperature?

45. Calculate A 5.0-g lead bullet is fired into a fence post. The initial speed of the bullet is 250 m/s, and when it comes to rest, half of its kinetic energy goes into heating it. How much does the bullet's temperature increase?

10.4 Phase Changes and Latent Heat

Vocabulary

- phase
- pressure
- equilibrium vapor pressure
- boiling point
- evaporation
- latent heat

Do you like chocolate? If so, you've probably noticed that your favorite chocolate bar is a solid. When you hold the bar in your hand too long, it melts into a liquid. Heating the liquid on a stove produces chocolate steam, a gas. These states of matter—solid, liquid, gas—are referred to as **phases**. In this lesson you'll learn more about phases and the heat needed to change from one to another.

Particles can move from one phase to another

On a cool autumn morning a lake often has small clouds of "steam" rising into the air from its surface. This is a common occurrence, even though the water isn't even close to its boiling temperature. What exactly is happening here?

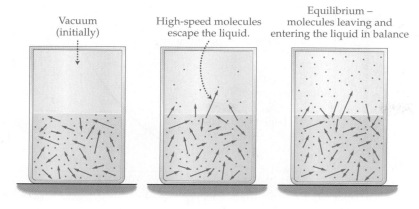

Vacuum (initially)

High-speed molecules escape the liquid.

Equilibrium – molecules leaving and entering the liquid in balance

◀ **Figure 10.17 A liquid in equilibrium with its vapor**
Initially, a liquid is placed in a container, and the volume above it is a vacuum. High-speed particles in the liquid are able to escape into the upper region of the container, forming a low-density gas. As the gas becomes more dense, the number of particles leaving the liquid is balanced by the number returning to the liquid. At this point the system is in equilibrium.

To understand this situation, consider a closed container that is partially filled with a liquid, as in **Figure 10.17**. Initially, the volume above the liquid is empty of particles—it is a vacuum. Soon, however, some of the more energetic particles in the liquid—which are moving fast enough to break free from their neighbors—escape into the vacuum and start to form a gas (or vapor). Similarly, fast-moving particles escaping from liquid water are what cause the "steam" rising from the lake.

This process of particles escaping from the liquid continues until the vapor is dense enough that a large number of gas particles hit the liquid and return to it. ⬡ **When the number of particles returning to the liquid equals the number leaving the liquid, we say that the system has reached** *phase equilibrium* **between the liquid and gas phases.** The same idea applies to equilibrium between other phases.

A useful way to characterize gas and liquid phases is with the physical quantity known as *pressure*. In general, **pressure** is the amount of force exerted on a given area. The symbol for pressure is P, and the pressure produced by a force F pushing on an area A is given by the following:

> **Definition of Pressure, P**
>
> $$\text{pressure} = \frac{\text{force}}{\text{area}}$$
>
> $$P = \frac{F}{A}$$
>
> SI units: N/m^2

A shorthand name for the combination of units N/m^2 is the **pascal** (Pa), named for the French scientist Blaise Pascal (1623–1662). Thus,

$$1\ Pa = 1\ N/m^2$$

Typical pressures are in the range of 1000 Pa. Thus, we often give pressures in terms of the kilopascal (kPa):

$$1\ kPa = 10^3\ Pa$$

For example, atmospheric pressure is 101 kPa (about 14.7 pounds per square inch).

⬡ *What is the key characteristic of phase equilibrium?*

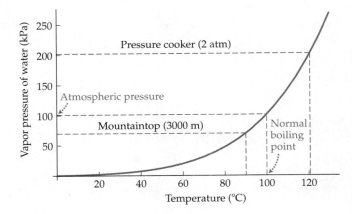

▲ **Figure 10.18 The equilibrium vapor-pressure curve for water**
The equilibrium vapor pressure of water increases with increasing temperature. In particular, at the boiling point of water, $T = 100\ °C$, the vapor pressure is equal to atmospheric pressure.

COOL PHYSICS
Pressure Cookers and Autoclaves

A home pressure cooker is basically a sealed pot that uses elevated pressures and temperatures to cook foods faster. An *autoclave* is an elaborate version of this device used by hospitals to sterilize surgical tools. If surgical tools were heated in boiling water open to the atmosphere, they would experience a temperature of 100 °C. In the autoclave the pressure rises to values significantly greater than atmospheric pressure. As a result, the water has a much higher boiling temperature, and the sterilization is more effective.

The pressure of a gas in equilibrium with a liquid is referred to as the **equilibrium vapor pressure**. In **Figure 10.18** we plot the equilibrium vapor pressure of water.

Equilibrium conditions depend on temperature

What happens if we increase the temperature in Figure 10.17? Well, increasing the temperature produces more high-speed particles in the liquid that can escape into the gas. Thus, to have an equal number of gas particles returning to the liquid, it's necessary for the pressure of the gas to be greater. Thus, the equilibrium vapor pressure increases with temperature. This is shown in Figure 10.18.

For each temperature there is just one equilibrium vapor pressure—just one pressure that creates a precise balance between the phases. Thus, when we plot the equilibrium vapor pressure versus temperature, as in Figure 10.18, the result is a curve—the vapor-pressure curve. The significance of this curve is that it determines the *boiling point* of a liquid. A liquid boils at its **boiling point**, which is the temperature at which its vapor pressure equals the external pressure. For example, you can see in Figure 10.18 that atmospheric pressure corresponds to the boiling temperature of water (100 °C), as expected.

> ### CONCEPTUAL Example 10.11 Boiling Temperature
>
> When water boils at the top of a mountain, is the boiling temperature greater than, less than, or equal to 100 °C?
>
> **Reasoning and Discussion**
> At the top of a mountain air pressure is less than it is at sea level. Therefore, according to Figure 10.18, the boiling temperature of water is less as well.
>
> **Answer**
> The boiling temperature of water is less than 100 °C on a mountaintop.

◀ **Figure 10.19 Extreme evaporation**
Although the body temperature of this athlete is well below the boiling point of water, the water contained in his sweat is evaporating rapidly from his skin. (Since water vapor is invisible, the "steam" we see in this photo is not the water vapor itself. Rather, it is a cloud of tiny droplets that form when the water vapor loses thermal energy to the cold air around it and condenses back to the liquid state.)

Evaporation keeps you cool

When you're hot and sweaty after a workout, "steam" rises from your skin. A good example of this is provided by the athlete in **Figure 10.19**. This is a bit odd, though, isn't it? After all, water boils at 100 °C, so how can a skin temperature of only about 35 °C result in "steam"?

Basically, the situation is similar to "steam" rising from a lake, as discussed earlier in this lesson. Recall that if a liquid is placed in a closed container with a vacuum above it, some of the fastest particles are able to break loose from the liquid and form a gas. When the gas becomes dense enough, the system attains equilibrium. But what happens if you open the container and let a breeze blow across it, removing much of the gas? In that case the release of particles from the liquid into the gas continues without reaching equilibrium. The process of particles leaving the liquid phase and going into the gas phase is referred to as **evaporation**. As the particles are continually removed, the liquid progressively evaporates until none is left.

Let's investigate how evaporation helps cool us when we exercise or work up a sweat. First, consider a droplet of sweat on the skin, as illustrated in **Figure 10.20**. As mentioned before, the high-speed particles in the droplet are the ones that escape from the liquid into the surrounding air. The moving air takes these particles away as soon as they escape.

Now, what does this mean for the particles that are left behind in the sweat droplet? Well, since the droplet tends to lose high-speed particles, the average kinetic energy of the remaining particles must decrease. As we know, this means that the temperature of the droplet must also decrease. Since the droplet is now cooler than its surroundings, including

High-speed molecules leave the droplet ...   ... reducing its temperature ...

... and causing thermal energy to be drawn from the warm skin.

◀ **Figure 10.20 A droplet of sweat resting on the skin**
High-speed particles in a droplet of sweat are able to escape the droplet and become part of the atmosphere. The average speed of the particles that remain in the droplet is reduced, so the temperature of the droplet is reduced as well.

▶ **Figure 10.21 Staying cool**
Humans keep cool by sweating. Because the most energetic molecules are the ones most likely to escape by evaporation, significant quantities of thermal energy are removed from the body when perspiration evaporates from the skin. Dogs, lacking sweat glands, nevertheless take advantage of the same mechanism to help regulate body temperature. In hot weather they pant to promote evaporation from the tongue. Of course, sitting in cool water can also help.

🔑 *Why does sweating help keep you cool?*

the skin on which it rests, it draws in thermal energy from the body. This warms the droplet, increasing the speed of its particles and continuing the evaporation process at more or less the same rate. 🔑 **Thus, sweat droplets are an effective means of drawing thermal energy from the body and shooting it off into the surrounding air in the form of high-speed water molecules.** Certainly, evaporation is helping to cool the jogger in **Figure 10.21 (a)**, and evaporation from its tongue helps to cool the dog in **Figure 10.21 (b)**.

Atmospheres also evaporate

The atmosphere of a planet or a moon can evaporate in much the same way a drop of sweat evaporates from your forehead. In the case of an atmosphere, however, it is the force of gravity that must be overcome.

For example, if a molecule of gas is to escape Earth, it must have a speed of at least 11,200 m/s—the same as the escape speed of a rocket. The speed of nitrogen and oxygen molecules (N_2 and O_2) in our atmosphere is roughly the speed of sound, 343 m/s, much less than Earth's escape speed. Thus, the odds of a nitrogen or oxygen molecule having enough speed to escape Earth are astronomically small. Good thing, too. That's why these molecules have persisted in our atmosphere for billions of years.

On the other hand, consider a lightweight molecule like hydrogen, H_2. Because of its low mass, a hydrogen molecule is very likely to have a speed on the order of a couple of thousand meters per second. Therefore, the probability that a hydrogen molecule has enough speed to escape Earth is about 300 orders of magnitude greater than the corresponding probability for an oxygen molecule. It's no surprise, then, that Earth's atmosphere contains essentially no hydrogen.

The Moon has a rather weak gravitational field and is unable to maintain any atmosphere at all. Whatever atmosphere it may have had early in its history has long since evaporated. You might say that the Moon's atmosphere is "lost in space."

Changing from one phase to another requires heat

🔑 *What happens to the temperature when heat is added to convert a substance from one phase to another?*

If you heat a substance, its temperature increases, right? Well, not always. When two phases coexist, something rather strange and surprising happens—the temperature remains the same even when you add thermal energy. How can that be?

To understand this odd behavior, let's start by considering an ice cube initially at the temperature $-10\ °C$. As you learned in the preceding lesson, adding thermal energy to the ice cube increases its temperature. When the temperature reaches 0 °C, however, additional thermal energy does not cause an additional increase in temperature. Instead, the added energy goes into converting some of the ice into water. On a microscopic level the added energy causes some of the molecules in the solid ice to break loose from neighboring molecules and become part of the liquid. Thus, the thermal energy added when the ice cube is at 0 °C goes into increasing the potential energy of the system rather than the kinetic energy—this is why the temperature doesn't change.

Thus, as long as any ice remains in a cup of water *and* the water and the ice are in equilibrium, you can be sure that both the ice and the water are at 0 °C. If thermal energy is added to the system, the amount of ice decreases but the temperature stays the same. If thermal energy is removed, the amount of ice increases but the temperature stays the same. In general, the **latent heat**, L, is the thermal energy required to change 1 kilogram of a substance from one phase to another. 🔒 **During the conversion process from one phase to another, the temperature of the system remains constant.** It follows that to convert a mass m from one phase to another requires the following heat, Q:

Heat Required to Change from One Phase to Another

heat to change phase $=$ mass $\times$ latent heat
$$Q = mL$$

Latent heat depends on which phases are present

The heat needed to melt ice into water is not the same as the heat needed to boil water into water vapor. In general, the latent heat depends on the phases involved. Here are some specific examples:

- The heat needed to freeze (fuse) a liquid into a solid is the *latent heat of fusion*, L_f.
- The heat needed to boil (vaporize) a liquid into a gas is the *latent heat of vaporization*, L_v.
- The heat needed to *sublimate* a solid directly into a gas is the *latent heat of sublimation*, L_s.

Values of the latent heats of fusion and vaporization for a variety of substances are given in **Table 10.3**.

Table 10.3 Latent Heats

Material	Latent heat of fusion, L_f (J/kg)	Latent heat of vaporization, L_v (J/kg)
Water	33.5×10^4	22.6×10^5
Ammonia	33.2×10^4	13.7×10^5
Copper	20.7×10^4	47.3×10^5
Benzene	12.6×10^4	3.94×10^5
Ethyl alcohol	10.8×10^4	8.55×10^5
Gold	6.28×10^4	17.2×10^5
Nitrogen	2.57×10^4	2.00×10^5
Lead	2.32×10^4	8.59×10^5
Oxygen	1.39×10^4	2.13×10^5

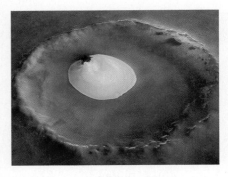

Figure 10.22 Sublimating ice on Mars
This recently discovered ice lake on the surface of Mars lies on the floor of an impact crater. It grows or shrinks with the Martian seasons. Because atmospheric pressure is so low on Mars, the ice does not melt during the Martian summer—instead, it sublimates directly to the vapor phase. On Mars, water ice behaves much like dry ice here on Earth.

The sublimation process (from solid directly to gas) is a little less familiar than the other processes. Ice doesn't sublimate on Earth, for example, though dry ice (solid CO_2) does. However, water ice does sublimate on Mars as a result of the thin atmosphere there, as indicated in **Figure 10.22**.

Heating Curves The temperature of a substance in response to heating is known as its *heating curve*. **Figure 10.23** shows the heating curve for water. Initially 1 kilogram of water is in the form of ice at $-20\ °C$. As the ice is heated, its temperature rises until it begins to melt at $0\ °C$. The temperature then remains constant until the latent heat of fusion has been supplied to the system. When all the ice has melted to water at $0\ °C$, continued heating results in a renewed increase in temperature. When the temperature of the water rises to $100\ °C$, boiling begins and the temperature again remains constant—this time until an amount of thermal energy equal to the latent heat of vaporization is added to the system. Finally, with the entire amount of water converted to steam, continued heating again produces an increasing temperature.

> **CONCEPTUAL Example 10.12 Which Is Worse?**
>
> Both water at $100\ °C$ and steam at $100\ °C$ can cause serious burns. Is a burn produced by steam likely to be more severe, less severe, or the same as a burn produced by water?
>
> **Reasoning and Discussion**
> As the water or steam comes into contact with the skin, it cools from $100\ °C$ to skin temperature, something like $35\ °C$. For water this means that a certain amount of thermal energy is transferred to the skin, which can cause a burn. The steam, on the other hand, must first give off the thermal energy required for it to condense to water at $100\ °C$. After that, the condensed water cools to body temperature, as before. Thus, the thermal energy transferred to the skin is larger in the case of steam, resulting in a more serious burn.
>
> **Answer**
> The steam burn is more severe.

Figure 10.23 Temperature versus heat added or removed
The temperature of 1 kilogram of water as thermal energy is added to or removed from the system. Notice that the temperature stays the same when the system is changing from one phase to another.

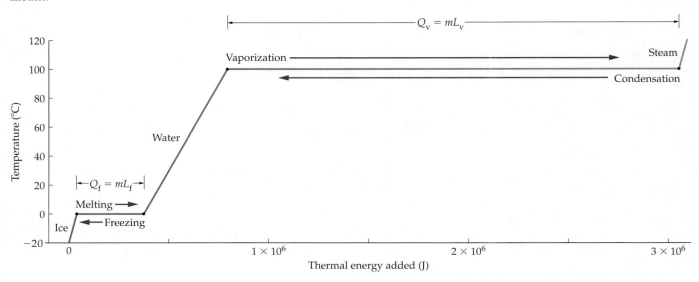

How much thermal energy must be removed from 0.72 kg of water at 0 °C to make ice cubes at 0 °C?

Solution

Since we are converting liquid water to solid water, we must use the latent heat of fusion for water. From Table 10.3 we see that this latent heat is $L_f = 33.5 \times 10^4$ J/kg. The amount of heat needed for $m = 0.72$ kg of water is

$$Q = mL_f$$
$$= (0.72 \text{ kg})(33.5 \times 10^4 \text{ J/kg})$$
$$= \boxed{2.4 \times 10^5 \text{ J}}$$

Practice Problem

46. How much heat is needed to convert 1.26 kg of water at 100 °C to steam at 100 °C?

Calculating the total heat is done step by step

As a example of using latent heat in a practical situation, let's calculate the thermal energy required to change 0.550 kg of ice at −20.0 °C to liquid water at 20.0 °C. This corresponds to going from point A to point D in **Figure 10.24**. The way to approach a problem like this is to consider each phase and each conversion from one phase to another, one at a time.

The first step (A to B in Figure 10.24) is to find the thermal energy necessary to warm the ice from −20.0 °C to 0 °C. Using the specific heat capacity of ice, $c_{ice} = 2090$ J/(kg·°C), we find

$$Q_1 = mc_{ice}\Delta T$$
$$= (0.550 \text{ kg})/\left(2090\frac{\text{J}}{\text{kg}\cdot°\text{C}}\right)(20.0 °\text{C})$$
$$= 23{,}000 \text{ J}$$

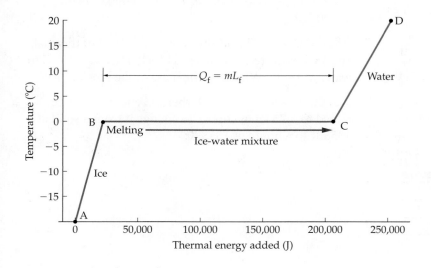

◄ **Figure 10.24 Thermal energy required for a given change in temperature**
The amount of thermal energy required to change 0.550 kg of H_2O from ice at −20.0 °C to water at 20.0 °C is the energy difference between points A and D. To calculate this thermal energy we sum the following three heats: (1) the heat to warm the ice from −20.0 °C to 0 °C; (2) the heat to melt all the ice; (3) the heat to warm the water from 0 °C to 20.0 °C.

COOL PHYSICS
Homemade Ice Cream

A pleasant application of latent heat is making homemade ice cream. As you may know, it is necessary to add salt to the ice–water mixture surrounding the container holding the ingredients of the ice cream. The dissolved salt impairs the ability of the water molecules to connect and freeze. Thus, a lower temperature is required to establish the ice-water equilibrium. As a result, the ice in the ice–water mixture begins to melt. As the ice melts, it draws the required heat from its surroundings—which includes the ice cream mixture. The ice cream quickly freezes and is soon ready to eat. Cool.

The second step in the process (B to C) is to melt the ice at 0 °C. The thermal energy required for this is found using the latent heat of fusion for water ($L_f = 33.5 \times 10^4$ J/kg):

$$Q_2 = mL_f$$
$$= (0.550 \text{ kg})(33.5 \times 10^4 \text{ J/kg})$$
$$= 184{,}000 \text{ J}$$

Finally, the third step (C to D) is to heat the water from 0 °C to 20.0 °C. This time we use the specific heat capacity for water, $c_{water} = 4186$ J/(kg·°C):

$$Q_3 = mc_{water}\Delta T$$
$$= (0.550 \text{ kg})\left(4186\frac{\text{J}}{\text{kg}\cdot°\text{C}}\right)(20.0 °\text{C})$$
$$= 46{,}000 \text{ J}$$

The total thermal energy required for this process, then, is

$$Q_{total} = Q_1 + Q_2 + Q_3$$
$$= 23{,}000 \text{ J} + 184{,}000 \text{ J} + 46{,}000 \text{ J}$$
$$= 253{,}000 \text{ J}$$

A step-by-step approach like this is the best way to calculate the total thermal energy.

10.4 LessonCheck ⓂⓅ

Checking Concepts

47. Determine How do you know if two phases are in equilibrium?

48. Explain Why is the evaporation of sweat from the skin a cooling process?

49. Describe What is latent heat, and how does it affect the temperature of a substance?

50. Identify What is force per area called?

51. Identify What do we call the pressure of a gas that is in equilibrium with a liquid phase?

52. Analyze Referring to the equilibrium vapor pressure curve in Figure 10.18, state whether water is a liquid or a gas at the following temperatures and pressures:

	Pressure	Temperature
(a)	pressure cooker	80 °C
(b)	pressure cooker	140 °C
(c)	mountaintop	100 °C
(d)	mountaintop	60 °C

Solving Problems

53. Calculate How much thermal energy must be removed from 0.96 kg of water at 0 °C to make ice cubes at 0 °C?

54. Calculate How much thermal energy must be added to 0.96 kg of water at 100 °C to make steam at 100 °C?

55. Calculate The addition of 9.5×10^5 J of thermal energy is required to convert a block of ice at −15 °C to water at 15 °C. What was the mass of the block of ice?

56. Calculate How much thermal energy must be added to 1.75 kg of copper to change it from a solid at 1358 K to a liquid at 1358 K?

Physics & You

Optical Pyrometer

What Is It? An *optical pyrometer* is a telescope-like instrument used to measure the temperature of very hot objects. It determines the temperature from a safe distance so that the operator does not have to make physical contact with the object.

A researcher monitors the temperature of a lava flow.

How Does It Work? An optical pyrometer uses the light emitted by very hot objects to determine their temperature. When the operator views the target object through the pyrometer, he or she also sees a thin, glowing filament. This filament is inside the pyrometer between lenses. The filament appears as a light or dark line superimposed on the image of the object.

The operator then adjusts the voltage that is applied to the filament. The voltage controls the brightness of the filament; when the brightness of the filament matches that of the object, the line disappears. Internal electronics determine the temperature that corresponds to the voltage setting. The displayed temperature indicates the temperature of the target object.

Because the filament and the target are likely made of different materials, they do not emit exactly the same color or light (you'll learn about light waves in Chapter 15). To correct for this, a red filter inside the pyrometer excludes all but a narrow range of colors, allowing the filament line to disappear completely with any target material. Calibration tables provide temperature corrections for various materials and conditions.

What Are Its Uses? Optical pyrometers are used in scientific, industrial, medical, and consumer applications. For example, volcanologists use them to measure the temperature of lava flows, and steel workers use them to monitor the temperature of molten steel. Cooler objects emit invisible infrared radiation that can be detected using an infrared-sensitive pyrometer. Infrared fever pyrometers are used in hospitals by pointing them into a patient's ear. Pyrometers called *radiometers* are used by radio astronomers to measure the surface temperatures of planets, the temperature of interstellar gas clouds, and the black body background temperature of the universe.

Take It Further

1. Research *The pyrometer makes use of stored calibration data. Use the library and the Internet to research how a device called a thermocouple is used to determine the calibration data.*

2. Reason *It is difficult to aim an infrared-sensitive pyrometer at a target because there is no visible image to view. Why is this not a problem for an infrared fever pyrometer?*

Physics Lab Investigating Specific Heat Capacity

This lab explores the property of specific heat capacity. You will determine the specific heat capacity of several metal samples and then use the data to identify an unknown metal.

Materials
- several objects of known composition
- one object of unknown composition
- two foam cups
- Celsius thermometer
- 250-mL beaker
- beaker tongs
- hot plate
- balance
- string

Procedure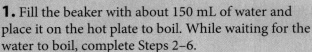

1. Fill the beaker with about 150 mL of water and place it on the hot plate to boil. While waiting for the water to boil, complete Steps 2–6.

2. Measure the mass of each metal object and record it in the data table.

3. Measure the mass of the empty foam cup and record it in the data table.

4. Fill the foam cup with enough cold tap water that all the metal objects will be submerged when they are placed in the cup.

5. Measure the mass of the cup plus the water and record it in the data table.

6. Place the thermometer in the cold water and allow it to come to thermal equilibrium with the water. Record this as the initial temperature. Remove the thermometer.

7. Tie a piece of string around each metal object. Once the water in the beaker begins to boil, use the string to carefully lower one of the known objects into the water. Leave the string hanging outside the beaker for easy retrieval.

8. Allow the beaker to return to a boil and then leave the object in the water for 2 min.

9. Use the tongs or string to very carefully lift the heated metal object out of the boiling water. Gently shake off any water that might be clinging to the metal and place it in the foam cup filled with cold water.

10. Place the second foam cup over the first and insert the thermometer through the top into the water.

11. Observe the temperature change on the thermometer. When the temperature remains constant for 1 min, record this as the final temperature.

12. Remove the metal from the cup and repeat Steps 3–11 for the other objects.

Data Table

Sample	Mass of Object (g)	Mass of Empty Cup (g)	Mass of Cup and Water (g)	Mass of Water Alone (g)	Initial Temperature of Object (°C)	Final Temperature of Object (°C)	Initial Temperature of Water (°C)	Final Temperature of Water (°C)
A								
B								
C								
D								

Analysis

1. For each trial, determine **(a)** the quantity of heat transferred to the water, **(b)** the quantity of heat transferred out of the metal object, and **(c)** the specific heat capacity of the metal.

2. Find out the type of metal that makes up each object from your teacher. Look up the specific heat capacity for each metal used. Calculate the percent error for each of your experimental values of the specific heat capacity.

3. What was your unknown metal?

Conclusions

1. Explain what the specific heat capacity tells you about the substance.

2. Why were foam cups used instead of clear plastic cups?

3. Why was it important to shake off the excess water before putting each object into the cold water?

4. Why did you wait 2 min before removing the metal from the boiling water?

5. Identify two possible sources of error in this experiment. Describe one way to reduce each of these errors.

10 Study Guide

Big Idea

Heat is a form of energy that is transferred because of temperature differences.

Energy that is transferred from one object to another because of a difference in temperature is referred to as *heat*. When energy is transferred as heat from object 1 to object 2, the thermal energy of object 1 decreases and the thermal energy of object 2 increases.

10.1 Temperature, Energy, and Heat

🔑 Temperature doesn't depend on the total amount of energy in an object; it depends on the average kinetic energy of the particles in the object.

🔑 An object has a certain amount of thermal energy, and the thermal energy it *exchanges* with other objects because of a temperature difference is called *heat*.

• The zeroth law of thermodynamics states that objects in contact with one another are in thermal equilibrium if they have the same temperature. Nothing else matters.

Key Equations

Use the following equations to convert between degrees Celsius (T_C), degrees Fahrenheit (T_F), and kelvins (T):

$$T_F = \tfrac{9}{5}T_C + 32$$
$$T_C = \tfrac{5}{9}(T_F - 32)$$
$$T = T_C + 273.15$$

10.2 Thermal Expansion and Energy Transfer

🔑 The thermal expansion of an object—its change in length—is proportional to *both* the initial length, L_i, and the temperature change, ΔT.

🔑 Conduction occurs as higher energy particles collide with and jostle neighboring lower energy particles, transferring kinetic energy from one particle to the next.

🔑 Convection differs from conduction in that it transfers thermal energy through the physical movement of particles from one place to another, rather than between particles that vibrate in place.

🔑 The warmth you feel from the Sun is due to light and infrared waves that it emits. These waves (referred to as *electromagnetic waves*) carry energy from the Sun across the vacuum of space to Earth.

Key Equation

When an object of length L_i changes it temperature by the amount ΔT, its length changes by the amount ΔL (where α is the coefficient of thermal expansion):

$$\Delta L = \alpha L_i \Delta T$$

10.3 Heat Capacity

🔑 The mechanical equivalent of heat is the precise amount of mechanical work that has the same effect as the transfer of a given amount of thermal energy.

🔑 A substance with a high specific heat capacity requires a lot of thermal energy to show a given change in temperature.

🔑 In calorimetry the final temperatures of the sample and the water are equal, since the system is in thermal equilibrium. The total energy of the system is conserved.

• One kilocalorie (kcal) is the thermal energy needed to raise the temperature of 1 kg of water from 14.5 °C to 15.5 °C.

• The specific heat capacity of a substance is the thermal energy required to change the temperature of 1 kg of the substance by 1 °C.

Key Equations

The mechanical equivalent of heat:

$$1 \text{ cal} = 4.186 \text{ J}$$

The specific heat capacity, c, is defined as follows:

$$c = \frac{Q}{m\Delta T}$$

10.4 Changes of Phase and Latent Heat

🔑 When the number of particles returning to a liquid equals the number leaving the liquid, we say that the system has reached *phase equilibrium* between the liquid and gas phases.

🔑 Sweat droplets are an effective means of drawing thermal energy from the body and shooting it off into the surrounding air in the form of high-speed water molecules.

🔑 During the conversion process from one phase to another, the temperature of the system remains constant.

• Pressure is the amount of force exerted on a given area.

• Latent heat is the thermal energy that must be added to or removed from 1 kilogram of a substance to convert it from one phase to another. Values of latent heat depend on the substance and the phase change involved.

10 Assessment

For instructor-assigned homework, go to www.masteringphysics.com

ANSWERS TO SELECTED ODD-NUMBERED PROBLEMS APPEAR IN APPENDIX A.

Lesson by Lesson

10.1 Temperature, Energy, and Heat

Conceptual Questions

57. How do temperature and thermal energy differ?

58. An ice cube is thrown into a swimming pool. Is this system in thermal equilibrium? Is the temperature of the water in the pool 0 °C? Explain.

59. Is a change in temperature of 20 °C greater than, less than, or equal to a change in temperature of 20 K?

60. Is a change in temperature of 20 °C greater than, less than, or equal to a change in temperature of 20 °F?

61. Which temperature scale has no negative temperatures?

62. Suppose someone makes the following comment: "Boiling and freezing water are opposites, and this is why the Fahrenheit temperature difference between them is 180°." How would you respond to this assertion?

63. Is it valid to say that a hot object contains more heat than a cold object?

Problem Solving

64. Normal body temperature for humans is 98.6 °F. What is the corresponding temperature in **(a)** degrees Celsius and **(b)** kelvins?

65. What is the temperature 1.0 K equivalent to on the Celsius scale?

66. The temperature in a freezer is set at 30 °F. What is this temperature on the Celsius scale?

67. The temperature of molten lava is about 1200 °C. Convert this temperature to **(a)** the Kelvin scale and **(b)** the Fahrenheit scale.

68. At what temperature is the reading on the Fahrenheit scale twice the reading on the Celsius scale?

69. **Greatest Temperature Change** A world record for the greatest change in temperature was set in Spearfish, South Dakota, on January 22, 1943. At 7:30 A.M. the temperature was −4.0 °F, and 2 minutes later the temperature was 45 °F. Find the average rate of temperature change during those 2 minutes in kelvins per second.

10.2 Thermal Expansion and Energy Transfer

Conceptual Questions

70. Sometimes the metal lid on a glass jar has been screwed on so tightly that it is very difficult to open. Explain why holding the lid under hot running water often loosens it enough for easy opening.

71. Why do you hear creaking and groaning sounds in a house, particularly at night as the air temperature drops?

72. Updrafts of air allow hawks and eagles to glide effortlessly, all the while gaining altitude. What causes the updrafts?

73. When a car is heated, say by taking it to the desert, it expands. Is the increase in length of the car greater than, less than, or equal to its increase in height? Explain.

74. Bimetallic strip A is made of copper and steel; bimetallic strip B is made of aluminum and steel. Referring to Table 10.1, which bimetallic strip bends more for a given change in temperature?

75. Rank The five plates in **Figure 10.25** are all at the same temperature and all made from the same metal. They are all placed in an oven and heated by the same amount. Rank the plates in order of increasing expansion in **(a)** the vertical and **(b)** the horizontal direction. Indicate ties where appropriate.

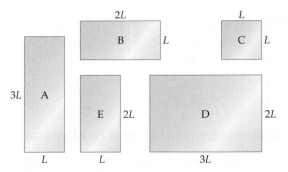

Figure 10.25

76. **Predict & Explain** A brass plate has a circular hole whose diameter is slightly smaller than the diameter of an aluminum ball. **(a)** If the ball and the plate are always kept at the same temperature, should the temperature of the system be increased or decreased in order for the ball to fit through the hole? **(b)** Choose the *best* explanation from among the following:

A. The aluminum ball changes its diameter more with temperature than the brass plate changes its dimensions, and therefore the temperature should be decreased.

B. Changing the temperature won't change the fact that the ball is larger than the hole.

C. Heating the brass plate makes its hole larger, which will allow the ball to pass through.

Problem Solving

77. A hole in an aluminum plate has a diameter of 1.178 cm at 23.00 °C. What is the diameter of the hole at 199.0 °C?

78. What temperature change is required to change the length of a 5.5-m steel beam by 0.0012 m?

79. A metal ball that is 1.2-m in diameter expands by 2.2 mm when the temperature is increased by 95 °C. **(a)** What is the coefficient of thermal expansion for the ball? **(b)** What substance is this ball likely made from? (Refer to Table 10.1.)

80. **Think & Calculate** It is desired to slip an aluminum ring over a steel bar, as shown in **Figure 10.26**. At 10.00 °C the inside diameter of the ring is 4.000 cm and the diameter of the rod is 4.040 cm. **(a)** In order for the ring to slip over the bar, should the ring be heated or cooled? Explain. **(b)** Find the temperature of the ring at which it fits over the bar. The bar remains at 10.00 °C.

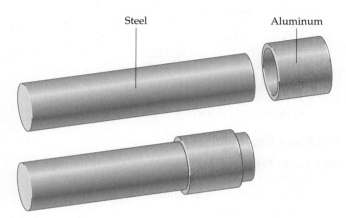

Steel   Aluminum

Figure 10.26

81. Some cookware has a stainless steel interior ($\alpha = 17.3 \times 10^{-6}\ \text{K}^{-1}$) and a copper bottom ($\alpha = 17.0 \times 10^{-6}\ \text{K}^{-1}$) for better heat distribution. Suppose a 21-cm-diameter pot of this type is heated to 610 °C on a stove. If the initial temperature of the pot was 22 °C, what is the difference in diameter change for the copper and the steel?

10.3 Heat Capacity
Conceptual Questions

82. Two different objects receive the same amount of thermal energy. Give at least two reasons why their temperature changes may not be the same.

83. What happens when 1 kilocalorie of energy is added to 1 kilogram of water?

84. If a lighted match is held beneath a balloon inflated with air, the balloon quickly bursts. If, instead, the lighted match is held beneath a balloon filled with water, the balloon remains intact, even if the flame comes in contact with the balloon. Explain.

85. **Predict & Explain** A certain amount of thermal energy is transferred to 2 kg of aluminum, and the same amount of thermal energy is transferred to 1 kg of ice. **(a)** Referring to Table 10.2, is the increase in temperature of the aluminum greater than, less than, or equal to the

increase in temperature of the ice? **(b)** Choose the *best* explanation from among the following:

A. Twice the specific heat capacity of aluminum is less than the specific heat capacity of ice, and hence the aluminum has the greater temperature change.

B. The aluminum has the smaller temperature change since its mass is less than that of the ice.

C. The same amount of thermal energy will cause the same change in temperature.

86. **Predict & Explain** The specific heat capacity of isopropyl alcohol is about half that of water. Suppose you have 0.5 kg of isopropyl alcohol at 20 °C in one container and 0.5 kg of water at 30 °C in a second container. **(a)** When these fluids are poured into the same container and allowed to come to thermal equilibrium, is the final temperature greater than, less than, or equal to 25 °C? **(b)** Choose the *best* explanation from among the following:

A. The low specific heat capacity of isopropyl alcohol means that it accepts more thermal energy, giving a final temperature that is less than 25 °C.

B. More thermal energy is required to change the temperature of water than to change the temperature of isopropyl alcohol. Therefore, the final temperature will be greater than 25 °C.

C. Equal masses are mixed together; therefore, the final temperature will be 25 °C, the average of the two initial temperatures.

Problem Solving

87. A 0.141-kg piece of iron is at an initial temperature of 26.5 °C. If 66.2 J of thermal energy is removed from the piece of iron, what is its final temperature?

88. How much thermal energy must be removed from a 0.21-kg chunk of ice to lower its temperature by 7.5 °C?

89. Estimate the thermal energy required to heat a 0.15-kg apple from 12 °C to 36 °C. (Assume that the apple is mostly water.)

90. **Sleeping Metabolic Rate** When people sleep, their metabolic rate is about $2.6 \times 10^{-4}\ \text{C}/(\text{s}\cdot\text{kg})$. How many nutritional Calories does a 75-kg person metabolize while getting a good night's sleep lasting 8.0 hr?

91. An exercise machine indicates that you have worked off 2.5 nutritional Calories in $1\frac{1}{2}$ min of running in place. What was your power output during this time? Give your answer in both watts and horsepower.

92. At the county fair you watch as a blacksmith drops a 0.50-kg iron horseshoe into a bucket containing 25 kg of water. If the initial temperature of the horseshoe is 450 °C and the initial temperature of the water is 23 °C, what is the equilibrium temperature of the system? Assume that no thermal energy is exchanged with the surroundings.

93. A 235-g lead ball at a temperature of 84.2 °C is placed in a calorimeter containing 177 g of water at 21.5 °C. Find the equilibrium temperature of the system.

94. Think & Calculate Silver pellets with a mass of 1.0 g and a temperature of 85 °C are added to 220 g of water at 14 °C. **(a)** How many pellets must be added to increase the equilibrium temperature of the system to 25 °C? Assume that no thermal energy is exchanged with the surroundings. **(b)** If copper pellets are used instead, does the required number of pellets increase, decrease, or stay the same? Explain. **(c)** Find the number of copper pellets that are required.

95. To determine the specific heat capacity of an object, a student heats it to 100 °C in boiling water. She then places the 38.0-g object in a 155-g aluminum calorimeter containing 103 g of water. The aluminum and water are initially at a temperature of 20.0 °C and are thermally insulated from their surroundings. If the final temperature is 22.0 °C, what is the specific heat capacity of the object? Referring to Table 10.2, identify the material that the object is made of.

96. The ceramic coffee cup in **Figure 10.27**, with $m = 116$ g and $c = 1090$ J/(kg·°C), is initially at room temperature (24.0 °C). If 225 g of 80.3 °C coffee and 12.2 g of 5.00 °C cream are added to the cup, what is the equilibrium temperature of the system? Assume that no thermal energy is exchanged with the surroundings and that the specific heat capacities of coffee and cream are the same as that of water.

Figure 10.27

10.4 Changes of Phase and Latent Heat
Conceptual Questions

97. Waving a fan back and forth can make you feel cooler on a hot day. Explain why.

98. Isopropyl alcohol is sometimes rubbed onto a patient's arms to lower the body temperature. Why is this effective?

99. Which of the following accompanies the formation of ice from water: **(a)** an absorption of thermal energy by the water, **(b)** an increase in temperature, **(c)** a decrease

in temperature, or **(d)** a removal of thermal energy from the water?

100. A system consists of an ice cube floating in a glass of water. The entire system is at 0 °C. Thermal energy is then added slowly to the system, causing half of the ice to melt. Is the resulting temperature of the system greater than, less than, or equal to 0 °C? Explain.

101. Rank Four liquids are at their freezing temperatures. Thermal energy is then removed until each solidifies. The amount of heat removed, Q, and the mass of each liquid, m, are as follows:

Liquid	A	B	C	D
Heat (J)	16,600	3150	3350	5400
Mass (kg)	0.05	0.025	0.01	0.05

Rank these liquids in order of increasing latent heat of fusion. Indicate ties where appropriate.

Problem Solving

102. **Vapor Pressure for Water** A portion of the vapor-pressure curve for water is shown in **Figure 10.28**. Referring to the figure, estimate the pressure (in kPa) required for water to boil at 30 °C.

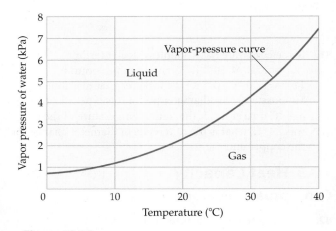

Figure 10.28

103. Using the vapor-pressure curve in Figure 10.28, estimate the temperature at which water boils when the pressure is 1.5 kPa.

104. How much thermal energy must be removed from 1.7 kg of water at 0 °C to make ice cubes at 0 °C?

105. How much thermal energy must be added to a 0.95-kg piece of lead to convert it from a solid to a liquid at 327.5 °C?

106. A block of ice at −12 °C is converted to water at 24 °C when 8.8×10^5 J of thermal energy is added to the system. What is the mass of the water?

107. A 1.1-kg block of ice is initially at a temperature of −5.0 °C. If 5.2×10^5 J of thermal energy is added to the ice, what is the final temperature of the system? Find the amount of ice, if any, that remains.

108. To help keep her barn warm on cold days, a farmer stores 865 kg of warm water in the barn. How many hours would a 2.00-kW electric heater have to operate to provide the same amount of thermal energy as is given off by the water as it cools from 20.0 °C to 0 °C and then freezes at 0 °C?

109. Suppose the 1.000 kg of water ice in Figure 10.23 *begins* melting at time zero. Thermal energy is added to this system at the rate of 12,250 J/s. How much time does it take for the system to reach a temperature of 15 °C?

110. A large punch bowl holds 3.99 kg of lemonade (which is essentially water) at 20.5 °C. A 0.0550-kg ice cube at −10.2 °C is placed in the lemonade. What is the final temperature of the system and the amount of ice (if any) remaining? Ignore any heat exchange with the bowl or the surroundings.

111. A 155-g aluminum cylinder is removed from a liquid nitrogen bath, where it has been cooled to −196 °C. The cylinder is immediately placed in an insulated cup containing 80.0 g of water at 15.0 °C. What is the equilibrium temperature of this system? If your answer is 0 °C, determine the amount of water that has frozen. The average specific heat capacity of aluminum over this temperature range is 653 J/(kg·°C).

Mixed Review

112. What form of thermal energy exchange describes the flow of heat through a vacuum from the Sun to the Earth?

113. A steel tape measure is marked in such a way that it gives accurate length measurements at a normal room temperature of 20 °C. If this tape measure is used outdoors on a cold day when the temperature is 0°C, are its measurements too long, too short, or still accurate?

114. (a) Referring to Figure 10.28, is water a liquid or a gas at a temperature of 20 °C and a pressure of 1 kPa? At 20 °C and 2 kPa? At 20 °C and 3 kPa? **(b)** What is the approximate boiling temperature of water at a pressure of 1 kPa? A pressure of 2 kPa? A pressure of 3 kPa?

115. A copper ring stands on edge with a metal rod placed inside it, as shown in **Figure 10.29**. As this system is heated, will the rod ever touch the top of the ring? Answer yes or no for a rod that is made of **(a)** copper, **(b)** aluminum, and **(c)** steel.

Figure 10.29

116. The Hottest Living Things In the surreal realm of deep-sea hydrothermal vents, 200 miles offshore from Puget Sound, lives a newly discovered *hyperthermophilic*—or extreme heat-loving—microbe that holds the record for existing in the hottest conditions known to science. This microbe has been tentatively named *Strain 121* for the temperature at which it thrives: 121 °C. (At sea level, water at this temperature would boil vigorously, but the extreme pressures at the ocean floor prevent boiling from occurring.) What is this temperature in degrees Fahrenheit?

117. Triple Choice Liquids A and B are at their freezing temperatures. Thermal energy is then removed until each solidifies. The amount of heat removed and the mass of each liquid are as follows:

	Heat Removed	Mass of Liquid
Liquid A	33,500 J	0.100 kg
Liquid B	166,000 J	0.500 kg

Is the latent heat of fusion of liquid A greater than, less than, or the same as the latent heat of fusion of liquid B?

118. Cooling Computers Researchers are developing heat exchangers for laptop computers that take thermal energy from the laptop—to keep it from being damaged by overheating—and use it to vaporize methanol. Given that 5100 J of thermal energy is removed from a laptop when 4.6 g of methanol is vaporized, what is the latent heat of vaporization for methanol?

119. Rank Four materials of differing masses are heated by different amounts and experience different changes in temperature, as shown below. Rank the materials in order of increasing specific heat capacity. Indicate ties where appropriate.

Material	A	B	C	D
Heat	Q	$2Q$	$3Q$	$4Q$
Mass	m	$3m$	$3m$	$4m$
Temperature Change	ΔT	$3\Delta T$	ΔT	$2\Delta T$

120. Thermal energy is added to 150 g of water at the rate of 55 J/s for 2.5 min. How much does the temperature of the water increase?

121. Thermal energy is added to 180 g of water at a constant rate for 3.5 min, resulting in an increase in temperature of 12 °C. What is the heating rate, in joules per second?

122. How much thermal energy must be extracted from 1.5 kg of steam at 110 °C to convert it to ice at 0.0 °C?

123. Mrs. Green uses solar power to heat water. Her solar collector has an area of 5.5 m² and receives sunlight that delivers a power of 520 W/m². How much time is required to heat 45 kg of water by 12 °C?

124. A system consists of a 0.130-kg chunk of ice floating in 1.12 kg of water, all at 0 °C. How much thermal energy must be added to the system to convert it to 1.25 kg of water at 15.0 °C?

125. Think & Calculate A sheet of aluminum has a circular hole with a diameter of 10.0 cm. A 9.99-cm-long steel rod is placed inside the hole, along a diameter of the circle, as shown in **Figure 10.30**. It is desired to change the temperature of this system until the steel rod just touches both sides of the circle. **(a)** Should the temperature be increased or decreased? Explain. **(b)** By how much should the temperature be changed?

Figure 10.30

126. Think & Calculate A student drops a 0.33-kg piece of steel at 42 °C into a container of water at 22 °C. The student also drops a 0.51-kg chunk of lead into the same container at the same time. The temperature of the water remains the same. **(a)** Was the temperature of the lead greater than, less than, or equal to 22 °C? Explain. **(b)** What was the temperature of the lead?

127. **The Cricket Thermometer** The rate of chirping of the snowy tree cricket (*Oecanthus fultoni* Walker) varies with temperature in a predictable way. A linear relationship provides a good match to the chirp rate, but an even more accurate relationship is the following:

$$N = (5.63 \times 10^{10})e^{-(6290 \text{ K})/T}$$

In this expression, N is the number of chirps in 13.0 s and T is the temperature in kelvins. If a cricket is observed to chirp 185 times in 60.0 s, what is the temperature in degrees Celsius?

Writing about Science

128. Research the making of homemade ice cream. What temperature should the ice-cream mixture be held at as it is stirred, and for how much time? How effective is adding salt to the ice-water mixture in lowering the temperature? (You might want to try this with a simple lab experiment.) What temperatures can be reached in this way? How much salt is needed? Why is rock salt used for this purpose?

129. Connect to the Big Idea Have you heard about the survival skill of rubbing two sticks together to make a fire? Discuss this in terms of thermal energy and energy conservation. Is it better to rub wooden sticks together or steel rods? Explain.

Read, Reason, and Respond

Faster than a Speeding Bullet The SR-71 Blackbird, shown in **Figure 10.31**, is 32.74 m long (107 ft 5 in) and is remarkable in many ways. For example, on July 28, 1976, it set an altitude record for sustained horizontal flight of 25,929 m (85,069 ft). The same day it set a closed-course speed record of 980.433 m/s (2193.167 mph). At this speed the airplane is—literally—faster than a speeding bullet.

Called the *Blackbird* because of its distinctive black paint job, the SR-71 becomes very hot because of air resistance when it flies at supersonic speeds. Its typical cruising speed is 3.2 times the speed of sound, referred to as Mach 3.2. The extreme heating at these speeds results in a number of interesting consequences, including the fact that the Blackbird is too hot to touch for about 30 min after it lands. In addition, pilots have been known to heat their lunch by holding it against the windshield, which reaches temperatures comparable to an oven.

Of course, temperatures this high also result in significant thermal expansion. For example, portions of the upper and lower inboard wing skin of the SR-71 are corrugated, allowing expansion during flight. In addition, when the Blackbird lands, it is 20 cm (~8.0 in) longer than it was when it took off. No wonder pilots say this is one "hot" airplane.

Figure 10.31 The SR-71 Blackbird reconnaissance aircraft can fly 3.2 times faster than the speed of sound.

130. How hot is the Blackbird when it lands, assuming that its coefficient of thermal expansion is $22 \times 10^{-6} \text{ K}^{-1}$ and its temperature at takeoff is 23 °C?

　A. 110 °C　　　　　　　C. 300 °C

　B. 280 °C　　　　　　　D. 560 °C

131. How long is the Blackbird when it is 120 °C? (The coefficient of thermal expansion is given in the previous problem.)

　A. 107 ft 2.5 in　　　　　C. 107 ft 7.5 in

　B. 107 ft 3.0 in　　　　　D. 107 ft 8.1 in

Standardized Test Prep

Multiple Choice

1. Energy transferred from a higher temperature object to a lower temperature object is
   - (A) heat.
   - (B) kinetic energy.
   - (C) thermal energy.
   - (D) internal energy.

2. Which is the sum of the kinetic and potential energies of an object?
   - (A) heat energy
   - (B) specific heat capacity
   - (C) thermal energy
   - (D) temperature

3. Which *best* describes temperature?
   - (A) It is equal to the heat difference between two objects.
   - (B) It is proportional to the total energy in an object.
   - (C) It is proportional to the average kinetic energy of particles in an object.
   - (D) It is equal to an object's specific heat capacity.

Use the graph below to answer Questions 4–6. The graph shows the temperatures of 1-kg samples of three materials, A, B, and C, as they are heated. Heating occurs at a rate of 30.0 J/s for 60.0 s.

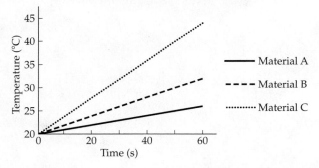

4. What is the approximate specific heat capacity of material A?
   - (A) 360 J/kg °C
   - (B) 500 J/kg °C
   - (C) 900 J/kg °C
   - (D) 720 J/kg °C

5. Material A starts to melt at 60 °C. At that point, the line on the graph will
   - (A) continue on as before.
   - (B) begin to curve upward.
   - (C) become steeper.
   - (D) become horizontal.

6. Which material has molecules with the highest average kinetic energy after 40 s of heating?
   - (A) material A
   - (B) material B
   - (C) material C
   - (D) impossible to determine from this graph

7. The diameter of a brass ball is such that it just fits through a brass ring. What happens if the ball and the ring are cooled to a much lower temperature?
   - (A) Both change in size at the same rate, and the ball still fits through the ring.
   - (B) The ball decreases more in size and easily fits through the ring.
   - (C) The diameter of the ball gets smaller while the diameter of the ring remains constant.
   - (D) The results depend on the specific temperature to which the ball and ring are cooled, and thus cannot be determined.

Extended Response

8. Explain why buildings' heating and cooling systems operate most efficiently if the heating vents are located near the floor and the cooling vents are located near the ceiling.

Test-Taking Hint

Make sure the units for each quantity in your solution are consistent with those of the specific heat capacity or the latent heat.

If You Had Difficulty With . . .

Question	1	2	3	4	5	6	7	8				
See Lesson	10.2	10.1	10.1	10.3	10.3	10.1	10.1	10.2				

11 Thermodynamics

Inside

These monkeys are relaxing in water that is warmer than their body temperature. As a result, thermal energy is transferred to their bodies, keeping them comfortably warm in a cold environment.

Big Idea

Energy conservation applies to thermal energy and heat.

In this chapter we present the laws of nature that relate to heat and thermal energy. As you will see, these laws of thermodynamics extend the principle of energy conservation to include thermal energy and its transfer. They also introduce the idea that processes in the real world proceed in one preferred direction. Thus, thermodynamics is more than just the study of thermal energy and temperature—it relates to the basic operating principles of the universe.

Inquiry Lab   How does doing work on an object affect it?

Explore 🤚

1. Take a large metal paper clip and unbend it so that it is straight.
2. Hold the ends of the paper clip and bend it back and forth several times.
3. Immediately after bending the clip, touch the spot where the bending occurred to your cheek.
4. Continue bending the clip several more times and once again hold the clip against your cheek.
5. Repeat Steps 1–4 using a plastic soda straw.

Think

1. **Observe** How were the paper clip and the soda straw affected by the bending? Did both objects respond in the same way? How did continuing to bend the objects affect them?
2. **Explain** Describe how work was done on each object. If each object is bent back and forth the same number of times, is the work done on each the same?
3. **Identify** What types of energy transformations occurred during this lab?
4. **Extend** You can start a fire by rubbing two sticks together. Explain why this process works.

11.1   The First Law of Thermodynamics

Have you ever rubbed your hands together to get warm? If so, you've experienced firsthand the connection between mechanical work and thermal energy. The first law of thermodynamics—which is the topic of this lesson—explains this experience. The law is a statement of energy conservation that specifically includes both mechanical work and thermal energy.

For example, suppose you push a notebook across a table to a friend at the other end. You do mechanical work on the notebook to give it an initial kinetic energy. As the notebook slides, it loses kinetic energy and eventually comes to rest. Does the initial energy of the notebook no longer exist? Does this system violate energy conservation?

Not at all. Friction converts the notebook's initial kinetic energy to a slight warming of the notebook and the tabletop. If you could measure the thermal energy that caused this warming, you would find that it *exactly* equals the initial kinetic energy of the notebook. Energy is conserved in this system—as it is in all systems—you just have to take thermal energy into account.

The thermal energy of a system is affected by heat

Recall from Chapter 10 that the thermal energy of a system is the sum of all its kinetic and potential energy. If energy is added to or taken away from a system in the form of heat or work, the thermal energy, E, of the system changes—just as you would expect. Let's consider the effects of heat first.

Vocabulary

- heat engine
- thermal reservoir
- efficiency

CONNECTINGIDEAS

In previous chapters you learned about temperature, thermal energy, heat, and work.

- All of these concepts play important roles throughout this chapter on thermodynamics.

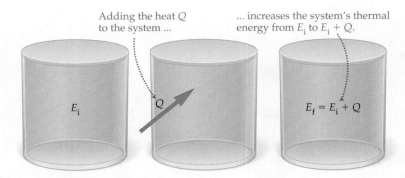

Adding the heat Q to the system ...

... increases the system's thermal energy from E_i to $E_i + Q$.

E_i

Q

$E_f = E_i + Q$

▲ Figure 11.1 **The thermal energy of a system**
A system initially has the thermal energy E_i (left). After the heat Q is added, the system's new thermal energy is $E_f = E_i + Q$ (right).

Consider the cylinder of gas shown in **Figure 11.1**. We'll call the initial value of the system's thermal energy E_i:

$$E_i = \text{initial thermal energy}$$

If the heat Q is added to the system, the thermal energy increases to the final value, E_f:

$$E_f = \text{final thermal energy}$$
$$= \text{initial thermal energy} + \text{heat}$$
$$= E_i + Q$$

Thus, the change in thermal energy is equal to the heat Q:

$$\Delta E = E_f - E_i = Q$$

To summarize:

- Adding heat to a system increases its thermal energy.
- Removing heat from a system decreases its thermal energy.

We can represent these observations by giving Q a *positive* value when the system *gains* heat and a *negative* value when it *loses* heat.

- Q is positive when heat is *added to* a system.
- Q is negative when heat is *removed from* a system.

This is consistent with our discussion in Chapter 10.

The thermal energy of a system is affected by work

As mentioned above, the thermal energy of a system can be changed by heat and by work. We've seen the effects of heat, so let's turn our attention to work.

Consider the system shown in **Figure 11.2**. Here we have a cylinder of gas with a moveable piston. The cylinder is insulated so that no heat can be added to or removed from the system. We start out by holding the piston down near the bottom of the cylinder. We then release it and let the gas expand upward. The gas does work as it lifts the piston and the weight sitting on the piston.

Let's suppose that the initial thermal energy is E_i:

$$E_i = \text{initial thermal energy}$$

As the gas forces the piston upward, it does the work W. The energy to do that work had to come from somewhere. The only place it can come from is

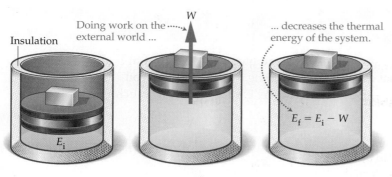

Insulation

Doing work on the
external world ...

W

... decreases the thermal
energy of the system.

$E_f = E_i - W$

E_i

▲ **Figure 11.2 Work and thermal energy**
A system initially has the thermal energy E_i (left). After the system does the work W on the external world, its remaining thermal energy is $E_f = E_i - W$ (right). The insulation guarantees that no thermal energy is transferred into or out of the system in the form of heat.

the thermal energy of the system. Therefore, the final thermal energy of the system is reduced by the amount W:

$$E_f = \text{final thermal energy}$$
$$= \text{initial thermal energy} - \text{work}$$
$$= E_i - W$$

This means that the change in thermal energy is the negative of the work done, $-W$:

$$\Delta E = E_f - E_i = -W$$

On the other hand, if work is done *on* the system, the system's thermal energy increases. To summarize:

- Thermal energy *decreases* when work is *done by* a system.
- Thermal energy *increases* when work is *done on* a system.

We can represent these observations by giving W a *positive* value when a system does work and a *negative* value when work is done on a system.

- W is positive when work is *done by* a system.
- W is negative when work is *done on* a system.

The sign conventions for heat and work are summarized in **Table 11.1**.

The first law of thermodynamics is a conservation law

Now that we've seen how heat and work affect the thermal energy of a system, let's combine those results into one statement. 🔲 **The change in a system's thermal energy is equal to the heat added to the system minus the work done by the system.**

First Law of Thermodynamics

change in thermal energy = heat − work
$$\Delta E = Q - W$$

Applying our sign conventions shows that adding heat to a system $(+Q)$ increases its thermal energy. Similarly, the work done by a system $(+W)$ decreases the system's thermal energy. In general, heat and work are two different means of transferring energy from one system to another.

Table 11.1 Signs of Q and W

Q positive	System *gains* heat
Q negative	System *loses* heat
W positive	Work done *by* system
W negative	Work done *on* system

🔲 *How is the first law of thermodynamics related to energy conservation?*

Jogging along the beach one day, you do 4.3×10^5 J of work and give off 3.8×10^5 J of heat. (**a**) What is the change in your thermal energy? (**b**) Switching over to walking, you give off 1.2×10^5 J of heat and your thermal energy decreases by 2.6×10^5 J. How much work have you done while walking?

Picture the Problem

Our sketch shows a person jogging along the beach. The fact that the person does work on the external world means that W is positive. As for the heat, the fact that heat is given off by the person means that Q is negative.

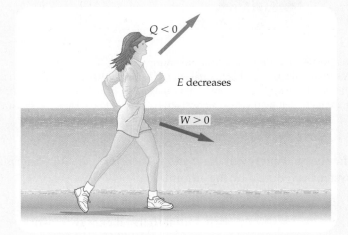

Strategy

The signs of W and Q have been determined in our sketch, and their magnitudes are given in the problem statement. To find the change in thermal energy for part (a), we simply use the first law of thermodynamics, $\Delta E = Q - W$. To find the work done for part (b), we rearrange the first law to solve for W, which gives $W = Q - \Delta E$.

Known	Unknown
(**a**) $W = 4.3 \times 10^5$ J	(**a**) $\Delta E = $?
$Q = -3.8 \times 10^5$ J	(**b**) $W = $?
(**b**) $Q = -1.2 \times 10^5$ J	
$\Delta E = -2.6 \times 10^5$ J	

Solution

Part (a)

1 Calculate ΔE, using $Q = -3.8 \times 10^5$ J and $W = 4.3 \times 10^5$ J:

$$\Delta E = Q - W$$
$$= (-3.8 \times 10^5 \text{ J}) - 4.3 \times 10^5 \text{ J}$$
$$= \boxed{-8.1 \times 10^5 \text{ J}}$$

Math HELP
Negative Numbers
See Math Review, Section I

Part (b)

2 Solve $\Delta E = Q - W$ for W:

$$W = Q - \Delta E$$

3 Substitute $Q = -1.2 \times 10^5$ J and $\Delta E = -2.6 \times 10^5$ J:

$$W = -1.2 \times 10^5 \text{ J} - (-2.6 \times 10^5 \text{ J})$$
$$= \boxed{1.4 \times 10^5 \text{ J}}$$

Insight

Notice the importance of using the correct signs for Q, W, and ΔE. When the proper signs are used, the first law is simply a way of keeping track of all the energy exchanges that occur in a system.

Practice Problems

1. **Follow-up** After walking for a few minutes, you begin to run, doing 5.1×10^5 J of work and decreasing your thermal energy by 8.8×10^5 J. How much heat did you give off while running?

2. A swimmer does 4.3×10^5 J of work and gives off 1.7×10^5 J of heat during a workout. Determine ΔE, W, and Q for the swimmer.

3. Give the change in thermal energy of a system if (**a**) $W = 50$ J, $Q = 50$ J; (**b**) $W = -50$ J, $Q = -50$ J; or (**c**) $W = 50$ J, $Q = -50$ J.

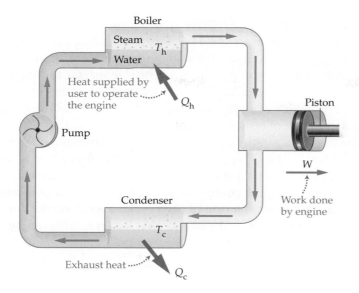

◀ **Figure 11.3 A schematic diagram of a steam engine**
The basic elements of a steam engine are a boiler—where thermal energy converts water to steam—and a piston that can be moved by the expanding steam. In some engines, the steam is simply exhausted into the atmosphere after it has expanded against the piston. More sophisticated engines send the exhaust steam to a condenser, where it is cooled and condensed back to liquid water, then recycled to the boiler.

Heat engines obey the first law of thermodynamics

If you've ever seen a puffing steam locomotive pull away from a train station, you've seen thermal energy converted to mechanical energy. In general, a **heat engine** is a device that converts thermal energy into work. Steam engines are the classic example, but heat engines are more common than you might think. Gas engines in cars turn thermal energy from burning fuel into forward motion. Even living creatures are heat engines. For example, you consume food and "burn" it during digestion. You then use that thermal energy to operate your body and mind.

Heat Engine Operation The basic elements of a steam engine are illustrated in **Figure 11.3**. First, some form of fuel (oil, wood, coal, etc.) is used to vaporize liquid water in the boiler. The resulting steam enters the engine, where it expands against a piston and does work. As the piston moves, it causes gears or wheels to rotate. After leaving the engine, the steam proceeds to a condenser. Here it gives off thermal energy to the cool air in the atmosphere and condenses to liquid water again. Two real-world examples of heat engines are shown in **Figure 11.4**.

CONNECTINGIDEAS

Here you learn that a heat engine converts thermal energy into mechanical work.
• Later on, in Chapter 23, you will learn how the mechanical work produced by a heat engine can be converted into electrical energy by a generator.

(a)

(b)

◀ **Figure 11.4 Heat engines**
(a) A modern version of Hero's engine, invented by the Greek mathematician and engineer Hero of Alexandria. In this simple heat engine the steam that escapes from a heated vessel of water is directed tangentially, causing the vessel to rotate. This converts the thermal energy supplied to the water into mechanical energy, in the form of rotational motion. (b) A steam engine of slightly more recent design hauls passengers up and down Mt. Washington in New Hampshire. Notice in the photo that the locomotive is belching two clouds, one black and one white. Can you explain their origins and the difference between them?

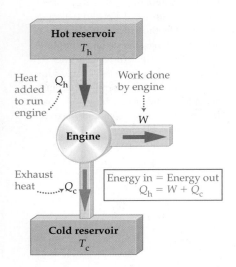

▲ **Figure 11.5 A schematic heat engine**
The engine absorbs the heat Q_h from the hot reservoir, performs the work W, and gives off the heat Q_c to the cold reservoir. Thus, the energy going into the engine is Q_h, and the energy going out of the engine is $W + Q_c$. Energy conservation requires that energy in equals energy out, and therefore $Q_h = W + Q_c$.

Representing a Heat Engine We represent heat engines schematically as shown in **Figure 11.5**. First, notice that a certain amount of heat is supplied to the engine from a high-temperature, or <u>hot</u>, *thermal reservoir*. This heat is denoted as Q_h. A **thermal reservoir** (or *reservoir*, for short) is an object, like a large body of water, that supplies or receives thermal energy with essentially no change in temperature. We can think of a reservoir as a source of heat at constant temperature.

Of the heat going to the engine in Figure 11.5 from the hot reservoir, a fraction appears as work, W. The rest is given off as waste heat at a relatively low temperature to a <u>c</u>old reservoir. This waste heat is denoted as Q_c.

There is no change in energy for a heat engine, because it returns to its initial state at the completion of each cycle. Therefore, the energy that enters the engine is equal to the energy that leaves it. This can be stated as follows:

$$\text{Energy in} = \text{Energy out}$$
$$Q_h = W + Q_c$$

Solving for the amount of work done by the engine, we find

$$W = Q_h - Q_c$$

This equation expresses energy conservation for the heat engine.

A heat engine is characterized by its efficiency

Of particular interest for any heat engine is how effectively it converts thermal energy to work. Fuel efficiency, for example, is an important consideration when buying a car. In general, the **efficiency**, e, of a heat engine is the fraction of the heat supplied to the engine that is converted to work. You can calculate the efficiency of a heat engine as follows:

Efficiency of a Heat Engine, e

$$\text{efficiency} = \frac{\text{work done by heat engine}}{\text{heat supplied by hot reservoir}}$$

$$e = \frac{W}{Q_h}$$

SI unit: dimensionless

Recalling that work is equal to the heat input to the engine minus the waste heat, $W = Q_h - Q_c$, we can also write the efficiency as

$$e = \frac{Q_h - Q_c}{Q_h}$$

$$= 1 - \frac{Q_c}{Q_h}$$

If the efficiency of a heat engine is $e = 0.20$, we say that the engine is 20% efficient. In this case, 20% of the input heat is converted to work, and 80% goes to waste heat that is exhausted into the air. Efficiency, then, can be thought of as the ratio of how much you receive (work) to how much you have to pay to run the engine (input heat).

A heat engine with an efficiency of 24.0% performs 1250 J of work. Find (**a**) the heat absorbed from the hot reservoir and (**b**) the heat given off to the cold reservoir.

Picture the Problem

Our sketch shows a schematic of the heat engine. We know the amount of work that is done and the efficiency of the engine. We want to find the heats Q_h and Q_c. Notice that an efficiency of 24.0% means that $e = 0.240$.

Strategy

(**a**) We can find the heat absorbed from the hot reservoir directly from the definition of efficiency, $e = W/Q_h$.
(**b**) We can find Q_c by rearranging the energy conservation equation, $W = Q_h - Q_c$, and substituting the values for Q_h and W.

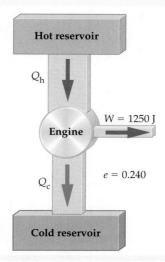

Known	Unknown
$e = 0.240$	(**a**) $Q_h = ?$
$W = 1250$ J	(**b**) $Q_c = ?$

Solution

1 (**a**) Rearrange $e = W/Q_h$ to solve for Q_h:

$$Q_h = \frac{W}{e}$$

$$= \frac{1250 \text{ J}}{0.240}$$

$$= \boxed{5210 \text{ J}}$$

2 (**b**) Rearrange the equation for energy conservation, $W = Q_h - Q_c$, to solve for Q_c:

$$Q_c = Q_h - W$$

$$= 5210 \text{ J} - 1250 \text{ J}$$

$$= \boxed{3960 \text{ J}}$$

Insight

Notice that when the efficiency of a heat engine is less than one-half (50%), as in this case, the amount of heat given off as waste to the cold reservoir is more than the amount of heat converted to work.

Practice Problems

4. |Follow-up| What is the efficiency of a heat engine that does 1250 J of work and gives off 5250 J of heat to the cold reservoir?

5. What is the efficiency of an engine that exhausts 870 J of heat in the process of doing 340 J of work?

6. An engine receives 690 J of heat from a hot reservoir and gives off 430 J of heat to a cold reservoir. What are (**a**) the work done by the engine and (**b**) the efficiency of the engine?

Engine A has an efficiency of 66%. Engine B absorbs the same amount of heat from the hot reservoir as engine A, but exhausts twice as much heat to the cold reservoir. Is the efficiency of engine B greater than, less than, or the same as the efficiency of engine A?

Reasoning and Discussion
Engine B takes in the same amount of heat as engine A. Part of this heat goes into work, and part is exhausted to the cold reservoir. Since engine B exhausts more heat to the cold reservoir than engine A, it has less energy left over to appear as work. Therefore, engine B produces less work for the same input of heat. As a result, engine B has the lower efficiency.

Answer
The efficiency of engine B is less than the efficiency of engine A.

11.1 LessonCheck (MP)

Checking Concepts

7. Explain How does the first law of thermodynamics extend the principle of conservation of energy?

8. Describe How does the energy of a heat engine change as it goes through one cycle?

9. Assess Which of the following indicates that the efficiency of a heat engine has increased? Explain your reasoning.

Case A: Adding more heat to the engine produces the same amount of work.

Case B: Adding the same amount of heat to the engine produces more work.

10. Big Idea If an object slides across a floor and comes to rest, what has become of its kinetic energy? Is this a violation of energy conservation?

Solving Problems

11. Rank Determine the thermal energy changes of systems A through D, described below, and then rank them in order of increasing change in thermal energy. Indicate ties where appropriate.

System	A	B	C	D
W (J)	10	−10	30	−20
Q (J)	20	−20	−50	−10

12. Rank Determine the efficiencies of the engines A through D, described below, and then rank them in order of increasing efficiency. Indicate ties where appropriate.

Engine	A	B	C	D
Q_h (J)	40	140	80	240
Q_c (J)	20	120	40	220

13. Calculate A system's thermal energy decreases by 20 J while the system performs 10 J of work. How much heat was added to the system?

14. Calculate A gas does 100 J of work as it expands. How much heat must be added to this gas for its thermal energy to decrease by 40 J?

15. Calculate A horse pulling a sled does 6.7×10^5 J of work and gives off 4.1×10^5 J of heat. Determine ΔE, W, and Q for the horse.

16. Calculate A heat engine takes in 1220 J of heat from the hot reservoir and exhausts 680 J of heat to the cold reservoir.
(a) How much work is done by the engine?
(b) What is the efficiency of the heat engine?

11.2 Thermal Processes

The thermal energy of a system can be changed in a number of ways. In general, a process that changes a system's thermal energy is referred to as a *thermal process*. In this lesson you'll learn about the most common and important types of thermal processes.

Constant-volume processes do no work

Suppose you pick up a can of hair spray and cup it in your hands. Thermal energy flows from your warm hands to the contents of the can. As a result, both the pressure and the temperature of the gas in the can increase. The volume of the can remains the same, however. A thermal process in which the volume of a system remains constant is referred to as a *constant-volume process*.

For example, suppose thermal energy is added to a gas in a container of fixed volume, as illustrated in **Figure 11.6**. Since the volume of the container doesn't change, there is no displacement of any of its walls. We know that work is force times displacement, so it follows that the gas does no work. This is true for *any* constant-volume process.

If we set the work equal to zero ($W = 0$) in the first law of thermodynamics, $\Delta E = Q - W$, we find

$$\Delta E = Q \quad \text{(for a constant-volume process)}$$

A constant-volume process does zero work, and the change in thermal energy is equal to the heat.

The volume changes in constant-pressure processes

If you've ever seen a large hot-air balloon being inflated, then you know that the gas in the balloon expands as it is heated. The balloon is surrounded by the atmosphere as it is heated, and the atmosphere is at a constant pressure. Therefore, the heating and expansion of the balloon occur with no change in pressure. Any process that occurs while pressure remains steady is referred to as a *constant-pressure process*.

Vocabulary

- isothermal
- adiabatic

What are the key characteristics of thermal processes?

◀ **Figure 11.6 Adding thermal energy to a system of constant volume**
Adding thermal energy to a system of constant volume increases the pressure. Because there is no displacement of the walls, however, no work is done in this process.

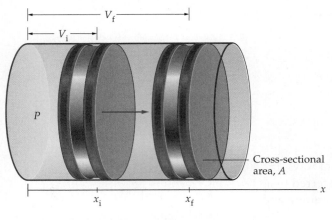

Figure 11.7 Work done by an expanding gas
A gas in a cylinder of cross-sectional area A expands with a constant pressure P from an initial volume $V_i = Ax_i$ to a final volume $V_f = Ax_f$. As it expands, it does the work $W = P(V_f - V_i)$.

As an example, consider a gas that has the pressure P and is held in a cylinder of cross-sectional area A, as shown in **Figure 11.7**. The gas is contained in the cylinder by a piston, and the piston end of the cylinder is open to the atmosphere. This ensures that the gas in the cylinder is always at a constant pressure.

If the gas in the cylinder is heated, it expands just like the gas in a hot-air balloon. As it expands, the gas does work on the piston. Recall that the force exerted by a gas is its *pressure* times the *area*. Since the pressure of the gas is P and the piston has the area A, the force the gas exerts on the piston is

$$F = PA$$

Next, recall that work is force times distance. In this case the gas moves the piston from the initial position x_i to the final position x_f. Therefore, the work done by the gas is

$$W = F(x_f - x_i) = PA(x_f - x_i) = P(Ax_f - Ax_i)$$

The initial volume of the gas is $V_i = Ax_i$, and its final volume is $V_f = Ax_f$, as you can see in Figure 11.7. Therefore, the work done by the gas is

$$W = P(V_f - V_i) = P\Delta V \qquad \text{(for a constant-pressure process)}$$

🔑 **The work done in a constant-pressure process equals the pressure times the change in volume.**

> ### QUICK Example 11.4 How Much Work?
>
> A gas with a constant pressure of 150 kPa expands from a volume of $0.76\ \text{m}^3$ to a volume of $0.92\ \text{m}^3$. How much work does the gas do?
>
> **Solution**
> The process is at constant pressure. Thus, the work done is the pressure times the change in volume. This gives
>
> $$\begin{aligned} W &= P\,\Delta V \\ &= (150\ \text{kPa})(0.92\ \text{m}^3 - 0.76\ \text{m}^3) \\ &= \boxed{24{,}000\ \text{J}} \end{aligned}$$
>
> This much energy could light a 100-W lightbulb for 4 minutes.

Practice Problems

17. As a gas expands at constant pressure from a volume of $0.74\ \text{m}^3$ to a volume of $2.3\ \text{m}^3$, it does 93 J of work. What is the pressure of the gas during this process?

18. A gas with a constant pressure of 270 kPa does 36,000 J of work as it expands. What was the change in volume of the gas?

Work can be calculated from a pressure-volume graph

The constant-pressure process illustrated in Figure 11.7 is represented graphically in **Figure 11.8**. A graph like this, of pressure versus volume, is called a *pressure-volume graph*, or *PV graph*. The process of expansion from the volume V_i to the volume V_f at the constant pressure P is indicated by the horizontal line in the PV graph.

Notice that the shaded region below the horizontal line is a rectangle of height P and width $\Delta V = V_f - V_i$. Therefore, the area of the rectangle is equal to the work. Though this result was obtained for the special case of constant pressure, it applies to any process.

$$\text{work} = \text{area under curve in a } PV \text{ graph} \quad \text{(for an expanding gas)}$$

🔑 **The work done by an expanding gas is equal to the corresponding area under the curve in a pressure-volume graph.** This result is applied in the next Guided Example.

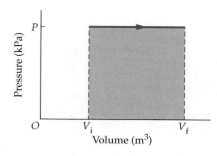

▲ **Figure 11.8 Graph of a constant-pressure process**
The area of the shaded region, $P(V_f - V_i)$, is equal to the work done by the expanding gas.

GUIDED Example 11.5 | Work Area   Pressure-Volume Graph

A gas expands from an initial volume of 0.40 m^3 to a final volume of 0.62 m^3 as the pressure increases linearly from 110 kPa to 230 kPa. Find the work done by the gas.

Picture the Problem

The sketch shows the process for this problem. As the volume increases from $V_i = 0.40 \text{ m}^3$ to $V_f = 0.62 \text{ m}^3$, the pressure increases linearly from $P_i = 110 \text{ kPa}$ to $P_f = 230 \text{ kPa}$.

Strategy

The work done by this gas is equal to the shaded area in the sketch. We can calculate this area as the sum of the area of a rectangle and the area of a triangle, as indicated in the sketch.

In particular, the rectangle has the height $P_i = 110 \text{ kPa}$ and the width $\Delta V = (V_f - V_i)$. Similarly, the triangle has the height $(P_f - P_i)$ and the base $\Delta V = (V_f - V_i)$.

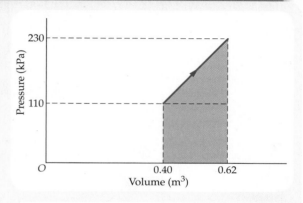

Known		Unknown
$P_i = 110 \text{ kPa}$	$P_f = 230 \text{ kPa}$	$W = ?$
$V_i = 0.40 \text{ m}^3$	$V_f = 0.62 \text{ m}^3$	

Solution

1 Calculate the area of the rectangular portion of the shaded area:

$$\begin{aligned} A_{\text{rectangle}} &= P_i(V_f - V_i) \\ &= (110 \text{ kPa})(0.62 \text{ m}^3 - 0.40 \text{ m}^3) \\ &= 2.4 \times 10^4 \text{ J} \end{aligned}$$

2 Next, calculate the area of the triangular portion of the shaded area. Recall that the area of a triangle is one-half the base times the height:

$$\begin{aligned} A_{\text{triangle}} &= \tfrac{1}{2}(P_f - P_i)(V_f - V_i) \\ &= \tfrac{1}{2}(230 \text{ kPa} - 110 \text{ kPa}) \\ &\quad (0.62 \text{ m}^3 - 0.40 \text{ m}^3) \\ &= 1.3 \times 10^4 \text{ J} \end{aligned}$$

3 Add the areas found in Steps 1 and 2 to find the work done by the gas:

$$\begin{aligned} W &= A_{\text{rectangle}} + A_{\text{triangle}} \\ &= 2.4 \times 10^4 \text{ J} + 1.3 \times 10^4 \text{ J} \\ &= \boxed{3.7 \times 10^4 \text{ J}} \end{aligned}$$

> **Math HELP**
> Calculating Area
> See Math Review, Section V

Insight

We could also have solved this problem by noting that the average pressure for the process is $P_{\text{av}} = \tfrac{1}{2}(P_f + P_i) = 170 \text{ kPa}$. Multiplying the average pressure by the volume change gives the work: $W = P_{\text{av}} \Delta V = (170 \text{ kPa})(0.22 \text{ m}^3) = 3.7 \times 10^4 \text{ J}$.

19. **Follow-up** Suppose the pressure in Guided Example 11.5 varies linearly from 110 kPa to 260 kPa. How much work does the gas do in this case?

20. **Rank** Figure 11.9 shows three different multistep processes, labeled A, B, and C. Rank these processes in order of increasing work done by a gas that undergoes the process. Indicate ties where appropriate.

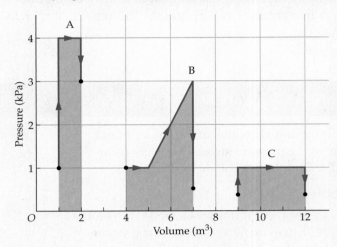

▲ Figure 11.9

Isothermal processes have a constant temperature

Figure 11.10 shows a weatherglass, a useful device for determining atmospheric pressure and predicting changes in the weather. The pocket of air inside the weatherglass presses on one surface of the colored water, and atmospheric pressure presses on the other surface. If the two pressures are equal, the water levels are equal as well. If the atmospheric pressure drops—with the advance of a storm, let's say—the air in the weatherglass expands and changes the water levels. The expansion of the air in the weatherglass occurs at the constant temperature in the room. A process that occurs at a constant temperature is referred to as an **isothermal** process.

In general, constant temperature means constant thermal energy. Thus, we can say that there is no change in the thermal energy of a system in an isothermal process. Setting $\Delta E = 0$ in the first law of thermodynamics, $\Delta E = Q - W$, yields

$$Q = W \qquad \text{(for an isothermal process)}$$

🔑 **In general, the work done by an isothermal process is equal to the heat.**

In the case of the weatherglass, the air in the glass draws heat from the room as it expands. It takes in just enough heat to keep its temperature constant, and that heat is exactly equal to the work the air does in pushing down on the water.

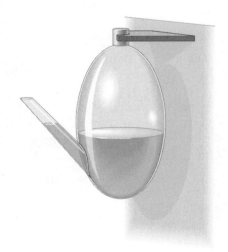

◀ **Figure 11.10 A weatherglass**
A weatherglass indicates changes in atmospheric pressure. The expansion or contraction of the air inside the weatherglass is a constant-temperature (isothermal) process.

Adiabatic processes do not exchange thermal energy

We've saved the most interesting thermal process—one with no transfer of heat—for last. A thermal process with no exchange of thermal energy ($Q = 0$) is referred to as an **adiabatic** process. A process is typically adiabatic when the system is well insulated, to keep heat from flowing. A process might also be adiabatic because it occurs so rapidly that there is no time for heat to flow.

You can experience an adiabatic process right now, as you read this book. Open your mouth wide and exhale. Hold your hand in the outgoing air. It's warm, isn't it? This is to be expected, since the air is coming from your warm body. Now purse your lips, and exhale again. This time the very same air feels cool. Why? The air cools as it leaves your mouth because it has expanded rapidly. This is an example of adiabatic cooling.

A similar example is the adiabatic cooling that occurs when air is let out of a bicycle tire. The escaping air does work on the atmosphere as it expands, just like the air exhaled through pursed lips. This produces a cooling effect that can be quite noticeable. In extreme cases the cooling is great enough to create frost on the valve stem of the tire.

Physics & You: Technology Adiabatic processes can produce heating as well as cooling. An example of adiabatic heating occurs when you pump air into a bicycle tire. In this case the work done on the pump appears as an increased temperature. An extreme example of adiabatic heating is shown in **Figure 11.11**. Here, a piston is pushed downward rapidly in a cylinder that contains bits of paper in the bottom. The temperature rises dramatically, and the paper bursts into flames. This is how gasoline is ignited in a *diesel engine*—no spark plugs are needed. The pistons in a diesel engine compress an air-fuel mixture so rapidly that it ignites explosively, which powers the engine.

Reading Support ✓

Word Origin of **adiabatic**

(adjective) occurring without loss or gain of heat

from the Greek *adiabatos*, "not to be crossed"; from *a-* ("not") + *diabatos* ("passable"); from *dia-* ("across") + *bainein* ("to go")

▲ **Figure 11.11 Adiabatic heating**
Pushing down rapidly on this plunger is an adiabatic process, because there is no time for thermal energy to flow into or out of the system. Therefore, the work done on the plunger appears as an increased thermal energy in the system, and the resulting increase in temperature is enough to ignite bits of paper at the bottom of the tube.

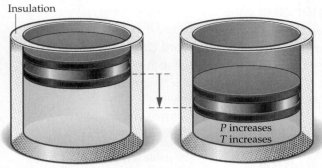

(a) Compressing with no flow of thermal energy increases P and T.

(b) Expanding with no flow of thermal energy decreases P and T.

▲ **Figure 11.12 An adiabatic process**
In an adiabatic process no thermal energy flows into or out of the system. In the cases shown in this figure, the flow of thermal energy is prevented by insulation. **(a)** An adiabatic compression increases both the pressure and the temperature. **(b)** An adiabatic expansion results in a decrease in pressure and temperature.

Calculating Energy Change How do you calculate the change in thermal energy for an adiabatic process? It's really quite simple. **Figure 11.12 (a)** shows a cylinder that is insulated so that no thermal energy can go into or out of the system. When the piston is pushed downward in the cylinder—decreasing the volume—the gas heats up and its pressure increases. Similarly, in **Figure 11.12 (b)** an adiabatic expansion causes the temperature and pressure of the gas to decrease. Because no thermal energy flows through the insulated walls, we conclude that $Q = 0$. From the first law of thermodynamics, $\Delta E = Q - W$, it follows that

$$\Delta E = -W \qquad \text{(for an adiabatic process)}$$

🔑 **If work is done on a system ($W < 0$) in an adiabatic process, the system's thermal energy increases ($\Delta E > 0$). Similarly, if a system does work ($W > 0$) in an adiabatic process, its thermal energy decreases ($\Delta E < 0$).**

GUIDED Example 11.6 | Work into Energy   **Adiabatic Process**

A gas is compressed adiabatically, and the amount of work done on it is 640 J. Find the change in thermal energy of the gas.

Picture the Problem

Our sketch shows a piston being pushed downward, compressing a gas in an insulated cylinder. The insulation ensures that no thermal energy can flow—as required in an adiabatic process.

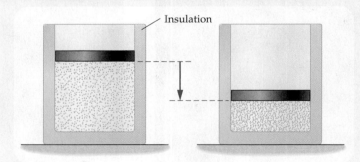

Strategy

We know that 640 J of work is done on the gas and that no thermal energy is exchanged ($Q = 0$) in an adiabatic process. Thus, we can find ΔE by substituting Q and W into the first law of thermodynamics, $\Delta E = Q - W$. One note of caution: Be careful to use the correct sign for the work. In particular, recall that work done *on* a system is negative.

Known

$W = -640\ \text{J}$ (work done on system is negative)

$Q = 0$

Unknown

$\Delta E = ?$

Solution

1 Identify the work and heat for this process:

$W = -640 \text{ J}$

$Q = 0$

2 Substitute Q and W into the first law of thermodynamics to find the change in internal energy, ΔE:

$\Delta E = Q - W$

$\quad = 0 - (-640 \text{ J})$

$\quad = \boxed{640 \text{ J}}$

> **Math HELP**
> Negative Numbers
> See Math Review, Section I

Insight

Because no energy can enter or leave the system in the form of heat, all the work done on the system goes into increasing its thermal energy. As a result, the temperature of the gas increases. This is adiabatic heating.

Practice Problems

21. **Follow-up** If a system's thermal energy decreases by 470 J in an adiabatic process, how much work was done by the system?

22. A gas expands adiabatically and does 520 J of work. What is the change in thermal energy of the gas?

Adiabatic processes affect the weather

Adiabatic heating and cooling can have important effects on the climate of a region. For example, moisture-laden winds blowing from the Pacific Ocean into western Oregon are deflected upward when they encounter the Cascade Mountains. As the air moves up the mountain slopes, the atmospheric pressure decreases, which allows the air to expand and undergo adiabatic cooling. The result is that the moisture in the air condenses to form clouds and precipitation falls on the western side of the mountains.

When the winds reach the eastern side of the mountains, they have little moisture remaining. Thus, eastern Oregon is in the *rain shadow* of the Cascade Mountains. In addition, as the moving air descends on the eastern side of the mountains, it undergoes adiabatic heating. These are the primary reasons why the summers in western Oregon are moist and mild, while the summers in eastern Oregon are hot and dry.

The key characteristics of constant-pressure, constant-volume, isothermal, and adiabatic processes are summarized in **Table 11.2**.

CⵘOL PHYSICS
Lenticular Clouds

The unusual lens-shaped cloud shrouding the summit of Oregon's Mt. Hood is called a *lenticular cloud*. Lenticular clouds, or "lennies" as they are called by cloud enthusiasts, are some of the most interesting and beautiful clouds in nature. They are formed by adiabatic cooling. As moisture-laden wind is deflected upward by a mountain, the moisture in the moving air cools due to adiabatic expansion and condenses to form a cloud. Some lennies look like flying saucers. In fact, the first report of a UFO was from a pilot flying near Mt. Rainier in 1947. Was he seeing a lennie instead?

Table 11.2 Thermodynamic Processes and Their Characteristics

Constant pressure	$W = P\Delta V$	$Q = \Delta E + P\Delta V$
Constant volume	$W = 0$	$Q = \Delta E$
Isothermal (constant temperature)	$W = Q$	$\Delta E = 0$
Adiabatic (no heat)	$W = -\Delta E$	$Q = 0$

11.2 LessonCheck (MP)

Checking Concepts

23. 🔍 **Scrutinize** State whether each of the following statements is true or false. If a statement is false, revise it so that it becomes true.
(**a**) The change in thermal energy in a constant-volume process is zero.
(**b**) The work done in a constant-pressure process is zero.
(**c**) The area under the curve on a pressure-volume graph is equal to the work.
(**d**) Work and heat are equal in an isothermal process.
(**e**) The thermal energy of a system increases when work is done on it in an adiabatic process.

24. Assess The temperature of a system is held fixed. Is it possible for thermal energy to flow into the system? Give an explanation if your answer is no. If your answer is yes, give a specific example.

25. Relate How is the change in thermal energy related to the work in an adiabatic process?

26. [Triple Choice] A gas does 50 J of work as it expands adiabatically. Is the change in thermal energy of this gas 50 J, 0 J, or −50 J? Explain.

Solving Problems

27. Calculate How much heat must be added to a gas that does 10 J of work in a constant-temperature (isothermal) process?

28. Calculate A gas at constant volume has 25 J of heat added to it. What is the work done by the gas?

29. Calculate A gas at a constant pressure of 95 kPa does 110 J of work as it expands from the volume V to the volume $2V$. What is the volume V?

30. Calculate An ideal gas is compressed at a constant pressure of 120 kPa to one-half of its initial volume. The work done on the gas is 790 J. What was the initial volume of the gas?

11.3 The Second and Third Laws of Thermodynamics

Vocabulary

• entropy

🔑 *How is the second law of thermodynamics more restrictive than the first law?*

When you stop to think about it, events in nature proceed in a certain direction. For example, the dropped egg in **Figure 11.13** splatters on the floor, but it never reassembles itself and flies back into your hand. Similarly, a movie running backward is immediately recognized as unnatural. In this lesson we explore the physics behind the directionality of nature.

Thermal energy flows from hot to cold

Have you ever warmed your hands by pressing them against a block of ice? Probably not. But if you think about it, you might wonder why it doesn't work. After all, the first law of thermodynamics is satisfied if energy flows from the ice to your hands. The ice gets colder while your hands get warmer, and the energy of the universe remains the same.

As we know, however, this sort of thing just doesn't happen—thermal energy *always* flows from warmer objects to cooler objects. This simple observation, in fact, is one of many ways of expressing the *second law of thermodynamics*:

> ### Second Law of Thermodynamics
> When objects of different temperatures are brought into contact, the flow of thermal energy is always from the higher-temperature object to the lower-temperature object.

The second law of thermodynamics is more restrictive than the first law. **The second law says that of all the processes that conserve energy (and satisfy the first law) only those that proceed in a certain direction actually occur.** In a sense, the second law implies directionality for the behavior of nature. For this reason, it is sometimes referred to as the "arrow of time."

The second law is very simple, but it has profound effects on nature and technology. For example, it determines the efficiency of engines, as we will see.

▲ **Figure 11.13 Directionality in nature** Events in nature occur in a certain direction. For example, an egg drops to the ground and splatters. The reverse process never occurs, even though it would conserve energy. This is an example of the second law of thermodynamics in action.

Heat Engine Efficiency

If you design an engine, you want it to be as efficient as possible. You want to maximize the amount of work you get from the engine for a given input of energy. For example, if you design an efficient car engine like the one in **Figure 11.14**, you want it to get as many miles per gallon as possible.

Heat engines are limited by their Carnot efficiency

It turns out that the second law of thermodynamics sets a limit on just how efficient an engine can be. No matter how advanced technology may become, efficiency can never reach 100%. Even a frictionless engine is not completely efficient. This connection was first discovered by the French engineer Sadi Carnot (1796–1832).

Carnot Efficiency What is the maximum efficiency of an engine? To answer this question, let's consider a heat engine that operates between a hot reservoir at the Kelvin temperature T_h and a cold reservoir at the Kelvin temperature T_c. (Recall that a thermal reservoir supplies or receives heat without changing its temperature.) The hot reservoir might be a boiler in a steam engine, and the cold reservoir the air in the atmosphere. *Carnot's theorem* states that a heat engine operating between the Kelvin temperatures T_h and T_c has a maximum efficiency, e_{max}, given by the following:

> ### Carnot's Theorem of Maximum Efficiency
> $$\text{maximum efficiency} = 1 - \frac{\text{temperature of cold reservoir}}{\text{temperature of hot reservoir}}$$
> $$e_{max} = 1 - \frac{T_c}{T_h}$$

Carnot's theorem is remarkable for a number of reasons. First, it means that no engine, no matter how sophisticated or technologically advanced, can ever exceed the maximum efficiency, e_{max}. We can strive to improve the technology of heat engines, to improve the methods of manufacture, but there is an upper limit to engine efficiency.

▲ **Figure 11.14 An efficient car engine** Modern technology has produced engines that are highly efficient, like the car engine shown here. The maximum efficiency of an engine is limited by the second law of thermodynamics, however, and no amount of technological innovation can ever exceed the limit.

Second, the theorem is just as remarkable for what it does not say. It says nothing about the type of heat engine, what the engine is made of, or how it is constructed. All that *does* matter are the two temperatures, T_c and T_h.

Ideal Engines   Suppose for a moment that we *could* construct an ideal engine, perfectly free from all forms of friction. Surely an ideal engine like this would have 100% efficiency, wouldn't it? Well, no, it wouldn't. 🔑 **No matter how perfect the engine, some of the input heat is always wasted—given off as Q_c—rather than converted to work.**

Recall that the efficiency is defined to be $e = W/Q_h$. Solving this equation for maximum work, W_{max}, yields

$$W_{max} = e_{max}Q_h$$

Substituting from our earlier equation for e_{max} gives

$$W_{max} = \left(1 - \frac{T_c}{T_h}\right)Q_h$$

This is the maximum amount of work an ideal engine can produce.

Efficiency and Reservoir Temperature   The temperatures of the hot and cold thermal reservoirs dictate an engine's efficiency. An extreme case proves the point. What happens when the hot and cold reservoirs have the same temperature, that is, when $T_c = T_h$? In this case the maximum efficiency is zero. As a result, the amount of work that such an engine can do is also zero. Heat engines always require a temperature difference to operate.

CONCEPTUAL Example 11.7 Comparing Efficiencies

Suppose you have an ideal heat engine that can operate in one of two different modes. In mode 1 the temperatures of the two reservoirs are $T_c = 200$ K and $T_h = 400$ K. In mode 2 the temperatures are $T_c = 400$ K and $T_h = 600$ K. Is the efficiency of mode 1 greater than, less than, or equal to the efficiency of mode 2?

Reasoning and Discussion
At first, you might think that because the temperature difference is the same in the two modes, the efficiency is the same as well. This is not the case. In fact, efficiency depends on the *ratio* of the two temperatures rather than on their difference.

The efficiency of mode 1 is

$$e_1 = 1 - \frac{200}{400} = \frac{1}{2}$$

The efficiency of mode 2 is

$$e_2 = 1 - \frac{400}{600} = \frac{1}{3}$$

Thus, mode 1 is more efficient, even though it operates at lower temperatures.

Answer
The efficiency of mode 1 is greater than the efficiency of mode 2.

🔑 *What does Carnot's theorem say about the efficiency of an ideal engine?*

A heat engine is operating at its maximum efficiency of $e_{max} = 0.288$. The cold reservoir of the engine is at a temperature of 295 K. What is the temperature of the engine's hot reservoir?

Solution (*Perform the calculations indicated in each step.*)

1. Write the maximum efficiency, e_{max}, in terms of the hot and cold temperatures:

$$e_{max} = 1 - T_c/T_h$$

2. Solve for the hot temperature, T_h:

$$T_h = T_c/(1 - e_{max})$$

3. Substitute the known values of T_c and e_{max} and calculate T_h:

$$T_h = 414 \text{ K}$$

Math HELP
Solving Simple
Equations
**See Math
Review,
Section III**

Insight

Though an efficiency of 28.8% may seem low, it is characteristic of many real engines.

Entropy

The messy room in **Figure 11.15 (a)** is more disordered than the neat room in **Figure 11.15 (b)**. Similarly, a pile of bricks is more disordered than a building constructed from the bricks. A puddle of water is more disordered than the block of ice from which it melted. Though it might not seem like it, these examples are related to the second law of thermodynamics.

Entropy is a measure of the disorder in a system

In physics, the quantity called **entropy**, S, measures the amount of disorder in a system. Entropy, which is related to the second law of thermodynamics, is defined in terms of its change. In particular, the change in entropy, ΔS, is defined as the heat divided by the temperature:

(a)

Definition of Entropy Change, ΔS

change in entropy $= \dfrac{\text{heat}}{\text{temperature}}$

$$\Delta S = \frac{Q}{T}$$

SI units: J/K

(b)

For this definition to be valid, Kelvin temperatures must be used.

When thermal energy is added to a system ($Q > 0$), the entropy of the system increases. For example, heating water and boiling it until it is converted to steam increases the disorder of the water molecules. The molecules are flying randomly around the room rather than being confined to the pot.

Similarly, if thermal energy is removed from a system ($Q < 0$), the system's entropy, or disorder, decreases. An example is freezing water to form ice. The molecules in liquid water move all around the container, but in ice they are confined to definite positions within the orderly crystal structure.

▲ **Figure 11.15 A messy room versus a neat room**

(a) A messy room just seems to happen—almost like a law of nature. In fact, nature tends to move in the direction of increasing disorder (messiness). It's another aspect of the second law of thermodynamics.
(b) Keeping a room neat requires a considerable amount of work.

Calculate the change in entropy when a 0.125-kg chunk of ice melts at 0 °C.

Picture the Problem

In our sketch we show a 0.125-kg chunk of ice at the temperature 0 °C. As the ice absorbs the thermal energy Q from its surroundings, it melts to water at 0 °C. Because the system absorbs thermal energy, its entropy increases.

Strategy

The entropy change is $\Delta S = Q/T$, where $T = 0\,°C = 273$ K. To find the heat Q, we note that to melt the ice we must add to it the latent heat of fusion, L_f. Thus, the heat is $Q = mL_f$, where $L_f = 33.5 \times 10^4$ J/kg.

Known	Unknown
$m = 0.125$ kg	$\Delta S = ?$
$T = 0\,°C = 273$ K	

Solution

1 Find the heat needed to melt the ice:

$$Q = mL_f$$
$$= (0.125 \text{ kg})(33.5 \times 10^4 \text{ J/kg})$$
$$= 4.19 \times 10^4 \text{ J}$$

2 Calculate the corresponding change in entropy:

$$\Delta S = \frac{Q}{T}$$
$$= \frac{4.19 \times 10^4 \text{ J}}{273 \text{ K}}$$
$$= \boxed{153 \text{ J/K}}$$

> **Math HELP**
> Significant Figures
> See Lesson 1.4

Insight

Notice that we were careful to convert the temperature of the system from 0 °C to 273 K before we applied the equation $\Delta S = Q/T$. This must always be done when calculating an entropy change. If we had neglected to do the conversion in this case, we would have found an infinite increase in entropy—which is impossible.

Practice Problems

31. Follow-up Find the mass of ice that would have to melt to give an entropy change of 275 J/K.

32. Heat is added to a 0.14-kg block of ice at 0 °C, increasing its entropy by 98 J/K. How much ice melts?

◄ What is the direction of entropy change in the universe?

The entropy of the universe never decreases

One of the key features of entropy is that no process in nature can decrease the total amount of entropy in the universe. The best any process can do is to leave the entropy unchanged.

- The total entropy of the universe either *stays the same or increases* during any process.

◄ In terms of entropy the universe moves in only one direction—toward ever-increasing entropy. This is quite different from the behavior with regard to the universe's energy, which remains constant no matter what type of process occurs. This statement about entropy is equivalent to the second law of thermodynamics—it is simply a different way of saying the same thing.

When certain processes occur, it sometimes appears as if the entropy of the universe has decreased. On closer examination, however, it always turns out that a larger increase in entropy has occurred elsewhere, resulting in a net increase.

For example, suppose you put an ice-cube tray filled with water in the freezer, and some time later the water has turned to ice, as shown in **Figure 11.16**. It might seem that the entropy of the universe has decreased. After all, thermal energy is removed from the water to freeze it, and, as we know, removing thermal energy from an object lowers its entropy. On the other hand, we also know that the freezer does work to draw thermal energy from the water. In fact, the refrigerator exhausts more thermal energy into the air in the kitchen than it absorbs from the water. Detailed calculations always show that the entropy of the air that is heated in the kitchen increases by more than the entropy of the water that is frozen into ice cubes decreases. Therefore, the total entropy of the universe increases—as it must—for *any* real process.

▲ **Figure 11.16 In the freezer**
These ice cubes entered the freezer as water, but now they are solid ice. Though the entropy (disorder) of the water has decreased, the entropy of the air in the kitchen has increased. The net result is an increase in the entropy of the universe.

Absolute zero can never be reached

Just one law of thermodynamics remains to be considered. The *third law of thermodynamics* states that there is no temperature lower than absolute zero and that absolute zero is unattainable. It is possible to cool objects to temperatures extremely close to absolute zero—experiments have reached temperatures as low as 0.00000000045 K—but no object can ever be cooled to precisely 0 K.

As an analogy to cooling toward absolute zero, imagine walking toward a wall. Let's suppose that each step you take is half the distance between you and the wall. Even if you take an infinite number of steps, you never reach the wall. You can get as close as you like, of course, but you never get all the way there.

The same sort of thing happens when cooling. Suppose you have a collection of objects at 0 K to use for cooling. You put an object in contact with one of the 0 K objects and it cools, while the 0 K object warms up. You continue this process, throwing away each "warmed up" 0 K object and using a new one. Each time you cool your object, its temperature gets closer to 0 K but never gets all the way there. So, even if you could use 0 K objects for cooling (which isn't possible), you could still never cool another object to 0 K.

In light of this discussion, we can express the third law of thermodynamics as follows:

> **The Third Law of Thermodynamics**
>
> It is impossible to lower the temperature of an object all the way to absolute zero.

Thus, absolute zero is the limiting temperature. It can be approached very closely, but it can never be attained.

Thermodynamic laws limit the behavior of nature

We can summarize our conclusions regarding the laws of thermodynamics as follows:

- The first law of thermodynamics states that you can't get something for nothing.

To be specific, you can't get more work out of a heat engine than the amount of thermal energy you put in. The best you can do is break even. This is energy conservation.

The second law of thermodynamics is even more restrictive:

- The second law of thermodynamics states that you can't even break even—some of the input heat must be wasted.

It might not seem fair, but it's a law of nature. An engine can be only so efficient, no matter how much we improve its design.

For example, if an engine took energy from a hot reservoir and converted it completely to work, the entropy of the hot reservoir—and the universe—would decrease. But exhausting some of the energy to the cold reservoir—as waste heat—increases the entropy of the cold reservoir enough to offset the entropy decrease in the hot reservoir. This ensures that the total entropy of the universe either stays the same or increases. Thus, entropy plays a direct role in determining the efficiency of engines.

Finally, the third law restricts our ability to cool an object:

- The third law of thermodynamics states that no object can be cooled all the way to absolute zero.

Nature shows incredible variety and versatility. Still, its behavior is limited by these fundamental laws of thermodynamics.

11.3 LessonCheck (MP)

Checking Concepts

33. Identify Which law of thermodynamics says that thermal energy flows from hot objects to cold objects?

34. Explain What does Carnot's theorem say about the feasibility of a 100% efficient heat engine?

35. Decide The molecules in ice are in a more orderly and structured state than the molecules in liquid water. Does freezing water to form ice decrease the entropy of the universe?

36. Explain If the entropy of a system increases, what can you say about its randomness?

37. Decide Which has more entropy: (a) popcorn kernels or the resulting popcorn, (b) two eggs in a carton or an omelet made from the eggs, (c) a pile of bricks or a house made from them, (d) a piece of paper or the ash after the paper has been burned?

Solving Problems

38. Rank The reservoir temperatures for heat engines A through D are given below. Rank the engines in order of increasing efficiency. Indicate ties where appropriate.

Engine	A	B	C	D
T_h (K)	400	440	800	1240
T_c (K)	200	420	600	1020

39. Calculate A heat engine has a high-temperature reservoir at 330 K and a low-temperature reservoir at 260 K. What is the maximum efficiency of this engine?

40. Calculate A heat engine has a high-temperature reservoir at 410 K and operates at a maximum efficiency of 0.24. What is the temperature of this engine's low-temperature reservoir?

41. Calculate What is the efficiency of a heat engine that exhausts 870 J of heat in the process of doing 340 J of work?

42. Calculate An ideal heat engine operates between the temperatures 390 K and 240 K.
(a) How much heat must be given to the engine to produce 1200 J of work?
(b) How much heat is discarded to the cold reservoir as this work is done?

43. Calculate Determine the change in entropy that occurs when 3.1 kg of water freezes at 0 °C.

Cryogenics

What Is It? Cryogenics is the branch of physics that deals with the production of super-cold temperatures and the observation of their effects. The history of cryogenics describes the scientific journey down the road to absolute zero.

How Was the Field Established? The story begins with the liquefaction of gases. The adiabatic expansion of air can transfer enough thermal energy to cause the air to liquefy, producing a temperature of 78 K (at atmospheric pressure). The liquefied air is easily separated into liquid nitrogen (77 K) and liquid oxygen (50.5 K). Industry uses large quantities of these refrigerants, which are delivered by insulated tanker trucks. Liquid hydrogen (20.28 K) was first produced in 1898. In 1908 the Dutch physicist Heike Onnes produced an even colder substance, liquid helium at 4.2 K.

Three years later, in 1911, Onnes discovered *superconductivity* when he observed the electrical resistance of mercury drop to *zero* at temperatures below 4.19 K. Mercury and many other metals are *superconductors* at extremely low temperatures (see Chapter 21). Temperatures below 4.2 K can be produced using a vacuum pump to reduce the pressure exerted on liquid helium. The reduced pressure results in temperatures around 1 K.

How Has the Field Advanced? Over the past century the record low temperature has been reduced by a factor of 10 approximately every ten years. One method involves *adiabatic demagnetization*, in which magnetic dipoles of a super-cooled solid are first aligned by a strong external magnetic field. As the field is slowly turned off, energy is transferred out of the substance. The loss of energy reduces vibrations in the solid and cools it further. A newer method called *laser cooling* uses light beams to trap and slow down a bundle of several thousand atoms. The current record low temperature is about 100 picokelvins (pK), that is, about $1 \cdot 10^{-10}$ K.

Every time the record low temperature is lowered, the chance of observing a new low-energy phenomenon exists. In this sense, the road to absolute zero is an open-ended journey and not a dead end.

Interesting things happen at extremely cold temperatures. This rose, which was dipped in liquid nitrogen, shatters on impact.

Take It Further

1. Write *Several interesting phenomena occur at very low temperatures. Use the library and the Internet to research superfluidity, the quantum Hall effect, and Bose-Einstein condensation. Write a paragraph in your own words about one of these phenomena.*

Physics Lab The Mechanical Equivalent of Heat

This lab explores the mechanical equivalent of heat. You will raise the temperature of a metal sample by repeatedly subjecting it to energy conversions. Analysis of your data will allow you to approximate the quantitative relationship between heat and mechanical energy.

Materials
- 1-m-long cardboard or PVC tube
- two rubber stoppers
- 0.5 kg of copper shot
- plastic foam cup (calorimeter)
- thermometer

Procedure

Overview The copper shot is placed in a tube and sealed in with rubber stoppers. The shot is made to fall vertically from one end of the tube to the other a large number of times. This process converts potential energy to internal energy (of the shot). The temperature of the shot is measured before and after the process. Analysis of the data allows the mechanical equivalent of heat to be calculated.

1. Determine the distance the shot will fall from one end of the tube to the other, taking into account the penetration depth of the two stoppers. Record the end-to-end distance. Then calculate and record the total distance for 100 end-to-end drops.

2. Carefully pour the shot into the foam cup and measure the mass of the shot. (The mass of the foam cup is considered negligible.) Record the mass.

3. Carefully insert the thermometer into the shot. Record the temperature in the data table. Record your data in the row for your group number.

4. Firmly insert a stopper into one end of the tube and then add the shot. Seal the tube by firmly inserting the second stopper.

5. Invert the tube 100 times in rapid succession. Be sure to allow the shot fall through the entire length of the tube.

6. After completing Step 5, quickly (and carefully) pour the shot into the foam cup and measure the temperature. Record the temperature in the data table.

Data Table

Group	Initial Temperature of Shot (°C)	Final Temperature of Shot (°C)	Temperature Change of Shot (°C)	Change in Potential Energy Due to Drop (J)	Thermal Energy Gained by Shot (cal)	Mechanical Equivalent of Heat (J/cal)
1						

Analysis

Perform Steps 1–4 for each trial and record the results in the data table.

1. Use the equation $\Delta t = t_f - t_i$ to calculate the temperature change (Δt) of the shot.

2. Use the equation $\Delta PE = mgh$ to calculate the change in potential energy of the shot based on the total distance it drops through.

3. Use the equation $Q = mc\Delta t$ to calculate the thermal energy gained by the shot. The specific heat of copper is $c_{copper} = 0.092 \ cal/g°C$.

4. Calculate the mechanical equivalent of heat by dividing the change in potential energy (in joules) by the thermal energy (in calories) needed to bring about the same change in temperature.

5. Collect the experimental results from other groups and calculate the average value of the mechanical equivalent of heat. Record the average value.

Conclusions

1. What is the percent error in the average value of the mechanical equivalent of heat? The accepted value is 4.19 J/cal.

2. Identify the main sources of error in this experiment. What could be done to reduce these errors?

11 Study Guide

Big Idea

Energy conservation applies to thermal energy and heat.

Energy is always conserved. Friction and air resistance often convert the energy of motion to thermal energy, but the total amount of energy is still the same.

11.1 The First Law of Thermodynamics

🔑 The change in a system's thermal energy is equal to the heat added to the system minus the work done by the system.

🔑 There is no change in energy for a heat engine, because it returns to its initial state at the completion of each cycle. Therefore, the energy that enters the engine is equal to the energy that leaves it.

🔑 If the efficiency of a heat engine is $e = 0.20$, we say that the engine is 20% efficient. In this case, 20% of the input heat is converted to work, and 80% goes to waste heat that is exhausted into the air.

• The first law of thermodynamics is a statement of energy conservation that specifically includes both mechanical work and heat.

• A heat engine is a device that converts heat into work.

Key Equations

The first law of thermodynamics can be written as follows:

$$\Delta E = Q - W$$

The efficiency of a heat engine is

$$e = \frac{W}{Q_h} \quad \text{or} \quad e = 1 - \frac{Q_c}{Q_h}$$

11.2 Thermal Processes

🔑 A constant-volume process does zero work, and the change in thermal energy is equal to the heat.

🔑 The work done in a constant-pressure process equals the pressure times the change in volume.

🔑 The work done by an expanding gas is equal to the corresponding area under the curve in a pressure-volume graph.

🔑 In general, the work done by an isothermal process is equal to the heat.

🔑 If work is done on a system ($W < 0$) in an adiabatic process, the system's thermal energy increases ($\Delta E > 0$). Similarly, if a system does work ($W > 0$) in an adiabatic process, its thermal energy decreases ($\Delta E < 0$).

• A process that changes a system's thermal energy is a thermal process.

• A thermal process that occurs at constant temperature is an isothermal process.

• A thermal process in which there is no exchange of thermal energy ($Q = 0$) is an adiabatic process.

11.3 The Second and Third Laws of Thermodynamics

🔑 The second law says that of all the processes that conserve energy (and satisfy the first law) only those that proceed in a certain direction actually occur.

🔑 No matter how perfect the engine, some of the input heat is always wasted—given off as Q_c—rather than converted to work.

🔑 In terms of entropy the universe moves in only one direction—toward ever-increasing entropy.

• According to the second law of thermodynamics, the flow of thermal energy is always from the higher-temperature object to the lower-temperature object.

• Entropy is a measure of the disorder of a system. As entropy increases, a system becomes more disordered.

• The third law of thermodynamics states that there is no temperature lower than absolute zero and that absolute zero is unattainable.

Key Equations

The maximum efficiency of a heat engine operating between the Kelvin temperatures T_h and T_c is

$$e_{max} = 1 - \frac{T_c}{T_h}$$

The change in entropy corresponding to the heat Q at the Kelvin temperature T is

$$\Delta S = \frac{Q}{T}$$

11 Assessment

For instructor-assigned homework, go to www.masteringphysics.com

Lesson by Lesson

11.1 The First Law of Thermodynamics

Conceptual Questions

44. Why do heat and work have opposite signs in the equation $\Delta E = Q - W$?

45. A system receives 100 J of heat. If the thermal energy of the system remains constant, how much work does the system do?

46. Engine 1 takes in 100 J of heat from a hot reservoir and does 20 J of work. Engine 2 takes in the same amount of heat from the hot reservoir and does 25 J of work. Is the efficiency of engine 1 greater than, less than, or equal to the efficiency of engine 2? Explain.

47. Engine 1 takes in 100 J of heat from a hot reservoir and does 20 J of work. Engine 2 takes in 600 J of heat from the hot reservoir and does 60 J of work. Is the efficiency of engine 1 greater than, less than, or equal to the efficiency of engine 2? Explain.

Problem Solving

48. A runner does 8.2×10^5 J of work and gives off 3.3×10^5 J of heat. Determine ΔE, W, and Q for the runner.

49. Find the heat associated with each of the following processes: **(a)** $W = 50$ J, $\Delta E = 50$ J; **(b)** $W = -50$ J, $\Delta E = -50$ J; **(c)** $W = 50$ J, $\Delta E = 150$ J.

50. An engine receives 770 J of heat from a hot reservoir and does 160 J of work. What is **(a)** the efficiency of this engine and **(b)** the heat given off to the cold reservoir?

51. What is the efficiency of an engine that exhausts 440 J of heat to a cold reservoir and receives 570 J of heat from a hot reservoir?

52. What is the efficiency of the heat engine shown in **Figure 11.17**?

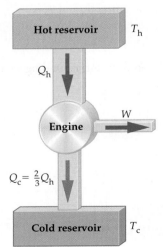

Figure 11.17

53. If the heat added to the engine in Figure 11.17 is $Q_h = 280$ J, how much work does the engine do?

54. A basketball player does 2.43×10^5 J of work during her time in the game, and 0.110 kg of water evaporates from her skin. Assuming a latent heat of 2.26×10^6 J/kg for the evaporation of sweat (the same as for water), determine the change in the player's thermal energy.

55. Three different processes act on a system. **(a)** In process A, 42 J of work are done on the system and 77 J of heat are added to the system. Find the change in the system's thermal energy. **(b)** In process B, the system does 42 J of work and 77 J of heat are added to the system. What is the change in the system's thermal energy? **(c)** In process C, the system's thermal energy decreases by 120 J while the system performs 120 J of work on its surroundings. How much heat was added to the system?

11.2 Thermal Processes

Conceptual Questions

56. Which of the physical quantities, Q, W, or ΔE, is zero in a constant-volume process?

57. Which of the physical quantities, Q, W, or ΔE, is zero in an isothermal process?

58. Which of the physical quantities, Q, W, or ΔE, is zero in an adiabatic process?

59. How is the work done in a constant-pressure process determined from a PV graph?

60. The pressure of a system is held fixed. Is it possible for the thermal energy of the system to change? Give an explanation if your answer is no. If your answer is yes, give a specific example.

61. A gas does a certain amount of work as it expands by a volume ΔV at a pressure P. If the pressure of the gas is increased, but the volume change remains the same, does the work done by the gas increase, decrease, or stay the same? Explain.

Problem Solving

62. A fluid expands by 0.42 m³ at a pressure of 121 kPa. How much work is done by the fluid?

63. A gas is contained in a cylinder with a pressure of 140 kPa and an initial volume of 0.66 m³. How much work is done by the gas as it expands at constant pressure to twice the initial volume?

64. A gas expands at a constant pressure of 190 kPa and does 82 J of work. What was the change in volume?

65. Think & Calculate A gas is at a constant pressure of 115 kPa. **(a)** If work is done on the gas, does its volume increase or decrease? Explain. **(b)** Find the change in volume of the gas if 62 J of work is done on it.

66. What is the pressure of a gas that does 22 J of work as it expands by 0.75 m^3? Assume constant pressure.

67. Think & Calculate The thermal energy of a system increases as the result of an adiabatic process. **(a)** Is work done on the system or by the system? **(b)** Calculate the work done in part (a) if the system's thermal energy increases by 890 J.

68. A system expands by 0.75 m^3 at a constant pressure of 125 kPa. Find the heat that flows into or out of the system if its thermal energy **(a)** increases by 65 J or **(b)** decreases by 1850 J. In each case, give the direction of the heat flow.

11.3 The Second and Third Laws of Thermodynamics

Conceptual Questions

69. If you clean up a messy room, putting things back where they belong, you decrease the room's entropy. Does this violate the second law of thermodynamics? Explain.

70. A heat engine operates between a hot reservoir at the Kelvin temperature T_h and a cold reservoir at the Kelvin temperature T_c. If both temperatures are doubled, does the efficiency of the engine increase, decrease, or stay the same? Explain.

71. A heat engine operates between a hot reservoir at the Kelvin temperature T_h and a cold reservoir at the Kelvin temperature T_c. If both temperatures are increased by 50 K, does the efficiency of the engine increase, decrease, or stay the same? Explain.

72. Predict & Explain **(a)** If you rub your hands together, does the entropy of the universe increase, decrease, or stay the same? **(b)** Choose the *best* explanation from among the following:

A. Rubbing the hands together draws heat from the surroundings and therefore lowers the entropy.

B. No mechanical work is done by the rubbing and hence the entropy does not change.

C. The heat produced by the rubbing raises the temperature of the hands and the air, which increases the entropy.

73. Rank Four heat engines operate with the following sets of reservoir temperatures:

Engine	A	B	C	D
T_h (K)	800	600	1200	1000
T_c (K)	400	400	800	800

Rank these engines in order of increasing efficiency. Indicate ties where appropriate.

Problem Solving

74. What is the efficiency of an engine that takes in 610 J of heat in the process of doing 230 J of work?

75. An engine receives 780 J of heat from a hot reservoir and does 110 J of work. What is **(a)** the heat given off to the cold reservoir and **(b)** the efficiency of this engine?

76. A nuclear power plant has a reactor that produces heat at the rate of 835 MW. This heat is used to produce 250 MW of mechanical power to drive an electrical generator. **(a)** At what rate is heat discarded to the environment by this power plant? **(b)** What is the thermal efficiency of the plant?

77. At a coal-burning power plant a steam turbine operates with a power output of 550 MW. The thermal efficiency of the power plant is 32%. **(a)** At what rate is heat discarded to the environment by this power plant? **(b)** At what rate must heat be supplied to the power plant by burning coal?

78. Find the change in entropy when 1.85 kg of water at 100 °C is boiled away to steam at 100 °C.

79. Think & Calculate A heat engine does 2700 J of work with an efficiency of 0.18. What is **(a)** the heat taken in from the hot reservoir and **(b)** the heat given off to the cold reservoir? **(c)** If the efficiency of the engine is increased, does your answer to part (a) increase, decrease, or stay the same? Explain.

80. The efficiency of a heat engine with a cold reservoir at a temperature of 265 K is 21%. Assuming that the temperature of the hot reservoir remains the same, find the temperature the cold reservoir must have for the engine's efficiency to be 25%.

81. Think & Calculate The efficiency of a particular heat engine is 0.300. **(a)** If the high-temperature reservoir is at a temperature of 545 K, what is the temperature of the low-temperature reservoir? **(b)** To increase the efficiency of this engine to 0.400, must the temperature of the low-temperature reservoir be increased or decreased? Explain. **(c)** Find the temperature of the low-temperature reservoir that gives an efficiency of 0.400.

82. A heat engine takes 2500 J of heat from a high-temperature reservoir and performs 2200 J of work. **(a)** What is the efficiency of this engine? **(b)** How much heat is exhausted to the low-temperature reservoir during each cycle? **(c)** What is the ratio, T_h/T_c, of the two reservoir temperatures?

83. On a cold winter's day thermal energy leaks slowly out of a house at the rate of 25 kW. If the inside temperature is 22 °C and the outside temperature is −14 °C, what is the rate of the entropy increase of the universe? (*Hint*: Include both the entropy decrease inside the house and the entropy increase outside the house.)

84. An 88-kg parachutist descends through a vertical height of 360 m with constant speed. Find the increase in entropy produced by the parachutist, assuming that the air temperature is 21 °C.

Mixed Review

85. Which thermodynamic law is violated by heat flowing spontaneously between two objects of equal temperature?

86. Which law of thermodynamics is most pertinent to the fact that rubbing your hands together makes them warmer?

87. Which law of thermodynamics is most pertinent to the statement that "all the king's horses and all the king's men couldn't put Humpty Dumpty back together again"?

88. What do we call a thermal process in which no heat is exchanged?

89. A gas follows the process from point A to point B in **Figure 11.18**. What is the work done by the gas?

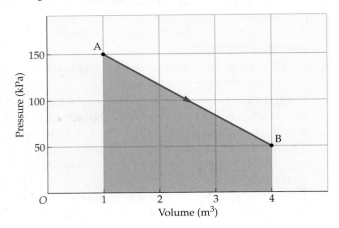

Figure 11.18

Writing about Science

90. Research the adiabatic processes involved in the operation of a diesel engine. By what factor is the air-fuel mixture in the cylinders of the engine compressed to produce ignition? What are some of the advantages of a diesel engine? What are some of its disadvantages?

91. **Connect to the Big Idea** If you ride a bicycle down a steep hill at constant speed, your kinetic energy remains constant. Your gravitational potential energy is continuously decreasing, however. Where is the potential energy going? Explain in detail.

Read, Reason, and Respond

Energy from the Ocean Whenever two objects are at different temperatures, thermal energy can be extracted with a heat engine. A case in point is the ocean, where one "object" is the warm water near the surface, and the other "object" is the cold water at considerable depth. Tropical seas, in particular, can have a significant temperature difference between the sun-warmed surface water and the cold, dark water 1000 m or more below the surface. A typical tropical ocean temperature profile is shown in **Figure 11.19**, where we see a rapid change in temperature—a *thermocline*—between depths of approximately 250 m and 900 m.

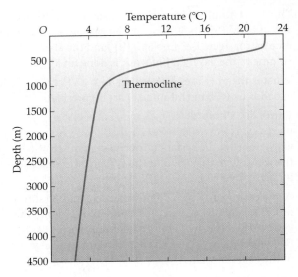

Figure 11.19 Ocean water temperature versus depth.

The idea of tapping this potential source of energy has been around for a long time. In 1870, for example, Captain Nemo in Jules Verne's *Twenty Thousand Leagues Under the Sea* said, "I owe all to the ocean; it produces electricity, and electricity gives heat, light, motion, and, in a word, life to the *Nautilus*." Just 11 years later, the French physicist Jacques Arsene d'Arsonval proposed a practical system referred to as Ocean Thermal Energy Conversion (OTEC), and in 1930 Georges Claude, one of d'Arsonval's students, built and operated the first experimental OTEC system off the coast of Cuba.

OTEC systems, which are potentially low-cost and carbon neutral, can provide not only electricity but also desalinated water. In fact, an OTEC plant generating 2 MW of electricity is expected to produce over 14,000 cubic feet of desalinated water a day. The governments of Hawaii, Japan, and Australia are actively pursuing plans for OTEC systems. The National Energy Laboratory of Hawaii Authority (NELHA), for example, operated a test facility near Kona from 1992 to 1998 and plans further tests in the future.

92. Suppose an OTEC system operates with surface water at 22 °C and deep water at 4.0 °C. What is the maximum efficiency this system could have?

A. 6.1%   C. 9.4%

B. 8.2%   D. 18%

93. If 1500 kg of water at 22 °C is cooled to 4.0 °C, how much energy is released? (For comparison, the energy released in burning a gallon of gasoline is 1.3×10^8 J.)

A. 2.5×10^7 J   C. 1.4×10^8 J

B. 1.1×10^8 J   D. 1.6×10^8 J

94. If an OTEC system goes deeper for cold water, to where the temperature is only 2.0 °C, what is its maximum efficiency?

A. 6.8%   C. 9.3%

B. 9.1%   D. 19%

Standardized Test Prep

Multiple Choice

Use the pressure-volume graph below to answer Questions 1–3. The graph shows a four-step thermodynamic process for a gas.

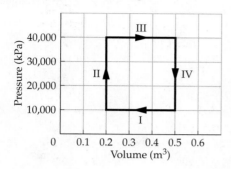

1. During which step of the process is work done *on* the gas?
   (A) I
   (B) II
   (C) III
   (D) IV

2. During which step of the process must the amount of heat transferred to the gas be equal to the increase in thermal energy of the gas?
   (A) I
   (B) II
   (C) III
   (D) IV

3. What is the net work done by the gas during one complete cycle?
   (A) 4,000 J
   (B) 9,000 J
   (C) 12,000 J
   (D) 30,000 J

4. Which equation correctly expresses the first law of thermodynamics?
   (A) $\Delta E = Q = W$
   (B) $\Delta E = W - Q$
   (C) $\Delta E = 0$
   (D) $\Delta E = Q - W$

5. Which is associated with a vertical line on a pressure-volume graph?
   (A) constant pressure
   (B) adiabatic process
   (C) maximum work
   (D) zero work

6. A heat engine will increase its efficiency if it produces
   (A) greater work output for the same heat input.
   (B) the same work output for increased heat input.
   (C) less work output for increased heat input.
   (D) zero work output for any amount of heat input.

7. Which *best* describes a process in which the work done equals the heat?
   (A) constant-volume
   (B) isothermal
   (C) adiabatic
   (D) thermal

8. During an isothermal process 350 J of heat is added to a gas. Which is true?
   (A) The temperature of the gas increases.
   (B) This process could be at constant volume.
   (C) The gas does 350 J of work.
   (D) The volume of the gas decreases.

Extended Response

9. The equation below shows the maximum work done by a heat engine operating between a hot reservoir at T_h and a cold reservoir at T_c. Explain **(a)** how increasing T_h affects the work done, **(b)** how decreasing T_c affects the work done, **(c)** how increasing the thermal energy flowing into the engine affects the work done, and **(d)** why the engine can never operate at 100% efficiency.

$$W_{max} = \left(1 - \frac{T_c}{T_h}\right)Q_h$$

Test-Taking Hint

The prefix *iso-* means "the same"; it is used to name quantities that do not change during a process.

If You Had Difficulty With . . .

Question	1	2	3	4	5	6	7	8	9			
See Lesson(s)	11.1	11.1	11.2	11.1	11.2	11.2, 11.3	11.3	11.2	11.3			

12

Gases, Liquids, and Solids

Inside

Hot-air balloons rise because of the buoyant force exerted on them by the surrounding air. The buoyant force is great enough to lift these balloons because hot air is less dense than cool air.

Big Idea

Fluids flow and change shape easily, whereas solids maintain a definite shape unless acted on by a force.

The substances we come into contact with every day are in one of three forms: solid, liquid, or gas. The air we breathe is a gas, the water we drink is a liquid, and the salt on our popcorn is a solid. In this chapter we consider these three phases of matter in detail.

Inquiry Lab   What makes water move through a straw?

Explore

1. Cut a straw into two pieces. One piece should be slightly longer than the other.
2. Place the longer straw into a water-filled cup, and then place your finger over the top of the straw. Lift the straw just above the water while keeping your finger over the top. Observe what happens, and then lower the straw back into the water.
3. Repeat Step 2, this time removing your finger from the top of the straw *after* lifting the straw above the water. Observe what happens.

4. Hold the longer straw vertically so that part of it is in the water. Then have your lab partner use the shorter straw to blow air horizontally over the top of the longer straw. Be careful to blow air *across* the open end, not into it. Observe what happens.

Think

1. **Describe** Imagine drinking water through a straw. Describe how this is done, and try to explain why it works.
2. **Identify** The weight of the air in the atmosphere exerts force on everything it is in contact with.

Use this fact to describe how atmospheric pressure exerts force on the water in the longer straw in Steps 2, 3, and 4.
3. **Apply** A resting object or fluid (such as water) remains at rest unless a net force acts on it. Which steps subjected the water in the straw to a net force? What caused the net force?
4. **Discuss** Consult with your lab partner and come up with an explanation for each of your observations in Steps 2, 3, and 4. Discuss your reasoning with your classmates.

12.1 Gases

Have you been breathing today? Good. Just think of how much air has gone in and out of your lungs during the day. You might say that the topic of this lesson is like a breath of fresh air, because air is one of the best examples of a gas.

Gases and liquids are examples of fluids

Air is a mixture of gases, and water is a liquid. One of the key characteristics of both gases and liquids is that they don't have a specific shape—you can't make a sculpture out of a gas or a liquid. In general, a substance that can flow from one location to another and has no set shape of its own is referred to as a **fluid**. Thus, both gases and liquids are fluids. They differ in that gases expand to fill their container, whereas liquids have a definite volume and may only partially fill their container. Liquids are discussed in the next lesson.

Gases exert pressure

Suppose you hold your hand in a horizontal position, palm down. The pressure of the atmosphere pushes down on the top of your hand with a significant force. Your hand doesn't move downward, though. Why? Because atmospheric pressure also pushes up on the palm of your hand with essentially the same force.

Vocabulary

- fluid
- ideal gas
- mole
- molar mass
- atomic mass

CONNECTING IDEAS

Chapter 10 introduced the concept of pressure. Pressure, which is the amount of force exerted on a given area, can also be thought of as the application of force by a fluid.

- Pressure plays a key role throughout much of this chapter.

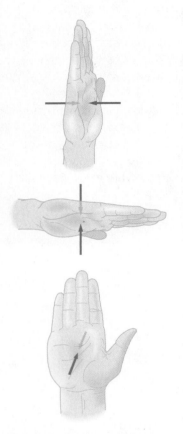

What are the key characteristics of gas pressure?

The same is true regardless of the orientation of your hand, as illustrated in **Figure 12.1**. This observation leads to an important conclusion: The pressure in a gas (or any fluid) acts equally in all directions. In addition, the pressure is always at right angles to any surface it acts on. The second point is particularly clear in **Figure 12.2**, where we see that a stream of water exiting from a hole in a container always starts out perpendicular to the surface of the container.

Gauge Pressure Suppose a car runs over a nail and a tire goes flat. Is the pressure inside the flat tire equal to zero? It might seem so at first, but it isn't. The hole in the flat tire allows air to pass freely from inside the tire to the atmosphere. As a result, the pressure in the tire is equal to atmospheric pressure.

Now, let's say you patch the tire and inflate it to a typical value of 241 kPa (about 35 lb/in^2). What this means is that the pressure inside the tire is *greater* than atmospheric pressure by 241 kPa. Recalling that atmospheric pressure is 101 kPa, we see that the pressure in the inflated tire is

$$P = 241 \text{ kPa} + P_{\text{atmospheric}}$$
$$= 241 \text{ kPa} + 101 \text{ kPa}$$
$$= 342 \text{ kPa}$$

(Recall that 1 Pa = 1 N/m^2 and 1 kPa = 10^3 N/m^2.)

The increased pressure inside the tire pushes outward on the walls of the tire, making it firm enough to support the car. We describe pressures like this in terms of *gauge pressure*. The gauge pressure, P_{gauge}, is defined as follows:

$$P_{\text{gauge}} = P - P_{\text{atmospheric}}$$

In this equation the pressure P is the actual pressure inside the tire, and the pressure that you read on a tire **gauge** is—appropriately enough—the gauge pressure. Many problems in this chapter refer to the gauge pressure. Remember that the gauge pressure is always less than the actual pressure by the amount $P_{\text{atmospheric}}$.

▲ **Figure 12.1 Pressure is the same in all directions**
The forces exerted on the two sides of a hand cancel, regardless of the hand's orientation. Thus, atmospheric pressure acts equally in all directions, as does the pressure due to any fluid.

Water shoots out perpendicular to the surface of the vase.

▲ **Figure 12.2 Pressure is perpendicular to a surface**
Pressure due to a fluid always exerts a force that is perpendicular to the surface in contact with the fluid. This is why the water shooting out from the leaks in this vase starts out in a direction that is perpendicular to the surface of the vase.

To determine the gauge pressure in a basketball, you push down on it and note the area of contact it makes with the floor. When you push down with a force of 22 N, the ball flattens out, and its bottom forms a circular area with a diameter of 2.0 cm. What is the gauge pressure of the basketball?

Picture the Problem

Our sketch shows the basketball both in its original state and with a force $F = 22$ N acting downward on it. In the latter case the ball has a circular area of contact with the floor of diameter $d = 2.0$ cm.

Strategy

The gauge pressure of the basketball is the force F divided by the circular area, which is $A = \pi r^2$. The circular area has a diameter of 2.0 cm, which corresponds to a radius of $r = 0.010$ m.

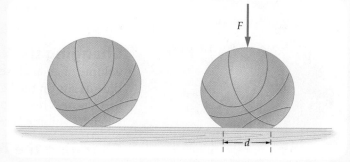

Known	Unknown
$F = 22$ N	$P_{gauge} = ?$
$d = 2.0$ cm ($r = 0.010$ m)	

Solution

1 The gauge pressure is force divided by area. Substitute the numerical values for force and area:

$$P_{gauge} = \frac{F}{A} = \frac{F}{\pi r^2}$$

$$= \frac{22 \text{ N}}{\pi (0.010 \text{ m})^2}$$

$$= \boxed{7.0 \times 10^4 \text{ Pa}}$$

Math HELP
Calculating Area
See Math Review, Section V

Insight

Thus, the basketball is inflated to a gauge pressure of 70 kPa, which corresponds to about 10 lb/in^2.

Practice Problems

1. Follow-up What is the diameter of the circular area of contact if you push down with a force of 44 N on a basketball whose gauge pressure is 75 kPa?

2. An inflated basketball has a gauge pressure of 74 kPa. What is the actual pressure inside the ball?

3. Your family's 1420-kg car is supported equally by its four tires, each inflated to a gauge pressure of 242 kPa. What is the area of contact each tire makes with the road?

Real gases approximate ideal gases

In physics we sometimes talk about frictionless surfaces and projectiles with no air resistance. These idealized cases are well approximated by many real-world situations. A similar idealization is used when studying gases.

Consider the air around you. It moves very freely—so freely, in fact, that you often forget it's there. The fact that it moves with such ease means that the particles within it have little effect on one another. We consider this

Reading Support ✓
Multiple Meanings of gauge

scientific meanings: a measurement device, usually with some sort of visual display; the diameter of a wire, rod, string, etc.

The air pressure gauge indicated that the tire was underinflated.

The electrician used a 12-gauge wire, as required by local building codes.

common meaning: to estimate or form a judgment about something

The lawyer was unable to gauge the mood of his client.

▲ **Figure 12.3 Inflating a basketball**
A hand pump can be used to increase the number of gas particles inside a basketball. This increases the pressure of the gas in the ball, causing it to inflate.

The pump forces more molecules into the ball ...

... which increases the pressure.

behavior *ideal*. An **ideal gas** is one in which the particles have no effect on one another. Like the air around us, most real gases are good approximations to an ideal gas.

Pressure is the key to understanding the behavior of an ideal gas. Let's explore three common ways of changing the pressure exerted by a gas.

Changing the Number of Particles
Consider the basketball in **Figure 12.3**. The pressure inside the ball increases as more gas particles are pumped into it. In general, increasing the number of gas particles in an enclosed space increases the pressure.

Changing the Volume
An increase in pressure also occurs when the volume of an enclosed gas decreases. That's what's happening in **Figure 12.4**; the girl sits on the basketball, and her weight distorts the ball and decreases its volume. The gas particles strike the inside of the ball more frequently when its volume is reduced, and this results in an increase in pressure.

Changing the Temperature
Finally, the pressure exerted by a gas depends on its temperature. Heating an enclosed gas increases the average kinetic energy of its particles. These more energetic particles hit the walls of the container at higher speeds, causing an increase in pressure. Thus, leaving a basketball out in the sun on a hot summer day increases the pressure inside it.

Ideal Gas Law
We can combine all of these observations into one simple equation for the pressure of an ideal gas:

Ideal Gas Equation

$$\text{pressure} = k\left(\frac{\text{number of gas particles} \times \text{temperature of gas}}{\text{volume of gas}}\right)$$

$$P = k\frac{NT}{V}$$

In this equation, N is the number of gas particles, T is the Kelvin temperature, and V is the volume of the gas. The constant k is known as the *Boltzmann constant*. It is named for the Austrian physicist Ludwig Boltzmann (1844–1906), and its numerical value is as follows:

Boltzmann Constant, k

$$k = 1.38 \times 10^{-23} \text{ J/K}$$

SI units: J/K

You may also see the ideal gas equation rearranged into the following equivalent form:

Ideal Gas Equation (alternative form)

$$PV = NkT$$

▲ **Figure 12.4 Increasing pressure by decreasing volume**
Sitting on a basketball reduces its volume and increases the pressure of the gas it contains.

The ideal gas equation shows that gas pressure increases if the number of gas particles increases, the temperature of the gas increases, or the volume of the gas decreases. The next Guided Example shows how to use this equation.

A person's lungs can hold 6.0 L of air at 310 K (normal body temperature) and 101 kPa (normal atmospheric pressure). Given that air is 21% oxygen, find the number of oxygen molecules in the person's lungs.

Picture the Problem

Our sketch shows a person's lungs, with their combined volume of $V = 6.0$ L. The sketch also indicates that the pressure in the lungs is $P = 101$ kPa and the temperature is $T = 310$ K.

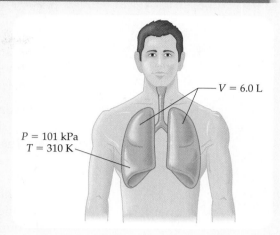

$V = 6.0$ L

$P = 101$ kPa
$T = 310$ K

Strategy

We will treat the air in the lungs as an ideal gas. Given the volume, temperature, and pressure of the gas, we can rearrange $PV = NkT$ to solve for the number of gas particles, N. Finally, because only 21% of air is oxygen, we multiply by 0.21 to find the number of oxygen molecules.

Known

$V = 6.0$ L $T = 310$ K $P = 101$ kPa

Air is 21% oxygen.

Unknown

$N = ?$

Solution

1 Rearrange $PV = NkT$ to solve for the number of gas particles:

$$PV = NkT$$

$$N = \frac{PV}{kT}$$

> **Math HELP**
> Solving Simple Equations
> **See Math Review, Section III**

2 Substitute the numerical values to find the number of gas particles in the lungs:

$$N = \frac{(1.01 \times 10^5 \, \text{Pa})(6.0 \times 10^{-3} \, \text{m}^3)}{(1.38 \times 10^{-23} \, \text{J/K})(310 \, \text{K})}$$

$$= 1.4 \times 10^{23} \, \text{particles}$$

3 Multiply the number of gas particles by 0.21 to find the number of oxygen molecules:

$$0.21(1.4 \times 10^{23} \, \text{particles})$$

$$= \boxed{2.9 \times 10^{22} \, \text{particles}}$$

Insight

As you might expect, the number of molecules in an average-size pair of lungs is enormous. In fact, a number this large is difficult to compare to anything familiar. For example, the number of stars in the Milky Way galaxy is estimated to be "only" about 10^{11}. Thus, if each star in the Milky Way were a galaxy of 10^{11} stars, then all of those stars combined would begin to approach the number of oxygen molecules in your lungs at this very moment!

Finally, one note about units: The number of particles, N, is dimensionless—it is simply a number. If you substitute 1 Pa $= 1$ N/m^2 and 1 J $= 1$ N·m in Step 2, you will see that all of the units cancel, leaving N dimensionless.

Practice Problems

4. Follow-up If the person takes a particularly deep breath, so that the lungs hold a total of 1.5×10^{23} gas particles, what is the new volume of the lungs?

5. Follow-up If the person climbs to the top of a mountain, where the air pressure is considerably less than 101 kPa, does the number of molecules in the lungs increase, decrease, or stay the same?

CONCEPTUAL Example 12.3 Does the Number of Molecules Change?

Feeling a bit cool, you turn up the thermostat in your living room. A short time later the air is warmer. Assuming that the air in the room is always at atmospheric pressure, is the number of molecules in the room greater than, less than, or the same as it was before you turned up the heat?

Reasoning and Discussion
Increasing the temperature increases the speed of the molecules in the air. Therefore, fewer molecules are needed to give the same pressure, since each molecule packs more of a "punch" when it hits the walls of the room.

This conclusion is consistent with the ideal gas equation, which becomes $N = (\text{constant})/T$ when everything in it is constant except for the number of particles and the temperature. It follows that increasing T decreases N.

Answer
The number of molecules is less than it was before you turned up the heat.

The ideal gas equation can be expressed in moles

When you buy salt, you don't pay for it grain by grain. Instead, you buy it in terms of weight, since even a modest weight of salt contains an enormous number of grains.

A similar distinction applies to the particles of a gas. We've seen that the number of particles in a modest volume of gas (like the volume of your lungs) is enormous. In many cases it's more convenient to talk about the amount of gas in terms of weight rather than in terms of the number of particles.

This way of measuring the amount of a gas uses *moles*. A **mole** (mol) is the amount of a substance that contains as many particles as there are atoms in 12 grams of carbon-12. Experiments show that the number of atoms in a mole of carbon-12 is 6.022×10^{23}. This number is known as *Avogadro's number*, N_A, named for the Italian physicist and chemist Amedeo Avogadro (1776–1856):

> **Avogadro's Number, N_A**
>
> $N_A = 6.022 \times 10^{23}$ molecules/mol
>
> SI unit: mol^{-1}
>
> (The number of molecules is dimensionless.)

Examples of a mole of various substances are shown in **Figure 12.5**.

◀ **Figure 12.5 A mole of various substances**
Counting atoms or molecules is hard to do, but the mole concept provides a useful way of dealing with this difficulty. A mole of any substance contains Avogadro's number of particles; thus, measuring out a mole of a substance is equivalent to counting 6.022×10^{23} particles. This photo shows a mole of four different substances: helium (in the balloon), sulfur (lower left), copper (lower center), and mercury (lower right).

Let's use the symbol n to stand for the number of moles in an amount of gas. If an amount of gas consists of n moles, then the number of gas particles, N, is simply the number of moles times Avogadro's number:

number of particles = number of moles × Avogadro's number

$$N = nN_A$$

Using this expression for N in the ideal gas equation ($PV = NkT$) yields

$$PV = nN_A kT$$

We simplify things by replacing $N_A k$ with a constant called the *universal gas constant*, R. That is, $R = N_A k$. The value of R is defined below.

Universal Gas Constant, R

$R = N_A k$

$= 8.31 \ J/(mol \cdot K)$

SI units: $J/(mol \cdot K)$

Thus, the ideal gas equation in terms of moles is

Ideal Gas Equation (in terms of moles)

$$\text{pressure} \times \text{volume} = \frac{\text{number}}{\text{of moles}} \times \frac{\text{universal}}{\text{gas constant}} \times \text{temperature}$$

$$PV = nRT$$

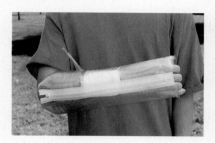

COOL PHYSICS
Air Splint

Backpackers and hikers often take an inflatable air splint on long hikes as a safety measure. An air splint is about the size of a credit card before use. When inflated, it can immobilize a broken arm, as shown in the photo. It can also slow bleeding from an injured area. Only a modest gauge pressure is required to inflate an air splint, and this can be supplied with a blow tube.

ACTIVE Example 12.4 Calculate the Amount of Air

How many moles of air are in an inflated basketball? Assume that the pressure in the ball is 171 kPa, the temperature is 293 K, and the diameter of the ball is 30.0 cm.

Solution *(Perform the calculations indicated in each step.)*

1. Solve $PV = nRT$ for the number of moles, n:

   $n = PV/RT$

2. Calculate the volume of the ball:

   $V = \frac{4}{3}\pi r^3 = 0.0141 \ m^3$

3. Substitute the numerical values:

   $n = 0.990 \ mol$

Insight

Thus, an inflated basketball contains approximately 1 mole of air.

> **Math HELP**
> Calculating Volume
> **See Math Review, Section V**

Practice Problems

6. A balloon is filled with helium at a pressure of 3.1×10^5 Pa. The balloon is at a temperature of 22 °C and has a radius of 0.35 m. How many helium atoms are contained in the balloon?

7. A tire contains 1.5 mol of air at a gauge pressure of 245 kPa and a temperature of 282 K. What is the volume of the tire?

■ How are the concepts of moles and molar mass useful when measuring quantities of matter?

Molar mass is the mass of a mole of a substance

A mole of anything has precisely the same number of particles, N_A. What differs from substance to substance is the *mass* of 1 mole. For example, 1 mole of helium atoms has a mass of 4.00260 g, and 1 mole of copper atoms has a mass of 63.546 g. In general, the **molar mass**, M, of a substance is the mass *in grams* of 1 mole of that substance. Thus, the molar mass of helium is $M = 4.00260$ g/mol, and that of copper is $M = 63.546$ g/mol. The periodic table in Appendix C gives the molar masses for all the elements. In contrast, the **atomic mass** of an element is the mass of one atom of that element.

■ Molar mass provides a convenient bridge between the macroscopic world, where we measure the mass of a substance in grams, and the microscopic world, where the number of particles in a sample of a substance is typically 10^{23} or more. As we've seen, if you measure out a mass of copper equal to 63.546 g, you have, in effect, counted out $N_A = 6.022 \times 10^{23}$ atoms of copper. Each of the copper atoms has a mass of $(63.546 \text{ g})/(6.022 \times 10^{23}) = 1.055 \times 10^{-22}$ g.

Kinetic theory relates the temperature of a gas to the pressure it exerts

As you read this book, innumerable gas molecules are bouncing off your body with speeds close to the speed of sound. In fact, the impact from these collisions is what creates atmospheric pressure. Each collision results in a change of momentum for a gas particle, just like throwing a ball at a wall and having it bounce back. The total change in momentum of the particles per time is the force they exert. The amount of force per area is the pressure of the gas.

This picture of a gas consisting of innumerable particles flying about randomly at high speeds is known as the *kinetic theory of gases*. Experiments show that the average kinetic energy of the particles in a gas is directly proportional to the Kelvin temperature. Thus, when we heat a gas, its particles move faster. This is why heating a gas increases its pressure—the heated gas particles move faster when they collide with the walls of the container, and hence they exert a greater outward force.

Chemical reactions usually take place only when molecules collide. In general, higher temperatures mean higher molecular speeds and increased rates of chemical reactions. This is why foods cook faster at higher temperatures and why the chirping rate of a cricket increases with temperature.

Similarly, phase transitions can be understood in terms of the kinetic behavior of molecules—this explanation is sometimes referred to as the *kinetic molecular theory*. For example, as the temperature of a solid is increased, the molecules oscillate about their fixed positions with more and more energy. When the temperature is high enough, the molecules have enough energy to break free of one another and move about more or less freely in the liquid phase. Increasing the temperature even more gives the molecules enough energy to break out of the liquid phase, forming the gas phase.

12.1 LessonCheck (MP)

Checking Concepts

8. ▭ **Identify** What is the direction of the pressure exerted by a gas?

9. ▭ **List** A classmate correctly states that there are three ways to increase the pressure of an enclosed gas. List the three ways.

10. ▭ **Apply** A 25.0-g sample of a known gas is sealed in a flask. Explain how you know the number of gas particles contained in the sample.

11. Identify What do we call an amount of a substance that contains the same number of particles as 12 g of carbon-12?

12. Describe How can the pressure exerted by a gas be understood on the molecular level?

13. Rank Four ideal gases, A through D, have the pressures, volumes, and amounts shown below. Rank the gases in order of increasing temperature. Indicate ties where appropriate.

Gas	A	B	C	D
P (kPa)	100	200	50	50
V (m³)	1	2	1	4
n (mol)	10	20	50	5

14. Apply Two containers hold ideal gases at the same temperature. Container A has twice the volume and half the number of particles as container B. What is the ratio P_A/P_B, where P_A is the pressure in container A and P_B is the pressure in container B?

Solving Problems

15. Calculate What is the pressure of an ideal gas if there are 2.2×10^{23} particles in a volume of 0.026 m³ at a temperature of 290 K?

16. Calculate What is the volume of an ideal gas if there are 4.10×10^{23} particles at a temperature of 262 K and a pressure of 195 kPa?

17. Calculate A tire contains 1.5 mol of air at a gauge pressure of 205 kPa. If the volume of the air in the tire is 0.012 m³, what is its temperature?

18. Calculate In the morning, when the temperature is 286 K, a bicyclist finds that the gauge pressure in her tires is 401 kPa. That afternoon she finds that the gauge pressure in the tires has increased to 419 kPa. What is the afternoon temperature?

12.2 Fluids at Rest

Vocabulary

- density
- buoyant force

🔑 *What does it mean to say that one substance is denser than another substance?*

Have your ears ever "popped" in an airplane or automobile as its elevation changed? Have you ever taken a deep breath and floated motionless in a pool of water? If so, you already have some understanding of the interesting behavior of fluids at rest. You'll learn more about these situations and others in this lesson.

Fluids are characterized by their density

The properties of a fluid can be hard to pin down. After all, fluids flow from place to place and can even change their shape. One of the best ways to describe a fluid is in terms of the amount of mass it has per volume. In general, the **density** of a substance (fluid or not) is the mass m of the substance divided by its volume, V. We use the Greek letter rho, ρ (pronounced "row"), to stand for the density.

Definition of Density, ρ

$$\text{density} = \frac{\text{mass}}{\text{volume}}$$

$$\rho = \frac{m}{V}$$

SI units: kg/m^3

🔑 **The denser a substance, the more mass it has in any given volume.** Figure 12.6 shows some of the effects of different densities.

Typical Densities To get a feeling for the densities of common substances, let's start with water. Visualize a container 1 meter on a side. The volume of the container is thus 1 cubic meter (1 m^3). It takes 1000 kilograms of water to fill the container, and therefore the density of water is

$$\rho_{\text{water}} = \frac{1000 \text{ kg}}{1 \text{ m}^3} = 1000 \text{ kg/m}^3$$

▶ **Figure 12.6 Density differences**
The cylindrical flask (left) contains three differently colored liquids with different densities. The most dense liquid (green) has settled to the bottom, and the least dense liquid (red) floats on top. The small helium-filled blimp (right) floats because helium is less dense than air.

For comparison, the density of the air in your physics classroom is roughly 1.29 kg/m³. The density of helium in a helium-filled balloon is even less—only about 0.179 kg/m³. Further examples of densities for a variety of solids, liquids, and gases are given in **Table 12.1.**

Solids and liquids are virtually incompressible, so their densities are practically constant. Gases, on the other hand, expand (become less dense) with heating or a decrease in pressure and compress (become more dense) with cooling or an increase in pressure. Therefore, the values for gas densities in Table 12.1 are given at standard temperature and pressure (STP): 0 °C and 1 atm.

Mass and Volume for a Given Density

The density equation can also be used to solve for the mass of a given volume—or the volume of a given mass—if the density of the substance is known. For example, the mass of 3.79 liters (3.79×10^{-3} m³, or 1 gallon) of water is

$$m = \rho_{\text{water}} V$$
$$= (1000 \text{ kg/m}^3)(3.79 \times 10^{-3} \text{ m}^3)$$
$$= 3.79 \text{ kg}$$

As a rule of thumb, a gallon of water weighs about 36 N (~8 lb). Similarly, the volume of 10 kilograms of water is

$$V = \frac{m}{\rho_{\text{water}}} = \frac{10 \text{ kg}}{1000 \text{ kg/m}^3} = 0.010 \text{ m}^3$$

QUICK Example 12.5 What's the Mass?

One day you look in your refrigerator and find nothing but a dozen eggs (mass of 44 g each). A quick measurement shows that the inside of the refrigerator is 1.0 m by 0.60 m by 0.75 m. How does the mass of the air in the refrigerator compare with the mass of the eggs?

Solution
At first it might seem that the "thin air" in the refrigerator would have much less mass than a dozen eggs. A brief calculation shows that this is not the case. For the mass of the eggs, we have

$$m_{\text{eggs}} = 12 \text{ eggs} \times \frac{0.044 \text{ kg}}{\text{egg}} = 0.53 \text{ kg}$$

For the mass of the air, we first note that the volume of the refrigerator is $V = 1.0 \text{ m} \times 0.60 \text{ m} \times 0.75 \text{ m}$. Thus, the mass of the air is

$$m_{\text{air}} = \rho_{\text{air}} V$$
$$= (1.29 \text{ kg/m}^3)(1.0 \text{ m} \times 0.60 \text{ m} \times 0.75 \text{ m})$$
$$= \boxed{0.58 \text{ kg}}$$

The air in the refrigerator has more mass than the dozen eggs.

Math HELP
Calculating Volume
See Math Review, Section V

Table 12.1 Densities of Common Substances

Substance	Density (kg/m³)
Solids	
Gold	19,300
Mercury	13,600
Lead	11,300
Silver	10,500
Iron	7860
Aluminum	2700
Ebony (wood)	1220
Ice	917
Cherry (wood)	800
Balsa (wood)	120
Fluids	
Ethylene glycol (antifreeze)	1114
Whole blood (37 °C)	1060
Seawater	1025
Freshwater	1000
Olive oil	920
Ethyl alcohol	806
Oxygen	1.43
Air	1.29
Helium	0.179

Practice Problems

19. A certain gas occupies a volume of 0.28 m³ and has a mass of 0.40 kg. What is the density of the gas? (Refer to Table 12.1 to identify the gas.)

20. What volume of olive oil has a mass of 12 kg? (Refer to Table 12.1.)

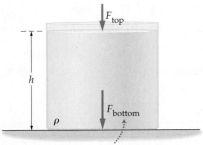

The force on the bottom is equal to the force on the top plus the weight of the fluid in the container.

▲ **Figure 12.7 Pressure and the weight of a fluid**
The force pushing down on the bottom of the container is greater than the force pushing down on the surface of the fluid. The difference in force is the weight of the fluid in the container.

The pressure in a fluid increases with depth

Countless movies have educated us about the perils of taking a submarine too deep. The hull creaks and groans, rivets start to pop, water begins to spray into the ship, and the captain sweats while watching the depth gauge. What causes the pressure to increase as a submarine dives? As you might expect, the increase in pressure is due to the added weight of water pressing on the submarine as it goes deeper.

To see how this works, consider a fluid-filled cylindrical container, like the one in **Figure 12.7**. The top surface of the fluid is open to the atmosphere, which has the pressure $P_{atmospheric}$. If the cross-sectional area of the container is A, the downward force exerted on the top surface by the atmosphere is

$$F_{top} = P_{atmospheric}A$$

At the bottom of the container, the downward force is F_{top} *plus* the weight of the fluid. The mass of the fluid in the container is $m = \rho V$, where ρ is the density of the fluid and $V = hA$ is its volume. Thus, the weight, W, of the fluid is

$$W = mg = \rho Vg = \rho(hA)g$$

It follows that the total force at the bottom of the container is

$$F_{bottom} = F_{top} + W$$
$$= P_{atmospheric}A + \rho(hA)g$$

Dividing the force by the area yields the pressure:

$$P_{bottom} = \frac{F_{bottom}}{A}$$
$$= P_{atmospheric} + \rho gh$$

This equation holds not only for the bottom of the container, but for *any depth below the surface.* Thus, the pressure at a depth h below the surface of any fluid with a density ρ and a pressure $P_{atmospheric}$ at its upper surface is given by the following equation:

Dependence of Pressure on Depth

$$\begin{pmatrix} \text{pressure} \\ \text{at depth } h \end{pmatrix} = \begin{pmatrix} \text{atmospheric} \\ \text{pressure} \end{pmatrix} + \left(\text{density} \times \begin{array}{c} \text{acceleration} \\ \text{due to gravity} \end{array} \times \text{depth} \right)$$

$$P = P_{atmospheric} + \rho gh$$

QUICK Example 12.6 What's the Pressure?

The *Titanic* was found in 1985 on the bottom of the North Atlantic Ocean at a depth of about 4000 m (~2.5 mi). What is the pressure at this depth?

Solution
Applying $P = P_{atmospheric} + \rho gh$ with $\rho = 1025$ kg/m³ for seawater, we have

$$P = P_{atmospheric} + \rho gh$$
$$= 1.01 \times 10^5 \text{ Pa} + (1025 \text{ kg/m}^3)(9.81 \text{ m/s}^2)(4000 \text{ m})$$
$$= \boxed{4.0 \times 10^7 \text{ Pa}}$$

This is about 400 atm, or roughly 6000 lb/in².

21. **Rank** Four containers are shown in **Figure 12.8**, each filled with water to the same level. Rank the containers in order of increasing pressure at the indicated depth h. Indicate ties where appropriate.

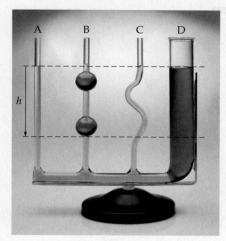

▲ Figure 12.8

The equation $P = P_{\text{atmospheric}} + \rho g h$ can be applied to any two points in a fluid. For example, suppose the pressure at one point is P_1. The pressure P_2 at a depth h below that point is greater than P_1; in fact, it is greater by the amount $\rho g h$. As one would expect, the increase in pressure depends on the density of the fluid (ρ), the acceleration due to gravity (g), and the increased depth (h). It follows that the dependence of pressure on depth is given by the following:

Dependence of Pressure on Depth

$$P_2 = P_1 + \rho g h$$

This relationship is illustrated in **Figure 12.9** and utilized in the next Guided Example.

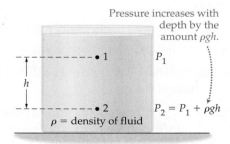

▲ **Figure 12.9 Pressure variation with depth**
If point 2 is deeper than point 1 by the amount h, the pressure at point 2 is greater than that at point 1 by the amount $\rho g h$.

A cubical box of wood 20.00 cm on a side is completely immersed in a fluid. At the top of the box, the pressure is 105.0 kPa. At the bottom of the box, the pressure is 106.8 kPa. What is the density of the fluid?

Picture the Problem

Our sketch shows the box at an unknown depth d below the surface of the fluid. The important dimension for this problem is the height of the box, which is 20.00 cm. We have also been given the pressures at the top and bottom of the box: $P_1 = 105.0$ kPa and $P_2 = 106.8$ kPa, respectively.

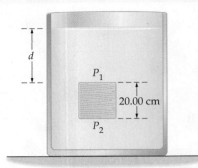

Strategy

The pressures at the top and bottom of the box are related by the equation $P_2 = P_1 + \rho g h$. Since the pressures and the height of the box are given, this relation can be solved for the unknown density, ρ.

Known	Unknown
$P_1 = 105.0$ kPa	$\rho = ?$
$P_2 = 106.8$ kPa	
$h = 20.00$ cm	

Solution

1 Rearrange $P_2 = P_1 + \rho g h$ to solve for the density:

$$P_2 - P_1 = \rho g h$$

$$\rho = \frac{P_2 - P_1}{g h}$$

> **Math HELP**
> Solving Simple Equations
> **See Math Review, Section III**

2 Substitute the numerical values and calculate the density:

$$\rho = \frac{1.068 \times 10^5 \, \text{Pa} - 1.050 \times 10^5 \, \text{Pa}}{(9.81 \, \text{m/s}^2)(0.2000 \, \text{m})}$$

$$= \boxed{920 \, \text{kg/m}^3}$$

Insight

From Table 12.1, it appears that the fluid in question is olive oil. If the box had been immersed in water, with its greater density, the difference in pressure between the top and the bottom of the box would have been greater.

Practice Problems

22. **Follow-up** Given that the density of the fluid is 920 kg/m³, what is the depth, d, to the top of the box? (For this problem, let $P_{\text{atm}} = 101.3$ kPa.)

23. Water in the lake behind Hoover Dam is 221 m deep. What is the water pressure at the base of the dam?

24. A cylindrical container is open to the atmosphere and holds a fluid of density 796 kg/m³. At the bottom of the container the pressure is 121 kPa. What is the depth of the fluid?

While swimming below the surface of the ocean, you let out a stream of air bubbles from your mouth. As the bubbles rise toward the surface, do their diameters increase, decrease, or stay the same?

Reasoning and Discussion

As the bubbles rise, the pressure in the surrounding water decreases. This allows the air in each bubble to expand and occupy a larger volume.

Answer

The diameters of the bubbles increase.

Barometers measure atmospheric pressure

An interesting application of the change of pressure with depth is the *barometer*. The simplest type of barometer was first proposed by the Italian physicist Evangelista Torricelli (1608–1647) in 1643. It's easy to construct a barometer like his. First, fill a long glass tube—open at one end and closed at the other—with a fluid. Next, invert the tube and place its open end below the surface of the same fluid in a bowl, as shown in **Figure 12.10**. Some of the fluid in the tube will flow into the bowl, leaving an empty space (vacuum) at the top of the tube. Enough will remain, however, to create a difference between the level of the fluid in the bowl and the height of the fluid in the tube.

How a Barometer Works The basic idea of the barometer is that the height difference is directly related to the atmospheric pressure that pushes down on the fluid in the bowl. To see the connection, first note that the pressure in the vacuum at the top of the tube is zero. Thus, the pressure in the tube at a depth h below the vacuum is

$$0 + \rho g h = \rho g h$$

Now, at the level of the fluid in the bowl, we know that the pressure is 1 atmosphere, or $P_{\text{atmospheric}}$. Therefore, it follows that

$$P_{\text{atmospheric}} = \rho g h$$

Thus, a measurement of the height difference (h) immediately gives atmospheric pressure.

Mercury is a fluid that is often used in such a barometer; the density of mercury is $\rho = 1.3595 \times 10^4 \text{ kg/m}^3$. The height of a column of mercury at normal atmospheric pressure is

$$h = \frac{P_{\text{atmospheric}}}{\rho g} = \frac{1.013 \times 10^5 \text{ Pa}}{(1.3595 \times 10^4 \text{ kg/m}^3)(9.81 \text{ m/s}^2)} = 760 \text{ mm}$$

Units of Atmospheric Pressure The unit in the above result, millimeters of mercury (mmHg), is, in fact, used to define normal atmospheric pressure:

$$1 \text{ atmosphere} = P_{\text{atmospheric}} = 760 \text{ mmHg}$$

Table 12.2 summarizes the various units in which atmospheric pressure can be expressed.

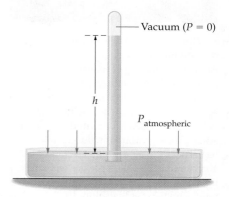

▲ **Figure 12.10 A simple barometer** Atmospheric pressure is related to the height of the fluid in the tube by the equation $P_{\text{atmospheric}} = \rho g h$.

Table 12.2 Atmospheric Pressure

1 atmosphere =	$P_{\text{atmospheric}}$
	= 760 mmHg (definition)
	= 14.7 lb/in^2
	= 101 kPa
	= 101 kN/m^2
~1 bar =	100 kPa

▲ Figure 12.11 A novel way to determine altitude
At sea level this bag of snacks seems to be only about half inflated. This is because atmospheric pressure pushes against the outside of the bag, keeping its volume smaller. At high altitude, however, atmospheric pressure is reduced, and the bag expands to a fully inflated condition. You can imagine that if the bag were fully inflated at sea level, it might expand too much and burst at high altitude.

🔑 *How does an external applied pressure affect an enclosed fluid?*

Thus, atmospheric pressure is due to the weight of air above our heads pushing downward. At higher elevations there is less air above us, and the air pressure is lower. If we change elevation rapidly, as in a car or an airplane, the result is a rapid change in air pressure. This often results in a "popping" in our ears as they adjust to the change.

You can see the effect of different air pressures by looking at a bag of snack food. **Figure 12.11** shows that at sea level, where air pressure is high, the bag is partially inflated. At high altitude, where air pressure is low, the air inside the bag expands—inflating the bag almost to the point of bursting.

Pascal's principle describes pressure within a fluid

Suppose you have a long, thin balloon, like the ones used to make balloon animals. If you squeeze the balloon in the middle, you increase the pressure in the balloon. The increase in pressure occurs everywhere in the balloon, not just at the point where you're squeezing. This is an example of *Pascal's principle.* 🔑 **An external pressure applied to an enclosed fluid is transmitted unchanged to every point within the fluid.** In our example the "external pressure" is the pressure you produce by squeezing the balloon, and the "enclosed fluid" is the air inside the balloon.

A classic example of Pascal's principle at work is the *hydraulic lift*, which is shown schematically in **Figure 12.12**. Here we see two cylinders, one of cross-sectional area A_1 and the other of cross-sectional area $A_2 > A_1$. The cylinders, each of which is fitted with a piston, are connected by a tube and filled with a fluid. Initially the pistons are at the same level and exposed to the atmosphere.

Now, suppose we push down on piston 1 with the force F_1. This increases the pressure in that cylinder by the amount

$$\Delta P = F_1/A_1$$

By Pascal's principle the pressure in cylinder 2 increases by the *same* amount. Therefore, the increased upward force on piston 2 due to the increased pressure of the fluid is

$$F_2 = (\Delta P)A_2$$

Substituting for the increase in pressure from $\Delta P = F_1/A_1$, we find

$$F_2 = (\Delta P)A_2 = \left(\frac{F_1}{A_1}\right)A_2 = F_1\left(\frac{A_2}{A_1}\right)$$

To be specific, let's assume that A_2 is 100 times greater than A_1. Then, by pushing down on piston 1 with a force F_1, we push upward on piston 2 with a force $F_2 = 100F_1$. Our force has been magnified 100 times!

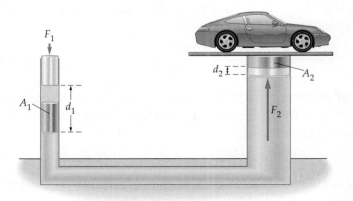

▶ Figure 12.12 A hydraulic lift
A small force, F_1, exerted on the small piston causes a much larger force, F_2, on the large piston.

Hydraulic lifts don't give something for nothing

If the force magnification of a hydraulic lift sounds too good to be true, rest assured, you are not getting something for nothing. Just as with a lever, there's a trade-off between the distance through which a force must be applied and the magnification of the force. This is illustrated in Figure 12.12, where we show piston 1 being pushed down through a distance d_1. This displaces a volume of fluid equal to $A_1 d_1$. The same volume flows into cylinder 2, where it causes piston 2 to rise through a distance d_2. Equating the two volumes, we have

$$A_1 d_1 = A_2 d_2$$

or, rearranging to solve for d_2:

$$d_2 = d_1 \left(\frac{A_1}{A_2} \right)$$

Thus, in the example just given, if we move piston 1 down a distance d_1, piston 2 rises a distance $d_2 = d_1 / 100$. Our *force* has been *magnified* 100 times at piston 2, but the *distance* has been *reduced* 100 times. Force times distance is exactly the same, however, as it must be to conserve energy.

QUICK Example 12.9 What's the Force?

A hydraulic lift is used to raise a 14,500-N car. If the radius of the small piston is 4.0 cm and the radius of the large piston is 17 cm, find the force that must be exerted on the small piston to lift the car.

Solution
Solving $F_2 = F_1(A_2/A_1)$ for F_1 and noting that the cross-sectional area of each piston is given by πr^2, we find

$$F_1 = F_2 \left(\frac{A_1}{A_2} \right)$$

$$= (14{,}500 \text{ N}) \left[\frac{\pi (0.040 \text{ m})^2}{\pi (0.17 \text{ m})^2} \right] = \boxed{800 \text{ N}}$$

> **Math HELP**
> Direct Proportion
> See Math Review, Section IV

Practice Problems

25. The hydraulic lift in Figure 12.12 can lift a 12,200-N car when a force of 715 N is applied to the small piston. If the radius of the large piston is 19 cm, what is the radius of the small piston?

26. A hydraulic lift like that in Figure 12.12 magnifies a force by a factor of 72. If piston 1 moves through a distance of 24 cm, through what distance does piston 2 move?

A buoyant force acts on any object in a fluid

Wouldn't it be fun to float with no effort? Just look at the person in **Figure 12.13**. What makes it so easy for her to float? Well, her ability to float is due to a relatively large upward force exerted on her by the dense water of the Dead Sea. In general, an upward force due to a surrounding fluid is referred to as a **buoyant force**.

▲ **Figure 12.13 Floating with ease**
The water of the Dead Sea is unusually dense because of its large concentration of dissolved salts. As a result, swimmers float high in the water and can engage in recreational activities that aren't usually associated with a dip in the ocean—like reading a magazine.

A fluid surrounding an object exerts a buoyant force in the upward direction. The direction of the buoyant force is due to the fact that pressure increases with depth, and hence the upward force on the object, F_2, is greater than the downward force, F_1. Forces acting to the left and to the right cancel out.

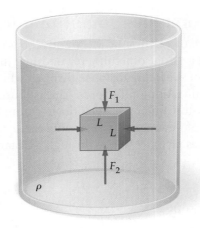

🔑 *What determines the buoyant force acting on an object in a fluid?*

To understand the origin of buoyancy, consider a block immersed in a fluid, as in **Figure 12.14**. Notice that the downward force exerted on the top of the block is less than the upward force exerted on the bottom of the block. This is because the pressure is greater at the bottom of the block. The difference in the forces gives rise to a net upward force on the block—the buoyant force. 🔑 **It can be shown that the buoyant force acting on an object is equal to the *weight of fluid* that the object displaces.** This is referred to as *Archimedes' principle.*

Archimedes' Principle

An example of Archimedes' principle is illustrated in **Figure 12.15**. We start with an object hanging from a scale in **Figure 12.15 (a)**. The weight of the object is 30 N. Next we immerse the object in a fluid, as shown in **Figure 12.15 (b)**. This causes liquid to spill from the main container into a smaller side container. The weight of the spilled liquid is found to be 10 N. This means that the buoyant force exerted on the object is 10 N. Sure enough, the apparent weight of the object, as read on the scale, is reduced to 30 N − 10 N = 20 N.

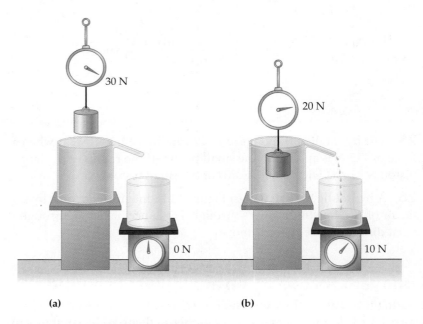

(a)   (b)

▲ **Figure 12.15 Archimedes' principle in action**
Dipping a block into water decreases the reading on a scale. The amount of the decrease is equal to the weight of the water that the block displaces. In this case 10 N of water is displaced, and the apparent weight of the block decreases by 10 N.

A flask of water rests on a scale. You dip your finger into the water, without touching the flask. Does the reading on the scale increase, decrease, or stay the same?

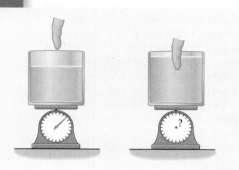

Reasoning and Discussion

Your finger experiences an upward buoyant force when it is dipped into the water. By Newton's third law, the water experiences an equal and opposite reaction force acting downward. This downward force is transmitted to the scale, which in turn gives a higher reading.

   Another way to look at this result is to note that when you dip your finger into the water, its depth increases. This results in a greater pressure at the bottom of the flask, and hence a greater downward force on the flask. The scale reading reflects this increased downward force.

Answer

The reading on the scale increases.

Objects float when buoyant force equals weight

When a person floats in water, the person displaces an amount of water equal to his or her weight. This means that the upward buoyant force is equal in magnitude to the downward force of gravity. The forces cancel, and the person is suspended by the water. If the water is quite dense, like that in the Dead Sea or the Great Salt Lake in Utah, less water than normal must be displaced. This is why people can float with ease in those bodies of water.

A cup is filled to the brim with water and a floating ice cube. When the ice melts, does the water overflow the cup, the water level become lower, or the water level remain the same?

Reasoning and Discussion

Since the ice cube floats, it displaces a volume of water equal to its weight. When it melts, it becomes water, and its weight is the same. Thus, the melted water from the ice cube fills exactly the same volume that the ice cube displaced when floating. As a result, the water level is unchanged.

Answer

The water level remains the same.

High-density objects can also float

If you take a steel block and drop it into a container of water, it *sinks* to the bottom. In general, an object sinks when the buoyant force is less than the weight of the object. In the case of the steel block, the weight of the water displaced by the block is less than the weight of the block itself. This is because the density of the block is greater than the density of water—the block contains more mass in a given volume than water does.

▶ **Figure 12.16 Floating an object that is more dense than water**
(a) A container of water. **(b)** A steel block is placed in the water. The block sinks because the buoyant force acting on it is less than its weight. That is, the weight of water displaced by the block is less than the weight of the block. **(c)** If the steel block is molded into the shape of a bowl, it can displace more water than the volume of the steel itself. In fact, it can displace enough water to float.

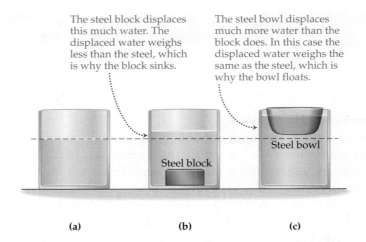

The steel block displaces this much water. The displaced water weighs less than the steel, which is why the block sinks.

The steel bowl displaces much more water than the block does. In this case the displaced water weighs the same as the steel, which is why the bowl floats.

Steel bowl

Steel block

(a) (b) (c)

The result of dropping a metal block into a container of water is shown in **Figure 12.16 (a)** and **(b)**. But there's more to the story. What if you take the steel block and mold it into the shape of a bowl, as shown in **Figure 12.16 (c)**? In this case the steel object floats nicely, even though it still has a greater density than water. Why? Well, the bowl shape displaces more water than the block of steel. In fact, the bowl displaces a volume of water whose weight is the same as the weight of the steel—and thus the steel can now float. This is the principle that allows ships to be made of steel and iron.

12.2 LessonCheck (MP)

Checking Concepts

27. ⬛ **Describe** What are the two key characteristics of the pressure exerted by a gas?

28. ⬛ **Give Examples** Explain Archimedes' principle in your own words, and give an example of a situation to which it applies.

29. Explain What holds a suction cup in place?

30. Explain Why is it possible for people to float without effort in Utah's Great Salt Lake?

31. | Triple Choice | A cup is filled to the brim with water. An ice cube floats in the water, and resting on top of the ice cube is a small pebble, as shown in **Figure 12.17**. When the ice melts and the pebble drops to the bottom, does the water level rise (overflowing the cup), fall, or stay the same. Explain.

▲ **Figure 12.17**

32. | Triple Choice | Lead is more dense than aluminum. Is the buoyant force on a solid lead sphere submerged in water greater than, less than, or equal to the buoyant force on a solid aluminum sphere of the same diameter? Explain.

Solving Problems

33. Approximate Use what you have learned in this lesson to estimate the mass of air in your physics classroom.

34. Calculate What is the downward force exerted by the atmosphere on a football field whose dimensions are 110 m by 49 m?

35. Calculate A raft is 4.2 m wide and 6.5 m long. When a horse is loaded onto the raft, it sinks 2.7 cm deeper into the water. What is the weight of the horse?

36. Calculate The force exerted on the small piston of a hydraulic lift is 750 N. If the area of the small piston is 0.0075 m² and the area of the large piston is 0.13 m², what is the force exerted by the large piston?

12.3 Fluids in Motion

In Florida building codes specify that the roof of a house must be strapped down to posts that are fixed in the ground. Why strap the roof down like that? The answer is that the high winds of hurricanes tend to lift roofs upward. The physics behind this phenomenon, as well as other aspects of fluids in motion, is the subject of this lesson.

Fluid speed depends on cross-sectional area

Suppose you want to water the yard, or shoot water on a friend, but you don't have a spray nozzle for the end of the hose. Without a nozzle the water flows rather slowly from the hose and hits the ground within half a meter. If you place your thumb over the end of the hose, however, and narrow the opening to a fraction of its original size, as shown in **Figure 12.18**, the water sprays out with a high speed and a large range. Why does decreasing the size of the opening make the water shoot out faster and farther?

To answer this question, consider a simple system that shows the same behavior. Imagine a fluid flowing with a speed v_1 through a cylindrical pipe of cross-sectional area A_1, as in the left-hand portion of **Figure 12.19**. If the pipe narrows to a cross-sectional area A_2, as in the right-hand portion of Figure 12.19, the fluid will flow with a new speed, v_2.

The speed in the narrow section of the pipe is found by noting that any amount of fluid that passes point 1 in a given time must also flow past point 2 in the same time. Otherwise, there would be a buildup or a loss of fluid between points 1 and 2—that is, conservation of mass would be violated. To satisfy conservation of mass, the product of speed v and cross-sectional area A must be the same at points 1 and 2. This leads to the *equation of continuity*:

Equation of Continuity

$$\begin{array}{c}\text{cross-sectional} \\ \text{area 1}\end{array} \times \text{velocity 1} = \begin{array}{c}\text{cross-sectional} \\ \text{area 2}\end{array} \times \text{velocity 2}$$

$$A_1 v_1 = A_2 v_2$$

The equation of continuity shows that if the cross-sectional area through which a fluid is flowing is reduced, the speed of the fluid increases. For example, if the area is halved, the fluid speed is doubled.

We apply this equation to the case of water flowing through the nozzle of a fire hose in the next Guided Example.

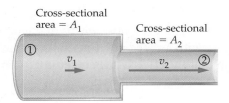

Cross-sectional area = A_1

Cross-sectional area = A_2

① v_1

v_2 ②

Vocabulary

- Bernoulli's principle
- lift
- viscosity
- surface tension

▲ **Figure 12.18 Using a thumb to increase water speed**
Narrowing the opening in a hose with a nozzle (or a thumb) increases the velocity of flow.

How does a change in the cross-sectional area through which a fluid is flowing affect its speed?

◄ **Figure 12.19 Fluid flow through a pipe of varying diameter**
As a fluid flows from a large pipe to a small pipe, the same mass of fluid passes a given point in a given amount of time. Thus, the fluid's speed in the small pipe is greater than it is in the large pipe.

Water flows through a 9.6-cm-diameter fire hose with a speed of 1.3 m/s. At the end of the hose, the water flows out through a nozzle whose diameter is 2.5 cm. What is the speed of the water coming out of the nozzle?

Picture the Problem

In our sketch we label the speed of the water in the hose v_1 and the speed of the water coming out the nozzle v_2. We are given $v_1 = 1.3$ m/s. We also know that the diameter of the hose is $d_1 = 9.6$ cm and the diameter of the nozzle is $d_2 = 2.5$ cm.

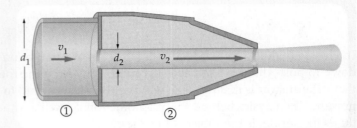

Strategy

We can find the speed of the water in the nozzle by applying the equation of continuity, $A_1 v_1 = A_2 v_2$. The hose and the nozzle are circular in cross section, and hence their cross-sectional areas are given by $A = \pi r^2 = \pi d^2 / 4$, where d is the diameter.

Known	Unknown
$v_1 = 1.3$ m/s	$v_2 = ?$
$d_1 = 9.6$ cm	
$d_2 = 2.5$ cm	

Solution

1 Solve the continuity equation for v_2, the speed of the water in the nozzle:

$$A_1 v_1 = A_2 v_2$$
$$v_2 = \frac{A_1 v_1}{A_2}$$
$$= v_1 \left(\frac{A_1}{A_2} \right)$$

> **Math HELP**
> Calculating Area
> **See Math Review, Section V**

2 The two cross-sectional areas are circular, and therefore $A = \pi d^2 / 4$. Substitute this expression for A_1 and A_2 and simplify:

$$v_2 = v_1 \left(\frac{\pi d_1^2 / 4}{\pi d_2^2 / 4} \right)$$
$$= v_1 \left(\frac{d_1^2}{d_2^2} \right)$$

3 Substitute the numerical values:

$$v_2 = (1.3 \text{ m/s}) \left(\frac{9.6 \text{ cm}}{2.5 \text{ cm}} \right)^2$$
$$= \boxed{19 \text{ m/s}}$$

Insight

Notice that a small-diameter nozzle can produce a very high speed. In fact, the speed depends inversely on the diameter squared. In this case the speed of the water increased by a factor of more than 14.

Practice Problems

37. **Follow-up** What nozzle diameter would be required to give the water a speed of 21 m/s?

38. To water the yard you use a hose with a diameter of 3.4 cm. Water flows from the hose with a speed of 1.1 m/s. If you partially block the end of the hose so that the effective diameter is 0.57 cm, with what speed does water come out of the hose?

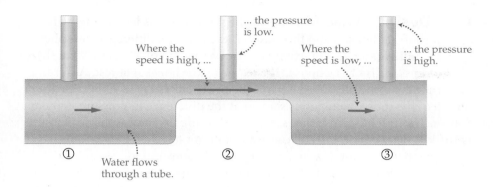

A change in pressure accompanies a change in speed

When a fluid flows at high speed, like the water shooting from a small nozzle, is its pressure greater than, less than, or the same as when it flows at low speed? The correct answer might surprise you. 🔑 **The pressure of water—or any other fluid—is *reduced* when it flows at a *higher* speed.** This fact is known as **Bernoulli's principle**.

As a demonstration of this behavior, consider the experiment shown in **Figure 12.20**. Here we see a fluid flowing through a pipe that starts out wide, narrows down, then widens out again. From the equation of continuity, we know that the speed of the fluid is greater at point 2 than at point 1. How about the pressure?

A convenient way to measure the pressure in this system is with a vertical tube containing a column of fluid, as shown in Figure 12.20. The greater the pressure in the fluid, the higher up it pushes the column of fluid in the vertical tube. Notice that the pressure is high at point 1, drops to a lower value at point 2 (where the fluid speed is greater), and then rises again at point 3 (where the fluid slows down again). Indeed, we can clearly see that higher speed produces lower pressure.

Bernoulli's principle has practical applications

Perhaps the easiest way to demonstrate Bernoulli's principle is to blow across the top of a piece of paper. If you hold the paper as shown in **Figure 12.21** and then blow over the top surface, the paper will lift upward. The reason is that there is a difference in air speed between the top and the bottom of the paper, with the higher speed on top. According to Bernoulli's principle, the pressure above the paper is therefore less than the pressure below the paper. This pressure difference, in turn, results in a net upward force, referred to as **lift**, and the paper rises.

Airplane Wings A similar example of the relation of pressure and speed is provided by an airplane wing. A cross section of a typical wing is shown in **Figure 12.22**. The shape of the wing is designed so that air flows more rapidly over the upper surface than the lower surface. As a result, Bernoulli's principle states that the pressure is less on the top of the wing. As with the piece of paper, the pressure difference results in a net upward force (lift) on the wing.

Notice that lift is a dynamic effect; it requires a flow of air. Airplane wings provide lift only when the plane is moving through the air. This is why planes have to reach a certain takeoff speed in order to become airborne.

🔑 *How does a change in the speed of a fluid affect its pressure?*

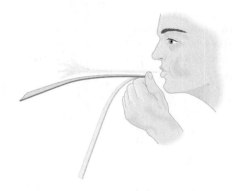

▲ **Figure 12.21 The Bernoulli effect on a sheet of paper**
If you hold a piece of paper by its end, it will bend downward. Blowing across the top of the paper reduces the pressure there, resulting in a net upward force that lifts the paper to a nearly horizontal position.

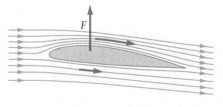

▲ **Figure 12.22 Airflow and lift in an airplane wing**
An airplane wing is shaped so that air flows more rapidly over the top than along the bottom. As a result, the pressure on the top of the wing is reduced, and a net upward force (lift) is generated.

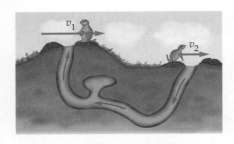

▲ **Figure 12.23** **Air circulation in a prairie dog burrow**
A prairie dog burrow typically has a high mound on one end and a low mound on the other. Since the wind speed increases with height above the ground, the pressure is smaller at the high-mound end of the burrow. The result is a beneficial circulation of fresh air through the burrow.

Roof Design As mentioned at the beginning of this lesson, high winds can lift roofs off houses, and the reason is the pressure difference described by Bernoulli's principle. For example, suppose a wind of 35 m/s (78 mi/h) blows across the roof of a house. The air inside the house is at rest. According to Bernoulli's principle, this difference in air speed results in a lower pressure on top of the roof. Calculations show that the difference in pressure is about 800 Pa. This might seem rather small, considering that atmospheric pressure is 101 kPa. However, it can still cause a significant force over a relatively large area, like that of a roof. If a roof has an area of about 150 m², for example, a pressure difference of 800 Pa results in an upward force of over 120,000 N (about 27,000 lb)! This is why roofs are often torn from houses during severe windstorms and why Florida mandates that roofs be strapped down to solid ground.

Bernoulli's Principle in Nature On a lighter note, prairie dogs benefit from the effects of Bernoulli's equation. A schematic prairie dog burrow is pictured in **Figure 12.23**. Notice that one of the entrance/exit mounds is higher than the other. This is significant because the speed of air flow due to the incessant prairie winds varies with height. In particular, the speed is zero right at ground level and increases to its maximum value within a few feet above the surface. As a result, the speed of air over the higher mound is greater than that over the lower mound. This causes the pressure to be less over the higher mound. The pressure difference between the two mounds causes air to be drawn through the burrow, giving a form of natural air conditioning.

CONCEPTUAL Example 12.13 **What Happens to the Ragtop Roof?**

A small ranger vehicle has a soft, ragtop roof. When the car is at rest, the roof is flat. When the car is cruising at highway speed with its windows rolled up, does the roof bow upward, remain flat, or bow downward?

Reasoning and Discussion
When the car is in motion, air flows over the top of the roof. Meanwhile, the air inside the car is at rest—since the windows are closed. Thus, there is less pressure over the roof than under it. As a result, the roof bows upward.

Answer
The roof bows upward.

🗝 *How is fluid viscosity related to work?*

Real fluids have viscosity and surface tension

When a block slides across a rough floor, it experiences a frictional force opposing the motion. Similarly, a fluid flowing past a stationary surface experiences a force opposing the flow. The tendency of a fluid to resist flow is referred to as the **viscosity** of the fluid. Fluids like air have low viscosities, thicker fluids like water are more viscous, and fluids like honey and molasses are characterized by high viscosity.

Viscosity becomes most apparent when a fluid flows through a tube or a pipe. Examples include water flowing through a pipe in a house and blood flowing through an artery or a vein. If the fluid is ideal, it flows through the tube with no resistance. 🔵 **Real fluids always have some viscosity, and hence work must be done to force a fluid to flow through a tube.** This is the reason the blood in our arteries has a relatively high pressure—the pressure is needed to push the blood against the resistance due to its viscosity. If a blood vessel is partially blocked, the resistance to the blood flow increases, resulting in a higher than normal blood pressure.

Another way real fluids differ from ideal fluids is in the behavior of their surface. Examine the spider resting on the surface of a pond in **Figure 12.24**. If you look carefully, you can see that the spider creates tiny dimples in the water's surface. It's almost as if the spider is supported by a thin sheet of rubber that tries to resist being stretched. In general, the force that tries to minimize the surface area of a fluid is referred to as the **surface tension**. It is surface tension that makes the surface behave like an elastic sheet. Even a needle or a razor blade can be supported by surface tension if it is put into place gently.

▲ **Figure 12.24 Visualizing surface tension**
Surface tension causes the surface of a liquid to behave like an elastic skin or membrane. When a small force is applied to the liquid surface, it tends to stretch. This enables a fishing spider, like the one shown here, to walk on the surface of a pond.

12.3 LessonCheck 🔵

Checking Concepts

39. 🔵 **Choose** You put your thumb over the end of a garden hose, partially blocking the opening. Is the speed of the water coming out of the hose greater than, less than, or the same as the speed of the water within the hose? Explain.

40. 🔵 **Explain** If you blow air across the top of a sheet of paper, the paper rises. Use Bernoulli's principle to explain why this happens.

41. 🔵 **Explain** Why is work required to get a real fluid to flow through a tube?

42. Explain Why does a stream of water emerging from a water faucet become narrower as it falls?

43. Choose A fluid flows through a pipe with a speed v. If the diameter of the pipe increases by a factor of 3, is the new speed of the fluid equal to $v/9$, $v/3$, v, $3v$, or $9v$? Explain.

Solving Problems

44. Calculate Water flows with a speed of 1.3 m/s through a section of hose with a cross-sectional area of 0.0075 m². If the cross-sectional area narrows down to 0.0033 m², what is the new speed of the water?

45. ⎹Think & Calculate⎸ Water flows with a speed of 2.5 m/s through a section of hose with a cross-sectional area of 0.0084 m². Further along the hose the cross-sectional area changes, and the water speed is reduced to 1.1 m/s.
(**a**) Is the new cross-sectional area greater than, less than, or equal to 0.0084 m²?
(**b**) Calculate the new cross-sectional area.

46. Calculate A fluid flows through a pipe with a speed of 1.8 m/s. The diameter of the pipe is 2.7 cm. Further along, the diameter of the pipe changes, and the fluid flowing in this section has a speed of 3.1 m/s. What is the new diameter of the pipe?

12.4 Solids

Vocabulary

- Hooke's law
- elastic

Solids maintain their shape. When Michelangelo (1475–1564), an Italian Renaissance painter, sculptor and architect, carved his statue of David from a solid block of marble, he was confident it would retain its shape long after his work was done. Even so, the shape of a solid *can* be changed—though usually only slightly—if a force is applied. In this lesson you'll learn about some of the deformations that can occur in solids.

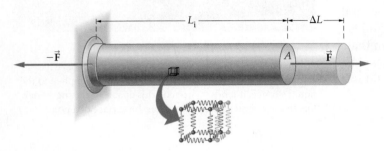

▲ **Figure 12.25 Stretching a rod**
Equal and opposite forces applied to the ends of a rod cause it to stretch. On the atomic level the forces stretch the "intermolecular springs" in the solid, resulting in an overall increase in length.

A force can stretch or compress a solid

A useful way to think about a solid is as a collection of small balls representing molecules. The molecules are connected to one another by springs, as shown in **Figure 12.25**. The springs represent the forces that the molecules exert on one another. When you pull on a solid rod with a force, you stretch each of the "intermolecular springs" in the direction of the force. The net result is that the entire solid increases in length by an amount proportional to the force, as indicated in Figure 12.25.

Hooke's law describes length change in a solid

If you've ever pulled on a spring, you know that the amount of stretch varies with the force you apply. The greater the force, the greater the stretch. In fact, this is the principle behind a spring scale, as illustrated in **Figure 12.26**. The greater the weight you place on the scale, the greater the stretch of the spring and the higher the reading.

These observations can be summarized with the following equation:

$$F = k\,\Delta L$$

This equation is known as Hooke's law. **Hooke's law** states that the change in length of a solid object is proportional to the applied force. The change in length can be either a stretch or a compression. The constant k is called the *spring constant*, and its units are newtons per meter (N/m).

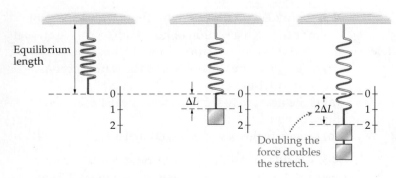

▲ **Figure 12.26 A spring scale**
A spring scale is used to measure the weight of an object. The heavier the object, the more the spring stretches.

We generally represent the change in length of a spring with the symbol x, as shown in **Figure 12.27**. When $x = 0$, the spring is relaxed and there is no change in length. When the spring is stretched or compressed, x represents the distance from equilibrium. Replacing ΔL with x yields Hooke's law for the specific case of a spring:

> **Hooke's Law for a Spring**
>
> force = spring constant $\times$ change in length
> $$F = kx$$

The distance x is always positive, and hence $F = kx$ is the magnitude of the spring force.

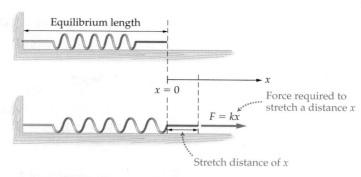

▲ **Figure 12.27 Stretching a spring**
The equilibrium position of the spring is indicated by $x = 0$. The magnitude of force required to stretch (or compress) the spring by a distance x is $F = kx$.

QUICK Example 12.14 **What's the Force?**

A spring with a spring constant of 21 N/m is stretched 3.4 cm. What force is required to cause this amount of stretching?

Solution
Using $k = 21$ N/m and $x = 0.034$ m, we find

$$F = kx = (21 \text{ N/m})(0.034 \text{ m}) = \boxed{0.71 \text{ N}}$$

Practice Problems

47. Pulling on a spring with a force of 1.2 N causes it to stretch by 6.4 cm. What is the spring constant for this spring?

48. What is the stretch distance when you pull with a force of 2.3 N on a spring with a spring constant of 19 N/m?

Solids can become permanently deformed

If you pull lightly on a spring, it stretches. When you let go, the spring returns to its original length. Objects that return to their original shape and size after being deformed are said to be **elastic**.

Stretching a spring too far, however, causes permanent deformation. In this case the spring never returns to its original state, even when you release it. You have exceeded the *elastic limit* of the spring.

These behaviors are illustrated by the stretch-force graph in **Figure 12.28**. For small forces the change in length is proportional to the force (as given by Hooke's law). The straight line that begins at the origin shows this relationship. Hooke's law, however, does not apply to somewhat larger forces. The spring is still elastic—it still returns to its original length when the force goes to zero—but it deviates from the behavior predicted by Hooke's law.

If the stretching force is increased even further, the elastic limit is reached. Stretching beyond the elastic limit causes permanent deformation of the object. This is what happens to a spring that is "sprung" by being stretched too far or a car fender that has been dented. Reducing the force to zero does not return the system to its original length or shape.

What causes the permanent deformation of a solid?

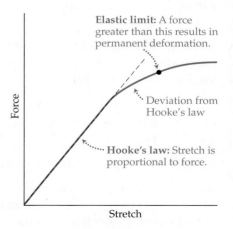

▲ **Figure 12.28 Stretch versus force**

Springs are coiled to increase their length

A long rope stretches considerably more for a given force than a short section of the same rope. In general, the amount an object stretches is proportional to its initial length.

For example, the wire shown in **Figure 12.29 (a)** stretches very little when a weight is attached to it. The longer wire in **Figure 12.29 (b)** stretches a little more with the same weight, even though it's made of the same material. To get a really useful amount of stretch, however, requires a considerably greater initial length. But how can you increase the length of a wire by a large amount without taking up a lot of extra space? The answer is to coil the wire into the shape of a helix; that is, to make it into a spring. This greatly increases the length, and also the amount of stretch, as we see in **Figure 12.29 (c)**. Thus, the coiled shape of a spring is functional—it's part of what makes a spring useful.

▶ **Figure 12.29 Stretch distance is proportional to initial length**
The amount that a wire, or any object, will stretch is proportional to the initial length of the wire. A convenient way to increase the initial length of a wire is to wrap it into a coil of many turns. This is how a spring is made.

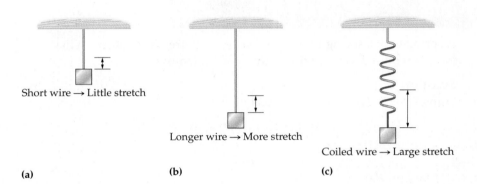

Short wire → Little stretch

Longer wire → More stretch

Coiled wire → Large stretch

(a)

(b)

(c)

12.4 LessonCheck (MP)

Checking Concepts

49. ▱ Compare How do the final lengths of two identical rods compare if rod 1 is stretched by a force that is less than its elastic limit, and rod 2 is stretched by a force that is greater than its elastic limit but less than its breaking strength?

50. Explain Hooke's law says that there is a definite relationship between the stretching of an object and the force exerted on it. What is this relation?

51. Describe An elastic object is stretched by an applied force. What happens after the applied force is removed?

52. Big Idea Describe, on a molecular level, how an applied force can change the length of a solid object.

53. | Triple Choice | Two identical springs are connected end to end. Is the spring constant of the resulting compound spring greater than, less than, or equal to that of each single spring? Explain.

Solving Problems

54. Calculate A force of 5.2 N stretches a certain spring by 14 cm. What is the spring constant of the spring?

55. Calculate A spring has a spring constant of 56 N/m. How much is the spring compressed by a force of 6.1 N?

56. Calculate A spring with a spring constant of 62 N/m is stretched 19 cm by an applied force. What is the magnitude of the applied force?

57. Calculate A spring with a spring constant of 42 N/m is attached end to end to a spring whose spring constant is 66 N/m. The two springs together act as one compound spring. What is the spring constant of this compound spring?

Physics & You

Meteorologist

A *meteorologist* is an atmospheric scientist who studies the physical and chemical processes of the atmosphere. Meteorologists use information about these processes to predict weather and climate patterns. They also track powerful storms, especially hurricanes, to protect coastal communities, limit structural damage, and ensure economic continuity.

Meteorologists track severe storms using direct and indirect measurements. Direct measurements are taken from reconnaissance aircraft and buoys in the ocean. Indirect measurements are collected from satellites and Doppler radar. All these date are analyzed in order to make predictions about a storm's path, wind speeds, and precipitation potential. Advances in technology continue to reduce the *error cone* associated with a storm's predicted path. Severe storm information is conveyed to coastal communities so that citizens can prepare their homes, stockpile supplies, and, when necessary, evacuate the area.

Becoming a meteorologist generally requires a 4-year college degree in meteorology or a degree in another science with a certain amount of coursework in meteorology. Employers look for job candidates with strong backgrounds in mathematics, computer science, and physics, along with good communication skills. Many meteorologists work for the National Oceanic and Atmospheric Administration (NOAA) or the United States armed services. Other job opportunities exist at colleges and universities, airlines, government agencies, and television networks.

▲ Meteorologists consider atmospheric dynamics when tracking tropical storms that might grow into hurricanes.

▲ A small number of meteorologists work for local and national media companies.

Take It Further

1. Research Use the Internet to research and compare the qualifications and pay of different jobs in meteorology. Be sure to consider a range of jobs, from those in the military and government to those at local television stations and national networks.

2. Apply Write a short report on the weather systems known as warm fronts and cold fronts. Use the physics principles from this chapter and previous ones to explain their characteristics.

Physics Lab Investigating Hooke's Law

This lab explores Hooke's law and the elastic nature of solids. You will use force and displacement data to determine two spring constants and also test to see if a gummy worm obeys Hooke's Law.

Materials
- spring scale (or force probe)
- two springs of different stiffnesses
- meterstick
- masking tape
- gummy worm
- ring stand

Procedure

1. Select a spring and secure one of its ends to a ring stand on the tabletop.

2. Lay the spring in a straight line across the tabletop. Place the meterstick in line with the spring, with its zero end at the end of the spring. Secure the meterstick in place with masking tape.

3. Use the spring scale (or force probe) to stretch the spring 5.0 cm longer than its unstretched length. Record the force required to do this in Data Table 1.

4. Repeat Step 3 for stretch distances of 10.0 cm, 15.0 cm, and 20.0 cm.

5. Remove the first spring, and then repeat Steps 1–4 with the second spring.

6. Use the spring scale (or force probe) to measure the force needed to stretch a gummy worm by three different amounts. Use care when choosing your stretch amounts—don't overstretch your worm! Record your data in Data Table 2.

Data Table 1: Spring Data

Stretch Distance (cm)	Spring 1	Spring 2
	Force (N)	
5.0		
10.0		
15.0		
20.0		

Data Table 2: Gummy Worm Data

Stretch Distance (cm)	Force (N)

Analysis

1. Plot a force-stretch graph using the data in Data Table 2. Plot the data for both springs on the same graph.

2. Draw a best-fit straight line through the data points for each spring. Use a different color for each best-fit line.

3. Calculate the slope of the best-fit line for each spring. What are the units of the slope? What does the slope represent?

4. Plot a force-stretch graph for the gummy worm data in Data Table 2.

5. Can you draw a best-fit straight line through the gummy worm data points? If so, do so. If not, explain why not.

Conclusions

1. What mathematical term describes the force-stretch relationship shown by the spring data?

2. Look carefully at your data for the gummy worm. Does the gummy worm obey Hooke's law? Explain why or why not.

12 Study Guide

Big Idea

Fluids flow and change shape easily, whereas solids maintain a definite shape unless acted on by a force.

Fluids are characterized by their pressure, which depends on volume, temperature, and the number of particles. Solids have a definite shape, but the length of a solid object can be changed by applying a force.

12.1 Gases

🔑 The pressure in a gas (or any fluid) acts equally in all directions. In addition, the pressure is always at right angles to any surface it acts on.

🔑 The ideal gas equation shows that gas pressure increases if the number of gas particles increases, the temperature of the gas increases, or the volume of the gas decreases.

🔑 Molar mass provides a convenient bridge between the macroscopic world, where we measure the mass of a substance in grams, and the microscopic world, where the number of particles in a sample of a substance is typically 10^{23} or more.

• An ideal gas is one in which the particles have no effect on one another.

• A mole (mol) is the amount of a substance that contains as many particles as there are atoms in 12 grams of carbon-12. The number of molecules in a mole of any substance is Avogadro's number, $N_A = 6.022 \times 10^{23}$.

Key Equations

The pressure of a gas is described by the ideal gas equation:

$$PV = NkT$$

The ideal gas equation can also be expressed in terms of moles:

$$PV = nRT$$

Gauge pressure is always less than the actual pressure by the amount $P_{atmospheric}$:

$$P_{gauge} = P - P_{atmospheric}$$

12.2 Fluids at Rest

🔑 The denser a substance, the more mass it has in any given volume.

🔑 An external pressure applied to an enclosed fluid is transmitted unchanged to every point within the fluid. This is Pascal's principle.

🔑 The buoyant force acting on an object is equal to the *weight of fluid* that the object displaces. This is Archimedes' principle.

• An upward force due to a surrounding fluid is referred to as a *buoyant force*.

• The pressure of a fluid at rest increases with depth.

Key Equations

The density, ρ, of a substance is given by

$$\rho = \frac{m}{V}$$

If the pressure at one point in a fluid is P_1, the pressure at a depth h below that point is

$$P_2 = P_1 + \rho g h$$

12.3 Fluids in Motion

🔑 The equation of continuity shows that if the cross-sectional area through which a fluid is flowing is reduced, the speed of the fluid increases.

🔑 The pressure of water—or any other fluid—is *reduced* when it flows at a *higher* speed. This is Bernoulli's principle.

🔑 Real fluids always have some viscosity, and hence work must be done to force a fluid to flow through a tube.

• Viscosity is the tendency of a fluid to resist flow.

• Surface tension is the force that tries to minimize the surface area of a fluid.

Key Equation

The equation of continuity relates the cross-sectional area through which a fluid is flowing to its flow speed:

$$A_1 v_1 = A_2 v_2$$

12.4 Solids

🔑 Stretching beyond the elastic limit causes permanent deformation of the object.

• Hooke's law states that the change in length of a solid object is proportional to the applied force.

• Objects that return to their original shape and size after being deformed are said to be elastic.

Key Equation

The stretch or compression of a spring is given by Hooke's law:

$$F = kx$$

12 Assessment

For instructor-assigned homework, go to www.masteringphysics.com

ANSWERS TO SELECTED ODD-NUMBERED PROBLEMS APPEAR IN APPENDIX A.

Lesson by Lesson

12.1 Gases
Conceptual Questions

58. At the beginning of an airline flight, you are instructed about the proper use of the oxygen masks that will fall from the ceiling if the cabin pressure suddenly drops. You are advised that the oxygen masks will work properly, even if the bags do not fully inflate. In fact, the bags expand to their fullest if cabin pressure is lost at high altitude but expand only partially if the plane is at low altitude. Explain.

59. **High-Elevation Airport** One of the highest airports in the world is located in La Paz, Bolivia. Pilots prefer to take off from this airport in the morning or the evening, when the air is quite cold. Explain.

60. Triple Choice Is the number of atoms in 1 mole of helium greater than, less than, or equal to the number of atoms in 1 mole of oxygen? Helium consists of individual atoms, He, and oxygen is a diatomic gas, O_2.

61. Predict & Explain (a) If you put a helium-filled balloon in the refrigerator, will its volume increase, decrease, or stay the same? (b) Choose the *best* explanation from among the following:

A. Lowering the temperature of an ideal gas at constant pressure results in a reduced volume.

B. The same amount of gas is in the balloon; therefore, its volume remains the same.

C. The balloon can expand more in the cool air of the refrigerator, giving it an increased volume.

Problem Solving

62. A 79-kg person sits on a 3.7-kg chair. Each leg of the chair makes contact with the floor on a circle that is 1.3 cm in diameter. Find the pressure exerted on the floor by each leg of the chair, assuming that the weight is evenly distributed.

63. To prevent damage to floors (and to increase friction), a crutch often has a rubber tip attached to its end. If the end of the crutch is a circle of radius 1.2 cm without the tip, and the rubber tip is a circle of radius 2.5 cm, by what factor does the tip reduce the pressure exerted by the crutch?

64. An inflated football has a gauge pressure of 68 kPa. What is the actual pressure inside the ball?

65. An automobile tire has a volume of 0.0185 m³. At a temperature of 294 K the pressure in the tire is 212 kPa. How many moles of air must be pumped into the tire to increase its pressure to 252 kPa, given that the temperature and volume of the tire remain constant?

66. **Helium-Filled Blimp** The blimp *Spirit of Akron* is 62.6 m long and contains 7023 m³ of helium. When the temperature of the helium in the blimp is 285 K, its pressure is 112 kPa. Find the mass of the helium in the blimp.

67. Think & Calculate The weight of a 1420-kg car is supported equally by its four tires, each inflated to a gauge pressure of 351 kPa. (a) What is the area of contact each tire makes with the ground? (b) If the gauge pressure is increased, does the area of contact increase, decrease, or stay the same? Explain.

68. A cylindrical container is fitted with an airtight piston that is free to slide up and down, as shown in **Figure 12.30**. A mass rests on top of the piston. The initial temperature of the system is 313 K, and the pressure of the gas is held constant at 137 kPa. The temperature is then increased until the height of the piston rises from 23.4 cm to 26.0 cm. What is the final temperature of the gas?

23.4 cm

Figure 12.30

12.2 Fluids at Rest
Conceptual Questions

69. Suppose you drink a liquid through a straw. Explain why the liquid moves upward, against gravity, into your mouth.

70. Considering your answer to Problem 69, is it possible to sip liquid through a straw on the surface of the Moon? Explain.

71. **Movie Physics** A science fiction movie from the 1960s shows Earth experiencing a rapid warming. In one scene large icebergs break up into small, car-size chunks that drop downward through the water and bounce off the hull of a submerged submarine. Is this an example of good, bad, or ugly physics? Explain.

72. Using a Hydrometer A *hydrometer* is a device for measuring fluid density. It is constructed as shown in **Figure 12.31**. If the hydrometer pulls a sample of fluid 1 into it, the small float inside the tube is submerged to the level 1. When fluid 2 is sampled, the float is submerged to level 2. Is the density of fluid 1 greater than, less than, or equal to the density of fluid 2? (This is how a car mechanic tests your car's antifreeze level. Since antifreeze [ethylene glycol] is more dense than water, the higher the density of coolant in your radiator, the more antifreeze protection your car has.)

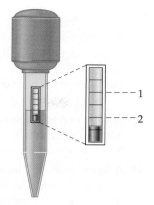

Figure 12.31

73. **Predict & Explain** A block of wood has a steel ball glued to one surface. The block can float in water with the ball "high and dry" on its top surface. **(a)** When the block is inverted and the ball is immersed in the water, does the volume of wood that is submerged increase, decrease, or stay the same? **(b)** Choose the *best* explanation from among the following:

A. When the block is inverted, the ball pulls it downward, causing more of the wood to be submerged.

B. The same amount of mass is supported in either case; therefore, the amount of wood that is submerged is the same.

C. When the block is inverted, the ball experiences a buoyant force, which reduces the buoyant force that must be provided by the wood.

74. A block of wood floats on water. A layer of oil is then poured on top of the water to a depth that more than covers the block, as shown in **Figure 12.32**. Is the volume of wood submerged in the water greater than, less than, or the same as before the oil was added?

Oil

Water

Figure 12.32

Problem Solving

75. In a classroom demonstration the pressure inside a soft drink can is suddenly reduced to essentially zero. Assuming the can to be a cylinder with a height of 12 cm and a diameter of 6.5 cm, find the total inward force exerted on the vertical sides of the can due to atmospheric pressure.

76. As a storm front moves in, you notice that the column of mercury in a barometer rises to only 736 mm. What is the air pressure in kPa?

77. A cylindrical container with a cross-sectional area of 65.2 cm² holds a fluid of density 806 kg/m³. The top surface of the fluid is open to the atmosphere. At the bottom of the container the pressure is 116 kPa. What is the depth of the fluid?

78. **Think & Calculate** A rooftop water storage tower, like the one shown in **Figure 12.33**, is filled with fresh water to a depth of 6.4 m. **(a)** What is the pressure 4.5 m below the surface of the water? **(b)** Why are the circular metal bands more closely spaced near the base of the tank?

Figure 12.33

79. **Think and Calculate** A new submarine is being developed to take tourists on sightseeing trips to tropical coral reefs. According to guidelines of the American Society of Mechanical Engineers (ASME), to be safe for human occupancy the submarine's hull must be able to withstand a pressure of 10.0 N/mm². **(a)** To what depth can the submarine safely descend in seawater? **(b)** If the submarine is used in freshwater, is its maximum safe depth greater than, less than, or the same as in seawater? Explain.

80. **Think & Calculate** The patient in **Figure 12.34** is to receive an intravenous injection of medication. For the injection to be administered properly, the pressure of the fluid containing the medication must be 109.0 kPa at the injection point. **(a)** If the fluid has a density of 1020 kg/m³, find the height, h, at which the bag of fluid must be suspended above the patient. Assume that the pressure inside the bag is 1 atm. **(b)** If a less dense fluid is used instead, must the suspension height be increased or decreased? Explain. (For this problem, let $P_{atm} = 101.3$ kPa.)

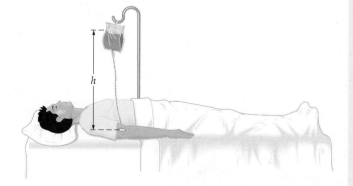

h

Figure 12.34

81. An air mattress is 2.3 m long, 0.66 m wide, and 14 cm deep. If the air mattress itself has a mass of 0.22 kg, what is the maximum mass it can support in freshwater?

82. In the hydraulic lift shown in **Figure 12.35**, the piston on the left has a diameter of 4.4 cm and a mass of 1.8 kg. The piston on the right has a diameter of 12 cm and a mass of 3.2 kg. If the density of the fluid is 750 kg/m³, what is the height difference, h, between the two pistons?

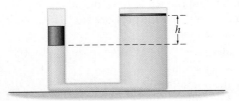

Figure 12.35

12.3 Fluids in Motion
Conceptual Questions

83. A sheet of water passing over a waterfall is thicker near the top than near the bottom. Explain.

84. It is a common observation that smoke rises more rapidly through a chimney when wind is blowing outside. Explain.

85. Is it best for an airplane to take off into the wind or with the wind? Explain.

86. If you have a hair dryer and a small plastic ball at home, try this demonstration. Direct the air from the dryer in a direction just above horizontal. Next, place the ball in the stream of air. If done just right, the ball will remain suspended in midair. Use Bernoulli's principle to explain this behavior.

87. A person prepares to blow through a horizontal drinking straw. The air from the horizontal straw will be directed across the top of a vertical straw whose lower end is submerged in a pan of water, as shown in **Figure 12.36**. When the person blows air through the horizontal straw, does the water level in the vertical straw rise, fall, or stay the same? Explain.

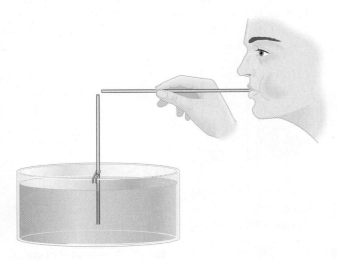

Figure 12.36

Problem Solving

88. Water flows with a speed of 0.43 m/s through a hose with a diameter of 3.2 cm. If the hose is attached to a nozzle with a diameter of 0.732 cm, what is the speed of the water in the nozzle?

89. A garden hose with a diameter of 1.7 cm has water flowing in it with a speed of 0.58 m/s. At the end of the hose is a nozzle. If the speed of water in the nozzle is 3.8 m/s, what is the diameter of the nozzle?

90. Arterial Plaque The buildup of plaque on the walls of an artery may decrease its diameter from 1.1 cm to 0.75 cm. If the blood flows with a speed of 15 cm/s before reaching the region of plaque buildup, find the speed of blood flow within that region.

91. Think and Calculate   A horizontal pipe contains water at a pressure of 110 kPa flowing with a speed of 1.6 m/s. **(a)** When the pipe narrows to half its original diameter, what is the speed of the water? **(b)** Is the pressure of the water in the narrower section of pipe greater than, less than, or equal to 110 kPa? Explain.

12.4 Solids
Conceptual Questions

92. What does it mean to say that an object is elastic?

93. What happens when a solid is stretched beyond its elastic limit?

94. If the force exerted on a spring is doubled, by what factor does the stretch distance increase?

95. Triple Choice   Two identical springs are connected parallel to one another; that is, they lie side by side. Is the spring constant of the resulting compound spring greater than, less than, or equal to the spring constant of a single spring? Explain.

96. Triple Choice   A spring is cut in half. Is the spring constant of each half greater than, less than, or equal to the spring constant of the original spring? Explain.

Problem Solving

97. How much force is required to stretch a spring 9.7 cm if the spring constant of the spring is 61 N/m?

98. How much force is required to compress a spring 12 cm if the spring constant of the spring is 57 N/m?

99. What spring constant must a spring have if it is to stretch 11 cm when a force of 5.2 N is applied?

100. The spring in a mattress has a spring constant of 38 kN/m. How much does the spring compress when a force of 51 N pushes on it?

101. A 25-kg chimpanzee hangs from the end of a horizontal branch, as shown in **Figure 12.37**. The branch sags downward through a vertical distance of 15 cm. Treating the branch as a spring satisfying Hooke's law, what is its spring constant?

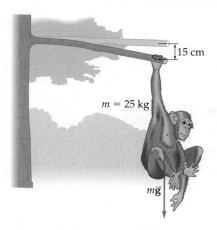

Figure 12.37

102. **Spiderweb** An orb weaver spider hangs vertically from one of its threads, which has a spring constant of 7.4 N/m. If the spider stretches the thread by 0.33 mm, what is the spider's mass?

103. A steel wire 5.1 m long stretches 0.13 cm when it is given a tension of 380 N. What is the spring constant of the wire?

Mixed Review

104. **Predicting the Weather** A weatherglass, as shown in **Figure 12.38**, is used to give an indication of a change in the weather. Does the water level in the neck of the weather glass move up, or move down when a low-pressure system approaches? Explain.

Figure 12.38

105. Rank The three containers in **Figure 12.39** are open to the air and filled with water to the same level. A block of wood floats in container A. An identical block of wood floats in container B and supports a small lead weight. Container C holds only water. Rank the three containers in order of increasing weight of the water they contain. Indicate ties where appropriate.

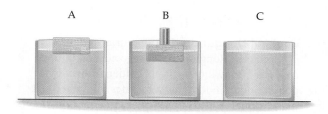

Figure 12.39

106. **Blood Pressure** When a person's blood pressure is taken, it is measured on an arm at approximately the same level as the heart. If the measurement were taken on the person's leg instead, would the reading be greater than, less than, or the same as a measurement made on the arm?

107. **Pain Threshold** A useful instrument for evaluating fibromyalgia and trigger-point tenderness is the *doloriometer* (or *algorimeter*). This device consists of a force meter attached to a circular probe that is pressed against the skin until pain is experienced. If the reading on the force meter is 14.5 N (3.25 lb) and the diameter of the circular probe is 1.39 cm, what is the pressure applied to the skin? Give your answer in pascals.

108. A wooden block with a density of 710 kg/m^3 and a volume of 0.012 m^3 is attached to the bottom of a vertical spring whose force constant is $k = 540$ N/m. Find the amount by which the spring is stretched.

109. **Airburst over Pennsylvania** On the evening of July 23, 2001, a meteor streaked across the skies of Pennsylvania, creating a spectacular fireball before exploding in the atmosphere with an energy release equivalent to 3 kilotons of TNT. The pressure wave from the airburst caused an increase in pressure of 0.50 kPa, enough to shatter some windows. Find the force that this "overpressure" would exert on a window whose dimensions are 86 cm × 115 cm. Give your answer in newtons and in pounds.

110. A person floats in a boat in a small backyard swimming pool. Inside the boat with the person are several blocks of wood. Suppose the person throws the blocks of wood into the pool, where they float. **(a)** Does the boat float higher, lower, or at the same level relative to the water? **(b)** Does the water level in the pool increase, decrease, or stay the same?

Writing about Science

111. Write a report on applications of Bernoulli's principle. Discuss how the principle explains the functioning of airplane wings, helicopter rotors, and atomizers, which are used to spray a mist of perfume onto people's necks (or to create an air-gas mist to be ignited in car engines).

112. **Connect to the Big Idea** Write a paragraph in which you compare and contrast the ability of particles in fluids and solids to move freely. Explain how freedom of movement results in the basic properties of each state.

Read, Reason, and Respond

Cooking Doughnuts Doughnuts are cooked by dropping the dough into hot vegetable oil until it changes from white to a rich, golden brown. One popular doughnut outlet automates this process; it makes doughnuts on an assembly line that customers can view in operation as they wait to order. Watching the doughnuts cook gives the customers time to develop an appetite as they ponder the physics of the process.

First, the uncooked doughnut is dropped into hot vegetable oil, whose density is $\rho = 919$ kg/m^3. There it browns on one side as it floats on the oil. After the doughnut has cooked for the proper amount of time, a mechanical lever flips it over so that it can cook on the other side. The doughnut floats fairly high in the oil, with less than half of its volume submerged. As a result, the final product has a characteristic white stripe around the middle, where the dough remains out of the oil, as indicated in **Figure 12.40**.

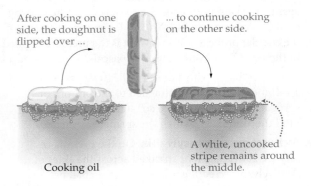

Figure 12.40 Steps in cooking a doughnut.

The connection between the density of the doughnut and the height of the white stripe is illustrated in **Figure 12.41**. On the horizontal axis we plot the density of the doughnut as a fraction of the density of the vegetable cooking oil. The vertical axis shows the height of the white stripe as a fraction of the total height of the doughnut. Notice that the height of the white stripe is plotted for both positive and negative values.

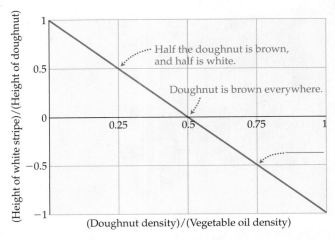

Figure 12.41 A finished doughnut has a stripe around the middle, whose height is related to the density of the doughnut by the straight line shown here.

113. Figure 12.41 describes the doughnut at the points where the height of the white stripe is $0.5H$ and 0. The description for $-0.5H$ has been left blank. Which of the following statements is most appropriate for this point?

   A. The doughnut sinks.

   B. The white stripe has a negative height.

   C. Half the doughnut is light brown; half is dark brown.

   D. The top and bottom of the doughnut are white, the middle one-half is brown.

114. Assuming that the doughnut has a cylindrical shape of height H and diameter D and that the height of the white stripe is $0.22H$, what is the density of the doughnut?

   A. 260 kg/m^3 C. 720 kg/m^3

   B. 360 kg/m^3 D. 820 kg/m^3

115. Suppose a new kind of doughnut whose density will be 330 kg/m^3 is going to be produced. If the height of the doughnut will be H, what will be the height of the white stripe?

   A. $0.14H$ C. $0.28H$

   B. $0.24H$ D. $0.64H$

Standardized Test Prep

Multiple Choice

1. A ball floats, submerged halfway, in a liquid. Which statement is true?
   - (A) The ball's density is the same as the liquid's density.
   - (B) The buoyant force on the ball is greater than the weight of the ball.
   - (C) The buoyant force on the ball is less than the weight of the ball.
   - (D) The ball's weight is equal to the weight of the fluid displaced.

2. An air bubble that has a volume of 0.001 m^3 is released at a depth of 21 m in a freshwater lake. The volume of the bubble when it reaches the surface is closest to
   - (A) 0.001 m^3.
   - (B) 0.002 m^3.
   - (C) 0.003 m^3.
   - (D) 0.004 m^3.

3. Which relates the speed of a fluid moving through a pipe to the cross-sectional area of the pipe?
   - (A) Pascal's principle
   - (B) Bernoulli's principle
   - (C) equation of continuity
   - (D) conservation of velocity

4. Which statement about an ideal gas held in a sealed, rigid container is true?
   - (A) The pressure of the gas increases as the temperature increases.
   - (B) The average kinetic energy of the particles in the gas decreases as the pressure increases.
   - (C) The internal energy of molecules in the gas decreases as temperature increases.
   - (D) The pressure of the gas remains constant as the temperature increases.

5. Water flows at 4.0 m/s through a horizontal section of pipe with a diameter of 6.0 cm. Further on, the pipe narrows, and the water flows at 8.0 m/s. What happens to the water pressure in the narrower section of the pipe?
   - (A) The pressure remains the same.
   - (B) The pressure doubles.
   - (C) The pressure quadruples.
   - (D) The pressure decreases.

6. A sample of helium gas with a temperature of 25 °C and a pressure of 310 kPa is held in a container with a volume of 0.022 m^3. Approximately how many helium atoms are inside the container? Assume that helium as an ideal gas.
   - (A) 1.7×10^{24}
   - (B) 2.3×10^{25}
   - (C) 7.5×10^{26}
   - (D) 4.9×10^{27}

7. An object with volume V floats in a fluid with density ρ, and 75% of the object is submerged in the fluid. Which is the correct expression for the density of the object?
   - (A) $\dfrac{4\rho}{3}$
   - (B) $\dfrac{3\rho}{4}$
   - (C) $\dfrac{\rho}{3}$
   - (D) $\dfrac{\rho}{4}$

8. A spring with spring constant k is cut in half, producing a new spring half as long. What is the spring constant of the new spring?
   - (A) $k/2$
   - (B) k
   - (C) $2k$
   - (D) $4k$

Extended Response

9. The cylindrical container shown in the figure below contains an ideal gas and is closed at the top by a 2.0-kg piston that is free to slide up and down. The space below the piston holds an initial volume of gas, V_i, at an initial pressure, P_i. Discuss what happens to the pressure and the volume of the gas when a 1.0-kg mass is placed on top of the piston. Assume that the temperature of the gas remains constant.

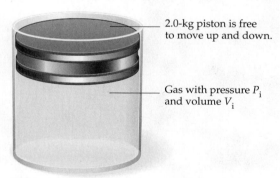

2.0-kg piston is free to move up and down.

Gas with pressure P_i and volume V_i

> **Test-Taking Hint**
>
> The temperature used in the ideal gas equation must be in kelvins.

Question	1	2	3	4	5	6	7	8	9
If You Had Difficulty With . . .									
See Lesson	12.2	12.1	12.3	12.1	12.3	12.1	12.2	12.4	12.1

13 Oscillations and Waves

Inside

Though she may not be thinking about it, this girl is demonstrating the basic physics of oscillatory motion.

Big Idea

Waves are traveling oscillations that carry energy.

In Chapter 8 you studied objects that are at rest. Such objects are seldom left undisturbed for very long, it seems. Soon they are disturbed by a bump, a kick, or a nudge. When this happens, you may see a series of back-and-forth motions referred to as *oscillations*. The properties of oscillations are explored in this chapter.

You will also learn that oscillations can produce *waves*, a topic of central importance in physics. This chapter introduces you to the properties of waves and explains how waves interact with one another.

Inquiry Lab   How do waves move in water?

Explore

1. Fill a rectangular pan half full of water.
2. Hold a pencil horizontally between your thumb and forefinger. Gently dip the pencil into the water at one end of the pan, making sure that the whole length of the pencil strikes the water at the same time. Observe the result.
3. Repeat Step 2, this time dipping the pencil into the water more forcefully, but not so hard as to make a splash.
4. Repeat Step 2, this time continually dipping the pencil in the water about once per second for 10 s.

Think

1. **Observe** What causes the initial disturbance in the water? In which direction does the wave that forms move? What happens when the wave hits the end of the pan?
2. **Assess** Did anything affect the height of the wave or its speed? If so, identify each factor.
3. **Predict** Do you think the results would differ if the water were replaced by maple syrup? Explain your reasoning.

13.1 Oscillations and Periodic Motion

If you sit quietly in a rocking chair, it remains at rest. It is in equilibrium. If you adjust your position slightly, however, the chair begins to rock back and forth. This is just one example of an oscillating system. Oscillating systems are common and can range from the head on a bobble-head doll to water molecules that oscillate in a microwave oven. Even the universe may be an oscillator, with a series of "big bangs" followed by equally momentous "big crunches."

Periodic Motion

Perhaps the most familiar oscillating system is a simple pendulum, like the one that keeps time in a grandfather clock. Picture in your mind the swinging motion of the clock's pendulum as it moves back and forth along the same path. The motion of the pendulum repeats over time. Other repeating motions are the motions of a child on a swing and of race cars on an oval track. In general, any motion that repeats itself over and over is referred to as **periodic motion.**

Period and frequency describe periodic motion

Periodic motion has a *cycle* that repeats. For example, the electrocardiogram in **Figure 13.1** on the next page shows a repeating cycle. If you match similar points in the pattern, you will find that approximately three cycles (heartbeats) are shown.

Vocabulary

- periodic motion
- period
- frequency
- hertz
- simple harmonic motion
- restoring force
- amplitude

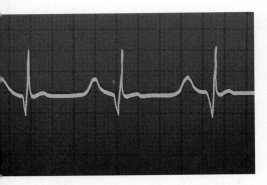

▲ **Figure 13.1 Electrocardiogram**
An electrocardiogram (also called an ECG or EKG) measures and displays the periodic electrical activity in a beating heart. A complete cycle of electrical activity occurs when the pattern repeats.

🔑 *How are the period and the frequency of periodic motion related?*

▲ **Figure 13.2 Radio frequencies**
Radio stations send out, or broadcast, their signals at specific frequencies. What is the frequency of your favorite station? What is the period of its signal?

Period   One of the key characteristics of periodic motion is its period. The **period**, T, is the time required to complete one full cycle of the motion. For example, the pendulum in a grandfather clock might take one second to swing from maximum displacement in one direction to maximum displacement in the opposite direction, and then another second to swing back to the starting point. In this case it takes two seconds to complete one cycle of oscillation, and the period is 2 s.

> **Definition of Period, T**
>
> period = time required for one cycle of a periodic motion
> $$T = \text{period}$$
>
> SI unit: seconds/cycle = s

Notice that a cycle is dimensionless.

Frequency   Closely related to the period is the **frequency**, f, which is the number of oscillations per unit of time. The higher the frequency, the more rapid the oscillations. As an example, your heart beats about 60 times per minute, or about once per second. This means that its frequency is 1 cycle (beat) per second. 🔑 **Frequency is calculated by taking the inverse of the period.** Thus, a pendulum with a period of 2 s has a frequency of $\frac{1}{2\text{ s}}$, or 0.5 s^{-1}.

> **Definition of Frequency, f**
>
> $$\text{frequency} = \frac{1}{\text{period}}$$
> $$f = \frac{1}{T}$$
>
> SI unit: cycle/second = 1/s = s^{-1}

A special unit called the **hertz** (Hz) is used to measure frequency. It is named for the German physicist Heinrich Hertz (1857–1894) in honor of his studies of radio waves. One hertz equals one cycle per second:

$$1\text{ Hz} = 1\,\frac{\text{cycle}}{\text{second}}$$

A pendulum whose frequency is 0.5 s^{-1} oscillates with a frequency of 0.5 Hz.

High frequencies are often measured in kilohertz (kHz), where 1 kHz = 10^3 Hz, or megahertz (MHz), where 1 MHz = 10^6 Hz. Consider the frequency of a radio signal, as shown in **Figure 13.2**. AM frequencies range from 530 kHz to 1710 kHz. This means, for example, that the broadcast signal from a 650-kHz AM radio station oscillates 650,000 times every second. Even more rapid are the oscillations for FM stations, which range from 88 MHz to 108 MHz. An FM signal at 101 MHz oscillates an amazing 101,000,000 times per second.

Period and frequency are reciprocals of one another. That is,

$$f = 1/T \qquad \text{and} \qquad T = 1/f$$

This means that when one is large, the other is small. For example, an oscillation with a large frequency has a small period, and vice versa. You can see this relationship in **Table 13.1**, which shows periods and frequencies for a wide range of periodic motions.

Table 13.1 Common Periods and Frequencies

System	Period (s)	Frequency (Hz)
Hour hand of a clock	43,200 (1 cycle per 12 hours)	2.3×10^{-5}
Minute hand of a clock	3600 (1 cycle per hour)	2.8×10^{-4}
Second hand of a clock	60 (1 cycle per minute)	0.017
Pendulum in a grandfather clock	2.0	0.50
Human heartbeat	1.0	1.0
Sound at lower range of human hearing	5.0×10^{-2}	20
Wing beat of a housefly	5.0×10^{-3}	200
Sound at upper range of human hearing	5.0×10^{-5}	20,000
Computer processor	3.1×10^{-10}	3.2×10^{9}

C⦿⦿LPHYSICS
Quartz Clock

Modern clocks, including those in computers, use a small vibrating quartz crystal to keep track of time. The crystal, which is often shaped like a miniature tuning fork, vibrates 32,768 times a second. An electronic circuit counts the vibrations and tracks the time. Quartz crystals keep time at least ten times more accurately than the pendulums or springs in old-fashioned clocks and watches.

QUICK Example 13.1 What Are the Frequency and the Period?

After walking up a flight of stairs, you take your pulse and observe 82 heartbeats in a minute. **(a)** What is the frequency of your heartbeat, in hertz? **(b)** What is the period of your heartbeat, in seconds?

Solution

(a) Convert the given frequency of 82 beats per minute to hertz:

$$f = \left(82 \, \frac{\text{beats}}{\text{minute}}\right)\left(\frac{1 \, \text{minute}}{60 \, \text{s}}\right) = 1.4 \, \frac{\text{beats}}{\text{s}} = \boxed{1.4 \, \text{Hz}}$$

(b) Take the reciprocal of the frequency to find the period:

$$T = \frac{1}{f} = \frac{1}{1.4 \, \text{Hz}} = \boxed{0.71 \, \text{s}}$$

Math HELP
Unit Conversion
See
Lesson 1.3

Practice Problems

1. A bird flaps its wings 5 times a second. What are the frequency and period of this motion?

2. Concept Check If the frequency of a motion is doubled, what happens to its period? Verify your answer by calculating the period of a bird's wing that flaps 10 times a second. Compare with the period you found in Problem 1.

3. Triple Choice The concert A string on a violin oscillates 440 times per second. If we increase the frequency of these oscillations, does the period increase, decrease, or stay the same? Explain.

4. A tennis ball is hit back and forth between two players. If it takes 2.3 s for the ball to go from one player to the other, what are the period and frequency of the ball's motion?

Simple Harmonic Motion

Periodic motion comes in many forms. A tennis ball going back and forth between players is a good example. Other examples include a basketball coach pacing the sidelines during a heated game, and a city mayor on a parade float waving her hand from side to side. One type of periodic motion is of particular importance. This is **simple harmonic motion,** and it occurs when the force pushing or pulling an object toward equilibrium is proportional to the displacement from equilibrium.

CONNECTINGIDEAS

In Chapter 5 you learned about Newton's second law; in Chapter 12 you learned about Hooke's law.
• Here these two laws are applied to a mass oscillating on a spring.

🔑 *What type of restoring force produces simple harmonic motion?*

Simple harmonic motion requires a restoring force

A classic example of simple harmonic motion is provided by a mass attached to a spring. In **Figure 13.3** we see a typical case, an air-track cart of mass m attached to a spring with a spring constant k. When the spring is neither stretched nor compressed, the cart is at the equilibrium position, $x = 0$. The cart remains at rest if it is left undisturbed.

Now, suppose you displace the cart away from equilibrium. Because of the displacement, the spring exerts a force back toward equilibrium. A force that acts to bring an object back to equilibrium is a **restoring force**. After all, it's a force that tries to "restore" an object to equilibrium.

🔑 **Simple harmonic motion occurs when the restoring force is** *proportional to the displacement* **from equilibrium.** Recall from Chapter 12 that the force exerted by a spring displaced a distance x from equilibrium is given by Hooke's law, $F = kx$. This force is proportional to x, and therefore a cart on a spring moves with simple harmonic motion.

Let's look at the motion of the cart in Figure 13.3. The cart is displaced to $x = A$ and released from rest. The stretched spring exerts a force on the cart to the left. This accelerates the cart toward the equilibrium position.

When the cart reaches $x = 0$ in Figure 13.3 (b), the spring is at its equilibrium position and the force acting on the cart is zero. Does the cart

▶ **Figure 13.3 Simple harmonic motion of a mass attached to a spring**
The simple harmonic motion of a mass attached to a spring is shown in **(a)** through **(e)**. The mass oscillates from one side of its equilibrium position ($x = 0$) to the other.

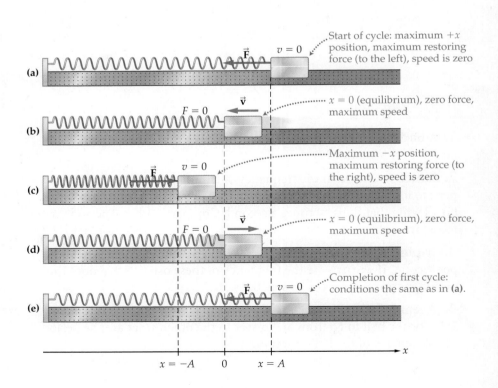

stop at this point? No. The cart's speed isn't zero, and no force acts to stop it. Therefore, it continues moving to the left.

As the cart moves to the left of equilibrium, it compresses the spring and produces a restoring force acting to the right. This force decelerates the cart and brings it to rest at $x = -A$, as shown in Figure 13.3 (c).

Finally, the cart begins to move to the right. It passes through equilibrium, as shown in Figure 13.3 (d), and comes to rest again at $x = A$, as shown in Figure 13.3 (e). At this point the cart has completed one oscillation of simple harmonic motion. The time required for one oscillation is the period, T.

Amplitude is a key characteristic of periodic motion

The motion in Figure 13.3 can be divided into four parts, each taking an equal amount of time (one-quarter of the period, or $T/4$).

- The cart moves from $x = A$ to $x = 0$ in the time $T/4$.
- The cart moves from $x = 0$ to $x = -A$ in the time $T/4$.
- The cart moves from $x = -A$ to $x = 0$ in the time $T/4$.
- The cart moves from $x = 0$ to $x = A$ in the time $T/4$.

The cart oscillates back and forth between $x = A$ and $x = -A$. Notice that both of these points are a distance A from equilibrium. The maximum displacement from equilibrium is the **amplitude** of motion. In this case the amplitude is A. As you might expect, the greater the amplitude of motion, the greater the energy of the oscillations.

A sine wave describes simple harmonic motion

Suppose you attach a pen to the cart in Figure 13.3 and let it trace the cart's motion on a strip of paper moving with constant speed, as indicated in **Figure 13.4 (a)**. This *strip chart* records the cart's motion as a function of time. Notice that the motion looks like a sine or cosine function.

The shape of the plot on the strip chart is shown in more detail in **Figure 13.4 (b)**. You can see that the time from one peak to the next is the period T. In general, whenever the time increases by the amount T, the motion repeats. In addition, the strip chart shows that the cart's motion is limited to displacements between $x = A$ and $x = -A$, where A is the amplitude.

▼ **Figure 13.4 Position-time graph for simple harmonic motion**
(a) The strip chart shows that the motion of the oscillating cart is like a sine or cosine function. Slightly more than one cycle has been completed in this sketch. **(b)** The motion of the cart is represented by a sine function if $t = 0$ is taken to be the point where the displacement is zero.

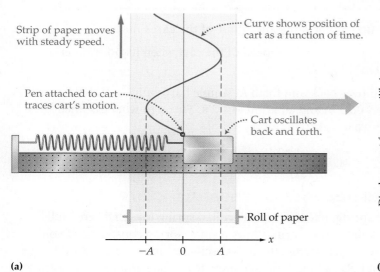

(a)

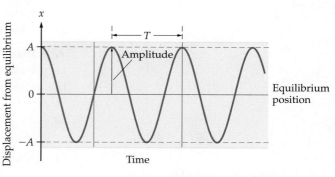

(b)

An air-track cart attached to a spring completes one oscillation every 2.4 s. At $t = 0$ the cart is released from rest at a distance of 0.10 m from its equilibrium position. What is the position of the cart at (**a**) 0.60 s and (**b**) 1.2 s?

Picture the Problem

We place the origin of the x-axis at the equilibrium position of the cart, with the positive direction pointing to the right. The cart is released from rest at $x = 0.10$ m, which means that the amplitude is $A = 0.10$ m. After it is released, the cart oscillates back and forth between $x = 0.10$ m and $x = -0.10$ m.

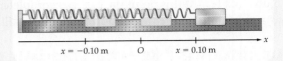

$x = -0.10$ m O $x = 0.10$ m

Known

$T = 2.4$ s $x = 0$ at $t = 0$

Unknown

(**a**) $x = ?$ at $t = 0.60$ s
(**b**) $x = ?$ at $t = 1.2$ s

Strategy

The period of oscillation is given as $T = 2.4$ s. Thus, we know that it takes the time $T/4 = 0.60$ s to go from $x = 0.10$ m to $x = 0$. Similarly, it takes the same time $(T/4)$ to go from $x = 0$ to $x = -0.10$ m, from $x = -0.10$ m to $x = 0$, and from $x = 0$ to $x = 0.10$ m. With this information we can determine the position x at the desired times.

Solution

1 (**a**) The time $t = 0.60$ s is equal to $T/4$. Therefore, the cart has moved from $x = 0.10$ m to $x = 0$:

$x = 0$

2 (**b**) The time $t = 1.2$ s is equal to $T/2$. At this time the cart has moved from $x = 0.10$ m to $x = -0.10$ m:

$x = -0.10$ m

Insight

At $t = 1.8$ s the cart is at $x = 0$ again, and at $t = T = 2.4$ s (one period) the cart has returned to its starting position, $x = 0.10$ m.

Practice Problems

5. **Follow-up** What is the amplitude of the cart's motion?

6. **Follow-up** What is the speed of the cart when its position is $x = -0.10$ m?

7. A mass moves back and forth in simple harmonic motion with an amplitude of 0.25 m and a period of 1.2 s. Through what distance does the mass move in 2.4 s?

The period depends on several factors

🔑 *How is the period of a mass on a spring related to the spring constant and the mass?*

Imagine a soft spring, like the popular coiled spring toys that can "walk" down a set of stairs. These springs have a small spring constant k. If you attach a mass to such a spring, the mass oscillates slowly. The period in this case is large—it takes a long time to complete an oscillation. On the other hand, a mass on a stiff spring (with a large spring constant k) oscillates rapidly. The period in this case is small. ⟨━⟩ **It follows that the period of a mass on a spring varies inversely with the spring constant.**

Similarly, the more mass you attach to a given spring, the slower it oscillates. ⟨━⟩ **Therefore, the period of a mass on a spring varies directly with the mass.** In fact, it can be shown with mathematical analysis or experiment that the period of a mass m on a spring with spring constant k is given by the following equation:

Period of a Mass on a Spring

$$\text{period} = 2\pi\sqrt{\frac{\text{mass}}{\text{spring constant}}}$$

$$T = 2\pi\sqrt{\frac{m}{k}}$$

SI unit: s

Notice that a larger mass m results in a larger period. On the other hand, a larger spring constant k results in a smaller period.

QUICK Example 13.3 What's the Period?

A 0.22-kg air-track cart is attached to a spring with a spring constant of 12 N/m and set into motion. What is the period of its oscillation?

Solution
Substituting for m and k in the period equation for a mass on a spring yields

$$T = 2\pi\sqrt{\frac{m}{k}}$$

$$= 2\pi\sqrt{\frac{0.22 \text{ kg}}{12 \text{ N/m}}}$$

$$= \boxed{0.85 \text{ s}}$$

PROBLEM-SOLVING NOTE

Recall that force is equal to mass times acceleration ($F = ma$). Substituting the SI units for each quantity in this equation shows that $N = (\text{kg})(\text{m/s}^2)$. Making this substitution for N in the calculation in this example yields the correct unit for the period.

Practice Problems

8. Concept Check If the mass attached to a given spring is increased by a factor of 4, by what factor does the period change?

9. Triple Choice If the mass and spring constant of a mass–spring system are both doubled, does the period of the system increase, decrease, or stay the same?

10. A 0.32-kg mass attached to a spring undergoes simple harmonic motion with a frequency of 1.6 Hz. What is the spring constant of the spring? (*Hint:* Rearrange $T = 2\pi\sqrt{m/k}$ to solve for the spring constant k.)

11. You have a spring with a spring constant of 22 N/m. What mass should you attach to this spring so that its motion has a period of 0.95 s? (*Hint:* Rearrange $T = 2\pi\sqrt{m/k}$ to solve for the mass m.)

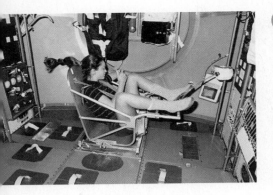

▲ Figure 13.5 Determining the mass of a weightless astronaut
The chair in which astronaut Tamara Jernigan sits is attached to a spring. Her mass can be determined by measuring the period of her oscillations as she rocks back and forth.

Engineers at NASA use the formula $T = 2\pi\sqrt{m/k}$ to measure the mass of astronauts in orbit. Astronauts are in free fall as they orbit the Earth, and therefore they are "weightless." As a result, they cannot simply step onto a bathroom scale to determine their mass. To get around this problem, NASA has developed a device called the *body mass measurement device (BMMD)*. Basically, the BMMD is a spring attached to a chair, into which an astronaut is strapped, as shown in **Figure 13.5**. As the astronaut oscillates back and forth, the period of oscillation is measured. Knowing the spring constant, k, and the period of oscillation, T, allows the astronaut's mass to be determined by rearranging $T = 2\pi\sqrt{m/k}$. Cool.

A change in amplitude does not affect the period

Suppose you have two identical mass–spring systems—identical masses and identical springs. Now suppose you give one mass twice the amplitude of the other. Does it take more time for the mass with the larger amplitude to complete one cycle than it does for the other mass? After all, the mass with the larger amplitude has to cover a greater distance during each cycle. Surprisingly, the answer is no—the two masses have the same period.

While it is true that a mass on a spring covers a greater distance when its amplitude is increased, it is also true that the larger amplitude causes a larger force to be exerted by the spring. The larger force, in turn, causes the mass to move more rapidly. In fact, the speed of the mass increases just enough to make it cover the greater distance in precisely the same time. That's why the formula for the period, $T = 2\pi\sqrt{m/k}$, does not contain the amplitude, A.

Figure 13.6 summarizes how the mass, the spring constant, and the amplitude affect the simple harmonic motion of a mass–spring system.

▼ Figure 13.6 Factors affecting the motion of a mass on a spring
The simple harmonic motion of a mass–spring system with spring constant k, mass m, and amplitude A is shown in the center. The effects of changing the values of k, m, and A are shown in **(a)** through **(d)**.

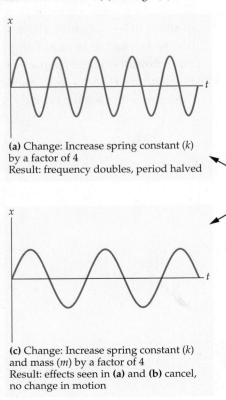

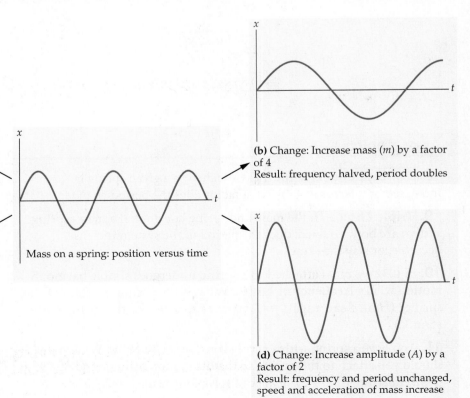

(a) Change: Increase spring constant (k) by a factor of 4
Result: frequency doubles, period halved

(b) Change: Increase mass (m) by a factor of 4
Result: frequency halved, period doubles

Mass on a spring: position versus time

(c) Change: Increase spring constant (k) and mass (m) by a factor of 4
Result: effects seen in **(a)** and **(b)** cancel, no change in motion

(d) Change: Increase amplitude (A) by a factor of 2
Result: frequency and period unchanged, speed and acceleration of mass increase

When a 0.10-kg mass is attached to a spring, the spring stretches by 0.20 m. Find the period of motion when this mass–spring system oscillates.

Solution (Perform the calculations indicated in each step.)

1. Solve Hooke's law for the spring constant, k:

$$k = \frac{F}{x}$$

2. Calculate the value of k:

$$k = \frac{F}{x} = \frac{mg}{x} = 4.9 \text{ N/m}$$

3. Substitute numerical values into $T = 2\pi\sqrt{m/k}$ and calculate the period, T:

$$T = 0.90 \text{ s}$$

Insight
Notice that the stretch distance of the spring was used to calculate the spring constant and that the force stretching the spring is the weight of the mass, $F = mg$.

Practice Problems

12. Triple Choice If the mass attached to a spring is increased and the spring constant is decreased, does the period of the oscillating motion increase, decrease, or stay the same?

13. When a 0.213-kg mass is attached to a vertical spring, it causes the spring to stretch a distance d. If the mass is displaced from equilibrium, it makes 102 oscillations in 56.7 s. Find the stretch distance, d.

13.1 LessonCheck 🅜🅟

Checking Concepts

14. 💬 **Determine** If the period increases, does the frequency increase, decrease, or stay the same? Explain.

15. 💬 **Describe** What characteristics of a restoring force lead to simple harmonic motion?

16. 💬 **Analyze** The spring constant of a spring is increased by a factor of 4, but the mass attached to the spring remains the same. How does the period of oscillation change?

17. Triple Choice If the mass attached to a spring is increased, does the frequency of the mass–spring system increase, decrease, or stay the same? Explain.

18. Determine You would like to increase the frequency of oscillations of a mass–spring system. Should you increase or decrease the mass? Explain.

Solving Problems

19. Rank Four mass–spring systems are described below. Rank the systems in order of increasing period. Indicate ties where appropriate.

	System A	System B	System C	System D
m	0.1 kg	0.4 kg	0.4 kg	0.1 kg
k	10 N/m	40 N/m	10 N/m	40 N/m

20. Calculate It takes a force of 12 N to stretch a spring 0.16 m. **(a)** What is the period of oscillation when a 2.2-kg mass is attached to the spring? **(b)** What is the frequency of oscillation in this case?

21. Challenge The processing speed of a computer refers to the number of binary operations it can perform in a second. Thus, the processing speed is actually a frequency. If the processor of a personal computer operates at 1.80 GHz (1 GHz = 10^9 Hz), how much time is required for one processing cycle?

13.2 The Pendulum

One Sunday in 1583, as Galileo Galilei attended services in a cathedral in Pisa, Italy, he suddenly realized something interesting about the chandeliers hanging from the ceiling. Air currents circulating through the cathedral had set them in motion, with small oscillations. Galileo noticed that chandeliers of equal length oscillated with equal periods, even if their masses were quite different. He verified this observation by timing the oscillations with the only timing device he had handy—his pulse.

Excited by his observations, Galileo rushed home to experiment. He constructed a series of pendulums (each one basically a mass suspended from a string) of different lengths and masses. Continuing to use his pulse as a stopwatch, he observed that the period of a pendulum varies with its length but is not affected by the mass attached to the string. Thus, in one exhilarating afternoon, the young Galileo discovered the key characteristics of a pendulum and launched himself into a career in science. Later, he would construct the first pendulum clock and a device to measure a patient's pulse.

Properties of Pendulums

Pendulums come in all shapes and sizes. A person on a swing is a pendulum. A hypnotist's pocket watch swinging back and forth ("You are getting sleepy") is a pendulum. A grandfather clock keeps time with a long pendulum consisting of a light rod with a heavy mass at the end. In general, we consider a **simple pendulum** to be a mass m suspended by a light string or rod of length L. The mass is often called the *bob*.

Pendulums move with simple harmonic motion

An example of a simple pendulum is shown in **Figure 13.7 (a)**. At rest, the pendulum hangs straight down in its equilibrium position. When displaced by a small angle from the vertical and released, the mass swings back and forth from one side of the equilibrium position to the other. If a stream of sand leaks from the mass onto a moving strip of paper, as in **Figure 13.7 (b)**, it produces a sine wave. Thus, the motion of the pendulum is very similar to that of a mass on a spring—both are examples of simple harmonic motion.

A pendulum moves with simple harmonic motion because a restoring force acts on it that is approximately proportional to the angle of displacement. As you can see in **Figure 13.7 (c)**, the weight of the bob, $m\vec{\mathbf{g}}$, acts downward, and a tension force, $\vec{\mathbf{T}}$, acts along the length of the string. The tension cancels part of the bob's weight, but a component of the weight ($mg \sin \theta$) acts in the direction of the equilibrium position. This is the restoring force. For small angles (about 15° or less) the restoring force is essentially proportional to the angle of displacement, θ, and this results in simple harmonic motion.

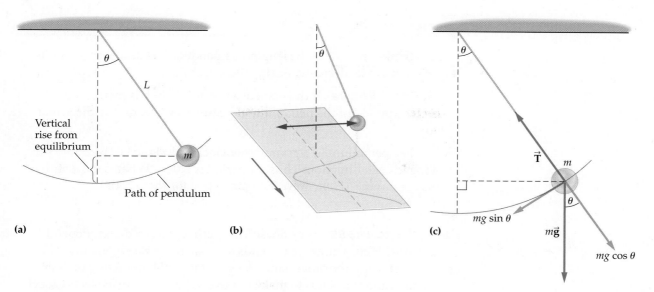

(a)   (b)   (c)

▲ Figure 13.7 **Motion of a pendulum**

▲ Figure 13.7 **Motion of a pendulum**
(a) A swinging pendulum oscillates back and forth from one side of the vertical position to the other. (b) Because the pendulum's angle of displacement, θ, varies with time like a sine function, the leaking sand produces a sine wave on the strip chart. (c) A component of the pendulum bob's weight acts as a restoring force. The strength of the restoring force is $mg \sin \theta$.

Different pendulums oscillate at different rates

The period of a pendulum—the time required for one full oscillation—can be determined mathematically or experimentally. For a pendulum of length L, the period is given by the following formula:

> **Period of a Pendulum**
>
> $$\text{period} = 2\pi\sqrt{\frac{\text{length of pendulum}}{\text{acceleration due to gravity}}}$$
>
> $$T = 2\pi\sqrt{\frac{L}{g}}$$
>
> SI unit: s

🔑 **The period of a pendulum depends on its length and on the acceleration due to gravity.** It's quite interesting, though, that the period doesn't depend at all on the amplitude of the motion or on the mass.

🔑 **What factors affect the period of a pendulum?**

QUICK Example 13.5 **What's the Period?**

A yo-yo has a string that is 0.75 m in length. What is the period of oscillation if the yo-yo is allowed to swing back and forth at the end of its string?

Solution
The length of the (yo-yo) pendulum is $L = 0.75$ m, and the acceleration due to gravity is $g = 9.81$ m/s^2. Substitute these values into the formula for the period of a pendulum:

$$T = 2\pi\sqrt{\frac{L}{g}}$$

$$= 2\pi\sqrt{\frac{0.75 \text{ m}}{9.81 \text{ m/s}^2}}$$

$$= \boxed{1.7 \text{ s}}$$

> **Math HELP**
> Significant
> Figures
> **See**
> **Lesson 1.4**

22. Triple Choice If the length of a pendulum is decreased, does its period increase, decrease, or stay the same? Explain.

23. Triple Choice If the mass of a pendulum bob is increased, does the period of the pendulum increase, decrease, or stay the same? Explain.

24. The pendulum in a grandfather clock is designed to take 1.00 s to swing in each direction. Thus, its period is 2.00 s. What is the length of this pendulum? (*Hint*: Solve $T = 2\pi\sqrt{L/g}$ for the length.)

Period and Mass

Why doesn't the period of a pendulum depend on the mass? Well, a larger mass tends to move more slowly because of its greater inertia. On the other hand, the gravitational force acting on it is also greater, and this tends to make it move faster. These two effects cancel exactly, just as they do in free fall. Thus, free-fall acceleration and the period of a pendulum are independent of mass for the same reason.

Period and Amplitude

Similarly, why doesn't a pendulum's period depend on the amplitude? To see this, consider the two pendulums shown in **Figure 13.8**. The pendulum with the larger amplitude has farther to go to get back to its equilibrium position. This would seem to indicate a larger period. However, this pendulum also has a greater restoring force acting on it. Because of this greater force, it covers the larger distance in exactly the same time as the pendulum with the smaller amplitude.

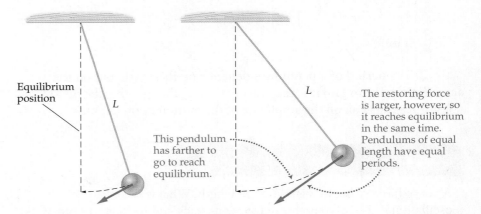

Equilibrium position

L

This pendulum has farther to go to reach equilibrium.

L

The restoring force is larger, however, so it reaches equilibrium in the same time. Pendulums of equal length have equal periods.

▲ **Figure 13.8 Pendulums with different amplitudes**
The pendulum with the larger amplitude has farther to go to reach the vertical position, but the restoring force acting on it is also greater. The two effects cancel, and thus the period doesn't depend on the amplitude.

Period and Length

You can also see why the period depends on the length of a pendulum. **Figure 13.9** shows two pendulums displaced by the same angle, θ. Both masses experience the same force of gravity pulling them back toward the vertical position. Therefore, they will have the same acceleration. Since the mass with the greater length must travel farther to reach the vertical, it will take longer for it to complete an oscillation. It follows that the period of a pendulum increases as its length increases.

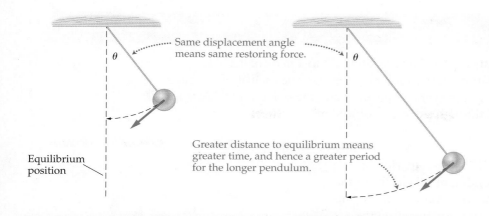

Same displacement angle means same restoring force.

Equilibrium position

Greater distance to equilibrium means greater time, and hence a greater period for the longer pendulum.

◀ **Figure 13.9 Effect of pendulum length on period**
The longer pendulum has farther to travel to reach equilibrium, but the force acting on it is the same as the force acting on the shorter pendulum. Therefore, the period of the longer pendulum is greater than the period of the shorter pendulum.

CONCEPTUAL Example 13.6 Raise or Lower the Weight?

If you look carefully at a grandfather clock, you'll notice that the weight at the bottom of the pendulum can be moved up or down by turning a small screw. Suppose you have a grandfather clock that runs slow. Should you raise or lower the weight?

Reasoning and Discussion

To make the clock run faster, you want it to take less time to go from *tick* to *tock*. In other words, the period of the pendulum should be decreased. From the period formula for a pendulum, $T = 2\pi\sqrt{L/g}$, you can see that shortening the pendulum (decreasing L) decreases the period. Therefore, you should raise the weight, which shortens the pendulum.

Answer

The weight should be raised. This shortens the period and makes the clock run faster.

The value of *g* affects the period of a pendulum

Physics & You: Technology The fact that the acceleration due to gravity varies from place to place on Earth was mentioned in Chapter 2. In fact, *gravity maps*, such as the one shown in **Figure 13.10**, are valuable tools for geologists attempting to understand the properties of the rock underlying a given region. The main instrument geologists use to make gravity maps is basically a very precise pendulum whose period can be accurately measured. Slight changes in the pendulum's period from one location to another, or from one elevation to another, correspond to slight changes in *g*. More common in recent decades is an electronic gravimeter that contains a mass on a spring and relates the force of gravity to the stretch of the spring. These gravimeters can measure *g* with an accuracy of 1 part in 100 million.

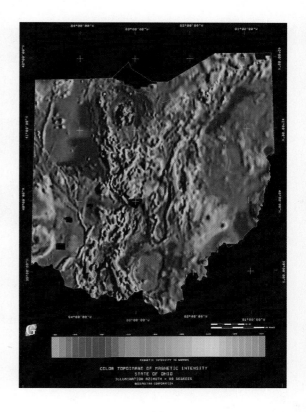

▶ **Figure 13.10 Map of gravitational strength**
Areas where the gravitational strength is greatest have denser rock underlying the surface.

A pendulum is constructed from a string 0.627 m long attached to a mass of 0.250 kg. When set in motion at a certain location, the pendulum completes one oscillation every 1.59 s. If the pendulum is held at rest in the vertical position and the string is cut, how long does it take for the mass to fall through a distance of 1.00 m?

Picture the Problem

The pendulum has a length $L = 0.627$ m and a period of oscillation $T = 1.59$ s. When the pendulum is held at rest in the vertical position, a pair of scissors is used to cut the string. The mass then falls straight downward through a distance $y = 1.00$ m. The acceleration of the mass is equal to the acceleration due to gravity at the pendulum's location.

Strategy

At first it might seem that the period of oscillation and the time of fall are unrelated. Recall, however, that the period of a pendulum depends on both its length *and* the acceleration due to gravity *at the location of the pendulum*.

To solve this problem, then, we first use the period T to find the local acceleration due to gravity, g_{local}. Once g_{local} is known, finding the time of fall is a straightforward kinematics problem, which we can solve with $y = \frac{1}{2}g_{local}t^2$.

Known

$L = 0.627$ m
$T = 1.59$ s
$y = 1.00$ m

Unknown

$t =$ time to fall 1.00 m

Solution

1 Use the formula for the period of a pendulum to solve for the acceleration due to gravity:

$$T = 2\pi\sqrt{L/g}$$

$$g = \frac{4\pi^2 L}{T^2}$$

Math HELP
Solving Simple Equations
See Math Review, Section III

2 Substitute numerical values to find g_{local}:

$$g_{local} = \frac{4\pi^2 L}{T^2}$$

$$= \frac{4\pi^2(0.627 \text{ m})}{(1.59 \text{ s})^2}$$

$$= 9.79 \text{ m/s}^2$$

3 Use kinematics to solve for the time to drop from rest through a distance y:

$$y = \frac{1}{2}gt^2 \quad \text{or} \quad t = \sqrt{2y/g}$$

4 Substitute $y = 1.00$ m and the value of g_{local} found in Step 2 to find the time:

$$t = \sqrt{2y/g_{local}}$$

$$= \sqrt{\frac{2(1.00 \text{ m})}{9.79 \text{ m/s}^2}}$$

$$= \boxed{0.452 \text{ s}}$$

Insight

This calculation shows a small local variation from the standard value of $g = 9.81 \text{ m/s}^2$.

25. Follow-up If a mass falls 1.00 m in a time of 0.451 s, what are **(a)** the local acceleration due to gravity and **(b)** the period of a pendulum of length 0.500 m at this location?

26. What is the frequency of a pendulum of length 1.25 m at a location where the acceleration due to gravity is 9.82 m/s^2?

27. Triple Choice The acceleration due to gravity is greater at location A than at location B. If identical pendulums are taken to both locations, is the period of the pendulum at location A greater than, less than, or equal to the period of the pendulum at location B? Explain.

Your legs act as pendulums when you walk

The next time you take a leisurely stroll, think about the way your legs pivot about the hip joint. Perhaps you hadn't thought about it before, but each leg is a large pendulum. In fact, it swings back and forth with each step, as illustrated in **Figure 13.11 (a)**. The length of your leg determines both the period of oscillation and the length of your stride. Thus, pendulum motion is important in understanding your walking speed and that of other animals, such as the elephants shown in **Figure 13.11 (b)**.

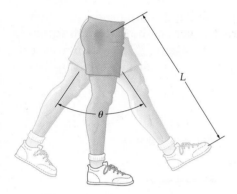

(a)

(b)

▲ Figure 13.11 **The leg acts like a pendulum**
(a) As you walk, each leg swings like a pendulum. The frequency of your steps is determined by the period of your leg. **(b)** The longer an animal's legs, the more time it takes to complete one step forward. The length of each step is also greater, however. The net effect is that animals with longer legs have greater walking speeds.

Resonance

External forces affect objects all the time, including those that oscillate. Surprising things can happen, however, if an external force is applied repeatedly at just the right frequency.

Resonance can produce large-amplitude oscillations

🔑 *What conditions produce resonance?*

If you stop pushing a friend on a swing, he will soon stop oscillating. The energy you add with each push is needed to *drive* the swing and keep it going. **Figure 13.12** shows a similar situation, a string from which a small weight is suspended. If the weight is set in motion and you hold your hand still, it will soon stop oscillating. If you move your hand back and forth in a horizontal direction, however, you can keep the weight oscillating indefinitely. The motion of your hand drives the motion of the weight. In general, *driven oscillations* are those caused by an applied force.

The response of the weight in Figure 13.12 depends on the frequency of the hand's back-and-forth motion. If possible, make a simple pendulum and verify this for yourself. When you move your hand very slowly, the weight simply tracks along with the motion of your hand. When you oscillate your hand rapidly, the weight exhibits only small oscillations. Oscillating your hand at an intermediate frequency, however, produces large-amplitude oscillations.

So, just what is the appropriate intermediate frequency to produce large oscillations? It turns out that large oscillations are produced when your hand drives the weight at its natural frequency. The **natural frequency** is the frequency at which an object oscillates by itself. 🔑 **A system driven at its natural frequency is in resonance**. Systems in resonance typically have rather large amplitudes.

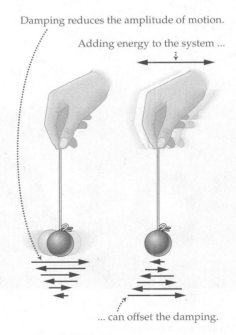

Damping reduces the amplitude of motion.

Adding energy to the system ...

... can offset the damping.

▲ **Figure 13.12 Driven oscillations**
On the left, friction quickly slows, or dampens, the motion of a pendulum when the hand is held still. On the right, back-and-forth oscillations of the hand "drive" the pendulum, and it continues to swing.

▲ Figure 13.13 **Resonance**
The collapse of the Tacoma Narrows Bridge is a famous example of the possible dangerous consequences of resonance.

Resonance affects how things are designed

Resonance plays an important role in a variety of physical systems. In Chapter 23 you'll learn how adjusting the tuning knob of a radio changes the resonance frequency of the electrical circuit in the tuner. When that resonance frequency matches the frequency being broadcast by a station, the radio picks up the broadcast.

Buildings and other structures can show resonance effects as well. One of the most dramatic and famous examples is the collapse of Washington's Tacoma Narrows Bridge in 1940. High winds through the narrows had often set the bridge into a gentle swaying motion, resulting in its affectionate nickname, "Galloping Gertie." During one windstorm, however, the bridge showed a resonance-like effect, and the amplitude of its swaying motion began to increase. Alarmed officials closed the bridge to traffic. A short time later the swaying motion became so great that the bridge broke apart and fell into the water below, as shown in **Figure 13.13**. Needless to say, bridges built since that time have been designed to prevent such catastrophic oscillations.

13.2 LessonCheck (MP)

Checking Concepts

28. ⬤ **Analyze** How does the period of a pendulum change if the pendulum is lengthened? Explain.

29. ⬤ **State** At what frequency should you drive a pendulum for it to be in resonance?

30. [Triple Choice] Two identical pendulums, A and B, are set into motion at the same location. Pendulum A has twice the amplitude of pendulum B. Is the average speed of the mass on pendulum A greater than, less than, or equal to the average speed of the mass on pendulum B? Explain.

Solving Problems

31. Calculate The chains of a swing on a playground swing set are 3.0 m long. What is the period of this swing?

32. Calculate How long must a pendulum be to have a period of 1.0 s? Assume the acceleration due to gravity is 9.81 m/s^2.

33. Calculate A simple pendulum of length 2.5 m makes 5.0 complete swings in 16 s. What is the acceleration due to gravity at the location of the pendulum?

13.3 Waves and Wave Properties

Vocabulary

- wave
- transverse wave
- longitudinal wave
- crest
- trough
- wavelength
- medium
- mechanical wave

Reading Support ✓

Multiple Meanings of propagate

scientific meaning: To transmit a disturbance (motion, light, sound, etc.) through a medium.

Sound waves from a loudspeaker propagate through the air.

common meaning: To spread a piece of information or promote an idea.

The company president propagated a rumor about a new product.

🔑 *How does the motion of a wave differ from the motion of the particles making up the wave?*

Waves are meant to travel. Just think of ocean waves rolling up on a white sand beach, the sound of distant thunder, or even the twinkling light from a faraway star. Waves are the ultimate expression of the ripple effect.

In this lesson you will learn how oscillations produce waves, how waves travel, and how to determine their speed.

Wave Formation and Wave Types

Consider a group of swings in a playground swing set. Each swing by itself behaves like a simple pendulum. That is, each swing is an oscillator. Now suppose we connect the swings to one another by tying a rope from the seat of the first swing to its neighbor, and then another rope from the second swing to the third swing, and so on. When the swings are at rest—in equilibrium—the connecting ropes have no effect. But, if you sit in the first swing and begin oscillating—thus disturbing the equilibrium—the connecting ropes cause the other swings in the set to start oscillating as well. You have created a traveling disturbance.

A wave is a traveling disturbance

In general, a **wave** is a disturbance that **propagates**, or is transmitted, from place to place, carrying energy as it travels. For example, when you feel the warmth of bright sunlight, you are experiencing some of the energy that was carried by the light waves from the Sun to the Earth. Similarly, when a wave hits you on the beach, you are experiencing some of the energy that is carried by waves moving through ocean water.

It is important to distinguish between the motion of the wave itself and the motion of the individual particles in the wave. For example, consider the surfers in **Figure 13.14 (a)**. Those who don't "catch the wave" simply bob up and down in the same location as the wave moves past them. Similarly, a "wave" at a ball game, shown in **Figure 13.14 (b)**, travels rapidly around the stadium. Even so, the individual people making up the wave simply stand up and sit down in one place. The people (the "particles" in this wave) don't run around the stadium. 🔑 **In general, waves travel from place to place, but the particles in a wave oscillate back and forth about one location.**

▶ **Figure 13.14 Wave motion versus particle motion**
A wave moves from one location to another, but the particles in a wave only move back and forth. Here the surfers and the fans are the "particles."

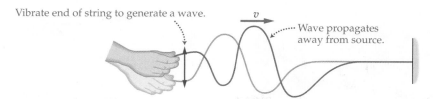

▲ **Figure 13.15 A wave on a string**
Vibrating one end of the string generates a wave that travels away from its source.

From these simple examples it is clear that waves come in a variety of different forms. Let's consider the most important types of waves.

Perpendicular motions occur in transverse waves

Waves can be categorized by the way their particles move. To see how this works, let's look at the easiest type of wave to visualize, a wave on a string, as shown in **Figure 13.15**. You can generate such a wave by tying a long string to a wall, pulling it taut, and moving your hand up and down. This creates a wave that travels along the string toward the wall. If your hand moves up and down in simple harmonic motion, the wave on the string will have the shape of a sine wave.

Notice that the wave travels in the horizontal direction, even though your hand oscillates vertically. In fact, every point on the string oscillates vertically, with no horizontal motion at all. This is shown in **Figure 13.16**, which tracks the motion of an individual point on a string as a wave travels past. Notice that the particles in the string move at right angles to the motion of the wave. In general, a wave in which the particles oscillate at right angles to the direction the wave travels is called a **transverse wave**.

Parallel motions occur in longitudinal waves

Not all waves are transverse waves. In particular, a wave in which the particles oscillate parallel to the direction of propagation is called a **longitudinal wave**.

A classic example of a longitudinal wave can be created using a spring toy, as shown in **Figure 13.17**. If you have such a toy handy, stretch it out on the floor and oscillate one end back and forth in line with the spring. The "disturbance" you produce at one end travels through the spring as a series of compressions and expansions. Each coil of the spring moves forward and backward horizontally, in the same direction as the wave itself. This is a longitudinal wave.

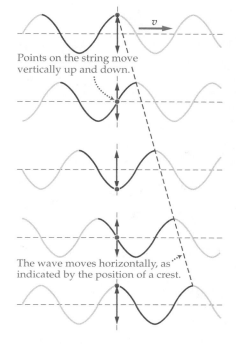

▲ **Figure 13.16 The motion of a wave on a string**
As a wave on a string moves horizontally, all points on the string oscillate in the vertical direction.

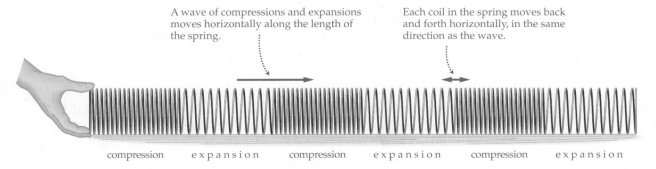

▲ **Figure 13.17 A longitudinal wave on a spring toy**
Oscillating one end of a spring toy back and forth produces a longitudinal wave.

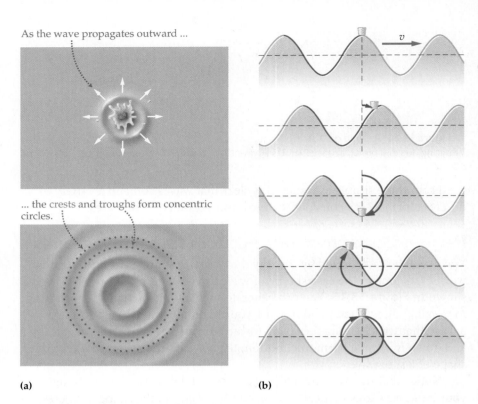

As the wave propagates outward ...

... the crests and troughs form concentric circles.

(a)   (b)

▲ **Figure 13.18 The motion of a water wave**
(a) A pebble dropped into a pool of water creates waves that move symmetrically away from the disturbance. **(b)** A cork floating in the water (or a water molecule) moves in a roughly circular path as a wave travels past.

Water waves have a combination of properties

If you drop a pebble into a pool of water, a series of concentric waves move away from the splash, as illustrated in **Figure 13.18 (a)**. A floating piece of cork traces the motion of the water itself as the waves travel outward. Notice that the cork moves in a roughly circular path, as shown in **Figure 13.18 (b)**, returning to its approximate starting point. Similarly, each molecule of water moves both vertically and horizontally as the wave travels in the horizontal direction. This pattern of motion indicates that a water wave is a combination of transverse and longitudinal waves.

Practice Problems

34. **Concept Check** A car brakes suddenly in heavy traffic, causing the cars behind it to bunch up. This bunching effect moves through the traffic like a wave. Is this wave transverse, longitudinal, or more like a water wave? Explain.

35. **Concept Check** A group of students want to produce waves by moving back and forth, up and down, and so on. How should the students move to make a transverse wave? How should they move to produce a longitudinal wave? Can you think of a way they could move to make something like a water wave?

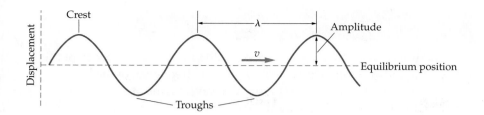

Figure 13.19 Wave characteristics
A wave repeats over a distance equal to the wavelength, λ. For example, the wavelength is the distance between two consecutive crests or troughs. The amplitude is the distance from the wave's equilibrium position to its maximum displacement, which occurs at a crest or a trough. The time a wave takes to move one wavelength is equal to its period, T.

Wave Properties and Behavior

A simple wave, like the one in **Figure 13.19**, is a regular, rhythmic disturbance that travels from one location to another. The wave repeats itself both in space and in time. The highest points on the wave are its **crests**, and the lowest points are its **troughs**. The size of the crests and troughs and the spacing between them determine several key wave properties.

Waves have characteristic properties

The distance from one crest to the next, or from one trough to the next, is the *repeat length*, or **wavelength**, λ, of the wave.

> **Definition of Wavelength, λ**
>
> λ = wavelength
>   = distance over which a wave repeats
>
> SI unit: m

Similarly, a wave can also be characterized by the time required for it to complete one cycle. Thus, the *repeat time*, or period, T, of a wave is the time required for one wavelength to pass a given point. Just as with oscillations, the frequency of a wave is the inverse of the period, $f = 1/T$.

Waves also have amplitude, just like oscillations. The amplitude of a wave is the greatest displacement of the wave from the equilibrium position. Thus, the distance from equilibrium to a crest or to a trough is the amplitude.

Wavelength and frequency determine wave speed

Combining our observations about repeat length and repeat time, we see that a wave travels a distance λ in the time T. Since speed is distance divided by time, it follows that the speed of a wave is

$$\text{speed} = \frac{\text{wavelength}}{\text{period}}$$
$$v = \frac{\lambda}{T}$$

Recalling that frequency is the inverse of the period, $f = 1/T$, we can write the equation for the wave speed as follows:

> **Speed of a Wave**
>
> speed = wavelength $\times$ frequency
> $v = \lambda f$
>
> SI units: m/s

This result applies to all waves, no matter what their type or how they are produced.

CONNECTING IDEAS

Earlier in this chapter you learned about the period, frequency, and amplitude of an oscillation.
• These same terms also describe wave characteristics. After all, a wave *is* a traveling oscillation.

Sound waves travel in air with a speed of 343 m/s. The frequency of the lowest sound you can hear is 20.0 Hz, and the frequency of the highest sound you can hear is 20.0 kHz. Find the wavelengths of sound waves having frequencies of 20.0 Hz and 20.0 kHz.

Solution

Solve $v = \lambda f$ for the wavelength, λ. This gives $\lambda = v/f$. Next, substitute the given values for v and f:

$$(20.0 \text{ Hz}) \quad \lambda = \frac{v}{f}$$
$$= \frac{343 \text{ m/s}}{20.0 \text{ s}}$$
$$= \boxed{17.2 \text{ m}}$$

$$(20.0 \text{ kHz}) \quad \lambda = \frac{v}{f}$$
$$= \frac{343 \text{ m/s}}{20,000 \text{ s}}$$
$$= \boxed{1.72 \text{ cm}}$$

Practice Problems

36. Concept Check How far does a wave travel in one period?

37. A wave oscillates 4.0 times a second and has a wavelength of 3.0 m. What are the **(a)** frequency, **(b)** period, and **(c)** speed of this wave?

38. You produce a wave by oscillating one end of a rope up and down 2.0 times a second.
(a) What is the frequency of this wave?
(b) What is the period of this wave?
(c) If the wave travels with a speed of 5.0 m/s, what is the wavelength?

Waves have different speeds in different materials

What determines the speed of a wave?

Waves travel at different speeds in different materials. **In fact, the speed of a wave is determined by the properties of the material, or *medium*, through which it travels.** A **medium** can be any form of matter, such as air, water, or steel.

In general, waves travel faster in a medium that is hard or stiff. A soft or squishy medium results in slow-moving waves. For example, sound waves in air have a speed of about 343 m/s. In water, which is not as compressible as air, the speed of sound waves is about four times faster, or 1400 m/s. Sound waves travel even faster in the solid medium steel, where their speed is about 5960 m/s. Similarly, waves in springs, strings, and ropes have speeds that depend on the tension force and on the mass per length.

Wave speeds also depend on other properties of a medium. For example, sound travels faster in warm air that it does in cool air. This is due to the higher speed of the warm-air molecules, as we saw in Chapter 12. The speed of water waves is greater in deep water than it is in shallow water. This is why waves approaching a beach slow down and produce surf.

Mechanical waves travel through matter. Sound waves and water waves are examples of mechanical waves. Other types of waves, such as radio waves and light waves, can travel through a vacuum. The properties of the vacuum determine the speed of these waves, which you'll study in upcoming chapters.

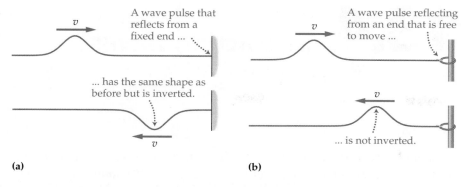

A wave pulse that reflects from a fixed end ...

A wave pulse reflecting from an end that is free to move ...

... has the same shape as before but is inverted.

... is not inverted.

(a)

(b)

◀ **Figure 13.20 A wave pulse reflecting from a fixed or a free end**
(a) A wave pulse on a string is inverted (turned upside-down) when it reflects from an end that is tied down. **(b)** A wave pulse on a string whose end is free to move is reflected right-side up.

A wave can reflect in different ways

Have you ever heard an echo? (Hello, hello, hello.) The echo is produced by sound waves bouncing, or reflecting, off a barrier (like a cliff face). In general, waves are reflected whenever they hit a barrier. How the reflection occurs depends on the type of barrier.

As an example, suppose a string is anchored firmly to a wall, as shown in **Figure 13.20 (a)**. When you flick the free end of the string, a brief disturbance, or pulse, travels toward the fixed end, where it exerts an upward force on the wall. The wall exerts an equal and opposite downward force to keep the end of the string at rest. Thus, the wall exerts a downward force on the string that is just the *opposite* of the upward force you exerted when you created the pulse. This causes the reflected pulse to be inverted, or upside-down, as indicated in Figure 13.20 (a).

Another way to set up the far end of the string is shown in **Figure 13.20 (b)**. In this case, the string is tied to a small ring that slides vertically with little friction on a vertical pole. This string still has a tension in it, but it is also free to move up and down.

When a pulse reaches the end of this string, it lifts the ring upward and then lowers it back down. In fact, the pulse flicks the far end of the string in the *same way* that you flicked the near end when you created the pulse. Therefore, the far end of the string simply creates a new pulse, identical to the first but traveling in the opposite direction.

Thus, when waves reflect from a barrier, they may or may not be inverted, depending on the characteristics of the barrier.

13.3 LessonCheck (MP)

Checking Concepts

39. ⬤ **Assess** A "wave" at a ballgame can be an exciting event. Is this wave transverse, longitudinal, or more like a water wave? Explain.

40. ⬤ **Predict** In which medium would you expect the speed of sound waves to be greater, a chocolate candy bar or a chocolate cake? Explain.

41. Big **Idea** Give everyday examples of energy carried by light waves, radio waves, water waves, and sound waves. In each case, describe how the wave's energy affects its surroundings.

Solving Problems

42. Calculate A wave oscillates 5.0 times a second and has a speed of 6.0 m/s. What are the **(a)** frequency, **(b)** period, and **(c)** wavelength of this wave?

43. Calculate As you sit in a fishing boat, you notice that 12 waves pass the boat every 45 s. If the distance from one crest to the next is 7.5 m, what is the speed of these waves?

44. Calculate A wave moves by you with a speed of 5.6 m/s. The distance from a crest of this wave to the next trough is 2.4 m. What is the frequency of the wave?

13.4 Interacting Waves

Vocabulary

- resultant wave
- principle of superposition
- constructive interference
- destructive interference
- standing wave
- node
- antinode

When air-track carts collide, they hit and bounce backward. When lumps of putty collide, they "smush" together into a big blob. What about waves? Well, when waves collide, they pass right through one another and keep going! In some ways, they're like a ghost passing through a wall. This lesson explores the interesting behavior of interacting waves.

Superposition and Interference

Whenever two or more individual waves overlap, they combine to form a **resultant wave**. The **principle of superposition** states that a resultant wave is simply the sum of the individual waves that make it up. This lesson explores these ideas.

Overlapping waves interact with one another

🔑 How do overlapping waves interact with one another?

Consider two waves on a string, traveling in opposite directions as shown in **Figure 13.21 (a)**. When the pulses arrive in the same region, they *superpose* and add to form a larger peak. 🔑 **Waves that occupy the same space combine in the simplest possible way—they just add together.**

A collision between waves doesn't affect the individual waves in any way. In fact, the waves pass right through one another and continue on as if nothing had happened. Think of listening to an orchestra with many different instruments playing simultaneously. Even though the sounds combine throughout the concert hall, you can still hear individual instruments. Each instrument makes its own sound as if the others were not present.

▼ **Figure 13.21 Superposition of waves**
Wave pulses combine (resulting in interference) as they pass through one another. Afterward, the pulses continue on unchanged. **(a)** Constructive interference occurs when waves add to give a larger amplitude. **(b)** Destructive interference occurs when waves combine to give a smaller amplitude.

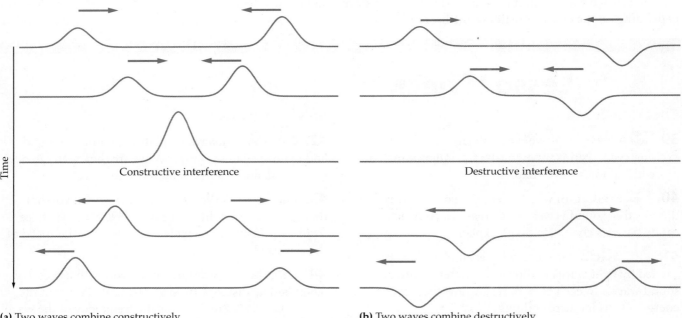

Constructive interference   Destructive interference

(a) Two waves combine constructively   **(b)** Two waves combine destructively

Since waves add, is the resultant wave always larger than the individual waves that combine?

Reasoning and Discussion

The resultant wave is the sum of the individual waves, so it would seem that the resultant should be larger. However, the individual waves are sometimes positive and sometimes negative. For example, if wave 1 is positive at a given time, and wave 2 is negative, the sum of the two waves can be small or even zero. Thus, the resultant wave can be smaller than the individual waves.

Answer

No. The resultant wave can be smaller than the individual waves.

Resultant waves can be larger or smaller

As simple as the principle of superposition is, it still leads to interesting consequences. For example, consider the wave pulses shown in Figure 13.21 (a). When they combine, the resulting pulse has a larger amplitude, equal to the sum of the amplitudes of the individual pulses. Whenever waves combine to form a larger wave, the result is referred to as **constructive interference**.

On the other hand, two pulses like those in **Figure 13.21 (b)** may combine. When this happens, the positive displacement of one wave adds to the negative displacement of the other to create a net displacement of zero. That is, the pulses momentarily cancel one another. When waves superpose to form a smaller wave, the result is referred to as **destructive interference**.

Recall that waves are not changed when they pass through one another. This is true in both constructive and destructive interference. For example, in Figure 13.21 (b) the wave pulses continue on unchanged after they interact. This makes sense from an energy point of view. Each wave carries energy, and that energy cannot simply vanish.

Practice Problems

45. Draw the resultant wave for **(a)** the two waves shown in **Figure 13.22 (a)** and **(b)** the two waves shown in **Figure 13.22 (b)**.

46. Two wave pulses on a string approach one another at the time $t = 0$, as shown in **Figure 13.23**. Each pulse moves with a speed of 1.0 m/s. Make a careful sketch of the resultant wave at the times $t = 1.0$ s, 2.0 s, 2.5 s, 3.0 s, and 4.0 s.

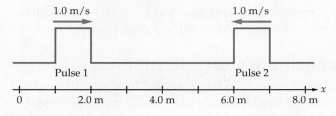

▲ Figure 13.23

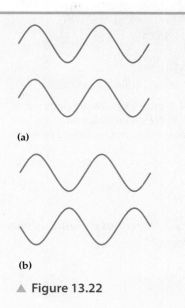

(a)

(b)

▲ Figure 13.22

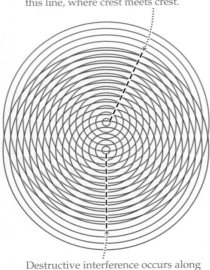

Constructive interference occurs along this line, where crest meets crest.

Destructive interference occurs along this line, where crest meets trough.

▲ **Figure 13.24 The interference of circular waves**
Two circular waves whose centers are offset produce an interference pattern. The light areas correspond to constructive interference, and the dark areas correspond to destructive interference.

▲ **Figure 13.25 Moiré patterns**
A moiré pattern is similar to an interference pattern formed by waves.

🔑 *How does a standing wave form?*

Interacting waves can form interference patterns

Interference effects are not limited to waves on a string. In fact, interference is one of the key characteristics that define waves. In general, when waves combine, they form interference patterns that include regions of constructive interference and regions of destructive interference.

An example of an interference pattern is shown in **Figure 13.24**. Here we see the superposition of two circular waves. Notice the light and dark "rays" radiating outward in this interference pattern. The light areas are where wave crests meet, which means the waves are *in phase*. This produces constructive interference. The dark areas are places where the waves are *out of phase*; that is, where a wave crest meets a wave trough. These are regions of destructive interference.

The interference pattern in Figure 13.24 is often referred to as a *moiré pattern*. You can create a moiré pattern by overlapping two repeating patterns. **Figure 13.25** shows how a slight displacement of one pattern from another creates an intriguing moiré pattern.

Physics & You: Technology Destructive interference is used to reduce noise in factories, busy offices, and even airplane cabins. The process, referred to as *active noise reduction (ANR)*, begins with a microphone that picks up the noise to be reduced. The signal from the microphone is then inverted and sent to a speaker. As a result, the speaker emits sound that is the opposite of the incoming noise—in effect, the speaker produces "anti-noise." In this way, the noise is *actively* canceled by destructive interference.

Standing Waves

If you've ever plucked a guitar string, you have produced a *standing wave*. In general, a **standing wave** is a wave that oscillates in a fixed location. The wave is said to be "standing" because its location does not change.

Specific points along a standing wave do not move

Consider a string of length L that is tied down at both ends, as shown in **Figure 13.26**. If you pluck this string in the middle, it vibrates and forms a standing wave, as shown in **Figure 13.26 (a)**. This is referred to as the *fundamental mode* of vibration for this string. It is also called the *first harmonic*. The string assumes a wavelike shape, but the wave stays in the same place. The fundamental mode corresponds to half a wavelength of a usual wave on a string.

Notice that the ends of the plucked string are fixed and do not move. Points on a standing wave that do not move are called **nodes**. Halfway between any two nodes is a point of maximum displacement known as an **antinode**. The first harmonic consists of two nodes (N) and one antinode (A) in the sequence N-A-N. This is illustrated in Figure 13.26 (a).

Certain frequencies produce standing waves

You can think of the first harmonic as being formed by a wave that reflects back and forth between the fixed ends of a string. 🔑 **When the frequency is just right, the reflected waves interfere constructively and the standing wave is formed. All standing waves are the result of interference.** If the frequency differs from the first-harmonic frequency, the reflections result in destructive interference and a standing wave does not form.

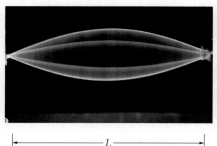

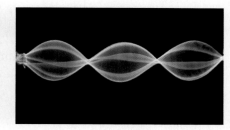

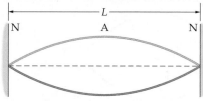

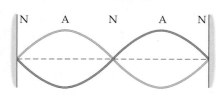

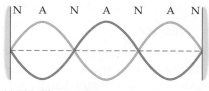

(a) First harmonic (fundamental) **(b) Second harmonic** **(c) Third harmonic**

▲ **Figure 13.26 Standing waves and harmonics**
(a) When a string is tied down at both ends and plucked in the middle, a standing wave results. This is the string's fundamental mode of oscillation, or first harmonic. The wavelength in this case is $2L$. **(b)** The second harmonic consists of two half-wavelength segments with a node in the middle. In this case the wavelength is L. **(c)** The third harmonic consists of three half-wavelength segments separated by two nodes. In this case the wavelength is $2L/3$. For each next higher harmonic, an additional half wavelength is added to the standing wave.

The wavelength and frequency of the first harmonic can be calculated from the length of the string, L, and the speed of the waves on the string, v, as follows:

> **First Harmonic of a Standing Wave on a String**
>
> wavelength $=$ 2(string length)
> $$\lambda_1 = 2L$$
>
> SI unit of λ_1: m
>
> $$\text{frequency} = \frac{\text{wave speed}}{2(\text{string length})}$$
> $$f_1 = \frac{v}{2L}$$
>
> SI unit of f_1: Hz $=$ cycles/s $=$ s^{-1}

However, the fundamental mode, described above, is not the only standing wave that can exist on a string.

Harmonics
A given string has an infinite number of standing wave modes—or *harmonics*. For example, the second harmonic is shown in **Figure 13.26 (b)**. Notice that its sequence of nodes and antinodes is N-A-N-A-N, which has one more antinode (A) and one more node (N) than the first harmonic. Adding one more antinode and node yields the third harmonic, N-A-N-A-N-A-N, as shown in **Figure 13.26 (c)**. The frequency of the second harmonic is twice the frequency of the first harmonic, and the frequency of the third harmonic is three times that of the first. This pattern continues for all higher harmonics.

A string 1.30 m in length is oscillating in its first harmonic mode. The frequency of oscillation is 7.80 Hz. (**a**) What is the wavelength of the first harmonic? (**b**) What is the speed of waves on this string?

Picture the Problem

Our sketch shows the string oscillating in its first harmonic mode. Also indicated is the length of the string.

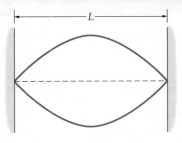

Strategy

(**a**) Only half a wavelength fits on a string oscillating in its first harmonic mode. Therefore, the wavelength of the first harmonic is twice the length of the string.

(**b**) The frequency, speed, and length are related by the formula $f_1 = v/2L$. We can rearrange this formula to solve for the speed, v. The other quantities in the formula are given in the problem statement.

Known	Unknown
$L = 1.30$ m	(**a**) $\lambda = ?$
$f_1 = 7.80$ Hz	(**b**) $v = ?$

Solution

Part (a)

 1 The wavelength of the first harmonic is twice the string length:

$$\lambda_1 = 2L$$
$$= 2(1.30 \text{ m})$$
$$= \boxed{2.60 \text{ m}}$$

Part (b)

2 Solve $f_1 = v/2L$ for the speed, v:

$$f_1 = v/2L$$
$$v = 2Lf_1$$

Math HELP
Solving Simple
Equations
**See Math
Review,
Section III**

3 Substitute the given numerical values:

$$v = 2Lf_1$$
$$= 2(1.30 \text{ m})(7.80 \text{ Hz})$$
$$= \boxed{20.3 \text{ m/s}}$$

Insight

Higher harmonics have higher frequencies, but the speed of the waves is the same for all harmonics.

Practice Problems

47. **Follow-up** Suppose the waves have a speed of 22.0 m/s rather than 20.3 m/s. What is the frequency of the first harmonic in this case?

48. **Concept Check** A guitar string plays flat—that is, its frequency of oscillation is lower than desired. To get the string in tune, should you increase or decrease the speed of waves on the string? Explain.

▲ Figure 13.27 **The shape of a piano**
The shape of a piano is a reflection of its function, not a stylistic expression. The long strings on the left side of a piano produce low notes, and the short strings on the right side produce high notes.

The length of a musical instrument determines its frequency range

When you pluck a guitar string, it vibrates primarily in its fundamental mode, the first harmonic. Higher harmonics make only small contributions to the sound produced. The same is true of the strings in a piano, a violin, and other string instruments. It follows that notes of different frequency, or pitch, can be produced by using strings of different length. To see the connection, recall that the first-harmonic frequency for a string of length L is $f_1 = v/2L$. Assuming that all other variables remain the same, longer strings (larger L) produce lower frequencies and shorter strings (smaller L) produce higher frequencies.

This fact accounts for the general shape of a piano, shown in **Figure 13.27**. Notice that the strings shorten toward the right side of the piano, where the notes are of higher frequency. Similarly, a violin is smaller than a cello, which is smaller than a double bass, as you can see in **Figure 13.28**. As a result, the frequency range of the violin is higher than that of the cello, and the double bass produces the lowest frequencies of all. Take a look at the instruments in your school band, and see if you can detect a connection between the size of the instrument and the pitch of the notes it plays. Physics is truly all around us!

▶ Figure 13.28 **A range of string instruments**
The frequency range of a musical instrument is directly related to the instrument's size. For example, the violin, the cello, and the double bass are similar in construction—they differ mainly in size. The smallest of these string instruments, the violin, has the highest frequency range, producing notes with frequencies from 196 Hz to 4200 Hz. Next in size is the cello, with a frequency range of 65 Hz to 988 Hz. The largest instrument, the double bass, produces frequencies in the range from 41 Hz to 245 Hz. Notice that the *highest* note of the double bass has only a slightly higher frequency than the *lowest* note of the violin.

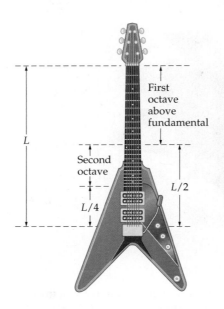

► **Figure 13.29 Frets on a guitar**
The thin metal bars that run across the neck of a guitar are the frets. Pressing a string down onto a fret changes the length of the string that is vibrating, which also changes the pitch of the sound. For example, to go up one octave from the fundamental mode, the effective length of a guitar string must be halved. To rise one more octave, it is necessary to halve the length of the string again. Thus, the distance between frets is not uniform; they are more closely spaced near the base of the neck.

First octave above fundamental

Second octave

L

$L/2$

$L/4$

The frets on a guitar are not equally spaced

Two full octaves of notes and more can be produced on a single guitar string. Different notes are produced by pressing the string down against frets to effectively change its length. It might seem surprising at first, but the separation between frets is not uniform, as you can see in **Figure 13.29**. There's a good reason for this. In order to produce the notes of a standard musical scale, it is necessary that each fret going down the neck of a guitar shorten the string by 6% compared with the previous fret. But 6% of a long string is a greater length than 6% of a short string. Therefore, the frets have a greater spacing at the top of the neck, where the string is longest.

13.4 LessonCheck Ⓜ

Checking Concepts

49. ⟡ Apply Two identical wave pulses with positive displacement head toward each other from opposite directions. Describe the resultant wave that forms when the pulses meet.

50. ⟡ Critique A classmate states that a standing wave involves both constructive and destructive interference. Is the classmate correct? Explain.

51. Cite Give two everyday examples of standing waves and describe how they form.

52. Assess Do higher harmonics always have greater frequencies than lower harmonics? Explain.

Solving Problems

53. Calculate A guitar string 66 cm long vibrates with a standing wave that has two antinodes.
(a) Which harmonic is this?
(b) What is the wavelength of this wave?

54. Evaluate A 0.33-m string is tied to a wall at either end. Which of the following wavelengths produces a standing wave on this string?

Wavelength 1	Wavelength 2	Wavelength 3	Wavelength 4
0.22 m	0.33 m	0.44 m	0.66 m

55. Calculate A string is tied to a wall at either end. If the wavelength of the first harmonic on this string is 0.65 m, what is the separation between the walls?

Physics & You

Tuned Mass Damper

What Is It? Installed near the top of a skyscraper, *a tuned mass damper (TMD)* is a large mass that oscillates horizontally with simple harmonic motion. Its purpose is to reduce, or dampen, the swaying motion caused by wind. Without a damper, the top of a building that was 500 m tall could sway back and forth up to 1 m in each direction.

The 509-m-tall Taipei 101 skyscraper.

The mass of the 5.5-m-diameter pendulum bob is 660,000 kg.

When Was It Invented? Naval engineers began using so-called resonant vibration absorbers in the early twentieth century to reduce the rolling motion of ships. Application of the technology to skyscrapers began in the 1960s when the reduced use of masonry building materials made new buildings more flexible.

How Does It Work? Winds cause skyscrapers to sway back and forth. The period of the swaying motion is determined by a building's shape, mass, and flexibility. Large oscillations can occur if the wind exerts a force at the building's natural frequency.

The 509-m-tall Taipei 101 skyscraper in Taiwan, Republic of China, is one of the world's tallest buildings. It contains a giant pendulum to oppose swaying motion. The period of the pendulum was "tuned" to match the building's natural frequency. When wind pushes the building in one direction, the pendulum swings in the opposite direction. Reaction forces from the pendulum's motion act on the building, opposing the wind force and thereby reducing the amplitude of the swaying.

The reaction forces are transmitted to the building through the pendulum pivot and several piston-type shock absorbers that connect the pendulum bob and the building. These shock absorbers absorb energy and control the amplitude of the pendulum. Softer shock absorbers allow for larger amplitudes and greater damping. Of course, the shock absorbers cannot be so soft as to allow the pendulum to batter the building. Even though the mass of the TMD is a tiny fraction of the building's total mass, it is able to reduce swaying by up to 40%.

Tuned mass damper

◄ The tuned mass damper is mounted near the top of the building.

Why Is It Important? Even a gentle swaying motion can cause people on the upper floors of a skyscraper to suffer motion sickness. Large-amplitude swaying motion over a long period of time could fatigue a building's steel structure and lead to a sudden collapse.

Take It Further

1. Analysis *The Taipei 101 building has 101 floors, and the height of the top floor is about 440 m. The mass of the 5.5-m-diameter pendulum bob is 660,000 kg. The pendulum extends down through four floors. Use the formula* $T = 2\pi\sqrt{L/g}$ *to estimate the period of this pendulum.*

2. Research *Look up* vortex shedding *to see how a steady constant wind can interact with a building to produce the time-varying force that drives resonant swaying.*

Physics Lab Standing Waves on a Coiled Spring

This lab explores the harmonic modes of vibration of a stretched coiled spring. You will use experimental data and observations to discover the properties of standing waves and make predictions about harmonics.

Materials
- coiled spring
- meterstick
- stopwatch
- masking tape

Procedure

1. Place the spring on the floor and stretch it between yourself and your lab partner. The length of the stretched spring should be between 2 and 3 m.

Caution: *Do not overstretch the spring or release it once it has been stretched.*

2. Place masking tape on the floor at each end of the stretched spring. Keep the spring stretched between these marks throughout the experiment. Measure and record the length of the stretched spring.

3. Generate waves by holding one end of the spring in place and moving the other end side to side at a constant rate. Adjust the frequency of oscillation until the first harmonic is produced. The first harmonic has a single antinode.

4. Use a stopwatch to measure the time needed to complete 20 waves (back-and-forth cycles) of the spring at the first harmonic. Record the time, and sketch the shape of the spring.

5. Repeat Steps 3 and 4 while increasing the frequency of oscillation until the second harmonic is produced. The second harmonic has two antinodes.

6. Repeat Step 5 for the third, fourth, and fifth harmonics.

Data Table

Length of stretched spring _____

Harmonic	Sketch of Wave Shape	Number of Half Wavelengths	Time for 20 Cycles (s)	Frequency (Hz)	Wavelength (m)	Wave Speed (m/s)
First						
Second						

Analysis

1. Describe how a standing wave forms on the spring.

2. Calculate the frequency of each harmonic. The frequency equals the number of complete cycles divided by the time needed to form them. Record your results in the data table.

3. Calculate the wavelength of each harmonic. The wavelength is determined using the length of the stretched spring and the number of half wavelengths. Record your results in the data table.

4. Use the wave speed equation to calculate the wave speed of each harmonic.

Conclusions

1. What effect does increasing the frequency have on the wavelength of standing waves?

2. How does the number of half wavelengths change between successive harmonics?

3. Will any frequency of oscillation produce a standing wave on a given length of spring? Explain your answer.

4. Predict the wavelength and frequency of the sixth harmonic (for the length of spring used in this experiment).

5. Predict what would happen to the wavelengths, frequencies, and wave speeds if the ends of the spring remained at the same place on the floor but the tension in the spring increased.

13 Study Guide

Big Idea
Waves are traveling oscillations that carry energy.

It takes energy to create an oscillation, like the push needed to start a person moving on a swing. In a wave, the oscillations that start at one location move to other locations with a speed determined by the properties of the medium through which the wave travels. The energy carried by a wave might heat the Earth (light), erode a beach (water waves), or shatter a drinking glass (sound).

13.1 Oscillations and Periodic Motion

🔑 Frequency is calculated by taking the inverse of the period.

🔑 Simple harmonic motion occurs when the restoring force is proportional to the displacement from equilibrium.

🔑 The period of a mass on a spring varies inversely with the spring constant. The period of a mass on a spring varies directly with the mass.

• In general, any motion that repeats itself over and over is referred to as *periodic motion*.

• The period of a mass on a spring depends on both the mass and the spring constant.

Key Equations

The period is the time required for a motion to repeat:

$$T = \text{time to complete a periodic motion}$$

The frequency is the inverse of the period:

$$f = 1/T$$

The period of a mass m on a spring with a spring constant k:

$$T = 2\pi\sqrt{\frac{m}{k}}$$

13.2 The Pendulum

🔑 The period of a pendulum depends on its length and on the acceleration due to gravity.

🔑 A system driven at its natural frequency is in resonance.

• A simple pendulum is a mass attached to the end of a string.

• The natural frequency of an oscillating system is the frequency at which it oscillates when free from external disturbances.

Key Equation

Period of a pendulum of length L:

$$T = 2\pi\sqrt{\frac{L}{g}}$$

13.3 Waves and Wave Properties

🔑 In general, waves travel from place to place, but the particles in a wave oscillate back and forth at one location.

🔑 The speed of a wave is determined by the properties of the material, or medium, through which it travels.

• A wave is a traveling oscillation that carries energy.

• The distance from one crest to the next, or from one trough to the next, is the wavelength, λ, of the wave.

$$\lambda = \text{wavelength} = \text{distance over which a wave repeats}$$

• In a transverse wave the individual particles move at right angles to the direction of the wave motion.

• In a longitudinal wave the individual particles move in the same direction as the wave motion.

Key Equation

Speed of a wave with wavelength λ and frequency f:

$$v = \lambda f$$

13.4 Interacting Waves

🔑 Waves that occupy the same space combine in the simplest possible way—they just add together.

🔑 When the frequency of waves reflecting back and forth between fixed ends of a string is just right, the waves constructively interfere and a standing wave is formed. All standing waves are the result of interference.

• Waves that add to give a larger resultant wave show constructive interference.

• Waves that add to give a smaller resultant wave show destructive interference.

• Standing waves oscillate in a fixed location.

Key Equations

The wavelength and frequency of the first harmonic for a standing wave on a string:

$$\lambda_1 = 2L \qquad f_1 = \frac{v}{2L}$$

13 Assessment

For instructor-assigned homework, go to www.masteringphysics.com

ANSWERS TO SELECTED ODD-NUMBERED PROBLEMS APPEAR IN APPENDIX A.

Lesson by Lesson

13.1 Oscillations and Periodic Motion

Conceptual Questions

56. Would the period of a mass–spring system be different on the Moon than it is on the Earth? Explain.

57. A mass on a spring moves back and forth in simple harmonic motion with amplitude A and period T. In terms of T, how long does it take for the mass to move through a total distance of $3A$?

58. Rank Rank the following mass–spring systems in order of increasing period of oscillation, indicating any ties. System A consists of a mass m attached to a spring with a spring constant k; system B has a mass $2m$ attached to a spring with a spring constant k; system C has a mass $3m$ attached to a spring with a spring constant $6k$; and system D has a mass m attached to a spring with a spring constant $4k$.

59. Predict & Explain An old car with worn-out shock absorbers oscillates with a given frequency when it hits a speed bump. **(a)** If the driver takes on two more passengers and hits another speed bump, is the car's frequency of oscillation greater than, less than, or equal to what it was before? **(b)** Choose the *best* explanation from among the following:

A. Increasing the mass on a spring increases its period and thus decreases its frequency.

B. The frequency depends on the spring constant of the spring but is independent of the mass.

C. Adding mass makes the spring oscillate more rapidly, which increases the frequency.

Problem Solving

60. A person in a rocking chair completes 12 cycles in 21 s. What are the period and frequency of the rocking motion?

61. While fishing for catfish, a fisherman suddenly notices that the bobber (a floating device) attached to his line is bobbing up and down with a frequency of 2.6 Hz. What is the period of the bobber's motion?

62. **Tuning Forks in Neurology** Tuning forks are used to diagnose a nervous system disorder that has reduced sensitivity to vibrations as a symptom. The medical tuning fork in **Figure 13.30** has a frequency of 128 Hz. What is its period of oscillation?

Figure 13.30 A neurological hammer with a tuning fork.

63. If you dribble a basketball with a frequency of 1.8 Hz, how long does it take for you to complete 12 dribbles?

64. You take your pulse and observe 74 heartbeats in a minute. What are the period and frequency of your heartbeat?

65. A plot of force versus length for a given spring is presented in **Figure 13.31**. **(a)** What is the spring constant for this spring? **(b)** What is the period of oscillation when a 0.10-kg mass is attached to this spring?

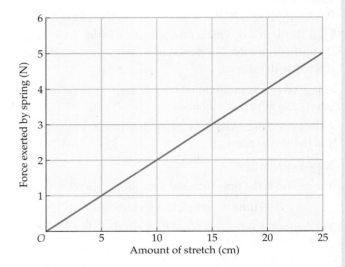

Figure 13.31

66. A 0.46-kg mass attached to a spring undergoes simple harmonic motion with a period of 0.77 s. What is the spring constant of the spring?

67. Think & Calculate **(a)** Your heart beats with a frequency of 1.45 Hz. How many beats occur in a minute? **(b)** If the frequency of your heartbeat increases, will the number of beats in a minute increase, decrease, or stay the same? **(c)** How many beats occur in a minute if the frequency increases to 1.55 Hz?

68. When a 0.50-kg mass is attached to a vertical spring, the spring stretches by 15 cm. How much mass must be attached to the spring to result in a 0.75-s period of oscillation?

13.2 The Pendulum

Conceptual Questions

69. A pendulum swings back and forth. How many times does it pass through the equilibrium position during one cycle of its motion? Assume the cycle begins when the pendulum is at maximum displacement from equilibrium.

70. A pendulum of length L has a period T. How must the length of the pendulum change in order to triple its period to $3T$? Give your answer in terms of L.

71. [Triple Choice] Pendulum A has a length L and a mass m. Pendulum B has a length L and a mass $2m$. Is the period of pendulum A greater than, less than, or equal to the period of pendulum B? Explain.

72. [Triple Choice] Billie releases the bob of her pendulum at an angle of 10° from the vertical. At the same time, Bobby releases the bob of his pendulum at an angle of 20° from the vertical. The two pendulums have the same length. Does Billie's bob reach the vertical position before, after, or at the same time as Bobby's bob? Explain.

73. Metronomes, such as the penguin shown in **Figure 13.32**, are useful devices for music students. If a student wants to have the penguin metronome tick with a greater frequency, should she move the penguin's bow tie weight upward or downward? Explain.

Figure 13.32
How do you like my tie?

74. [Predict & Explain] A grandfather clock keeps perfect time at sea level. **(a)** If the clock is taken to a ski lodge at the top of a nearby mountain, do you expect it to keep correct time, run slow, or run fast? **(b)** Choose the *best* explanation from among the following:

A. Gravity is weaker at the top of the mountain, leading to a greater period of oscillation.

B. The length of the pendulum is unchanged, and therefore its period remains the same.

C. The extra gravity from the mass of the mountain causes the period to decrease.

Problem Solving

75. A pendulum oscillates with a frequency of 0.24 Hz. What is its length?

76. A pendulum in the service shaft of a building is used by a physics class for experiments. The shaft is dark, and the top of the pendulum cannot be seen. The bob of the pendulum is visible, and it is observed to complete 13 oscillations in 110 s. What is the length of the pendulum? Assume $g = 9.81 \text{ m/s}^2$.

77. A simple pendulum of length 2.5 m makes 5.0 complete swings in 16 s. What is the acceleration due to gravity at the location of the pendulum?

78. Find the length of a simple pendulum that has a period of 2.00 s. Assume $g = 9.81 \text{ m/s}^2$.

79. A *seconds pendulum* is one that passes through its equilibrium position once every 1.000 s. **(a)** What is the period of a seconds pendulum? **(b)** A seconds pendulum in Chicago has a length of 0.9933 m. What is the acceleration due to gravity in Chicago?

80. **United Nations Pendulum** A large pendulum swings in the lobby of the United Nations building in New York City. The pendulum has a 91-kg gold-plated bob and a length of 22.9 m. How much time does it take for the bob to swing from its maximum displacement to its equilibrium position? Assume $g = 9.81 \text{ m/s}^2$.

81. A fan at a baseball game notices that the radio commentators have lowered a microphone from their booth to just a few centimeters above the ground, as shown in **Figure 13.33**. The field-level microphone slowly swings like a simple pendulum. If the microphone completes 10 oscillations in 60.0 s, how high above the field is the radio booth?

Figure 13.33

13.3 Waves and Wave Properties
Conceptual Questions

82. How much time is required for a wave to travel two wavelengths?

83. Red light has a longer wavelength than violet light, but the waves of both have the same speed. Which light waves have the greater frequency?

84. Is it possible for fans at baseball game to produce a longitudinal "wave"? If so, explain how the fans might move their bodies to accomplish this.

85. In a classic TV commercial, a group of cats feed from bowls of cat food that are lined up side by side. Initially there is one cat for each bowl. When an additional cat is added to the scene, it runs to a bowl at the end of the line and begins to eat. The cat that was there originally moves to the next bowl, displacing that cat, which moves to the next bowl, and so on down the line. What type of wave have the cats created? Explain.

86. Rank The three waves on strings, A, B, and C, shown in **Figure 13.34** move to the right with equal speeds. Rank the waves in order of (a) increasing frequency and (b) increasing wavelength, indicating any ties.

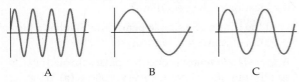

Figure 13.34

Problem Solving

87. You dip your finger into the water of a pond twice a second. The waves you produce have crests that are separated by 0.18 m. Determine the frequency, period, wavelength, and speed of these waves.

88. A wave travels along a stretched horizontal rope. The vertical distance from crest to trough of this wave is 13 cm and the horizontal distance from crest to trough is 28 cm. What are (a) the wavelength and (b) the amplitude of this wave?

89. A surfer floating beyond the breakers notes 14 waves per minute passing her position. If the wavelength of these waves is 34 m, what is their speed?

90. Tsunami A tsunami (tidal wave) traveling across deep water can have a speed of 750 km/h and a wavelength of 310 km. What is the frequency of such a wave?

91. The speed of water waves decreases as the water becomes shallower. Suppose waves travel across the surface of a lake with a speed of 2.0 m/s and a wavelength of 1.5 m. When these waves move into a shallower part of the lake, their speed decreases to 1.6 m/s, though their frequency remains the same. Find the wavelength of the waves in the shallower water.

92. While sitting on a dock of the bay, you notice a series of waves going past. You observe that 11 waves go past you in 45 s and that the distance from one crest to the next trough is 3.0 m. Find the (a) period, (b) frequency, (c) wavelength, and (d) speed of these waves.

13.4 Interacting Waves

Conceptual Questions

93. Which of the waves in **Figure 13.35**—blue, green, or red—is the resultant of the other two waves?

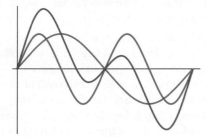

Figure 13.35

94. Sketch the resultant wave that is produced by the two waves in **Figure 13.36**.

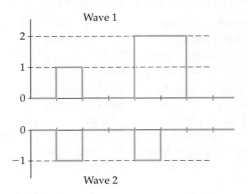

Figure 13.36

95. A string 1 m in length is tied down at both ends. The three lowest-frequency standing waves on this string have frequencies of 100 Hz, 200 Hz, and 300 Hz. (a) What is the fundamental frequency of this string? (b) What is the wavelength of the fundamental mode?

Problem Solving

96. The speed of waves on a string that is tied down at both ends is 32 m/s. If the fundamental frequency for this string is 65 Hz, what is the length of the string?

97. The first harmonic of a string tied down at both ends has a frequency of 26 Hz. If the length of the string is 0.83 m, what is the speed of waves on the string?

98. Think & Calculate A string tied down at both ends has a length of 0.64 m, and its first harmonic has a frequency of 41 Hz. (a) If the length of the string is increased while all other variables are kept the same, does the first-harmonic frequency increase, decrease, or stay the same? Explain. (b) Calculate the first-harmonic frequency for this string if its length is increased to 0.88 m.

99. Two wave pulses on a string approach one another at the time $t = 0$, as shown in **Figure 13.37**. Each pulse moves with a speed of 1.0 m/s. Make a careful sketch of the resultant wave at the times $t = 1.0$ s, 2.0 s, 2.5 s, 3.0 s, and 4.0 s.

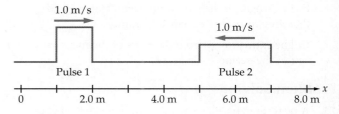

Figure 13.37

100. Suppose pulse 2 in Problem 99 is inverted so that it has a downward deflection. Sketch the resultant wave at the given times for this case.

101. A guitar string 66 cm long vibrates with a standing wave that has three antinodes. **(a)** Which harmonic is this? **(b)** What is the wavelength of this wave?

Mixed Review

102. An object undergoes simple harmonic motion with a period of 2.0 s. In 3.0 s the object moves through a total distance of 24 cm. What is the object's amplitude of motion?

103. Using a BMMD An astronaut uses a body mass measurement device (BMMD; see page 460 for more information) to determine her mass. What is the astronaut's mass, given that the spring constant of the BMMD is 2600 N/m and the period of oscillation is 0.85 s?

104. Sunspots Galileo made the first European observations of sunspots in 1610, and daily observations were begun in Zurich in 1749. The number of sunspots varies from year to year, but the counts exhibit a roughly 11-year cycle. What is the frequency of the sunspot cycle? Give your answer in hertz.

105. Helioseismology In 1962 physicists at Cal Tech discovered that the surface of the Sun vibrates as a result of the violent nuclear reactions within its core. This discovery led to a new field of science known as *helioseismology*. A typical vibration is shown in **Figure 13.38**; it has a period of 5.7 min. The blue regions are moving outward while the red regions are moving inward. Find the frequency of this vibration.

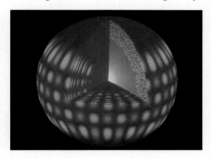

Figure 13.38 A typical vibration pattern of the Sun's surface.

106. Predict & Explain The two blocks in **Figure 13.39** have the same mass, *m*. All the springs have the same spring constant, *k*, and are at their equilibrium length. **(a)** When the blocks are set into oscillation, is the period of block 1 greater than, less than, or equal to the period of block 2? **(b)** Choose the *best* explanation from among the following:

A. Springs in parallel are stiffer than springs in series; therefore the period of block 1 is smaller than the period of block 2.

B. The two blocks experience the same restoring force for a given displacement from equilibrium, and thus they have equal periods of oscillation.

C. The forces exerted by the two springs on block 2 partially cancel one another, leading to a longer period of oscillation.

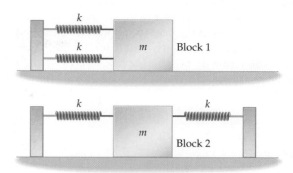

Figure 13.39

107. A standing wave with a frequency of 603 Hz is produced on a string that is 1.33 m long and fixed at both ends. If the speed of waves on this string is 402 m/s, how many antinodes are there in the standing wave?

108. Figure 13.40 shows a displacement-versus-time graph of the periodic motion of a 3.8-kg mass on a spring. **(a)** Referring to the figure, what is the period of motion? **(b)** What is the amplitude of motion? **(c)** Calculate the spring constant of this spring.

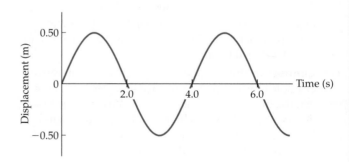

Figure 13.40

109. Think & Calculate A 3.8-kg mass on a spring oscillates as shown in the displacement-versus-time graph in Figure 13.39. **(a)** Referring to the graph, at what times between *t* = 0 and *t* = 6 s does the mass experience a force of maximum magnitude? Explain. **(b)** At what times shown in the graph does the mass experience zero force? Explain. **(c)** Calculate the magnitude of the maximum force exerted on the mass.

110. A 0.45-kg crow lands on a slender branch and bobs up and down with a period of 1.4 s. An eagle flies to the same branch, scaring the crow away, and lands. The eagle bobs up and down with a period of 3.6 s. Treating the branch as an ideal spring, find **(a)** the effective spring constant of the branch and **(b)** the mass of the eagle.

111. Think & Calculate The string of the pendulum shown in **Figure 13.41** is stopped by a peg when the bob swings to the left, but moves freely when the bob swings to the right. **(a)** Is the period of this pendulum greater than, less than, or the same as the period of the same pendulum without the peg? **(b)** Express the period of this pendulum in terms of L and ℓ. **(c)** Evaluate your expression from part (b) for the case where $L = 1.0$ m and $\ell = 0.25$ m.

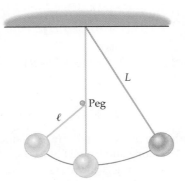

Figure 13.41

Writing about Science

112. Research the technology used in quartz digital watches. Your report should address questions such as the following: How big are the quartz crystals? Why are quartz crystals more accurate than pendulums and springs? When were the first quartz watches made, and who developed them?

113. Connect to the Big Idea A particularly interesting example of waves and resonance involves a singer who can use her voice to shatter a glass. Write a step-by-step description of how this happens. Be sure to discuss how the sound forms, travels, and is able to shatter the glass. What determines the "right note" for a particular glass?

Read, Reason, and Respond

A Cricket Thermometer Insects are ectothermic; that is, their body temperature is largely determined by the temperature of their surroundings. One of the most interesting effects of this characteristic is the temperature dependence of the chirp rate of certain insects. A particularly precise connection between chirp rate and temperature is found in the snowy tree cricket (*Oecanthus fultoni* Walker), shown in **Figure 13.42**. This type of cricket chirps at a rate that follows the formula $N = T - 39$. In this expression, N is the number of chirps in 13 s, and T is the numerical value of the temperature in degrees Fahrenheit. This formula, which is known as *Dolbear's law*, is plotted in **Figure 13.43** (green line) along with data points (blue dots) for the snowy tree cricket.

Figure 13.42 The snowy tree cricket.

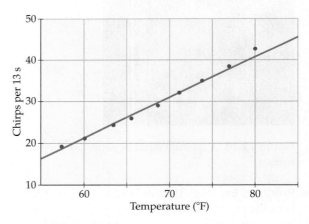

Figure 13.43

114. What is the temperature (in degrees Fahrenheit) if a cricket is observed to give 35 chirps in 13 s?

A. 13   C. 74

B. 35   D. 90

115. What is the frequency of the cricket's chirping (in hertz) when the temperature is 68 °F?

A. 0.45   C. 5.2

B. 2.2   D. 29

Standardized Test Prep

Multiple Choice

Use the graph below to answer Questions 1 and 2. The graph plots the position of an oscillator versus time.

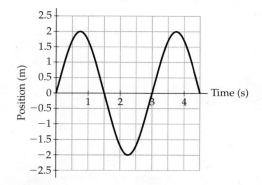

1. What are the frequency and the amplitude of the oscillator?
   - (A) 0.67 Hz and 4 m
   - (B) 0.67 Hz and 2 m
   - (C) 0.33 Hz and 4 m
   - (D) 0.33 Hz and 2 m

2. At what times is the oscillator moving at maximum speed?
   - (A) 0.75, 2.25, and 3.75 s
   - (B) 1.5, 3.0, and 4.5 s
   - (C) 1.5, 3.0, and 5.0 s
   - (D) 2.0, 4.0, and 6.0 s

3. A mass on an oscillating spring has its maximum speed when
   - (A) displacement is maximum and force is maximum.
   - (B) displacement is maximum and force is minimum.
   - (C) displacement is minimum and force is minimum.
   - (D) displacement is minimum and force is maximum.

4. A swinging pendulum can be considered a simple harmonic oscillator only if
   - (A) the pendulum length is large.
   - (B) the mass on the pendulum is small.
   - (C) the angle of release is small.
   - (D) the value of g is small.

5. Which statement about the first harmonic of a standing wave on a string is correct?
   - (A) The wave has a node at one end and an antinode at the other end.
   - (B) The wavelength of the wave is twice the length of the string.
   - (C) The wave has an antinode at both ends.
   - (D) The wavelength of the wave is equal to the length of the string.

6. A string with a length of 0.75 m is tied down at both ends. The frequency of the first harmonic on this string is 2.4 Hz. What are the period of the first harmonic and the speed of the waves on the string?
   - (A) period = 0.42 s, speed = 3.6 m/s
   - (B) period = 0.75 s, speed = 2.4 m/s
   - (C) period = 1.3 s, speed = 1.9 m/s
   - (D) period = 2.4 s, speed = 1.7 m/s

7. What happens when a mechanical wave travels from a "faster" medium (where the wave speed is greater) to a "slower" medium?
   - (A) Wavelength and frequency decrease.
   - (B) Wavelength and frequency increase.
   - (C) Wavelength increases, and frequency decreases.
   - (D) Wavelength decreases, and frequency does not change.

Extended Response

8. Describe how you would set up an experiment to determine a value of g using an oscillator, specifying (a) what type of oscillator you would use and (b) what variables you would manipulate to determine the best value.

> **Test-Taking Hint**
>
> Remember that the frequency and period of a harmonic oscillator do not change as the oscillator loses amplitude.

If You Had Difficulty With . . .

Question	1	2	3	4	5	6	7	8				
See Lesson	13.1	13.1	13.1	13.2	13.4	13.4	13.3	13.2				

14 Sound

Inside

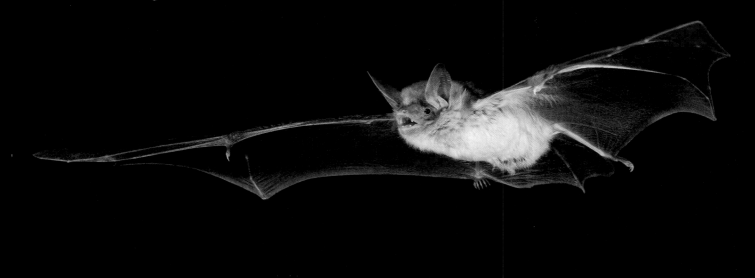

Bats live in a world defined by sound. This bat is emitting sounds, and its ears are positioned to pick up the resulting echoes. The timing and intensity of the echoes allow the bat to navigate in total darkness.

Big Idea

Sound carries energy in the form of a traveling wave of compressions and expansions.

Sound is an important part of how we experience the world. We use sound to communicate with one another. We use sound to enrich our lives with songs and to cheer our favorite sports team. Sound even allows us to hear things that happen around a corner—out of sight—and to peer inside the human body. This chapter explores the most important aspects of sound, including the way our sense of hearing allows us to appreciate the world around us.

14.1 Sound Waves and Beats

The first thing we do when we come into the world is make a sound. It's many years before we learn some of the scientific details about sound. In this lesson we study some of the basic properties of sound waves.

Properties of Sound Waves

Sound is a wave that travels through the air and other substances. It has all the wave characteristics you learned about in Chapter 13: frequency, period, wavelength, and wave speed.

Sound waves travel as compressions and expansions

As a child, you probably played with a coiled spring toy, like the one shown in **Figure 14.1 (a)** on the next page. If you oscillate one end of the spring back and forth, you will see a longitudinal wave moving away from you. The wave consists of regions where the coils of the spring are compressed alternating with regions where the coils are expanded.

Similarly, **Figure 14.1 (b)** shows how a vibrating tuning fork produces sound waves as its tines oscillate back and forth. Just as with the toy spring, a wave travels away from this vibrating source. 🔑 **In general, a sound wave is formed when an oscillating object creates alternating regions of compressed and expanded air. These alternating regions move away from the source as a longitudinal wave.**

Vocabulary

- pitch
- infrasonic
- ultrasonic
- beat
- beat frequency

🔑 *How is a sound wave produced?*

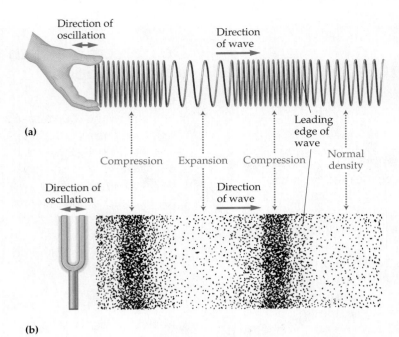

(a)

Compression Expansion Compression Normal density

Leading edge of wave

(b)

▲ **Figure 14.1 The production of sound waves**
An oscillating tuning fork makes sound waves in the air just like an oscillating hand makes waves in a coiled spring toy.

Sound waves can exhibit sine wave behavior

At first glance a sound wave seems different from a wave on a string. In particular, the sound wave doesn't seem to have the regular shape of a sine wave. If you plot the appropriate quantities, however, the classic wave shape emerges. For example, in **Figure 14.2 (a)** we show the compressions and **rarefactions** of a typical sound wave. In **Figure 14.2 (b)** we plot the corresponding density of the air. Clearly, the density oscillates in a wavelike fashion, just like a wave on a string. Similarly, the electrical output of a microphone that is picking up the sound from an oscillating tuning fork is shown in **Figure 14.3**. It also has the typical shape of a sine wave.

The speed of sound varies

Like a wave on a string, the *speed* of a sound wave is determined by the properties of the medium through which it moves.

Speed of Sound in Air   In air, under normal atmospheric pressure and temperature, the speed of sound is 343 m/s. This is pretty fast—about 770 mi/h in everyday terms. The reason for this high speed is that molecules in the air are constantly moving at roughly that speed. Thus, at this very moment air molecules are colliding with your body at about the speed of sound. As air is heated, the molecules in it move even faster. As a result, the speed of sound also increases with temperature.

You should use 343 m/s for the speed of sound throughout this text, unless specifically instructed otherwise.

Speed of Sound in Air (at room temperature, 20 °C)
$v = 343$ m/s   (approximately 1000 ft/s or 770 mi/h)
SI units: m/s

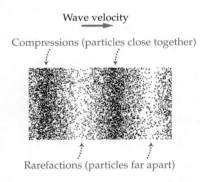

Wave velocity

Compressions (particles close together)

Rarefactions (particles far apart)

(a)

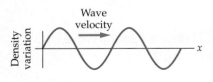

Wave velocity

Density variation

x

(b)

▲ **Figure 14.2 Wave properties of sound**
(a) A sound wave moving through the air.
(b) The density of the air within the sound wave. Notice that the density has the shape of a wave on a string.

The following example gives an easy-to-remember rule of thumb for calculating distances by making use of the speed of sound.

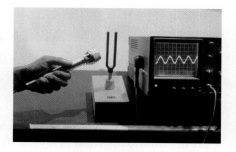

▲ **Figure 14.3 Electrical output of a microphone**
An oscilloscope connected to a microphone shows the wave form of a pure tone, created here by a tuning fork. The trace on the screen shows that the wave form is like that of a sine wave.

CONCEPTUAL **Example 14.1 How Far Away Is the Lightning?**

Three seconds after you see a brilliant flash of lightning, thunder shakes your house. Did the lightning strike about a kilometer away, much closer than a kilometer, or much farther away than a kilometer?

Reasoning and Discussion
As mentioned, the speed of sound in air is 343 m/s. Thus, in 3 s sound travels roughly 1000 m, or 1 km. A useful rule of thumb, then, is as follows: Divide the number of seconds for the sound of thunder to arrive by 3, and you have the distance in kilometers. (With non-SI units the rule is to divide the number of seconds by 5 to obtain the distance in miles.)

Answer
The lightning struck about a kilometer away.

Practice Problems

1. Sitting in the outfield bleachers, 170 m from home plate, you see a batter hit the ball for a home run. How long after you see the ball hit do you hear the crack of the bat?

2. At Zion National Park a loud shout produces an echo 1.75 s later from a colorful sandstone cliff. How far away from the shouter is the cliff? (*Hint*: The echo reflects off the distant cliff and returns to its source. Use the equation $d = vt$ to solve for distance.)

Speed of Sound in Other Materials   In general, the speed of sound depends on the "stiffness" of a material. The stiffer the material, the faster the sound wave. For example, air is quite compressible—not "stiff" at all—and the speed of sound is relatively low in air compared to liquids and solids. Water is not very compressible, and the speed of sound in it is about 4 times greater than in air. Sound travels even faster in solids than in liquids. In fact, the speed of sound in steel is about 17 times greater than in air. **Table 14.1** gives a sampling of the speed of sound for a range of materials.

Table 14.1 Speed of Sound in Various Materials

Material	Speed (m/s)
Aluminum	6420
Granite	6000
Steel	5960
Pyrex glass	5640
Copper	5010
Plastic	2680
Fresh water (20 °C)	1482
Fresh water (0 °C)	1402
Hydrogen (0 °C)	1284
Helium (0 °C)	965
Air (20 °C)	343
Air (0 °C)	331

You drop a stone from rest into a well in which the water is 7.35 m below the top. How long does it take before you hear the splash?

Picture the Problem

Our sketch shows the well into which the stone is dropped. Notice that the distance to the water is $d = 7.35$ m. After the stone falls the distance d, the sound from the splash rises the same distance d before it is heard.

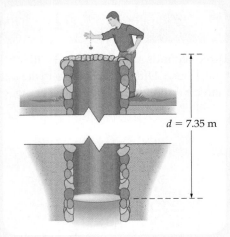

$d = 7.35$ m

Strategy

The time until the splash is heard is the sum of the time, t_1, for the stone to drop the distance d, and the time, t_2, for sound to travel upward the distance d. To find the drop time, use one-dimensional kinematics with an initial velocity $v_i = 0$ (since the stone is dropped from rest) and an acceleration of $g = 9.81$ m/s^2. The relationship between distance and time for the falling stone is $d = \frac{1}{2}gt_1^2$. For the sound wave, use $d = vt_2$, with $v = 343$ m/s.

Known	Unknown
$d = 7.35$ m	$t = ?$

Solution

1 Solve $d = \frac{1}{2}gt_1^2$ for time t_1 (the time for the stone to drop a distance d). Then, substitute the known values and calculate t_1:

$$d = \tfrac{1}{2}gt_1^2$$

$$t_1 = \sqrt{\frac{2d}{g}} = \sqrt{\frac{2(7.35 \text{ m})}{9.81 \text{ m/s}^2}}$$

$$= 1.22 \text{ s}$$

2 Solve $d = vt_2$ for t_2 (the time for sound to travel upward a distance d). Then substitute the numerical values and calculate t_2:

$$d = vt_2$$

$$t_2 = \frac{d}{v} = \frac{7.35 \text{ m}}{343 \text{ m/s}}$$

$$= 0.0214 \text{ s}$$

> **Math HELP**
> Solving Simple Equations
> **See Math Review, Section III**

3 Sum the times found in Steps 1 and 2:

$$t = t_1 + t_2$$
$$= 1.22 \text{ s} + 0.0214 \text{ s}$$
$$= \boxed{1.24 \text{ s}}$$

Insight

We use the same speed for a sound wave whether it is traveling horizontally, vertically upward, or vertically downward. The speed of sound is independent of its direction of motion. As a result, the waves from a source of sound move outward uniformly in all directions.

Practice Problems

3. **Follow-up** If the distance to the water level in the well is doubled, is the time until you hear the splash twice what it was before, more than twice, or less than twice? Explain.

4. **Follow-up** Suppose you throw the stone downward into the well with a speed of 2.1 m/s. How long does it take before you hear the splash?

5. **Concept Check** Is the speed of sound likely to be faster in a soft, squishy rubber ball or a hard, rigid steel ball?

6. **Challenge** A cannon 95 m away shoots a cannonball straight up in the air with an initial speed of 44 m/s. What is the speed of the cannonball when you hear the shot?

The speed of sound is the same for all frequencies

The speed of sound is the same for all directions of travel and for all frequencies. Thus, the speed v remains constant in the wave speed equation:

$$v = \lambda f$$
$$\text{speed} = \text{wavelength} \times \text{frequency}$$

For example, if the frequency of a wave is doubled, its wavelength is halved, and the speed v stays the same.

The fact that different frequencies travel with the same speed is evident when you listen to an orchestra in a large room. Different instruments are producing sounds of different frequencies. Even so, you hear all the sounds at the same time. Otherwise, listening to music from a distance would be quite a strange and inharmonious experience.

Sound can be described by its pitch

When we hear a sound, its frequency makes a great impression on us. How high or low we perceive a sound to be is known as its **pitch**. 🔑 **The pitch of a sound is simply the frequency of the corresponding sound wave.** For example, as you run your hand from left to right across the keys of a piano, you hear a series of notes with ever-increasing pitch, or frequency, from 55 Hz to 4187 Hz. Similarly, as you hum a song, you change the size and shape of your vocal chords slightly and this changes the pitch of the sound you produce.

Human Hearing and Beats

Did you know that your sense of hearing detects only a small portion of the sound waves that are created in nature? Did you know that two objects vibrating at slightly different frequencies create an interesting pulsating sound at a third frequency? Let's explore these phenomena.

Human hearing has a limited range

Earlier we described pitch in terms of the keys on a piano. Your hearing extends well beyond the range of a piano, however. As a rule of thumb, humans can hear sounds between 20 Hz on the low-frequency end and 20,000 Hz on the high-frequency end. Sounds with a frequency less than 20 Hz are referred to as **infrasonic**. Sounds with a frequency greater than 20,000 Hz are called **ultrasonic**. (The term *supersonic* refers to an object that moves faster than the speed of sound; it does not refer to frequency.)

Though we are unable to hear ultrasound and infrasound, these frequencies occur commonly in nature. For example, bats and dolphins produce ultrasound as they go about their daily lives. They send out ultrasonic waves that reflect back to them from objects in their vicinity. The reflected waves—the echoes—are used in a process known as *echolocation* to locate prey and to navigate. The bat in **Figure 14.4** is using echolocation to zero in on a juicy moth. Some insects can hear the ultrasonic sounds of a hunting bat and take evasive action. In fact, certain moths fold their wings in flight and drop into a steep dive when they hear an ultrasonic bat call.

It was also recently discovered that elephants communicate with one another using sounds with frequencies as low as 15 Hz. In fact, it may be that *most* elephant communication is infrasonic. These sounds, which humans feel as vibrations rather than hear, can carry over an area of about 30 square kilometers on the dry African savanna.

CONNECTING IDEAS

Frequency, a fundamental characteristic of all waves, was introduced in Chapter 13.
- This lesson explores how the frequency of a sound affects what we hear.
- Lesson 14.2 looks at the frequencies of standing waves that form in columns of air.
- Lesson 14.3 examines the frequency shift known as the *Doppler effect*.

🔑 *What determines the pitch of a sound?*

▲ **Figure 14.4 A bat hunting a moth**
Bats navigate in the dark and locate their prey using a system of biological sonar. They emit a continuous stream of ultrasonic sounds and detect the echoes from objects around them.

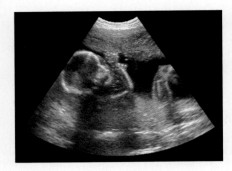

▲ **Figure 14.5 Ultrasound scan**
Ultrasound scans are created by beaming ultrasonic pulses into the body and measuring the time required for the echoes to return. This technique is commonly used to evaluate heart function (echocardiograms) and to visualize developing fetuses (sonograms).

🔑 *What causes beats?*

Medical applications of ultrasound are also common. Perhaps the most familiar is the ultrasound scan that is used to image a fetus in the womb, as illustrated in **Figure 14.5**. Sending bursts of ultrasound into the body and measuring the time delay of the resulting echoes—the technological equivalent of echolocation—makes it possible to map the structures that lie hidden beneath the skin.

Two sounds can result in beats

Imagine plucking—at the same time—two guitar strings that have slightly different frequencies. If you listen carefully, you'll notice that the sound produced by the two strings changes with time. In fact, the loudness increases then decreases, increases then decreases, over and over. In general, changes in loudness produced by sounds of different frequency are referred to as **beats**. 🔑 **Beats are the result of two waves interfering with one another, sometimes constructively (loud sound) and sometimes destructively (soft sound).**

Beat Formation To see how beats are produced, consider the two waves (one red, the other blue) shown in **Figure 14.6 (a)**. Initially the two waves interfere constructively, giving a large amplitude. The sound we hear at this time is loud, as indicated in **Figure 14.6 (b)**. A short time later the two waves interfere destructively, giving zero amplitude and no sound.

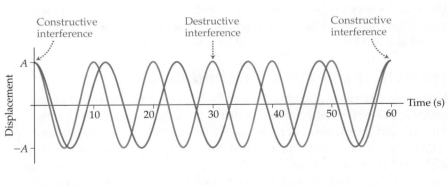

(a)

▶ **Figure 14.6 Interference of two waves with slightly different frequencies**
(a) A graph of two waves, one blue and the other red, with slightly different frequencies. (b) The resultant wave for the two waves shown in part (a). Notice the alternating constructive and destructive interference, resulting in beats.

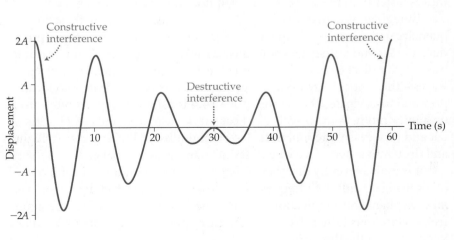

(b)

Beat Frequency   A key characteristic of a beat is its repeating pattern of loud and soft sounds. The frequency at which a beat repeats itself is known as the **beat frequency**. The beat frequency is simply the difference in frequency between the two original waves. If one wave has the frequency f_1 and the other has the frequency f_2, the beat frequency is as follows:

Definition of Beat Frequency

beat frequency = absolute value of the difference in frequency

$$f_{beat} = |f_1 - f_2|$$

SI unit: $1/s = s^{-1}$

Taking the absolute value of the difference simply ensures that the beat frequency is always positive.

As an example, suppose two guitar strings have the frequencies 438 Hz and 442 Hz. If you pluck them at the same time, you hear increasing and decreasing loudness with a beat frequency of 4 Hz. This means that you hear maximum loudness four times a second and minimum loudness four times a second. If the frequencies of the guitar strings are closer (say, 439 Hz and 441 Hz), the beat frequency is smaller, and fewer maxima and minima are heard each second (two times per second in this case).

CONCEPTUAL Example 14.3 Comparing String Frequencies

When guitar strings A and B are plucked at the same time, a beat frequency of 5 Hz is heard. If the frequency of string A is increased, the beat frequency increases to 6 Hz. Was the initial frequency of string A greater than, less than, or equal to the initial frequency of string B?

Reasoning and Discussion

Initially, one of the strings had a frequency 5 Hz greater than the frequency of the other string. If string A had the lower frequency initially, increasing its frequency would give it a frequency closer to that of string B. This would reduce the beat frequency. Since the beat frequency increases when the frequency of string A is increased, it follows that string A had the higher frequency initially. Increasing its frequency even more just makes the beat frequency larger.

Answer

The initial frequency of string A was greater than the initial frequency of string B.

Clearly, beats can be used to tune a musical instrument to a desired frequency. To tune a guitar string to 440 Hz, for example, the string can be played simultaneously with a 440-Hz tuning fork. Listening to the beats, the musician can then increase or decrease the tension in the guitar string until the beat frequency becomes practically zero.

To tune a guitar string to 440 Hz, you compare its frequency with a 440-Hz tuning fork. Initially, you hear a beat frequency of 4 Hz. Then you slowly tighten the guitar string, which increases its frequency. As a result, the beat frequency decreases steadily to 3 Hz. What are the initial and final frequencies of the guitar string?

Picture the Problem

Our sketch shows the before and after situations for this problem. The original beat frequency is 4 Hz. As the string is tightened, the beat frequency decreases to 3 Hz.

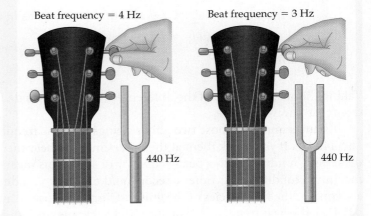

Beat frequency = 4 Hz   Beat frequency = 3 Hz

440 Hz   440 Hz

> **Known**
>
> frequency of tuning fork = 440 Hz
> change in beat frequency, f_{beat}: 4 Hz → 3 Hz
>
> **Unknown**
>
> initial frequency of guitar string = ?
> final frequency of guitar string = ?

Strategy and Solution

The initial beat frequency is 4 Hz, and hence the initial frequency of the string is either 436 Hz or 444 Hz. The final beat frequency is 3 Hz, and hence the final frequency of the string is either 437 Hz or 443 Hz. Only 437 Hz satisfies the condition that the frequency of the string increased as it was tightened. Therefore, the frequency of the string increases from 436 Hz to 437 Hz as it is tightened.

Insight

In this case, the final frequency is still too low. To tune the string properly, it would be necessary to tighten it a little more.

Practice Problems

7. **Follow-up** Suppose the initial beat frequency is 4 Hz and tightening the string causes the beat frequency to increase steadily to 6 Hz. What are the initial and final frequencies of the guitar string in this case?

8. You have three tuning forks with frequencies of 252 Hz, 256 Hz, and 259 Hz. What beat frequencies are possible with these tuning forks?

9. Two musicians are comparing their clarinets. The first clarinet produces a tone that is known to be 441 Hz. When the two clarinets are played together, they produce eight beats every 2.00 s. If the second clarinet produces a higher-pitched tone than the first clarinet, what is the second clarinet's frequency?

14.1 LessonCheck (MP)

Checking Concepts

10. ☐ **Explain** How is a sound wave produced?

11. ☐ **Choose** Which sound has the higher pitch, a sound at 400 Hz or a sound at 600 Hz?

12. ☐ **Describe** What causes beats to be heard?

13. Triple Choice The frequency of a sound is doubled. Does the wave speed of this sound increase, decrease, or stay the same? Explain.

14. Triple Choice The frequency of a sound is doubled. Does the wavelength of this sound increase, decrease, or stay the same? Explain.

15. Big Idea Describe a way in which sound waves are similar to waves on a string. Describe a way in which sound waves differ from waves on a string.

Solving Problems

16. Calculate How long does it take for an echo to return to you from a cliff that is 155 m away?

17. Calculate Two tuning forks have frequencies of 278 Hz and 292 Hz. What is the beat frequency if both tuning forks are sounded simultaneously?

18. Rank In the four cases described below, two sounds with frequencies f_1 and f_2 are played simultaneously. Rank the cases in order of increasing beat frequency. Indicate ties where appropriate.

Case A	Case B	Case C	Case D
$f_1 = 149\,\text{Hz}$	$f_1 = 12\,\text{Hz}$	$f_1 = 901\,\text{Hz}$	$f_1 = 332\,\text{Hz}$
$f_2 = 145\,\text{Hz}$	$f_2 = 22\,\text{Hz}$	$f_2 = 900\,\text{Hz}$	$f_2 = 338\,\text{Hz}$

14.2 Standing Sound Waves

Waves like to travel. Even so, it's possible to confine a wave—to make the wave stay in one location. In Chapter 13 we saw that this can be done with a wave on a string by tying the string down at either end—like a guitar string. It can also be done with sound waves by confining them to a pipe or a tube—as in a trumpet or trombone. We consider the case of sound waves in this lesson.

Standing waves can form in a pipe open at one end

If you've ever plucked a guitar string or blown into a flute, you've created a *standing sound wave*. The girl in **Figure 14.7** shows how to produce a standing sound wave with a bottle. As we saw in Chapter 13, a standing wave is one that oscillates with time but remains fixed in location. In this sense the wave is "standing." In addition, a standing wave always has *nodes*, where the amplitude of the wave is zero, and *antinodes*, where the amplitude of the wave is a maximum.

▲ **Figure 14.7 Producing a standing wave in a bottle**
Blowing across the mouth of a bottle sets the air column in the bottle vibrating, producing a delightful tone.

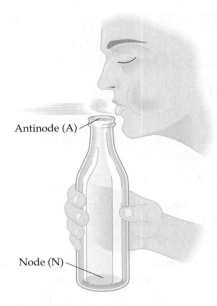

▲ **Figure 14.8 Swirling air creates a standing wave**
When air is blown across the open top of a pop bottle, the turbulent air flow can cause an audible standing wave. The standing wave has an antinode, A, at the top (where the air is moving) and a node, N, at the bottom (where the air cannot move).

As a first example, consider the case of a standing wave in a pop bottle. Blow across the open end of a pop bottle, as in **Figure 14.8**, and you hear a tone of a certain frequency. Now pour some water into the bottle and repeat the experiment. This time the sound has a higher frequency. Let's see how this comes about.

What are the conditions for a standing wave in a bottle or in a pipe that is open at one end?

Standing Wave Formation
When you blow across the opening in a bottle, as illustrated in Figure 14.8, the result is a swirling movement of air that produces rarefactions and compressions. Since any location where air is expanding and compressing is an antinode (A) of a sound wave, the opening of the bottle is an antinode. On the other hand, the bottom of the bottle is closed, preventing movement of the air—just like tying down the end of a string prevents it from moving. Therefore, the bottom is a node (N). **In general, a standing wave in a bottle must have a node at the bottom and an antinode at the top.**

Harmonics
The lowest-frequency standing wave that meets these conditions for a node and an antinode is shown in **Figure 14.9 (a)**. A graph of the air density shows that one-quarter of a wavelength fits inside a bottle of length L. Thus, the wavelength of the fundamental wave must be four times the length of the bottle; that is, $\lambda = 4L$. The corresponding frequency can be found using the formula for the speed of a wave, $v = \lambda f$. Letting f_1 stand for the fundamental (first harmonic) frequency, we find

$$f_1 = \frac{v}{\lambda} = \frac{v}{4L}$$

This frequency is inversely proportional to the length. Therefore, shortening the air column increases the frequency. This is why you hear a higher frequency when you add water to the bottle in Figure 14.8.

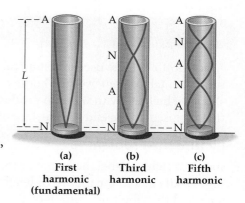

▶ **Figure 14.9 Standing waves in a pipe that is open at one end**
The first three harmonics for waves in a column of air of length L that is open at one end.
(a) $\lambda/4 = L$, $\lambda = 4L$; **(b)** $3\lambda/4 = L$, $\lambda = 4L/3$; **(c)** $5\lambda/4 = L$, $\lambda = 4L/5$.

(a)	(b)	(c)
First harmonic (fundamental)	Third harmonic	Fifth harmonic

The next harmonic is produced by adding half a wavelength to the standing wave, just as in the case of the string. Thus, if the first harmonic is represented by N-A, the next harmonic can be written as N-A-N-A. Since the distance from a node to an antinode is a quarter of a wavelength, we see that three-quarters (3/4) of a wavelength fits into the bottle for this harmonic. This is shown in **Figure 14.9 (b)**. Therefore, $3\lambda/4 = L$, or equivalently

$$\lambda = \tfrac{4}{3}L$$

As a result, the frequency of the next harmonic is

$$\frac{v}{\lambda} = \frac{v}{\frac{4}{3}L} = 3\frac{v}{4L} = 3f_1$$

Notice that this is the *third* harmonic of the pipe, since its frequency is *three* times f_1.

Similarly, the next higher harmonic is represented by N-A-N-A-N-A, as indicated in **Figure 14.9 (c)**. In this case, $5\lambda/4 = L$, and the frequency is

$$\frac{v}{\lambda} = \frac{v}{\frac{4}{5}L} = 5\frac{v}{4L} = 5f_1$$

This is the *fifth* harmonic of the pipe.

Notice that only the *odd* harmonics are present in this case, in contrast to waves on a string, for which all integer harmonics occur. To summarize, the progression of harmonics for a column of air that is closed at one end and open at the other end is described by the following frequencies and wavelengths:

Standing Waves in a Column of Air Closed at One End

Fundamental frequency: $f_1 = \dfrac{v}{4L}$

Harmonic frequencies and wavelengths:

$$\left.\begin{cases} f_n = nf_1 = n\dfrac{v}{4L} \\[2mm] \lambda_n = \dfrac{\lambda_1}{n} = \dfrac{4L}{n} \end{cases}\right\} \text{ for } n = 1, 3, 5, \ldots$$

Practice Problems

19. What is the fundamental frequency for a bottle that has a length of 0.22 m?

20. What is the third harmonic frequency for a bottle that has a height of 0.18 m?

21. The fundamental wavelength of a particular bottle is 0.88 m. What is the length of the bottle?

COOLPHYSICS
The Vibrating Sun

Standing waves have been observed on the surface of the Sun. Like an enormous, low-frequency musical instrument, the Sun vibrates once every 5 minutes, a result of the intense nuclear reactions that take place within its core. One of the goals of the Solar and Heliospheric Observatory (SOHO) is to study these solar vibrations in detail. By observing the variety of standing waves produced in the Sun, scientists hope to learn more about its internal structure and dynamics.

A pop bottle is to be used as a musical instrument. Suppose we would like the fundamental frequency of the bottle to be 525.0 Hz. If the bottle is 26.0 cm tall, how high should it be filled with water to produce the desired frequency? Treat the bottle as a pipe that is closed at one end (the surface of the water) and open at the other end.

Picture the Problem

In our sketch we label the height of the bottle with $H = 26.0$ cm and the unknown height of the water with h. The length of the vibrating column of air is $L = H - h$.

Strategy

Given the fundamental frequency ($f_1 = 525.0$ Hz) and the speed of sound in air ($v = 343$ m/s), we can use $f_1 = v/4L$ and solve for L, the length of the air column. The height of the water is then $h = H - L$.

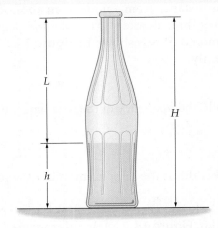

Known	Unknown
$f_1 = 525.0$ Hz	$h = ?$
$H = 26.0$ cm	

Solution

1 Solve the fundamental frequency equation ($f_1 = v/4L$) for the length of the air column, L:

$$f_1 = \frac{v}{4L}$$
$$L = \frac{v}{4f_1}$$

Math HELP
Solving Simple Equations
See Math Review, Section II

2 Substitute the known values for v and f_1 and calculate L:

$$L = \frac{343 \text{ m/s}}{4(525.0 \text{ Hz})}$$
$$= 0.163 \text{ m}$$

3 Use $h = H - L$ to find the height of the water:

$$h = H - L$$
$$= 0.260 \text{ m} - 0.163 \text{ m}$$
$$= \boxed{0.097 \text{ m}}$$

Insight

The height of the water is 0.097 m, or 9.7 cm. If more water were added to the bottle, the air column would shorten, and the fundamental frequency would be greater than 525.0 Hz. All higher harmonics would be increased in frequency as well.

Practice Problems

22. **Follow-up** What is the frequency of the next higher harmonic above the fundamental?

23. A pipe that is closed at one end has a fundamental frequency of 330 Hz. How long is the pipe?

24. The wavelength of the third harmonic in a bottle is 0.22 m. What is the length of the bottle?

Standing waves can form in a pipe open at both ends

Standing waves also form in pipes that are open at both ends. For example, have you ever noticed the metal pipes that hang down below the keys on a xylophone? These pipes, which are open at both ends, are referred to as *resonators*. Their purpose is to enhance and sustain the notes produced by the keys. They do this by producing standing waves.

Figure 14.10 illustrates the lowest three harmonics in a pipe open at both ends. **A standing wave in a pipe open at both ends must have an antinode at each end of the pipe.** Thus, the first harmonic, or fundamental, is A-N-A, as shown in **Figure 14.10 (a)**. Notice that half a wavelength fits into the pipe; thus,

$$f_1 = \frac{v}{2L}$$

This is the same as the corresponding result for a wave on a string.

The next harmonic is A-N-A-N-A, for which one complete wavelength fits in the pipe. This harmonic is shown in **Figure 14.10 (b)** and has the frequency

$$f_2 = \frac{v}{L} = 2f_1$$

This is the *second* harmonic of the pipe. Higher harmonics continue in integer steps ($3f_1$, $4f_1$, $5f_1$, and so on) just like the harmonics for waves on a string. Thus, the frequencies and wavelengths in a column of air open at both ends are as follows:

Standing Waves in a Column of Air Open at Both Ends

Fundamental frequency: $f_1 = \dfrac{v}{2L}$

Harmonic frequencies and wavelengths:

$$\left. \begin{array}{l} f_n = nf_1 = n\dfrac{v}{2L} \\[2mm] \lambda_n = \dfrac{\lambda_1}{n} = \dfrac{2L}{n} \end{array} \right\} \text{ for } n = 1, 2, 3, \ldots$$

What are the conditions for a standing wave in a pipe that is open at both ends?

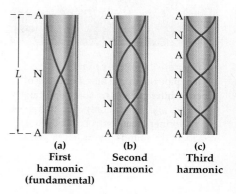

(a) First harmonic (fundamental)   **(b)** Second harmonic   **(c)** Third harmonic

▲ **Figure 14.10 Standing waves in a pipe that is open at both ends**
The first three harmonics for waves in a column of air of length L in a pipe that is open at both ends: **(a)** $\lambda/2 = L$, $\lambda = 2L$; **(b)** $\lambda = L$; **(c)** $3\lambda/2 = L$, $\lambda = 2L/3$.

QUICK Example 14.6 What's the Length?

The frequency of the standing wave in the open pipe shown below is 310 Hz. How long is the pipe?

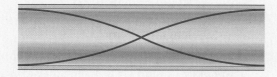

Solution
Comparing with Figure 14.10, we see that the standing wave is the first harmonic. Therefore, its frequency is given by $f_1 = v/2L$. Solving for the length, L, we find

$$L = \frac{v}{2f_1}$$

$$= \frac{343 \text{ m/s}}{2(310 \text{ Hz})}$$

$$= \boxed{0.55 \text{ m}}$$

Figure 14.11 Pipe organ
A large pipe organ can have hundreds of pipes of different lengths, some open at both ends and some at only one. This gives the performer great control over the tonal quality of the sounds produced, as well as the pitch.

Practice Problems

25. **Triple Choice** If the length of a pipe is increased, does the fundamental frequency increase, decrease, or stay the same? Does your answer depend on whether the pipe is open at both ends or closed at one end? Explain.

26. A pipe open at both ends is 1.2 m long. What is the fundamental frequency of this pipe? What is the fundamental frequency of a similar pipe that is 2.4 m long?

27. What is the wavelength of the second harmonic in a 2.5-m-long pipe that is open at both ends?

Physics & You: Technology A pipe organ uses a variety of pipes of different length, as we can see in **Figure 14.11**. Some are open at both ends, and some are open at one end only. Pressing a key on the organ forces air through a given pipe, and the length of the pipe determines the pitch of the sound produced. It's interesting to note, however, that the open and closed pipes have different harmonic frequencies. As a result, they sound distinctly different even if they have the same fundamental frequency. Thus, careful choices of both the length and type of a pipe give an organ a wide range of different sounds, allowing it to mimic a trumpet, a trombone, a clarinet, and so on.

14.2 LessonCheck ⓂⓅ

Checking Concepts

28. **List** What are the conditions necessary for a standing wave in a pipe that is open at one end?

29. **Describe** What conditions produce a standing wave in a pipe that is open at both ends?

30. **Compare** Which has more antinodes, the first harmonic of a bottle or that of a pipe that is open at both ends?

31. **Triple Choice** The fundamental frequency of a pipe that is open at both ends is 200 Hz. If you cut the pipe in half, will the fundamental frequency of each half be greater than, less than, or equal to 200 Hz? Explain.

Solving Problems

32. **Rank** Four standing waves, labeled A through D, are described below. Rank the standing waves in order of increasing frequency. Indicate ties where appropriate.
Wave A: first harmonic; pipe closed at one end; length = 1 m
Wave B: first harmonic; pipe open at both ends; length = 1 m
Wave C: second harmonic; pipe open at both ends; length = 3 m
Wave D: third harmonic; pipe closed at one end; length = 3 m

33. **Calculate** Find the fundamental frequency in Guided Example 14.5 if the height of the water, h, is decreased to 7.00 cm.

34. **Calculate** What is the wavelength of the third harmonic in a 2.7-m-long pipe that is closed at one end?

35. **Think & Calculate** A bottle has a standing wave with a frequency of 375 Hz. The next higher-frequency standing wave in the bottle is 625 Hz.
(a) Is 375 Hz the fundamental frequency for the bottle? Explain.
(b) What is the next higher-frequency standing wave after 625 Hz?

14.3 The Doppler Effect

Have you ever noticed the sudden change in pitch of a train whistle or a car horn as the vehicle moves by you? The physics of this behavior is the topic of this lesson.

The Doppler effect is caused by relative motion

The next time an emergency vehicle drives past you at high speed, listen carefully to its siren. Though the siren always plays the same note, what you hear is a "wow" effect, as the pitch changes from high to low. The change in pitch, due to the relative motion between a source of sound and the person hearing the sound, is called the **Doppler effect.** 🔑 **When a source of sound moves toward an observer, the frequency heard is higher than the frequency produced by the source. When a source of sound moves away from an observer, the frequency heard is lower than the frequency produced by the source.** We'll explore this effect in detail.

Christian Doppler (1803–1853) was the first to study the effect scientifically. His experiment involved a trumpeter on a moving train car. While standing on the ground, Doppler listened carefully as the moving trumpeter played a single note. Using his perfect pitch, Doppler was able to determine how much higher or lower the trumpeter's note sounded. From these observations, he was able to derive the basic properties of the effect that now bears his name.

The Doppler effect occurs with all kinds of waves, not just sound waves. A fascinating example involves light from distant galaxies. The relative motion between a galaxy (the source) and the Earth (the receiver) means that the light astronomers on Earth detect has a different frequency than the light that was emitted by the galaxy. As a result, a galaxy that is moving away from the Earth appears redder than normal, and a galaxy that is moving toward the Earth appears bluer. The connection between the frequency of light and its color is explored in detail in Chapter 15.

Waves from a moving source bunch up or spread out

When the occupant of a parked truck honks the horn, everyone standing nearby hears the same pitch. This is illustrated in **Figure 14.12**.

Vocabulary

- Doppler effect

🔑 *How is the frequency you hear related to the motion of the source?*

▼ Figure 14.12 **A stationary source of sound**
All observers hear the same frequency from a stationary source of sound.

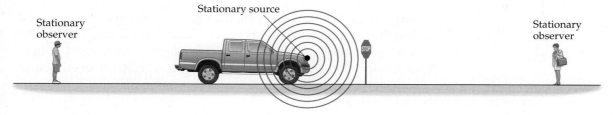

Stationary observer

Stationary source

Stationary observer

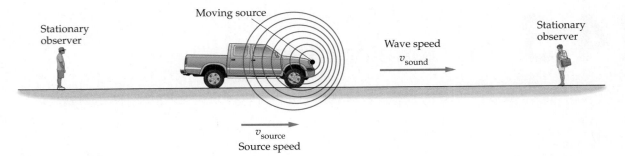

Moving source

Stationary observer

Wave speed
v_{sound}

Stationary observer

v_{source}
Source speed

▲ **Figure 14.13 The Doppler effect for a moving source**
Sound waves from a moving source are bunched up in the forward direction, causing a shorter wavelength and a higher frequency. The waves are spread out in the reverse direction, resulting in a lower frequency.

Now suppose the truck moves toward the right while sounding its horn, as shown in **Figure 14.13**. Notice that the wave crests are bunched up close together in the forward direction. This means that the observer ahead of the truck experiences more crests per second, and hence hears a higher-frequency sound. An observer behind the truck experiences wave crests that are more spread out, which results in a lower-frequency sound. As we noted above, the frequency is *higher* when a source approaches an observer and *lower* when a source moves away from an observer.

Modeling the Doppler Effect A simple experiment with a pan of water shows why the waves in Figure 14.13 bunch up in the forward direction. Imagine tapping your finger in a pan of water. Keeping your finger in the same location will produce ripples that form concentric circles centered at your finger. Now imagine moving your finger to the right as you tap. In this case your finger will almost keep up with the ripples that are moving to the right. Therefore, the separation between one wave crest and the next is smaller than normal in the forward direction. Similar considerations show why the waves crests are spread out in the reverse direction.

Calculating Frequency Shift To get a precise expression for the change in frequency, suppose a source emits sound with a frequency f_{source} and moves with a speed v_{source}. The speed of sound is v_{sound}. Careful measurements, or logical reasoning using mathematics, shows that the frequency heard by an observer is the following:

Doppler Effect for a Moving Source

$$f_{observer} = \frac{f_{source}}{\left(1 \pm \dfrac{v_{source}}{v_{sound}}\right)}$$

$+$ sign for source *moving away* from the observer
$-$ sign for source *moving toward* the observer

SI unit: $Hz = s^{-1}$

Using the plus sign ($+$) in the denominator gives an observer frequency that is less than the source frequency. Therefore, *the plus sign is used for a source that moves away from the observer*. Similarly, using the minus sign ($-$) in the denominator gives an observer frequency that is greater than the source frequency. Therefore, *the minus sign is used for a source that moves toward the observer*.

An express train sounds its whistle as it approaches a station. The whistle produces a tone of 655 Hz, and the train travels with a speed of 21.2 m/s. Find the frequency heard by an observer standing on the station platform as the train (**a**) approaches the station and (**b**) passes the station and moves away.

Picture the Problem

The train moves with the speed $v_{source} = 21.2$ m/s and emits sound that has the frequency $f_{source} = 655$ Hz. The observer on the station platform hears a higher frequency as the train approaches and a lower frequency as the train moves away.

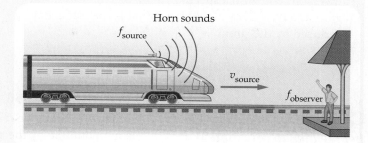

Horn sounds

f_{source}

v_{source}

$f_{observer}$

Strategy

(**a**) When the source moves toward the observer, the observed frequency is given by $f_{observer} = f_{source}/(1 - v_{source}/v_{sound})$.
(**b**) When the source moves away from the observer, the observed frequency is given by $f_{observer} = f_{source}/(1 + v_{source}/v_{sound})$.

Known

$f_{source} = 655$ Hz
$v_{source} = 21.2$ m/s

Unknown

(**a**) $f_{observer}$ for approaching train = ?
(**b**) $f_{observer}$ for receding train = ?

Solution

1 (**a**) Use the Doppler equation for a source moving toward the observer to find the Doppler shift from f_{source} to $f_{observer}$:

$$f_{observer} = \frac{f_{source}}{\left(1 - \dfrac{v_{source}}{v_{sound}}\right)}$$

$$= \frac{655 \text{ Hz}}{\left(1 - \dfrac{21.2 \text{ m/s}}{343 \text{ m/s}}\right)}$$

$$= \frac{655 \text{ Hz}}{(1 - 0.0618)}$$

$$= \boxed{698 \text{ Hz}}$$

2 (**b**) Use the Doppler equation for a source moving away from the observer to find the Doppler shift from f_{source} to $f_{observer}$:

$$f_{observer} = \frac{f_{source}}{\left(1 + \dfrac{v_{source}}{v_{sound}}\right)}$$

$$= \frac{655 \text{ Hz}}{\left(1 + \dfrac{21.2 \text{ m/s}}{343 \text{ m/s}}\right)}$$

$$= \frac{655 \text{ Hz}}{(1 + 0.0618)}$$

$$= \boxed{617 \text{ Hz}}$$

Insight

All that changed between part (a) and part (b) was the sign in the denominator.

36. Follow-up If the stationary observer in Guided Example 14.7 hears a frequency of 702 Hz as the train approaches the station, what is the speed of the train?

37. An emergency vehicle blowing its siren is moving away from you with a speed of 23 m/s. The sound you hear has a frequency of 590 Hz. What is the frequency produced by the siren?

38. Think & Calculate A person with perfect pitch sits on a park bench listening to the 450-Hz horn of a moving car. (**a**) If the person detects a frequency of 470 Hz, is the car approaching or moving away? Explain. (**b**) How fast is the car moving?

A moving observer also experiences a Doppler effect

🔑 *How is the frequency you hear related to your motion relative to the source of sound?*

In **Figure 14.14** we see a stationary source of sound on a park bench. The sound emitted by the source is represented by the circular pattern of wave crests moving outward in all directions with the speed v_{sound}. We also see an observer approaching the sound source with the speed $v_{observer}$. This observer will encounter more wave crests per second than if he had been at rest. Therefore, he hears a higher-frequency sound. Similarly, if the observer is moving away from the sound source, fewer wave crests are encountered per second and the observed frequency is lower. Thus, a moving observer also experiences a Doppler effect.

🔑 **A moving observer experiences a Doppler effect for sound from a stationary source. The observer hears a higher frequency when approaching the source and a lower frequency when moving away from the source.** It turns out, though, that the observed frequency is different when the observer is moving than it is when the source is moving. In fact, the equation for the frequency heard by a *moving observer* is as follows:

Doppler Effect for a Moving Observer

$$f_{observer} = f_{source}\left(1 \pm \frac{v_{observer}}{v_{sound}}\right)$$

+ sign for observer *moving toward* the source
− observer moving *away from* the source

SI unit: $\text{Hz} = \text{s}^{-1}$

Using the plus sign (+) gives an observer frequency that is greater than the source frequency. Therefore, *the plus sign is used for an observer that moves toward the source.* Similarly, using the minus sign (−) gives an observer frequency that is less than the source frequency. Therefore, *the minus sign is used for an observer that moves away from the source.*

▶ **Figure 14.14 The Doppler effect for a moving observer**
Sound waves from a stationary source form concentric circles moving outward with the speed v_{sound}. The observer, who moves toward the source with the speed $v_{observer}$, encounters more wave crests per second than if he had been at rest.

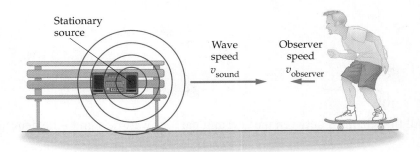

Stationary source

Wave speed v_{sound}

Observer speed $v_{observer}$

A street musician sounds the A string of his violin, producing a tone of 440 Hz. What frequency does a bicyclist hear as she **(a)** approaches and **(b)** moves away from the musician with a speed of 11.0 m/s?

(a)

(b)

Solution *(Perform the calculations indicated in each step.)*

Part (a)

1. Apply the Doppler effect equation for an observer moving toward a source, with $v_{observer} = 11.0$ m/s and $v_{sound} = 343$ m/s:

$f_{observer} = 454$ Hz

Part (b)

2. Apply the Doppler effect equation for an observer moving away from a source, with $v_{observer} = 11.0$ m/s and $v_{sound} = 343$ m/s:

$f_{observer} = 426$ Hz

Insight

As the bicyclist passes the musician, the observed frequency of the sound decreases, giving a "wow" effect. The difference in frequency is about 1 semitone, which is the frequency difference between adjacent notes on a piano. See Table 14.3 in Lesson 14.4 for semitones in the vicinity of middle C.

Practice Problems

39. Follow-up If the bicyclist hears a frequency of 451 Hz when approaching the musician, what is her speed?

40. A parked car is sounding its horn as a motorcyclist approaches with a speed of 25 m/s. The motorcyclist hears a frequency of 375 Hz. What is the frequency produced by the car's horn?

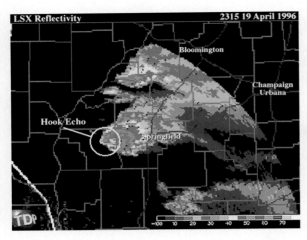

▲ **Figure 14.15 Doppler radar**
Doppler radar, widely used at airports and for weather forecasting, makes it possible to determine the speed and direction of winds in a distant storm by measuring the change in frequency they produce. The image shown here is the Doppler radar scan of a severe thunderstorm that struck the town of Ogden, Illinois, on April 19, 1996. Reddish colors indicate winds blowing toward the radar station; bluish colors indicate winds blowing away from it. The hook-shaped echo marked on the image is characteristic of a tornado in the making.

The Doppler effect has many useful applications

The Doppler effect is used in an amazing variety of technological applications, including weather analysis and forecasting, medical diagnostics, and sports measurements.

Physics & You: Technology Perhaps the most familiar Doppler effect application is the radar gun used to measure the speed of a speeding car or a pitched baseball. Though the radar gun uses radio waves rather than sound waves, the basic physical principle is the same. Thus, by measuring the Doppler-shifted frequency of waves reflected from a moving object, the radar gun can calculate its speed. Doppler radar, used in weather forecasting, applies this same principle to tracking the motion of precipitation caused by storm clouds, as we see in **Figure 14.15**. This is an invaluable tool for predicting violent weather, like tornados and hurricanes.

In medicine the Doppler effect is used to measure the speed of blood flow in an artery or in the heart itself. In this application a beam of ultrasound is directed toward an artery in a patient. Some of the sound is reflected back by red blood cells moving through the artery. The reflected sound is detected, and its frequency is used to determine the speed of blood flow. If this information is color coded, with different colors indicating different speeds and directions of flow, an impressive image of blood flow can be constructed.

You can imagine that when Doppler conducted his experiments with a trumpeter on a moving railroad car, he never anticipated all the ways in which the effect would be applied in the future. Even so, his simple experiments led to the development of technologies that we rely on today to save lives and to improve our health. Not bad for a trumpet and a railroad car!

14.3 LessonCheck ⓜ

Checking Concepts

41. ⚃ **Analyze** The sound you hear from a moving horn has a greater frequency than the sound produced by the horn. Is the horn moving toward you or away from you? Explain.

42. ⚃ **Apply** If you move away from a stationary source of sound, is the frequency you hear greater than, less than, or equal to the frequency produced by the source? Explain.

43. Triple Choice A northern mockingbird sings a single note with a frequency 220 Hz as it flies directly toward you. Is the frequency you hear greater than, less than, or equal to 220 Hz? Explain.

Solving Problems

44. Calculate A pedestrian waiting for the light to change at an intersection hears a car approaching with its horn blaring. The car's horn produces sound with a frequency of 381 Hz, but the pedestrian hears a frequency of 389 Hz. How fast is the car moving?

45. Calculate A bat approaches a stationary moth with a speed of 12 m/s. If the bat emits sound with a frequency of 28,000 Hz, what frequency does the moth hear?

46. Calculate Which do you think produces the higher observed frequency, a 110-Hz horn moving toward you at 12 m/s or a 220-Hz horn moving away from you with a speed of 24 m/s? Verify your answer by calculating the observed frequency in each case.

14.4 Human Perception of Sound

The noise made by a jackhammer is much louder than the song of a sparrow. On this we can all agree. But how can we make an observation like this more precise? What is it that actually determines the loudness of a sound? We address these questions in this lesson.

Loudness is determined by intensity

As you know, waves carry energy. The amount of sound energy passing through a given area in a given time is the loudness, or **intensity**, I, of a sound wave. This is illustrated in **Figure 14.16**. To be specific, suppose sound energy E passes through area A in time t. In this case the intensity of the sound wave equals the energy divided by the area and the time:

$$\text{intensity} = \frac{\text{energy}}{\text{area} \times \text{time}}$$

$$I = \frac{E}{At}$$

Recalling that power is energy per time, $P = E/t$, we can express the intensity as power per area:

Definition of Intensity, I

$$\text{intensity} = \frac{\text{power}}{\text{area}}$$

$$I = \frac{P}{A}$$

SI units: W/m^2

The units are those of power (watts, W) divided by those of area (meters squared, m^2).

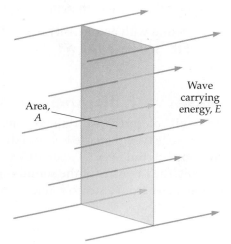

Area, A

Wave carrying energy, E

◄ **Figure 14.16 Intensity of a wave**
If a wave carries an energy E through an area A in the time t, the corresponding intensity is $I = E/At = P/A$, where $P = E/t$ is the power.

Table 14.2   Sound Intensities (W/m²)

Loudest sound produced in a laboratory	10^9
Saturn V rocket at 50 m	10^8
Rupture of the eardrum	10^4
Jet engine at 50 m	10
Threshold of pain	1
Rock concert	10^{-1}
Jackhammer at 1 m	10^{-3}
Heavy street traffic	10^{-5}
Conversation at 1 m	10^{-6}
Classroom	10^{-7}
Whisper at 1 m	10^{-10}
Normal breathing	10^{-11}
Threshold of human hearing	10^{-12}

The frequency of a sound wave determines its pitch. So what property determines its intensity? The property of a sound wave that determines the intensity is the amplitude. Recall that the amplitude of a wave on a string is the maximum height of the string above or below its equilibrium position. In the case of sound waves, the amplitude is the maximum difference in pressure between areas of compression and areas of expansion in the wave. The greater the difference in pressure, the louder the sound you hear.

Though we have introduced the concept of intensity in terms of sound, it applies to all types of waves. For example, the intensity of light from the Sun as it reaches the Earth's upper atmosphere is about 1380 W/m^2. If this intensity could be heard as sound, it would be painfully loud—roughly the equivalent of four jet planes taking off simultaneously. By comparison, the intensity of microwaves in a microwave oven is even greater, about 6000 W/m^2. On the other hand, the intensity of a whisper is an incredibly tiny 10^{-10} W/m^2. A sampling of interesting intensities is given in **Table 14.2**.

QUICK Example 14.9 What's the Intensity?

A large stadium loudspeaker puts out 0.15 W of sound through a square-shaped area 2.0 m on each side. What is the intensity of this sound?

Solution
Applying $I = P/A$, with $P = 0.15 \ W$ and $A = (2.0 \text{ m})^2$, we find

$$I = \frac{P}{A}$$

$$= \frac{0.15 \text{ W}}{(2.0 \text{ m})^2}$$

$$= \boxed{0.038 \text{ W/m}^2}$$

Practice Problems

47. **Triple Choice** If the power of a speaker is doubled and the area through which the sound is emitted is also doubled, does the intensity of the sound increase, decrease, or stay the same? Explain.

48. What area does a sound with an intensity of 0.067 W/m^2 pass through if the power of the source is 0.21 W?

Intensity decreases with distance from the source

Have you ever noticed that if you spread a certain amount of jam over a larger piece of bread, the intensity of the taste is reduced? It's much the same with sound. **Sound spreads out over a larger area as it moves away from its source. As a result, the intensity of the sound—and its loudness—is reduced.** That's why the music from a stereo sounds softer as you move farther away from it.

Why does the loudness of a sound decrease as you move away from it?

Figure 14.17 Echolocation
Two moths, at distances r_1 and r_2, hear the sonar signals sent out by a bat. The intensity of the signal decreases with the square of the distance from the bat. The bat, in turn, hears the echoes sent back from the moths. It can then use the direction and intensity of the returning echoes to locate its prey.

Comparing Intensities

In **Figure 14.17** we show a source of sound (a bat) and two observers (moths) listening at the distances r_1 and r_2. Let's assume that distance r_2 is twice that of r_1; that is, $r_2 = 2r_1$. How do the sound intensities heard by the two moths compare?

The sound waves from the bat move outward uniformly in all directions, and hence the wave crests are concentric spheres. Since the area of a sphere is $4\pi r^2$, it follows that the intensity detected by the first moth is

$$I_1 = \frac{P}{4\pi r_1^2}$$

In writing this expression we've designated the power of the sound emitted by the bat as P. Similarly, the second moth hears the same sound, but at a greater distance. Therefore, the intensity of the sound for the second moth is

$$I_2 = \frac{P}{4\pi r_2^2}$$

Given that $r_2 = 2r_1$, it's easy to see that the two intensities are related by $I_2 = \frac{1}{4}I_1$. You might want to do this calculation yourself to verify the result.

Notice that the power P is the same in each case—it simply represents the amount of sound produced by the bat. The sound has spread out over a considerably larger area for the second moth, however, and hence the intensity it experiences is less.

Calculating Intensity

To summarize, the intensity of a sound at a distance r from a point source of power P is as follows:

> **Intensity of a Sound from a Point Source at a Distance r**
>
> $$\text{intensity} = \frac{\text{power}}{\text{area of a sphere}}$$
>
> $$I = \frac{P}{4\pi r^2}$$
>
> SI units: W/m^2

This result assumes that no sound is reflected or absorbed. These assumptions are applied in the next Guided Example.

Two people relaxing on a deck listen to a songbird sing. One person, only 1.00 m from the bird, hears the sound with an intensity of 2.80×10^{-6} W/m². (**a**) What is the power output of the bird's song? (**b**) What is the intensity of the sound heard by the second person, who is 4.25 m from the bird? Assume that no sound is reflected or absorbed.

Picture the Problem

Our sketch shows the two observers, one at a distance of $r_1 = 1.00$ m from the bird and the other at a distance of $r_2 = 4.25$ m. The sound emitted by the bird is assumed to spread out spherically, with no reflection or absorption.

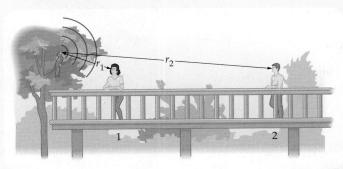

Strategy

(**a**) The intensity I_1 at the distance r_1 is known. Thus, we can solve the equation $I = P/4\pi r^2$ for the power P, where $r = r_1 = 1.00$ m and $I = I_1 = 2.80 \times 10^{-6}$ W/m². (**b**) Use the power P from part (a) to determine the intensity I_2 at the distance $r = r_2 = 4.25$ m.

Known	Unknown
$r_1 = 1.00$ m	(**a**) $P = ?$
$I_1 = 2.80 \times 10^{-6}$ W/m²	(**b**) $I_2 = ?$
$r_2 = 4.25$ m	

Solution

Part (a)

1 Solve $I = P/4\pi r^2$ for the power, P:

$$I = \frac{P}{4\pi r^2}$$

$$P = I(4\pi r^2)$$

2 Substitute the numerical values for I_1 and r_1 to calculate P:

$$P = (2.80 \times 10^{-6} \text{ W/m}^2)(4\pi)(1.00 \text{ m})^2$$

$$= \boxed{3.52 \times 10^{-5} \text{ W}}$$

Part (b)

3 Use $I = P/4\pi r^2$ to find the intensity at point 2. Let $r = r_2 = 4.25$ m and use the power found in part (a):

$$I_2 = \frac{P}{4\pi r_2^2}$$

$$= \frac{3.52 \times 10^{-5} \text{ W}}{4\pi (4.25 \text{ m})^2}$$

$$= \boxed{1.55 \times 10^{-7} \text{ W/m}^2}$$

Math HELP
Scientific Notation
See Math Review, Section I

Insight

The intensity at the first person is 18.1 times the intensity at the second person. Therefore, the sound is louder for the first person, as expected.

Practice Problems

49. **Follow-up** If a third person experiences an intensity of 7.40×10^{-7} W/m², how far is this observer from the songbird?

50. What is the power of a point source of a sound that has an intensity of 3.2×10^{-6} W/m² at a distance of 48 m?

51. **Rank** The power of a point source of sound and the distance at which the sound is heard are given below for four different cases. Rank cases A through D in order of increasing intensity. Indicate ties where appropriate.

Case A: $P = 10$ W, $r = 1$ m
Case B: $P = 20$ W, $r = 2$ m
Case C: $P = 50$ W, $r = 5$ m
Case D: $P = 100$ W, $r = 5$ m

The ear perceives loudness in an unexpected way

Hearing, like most of our senses, is incredibly versatile and sensitive. We can detect sounds that are about a million times fainter than a typical conversation. We can also listen to sounds that are a million times louder than conversations without experiencing pain.

Perceived Loudness Let's use a simple example to illustrate the unusual way in which we perceive loudness. Imagine hearing a sound of a certain intensity. Next, you hear a sound that seems twice as loud as the first sound. If the two intensities are compared, it turns out that the intensity of the second sound is about 10 times greater than the intensity of the first sound. Thus, our perception of sound is such that *doubling* the loudness corresponds to increasing the intensity *by a factor of 10*.

Let's go a little further with this. How do you think the intensity of a third sound—twice as loud as the second sound—is related to the intensity of the first sound? This third sound is 10 times louder than the second sound, which is 10 times louder than the first sound. Thus, the third sound has an intensity 100 times greater than the intensity of the first sound ($10 \times 10 = 100$).

Measuring Loudness The loudness of a sound is measured in a unit known as the *bel*, named after Alexander Graham Bell (1847–1922), inventor of the telephone. Since the bel is a fairly large unit, it's more common to use a unit that is one-tenth of a bel. Thus, the unit commonly used to measure loudness is the **decibel**, abbreviated dB.

On the decibel scale the faintest sound a human can hear is zero decibels, 0 dB. 🔊 **Doubling the *loudness* of a sound, which increases the intensity by a factor of 10, corresponds to an increase of 10 dB.** Doubling the *intensity* of a sound increases the loudness by about 3 dB. And finally, the *smallest* increase in loudness that can be detected is about 1 dB. Decibel levels for a variety of sounds are presented in **Figure 14.18**. The loudness of a sound given in decibels is referred to as the **intensity level**.

🔊 *How is our perception of loudness related to the decibel scale?*

COOLPHYSICS
Hearing Sensitivity

When detecting the faintest of sounds, our hearing is more sensitive than one would ever guess. For example, a faint sound with an intensity of about 10^{-11} W/m^2 causes a displacement of molecules in the air of only 10^{-10} m. This displacement is roughly the diameter of an atom!

QUICK Example 14.11 What's the Intensity?

A crying baby emits a sound with an intensity level of 69 dB. If 10 identical babies are in a nursery together, all crying with the same intensity, what is the intensity level of the sound they produce?

Solution

The 10 babies put out 10 times the sound intensity of a single baby. Therefore, the intensity level increases by 10 dB, from 69 dB to 79 dB. Even so, the 10 babies sound only twice as loud as a single baby.

Practice Problem

52. In a pig-calling contest a caller produces a sound with an intensity level of 100 dB. How many such callers would be required to reach the pain level of 120 dB?

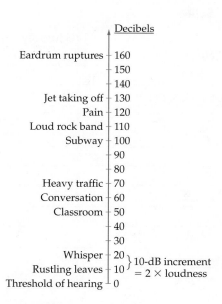

▲ **Figure 14.18 Intensity levels for common sounds**

A slightly louder sound has a much greater intensity

The decibel relationship between loudness and intensity means that a small change in loudness requires a large change in intensity. This fact produces some surprising results. For instance, suppose a large nursery in a hospital has so many crying babies that the loudness of the sound they produce is twice the safe level, as determined by OSHA (the Occupational Safety and Health Administration). Should you remove half of the babies to get back to a safe level? No. To reduce the loudness to a safe level requires reducing the sound level by 10 dB, which means the intensity must be reduced by a factor of 10. Thus, 90% of the babies must be removed from the nursery so that only 1/10, or 10%, of the original number remain.

> **CONCEPTUAL** Example 14.12 **Does the Intensity Change?**
>
> The 20 violinists in a local orchestra all playing together produce a sound with an intensity level of 60 dB. If 18 of the violinists stop playing, is the new intensity level greater than, less than, or equal to 50 dB?
>
> **Reasoning and Discussion**
> When the intensity is reduced by a factor of 10 (from 20 violinists to 2) the loudness of the sound, in decibels, drops by 10 dB. Thus, the new intensity level corresponds to 50 dB.
>
> **Answer**
> The new intensity level is equal to 50 dB.

The ear responds to frequency in an unusual way

Earlier you learned that healthy human ears sense sounds between 20 Hz and 20,000 Hz. The manner in which human hearing responds over this frequency range is similar to the way it responds to loudness.

Frequency and Musical Notes The human ear responds to frequency differently than you might think. For example, consider the C notes on a piano. Middle C has a frequency of 261.7 Hz. Moving up one octave to the next C doubles the frequency, to 523.3 Hz. Going up one more octave doubles the frequency again, and the next C is 1047 Hz. Thus, the frequencies that sound "equally spaced" to our ears—one octave to the next to the next—are actually increasing by a factor of 2. Since there are 12 semitones in one octave of the chromatic scale, the frequency increases from one semitone to the next by the multiplicative factor $2^{1/12}$. The frequencies for a full chromatic octave are given in **Table 14.3**.

The ear is particularly sensitive to certain frequencies

Also interesting is the fact that the human ear is particularly sensitive to certain frequencies. Though you might not have thought about it before, your ear canal is basically a column of air that is closed at one end (at the eardrum) and open at the other end. Standing waves in the ear canal can lead to resonance effects at certain frequencies. At these frequencies the amplitude of the sounds in your ear is enhanced by resonance, and your ability to hear these sounds is increased.

Table 14.3 Chromatic Musical Scale

Note	Frequency (Hz)
Middle C	261.7
C$^{\#}$ (C-sharp), D^{b} (D-flat)	277.2
D	293.7
D$^{\#}$, E^{b}	311.2
E	329.7
F	349.2
F$^{\#}$, G^{b}	370.0
G	392.0
G$^{\#}$, A^{b}	415.3
A	440.0
A$^{\#}$, B^{b}	466.2
B	493.9
C	523.3

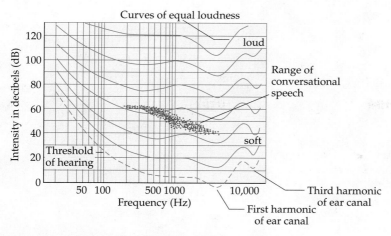

▲ **Figure 14.19 Human response to sound**
The human ear is more sensitive to some frequencies of sound than to others. Thus, every point on a curve of equal loudness seems just as loud to us as any other point, even though the corresponding physical intensities differ. For example, notice that the threshold of hearing is not equal to 0 dB for all frequencies. In fact, it is approximately 25 dB at 100 Hz, about 5 dB at 1000 Hz, and is even slightly negative near 3500 Hz. Regions where the curves dip downward correspond to increased sensitivity of the ear—in fact, near 3500 Hz we can hear sounds that are a thousand times less intense than sounds at 100 Hz. The two most prominent dips occur near 3500 Hz and 11,000 Hz, corresponding to standing waves in the ear canal analogous to those shown in Figure 14.9 (a) and (b), respectively.

This is illustrated in **Figure 14.19**, which shows *curves of equal loudness* as a function of frequency. At the frequencies where these curves dip downward, sounds of lower intensity seem just as loud as sounds of higher intensity at other frequencies. The two prominent dips near 3500 Hz and 11,000 Hz are due to standing waves in the ear canal. These standing waves correspond to the standing waves shown in Figures 14.9 (a) and (b), respectively. Thus, standing waves in a pipe closed at one end have an important effect on how we hear the world around us.

14.4 LessonCheck (MP)

Checking Concepts

53. ⟬⟭ **Explain** Why does a sound become louder as you get closer to the source?

54. ⟬⟭ **Apply** Imagine waking up to two different alarm clocks, one 20 dB louder than the other. How many times louder does the "loud" alarm sound to your ears?

55. Determine If you double your distance from a point source of sound, by what factor does the intensity change? Explain.

56. Big Idea Compare and contrast regions of compression and expansion (rarefaction) in a sound wave. Do these regions transfer energy from one place to another? Explain.

Solving Problems

57. Calculate A sound wave passes through an area of 1.8 m² with an intensity of 4.4×10^{-4} W/m². What is the power of this sound?

58. Calculate What is the intensity of a sound from a 25-W point source at a distance of 5.1 m?

59. Calculate A car horn blasts out sound with an intensity level of 68 dB. How many such car horns would be required to reach an intensity level of 78 dB?

60. Calculate One hundred violins combine to give an intensity level of 76 dB. What is the intensity level of just one violin by itself?

Physics & You

Sonar-generated map of the ocean floor.

Sonar Mapping

What Is It? *Sonar* (an acronym derived from *sound navigation and ranging*) is a technology that uses sound waves. Oceanographers apply this technology to measure the distance from the ocean surface to the ocean floor. With sonar data oceanographers have created maps that reveal mountain ranges, plateaus, and deep trenches on the ocean floor.

Who Invented It? Leonardo da Vinci first examined how sound could be used to detect objects in the water more than 500 years ago. In 1822 Daniel Colloden calculated the speed of sound in water using an underwater bell in Lake Geneva, Switzerland. Less than 100 years later an early sonar prototype was developed to aid ships navigating through iceberg-laden waters. During World War I (1914–1918) Paul Langevin developed a sonar device to detect submarines. The United States and Great Britain improved on this technology during World War II (1939–1945). Later the technology was released to the public. Civilian use of sonar led to significant discoveries in the field of oceanography.

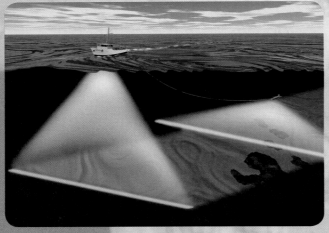

▲ Sonar signals can be broadcast from a boat or from an underwater "fish" towed by the boat. The underwater "fish" provides better images because it is isolated from surface waves and the motion of the boat.

How Does It Work? An active sonar measurement is taken when an electronic device creates a sound pulse, called a *ping*. The sound wave travels through the ocean water. When it hits the ocean floor, it bounces back to the surface. A receiver records the returning sound wave, which is known as an *echo*. Using the known speed of sound in salt water, the total travel time of the wave is converted into a distance.

Why Is It Important? Sonar provides scientists with access to a region of the ocean previously impossible to study. Using this technology scientists have developed high-resolution maps of the ocean floor. The maps show long mountain chains called *mid-ocean ridge systems*, deep valleys called *trenches*, and areas of rough ocean floor called *fracture zones*. Scientists used such maps and other data to develop the theory of plate tectonics. This theory explains how new crust is produced along a mid-ocean ridge system and recycled back into the Earth's interior along trenches. This concept explains how landmasses and oceans transform the surface of the planet over time.

Take It Further

1. Research *Use the Internet and other sources to learn about multi-beam sonar systems and computer programs used to map the ocean floor. Summarize your findings in a one-page written report.*

2. Critical Thinking *Explain why you think sonar is widely used in the fishing industry. Compare how sonar is used to map ocean floor features and how it could be used by the fishing industry.*

Physics Lab Determining the Speed of Sound in Air

In this lab you will use resonance, a wave phenomenon you learned about in Chapter 13, to determine the speed of sound in air.

Materials
- two tuning forks of different frequency
- 1000-mL graduated cylinder
- 1-m length of plastic pipe
- rubber hammer
- meterstick
- tap water

Procedure

1. Hold the plastic pipe upright in the graduated cylinder. Make sure the pipe rests on the bottom of the graduated cylinder.

2. Fill the graduated cylinder nearly to the top with water.

3. Strike the tuning fork with the rubber hammer. Hold the vibrating tuning fork over the mouth of the plastic pipe while your partner slowly moves the pipe upward until the sound becomes quite loud. This is the first resonance.

Caution: *Do not allow the tuning fork to touch the glass beaker.*

4. Using the meterstick, measure the length of the resonating column of air to the nearest centimeter. This length is the distance from the surface of the water to the top of the plastic pipe. Record the length in the Data Table 1.

5. Have your lab partner continue raising the plastic pipe until other resonances are detected. (You might need to strike the tuning fork again with the rubber hammer.) Measure the lengths of the air columns corresponding to these resonances. Record these lengths in Data Table 1.

6. Repeat Steps 1–5 using the other tuning fork. Record the data in a second data table just like Data Table 1.

Data Table 1: Tuning Fork 1

Frequency of tuning fork (Hz): _____

Resonance	Length of Air Column (m)	Harmonic (*n*)	Calculated Speed of Sound (m/s)
First			
Second			
Third			

Analysis

1. Determine the harmonic, *n*, of each resonance in Data Tables 1 and 2. Recall that *n* can only have integer values: 1, 2, 3, ...

2. Calculate the speed of sound in air for each resonance for each tuning fork. Use $v_{sound} = \dfrac{f_n 4L}{n}$ for this calculation. Record your results in the data tables.

3. Ask your teacher for the temperature (in Celsius degrees) of the air in the room and record it. Then calculate the theoretical speed of sound in air using the relationship $v_{sound} = 331\,\text{m/s} + \left(0.6\dfrac{\text{m/s}}{°\text{C}}\right)T$, where *T* is the temperature in Celsius degrees. Record your result.

4. Determine the percent error in the measured speed of sound compared to the theoretical value.

Conclusions

1. What is the relationship between the length of the closed air column for the first resonance and the wavelength of the sound produced by each tuning fork?

2. Describe the change in the wavelength of sound as the frequency of the tuning fork increased.

3. Is the speed of sound independent of the frequency? How do you know?

4. How would your measurements change if the room temperature were higher?

5. What difficulties would you face if you used a 100-Hz tuning fork?

6. What limited the number of resonances you found in this experiment?

14 Study Guide

Big Idea

Sound carries energy in the form of a traveling wave of compressions and expansions.

Sound is a type of wave. Like all waves it has certain characteristic properties, including wavelength, frequency, and speed. Another characteristic shared by all waves is that they carry energy. In the case of sound waves, the energy is carried in the form of compressions and expansions of the substance through which the waves travel.

14.1 Sound Waves and Beats

🔑 In general, a sound wave is formed when an oscillating object creates alternating regions of compressed and expanded air. These alternating regions move away from the source as a longitudinal wave.

🔑 The pitch of a sound is simply the frequency of the corresponding sound wave.

🔑 Beats are the result of two waves interfering with one another, sometimes constructively (loud sound) and sometimes destructively (soft sound).

Key Equation

Beat frequency:

$$f_{beat} = |f_1 - f_2|$$

14.2 Standing Sound Waves

🔑 A standing wave in a bottle must have a node at the bottom and an antinode at the top.

🔑 A standing wave in a pipe open at both ends must have an antinode at each end of the pipe.

• Standing waves oscillate in a fixed location.

Key Equations

Harmonics for a bottle:

$$f_n = nf_1 = n\frac{v}{4L} \qquad n = 1, 3, 5, \ldots$$

$$\lambda_n = \frac{\lambda_1}{n} = \frac{4L}{n}$$

Harmonics for a pipe open at both ends:

$$f_n = nf_1 = n\frac{v}{2L} \qquad n = 1, 2, 3, \ldots$$

$$\lambda_n = \frac{\lambda_1}{n} = \frac{2L}{n}$$

14.3 The Doppler Effect

🔑 When a source of sound moves toward an observer, the frequency heard is higher than the frequency produced by the source. When a source of sound moves away from an observer, the frequency heard is lower than the frequency produced by the source.

🔑 A moving observer experiences a Doppler effect for sound from a stationary source. The observer hears a higher frequency when approaching the source and a lower frequency when moving away from the source.

Key Equations

Doppler effect for a moving source:

$$f_{observer} = \frac{f_{source}}{\left(1 \pm \dfrac{v_{source}}{v_{sound}}\right)}$$

use + if source moves away from observer
use − if source moves toward observer

Doppler effect for a moving observer:

$$f_{observer} = f_{source}\left(1 \pm \frac{v_{observer}}{v_{sound}}\right)$$

use + if observer moves toward source
use − if observer moves away from source

14.4 Human Perception of Sound

🔑 Sound spreads out over a larger area as it moves away from its source. As a result, the intensity of the sound—and its loudness—is reduced.

🔑 Doubling the loudness of a sound, which increases the intensity by a factor of 10, corresponds to an increase of 10 dB.

• The decibel (dB) scale measures the loudness of sounds.

• The range of human hearing extends from 20 Hz to 20,000 Hz.

Key Equations

Intensity of a sound wave:

$$I = \frac{P}{A}$$

Intensity of a sound from a point source at a distance r:

$$I = \frac{P}{4\pi r^2}$$

14 Assessment

ANSWERS TO SELECTED ODD-NUMBERED PROBLEMS APPEAR IN APPENDIX A.

Lesson by Lesson

14.1 Sound Waves and Beats

Conceptual Questions

61. When guitar strings A and B are plucked at the same time, a beat frequency of 4 Hz is heard. If string A is tightened, the beat frequency decreases to 3 Hz. Which of the two strings had the lower frequency initially?

62. A person in the distance is hammering a nail into a board. You see the hammer strike the nail before you hear the sound. Explain.

63. Triple Choice The wavelength of sound coming from a loudspeaker is doubled. Does the frequency of the sound waves increase, decrease, or stay the same? Explain. Does the speed of the sound waves increase, decrease, or stay the same? Explain.

64. Predict & Explain (a) Is the beat frequency produced when a 245-Hz tone and a 240-Hz tone are played together greater than, less than, or equal to the beat frequency produced when a 140-Hz tone and a 145-Hz tone are played together? (b) Choose the *best* explanation from among the following:

A. The beat frequency is determined by the difference between the frequencies and is independent of their actual values.

B. The higher frequencies will produce a greater beat frequency.

C. The percent change in frequency for 240 Hz and 245 Hz is less than for 140 Hz and 145 Hz, resulting in a smaller beat frequency.

Problem Solving

65. **Dolphin Ultrasound** Dolphins of the open ocean are classified as Type II *Odontocetes* (toothed whales). These animals use ultrasonic "clicks" with a frequency of 55 kHz to navigate and find prey. (a) Suppose a dolphin sends out a series of clicks that are reflected back from the bottom of the ocean 75 m below. How much time elapses before the dolphin hears the echoes of the clicks? (The speed of sound in seawater is approximately 1530 m/s.) (b) What is the wavelength of a 55-kHz sound in the ocean?

66. Two tuning forks have frequencies of 268 Hz and 272 Hz. What is the beat frequency if both forks are sounded simultaneously?

67. **Tuning a Piano** To tune middle C on a piano, a tuner hits the key and at the same time sounds a 261-Hz tuning fork. If the tuner hears three beats per second, what are the possible frequencies of the piano key?

68. Think & Calculate A sound wave in air has a frequency of 425 Hz. (a) What is its wavelength? (b) If the frequency of the sound is increased, does its wavelength increase, decrease, or stay the same? Explain. (c) Calculate the wavelength for a sound wave with a frequency of 475 Hz.

69. Think & Calculate When you drop a rock into a well, you hear the splash 1.5 s later. (a) If the distance to the water in the well were doubled, would the time required to hear the splash be greater than, less than, or equal to 3.0 s? Explain. (b) How far down was the water originally?

70. Think & Calculate A tuning fork with a frequency of 320.0 Hz and a tuning fork with an unknown frequency produce beats with a frequency of 4.5 Hz. If the frequency of the 320.0-Hz fork is lowered slightly by placing a bit of putty on one of its tines, the new beat frequency is 7.5 Hz. (a) Which tuning fork has the lower frequency? Explain. (b) What is the final frequency of the 320.0-Hz tuning fork? (c) What is the frequency of the other tuning fork?

14.2 Standing Sound Waves

Conceptual Questions

71. Explain the function of the sliding part of a trombone.

72. Why do large animals generally produce sounds that are lower in frequency than the sounds produced by smaller animals?

73. Predict & Explain (a) When you blow across the opening of a partially filled 2-L plastic bottle, you hear a tone. After pouring some water out of the bottle, you blow across the opening again. Does this tone have a higher frequency, a lower frequency, or the same frequency as the first tone? (b) Choose the *best* explanation from among the following:

A. The same bottle produces the same frequency.

B. The greater distance from the top of the bottle to the level of the water results in a higher frequency.

C. A lower level of water results in a longer column of air and hence a lower frequency.

Problem Solving

74. An organ pipe that is open at both ends is 3.5 m long. What is its fundamental frequency?

75. An organ pipe that is closed at one end is 3.5 m long. What is its fundamental frequency?

76. An organ pipe is 1.5 m long and open at both ends. What are the first three harmonic frequencies of this pipe?

77. An organ pipe is 1.5 m long and closed at one end. What are the first three harmonic frequencies of this pipe?

78. The fundamental frequency in an organ pipe closed at one end is 275 Hz. **(a)** What is the length of this pipe? **(b)** What are the frequencies of the next two harmonics?

79. Some of the standing waves in an organ pipe open at both ends have the following frequencies: 100 Hz, 200 Hz, 250 Hz, and 300 Hz. It is also known that there are no standing waves with frequencies between 250 Hz and 300 Hz. **(a)** What is the fundamental frequency of this pipe? **(b)** What is the frequency of the third harmonic?

80. Think & Calculate The human ear canal is much like an organ pipe that is closed at one end (at the tympanic membrane, or eardrum) and open at the other (see **Figure 14.20**). A typical ear canal has a length of about 2.4 cm. **(a)** What are the fundamental frequency of the ear canal and the wavelength of that standing wave? **(b)** Find the frequency and wavelength of the ear canal's third harmonic. (Recall that the third harmonic in this case is the standing wave with the second-lowest frequency.) **(c)** Suppose a person has an ear canal that is shorter than 2.4 cm. Is the fundamental frequency of that person's ear canal greater than, less than, or the same as the value found in part (a)? Explain. [Notice that the frequencies found in parts (a) and (b) correspond closely to the frequencies of enhanced sensitivity in Figure 14.19.]

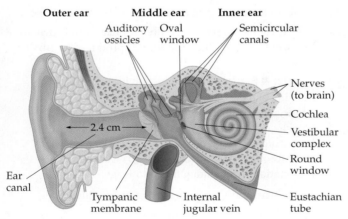

Figure 14.20

81. The organ pipe in **Figure 14.21** is 2.75 m long. It is closed at one end. **(a)** What is the frequency of the standing wave shown in the pipe? **(b)** What is the fundamental frequency of this pipe?

Figure 14.21

82. The frequency of the standing wave shown in **Figure 14.22** is 202 Hz. Notice that the pipe is open at both ends. **(a)** What is the fundamental frequency of this pipe? **(b)** What is the length of the pipe?

Figure 14.22

14.3 The Doppler Effect
Conceptual Questions

83. A radar gun is often used to measure the speed of a Major League pitch by reflecting a beam of radio waves off the moving ball. Describe how the Doppler effect can give the speed of the ball from a measurement of the frequency of the reflected beam.

84. A person travels toward a stationary source of sound. Is the observed frequency of the sound greater than, less than, or the same as when the person is at rest? Explain. Is the wavelength of the sound greater than, less than, or the same as when the person is at rest? Explain.

85. A source of sound travels toward a stationary person. Is the observed frequency of the sound greater than, less than, or the same as when the source is at rest? Explain. Is the wavelength of the sound greater than, less than, or the same as when the source is at rest? Explain.

86. You are heading toward an island in your speedboat when you see a friend standing on shore. You sound the boat's horn to get your friend's attention. Is the wavelength of the sound produced by the horn greater than, less than, or equal to the wavelength of the sound heard by your friend? Explain.

Problem Solving

87. A train moving with a speed of 31.8 m/s sounds a 136-Hz horn. What frequency is heard by an observer standing near the tracks as the train approaches?

88. In Problem 87, suppose the stationary observer sounds a horn that is identical to the one on the train. What frequency does the sound from this horn have for a passenger in the train?

89. Think & Calculate A bat moving with a speed of 3.25 m/s and emitting sound at 35.0 kHz approaches a moth at rest on a tree trunk. **(a)** What frequency is heard by the moth? **(b)** If the speed of the bat is increased, is the frequency heard by the moth higher or lower? **(c)** Calculate the frequency heard by the moth when the speed of the bat is 4.25 m/s.

90. A motorcyclist approaches a stationary police car with a speed of 13.0 m/s. The police car emits a sound with a frequency of 502 Hz. What frequency does the motorcyclist hear?

91. A particular jet engine produces a tone of 495 Hz. Suppose that one jet is at rest on the tarmac while a second identical jet flies overhead at 82.5% of the speed of sound. The pilot of each jet listens to the sound produced by the engine of the other jet. **(a)** Calculate the frequency heard by the pilot in the moving jet. **(b)** Calculate the frequency heard by the pilot in the stationary jet, assuming the flying jet is moving toward it.

14.4 Human Perception of Sound

Conceptual Questions

92. If the power of a sound source is quadrupled, and the area through which the sound passes is doubled, by what factor does the intensity of the sound change? Explain.

93. What are the units of the intensity of sound? What are the units of intensity times time?

94. Explain why the intensity of a point source of sound decreases with the square of the distance from the source.

95. A particular sound has an intensity level of 45 dB. What is the intensity level of a second sound that is perceived to be twice as loud as the first sound?

Problem Solving

96. A bird watcher is hoping to add the white-throated sparrow to her life list of observed species. How far could she be from the bird and still be able to hear it? Assume that there is no reflection or absorption of the sparrow's sound and that the power of the sound output is 3.15×10^{-5} W. (Recall that the minimum intensity of sound a human can hear is 10^{-12} W/m^2.)

97. Residents of Hawaii are warned of the approach of a tsunami by sirens mounted on top of towers. Suppose a siren produces a sound that has an intensity of 0.50 W/m^2 at a distance of 2.0 m. Treating the siren as a point source of sound and ignoring reflection and absorption, find the intensity of sound at a distance of **(a)** 12 m and **(b)** 21 m from the siren. **(c)** How far away can the siren be heard? (Recall that the minimum intensity of sound a human can hear is 10^{-12} W/m^2.)

98. The Eardrum The radius of a typical human eardrum is about 4.0 mm. Find the energy per second received by an eardrum when it experiences a sound that is **(a)** at the threshold of hearing and **(b)** at the threshold of pain.

99. Think & Calculate Ten violins playing simultaneously with the same intensity combine to give an intensity level of 70 dB. **(a)** What is the intensity level of each violin? **(b)** If the number of violins is increased to 100, will the intensity level be more than, less than, or equal to 80 dB? Explain.

Mixed Review

100. Triple Choice You stand near the tracks as a train approaches with constant speed. The train is blowing its horn continuously, and you listen carefully to the sound it makes. For each of the following properties of the sound, state whether it increases, decreases, or stays the same as the train gets closer: **(a)** intensity; **(b)** frequency; **(c)** wavelength; **(d)** wave speed.

101. When you drive a nail into a piece of wood, you hear a tone with each blow of the hammer. In fact, the tone increases in pitch as the nail is driven farther into the wood. Explain.

102. Crack of the Bat Physicist Robert Adair, appointed the official physicist to the National League by the baseball commissioner, believes that the crack of the bat can tell an outfielder how well the ball has been hit. According to Adair, a good hit makes a sound of 510 Hz, while a poor hit produces a sound of 170 Hz. What is the *difference* in wavelength of these two sounds?

103. Medical ultrasound often uses a frequency of 3.5 MHz. What is the wavelength of these ultrasound waves? Assume that the speed of sound waves in the human body is 1,500 m/s, the same as the speed of sound in salt water.

104. The fundamental frequency of an organ pipe that is closed at one end and open at the other end is 261.6 Hz (middle C). The second harmonic of an organ pipe that is open at both ends has the same frequency. What are the lengths of these two pipes?

105. A machine shop has 120 equally noisy machines that together produce an intensity level of 92 dB. If the intensity level must be reduced to 82 dB, how many machines must be turned off?

106. Two trains with 124-Hz horns approach one another. The slower of the two trains has a speed of 26 m/s. What is the speed of the fast train if an observer standing near the track between the trains hears a beat frequency of 4.4 Hz?

107. Think & Calculate Jim is speeding toward James Island with a speed of 24 m/s when he sees Betsy standing on shore at the base of a cliff (**Figure 14.23**). Jim sounds a 330-Hz horn. **(a)** What frequency does Betsy hear? **(b)** Jim can hear the echo of his horn reflected back to him by the cliff. Is the frequency of this echo greater than or equal to the frequency heard by Betsy? Explain. **(c)** Calculate the frequency Jim hears for the echo from the cliff.

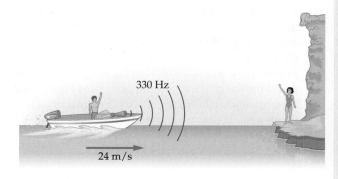

Figure 14.23

108. An organ pipe 2.5 m long is open at one end and closed at the other end. What is the linear distance between a node and the adjacent antinode for the third harmonic in this pipe?

Writing about Science

109. Write a report about the way bats use ultrasound to navigate and hunt. What frequencies do the bats use? What is the wavelength of their sounds, and how does the wavelength compare with the size of their insect prey? Also, report on the ways their prey try to escape capture.

110. Connect to the Big Idea A standing wave on a string produces a sound wave. How does the frequency of the sound wave compare with the frequency of the wave on the string? How does the wavelength of the sound wave compare with that of the wave on the string? Explain any differences or similarities.

Read, Reason, and Respond

The Sound of a Dinosaur Most animals alive today make extensive use of sounds in their interactions with others. Some sounds are meant primarily for members of the same species, like the cooing calls of a pair of doves, the long-range infrasound communication between elephants, and the songs of the hump-backed whale. Other sounds may be used as a threat to other species, such as the rattle of a rattlesnake and the roar of a lion or bear.

There is little doubt that extinct animals used sounds in much the same ways. But how can we ever hear the call of a long-vanished animal like a dinosaur since sounds don't fossilize? In some cases basic physics may have the answer.

Consider the long-crested, duck-billed dinosaur *Parasaurolophus walkeri*, shown in **Figure 14.24**, which roamed the Earth 75 million years ago. This dinosaur possessed the largest crest of any duck-billed dinosaur—so long, in fact, that it had a notch in its back to make room for the crest when it tilted its head backward. Many paleontologists believe that the air passages in the dinosaur's crest acted like bent organ pipes open at both ends and produced sounds that *P. walkeri* used to communicate with others of its kind. As air was forced through the passages, the predominant sound they produced would be that of the fundamental standing wave, with a small amount of higher harmonics mixed in as well. The frequencies of these standing waves can be determined with basic physical principles. **Figure 14.25** presents a graph of the lowest 10 harmonics of a pipe that is open at both ends as a function of the length of the pipe.

Figure 14.24
The long crest of *Parasaurolophus walkeri* played a key role in its communications with others.

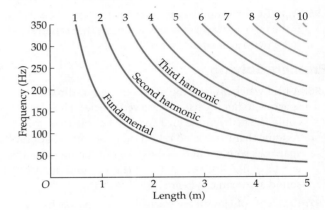

Figure 14.25
Standing wave frequencies as a function of length for a pipe open at both ends. The first ten harmonics ($n = 1, 2, \ldots, 10$) are shown.

111. Suppose the nasal passages in a certain *P. walkeri* crest corresponded to a bent tube 2.7 m long. What is the fundamental frequency of this tube, assuming that the bend has no effect on the frequency? (For comparison, a typical human hearing range is from 20 Hz to 20 kHz.)

  A. 0.0039 Hz   C. 64 Hz

  B. 32 Hz   D. 130 Hz

112. Paleontologists believe that the crest of a female *P. walkeri* was probably shorter than the crest of a male. If this was the case, would the fundamental frequency produced by a female be greater than, less than, or equal to the fundamental frequency from a male?

113. Suppose the fundamental frequency of a particular female was 74 Hz. What was the length of this female's nasal passages?

  A. 1.2 m   C. 2.7 m

  B. 2.3 m   D. 4.6 m

Standardized Test Prep

Multiple Choice

1. Two open plastic pipes produce the same 260-Hz sound when they are struck. After one of the pipes is shortened, a beat frequency of 4 beats per second is heard when the pipes are struck at the same time. What is the frequency of the shorter pipe?
   (A) 256 Hz
   (B) 260 Hz
   (C) 264 Hz
   (D) 1040 Hz

2. A musician plays his clarinet (a wind instrument) for several minutes in order to warm it up. As the temperature of the air in the clarinet increases, which of the following changes occur?
   (A) Both the fundamental frequency and the fundamental wavelength increase.
   (B) The fundamental frequency increases, and the fundamental wavelength decreases.
   (C) The wave speed increases, and the fundamental frequency increases.
   (D) The wave speed increases, and the fundamental frequency decreases.

3. You blow across the top of an open straw to produce a musical note. Then you cut the straw in half, close the bottom end, and blow across the top again. How does the pitch of the new note compare to the original note?
   (A) same pitch
   (B) an octave higher (twice the original frequency)
   (C) an octave lower (half the original frequency)
   (D) two octaves lower (one-fourth the original frequency)

4. A speaker is moved twice as far away from you as it had been, but the volume is turned up so that the intensity at the source doubles. What is the resultant intensity of the sound where you sit, compared to what it was before the speaker was moved?
   (A) one-fourth the original intensity
   (B) one-half the original intensity
   (C) twice the original intensity
   (D) four times the original intensity

5. What are the frequencies of the first and second harmonics for a standing wave in a 1.2-m-long tube that is open at both ends? (The speed of sound in air is 343 m/s.)
   (A) 71 Hz and 142 Hz
   (B) 71 Hz and 286 Hz
   (C) 143 Hz and 286 Hz
   (D) 286 Hz and 572 Hz

6. Which of the following does *not* increase the pitch of a guitar string, if all other factors remain the same?
   (A) tightening the string
   (B) shortening the string
   (C) replacing the string with a thinner string
   (D) plucking the string with more force

7. Where is the speed of sound the greatest?
   (A) in air
   (B) in a vacuum
   (C) in a rigid solid, such as steel
   (D) in a flexible solid, such as rubber

8. The intensity of a sound increases from 1×10^{-8} W/m^2 to 4×10^{-8} W/m^2. How does the loudness measured in decibels change?
   (A) It increases by 6 dB.
   (B) It increases by 10 dB.
   (C) It increases by 20 dB.
   (D) It increases by 40 dB.

Extended Response

9. Describe how the frequency heard by an observer moving toward a stationary sound source compares to the frequency heard by a stationary observer approached by a moving sound source. Then, calculate the observed frequency in each case, given that the speed of the moving observer and the moving sound source are both equal to 0.10 of the speed of sound.

> **Test-Taking Hint**
>
> The harmonics for a tube open on both ends follow the same pattern as the harmonics for a string fixed at both ends:
> $\lambda = \dfrac{2L}{n}$, where n is number of the harmonic.

If You Had Difficulty With . . .

Question	1	2	3	4	5	6	7	8	9				
See Lesson(s)	14.1	14.1, 14.2	14.2	14.4	14.3	14.2	14.1	14.4	14.3				

15

The Properties of Light

You know this simple flower as a yellow daisy. To a bee—with its ability to see ultraviolet light that is invisible to us—the very same daisy looks like a target, with the bull's-eye centered on the nectar.

Big Idea

Light is a small but important part of the electromagnetic spectrum. Everything you see either emits or reflects light.

This chapter delves into the mysteries and delights of light. Mysteries? Yes, light is intriguing and mysterious because it behaves like a wave and also like a particle. Delights? Absolutely. This chapter also discusses such wonders of nature as the clear blue sky and fiery red sunsets.

15.1 The Nature of Light

Perhaps you've seen the classic physics demonstration where a ringing bell is placed under a glass jar and then the air inside the jar is pumped out. Once the air is removed from the jar, leaving only a vacuum, the bell can no longer be heard. You can still *see* the bell, however. Thus, the waves that make up light can travel through a vacuum, unlike the mechanical waves of sound.

Light could travel around the world seven times in a second

Nothing can travel faster than light in a vacuum. For historical reasons its speed is not indicated with a symbol like v_{light}. Instead we use the symbol c, which comes from the word *celerity*, meaning "speed or swiftness." The approximate value of the speed of light in a vacuum is as follows:

> **Speed of Light in a Vacuum**
>
> $c = 3.00 \times 10^8 \text{ m/s}$

This is a large speed, corresponding to about 186,000 mi/s. At this speed light could circumnavigate Earth about seven times every second.

The speed of light varies depending on the material through which it travels. In air the speed of light is slightly less than it is in a vacuum. In denser materials, like glass and water, the speed of light is reduced to about two-thirds of its vacuum value.

Vocabulary

- photon

CONNECTING IDEAS

The speed of light applies to all electromagnetic waves moving through a vacuum.

- In Chapter 17 you'll learn how the reduced speed of light traveling in a material can lead to a change in direction called *refraction*.

The distance from the Sun to Earth is 1.50×10^{11} m. How long does it take for light to cover this distance?

Solution

Recall that speed is distance divided by time, $v = d/t$. This equation can be solved for the time, giving $t = d/v$. Using $v = c$ for the speed, we find

$$t = \frac{d}{c}$$

$$= \frac{1.50 \times 10^{11} \text{ m}}{3.00 \times 10^8 \text{ m/s}}$$

$$= \boxed{500 \text{ s}}$$

Because 500 s is $8\frac{1}{3}$ min, we say that Earth is about 8 light-minutes from the Sun.

> **Math HELP**
> Scientific Notation
> **See Math Review, Section I**

Practice Problem

1. When the rover *Sojourner* was deployed on the surface of Mars in July 1997, radio signals took about 12 min to travel from Earth to the rover. How far was Mars from Earth at that time?

Measuring the speed of light is difficult

Because the speed of light is so large, its value is difficult to determine. Let's look at some of the important milestones on the road to determining c.

Galileo's Experiment The first scientific attempt to measure the speed of light was a bit of a failure. In this experiment Galileo Galilei (1564–1642) and an assistant used two lanterns. Galileo opened the shutters of one lantern, and the assistant—who was positioned a considerable distance away—was instructed to open the shutter on the second lantern as soon as he observed the light from Galileo's lantern. Galileo then attempted to measure the time that elapsed before he saw the returning light from his assistant's lantern. Seeing no delay beyond normal human reaction time, Galileo concluded that the speed of light must be very large—too large to measure with such an experiment.

Romer's Observations The first to give a numerical value to the speed of light was the Danish astronomer Ole Romer (1644–1710). The ironic thing is that Romer didn't set out to measure the speed of light at all. He was working on the problem of determining longitude by measuring the times when the moons of Jupiter disappeared behind that planet. He was surprised to find that the time of these eclipses varied during the course of a year.

This behavior finally made sense to Romer when he realized that the eclipses occurred earlier when Earth was closer to Jupiter in its orbit and later when Earth was farther away. This difference is illustrated in **Figure 15.1**. From the results of Quick Example 15.1, we know that light requires just over 8 minutes to go from the Sun to the Earth, or about 16 minutes to travel from one side of Earth's orbit to the other. This is roughly the discrepancy in

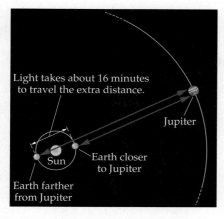

▲ **Figure 15.1 Using Jupiter to determine the speed of light**
When the Earth is at its greatest distance from Jupiter, light takes longer to travel between them. This time lag allowed Ole Romer to estimate the speed of light.

eclipse times observed by Romer, though his measurements weren't nearly as accurate as those made today. Romer's value for the speed of light was about 2×10^8 m/s, while the modern value is 3×10^8 m/s. Still, this was the first direct indication of a finite speed for light.

Fizeau's Experiment The first laboratory measurement of the speed of light was performed by the French scientist Armand Fizeau (1819–1896). The basic elements of his experiment are shown in **Figure 15.2**. Notice that light passes through one notch in a rotating wheel and travels to a mirror a considerable distance away. The light is reflected back and—if the rotational speed of the wheel is just right—passes through the *next notch* in the wheel. By measuring the rotational speed of the wheel and the distance from the wheel to the mirror, Fizeau was able to make an accurate measurement of the speed of light. His value was about 3.13×10^8 m/s.

Today, measurements of the speed of light have been refined to such a degree that we now use that speed to *define* the meter, as was mentioned in Chapter 1. Thus, by definition, the speed of light in a vacuum is

$$c = 299,792,458 \text{ m/s}$$

For most routine calculations, however, the value $c = 3.00 \times 10^8$ m/s is adequate.

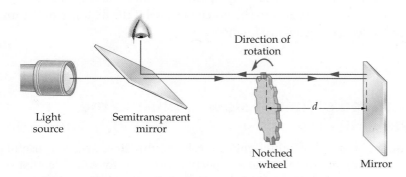

▲ **Figure 15.2 Fizeau's experiment to measure the speed of light**
If the time required for light to travel to the far mirror and back is equal to the time it takes the wheel to rotate from one notch to the next, light will pass through the wheel and on to the observer.

Consider a Fizeau experiment with a notched wheel. Light passing through one notch travels to the mirror and back just in time to pass through the next notch. If the distance from the wheel to the mirror is 9500 m (about 5.9 mi) and the time for the wheel to rotate from one notch to the next is 6.3×10^{-5} s, what is the speed of light obtained by this measurement?

Picture the Problem

Our sketch shows an experimental setup similar to Fizeau's. The notched wheel is $d = 9500$ m from a mirror.

Strategy

In general, speed is distance divided by time. In this case the distance traveled is $2d$ (to the mirror and back), and the time is $t = 6.3 \times 10^{-5}$ s.

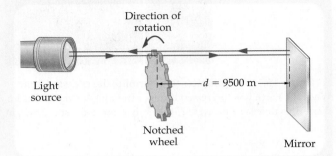

Known	Unknown
$d = 9500$ m	$c = ?$
$t = 6.3 \times 10^{-5}$ s	

Solution

1 Divide the round-trip distance by the time to find the speed of light:

$$c = \frac{2d}{t}$$
$$= \frac{2(9500 \text{ m})}{6.3 \times 10^{-5} \text{ s}}$$
$$= \boxed{3.0 \times 10^8 \text{ m/s}}$$

Math HELP
Significant Figures
See Lesson 1.4

Insight

Notice that even with a rather large distance for the round-trip (almost 12 mi), the travel time of the light is small, only 0.000063 s to be exact. This illustrates the difficulty of trying to make an accurate measurement of c.

Practice Problems

2. **Follow-up** If the distance from the notched wheel to the mirror is increased, should the rotational speed of the wheel be increased or decreased to let the returning light pass through the next notch?

3. **Follow-up** How long does it take for light to return to the notched wheel if the distance to the mirror is 10.5 km?

The speed of light is slow relative to the size of the universe

Although the speed of light is enormous by earthly standards, it is useful to look at it from an astronomical perspective. Imagine, for example, that you could shrink the Solar System to fit onto a football field. Have the Sun be at one end zone and Pluto at the other end zone. On this scale the Earth is a grain of sand located at the 2.5-yard line near the Sun. Light would take about 8 minutes to cover that small distance. To travel to Pluto, at the other end of the field, light would require about 5.5 hours. A caterpillar could walk from one end of a football field to the other in that amount of time.

Thus, on the scale of the Solar System, the speed of light is like the crawl of a caterpillar. When you recall that the Solar System is but a speck on the outskirts of the Milky Way galaxy and that the nearest major galaxy to our own—the Andromeda galaxy, shown in **Figure 15.3**—is about 2.2 million light-years away, the speed of light doesn't appear so great after all! Light literally crawls along as it crosses a galaxy or travels from one galaxy to another.

The Doppler effect applies to light

In Chapter 14 we discussed the Doppler effect for sound waves. There we saw that motion between a source of sound and an observer results in a change in frequency of the sound. 🔑 **The Doppler effect changes the frequency of light waves just as it does for sound waves.**

Consider a source of light with a frequency f_{source}. If this source has a speed v_{source} relative to an observer, the observed frequency, $f_{observed}$, is

$$f_{observed} = f_{source}\left(1 \pm \frac{v_{source}}{c}\right)$$

Because v_{source} and c are speeds, they are always positive. The appropriate sign in front of the term v_{source}/c is chosen for a given situation.

- The plus sign $(+)$ is used when the source is approaching the observer.

$$f_{observed} = f_{source}\left(1 + \frac{v_{source}}{c}\right)$$

- The minus sign $(-)$ is used when the source is moving away from the observer.

$$f_{observed} = f_{source}\left(1 - \frac{v_{source}}{c}\right)$$

This Doppler effect applies to all types of *electromagnetic waves*, including light waves, radio waves, microwaves, and so on. In addition, the equations given above apply regardless of whether the source or the observer is moving.

QUICK Example 15.3 What's the Change in Frequency?

An FM radio station broadcasts at a frequency of 88.5 MHz. If you drive your car toward the station at 32.0 m/s, what change in frequency do you observe?

Solution
We can find the change in frequency, $f_{observed} - f_{source}$, by rearranging the Doppler equation, $f_{observed} = f_{source}(1 + v_{source}/c)$. This gives

$$f_{observed} - f_{source} = f_{source}\frac{v_{source}}{c}$$

$$= (88.5 \times 10^6 \text{ Hz})\frac{32.0 \text{ m/s}}{3.00 \times 10^8 \text{ m/s}}$$

$$= \boxed{9.44 \text{ Hz}}$$

Thus, the frequency changes by 9.44 Hz, which is only 0.00000944 MHz. No need to retune your car's radio receiver!

🔑 **How does the Doppler effect apply to light?**

CONNECTING IDEAS

The Doppler effect for sound waves was introduced in Chapter 14.
- Here the Doppler effect is applied to light. The basic concepts are the same.

▲ **Figure 15.3 The Andromeda galaxy** Even traveling at 300 million meters per second, light from the Andromeda galaxy takes over 2 million years to reach us. Yet that galaxy is one of our nearest cosmic neighbors.

Math HELP
Solving Simple Equations
See Math Review, Section III

4. Concept Check What is the lowest frequency the Doppler effect for light can produce? What is the corresponding speed of the source?

5. A distant star is traveling directly toward Earth with a speed of 36,500 km/s. By what factor is the frequency of light emitted by this star changed?

The universe is expanding

American astronomer Edwin Hubble (1889–1953) devised a reliable method for determining the distances to remote galaxies in the 1920s. As he collected more and more data, he discovered that light from distant galaxies is Doppler shifted. In fact, he found that the greater the distance to a galaxy, the greater the *Doppler shift*. Thus, Hubble found that most galaxies are moving away from us and that their speed is directly proportional to their distance. This was a complete surprise at the time.

Hubble's observations gave strong support to the *Big Bang theory*. According to this theory, the universe started in a hot dense state and then expanded rapidly outward. As a result, objects that left the Big Bang with a higher speed are farther away today (as one would expect) than objects that had a lower speed. This is in perfect agreement with the results of Hubble's work. Since Hubble's time, the Big Bang theory has gained even more support from a wide variety of different experimental techniques, and it is the currently accepted model for the development of the universe.

Light behaves like a wave and a particle

What gives light its particle-like properties?

Light is certainly a wave—there's no doubt about that. Light displays all of the properties that define a wave. Even so, it also displays some of the properties associated with *particles*.

To understand this behavior you might say that light "bottles" its energy, rather than delivering it from a "tap." Suppose you want to get a drink of water. If you turn on the water faucet you can fill your glass to any desired level, as illustrated in **Figure 15.4 (a)**. In contrast, suppose you decide to get your water from a bottle, as in **Figure 15.4 (b)**. Bottled water comes in discrete packages— the individual bottles. You can still get some water to drink, but the amount of water you get will be either one bottle, two bottles, three bottles, or more. This is like the energy that is carried by a light wave. It doesn't come in any amount at all, but only in bundles of energy of a fixed amount—just like bottled water.

The "bottled-up" packets of energy in a beam of light are referred to as *photons*. You can think of a **photon** as a "particle" of light that carries energy but has no mass. **Photons are what give light its particle-like nature.** As we will see when we consider quantum physics in Chapter 24, this *wave-particle duality* not only applies to light but also plays a fundamental role in our understanding of modern physics.

The speed of light plays a key role in relativity theory

How does the speed of light depend on the speed of an observer?

One final mystery of light involves its speed as measured by different observers. Albert Einstein made the following bold prediction in 1905: **All observers measure the same speed for light, regardless of their speed relative to one another.** This is one of the basic principles of Einstein's theory of relativity, and it has survived every experimental test.

(a)

(b)

◀ **Figure 15.4 Water (energy) from the tap or in bottles**
(a) A wave with energy that can have any value is like water coming from a tap, which can provide any amount.
(b) Energy in a light wave is "bottled" in fixed amounts, like bottled water.

To see how "mysterious" this prediction really is, consider the following example. Suppose a friend zooms by you in her spaceship at 90% of the speed of light. A pretty fast spaceship! After she goes by, you shine a beam of light in her direction, as in **Figure 15.5**. The light catches up with her and passes her ship. As the beam goes by, she measures its speed.

Now, here's the interesting question: What is the speed of the light beam as measured by your friend moving at 90% of the speed of light? Naturally, most reasonable people will say that your friend sees the light beam moving at 10% of the speed of light. That certainly sounds right. But according to Einstein—and verified by experiment—your friend sees the beam of light going past her at 100% of the speed of light! Both you and your friend measure *exactly the same speed* for the light beam, even though she's moving very fast relative to you.

As we will see when we study relativity in Chapter 27, the fact that all observers measure the same speed of light leads to additional interesting consequences. Among these are the facts that clocks run slow at high speed and metersticks shrink in length. Observations like this make you realize that the universe is not only more amazing than we imagine, but more amazing than we can imagine.

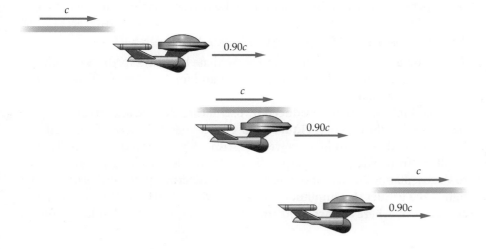

◀ **Figure 15.5 Light speed as viewed from a moving spaceship**
Even though the spaceship moves with a speed of 0.90c, a beam of light passing the ship is observed to be moving with the speed c.

15.1 LessonCheck (MP)

Checking Concepts

6. 🔑 **Explain** A source of light with a frequency f_{source} moves toward an observer. How does the frequency of the light as measured by the observer compare to the frequency of the source?

7. 🔑 **Identify** Which of the following properties of a light wave is "bottled" in particle-like lumps: (**a**) wavelength, (**b**) frequency, (**c**) speed, or (**d**) energy?

8. 🔑 **Explain** Imagine that you are moving in the same direction as a beam of light. As you increase your speed, how does your measurement of the speed of the light beam change?

9. Relate How is the speed of a distant galaxy related to its distance from Earth?

10. List What are the basic defining characteristics of a photon?

Solving Problems

11. ⬚ Triple Choice ⬚ A source of light moves in a circle with the speed $v_{source} = c/2$. When the source is moving directly toward you, its observed frequency is f_1. When the source is moving directly away from you, its observed frequency is f_2. Is the ratio f_1/f_2 equal to 1, 2, or 3? Explain.

12. Convert A light-year is a unit of distance; it is the distance that light travels in 1 year. How many kilometers are in 1 light-year?

13. Calculate Most of the galaxies in the universe are observed to be moving away from Earth. Suppose a particular galaxy emits orange light with a frequency of 5.000×10^{14} Hz. If the galaxy is moving away from Earth with a speed of 3325 km/s, what is the frequency of the light when it is observed on Earth?

15.2 Color and the Electromagnetic Spectrum

Vocabulary

- electromagnetic wave
- visible light
- electromagnetic spectrum
- primary colors
- additive primary colors
- subtractive primary colors

🔑 *How are light waves produced?*

When sunlight shines through a prism, it spreads out into a rainbow of colors. If you look carefully at a rainbow, you will see that red light is on one end and violet light is on the other end. All the other colors of the rainbow are spread out in between. In this lesson we explore the reasons for the different colors.

Light is an electromagnetic wave

A wave on a string causes the string to oscillate back and forth. A sound wave causes air molecules to oscillate back and forth. What oscillates back and forth in a light wave?

🔑 Light waves are produced by oscillating electric and magnetic fields. This is illustrated in **Figure 15.6**. Notice that the electric field oscillates up and down, just like a wave on a string. Similarly, the magnetic field oscillates from side to side, perpendicular to the electric field. The amplitude of the oscillating electric and magnetic fields determines the brightness of the light—the greater the amplitude, the greater the brightness. This is just like a sound wave, where an increased amplitude corresponds to a louder sound.

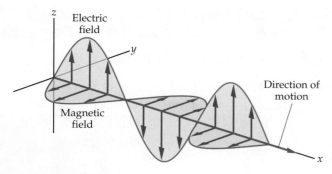

▲ **Figure 15.6 A light wave**
Light waves consist of oscillating electric and magnetic fields. Notice that the electric and magnetic fields are perpendicular to one another. Light is just one example of an electromagnetic wave.

Light is just one example of a large group of waves known as *electromagnetic waves*. In general, any wave produced by oscillating electric and magnetic fields is an **electromagnetic wave**. The electromagnetic waves that our eyes can detect are known as **visible light**. Everything you've ever seen was emitting or reflecting visible light to your eyes. Take a look at some of the objects around you right now. Can you identify those that are emitting light and those that are reflecting it?

Electromagnetic waves are produced in nature when an electron in an atom oscillates back and forth. This sends out an electromagnetic wave, just like shaking one end of a string sends out a wave. Electromagnetic waves can also be produced when oscillating electric currents are present in antennas, like those shown in **Figure 15.7**. Even shaking a bar magnet produces an electromagnetic wave.

Electromagnetic waves are characterized by their frequencies

The frequency of an electromagnetic wave is the key to its behavior. For example, different colors of visible light are produced by electromagnetic waves of different frequencies. But light isn't the only type of electromagnetic wave.

In principle, the frequency of an electromagnetic wave can have any value at all, and visible light corresponds to only a small range of possible frequencies. The full range of frequencies of electromagnetic waves is known as the **electromagnetic spectrum**. Later in this lesson we'll discuss other types of electromagnetic waves with frequencies different from those of visible light, including infrared, ultraviolet, and X-rays.

In Chapter 13 you learned that a wave's speed, frequency, and wavelength are related by the equation $v = f\lambda$. All electromagnetic waves in a vacuum have the same speed, c. Therefore, the frequency, f, and wavelength, λ, of an electromagnetic wave are related as follows:

$$c = f\lambda$$

🔑 The product of an electromagnetic wave's frequency and wavelength must equal c. Thus, if the frequency of an electromagnetic wave increases, its wavelength must decrease.

In the following Guided Example we calculate the frequency of red and violet light, given the corresponding wavelengths. The wavelengths are given in units of nanometers (nm), where $1 \text{ nm} = 10^{-9}$ m.

▲ **Figure 15.7 Antennas**
Electromagnetic waves are produced by (and detected as) oscillating electric currents in a wire. The actual antenna is often much smaller than is commonly imagined. For example, the bowl-shaped structures that we tend to think of as antennas, such as these microwave relay dishes, serve to focus the transmitted beam in a particular direction or concentrate the received signal on the actual detector.

🔑 *How are the frequency and the wavelength of an electromagnetic wave related?*

Find (**a**) the frequency of red light, with a wavelength of 700.0 nm, and (**b**) the frequency of violet light, with a wavelength of 400.0 nm.

Picture the Problem

The visible electromagnetic spectrum, along with representative wavelengths, is shown in our sketch. In addition to the wavelengths of 700.0 nm for red light and 400.0 nm for violet light, the diagram shows 600.0 nm for yellowish-orange light and 500.0 nm for greenish-blue light.

700.0 nm   600.0 nm   500.0 nm   400.0 nm

Known

$\lambda = 700.0$ nm (red)

$\lambda = 400.0$ nm (violet)

Unknown

$f = ?$

Strategy

We obtain the frequency by rearranging $c = f\lambda$ to yield $f = c/\lambda$.

Solution

1   (a) Substitute $\lambda = 700.0$ nm to calculate the frequency of red light:

$$f = \frac{c}{\lambda}$$

$$= \frac{3.00 \times 10^8 \text{ m/s}}{700.0 \times 10^{-9} \text{ m}}$$

$$= \boxed{4.29 \times 10^{14} \text{ Hz}}$$

Math HELP
Exponents
See Math
Review,
Section I

2   (b) Substitute $\lambda = 400.0$ nm to calculate the frequency of violet light:

$$f = \frac{c}{\lambda}$$

$$= \frac{3.00 \times 10^8 \text{ m/s}}{400.0 \times 10^{-9} \text{ m}}$$

$$= \boxed{7.50 \times 10^{14} \text{ Hz}}$$

Insight

The frequency of any visible light is extremely large. Even so, the *range* of frequencies for this type of electromagnetic wave is small when compared with other portions of the electromagnetic spectrum. (Recall that $1 \text{ Hz} = 1 \text{ s}^{-1}$.)

Practice Problems

14. [Follow-up] What is the wavelength of light with a frequency of 5.25×10^{14} Hz?

15. [Triple Choice] If the wavelength of a light wave is increased, does its frequency increase, decrease, or stay the same? Explain.

16. How many wavelengths of orange light ($\lambda = 620$ nm) make up the height of a person who is 1.8 m tall?

17. X-rays produced in a dentist's office typically have a wavelength of 0.30 nm. What is the frequency of these electromagnetic waves?

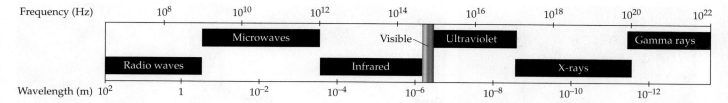

▲ Figure 15.8 **The electromagnetic spectrum**
Notice that the visible portion of the spectrum is relatively narrow. The boundaries between various bands of the spectrum are not sharp but are instead somewhat arbitrary.

Electromagnetic waves have a range of properties

Recall from Chapter 14 that sound waves can have any frequency, from those that are well below our hearing range to those that are well above what we can hear. The same idea applies to electromagnetic waves.

In the electromagnetic spectrum certain portions are given special names. This is indicated in **Figure 15.8**. For example, we just saw in Guided Example 15.4 that visible light occupies a relatively narrow band of frequencies from 4.29×10^{14} Hz to 7.50×10^{14} Hz. Let's discuss the most important regions of the electromagnetic spectrum in order of increasing frequency.

Radio Waves The lowest-frequency electromagnetic waves of practical importance are *radio waves*, in the frequency range from roughly 10^6 Hz to 10^9 Hz. These are the waves that are used in both radio and television broadcasting. In addition, molecules and accelerated electrons in space give off radio waves, and radio astronomers can detect these waves with large dish receivers like those shown in **Figure 15.9**. Radio waves are also produced as a piece of adhesive tape is slowly peeled from a plastic surface. You can confirm this yourself by holding a transistor radio near the tape and listening for pops and snaps coming from the speaker. Most commonly, the radio waves we pick up with our radios and televisions are produced by alternating currents in metal antennas.

Microwaves Electromagnetic radiation with frequencies from 10^9 Hz to about 10^{12} Hz are referred to as *microwaves*. Waves in this frequency range are versatile—they can cook your food in microwave ovens and also carry your long-distance telephone conversations. In fact, as shown in **Figure 15.10**, most modern communication, whether by cell phone or through a WiFi connection to the Internet, relies on microwave technology.

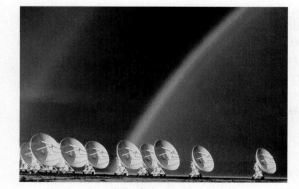

▲ Figure 15.9 **A radiotelescope**
Since the development of the first radiotelescopes in the 1950s, the radio portion of the electromagnetic spectrum has provided astronomers with a valuable window on the universe. These antennas are part of the Very Large Array (VLA), located in San Augustin, New Mexico.

▲ Figure 15.10 **WiFi communication**
The people at this café are connected wirelessly to the Internet, thanks to microwaves.

(a)

(b)

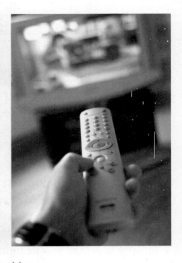

(c)

▲ **Figure 15.11 Infrared waves**
(a) Snakes called *pit vipers* can "see" infrared radiation with the pit organs located just in front of their eyes. **(b)** Photographs made with infrared radiation are often called *thermograms*. In this case the areas of the cat's head that are warmest (bright pink) and coolest (dark blue) are clearly identified. **(c)** We use infrared waves all the time, even though they are invisible to us. If you change the channel with a remote control, your signal is sent by an infrared beam.

▲ **Figure 15.12 The colors of the rainbow**
The dazzling colors of a rainbow illustrate the significance of the frequency of an electromagnetic wave. Each color is a different frequency.

Infrared Waves Electromagnetic waves with frequencies just below that of red light—roughly 10^{12} Hz to 4.3×10^{14} Hz—are known as *infrared* (IR) *waves*. These waves can be felt as heat on our skin but cannot be seen with our eyes—thus, they are often used for heating, as in the infrared lamps used to keep food warm at restaurants. Many animals, like the pit viper snake shown in **Figure 15.11 (a)**, have specialized infrared receptors (pits) that allow them to "see" the infrared radiation given off by a warm-blooded prey animal, even in total darkness. Humans can achieve the same result with the technology of night-vision devices that detect infrared radiation. These devices produce images that show where an object is warmest and coldest. For example, notice the cold nose (dark blue) and warm ears (bright pink) in the infrared image of a cat in **Figure 15.11 (b)**. In addition, remote controls like the one in **Figure 15.11 (c)** operate on a beam of infrared light, with a wavelength of about 1000 nm. This infrared light is so close to the visible spectrum and so low in intensity that it cannot be felt as heat.

Visible Light The portion of the electromagnetic spectrum most familiar to us is the spectrum of visible light, represented by the full range of colors seen in a rainbow. Each of the different colors in **Figure 15.12** is produced by an electromagnetic wave with a different frequency.

Notice that the visible part of the electromagnetic spectrum is actually the smallest of the frequency bands in Figure 15.8. This accounts for the fact that a rainbow produces only a narrow strip of color in the sky. If the visible part of the spectrum were wider, the rainbow would be wider as well. It should be re-membered, however, that there is nothing particularly special about the visible band. In fact, what makes up visible light is species dependent. For example, some bees and butterflies can see ultraviolet light that is invisible to us.

Ultraviolet Light When electromagnetic waves have frequencies just above that of violet light—from about 7.5×10^{14} Hz to 10^{17} Hz—they are called *ultraviolet (UV) rays*. Although these rays are invisible, they often make their presence known by causing suntans with moderate exposure. More prolonged or intense exposure to UV rays can have harmful consequences, from sunburn to an increased probability of developing skin cancer. Fortunately, most of the UV radiation that reaches Earth from the Sun is absorbed in the upper atmosphere by ozone (O_3) and other molecules. A significant reduction in the ozone concentration in the stratosphere could result in an unwelcome increase of UV radiation on Earth's surface. UV light is also given off by galaxies in regions of star formation, as shown in **Figure 15.13**.

X-Rays As the frequency of electromagnetic waves becomes even higher, in the range between about 10^{17} Hz and 10^{20} Hz, this part of the spectrum is known as *X-rays*. These energetic rays pass through our bodies rather freely, except when they encounter bones, teeth, or other relatively dense material. This property makes X-rays valuable for medical diagnosis, research, and treatment. Simple X-rays cast a shadow of bones or teeth onto a sheet of photographic film, as shown in **Figure 15.14**, but today's sophisticated CAT scanners send X-rays onto a part of the body from all directions to produce a three-dimensional image. Still, X-rays can cause damage to human tissue, and it is desirable to reduce unnecessary exposure to these rays as much as possible. That's why a dentist puts a lead-lined vest on your chest and steps behind a lead wall when a dental X-ray is taken.

Gamma Rays Finally, electromagnetic waves with frequencies above 10^{20} Hz are referred to as *gamma (γ) rays*. Gamma rays are highly energetic and can be destructive to living cells. It is for this reason that they are used to kill cancer cells and, more recently, microorganisms in food. Irradiated food, however, has yet to become popular with the general public, even though NASA has irradiated astronauts' food since the 1960s. Next time you see irradiated food in the grocery store, you will know that it has been exposed to gamma rays from cobalt-60 for 20 to 30 minutes. A comparison between food that has been treated with gamma rays and untreated food is presented in **Figure 15.15**.

Adding colors of light produces a spectrum

The human eye has three types of light-sensitive cells that detect red, green, and blue light, respectively. Because of this, the colors red, green, and blue are known as the **primary colors**. 🔑 **All of the colors we see in nature—from red to orange to yellow to green to blue—are produced in our eyes by different amounts of the primary colors.**

The specific way that primary colors combine to form other colors is illustrated in **Figure 15.16**. First, notice that this figure is produced by letting three beams of light overlap. One beam is red, one green, and one blue. Where the red and green beams overlap, the result is yellow light. Even though none of the beams of light is yellow, our eyes perceive the combination of red and green light as yellow. Similarly, notice that the overlap of red and blue light yields magenta, and the overlap of blue and green produces cyan. In the middle of the figure, where all three colors overlap, we see white light. Since red, green, and blue light *add together* to produce white light, we say that these colors are the **additive primary colors**.

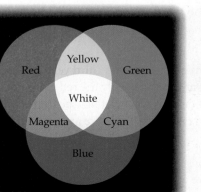

▲ **Figure 15.13 A galaxy viewed in ultraviolet light**
The bright spots in this image of a spiral galaxy are areas of intense star formation, populated by hot young stars that radiate heavily in the ultraviolet.

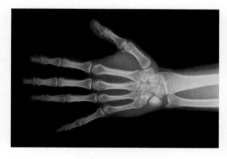

▲ **Figure 15.14 An X-ray image**

🔑 *How do primary colors relate to the colors we see?*

▲ **Figure 15.15 Food preservation**
The use of radiation to preserve food is quite effective, but still controversial. Both boxes of strawberries shown here were stored for about 2 weeks in a refrigerator. Before storage, the box at right was irradiated to kill microorganisms and mold spores.

◀ **Figure 15.16 Additive primary colors**
The additive primary colors are red, green, and blue. Combining two or more of these primaries results in other colors, or even white light.

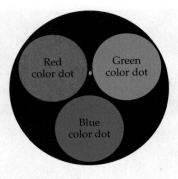

▲ **Figure 15.17** **A pixel from a color TV screen**
A picture element, or pixel, on a color TV screen consists of three color dots, one for each of the three additive primaries.

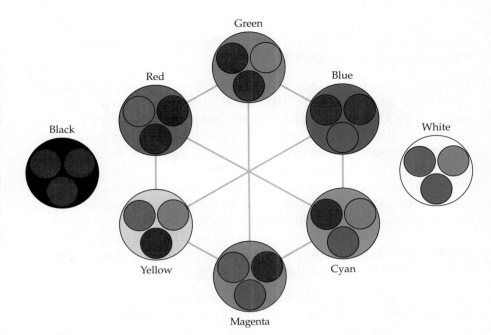

▲ **Figure 15.18** **Colors produced by a pixel**
Lighting various combinations of the color dots in a pixel produces a full range of colors.

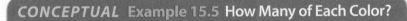

Physics & You: Technology Manufacturers of television and computer screens take advantage of the additive primaries. Each picture element—or *pixel*—on a TV screen consists of three color dots, as shown in **Figure 15.17**. As you might expect, these color dots produce the three additive primaries— red, green, and blue. Lighting combinations of the color dots and varying the brightness allows the screen to display any desired color. **Figure 15.18** shows what a pixel looks like as it produces various colors. Notice that the red and green color dots are lit to produce yellow, the red and blue color dots produce magenta, and the blue and green dots produce cyan. Lighting all three dots in a pixel produces white, and lighting none of them produces black.

Think about pixels the next time you see a beautiful image on TV. Even though the picture looks just like it would in real life, it is produced by just three colors. Amazing!

CONCEPTUAL **Example 15.5** **How Many of Each Color?**

A TV screen with 3,000,000 pixels has the image shown. How many pixels on this screen have **(a)** only the red color dot lit, **(b)** only the green color dot lit, **(c)** both the green and the red color dots lit, or **(d)** only the yellow color dot lit?

Reasoning and Discussion
One-third of the screen is red, one-third green, and one-third yellow (red plus green). The third of the pixels that are red have only the red color dot lit. The third that are green have only the green color dot lit. The third in the middle have both the red and the green color dots lit. There are no yellow color dots on the screen.

Answer
The number of pixels for each part are as follows: **(a)** 1,000,000; **(b)** 1,000,000; **(c)** 1,000,000; and **(d)** zero.

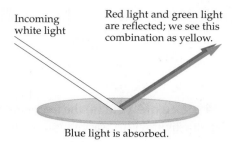

Incoming white light

Red light and green light are reflected; we see this combination as yellow.

Blue light is absorbed.

▲ **Figure 15.19 Yellow reflected light**
When white light shines on a yellow surface, red light and green light are reflected and blue light is absorbed.

Subtracting colors also produces a spectrum

Not all colors are produced by sources of light like the color dots on a TV screen. Sometimes color is produced by subtracting, or removing, some of the colors in white light. This is done with pigments, like those used in paints and dyes.

Consider the yellow paint used for lane lines on highways. 🔑 **Yellow paint (pigment) looks yellow because it reflects red and green light to our eyes. We see the combination of these two colors as "yellow." Other colors of paint work in a similar way.**

Notice that in order for a yellow pigment to reflect just red and green light, it must absorb blue light. This is illustrated in **Figure 15.19**. Similarly, a cyan pigment absorbs red light and reflects green and blue. A magenta pigment absorbs green light and reflects red and blue light.

These three color pigments—cyan, magenta, and yellow—are known as the **subtractive primary colors**. They are the colors that can combine to produce any desired color by subtracting light, just like the additive primaries produce different colors by adding light. If all three subtractive primaries are combined, they subtract all colors from light, leaving black. This is indicated in the center of **Figure 15.20**. All of the color images in magazines and books, including this one, are printed using what are called *CMYK inks*. CMY stands for "cyan, magenta, yellow"; K stands for "black," which is used to produce really deep black and darker shades of other colors.

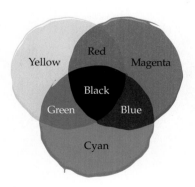

▲ **Figure 15.20 Subtractive primary colors**
The subtractive primary colors are yellow, cyan, and magenta. Combining two or more of these primaries results in new colors. Combining all three subtractive primaries produces black; that is, all light is absorbed in this case.

🔑 *How does a paint, or a pigment, produce its characteristic color?*

COOLPHYSICS
Matching Paint Colors

Have you ever gone to a home improvement store to have a paint sample matched? It's an interesting process with lots of cool physics. First, the matching machine shines a bright white light on the sample. Some colors are absorbed by the sample and others are reflected. The reflected light then passes through 31 different filters, each of which transmits just one color. A computer measures the intensity of light passing through each filter and uses that information to calculate the amount of each corresponding paint that is needed in the final mixture. The result is a color match that is much better than what could be produced by the eye alone.

15.2 LessonCheck (MP)

Checking Concepts

18. 🔘 **Identify** What oscillates in an electromagnetic wave?

19. 🔘 **Differentiate** Which waves have the smaller wavelength, infrared waves or ultraviolet waves?

20. 🔘 **Apply** What color do you see when red and blue light combine?

21. 🔘 **Analyze** Which subtractive primary color corresponds to the combination of green and blue additive primaries?

22. Determine A TV screen has 3,000,000 pixels. This means that it has 3,000,000 red color dots, 3,000,000 green color dots, and 3,000,000 blue color dots. When the screen shows the image in **Figure 15.21**, how many red color dots are lit? How many green color dots are lit? How many blue color dots are lit?

▲ Figure 15.21

23. Big Idea What part of the electromagnetic spectrum is just below the frequency of visible light? What part of the spectrum is just above the frequency of visible light? What is the name of the highest-frequency part of the electromagnetic spectrum?

Solving Problems

24. Calculate Find (**a**) the frequency and (**b**) the period of blue light with a wavelength of 460 nm.

25. Calculate A cell phone transmits at a frequency of 1.25×10^9 Hz. What is the wavelength of the electromagnetic wave used by this phone?

26. Calculate At this very moment you are giving off electromagnetic waves with a wavelength of about 9.0 microns (9.0×10^{-6} m).
(**a**) What is the frequency of these waves?
(**b**) To what portion of the electromagnetic spectrum do these waves belong?

15.3 Polarization and Scattering of Light

Our senses don't respond to everything the world has to offer. Dogs, for example, can hear sounds that are inaudible to us. Snakes can "see" infrared waves that are invisible to us. In this lesson we learn about another physical phenomenon that humans cannot sense but that some birds and insects respond to and use in their everyday lives.

Light can be described by its polarization

When looking into the blue sky of a crystal-clear day, humans see light that is uniform. However, for some animals, like honeybees and pigeons, the light in the sky is far from uniform. The reason is that these animals are sensitive to the direction of the electric field in a beam of light. **In general, the direction of the electric field in a light wave, or any other electromagnetic wave, is referred to as its polarization.** Bees use the polarization of light as an aid in navigating from flower to hive. Pigeons use it to help in navigating from one place to another during migration.

Polarization Direction To understand polarization more clearly, consider the electromagnetic waves pictured in **Figure 15.22**. Each of these waves has an electric field that points along a single line. For example, the electric field in Figure 15.22 (a) oscillates up and down in the vertical direction. We say that this wave is *linearly polarized* in the vertical direction. Similarly, the direction of polarization for the wave in Figure 15.22 (b) is at an angle of 30° relative to the vertical.

Vocabulary

- polarization
- polarizer

What determines the direction of polarization?

▼ **Figure 15.22 Polarization of electromagnetic waves**
The polarization of an electromagnetic wave is the direction along which its electric field points. The waves shown illustrate **(a)** polarization in the vertical direction and **(b)** polarization at an angle of 30° with respect to the vertical.

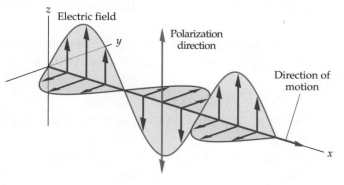

(a)

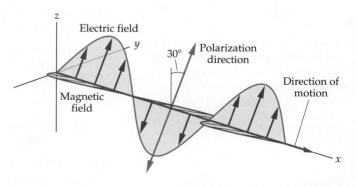

(b)

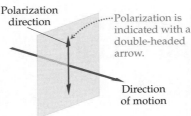

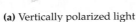

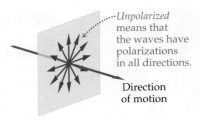

(a) Vertically polarized light

(b) Unpolarized light

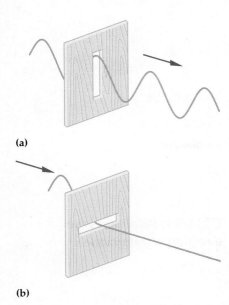

(a)

(b)

▲ **Figure 15.24 A mechanical polarizer**
(a) The polarization of this wave is in the same direction as the polarizer. As a result, the wave passes through unaffected.
(b) Here the polarization of the wave is at right angles to the direction of the polarizer. In this case the wave is absorbed.

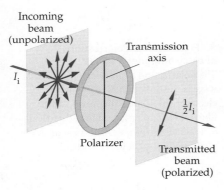

▲ **Figure 15.25 Transmission of unpolarized light through a polarizer**
Notice that the transmitted beam has an intensity of $\frac{1}{2}I_i$ and is polarized in the direction of the polarizer's transmission axis.

🔑 *How is a beam of light affected when it passes through a polarizer?*

Unpolarized Light

The polarization of light is indicated with a red double-headed arrow, as shown in **Figure 15.23**. Part (a) shows light polarized in the vertical direction. Part (b) shows light that is a combination of waves with polarizations in different, random directions. Light with random polarization directions is called *unpolarized*. A common incandescent lightbulb produces unpolarized light because each atom in the heated filament sends out a light wave that has random polarization. Similarly, the light from the Sun is unpolarized.

A polarizer produces polarized light

Unpolarized light can be polarized by passing it through a **polarizer**, a filter that transmits light waves with only one direction of polarization. A simple mechanical polarizer is shown in **Figure 15.24**. Here we see a wave that displaces a string in the vertical direction as it moves toward a slot cut into a block of wood. If the slot is vertical, as in Figure 15.24 (a), the wave passes through unhindered. If the slot is horizontal it stops the wave, as indicated in Figure 15.24 (b).

A polarizer performs a similar function on a beam of light. In general, a polarizer has a *transmission axis,* which is the direction of polarized light that it transmits. If a polarizer has a vertical transmission axis, like the vertical slot in Figure 15.24 (a), it passes vertically polarized light. A polarizer with a horizontal transmission axis, like Figure 15.24 (b), completely blocks vertically polarized light.

Effects of a Polarizing Filter

What happens when unpolarized light encounters a polarizer? This is illustrated in **Figure 15.25**, where we see an unpolarized beam of light passing through a polarizer with a vertical transmission axis. Some of the light in the unpolarized beam has a vertical polarization and passes right through the polarizer. Some of the light has a horizontal polarization and is blocked. Averaging over all possible polarization directions, we find that exactly half of the light passes through the polarizer.

- When an unpolarized light beam with initial intensity I_i passes through a polarizer, the transmitted, or final, intensity, I_f, is one-half of the initial intensity:

$$I_f = \tfrac{1}{2}I_i$$

Just as important as the change in intensity is what happens to the *polarization direction* of the transmitted light:

- When a beam of light is transmitted through a polarizer, it becomes polarized in the direction of the polarizer's transmission axis.

🔑 Thus, a polarizer affects both the intensity *and* the polarization of a beam of light.

Polarizers reduce light's intensity in a predictable way

Have you ever looked through a pair of polarizing sunglasses? If so, you know that the intensity of the light you see depends on how you tilt your head. Let's see how to calculate this intensity.

Law of Malus The transmission of polarized light through a polarizer is illustrated in **Figure 15.26**. Here we see light with a vertical polarization and initial intensity I_i passing through a polarizer whose transmission axis is at an angle θ to the vertical. In a case like this, the polarizer reduces the intensity of the light that passes through it according to the following law:

> **Law of Malus**
>
> $I_f = I_i \cos^2 \theta$

Notice that the intensity is unchanged ($I_f = I_i$) if $\theta = 0$, since $\cos 0° = 1$. Similarly, the transmitted intensity is zero ($I_f = 0$) if $\theta = 90°$, since $\cos 90° = 0$. The light that passes through the polarizer is polarized in the same direction as the polarizer's transmission axis.

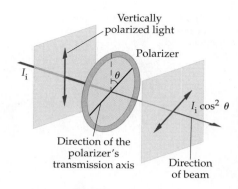

▲ **Figure 15.26 Transmission of polarized light through a polarizer**
A polarized beam of light, with intensity I_i, encounters a polarizer oriented at an angle θ relative to the light's polarization direction. The intensity of light transmitted through the polarizer is $I_f = I_i \cos^2 \theta$. After passing through the polarizer, the light is polarized in the same direction as the polarizer's transmission axis.

QUICK Example 15.6 What's the Intensity?

Vertically polarized light with an initial intensity of 515 W/m² passes through a polarizer oriented at an angle θ to the vertical. Find the transmitted (final) intensity of the light for **(a)** $\theta = 10.0°$, **(b)** $\theta = 45.0°$, and **(c)** $\theta = 90.0°$.

Solution
Applying $I_f = I_i \cos^2 \theta$, we obtain
(a) $I_f = (515 \text{ W/m}^2)(\cos 10.0°)^2 = \boxed{499 \text{ W/m}^2}$

(b) $I_f = (515 \text{ W/m}^2)(\cos 45.0°)^2 = \boxed{258 \text{ W/m}^2}$

(c) $I_f = (515 \text{ W/m}^2)(\cos 90.0°)^2 = \boxed{0}$

Practice Problems

27. A beam of horizontally polarized light passes through a polarizer whose transmission axis is at an angle of 35.0° with the vertical. If the intensity of the transmitted light is 0.55 W/m², what was the initial intensity of the beam?

28. Vertically polarized light encounters a polarizer. At what angle to the vertical should the polarizer be aligned if the transmitted light is to have half the intensity of the incident beam?

29. Triple Choice Is the intensity of unpolarized light transmitted through a vertical polarizer greater than, less than, or equal to the intensity of unpolarized light transmitted through a horizontal polarizer? Explain.

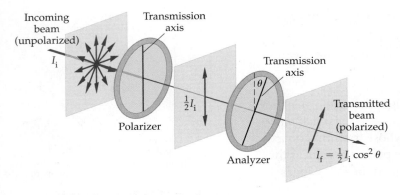

▲ Figure 15.27 A polarizer and an analyzer
An unpolarized beam of intensity I_i is polarized in the vertical direction by a polarizer with a vertical transmission axis. Next, it passes through another polarizer, the analyzer, whose transmission axis is at an angle θ relative to the transmission axis of the first polarizer. The final intensity of the beam is given by $I_f = \frac{1}{2} I_i \cos^2 \theta$.

Using Multiple Polarizers

A common type of polarization experiment is shown in **Figure 15.27**. Here an unpolarized beam is passed through a polarizer to give the light a specified polarization. The light then passes through a second polarizer, referred to as the *analyzer*, whose transmission axis is at an angle θ relative to that of the first polarizer. The orientation of the analyzer can be adjusted to give a beam of light of variable intensity and polarization. We consider this use of more than one polarizer in the next Guided Example.

GUIDED Example 15.7 | Analyze This   **Polarization**

In the polarization experiment shown in our sketch, the final intensity of the beam is $\frac{1}{5} I_i$. What is the angle θ between the transmission axes of the analyzer and the polarizer?

Picture the Problem

The experimental setup is shown in the sketch. As indicated, the intensity of the unpolarized incident beam, I_i, is reduced to $\frac{1}{2} I_i$ after passing through the first polarizer. The analyzer reduces the intensity further, to $\frac{1}{5} I_i$.

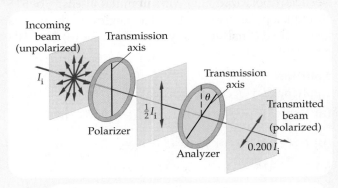

Strategy

Referring to Figure 15.27, we see that the final intensity of the light beam is $I_f = \frac{1}{2} I_i \cos^2 \theta$. This means that $\cos^2 \theta$ must be equal to $\dfrac{1}{2.50}$. Solve this relation for the angle θ.

Known

$I_f = \frac{1}{5} I_i$

Unknown

$\theta = ?$

Solution

1 Set $\cos^2 \theta$ equal to $1/2.50$ and solve for $\cos \theta$:

$$\cos^2 \theta = \frac{1}{2.50}$$

$$\cos \theta = \frac{1}{\sqrt{2.50}}$$

> **Math HELP**
> Solving Simple Equations
> **See Math Review, Section III**

2 Solve for the angle θ:

$$\theta = \cos^{-1}\left(\frac{1}{\sqrt{2.50}}\right) = \boxed{50.8°}$$

Insight

Since the analyzer absorbs part of the light as the beam passes through, it also absorbs energy. Therefore, the analyzer experiences a slight heating in this experiment. As always, energy must be conserved.

30. Follow-up If the angle θ in Guided Example 15.7 is increased slightly, does the final intensity of the light beam increase, decrease, or stay the same? Check your answer by finding the final intensity for 60.0°.

31. What angle θ is required if the transmitted intensity in Figure 15.27 is to be $0.250I_i$?

32. Unpolarized light passes through two polarizers whose transmission axes are at an angle of 30.0° with respect to each other. What fraction of the light's initial intensity is transmitted through the polarizers?

33. Vertically polarized light with an intensity of 1.2 W/m² passes through a polarizer whose transmission axis is at an angle of 35° to the vertical. It then passes through a second polarizer whose transmission axis is at an angle of 45° to the first polarizer. What is the final intensity of the light?

Polarizers have many practical applications

Polarizers with transmission axes at right angles to one another are referred to as *crossed polarizers*. The transmission through a pair of crossed polarizers is zero according to the Law of Malus, since $\theta = 90°$. Crossed polarizers are illustrated in **Figure 15.28** and are featured in the following Conceptual Example.

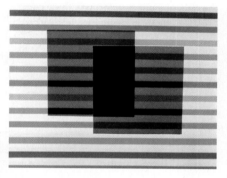

▲ **Figure 15.28 Crossed polarizers** When unpolarized light strikes a single layer of polarizing material, half of the light is transmitted. This is true regardless of how the transmission axis is oriented. However, no light at all can pass through a pair of polarizing filters with axes at right angles (crossed polarizers).

CONCEPTUAL Example 15.8 Is the Light Completely Blocked?

Consider a set of three polarizers. Polarizer 1 has a vertical transmission axis, and polarizer 3 has a horizontal transmission axis. Taken together, polarizers 1 and 3 are a pair of crossed polarizers. Polarizer 2, with a transmission axis at 45° to the vertical, is placed between polarizers 1 and 3, as shown below. A beam of unpolarized light shines on polarizer 1 from the left. Is any light transmitted through the three polarizers, or is transmission completely blocked?

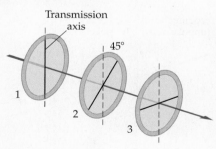

Reasoning and Discussion

Since polarizers 1 and 3 are crossed, it might seem that no light can be transmitted. When we recall that a polarizer causes a beam to have a polarization in the same direction as its transmission axis, however, it becomes clear that transmission is indeed possible.

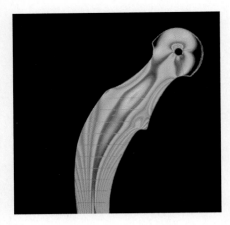

▲ Figure 15.29 Photoelastic stress analysis
In photoelastic stress analysis a plastic model of an object being studied is placed between crossed polarizers. Here the object is a prosthetic hip joint. If the polarization of the light is unchanged by the plastic, the light will not pass through the second polarizer. In areas where the plastic is stressed, however, it rotates the plane of polarization, allowing some of the light to pass through.

To be specific, some of the light that passes through polarizer 1 will also pass through polarizer 2, since the angle between their transmission axes is less than 90°. After passing through polarizer 2, the light is polarized at 45° to the vertical. As a result, some of it can pass through polarizer 3 because, again, the angle between the polarization direction and the transmission axis is less than 90°.

Answer
Some light is transmitted through the three polarizers.

Physics & You: Technology There are many practical uses for crossed polarizers. For example, engineers often construct a plastic replica of a building, bridge, or similar structure to study the stress in its various parts with a technique known as *photoelastic stress analysis*. Dentists use the same technique to study stresses in teeth, and doctors use it when they design prosthetic joints. In this technique the plastic replica plays the role of polarizer 2 in Conceptual Example 15.8. In those regions of the structure where the stress is high, the plastic acts to rotate the plane of polarization and—just as with polarizer 2 in Conceptual Example 15.8—this allows light to pass through the system. An example is shown in **Figure 15.29**. By examining such models with crossed polarizers, engineers can gain valuable insight into the safety of the structures they plan to build, dentists can determine where a tooth is likely to break, and doctors can see where an artificial hip joint needs to be strengthened.

Scattering causes polarization

Nature has been polarizing light since long before people invented polarizing filters. For example, when unpolarized light is scattered, it can become polarized. This is illustrated in **Figure 15.30**, where we see an unpolarized beam of light being scattered by a molecule. An observer in the forward direction, at point A, sees light of all polarizations—that is, unpolarized light. An observer at point B, however, sees vertically polarized light. An observer at an intermediate angle, like point C, sees light with an intermediate amount of polarization.

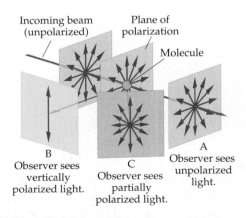

▲ Figure 15.30 Unpolarized light scattering from a molecule
In the forward direction (A) the scattered light is unpolarized. At right angles to the initial beam of light (B) the scattered light is polarized. Along other directions the light is only partially polarized.

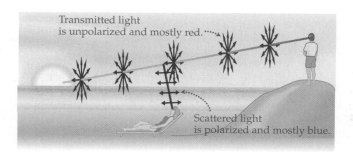

▲ **Figure 15.31 The effects of scattering on sunlight**
The scattering of sunlight by the atmosphere produces polarized light for an observer looking at a right angle to the direction in which the Sun lies. This observer also sees more blue light than red. An observer looking toward the Sun sees unpolarized light that contains more red light than blue.

This mechanism produces polarization in the light coming from the sky. In particular, maximum polarization is observed in a direction at right angles to the Sun, as can be seen in **Figure 15.31**. Thus, to a creature that is sensitive to the polarization of light, like a bee or a bird, the light from the sky varies with the direction relative to the Sun. An example of the sky seen with and without a polarizing filter is shown in **Figure 15.32**. The surface of a lake can also act as a polarizer, as we see in **Figure 15.33**. We discuss polarization due to reflection next.

▲ **Figure 15.32 A polarizing filter darkens the sky**
The top photo was taken without a polarizing filter. The sky is light blue. The lower photo was taken with a polarizing filter. This filter blocked much of the polarized light from the sky, resulting in a much darker blue sky.

▲ **Figure 15.33 The surface of a lake is like a polarizing filter**
A smooth surface such as a lake can act as a polarizing filter. That is why the sky is darker and the clouds are more visible in the reflected image than in the direct view.

··· Reflected light is
horizontally polarized.

▲ **Figure 15.34 Polarization by reflection**
Light reflected from a horizontal surface is polarized horizontally. Polarizing sunglasses with a vertical transmission axis reduce this kind of reflected glare.

Reflection also causes polarization

Light is also polarized when it reflects from a smooth surface, like the top of a table or the surface of a calm lake. **Figure 15.34** shows a typical situation, with unpolarized light from the Sun reflecting from the surface of a lake. The reflected light from the lake is polarized *horizontally*. Polarizing sunglasses take advantage of this effect by using sheets of polarizing material with a *vertical* transmission axis. With this orientation, the horizontally polarized reflected light—the glare—is not transmitted.

A person wearing polarized sunglasses presents a potential problem for the makers of digital watches and electronic calculators with LCD displays. The light emerging from an LCD display is linearly polarized. If the polarization direction is vertical, the light will pass through the sunglasses and the display can be read as usual. On the other hand, if the polarization direction of the display is horizontal, it will appear completely black through a pair of polarizing sunglasses—no light will pass through at all.

Blue sky is caused by light scattering from air molecules

One of humanity's oldest questions is the following: Why is the sky blue? The answer has to do with the way light scatters, and the fact that it scatters most effectively when its wavelength is comparable to the size of the scatterer. The molecules in the atmosphere are generally much smaller than the wavelength of visible light. But blue light, with its relatively short wavelength, is scattered more effectively by air molecules than red light, with its longer wavelength. Similarly, microscopic particles of dust in the upper atmosphere also scatter the short-wavelength blue light more effectively. This is why we see a blue sky.

Taking this phenomenon one step further, a sunset appears red because you are looking directly at the Sun through a long expanse of the atmosphere. Most of the Sun's blue light has been scattered off in other directions. This leaves you with red light. In fact, the blue light that is missing from your red sunset is the blue light of someone else's blue sky, as indicated in Figure 15.31. Next time you admire the clear blue sky, just remember that all the blue light you see is missing from someone else's beautiful red sunset.

C⊙OL *PHYSICS*
LCD Displays

Here's an experiment you can try at home. Take a look at an LCD display through a pair of polarizing sunglasses. You might try a calculator display, a digital watch, or an LCD television or computer screen. Now, slowly rotate the sunglasses as you keep an eye on the display. What do you see? The display should darken as you rotate the sunglasses, and eventually, at 90° of rotation, the display should go black, as shown below for an LCD computer screen. Clearly, it would be unwise for a manufacturer to make an LCD display with a horizontal polarization direction—a person wearing polarizing sunglasses would think the display didn't work.

15.3 LessonCheck (MP)

Checking Concepts

34. 🔟 **Identify** The electric field in a given electro-magnetic wave is vertical and the magnetic field is horizontal. What is the direction of polarization of this wave?

35. 🔟 **Analyze** Vertically polarized light passes through a polarizer whose transmission axis is 30° from the vertical. What is the polarization direction of light that passes through this filter?

36. Explain What is the key factor in explaining the blue sky, the wavelength of light or its speed?

37. Explain While wearing your polarizing sunglasses at the beach, you notice that they reduce the glare from the water better when you are sitting upright than when you are lying on your side. Explain.

38. Triple Choice The transmission axis of a polarizing filter is at an angle of 45° relative to the vertical. Does more light pass through the polarizer if the light is polarized vertically or horizontally, or is the transmission the same in either case? Explain.

Solving Problems

39. Rank A beam of vertically polarized light encounters two polarizing filters, as shown in **Figure 15.35**. Rank the three cases (A, B, and C) in order of increasing transmitted intensity. Indicate ties where appropriate.

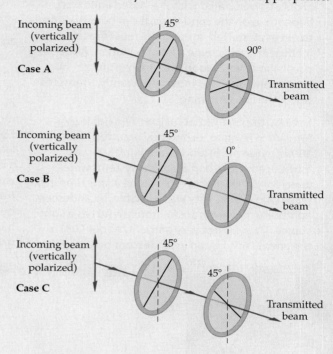

▲ Figure 15.35

40. Calculate Find the transmitted intensity for each of the cases shown in Figure 15.35, assuming that the initial intensity is 37.0 W/m².

41. Apply An unpolarized light beam with an intensity of 25.5 W/m² passes through three polarizers as shown in **Figure 15.36**.
(a) Determine the transmitted intensity.
(b) If the 45° angle of the middle polarizer is slowly reduced to zero, does the transmitted intensity increase, decrease, or stay the same? Explain.

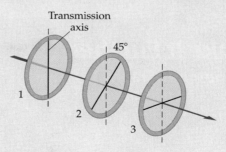

▲ Figure 15.36

Physics & You

Lighting Technologies and Energy Usage

Fire The original lighting source was fire. Later candles, kerosene lanterns, and gaslights were used, all of which are fire-based sources. There are major problems associated with fire-based light. First, depending on the conditions, it can be very difficult to get a fire started. In addition, most fires burn inefficiently and create a lot of pollution. Airborne particulates from wood fires are harmful to your health. And, perhaps most importantly, there is the risk of the fire spreading.

Incandescent Lightbulbs A major leap forward in lighting technology occurred in the 1890s, when the incandescent lightbulb was introduced and urban electrical systems were developed. Now lights could shine at any time in any place where electricity was available. Incandescent lightbulbs, however, are far from perfect as a light source. They are not very efficient. About 90% of the energy used by an incandescent bulb is given off as heat, not light. Considering the number of lightbulbs used around the world, it is staggering to think of the amount of energy wasted by incandescent bulbs.

Newer Technologies Today, a trip to the lighting section of a home store reveals the extent of major advances in lighting. The shelves are stocked with halogen bulbs, fluorescent tubes, compact fluorescent (CFL) bulbs, light emitting diode (LED) bulbs, and others. Most of these bulbs are much more efficient than a standard incandescent bulb. Each type of bulb, however, has its own strengths and weaknesses. Some bulbs are inexpensive, some last a very long time, some pose a disposal hazard, and each emits its own unique range of frequencies. The frequencies emitted by a bulb give the light certain qualities, often described in terms such as cool, warm, harsh, natural, and so on. Choosing the best bulb for a given situation involves considering a lot of factors.

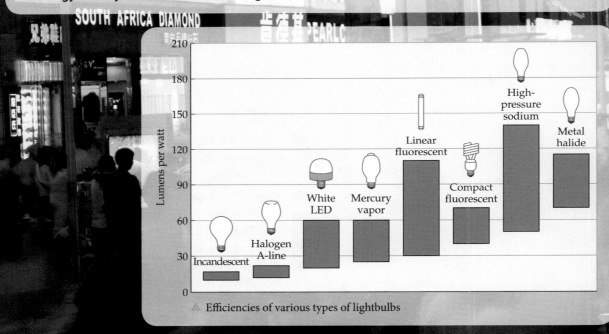

Efficiencies of various types of lightbulbs

Take It Further

1. Interpret Graphs *What is the most efficient type of lightbulb shown in the graph? How many times more efficient is the most efficient bulb than the least efficient bulb?*

2. Analyze Costs and Benefits *Visit a local store and survey the lightbulbs it sells or use the Internet to search the offerings of a national retailer. List the pros and cons of three different lightbulbs that emit the same amount of light as a 60-W incandescent bulb. Decide which bulb you would purchase for a desk lamp to use at home. Use factors such as cost, life expectancy, light output, disposal issues, and light quality to support your decision.*

Physics Lab Polarization

This lab explores the polarization of light. You will use polarizing filters to observe light from several different sources and will see how light is affected when it reflects from a surface or passes through certain materials.

Materials

- clear plastic object (ruler, protractor, or fork)
- incandescent or fluorescent light source
- beaker containing corn syrup
- sheet of aluminum foil
- two polarizing filters

Procedure

Complete each of the following activities. Record your observations on a sheet of paper or in your notebook.

1. Polarization by Selective Absorption: View a light source, such as a lightbulb or fluorescent lamp, through a polarizing filter. Observe what happens as you rotate the filter while viewing the source.

2. Overlapping Polarizers: Overlap two polarizing filters. Look at a source of light through the filters while rotating one of the filters relative to the other. Describe what you see.

3. Polarization by Reflection: Find a surface, such as a shiny tabletop or a polished tile floor, from which reflected light produces glare. View the glare through a rotating polarizing filter. Repeat this procedure with light reflected from a metallic surface such as a piece of aluminum foil.

4. Polarization by Scattering: If the weather permits, go outside and investigate light from the sky with a polarizing filter.

Caution: *Never look directly at the Sun!*

Slowly rotate the filter as you view a portion of the sky. Next, use the filter to examine other regions of the sky. Observe the appearance of clouds when viewed through the rotating filter.

5. Birefringence: *Birefringence* is an effect produced by certain types of crystalline structures in which a ray of light passing through the crystal is broken into two rays that pass through at different speeds. Plastic and glass can produce birefringence when stressed by a force. Place a plastic fork or other plastic object between crossed polarizing filters. Stress the object and describe what you see.

6. Liquid Crystal Display: Examine a liquid crystal display (LCD) on a watch, calculator, television, or laptop through a single polarizing filter. Rotate the filter and note the effect.

7. Optically Active Substances: Optically active substances, such as corn syrup, change the plane of polarization of a beam of light and can, in the process, separate white light into its component colors. Place polarizing filters on the top and bottom of a beaker containing corn syrup, and shine light upward through the lower filter. Rotate the top filter and observe the color of the light.

Analysis

1. Describe how the intensity of light passing through a single polarizing filter is affected when the filter is stationary and when it is rotated.

2. Describe what happens to the intensity of the light passing through two overlapping polarizing filters when one of the filters is rotated.

3. Compare the reflection of unpolarized light from metallic and nonmetallic surfaces.

4. Describe what you observed when you viewed the sky through a rotating polarizing filter. Does the light in one portion of the sky seem to be more strongly polarized than the light in others? Describe the appearance of clouds when viewed through a polarizing filter.

5. Describe what you observed when you viewed an LCD through a polarizing filter.

Conclusions

1. Describe three ways in which light can be polarized.

2. Explain how you could determine whether a pair of sunglasses had polarizing lenses or just tinted glass.

3. What did your observations when viewing an LCD through a polarizing filter tell you about the light it emits?

15 Study Guide

Big Idea

Light is a small but important part of the electromagnetic spectrum. Everything you see either emits or reflects light.

Light is an electromagnetic wave, a wave produced by oscillating electric and magnetic fields. The color of light is like the pitch of a sound, determined by the frequency of the wave. Frequencies above and below the range of visible light are characteristic of other types of electromagnetic waves, like infrared waves, microwaves, and X-rays. The full range of frequencies is referred to as the *electromagnetic spectrum*.

15.1 The Nature of Light

🔑 The Doppler effect changes the frequency of light waves in the same way that it does for sound waves.

🔑 Photons are what give light its particle-like nature.

🔑 All observers measure the same speed for light, regardless of their speed relative to one another.

• The speed of light in a vacuum is $c = 3.00 \times 10^8$ m/s. The speed is related to the wavelength and the frequency by $c = f\lambda$.

• The Doppler effect reduces the frequency of light from distant galaxies, indicating that the galaxies are moving away from us.

• The frequency of light is increased when the source is moving toward the observer. The frequency is decreased when the source is moving away from the observer.

Key Equations

Doppler effect for light:

$$f_{\text{observed}} = f_{\text{source}}\left(1 \pm \frac{v_{\text{source}}}{c}\right)$$

15.2 Color and the Electromagnetic Spectrum

🔑 Light waves are produced by oscillating electric and magnetic fields.

🔑 The product of an electromagnetic wave's frequency and wavelength must equal c. Thus, if the frequency of an electromagnetic wave increases, its wavelength must decrease.

🔑 All of the colors we see in nature—from red to orange to yellow to green to blue—are produced in our eyes by different amounts of the primary colors.

🔑 Yellow paint (pigment) looks yellow because it reflects red and green light to our eyes. We see the combination of these two colors as "yellow." Other colors of paint work in similar ways.

• The electromagnetic spectrum is the full range of electromagnetic waves with all possible frequencies.

15.3 Polarization and Scattering of Light

🔑 The direction of the electric field in a light wave, or any other electromagnetic wave, is referred to as its *polarization*.

🔑 A polarizer affects both the intensity *and* the polarization of a beam of light.

• The polarization of a beam of light is the direction along which its electric field points. An unpolarized beam has waves with polarizations in random directions.

• The transmission axis of a polarizer determines the direction of polarization of light that is transmitted through the polarizer.

• When light reflects from a horizontal surface, like a tabletop or the surface of a lake, it is polarized in the horizontal direction.

• The sky is blue because light is scattered by molecules and dust particles in the air. Since the molecules and dust particles are small, they scatter short-wavelength light (blue light) most effectively, giving a blue color to the sky.

Key Equations

The intensity of unpolarized light transmitted through a polarizer is given by

$$I_f = \tfrac{1}{2} I_i$$

The intensity of polarized light transmitted through a polarizer is given by the Law of Malus:

$$I_f = I_i \cos^2 \theta$$

15 Assessment

ANSWERS TO SELECTED ODD-NUMBERED PROBLEMS APPEAR IN APPENDIX A.

Lesson by Lesson

15.1 The Nature of Light

Conceptual Questions

42. Explain why an "invisible man" would be unable to see.

43. Is a photon more like a particle or a wave?

44. How did Romer determine the speed of light?

45. **Triple Choice** When a source of light moves toward you, is the frequency of the light you observe greater than, less than, or equal to the frequency of light that the source emits? Explain.

46. A source of light moves with a constant speed, v_{source}, in a circular path. When the source is moving directly toward you, its observed frequency is f_1. When the source is moving directly away from you, its observed frequency is f_2. If the ratio f_1/f_2 is equal to 2, is the speed of the source $c/4$, $c/3$, or $c/2$? Explain.

47. **Apply** The magnitude of the Doppler effect tells how rapidly a source of light is moving. What determines whether the source is approaching or moving away from the observer?

Problem Solving

48. Alpha Centauri, the closest star to the Sun, is 4.3 light-years away. How far is this in meters?

49. **Spacecraft Communication** When *Voyager I* and *Voyager II* were exploring the outer planets, NASA flight controllers had to plan the movements of the spacecraft well in advance. How many seconds elapse between the time a command is sent from Earth and the time the command is received by *Voyager* at Neptune? Assume the distance from Earth to Neptune is 4.5×10^{12} m.

50. Baseball scouts use a radar gun to measure the speed of pitches. One particular model of radar gun emits a microwave signal at a frequency of 10.525 GHz. What is the increase in frequency of these waves as seen by a 40.2-m/s (90.0-mi/h) fastball headed straight toward the gun?

51. **Think & Calculate** A distant star is traveling directly away from Earth with a speed of 36,500 km/s. **(a)** When the wavelengths in this star's spectrum are measured on Earth, are they greater than, less than, or the same as the wavelengths that would be observed if the star were at rest relative to Earth? Explain. **(b)** By what fraction are the frequencies in this star's spectrum shifted?

52. **Think & Calculate** The frequency of light reaching Earth from a particular galaxy is 15% lower than the frequency the light had when it was emitted. **(a)** Is this galaxy moving toward or away from Earth? Explain. **(b)** What is the speed of this galaxy relative to Earth? Give your answer as a fraction of the speed of light.

53. **Measuring the Speed of Light** Galileo attempted to measure the speed of light by measuring the time elapsed between his opening a lantern and his seeing the light return from his assistant's lantern. The experiment is illustrated in **Figure 15.37**. What distance, d, must separate Galileo and his assistant in order for the human reaction time, $\Delta t = 0.2$ s, to introduce no more than a 15% error in this measurement of the speed of light?

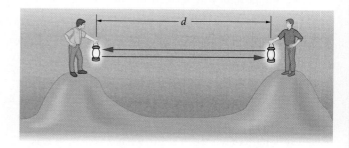

Figure 15.37

54. **Michelson's Speed Measurement** In 1926 Albert Michelson measured the speed of light with a technique similar to that used by Fizeau. In place of a toothed wheel, Michelson used an eight-sided mirror rotating at 528 rev/s, as illustrated in **Figure 15.38**. The distance from the rotating mirror to the fixed reflector was 35.5 km. If the light completed the 71.0-km round-trip in the time it took the mirror to complete one-eighth of a revolution, what is the speed of light?

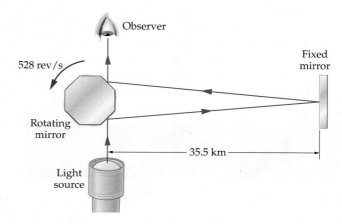

Figure 15.38

55. **Think & Calculate** **(a)** How fast would a motorist have to be traveling for a yellow ($\lambda = 590$ nm) traffic light to appear green ($\lambda = 550$ nm) because of the Doppler shift? **(b)** Should the motorist be traveling toward or away from the traffic light to see this effect? Explain.

56. A father and his daughter are interested in the same baseball game. The father sits next to his radio at home and listens to the game; his daughter attends the game and sits in the outfield bleachers. In the bottom of the ninth inning, a home run is hit. If the father's radio is 132 km from the radio station and the daughter is 115 m from home plate, who hears the home run first? (Assume that there is no time delay between the baseball being hit and its sound being broadcast by the radio station and that the speed of sound in the stadium is 343 m/s.)

57. Most of the galaxies in the universe are observed to be moving away from Earth. Suppose a particular galaxy emits orange light with a frequency of 5.000×10^{14} Hz. If the galaxy is receding from Earth with a speed of 3325 km/s, what is the frequency of the light when it reaches Earth?

58. Two starships, the *Enterprise* and the *Constitution,* are approaching each other head-on from a great distance. The separation between them is decreasing at a rate of 722.5 km/s. The *Enterprise* sends a laser signal toward the *Constitution*. If the *Constitution* observes a wavelength of $\lambda = 670.3$ nm, what wavelength was emitted by the *Enterprise*?

59. A radar unit in a highway patrol car uses a frequency of 8.00×10^9 Hz. What frequency difference will the unit detect from a car receding at a speed of 44.5 m/s from the stationary patrol car? (*Hint:* The car reflects Doppler-shifted waves back to the patrol car. Thus, the radar unit observes the effect of two successive Doppler shifts.)

60. Consider a spiral galaxy that is moving directly away from Earth with a speed $V = 3.600 \times 10^5$ m/s as measured from its center, as shown in **Figure 15.39**. The galaxy is also rotating about its center, so that points in its spiral arms are moving with a speed $v = 6.400 \times 10^5$ m/s relative to the center. If light with a frequency of 8.230×10^{14} Hz is emitted in both arms of the galaxy, what frequency is detected by astronomers observing the arm that is moving **(a)** toward and **(b)** away from Earth? (Measurements of this type are used to map the speeds of various regions in distant, rotating galaxies.)

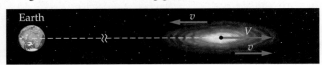

Figure 15.39

15.2 Color and the Electromagnetic Spectrum

Conceptual Questions

61. What is an electromagnetic wave? How is one generated?

62. Triple Choice Is the wavelength of infrared waves greater than, less than, or equal to the wavelength of ultraviolet waves? Explain.

63. Triple Choice Is the frequency of red light greater than, less than, or equal to the frequency of blue light? Explain.

64. Triple Choice Suppose the wavelength of a given light source is doubled. Does the frequency of the light increase by a factor of 2, decrease by a factor of 2, or remain the same? Explain.

65. What are the three additive primary colors?

66. What are the three subtractive primary colors?

67. What color do you get by combining blue and yellow light? By combining blue and green light?

68. What color do you get by combining cyan and yellow pigment? By combining cyan and red pigment?

69. A TV screen is half blue, half white. The screen has 3,000,000 pixels, each with one red color dot, one green color dot, and one blue color dot. How many red dots are lit? How many green dots are lit? How many blue dots are lit?

70. A TV screen is half magenta, half green. The screen has 3,000,000 pixels, each with one red color dot, one green color dot, and one blue color dot. How many red dots are lit? How many green dots are lit? How many blue dots are lit?

Problem Solving

71. Find the frequency of blue light with a wavelength of 460 nm.

72. Yellow light has the wavelength $\lambda = 590$ nm. How many of these waves would span the 1.0-mm thickness of a dime?

73. A cell phone transmits at a frequency of 1.25×10^8 Hz. What is the wavelength of the electromagnetic waves emitted by this phone?

74. **UV Radiation** Ultraviolet light is typically divided into three categories. UV-A, with wavelengths between 400 nm and 320 nm, has been linked with malignant melanomas. UV-B radiation, which is the primary cause of sunburn and skin cancers other than malignant melanomas, has wavelengths between 320 nm and 280 nm. Finally, the region known as UV-C extends to wavelengths of 100 nm. **(a)** Find the range of frequencies for UV-B radiation. **(b)** In which of these three categories does radiation with a frequency of 7.9×10^{14} Hz belong?

75. **Submarine Communication** Normal radiofrequency waves cannot penetrate more than a few meters below the surface of the ocean. One method of communicating with submerged submarines uses very low frequency (VLF) radio waves. What is the wavelength (in air) of a 10.0-kHz VLF radio wave?

76. Think & Calculate When an electromagnetic wave travels from one medium into another where it has a different speed, the frequency of the wave remains the same. Its wavelength, however, changes. **(a)** If the wave speed decreases, does the wavelength increase or decrease? Explain. **(b)** Consider a case where the wave speed decreases from c to $\frac{3}{4}c$. By what factor does the wavelength change?

77. **Think & Calculate** **(a)** Which color of light has the higher frequency, red or violet? **(b)** Calculate the frequencies of blue light with a wavelength of 470 nm and red light with a wavelength of 680 nm.

78. ULF (ultra low frequency) electromagnetic waves, produced in the depths of outer space, have been observed to have a wavelength of 29 million kilometers. What is the period of such a wave?

79. A television is tuned to a station broadcasting at a frequency of 6.60×10^7 Hz. For best reception the set's rabbit-ear antenna should be adjusted to have a tip-to-tip length equal to half a wavelength of the broadcast signal. Find the optimum length of the antenna.

80. An AM radio station's antenna is constructed to be $\lambda/4$ tall, where λ is the wavelength of the radio waves. How tall should the antenna be for a station broadcasting at a frequency of 880 kHz?

81. As you drive by an AM radio station, you notice a sign saying that its antenna is 112 m high. If this height represents one-quarter of the wavelength of its signal, what is the frequency of the station?

15.3 Polarization and Scattering of Light

Conceptual Questions

82. You want to check the time while wearing your polarizing sunglasses. If you hold your forearm horizontally, you can read the time easily. If you hold your forearm vertically, however, so that you are looking at your watch sideways, you notice that the display is black. Explain.

83. You are given a sheet of polarizing material. Describe how to determine the direction of its transmission axis if none is indicated on the sheet.

84. Can sound waves be polarized? Explain.

85. **Predict & Explain** Consider the two polarization experiments shown in **Figure 15.40**. **(a)** If the incident light is unpolarized, is the transmitted intensity in case A greater than, less than, or the same as the transmitted intensity in case B? **(b)** Choose the *best* explanation from among the following:

A. The transmitted intensity is the same in either case; the first polarizer lets through one-half of the incident intensity, and the second polarizer is at the angle θ relative to the first.

B. Case A has a smaller transmitted intensity than case B because the first polarizer is at the angle θ relative to the incident beam.

C. Case B has a smaller transmitted intensity than case A because the direction of polarization is rotated in the clockwise direction by the angle θ in case B.

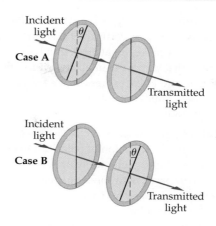

Figure 15.40

86. **Predict & Explain** Consider the two polarization experiments shown in Figure 15.40. **(a)** If the incident light is polarized in the horizontal direction, is the transmitted intensity in case A greater than, less than, or the same as the transmitted intensity in case B? **(b)** Choose the *best* explanation from among the following:

A. The two cases have the same transmitted intensity because the angle between the polarizers is θ in each case.

B. The transmitted intensity is greater in case B because all of the initial beam gets through the first polarizer.

C. The transmitted intensity is smaller in case B than in case A; in fact, the transmitted intensity in case B is zero because the first polarizer is oriented vertically.

87. Suppose linearly polarized light is incident on the polarization experiments shown in Figure 15.40. In what direction, relative to the vertical, must the incident light be polarized if the transmitted intensity is to be the same in both experiments? Explain.

Problem Solving

88. Vertically polarized light with an intensity of 0.55 W/m^2 passes through a polarizer whose transmission axis is at an angle of $35.0°$ with the vertical. What is the intensity of the transmitted light?

89. A person riding in a boat observes that the sunlight reflected by the water is polarized parallel to the surface of the water. The person is wearing polarizing sunglasses with a vertical transmission axis. If the person leans at an angle of $27.5°$ to the vertical, what fraction of the reflected light intensity will pass through the sunglasses?

90. Unpolarized light passes through two polarizers whose transmission axes are at an angle of $50.0°$ with respect to each other. What fraction of the incident intensity is transmitted through the polarizers?

91. In Problem 90, what should be the angle between the transmission axes of the polarizers for one-tenth of the incident intensity to be transmitted?

92. Unpolarized light with intensity I_i falls on a polarizing filter whose transmission axis is vertical. The axis of a second polarizing filter makes an angle θ with the vertical. Plot a graph that shows the intensity of the light transmitted by the second filter (expressed as a fraction of I_i) as a function of θ. Your graph should cover the range $\theta = 0$ to $\theta = 360°$.

93. **Think & Calculate** Optically active molecules have the property of rotating the direction of polarization of linearly polarized light. Many biologically important molecules have this property, with some causing a counterclockwise rotation (negative rotation angle) and others causing a clockwise rotation (positive rotation angle). For example, a 5.00 g per 100 mL solution of *l*-leucine causes a rotation of $-0.550°$; the same concentration of *d*-glutamic acid causes a rotation of $0.620°$. **(a)** If placed between crossed polarizers, which of these solutions transmits the greater intensity? Explain. **(b)** Find the transmitted intensity for each of these solutions when placed between crossed polarizers. The incident beam is unpolarized and has an intensity of $12.5 \ \mathrm{W/m^2}$.

94. A helium-neon laser emits a beam of unpolarized light that passes through three polarizing filters, as shown in **Figure 15.41**. The intensity of the laser beam is I_i. **(a)** What is the intensity of the beam at point A? **(b)** What is the intensity of the beam at point B? **(c)** What is the intensity of the beam at point C? **(d)** If filter 2 is removed, what is the intensity of the beam at point C?

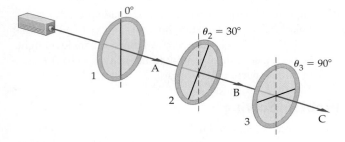

Figure 15.41

95. Referring to Figure 15.41, suppose that filter 3 is at a general angle θ with the vertical, rather than the angle 90°. **(a)** Find an expression for the transmitted intensity as a function of θ. **(b)** Plot your result from part (a), and determine the maximum transmitted intensity. **(c)** At what angle θ does the maximum transmission occur?

Mixed Review

96. **Rank** Four types of electromagnetic waves are listed below. Rank the waves in order of increasing wavelength. Indicate ties where appropriate.

A. infrared waves   **C.** green light

B. X-rays   **D.** radio waves

97. **Triple Choice** If the frequency of a light wave is increased, does its speed increase, decrease, or stay the same? Explain.

98. Common laser pointers emit light with a wavelength of 650 nm. **(a)** What is the color of this light? **(b)** What is the frequency of this light?

99. Name two objects in your physics classroom that **(a)** reflect light and **(b)** emit light.

100. A musical octave represents a doubling of the frequency of the waves. Compare the number of octaves in the human hearing range to the number of octaves in the human vision range.

101. Let's say you have cyan, yellow, and magenta pigments. How can you combine these pigments to produce a green pigment? A blue pigment?

102. At a garage sale you find a pair of sunglasses that are priced to sell and are claimed to have polarizing lenses. You are not sure, however, if the lenses are truly polarizing or are simply tinted. How can you tell which is the case? Explain.

103. A typical medical X-ray has a frequency of 1.50×10^{19} Hz. What is the wavelength of such an X-ray?

104. How many hydrogen atoms, 0.10 nm in diameter, must be placed end to end to fit into one wavelength of 410-nm violet light?

105. An incident beam of light with an intensity I_i passes through a polarizing filter whose transmission axis is at an angle θ to the vertical. As the angle is changed from $\theta = 0$ to $\theta = 90°$, the intensity as a function of the angle is given by one of the curves in **Figure 15.42**. Give the color of the curve corresponding to an incident beam that is **(a)** unpolarized, **(b)** vertically polarized, and **(c)** horizontally polarized.

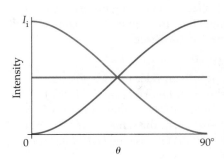

Figure 15.42

106. **3-D Movies** Modern-day 3-D movies are produced by projecting two different images onto the screen, with polarization directions that are at 90° relative to one another. Viewers must wear headsets with polarizing filters to experience the 3-D effect. Explain how this works.

107. Radiofrequency Ablation In *radiofrequency (RF) ablation*, a small needle is inserted into a cancerous tumor. When radiofrequency oscillating currents are sent into the needle, ions in the neighboring tissue respond by vibrating rapidly, causing local heating to temperatures as high as 100°C. This kills the cancerous cells but, because of the small size of the needle, relatively few of the surrounding healthy cells. A typical RF ablation treatment uses a frequency of 750 kHz. What is the wavelength that such radio waves would have in a vacuum?

108. Figure 15.43 shows four polarization experiments (A, B, C, and D) in which unpolarized incident light passes through two polarizing filters with different orientations. Rank the four cases in order of increasing amount of transmitted light. Indicate ties where appropriate.

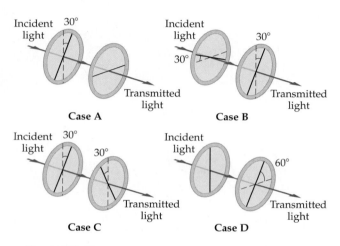

Figure 15.43

109. Moon-Based Reflector One experimental apparatus placed on the Moon's surface by *Apollo 11* astronauts was a reflector that is used to measure the Earth–Moon distance with high accuracy. A laser beam on Earth is bounced off the reflector, and its round-trip travel time is recorded. If the travel time can be measured to within an accuracy of 0.030 ns, what is the uncertainty in the measured Earth–Moon distance?

110. The H_β line of the hydrogen atom's spectrum has a normal wavelength of $\lambda_\beta = 486$ nm. This same line is observed in the spectrum of a distant quasar, but lengthened by 20.0 nm. What is the speed of the quasar relative to Earth, assuming that it is moving along our line of sight?

111. Think & Calculate Suppose the distance to the fixed mirror in Figure 15.38 is decreased to 20.5 km. **(a)** Should the angular speed of the rotating mirror be increased or decreased to ensure that the experiment works? **(b)** Find the required angular speed, assuming that the speed of light is 3.00×10^8 m/s.

112. Polarized Spider Vision The jumping spider *Drassodes cupreus* has a pair of eyes that behave like polarization filters. The transmission axes of the two eyes have an angle of 90° between them. Suppose linearly polarized light with an intensity of 825 W/m² shines from the sky onto the spider and that the intensity transmitted by one of the polarizing eyes is 212 W/m². **(a)** For this eye, what is the angle between its transmission axis and the polarization direction of the incident light? **(b)** What is the intensity transmitted by the other polarizing eye?

113. A radar unit in a highway patrol car uses a frequency of 9.00×10^9 Hz. What frequency difference will the unit detect from a car approaching the parked patrol car with a speed of 35.0 m/s? (*Hint*: The car reflects Doppler shifted waves back to the patrol car. Thus, the radar unit observes the effect of two successive Doppler shifts.)

114. Three polarizers are arranged as shown in Figure 15.41. If the incident beam of light is unpolarized and has an intensity of 1.60 W/m², find the transmitted intensity **(a)** when $\theta_2 = 25.0°$ and $\theta_3 = 50.0°$, and **(b)** when $\theta_2 = 50.0°$ and $\theta_3 = 25.0°$.

115. A typical home may require a total of 2.00×10^3 kWh of energy per month. Suppose you would like to obtain this energy from sunlight, which has an average daily intensity of 1.00×10^3 W/m². Assuming that sunlight is available 8.0 hours per day, 25 days per month (accounting for cloudy days) and that you have a way to store energy from your collector when the Sun isn't shining, determine the smallest collector size that will provide the needed energy, given a conversion efficiency of 25%.

Writing about Science

116. Write a report on how paint color is matched. Address such questions as these: How many filters are used in the spectrophotometer? What kind of light is used to illuminate the sample? How accurate are the color matches? How much does your local home improvement store charge to do a color match?

117. Connect to the Big Idea Is the speed of a microwave greater than, less than, or equal to the speed of an X-ray? Is the wavelength of a microwave greater than, less than, or equal to the wavelength of an X-ray? Explain.

Read, Reason, and Respond

Visible-Light Curing in Dentistry An essential part of modern dentistry is visible-light curing (VLC), a procedure that hardens the materials used in fillings, veneers, and other applications. The curing lights work by activating molecules known as *photoinitiators*. The photoinitiators, in turn, start a process of polymerization that causes molecules to link together to form a tough, solid polymer network. Thus, with VLC a dentist can apply and shape a soft material as desired, shine a bright light on the result, and in 20 seconds have a completely hardened—or cured—final product. The process is illustrated in **Figure 15.44**.

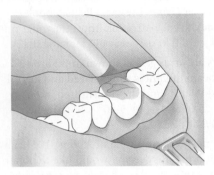

Figure 15.44 An intense beam of light cures, or hardens, the restorative material used to fill a cavity.

The most common photoinitiator is camphoroquinone (CPQ). To cure CPQ in the least time, a dentist needs to illuminate it with light having a wavelength of approximately 465 nm. Many VLC units use a halogen light, but more recent models have begun to use LEDs as their light source. The LEDs produce light with an intensity as high as 1000 mW/cm^2, which is about 10 times the intensity of sunlight on the surface of the Earth.

118. What is the color of the light that is most effective at activating the photoinitiator CPQ?
    - A. red
    - B. yellow
    - C. green
    - D. blue

119. What is the frequency of the light that is most effective at activating CPQ?
    - A. 140 Hz
    - B. 1.00×10^{14} Hz
    - C. 6.45×10^{14} Hz
    - D. 1.55×10^{15} Hz

120. How much energy does the LED in a VLC unit deliver in 30 s to a filling with an area of 0.50 cm^2?
    - A. 0.50 J
    - B. 15 J
    - C. 465 J
    - D. 1000 J

Standardized Test Prep

Multiple Choice

1. Which describes the polarization of the light wave shown in the figure below?

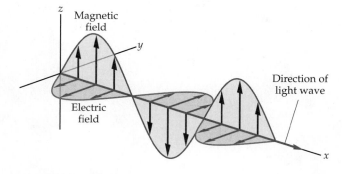

   (A) vertically polarized
   (B) horizontally polarized
   (C) unpolarized
   (D) randomly polarized

2. What determines the color of visible electromagnetic radiation we see?
   (A) speed
   (B) amplitude
   (C) frequency
   (D) polarization

3. An object looks blue when viewed under a white light source. What color will the object appear when it is illuminated by the white light and viewed through a yellow filter?
   (A) orange
   (B) black
   (C) red
   (D) green

4. Which of the following correctly ranks electromagnetic waves from lowest to highest frequency?
   (A) red, green, infrared, gamma
   (B) infrared, blue, ultraviolet, x-ray
   (C) yellow, red, infrared, radio
   (D) ultraviolet, green, red, infrared

5. Which statement regarding the polarization of light is correct?
   (A) When unpolarized light passes through a polarizer, its intensity is reduced by half.
   (B) A polarizer has no effect on unpolarized light.
   (C) When polarized light passes through a polarizer, its intensity is reduced by half.
   (D) The intensity of polarized light is not reduced if it passes through crossed polarizers.

6. The hydrogen spectrum from a star is observed to have shifted from 470.8 nm to 537.4 nm due to the Doppler effect. Which statement is true?
   (A) The star is moving toward Earth.
   (B) The star is moving away from Earth.
   (C) The star is moving at $1.3c$.
   (D) The star is moving at $0.9c$.

7. Wavelengths for visible light range from about 400 nm to 700 nm. What is the visible frequency range?
   (A) 4.3×10^5 Hz to 7.5×10^5 Hz
   (B) 2.4×10^5 Hz to 5.3×10^5 Hz
   (C) 4.3×10^{14} Hz to 7.5×10^{14} Hz
   (D) 4.3×10^{15} Hz to 7.5×10^{15} Hz

Extended Response

8. An experiment involves three flashlights and three color filters—red, blue, and green. Describe what happens in each case: (a) The filters are placed separately on the three flashlights, and the colored beams overlap on a white screen. (b) All three color filters are placed on one flashlight, and the beam is directed toward a white screen. Explain your answers.

> **Test-Taking Hint**
>
> The sequence of colors of visible light, from lowest frequency to highest, is given by the mnemonic ROY G BIV (red, orange, yellow, green, blue, indigo, violet).

If You Had Difficulty With . . .

Question	1	2	3	4	5	6	7	8					
See Lesson	15.3	15.2	15.2	15.2	15.3	15.1	15.2	15.2					

16

Reflection and Mirrors

Inside

Each of the roughly 2000 mirrors of this solar power plant can be aimed independently, allowing each one to follow the Sun and focus the reflected sunlight directly on the top of the heating tower. The intense heat that results is used to generate electricity.

Big Idea

Mirrors are particularly good at reflecting light; a mirror's shape determines the size, location, and orientation of the reflected image.

When you look into a mirror, you see images of yourself and the objects nearby. If the surface of the mirror is flat, the images look just like those in the real world—except with right and left reversed. If the surface of the mirror is curved, the images can be larger or smaller than life size, or even upside-down. We explore all of these interesting possibilities in this chapter.

16.1 The Reflection of Light

Look around. Is there something shiny in your classroom? Maybe a mirror on the wall or the metal legs of a chair. What makes these objects shiny? Well, if you think about it, what these object have in common is that they *reflect* a lot of light. In general, **reflection** occurs whenever a wave (any type of wave) hits a surface or a boundary and bounces off in a different direction. Objects that reflect most of the light waves that hit them look "bright" and "shiny." In this lesson we'll investigate the basic physics of reflection.

Reflection occurs at boundaries

If you throw a ball at a wall, it bounces back. Sound and light waves also bounce (reflect) off a wall. That's why you hear echoes from a wall and can see a wall even though it produces no light of its own.

In general, waves are reflected—at least partially—any time they encounter a boundary between two different materials. In the case of a wall, the two materials are the air in the room and the substance of the wall. Other examples include the boundaries between substances like air and water, water and glass, and air and glass. When light hits the boundary between air and glass, for example, some of the light passes into the glass, and some reflects back into the air. The reflected light stays in the original substance and travels in a different direction. *Mirrors* are simply objects that are particularly good at reflecting light waves.

Vocabulary

- reflection
- ray
- plane wave
- normal
- specular reflection
- diffuse reflection

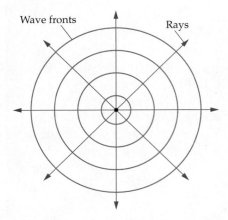

▲ **Figure 16.1 Wave fronts and rays**
The wave fronts shown here represent the crests of water waves moving outward from a splash. The rays indicate the direction of motion at each location.

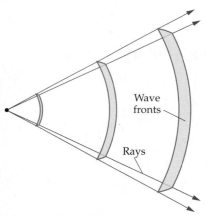

(a) Spherical wave fronts

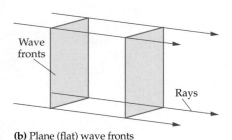

(b) Plane (flat) wave fronts

▲ **Figure 16.2 Spherical and plane (flat) wave fronts**

Light can be represented by wave fronts and rays

To study the reflection of light, we need a simple way to draw the situations we are interested in. A convenient method is to use *rays*. Simply put, a **ray** is an arrow that points in the direction that light travels.

Consider the waves created by a rock dropped into a pool of water. As we know, these waves form concentric outward-moving circles. A simplified version of this situation is shown in **Figure 16.1**. Here the circles indicate the crests of the outgoing waves. We refer to these circles as *wave fronts*. The outward motion of the waves is indicated by the outward-pointing arrows—the rays. **Rays are always at right angles to the wave fronts.**

A similar situation applies to light and other electromagnetic waves, as illustrated in **Figure 16.2 (a)**. In this case the waves move outward in three dimensions, giving rise to spherical wave fronts. As expected, *spherical wave fronts* such as these have rays that point radially outward. **Figure 16.2 (b)** shows that as they move farther from a source, spherical wave fronts become flat planes, and the rays become parallel. In general, **plane waves** have flat wave fronts and parallel rays all pointing in the same direction.

As you will see, plane waves and their corresponding rays are useful when investigating the properties of mirrors. It can be a bit messy, though, to draw both wave fronts and rays. Thus, we usually simplify our representation of light beams by omitting the wave fronts and plotting only one, or a just few, rays.

Light waves reflect in the simplest way possible

Consider a beam of light that reflects from a mirror. To study this situation we begin by drawing the *normal* to the surface of the mirror. The **normal** to a reflecting surface is a line drawn *perpendicular* to the surface. An example is shown as a dashed line in **Figure 16.3**.

The incident and reflected beams of light in Figure 16.3 are each represented by a single ray. Notice that the incident ray hits the surface of the mirror at the angle θ_i to the normal. We call θ_i the *angle of incidence*. Similarly, the *angle of reflection*, θ_r, is the angle that the reflected ray makes with the normal. **The relationship between the angle of reflection and the angle of incidence is very simple—they are equal.** Thus, a beam of light reflects from a mirror at the same angle as it strikes the mirror.

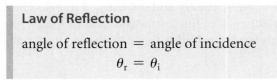

Law of Reflection

angle of reflection $=$ angle of incidence
$$\theta_r = \theta_i$$

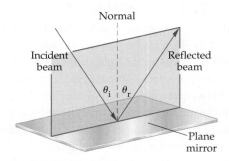

▲ **Figure 16.3 Reflection from a smooth surface**
In this simple diagram of reflection, the incident and reflected beams are indicated by a single ray pointing in the direction of travel. The angle of reflection, θ_r, is equal to the angle of incidence, θ_i.

(a) Specular reflection

(b) Mirror image due to specular reflection

▲ **Figure 16.4 Reflection from a smooth surface**
(a) A smooth surface produces specular reflection, with all of the reflected light traveling in a single direction. **(b)** The smooth surface of this lake produces specular reflection.

🔖 *What type of reflection produces clear images?*

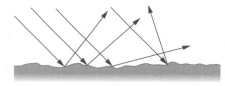

(a) Diffuse reflection

(b) No image with diffuse reflection

▲ **Figure 16.5 Reflection from a rough surface**
(a) A rough surface produces reflected light traveling in a range of different directions. This is diffuse reflection. **(b)** The rough surface of this lake reflects light in many different directions, which prevents the reflected light from forming an image.

CONCEPTUAL Example 16.1 How Does the Direction Change?

In case A in the figure below, a beam of light with an angle of incidence of 10° is reflected from a mirror. The change in direction of this beam is 160°, as shown. In case B a beam of light is incident on a mirror at an angle of 45°. Is the change in direction of this beam greater than, less than, or equal to 160°? Explain.

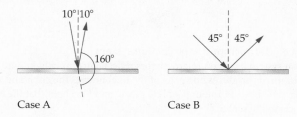

Case A   Case B

Reasoning and Discussion
In case B the change in direction of the beam is only 90°, as shown below.

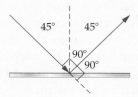

Thus, a beam of light with a more glancing angle of reflection (case B) changes its direction less than a beam of light that is practically turned around by the mirror (case A).

Answer
The change in direction of the beam in case B is less than 160°.

Reflection can be specular or diffuse

When light reflects from a surface, the texture of the surface determines its appearance. For example, a smooth surface looks shiny because the reflected light is "beamed" in one direction. This is illustrated in **Figure 16.4 (a)**. Reflection from a smooth surface, with all the reflected light moving in a single direction, is referred to as **specular reflection**. 🔖 **Specular reflection is responsible for the sharp, clear images seen in mirrors.** Such an image appears in the mirror-like surface of a lake in **Figure 16.4 (b)**.

Reflection from rough surfaces is quite different. Think of light reflecting from the surface of a bathroom towel. The rough surface of the towel reflects light in all directions. Reflection that sends light off in a variety of directions is referred to as **diffuse reflection**. An example is shown in **Figure 16.5**. Any surface in your classroom that is not shiny, like the pages in this book or the fabric of your shirt, is causing diffuse reflection of light.

The surface of a road provides a good illustration of the difference between specular and diffuse reflection. When the road is wet, the water creates a smooth surface. Headlights reflecting from the wet road undergo specular reflection, producing an intense glare. When the same road is dry, its surface is rough, and the headlights are reflected in many different directions. There's no bright glare in this case. The law of reflection is obeyed in both cases, of course—it's the texture of the surface that differs.

▲ **Figure 16.6 Micromirrors**
An ant's leg provides a sense of scale in this photo of the array of mirrors in a digital micromirror device (DMD).

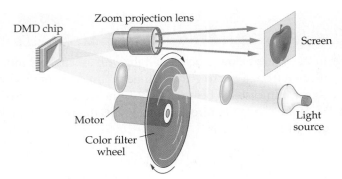

▲ **Figure 16.7 A digital micromirror projection system**
A digital projection system based on micromirrors reflects incoming light onto a distant screen. Each micromirror produces one pixel of the final image. The color and intensity of a given pixel is determined by the amounts of red, green, and blue light that the corresponding micromirror reflects to the screen.

Physics & You: Technology A clever application of specular and diffuse reflection occurs in an electronic chip known as a *digital micromirror device (DMD)*. These small devices consist of as many as 1.3 million microscopic plane mirrors. Each micromirror, though smaller than the diameter of a human hair, can be oriented independently in response to electrical signals. For scale, **Figure 16.6** shows an ant's leg in front of a DMD. The reflection from each micromirror is specular, and if all 1.3 million micromirrors are oriented in the same direction, the DMD acts like a small plane mirror. Conversely, if the micromirrors are oriented randomly, the reflection from the DMD is diffuse.

When a DMD is used to project a movie, each micromirror plays the role of a single pixel in the projected image. In such a projection system the light directed onto the DMD cycles rapidly from red to green to blue, and each micromirror reflects only the appropriate colors for that pixel onto the screen, as shown in **Figure 16.7**. The result is a projected image of great brilliance and vividness that eliminates the need for film.

Light travels along the path of least time

Light gets where it's going fast. The speed of light is greater than the speed of anything else in the universe. Not only that, but light travels along the path that gives the shortest possible travel time.

As an example, when light travels from point A to point B in **Figure 16.8**, it travels along a straight line from A to B. As we know, a straight line is the shortest distance between two points. Therefore, the path that light takes is the path of least time.

Suppose, instead, that light travels from point A to a mirror, reflects from the mirror, and then continues to point B. This is shown in **Figure 16.9**. Which path should the light take if it is to get to B in the least time? That is, from what point on the mirror should the light reflect?

It turns out that the travel time is least when the light follows path 2. This path obeys the law of reflection, with the angle of reflection equal to the angle of incidence. This path is also the *shortest possible reflecting path* from A to B. Light chooses this path over all others! The distances (and travel times) along paths 1 and 3 are greater, as we show in **Figure 16.10**. You can always count on light to get to its destination quickly.

▲ **Figure 16.8 The least time between two points**
When light travels from point A to point B, it follows a straight-line path. Since a straight line is the shortest path between A and B, the light travels along the path of least time.

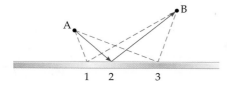

▲ **Figure 16.9** **The least time with a reflection**
Light travels from point A to the mirror, reflects, then continues to point B. The path that takes the least time for light to travel is path 2, which obeys the law of reflection—with the angle of reflection equal to the angle of incidence. Paths 1 and 3 are longer than path 2 and would take more time to travel.

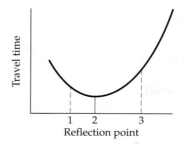

▲ **Figure 16.10** **Travel time**
This graph plots the travel time (vertical axis) versus the reflection point (horizontal axis) corresponding to different paths for reflection from the mirror shown in Figure 16.9. The travel times for paths 1 and 3 are greater than the travel time for path 2, which obeys the law of reflection. Thus, light chooses the path of least time.

16.1 LessonCheck (MP)

Checking Concepts

1. ⚏ **Compare** How are rays and wave fronts similar? How are they different?

2. Analyze A beam of light reflects from a mirror as shown in **Figure 16.11**. If the angle of incidence of the beam is increased, does the reflected beam shift in direction 1, shift in direction 2, or stay in the same direction? Explain.

▲ Figure 16.11

3. ⚏ **Identify** Which type of surface, smooth or rough, produces clear reflected images?

4. Triple Choice Figure 16.12 shows a horizontal incident beam of light reflecting from a mirror inclined at an angle θ above the horizontal. The reflected beam of light makes an angle ϕ with the horizontal. If θ is increased, does ϕ increase, decrease, or stay the same? Explain.

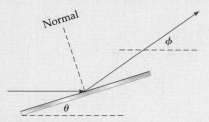

▲ Figure 16.12

Solving Problems

5. Calculate Suppose the angle of the mirror relative to the horizontal in Figure 16.12 is $\theta = 30°$. What is the angle of incidence?

6. Calculate Suppose the angle of the mirror relative to the horizontal in Figure 16.12 is $\theta = 20°$. What angle ϕ does the reflected beam make with the horizontal?

16.2 Plane Mirrors

Vocabulary

- plane mirror
- virtual image
- corner reflector

🔑 How is the image's location related to the object's location for a plane mirror?

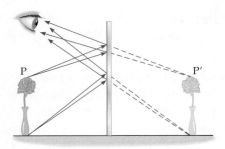

(a) Image formed by a plane mirror

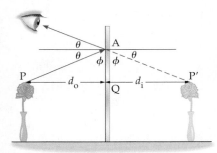

(b) Image appears as far behind the mirror as object is in front.

▲ **Figure 16.13 Locating a mirror image**
(a) Rays of light from point P at the top of the flower appear to originate from point P′ behind the mirror. **(b)** This construction shows that the length of the line PQ (object to mirror) is the same as the length of the line QP′ (mirror to image).

Have you looked in a mirror recently? If so, you probably used a mirror that was perfectly flat, a **plane mirror**. When you look into a plane mirror, you see an upright image (right side up) of yourself that appears to be as far behind the mirror as you are in front of it. In addition, the image is reversed right to left. If you raise your right hand, the mirror image raises its left hand. In this lesson we use the simple law of reflection to investigate mirror images.

A plane mirror produces an image behind the mirror

Before going into the details of mirror images, let's consider how objects produce images in our eyes. Any nearby object is bathed in light coming at it from all directions. As the object reflects the light back into the room, each point on it acts like a source of light. When you view the object, the light coming from a point on the object enters your eyes and is focused to a point on your retina. This is the case *for every point* that you can see on the object. Therefore, each point on the object is detected by a corresponding point on the retina. This results in a one-to-one connection between the physical object and the image on the retina.

The formation of a mirror image occurs in the same way, except the light from an object reflects off a mirror before it enters the eyes. This is illustrated in **Figure 16.13 (a)**. Here we see an object—a small flower in a vase—placed in front of a plane mirror. Rays of light leaving the top of the flower at point P reflect from the mirror and enter the eye of an observer. To the observer it appears that the rays are coming from the point P′ *behind the mirror*. Similar remarks apply to rays of light coming from the base of the flower vase.

In **Figure 16.13 (b)** we trace a ray from the flower to the mirror—where it reflects—and then to the eye. We indicate the location of the object with d_o and the location of the image with d_i. One ray from the top of the flower is shown reflecting from the mirror and entering the observer's eye. Extending this ray back to the image produces two triangles, PAQ and P′AQ. Geometry shows that these triangles are equal to one another. Therefore, the length of side PQ is the same as the length of the side QP′. 🔑 **In other words, the image formed by a plane mirror appears as far behind the mirror as the object is in front of the mirror.** We can write this conclusion in the form of an equation as follows:

> **Image Distance for a Plane Mirror**
>
> image distance = −(object distance)
> $$d_i = -d_o$$

For example, if the object distance is 5 cm, the image distance is −5 cm. The *negative* image distance means that the image is *behind* the mirror, as indicated in Figure 16.13. In general, an image that is behind a mirror is known as a **virtual image**. The term *virtual* is used to indicate that no light passes through the image and that it cannot be projected onto a screen. A virtual image looks just as real to your eye as any physical object, however.

One last point about Figure 16.13. Notice that the height of the image is the same as the height of the object. This is always the case for plane mirrors. If we let h_i denote the image height and h_o the object height, we can express this result with the following simple equation:

Image Height for a Plane Mirror

image height = object height

$$h_i = h_o$$

GUIDED Example 16.2 | Reflecting on a Flower   Reflection

An observer is at table level, a distance d to the left of a flower of height h. The flower is a distance d to the left of a mirror, as shown below. A ray of light traveling from the top of the flower to the observer's eye reflects from the mirror at a height y above the table. Find y in terms of the height of the flower, h.

Picture the Problem

Our sketch shows a ray from the top of the flower to the eye of the observer. The point where the ray hits the mirror is a height y above the table. The flower is a distance d to the left of the mirror, and its image is a distance d to the right of the mirror. The observer's eye is a distance $2d$ to the left of the mirror.

Strategy

We draw a single ray from the top of the flower to the mirror and then to observer's eye. When this ray is extended behind the mirror a distance d, it is at the top of the flower's image, which is a height h above the table.

The straight line from the observer's eye to the top of the flower's image slants steadily upward. At a horizontal distance $3d$ from the eye, the height of this line is h. Thus, the height increases by an amount $h/3$ for each change in horizontal distance of d. The mirror is a horizontal distance $2d$ from the eye, and therefore the height of the ray at the mirror is $y = 2h/3$.

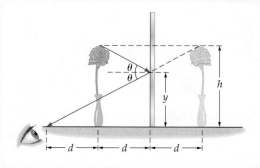

Known

distance from mirror to observer = $2d$
object distance = d
image distance = $-d$

Unknown

$y = ?$

Solution

1 The height, y, at which the ray reflects (in terms of h) is given by

$$y = \boxed{2h/3}$$

Insight

The observer sees the entire image of the flower in a section of mirror that is only two-thirds the height of the flower.

7. **Follow-up** If the observer in Guided Example 16.2 moves farther from the base of the flower, does the height of the point of reflection on the mirror increase, decrease, or stay the same? As a check on your answer, calculate the height of the point of reflection for the case where the distance from the observer to the base of the flower is $2d$.

8. A student walks toward a full-length mirror on a wall with a speed of 1.1 m/s. How fast are the student and her image approaching one another?

Plane mirrors reverse right and left

The basic features of images formed by a plane mirror are as follows:

- The mirror image appears to be the same distance behind the mirror as the object is in front of the mirror.
- The mirror image is the same size as the object.
- The mirror image is upright, but reversed right to left.

According to the last property, any writing or text reflected in a mirror is reversed right to left. This is the reason ambulances and other emergency vehicles have mirror-image labels on the front. When such a label is viewed in the rear-view mirror of a car, it looks normal, making it easy to read and understand. Leonardo da Vinci used mirror-image writing to make it hard for others to decipher his notes. An example is shown in **Figure 16.14**. When the writing in his notebooks is viewed in a mirror, it appears normal.

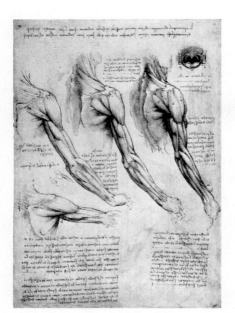

▲ **Figure 16.14 Leonardo da Vinci's mirror writing**
Leonardo da Vinci (1452–1519), the **quintessential** Renaissance man, kept the contents of his notebooks private by setting down his observations in mirror-image writing.

CONCEPTUAL Example 16.3 How Tall Is the Mirror?

To save money you would like to buy the shortest mirror that will allow you to see your entire body. Should the mirror's height be half of your height, two-thirds of your height, or equal to your height?

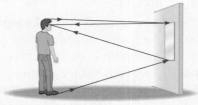

Reasoning and Discussion
First, in order for you to see your feet, the mirror must extend from your eyes downward to a point halfway between your eyes and feet, as shown in the sketch.

Similarly, the mirror must extend upward from your eyes half the distance to the top of your head. Altogether, then, the mirror must have a height equal to half of your total height.

Answer
The mirror's height needs to be half of your height.

Mirror images have practical applications

A mirror is a simple device, yet it is able to do impressive things—like changing the direction of a beam of light with little loss of energy. Not surprisingly, mirrors are used in many devices, both commercial and scientific.

▲ **Figure 16.15 Heads-up display**
A heads-up display in an airplane cockpit displays important flight information by reflecting it on a transparent screen near the windshield. This lets the pilot view the data without looking away from the scene ahead.

Physics & You: Technology An interesting application of mirror images is a *heads-up display*. An example from an airplane is shown in **Figure 16.15**. Some automobiles also use heads-up displays. In such a car a small illuminated display screen is recessed in the dashboard, out of direct sight of the driver. The screen shows important information, like the speed of the car, in mirror image. The driver sees the information by looking at its reflection in the windshield. Thus, while still looking straight ahead (heads up), the driver can see both the road and the reading of the speedometer.

A similar device is used in movie theaters to provide subtitles for people who are hearing impaired. In this case a transparent plastic screen is mounted on the arm of a person's chair. This screen is adjusted so that the person can look through it to see the movie and, at the same time, see the reflection of a text screen in the back of the theater. The text screen presents the movie's subtitles in mirror-image writing.

GUIDED Example 16.4 | Two-Dimensional Corner Reflector   Reflection

Two plane mirrors are placed at right angles, as shown below. A ray of light comes from the right and reflects from the lower mirror. The angle this ray makes with the horizontal is 30°. Find the angle the outgoing ray makes with the horizontal after reflecting once from each mirror.

Picture the Problem

The physical system is shown in our sketch. Notice that the normal for the first reflection is vertical. The normal for the second reflection is horizontal.

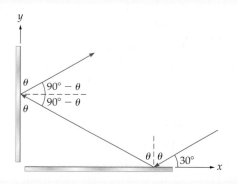

Strategy

We apply the law of reflection to each of the reflections in turn.

First reflection: The normal is vertical, and the angle of incidence is $\theta = 90° - 30° = 60°$. Therefore, the reflected ray also makes an angle of $\theta = 60°$ with the vertical.

Second reflection: The normal is horizontal in this case. It follows that the angle of incidence is $90° - \theta = 90° - 60° = 30°$. The angle of reflection is equal to the angle of incidence; therefore, the reflected ray is at angle of 30° above the horizontal.

Known

angle of incidence $= 30°$

Unknown

angle of outgoing ray above the horizontal $= ?$

Solution

1 The angle of the outgoing ray (relative to the horizontal) is equal to $90° - \theta$:

$$90° - \theta = \boxed{30°}$$

Insight

The outgoing ray travels parallel to the incoming ray but in exactly the opposite direction. This is true for any angle of incidence when plane mirrors are arranged like these two.

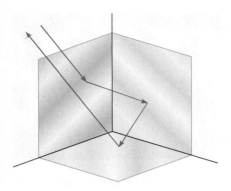

▲ Figure 16.16 A corner reflector
A three-dimensional corner reflector constructed from three plane mirrors at right angles to one another. A ray entering the reflector is sent back in the direction from which it came.

Practice Problems

9. |**Follow-up**| If the incoming ray in Guided Example 16.4 hits the horizontal mirror farther to the right, is the angle the outgoing ray makes with the horizontal greater than, less than, or equal to 30°?

10. |**Follow-up**| If the incoming ray in Guided Example 16.4 makes an angle of 25° with the surface of the horizontal mirror, what is its angle of incidence with the vertical mirror?

Physics & You: Technology If three plane mirrors are joined at right angles, as in **Figure 16.16**, the result is referred to as a **corner reflector**. A corner reflector is the three-dimensional version of the two-mirror arrangement considered in Guided Example 16.4. Specifically, a ray incident on a corner reflector is sent back in the same direction from which it came.

Corner reflectors are used on ships, and especially lifeboats, where they reflect radar waves directly back to the source. For this reason they are often referred to as *retroreflectors*. Retroreflectors make the vessel show up bright and clear on a radar scan of the area. Similar retroreflectors are common on the taillights of cars and bicycles—and even on the backs of running shoes.

16.2 LessonCheck ⓜ

Checking Concepts

11. 🔗 **Apply** If an object is 1 m in front of a plane mirror, how far is the image behind the mirror?

12. |**Triple Choice**| As you walk toward a plane mirror, does the distance between you and your image increase, decrease, or stay the same? Explain.

13. Diagram Two plane mirrors meet at right angles at the origin of a coordinate system, as shown in **Figure 16.17**. A long object extends from $x = 1$ m and $y = 1$ m to $x = 4$ m and $y = 2$ m. Draw *all three* of the images formed by the two mirrors. (*Hint:* One of the three images is due to reflections from both mirrors.)

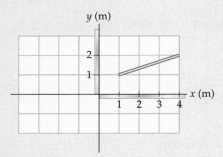

▲ Figure 16.17

Solving Problems

14. Calculate You stand 2.3 m in front of a plane mirror. Your little brother is 1.2 m in front of you, directly between you and the mirror. What is the distance from you to your brother's image?

15. Calculate A nearsighted python with a length of 4.0 m is stretched out perpendicular to a plane mirror, admiring its reflected image. If the greatest distance to which the snake can see clearly is 9.6 m, how close must its head be to the mirror for it to see a clear image of its tail?

16. Calculate Two plane mirrors are placed with an angle of 60° between them, as shown in **Figure 16.18**. A horizontal ray of light strikes the slanted mirror. After the ray reflects from both mirrors, what angle θ does it make with the horizontal?

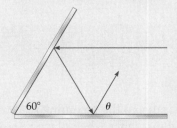

▲ Figure 16.18

16.3 Curved Mirrors

Have you ever been in a car that had a mirror with the message "OBJECTS IN THE MIRROR ARE CLOSER THAN THEY APPEAR"? If so, you can be sure that the mirror was curved. Curved mirrors produce all sorts of interesting effects, like enlarging an object, shrinking an object, or even turning an object upside down. In this lesson we'll look at all of these behaviors.

Types of Curved Mirrors and Their Characteristics

There are two basic types of curved mirrors, concave and convex. A **concave mirror** is one that curves inward, forming a sort of "cave" within the mirror. In contrast, a **convex mirror** has the opposite shape—it bulges outward like the surface of a ball.

Most curved mirrors have a spherical shape, as indicated in **Figure 16.19 (a)**, and are referred to as *spherical mirrors*. A typical spherical mirror is just a portion of a spherical shell of radius R. If the inside of this spherical section is a reflecting surface, the result is a concave spherical mirror. If the outside surface is reflecting, the result is a convex spherical mirror. These two situations are illustrated in **Figure 16.19 (b)** and **(c)**.

Vocabulary

- concave mirror
- convex mirror
- center of curvature
- principal axis
- focal point
- focal length
- principal rays
- real image
- magnification

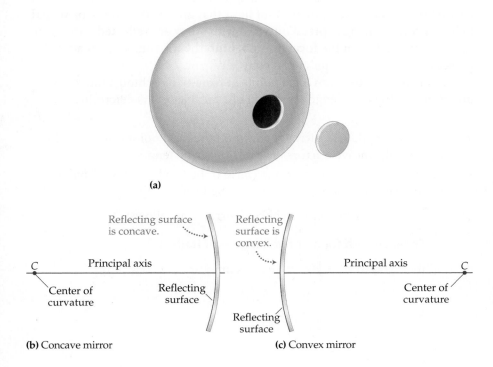

(a)

(b) Concave mirror

Reflecting surface is concave.

Reflecting surface is convex.

Principal axis

C

Center of curvature

Reflecting surface

Reflecting surface

(c) Convex mirror

Principal axis

Center of curvature

C

◀ **Figure 16.19 Spherical concave and convex mirrors**
(a) A spherical mirror has the same shape as a section of a spherical shell. (b) If the inside surface of a spherical mirror is reflecting, the mirror is concave—like a cave. The center of curvature, *C*, is on the same side of the mirror as the reflecting surface in this case. (c) If the outside surface reflects, the mirror is convex. In this case the center of curvature, *C*, is on the opposite side of the mirror from its reflecting surface. This means that the center of curvature is behind the mirror.

Figure 16.19 also shows the *center of curvature* and the *principal axis* for each type of mirror. The **center of curvature**, C, is the center of the spherical shell with radius R of which the curved mirror is a section. The **principal axis** is a straight line drawn through the center of curvature and the midpoint of the mirror. Notice that the principal axis intersects the mirror at right angles.

A concave mirror converges rays in front of the mirror

Consider the concave mirror shown in **Figure 16.20**. A beam of light is directed toward the mirror along its principal axis. This beam is represented in the figure by several parallel rays. The rays reflect from the surface of the mirror and converge—or *focus*—at the **focal point**, F.

From Figure 16.20 we can see that the focal point, F, is halfway between the center of curvature, C, and the surface of the mirror. Since the center of curvature is a distance R from the surface, it follows that the distance from the mirror to the focal point is $R/2$. In general, the **focal length**, f, of a concave mirror is the distance from the surface of the mirror to the focal point:

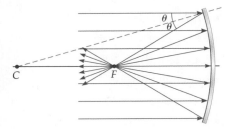

▲ Figure 16.20 Parallel rays reflecting from a concave mirror
Parallel rays reflecting from a concave mirror pass through the focal point, F, which is halfway between the surface of the mirror and the center of curvature.

> **Focal Length for a Concave Mirror of Radius R**
> focal length $= \frac{1}{2} \times$ radius of curvature
> $$f = \frac{1}{2}R$$
> SI unit: m

With a concave mirror, incoming rays of light that are parallel to the principal axis are reflected through the focal point.

A convex mirror spreads rays outward

Several parallel rays of light are shown approaching a convex mirror in **Figure 16.21**. Incoming rays of light that are parallel to the principal axis of a convex mirror spread outward when they are reflected—just as if they had started from the focal point behind the mirror. No light actually passes through the focal point of a convex mirror.

To distinguish between focal points in front of or behind a mirror, we give a sign to the focal length. The sign of a focal length is determined as follows:

- The focal length of a mirror is *positive* if the focal point is in *front* of the mirror. All concave mirrors have positive focal lengths.
- The focal length of a mirror is *negative* if the focal point is *behind* the mirror. All convex mirrors have negative focal lengths.

Therefore, the focal length for a convex mirror is as follows:

What point do light rays pass through after reflecting from a concave mirror?

Where do the light rays reflected by a convex mirror appear to come from?

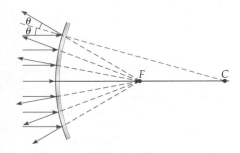

▲ Figure 16.21 Parallel rays reflecting from a convex mirror
When parallel rays of light reflect from a convex mirror, they diverge as if coming from the focal point, which is behind the mirror.

> **Focal Length for a Convex Mirror of Radius R**
> focal length $= -\frac{1}{2} \times$ radius of curvature
> $$f = -\frac{1}{2}R$$
> SI unit: m

Suppose you would like to use the Sun to start a fire in the wilderness. Which type of mirror would work best, a concave mirror or a convex mirror?

Reasoning and Discussion

If you used a convex mirror, the situation would be like that shown in Figure 16.21. In this case the rays of sunlight would spread out after reflection. On the other hand, a concave mirror would bring the rays together at a point, as in Figure 16.20. It follows that the concave mirror is the one to use, since it concentrates the sunlight.

Answer

The concave mirror would work best.

Ray Tracing

The easiest way to find the image formed by a mirror is to draw a few rays and see how they reflect. In this method, known as *ray tracing*, we draw the paths of rays of light as they reflect from a mirror and use them to find the location of the image. This was done for the plane mirror in Figure 16.13. Ray tracing for spherical mirrors applies the same idea.

Principal rays show how mirrors reflect light

Three rays that show simple behavior, known as the **principal rays**, are used in ray tracing with spherical mirrors. These rays are illustrated in **Figure 16.22** for a concave mirror and in **Figure 16.23** for a convex mirror.

Parallel Ray We start with the parallel ray (*P* ray), which, as its name implies, is parallel to the principal axis of the mirror. A *P* ray is reflected through the focal point of a concave mirror, as shown by the purple ray in Figure 16.22. Similarly, a *P* ray reflects from a convex mirror along a line that *extends back* through the focal point, like the purple ray in Figure 16.23. *P* rays are always drawn with purple arrows.

Focal-Point Ray Next, a ray that passes through the focal point of a concave mirror is reflected parallel to the principal axis, as indicated by the green ray in Figure 16.22. Thus, in a sense, a focal-point ray (*F* ray) for a concave mirror is the reverse of a *P* ray. The *F* ray for a convex mirror is shown in green in Figure 16.23. *F* rays are always drawn with green arrows.

Center-of-Curvature Ray Finally, any straight line drawn from the center of curvature intersects the mirror at right angles. Thus, a ray moving along such a path is reflected back along the same path. Center-of-curvature rays (*C* rays) are illustrated in red in Figures 16.22 and 16.23. *C* rays are always drawn with red arrows.

Convex mirrors form virtual images

To see how principal rays can be used to obtain an image, consider the convex mirror shown in **Figure 16.24** on the next page. In front of the mirror is an object, represented symbolically by the red arrow. Also

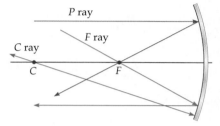

▲ **Figure 16.22 Principal rays used in ray tracing for a concave mirror**
The parallel ray (*P* ray) reflects through the focal point. The focal-point ray (*F* ray) reflects parallel to the principal axis, and the center-of-curvature ray (*C* ray) reflects back along its incoming path.

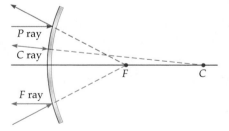

▲ **Figure 16.23 Principal rays used in ray tracing for a convex mirror**
The parallel ray (*P* ray) reflects along a direction that extends back to the focal point. Similarly, the focal-point ray (*F* ray) moves toward the focal point until it reflects from the mirror, after which it moves parallel to the axis. The center-of-curvature ray (*C* ray) is directed toward the center of curvature. It reflects from the mirror and goes back along its incoming path.

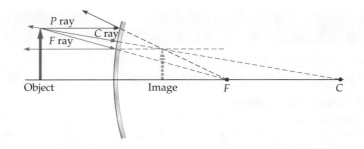

▶ Figure 16.24 Image formation with a convex mirror
Ray diagram showing an image formed by a convex mirror. The three outgoing rays (*P*, *F*, and *C*) extend back to a single point at the top of the image.

indicated in the figure are the principal rays. Notice that these rays diverge from the mirror as if they had originated from the tip of the dashed orange arrow behind the mirror. This arrow is the image of the object.

Recall that an image formed behind a mirror (with no light passing through the image) is a virtual image. For example, the orange arrow in Figure 16.24 is a virtual image of the red arrow. We indicate this on the diagram by showing the virtual image as a dashed arrow. As you can see from the diagram, the virtual image is upright, smaller than the object, and located between the mirror and the focal point, *F*.

Even though we drew three rays in Figure 16.24, any two would have given the intersection point at the tip of the virtual image. This is commonly the case with ray diagrams. When possible, it is useful to draw all three rays as a check on your results.

If an object is very close to a convex mirror, the mirror is essentially flat and behaves like a plane mirror. Thus, the virtual image is about the same distance behind the mirror as the object is in front and about the same size as the object. On the other hand, if an object is far from a convex mirror, the image is very small and practically at the focal point. These limits are illustrated in **Figure 16.25**.

Concave mirrors produce a variety of images

⊙ How do real images and virtual images differ?

Let's now consider the images formed by concave mirrors. The *F* and *P* rays for the case where the object is farther from the mirror than the center of curvature are shown in **Figure 16.26 (a)**. The *C* ray isn't needed in this case and has been omitted for clarity. Notice that the image is inverted (upside-down), closer to the mirror, and smaller than the object.

The image in Figure 16.26 (a) is formed by converging rays. When light rays pass through an image, as in this case, it is known as a **real image**. Real images are shown without dashing. **⊙ A real image (with light rays passing through it) can be projected onto a screen; a virtual image (with no light rays passing through it) cannot.**

▶ Figure 16.25 Image size and location with a convex mirror
(a) When an object is close to a convex mirror, the image is practically the same size and distance from the mirror. **(b)** If the object is far from the mirror, the image is small and close to the focal point.

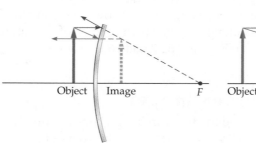
(a) An object close to a convex mirror

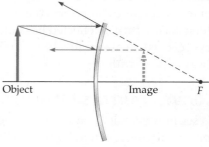
(b) An object far from a convex mirror

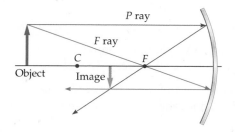

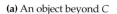

(a) An object beyond *C*

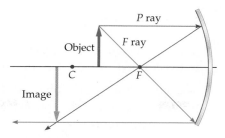

(b) An object between *C* and *F*

◀ **Figure 16.26 Image formation with a concave mirror**
Ray diagrams showing the image formed by a concave mirror when the object is **(a)** beyond the center of curvature and **(b)** between the center of curvature and the focal point.

The ray diagram for the case where the object is between the center of curvature and the focal point is shown in **Figure 16.26 (b)**. Again, the *C* ray has been omitted for clarity. The image is real and inverted, but it is now farther from the mirror and larger than the object.

The case in which the object is between the mirror and the focal point is considered in the next Guided Example.

GUIDED Example 16.6 | Image Formation   **Concave Mirrors**

Use a ray diagram to find the location, orientation, and size of the image formed by a concave mirror when the object is between the mirror and its focal point.

Picture the Problem

The situation is shown in our ray diagram, along with the three principal rays. Notice that after reflection these three rays diverge from one another, just as if they had originated at a point behind the mirror.

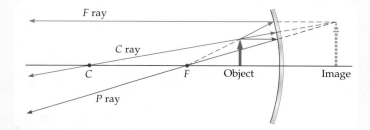

Strategy

Analysis of the principal rays for this case gives us the result. Let's consider these three rays one at a time.

P ray: The *P* ray is the simplest of the three. It starts parallel to the axis, then reflects through the focal point.

F ray: The *F* ray does not go through the focal point, as is usually the case. Instead, it starts on a line that *extends back* to the focal point and then reflects parallel to the axis.

C ray: The *C* ray starts at the top of the object, is incident on the mirror at right angles, and then reflects back along its initial path and through the center of curvature.

Notice that the extensions of the three reflected principal rays (*P*, *F*, and *C*) form a virtual image behind the mirror. Analyzing the ray diagram yields a description of the image.

Solution

The image has these characteristics:

behind the mirror
upright (same orientation as object)
enlarged

Insight

Makeup mirrors are concave mirrors with fairly large focal lengths. The person applying makeup is between the mirror and its focal point, just like the object in this Guided Example. Therefore, the image of the person's face is upright and enlarged, as desired.

17. Follow-up If the object in Guided Example 16.6 is moved closer to the concave mirror, does the image increase or decrease in size?

18. Follow-up If the object in Guided Example 16.6 is moved closer to the focal point of the concave mirror, does the image increase or decrease in size?

The imaging characteristics of convex and concave mirrors are summarized in **Table 16.1**.

Table 16.1 Imaging Characteristics of Convex and Concave Spherical Mirrors

	Object Location	Image Orientation	Image Size	Image Type
CONVEX MIRROR				
	arbitrary	upright	reduced	virtual
CONCAVE MIRROR				
	beyond C	inverted	reduced	real
	C	inverted	same as object	real
	between F and C	inverted	enlarged	real
	just beyond F	inverted	approaching infinity	real
	just inside F	upright	approaching infinity	virtual
	between mirror and F	upright	enlarged	virtual

CONCEPTUAL Example 16.7 Concave or Convex?

The passenger-side rear-view mirrors in newer cars often have a warning label that reads "OBJECTS IN THE MIRROR ARE CLOSER THAN THEY APPEAR." Are these rear-view mirrors concave or convex?

Reasoning and Discussion
Objects in such a rear-view mirror are closer than they appear because the mirror produces an image that is reduced in size, which makes the object look as if it is farther away. In addition, we know that the rear-view mirror always gives an upright image, no matter how close or far away the object. The type of mirror that always produces upright and reduced images is a convex mirror.

Answer
The rear-view mirrors are convex.

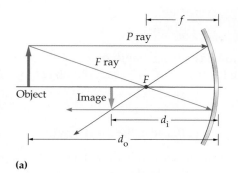

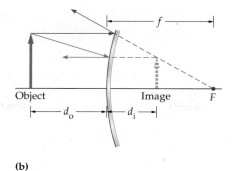

◀ **Figure 16.27 Applying the mirror equation**
Examples of object distance, d_o, image distance, d_i, and focal length, f, for **(a)** a concave mirror and **(b)** a convex mirror. The image distance and focal length are positive for the concave mirror. The image distance and focal length are negative for the convex mirror.

The Mirror Equation

While ray tracing is very useful, images can be located more precisely with an equation. 🔑 The *mirror equation* **is a precise mathematical relationship between object distance, image distance, and focal length for a given mirror.**

For example, suppose an object is a distance d_o from a mirror, as in **Figure 16.27**. The image is a distance d_i from the mirror, and the focal point is a distance f from the mirror. These three quantities, d_o, d_i, and f, are related by the following equation:

🔑 *What quantities are related by the mirror equation?*

> **The Mirror Equation**
>
> $$\frac{1}{\text{object distance}} + \frac{1}{\text{image distance}} = \frac{1}{\text{focal length}}$$
>
> $$\frac{1}{d_o} + \frac{1}{d_i} = \frac{1}{f}$$

If any two of these quantities are known, the mirror equation yields the third.

Correct signs must be used in the mirror equation

It is important to identify and use the correct sign for each term in the mirror equation. First, recall that f is positive for concave mirrors and negative for convex mirrors. Next, note that the image distance d_i is positive for images in front of a mirror (real images) and negative for images behind a mirror (virtual images). Finally, the object distance d_o is positive for all real objects (the only situation considered in this book). These simple rules are summarized below:

> **Focal Length**
>
> f is positive for concave mirrors.
>
> f is negative for convex mirrors.
>
> **Image Distance**
>
> d_i is *positive* for images *in front* of a mirror (real images).
>
> d_i is *negative* for images *behind* a mirror (virtual images).
>
> **Object Distance**
>
> d_o is *positive* for real objects.

Try applying these rules to Figure 16.27 (a) and (b). Do you see that f and d_i are negative in Figure 16.27 (b)?

PROBLEM-SOLVING NOTE

The situation in Quick Example 16.8 corresponds to Figure 16.26 (a). Similarly, part (a) of Problem 19 corresponds to Figure 16.26 (b), and part (b) of Problem 19 corresponds to Guided Example 16.6.

QUICK Example 16.8 Where's the Image?

The concave side of a shiny metal spoon has a focal length of $f = 5.00$ cm. Find the image distance for this "mirror" when the object distance is 25.0 cm.

Solution
Start with the mirror equation $\frac{1}{d_o} + \frac{1}{d_i} = \frac{1}{f}$. Because the focal length and the object distance are known, the only unknown is the image distance. Rearrange the mirror equation to solve for d_i:

$$\frac{1}{d_i} = \frac{1}{f} - \frac{1}{d_o}$$

$$\frac{1}{d_i} = \frac{1}{5.00 \text{ cm}} - \frac{1}{25.0 \text{ cm}}$$

$$\frac{1}{d_i} = 0.200 \text{ cm}^{-1} - 0.0400 \text{ cm}^{-1}$$

$$\frac{1}{d_i} = 0.160 \text{ cm}^{-1}$$

$$d_i = \frac{1}{0.160 \text{ cm}^{-1}} = \boxed{6.25 \text{ cm}}$$

Math HELP
Solving Simple Equations
See Math Review, Section III

Practice Problems

19. Follow-up Find the image distance in Quick Example 16.8 when the object distance is (**a**) 9.00 cm and (**b**) 2.00 cm.

20. A concave mirror produces an image that is 4.1 cm behind the mirror when the object is 1.8 cm in front of the mirror. What is the focal length of the mirror?

21. A concave mirror with a focal length of 4.8 cm produces an image 8.9 cm in front of the mirror. What is the object distance?

The mirror equation applies equally well to a *convex* mirror, as long as we recall that the focal length in this case is *negative*.

QUICK Example 16.9 Where's the Image?

The convex mirror in Figure 16.24 has a 20.0-cm radius of curvature. Find the image distance for this mirror when the object distance is 6.33 cm.

Solution
Recall that $f = -\frac{1}{2}R$ for a convex mirror. Therefore, $f = -10.0$ cm and $d_o = 6.33$ cm in this case. The image distance is calculated as follows:

$$\frac{1}{d_o} + \frac{1}{d_i} = \frac{1}{f}$$

$$\frac{1}{d_i} = \frac{1}{-10.00 \text{ cm}} - \frac{1}{6.33 \text{ cm}}$$

$$d_i = \boxed{-3.88 \text{ cm}}$$

Math HELP
Solving Simple Equations
See Math Review, Section III

Practice Problems

22. A convex mirror produces an image that is 3.7 cm behind the mirror when the object is 5.9 cm in front of the mirror. What is the focal length of the mirror?

23. A convex mirror with a focal length of -9.6 cm produces an image 4.5 cm behind the mirror. What is the object distance?

An image can be larger or smaller than the object

Curved mirrors typically produce images that are either larger or smaller than the objects. For example, **Figure 16.28** shows an image that is reduced in size. In general, the ratio of the height of the image, h_i, to the height of the object, h_o, is defined as the **magnification**, m. The magnification can be determined in terms of the object and image distances as follows:

Magnification, m

$$\text{magnification} = \frac{\text{image height}}{\text{object height}} = -\left(\frac{\text{image distance}}{\text{object distance}}\right)$$

$$m = \frac{h_i}{h_o} = -\frac{d_i}{d_o}$$

The sign of the magnification tells whether the image is upright or inverted:

Sign Convention for Magnification

m is positive for upright images.

m is negative for inverted images.

For example, if both d_o and d_i are positive, as in Figure 16.27 (a), the magnification is negative, and the image is inverted. Conversely, if the image is behind the mirror, we know that d_i is negative. The magnification is positive in this case, and the image is upright, as we see in Guided Example 16.6 and Figure 16.27 (b).

What about the magnitude of the magnification? Well, the magnitude gives the factor by which the size of the image is increased or decreased compared with the object. For example, if the magnification is $m = 0.5$, the image is upright and one-half the size of the object. If the magnification is $m = -3$, the image is inverted and three times the size of the object.

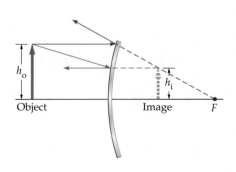

▶ **Figure 16.28 Magnification of a curved mirror**
The magnification of a mirror is the ratio of the image height, h_i, to the object height, h_o. The magnification in this case is positive. A negative magnification means that the image is inverted.

After leaving some presents under the tree, Santa notices his image in a shiny, spherical Christmas ornament. The ornament is 8.50 cm in diameter and 1.10 m away from Santa. Curious to know (**a**) the location and (**b**) the size of his image, Santa consults a book on physics. Knowing that Santa likes to "check it twice," what results should he obtain, assuming that his height is 1.75 m?

Picture the Problem

The physical situation is illustrated in our sketch, which also shows the image Santa sees in the ornament. Because the spherical ornament is a convex mirror, the image it forms is upright and reduced.

Strategy

The ornament is a convex mirror of radius $R = \frac{1}{2}(8.50 \text{ cm}) = 4.25 \text{ cm} = 0.0425 \text{ m}$. We can find the location of the image by using the mirror equation with the focal length given by $f = -\frac{1}{2}R$. Once we've determined the location of the image, we can find its magnification and size using the equation $m = h_i/h_o = -(d_i/d_o)$.

Known	Unknown
$R = (8.50 \text{ cm})/2$	$d_i = ?$
$\quad = 4.25 \text{ cm} = 0.0425 \text{ m}$	$h_i = ?$
$d_o = 1.10 \text{ m}$	
$h_o = 1.75 \text{ m}$	

Solution

Part (a)

1   Calculate the focal length for the ornament (remembering to use a negative sign since the surface of the ornament is a convex mirror):

$$f = -\tfrac{1}{2}R$$
$$= -\tfrac{1}{2}(0.0425 \text{ m})$$
$$= -0.0213 \text{ m}$$

> **Math HELP**
> Significant Figures
> See
> Lesson 1.4

2   Rearrange the mirror equation to find the image distance, d_i:

$$\frac{1}{d_i} = \frac{1}{f} - \frac{1}{d_o}$$
$$= \frac{1}{(-0.0213 \text{ m})} - \frac{1}{1.10 \text{ m}}$$
$$= -47.9 \text{ m}^{-1}$$
$$d_i = \frac{1}{(-47.9 \text{ m}^{-1})}$$
$$= -0.0209 \text{ m}$$
$$= \boxed{-2.09 \text{ cm}}$$

Part (b)

3 Determine the magnification of the image using $m = -(d_i/d_o)$:

$$m = -\left(\frac{d_i}{d_o}\right)$$

$$= -\left(\frac{-0.0209 \text{ m}}{1.10 \text{ m}}\right)$$

$$= 0.0190$$

4 Determine the height of the image by rearranging $m = h_i/h_o$:

$$m = \frac{h_i}{h_o}$$

$$h_i = mh_o$$

$$= (0.0190)(1.75 \text{ m})$$

$$= \boxed{3.33 \text{ cm}}$$

Insight

Santa's image is 2.09 cm behind the surface of the ornament—about halfway between the surface and the center. The height of his image is 3.33 cm. Thus, Santa's image fits on the surface of the ornament with room to spare.

Practice Problems

24. [Follow-up] How far from the ornament must Santa stand if his image is to be 1.75 cm behind its surface? What is the height of his image in this case?

25. What is the focal length of a spherical mirror if the image distance is –3.1 cm when the object distance is 1.5 m? What is the magnification in this case?

ACTIVE Example 16.11 Determine the Magnification and the Focal Length

A dentist uses a small mirror attached to a thin rod to examine one of your teeth. When the tooth is 1.20 cm in front of the mirror, the image it forms is 9.25 cm behind the mirror. Find **(a)** the magnification of the image and **(b)** the focal length of the mirror.

Solution *(Perform the calculations indicated in each step.)*

1. From the information given, identify the object and image distances:

$d_o = 1.20 \text{ cm}, d_i = -9.25 \text{ cm}$

2. For part (a), use $m = -(d_i/d_o)$ to find the magnification:

$m = 7.71$

3. For part (b), use the mirror equation to calculate the focal length:

$f = 1.38 \text{ cm}$

Insight

Because the focal length is positive and the magnification is greater than 1, we conclude that the dentist's mirror is concave. It will make the tooth look 7.71 times larger than life. A convex mirror, in contrast, always produces a reduced image with a magnification less than 1.

16.3 LessonCheck (MP)

Checking Concepts

26. **Explain** Where is the focal point of a concave mirror whose radius of curvature is R?

27. **Explain** Where is the focal point of a convex mirror whose radius of curvature is R?

28. **Identify** To use the mirror equation you must know two of the three quantities that the equation relates. List the three quantities.

29. **Compare** What are the key differences between real images and virtual images?

30. **Big Idea** Describe situations in which mirrors produce real images and situations in which they produce virtual images.

31. Apply Many homes have dish receivers for satellite TV. Which surface of the dish is pointed toward the satellite, the concave surface or the convex surface? Explain.

32. Apply You hold a shiny tablespoon at arm's length and look at the convex back surface.
(a) Is the image you see of yourself upright or inverted?
(b) Is the image enlarged or reduced?
(c) Is the image real or virtual?

Solving Problems

33. Calculate A section of a spherical shell has a radius of curvature of 0.86 m. If this section is painted with a reflective coating on both sides, what is the focal length of (a) the convex side and (b) the concave side?

34. Calculate Sunlight reflects from a concave mirror and converges to a point 15 cm from the mirror's surface. What is the radius of curvature of the mirror?

35. Diagram An object with a height of 42 cm is placed 2.0 m in front of a concave mirror with a focal length of 0.50 m.
(a) Determine the approximate location and size of the image using a ray diagram.
(b) Is the image upright or inverted?

36. Calculate Use the mirror equation and the magnification equation to find the precise location and magnification of the image produced by the mirror in Problem 35. What is the height of the image?

The Hubble Space Telescope (HST)

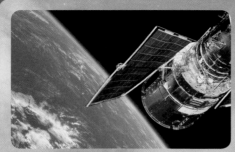

The HST is about the size of a large school bus and moves at approximately 28,000 km/h in an orbit 559 km above Earth's surface.

This image from the HST shows that thousands of stars and planets are forming in the Orion Nebula.

What Is It? The Hubble Space Telescope (HST) is a space-based, solar-powered telescope named for the astronomer Edwin Hubble. Two highly polished curved mirrors give it an extraordinary ability to capture images from the vastness of space. The curves of the mirrors are virtually flawless, deviating by only about 0.00003 mm from a perfect curve. The HST has extremely advanced optics and the advantage of operating in the vacuum of space. Ground-based telescopes are handicapped by having to view space through Earth's thick atmosphere, which is polluted by particles and ambient light.

How Does It Work? Light enters though the open end of the telescope, and the concave primary mirror reflects the light to the convex secondary mirror. That mirror focuses the light on a focal point behind the primary mirror. Instruments digitize this information and send it to Earth, where it is converted into images.

What Has Been Discovered? The HST began its career in 1990 and has been unveiling stunning secrets of the universe and inspiring our imaginations ever since. The telescope has shown us a breathtaking, beautiful, and diverse universe. It has documented stars and planets forming in the Orion nebula, a "nearby" gigantic cloud of dust, gas, and concentrated energy that is trillions of kilometers away. The telescope has also revealed countless planets orbiting distant stars. This discovery was a revelation to many astronomers, as it meant that there were a tremendous number of planets "out there." Thus, the HST may help us to answer an age-old question: Are we the only island of life in the universe? Even if extraterrestrial life is extremely uncommon, the sheer number of planets that the HST has revealed hints that we may not be alone. Perhaps alien civilizations using Hubble-like telescopes are making the same astonishing discovery.

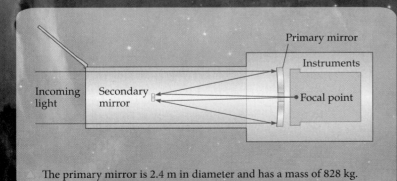

The primary mirror is 2.4 m in diameter and has a mass of 828 kg.

Take It Further

1. Explain *Why does the HST use curved mirrors instead of plane mirrors?*

2. Reason *How could the HST be used in the search for extraterrestrial life?*

Physics Lab Focal Length of a Concave Mirror

This lab explores the focusing characteristics of a concave mirror. You will use two different techniques to determine the focal length of a concave mirror.

Materials
- concave mirror
- index card
- meterstick
- light source (lightbulb or candle)

Procedure

Read the entire procedure and prepare a data table.

Part I: Focusing a Distant Object
A concave mirror should focus light rays from a very distant object at its focal point.

1. Place a light source as far away from the concave mirror as possible. Place the mirror so that it faces the light source.

2. Slowly move the index card away from the mirror until a clear image of the light source comes into focus on the card.

3. Measure the distance from the mirror to the card, to the nearest 0.1 cm. This is the focal length. Record the value in a data table.

4. Repeat Steps 1–3 two more times, using other members of the group.

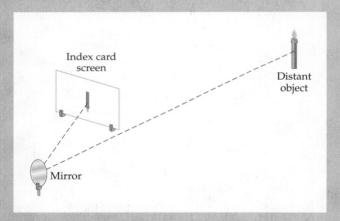

Part II: Using the Mirror Equation
A concave mirror should obey the mirror equation, $1/d_o + 1/d_i = 1/f$. By placing the object at various distances, d_o, and locating the image at d_i, the focal length, f, of the mirror can be calculated.

Important: *For each trial in Part II you must place the light source (the object) at a distance from the mirror that is greater than the average focal length determined in Part I—see Step 5 below.*

5. Compute the average focal length obtained in Part I, and record it in your data table.

6. Add about 10 cm to the average focal length calculated in Step 5 (you want to obtain a whole number, such as 45 cm), and place the mirror and the light source (the object) this distance apart. Record the object distance, d_o, in your data table.

7. Move the index card back and forth until a clear image of the light source comes into focus on the card, as shown. Measure the image distance, d_i, and record it in your data table.

8. Repeat Steps 6–7 three more times, adding an additional 10 cm to the object distance each time.

Analysis

1. Use the mirror equation to compute the focal length for each trial in Part II. Record the values in your data table.

2. Compute the average focal length for Part II, and record the value in your data table.

3. Obtain the focal length of the mirror from your teacher. Compute the percent error in the focal lengths you found in Parts I and II.

4. Use the mirror equation to predict the image location when the object is located at $0.75f$. How will this image differ from those produced in this lab?

Conclusions

1. Which method was better at determining the focal length of your mirror? Why was this method better?

2. What changes could you make to this experiment to reduce the sources of error in the less accurate method?

3. All the images in this lab were real images. What characteristics did they have in common?

16 Study Guide

Big Idea

Mirrors are particularly good at reflecting light; a mirror's shape determines the size, location, and orientation of the reflected image.

The image produced by a plane mirror is the same size as the object and cannot be projected onto a screen. Curved mirrors can produce enlarged, reduced, or same-size images, some of which can be projected onto a screen. All mirrors, whether flat or curved, reflect light according to the law of reflection: The angle of reflection is equal to the angle of incidence.

16.1 The Reflection of Light

🔑 Light rays are always at right angles to the wave fronts.

🔑 The relationship between the angle of incidence and the angle of reflection is very simple—they are equal.

🔑 Specular reflection is responsible for the sharp, clear images seen in mirrors.

• Smooth surfaces produce specular reflection; rough surfaces produce diffuse reflection.

Key Equation

The law of reflection states that the angle of reflection, θ_r, is equal to the angle of incidence, θ_i:

$$\theta_r = \theta_i$$

16.2 Plane Mirrors

🔑 The image formed by a plane mirror appears as far behind the mirror as the object is in front of the mirror.

• The image formed by a plane mirror has the following characteristics:

The image appears to be the same distance behind the mirror as the object is in front of the mirror.

The image is the same size as the object.

The image is upright, but reversed right to left.

• An image that is formed behind a mirror is a virtual image.

• Virtual images are formed by diverging light rays. Because no light rays pass through a virtual image, it cannot be projected onto a screen.

• All images formed by plane mirrors are virtual images.

16.3 Curved Mirrors

🔑 With a concave mirror, incoming rays of light that are parallel to the principal axis are reflected through the focal point.

🔑 Incoming rays of light that are parallel to the principal axis of a convex mirror spread outward when they are reflected—just as if they had started from the focal point behind the mirror.

🔑 A real image (with light rays passing through it) can be projected onto a screen; a virtual image (with no light rays passing through it) cannot.

🔑 The *mirror equation* is a precise mathematical relationship between object distance, image distance, and focal length for a given mirror.

• Ray tracing involves drawing the paths of rays of light as they reflect from a mirror and using them to find the location of the image.

• A spherical mirror has a reflecting surface that is a section of a spherical shell. A convex spherical mirror has a reflecting surface that bulges outward. A concave spherical mirror has a hollowed-out reflecting surface.

• The focal length is positive for a concave mirror and negative for a convex mirror. Similarly, the image distance is positive for an image in front of the mirror and negative for an image behind the mirror.

Key Equations

The focal length for a concave mirror with a radius of curvature R is

$$f = \tfrac{1}{2}R$$

The focal length of a convex mirror is

$$f = -\tfrac{1}{2}R$$

The mirror equation relates the object distance, d_o, image distance, d_i, and focal length, f, as follows:

$$\frac{1}{d_o} + \frac{1}{d_i} = \frac{1}{f}$$

The magnification of an image is given by

$$m = \frac{h_i}{h_o} = -\frac{d_i}{d_o}$$

For instructor-assigned homework, go to www.masteringphysics.com

Lesson by Lesson

16.1 The Reflection of Light

Conceptual Questions

37. Triple Choice Two cases where a beam of light reflects from a mirror are shown in **Figure 16.29**. Is the angle of incidence in case A greater than, less than, or equal to the angle of incidence in case B? Explain.

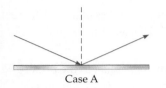
Case A

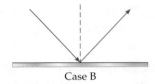
Case B

Figure 16.29

38. Triple Choice If the angle of incidence of a beam of light is increased, does the angle of reflection increase, decrease, or stay the same? Explain.

39. A beam of light is incident on a mirror. If the angle of incidence of this beam is increased, does the beam shift closer to the normal or farther from the normal? Explain.

40. What is the largest possible angle of incidence? Explain.

41. What is the smallest possible angle of incidence? Explain.

Problem Solving

42. A laser beam is reflected by a plane mirror. It is observed that the angle between the incident and reflected beams is 28°. What is the angle of incidence?

43. A beam of light strikes a mirror at an angle of 23° to the normal. What is the angle between the incident and reflected beams?

44. A ray of light reflects from a plane mirror with an angle of incidence of 37°. What is the angle between the plane of the mirror and the reflected beam?

45. How many times does the light beam shown in **Figure 16.30** reflect from **(a)** the top and **(b)** the bottom mirror?

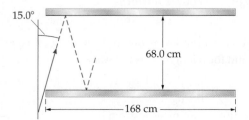
15.0°
68.0 cm
168 cm

Figure 16.30

46. Think & Calculate Standing 2.0 m in front of a very small mirror, you see the reflection of your belt buckle, which is 0.70 m below your eyes. **(a)** What is the vertical location of the mirror relative to the level of your eyes? **(b)** What angle do your eyes make with the horizontal when you look at the buckle? **(c)** If you now move backward until you are 6.0 m from the mirror, will you still see the buckle, or will you see a point on your body that is above or below the buckle? Explain.

47. Sunlight enters a room at an angle of 32° above the horizontal and reflects from a small mirror lying flat on the floor. The reflected light forms a spot on a wall that is 2.0 m behind the mirror, as shown in **Figure 16.31**. If you place a pencil under the edge of the mirror nearer the wall, tilting it upward by 5.0°, how much higher on the wall (Δy) will the spot of light appear?

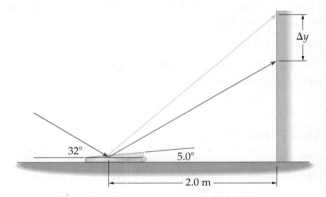
Δy
32°
5.0°
2.0 m

Figure 16.31

16.2 Plane Mirrors

Conceptual Questions

48. Do plane mirrors form real or virtual images?

49. If you stand 1 m in front of a plane mirror, what is the distance between you and your mirror image?

50. If you view a clock in a mirror, as shown in **Figure 16.32**, do the mirror-image hands rotate clockwise or counterclockwise?

Figure 16.32

51. Two plane mirrors meet at right angles at the origin of a coordinate system, as indicated in **Figure 16.33**. **(a)** Draw the locations and orientations of *all* three images of L-shaped object A formed by the two mirrors. **(b)** Draw the locations and orientations of *all* three images of L-shaped object B formed by the two mirrors. (*Hint:* One of the three images in each case is due to reflections from both mirrors.)

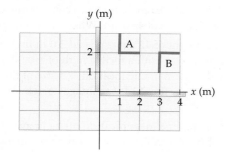

Figure 16.33

Problem Solving

52. How rapidly does the distance between you and your mirror image decrease if you walk directly toward a mirror with a speed of 2.6 m/s?

53. You stand 3.5 m in front of a large mirror, and your little sister stands 2.0 m directly in front of you. At what distance should you focus your camera if you want to take a picture of your sister in the mirror?

54. The rear window in a car is approximately a rectangle, 1.3 m wide and 0.30 m high. The inside rear-view mirror is 0.50 m from the driver's eyes and 1.50 m from the rear window. What are the minimum dimensions that the rear-view mirror should have if the driver is to be able to see the entire width and height of the rear window in the mirror without moving her head?

55. You hold a small plane mirror 0.50 m in front of your eyes, as shown in **Figure 16.34**. The mirror is 0.32 m high, and in it you see the image of a tall building behind you. If the building is 95 m behind you, what vertical height of the building, *H*, can be seen in the mirror at any one time?

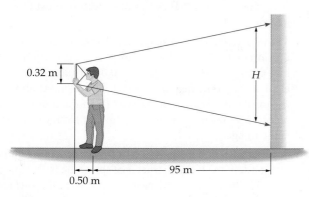

Figure 16.34

56. Two rays of light converge toward each other, as shown in **Figure 16.35**, forming an angle of 27°. Before they intersect, however, they are reflected from a plane mirror with a width of 11 cm. If the mirror can be moved horizontally to the left or right, what is the greatest possible distance *d* from the mirror to the point where the reflected rays meet?

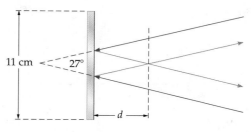

Figure 16.35

16.3 Curved Mirrors
Conceptual Questions

57. Explain the difference between a positive and negative image distance.

58. Explain the difference between a positive and negative magnification.

59. What kind of mirror can produce a real image?

60. Suppose you would like to start a fire by focusing sunlight onto a piece of paper. Conceptual Example 16.5 showed that a concave mirror would work better than a convex mirror for this purpose. At what distance from the mirror should the paper be held for best results?

61. Astronomers often use large mirrors in their telescopes to gather as much light as possible from faint, distant objects. Should the mirror in such a telescope be concave or convex? Explain.

62. You hold a shiny tablespoon at arm's length and look at the concave front side of the spoon. **(a)** Is the image you see of yourself upright or inverted? **(b)** Is the image enlarged or reduced? **(c)** Is the image real or virtual?

63. An object is placed in front of a convex mirror whose radius of curvature is *R*. What is the greatest distance behind the mirror that the image can be located?

64. An object is placed to the left of a convex mirror. In which direction will the image move (left or right) when the object is moved farther to the left?

65. Which of the four points (1, 2, 3, or 4) in **Figure 16.36** best represents the location of **(a)** the focal point, *F*, and **(b)** the center of curvature, *C*, of the concave mirror?

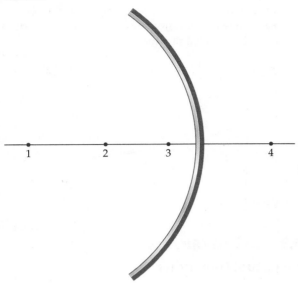

Figure 16.36

66. An object (red arrow) is placed in front of a convex spherical mirror, as shown in **Figure 16.37**. Which of the four green arrows (1, 2, 3, or 4) best represents the image formed by the mirror?

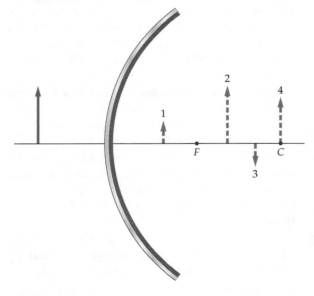

Figure 16.37

Problem Solving

67. A reflecting globe in a garden is 31.5 cm in diameter. What is the focal length of the globe?

68. A concave mirror has a focal length of 12 cm. What is its radius of curvature?

69. A convex mirror has a radius of curvature of 28 cm. What is its focal length?

70. When an object is placed 33 cm in front of a convex mirror, it produces a virtual image 11 cm from the mirror. What is the magnification of the image? Is the image upright or inverted?

71. When an object is placed 24 cm in front of a concave mirror, it produces a real image 12 cm from the mirror. What is the magnification of the image? Is the image upright or inverted?

72. A small object is located 30.0 cm in front of a concave mirror with a radius of curvature of 40.0 cm. Where is the image?

73. An object with a height of 42 cm is placed 2.0 m in front of a convex mirror with a focal length of −0.50 m. **(a)** Determine the approximate location and size of the image using a ray diagram. **(b)** Is the image upright or inverted?

74. Find the location and the magnification of the image produced by the mirror in Problem 73 using the mirror equation and the magnification equation.

75. During a daytime football game you notice that a player's reflective helmet forms an image of the Sun 4.8 cm behind the surface of the helmet. What is the radius of curvature of the helmet, assuming it to be roughly spherical?

76. **Figure 16.38** is a graph of image distance versus object distance for a mirror. **(a)** Is the mirror convex, plane, or concave? Explain. **(b)** What is the focal length of the mirror?

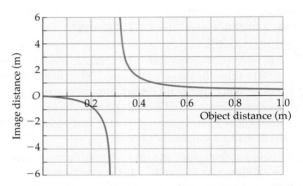

Figure 16.38

77. It is desired that a convex mirror produce a magnification of 0.25 when an object is placed 32 cm from the mirror. What radius of curvature must the mirror have?

78. Think & Calculate A magician wishes to create the illusion of a 2.74-m-tall elephant. He plans to do this by forming a virtual image of a 50.0-cm-tall model elephant with the help of a spherical mirror. **(a)** Should the mirror be concave or convex? **(b)** If the model is placed 3.00 m from the mirror, how far from the mirror is the image?

79. A person 1.6 m tall stands 3.6 m from a reflecting globe in a garden. **(a)** If the diameter of the globe is 18 cm, where is the image of the person, relative to the surface of the globe? **(b)** How large is the person's image?

80. Shaving and makeup mirrors typically have one flat and one concave (magnifying) surface. You find that you can project a magnified image of a lightbulb onto the wall of your bathroom if you hold such a mirror 1.8 m from the bulb and 3.5 m from the wall. **(a)** What is the magnification of the image? **(b)** Is the image upright or inverted? **(c)** What is the focal length of the mirror?

81. Hale Telescope The 5.08-m-diameter concave mirror of the Hale Telescope on Mount Palomar has a focal length of 16.9 m. An astronomer stands 20.0 m in front of this mirror. **(a)** Where is her image located? Is it in front of or behind the mirror? **(b)** Is her image real or virtual? How do you know? **(c)** What is the magnification of her image?

82. A concave mirror produces a virtual image that is three times as tall as the object. **(a)** If the object is 22 cm in front of the mirror, what is the image distance? **(b)** What is the focal length of this mirror?

83. You view a nearby tree in a concave mirror. The inverted image of the tree is 3.5 cm high and is located 7.0 cm in front of the mirror. If the tree is 21 m from the mirror, what is its height?

84. A shaving mirror produces an upright image that is magnified by a factor of 2.0 when your face is 25 cm from the mirror. What is the mirror's radius of curvature?

Mixed Review

85. A beam of light is incident on a mirror along the normal. What is the angle of reflection? Explain.

86. An object is placed to the left of a concave mirror, beyond the focal point. In which direction (left or right) will the image move when the object is moved farther to the left?

87. Use ray diagrams to show whether the image formed by a convex mirror increases or decreases in size as an object is brought closer to the mirror's surface.

88. Figure 16.39 is a graph of image distance versus object distance for a mirror. Is the mirror concave, plane, or convex? Explain.

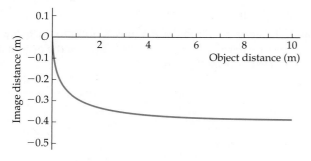

Figure 16.39

89. An object (red arrow) is placed in front of a concave spherical mirror, as shown in **Figure 16.40**. Which of the four dashed green arrows (1, 2, 3, or 4) best represents the image formed by the mirror?

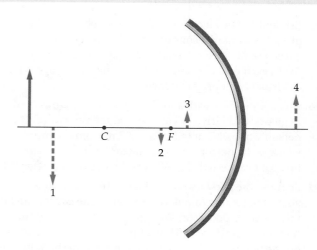

Figure 16.40

90. A boy sees his reflection in a reflecting globe 0.55 m away in a garden. If the reflecting globe has a diameter of 45 cm and the boy is 1.7 m tall, what is the height of his image in the globe?

91. An object is at the center of curvature of a concave mirror. **(a)** Is the resulting image real or virtual? **(b)** What is the location of the image? **(c)** Is the image upright or inverted? **(d)** What is the magnification of the image?

92. Show that the mirror equation for a curved mirror reduces to the mirror equation for a plane mirror ($d_i = -d_o$) when the focal length becomes infinite. (This makes sense, because the surface of a sphere with a large radius of curvature appears almost flat, like a plane mirror.)

93. An object is placed in front of a concave mirror, as shown in **Figure 16.41**. Find the location and height of the image.

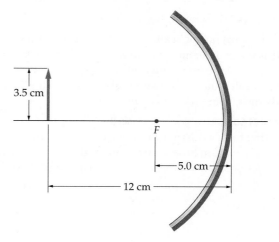

Figure 16.41

94. ⎡**Think & Calculate**⎤ Suppose the separation between the two mirrors in Figure 16.30 is increased by moving the top mirror upward. **(a)** Will this change affect the number of reflections made by the beam of light? If so, how? **(b)** What is the total number of reflections made by the beam of light when the separation between the mirrors is 125 cm?

95. (a) Find the two locations at which an object can be placed in front of a concave mirror with a radius of curvature of 38 cm and have its image be twice its size. **(b)** In each of these cases, state whether the image is real or virtual and upright or inverted.

96. A convex mirror with a focal length of −75 cm is used to give a truck driver a view behind the vehicle. **(a)** If a person who is 1.7 m tall stands 2.2 m from the mirror, where is the person's image located? **(b)** Is the image upright or inverted? **(c)** What is the size of the image?

97. Suppose the person's eyes in Figure 16.34 are 1.6 m above the ground and that the small plane mirror can be moved up or down. **(a)** Find the height of the bottom of the mirror at which the lowest point the person can see on the building is 19.6 m above the ground. **(b)** With the mirror held at the height found in part (a), what is the highest point on the building the person can see?

98. A concave mirror produces a real image that is three times as large as the object. **(a)** If the object is 22 cm in front of the mirror, what is the image distance? **(b)** What is the focal length of this mirror?

99. The virtual image produced by a convex mirror is one-quarter the size of the object. **(a)** If the object is 32 cm in front of the mirror, what is the image distance? **(b)** What is the focal length of this mirror?

100. **Think & Calculate** A 1.7-m-tall shopper is 5.2 m from a convex security mirror in a department store. The shopper notices that his image in the mirror appears to be only 16.25 cm tall. **(a)** Is the shopper's image upright or inverted? Explain. **(b)** What is the mirror's radius of curvature?

Writing about Science

101. Write a report on applications of the heads-up display. Find as many different applications as you can, including airplanes, automobiles, and captioning systems in a movie theater. Include a diagram showing how the heads-up system produces an image.

102. **Connect to the** Big Idea Describe how light reflecting from a mirror can produce an image. In particular, explain how mirrors can produce images that are larger or smaller than life size, as well as upright or inverted.

Read, Reason, and Respond

A Sizzling Hotel Las Vegas is known for glamorous attractions and sizzling entertainment. The Vdara Hotel, shown in **Figure 16.42**, seems to have taken this idea a bit too literally. The front of the hotel is covered in glass, making it a giant mirror. Light from the sun reflected onto the pool area has been reported to be hot enough to melt plastic.

Figure 16.42 Vdara Hotel in Las Vegas.

103. Based on the physics of mirrors, what is the likely shape of the front of the hotel?

A. convex   C. plane

B. concave   D. none of the above

104. If the front of the hotel curves with a radius of 160 m, at what distance from the hotel do you expect the greatest heating due to focused sunlight?

A. 40 m   C. 160 m

B. 80 m   D. 320 m

105. A person stands in front of the hotel, 25 m from its reflecting surface. What is the image distance for this person? (The radius of the front of the hotel is 160 m.)

A. −36 m   B. 25 m

B. −19 m   D. 80 m

106. If the person in Problem 105 moves closer to the hotel, does the person's image move (A) closer to the person or (B) farther from the person?

Standardized Test Prep

Multiple Choice

Use the graph below to answer Questions 1–4. The graph plots the image distance versus the object distance for a concave mirror.

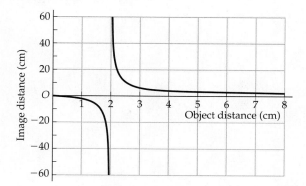

1. What is the focal length of the mirror?
   (A) zero
   (B) 2 cm
   (C) 4 cm
   (D) more than 8 cm

2. Which set of phrases describes the image distance and the type of image formed for an object placed 4 cm from the mirror?
   (A) about 4 cm, real image
   (B) about 4 cm, virtual image
   (C) about 10 cm, real image
   (D) about 10 cm, real image

3. At which of the following object distances is the image formed by the mirror a real image?
   (A) 0.5 cm
   (B) 1.0 cm
   (C) 1.5 cm
   (D) 2.5 cm

4. At which of the following object distances is the image formed by the mirror an upright image?
   (A) 1.5 cm
   (B) 2.5 cm
   (C) 3.5 cm
   (D) 4.5 cm

5. Which principle or concept explains why a surface might appear dull when illuminated by a beam of light?
   (A) law of reflection
   (B) diffuse reflection
   (C) specular reflection
   (D) paraxial reflection

6. Describe the image formed when an object is placed 10 cm in front of a convex mirror that has a radius of curvature of 20 cm.
   (A) virtual, upright, twice as large as the object
   (B) real, inverted, twice as large as the object
   (C) virtual, upright, half as large as the object
   (D) real, inverted, half as large as the object

7. Which is the *best* description of the image formed by a plane mirror?
   (A) real image, upright
   (B) real image, magnification of 1
   (C) virtual image, magnification of 1
   (D) real image, inverted left to right

8. An object is placed 3 cm in front of a concave mirror with a focal length of 6 cm. Which set of phrases describes the type of image formed and its magnification?
   (A) virtual image, twice as large as object
   (B) virtual image, half as large as object
   (C) real image, twice as large as object
   (D) real image, half as large as object

Extended Response

9. You are asked to determine the focal length and radius of curvature of a convex mirrored surface. Describe the procedure you would use to accomplish this and the expected results.

> **Test-Taking Hint**
>
> Convex, or diverging, mirrors produce virtual images. Concave, or converging, mirrors produce real images only when the object is outside the focal length.

If You Had Difficulty With . . .

Question	1	2	3	4	5	6	7	8	9			
See Lesson	16.3	16.3	16.3	16.3	16.1	16.3	16.2	16.3	16.3			

17

Refraction and Lenses

Inside

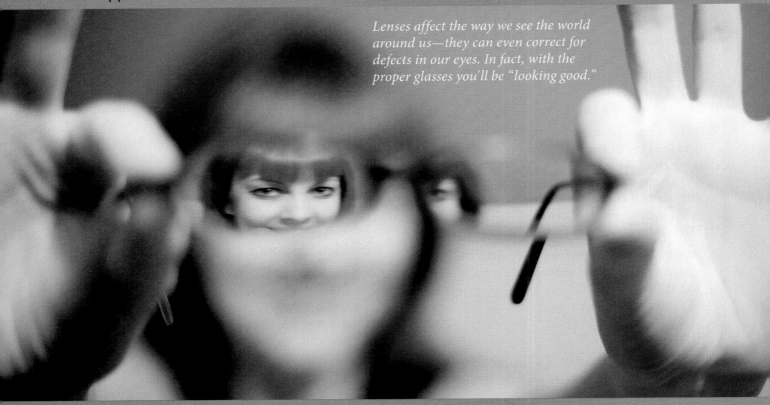

Lenses affect the way we see the world around us—they can even correct for defects in our eyes. In fact, with the proper glasses you'll be "looking good."

Big Idea

Lenses take advantage of refraction to bend light and form images.

When you look through a pair of binoculars, you see enlarged images of the world around you. These images are formed by lenses in the binoculars that bend rays of light. By simply changing the direction of light rays, lenses can create images of various sizes and orientations. In this chapter you'll learn how lenses bend light and how they produce images.

17.1 Refraction

Have you ever noticed that you can walk faster on a concrete sidewalk than on a sandy beach? Light behaves in the same way—it travels quickly through some materials and more slowly through others.

The speed of light is different in different materials

As you know, the speed of light in a vacuum is $c = 3.00 \times 10^8$ m/s. This is as fast as light (or anything else) can go. A vacuum is the easiest "material" for light to travel through—there's nothing to slow it down. When light travels through a dense material like water, however, its speed is reduced. In fact, **the speed of light through any material is slower than its speed in a vacuum.** Water, or any other material, is "harder" for light to travel through. It's just like a muddy field being harder to walk across than a paved sidewalk, as illustrated in **Figure 17.1**.

Vocabulary

- index of refraction
- refraction
- mirage

How does the speed of light in a vacuum compare with its speed through a material?

(a)

(b)

◀ **Figure 17.1 Walking speed is different on different surfaces**
(**a**) A person on a paved surface makes swift progress, like light traveling through a vacuum. (**b**) A person makes slow progress on a muddy surface, like light traveling through water or glass.

CONNECTING IDEAS

This lesson introduces you to the concept of the index of refraction.

• The index of refraction also plays an important role in determining how waves behave when they reflect from a material with a different index of refraction than the material in which they are traveling—a topic covered in the next lesson.

Measurements show that the speed of light in water is smaller than the speed of light in a vacuum by a factor of 1.33:

$$\text{speed of light in water} = \frac{c}{1.33}$$

The **index of refraction** of a material is the factor by which it reduces the speed of light. Therefore, the index of refraction of water is 1.33.

In general, if the speed of light in a material is v, its index of refraction n is defined as follows:

Definition of the Index of Refraction, n

$$\text{speed of light in a material} = \frac{\text{speed of light in a vacuum}}{\text{index of refraction of the material}}$$

$$v = \frac{c}{n}$$

Notice that the larger the index of refraction, the smaller the speed of light. Values of the index of refraction for a variety of materials are given in **Table 17.1**.

Table 17.1 Index of Refraction for Common Substances

	Substance	Index of refraction, n
SOLIDS	Diamond	2.42
	Flint glass	1.66
	Crown glass	1.52
	Fused quartz (glass)	1.46
	Ice	1.31
LIQUIDS	Benzene	1.50
	Ethyl alcohol	1.36
	Water	1.33
GASES	Carbon dioxide	1.00045
	Air	1.000293

QUICK Example 17.1 What's the Travel Time?

How much time does it take for light to travel 2.50 m in water?

Solution
The time t needed to travel a distance d with a velocity v is given by $t = d/v$. The velocity of light in water is $v = c/n$, where $n = 1.33$. Therefore, the time required to cover a distance $d = 2.50$ m is

$$t = \frac{d}{v} = \frac{d}{(c/n)}$$

$$= \frac{2.50 \text{ m}}{\left(\dfrac{3.00 \times 10^8 \text{ m/s}}{1.33}\right)}$$

$$= \boxed{1.11 \times 10^{-8} \text{ s}}$$

Math HELP
Scientific Notation
See Math Review, Section I

1. Triple Choice Is the speed of light in diamond greater than, less than, or equal to the speed of light in water? Explain. (Refer to Table 17.1.)

2. In a certain material light travels a distance of 1.56 m in 6.76×10^{-9} s. What is the index of refraction of this material?

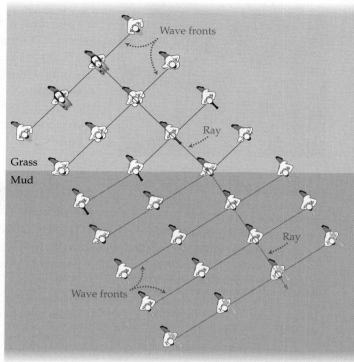

▲ **Figure 17.2 An analogy for refraction**
As a marching band moves from an area where the ground is solid to an area where it is soft and muddy, the direction of the band's motion changes.

Changing the speed of light can change its direction

How does a change in speed affect the direction of travel? To see the connection, consider a high-school marching band as it parades across a field. **Figure 17.2** shows an overhead view of the band. The band consists of several rows, each like a wave front in a beam of light. The direction of marching is indicated by a "ray" that is perpendicular to the "wave fronts."

Now suppose the band encounters a section of the field that is muddy. As band members on one end of each row encounter the mud, they slow down. The members still in the grass continue with their usual speed. This difference in speed causes the "wave fronts" to bend as they enter the muddy area. As a result, the band is traveling in a different direction in the mud. In general, a change in direction due to a change in speed is referred to as **refraction**. For light, refraction generally occurs when it passes from one material to another. The exception is when light is perpendicular to the boundary between two materials, as we will see later in this chapter.

Snell's Law To describe the new direction of travel, consider the simplified situation shown in **Figure 17.3**. Here we show just the "rays" for the band in the grass (material 1) and in the mud (material 2). We also show the normal to the boundary between the grass and the mud with a dashed line. The ray in material 1 makes an angle θ_1 with the normal. The ray in material 2 makes an angle θ_2 with the normal. If the index of refraction of material 1 is n_1 and the index of refraction of material 2 is n_2, the angles in these two materials are related by Snell's law:

Snell's Law	
in material 1	**in material 2**
index of refraction $\times$ sine of angle $=$	index of refraction $\times$ sine of angle
$n_1 \sin \theta_1 = n_2 \sin \theta_2$	

We refer to θ_1 as the *angle of incidence* and θ_2 as the *angle of refraction*.

▼ **Figure 17.3 Change in direction of travel**
A change in speed from one surface to another results in a change in direction of travel.

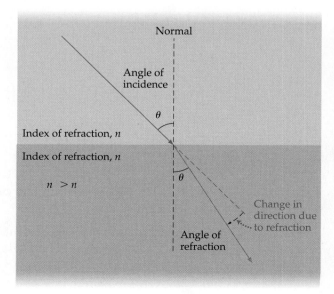

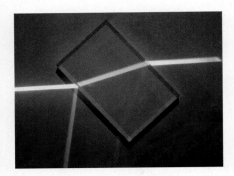

▲ Figure 17.4 Light reflects and refracts at an air-glass boundary
When a beam of light encounters the boundary between air and glass, part of it is reflected back into the air and part is transmitted (refracted) into the glass. The direction of the reflected beam is given by the law of reflection (Chapter 16), and the direction of the refracted beam is given by Snell's law.

In general, whenever light—or any wave—encounters the boundary between two different materials, some of the light is reflected and some is refracted (a small amount may also be absorbed). The reflected light travels in the same medium as the incoming light, though in a different direction—as discussed in Chapter 16. The refracted (transmitted) light is bent with respect to its original direction as it enters the new medium. The change in direction of the refracted light is directly related to its change in speed as it travels in the new medium. An example of light reflecting and refracting as it passes from air to glass is shown in **Figure 17.4**.

> **QUICK Example 17.2 What's the Angle of Refraction?**

The diagrams below show a beam of light passing from air into either **(a)** water or **(b)** diamond. Some of the light is *reflected* from the boundary between the two materials, with the angle of reflection equal to the angle of incidence, as expected. The remainder of the light is *refracted* as it enters the second material. Find the angle of refraction for the case where the beam of light enters the water, given that the angle of incidence is 60.0°.

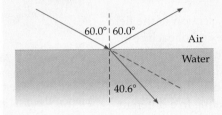

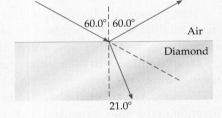

(a) Light passes from air into water.   **(b)** Light passes from air into diamond.

Solution
The beam of light starts in air. Referring to Table 17.1, we see that the corresponding index of refraction is $n_1 = 1.000293$, or simply $n_1 = 1.00$ to three significant figures. The angle of incidence is $\theta_1 = 60.0°$. With $n_2 = 1.33$ for water (Table 17.1), we find for case (a)

$$n_1 \sin \theta_1 = n_2 \sin \theta_2$$
$$\sin \theta_2 = \frac{n_1 \sin \theta_1}{n_2}$$
$$\theta_2 = \sin^{-1}\left(\frac{n_1 \sin \theta_1}{n_2}\right)$$
$$= \sin^{-1}\left(\frac{1.00}{1.33} \sin 60.0°\right)$$
$$= \boxed{40.6°}$$

> **CONNECTING** IDEAS
>
> **Trigonometric functions were introduced in Chapter 4 when we calculated vector components.**
> • Trigonometric functions are used extensively in this chapter to describe the behavior of refracted light.

Practice Problems

3. Follow-up What is the angle of refraction in case (b), where the light beam enters diamond at 60.0°?

4. Triple Choice Light traveling in air is incident on either ice (case A) or glass (case B) with an angle of 30°. Is the change in direction of light in case A greater than, less than, or equal to the change in direction in case B? Explain. (Refer to Table 17.1.)

5. The angle of refraction of a ray of light traveling through an ice cube is 31°. Find the angle of incidence, assuming that the incident ray was traveling in air.

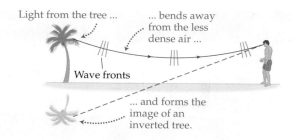

Light from the tree ...

... bends away from the less dense air ...

Wave fronts

... and forms the image of an inverted tree.

◀ **Figure 17.5 A mirage**
A mirage is produced when light bends upward due to the low index of refraction of heated air near the ground.

Refraction is responsible for mirages

A hiker in a hot, dry desert sometimes "sees" a glimmering pool of water in the distance. Often, this "pool of water" is nothing more than a *mirage*. A **mirage** is an optical illusion caused by the refraction of light. To see the connection between refraction and a mirage, consider **Figure 17.5**. Here we see a ray of light from the top of a tree heading toward the ground. Also shown are portions of the wave fronts associated with the ray.

Near the hot ground the air is less dense (more like a vacuum) than it is higher up. As a result, the speed of light is greater near the ground. Therefore, the bottom of a wave front moves farther in a given amount of time than the top of the wave front. This causes the wave fronts to rotate, changing the direction of the light rays. Eventually, the wave fronts rotate so much that the rays of light are heading upward, away from the ground.

The upward-moving light rays can then enter the eyes of an observer, as shown in Figure 17.5. To the observer, it looks like the light has come from the ground—exactly as if it had been reflected from a pool of water. A similar example of a mirage is shown in **Figure 17.6**, where a hot road surface appears to be covered with shimmering pools of water.

The index of refraction determines how light is bent

In general, light bends toward the normal when its speed decreases and away from the normal when its speed increases. 🔋 However, refraction only occurs if the light enters the new material at an angle *other than* 90°. In other words, light entering a new material at a right angle speeds up or slows down (depending on *n*), but does not change direction.

This is consistent with Snell's law, because the angle of incidence is zero for light entering a material at a right angle. The sine of zero is zero, and hence Snell's law implies that the angle of refraction must also be zero. With both angles equal to zero, both sides of Snell's law are zero and the equality is satisfied.

The basic features of refraction are summarized below.

- Light is bent *toward* the normal when it slows down because it has entered a material with a higher index of refraction.

- Light is bent *away* from the normal when it speeds up because it has entered a material with a lower index of refraction.

- The greater the change in the index of refraction, the greater the change in the direction of the light.

- If a ray of light goes from one material to another along the normal (perpendicular to the boundary between the materials), it does not change direction.

▲ **Figure 17.6 Refraction can make a dry road look wet**
One of the most common mirages, often seen in hot weather, makes a stretch of road look like the surface of a lake. The blue color that so resembles water to our eyes is actually an image of the sky, refracted by the hot, low-density air just above the road.

🔋 *What are the conditions required for refraction to occur?*

One night, while on vacation in the Caribbean, you walk to the end of a dock and, for no particular reason, shine your laser pointer into the water. When you shine the beam of light on the water a horizontal distance of 2.4 m from the dock, you see a glint of light from a shiny object on the sandy bottom—perhaps a gold doubloon. If the angle of incidence is 53° and the water is 5.5 m deep, what is the horizontal distance from the end of the dock to the shiny object?

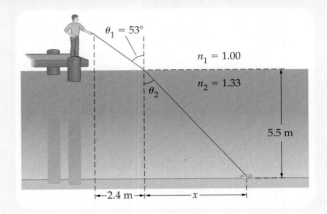

Picture the Problem

The person at the end of the dock and the shiny object on the sandy bottom are shown. All of the known distances are indicated, along with the angle of incidence, θ_1, the angle of refraction, θ_2, and the appropriate indices of refraction from Table 17.1.

Strategy

We can use Snell's law and basic trigonometry to find the horizontal distance to the shiny object. First, Snell's law ($n_1 \sin \theta_1 = n_2 \sin \theta_2$) can be solved for the angle of refraction, θ_2. Next, the sketch shows that the horizontal distance to the shiny object is 2.4 m + x. We can find the distance x from the angle of refraction, θ_2, since $\tan \theta_2 = x/(5.5 \text{ m})$.

Known

angle of incidence: $\theta_1 = 53°$

water depth = 5.5 m
(vertically downward from dock)

entry of beam into water = 2.4 m
(horizontally from dock)

Unknown

horizontal distance from end of dock to shiny object = ?

Solution

1   Use Snell's law to calculate the angle of refraction:

$$n_1 \sin \theta_1 = n_2 \sin \theta_2$$

$$\theta_2 = \sin^{-1}\left(\frac{n_1}{n_2} \sin \theta_1\right)$$

$$= \sin^{-1}\left[\left(\frac{1.00}{1.33}\right) \sin 53°\right]$$

$$= 37°$$

2   Calculate x using $\tan \theta_2 = \dfrac{x}{5.5 \text{ m}}$:

$$\tan \theta_2 = \frac{x}{5.5 \text{ m}}$$

$$x = (5.5 \text{ m}) \tan \theta_2$$

$$= (5.5 \text{ m}) \tan 37°$$

$$= 4.1 \text{ m}$$

Math HELP
Solving Simple Equations
See Math Review, Section III

3   Add 2.4 m to x to find the total horizontal distance to the shiny object:

$$\text{horizontal distance} = 2.4 \text{ m} + x$$

$$= 2.4 \text{ m} + 4.1 \text{ m}$$

$$= \boxed{6.5 \text{ m}}$$

Insight

If you were to simply stand at the end of the dock and look out into the water at the glint of gold on the bottom, you would be looking in the direction determined by θ_1, giving a line of sight that was too high. The gold object would be below your line of sight and closer than you thought.

6. Follow-up If the index of refraction for water were 1.35 instead of 1.33, would the shiny gold object be farther from the dock, nearer the dock, or at the same distance as calculated above? Check your answer by calculating the distance with $n_2 = 1.35$.

7. Light passes from air into benzene with an angle of incidence of 43°. The refracted beam makes an angle of 27° with the normal. Calculate the index of refraction of benzene.

8. The angle of refraction of a ray of light traveling from air into a slab of flint glass is 31°. What is the angle of incidence?

Refraction causes objects to look bent

You've probably noticed that a pencil placed in a glass of water appears to be bent. The cause of this illusion is shown in **Figure 17.7**. The figure shows that rays leaving the water bend away from the normal and make the pencil appear to be above its actual position.

Apparent Depth The phenomenon in Figure 17.7 is an example of what is known as *apparent depth*, in which an object appears to be closer to the water's surface than it really is. If you have side-by-side sinks in your kitchen, you might want to try the following experiment. Fill one sink halfway full with water and let the other one remain empty. Now look at the bottom of each sink. Does one seem closer to you and the other farther away? In fact, the one with water in it looks shallower because refraction has caused the image of the bottom to seem closer to the surface than it actually is—just like the pencil in Figure 17.7.

Displaced Light Another example of refraction bending light is shown in **Figure 17.8 (a)**. In this figure we see light passing through a slab of glass. The light is refracted twice—once at each surface of the slab. The first refraction bends the light rays closer to the normal, and the second refraction bends the rays away from the normal. As can be seen in the figure, the two changes in direction cancel, so the final direction of the light is the same as its original direction. The light has been displaced, however, by an amount proportional to the thickness of the slab. This sideways displacement of the light produces a disjointed image of an object behind the slab, as shown in **Figure 17.8 (b)**.

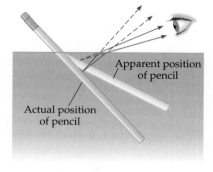

▲ Figure 17.7 **Refraction and the "bent" pencil**
Refraction causes a pencil to appear bent when placed in water. Rays leaving the water are bent away from the normal and hence extend back to a point that is higher than the actual position of the pencil.

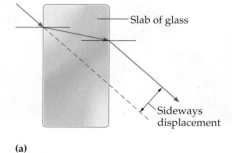

(a)

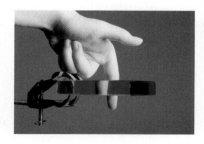

(b)

◄ Figure 17.8 **Light traveling through a glass slab**
(a) When a ray of light passes through a glass slab, it refracts first toward the normal, then away from the normal. The net result is that the ray continues in its original direction but is displaced sideways. (b) The finger behind this slab of glass appears to be disjointed, because light is refracted as it passes through the slab.

A horizontal ray of light encounters a prism, as shown below. After passing through the prism, is the ray deflected upward, deflected downward, or still horizontal?

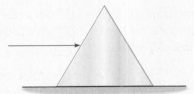

Reasoning and Discussion

When the ray enters the prism, it is bent toward the normal, which deflects it *downward*, as shown below. When it leaves through the opposite side of the prism, it is bent away from the normal. Because the sides of a prism are angled in opposite directions, however, bending away from the normal in the second refraction also causes a *downward* deflection.

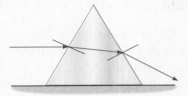

The net result, then, is a downward deflection of the ray.

Answer

The ray deflects downward.

Figure 17.9 The role of refraction in the sunset

(a) The Sun in this beautiful sunset was actually below the horizon when this picture was taken. (b) The atmosphere, which is thicker at the bottom than at the top, acts very much like a prism. The result is that light passing through the atmosphere is bent, as indicated. This makes the Sun appear to be above the horizon even though it is in fact below the horizon.

Sunsets

Sunsets are always beautiful. A good example is shown in **Figure 17.9 (a)**. The next time you see the Sun just about to set, mention to your friends or family members that even though the Sun is visible, it's already below the horizon! Their first reaction will be that you're nuts—how can the Sun be below the horizon when we can still see it? The answer—refraction.

Consider **Figure 17.9 (b)**, which shows a person observing a sunset. At the moment shown, a direct line from the person to the Sun goes below the

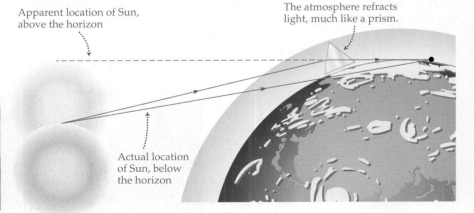

Apparent location of Sun, above the horizon

The atmosphere refracts light, much like a prism.

Actual location of Sun, below the horizon

(a)

(b)

horizon. Without an atmosphere, the Sun would indeed be invisible at this time. The atmosphere, however, refracts the sunlight like a giant prism. The refracted light is above the horizon, as shown in the figure, and this is where you see the Sun. In reality, the Sun is already below the horizon, and can only be seen because the atmosphere is refracting its light. Now that's cool physics.

Refraction obeys the principle of least time

Imagine you're a lifeguard at a beach. You're on the sandy beach at point A in **Figure 17.10** when you suddenly see a swimmer who needs help in the water at point B. You want to get to the swimmer as quickly as possible, so what route should you take to get there?

Path 1 in Figure 17.10 seems to be the logical choice, since it is the shortest route. The problem is, you can swim only half as fast as you can run on the beach. Therefore, path 1 takes a long time because you have to swim for a considerable distance.

Well, maybe path 3 is the one to take. This route minimizes the distance you have to swim. The problem here is that the total length of travel is quite bit longer. Path 3 also takes too long.

The fastest route to the swimmer is path 2. This path cuts down on the distance you have to swim compared to path 1, but it is also shorter than path 3. The interesting thing about path 2 is that it is precisely the path that obeys Snell's law—a beam of light would follow this path. Once again, we see that light follows the path that gets it to its destination in the least possible time—and this is also the path the wise lifeguard chooses.

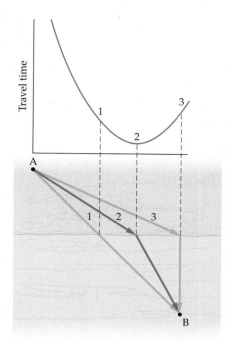

▲ **Figure 17.10 Path of least time**
From point A a lifeguard who runs on the beach and then swims in the water can reach a weak swimmer at point B in the least time by following path 2. This path obeys Snell's law and is the one that light follows in a similar situation.

17.1 LessonCheck ⓂⓅ™

Checking Concepts

9. 🔲 **Compare** Light travels through a material with an index of refraction equal to 1.20. How does the speed of light through this material compare to its speed through a material with an index of refraction equal to 1.33? Explain.

10. 🔲 **Explain** The speed of light increases as it passes from one material into another, and yet its direction might not change. Why?

11. ⬛ **Triple Choice** When a ray of light enters a piece of glass that is surrounded by air, it slows down. When it leaves the glass, does its speed increase, decrease, or stay the same? Explain.

12. Assess A classmate states that when light travels from one material to another, it always bends toward the normal. Is this correct? If not, give an example to prove your point.

Solving Problems

13. Calculate A pond covered by a transparent layer of ice has a total depth (ice plus water) of 4.00 m. The thickness of the ice is 0.32 m. Find the time required for light to travel vertically from the top surface of the ice to the bottom of the pond.

14. Calculate A beam of light travels from air into a transparent material. The angle of incidence is 25° and the angle of refraction is 15°. What is the index of refraction of the material?

15. Calculate A beam of light travels from air into a transparent material. The angle of incidence is 35° and the index of refraction of the material is 1.38. What is the angle of refraction of the beam of light?

16. ⬛ **Think & Calculate** The angle of refraction of a beam of light passing from a material with an index of refraction of 2.41 into a material with an index of refraction of 1.33 is 42°.
(**a**) Is the angle of incidence greater than, less than, or equal to 42°? Explain.
(**b**) Find the angle of incidence.

17.2 Applications of Refraction

Vocabulary
- total internal reflection
- dispersion

Refraction plays a key role in many technological applications. It is also responsible for the beautiful rainbows that inspire poets. In this lesson you'll learn about some of the more interesting aspects of refraction.

Light rays can become "trapped" inside a material

If you've ever looked upward from the bottom of a swimming pool, you've probably noticed an interesting effect. Directly overhead you see the ceiling or the sky. As you look farther away from the vertical, however, you can no longer see out of the pool. Instead, you see a reflection of the bottom of the pool. Let's see how this comes about.

Figure 17.11 (a) shows a ray of light in water encountering a water-air boundary. Part of the light is reflected back into the water at the boundary—as if from the surface of a mirror. The rest of the light emerges into the air traveling along a direction that is bent away from the normal, according to Snell's law.

What condition is required for total internal reflection to occur?

Total Internal Reflection If the angle of incidence is increased, as in Figure 17.11 (b), the angle of refraction increases as well. At a *critical angle* of incidence, θ_c, the refracted beam no longer enters the air but instead is parallel to the water-air boundary. This is shown in Figure 17.11 (c). In this case the angle of refraction is 90°. For angles of incidence greater than the critical angle, as in Figure 17.11 (d), all of the light is reflected back into the water. When light is completely reflected back into the original material in which it was traveling, we say that it has undergone **total internal reflection**.

Total internal reflection can occur only when light is trying to enter a material with a lower index of refraction. In our example light traveling in water ($n = 1.33$) is totally reflected when it encounters a boundary with air ($n = 1.00$). An example of total internal reflection for light going from water into air is shown in Figure 17.12. This figure shows that an observer at the bottom of the tank could see the top of the tank by looking in a direction close to the vertical, but would see a reflection of the bottom of the tank along lines of sight at angles farther away from the vertical, just like a swimmer at the bottom of a pool.

▼ **Figure 17.11 Total internal reflection** Total internal reflection can occur when light travels from a material with a high index of refraction to one with a lower index of refraction. The critical angle for total internal reflection occurs when the angle of refraction is equal to 90°.

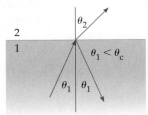
(a) Small angle of incidence

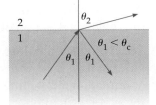
(b) Larger angle of incidence

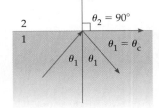
(c) Refracted beam parallel to interface

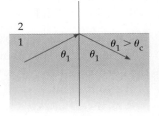
(d) Total internal reflection

Calculating the Critical Angle

The critical angle, θ_c, for total internal reflection is obtained by setting $\theta_2 = 90°$ in Snell's law. Noting that $\sin 90° = 1$, we find

$$n_1 \sin \theta_c = n_2$$

Solving this for the critical angle, θ_c, gives the following equation:

Critical Angle for Total Internal Reflection, θ_c

$$\text{critical angle} = \text{inverse sine} \left(\frac{\begin{array}{c}\text{index of refraction}\\ \text{of new material}\end{array}}{\begin{array}{c}\text{index of refraction}\\ \text{of original material}\end{array}} \right)$$

$$\theta_c = \sin^{-1}\left(\frac{n_2}{n_1}\right)$$

We apply this equation in the following Guided Example.

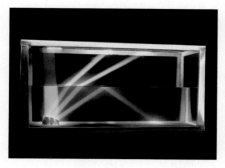

▲ **Figure 17.12 Total internal reflection** A powerful beam of light enters a tank of water from above and is reflected by mirrors on the bottom of the tank. Most of the light in the first two beams passes through the water-air boundary, undergoing refraction as it leaves the water. The third beam, however, strikes the boundary at an angle of incidence greater than the critical angle. As a result, all of the beam is reflected, as if the surface of the water were a mirror.

GUIDED Example 17.5 | Light Totally Reflected   **Total Internal Reflection**

Find the critical angle for light passing from glass ($n = 1.50$) into (**a**) air ($n = 1.00$) and (**b**) water ($n = 1.33$).

Picture the Problem

Our sketch shows the two cases being considered. Notice that in each case the original medium is glass. It follows that $n_1 = 1.50$. For part (a) we have $n_2 = 1.00$, and for part (b) we have $n_2 = 1.33$.

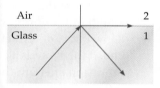

Strategy

The critical angle is defined by $\theta_c = \sin^{-1}(n_2/n_1)$. We obtain θ_c by substituting the appropriate indices of refraction for each case.

Known

$n_1 = 1.50$ (glass)
(**a**) $n_2 = 1.00$ (air)
(**b**) $n_2 = 1.33$ (water)

Unknown

(**a**) and (**b**) critical angle: $\theta_c = ?$

Solution

1 (a) Solve for the critical angle using $n_1 = 1.50$ and $n_2 = 1.00$:

$$\theta_c = \sin^{-1}\left(\frac{n_2}{n_1}\right) = \sin^{-1}\left(\frac{1.00}{1.50}\right) = \boxed{41.8°}$$

2 (b) Solve for the critical angle using $n_1 = 1.50$ and $n_2 = 1.33$:

$$\theta_c = \sin^{-1}\left(\frac{n_2}{n_1}\right) = \sin^{-1}\left(\frac{1.33}{1.50}\right) = \boxed{62.5°}$$

Insight

For water and glass the two indices of refraction are relatively close in value. Therefore, light escapes from glass to water over a wider range of incident angles (0° to 62.5°) than it does from glass to air (0° to 41.8°). In general, if the indices of refraction of two media are close in value, only light rays with very large angles of incidence will undergo total internal reflection.

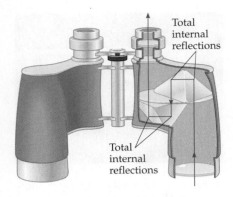

▲ **Figure 17.13 Binoculars with Porro prisms**

Prisms are used to "fold" the light path within a pair of binoculars. This design makes the binoculars easier to handle.

Reading Support ✔

Word Origin of **binoculars**

(noun) an optical device that has a lens for each eye and is used to view distant objects

prefix *bi-* ("two or twice") + Latin noun *oculus* ("eye")

🔑 *What colors of light undergo the most refraction?*

17. Follow-up Suppose the incident ray in Guided Example 17.5 is traveling in a different type of glass, with a different index of refraction. At this glass-air boundary the critical angle is 40.0°. Is the index of refraction of this glass greater than, less than, or equal to 1.50? Verify your answer with a calculation.

18. Concept Check In which cases can total internal reflection occur?
Case A: Light travels from water into air.
Case B: Light travels from air into water.
Case C: Light travels from water into ice.
Case D: Light travels from ice into water.

19. The critical angle for light passing from a certain material into air is 46.5°. What is the index of refraction of this material?

Total internal reflection has many applications

Physics & You: Technology Total internal reflection is frequently put to practical use. For example, many **binoculars** contain a set of prisms—called *Porro prisms*—that use total internal reflection to "fold" a relatively long light path into the short length of the binoculars, as shown in **Figure 17.13**. Thus, the user of binoculars handles a relatively short device that has the same optical behavior as a set of long, unwieldy telescopes. Thus, the characteristic zigzag shape of binoculars is not a fashion statement, but a reflection of the internal optical construction.

Optical fibers are another important application of total internal reflection. These thin fibers are generally composed of a glass or plastic core with a high index of refraction surrounded by an outer coating, or *cladding*, with a low index of refraction. Light is introduced into the core of the fiber at one end. It then travels along the fiber in a zigzag path, undergoing one total internal reflection after another, as illustrated in **Figure 17.14 (a)**. The total internal reflections allow the fiber to go around corners, and even to be tied into knots, and still deliver the light to the other end. An example of an optical fiber in action is shown in **Figure 17.14 (b)**.

The index of refraction depends on the light's color

Different materials—like air, water, and glass—have different indices of refraction. There is more to the story, however. The index of refraction for a given material also depends on the color of the light being refracted.

▶ **Figure 17.14 Sending light through an optical fiber**

(**a**) An optical fiber channels light along its core by means of a series of total internal reflections between the core and the cladding. (**b**) Total internal reflection makes it possible to send light through an optical fiber, as if it were a "light pipe."

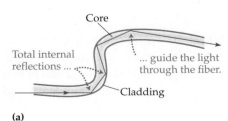

Core

Total internal reflections ...

... guide the light through the fiber.

Cladding

(a)

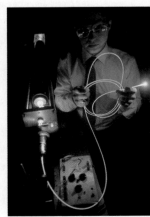

(b)

In general, a material has a higher index of refraction for light toward the blue end of the visible spectrum. This means that blue light bends more when refracted than red light does. The result is that white light, as a mixture of all colors, is spread out by refraction. This is why different colors of light travel in different directions after passing through a prism. The spreading out of refracted light according to color is known as **dispersion**.

GUIDED Example 17.6 | Prismatics   **Dispersion**

A flint-glass prism has a cross section in the shape of a 30°-60°-90° triangle, as shown in the diagram. Red and violet light are incident on the prism at right angles to its vertical side. Given that the index of refraction of flint glass is 1.66 for red light and 1.70 for violet light, find the difference in the refraction angles as the rays emerge from the prism.

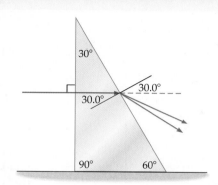

Picture the Problem

The prism and the red and violet rays are shown in our sketch. Notice that the angle of incidence on the vertical side of the prism is 0°. Therefore, the angle of refraction is also 0° for both rays. On the slanted side of the prism, the rays have an angle of incidence equal to 30.0°. Their angles of refraction are different, however.

Strategy

To find the final angle of refraction for each ray, we apply Snell's law with the appropriate index of refraction. We then subtract the angles to find the difference.

Known

angles for the triangle: 30°, 60°, 90°
$n = 1.66$ (red light)
$n = 1.70$ (violet light)

Unknown

difference in refraction angles $= ?$

Solution

1 Solve Snell's law ($n_1 \sin \theta_1 = n_2 \sin \theta_2$) for the angle of refraction, θ_2. Next, substitute the known values of $n_1 = 1.66$, $\theta_1 = 30.0°$, and $n_2 = 1.00$ to calculate θ_2 for red light:

> **Math HELP**
> Trigonometric Functions
> **See Math Review, Section VI**

$$n_1 \sin \theta_1 = n_2 \sin \theta_2$$

$$\theta_2 = \sin^{-1}\left(\frac{n_1}{n_2} \sin \theta_1\right)$$

$$= \sin^{-1}\left(\frac{1.66}{1.00} \sin 30.0°\right)$$

$$= 56.1°$$

2 Repeat Step 1 for violet light, with $n_1 = 1.70$, $\theta_1 = 30.0°$, and $n_2 = 1.00$:

$$\theta_2 = \sin^{-1}\left(\frac{n_1}{n_2} \sin \theta_1\right)$$

$$= \sin^{-1}\left(\frac{1.70}{1.00} \sin 30.0°\right)$$

$$= 58.2°$$

3 Subtract 56.1° from 58.2° to find the difference in the refraction angles:

$$58.2° - 56.1° = \boxed{2.1°}$$

Insight

This kind of difference in refraction angles is the reason for the familiar rainbow of colors seen with a prism.

20. | Follow-up | If green light emerges from the prism in Guided Example 17.6 with a refraction angle of 57.0°, what is the index of refraction of flint glass for this color of light?

21. | Triple Choice | Yellow light is bent less when it passes from air into glass than blue light is. Is the index of refraction of the glass for yellow light greater than, less than, or equal to the index of refraction for blue light? Explain.

22. The index of refraction for red light in a certain liquid is 1.320. The index of refraction for violet light in the same liquid is 1.332. Find the difference in refraction angles ($\theta_{violet} - \theta_{red}$) for red and violet light when both are incident from air on the flat surface of the liquid at an angle of 45.00° to the normal.

Dispersion is responsible for rainbows

The most famous and striking example of dispersion is the rainbow. As you know, you can't have a rainbow without rain—droplets of rain, in fact, that disperse the sunlight into its component colors. The physical situation is shown in **Figure 17.15 (a)**. Here you see a single drop of rain and an incident beam of sunlight. When sunlight enters the drop, it is separated into its red and violet components by dispersion. The light then reflects from the back of the drop, and finally refracts and undergoes additional dispersion as it leaves the drop.

The direction of the light as it emerges from the water drop is almost opposite to its incident direction. The difference is only 40° to 42°, depending on the color of the light. To be specific, violet light corresponds to an angle of 40°, and red light corresponds to an angle of 42°.

▼ **Figure 17.15 Dispersion in raindrops produces rainbows**
(a) White light entering a raindrop is spread out by dispersion into its various color components—including red and violet, which are shown here. This is the basic mechanism responsible for the formation of a rainbow. (The angles in this figure have been exaggerated for clarity.) **(b)** As a single drop of rain falls toward the ground, it sends all the colors of the rainbow to an observer. Notice that the top of the rainbow is red, and the bottom is violet.

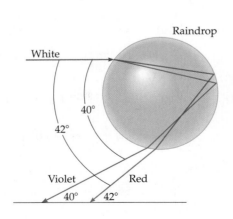

(a)

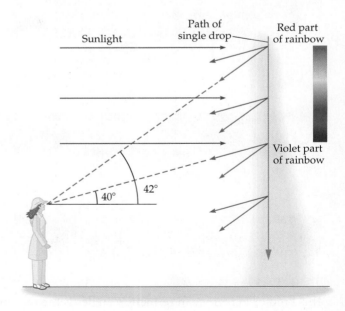

(b)

To see how the rainbow is formed in the sky, imagine standing with your back to the Sun, looking toward an area where rain is falling. Consider a single drop as it falls toward the ground. This drop, like all the other drops in the area, is sending out light of all colors in different directions. When the drop is at an angle of 42° above the horizontal, you see the red light coming from it, as indicated in **Figure 17.15 (b)**. As the drop continues to fall, its angle above the horizontal decreases. Eventually, it reaches a height where this angle is 40°. At this point the violet light from the drop reaches your eyes. In between, the drop sends out all the colors of the rainbow for your enjoyment and inspiration.

An impressive *double* rainbow over Isaac Newton's childhood home is shown in **Figure 17.16**. A double rainbow is caused by two internal reflections within each water droplet.

◄ **Figure 17.16 A double rainbow at the Newton home**
A rainbow over Isaac Newton's childhood home, the manor house of Woolsthorpe, near Grantham, Lincolnshire, England. (Notice the apple tree near the right side of the house.) Rays of light that reflect twice inside a raindrop before exiting produce the secondary rainbow, which is above the normal, or primary, rainbow. The sequence of colors in a secondary rainbow is reversed from that in the primary rainbow.

17.2 LessonCheck (MP)

Checking Concepts

23. Determine Which might result in total internal reflection, light traveling from a material with $n = 1.1$ to one with $n = 1.3$ or light traveling from a material with $n = 1.3$ to one with $n = 1.1$? Explain.

24. Determine A beam of white light strikes a prism. Which color of light is refracted the most? Which is refracted the least?

25. Identify What is the angle of refraction when the angle of incidence is equal to the critical angle for total internal reflection? Explain.

26. Triple Choice Is the critical angle for total internal reflection of red light passing from glass into air greater than, less than, or equal to the critical angle for blue light? Explain.

27. Apply What effect does dispersion have on a beam of light consisting of a single color? Explain.

Solving Problems

28. Calculate What is the critical angle for total internal reflection for light passing from glass with $n = 1.65$ into water with $n = 1.33$?

29. Calculate The critical angle for total internal reflection for light traveling from a particular type of glass to air is 39.1°. What is the index of refraction of this glass?

30. Calculate A horizontal beam of white light is incident on a right-angle prism, similar to the one in Guided Example 17.6. What is the difference in the angles of refraction of the outgoing beams of red and violet light? The prism's index of refraction is $n_{\text{violet}} = 1.505$ for violet light and $n_{\text{red}} = 1.421$ for red light.

17.3 Lenses

Vocabulary

- lens

When you look through a magnifying glass, things appear larger than life. Why? How does a simple piece of glass magnify objects? The wonderful behavior of lenses is the subject of this lesson.

Lens Types and Their Characteristics

A ray of light is refracted, or bent, as it passes from one material to another. A device that takes advantage of refraction and uses it to focus light is referred to as a **lens**. Let's look at the basic types of lenses and their characteristics.

The shape of a lens allows it to redirect light

How is the shape of a lens related to how it bends light?

Typically, a lens is a thin piece of glass with a curved surface. Converging lenses take parallel rays of light and bring them together at a focus, as shown in **Figure 17.17 (a)**. Diverging lenses cause parallel rays to spread out as if diverging from a point, as shown in **Figure 17.17 (b)**. A variety of converging and diverging lenses are illustrated in **Figure 17.17 (c)**. In general, a lens that is thicker in the middle converges light, and a lens that is thinner in the middle diverges light.

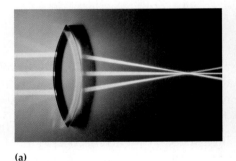
(a)

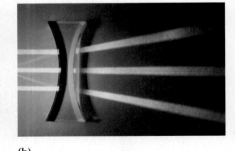
(b)

▶ **Figure 17.17 Converging and diverging lenses**
The paths of light rays through **(a)** a convex (converging) lens and **(b)** a concave (diverging) lens. **(c)** Converging and diverging lenses come in a variety of shapes. Generally speaking, converging lenses are thicker in the middle than at the edges, and diverging lenses are thinner in the middle.

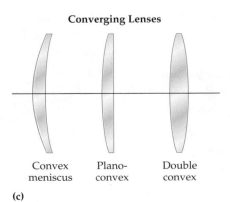
Converging Lenses

Convex meniscus   Plano-convex   Double convex

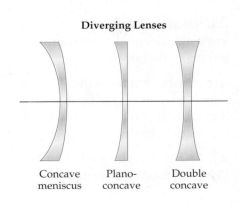
Diverging Lenses

Concave meniscus   Plano-concave   Double concave

(c)

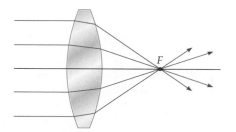

(a) A double-convex lens

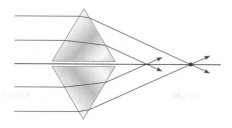

(b) Two prisms placed back to back

(c) Dewdrops acting as convex lenses

▲ **Figure 17.18 A convex lens compared with a pair of prisms**
The behavior of a convex lens **(a)** is similar to that of two prisms placed back to back **(b)**. In both cases light rays parallel to the axis are made to converge. The lens brings light to a focus at the focal point, *F*. **(c)** Drops of dew can act as convex lenses, with each drop producing its own image of a nearby object.

Convex Lenses Let's start by considering the convex (bulging outward) lens shown in **Figure 17.18 (a)**. To see why such a convex lens is converging, notice that it is similar to two prisms placed back to back, as shown in **Figure 17.18 (b)**. Recalling the bending of light described for a prism in Conceptual Example 17.4, we expect parallel rays of light to be brought together by this lens. In fact, convex lenses are shaped so that they bring parallel light rays to a focus at a focal point, *F*, along their center line, or axis, as indicated in the figure. A drop of water also acts like a convex lens, as you can see in **Figure 17.18 (c)**.

Concave Lenses A concave (curved inward like a cave) lens like the one in **Figure 17.19 (a)** is similar to two prisms placed point to point, as shown in **Figure 17.19 (b)**. In this case parallel rays are bent away from the axis of the lens. When the diverging rays from such a lens are extended back, they appear to originate at the focal point, *F*, on the axis of the lens.

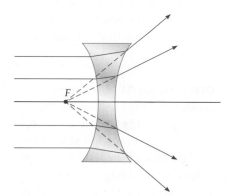

(a) A double-concave lens

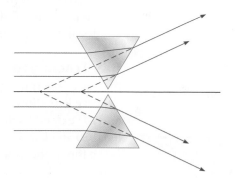

(b) Two prisms placed point to point

◀ **Figure 17.19 A concave lens compared with a pair of prisms**
A concave lens **(a)** is similar to two prisms placed point to point **(b)**. In both cases parallel light rays are made to diverge. In the case of the concave lens, the diverging rays appear to originate from the focal point, *F*.

Ray Tracing

Ray tracing is a simple and useful way to study the behavior of a lens. You can use ray tracing to find the location, size, and orientation of an image produced by a lens, just as you did in Chapter 16 for mirrors.

Principal rays show how lenses redirect light

There are three *principal rays for lenses*, and they are very similar to the three principal rays used with mirrors. The principal rays for lenses are shown in **Figure 17.20 (a)** (for a convex lens) and **Figure 17.20 (b)** (for a concave lens) on the next page.

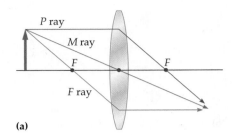

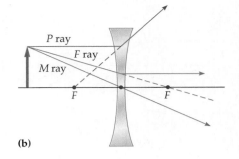

(a)　　　　　　　　　　　　　　**(b)**

▼ **Figure 17.21 Ray tracing for a concave lens**
Ray tracing can be used to find the image produced by a concave lens. Notice that the *P*, *F*, and *M* rays all extend back to the top of the virtual image, which is upright and reduced in size.

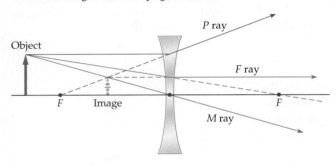

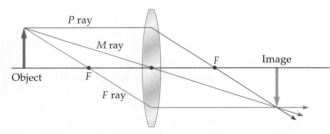

(a) Object beyond the focal point

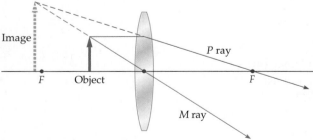

(b) Object between the focal point and the lens

▲ **Figure 17.22 Ray tracing for a convex lens**
(a) The object is beyond the focal point. The image in this case is real and inverted. **(b)** The object is between the lens and the focal point. In this case the image is virtual, upright, and enlarged.

The properties of the principal rays are as follows:

- **Midpoint Ray** The midpoint ray, or *M* ray, goes through the middle of a lens. The *M* ray continues in its original direction straight through the middle of the lens. In this book the midpoint ray is shown in red.
- **Parallel Ray** The parallel ray, or *P* ray, approaches a lens parallel to its axis. The *P* ray is bent so that it passes through the focal point, *F*, of a convex lens (Figure 17.20 (a)). The *P* ray extends back to the focal point, *F*, with a concave lens (Figure 17.20 (b)). In this book the parallel ray is shown in purple.
- **Focal-Point Ray** The focal-point ray, or *F* ray, for a convex lens is drawn through the focal point and then to the lens, as pictured in Figure 17.20 (a). For a concave lens the *F* ray is drawn toward the focal point *on the other side* of the lens, as shown in Figure 17.20 (b). In both cases, the lens bends the ray so that it is parallel to the lens's axis. In this book the focal-point ray is shown in green.

Ray Tracing for a Concave Lens To illustrate the use of ray tracing, consider the image formed by the concave lens shown in **Figure 17.21**. The three rays (*P*, *F*, and *M*) extend back to a single point on the left side of the lens. To an observer on the right side of the lens, this point is the top of the image. Notice that the image is upright and reduced in size. In addition, the image is virtual, since it is on the same side of the lens as the object. It is not possible to project this image onto a screen.

Ray Tracing for a Convex Lens The behavior of a convex lens is more interesting than that of a concave lens in that the type of image it forms depends on the location of the object. For example, if the object is placed beyond the focal point, as in **Figure 17.22 (a)**, the image is on the opposite side of the lens and upside-down. Light passes through the image, and so it is a real image that can be projected onto a screen.

If the object is placed between the lens and the focal point, the result is shown in **Figure 17.22 (b)**. The image is virtual (on the same side of the lens as the object) and upright, and cannot be projected on a screen.

Table 17.2 Imaging Characteristics of Concave and Convex Lenses

	Object location	Image orientation	Image size	Image type
CONCAVE LENS	arbitrary	upright	reduced	virtual
CONVEX LENS	beyond F	upside down	reduced or enlarged	real
	just beyond F	upside down	approaching infinity	real
	just inside F	upright	approaching infinity	virtual
	between lens and F	upright	enlarged	virtual

The imaging characteristics of concave and convex lenses are summarized in **Table 17.2**.

The material around a lens affects the focal point

The location of the focal point depends on the index of refraction of the lens, as well as the index of refraction of the surrounding material. This effect is considered in the next Conceptual Example.

CONCEPTUAL Example 17.7 Is the Focal Length Affected?

The lens shown in the diagram below is generally used in air. If it is placed in water instead, does its focal length increase, decrease, or stay the same?

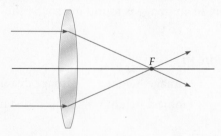

Reasoning and Discussion

The difference in index of refraction between the lens and its surroundings is less when the lens is in water than when it is in air. Therefore, light is bent less by the lens when it is in water. This is illustrated in the diagram below.

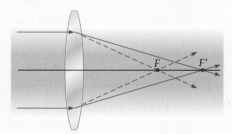

As a result, the focal length of the lens is increased when it is in water.

Answer

The focal length increases when the lens is placed in water.

The effect considered in Conceptual Example 17.7 explains why you can't focus when your eyes are underwater. The difference in the indices of refraction between the water and your eyes is so little that light passes into your eyes with almost no deflection. On the other hand, if you wear goggles, so that your eyes are in contact with air, your vision is normal.

The Thin-Lens Equation

While ray tracing is very useful, images can be located more precisely with an equation. The *thin-lens equation* is a precise mathematical relationship between object distance, image distance, and focal length for a given lens.

The thin-lens equation relates the object distance and the image distance

🔑 *How is the thin-lens equation related to the mirror equation?*

🔑 To calculate the precise location and size of an image formed by a lens, we use the thin-lens equation, which is identical in form to the mirror equation (presented in Chapter 16).

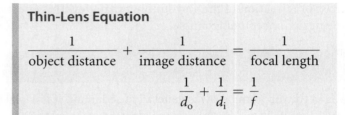

Thin-Lens Equation

$$\frac{1}{\text{object distance}} + \frac{1}{\text{image distance}} = \frac{1}{\text{focal length}}$$

$$\frac{1}{d_o} + \frac{1}{d_i} = \frac{1}{f}$$

The magnification, *m*, of an image is found in exactly the same way as for mirrors:

Magnification, m

$$\text{magnification} = \frac{\text{image height}}{\text{object height}} = -\frac{\text{image distance}}{\text{object distance}}$$

$$m = \frac{h_i}{h_o} = -\frac{d_i}{d_o}$$

As with mirrors, the sign of the magnification indicates the orientation of the image—positive for upright and negative for upside-down. The magnitude of the magnification gives the amount by which the image is enlarged or reduced compared with the object.

The sign conventions for lenses are summarized below.

Focal Length

f is positive for converging (convex) lenses.

f is negative for diverging (concave) lenses.

Magnification

m is positive for upright images.

m is negative for inverted images.

Image Distance

d_i is positive for images on the opposite side of the lens from the object. These are real images.

d_i is negative for images on the same side of the lens as the object. These are virtual images.

Object Distance

d_o is positive for all real objects. This text considers only real objects.

We apply the thin-lens equation and the definition of magnification in the next Guided Example.

GUIDED Example 17.8 | Object Distance and Focal Length — Convex Lenses

An object is placed 7.5 cm in front of a convex lens with a focal length of 5.0 cm. Find the location and magnification of the image.

Picture the Problem

Our sketch shows the ray diagram for this case. The focal length is $f = 5.0$ cm, and the object distance is $d_o = 7.5$ cm. Notice from the ray diagram that the image is upside-down, which means that the magnification is negative.

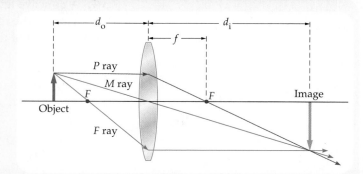

Strategy

We can find the image distance, d_i, with the thin-lens equation, $1/d_o + 1/d_i = 1/f$. Once the object and image distances are known, we can find the magnification with $m = -d_i/d_o$.

Known	Unknown
$d_o = 7.5$ cm	$d_i = ?$
$f = 5.0$ cm	$m = ?$

Solution

1 Apply the thin-lens equation to this system:

$$\frac{1}{d_o} + \frac{1}{d_i} = \frac{1}{f}$$

$$\frac{1}{7.5 \text{ cm}} + \frac{1}{d_i} = \frac{1}{5.0 \text{ cm}}$$

2 Rearrange the thin-lens equation to solve for the image distance, d_i:

$$\frac{1}{d_i} = \frac{1}{5.0 \text{ cm}} - \frac{1}{7.5 \text{ cm}}$$

$$\frac{1}{d_i} = \frac{1}{15 \text{ cm}}$$

$$d_i = \boxed{15 \text{ cm}}$$

3 Find the magnification with the relation $m = -d_i/d_o$:

$$m = -\frac{d_i}{d_o}$$

$$= -\frac{15 \text{ cm}}{7.5 \text{ cm}} = \boxed{-2.0}$$

Insight

The image in this case is twice as large as the object, but upside-down. The fact that the image distance is positive means that the image is real (it can be projected onto a screen) and that it is located on the opposite side of the lens from the object.

31. Follow-up Suppose you would like the image in Guided Example 17.8 to have a magnification of −3 instead of −2. Should the object be moved closer to the lens or farther from it? Calculate the object and image distances for a magnification of −3.

32. Use a ray diagram to determine the approximate location of the image produced by a convex lens when the object is at a distance of $\frac{1}{2}f$ from the lens. Is the image upright or inverted? Is the image real or virtual?

33. Use a ray diagram to determine the approximate location of the image produced by a concave lens when the object is at a distance of $\frac{1}{2}|f|$ from the lens. Is the image upright or inverted? Is the image real or virtual?

34. A concave lens has a focal length of −32 cm. Find the image distance and the magnification that result when an object is placed 23 cm in front of the lens.

17.3 LessonCheck (MP)

Checking Concepts

35. Describe You want to focus the Sun's rays on a point in order to start a campfire. Describe the lens shape that will accomplish this task.

36. Assess A classmate tells you that you solved a homework problem incorrectly because you applied the mirror equation to a lens, instead of using the thin-lens equation. Is your classmate right? Explain.

37. Big Idea Compare and contrast the way in which lenses form images to the way in which mirrors form images. What determines whether an image formed by a lens is real or virtual? Explain.

38. Analyze If you would like to increase the magnification of the image in Figure 17.21, should the object be moved closer to the lens or farther from it? Explain.

39. Analyze Suppose you move the object in Figure 17.22 (a) farther from the lens.
(**a**) Does the image move closer to the lens or farther from the lens? Explain.
(**b**) Does the image increase or decrease in size? Explain.

Solving Problems

40. Calculate A convex lens is held 26 cm above a piece of paper on a sunny day. The sunlight is focused to a point on the paper, which ignites. What is the focal length of the lens?

41. Calculate When an object is located 32 cm to the left of a lens, the image is formed 17 cm to the right of the lens. What is the focal length of the lens?

42. Calculate An object is placed 29 cm to the left of a lens with a focal length of 15 cm. What is the image distance?

43. Calculate An image is formed 12 cm to the left of a lens with a focal length of −23 cm. What is the object distance?

17.4 Applications of Lenses

Do you or someone in your family wear glasses? Do you or a friend own a camera or a pair of binoculars? If so, you're familiar with some of the many everyday applications of lenses. This lesson explores the workings of common optical instruments, as well as the marvelously sophisticated human eye.

Optical Devices

Modern technology makes use of lenses in many practical situations. Let's consider some of the most common of these.

The camera is a simple application of a lens

Physics & You: Technology The basic elements of a camera are shown in **Figure 17.23**. The lens forms a real, upside-down image on an optical sensor—usually a *charge-coupled device (CCD)* in a digital camera.

To focus a camera, the lens is moved either toward or away from the CCD. The lens is moved toward the CCD to focus on distant objects or away from the CCD to focus on close objects. The distances involved in focusing a camera are typically rather small, as you'll see in the next Active Example.

ACTIVE Example 17.9 Find the Displacement of the Lens

A simple camera uses a lens with a focal length of 50.0 mm. How far and in what direction must the lens be moved to change the focus of the camera from a person 20.0 m away to a person only 3.00 m away?

Solution (*Perform the calculations indicated in each step.*)

1. Use the thin-lens equation to calculate the image distance for an object distance of 20.0 m:

$d_i = 5.01$ cm

2. Calculate the image distance for an object distance of 3.00 m:

$d_i = 5.08$ cm

3. Find the difference in the image distances:

0.07 cm

Insight
Since the image distance is greater for the person who is 3.00 m away, it follows that the lens must be moved *away* from the CCD by just 0.07 cm to change the focus the desired amount.

Vocabulary

- spherical aberration
- chromatic aberration
- nearsighted
- farsighted

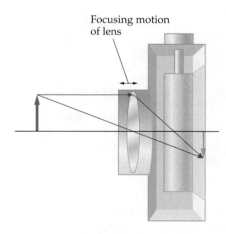

Focusing motion of lens

▲ **Figure 17.23 Basic elements of a camera**
A camera forms a real, upside-down image on photographic film or an electronic sensor, and the image is brought into focus by moving the lens back and forth. Unlike the adjustable shape of the human eye, the shape of the camera lens does not change.

A magnifying glass is a convex lens

A *magnifying glass* is nothing more than a simple convex lens. Even so, a *magnifier* can make objects appear many times larger than their actual size, as shown in **Figure 17.24**. Typically, magnifiers produce images that are upright, enlarged, and virtual (on the same side of the lens as the object).

If you have a magnifying glass at home, you can easily determine its magnification. Hold the magnifier over a page of ruled paper, as indicated in **Figure 17.25**. Notice that the rules on the page appear farther apart when viewed through the magnifier. In fact, we can see that two spaces on the page in Figure 17.25 match up with a single space as seen through the magnifier. Thus, this magnifier has a magnification of 2. This is a typical value for home magnifying glasses.

A simple microscope has two lenses

Physics & You: Technology Although a magnifying glass is a useful device, higher magnifications and improved optical quality can be obtained with a microscope. The simplest microscope, referred to as a *compound microscope*, consists of two converging lenses fixed at either end of a tube. An example is shown in **Figure 17.26 (a)**.

The basic optical elements of a microscope are the *objective* and the *eyepiece*. The objective is a converging lens with a relatively short focal length that is placed near the object to be viewed. It forms a real, upside-down, and enlarged image, as shown in **Figure 17.26 (b)**. To focus the microscope the precise location of this image is adjusted by moving the tube containing the eyepiece and the objective up or down. This image serves as the object for the second lens in the microscope—the eyepiece. In fact, the eyepiece is simply a magnifier that further enlarges the image formed by the objective.

The final magnification of the microscope is the product of the magnification of the objective and the magnification of the eyepiece. For example, a microscope might have a 10× eyepiece (meaning it magnifies 10 times) and a 50× objective. With these two lenses used together, the magnification of the microscope is 500.

▲ **Figure 17.24 Looking at nature**
A magnifying glass can be very helpful in studying an object of interest.

▲ **Figure 17.25 Determining the magnification**
The magnified rules have twice as much space between them as the rules on the sheet of paper. It follows that this is a two-power (2×) magnifying glass.

▶ **Figure 17.26 The elements and operation of a compound microscope**
(a) A compound microscope consists of two lenses—an objective and an eyepiece—fixed at either end of a movable tube.
(b) The object is placed just outside the focal point of the objective. The resulting enlarged image is then enlarged further by the eyepiece, which is basically a magnifying glass.

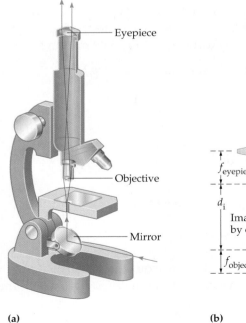

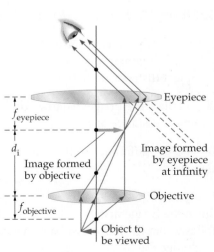

(a)

(b)

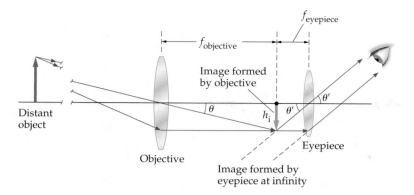

A telescope also has two lenses

Physics & You: Technology A telescope is similar in many respects to a microscope. Both instruments use two converging lenses to produce a magnified image of an object. In the case of a microscope, the object is small and close at hand. In the case of the telescope, the object is large—a planet or a galaxy perhaps—but its apparent size can be very small because of its great distance. The major difference between these instruments is that the telescope must deal with an object that is essentially infinitely far away.

Because the object is at infinity, the light entering the objective of a telescope is focused at the focal point of the objective, as shown in **Figure 17.27**. **As in a microscope, the image formed by a telescope's objective lens is the object for the eyepiece, which is basically a magnifier.** Thus, if the image from the objective is placed at the focal point of the eyepiece, it will form an image that is at infinity, as indicated in Figure 17.27. In this configuration the observer can view the final image of the telescope with a completely relaxed eye.

The magnification of a telescope is the ratio of the focal lengths of the objective and the eyepiece. For example, a telescope with an objective whose focal length is 1500 mm and an eyepiece whose focal length is 10 mm produces a magnification of $1500/10 = 150$.

Telescopes using two or more lenses, like that in Figure 17.27, are referred to as *refractors*. In fact, the first telescopes constructed for astronomical purposes, made by Galileo starting in 1609, were refractors. By the end of 1609 Galileo had produced a telescope whose magnification was about 20. This was more than enough to enable him to see—for the first time in human history—mountains on the Moon, stars in the Milky Way, the phases of Venus, and moons orbiting Jupiter. As a result of his telescopic observations, Galileo became a firm believer in the Copernican model of the Solar System.

Lenses are affected by aberrations

An ideal lens brings all parallel rays of light that strike it together at a single focal point. Real lenses, however, never quite live up to the ideal. Instead, a real lens blurs the focal point into a small but finite region of space. This, in turn, blurs the image the lens forms. The deviation of a lens from ideal behavior is referred to as an *aberration*.

Some lens shapes cause abberations. **Spherical aberration** occurs when a lens has a surface that is a section of a sphere. A lens with a spherical surface fails to focus parallel light rays at a single focal point. This is shown in **Figure 17.28 (a)** on the next page. To prevent spherical aberration a lens must be very precisely ground and polished to a nonspherical shape.

How is the functioning of a telescope similar to that of a microscope?

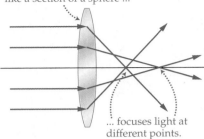

A lens whose surface is shaped like a section of a sphere ...

... focuses light at different points.

(a) Spherical aberration

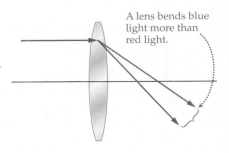

A lens bends blue light more than red light.

(b) Chromatic aberration

▲ **Figure 17.28 Types of aberration**
(a) In a lens with spherical aberration, light striking the lens at different locations comes together at different focal points. **(b)** Chromatic aberration is caused by the fact that blue light bends more than red light as it passes through a lens. As a result, different colors of light have different focal points.

▲ **Figure 17.29 Chromatic aberration in a simple lens**
The lens shows colored fringes, even though the object isn't colored.

🔑 *How does the way your eyes focus on an object differ from the way a camera does?*

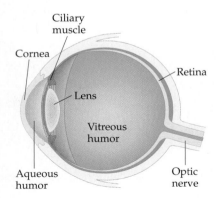

▲ **Figure 17.30 Basic elements of the human eye**
Light enters the eye through the cornea and the lens. It is focused onto the retina by the ciliary muscles, which change the shape of the lens.

Another common type of aberration is due to the basic properties of refraction. In general, **chromatic aberration** occurs when a lens bends light of different colors by different amounts. An example is shown in **Figure 17.28 (b)**. Just as a prism splits white light into a spectrum of colors, so too does a lens. The result is that white light passing through a lens does not focus to a single point. This is why you sometimes see a fringe of color around an image seen through a simple lens, as shown in **Figure 17.29**. Chromatic aberration can be corrected by combining two or more lenses to form a compound lens. For example, the "lens" in a 35-mm camera may actually consist of five or more individual lenses to correct for aberrations.

Human Vision

The eye is a marvelously sensitive and versatile optical instrument. It allows us to observe objects as distant as the stars and as close as a book in our hands. Perhaps most amazing is that the eye can accomplish all this even though its structure is essentially that of a spherical bag of water 2.5 cm in diameter. The slight differences that set the eye apart from a bag of water make all the difference in terms of optical performance, however.

The eye focuses over a wide range of distances

The key elements of the eye are shown in **Figure 17.30**. Light enters the eye through the transparent outer coating of the eye, the *cornea*. It then passes through the *aqueous humor*, the adjustable *lens*, and the jellylike *vitreous humor* before reaching the light-sensitive *retina* at the back of the eye, as shown in **Figure 17.31**. The retina is covered with millions of small structures known as *rods* and *cones*, which, when stimulated by light, send electrical impulses along the *optic nerve* to the brain.

🔑 **The human eye focuses by changing the shape of its lens, which changes the focal length, rather than by moving the lens back and forth as in a camera.** The image formed by the lens is real and inverted. How our nervous system interprets the upside-down image on the retina as a right-side-up object is another story altogether. Here we concentrate on the optical properties of the eye.

Most of the refraction in an eye occurs at the cornea

Most of the refraction needed to produce an image occurs at the cornea, as light first enters the eye. The reason is that the difference in indices of refraction is greater at the air-cornea boundary than at any other boundary within the eye. The lens itself accounts for only about a quarter of the total refraction needed for focusing. Still, the contribution made by the lens is crucial. By altering the shape of the lens, the *ciliary muscles* are able to change the precise amount of refraction the lens produces.

Focusing the Eye   When we view a distant object, our ciliary muscles are *relaxed*, as shown in **Figure 17.32 (a)**, allowing the lens to be relatively flat. As a result, it causes little refraction and its focal length is at its greatest. When we view a nearby object, the lens must shorten its focal length and cause more refraction, as shown in **Figure 17.32 (b)**. Thus, the ciliary muscles *tense* to give the lens a greater curvature.

The fact that the ciliary muscles must be tensed to focus on nearby objects means that our eyes can tire from muscular strain. That's why it's beneficial to pause occasionally from reading and look off into the distance. Viewing distant objects allows the ciliary muscles to relax, thus reducing the strain on our eyes.

Focusing Limits   The lenses in our eyes can be distorted only so much, however. In fact, there is a limit to how close the eyes can focus. The shortest distance at which a sharp focus can be obtained is the *near point*—anything closer will appear fuzzy no matter how hard we try to focus on it. For young people the near point is typically about 25 cm from the eye. A 40-year-old person often has a near point of 40 cm. In old age the near point may move to 500 cm or more. This increase is due to the lens becoming less flexible. Thus, as we age, it is not uncommon to have to move a piece of paper away from our eyes in order to read it. Eventually, reading glasses may be necessary.

At the other end of the scale, the *far point* is the greatest distance an object can be from the eyes and still be in focus. Since we can focus on the Moon and stars, it is clear that the normal far point is essentially infinity.

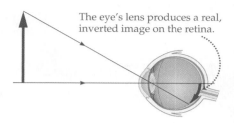

▲ **Figure 17.31 Image production in the eye**
As in a camera, the lens in an eye produces a real, upside-down image. Fortunately, the brain processes the upside-down images to give us a right-side-up view of the world.

C**OOL***PHYSICS*
Eye Floaters

Have you ever looked at a light-colored background or a clear blue sky and seen one or more "spots" float across your field of vision? Most people have these *floaters*. In fact, people are occasionally fooled by one of them into thinking a fly is buzzing about their head. For this reason floaters are also called *muscae volitantes*, which literally means "flying flies." Muscae volitantes are loose cells and other small impurities suspended in the vitreous humor of the eye or in the lens itself. They are usually quite harmless.

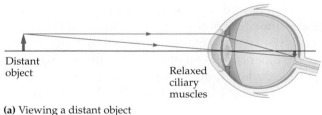

(a) Viewing a distant object

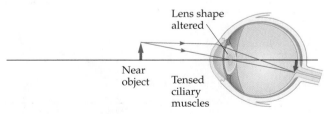

(b) Viewing a near object

▲ **Figure 17.32 Focusing in the human eye**
(a) When the eye is viewing a distant object, the ciliary muscles are relaxed and the focal length of the lens is at its greatest. **(b)** When the eye is focusing on a near object, the ciliary muscles are tensed, changing the shape and reducing the focal length of the lens.

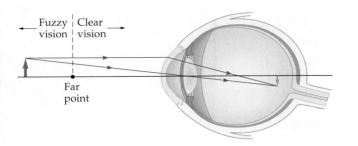

(a)

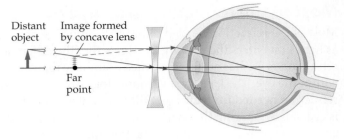

(b)

▲ **Figure 17.33 Nearsightedness and its correction**
(**a**) An eye that is elongated can cause nearsightedness. In this case an object at infinity comes to a focus in front of the retina. (**b**) A diverging lens in front of the eye can correct for nearsightedness. The concave lens focuses light from an object beyond the far point to produce an image that is at the far point. The eye can now focus on the image of the object.

A diverging lens corrects nearsightedness

Normally, the ciliary muscles of the eyes are relaxed when an object at infinity is in focus. If you are **nearsighted** (myopic), however, your relaxed eyes do not focus at infinity as they should. Instead they focus at a finite distance—the far point. This condition is known as *nearsightedness* because only objects *near* the eyes can be focused. Objects beyond the far point are fuzzy. This is illustrated in **Figure 17.33 (a)**.

The problem is that a nearsighted eye converges light in too short a distance. For example, the distant object in Figure 17.33 (a) forms an image in front of the retina. One cause of nearsightedness is an elongation of the eye, as indicated in the figure. The effect need not be large—an elongation of only a millimeter is enough to cause a problem.

Correcting nearsightedness requires "undoing" some of the excess convergence produced by the eye so that images fall on the retina. This correction can be achieved by placing a *diverging* lens in front of the eye. For example, consider an object at infinity. If a concave lens with the proper focal length produces a virtual image of this object at the nearsighted person's far point, as in **Figure 17.33 (b)**, the person's relaxed eye can focus on the object.

A converging lens corrects farsightedness

A person who is **farsighted** (hyperopic) can see clearly beyond a certain distance—the near point—but cannot focus on closer objects. Farsightedness is illustrated in **Figure 17.34 (a)**. Basically, the vision of a farsighted person differs from that of a person with normal vision by having a near point that is farther from the eye than the usual 25 cm. As a result, a farsighted person is typically unable to read clearly, since a book is too close to be brought into focus.

▼ **Figure 17.34 Farsightedness and its correction**
(**a**) An eye that is shorter than normal can cause farsightedness. Notice that an object inside the near point comes to a focus behind the retina. (**b**) A converging lens in front of the eye can correct for farsightedness. The convex lens focuses light from an object inside the near point to produce an image that is beyond the near point. The eye can then focus on the image of the object.

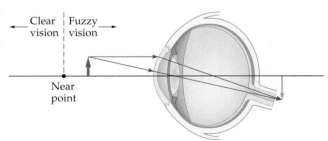

(a)

(b)

Farsightedness can be caused by an eyeball that is shorter than normal. It can also be caused by a lens that becomes stiff with age. In either case rays from an object inside the near point are brought to a focus behind the retina, as shown in Figure 17.34 (a). Thus, a farsighted eye does not converge light enough to focus it on the retina.

This problem can be corrected by "preconverging" the light—that is, by using a converging lens in front of the eye. For example, suppose an object is inside a person's near point, as in **Figure 17.34 (b)**. If a converging lens placed in front of the eye produces an image that is far away, the farsighted person can view the object with ease.

CONCEPTUAL Example 17.10 Which Glasses Should They Use?

Bill and Ted are on an excellent camping trip when they decide to start a fire by focusing sunlight with a pair of eyeglasses. If Bill is nearsighted and Ted is farsighted, should they use Bill's glasses or Ted's glasses?

Reasoning and Discussion
To focus the parallel rays of light from the Sun to a point requires a converging lens. As we have seen, farsightedness is corrected with a converging lens. Therefore, Ted's eyeglasses are converging and should be used to start the fire.

Answer
Ted's eyeglasses are the more excellent choice.

▲ Figure 17.35 **Glasses for nearsighted and farsighted people**
The eyes of a nearsighted person appear smaller than life size when viewed through the eyeglasses, as seen in the upper photo. The lower photo shows a farsighted person. Notice that the glasses make the person's eyes appear larger than life size.

In general, you can tell whether people are nearsighted or farsighted by looking at their eyes when they're wearing their glasses. Nearsighted people wear diverging lenses that make their eyes appear smaller. Farsighted people wear converging lenses that make their eyes look larger. This is shown in **Figure 17.35**.

17.4 LessonCheck (MP)

Checking Concepts

44. 🔧 **Compare** How does the way the human eye focuses light compare to the way a camera lens does?

45. 🔧 **Compare** How are the functions of the objective lens and the eyepiece lens similar in a microscope and a telescope?

46. Explain Why are convex lenses used to correct farsightedness?

47. Explain When you open your eyes underwater, everything looks blurry. Can this be thought of as an extreme case of nearsightedness or of farsightedness?

48. Infer Does chromatic aberration occur with mirrors? Explain.

Solving Problems

49. Calculate A camera lens is focused on an object 15.0 m away. If the focal length of the lens is 45.5 mm, how far must the lens be moved to focus on an object only 5.32 m away?

50. Calculate What is the distance from a camera lens to the CCD sensor if the lens has a focal length of 105 mm and the distance from the lens to the object is 0.750 m?

51. Calculate A lens is 6.2 cm from the CCD sensor in a camera. If the lens is focused on an object 3.8 m away, what is the focal length of the lens?

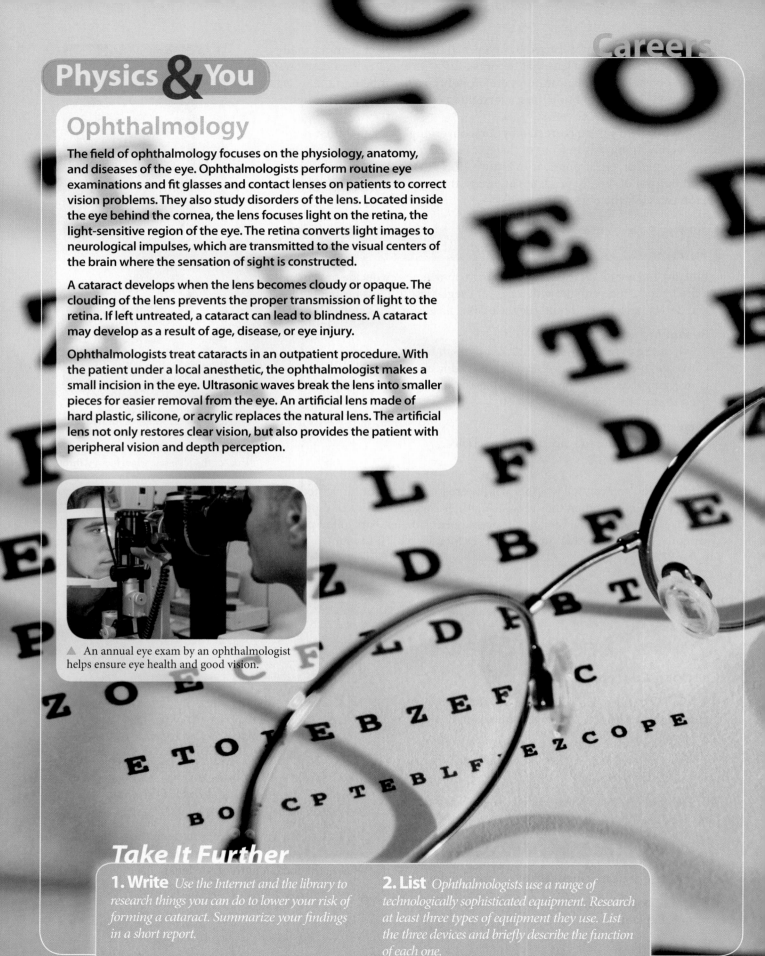

Physics & You

Ophthalmology

The field of ophthalmology focuses on the physiology, anatomy, and diseases of the eye. Ophthalmologists perform routine eye examinations and fit glasses and contact lenses on patients to correct vision problems. They also study disorders of the lens. Located inside the eye behind the cornea, the lens focuses light on the retina, the light-sensitive region of the eye. The retina converts light images to neurological impulses, which are transmitted to the visual centers of the brain where the sensation of sight is constructed.

A cataract develops when the lens becomes cloudy or opaque. The clouding of the lens prevents the proper transmission of light to the retina. If left untreated, a cataract can lead to blindness. A cataract may develop as a result of age, disease, or eye injury.

Ophthalmologists treat cataracts in an outpatient procedure. With the patient under a local anesthetic, the ophthalmologist makes a small incision in the eye. Ultrasonic waves break the lens into smaller pieces for easier removal from the eye. An artificial lens made of hard plastic, silicone, or acrylic replaces the natural lens. The artificial lens not only restores clear vision, but also provides the patient with peripheral vision and depth perception.

▲ An annual eye exam by an ophthalmologist helps ensure eye health and good vision.

Take It Further

1. Write *Use the Internet and the library to research things you can do to lower your risk of forming a cataract. Summarize your findings in a short report.*

2. List *Ophthalmologists use a range of technologically sophisticated equipment. Research at least three types of equipment they use. List the three devices and briefly describe the function of each one.*

Physics Lab Investigating Refraction

In this lab you will determine the relationship between the angle of incidence and the angle of refraction when light passes from one transparent medium into another.

Materials
- ray box (or narrow-beam light source)
- semicircular Lucite block
- semicircular plastic dish
- polar graph paper

Procedure

1. Carefully align the flat side of the semicircular Lucite block with the horizontal axis of the polar graph paper. The center of the block should lie on the vertical axis, which will become the normal from which the angles of incidence and refraction are measured.

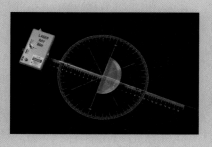

2. Using the ray box, send a ray of light at an angle of 0° to the flat side of the Lucite block. Notice that an angle of 0° is *normal* to the flat surface of the block. Observe and record the corresponding angle of refraction in Data Table 1.

3. Repeat Step 2 while increasing the angle of incidence in 10° increments. Continue until you reach an angle of incidence of 80°.

4. Replace the semicircular block of Lucite with a water-filled semicircular plastic dish. Repeat Steps 1–3, recording your data in a second data table that has the same column heads and rows as Data Table 1 but is titled "Data Table 2: Water."

Data Table 1: Lucite

Angle of Incidence, θ_i	Angle of Refraction, θ_r	$\theta_i - \theta_r$	$\dfrac{\theta_i}{\theta_r}$	$\dfrac{\sin \theta_i}{\sin \theta_r}$
0°				
10°				
20°				
30°				

Analysis

1. Perform the steps below for each angle of incidence and record the results in Data Tables 1 and 2.

- Calculate the difference between the angle of incidence and the angle of refraction.

- Calculate the ratio of the angle of incidence to the angle of refraction.

- Calculate the ratio of the sine of the angle of incidence to the sine of the angle of refraction.

2. Describe the path of a light ray after it enters the Lucite block when its angle of incidence is greater than 0°. Does the refracted ray bend toward or away from the normal?

3. Describe the path of a light ray after it enters the water when its angle of incidence is greater than 0°. Does the refracted ray bend toward or away from the normal? How does the amount of refraction in this case compare with the refraction in Lucite?

Conclusions

1. Does the light ray always refract? What condition is required in order for the light ray to refract when entering a new material?

2. Review the data in Data Tables 1 and 2. Do any of the calculated quantities remain constant as the angle of incidence changes? Does this suggest a mathematical relationship? If so, state the relationship.

17 Study Guide

Big Idea

Lenses take advantage of refraction to bend light and form images.

Different lens shapes form images with a variety of orientations and sizes. Light refracts (changes speed and direction) when it enters a lens at an angle other than 90°. The amount of refraction that occurs depends on the index of refraction of the lens and the color of the light.

17.1 Refraction

🔑 The speed of light through any material is slower than its speed in a vacuum.

🔑 Refraction only occurs if the light enters the new material at an angle *other than* 90°.

• The index of refraction of a material is the factor by which it reduces the speed of light.

• Refraction is a change in direction due to a change in the speed at which light is traveling.

• Refracted light is bent closer to the normal in a medium where its speed is reduced and away from the normal in a medium where its speed is increased.

Key Equations

The index of refraction, n, is defined as follows:

$$v = \frac{c}{n}$$

Snell's law relates the angles of incidence and refraction to the indices of refraction as follows:

$$n_1 \sin \theta_1 = n_2 \sin \theta_2$$

17.2 Applications of Refraction

🔑 Total internal reflection can occur only when light is trying to enter a material with a lower index of refraction.

🔑 In general, a material has a higher index of refraction for light toward the blue end of the visible spectrum. This means that blue light bends more when refracted than red light does.

• Total internal reflection occurs when light encounters a boundary with a new material and is completely reflected back into the original material.

• The spreading out of refracted light according to color is known as *dispersion*. Rainbows are caused by the dispersion of sunlight by raindrops.

Key Equation

The critical angle, θ_c, for total internal reflection is

$$\sin \theta_c = \frac{n_2}{n_1}$$

17.3 Lenses

🔑 In general, a lens that is thicker in the middle converges light, and a lens that is thinner in the middle diverges light.

🔑 To calculate the precise location and size of an image formed by a lens, we use the thin-lens equation, which is identical to the mirror equation (presented in Chapter 16).

• A lens is a device that takes advantage of refraction and uses it to focus light.

• The thin-lens equation relates object distance, image distance, and focal length.

• Ray tracing is a convenient way to determine the basic features of an image formed by a lens.

Key Equations

The thin-lens equation relates the object distance, d_o, the image distance, d_i, and the focal length, f, for a lens:

$$\frac{1}{d_o} + \frac{1}{d_i} = \frac{1}{f}$$

The magnification, m, of an image formed by a lens is given by the same equation used for mirrors:

$$m = -\frac{d_i}{d_o}$$

17.4 Applications of Lenses

🔑 As in a microscope, the image formed by a telescope's objective lens is the object for the eyepiece, which is basically a magnifier.

🔑 The human eye focuses by changing the shape of its lens, which changes the focal length, rather than by moving it back and forth as in a camera.

• The lens in a human eye forms a real, but upside-down, image on the retina.

• The near point is the closest distance to which the eye can focus. A typical value for the near point is 25 cm.

• The far point is the greatest distance at which the eye can focus. For a normal eye the far point is infinity.

• Any deviation of a lens from ideal behavior is an aberration.

17 Assessment

ANSWERS TO SELECTED ODD-NUMBERED PROBLEMS APPEAR IN APPENDIX A.

Lesson by Lesson

17.1 Refraction

Conceptual Questions

52. A swimmer at point B in **Figure 17.36** needs help. Two lifeguards depart simultaneously from their stand at point A, but they follow different paths. Although both lifeguards run with equal speed on the sand and swim with equal speed in the water, the lifeguard who follows the longer path, ACB, arrives at point B before the lifeguard who follows the shorter, straight-line path from A to B. Explain.

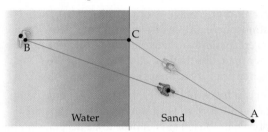

Figure 17.36

53. Sitting on a deserted beach one evening, you watch as the last bit of the Sun approaches the horizon. Just before the Sun disappears from sight, is the top of the Sun actually above or below the horizon? That is, if Earth's atmosphere could be instantly removed just before the Sun disappeared, would the Sun still be visible, or would it be below the horizon? Explain.

54. Can you see a virtual image? What is the difference between a real image and a virtual image?

55. **Triple Choice** Substance 1 has an index of refraction equal to 1.22; substance 2 has an index of refraction equal to 1.45. Is the speed of light in substance 1 greater than, less than, or equal to the speed of light in substance 2? Explain.

56. Two identical eyedroppers partially immersed in oil (left) and water (right) are shown in **Figure 17.37**. Explain why the dropper is invisible in the oil.

Figure 17.37 What happened to the dropper?

57. **Being Invisible** A common science fiction plot device is for a character to become invisible. Suppose a character becomes invisible by altering his index of refraction to match that of air. If someone could actually do this, would the person be able to see? Explain.

58. **Fishing by Hand** A young samurai wades into a small stream and plucks a fish from it for his dinner, as shown in **Figure 17.38**. **(a)** As he looks through the water at the fish, does he see it in the general vicinity of point 1 or point 2? Explain. **(b)** If the fish looks up at the man, does it see the samurai's head in the general vicinity of point 3 or point 4? Explain.

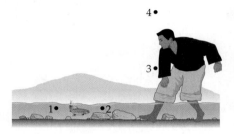

Figure 17.38

Problem Solving

59. What is the speed of light in a substance whose index of refraction is 1.62?

60. The speed of light in a substance is 2.14×10^8 m/s. What is the index of refraction of this substance?

61. Light travels a distance of 0.960 m in 4.00 ns in a given substance. What is the index of refraction of this substance?

62. A beam of light in substance 1 (index of refraction = 1.25) has an angle of incidence of 43° at the boundary with substance 2 (index of refraction = 1.62). What is the angle of refraction?

63. When a beam of light passes from substance 1 (index of refraction = 1.55) to substance 2, it has an angle of incidence of 33° and an angle of refraction of 24°. What is the index of refraction of substance 2?

64. A beam of light traveling in air enters a pool of water. If the angle of refraction is 41° and the index of refraction of water is 1.33, what is the angle of incidence?

65. A beam of light traveling in air enters a substance. If the angle of incidence is 39° and the angle of refraction is 21°, what is the index of refraction of the substance?

66. **Ptolemy's *Optics*** One of the many works published by the Greek astronomer Ptolemy (A.D. circa 90–170) was *Optics*. In this book Ptolemy reported the results of refraction experiments he conducted by observing light passing from air into water. Two of his results are as follows: (1) angle of incidence = 10.0°, angle of refraction = 8.00°; (2) angle of incidence = 20.0°, angle of refraction = 15.5°. Find the percent error for each of Ptolemy's measurements, assuming that the index of refraction of water is 1.33.

67. **Think & Calculate** A beam of light in substance 1 (index of refraction = 1.33) has an angle of incidence of 51° at the boundary with substance 2 (index of refraction = 1.66). **(a)** Is the angle of refraction greater than, less than, or equal to 51°? Explain. **(b)** What is the angle of refraction?

68. You have a semicircular disk of glass with an index of refraction of $n = 1.52$. Find the incident angle θ for which the beam of light in **Figure 17.39** will hit the indicated point on the screen.

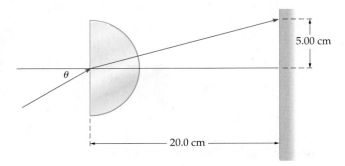

5.00 cm

θ

20.0 cm

Figure 17.39

69. The observer in **Figure 17.40** is positioned so that the far edge of the bottom of the empty glass is just visible. When the glass is filled to the top with water ($n = 1.33$), the center of the bottom of the glass is visible to the observer. Find the height, H, of the glass, given that its width is $W = 6.2$ cm. (Note that the dimensions in the figure are exaggerated for clarity. Also, assume that the viewing angle of the observer is unchanged.)

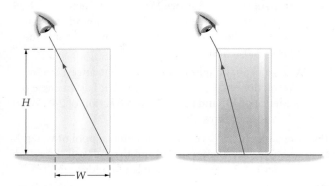

H

W

Figure 17.40

17.2 Applications of Refraction
Conceptual Questions

70. When you observe a mirage on a hot day, what are you actually seeing when you gaze at the "pool of water" in the distance?

71. **What's the Secret?** At the top of **Figure 17.41** are the words "SECRET CODE" written in different colors. If you placed a cylindrical rod of glass on top of the words, you would find that "SECRET" appears inverted, but "CODE" does not. Explain.

SECRET   CODE

ƧECЯEƚ   CODE

Figure 17.41

72. When light of color A and light of color B are sent through a prism, color A is bent more than color B. Which color of light travels more rapidly in the prism? Explain.

73. **Triple Choice** Imagine for a moment that Earth has no atmosphere. **(a)** Over the period of a year, would the average number of daylight hours at your home be greater than, less than, or equal to 12? **(b)** Does your answer change if Earth has an atmosphere? Explain.

74. A kitchen has twin side-by-side sinks of equal depth. The left sink is filled with water, and the right sink is empty. Does the left sink appear to be deeper, shallower, or the same depth as the right sink? Explain.

75. A light beam undergoes total internal reflection at the boundary between material A, in which it is traveling, and material B, on the other side of the boundary. Which material has the greater index of refraction? Explain.

Problem Solving

76. What is the critical angle for total internal reflection for a beam of light encountering a boundary between water ($n = 1.33$) and air?

77. A beam of light traveling in water ($n = 1.33$) encounters a boundary with another material. If the critical angle for total internal reflection is 81°, what is the index of refraction of the other material?

78. What is the critical angle for total internal reflection for a boundary between substance 1 with $n_1 = 1.42$ and substance 2 with $n_2 = 1.31$? In which substance does the total internal reflection occur?

79. The index of refraction for blue light in a certain type of glass is 1.670, and the index of refraction for red light in the same glass is 1.650. What is the difference in the refraction angles ($\theta_{blue} - \theta_{red}$) when light of both colors strikes the glass from air at an angle of incidence of 32.25°?

80. A ray of light enters the long side of a 45°-90°-45° prism and undergoes two total internal reflections, as diagrammed in **Figure 17.42**. The result is a reversal of the ray's direction of travel. Find the minimum value of the prism's index of refraction, n, for these internal reflections to be total.

Figure 17.42

81. **Think & Calculate** A beam of light traveling in fused quartz ($n = 1.46$) encounters a boundary with an unknown material. The critical angle for total internal reflection is 44°. **(a)** Is the index of refraction of the unknown material greater than, less than, or equal to 1.46? Explain. **(b)** What is the index of refraction of the unknown material?

82. **Think & Calculate** A glass paperweight with an index of refraction n rests on a desk, as shown in **Figure 17.43**. A ray of light enters the horizontal top surface of the paperweight at an angle of $\theta = 77.5°$ with the vertical. **(a)** Find the minimum value of n for which the internal reflection at the vertical surface of the paperweight is total. **(b)** If θ is decreased, does the minimum value of n increase or decrease? Explain.

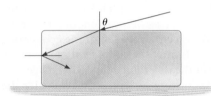

Figure 17.43

83. A horizontal beam of light enters a 45°-90°-45° prism at the center of its long side, as shown in **Figure 17.44**. The emerging ray travels in a direction that is 34° below the horizontal. What is the index of refraction of this prism?

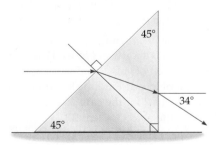

Figure 17.44

17.3 Lenses
Conceptual Questions

84. **Triple Choice** An object is moved closer to a lens. Does the focal length of the lens increase, decrease, or stay the same? Explain.

85. **Triple Choice** An object is moved closer to a concave lens. Does the image get larger, smaller, or stay the same size? Explain.

86. Is the image produced by a concave lens real or virtual? Is it larger or smaller than the object? Explain.

87. An object is between a convex lens and its focal point. Is the image real or virtual? Is it larger or smaller than the object? Explain.

88. The object distance for a convex lens is greater than the focal length. Is the image real or virtual? Is it upright or upside-down? Explain.

89. You have two lenses at your disposal, one with a focal length of 45 cm and the other with a focal length of −45 cm. Which of these two lenses would you use to project an image of a lightbulb onto a wall that is far away?

Problem Solving

90. **(a)** Trace the lens system shown in **Figure 17.45** onto a piece of paper. Find the image by drawing a ray diagram with all three principal rays on your tracing. **(b)** Is the image upright or inverted? **(c)** What is the approximate size and location of the image?

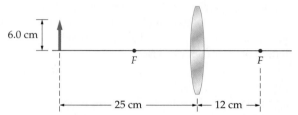

Figure 17.45

91. **(a)** Trace the lens system shown in **Figure 17.46** onto a piece of paper. Find the image by drawing a ray diagram with all three principal rays on your tracing. **(b)** Is the image upright or inverted? **(c)** What is the approximate size and location of the image?

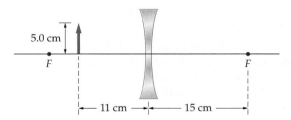

Figure 17.46

92. **(a)** Use a ray diagram to determine the approximate location of the image produced by a concave lens when the object is at a distance $2|f|$ from the lens. **(b)** Is the image upright or inverted? **(c)** Is the image real or virtual? Explain.

93. An object is a distance $2f$ from a convex lens. **(a)** Use a ray diagram to find the approximate location of the image. **(b)** Is the image upright or inverted? **(c)** Is the image real or virtual? Explain.

94. An object is a distance $4f$ from a convex lens. **(a)** Use a ray diagram to find the approximate location of the image. **(b)** Is the image upright or inverted? **(c)** Is the image real or virtual? Explain.

95. A concave lens has a focal length of -32 cm. Find the image distance and the magnification that result when an object is placed 23 cm in front of the lens.

96. When an object is located 32 cm to the left of a lens, the image is formed 17 cm to the right of the lens. What is the focal length of the lens?

97. A lens for a 35-mm camera has a focal length given by $f = 45.5$ mm. How close to the CCD sensor should the lens be placed to form a sharp image of an object that is 5.00 m away?

98. Think & Calculate An object is located to the left of a convex lens whose focal length is 36 cm. The magnification produced by the lens is $m = 3.0$. **(a)** To increase the magnification to 4.0, should the object be moved closer to the lens or farther away? Explain. **(b)** Find the object distance that gives a magnification of 4.0.

99. Think & Calculate A friend tells you that when he takes off his eyeglasses and holds them 21 cm above a printed page, the image of the print is upright but reduced to 0.67 of its actual size. **(a)** Are the lenses in the glasses concave or convex? Explain. **(b)** What is the focal length of your friend's glasses?

100. Think & Calculate A friend tells you that when she takes off her eyeglasses and holds them 21 cm above a printed page, the image of the print is upright but enlarged to 1.5 times its actual size. **(a)** Are the lenses in the glasses concave or convex? Explain. **(b)** What is the focal length of your friend's glasses?

17.4 Applications of Lenses
Conceptual Questions

101. Why is it restful to your eyes to gaze off into the distance?

102. Is the final image produced by a telescope real or virtual? Explain.

103. You are stranded on a deserted island and have in your possession two lenses, one with a positive focal length and the other with a negative focal length. **(a)** Which lens should you use to focus the rays of the Sun on dry grass to start a fire, or would either one work? **(b)** Choose the best explanation from among the following:

A. Either lens will work since all lenses focus light.

B. The lens with the positive focal length is a convex, or converging, lens—it focuses the Sun's rays into an intense real image that can start a fire.

C. The lens with the negative focal length is a concave, or diverging, lens—the virtual image it forms can light a fire.

104. A clerk at the local grocery store wears glasses that make her eyes look larger than they actually are. Is the clerk nearsighted or farsighted? Explain.

105. The umpire at a baseball game wears glasses that make his eyes look smaller than they actually are. Is the umpire nearsighted or farsighted? Explain.

Problem Solving

106. Your friend is 1.9 m tall. When she stands 3.2 m from you, what is the height of her image formed on the retina of your eye? Assume that the lens of your eye is 2.5 cm from the retina.

107. The lens of your Uncle Albert's eye is 2.60 cm from his retina. Find the near point for Uncle Albert if the smallest focal length his eye can produce is 2.20 cm.

108. To construct a telescope you are given a lens with a focal length of 32 mm for the eyepiece and a lens with a focal length of 1600 mm for the objective. What magnification does this telescope produce?

109. An object is placed 30.0 cm to the left of a converging lens with a focal length of $f_1 = 20.5$ cm. A diverging lens, with a focal length of $f_2 = -42.5$ cm, is placed 70.0 cm to the right of the first lens. What is the location of the image produced by the diverging lens? Give your answer relative to the position of the diverging lens. (The image produced by the converging lens is the object for the diverging lens.)

110. A converging lens of focal length 8.00 cm is 35.0 cm to the left of a diverging lens of focal length -6.00 cm. A coin is placed 12.0 cm to the left of the converging lens. Find the location of the coin's final image relative to the location of the diverging lens. (The image produced by the converging lens is the object for the diverging lens.)

111. Roughing It with Science A physics professor shipwrecked on Hooligan's Island decides to build a telescope from his eyeglasses and some coconut shells. Fortunately, the professor's eyes require different prescriptions, with the left lens having a focal length of 21 cm and the right lens having a focal length of 55 cm. **(a)** Which lens should he use as the objective? **(b)** What is the magnification of the professor's telescope?

112. Two lenses, with $f_1 = 20.0$ cm and $f_2 = 30.0$ cm, are placed on the x axis, as shown in **Figure 17.47**. An object is 50.0 cm to the left of lens 1, and lens 2 is a distance x to the right of lens 1. Find the location of the final image relative to lens 2 for the case $x = 115$ cm. (The image of lens 1 is the object for lens 2.)

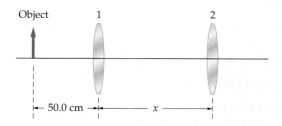

Figure 17.47

113. Two lenses that are 35 cm apart are used to form an image, as shown in **Figure 17.48**. Lens 1 is converging and has the focal length $f_1 = 14$ cm; lens 2 is diverging and has the focal length $f_2 = -7.0$ cm. An object is placed 24 cm to the left of lens 1. **(a)** Use a ray diagram to find the approximate location of the image produced by the two lenses. (The image of lens 1 is the object for lens 2.) **(b)** Is the final image upright or inverted? **(c)** Is the final image real or virtual? Explain.

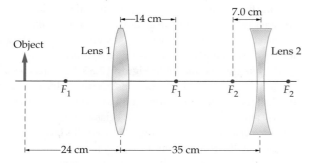

Figure 17.48

Mixed Review

114. Triple Choice   As a beam of light passes from substance 1 into substance 2, it bends closer to the normal. Is the index of refraction of substance 1 greater than, less than, or equal to the index of refraction of substance 2? Explain.

115. Predict & Explain   An *intracorneal ring* is a small plastic device implanted in a person's cornea to change its curvature. By changing the shape of the cornea, the intracorneal ring can correct a person's vision. **(a)** If a person is nearsighted, should the ring increase or decrease the cornea's curvature? **(b)** Choose the *best* explanation from among the following:

A. The intracorneal ring should increase the curvature of the cornea so that it bends light more. This will allow it to focus light coming from far away.

B. The intracorneal ring should decrease the curvature of the cornea so that it is flatter and bends light less. This will allow parallel rays from far away to be focused.

116. A large, empty coffee mug sits on a table. From your vantage point the bottom of the mug is not visible. When the mug is filled with water, however, you *can* see the bottom of the mug. Explain.

117. Your favorite aunt can read a newspaper only if it is within 15.0 cm of her eyes. **(a)** Is your aunt nearsighted or farsighted? Explain. **(b)** Should your aunt wear glasses that are converging or diverging to improve her vision? Explain.

118. The image distance for a lens is plotted versus its object distance in **Figure 17.49**. **(a)** Is the lens concave or convex? Explain. **(b)** What is the focal length of the lens?.

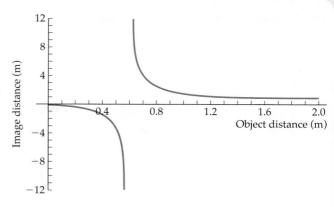

Figure 17.49

119. A person needs glasses with a focal length of -74 cm to be able to focus on distant objects. Is this person nearsighted or farsighted? Explain.

120. **Pricey Stamp** A rare misprinted 1918 "Jenny" stamp, depicting an upside-down Curtiss JN-4 airplane (known as a "Jenny"), sold at auction for $525,000. A collector uses a simple magnifying glass to examine the "Jenny," obtaining a magnification of 2.5 when the stamp is held 2.76 cm from the lens. What is the focal length of the magnifying glass?

121. **(a)** Trace the lens system shown in **Figure 17.50** onto a piece of paper. Find the image by drawing a ray diagram with all three principal rays on your tracing. **(b)** Is the image upright or inverted? **(c)** What is the approximate size and location of the image?

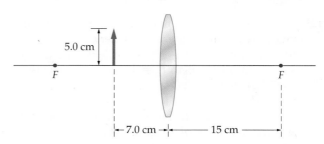

Figure 17.50

122. **(a)** Trace the lens system shown in **Figure 17.51** onto a piece of paper. Find the image by drawing a ray diagram with all three principal rays on your tracing. **(b)** Is the image upright or inverted? **(c)** What is the approximate size and location of the image?

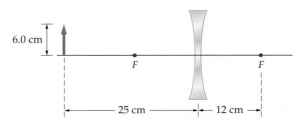

Figure 17.51

123. Figure 17.52 shows an arrow with height $h_o = 2.00$ cm located 75.0 cm from a lens with a focal length of $f = 30.0$ cm. What is the height, h_i, of the image?

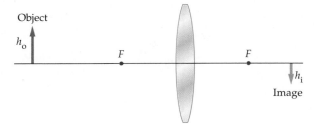

Figure 17.52

124. An arrow 2.00 cm long points toward a lens with the focal length $f = 30.0$ cm. The arrow lies *along* the lens's axis, extending from 74.0 cm to 76.0 cm from the lens, as indicated in **Figure 17.53**. What is the length of the arrow's image? (*Hint:* Use the thin-lens equation to locate the image of each end of the arrow.)

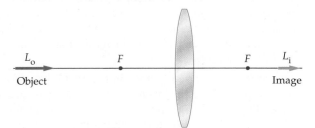

Figure 17.53

125. Two lenses that are 35 cm apart are used to form an image, as shown in **Figure 17.54**. Lens 1 is diverging and has the focal length $f_1 = -7.0$ cm; lens 2 is converging and has the focal length $f_2 = 14$ cm. The object is placed 24 cm to the left of lens 1. **(a)** Use a ray diagram to find the approximate location of the image produced by the two lenses. (The image of lens 1 is the object for lens 2.) **(b)** Is the final image upright or inverted? **(c)** Is the final image real or virtual? Explain.

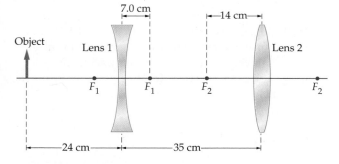

Figure 17.54

Writing about Science

126. Write a report on the lenses used in digital cameras. Address the following questions: How many individual lenses are in a typical camera "lens"? How are the lenses arranged? How do zoom lenses work?

127. Connect to the Big Idea Describe how light refracting through a lens can produce an image. Illustrate your answer with ray diagrams for both a concave and a convex lens.

Read, Reason, and Respond

Cataracts and Intraocular Lenses A *cataract* is an opacity or cloudiness that develops in the lens of an eye. Cataracts are the leading cause of blindness worldwide, and in the United States 60% of the population between the ages of 65 and 74 have cataracts to some extent.

Cataracts are generally treated by removing the cloudy lens and replacing it with an artificial *intraocular lens (IOL)*, like the one shown in **Figure 17.55**. In many cases the IOL is rigid; neither its focal length nor its location can be changed. Such a lens is designed to allow the eye to see clearly at infinity, but corrective eyeglasses must be worn for close vision. More recently, adaptive IOLs have been developed that move when the ciliary muscles (the eye muscles that change the shape of a natural lens) contract, thus providing a degree of accommodation. The focal length of an adaptive IOL is fixed, just like that of a normal IOL, but the ciliary muscles can move an adaptive IOL forward to focus on nearby objects.

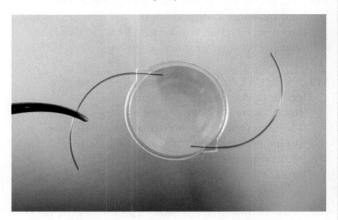

Figure 17.55 A spring-loaded intraocular lens.

128. A patient receives a normal IOL whose focus cannot be changed—it is designed to provide clear vision of objects at infinity. The patient will use corrective eyeglasses for close vision. Should the lenses of the glasses be converging or diverging?

129. Suppose a flexible, adaptive IOL has a focal length of 3.00 cm. How far forward must the IOL move to change the focus of the eye from an object at infinity to an object at a distance of 50.0 cm?

A. 1.9 mm   C. 3.1 mm

B. 2.8 mm   D. 3.2 mm

130. What focal length must an IOL have if it is to focus an object at a distance of 45 cm on the retina of the eye, which is 2.9 cm from the IOL?

A. 0.021 cm   C. 2.7 cm

B. 0.37 cm   D. 3.1 cm

Standardized Test Prep

Multiple Choice

Use the graph below to answer Questions 1–3. The graph is a plot of image distance versus object distance for a convex lens with a focal length of 3 cm.

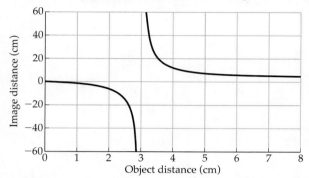

1. Which is a valid conclusion based on the data plotted in the graph?
   (A) As the object moves farther from the lens, the image gets larger.
   (B) A real image forms when the object is located beyond the focal point.
   (C) A real image forms when the object is located between the lens and the focal point.
   (D) As the object distance increases, the image distance always increases.

2. Which statement describes the image formed when an object is placed 2 cm in front of the lens?
   (A) The image is upright and on the opposite side of the lens from the object.
   (B) The image is inverted and located about 4 cm from the lens.
   (C) The image is upright and on the same side of the lens as the object.
   (D) The image is inverted and located on the opposite side of the lens from the object.

3. Which statement describes the image formed when an object is placed 4 cm in front of the lens?
   (A) The image is real and located about 10 cm from the lens.
   (B) The image is virtual and located about 10 cm from the lens.
   (C) The image is real and located about 4 cm from the lens.
   (D) The image is virtual and located about 4 cm from the lens.

Use the diagram below to answer Question 4. The diagram shows an object (upright arrow) located on the right side of a concave lens. Both focal points are also shown.

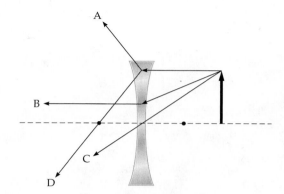

4. Which of the rays labeled A through D is not correct?
   (A) A   (C) C
   (B) B   (D) D

5. Which statement describes what happens when green light passes from air into a new material with an index of refraction of 1.5?
   (A) Its speed increases by a factor of 1.5.
   (B) Its speed decreases by a factor of 1.5.
   (C) Its direction must change.
   (D) Its color changes.

6. Which statement is true about white light traveling from air into a glass prism and being dispersed into colors?
   (A) Blue light is refracted through a smaller angle than red light.
   (B) All colors of light travel at the same speed.
   (C) Red light changes speed more than blue light.
   (D) Blue light changes speed more than red light.

Extended Response

7. Use the fact that light travels faster in plastic than it does in glass to explain why optical fibers are *not* made with a plastic core that is coated in glass.

> **Test-Taking Hint**
>
> The speed and the wavelength of light change when it enters a new material, but its frequency does not.

If You Had Difficulty With . . .

Question	1	2	3	4	5	6	7					
See Lesson(s)	17.3	17.3	17.3	17.3	17.1	17.1, 17.2	17.2					

18 Interference and Diffraction

Many of the brilliant colors in nature are due not to pigments, but to iridescence—a form of light interference. The blues and greens of this hummingbird are examples of iridescent colors.

Big Idea

Like all waves, light waves show the effects of superposition and interference.

To this point we've represented light as a ray that travels in a straight line. In this chapter you'll discover that light is also a wave and that its wave properties are of great importance. For example, because light is a wave, it can bend around obstacles. In addition, the wave properties of light are crucial to the operation of DVD players, the appearance of images on TV screens, and the brilliant iridescent colors of butterfly wings. You'll learn about all of these topics in this chapter.

Inquiry Lab   What is thin-film interference?

Explore 🔧 🧪

1. Thoroughly clean and dry two microscope slides.
2. After placing the slides on top of one another, lay them on a dark surface.
3. Illuminate the slides with white light. Tilt the slides so that you are able to see an image of the light source. What do you notice about the appearance of the surface of the slide?
4. Gently apply pressure to the top slide and describe how its appearance changes.

Think

1. **Observe** What did you observe when viewing the illuminated slides in Step 3?
2. **Describe** What happened when pressure was applied to the top slide in Step 4?
3. **Predict** How do you think your observations would change if the slides were illuminated with monochromatic light (light of a single wavelength) instead of white light?

18.1 Interference

As you learned in Chapter 13, waves can *interfere* with one another. Interfering waves can add to produce a larger amplitude, subtract to produce a smaller amplitude, or even cancel one another. This lesson explores the various effects that occur when light waves interfere.

Interference is caused by the superposition of waves

One fine summer day you watch boats zip across a quiet lake. The waves they make travel outward and overlap. If you look closely, you'll see that the waves formed by the overlapping are sometimes higher and sometimes lower than the original waves. This is an example of *superposition,* where the displacement of two or more waves is the sum of the displacements of the individual waves.

When waves combine to cause a larger displacement, they interfere *constructively*; when they combine to cause a smaller displacement, they interfere *destructively*. Interference between light waves results in an increase in brightness for constructive interference and a decrease in brightness for destructive interference.

🔑 Light wave interference is most noticeable when the light sources are *coherent* and *monochromatic*. **Monochromatic light** is light of a single color, or frequency. **Coherent light** sources maintain a constant *phase* relative to one another. Because a *laser* emits light that is both monochromatic and coherent, it is perfect for showing interference.

The phase difference between **incoherent light** sources varies randomly with time. Incoherent light sources—which include incandescent lightbulbs, fluorescent lights, and the Sun—do not form noticeable interference patterns.

Vocabulary

- monochromatic light
- coherent light
- incoherent light
- Huygens's principle

🔑 *Under what conditions is the interference of light most noticeable?*

CONNECTING IDEAS

Superposition and interference of waves were introduced in Chapter 13 for waves on a string.
- The concepts were extended to sound waves in Chapter 14.
- Here we apply the same concepts—superposition and interference—to light waves.

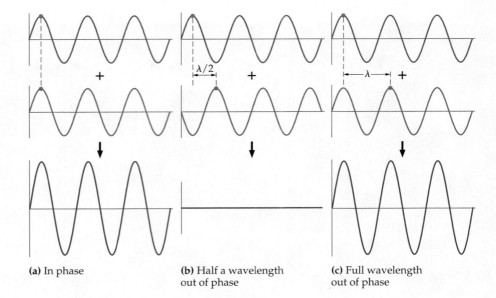

▶ Figure 18.1 Constructive and destructive interference
(a) Waves that are in phase add to give a larger displacement. This is constructive interference. (b) Waves that are half a wavelength out of phase interfere destructively. If the individual waves have equal amplitudes, as here, their sum will have zero displacement. (c) When waves are one wavelength out of phase, the result is again constructive interference, exactly the same as when the waves are in phase.

(a) In phase

(b) Half a wavelength out of phase

(c) Full wavelength out of phase

Wave interference depends on the phase difference

The basic principle that determines whether waves interfere constructively or destructively is their phase relative to one another. For example, two waves are *in phase* if their crests and troughs align with each other. In other words, waves that are in phase have a *phase difference* of zero. Such waves add constructively, and the result is an increased displacement, as indicated in **Figure 18.1 (a)**.

When two waves are 180° out of phase, the crest of one wave aligns with the trough of the other wave. Thus, the displacements are in opposite directions. These waves interfere destructively, as shown in **Figure 18.1 (b)**. Notice that a 180° difference in phase corresponds to waves being out of step by *half a wavelength*.

Similarly, if the phase difference between two waves is 360°, they are out of step by one full wavelength. In this case the interference is constructive. In fact, being 360° out of phase is exactly the same as being in phase, since 0° and 360° are two ways of saying the same thing. An example of waves that are out of phase by 360° is shown in **Figure 18.1 (c)**.

Interference occurs with all electromagnetic waves

Consider two radio antennas that send out electromagnetic waves of frequency f and wavelength λ, as illustrated in **Figure 18.2**. The antennas are connected to the same transmitter, and hence they emit waves that are in phase and coherent.

In-Phase Waves When waves from the antennas reach point M in Figure 18.2, they have traveled the same distance—that is, they have traveled the same number of wavelengths. Therefore, the waves are still in phase at M. As a result, they interfere constructively at that point, and the radio signal is strong. This is like the situation shown in Figure 18.1 (a).

Next, consider point P in Figure 18.2. To reach P the waves from the two antennas must travel different distances, ℓ_1 and ℓ_2. Let's assume that the difference between these distances is one wavelength:

$$\ell_2 - \ell_1 = \lambda$$

This means the waves are 360° out of phase at point P. It follows that they experience constructive interference there, just like the waves in Figure 18.1 (c).

🔑 How is the path-length difference from two wave sources related to interference?

📡 In general, if the difference in path length at a given point is an integer number of wavelengths $(0\lambda, 1\lambda, 2\lambda, 3\lambda, \ldots)$, then that point is a location of constructive interference.

> **Condition for Constructive Interference**
>
> $$\overbrace{\text{length of path 2} - \text{length of path 1}}^{\text{path-length difference}} = \text{integer} \times \text{wavelength}$$
>
> $$\ell_2 - \ell_1 = m\lambda \qquad m = 0, 1, 2, \ldots$$

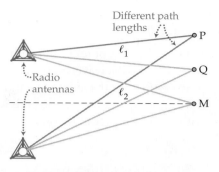

▲ **Figure 18.2 Two radio antennas transmitting the same signal**
Midway between the two antennas, at point M, the waves travel the same distance, and hence they interfere constructively. At point P the distance ℓ_2 is greater than the distance ℓ_1 by one wavelength. Therefore, P is also a point of constructive interference. At point Q, however, the distance ℓ_2 is greater than the distance ℓ_1 by half a wavelength. The waves interfere destructively at that point.

Out-of-Phase Waves

On the other hand, consider the situation at point Q in Figure 18.2. The difference in path length to point Q is half a wavelength:

$$\ell_2 - \ell_1 = \tfrac{1}{2}\lambda$$

Therefore, the waves cancel there, just like the waves in Figure 18.1 (b). Destructive interference also occurs when the difference in path length is *one and a half* wavelengths, *two and a half* wavelengths, and so on. 📡 In general, if the difference in path length at a given point is an integer plus one half wavelength $[\tfrac{1}{2}\lambda, (1 + \tfrac{1}{2})\lambda, (2 + \tfrac{1}{2})\lambda, \ldots]$, then that point is a location of destructive interference.

> **Condition for Destructive Interference**
>
> $$\overbrace{\text{length of path 2} - \text{length of path 1}}^{\text{path-length difference}} = (\text{integer} + \tfrac{1}{2}) \times \text{wavelength}$$
>
> $$\ell_2 - \ell_1 = (m + \tfrac{1}{2})\lambda \quad m = 0, 1, 2, \ldots$$

GUIDED Example 18.1 | Two May Not Be Better Than One　　　　　　**Interference**

Two friends tune their radios to the same frequency and pick up a signal transmitted simultaneously by a pair of antennas. The friend who is equidistant from the antennas, at point M, receives a strong signal. The friend at point Q receives a very weak signal. Find the wavelength of the radio waves if $d = 3.75$ km, $L = 14.0$ km, and $y = 1.88$ km in the figure below. Assume that Q is the first point of minimum signal as one moves away from M. (This means that the path-length difference at Q is half a wavelength.)

Picture the Problem

Our sketch shows the radio antennas and the two locations mentioned in the problem statement. Notice that the radio antennas are a distance $d = 3.75$ km above and below the dashed center line and that the points M and Q have a vertical separation of $y = 1.88$ km. The horizontal distance to both M and Q is $L = 14.0$ km.

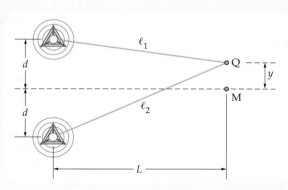

Strategy

We know that at point Q the path-length difference, $\ell_2 - \ell_1$, is half a wavelength. Thus, we can determine λ by calculating the lengths ℓ_2 and ℓ_1 and setting their difference equal to $\lambda/2$. The path lengths are calculated with the Pythagorean theorem.

Known	Unknown
$d = 3.75$ km	$\lambda = ?$
$L = 14.0$ km	
$y = 1.88$ km	

Solution

1 Use the Pythagorean theorem to calculate the path length ℓ_1:

$$\ell_1 = \sqrt{L^2 + (d - y)^2}$$
$$= \sqrt{(14.0 \text{ km})^2 + (3.75 \text{ km} - 1.88 \text{ km})^2}$$
$$= \sqrt{199 \text{ km}^2}$$
$$= 14.1 \text{ km}$$

2 Use the Pythagorean theorem to calculate the path length ℓ_2:

$$\ell_2 = \sqrt{L^2 + (d + y)^2}$$
$$= \sqrt{(14.0 \text{ km})^2 + (3.75 \text{ km} + 1.88 \text{ km})^2}$$
$$= \sqrt{228 \text{ km}^2}$$
$$= 15.1 \text{ km}$$

3 Set $\ell_2 - \ell_1$ equal to $\lambda/2$ and solve for the wavelength:

$$\ell_2 - \ell_1 = \tfrac{1}{2}\lambda$$
$$\lambda = 2(\ell_2 - \ell_1)$$
$$= 2(15.1 \text{ km} - 14.1 \text{ km})$$
$$= \boxed{2.0 \text{ km}}$$

> **Math HELP**
> Significant Figures
> See Lesson 1.4

Insight

Notice that radio waves have a rather large wavelength. In fact, the distance from one crest to the next (the wavelength) for these waves is 2.0 km (about 1.2 mi). Since radio waves travel at the speed of light, the corresponding frequency is $f = c/\lambda = 150$ kHz.

Practice Problems

1. **Follow-up** Suppose the wavelength broadcast by the two antennas is changed and that the vertical distance between M and Q increases as a result. Is the new wavelength greater than or less than 2.0 km? Find the wavelength for the case where $y = 2.91$ km.

2. Two sources emit waves that are coherent and in phase and have a wavelength of 26.0 m. Do the waves interfere constructively or destructively at a point 78.0 m from one source and 143 m from the other source?

3. Two sources emit waves that are in phase with each other. What is the largest wavelength that will give constructive interference at a point 161 m from one source and 295 m from the other source?

A two-slit experiment shows that light waves interfere with one another

We now consider a classic physics experiment referred to as the *two-slit experiment*. This experiment clearly demonstrates the wave nature of light. It also allows us to determine the wavelength of light, just as we did for radio waves in Guided Example 18.1.

The experiment was first performed in 1801 by the English physician and physicist Thomas Young (1773–1829). His experiment consists of a beam of *monochromatic* (single-color) light that passes through a small slit in a screen and then illuminates two slits, S_1 and S_2 in a second screen. After the light passes through the two slits, it shines on a distant screen, as shown in **Figure 18.3**. An *interference pattern* with bright and dark *fringes* (narrow bands) forms on the distant screen.

The slit in the first screen serves only to produce a small source of light. Because the slit is small, the light coming through it is essentially coherent. This was the best Young could do to produce a coherent beam of light in the days before lasers.

The key elements in the experiment are the two thin slits in the second screen. Since these slits are equidistant from the single slit, as shown in Figure 18.3, the light waves passing through them are in phase. Thus, the two slits act as monochromatic, coherent sources of light, just like the two radio antennas in Guided Example 18.1. With an idealized system like this, interference effects are easy to observe. Still, you might wonder why light passing through the slits spreads out on the distant screen, rather than just making two thin streaks of light. We address this question next.

Huygens's principle explains the interference of light

Have you ever noticed that water waves spread out when they pass through a small opening in an obstruction? This effect is shown in **Figure 18.4** for water waves. Huygens's principle explains this behavior. **Huygens's principle** states that every point on a wave front acts like a point source for new waves. Examples of how the principle applies to circular and plane wave fronts are shown in **Figure 18.5** on the next page. Notice that the waves from each point on a wave front add to form a new wave front of the same shape.

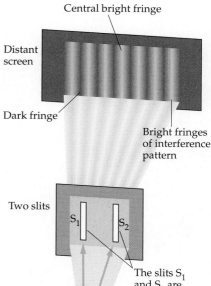
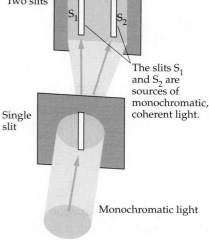

▲ **Figure 18.3 Young's two-slit experiment**
The first screen produces a small source of light that illuminates the two slits, S_1 and S_2. After passing through these slits, the light spreads out and produces an interference pattern of alternating bright and dark fringes on a distant screen.

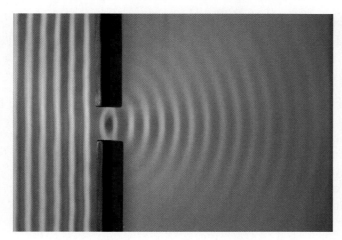

▲ **Figure 18.4 Water waves spreading outward**
When waves pass through a small opening, they spread outward. This is true of all waves, including water waves and light waves. In the case shown here, water waves initially traveling toward the right spread out in all directions after passing through an opening.

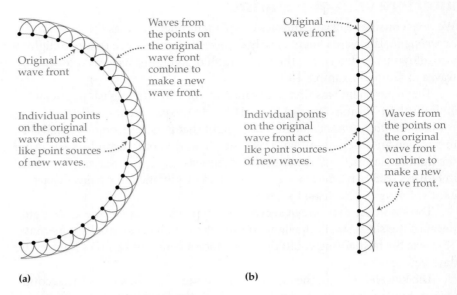

Figure 18.5 Huygens's principle
According to Huygens's principle, each point on a wave front acts like a point source of new waves. These new waves combine to form a new wave front. Here we see Huygens's principle applied to **(a)** circular and **(b)** plane wave fronts.

Why do waves spread out after passing through a small opening?

What happens when a wave passes through a small opening and most of the wave front is cut off? **The section of wave front that passes through the small opening acts almost like a point source of waves. This is why waves—whether they are water waves or light waves—spread out after they pass through a small opening.** This spreading is seen in **Figure 18.6**.

Huygens's principle is applied to the two slits in Young's experiment in **Figure 18.7**. Notice that the light radiates away from the slits in all forward directions—not just in the direction of the incoming light. The result is that light is spread out over a large area on the distant screen. The light does not just illuminate areas in direct line with the slits, as you might expect.

Interference of light waves forms fringes

The two-slit experiment produces a series of alternating bright and dark fringes, as illustrated in Figure 18.3. These fringes are the direct result of constructive and destructive interference.

Bright Fringes Notice that the central bright fringe in Figure 18.3 is midway between the two slits. This is a region where constructive interference occurs—just like at point M in Figure 18.2. The next bright fringe occurs when the difference in path length from the two slits is equal to one wavelength of light, as with point P in Figure 18.2.

Figure 18.8 shows that the path-length difference for two rays in a two-slit experiment is

$$d \sin \theta$$

As a result, the bright fringe closest to the midpoint, with a path-length difference of one wavelength, occurs at the angle θ given by this condition:

$$d \sin \theta = \lambda$$

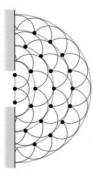

Figure 18.6 A wave passing through an opening
As a wave passes through an opening, the small section of wave front that emerges produces new wave fronts that spread out.

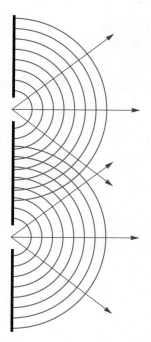

▲ **Figure 18.7 Huygens's principle in the two-slit experiment**
According to Huygens's principle, each of the two slits in Young's experiment acts as a source of light waves that travel in all forward directions. It follows that light waves from the two sources can overlap, resulting in an interference pattern.

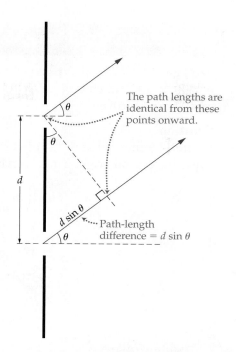

The path lengths are identical from these points onward.

Path-length difference = $d \sin \theta$

▲ **Figure 18.8 Path-length difference in the two-slit experiment**
Light traveling from two slits to a distant screen along parallel paths. The paths make an angle θ relative to the normal to the slits. The difference in path length is $d \sin \theta$, where d is the slit separation.

In general, a bright fringe occurs whenever the path-length difference is equal to an integer number of wavelengths ($0\lambda, 1\lambda, 2\lambda, 3\lambda, \ldots$). Therefore, bright fringes satisfy the following conditions:

> **Conditions for Bright Fringes in a Two-Slit Experiment**
>
> path-length difference = integer × wavelength
> $$d \sin \theta = m\lambda \qquad m = 0, \pm 1, \pm 2, \ldots$$

The value $m = 0$ corresponds to the central bright fringe, as shown in **Figure 18.9**. Positive values of m indicate fringes above the central bright fringe; negative values of m indicate fringes below the central bright fringe. The absolute value of m is referred to as the *order*. Thus, a first-order bright fringe corresponds to $m = \pm 1$, a second-order bright fringe corresponds to $m = \pm 2$, and so on.

Dark Fringes
Approximately halfway between any two bright fringes in Figure 18.9 is a dark fringe, where destructive interference occurs. The dark fringes correspond to differences in path length of half a wavelength, one and a half wavelengths, and so on. The path-length differences for bright and dark fringes are shown in Figure 18.9.

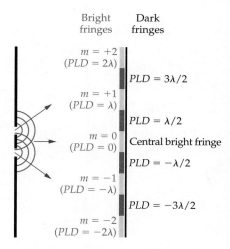

Bright fringes   Dark fringes

$m = +2$
$(PLD = 2\lambda)$

$PLD = 3\lambda/2$

$m = +1$
$(PLD = \lambda)$

$PLD = \lambda/2$

$m = 0$
$(PLD = 0)$ Central bright fringe

$PLD = -\lambda/2$

$m = -1$
$(PLD = -\lambda)$

$PLD = -3\lambda/2$

$m = -2$
$(PLD = -2\lambda)$

▲ **Figure 18.9 The two-slit pattern**
Bright and dark fringes in the two-slit experiment. The path-length difference, or *PLD*, is indicated in parentheses for each fringe, and the values of m are given for the bright fringes.

QUICK Example 18.2 What Are the Angles?

Red light ($\lambda = 752$ nm) passes through a pair of slits with a separation of 6.20×10^{-5} m. Find the angle for the first bright fringe above the central bright fringe.

Solution

Figure 18.9 shows that $m = +1$ corresponds to the first bright fringe above the central bright fringe. Solving $d \sin \theta = m\lambda$ for the angle gives

$$\theta = \sin^{-1}\left(m\frac{\lambda}{d}\right)$$

$$= \sin^{-1}\left[(1)\frac{7.52 \times 10^{-7} \, \cancel{m}}{6.20 \times 10^{-5} \, \cancel{m}}\right]$$

$$= \boxed{0.695°}$$

Math HELP
Solving Simple Equations
See Math Review, Section III

Practice Problems

4. Concept Check How would you expect the interference pattern of a two-slit experiment to change if white light were used instead of monochromatic light?

5. Laser light with a wavelength of $\lambda = 670$ nm is incident on a pair of slits along the normal. What slit separation will produce a first-order ($m = 1$) bright fringe at an angle of 35°?

Distance from the Central Fringe

A convenient way to characterize the location of interference fringes is in terms of their vertical distance from the central bright fringe, as indicated in **Figure 18.10**. If the horizontal distance to the screen is L, it follows that the vertical distance, y, is given by the following expression:

Linear Distance from the Central Bright Fringe

$$y = L \tan \theta$$

The following Guided Example shows how a measurement of the vertical distance between fringes can be used to determine the wavelength of light.

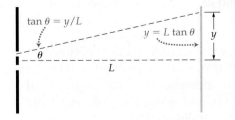

▲ **Figure 18.10 Vertical distance in an interference pattern**
If light travels at an angle θ relative to the normal to the slits, it is displaced a vertical distance $y = L \tan \theta$ on the distant screen.

Two slits with a separation of 8.5×10^{-5} m create an interference pattern on a screen 2.3 m away. If the tenth bright fringe above the central bright fringe is 12 cm above it, what is the wavelength of the light used in the experiment?

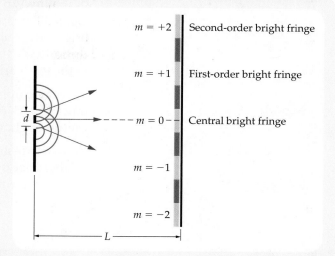

Picture the Problem

Our sketch shows that the first bright fringe above the central bright fringe corresponds to $m = +1$, the second bright fringe corresponds to $m = +2$, and so on. Therefore, the tenth bright fringe corresponds to $m = +10$.

The separation of the slits is $d = 8.5 \times 10^{-5}$ m, the distance to the screen is $L = 2.3$ m, and the vertical distance to the tenth bright fringe is $y = 12$ cm $= 0.12$ m.

Strategy

To find the wavelength we first determine the angle for the tenth bright fringe using the relation $y = L \tan \theta$. Once we know θ, we use the condition for bright fringes to determine the wavelength. That is, we set $m = +10$ in $d \sin \theta = m\lambda$ and solve for λ.

Known	Unknown
$d = 8.5 \times 10^{-5}$ m	$\lambda = ?$
$L = 2.3$ m	
$y = 0.12$ m	
$m = +10$	

Solution

1 Rearrange $y = L \tan \theta$ and calculate the angle for the tenth bright fringe (refer to Figure 18.10):

$$y = L \tan \theta$$
$$\theta = \tan^{-1}\left(\frac{y}{L}\right)$$
$$= \tan^{-1}\left(\frac{0.12 \text{ m}}{2.3 \text{ m}}\right)$$
$$= 3.0°$$

Math HELP
Solving Simple Equations
See Math Review, Section III

2 Rearrange $d \sin \theta = m\lambda$ to solve for the wavelength:

$$d \sin \theta = m\lambda$$
$$\lambda = \frac{d}{m} \sin \theta$$
$$= \left(\frac{8.5 \times 10^{-5} \text{ m}}{10}\right) \sin(3.0°)$$
$$= 4.4 \times 10^{-7} \text{ m}$$
$$= \boxed{440 \text{ nm}}$$

Insight

We have expressed the wavelength of the light in nanometers (nm), a common unit for wavelengths of light. Light with a wavelength of 440 nm has a dark blue color.

6. Follow-up (a) If the wavelength of light used in the experiment in Guided Example 18.3 is increased, does the linear distance to the tenth bright fringe above the central bright fringe increase, decrease, or stay the same? (b) Check your reasoning by calculating the linear distance to the tenth bright fringe for a wavelength of 550 nm.

7. Monochromatic light passes through two slits separated by a distance of 0.0334 mm. If the angle for the third bright fringe above the central bright fringe is 3.21°, what is the wavelength of the light?

8. In a two-slit experiment the first bright fringe above the central bright fringe occurs at an angle of 0.21°. If the wavelength of the light in the experiment is 520 nm, what is the slit separation?

18.1 LessonCheck (MP)

Checking Concepts

9. **Explain** Why don't you notice interference when the beams from two flashlights overlap?

10. **Identify** Two coherent light sources are half a wavelength out of phase at a certain point. Do they interfere? If so, how?

11. **Describe** Use Huygens's principle to explain how waves spread out after passing through a small opening.

12. **Big Idea** What is meant by the superposition of waves? What is the difference between constructive and destructive interference?

13. **Triple Choice** A two-slit experiment with blue light produces a set of bright fringes. Will the spacing between the fringes increase, decrease, or stay the same if red light is used instead? Explain.

Solving Problems

14. Calculate A two-slit experiment uses light with a wavelength of 410 nm and a slit separation of 3.4×10^{-5} m. What is the angle to the first bright fringe above the central bright fringe?

15. Calculate A two-slit experiment with slits separated by 48.0×10^{-5} m produces a second-order ($m = 2$) bright fringe at an angle of 0.0990°. Find the wavelength of the light used in this experiment.

16. Calculate Light with a wavelength of 546 nm passes through two slits and forms an interference pattern on a screen 8.75 m away. If the vertical distance from the central bright fringe to the first bright fringe above it on the screen is 5.36 cm, what is the separation of the slits?

18.2 Interference in Thin Films

You've probably seen a film of oil floating on water. If so, you know it's very colorful. In fact, you can see all the colors of the rainbow in the film. Where do all those colors come from? The short answer is that they are produced by a phenomenon known as *thin-film interference.* We'll fill in the details in this lesson.

Reflected rays of light can interfere with one another

In Lesson 18.1 you saw how light from two slits can interfere to form bright and dark fringes. Something similar happens when you shine light on a film of oil. In this case, instead of *two slits* there are *two surfaces* of the film (top and bottom), each causing a reflection, as shown in **Figure 18.11**. Reflected light from these two sources can interfere, just like light from two slits.

In the two-slit experiment the light from the two sources interferes because it travels different distances, which changes the relative phase of the light waves. This also happens with light reflected from a thin film. The interesting thing about reflection, however, is that the reflection process itself can change the phase. In fact, in the right situations, reflection causes a phase change of 180°. Let's see how this works.

The reflection of light can change its phase

You know that a wave on a string reflects differently depending on whether the end of the string is tied to a solid support or is free to move up and down. Specifically, a wave on a string with a loose end is reflected back exactly as it approached the end—there is no phase change. On the other hand, a wave on a string that is tied down is flipped upside-down when it reflects. This is equivalent to changing the phase of the wave by 180°, or half a wavelength.

CONNECTING IDEAS

In Chapter 13 you learned that a wave on a string can undergo a change in phase when it is reflected.
• Here you learn that a light wave can also experience a phase change when it is reflected.

(a)

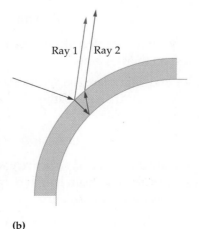

Ray 1   Ray 2

(b)

◀ **Figure 18.11 Light reflecting from a thin film**
When light reflects from a thin film, the reflections from the top (ray 1) and the bottom (ray 2) surfaces can interfere, just like light from two different slits in the two-slit experiment.

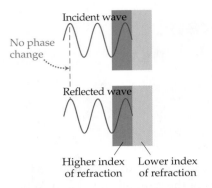 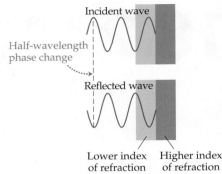

▶ **Figure 18.12 Phase change with reflection**
(a) A light wave reflects with no phase change when it encounters a medium with a lower index of refraction. **(b)** A light wave reflects with a half-wavelength (180°) phase change when it encounters a medium with a higher index of refraction.

Incident wave

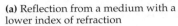 No phase change

Reflected wave

Higher index of refraction   Lower index of refraction

(a) Reflection from a medium with a lower index of refraction

Incident wave

Half-wavelength phase change

Reflected wave

Lower index of refraction   Higher index of refraction

(b) Reflection from a medium with a higher index of refraction

🔑 *What happens to the phase of light when it reflects?*

Since light is a wave, it undergoes a similar phase change when it reflects. 🔑 **When light encounters a medium with a lower index of refraction, it is reflected with no phase change.** This is shown in **Figure 18.12 (a)**. This reflection is just like that of a wave on a string whose end is free to move. 🔑 **In contrast, when light encounters a medium with a higher index of refraction, it is reflected with a phase change of half a wavelength.** This case, which is just like the reflection of a wave on a string whose end is fixed, is shown in **Figure 18.12 (b)**. Let's apply these observations to a specific case.

A wedge-shaped volume of air causes interference

An interesting example of interference due to reflection is provided by two plates of glass that touch at one end and have a small separation at the other end, as shown in **Figure 18.13 (a)**. The air between the plates occupies a thin, wedge-shaped region. As a result, this type of arrangement is referred to as an *air wedge*.

An air wedge causes interference between light reflected from the bottom surface of the upper glass plate (ray 1) and light reflected from the top surface of the lower plate (ray 2). Ray 1 reflects at a glass-to-air boundary. Air has a lower index of refraction than glass, and hence this ray reflects with no phase change.

Ray 2 travels a distance d through the air, reflects from the air-to-glass boundary, and then travels essentially the same distance d in the opposite direction before rejoining ray 1. Since glass has a higher index of refraction than air, ray 2 reflects with a phase change of 180°—the same as if it had traveled an extra half-wavelength, or ½ λ. Taking this phase change into account gives the *effective* path length for ray 2:

$$\text{effective path length of ray 2} = \tfrac{1}{2}\lambda + 2d$$

Because ray 2 has a greater effective path length than ray 1, the rays can interfere when they recombine:

- Rays 1 and 2 interfere *constructively* when the effective path length of ray 2 is greater than the path length of ray 1 by an integer number of wavelengths:

$$\tfrac{1}{2}\lambda + 2d = m\lambda \qquad m = 1, 2, 3, \ldots$$

- Rays 1 and 2 interfere *destructively* when the effective path length of ray 2 is greater than the path length of ray 1 by half a wavelength, one and a half wavelengths, and so on:

$$\tfrac{1}{2}\lambda + 2d = (m + \tfrac{1}{2})\lambda \qquad m = 0, 1, 2, \ldots$$

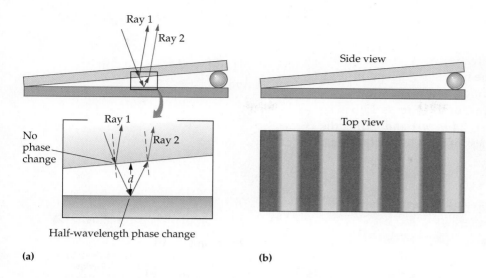

Ray 1

Ray 2

Side view

Top view

(a)

No phase change

Ray 1

Ray 2

d

Half-wavelength phase change

(b)

◀ **Figure 18.13 An air wedge**
(a) In an air wedge interference occurs between the light reflected from the bottom surface of the top plate of glass (ray 1) and the light reflected from the top surface of the bottom plate of glass (ray 2). The two rays are shown widely separated for clarity; in reality, they would be almost on top of one another. **(b)** When the rays recombine, they form interference fringes that are evenly spaced, as shown in the top view of the wedge.

Since the distance between the plates, d, increases steadily in an air wedge, it follows that the dark and bright interference fringes are evenly spaced, as shown in **Figure 18.13 (b)**.

CONCEPTUAL Example 18.4 Is the Fringe Dark or Bright?

Is there a dark fringe or a bright fringe where the glass plates touch in an air wedge?

Reasoning and Discussion

Recall that ray 1 undergoes no phase change on reflection. On the other hand, ray 2 experiences a 180° phase change due to reflection. When the path-length difference, $2d$, approaches zero, rays 1 and 2 will cancel each other because of destructive interference since they are 180° out of phase.

Answer

A dark fringe occurs where the two glass plates touch.

The next Guided Example uses the number of fringes observed in an air wedge to measure the thickness of a human hair.

GUIDED Example 18.5 | Splitting Hairs   Air Wedge

Consider an air wedge that is formed by placing a human hair between two glass plates on one end, and allowing them to touch on the other end. When this wedge is illuminated with red light ($\lambda = 741$ nm), it has 179 bright fringes. How thick is the hair?

Picture the Problem

The air wedge used in this experiment is shown in our sketch. The separation of the plates at the end with the hair is equal to the thickness of the hair, t. It is at this end that the 179th bright fringe is observed.

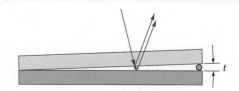

t

Strategy

Recall that the condition for bright fringes is $\frac{1}{2}\lambda + 2d = m\lambda$. The value $m = 1$ corresponds to the first bright fringe, $m = 2$ corresponds to the second bright fringe, and so on. Clearly, then, the 179th bright fringe corresponds to $m = 179$. Substituting this value for m in $\frac{1}{2}\lambda + 2d = m\lambda$ and setting the plate separation equal to the thickness of the hair, t, allows us to solve for t.

Known

$\lambda = 741 \text{ nm} = 7.41 \times 10^{-7} \text{ m}$

$m = 179$

Unknown

$t = \text{?}$

Solution

1 Rearrange the bright-fringe condition, $\frac{1}{2}\lambda + 2d = m\lambda$, to solve for the plate separation, d:

$$\frac{1}{2}\lambda + 2d = m\lambda$$
$$2d = m\lambda - \frac{1}{2}\lambda$$
$$d = \frac{\lambda}{2}\left(m - \frac{1}{2}\right)$$

> **Math HELP**
> Solving Simple Equations
> See Math Review, Section III

2 The plate separation is the thickness of the hair. Thus, d can be replaced with t. Next, substitute the given numerical values for m and λ and solve for the thickness.

$$t = \frac{\lambda}{2}\left(m - \frac{1}{2}\right)$$
$$= \frac{(7.41 \times 10^{-7} \text{ m})}{2}\left(179 - \frac{1}{2}\right)$$
$$= \boxed{6.61 \times 10^{-5} \text{ m}}$$

Insight

Thus, the thickness of the hair is roughly 66 micrometers. You can verify this conversion yourself using the following relationship:

$$1 \text{ micrometer} = 1 \text{ } \mu\text{m} = 1 \times 10^{-6} \text{ m}$$

Notice that to measure the thickness of a hair we have used a "ruler" (the red light) with a length scale that is comparable to that thickness. The hair has a thickness about 100 times larger than the wavelength of the red light. Thus, measuring the thickness of the hair with this light is like measuring the length of a football field with a meterstick.

Practice Problems

17. **Follow-up** If a thicker hair is used in this air-wedge experiment, will the number of bright fringes increase, decrease, or stay the same? How many bright fringes will be observed if the hair has a thickness of 80.0 μm?

18. **Follow-up** If blue light is used in the air-wedge experiment instead of red light, does the number of bright fringes increase, decrease, or stay the same? Explain.

19. Light with a wavelength of 550 nm shines on an air wedge. At a point where the distance between the two plates of glass is $d = 7.7 \times 10^{-5}$ m, do you expect to see a bright fringe or a dark fringe?

▲ **Figure 18.14 Colorful interference in a soap bubble**
The swirling colors in soap bubbles are created by interference, which eliminates certain wavelengths from the reflected light and enhances others. Which colors are removed or enhanced at a given point depends on the thickness of the film in that region.

Thin films produce colorful interference effects

We've all enjoyed the beautiful swirling patterns of color on the surface of a bubble. A good example is shown in **Figure 18.14**. 💬 **The colors in a thin film are the result of constructive interference and destructive interference that occur when white light reflects from the film's top and bottom surfaces.** Some colors experience destructive interference and are *eliminated* from the reflected light. Others colors are *enhanced* by constructive interference.

We analyze a thin film surrounded by air, like a soap bubble, in the same way as we analyzed the air wedge earlier in this lesson. The condition for destructive interference is found to be the following:

💬 *Why do thin films produce swirling patterns of color?*

Condition for Destructive Interference in a Thin Film

effective path length $=$ integer $\times$ wavelength
$$2nt = m\lambda \qquad\qquad m = 1, 2, 3, \ldots$$

In this equation, t is the thickness of the thin film, n is the index of refraction of the soap film, and λ is the wavelength of the light. Similarly, constructive interference requires an extra half-wavelength difference in the effective path length. This results in the following condition:

Condition for Constructive Interference in a Thin Film

effective path length $=$ $\left(\text{integer} + \frac{1}{2}\right) \times$ wavelength
$$2nt = \left(m + \tfrac{1}{2}\right)\lambda \qquad m = 0, 1, 2, \ldots$$

These conditions apply to any thin film suspended between materials whose indices of refraction are lower than that of the film.

A beam consisting of red light ($\lambda = 662$ nm) and blue light ($\lambda = 465$ nm) shines on a thin soap film. If the film has the index of refraction $n = 1.33$ and is suspended in air ($n = 1.00$), find the smallest thickness for which the blue light is eliminated by destructive interference. At this thickness the film appears red in reflected light.

Picture the Problem

Our sketch shows a soap film of thickness t and index of refraction $n = 1.33$ suspended in air ($n = 1.00$). At a certain thickness the reflected blue light will be eliminated by destructive interference, as indicated. The red light does not cancel.

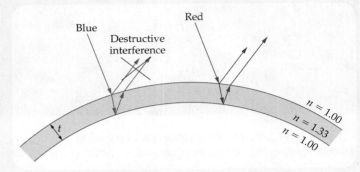

Strategy

As mentioned above, the desired thickness of the film is such that the blue light satisfies $2nt = m\lambda$, where m is equal to 1, 2, 3, and so on. The smallest film thickness corresponds to $m = 1$.

Known	Unknown
$\lambda = 465$ nm	$t = ?$
$n = 1.33$	
$m = 1$	

Solution

1 Solve $2nt = m\lambda$ for the thickness, t:

$$2nt = m\lambda$$
$$t = \frac{m\lambda}{2n}$$

2 Substitute the numerical values and calculate t:

$$t = \frac{(1)(465 \text{ nm})}{2(1.33)}$$
$$= \boxed{175 \text{ nm}}$$

Insight

Although blue light is eliminated at this thickness, red light is not. Thus, the film appears red.

Practice Problems

20. **Follow-up** What is the minimum film thickness that produces destructive interference for the red light in this system?

21. White light is incident on a soap film ($n = 1.30$) in air. The reflected light looks bluish because red light ($\lambda = 670$ nm) has been eliminated by destructive interference. What is the minimum thickness of the soap film?

22. A beam of monochromatic light experiences constructive interference when it shines on a soap film with a thickness of 183 nm and an index of refraction of 1.35. What is the wavelength of the light? (Assume that $m = 0$.)

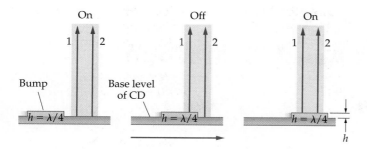

On   Off   On

1 2   1 2   1 2

Bump   Base level
of CD

$h = \lambda/4$ $h = \lambda/4$ $h = \lambda/4$

h

◀ **Figure 18.15 Reading information from a CD or DVD**
As a laser beam sweeps across the surface of a CD, the detector receives a weak "off" signal when a bump is entering or leaving the beam. When the beam reflects entirely from the top of a bump or from the base level of the CD, the detector receives a strong "on" signal. These on and off signals transmit information in much the same way that dots and dashes do in Morse code.

Physics & You: Technology Destructive interference plays a key role in the operation of a CD or DVD player. The basic idea behind these devices is that information is encoded in the form of a series of bumps on an otherwise smooth reflecting surface. A laser beam directed onto the surface is reflected back to a detector, and as the intensity of the reflected beam varies because of the bumps, the information on the CD is decoded. The basic idea is similar to using dots and dashes to send information in Morse code.

Imagine a laser beam shining on the surface of a CD. When a bump comes under the beam, as in **Figure 18.15**, the reflected beam has two parts. One part comes from the top of the bump, and the other part comes from the bottom. If these two parts are half a wavelength out of phase, they will be affected by destructive interference, and the detector will receive a weak signal. Thus, the bumps send the detector a series of "on" and "off" signals that can be converted to sound, pictures, or other types of information. The bumps are very tiny, with a height equal to a quarter of the wavelength of the laser light. This means that their height is only about 1/400 of the thickness of a human hair.

18.2 LessonCheck (MP)

Checking Concepts

23. 🔲 **State** If light reflects from a material with a lower index of refraction than the material in which it is traveling, what happens to its phase?

24. 🔲 **Explain** Does the thickness of a thin film affect the colors you see?

25. Triple Choice The ray shown in **Figure 18.16** undergoes a 180° phase change as it reflects. Is the index of refraction of material 2 higher than, lower than, or equal to 1.35? Explain.

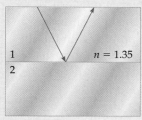

1
2    $n = 1.35$

▲ Figure 18.16

26. Triple Choice Is the phase change due to reflection of ray 1 in **Figure 18.17** greater than, less than, or equal to the phase change due to reflection of ray 2? Explain.

▶ Figure 18.17

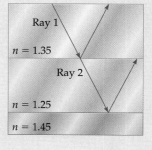

Ray 1

$n = 1.35$

Ray 2

$n = 1.25$

$n = 1.45$

Solving Problems

27. Calculate Two plates of glass are separated by an air-filled space with a thickness of 5.1×10^{-5} m. What wavelength of light experiences constructive interference for $m = 250$?

28. Calculate A beam of monochromatic light with a wavelength of 523 nm experiences destructive interference when it shines on a soap film with a thickness of 195 nm. Assuming that $m = 1$, what is the index of refraction of the soap film?

29. Calculate A film of oil ($n = 1.38$) floats on a puddle of water ($n = 1.33$). You notice that green light ($\lambda = 521$ nm) is absent in the reflection. What is the minimum thickness of the oil film?

18.3 Diffraction

Vocabulary

- diffraction
- resolution

Why do sound waves diffract more readily than light waves?

CONNECTINGIDEAS

This lesson explores the pattern of light and dark fringes formed when light diffracts through an opening.

- The Heisenberg uncertainty principle—which is fundamental to quantum physics and is discussed in Chapter 24—is similar in many respects to the diffraction of light.

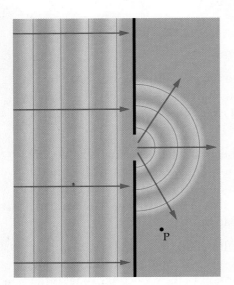

When you sit in a classroom waiting for the bell to ring, you can sometimes hear a friend down the hall talking and laughing. How can you hear your friend when the sound she is producing has to go around corners to get to you? Well, one of the key characteristics of all waves, including sound and light, is that they can bend around corners. This type of behavior is the topic of this lesson.

All waves bend around obstructions

Since light is a wave, it must exhibit behavior similar to that of the water waves in **Figure 18.18**. Notice that the waves are initially traveling directly to the right. After passing through the gap in the barrier, however, they spread out. This is as expected from Huygens's principle. Thus, an observer at point P detects waves even though she is not on a direct line with the initial waves and the gap. In general, waves spread out around barriers and through openings.

The bending of a wave around a barrier or through an opening is known as **diffraction**. A common result of diffraction is the ability to hear a person talking even when the person is out of sight around a corner. The sound waves from the person bend around the corner, just like the water waves in Figure 18.18.

It might seem that light cannot be a wave, since it does not bend around a corner along with the sound. There is a significant difference between sound waves and light waves, however. The wavelength of sound waves is about a meter, but the wavelength of light waves is about 10^{-7} m. **The huge difference in wavelength between sound waves and light waves greatly affects their behavior, because long waves diffract more than short waves. It follows that long-wavelength sound waves diffract much more than short-wavelength light waves.**

Light diffracts as it passes through a single slit

To investigate the diffraction of light, we start by considering the behavior of a beam of light as it passes through a single slit in a screen. Consider monochromatic light of wavelength λ passing through a slit of width W, as shown in **Figure 18.19**. After passing through the slit the light shines on a distant screen. According to Huygens's principle, each point within the slit can be considered as a source of new waves that travel outward toward the screen. The interference of these waves with one another generates a series of bright and dark fringes referred to as a *diffraction pattern*.

◄ Figure 18.18 **Diffraction of water waves**
When water waves pass through an opening, they spread outward, or *diffract*. Thus, an observer at point P detects waves even though this point is not on a line with the original direction of the waves through the opening. All waves exhibit similar behavior.

Analyzing the Diffraction Pattern

We can understand single-slit diffraction by referring to **Figure 18.20**. Here you see light waves traveling to a screen from various points in a slit. Consider waves from points 1 and 1′ traveling to the screen at an angle θ relative to the initial direction of the light. The path-length difference for these waves is $(W/2) \sin \theta$. This same path-length difference applies to the wave pairs 2 and 2′, 3 and 3′, and so on through all points in the slit. Each of these wave pairs acts like it came from the two slits in a two-slit experiment. Thus, we can apply the analysis of that simplified system to this more complex case.

The path-length difference for waves traveling in the forward direction is zero. As a result, all wave pairs interfere constructively, giving maximum intensity at the center of the diffraction pattern. Things change, however, when the angle θ does not equal zero. If the angle is increased until the path-length difference is half a wavelength, $\lambda/2$, then each wave pair experiences destructive interference. Thus, the first dark fringe in the diffraction pattern occurs when $(W/2) \sin \theta = \lambda/2$. Canceling the 2's on either side of this equation yields the condition for the first dark fringe:

$$W \sin \theta = \lambda$$

Similar analysis shows that the second dark fringe occurs at the angle given by

$$W \sin \theta = 2\lambda$$

In general, dark fringes satisfy the following conditions:

Conditions for Dark Fringes in Single-Slit Interference
$\dfrac{\text{path-length}}{\text{difference for slit}} = \text{integer} \times \text{wavelength}$
$W \sin \theta = m\lambda \qquad m = \pm 1, \pm 2, \pm 3, \ldots$

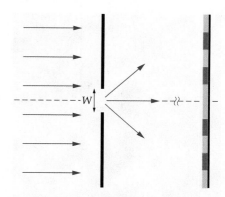

▲ Figure 18.19 Single-slit diffraction When light passes through a slit, a diffraction pattern of bright and dark fringes is formed.

PROBLEM-SOLVING NOTE

Using positive and negative values for m accounts for dark fringes both above and below the midpoint of the diffraction pattern.

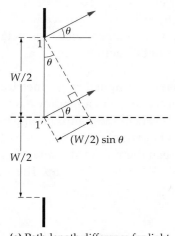

(a) Path-length difference for light waves from points 1 and 1′

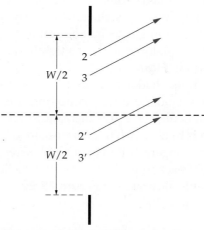

(b) Same path-length difference for other wave pairs

▲ Figure 18.20 Locating the first dark fringe in single-slit diffraction
The location of the first dark fringe in a single-slit diffraction pattern can be determined by considering pairs of waves traveling outward from the top and bottom half of a slit. **(a)** A wave pair originating at points 1 and 1′ has a path-length difference of $(W/2) \sin \theta$. These waves interfere destructively if the path-length difference is equal to half a wavelength. **(b)** The rest of the light coming from the slit can be considered to consist of additional wave pairs, like 2 and 2′, 3 and 3′, and so on. Each wave pair has the same path-length difference.

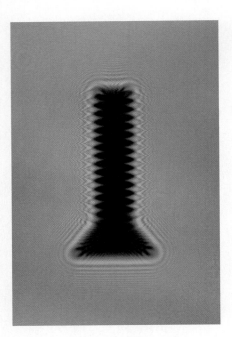

▲ Figure 18.21 Fringes around a shadow
On close inspection, the shadow of a sharp edge is seen to consist of numerous fringes produced by diffraction. The shadow cast by this metal screw is an example.

Monochromatic light passes through a slit of width $W = 1.2 \times 10^{-5}$ m. If the first dark fringe above the midpoint of the resulting diffraction pattern is at the angle $\theta = 3.25°$, what is the wavelength of the light?

Solution
Solving $W \sin \theta = m\lambda$ for the wavelength and using $m = 1$ for the first dark fringe gives

$$\lambda = \frac{W \sin \theta}{m}$$
$$= \frac{(1.2 \times 10^{-5} \text{ m}) \sin (3.25°)}{1}$$
$$= \boxed{6.80 \times 10^{-7} \text{ m}}$$

Practice Problems

30. **Concept Check** If the wavelength of the light used in a single-slit experiment is increased, does the angle to the first dark fringe increase, decrease, or stay the same? Explain.

31. What slit width produces first-order diffraction minima ($m = \pm 1$) at angles of $\pm 23°$ with 690-nm light?

Shadows are not as sharp as they seem

As mentioned earlier, waves diffract when they encounter a barrier or an opening. It follows, then, that the shadow cast by an object is not really as sharp as it might seem. For example, the shadow of an object such as a metal screw actually consists of a tightly spaced series of diffraction fringes, as shown in **Figure 18.21**.

Thus, shadows do not have the sharp boundaries implied by geometrical optics. Under ordinary conditions the diffraction pattern in a shadow may be difficult to see. It is there, however. A surefire way to see a diffraction pattern is to hold two fingers close together in front of your eyes. Try it. Don't let the fingertips touch, but look carefully between them toward a smooth, bright background. You should be able to see dark fringes in the gap. The effect is illustrated in **Figure 18.22**.

▶ Figure 18.22 Fingertip diffraction
If you hold two fingertips close to one another, you'll see dark fringes in the gap between them. This is an example of diffraction.

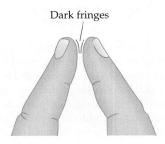

Dark fringes

Light with a wavelength of 511 nm forms a diffraction pattern after passing through a single slit of width 2.20×10^{-6} m. Find the angle associated with (**a**) the first and (**b**) the second dark fringe above the central bright fringe.

Picture the Problem

In our sketch we identify the first and second dark fringes above the central bright fringe. Notice that the first dark fringe corresponds to $m = 1$, and the second corresponds to $m = 2$. The width of the slit is $W = 2.20 \times 10^{-6}$ m.

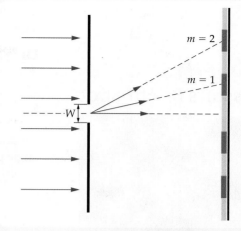

Strategy

We can find the desired angles by using the condition for dark fringes, $W \sin \theta = m\lambda$. As mentioned above, we use $m = 1$ for part (a), and $m = 2$ for part (b). The values of λ and W are given in the problem statement.

Known	Unknown
$\lambda = 511$ nm	$\theta = ?$
$W = 2.20 \times 10^{-6}$ m	

Solution

1 (a) Solve for θ using $m = 1$:

$$\theta = \sin^{-1}\left(\frac{m\lambda}{W}\right)$$

$$= \sin^{-1}\left[\frac{(1)(511 \times 10^{-9}\ \text{m})}{2.20 \times 10^{-6}\ \text{m}}\right]$$

$$= \boxed{13.4°}$$

> **Math HELP**
> Rules for Exponents
> See Math Review, Section I

2 (b) Solve for θ using $m = 2$:

$$\theta = \sin^{-1}\left(\frac{m\lambda}{W}\right)$$

$$= \sin^{-1}\left[\frac{(2)(511 \times 10^{-9}\ \text{m})}{2.20 \times 10^{-6}\ \text{m}}\right]$$

$$= \boxed{27.7°}$$

Insight

Notice that the angle to the second dark fringe is *not* simply twice the angle to the first dark fringe—though it's close. This is the case because the angle θ depends on the sine function, which is not linear.

Practice Problems

32. **Follow-up** Suppose the wavelength of the light in this single-slit experiment is changed to give the first dark fringe at an angle greater than 13.4°. Is the required wavelength greater than or less than 511 nm? Check your answer by calculating the wavelength required to give the first dark fringe at $\theta = 15.0°$.

33. Green light ($\lambda = 546$ nm) passes through a single slit. What slit width produces the first dark fringe at an angle of 16.0°?

34. Diffraction also occurs with sound waves. Consider 1300-Hz sound waves diffracted when they pass through a door that is 84 cm wide. What is the angle to the first-order ($m = 1$) diffraction minima?

Diffraction limits the way you see the world. For example, how sharply a scene appears to your eyes depends on the size of your pupils. In general, larger pupils produce sharper vision. This is why a bald eagle, with pupils that are even larger than yours, can see a small creature on the ground from a great height.

Diffraction through a Circular Opening

Let's take a look at the diffraction pattern created by a circular opening—such as the pupil of your eye. This pattern is slightly different from that produced by the linear slits studied earlier. The circular opening creates a circular diffraction pattern of alternating bright and dark regions, as shown in **Figure 18.23**. More specifically, a circular opening of diameter D produces a dark fringe at an angle given by the following condition:

First Dark Fringe for the Diffraction Pattern of a Circular Opening

$$\text{sine of angle to first dark fringe} = 1.22 \times \frac{\text{wavelength of light}}{\text{diameter of opening}}$$

$$\sin \theta = 1.22 \frac{\lambda}{D}$$

This equation shows that even if you could focus perfectly on a point source of light, it *would not* form a point image on your retina. Instead, diffraction blurs the image by spreading the light from any point out into a circular pattern, as in Figure 18.23. No matter how sharply your eye focuses, it can't avoid this blurring effect—there will always be a fuzzy circle of light rather than a sharp point on your retina. This is how diffraction limits your vision.

Sharpness and Resolution

The sharpness of vision—in particular, the ability to visually separate closely spaced objects—is referred to as **resolution**. Because diffraction through the pupil blurs images, it can be difficult to visually separate objects that are very close together. In other words, diffraction limits the *resolving power* of your vision. This is illustrated in **Figure 18.24 (a)**. Notice how the smearing of the light (due to diffraction) makes two separate light sources appear as one.

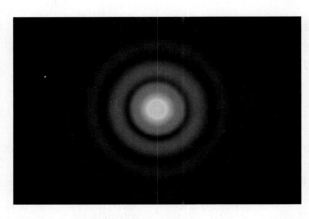

▲ Figure 18.23 **Diffraction from a circular opening**
Light passing through a circular aperture creates a circular diffraction pattern of alternating bright and dark regions.

COOLPHYSICS
Poisson's Bright Spot

When a penny is placed in a beam of light, a shadow is formed. As the light passes around the penny, each point on its rim acts like a point source of light. These sources are all the same distance from the center of the penny's shadow, and hence the center is a point of constructive interference. This results in a bright spot there, known as *Poisson's bright spot*, for the French scientist Siméon D. Poisson (1781–1840), who predicted its existence. The irony is that Poisson didn't believe that light is a wave. He used the prediction of a bright spot in the darkest part of a shadow to support his argument that the wave model must be wrong. The wave model of light gained almost universal acceptance, however, after experiments quickly showed that the bright spot does indeed exist—just as Poisson predicted.

(a)

(b)

▲ **Figure 18.24 Separating two point sources**
(a) If the angular separation between two sources is not great enough (left), their diffraction patterns overlap to the point where they appear as a single elongated source. With greater angular separation (right), the individual sources can be distinguished as separate. (b) As a car approaches, the angle between its headlights increases. This makes it easier to distinguish the headlights as separate sources of light.

An everyday example of diffraction-limited resolution is shown in **Figure 18.24 (b)**. In the first photo we see a brilliant light in the distance that may be the single headlight of an approaching motorcycle or the unresolved images of two headlights on a car. If we are seeing the headlights of a car, the angular separation between them will increase as the car approaches. When the angular separation exceeds the angle to the first dark fringe, as in the second photo, we are able to distinguish the two headlights as separate sources of light.

QUICK Example 18.9 **What's the Angle?**

Find the angle to the first dark fringe for yellow light (551 nm) passing through a circular opening with a diameter of 5.00 mm. (This is roughly the diameter of a human pupil.)

Solution
Substituting the known values into $\sin \theta = 1.22 \dfrac{\lambda}{D}$ and solving for θ gives

$$\sin \theta = 1.22 \frac{\lambda}{D}$$

$$\theta = \sin^{-1}\left[1.22 \times \frac{551 \times 10^{-9}\ \text{m}}{5.00 \times 10^{-3}\ \text{m}}\right]$$

$$= \boxed{0.00770°}$$

The small value of θ indicates that human vision has the potential for relatively high resolution.

◄ **Figure 18.25 Pointillism**
Georges Seurat's *A Sunday Afternoon on the Island of La Grande Jatte* (1884), an example of pointillism.

(**Practice Problems**)

35. [**Concept Check**] Is resolution greater with blue light or red light, all other factors being equal? Explain.

36. A spy camera is designed to read a car's license plate from an orbiting satellite. To make this possible, the angle to the first dark fringe of the camera's diffraction pattern has to be 0.000018°. What must the diameter of the camera's aperture (opening) be to accomplish this, assuming light with a wavelength of 550 nm?

Pointillism Have you ever seen a painting done in the artistic style known as *pointillism*? In this type of painting the artist applies paint to a canvas in the form of small dots of color, as in the example in **Figure 18.25**. When viewed from a distance, the individual dots blur together—due to diffraction—and the painting appears to be painted with continuous colors.

(**Physics & You: Technology**) Did you know that watching television is a lot like viewing a pointillist painting? Although a TV screen appears to show all the colors of the rainbow, it actually produces only three colors— red, green, and blue—the additive primaries. These three colors are grouped together in small points known as *pixels*, as shown in **Figure 18.26**. From a distance the three individual color spots blur together, and the eyes see the net effect of the various combinations of the three colors. Since any color can be created with the proper amounts of the three additive primary colors, the TV screen can reproduce any picture.

To see this effect in action, try the following experiment: Look for a region on a TV screen or a computer monitor where the picture is yellow. Since yellow is created by mixing red and green light equally, you will see on close examination (perhaps with the aid of a magnifying glass) that pixels in the yellow region of the screen have both the red and green dots illuminated, but the blue dots are dark. As you slowly move away from the screen, the red and green dots merge, leaving the brain with the sensation of yellow light, even though there are no yellow dots on the screen.

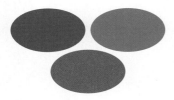

▲ **Figure 18.26 Pixels on a TV screen**
A typical pixel on the screen of a color television consists of three closely spaced color spots: one red, one blue, and one green. These are the only colors the screen actually produces.

18.3 LessonCheck (MP)

Checking Concepts

37. 🗨 **Explain** Why is diffraction more noticeable with sound waves than with light waves?

38. 🗨 **Describe** Why do two distant objects often appear as one?

39. Identify What do we call the ability to visually distinguish between closely spaced objects?

40. Apply Spy cameras use lenses with large apertures. Why are large apertures advantageous in such applications?

Solving Problems

41. Calculate Light with a wavelength of 676 nm passes through a slit 7.64 μm wide. What is the angle to the first dark fringe above the central bright fringe?

42. Calculate Light passes through a slit with a width of 2.7×10^{-6} m. If the angle to the first dark fringe above the central bright fringe is 12°, what is the wavelength of the light?

43. Calculate In a single-slit experiment light with a wavelength of 592 nm passes through the slit. The first dark fringe above the central bright fringe occurs at a angle of 21°. What is the width of the slit?

44. ⬛ **Think & Calculate** The diffraction pattern shown in **Figure 18.27** is produced by passing light from a helium-neon laser ($\lambda = 632.8$ nm) through a single slit. The angle from the central bright fringe to the first dark fringe is 17°.
(a) What is the width of the slit?
(b) If monochromatic yellow light with a wavelength of 591 nm is used instead, will the angle to the first dark fringe be greater than, less than, or equal to 17°? Explain.

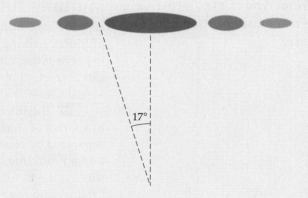

▲ Figure 18.27

45. Calculate Two point sources of light with a wavelength of 690 nm are viewed through the 5.1-mm-diameter pupil of an eye. What is the minimum angle of separation between the two sources if they are to be seen as being separate? (*Hint:* The desired minimum angle is equal to the angle to the first dark fringe of the diffraction pattern.)

18.4 Diffraction Gratings

Vocabulary

- diffraction grating
- iridescence

🔑 *What type of interference pattern does a diffraction grating produce?*

At first glance a CD looks like a shiny mirror. On closer examination you notice that it reflects all the colors of the rainbow as you turn it one way and another. What causes these colors? This lesson explores what's happening with the CD and similar devices known as *diffraction gratings*.

Multiple slits diffract light like a prism

As you know from the previous lessons, a screen with one or two slits produces striking patterns of interference fringes. It's natural to wonder what happens if the number of slits is increased. A screen with a large number of slits is referred to as a **diffraction grating**. In some cases a grating can have as many as 40,000 slits—or *lines*, as they are often called—per centimeter.

🔑 **The interference pattern formed by a diffraction grating consists of a series of sharp, widely spaced bright fringes called *principal maxima*, separated by relatively dark regions, which contain a number of weak *secondary maxima*.** Such a pattern is shown in **Figure 18.28** for the case of five slits. As the number of slits becomes larger, the principal maxima become brighter and the secondary maxima become insignificant.

Analyzing the Diffraction Pattern

As you might expect, the angle at which a principal maximum occurs depends on the wavelength of the light. This is why a grating acts like a prism—it sends the various colors (wavelengths) of white light off in different directions. In fact, a typical grating spreads the colors of light even more effectively than does a prism.

To see how to determine the angles at which principal maxima are found, consider a grating with a large number of slits. Each slit is separated from the next by a distance d, as shown in **Figure 18.29**. A beam of light with wavelength λ approaches the grating from the left and is diffracted onto a distant screen. At an angle θ to the incident direction, the path-length difference between successive slits is $d \sin \theta$, as shown in Figure 18.29. As usual, constructive interference (and hence a principal maximum) occurs when the path-length difference is an integer number of wavelengths, $m\lambda$.

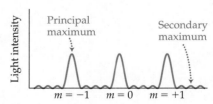

▲ Figure 18.28 **Diffraction pattern for five slits**
Light intensity in the interference pattern produced by a diffraction grating with five slits. The large principal maxima are sharper and brighter than the bright fringes in the two-slit experiment. The small secondary maxima are insignificant compared to the principal maxima.

Constructive Interference by a Diffraction Grating

path-length difference = integer × wavelength

$$d \sin \theta = m\lambda \qquad m = 0, \pm 1, \pm 2, \ldots$$

A grating with more lines per centimeter (that is, smaller spacing between slits, d) spreads light out over a wider range of angles. This occurs because a small value of d and a large value of θ have the same effect as a large d and a small θ:

$$d \sin \theta = d \sin \theta$$

QUICK Example 18.10 What's the Spacing?

Find the slit spacing necessary for 450-nm light to produce a first-order ($m = 1$) principal maximum at 15°.

Solution

First, solve the condition for constructive interference ($d \sin \theta = m\lambda$) for the spacing d. Next, substitute the numerical values for m, λ, and θ.

$$d = \frac{m\lambda}{\sin \theta}$$

$$= \frac{(1)(450 \times 10^{-9} \text{ m})}{\sin 15°}$$

$$= \boxed{1.7 \times 10^{-6} \text{ m}}$$

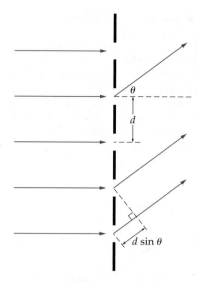

▲ **Figure 18.29 Path-length difference in a diffraction grating**
A diffraction grating consists of a number of slits with a spacing d. The difference in path length for waves from neighboring slits is $d \sin \theta$.

Practice Problems

46. Light shines on a diffraction grating with a slit spacing of 4.5×10^{-6} m. If the angle to the first-order ($m = 1$) principal maximum is 8.2°, what is the wavelength of the light?

47. A diffraction grating has a slit spacing of 1.3×10^{-5} m. What is the angle to the first-order ($m = 1$) principal maximum when light of 670 nm shines on the grating? (*Hint:* Solve for θ using the inverse sine function.)

Several principal maxima can be produced

The first-order principal maximum for the grating in Quick Example 18.10 is at 15°. What about the higher-order principal maxima? Well, the second-order principal maximum occurs at 32° and the third-order principal maximum is at 53°. When you try to solve for the fourth-order maximum, you find that $\sin \theta > 1$, which isn't possible. Therefore, there is no fourth-order principal maximum for this grating, and no maxima higher than that.

CDs typically produce three reflected beams, as shown in **Figure 18.30**. One beam reflects at an angle equal to the incident angle. This is referred to as the *specular beam*, since it is like the reflected beam that occurs with a plane mirror (*specular* means "like a mirror"). In addition, you can see two more beams corresponding to the first-order and second-order principal maxima. These beams wouldn't occur with a mirror—the microscopic lines on the CD's surface produce them.

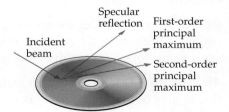

▲ **Figure 18.30 Reflection from a CD**
When a laser beam shines on a CD, a number of reflected beams are observed. The most intense of these is the specular beam, whose angle of reflection is equal to its angle of incidence—the same as if the CD were a mirror. Additional reflected beams are observed at angles corresponding to the principal maxima of the grating.

▶ **Figure 18.31 The blue morpho butterfly**
(a) The beautiful colors of the blue morpho butterfly are the result of iridescence, that is, the interference of light waves. (b) A butterfly's wings are covered with tiles, much like the shingles on a roof. The tiles themselves are lined with ridges, making them act just like a diffraction grating.

(a)

(b)

Diffraction gratings are also found in nature

Diffraction gratings are common in nature. Bird feathers and butterfly wings are familiar examples. **Figure 18.31 (a)** shows the beautiful wing of a blue morpho butterfly. A microscopic examination of a butterfly wing shows that it is covered by thousands of tiny tiles, like the shingles on a roof, as shown in **Figure 18.31 (b)**. Each of these tiles has a series of closely spaced ridges. These ridges act just like the lines on a CD, producing reflected light of different colors in different directions.

Coloration that is produced by the interference of light is referred to as **iridescence**. In fact, many of the colors you see in nature are produced by iridescence rather than by pigments. An example is shown in **Figure 18.32**. The next time you examine an iridescent object, notice how its color changes as you change your viewing angle. This is a key feature of iridescent colors. In contrast, colors produced by pigments look the same from any angle.

◀ **Figure 18.32 Iridescence in nature**
Feathers, seashells, and insects—such as this beetle—often exhibit iridescence.

18.4 LessonCheck ⓂⓅ

Checking Concepts

48. 🔌 **Describe** Use the words *light*, *dark*, *strong*, and *weak* to describe the principal maxima and secondary maxima formed by a diffraction grating.

49. Identify Does a diffraction grating have many slits or only one or two?

50. Describe What is iridescence?

51. Analyze If the spacing between the lines on a diffraction grating is increased, does the grating spread colors of light through a greater angle or a smaller angle? Explain.

52. Triple Choice If the wavelength of light used with a diffraction grating is increased, does the angle to the first principal maximum increase, decrease, or stay the same? Explain.

Solving Problems

53. Calculate The yellow light from a helium discharge tube has a wavelength of 587.5 nm. When this light shines on a certain diffraction grating, it produces a first-order principal maximum ($m = 1$) at an angle of 1.250°. Calculate the slit spacing for this grating.

54. Calculate Light with a wavelength of 692 nm shines on a diffraction grating with a slit spacing of 1.92×10^{-6} m. What is the angle to the second-order principal maximum ($m = 2$)?

55. Calculate The first-order principal maximum ($m = 1$) for a diffraction grating with a slit spacing of 2.2×10^{-6} m is at an angle of 21°. What is the wavelength of the light that is shining on this grating?

Physics & You

X-ray Diffraction

What Is It? Pure solids have a definite crystal structure. The crystals are arranged in repeatable patterns. X-ray diffraction is a nondestructive analytical technique that provides insight into the pattern of atoms—the basic building blocks of matter—in a solid. Scientists and engineers use this technique to understand the chemical and physical properties of materials.

X-ray diffraction pattern from an iridium crystal.

Who Invented It? In the seventeenth century Nicolas Steno determined that particular crystals have faces at specific angles. A century later William Hallowes Miller noted that a pure solid crystal is composed of "blocks," all having the same three-dimensional size and shape. Then, in 1895, Wilhelm Conrad Röntgen discovered X-rays. When an X-ray beam strikes a crystal, it is scattered in many directions. Almost 20 years later, Max Von Laue developed a law that connected the angle of the scattered X-rays to the orientation of the blocks in the crystal. This led to an understanding of the arrangement of atoms in matter.

How Does It Work? The process is fairly simple. A focused beam of X-rays is aimed at a sample. The outermost electrons in the atoms making up the sample diffract and scatter the X-ray beam. A detector records the angle and intensity of the diffracted rays. Finally, a computer program uses these data to construct a three-dimensional picture of the sample.

Why Is It Important? Scientists have used X-ray diffraction to uncover the structure of organic compounds, such as fatty acids, proteins, nucleic acids, penicillin, insulin, and chlorophyll. The pharmaceutical industry also uses this analytical technique to develop medications that work more efficiently in the human body to alleviate diseases and speed recovery.

▲ Professor Ada Yonath poses next to an X-ray diffraction image of a ribosome. She was awarded the Nobel Prize in chemistry in 2009.

Take It Further

1. Research *Use the Internet and other sources to learn how Ada Yonath used X-ray diffraction in her studies of the structure of the ribosome. Summarize your findings in a one-page written report.*

2. Apply *Nanotechnology is a field of engineering and materials science that focuses on the construction of materials using building blocks with dimensions of 100 nanometers or less. How might X-ray diffraction be used to improve nanoengineered products?*

Physics Lab An Application of Diffraction

In this lab you will use the diffraction of laser light to determine the width of a human hair.

Materials
- laser
- human hair
- thin wire
- meterstick
- viewing screen
- tape

Procedure

Caution: *Never direct the laser toward your eyes or the eyes of others!*

1. Obtain the wavelength of the laser from your teacher and record it.

> **Wavelength of laser, $\lambda = $ _____ m**

2. Use tape to fasten a human hair across the opening of the laser so that the hair is illuminated when the laser is turned on. The laser should only pass through the hair.

3. Darken the room and point the laser toward a white screen or wall. Turn the laser on and observe the diffraction pattern produced.

4. Measure the distance L from the hair to the diffraction pattern on the screen. Record the distance, to the nearest millimeter, in Data Table 1. (*Note*: for best results, L should be at least 2 m.)

5. Measure the distance x from the center of the central bright fringe to the either the first-order minimum ($m = 1$), the second-order minimum ($m = 2$), or the third-order minimum ($m = 3$). Record the distance x, to the nearest millimeter, and the order of the minimum, m, in Data Table 1.

6. Repeat Steps 4 and 5 four more times, each time changing either the distance between the hair and the screen or the order of the minimum.

7. Repeat Steps 2–6 using a fine wire in place of the hair. Record these results in Data Table 2 (not shown, but identical to Data Table 1 except for its title).

Data Table 1: Human Hair

Trial	L (m)	x (m)	m	d (m)
1				
2				
3				
4				
5				
Average	—	—	—	

Analysis

1. For each trial in Data Tables 1 and 2, use the relationship $d = m\lambda L/x$ to calculate the width d. Note that d is the width of the hair (or wire), m is the order of the minimum, λ is the wavelength of the laser light, L is the distance between the hair (or wire) and the screen, and x is the distance from the central maximum to the selected minimum. Record the calculated values for width in the appropriate data table.

2. Calculate the average of the hair width and wire width. Record the results in the appropriate data table.

Conclusions

1. Describe how the distance to the screen affects the pattern.

2. How does the width of the object inserted in the path of the laser beam affect the separation of the minima in the diffraction pattern?

3. Research the known values of the thickness of a human hair. How does the value you obtained compare to the range of known values?

4. Predict the pattern that would be produced by shining the laser through a fine wire mesh.

18 Study Guide

Big Idea

Like all waves, light waves show the effects of super-position and interference.

When two beams of light overlap, the waves add together. This is superposition. The result can be constructive interference (brighter light) or destructive interference (dimmer light).

18.1 Interference

🔑 Light wave interference is most noticeable when the light sources are *coherent* and *monochromatic*.

🔑 In general, if the difference in path length at a given point is an integer number of wavelengths $(\lambda, 2\lambda, 3\lambda, \ldots)$, then that point is a location of constructive interference. Likewise, if the difference in path length at a given point is an integer plus half a wavelength $\left(\dfrac{\lambda}{2}, \dfrac{3\lambda}{2}, \dfrac{5\lambda}{2}, \ldots\right)$, then that point is a location of destructive interference.

🔑 A section of wave front that passes through a small opening acts almost like a point source of waves. This is why waves—whether they are water waves or light waves—spread out after they pass through a small opening.

Key Equations

Bright fringes in a two-slit experiment occur at angles θ given by

$$d \sin \theta = m\lambda \qquad m = 0, \pm 1, \pm 2, \ldots$$

If a two-slit interference pattern is projected on a screen a distance L from the slits, the linear distance to a given bright fringe is

$$y = L \tan \theta$$

18.2 Interference in Thin Films

🔑 When a light encounters a medium with a lower index of refraction, it is reflected with no phase change. In contrast, when light encounters a medium with a higher index of refraction, it is reflected with a phase change of half a wavelength.

🔑 The colors in a thin film are the result of constructive interference and destructive interference that occur when white light reflects from the film's top and bottom surfaces.

Key Equations

When light of wavelength λ shines on an air wedge, bright fringes occur where the separation between the plates, d, is such that

$$\tfrac{1}{2}\lambda + 2d = m\lambda \qquad m = 1, 2, 3, \ldots$$

Similarly, dark fringes occur on an air wedge where the following condition is satisfied:

$$\tfrac{1}{2}\lambda + 2d = (m + \tfrac{1}{2})\lambda \qquad m = 0, 1, 2, \ldots$$

Destructive interference in a thin film with thickness t and index of refraction n satisfies the following condition:

$$2nt = m\lambda \qquad m = 1, 2, 3, \ldots$$

Constructive interference in a thin film satisfies the following condition:

$$2nt = (m + \tfrac{1}{2})\lambda \qquad m = 0, 1, 2, \ldots$$

18.3 Diffraction

🔑 The huge difference in wavelength between sound waves and light waves greatly affects their behavior, because long waves diffract more than short waves. It follows that long-wavelength sound waves diffract much more than short-wavelength light waves.

🔑 Because diffraction through the pupil blurs images, it can be difficult to visually separate objects that are very close together. In other words, diffraction limits the *resolving power* of your vision.

Key Equations

Dark fringes in a single-slit diffraction pattern are given by

$$W \sin \theta = m\lambda \qquad m = \pm 1, \pm 2, \pm 3, \ldots$$

The first dark fringe for a circular opening occurs at the angle θ given by

$$\sin \theta = 1.22 \dfrac{\lambda}{D}$$

18.4 Diffraction Gratings

🔑 The interference pattern formed by a diffraction grating consists of a series of sharp, widely spaced bright fringes called *principal maxima*, separated by relatively dark regions, which contain a number of weak *secondary maxima*.

Key Equation

The principal maxima produced by a diffraction grating occur at the angles given by

$$d \sin \theta = m\lambda \qquad m = 0, \pm 1, \pm 2, \ldots$$

where d is the distance between successive slits and λ is the wavelength of the light.

18 Assessment

For instructor-assigned homework, go to www.masteringphysics.com

ANSWERS TO SELECTED ODD-NUMBERED PROBLEMS APPEAR IN APPENDIX A.

Lesson by Lesson

18.1 Interference

Conceptual Questions

56. What happens to the two-slit interference pattern if the separation between the slits is reduced?

57. What happens to the two-slit interference pattern if the wavelength of the light is reduced?

58. **Triple Choice** Two coherent sources of light are three wavelengths out of phase. Does light from these sources experience constructive interference, destructive interference, or no interference at all? Explain.

59. If a radio station broadcasts its signal through two different antennas simultaneously, does this guarantee that the signal you receive will be stronger than the signal from a single antenna? Explain.

60. **Predict & Explain** A two-slit experiment with red light produces a set of bright fringes. **(a)** Will the spacing between the fringes increase, decrease, or stay the same if the color of the light is changed to blue? **(b)** Choose the *best* explanation from among the following:

A. The spacing will increase because blue light has a greater frequency than red light.

B. The spacing will decrease because blue light has a smaller wavelength than red light.

C. The spacing will stay the same because only the wave nature of light is important, not the color of the light.

61. When green light ($\lambda = 505$ nm) passes through a pair of slits, the interference pattern shown in **Figure 18.33 (a)** is observed. When light of a different color passes through the same pair of slits, the pattern shown in **Figure 18.33 (b)** is observed. Is the wavelength of the second color of light greater than, less than, or equal to 505 nm? Explain.

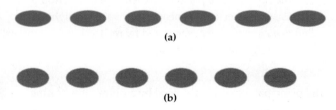

(a)

(b)

Figure 18.33

Problem Solving

62. Two sources emit waves that are coherent and in phase and have a wavelength of 26.0 m. Do the waves interfere constructively or destructively at a point 85.0 m from one source and 124 m from the other source?

63. Light with a wavelength of 520 nm is used in a two-slit experiment. If the angle to the first bright fringe above the central fringe is 0.067°, what is the separation of the slits?

64. The third bright fringe ($m = 3$) in a two-slit experiment occurs at an angle of 0.26°. If the slit separation is 4.1×10^{-4} m, what is the wavelength of the light?

65. A person stands at point P in **Figure 18.34**. The two transmitters (A and B) emit identical signals in phase with one another, which the person receives on a portable radio. If the radio picks up a maximum-strength signal at point P, what is the longest possible wavelength of the radio waves?

Figure 18.34

66. Suppose the portable radio in Figure 18.34 picks up a minimum-strength signal at point P. What is the largest possible value for the wavelength of the radio waves? Assume that the transmitters are in phase and that they emit waves with the same wavelength.

67. At point P in Figure 18.34, a woman's radio picks up a maximum signal from transmitters A and B, which emit waves with a wavelength of 75 m. If she then walks toward transmitter A with a speed of 1.1 m/s, how much time elapses before she reaches the next location with a maximum signal?

68. What is the angle to the first bright fringe above the central bright fringe in a two-slit experiment with a slit separation of 5.1×10^{-6} m and light of wavelength 550 nm?

69. Two students in a dorm room listen to a pure tone (single wavelength) produced by two loudspeakers that are in phase. The situation is shown in **Figure 18.35**. Both students hear a maximum-amplitude sound. What is the longest possible wavelength produced by the loudspeakers?

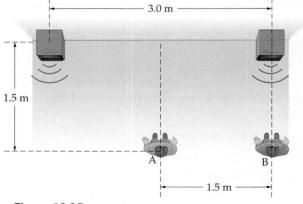

Figure 18.35

70. Moe, Larry, and Curly stand in a line with a spacing of 1.00 m. The boys are 3.00 m in front of a pair of stereo speakers 0.800 m apart, as shown in **Figure 18.36**. The speakers produce a single-frequency tone, vibrating in phase with each other. What is the longest wavelength that allows Larry to hear a loud tone while Moe and Curly hear very little?

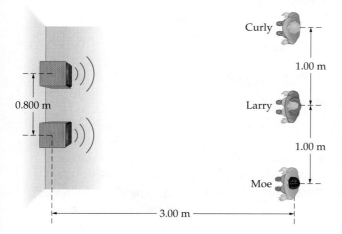

Figure 18.36

71. Think & Calculate A two-slit experiment with slits separated by 48.0×10^{-5} m produces a second-order bright fringe at an angle of 0.0990°. **(a)** Find the wavelength of the light used in this experiment. **(b)** If the slit separation is increased, does the angle for the second-order bright fringe increase, decrease, or stay the same? Explain.

72. Light from a helium-neon laser $(\lambda = 632.8 \text{ nm})$ passes through a pair of slits and forms an interference pattern on a screen located 1.40 m from the slits. **Figure 18.37** shows the interference pattern observed on the screen. What is the slit separation?

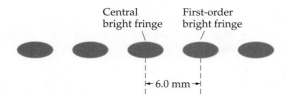

Figure 18.37

73. Think & Calculate Suppose the interference pattern shown in Figure 18.37 is produced by monochromatic light passing through two slits, with a separation of 132 μm, and onto a screen 1.20 m away. **(a)** What is the wavelength of the light? **(b)** If the wavelength of the light is increased, will the bright spots of the pattern move closer together or farther apart? Explain.

74. A physics instructor wants to produce a two-slit interference pattern large enough for her class to see, so she projects the pattern onto a screen. She would like the distance on the screen between the central bright

fringe and the bright fringe just above it to be 2.50 cm. If the slits have a separation of $d = 0.0220$ mm, what is the required distance from the slits to the screen when 632.8-nm light from a helium-neon laser is used?

18.2 Interference in Thin Films

Conceptual Questions

75. Triple Choice A ray of light travels in material 1 until it encounters material 2. The ray reflects with no phase change at the boundary between the two materials. Is the index of refraction of material 1 greater than, less than, or the same as the index of refraction of material 2? Explain.

76. Thin film 1 has a thickness that is just right to eliminate green light by destructive interference. Thin film 2 is slightly thicker than film 1. Is the light that film 2 eliminates closer to the red end or the blue end of the visible spectrum? Explain.

77. A thin film of oil $(n = 1.2)$ coats the surface of a glass lens $(n = 1.3)$. What is the phase change of light that reflects from **(a)** the air-to-oil boundary and **(b)** the oil-to-glass boundary? Explain in each case.

78. An oil film floating on water appears dark near the edges, where it is thinnest. Is the index of refraction of the oil greater than or less than that of water? Explain.

79. **Figure 18.38** shows four different cases where light of wavelength λ reflects from both the top and the bottom of a thin film of thickness d. The indices of refraction of the film and the materials above and below it are indicated in the figure. In which of the four cases will light reflected from the top and bottom of the film interfere constructively if $d = \lambda/4$?

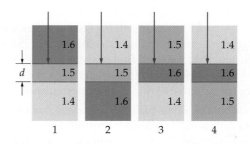

Figure 18.38

80. Repeat Problem 79, this time with $d = \lambda/2$.

Problem Solving

81. Light with a wavelength of 615 nm experiences constructive interference when it shines on a soap film whose thickness is 117 nm. What is the index of refraction of the soap film? (Assume that $m = 0$.)

82. What is the minimum thickness of a soap film $(n = 1.33)$ that will produce constructive interference of red $(\lambda = 652 \text{ nm})$ light that is traveling in air and reflects from the film?

83. Light experiences destructive interference when it shines on two glass plates separated by an air-filled space with a thickness of 4.7×10^{-5} m. What is the wavelength of light for the case where $m = 150$?

84. A thin film of oil ($n = 1.35$) floats on a puddle of water ($n = 1.33$). If light with a wavelength of 590 nm experiences destructive interference when it reflects from the film, what is the minimum thickness of the oil?

85. A soap bubble with walls 401 nm thick floats in air. If this bubble is illuminated with sunlight (wavelengths from 400 nm to 700 nm), what wavelength will be absent in the reflected light? Assume that the index of refraction of the soap film is 1.33.

86. Find the minimum thickness required for a soap film suspended in air to eliminate light with a wavelength of 495 nm. Assume that the index of refraction of the soap film is 1.33.

87. A soap film ($n = 1.33$) is 825 nm thick. What wavelengths in the range from 400 nm to 700 nm (visible light) will experience destructive interference if the film is surrounded by air on both sides?

88. **Think & Calculate** Two glass plates have a uniform separation of 4.9×10^{-5} m. **(a)** What is the value of m for which light with a wavelength of 695 nm is eliminated by destructive interference? **(b)** If the separation between the glass plates is increased, does your answer to part (a) increase, decrease, or stay the same? Explain.

89. A 742-nm-thick soap film ($n_{film} = 1.33$) rests on a glass plate ($n_{glass} = 1.52$). What wavelengths in the range from 400 nm to 700 nm (visible light) will constructively interfere when reflected from the film?

90. Light is incident from above on two plates of glass, which are separated on both ends by thin wires of diameter $d = 0.600$ μm. Considering only interference between light reflected from the bottom surface of the upper plate and light reflected from the upper surface of the lower plate, does light of wavelength 343 nm experience constructive or destructive interference?

91. **Think & Calculate** Glass plate 1 is separated from glass plate 2 by a distance d. **(a)** Considering only interference between light reflected from the bottom surface of the upper plate (plate 1) and light reflected from the upper surface of the lower plate (plate 2), what minimum separation is required for light with a wavelength of 412 nm to experience constructive interference? **(b)** Does your answer to part (a) depend on the indices of refraction of plates 1 and 2? Explain.

92. Two glass plates are separated by fine wires with diameters $d_1 = 0.0500$ mm and $d_2 = 0.0520$ mm, as shown in **Figure 18.39**. The wires are parallel and separated by a distance of 7.00 cm. If monochromatic light with $\lambda = 589$ nm is incident on the top plate from above, what is the distance (in centimeters) between adjacent dark bands in the pattern of the reflected light?

(Consider interference only between light reflected from the bottom surface of the upper plate and light reflected from the upper surface of the lower plate.)

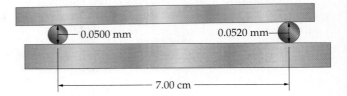

Figure 18.39

18.3 Diffraction
Conceptual Questions

93. Triple Choice A single-slit diffraction pattern is formed on a distant screen with red light. If blue light is used instead, does the width of the central fringe increase, decrease, or stay the same? Explain.

94. Triple Choice If the width of a single slit is increased, does the width of the central bright fringe increase, decrease, or stay the same? Explain.

95. Do you expect more diffraction with infrared light or ultraviolet light, all other factors being the same? Explain.

96. Triple Choice Some sunglasses claim improved vision because their lenses are tinted blue. What is the basis for this claim? Do you expect blue light to allow a resolution that is greater than, less than, or equal to the resolution with red light? Explain.

97. Explain why narrowing a slit causes light that passes through it to spread out more widely.

98. Predict & Explain **(a)** In principle, do your eyes have better resolution on a dark cloudy day or on a bright sunny day? **(b)** Choose the *best* explanation from the following:

A. Your eyes have better resolution on a cloudy day because your pupils are open wider.

B. Your eyes have better resolution on a sunny day because the bright light causes your pupils to narrow to a smaller opening.

Problem Solving

99. Light with a wavelength of 625 nm passes through a single slit in a screen. What is the angle to the first-order dark fringe ($m = 1$) if the slit width is 42×10^{-5} m?

100. What is the wavelength of light that passes through a slit of width 2.2×10^{-6} m and produces a first-order dark fringe ($m = 1$) at 18°?

101. When light with a wavelength of 562 nm passes through a slit, it forms a first-order dark fringe ($m = 1$) at 22°. What is the width of the slit?

102. Light from a helium-neon laser ($\lambda = 632.8$ nm) passes through a single slit. The angle to the second-order dark fringe ($m = 2$) of the diffraction pattern is 18.5°. What is the width of the slit?

103. What is the angle to the third-order dark fringe ($m = 3$) for a single slit of width 37×10^{-5} m if the wavelength of light is 589 nm?

104. **Think & Calculate** A single slit is illuminated with 610-nm light, and the resulting diffraction pattern is viewed on a screen 2.3 m away. **(a)** If the linear distance between the first-order and second-order dark fringes of the pattern is 12 cm, what is the width of the slit? **(b)** If the slit is made wider, will the distance between the first-order and second-order dark fringes increase or decrease? Explain.

105. A screen is placed 1.00 m behind a single slit. The central bright fringe in the resulting diffraction pattern on the screen is 1.60 cm wide—that is, the two first-order dark fringes are separated by 1.60 cm. What is the distance between the two second-order dark fringes?

106. **Splitting Binary Stars** As seen from Earth, the red dwarfs Krüger 60A and Krüger 60B form a binary star system with an angular separation of 2.5 arc seconds (1 arc second = $1/3600°$). What is the diameter of the circular opening of a telescope that would have the first-order dark fringe ($m = 1$) at an angle of 2.5 arc seconds? Assume that the light from the stars has a wavelength of 450 nm.

107. Find the diameter of the aperture (opening) of a camera that can resolve detail on the ground the size of a person (2.0 m) from an SR-71 Blackbird flying at an altitude of 27 km. Assume that the light used to form the image has a wavelength of 450 nm. (*Hint*: Set the angle for the first-order dark fringe of the diffraction pattern equal to the angle subtended by a person at 27 km.)

108. **The Resolution of HST** The Hubble Space Telescope (HST) orbits Earth at an altitude of 559 km. It has a circular objective mirror that is 2.4 m in diameter. If the HST were to look down on Earth's surface (rather than up at the stars), what is the minimum separation of two objects that could be resolved using 550-nm light? (*Note:* The HST is used only for astronomical work, but a (classified) number of similar telescopes are in orbit for spy purposes.) (*Hint:* Set the angle for the first-order dark fringe in the diffraction pattern equal to the angle between two objects at a distance of 559 km.)

109. Early cameras were little more than a box with a pinhole on the side opposite the film. **(a)** What is the angle to the first-order dark fringe in the diffraction pattern of a pinhole with a 0.50-mm diameter? **(b)** What is the greatest distance from the camera at which two point objects 15 cm apart can be resolved? Assume that the light has a wavelength of 520 nm. (*Hint*: Set the angle for the first-order dark fringe in the diffraction pattern equal to the angle between two objects 15 cm apart at a given distance.)

18.4 Diffraction Gratings
Conceptual Questions

110. The color of an iridescent object, like a butterfly wing or a feather, appears to be different when viewed from different angles. The color of a painted surface appears the same from all viewing angles. Explain the difference.

111. What is the difference between a diffraction grating and a screen with one or two slits in it?

112. What is the path-length difference between rays of light that form the first-order principal maximum of a diffraction grating?

113. Some diffraction gratings have only first-order and second-order principal maxima. Others have third-order or higher-order principal maxima as well. Explain the difference.

114. **Triple Choice** If the spacing between lines on a grating is increased, does the angle to the first-order principal maximum increase, decrease, or stay the same? Explain.

115. **Triple Choice** Is the angle to the first-order principal maximum for red light greater than, less than, or equal to the angle to the first-order principal maximum for blue light? Explain.

116. **Triple Choice** If the number of lines on a diffraction grating is increased, does the intensity of the principal maxima increase, decrease, or stay the same?

Problem Solving

117. What is the spacing (in meters) between lines on a diffraction grating with 2500 lines per centimeter?

118. A diffraction grating has lines that are separated by 1.3×10^{-5} m. How many lines per centimeter does this grating have?

119. What is the angle to the first-order principal maximum when light with a wavelength of 626 nm shines on a diffraction grating with a spacing between slits of 0.88×10^{-5} m?

120. What is the spacing between slits on a diffraction grating that produces a first-order principal maximum at 24.5° with light of wavelength 562 nm?

121. The second-order principal maximum ($m = 2$) is at an angle of 43° for a diffraction grating with a slit spacing of 2.7×10^{-6} m. What is the wavelength of light being used with this grating?

122. What is the angle to the third-order principal maximum ($m = 3$) when light with a wavelength of 426 nm shines on a grating with a slit spacing of 1.3×10^{-5} m?

123. **Think & Calculate** The second-order principal maximum produced by a diffraction grating with a slit spacing of 1.8×10^{-5} m is at an angle of 3.1°. **(a)** What is the wavelength of the light that shines on the grating? **(b)** If a grating with a smaller slit spacing is used with this light, is the angle of the second-order principal maximum greater than or less than 3.1°? Explain.

124. White light (wavelengths from 400 nm to 700 nm) strikes a diffraction grating with a slit spacing of 1.3×10^{-6} m. How many complete visible spectra will be formed on either side of the central maximum?

125. Monochromatic light strikes a diffraction grating before illuminating a screen 2.10 m away. If the two first-order principal maxima are separated by 1.53 m on the screen, what is the distance between the two second-order principal maxima?

Mixed Review

126. Monochromatic light passes through a single slit and forms a diffraction pattern of alternating bright and dark fringes. If the width of the slit is decreased, do the dark fringes move farther away from or closer to the center of the pattern? Explain.

127. If the wavelength of the light used in a two-slit experiment is decreased, does the spacing between the bright fringes increase, decrease, or stay the same? Explain.

128. If the diameter of the pupil increases, does the resolution of the eye increase, decrease, or stay the same? Explain.

129. Diffraction gratings are often rated by the number of lines (slits) they have per centimeter. If the number of lines per centimeter is increased in a grating, do the principal maxima move farther away from or closer to the center of the diffraction pattern? Explain.

130. Light whose wavelength is 561 nm passes through two slits in a screen. If the separation between the slits is 12×10^{-5} m, what is the angle to the second-order bright fringe above the central bright fringe?

131. **Predict & Explain** **(a)** If a thin film of liquid floating on water has an index of refraction less than that of water, will the film appear bright or dark in reflected light as its thickness goes to zero? **(b)** Choose the *best* explanation from the following:

A. The film will appear bright because as its thickness goes to zero, the path-length difference for the reflected rays goes to zero.

B. The film will appear dark because there is a phase change at both interfaces, which will cause destructive interference of the reflected rays.

132. Light passes through a single slit and produces a pattern of light and dark fringes on a wall 1.3 m from the slit. What is the linear distance from the center of the diffraction pattern to the first-order dark fringe if the light has a wavelength of 589 nm and the width of the slit is 9.7×10^{-5} m?

133. **Think & Calculate** Figure 18.40 shows a single-slit diffraction pattern formed by light passing through a slit of width $W = 11.2 \ \mu m$ and illuminating a screen 0.865 m behind the slit. **(a)** What is the wavelength of the light? **(b)** If the width of the slit is decreased, will the

← 15.2 cm →

Figure 18.40

distance indicated in Figure 18.41 be greater than or less than 15.2 cm? Explain.

Writing about Science

134. Write a report on iridescence. Find several examples from the natural world. Can some of these be found in your area? In addition, find several examples of iridescence in manufactured objects. Are there some iridescent objects in your home?

135. **Connect to the** Big Idea Describe how the pixels on a TV screen give the sensation of all possible colors. Also, take a look at a comic strip with a magnifying glass. Discuss what you see there in the context of the diffraction of light.

Read, Reason, and Respond

Resolving Lines on an HDTV The American Television Systems Committee (ATSC) sets the standards for high-definition television (HDTV). One of the approved HDTV formats is 1080p, which means 1080 horizontal lines scanned progressively (p)—that is, one line after another in sequence from top to bottom. Another standard is 1080i, or 1080 interlace. With this format it takes two scans of the screen to show a complete picture—the first scan shows the "even" horizontal lines, the second scan shows the "odd" horizontal lines.

For the following problems, assume that 1080 horizontal lines are displayed on a television with a screen that is 39.9 cm high (32-in diagonal) and that the light coming from the screen has a wavelength of 474 nm inside your eyes. Also, assume that the pupils of your eyes have a diameter of 5.50 mm.

136. What is the minimum angle (in degrees) your eye can resolve for the given wavelength and pupil diameter? (This is the angle to the first-order dark fringe of the diffraction pattern.)

A. 0.240×10^{-3} C. 4.94×10^{-3}

B. 4.05×10^{-3} D. 6.02×10^{-3}

137. What is the linear separation between horizontal lines on the screen?

A. 0.0235 mm   C. 0.369 mm

B. 0.145 mm   D. 0.926 mm

138. What is the angular separation (in degrees) of the horizontal lines as viewed from a distance of 3.66 m?

A. 5.79×10^{-3} C. 14.7×10^{-3}

B. 14.5×10^{-3} D. 69.3×10^{-3}

Standardized Test Prep

Multiple Choice

Use the diagram below to answer Questions 1–3. The screen on the left has two small slits spaced 1 μm apart. In-phase light waves pass through the slits and produce an interference pattern on the screen on the right. The central bright fringe is located at point B, and the first-order bright fringe is located at point A.

1. If point P is at the midpoint between the slits, point B is 10 cm from point P, and point A is 2 mm from point B, what is the wavelength of the light?
   (A) 20 nm
   (B) 10 μm
   (C) 200 mm
   (D) 10 cm

2. Which of the following describes a possible approximate location of another bright fringe?
   (A) halfway between points A and B
   (B) 4 mm from point B on the line between A and B
   (C) halfway between points P and A
   (D) halfway between points P and B

3. If the two slits are moved farther apart, the location of point A (the first-order bright fringe) will
   (A) remain the same distance from point B.
   (B) move closer to point B.
   (C) move farther from point B.
   (D) move closer to point P.

4. Two waves that differ in phase by 360° will interact to produce
   (A) constructive interference.
   (B) destructive interference.
   (C) reflection.
   (D) refraction.

5. A beam of blue light ($\lambda = 462$ nm) shines on a thin soap film with the index of refraction $n = 1.29$. What is the minimum thickness of the film that that will cause destructive interference of the blue light?
   (A) 179 nm   (C) 298 nm
   (B) 231 nm   (D) 358 nm

6. A thin soap film has the index of refraction $n = 1.31$ and a thickness $t = 108$ nm. What is the maximum wavelength of light that experiences constructive interference by this film?
   (A) 216 nm   (C) 432 nm
   (B) 283 nm   (D) 566 nm

7. Light changes phase by 180° when
   (A) passing from a medium with a lower index of refraction into a medium with a higher index of refraction.
   (B) passing from a medium with a higher index of refraction into a medium with a lower index of refraction.
   (C) reflecting from a boundary between a material with a higher index of refraction and one with a lower index of refraction.
   (D) reflecting from a boundary between a material with a lower index of refraction and one with a higher index of refraction.

Extended Response

8. A diffraction grating has 750 lines/mm. Laser light is used to create a diffraction pattern on a screen 1.00 m from the grating. The first-order principal maximum is located 54.0 cm from the midpoint of the pattern. Describe what happens to the diffraction pattern if laser light with a shorter wavelength is used instead. Use the appropriate equation to support your answer.

> **Test-Taking Hint**
> Interference between two light sources is determined mathematically in a way quite similar to the way interference between two sound sources is determined.

If You Had Difficulty With . . .

Question	1	2	3	4	5	6	7	8				
See Lesson(s)	18.1, 18.3	18.1, 18.3	18.1, 18.3	18.1	18.2	18.2	18.1	18.4				

19

Electric Charges and Forces

Inside

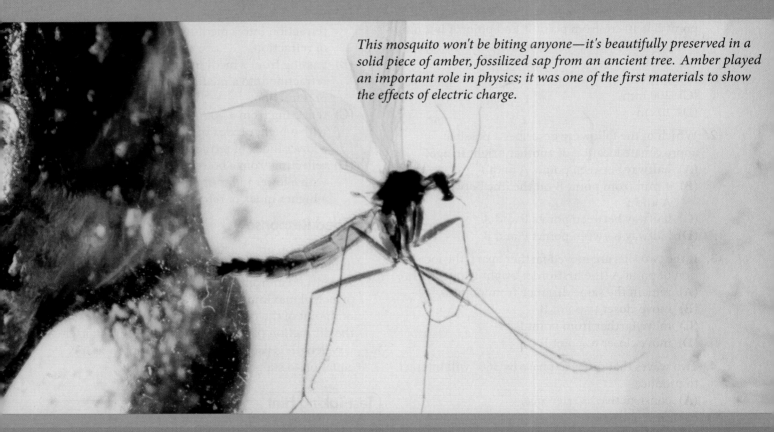

This mosquito won't be biting anyone—it's beautifully preserved in a solid piece of amber, fossilized sap from an ancient tree. Amber played an important role in physics; it was one of the first materials to show the effects of electric charge.

Big Idea

Matter is made of electric charges, and electric charges exert forces on one another.

We are all made of electric charges. Every atom in every human body contains positive and negative electric charges held together by an attractive force that is similar to gravity—only vastly stronger. In this chapter you'll learn about the basic properties of electric charges and the force that acts between them.

19.1 Electric Charge

You walk across a carpet, touch a doorknob, and—bazinga!—you get an electric shock. That shock is actually a small-scale lightning bolt. In this lesson we discuss the basic physics of electric charge and fill in the details behind some of the shocking experiences of everyday life.

Rubbing objects together can charge them

The effects of electric charge have been known since at least 600 B.C. About that time the Greeks noticed that amber—a solid, translucent material formed from the fossilized resin of extinct trees—has a peculiar property. If you rub a piece of amber with animal fur, it attracts small, lightweight objects. This phenomenon is illustrated in **Figure 19.1** on the next page.

For some time it was thought that amber was unique in its ability to become charged. Much later it was discovered that other materials behave in this way as well. For example, if you rub glass with a piece of silk, it too can attract small objects. In this respect, glass and amber seem to be the same. It turns out, however, that these two materials have different types of charges.

Vocabulary

- positive charge
- negative charge
- neutral
- coulomb
- charge quantization
- ion
- insulator
- conductor
- semiconductor

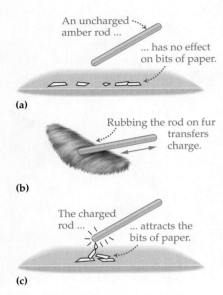

▲ Figure 19.1 Charging an amber rod
(a) An uncharged amber rod exerts no force on scraps of paper. (b) When the rod is rubbed against a piece of fur, it becomes charged. (c) The charged rod attracts the scraps of paper.

Electric charge comes in two distinct types

To see that the charges on amber and on glass are different, imagine suspending a small, charged rod of amber from a thread, as in **Figure 19.2**. If a second piece of charged amber is brought near the rod, as shown in Figure 19.2 (a), the rod rotates away. This indicates a repulsive force between the two pieces of amber. Thus, we conclude that like charges repel.

On the other hand, if a piece of charged glass is brought near the amber rod, the amber rotates toward the glass, as shown in Figure 19.2 (b). This indicates an attractive force. It follows that the *different* charges on glass and amber attract one another. We refer to different charges as being the *opposite* of one another, as in the familiar expression "opposites attract."

Positive and Negative Charge We know today that the two types of charge found on amber and glass are, in fact, the only types of electric charge that exist. We still use the names proposed by Benjamin Franklin (1706–1790) in 1747, **positive charge** (+) for the charge on glass and **negative charge** (−) for the charge on amber.

Calling the different charges positive and negative is arbitrary. We could just as well call them black and white or Sheldon and Penny. Actually, Franklin's names positive and negative turn out to be quite useful mathematically. For example, an object that contains equal amounts of positive charge and negative charge has a total charge of zero—the positive and negative charges add up to zero. Objects with zero total charge are said to be electrically **neutral**. Most everyday objects are electrically neutral because they contain equal amounts of positive and negative charge.

Atoms have positive, negative, and neutral particles

A familiar example of an electrically neutral object is the atom. Each atom contains a small, dense nucleus with a positive charge that is surrounded by a "cloud" of *electrons* (from the Greek word for amber, *elektron*) with an equal negative charge. Two types of particles are found in the nucleus; one is positively charged, and the other is electrically neutral. A simplified representation of an atom is shown in **Figure 19.3**.

▶ Figure 19.2 Likes repel and opposites attract
A charged amber rod is suspended by a string. According to the convention introduced by Benjamin Franklin, the charge on the amber is designated as negative. (a) When another charged amber rod is brought near the suspended rod, it rotates away, indicating a repulsive force between like charges. (b) When a charged glass rod is brought close to the suspended amber rod, the amber rotates toward the glass, indicating an attractive force and the existence of a second type of charge, which we designate as positive.

(a) Like charges repel.

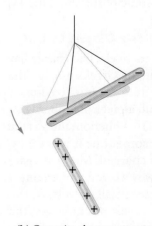

(b) Opposite charges attract.

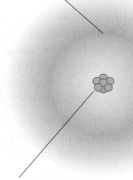

Negatively charged electron

Negatively charged electron cloud

Positively charged nucleus

Positively charged nucleus

(a) Simplified picture of an atom

(b) More accurate picture of an atom

◀ Figure 19.3 **The structure of an atom** (a) A crude representation of an atom, showing the positively charged nucleus at its center and the negatively charged electrons orbiting the nucleus. (b) More accurately, the electrons should be thought of as forming a "cloud" of negative charge surrounding the nucleus. The density of the cloud at a given location is proportional to the probability that an electron will be found there.

Electrons

All electrons have exactly the same electric charge. This charge is very small and is defined to have magnitude e.

Magnitude of an Electron's Charge, e

$e = 1.60 \times 10^{-19}$ C

SI unit: coulomb (C)

In this expression, C is a unit of charge referred to as the **coulomb**, after the French physicist Charles Augustin de Coulomb (1736–1806). Since electrons have negative charge, the charge on an electron is $-e$. This is one of the defining properties of the electron. Another defining property of the electron is its mass, m_e:

$$m_e = 9.11 \times 10^{-31} \text{ kg}$$

Protons

In contrast, the charge on a *proton*—one of the main constituents of the nucleus—is *exactly* $+e$. Therefore, the total charge on atoms, which have equal numbers of electrons and protons, is precisely zero. The mass of the proton is

$$m_p = 1.673 \times 10^{-27} \text{ kg}$$

This is about 2000 times larger than the mass of the electron.

Neutrons

The other main constituent of the nucleus is the *neutron*, which, as its name implies, has zero charge. Its mass is slightly larger than that of the proton:

$$m_n = 1.675 \times 10^{-27} \text{ kg}$$

Electric charge is quantized

Since electrons always have the charge $-e$ and protons always have the charge $+e$, it follows that all objects must have a total charge that is an integer multiple of e. This conclusion was confirmed early in the twentieth century by the American physicist Robert A. Millikan (1868–1953). In a classic series of experiments, Millikan found that the charge on an object (he used oil drops) can be $\pm e$, $\pm 2e$, $\pm 3e$, and so on, but never $1.5e$ or $-9.3847e$, for example. The fact that electric charge comes in integer multiples of e is referred to as **charge quantization**.

What amount of charge can an object have?

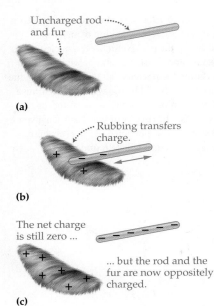

Uncharged rod and fur

(a)

Rubbing transfers charge.

(b)

The net charge is still zero ...

... but the rod and the fur are now oppositely charged.

(c)

▲ **Figure 19.4 Charge transfer**
(a) Initially, an amber rod and a piece of fur are electrically neutral; that is, they each contain equal amounts of positive and negative charge. **(b)** As they are rubbed together, charge is transferred from one to the other. **(c)** In the end, the fur and the rod have total charges of equal magnitude but opposite sign.

Math HELP
Dimensional Analysis
See Lesson 1.3

The quantization of electric charge is not noticeable in everyday situations because the unit of charge (e) is so small. Charge is like the air around us—which seems like a continuous fluid even though it's made of small individual molecules. Charge quantization is key to understanding the behavior of atoms and molecules, however. After all, the addition or removal of even a single electron is a significant event for an atom or a molecule.

A coulomb is a large amount of charge

It takes a lot of electrons to make 1 coulomb of charge. In fact, since the charge of an electron has a magnitude of only 1.60×10^{-19} C/electron, the number of electrons in a coulomb is

$$\frac{1 \text{ C}}{1.60 \times 10^{-19} \text{ C/electron}} = 6.25 \times 10^{18} \text{ electrons}$$

A powerful lightning bolt can deliver about 20–30 coulombs of charge. A more common everyday unit of charge is the microcoulomb, μC, where $1 \text{ μC} = 10^{-6}$ C. This smaller unit still corresponds to 6.25 trillion electrons.

QUICK Example 19.1 How Much Charge?

Find the amount of positive electric charge in 1 mole of helium atoms. (*Hint*: The nucleus of a helium atom consists of two protons and two neutrons.)

Solution

Each helium atom contains two protons, each with a positive charge of magnitude e. Therefore, the total positive charge in a mole of helium is

$$1 \text{ mole} \left(\frac{6.022 \times 10^{23} \text{ atoms}}{\text{mole}} \right) \left(\frac{2 \text{ protons}}{\text{atom}} \right) \left(\frac{1.60 \times 10^{-19} \text{ C}}{\text{proton}} \right)$$

$$= \boxed{1.93 \times 10^5 \text{ C}}$$

Thus, a mere 4 g of helium (1 mole) contains almost 200,000 C of positive charge. Of course, it contains the same amount of negative charge, as well.

Practice Problems

1. Find the total charge of a system consisting of 3.9×10^7 electrons.

2. How much negative electric charge is contained in 2 moles of carbon?

Electric charge can be transferred between objects

How is it that rubbing a piece of amber with fur gives the amber an electric charge? An early—but **erroneous**—idea was that the friction of rubbing *created* the charge. We now know, however, that rubbing the fur across the amber simply results in a *transfer* of charge from the fur to the amber. The total amount of charge remains the same in this and all other processes.

This transfer of charge is illustrated in **Figure 19.4**. Before charging, the fur and the amber are both neutral. During the rubbing process some electrons are transferred from the fur to the amber, giving the amber a negative charge. At the same time the fur acquires a positive charge. At no time during this process is charge ever created or destroyed. This, in fact, is an example of one of the fundamental conservation laws of physics:

🔋 **Electric charge is *conserved*. This means that the total electric charge in the universe is constant.** No physical process can result in an increase or decrease in the total amount of electric charge in the universe.

When charge is transferred from one object to another, it is generally due to the movement of electrons. In a typical solid the nuclei of the atoms are fixed in position. The outer electrons of these atoms, however, are weakly bound and easily separated. As a piece of fur rubs across amber, for example, some of the electrons that were originally part of the atoms in the fur are separated from those atoms and deposited onto atoms in the amber. An atom that gains or loses electrons is called an **ion**. More specifically, atoms that lose electrons become *positive ions* and atoms that gain electrons become *negative ions*. This charge-transfer process is referred to as *charging by separation*.

Charging by contact varies with the material

When two materials are rubbed together, the magnitude *and* the sign of the charge each material acquires depends on how strongly it holds onto its electrons. For example, if silk is rubbed against glass, the silk acquires a *negative* charge. If silk is rubbed against amber, however, the silk becomes *positively* charged.

Triboelectric Charging Transferring charge by rubbing objects together is a type of charging by separation known as *triboelectric charging*. The process can be understood by referring to **Table 19.1**. The larger the number of plus signs associated with a material in the table, the more readily it gives up electrons and becomes positively charged. Similarly, the larger the number of minus signs associated with a material, the more readily it acquires electrons and becomes negatively charged.

In general, when two materials in Table 19.1 are rubbed together, the one higher in the list becomes positively charged, and the one lower in the list becomes negatively charged. The greater the separation on the list, the greater the magnitude of the charge. For example, Table 19.1 shows that rubbing silk on amber gives the silk a positive charge. It also shows that rubbing silk on glass gives the silk a negative charge, since silk is lower on the list than glass. The table also shows that silk becomes negatively charged when rubbed against nylon, though the magnitude of the charge is less than when silk is rubbed on glass.

Charging by Collision Charge separation occurs not only when one object is rubbed against another, but also when objects collide. For example, collisions of crystals of ice in a rain cloud cause charge separation that can result in bolts of lightning that bring the charges together again. The rotating blades of a helicopter become charged due to the collisions between the blades and dust particles in the air. The charged blades give off sparks that are visible at night, as shown in **Figure 19.5** on the next page.

Similarly, particles in the rings of Saturn are constantly undergoing collisions and becoming charged. In fact, when the *Voyager* spacecraft examined the rings of Saturn, it observed electric discharges, similar to lightning bolts on Earth. In addition, the faint radial lines, or *spokes*, that extend across the rings of Saturn—which cannot be explained by gravitational forces alone—are the result of electric forces between the charged particles. Some of these spokes can be seen in **Figure 19.6** on the next page.

🔋 *How does the total amount of charge in the universe change over time?*

CONNECTINGIDEAS

The concept of a conserved quantity was introduced for energy in Chapter 6 and then applied to momentum in Chapter 7.
• Here we see that electric charge is also a conserved quantity.

Table 19.1	Triboelectric Charging
Material	**Relative Charging with Rubbing**
Rabbit fur	++++++
Glass	+++++
Human hair	++++
Nylon	+++
Silk	++
Paper	+
Cotton	−
Wood	−−
Amber	−−−
Rubber	−−−−
PVC	−−−−−
Teflon	−−−−−−

▲ **Figure 19.5 Sparking helicopter blades**
Rapidly rotating helicopter blades become charged as they collide with dust particles in the air. When the charge builds up enough, the blades begin to give off sparks.

▲ **Figure 19.6 Spokes in the rings of Saturn**
The dark radial lines in this photo of Saturn's rings are referred to as *spokes*. The spokes arise because the ice particles that make up the rings collide with one another and become charged.

CONCEPTUAL **Example 19.2 Does the Mass Change?**

Is the mass of an amber rod after charging it by rubbing it with fur greater than, less than, or the same as its mass before charging?

Reasoning and Discussion
Since an amber rod becomes negatively charged, it acquires electrons from the fur. Each electron has a small, but nonzero, mass. Therefore, the mass of the rod increases ever so slightly when it is charged.

Answer
The mass of the amber rod is greater after it has been charged.

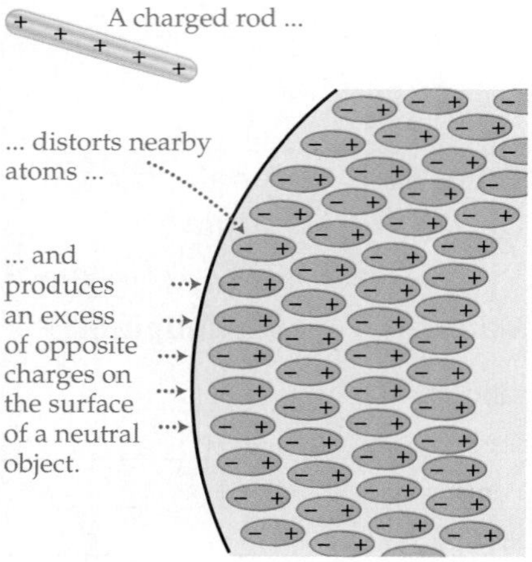

A charged rod ...

... distorts nearby atoms ...

... and produces an excess of opposite charges on the surface of a neutral object.

▲ **Figure 19.7 Electrical polarization**
When a positively charged rod is brought close to a neutral object, the atoms at the surface of the object distort, producing an excess of negative charge on the surface. The induced charge is referred to as a *polarization charge*. Because the polarization charge is opposite that on the rod, there is an attractive force between the rod and the object.

An electric charge can attract a neutral object

We know that charges of opposite sign attract. It is also possible, however, for a charged rod to attract small objects that have zero total charge. The mechanism responsible for this attraction is called *polarization*.

The Polarization Process To see how polarization works, consider **Figure 19.7**. Here we show a positively charged rod held close to an enlarged view of a neutral object. An atom at the surface of the neutral object is elongated because the negative electrons in it are attracted to the rod while the positive protons are repelled. As a result, a net negative charge—called a polarization charge—develops on the object's surface near the rod. The attractive force between the rod and this *induced* polarization charge produces an attraction between the rod and the neutral object.

Of course, the same conclusion is reached if we consider a negative rod held near a neutral object—except in this case the polarization charge is positive. Thus, the effect of polarization is to give rise to an attractive force regardless of the sign of the charged object. It is for this reason that both charged amber and charged glass attract neutral objects—even though their charges are opposite. A similar example is shown in **Figure 19.8**. The stream of water is uncharged, but the charged balloon attracts the stream by polarizing it.

▲ Figure 19.8 Deflecting a stream of water
Rubbing a balloon against a cloth surface gives the balloon a negative electrical charge. The balloon can then attract a stream of water, even though water molecules are electrically neutral.

Physics & You: Technology A potentially dangerous, and initially unsuspected, consequence of polarization arises in hospitals during endoscopic surgery. In these procedures a tube carrying a small video camera is inserted into the body. The resulting video image is produced by electrons striking the inside surface of a computer monitor's screen, which is kept positively charged to attract the electrons. Small airborne particles in the operating room—including dust, lint, and skin cells—are polarized by the positive charge on the screen and are attracted to its exterior surface.

The problem arises when a surgeon touches the screen to point out an important feature to other medical personnel. Even the slightest touch can transfer particles—many of which carry bacteria—from the screen to the surgeon's finger and from there to the patient. In fact, the surgeon's finger doesn't even have to touch the screen. As the finger approaches the screen, it too becomes polarized and attracts particles from the screen. Incidents like these have resulted in infections. Surgeons are now cautioned to keep their fingers away from the surface of the video monitor.

Not all materials are good conductors

Suppose you rub one end of an amber rod with fur. The result is that the rubbed portion becomes charged, and the other end remains neutral. The charge does not move about from one end of the rod to the other. Materials like amber, in which charges are not free to move, are called **insulators**. Most electrical insulators are nonmetallic substances, and most are also good thermal insulators. In contrast, a **conductor** is a material that allows charges to move freely from one location to another. Most metals are good conductors.

As an example of insulators and conductors, suppose an uncharged metal sphere is supported on an insulating base. If a charged rod is brought into contact with the sphere, as shown in **Figure 19.9 (a)**, some charge is transferred to the sphere at the point of contact. The charge does not stay put, however. Since the metal is a good conductor, the charges are free to move about the sphere, which they do because of their mutual repulsion. The result is a uniform distribution of charge over the surface of the sphere, as shown in **Figure 19.9 (b)**. The insulating base prevents charge from flowing from the sphere into the ground.

Free Electrons On a microscopic level the difference between conductors and insulators is that the atoms in conductors allow one or more of their outermost electrons to become detached. These detached electrons, often referred to as *conduction electrons*, can move freely throughout the conductor. In a sense, the conduction electrons behave almost like gas molecules moving about within a container. Insulators, in contrast, have very few, if any, free electrons. In an insulator the electrons are bound to their atoms and cannot move from place to place within the material. Since the flow of electric charge can be dangerous to people, insulating gloves like those in **Figure 19.10** on the next page are important to the safety of electrical workers.

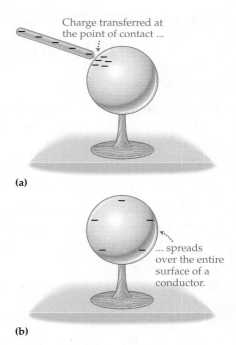

Charge transferred at the point of contact ...

(a)

... spreads over the entire surface of a conductor.

(b)

▲ Figure 19.9 Charging a conductor
(a) When an uncharged metal sphere is touched by a charged rod, some charge is transferred at the point of contact. **(b)** Because like charges repel and because charges move freely through a conductor, the transferred charge quickly spreads out and covers the entire surface of the sphere.

COOL PHYSICS
Photocopiers

Selenium is a *photoconductive* material that conducts electricity when light shines on it, but is an insulator when in the dark. This property makes selenium useful in photocopiers. Copiers have a selenium-coated drum on which an image of the document to be copied is projected. Negatively charged toner powder is then wiped across the drum. The toner sticks to those positively charged portions of the drum that were not illuminated. Finally, the drum is brought into contact with paper and the toner is fused to the paper by heat.

▲ Figure 19.10 **Insulating gloves**
Electric charge flowing through a person can be dangerous. To avoid this hazard electrical workers wear insulating gloves that prevent the flow of electricity.

Materials that have properties intermediate between those of a good conductor and a good insulator are referred to as **semiconductors**. These materials can be fine-tuned to display almost any desired degree of conductivity. The great versatility of semiconductors is one reason they have found such wide areas of application in electronics and computers.

19.1 LessonCheck MP

Checking Concepts

3. ⊂⊃ **Explain** If an object is electrically neutral, what can you say about the amounts of positive and negative charge it contains?

4. ⊂⊃ **Explain** What does charge quantization tell us about the amount of charge an object can have?

5. ⊂⊃ **Explain** Is the total amount of charge in the universe changed when we charge an object?

6. Predict & Explain An electrically neutral object is given a positive charge.
(**a**) Does the object's mass increase, decrease, or stay the same as a result of being charged?
(**b**) Choose the *best* explanation from among the following:

A. To give the object a positive charge means removing some of its electrons. This will reduce its mass.

B. Since electric charges have mass, giving the object a positive charge will increase its mass.

C. Charge is conserved, and therefore the mass of the object will remain the same.

7. **Apply** If rabbit fur is rubbed against glass, what is the sign of the charge each object acquires? Explain. (Refer to Table 19.1.)

Solving Problems

8. Think & Calculate A system consists of 55 electrons and 43 protons.
(**a**) Is the total charge of the system positive or negative? Explain.
(**b**) What is the total charge of the system?

9. **Calculate** A system consists of electrons and protons only. It contains 150 electrons and has a total charge of $+22e$. What is the mass of the system?

10. **Calculate** A system consists of electrons and protons only. It contains 320 protons and has a total charge of $-51e$. What is the mass of the system?

11. **Calculate** How much positive charge is contained in 3 moles of argon?

19.2 Electric Force

You've heard the expression "opposites attract and likes repel." This saying comes directly from the behavior of electric charges. Not only do opposite charges attract and like charges repel, but the strength of the attraction or repulsion depends on the magnitudes of the charges. The force between electric charges, which we refer to as the *electric or electrostatic force*, also depends on the separation between the charges. We explore the details of the electric force in this lesson.

Vocabulary
• Coulomb's law

Coulomb's law describes the electric force

Let's consider two electric charges, q_1 and q_2. The strength (or magnitude) of the force between these charges depends on both their magnitudes, $|q_1|$ and $|q_2|$, and their separation, r. What Coulomb discovered is that if you double charge q_1, the force doubles. If you double charge q_2, the force again doubles. **The electrostatic force depends on the product of the magnitude of the charges.**

$$F \text{ depends on } |q_1||q_2|$$

Thus, doubling either charge doubles the force.

Coulomb also discovered that the electric force becomes weaker as the charges are moved farther apart, as you might expect. In fact, he found that if you double the separation between the charges, the force drops by a factor of 4. **The electrostatic force also depends on the inverse square of the distance between the charges.**

Coulomb combined these observations into a law. **Coulomb's law** relates the strength of the electrostatic force between point charges to the magnitude of the charges and the distance between them:

> **Coulomb's Law for the Magnitude of the Electrostatic Force between Point Charges**
>
> $$\text{electrostatic force} = k \frac{\left(\begin{array}{c}\text{magnitude} \\ \text{of charge } q_1\end{array}\right) \times \left(\begin{array}{c}\text{magnitude} \\ \text{of charge } q_2\end{array}\right)}{(\text{distance between the charges})^2}$$
>
> $$F = k\frac{|q_1||q_2|}{r^2}$$
>
> SI unit: newton (N)

In this equation the constant k has the following value:

$$k = 8.99 \times 10^9 \text{ N} \cdot \text{m}^2/\text{C}^2$$

The units of k are those required for the electrostatic force to have the newton as its unit.

How is the electrostatic force between two charges related to the magnitude of the charges and the distance between them?

CONNECTING IDEAS

The inverse-square relationship was introduced in Chapter 9, where it occurred in Newton's law of universal gravitation.

• Here we see that an inverse-square relationship also appears in Coulomb's law of electrostatic force.

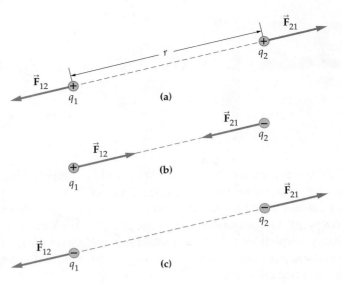

▲ **Figure 19.11 Forces between point charges**
The forces exerted by two point charges on one another always act along the line connecting the charges. If the charges have the same sign, as in **(a)** and **(c)**, the forces are repulsive; that is, each charge experiences a force that points away from the other charge. Charges of opposite sign, as in **(b)**, experience attractive forces. Notice that in all cases the forces exerted on the two charges form an action-reaction pair, where the one force is equal in magnitude and opposite in direction to the other force.

◀ **Figure 19.12 Bad hair day**
The Van de Graaff generator charges this student's hair, giving each strand of hair the same type of charge. The like charges repel, creating the ultimate bad hair day.

Electric force acts along a line between the charges

How is the direction of the electric force determined?

The magnitude of the electric force is given by Coulomb's law. The direction of this force is easy to determine. **The electric force acts along the line connecting the two charges. In addition, we know that like charges repel and opposite charges attract.** These properties are illustrated in **Figure 19.11**, where force vectors are shown for pairs of charges of various signs. Thus, to find the electric force we first calculate the magnitude of the force with Coulomb's law and then determine its direction with the "opposites attract, likes repel" rule.

Notice that Newton's third law, about paired action-reaction forces, applies to each of the cases shown in Figure 19.11. For example, the force exerted on charge 1 by charge 2 is always equal in magnitude and opposite in direction to the force exerted on charge 2 by charge 1. Expressed mathematically, this is $\vec{F}_{21} = -\vec{F}_{12}$.

Figure 19.12 illustrates the "opposites attract, likes repel" rule in a dramatic way. The student is touching the metal ball on the top of an electrostatic generator (known as a Van de Graaff generator). The generator gives the student a charge—literally! Each of his hairs is charged with charges of the same sign (like charges), and therefore the hairs repel one another. That is why the student's hair is standing on end—Coulomb's law in action. (The student is kept safe from electrical shock by standing on an insulating pad.)

An electron and a proton are separated and then released from rest simultaneously. The two particles are free to move. When they collide, are they at the midpoint of their initial separation, closer to the initial position of the proton, or closer to the initial position of the electron?

Reasoning and Discussion

Because of Newton's third law, the forces exerted on the electron and the proton are equal in magnitude and opposite in direction. For this reason, it might seem that the particles meet at the midpoint.

The masses of the particles are quite different, however. In fact, recall that the mass of the proton is about 2000 times greater than the mass of the electron. Therefore, the proton's acceleration $(a = F/m)$ is about 2000 times less than the electron's acceleration. As a result, the particles collide near the initial position of the proton. More specifically, they collide at the location of the center of mass of the system, which remains at the same point throughout the process.

Answer

The particles collide closer to the initial position of the proton.

Electric force is similar to gravitational force

It's interesting to compare Coulomb's law for the electric force and Newton's law for the force of gravity. The equations are as follows:

Coulomb's law $\qquad F = k\dfrac{|q_1||q_2|}{r^2}$

Newton's law of gravity $\qquad F = G\dfrac{m_1 m_2}{r^2}$

Similarities and Differences In each case the force decreases as the square of the distance between the objects. In addition, each force depends on the product of two magnitudes of a physical quantity. For electric force the physical quantity is the charge; for gravity it is the mass.

Equally significant, however, are the differences. For example, the force of gravity is always attractive, whereas the electric force can be attractive or repulsive. This difference has important consequences.

Astronomical Systems One consequence is that gravity dominates in astronomical systems, where the electric force plays essentially no role. A little thought shows why. **Because the electric force can be attractive or repulsive, the total electric force between neutral objects, such as the Earth and the Moon, is essentially zero. Basically, the attractive and repulsive forces cancel one another.**

Not so with gravity, however. Gravity is *always* attractive, exerting a larger total force on larger astronomical bodies. Thus, the total gravitational force between the Earth and the Moon is not zero. Gravity is also behind the formation of black holes—objects whose gravity is so strong that not even light can escape from them. If a star or other object comes too close to a black hole it will be pulled into the hole, as shown in **Figure 19.13** on the next page. This addition of matter makes the black hole's gravity even stronger. In astronomy it is truly gravity that rules; the electric force is unimportant.

How much electric force acts between neutral objects?

► **Figure 19.13 Black-hole gravity**
Black holes are formed by gravity, not by electric forces. Here we see an artist's conception of a star orbiting a black hole, with the strong gravity of the black hole pulling matter away from the star. The matter disappears inside the black hole, never to be seen again. This makes the black hole's gravity even stronger.

Atomic Systems On the other hand, the electric force rules at the atomic level, where gravity plays essentially no role. To see why, let's compare the electric and gravitational forces between a proton and an electron in a hydrogen atom. If we assume the distance between the particles is equal to the radius of hydrogen, $r = 5.29 \times 10^{-11}$ m, then the magnitude of the gravitational force is

$$F_g = G\frac{m_e m_p}{r^2}$$

$$= 6.67 \times 10^{-11}\ \mathrm{N \cdot m^2/kg^2} \left(\frac{(9.11 \times 10^{-31}\ \mathrm{kg})(1.673 \times 10^{-27}\ \mathrm{kg})}{(5.29 \times 10^{-11}\ \mathrm{m})^2} \right)$$

$$= 3.63 \times 10^{-47}\ \mathrm{N}$$

Calculating the magnitude of the electric force between the particles gives

$$F_e = k\frac{|q_1||q_2|}{r^2}$$

$$= 8.99 \times 10^9\ \mathrm{N \cdot m^2/C^2} \left(\frac{|-1.60 \times 10^{-19}\ \mathrm{C}||1.60 \times 10^{-19}\ \mathrm{C}|}{(5.29 \times 10^{-11}\ \mathrm{m})^2} \right)$$

$$= 8.22 \times 10^{-8}\ \mathrm{N}$$

Taking the ratio of these two forces, we find that the electric force is greater than the gravitational force by a staggeringly huge factor:

$$\frac{F_e}{F_g} = \frac{8.22 \times 10^{-8}\ \mathrm{N}}{3.63 \times 10^{-47}\ \mathrm{N}}$$

$$= 2.26 \times 10^{39}$$

$$= 2{,}260{,}000{,}000{,}000{,}000{,}000{,}000{,}000{,}000{,}000{,}000{,}000$$

This enormous factor explains why a small piece of charged amber can lift bits of paper off the ground, even though the entire mass of the Earth is pulling downward on the paper. It also explains why the force of gravity is unimportant in atomic systems.

In an effort to better understand the behavior of atomic systems, the Danish physicist Niels Bohr (1885–1962) introduced a simple model for the hydrogen atom. In the Bohr model the electron is imagined to move in a circular orbit about a stationary proton. The force responsible for the electron's circular motion is the electric force of attraction between the electron and the proton. Given that the radius of the electron's orbit is 5.29×10^{-11} m and its mass is $m_e = 9.11 \times 10^{-31}$ kg, find the electron's speed.

Picture the Problem

Our sketch shows the electron moving with a speed v in its orbit of radius r. Because the proton is so much more massive than the electron, it is essentially stationary at the center of the orbit. The electron has the charge $-e$ and the proton has the charge $+e$.

Strategy

The idea behind this model is that the force required to make the electron move in a circular path is provided by the electric force of attraction between the electron and the proton. Thus, as with any circular motion, we set the force acting on the electron equal to its mass times its centripetal acceleration. This allows us to solve for the centripetal acceleration, $a_{cp} = v^2/r$, which in turn gives us the speed, v.

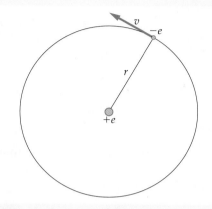

Known

$r = 5.29 \times 10^{-11}$ m

$m_e = 9.11 \times 10^{-31}$ kg

Unknown

$v = ?$

Solution

1 Set the Coulomb force between the electron and the proton equal to the centripetal force required for the electron's circular orbit. Then, substitute e for the magnitude of each charge and v^2/r for a_{cp}:

$$k\frac{|q_1||q_2|}{r^2} = m_e a_{cp}$$

$$k\frac{e^2}{r^2} = m_e\frac{v^2}{r}$$

Math HELP
Solving Simple Equations
See Math Review, Section III

2 Solve for the speed of the electron, v:

$$v = e\sqrt{\frac{k}{m_e r}}$$

3 Substitute the known values and calculate v (recall that $1\,\text{N} = 1\,\text{kg} \cdot \text{m/s}^2$):

$$v = (1.60 \times 10^{-19}\,\text{C}) \times$$

$$\sqrt{\frac{8.99 \times 10^9\,\text{N} \cdot \text{m}^2/\text{C}^2}{(9.11 \times 10^{-31}\,\text{kg})(5.29 \times 10^{-11}\,\text{m})}}$$

$$= \boxed{2.19 \times 10^6\,\text{m/s}}$$

Insight

If you could travel around the world at this speed, your trip would take only about 18 s. It would be a lethal trip, though. At this speed your centripetal acceleration would be 75,000 times the acceleration due to gravity. The centripetal acceleration of the electron in this first Bohr orbit around the proton is about 10^{22} times greater than the acceleration due to gravity at the surface of the Earth.

12. Follow-up The second Bohr orbit in Guided Example 19.4 has a radius that is four times the radius of the first orbit. What is the speed of an electron in this orbit?

13. At what separation distance is the electrostatic force between a $+11.2$-μC point charge and a $+29.1$-μC point charge equal in magnitude to 1.77 N?

14. The repulsive electric force between the point charges -6.36×10^{-6} C and q has a magnitude of 1.32 N when the separation between the charges is 1.07 m. Find the sign and the magnitude of the charge q.

Electric forces keep most objects close to neutral

Another indication of the strength of the electric force is given in the following Quick Example.

> QUICK Example 19.5 What's the Force?
>
> Find the electric force between two 1.00-C charges separated by 1.00 m.
>
> **Solution**
> Substituting $q_1 = q_2 = 1.00$ C and $r = 1.00$ m in Coulomb's law, we find
>
> $$F = k\frac{|q_1||q_2|}{r^2}$$
>
> $$= (8.99 \times 10^9 \, \text{N} \cdot \text{m}^2/\text{C}^2)\frac{(1.00 \, \text{C})(1.00 \, \text{C})}{(1.00 \, \text{m})^2}$$
>
> $$= \boxed{8.99 \times 10^9 \, \text{N}}$$
>
> This is a very large force in everyday terms.

The calculation above shows that two 1-coulomb charges separated by 1 meter exert about a million tons of force on each other. If the charge in your textbook could be separated into a pile of positive charge on one side of the room and a pile of negative charge on the other side, the force needed to hold them apart would be roughly 10^{10} tons! Thus, everyday objects are never far from electrical neutrality, because disturbing that neutrality would require tremendous forces.

19.2 LessonCheck (MP)

Checking Concepts

15. ⬤ **Apply** A system consists of two charges separated by 1 m. How does the magnitude of the electric force change when each charge and the separation distance are doubled? Explain.

16. ⬤ **Describe** What is the direction of the electric force between two positive charges?

17. ⬤ **Describe** What kind of object experiences zero electric force?

18. **Big Idea** What is the force that holds electrons and nuclei together in an atom?

19. **Rank** Four systems are described below. Rank the systems in order of increasing magnitude of the electric force. Indicate ties where appropriate.

System	q_1 (C)	q_2 (C)	r (m)
A	1	2	1
B	−2	4	3
C	5	−5	5
D	16	3	4

20. **Triple Choice** A system consists of two charges, q and $10q$. The force exerted on charge q has a magnitude of F. Does the force exerted on the charge $10q$ have a magnitude that is greater than, less than, or equal to F? Explain.

Solving Problems

21. **Think & Calculate** **(a)** What is the magnitude of the electric force between charges of 0.25 C and 0.11 C at a separation of 0.88 m? **(b)** If the separation between the charges is increased, does the magnitude of the force increase, decrease, or stay the same? Explain.

22. **Calculate** The attractive electric force between the point charges q and $-2q$ has a magnitude of 2.2 N when the separation between the charges is 1.4 m. What is the magnitude of charge q?

23. **Calculate** A point charge, $q = -0.35$ nC, is fixed at the origin. Where must a proton be placed in order for the electric force acting on it to be exactly opposite to its weight? (Let the y axis be vertical.)

24. **Calculate** When two identical ions are separated by a distance of 6.2×10^{-10} m, the electrostatic force each exerts on the other is 5.4×10^{-9} N. How many electrons are missing from each ion?

19.3 Combining Electric Forces

The electric force, like all forces, is a vector quantity. So when a charge experiences forces due to two or more other charges, the total force on it is the *vector* sum of the individual forces. You'll see how to apply this idea in this lesson.

Electric forces add by vector addition

How is the total electric force on a charge determined?

Finding the total force on an individual charge due to several other charges is straightforward. **The total force acting on a given charge is the *sum* of the individual forces between just *two* charges at a time, with the force between each pair of charges given by Coulomb's law.**

As a first example of combining electric forces, consider the system shown in **Figure 19.14**. In this case the total force on charge 1, $\vec{F}_1$, is the vector sum of the forces due to charge 2 and charge 3:

$$\vec{F}_1 = \vec{F}_{12} + \vec{F}_{13}$$

Thus, electric forces combine by vector addition to give the total force. This addition process is referred to as the *superposition* of forces.

In the following Guided Example, we apply superposition to three charges in a line.

CONNECTINGIDEAS

Adding vector quantities was initially applied to displacements and velocities in Chapter 4. Chapter 5 extended the concept to forces.
- Vector addition applies here as well, because the electric force—like all forces—is a vector quantity.

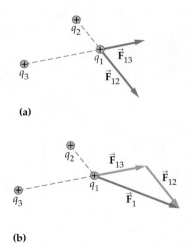

(a)

(b)

▲ **Figure 19.14 Superposition of forces**
(a) Forces are exerted on charge q_1 by the charges q_2 and q_3. These forces are $\vec{F}_{12}$ and $\vec{F}_{13}$, respectively. (b) The total force acting on q_1, which we label $\vec{F}_1$, is the vector sum of $\vec{F}_{12}$ and $\vec{F}_{13}$.

A charge $q_1 = -5.4$ μC is at the origin, and a charge $q_2 = -2.2$ μC is on the x axis at $x = 1.00$ m. Find the total force acting on a charge $q_3 = +1.6$ μC located on the x axis at $x = 0.75$ m.

Picture the Problem

The situation is shown in our sketch, with each charge at its appropriate location. Notice that the forces exerted on charge q_3 by the charges q_1 and q_2 are in opposite directions. We give the force on q_3 due to q_1 the label $\vec{\mathbf{F}}_{31}$ and the force on q_3 due to q_2 the label $\vec{\mathbf{F}}_{32}$.

Strategy

The total force on q_3 is the vector sum of the forces due to q_1 and q_2. In particular, notice that $\vec{\mathbf{F}}_{31}$ points in the negative x direction and $\vec{\mathbf{F}}_{32}$ points in the positive x direction. The magnitude of $\vec{\mathbf{F}}_{31}$ is $k|q_1||q_3|/r^2$, with $r = 0.75$ m. Similarly, the magnitude of $\vec{\mathbf{F}}_{32}$ is $k|q_2||q_3|/r^2$, with $r = 0.25$ m.

Known

$q_1 = -5.4$ μC

$q_2 = -2.2$ μC

$q_3 = +1.6$ μC

locations of all charges

Unknown

$F_3 = ?$

Solution

1 Find the force acting on q_3 due to q_1. This force is in the negative x direction and has a negative sign:

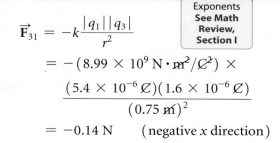

$$\vec{\mathbf{F}}_{31} = -k\frac{|q_1||q_3|}{r^2}$$

$$= -(8.99 \times 10^9 \text{ N} \cdot \text{m}^2/\text{C}^2) \times$$

$$\frac{(5.4 \times 10^{-6} \text{ C})(1.6 \times 10^{-6} \text{ C})}{(0.75 \text{ m})^2}$$

$$= -0.14 \text{ N} \quad (\text{negative } x \text{ direction})$$

2 Find the force acting on q_3 due to q_2. This force is in the positive x direction and has a positive sign:

$$\vec{\mathbf{F}}_{32} = k\frac{|q_2||q_3|}{r^2}$$

$$= (8.99 \times 10^9 \text{ N} \cdot \text{m}^2/\text{C}^2) \times$$

$$\frac{(2.2 \times 10^{-6} \text{ C})(1.6 \times 10^{-6} \text{ C})}{(0.25 \text{ m})^2}$$

$$= 0.51 \text{ N} \quad (\text{positive } x \text{ direction})$$

3 Superpose (add) these two forces to find the total force, $\vec{\mathbf{F}}_3$, acting on q_3:

$$\vec{\mathbf{F}}_3 = \vec{\mathbf{F}}_{31} + \vec{\mathbf{F}}_{32}$$

$$= -0.14 \text{ N} + 0.51 \text{ N}$$

$$= \boxed{0.37 \text{ N}} \quad (\text{positive } x \text{ direction})$$

Math HELP

Exponents
See Math
Review,
Section I

Insight

The total force acting on q_3 has a magnitude of 0.37 N, and it points in the positive x direction. Notice that the force due to q_2 is greater than that due to q_1, even though q_1 has the greater magnitude. The reason is that q_2 is three times closer to q_3 than q_1 is.

25. Follow-up Find the total force on q_3 in Guided Example 19.6 if it is at the location $x = 0.25$ m.

26. Given that $q = +12$ μC and $d = 16$ cm, find the direction and the magnitude of the total electrostatic force exerted on the point charge q_1 in **Figure 19.15**.

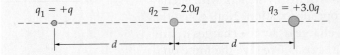

$q_1 = +q$ $q_2 = -2.0q$ $q_3 = +3.0q$

d d

▲ **Figure 19.15**

27. Think & Calculate (a) Given that $q = +12$ μC and $d = 16$ cm, find the direction and the magnitude of the total electrostatic force exerted on the point charge q_2 in Figure 19.15. (b) How would your answers to part (a) change if the distance d were tripled?

Electric forces can be added graphically

You may be wondering how we handle a situation where the individual forces do not act along the same line. In such cases we start by drawing arrows representing the individual force vectors. The sum of these vectors gives the total force. The sum can be found using components or graphically by placing the individual force vectors head to tail, head to tail, and so on, as in Figure 19.14.

CONCEPTUAL Example 19.7 Comparing Forces

A charge $-q$ is to be placed at either point A or point B in the accompanying figure. Assume that points A and B lie on a line that is midway between the two positive charges. Is the total force on $-q$ at point A greater than, less than, or equal to the total force on $-q$ at point B?

Reasoning and Discussion

Point A is closer to the two positive charges than is point B. As a result, the force exerted by each positive charge is greater when the charge $-q$ is placed at A. The *total* force, however, is zero at point A. This is because the equal attractive forces due to the two positive charges cancel, as shown in the diagram. At point B, on the other hand, the attractive forces combine to give a total force that is downward. Thus, the charge $-q$ experiences a smaller total force at point A than at point B.

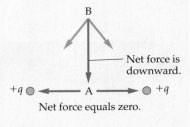

Answer

The total force at point A is less than the total force at point B.

A spherical distribution of charge acts just like a point charge

Although Coulomb's law is stated in terms of point charges, it can be applied to spherical charge distributions as well. For example, suppose a sphere has a charge Q spread evenly over its surface. If a point charge q is outside the sphere, a distance r from its center, the force between the point charge and the sphere is simply

$$F = k\frac{|q||Q|}{r^2}$$

Thus, for charges outside a sphere, a spherical charge distribution behaves as if all its charge were concentrated in a point at its center. Charges inside the sphere experience zero force. In general, the electrical behavior of spherical *charge* distributions is similar to the gravitational behavior of spherical *mass* distributions.

Surface Charge Density

In the next Active Example, we consider a system where a charge Q is spread evenly over the surface of a sphere. In such a case it is convenient to specify the amount of *charge per area* on the sphere. The charge per area is referred to as the *surface charge density, σ*. (The symbol σ is the lowercase Greek letter sigma.)

The surface charge density is defined by the following equation:

$$\sigma = \frac{Q}{A}$$

It follows that if a sphere has an area A and a surface charge density σ, its total charge is

$$Q = \sigma A$$

The dimensions of σ are charge per area, so its SI units are C/m^2. Since the amount of charge in everyday situations is generally much less than a coulomb, it's common to see the surface charge density expressed in units of $\mu C/m^2$.

ACTIVE Example 19.8   Find the Force Exerted by a Charged Sphere

A sphere with a surface area of $0.13 \ m^2$ has a uniform surface charge density equal to $5.9 \ \mu C/m^2$. A point charge of magnitude $0.71 \ \mu C$ is outside the sphere, at a distance of 0.45 m from its center. Find the magnitude of the force exerted by the sphere on the point charge.

Solution (*Perform the calculations indicated in each step.*)

1. Calculate the total charge on the sphere: $\qquad\qquad Q = 0.77 \ \mu C$

2. Use Coulomb's law to calculate the magnitude of the force between the sphere and the point charge: $\qquad F = 0.024 \ N$

Insight

As long as the point charge is outside the charged sphere, as it is in this case, the charge on the sphere can be treated as though it is concentrated at the center.

19.3 LessonCheck (MP)

Checking Concepts

28. (icon) **Explain** An object is acted on by more than one electric force. How do you determine the total electric force acting on the object?

29. Describe How can the analysis of a sphere having an evenly distributed electric charge on its surface be simplified?

30. Apply If the magnitude of each charge in Guided Example 19.6 is doubled, how does the total force exerted on charge q_3 change? Explain.

31. Triple Choice Suppose the sphere in Active Example 19.8 is replaced by one with half the radius, but with the same surface charge density. Is the force exerted by this sphere on the point charge greater than, less than, or equal to the force exerted by the original sphere? Explain.

Solving Problems

32. Calculate Three point charges lie on the x axis. Charge 1 ($+9.9$ μC) is at the origin, charge 2 (-5.1 μC) is at $x = 12$ cm, and charge 3 ($+4.4$ μC) is at $x = 15$ cm. What are the direction and magnitude of the total force exerted on charge 3?

33. Calculate Referring to Problem 32, what are the direction and the magnitude of the total force exerted on charge 2?

34. Calculate A sphere with a surface area of 0.0224 m^2 and a surface charge density of $+12.1$ μC/m^2 exerts an electrostatic force of magnitude 46.9×10^{-3} N on a point charge of $+1.75$ μC. Find the separation between the point charge and the center of the sphere.

35. Challenge In Guided Example 19.6, the total force acting on the charge q_3 is toward the right. To what value of x should q_3 be moved for the total force on it to be zero?

Electrocardiogram Technician

On a typical ECG of a healthy heart, the waves have distinct shapes.

What Is an Electrocardiogram? Electrocardiograms (ECGs) are graphs that provide a record of the electrical impulses produced by contractions of heart muscle. Each graph provides information about the health of a heart.

What Does an ECG Technician Do? Electrocardiogram technicians operate and maintain ECG equipment. They also prepare patients by placing electrodes on skin surfaces. The electrodes detect heart-generated electrical impulses and transmit them to the ECG machine that produces the graph.

ECG technicians work in a variety of medical locations, such as emergency rooms, cardiology clinics, and private practices. They should have outgoing personalities and good communication skills because they interact with patients on a very personal basis. An ECG technician is often one of the first people seen by a patient who is suffering from a serious heart problem. Therefore, these technicians must be able to work in tense situations. Helping to calm the patient is often the one of the first things the technician does. ECG technicians also interact with nurses, doctors, and other health-care professionals.

How Do You Become an ECG Technician? Unlike many other medical positions, the position of ECG technician does not require a bachelor's degree. Training can occur on the job, and many community colleges offer courses. Some programs offer additional training in taking samples of blood (phlebotomy), cardiopulmonary resuscitation (CPR), medical transcription, and first aid. This additional training can lead to better job opportunities.

▲ In a stress test an ECG monitors a person's heart during physical exertion.

Take It Further

1. Infer ECG electrodes are attached to a patient's skin with a gel. Should this gel be a good conductor or a good insulator? Explain why.

2. Research Use the Internet to research ECG technician training courses offered in your area. What are the prerequisites for such a course? Summarize your findings in a brief report.

Physics Lab Investigating Coulomb's Law

In this lab you will use a device known as a *Coulomb's law apparatus* to investigate how the electrostatic force depends on the distance between charged objects.

Materials
- Coulomb's law apparatus
- hard rubber rod
- fur

Procedure

Note: *All scale readings should be made when the mirror image of the ball is hidden behind the suspended ball.*

1. Adjust the suspended ball so that it hangs in the middle of the viewing area in front of the mirror, approximately centered on the scale markings. It should also be level with the fixed ball attached to the wooden guide block. Record the equilibrium position of the suspended ball in Data Table 1.

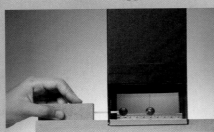

2. Charge the rubber rod by rubbing it with the fur.

3. Remove the guide block from the tower. Touch the charged rubber rod to the ball that is attached to the guide block. This transfers charge to the ball.

Note: *After the charge is transferred, the rest of the lab should be completed as quickly as possible.*

4. Slide the guide block back into the tower.

5. Gently slide the charged ball toward the suspended ball until the two balls touch. After touching, the two balls will have the same charge and will repel each other.

6. Move the guide block to the 1.5-cm position. Observe and record the new position of the suspended ball in Data Table 1.

7. Repeat Step 6 as needed in order to compete Data Table 1. Notice that the guide block moves to the right in 0.5-cm increments.

Data Table 1

Equilibrium Position of Suspended Ball: _____ cm				
Position of Ball on Guide Block (cm)	Position of Suspended Ball (cm)	d (cm)	r (cm)	$1/r^2$ (cm^{-2})
1.5				
2.0				
2.5				
3.0				
3.5				

Analysis

1. For each position of the guide block, calculate the following quantities and record them in Data Table 1:
- d, the suspended ball's distance from its equilibrium position
- r, the separation between the two balls
- $1/r^2$, the inverse square of the separation between the two balls

2. Plot two graphs to determine the dependence of the force on the distance between charges.
- Graph 1 plots the force (or d) versus r.
- Graph 2 plots the force (or d) versus $1/r^2$.

Assume that the separation distance, d, is directly proportional to the electrostatic force that acts. That is, use the values of d as a substitute for the magnitudes of the force when plotting the graphs.

Conclusions

1. Why were readings taken as quickly as possible?

2. Why were you instructed to make all readings of the suspended ball's position with the mirror image of the ball hidden behind the suspended ball?

3. What do the two graphs tell you about the relationship between electrostatic force and the distance between charged objects? Explain.

4. Cite two sources of experimental error, and discuss how they may have affected your results.

19 Study Guide

Big Idea

Matter is made of electric charges, and electric charges exert forces on one another.

All atoms contain both positive and negative electric charges. These charges exert strong forces on one another—much stronger, in fact, than the force of gravity. Like charges attract each other, while opposite charges repel each other.

19.1 Electric Charge

🔑 Most everyday objects are electrically neutral because they contain equal amounts of positive and negative charge.

🔑 Since electrons always have the charge $-e$ and protons always have the charge $+e$, it follows that all objects must have a total charge that is an integer multiple of e.

🔑 Electric charge is *conserved*. This means that the total electric charge in the universe is constant.

• Objects with zero total charge are said to be electrically *neutral*.

• The fact that electric charge comes in amounts that are always integer multiples of e is referred to as *charge quantization*.

• Conductors allow the flow of electric charge; insulators prevent the flow of electric charge.

• The SI unit of charge is the coulomb (C).

Key Equation

The charge on an electron has the following magnitude:

$$e = 1.60 \times 10^{-19}\,\text{C}$$

19.2 Electric Force

🔑 The electrostatic force depends on the product of the magnitude of the charges. It also depends on the inverse square of the distance between them.

🔑 The electrostatic force acts along the line connecting the two charges. In addition, like charges repel and opposite charges attract.

🔑 Because the electric force can be attractive or repulsive, the total electric force between neutral objects, such as the Earth and the Moon, is essentially zero. Basically, the attractive and repulsive forces cancel one another.

Key Equation

Coulomb's law gives the magnitude of the electrostatic force between two point charges, q_1 and q_2, separated by a distance r:

$$F = k\frac{|q_1|\,|q_2|}{r^2}$$

The constant k in Coulomb's law is

$$k = 8.99 \times 10^9\,\text{N} \cdot \text{m}^2/\text{C}$$

19.3 Combining Electric Forces

🔑 The total force acting on a given charge is the *sum* of the individual forces between just *two* charges at a time, with the force between each pair of charges given by Coulomb's law.

• Electric forces combine by vector addition. This addition process is referred to as *superposition* of forces.

• A spherical charge distribution behaves as if all its charge were concentrated in a point at its center.

19 Assessment
For instructor-assigned homework, go to www.masteringphysics.com

ANSWERS TO SELECTED ODD-NUMBERED PROBLEMS APPEAR IN APPENDIX A.

Lesson by Lesson

19.1 Electric Charge

Conceptual Questions

36. When an object that was neutral becomes charged, does the total charge of the universe change? Explain.

37. Explain why a comb that has been rubbed through your hair attracts small bits of paper, even though the paper is uncharged.

38. **Predict & Explain** An electrically neutral object is given a negative charge. **(a)** In principle, does the object's mass increase, decrease, or stay the same as a result of being charged? **(b)** Choose the *best* explanation from among the following:

A. To give an object a negative charge requires giving it more electrons, which will increase its mass.

B. A positive charge increases an object's mass; a negative charge decreases its mass.

C. Charge is conserved, and therefore the mass of the object will remain the same.

39. Use Table 19.1 to answer the following questions. **(a)** Is the charge on the rubber balloon shown in Figure 19.8 more likely to be positive or negative? Explain. (The balloon was rubbed on cloth to charge it.) **(b)** If the charge on the balloon is reversed, will the stream of water deflect toward or away from the balloon? Explain.

40. Use Table 19.1 to answer the following questions. **(a)** If glass is rubbed against nylon, what is the sign of the charge each acquires? Explain. **(b)** Repeat part (a) for the case of glass and rubber. **(c)** For the cases described in parts (a) and (b), in which one is the magnitude of the charge greater? Explain.

Problem Solving

41. How much positive charge is in 2 moles of carbon?

42. Find the total charge of a system consisting of 212 electrons and 165 protons.

43. What is the total electric charge of 1.5 kg of electrons?

44. What is the total electric charge of 1.5 kg of protons?

45. A system consists of electrons, protons, and neutrons. It contains 250 neutrons and 150 protons. If the total charge of the system is $-35e$, what is its mass?

46. A container holds a gas consisting of 1.75 moles of oxygen molecules. One in a million of these molecules has lost a single electron. What is the total charge of the gas?

47. **Adhesive Tape** When adhesive tape is pulled from a dispenser, the detached tape acquires a positive charge and the tape remaining in the dispenser acquires a negative charge. If the tape pulled from the dispenser has 0.14 µC of charge per centimeter, what length of tape must be pulled to transfer 1.8×10^{13} electrons to the remaining tape?

48. A system of 1525 particles, each of which is either an electron or a proton, has a total charge of -5.456×10^{-17} C. **(a)** How many electrons are in this system? **(b)** What is the mass of this system?

49. A system consisting solely of electrons and protons has a total electric charge of 2.4×10^{-16} C and a total mass of 4.5×10^{-24} kg. How many protons are in this system?

19.2 Electric Force

Conceptual Questions

50. A charged rod is brought near a suspended object. The object is repelled by the rod. Can we conclude that the suspended object is charged? Explain.

51. A charged rod is brought near a suspended object. The object is attracted to the rod. Can we conclude that the suspended object is charged? Explain.

52. Two charged objects exert equal but opposite forces on one another. Why don't these forces cancel, leaving zero total force on the objects?

53. **Rank** Four systems are described below. Rank the systems in order of increasing separation between the charges. Indicate ties where appropriate.

System	q_1 (C)	q_2 (C)	F (N)
A	16	3	3
B	−2	4	2
C	5	−5	5
D	1	2	2

54. **Triple Choice** In case 1 a charge q is at the origin, and a charge $5q$ is 1 m away. In case 2 a charge q is at the origin, and a charge $-5q$ is 1 m away. Is the magnitude of the force exerted on the charge at the origin in case 1 greater than, less than, or equal to the magnitude of the force exerted on that charge in case 2? Explain.

55. Describe some of the similarities and differences between Coulomb's law and Newton's law of gravity.

Problem Solving

56. Find the magnitude of the electric force between the charges 0.12 C and 0.33 C at a separation of 2.5 m. Is the force attractive or repulsive?

57. Two identical charges experience a repulsive force of magnitude 1.9 N when their separation is 1.2 m. What is the magnitude of each charge?

58. The attractive electrostatic force between the point charges $+8.44 \times 10^{-6}$ C and Q has a magnitude of 0.975 N when the separation between the charges is 1.31 m. Find the sign and magnitude of the charge Q.

59. **Think & Calculate** Two point charges, the first with a charge of $+3.13 \times 10^{-6}$ C and the second with a charge of -4.47×10^{-6} C, are separated by 0.255 m. **(a)** Find the magnitude of the electrostatic force experienced by the positive charge. **(b)** Is the magnitude of the force experienced by the negative charge greater than, less than, or the same as that experienced by the positive charge? Explain.

60. At what separation will the electric force between charges of 2.1 μC and 5.0 μC have a magnitude of 0.25 N?

61. Two identical ions are each missing two electrons. What is the magnitude of the force between the ions when their separation is 2.1×10^{-10} m?

62. If the speed of the electron in Guided Example 19.4 were 7.3×10^5 m/s, what would be the corresponding orbital radius?

63. A point charge $q = -0.51$ nC is fixed at the origin. Where must an electron be placed in order for the electric force acting on it to be exactly opposite to its weight? (Let the y axis be vertical.)

64. **Think & Calculate** **(a)** If the nucleus in Guided Example 19.4 had a charge of $+2e$ (as would be the case for the nucleus of a helium atom), would the speed of the electron be greater than, less than, or the same as that found in the example? Explain. (Assume that the radius of the electron's orbit is the same.) **(b)** Find the speed of the electron for a nucleus with the charge $+2e$.

65. A system consists of two positive point charges, q_1 and $q_2 > q_1$. The total charge of the system is $+62.0$ μC, and each charge experiences an electrostatic force of magnitude 75.0 N when the separation between them is 0.270 m. Find q_1 and q_2.

66. **Think & Calculate** Two identical point charges are connected by a string 7.6 cm long. The tension in the string is 0.21 N. **(a)** Find the magnitude of the charge on each of the point charges. **(b)** Using the information given in the problem statement, is it possible to determine the signs of the charges? Explain. **(c)** Find the tension in the string if $+1.0$ μC of charge is transferred from one point charge to the other. Compare with your result from part (a).

19.3 Combining Electric Forces

Conceptual Questions

67. **Rank** Consider the three electric charges, A, B, and C, shown in **Figure 19.16**. Rank the charges in order of increasing magnitude of the total force they experience. Indicate ties where appropriate.

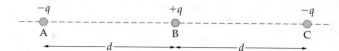

Figure 19.16

68. **Rank** Consider the three electric charges, A, B, and C, shown in **Figure 19.17**. Rank the charges in order of increasing magnitude of the total force they experience. Indicate ties where appropriate.

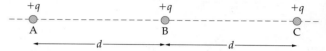

Figure 19.17

69. Suppose the magnitude of the total force experienced by charge C in **Figure 19.18** is 1.00 N. What is the magnitude of the total force experienced by **(a)** charge A and **(b)** charge B?

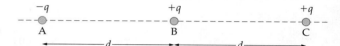

Figure 19.18

70. Which charge in **Figure 19.19** experiences **(a)** the greatest total force and **(b)** the smallest total force?

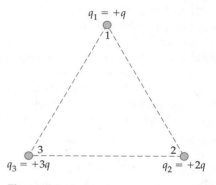

Figure 19.19

Problem Solving

71. A sphere has a surface area of 0.42 m² and a uniform surface charge density of $+7.1$ μC/m². A -6.6-μC point charge is 0.68 m from the center of the sphere. What is the magnitude of the force between the sphere and the charge?

72. Three point charges lie on the x axis. Charge 1 $(-2.1$ μC) is at the origin, charge 2 $(+3.2$ μC) is at $x = 7.5$ cm, and charge 3 $(-1.8$ μC) is at $x = 11$ cm. What are the direction and the magnitude of the total force exerted on charge 1?

73. Referring to Problem 72, what is the direction and magnitude of the total force exerted on charge 2?

74. A sphere has a surface area of 0.056 m² and a surface charge density of +6.2 µC/m² . If the sphere exerts an electrostatic force of magnitude 2.9×10^{-3} N on a point charge of −3.7 µC, find the separation between the point charge and the center of the sphere.

75. The force between a point charge of +6.7 µC and a charged sphere is 3.4×10^{-3} N. If the surface area of the sphere is 0.042 m² and the separation between the center of the sphere and the point charge is 0.27 m, what is the surface charge density of the sphere?

76. Find the magnitude of the total electric force exerted on the charge q_1 in **Figure 19.20**. Let $q = +2.4$ µC and $d = 33$ cm.

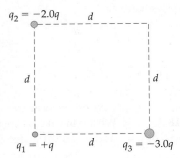

Figure 19.20

77. **Think & Calculate** **(a)** Find the magnitude of the total electric force exerted on charge q_1 in Figure 19.20 for the case where charge q_3 is $+3.0q$. **(b)** How would your answer to part (a) change if the distance d were doubled?

78. Four point charges are located at the corners of a square with sides of length a. Two of the charges are $+q$, and two are $−q$. Find the magnitude and the direction of the total electric force exerted on a charge $+Q$ located at the center of the square, for each of the following two arrangements of charge: **(a)** The charges alternate in sign $(+q, −q, +q, −q)$ around the square; **(b)** The two positive charges are on the top corners of the square, and the two negative charges are on the bottom corners.

79. **Challenge** The point charges in **Figure 19.21** have the following values: $q_1 = +2.1$ µC, $q_2 = +6.3$ µC, and $q_3 = −0.89$ µC. Given that the distance d is 4.35 cm, find the magnitude of the total electric force exerted on charge q_3.

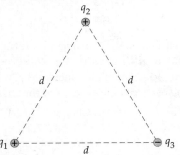

Figure 19.21

80. **Challenge** Two spheres with uniform surface charge density are separated by a center-to-center distance of 33 cm. The spheres have a combined charge of +55 µC and repel one another with a force of 0.75 N. What is the surface charge density on each sphere, given that one has a radius of 7.2 cm and the other has a radius of 4.7 cm?

Mixed Review

81. If glass is rubbed on nylon, what is the sign of the charge the nylon acquires? Explain.

82. Four systems are described below. The magnitude of the electric force in system D is 1 N. What is the magnitude of the electric force in systems **(a)** A, **(b)** B, and **(c)** C?

System	q_1 (C)	q_2 (C)	r (m)
A	$2q$	$−4q$	1
B	$−16q$	$4q$	4
C	$6q$	$−6q$	6
D	q	q	1

83. **Triple Choice** Is the charge transferred when silk is rubbed on glass greater than, less than, or equal to the charge transferred when cotton is rubbed on glass? Assume that all other factors are equal. Explain.

84. **Rank** Five point charges, $q_1 = +q$, $q_2 = +2q$, $q_3 = −3q$, $q_4 = −4q$, and $q_5 = −5q$, are placed in the vicinity of a hollow spherical shell with a charge $+Q$ spread evenly over its surface, as indicated in **Figure 19.22**. Rank the point charges in order of increasing magnitude of the force exerted on them by the sphere. Indicate ties where appropriate.

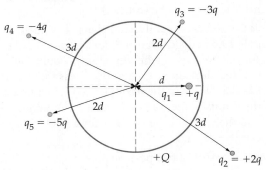

Figure 19.22

85. A charge $+q$ and a charge $−q$ are placed at opposite corners of a square. Will a third point charge experience a greater total force if it is placed at one of the empty corners of the square or at the center of the square? Explain.

86. **Ventricular Fibrillation** If a charge of 0.30 C passes through a person's chest in 1.0 s, the heart can go into *ventricular fibrillation*—a nonrhythmic "fluttering" of the ventricles that results in little or no blood being pumped to the body. If this rate of charge transfer persists for 4.5 s, how many electrons pass through the chest?

87. A system consisting entirely of electrons and protons has a total charge of 1.84×10^{-15} C and a total mass of 4.56×10^{-23} kg. How many **(a)** electrons and **(b)** protons are in this system?

88. **Predict & Explain** An electron and a proton are released from rest in space, far from any other objects. The particles move toward each other, as a result of their mutual electrical attraction. **(a)** When they meet, is the kinetic energy of the electron greater than, less than, or equal to the kinetic energy of the proton? **(b)** Choose the *best* explanation from among the following:

A. The proton has the greater mass. Since kinetic energy is proportional to mass, it follows that the proton will have the greater kinetic energy.

B. The two particles experience the same force, but the light electron moves farther than the massive proton. Therefore, the work done on the electron, and hence its kinetic energy, is greater.

C. The same force acts on the two particles. Therefore, they will have the same kinetic energy, and energy will be conserved.

89. **Predict & Explain** In Conceptual Example 19.7, suppose the charge that is to be placed at point A or point B is $+q$ rather than $-q$. **(a)** Is the magnitude of the total force acting on $+q$ at point A greater than, less than, or equal to the magnitude of the total force acting on it at point B? **(b)** Choose the *best* explanation from among the following:

A. Point B is farther from the two fixed charges. As a result, the total force on $+q$ is less at point B than at point A.

B. The total force at point A cancels, just as in Conceptual Example 19.7. Therefore, the nonzero total force at point B is greater than the zero total force at point A.

C. The total force is greater at point A because at that location the charge $+q$ experiences a net repulsion from each of the fixed charges.

90. **Rank** Three charges, $q_1 = +q$, $q_2 = -q$, and $q_3 = -q$, are at the vertices of an equilateral triangle, as shown in **Figure 19.23**. Rank the three charges in order of increasing magnitude of the total electric force they experience. Indicate ties where appropriate.

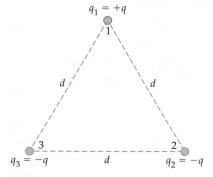

Figure 19.23

91. **Triple Choice** An electron $(\text{charge} = -e)$ orbits a helium nucleus $(\text{charge} = +2e)$. Is the force exerted on the helium nucleus by the electron greater than, less than, or the same as the force exerted on the electron by the helium nucleus? Explain.

92. In an operating room technicians and doctors must take care not to create an electric spark, since the oxygen gas used during operations increases the risk of a deadly fire. Should the operating-room personnel wear shoes that are conducting or nonconducting? Explain.

93. Suppose a charge $+Q$ is placed on the Earth, and another charge $+Q$ is placed on the Moon. Find the value of Q needed to "balance" the gravitational attraction between the Earth and the Moon.

94. Four lightweight, plastic spheres, labeled A, B, C, and D, are suspended from threads in various combinations, as illustrated in **Figure 19.24**. The total charge on sphere D is $+Q$, and the other spheres have a total charge of $+Q$, $-Q$, or 0. From the results of the four experiments shown and the fact that the spheres have equal masses, determine the total charge of **(a)** sphere A, **(b)** sphere B, and **(c)** sphere C.

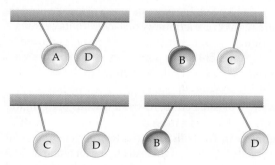

Figure 19.24

95. Find the magnitude of the force acting on charge 1 in Figure 19.23. Give your answer as a formula.

96. A square with sides of length L has a point charge at each of its four corners. Two corners that are diagonally opposite have charges equal to $+2.25$ μC; the other two corners have charges equal to Q. Find the magnitude and the sign of the charge Q such that each of the $+2.25$-μC charges experiences zero total force.

97. Twelve identical point charges, q, are equally spaced around the circumference of a circle of radius R. The circle is centered at the origin. One of the twelve charges, which happens to be on the positive x axis, is moved to the center of the circle. Find **(a)** the direction and **(b)** the magnitude of the total electric force exerted on this charge.

98. Four identical charges, $+Q$, occupy the corners of a square with sides of length d. A fifth charge, q, can be placed at any location. Find the location and the magnitude and sign of the fifth charge such that the total electric force acting on each of the original four charges, $+Q$, is zero.

99. **Think & Calculate** Three charges are placed at the vertices of an equilateral triangle of side $d = 0.63$ m, as shown in **Figure 19.25**. Charges 1 and 3 are $+7.3$ µC; charge 2 is -7.3 µC. **(a)** Find the magnitude and the direction of the total force acting on charge 3. **(b)** If charge 3 is moved to the origin, will the total force acting on it there be greater than, less than, or equal to the total force found in part (a)? Explain. **(c)** Find the total force on charge 3 when it is at the origin.

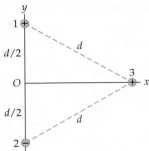

Figure 19.25

100. A small sphere with a charge of $+2.44$ µC is attached to a relaxed horizontal spring whose force constant is 89.2 N/m. The spring extends along the x axis, and the sphere rests on a frictionless surface with its center at the origin. A point charge $Q = -8.55$ µC is moved slowly from infinity to a point $x = d > 0$ on the x axis. This causes the small sphere to move to the position $x = 0.124$ m. Find d.

101. Two small plastic balls hang from threads of negligible mass. Each ball has a mass of 0.14 g and a charge of magnitude q. The balls are attracted to each other, and the threads attached to the balls make an angle of $20.0°$ with the vertical, as shown in **Figure 19.26**. Find **(a)** the magnitude of the electric force acting on each ball, **(b)** the tension in each of the threads, and **(c)** the magnitude of the charge on the balls.

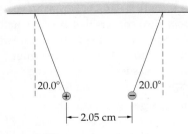

Figure 19.26

Writing about Science

102. Write a report about the electrical charging that occurs when you walk across a carpet. What time of year and what atmospheric conditions produce a greater charge buildup? How much charge is transferred when you touch a doorknob and get a shock?

103. **Connect to the** Big **Idea** Discuss the role of electric forces within an atom. If electric forces were weaker than they actually are, would atoms be larger or smaller?

Read, Reason, and Respond

Bumblebees and Static Cling Ben Franklin showed that the same kind of spark we experience when touching a doorknob after walking on a carpet is a bolt of lightning if it is scaled up in size. His insight led to the invention of lightning rods to conduct electricity safely out of a building and into the ground. Today, we employ static electricity in many technological applications, ranging from photocopiers to electrostatic precipitators that clean emissions from smokestacks.

Static electricity plays an important role in the pollination process. Imagine a bee busily flitting from flower to flower. As air rushes over its body, it acquires an electric charge—just as you do when your shoes rub against a carpet. A bee might have only 93.0 pC of charge, but that's more than enough to attract grains of pollen from a distance, like a charged comb attracting bits of paper. The result, shown in **Figure 19.27**, is a bee covered with grains of pollen as she travels from one flower to another. So, the next time you experience annoying static cling in your clothes, just remember that the same force helps pollinate the plants that all animals on Earth, including humans, rely on.

Figure 19.27 A bee with static cling.

104. How many electrons must be transferred away from a bee to produce a charge of $+93.0$ pC?

A. 1.72×10^{-9} C. 1.02×10^{20}

B. 5.81×10^{8} D. 1.49×10^{29}

105. Suppose two bees, each with a charge of 93.0 pC, are separated by a distance of 1.20 cm. Treating the bees as point charges, what is the magnitude of the electrostatic force experienced by the bees?

A. 6.01×10^{-17} N   C. 5.40×10^{-7} N

B. 6.48×10^{-9} N   D. 5.81×10^{-3} N

106. The force required to detach a grain of pollen from a stigma of an avocado flower is approximately 4.0×10^{-8} N. What is the maximum distance at which the electrostatic force between a bee and a grain of pollen is sufficient to detach the grain? Treat the bee and the grain of pollen as point charges, and assume that the pollen has a charge opposite in sign and equal in magnitude to that on the bee.

A. 4.7×10^{-7} m   C. 4.4 cm

B. 1.9 mm   D. 220 m

Standardized Test Prep

Multiple Choice

Use the diagram below to answers Questions 1–3. The diagram shows two point charges, a $+3$-μC charge at $x = 1$ m and a -4-μC charge at $x = 6$ m.

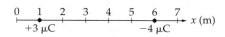

1. What is the force exerted on the -4-μC charge by the $+3$-μC charge?
   (A) 0.0043 N to the right
   (B) 0.022 N to the right
   (C) 0.0043 N to the left
   (D) 0.022 N to the left

2. What is the total force that would be exerted by the two charges on a proton placed at $x = 3$ m?
   (A) 4.4×10^{-16} N to the right
   (B) 4.4×10^{-16} N to the left
   (C) 1.7×10^{-15} N to the right
   (D) 1.7×10^{-15} N to the left

3. Where should a $+2$-μC charge be placed so that it would experience zero total force from the other two charges?
   (A) at some point between the two charges
   (B) at some point to the left of $x = 1$ m
   (C) at some point to the right of $x = 6$ m
   (D) There is no such position.

4. A point charge Q exerts a force on a second charge q when the two charges are a distance d apart. How does the magnitude of the force on charge q change when the charge Q and the distance d are doubled?
   (A) The force decreases by a factor of 2.
   (B) The force is unchanged.
   (C) The force doubles.
   (D) The force quadruples.

5. A hollow metal sphere carries a total charge of 10 μC on its surface. The sphere has a diameter of 2.0 m and is located with its center at $x = 2$ m.

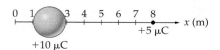

   What force does the sphere exert on a point charge of $+5$ μC located at $x = 8$ m?
   (A) 0.090 N to the right
   (B) 0.075 N to the right
   (C) 0.0125 N to the right
   (D) zero

6. A proton and an electron are released when they are at a distance of 1.0 m from each other. What is the magnitude of the acceleration of the proton at the moment it is released?
   (A) 250 m/s^2 (C) 51 m/s^2
   (B) 100 m/s^2 (D) 0.14 m/s^2

7. Referring to Question 6, which of the following describes the initial acceleration of the electron relative to that of the proton?
   (A) It is smaller than that of the proton.
   (B) It is equal to that of the proton.
   (C) It is larger than that of the proton.
   (D) It is in the same direction as that of the proton.

Extended Response

8. A rubber rod is rubbed with fur, transferring electrons to the rod. The rod is then brought near the 100-cm end of an uncharged metal meterstick that is suspended from its center by a nonconducting string. The meterstick begins to rotate toward the rod. **(a)** Explain why the meterstick begins to rotate, and **(b)** find the charge on the meterstick before, during, and after the rod is brought near one end.

> **Test-Taking Hint**
>
> Electric forces exerted by charged objects on each other are equal in magnitude and opposite in direction.

If You Had Difficulty With . . .

Question	1	2	3	4	5	6	7	8				
See Section	19.2	19.3	19.3	19.3	19.3	19.2	19.2	19.1				

20 Electric Fields and Electric Energy

Inside

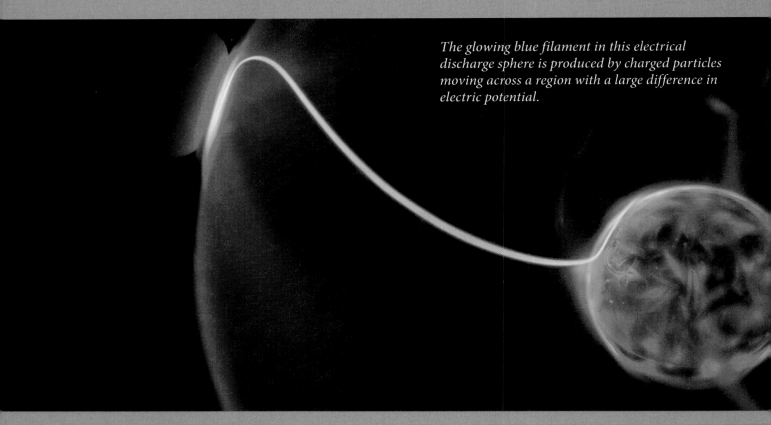

The glowing blue filament in this electrical discharge sphere is produced by charged particles moving across a region with a large difference in electric potential.

Big Idea

Electric charges produce fields that exert forces and store energy.

In this chapter you'll discover how a charged object produces a field that exerts forces on nearby objects. You'll also find out that the field stores energy. The ability to store electrical energy is important in everyday electronic devices, like MP3 players, TV sets, and cell phones. More importantly, storing electrical energy—and then delivering a jolt of it at the proper time—can save a life!

Inquiry Lab   How can an electric field be made stronger?

Explore 🦫

1. Place small bits of paper on a table.
2. Blow up a balloon and pinch the end shut with your fingers.
3. Give the balloon an electrical charge by rubbing it over your hair. This charge produces an electric field around the balloon.
4. Hold the balloon as close to the bits of paper as possible without attracting them.

5. Allow air to slowly escape from the balloon while keeping the distance between it and the paper constant. Observe what happens.

Think

1. **Describe** What happened to the bits of paper as the balloon deflated?

2. **Infer** What do you think happened to the charge on the balloon as it deflated? What do you think happened to the amount of charge per surface area of the balloon as it deflated? How might the answers to these questions explain what you observed?

20.1 The Electric Field

You've probably encountered the notion of a force field in various science fiction movies. Force fields do exist in nature. An electrically charged object sets up a force field around it; this force field is known as an **electric field**. You'll learn about the properties of the electric field in this lesson.

Vocabulary

- electric field
- electric dipole
- charging by induction

The electric field can be visualized

To begin, what does an electric field look like? To help visualize an electric field, look at the group of grass seeds suspended in a fluid in **Figure 20.1**. In Figure 20.1 (a) there is no net electric charge, and hence there is no electric field. The seeds point in random directions. If, on the other hand, the object at the center is given an electric charge, as in Figure 20.1 (b), the seeds line up in the direction of the electric field. Each seed experiences an electric force, and the force causes it to align with the field.

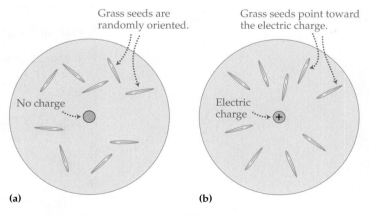

Grass seeds are randomly oriented.

Grass seeds point toward the electric charge.

No charge

Electric charge

(a)   (b)

▲ Figure 20.1 **Visualizing an electric field**
(a) When no electric charge is present, grass seeds in a fluid point in random directions. (b) An electric charge produces an electric field that aligns the seeds in such a way that they tend to point toward the charge.

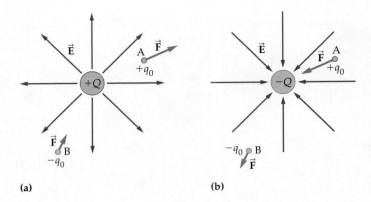

The electric field points **(a)** away from a positive charge and **(b)** toward a negative charge. The force on a positive charge is in the direction of the electric field and the force on a negative charge is in the opposite direction from the electric field.

🔑 How is the sign of an electric charge related to the electric field it produces?

The standard way to draw electric fields is shown in **Figure 20.2**. Here you see a positive charge $+Q$ at the center of Figure 20.2 (a) and a negative charge $-Q$ at the center of Figure 20.2 (b). 🔑 **The direction of an electric field is away from a positive charge and toward a negative charge.**

The direction of the force exerted on a charge in an electric field depends on the field and the sign of the charge. For example, a small positive test charge $(+q_0)$ at location A in Figure 20.2 experiences a force that is in the *same* direction as $\vec{E}$. A small negative test charge $(-q_0)$ at location B experiences a weaker force (since it's farther from the central charge) that is in the *opposite* direction from $\vec{E}$. To summarize,

- The electric field $\vec{E}$ points away from positive charges and toward negative charges.
- The force on a positive charge $(+)$ is *in the direction* of the electric field.
- The force on a negative charge $(-)$ is *in the opposite direction* from the electric field.

Because the force on a positive charge is in the same direction as the electric field, we always use positive test charges to determine the direction of $\vec{E}$.

The electric field is the amount of force per charge

You've just seen the connection between the direction of the electric field and the direction of the electric force. How do you determine the magnitude of the electric field? Well, by definition, the magnitude of the electric field is the electric force per charge:

Definition of the Magnitude of the Electric Field, *E*

If a small positive test charge q_0 experiences a force of magnitude F at a given location, the magnitude of the electric field E at that location is

$$\text{magnitude of electric field} = \frac{\text{force on positive charge}}{\text{amount of charge}}$$

$$E = \frac{F}{q_0}$$

SI units: N/C

In this definition it is assumed that the test charge is small enough that it does not disturb the position of any other charges in the system. That's why we always talk about "small" test charges.

Determining the Force on a Charge
In many practical situations you will be given the electric field $\vec{E}$ at a given location and asked to determine the force a charge q experiences at that location. This can be done as follows:

> **The Electric Force due to an Electric Field**
>
> A charge q in an electric field $\vec{E}$ experiences a force $\vec{F}$ given by
>
> force = amount of charge × electric field
>
> $$\vec{F} = q\vec{E}$$
>
> SI unit: N

This equation gives both the magnitude and the direction of the force. The magnitude is given by

$$F = |q|E$$

The direction is as mentioned before: A positive charge experiences a force $\vec{F}$ in the same direction as $\vec{E}$, and a negative charge experiences a force $\vec{F}$ in the opposite direction from $\vec{E}$.

We apply these results in the following Guided Example.

In a certain region of space a uniform electric field has a magnitude of 4.60×10^4 N/C and points in the positive x direction. Find the magnitude and the direction of the force this field exerts on a charge of (**a**) $+2.80$ μC and (**b**) -9.30 μC.

Picture the Problem

In our sketch we show the uniform electric field and the two charges mentioned in the problem statement. The positive charge experiences a force in the positive x direction (the direction of $\vec{E}$), and the negative charge experiences a force in the negative x direction (opposite to $\vec{E}$).

Strategy

Determine the magnitude of each force using $F = |q|E$. To use the equation we have to convert the charges from microcoulombs to coulombs. The relationship $1\ \mu C = 1 \times 10^{-6}$ C allows us to do this. The direction is indicated in our sketch.

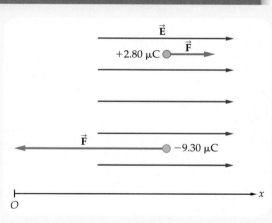

Known

$E = 4.60 \times 10^4$ N/C

(**a**) $q = +2.80\ \mu C = +2.80 \times 10^{-6}$ C

(**b**) $q = -9.30\ \mu C = -9.30 \times 10^{-6}$ C

Unknown

$F = ?$

Solution

1 (a) Calculate the magnitude of the force exerted on the +2.80-μC charge:

$$F = |q|E$$
$$= (2.80 \times 10^{-6}\,\cancel{C})(4.60 \times 10^4\,\text{N}/\cancel{C})$$
$$= \boxed{0.129\,\text{N}}$$

2 (b) Calculate the magnitude of the force exerted on the −9.30-μC charge:

$$F = |q|E$$
$$= (9.30 \times 10^{-6}\,\cancel{C})(4.60 \times 10^4\,\text{N}/\cancel{C})$$
$$= \boxed{0.428\,\text{N}}$$

Math HELP
Exponents
See Math Review, Section I

Insight

To summarize, the force on the +2.80-μC charge has the magnitude 0.129 N and points in the positive x direction. The force on the −9.30-μC charge has the magnitude 0.428 N and points in the negative x direction.

Practice Problems

1. **Follow-up** If the +2.80-μC charge in Guided Example 20.1 experiences a force of 0.25 N, what is the magnitude of the electric field?

2. An object with a charge of −3.6 μC and a mass of 0.012 kg experiences an upward electric force equal in magnitude to its weight. What are the direction and the magnitude of the electric field?

3. A charge in an electric field of magnitude 6.1×10^4 N/C experiences a force of magnitude 0.45 N in the same direction as the electric field. What are the sign and magnitude of the charge?

Point charge electric fields decrease with distance

Perhaps the simplest example of an electric field is the field produced by a point charge. Let's suppose a positive point charge q is at the origin, as shown in **Figure 20.3 (a)**. If a small positive test charge q_0 is placed a distance r from the origin, the force it experiences is directed away from the origin and has a magnitude given by Coulomb's law:

$$F = k\frac{qq_0}{r^2}$$

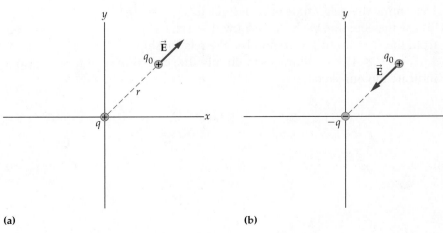

▶ **Figure 20.3** **The electric field of a point charge**
(a) The electric field $\vec{E}$ due to a positive charge $+q$ at the origin is directed radially outward. Its magnitude is $E = k|q|/r^2$.
(b) The electric field points toward a negative charge $-q$ at the origin. The magnitude of the field is the same as in part (a).

(a)

(b)

Applying the definition of the electric field, $E = F/q_0$, we find that the magnitude of the field is

$$E = \frac{F}{q_0} = k\frac{q}{r^2}$$

As you can see, the electric field due to a point charge decreases with the inverse square of the distance—that is, with $1/r^2$. In general, then, the electric field a distance r from a point charge q has the following magnitude:

> ### Magnitude of the Electric Field due to a Point Charge
>
> $$\text{magnitude of electric field} = k\frac{|\text{charge}|}{(\text{distance})^2}$$
>
> $$E = k\frac{|q|}{r^2}$$

The field points away from a positive point charge. **Figure 20.3 (b)** shows that the electric field points toward a negative point charge.

QUICK Example 20.2 What's the Electric Field?

Find the magnitude of the electric field produced by a 1.0-μC point charge at a distance of 0.75 m.

Solution

Applying $E = k\frac{|q|}{r^2}$ with $q = 1.0$ μC and $r = 0.75$ m yields

$$\begin{aligned}
E &= k\frac{|q|}{r^2} \\
&= (8.99 \times 10^9\,\text{N} \cdot \text{m}^2/\text{C}^2)\frac{(1.0 \times 10^{-6}\,\text{C})}{(0.75\,\text{m})^2} \\
&= \boxed{1.6 \times 10^4\,\text{N/C}}
\end{aligned}$$

Math HELP
Significant
Figures
See Lesson 1.4

Since E depends on $1/r^2$, it follows that doubling the distance results in a reduction in the electric field by a factor of 4.

Practice Problems

4. A charge at the origin produces an outward-pointing electric field of magnitude 2.7×10^4 N/C at a distance of 0.44 m. What are the sign and magnitude of the charge?

5. A charge $q = -3.1$ μC produces an electric field of magnitude 0.78×10^4 N/C at a distance r from the charge. Find the distance r.

Notice that the electric field due to a point charge decreases rapidly as the distance from the charge increases. The field never actually goes to zero, however. On the other hand, the electric field increases as the distance gets closer to zero. Thus, the closer you get to an electric charge, the stronger its electric field.

The total electric field at the point P is the vector sum of the fields due to the charges q_1 and q_2. Notice that $\vec{E}_1$ and $\vec{E}_2$ point away from the charges q_1 and q_2, respectively. This is as expected, since both of these charges are positive.

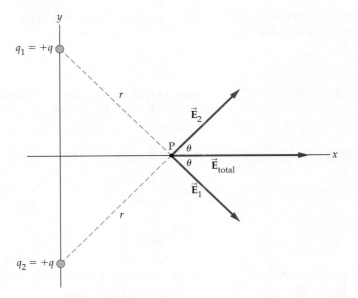

How do you calculate the electric field for a system of two or more charges?

Electric fields add together

When a system consists of several charges, the total electric field is found by *superposition*—that is, by calculating the vector sum of the electric fields due to the individual charges. This is just like what you've done before to calculate the total force due to a system of charges. If the charges in a system remain in fixed positions, the electric field they produce is referred to as a *static* electric field. All of the situations discussed in this chapter deal with static electric fields.

Applying Superposition As an example, let's calculate the total electric field at point P in **Figure 20.4** due to two charges. To simplify the calculation we'll use symbols for the charges and distances, rather than specific numbers.

First we sketch the directions of the fields $\vec{E}_1$ and $\vec{E}_2$ due to the charges $q_1 = +q$ and $q_2 = +q$, respectively. In particular, if a positive test charge is at point P, the force due to q_1 is down and to the right. Similarly, the force due to q_2 is up and to the right. From the geometry of the figure, we see that $\vec{E}_1$ is at an angle θ below the x axis and—by symmetry—$\vec{E}_2$ is at the same angle θ above the x axis. Since the two charges have the same magnitude and the distances from P to the charges are the same, it follows that $\vec{E}_1$ and $\vec{E}_2$ have the same magnitude, E:

$$E = k\frac{q}{r^2}$$

To find the total electric field, $\vec{E}_{total}$, we use components. First we consider the y direction. In this case we have $E_{1,y} = -E \sin \theta$ and $E_{2,y} = +E \sin \theta$. It follows that the y component of the total electric field is zero:

$$E_{total,y} = E_{1,y} + E_{2,y}$$
$$= -E \sin \theta + E \sin \theta$$
$$= 0$$

Similarly, we determine the x component of E_{total}:

$$E_{net,x} = E_{1,x} + E_{2,x}$$
$$= E \cos \theta + E \cos \theta$$
$$= 2E \cos \theta$$

Thus, the total electric field at P is in the positive x direction, as shown in Figure 20.4. The magnitude of the total electric field is equal to $2E \cos \theta$.

Two charges, q_1 and q_2, have equal magnitudes q and are placed as shown in the figure below. The total electric field at point P points vertically upward. Which of the following statements about the signs of q_1 and q_2 is correct? **(a)** The charge q_1 is positive, and q_2 is negative. **(b)** The charge q_1 is negative, and q_2 is positive. **(c)** The charges q_1 and q_2 have the same sign.

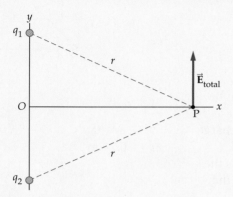

Reasoning and Discussion

If the total electric field at P points vertically upward, the x components of $\vec{E}_1$ and $\vec{E}_2$ must cancel, and the y components must both be in the positive y direction. The only way this can occur is with q_1 negative and q_2 positive, as shown in the following diagram.

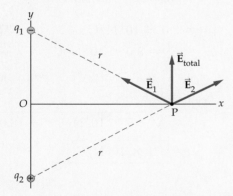

With this charge assignment a positive test charge at P is attracted to q_1 (so that $\vec{E}_1$ is upward and to the left) and repelled from q_2 (so that $\vec{E}_2$ is upward and to the right).

Answer

(b) The charge q_1 is negative, and q_2 is positive.

The next Guided Example uses specific numerical values for the superposition of electric forces.

An electric charge $-q$ is placed on the y axis at $y = 0.75$ m. An identical charge is placed on the x axis at $x = 0.75$ m. The magnitude of each charge is $q = 5.0$ µC. Find the magnitude of the total electric field at the origin due to the two charges.

Picture the Problem

The positions of the two charges are shown in our sketch. We also show the electric field produced by each charge. Notice that the charge on the y axis produces the y component of the total electric field; the charge on the x axis produces the x component of the total electric field. Both components are positive, because the electric field points toward a negative charge.

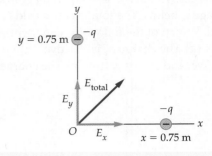

Strategy

The magnitude of the electric field produced at the origin by each charge individually is given by $E = kq/r^2$. In this case, $q = 5.0$ µC and $r = 0.75$ m. The field produced by the charge on the x axis is the x component of the total electric field. Similarly, the field produced by the charge on the y axis is the y component of the total electric field. Therefore, we use the Pythagorean theorem to calculate E_{total}.

Known	Unknown
$q = 5.0$ µC	$E_{total} = ?$
$r = 0.75$ m	

Solution

1 Find the x and y components of the total electric field:

$$E_x = k\frac{q}{r^2}$$

$$= (8.99 \times 10^9 \, \text{N} \cdot \text{m}^2/\text{C}^2)\frac{(5.0 \times 10^{-6} \, \text{C})}{(0.75 \, \text{m})^2}$$

$$= 8.0 \times 10^4 \, \text{N/C}$$

$$E_y = E_x = 8.0 \times 10^4 \, \text{N/C}$$

> **Math HELP**
> Pythagorean Theorem
> See Math Review, Section VI

2 Use the Pythagorean theorem to find the magnitude of the total electric field:

$$E_{total} = \sqrt{E_x^2 + E_y^2}$$

$$= \sqrt{(8.0 \times 10^4 \, \text{N/C})^2 + (8.0 \times 10^4 \, \text{N/C})^2}$$

$$= \boxed{1.1 \times 10^5 \, \text{N/C}}$$

Insight

Because of the symmetry of the system, the total electric field points in a direction that is 45° above the x axis.

Practice Problems

6. **Follow-up** Suppose the charge on the x axis is changed from $-q$ to $+q$. Is the magnitude of the total electric field in this case greater than, less than, or equal to 1.13×10^5 N/C? Explain.

7. **Follow-up** Suppose the charge on the x axis is changed from $-q$ to $+2q$. What is the magnitude of the total electric field in this case?

In general, the interaction between charges can be thought of as a two-step process. In the first step one or more charges produce an electric field at a given location. We've seen how this electric field can be calculated. In the next step, the electric field exerts a force on another charge at that location. In a sense, the electric field is what "connects" charges to one another.

Some animals can produce and detect electric fields

Many aquatic creatures are capable of producing electric fields. For example, some freshwater fish in Africa can use their specialized tail muscles to generate an electric field. They are also able to detect variations in this field as they move through their environment. Thus, these nocturnal feeders have an electrical guidance system that assists them in locating obstacles, enemies, and food. Much stronger electric fields are produced by electric eels and electric skates. In particular, the electric eel *Electrophorus electricus* generates an electric field strong enough to kill small animals and to stun larger animals, including humans.

Sharks are well known for their sensitivity to weak electric fields in their surroundings. In fact, they possess specialized organs for this purpose, known as *ampullae of Lorenzini*. These organs assist sharks in the detection of prey.

Simple rules are used for drawing electric field lines

When you look at diagrams like those in Figure 20.2, it's tempting to imagine a pictorial representation of the electric field. In fact, the following set of rules provides a consistent method of drawing electric field lines:

> ### Rules for Drawing Electric Field Lines
>
> 1. Electric field lines point in the *direction* of the electric field vector $\vec{E}$ at every point.
>
> 2. Electric field lines *start* at positive (+) charges or at infinity.
>
> 3. Electric field lines *end* at negative (−) charges or at infinity.
>
> 4. Electric field lines are *closer together* where $\vec{E}$ has a greater magnitude.
>
> 5. The number of electric field lines entering or leaving a charge is proportional to the magnitude of the charge.

Let's see how these rules are applied. As a first example, notice that the electric field lines in **Figure 20.5** all start at the positive charge. and go to infinity. This is in agreement with rule 2. In addition, the lines are closer together near the charge, where the field is large. This agrees with rule 4.

▶ Figure 20.5 **Electric field lines for a positive point charge**
Near a positive charge the field lines point radially away from the charge. The lines start on the positive charge and end at infinity.

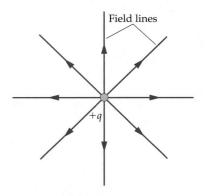
Electric field lines point away from positive charges.

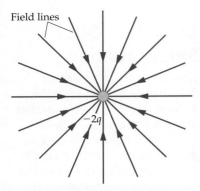

Field lines

−2q

Electric field lines point toward negative charges.

▲ **Figure 20.6 Electric field lines for a negative point charge**
Near a negative charge the field lines point radially inward. They start at infinity and end on the negative charge. They are also more dense where the field is more intense. The number of lines shown here is twice the number in Figure 20.5, a reflection of the relative magnitudes of the charges.

Similar considerations apply to **Figure 20.6**, where the charge is $-2q$. In this case the direction of the field lines is reversed—they end on the negative charge. In addition, the number of lines is doubled, since the magnitude of the charge has been doubled. This illustrates rule 5.

CONCEPTUAL Example 20.5 **Do They Intersect?**

Which of the following statements is correct? **(a)** Electric field lines can intersect. **(b)** Electric field lines cannot intersect.

Reasoning and Discussion
By definition, electric field lines always point in the direction of the electric field. Since the electric force, and hence the electric field, can point in only one direction at any given location, it follows that field lines cannot intersect. If they did, the field at the intersection point would have two conflicting directions.

Answer
(b) Electric field lines cannot intersect.

Electric fields have distinctive shapes and patterns

As you study electric fields, you will notice that the fields tend to form specific patterns depending on the charges involved. A few such patterns, for various combinations of charges, are shown in **Figure 20.7**. When examining these patterns, recall that the magnitude of $\vec{E}$ is greater in regions where the field lines are closer together. For example, the electric field is intense between the charges in Figure 20.7 (a). In contrast, the field is weak between the charges in Figure 20.7 (b), where the field lines are widely spaced. (In fact, the field vanishes midway between the two charges.)

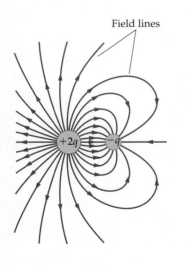

(a)

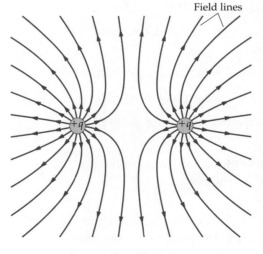

(b)

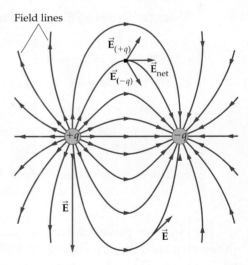

(c) Electric dipole

▲ **Figure 20.7 Electric field lines for systems of charges**
(a) In a system with a net charge, some field lines extend to infinity. If the charges have opposite signs, some field lines start on one charge and terminate on the other charge. At each point in space the electric field vector, $\vec{E}$, is tangent to the field lines. **(b)** All of the field lines in a system with charges of the same sign extend to infinity. **(c)** The electric field lines for an electric dipole form closed loops that become more widely spaced as the distance from the charges increases.

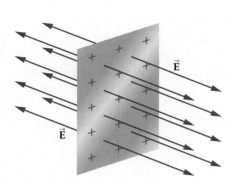

▲ **Figure 20.8 The electric field of a charged plate**
The electric field near a large charged plate is uniform in direction and magnitude.

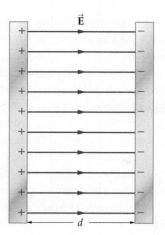

▲ **Figure 20.9 A parallel-plate capacitor**
In the ideal case the electric field is uniform between the plates of a parallel-plate capacitor and zero outside the plates.

Electric Dipoles The charge combination of $+q$ and $-q$ in Figure 20.7 (c) is known as an **electric dipole**. The total charge of a dipole is zero, but because the positive and negative charges are separated, the electric field does not vanish. Instead, the field lines form loops that are characteristic of a dipole. Dipoles are common in nature. Perhaps the most familiar example is the water molecule, which is positively charged at one end and negatively charged at the other.

Infinite Plates and Parallel-Plate Capacitors A simple but particularly important field picture results when charge is spread uniformly over a very large (essentially infinite) plate, as illustrated in **Figure 20.8**. The electric field is uniform in this case, in both direction and magnitude. The field points in a single direction—perpendicular to the plate. Most remarkably, the magnitude of the electric field doesn't depend on the distance from the plate—it's the same for all distances.

If two plates with opposite charge are placed parallel to one another and separated by a finite distance, the result is a *parallel-plate capacitor*. An example is shown in **Figure 20.9**. The field in this case is uniform between the plates and zero outside the plates. This case is the ideal, which is exactly true for infinite plates and a good approximation for large plates.

Excess charge moves to a conductor's outer surface

Conductors contain an enormous number of electrons that are free to move about. This simple fact has some rather interesting consequences. For example, any excess charge placed on a conductor moves to its outer surface, as indicated in **Figure 20.10**. In this way the individual charges are spread as far apart from one another as possible.

◄ **Figure 20.10 Charge distribution on a conducting sphere**
An excess charge placed on a conductor distributes itself uniformly on the surface. None of the excess charge is within the volume of the conductor.

What is the electric field inside a charged conductor?

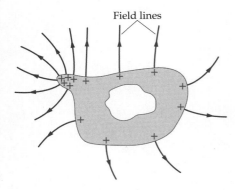

▲ **Figure 20.11 Intense electric field near a sharp point**
Electric charges and field lines are densely packed near a sharp point. This means that the electric field is intense in such regions.

The distribution of charge on the surface of a conductor guarantees that the electric field within the conductor is zero. This effect is referred to as *shielding*. Shielding occurs whether the conductor is solid or hollow. In fact, even a thin sheet of metal foil formed into the shape of a box shields its interior from external electric fields. This effect is put to use in numerous electrical devices, which often have a metal foil or wire mesh enclosure surrounding the sensitive electrical circuits. With such an enclosure a device can be isolated from the effects of other nearby devices that might otherwise interfere with its operation.

Related to shielding is the fact that electric field lines always contact a conductor at right angles to its surface. In addition, the field lines crowd together where a conductor has a point or a sharp projection, as illustrated in **Figure 20.11**. The result is an intense electric field at a sharp metal point.

Physics & You: Technology The crowding of field lines at a point is the basic principle behind the operation of lightning rods. All lightning rods have a pointed tip. During an electrical storm the electric field at the tip becomes so intense that electric charge is given off into the atmosphere. In this way a lightning rod discharges the area near a house—by giving off a steady stream of charge. This discharging can prevent lightning from striking the house, which would transfer a large amount of charge in one sudden blast. Similarly, sharp points on the rigging of a sailing ship at sea can give off streams of charge during a storm. This produces glowing lights referred to as *Saint Elmo's fire*.

An object can be charged without ever touching it

A straightforward way to charge an object is to touch it with a charged rod. Electric forces act at a distance, however, and therefore it is also possible to charge an object without touching it. The charging of an object without direct contact is referred to as **charging by induction**.

To see how this works, consider an uncharged metal sphere on an insulating base. If a negatively charged rod is brought close to the sphere, as in **Figure 20.12 (a)**, electrons in the sphere are repelled. An induced positive charge is produced on the near side of the sphere, and an induced negative charge on the far side. At this point the sphere is still electrically neutral.

The key step in the process of charging by induction, shown in **Figure 20.12 (b)**, is to connect the sphere to the ground using a conducting wire. This is referred to as *grounding* the sphere and is indicated by the symbol ⏚. (A table of electrical symbols can be found in Appendix C.) Since the ground is a good conductor of electricity, and since the Earth can receive or give up practically unlimited numbers of electrons, the effect of grounding the sphere is that the electrons repelled by the charged rod enter the ground. Now the sphere has a net positive charge. With the rod kept in place, the grounding wire is removed, as shown in **Figure 20.12 (c)**, trapping the net positive charge on the sphere. The rod can now be pulled away, as shown in **Figure 20.12 (d)**.

Notice that the *induced* charge on the sphere is opposite in sign to the charge on the rod. In contrast, when an object is charged by *contact*, it acquires a charge with the same sign as the charge on the rod.

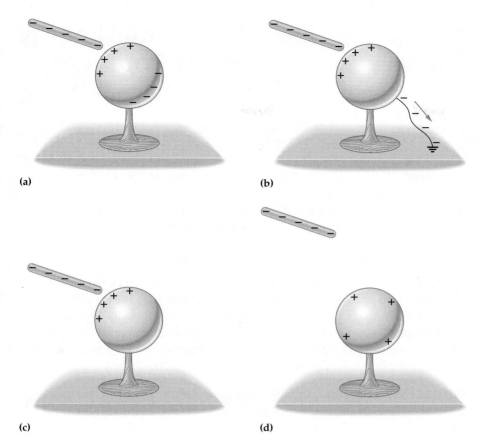

(a)

(b)

(c)

(d)

◄ Figure 20.12 Charging by induction
(a) A charged rod induces positive and negative charges on opposite sides of a conductor. **(b)** When the conductor is grounded, charges that are repelled by the rod enter the ground. There is now a net positive charge on the conductor. **(c)** Removing the grounding wire, with the charged rod still in place, traps the net charge on the conductor. **(d)** The charged rod can then be removed, and the conductor will retain a charge that is opposite in sign to that on the charged rod.

20.1 LessonCheck (MP)

Checking Concepts

8. Identify An electric field points away from a charge. What is the sign of the charge?

9. Explain How are the electric fields produced by individual charges combined to obtain the total electric field?

10. Describe What is the magnitude of the electric field inside a conductor?

11. Determine Two electric charges are separated by a finite distance. Somewhere between the charges, on the line connecting them, the total electric field is zero. Do the charges have the same or opposite signs? Explain.

12. Deduce An electron moves in a region of constant electric field. Is the electron's acceleration in the same direction as the electric field? Explain.

13. Deduce An electron moves in a region of constant electric field. Does it follow that the electron's velocity is parallel to the electric field? Explain.

Solving Problems

14. Calculate What is the magnitude of the electric field produced by a charge of magnitude 5.00 μC at a distance of (**a**) 1.00 m and (**b**) 2.00 m?

15. Calculate At what distance does the electric field produced by a charge of −11 μC have a magnitude equal to 9.3×10^4 N/C?

16. Think & Calculate Two point charges of equal magnitude are 8.0 cm apart. At the midpoint of the line connecting them, the total electric field has a magnitude of 45 N/C.
(**a**) Do the charges have the same or opposite signs? Explain.
(**b**) What is the magnitude of the charges?

17. Diagram Sketch the electric field lines produced by two equal negative charges separated by a distance d.

Vocabulary

- electric potential energy
- electric potential (voltage)
- volt

When is mechanical work done in an electrical system?

Both electric and gravitational forces can store mechanical work in the form of potential energy. Mechanical work that is stored as electrical energy is referred to as **electric potential energy**. We explore the energy associated with electric fields and electric charges in this lesson.

Electric fields store mechanical work

You know that mechanical work is done when you lift a box onto a shelf. You push upward against the force of gravity as you raise the box. The work you do isn't lost, however—it's stored as gravitational potential energy (recall that gravitational potential energy was introduced in Chapter 6). If the box falls from the shelf, the gravitational potential energy is converted to kinetic energy as the box falls. The same idea applies to electric forces.

Opposite Charges Suppose you have a positive charge in one hand and a negative charge in the other, as shown in **Figure 20.13**. The charges attract one another. To pull your hands apart you have to exert a force and do work. (It's like stretching a spring—you have to do work to increase the amount of stretch.) **Separating two charges that attract each other requires mechanical work. This work is stored in the electric field as *electric potential energy*.** If you release the charges, they speed up as they race toward each other, converting their electric potential energy into kinetic energy.

Like Charges Similarly, suppose you have a positive charge in each hand, as shown in **Figure 20.14**. Now the charges *repel* one another. **Moving two charges that repel each other closer together requires mechanical work. This work is stored in the electric field as *electric potential energy*.**

Work must be done to separate unlike charges, just like stretching a spring.

The work is stored as potential energy.

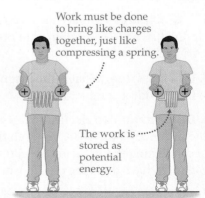

Work must be done to bring like charges together, just like compressing a spring.

The work is stored as potential energy.

▲ Figure 20.13 **Moving unlike charges farther apart**
The work done to separate these charges is stored in the form of electric potential energy. The energy can be recovered by releasing the charges and letting them speed toward one another.

▲ Figure 20.14 **Moving like charges closer together**
The work done to move two like charges together is stored as electric potential energy. The energy can be recovered by releasing the charges and letting them speed away from each other.

If you release these charges, they fly apart from one another, converting the electric potential energy to kinetic energy. (It's like trying to close an overstuffed suitcase. You have to push hard as you try to close it, but if you let go, it pops back open.)

Calculating Work Done in an Electric Field As a specific example, consider a uniform electric field, $\vec{E}$, as shown in **Figure 20.15 (a)**. A positive charge q is placed in this field, where it experiences a downward electric force of magnitude $F = qE$. If the charge is moved upward through a distance d, the electric force and the displacement are in opposite directions. Therefore, the work done by the electric force is negative and equal in magnitude to force times distance:

$$W = -qEd$$

The equation from Chapter 6 that relates work and potential energy, $\Delta PE = -W$, gives a direct connection between the work done to move charges in an electric field and the system's change in electric potential energy:

$$\Delta PE = -W \quad \text{or} \quad \Delta PE = qEd$$

Notice that the electric potential energy increases in this case. This is like increasing the gravitational potential energy by raising a ball against the force of gravity, as indicated in **Figure 20.15 (b)**.

Voltage is the electric potential energy per charge

In Lesson 20.1 we defined the electric field, $\vec{E}$, as the force for a given amount of charge. A similar definition in terms of electrical energy is also useful.

To be specific, suppose the electric potential energy of a charge q changes by the amount ΔPE. By definition, we say that the **electric potential**, V, of the charge changes by the amount $\Delta PE/q$.

> **Definition of Electric Potential, V**
>
> $$\text{change in electric potential} = \frac{\text{change in electric potential energy}}{\text{charge}}$$
>
> $$\Delta V = \frac{\Delta PE}{q}$$
>
> SI unit: joule/coulomb = volt (V)

$\Delta PE = qEd$

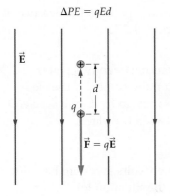

(a) Moving a charge in an electric field

$\Delta PE = mgd$

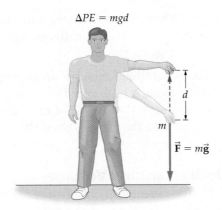

(b) Moving a mass in a gravitational field

◀ **Figure 20.15 Change in electric potential energy**
(a) A positive test charge q experiences a downward force due to the electric field $\vec{E}$. If the charge is moved upward a distance d, the work done by the electric field is $-qEd$. At the same time the electric potential energy of the system increases by qEd. The situation is analogous to that of an object in a gravitational field. **(b)** If a ball is lifted against the force exerted by gravity, the gravitational potential energy of the system increases.

Thus, electric potential is as important as energy, because it is basically electric potential energy per charge. The electric potential is generally referred to as *voltage* because it is measured in a unit called the *volt*, as we will see next.

Units of Voltage The unit of electric potential, or voltage, is the **volt** (V). You're probably familiar with voltage in the form of 120-V electricity in your home or 1.5-V batteries for your camera. The volt is named in honor of Alessandro Volta (1745–1827), who invented a predecessor to the modern battery. To be specific, the volt has the units of energy (J) per charge (C):

$$1 \text{ V} = 1 \text{ J/C}$$

Equivalently, 1 joule of energy is equal to 1 coulomb times 1 volt:

$$1 \text{ J} = (1 \text{ C})(1 \text{ V})$$

It follows that a 1.5-V battery does 1.5 J of work for every coulomb of charge that flows through it; that is, $(1 \text{ C})(1.5 \text{ V}) = 1.5 \text{ J}$. In general, the change in electric potential energy, ΔPE, as a charge q moves through an electric potential (voltage) difference ΔV is

$$\Delta PE = q\Delta V$$

This equation is simply a rearrangement of the definition of electric potential given above.

QUICK Example 20.6 **What's the Change in Electric Potential Energy?**

Find the change in electric potential energy, ΔPE, as a charge of 2.20×10^{-6} C moves from point A to point B, given that the change in electric potential between these points is $\Delta V = 24.0$ V.

Solution
Using $\Delta PE = q\Delta V$, we find

$$\begin{aligned} \Delta PE &= q\Delta V \\ &= (2.20 \times 10^{-6} \text{ C})(24.0 \text{ V}) \\ &= \boxed{5.28 \times 10^{-5} \text{ J}} \end{aligned}$$

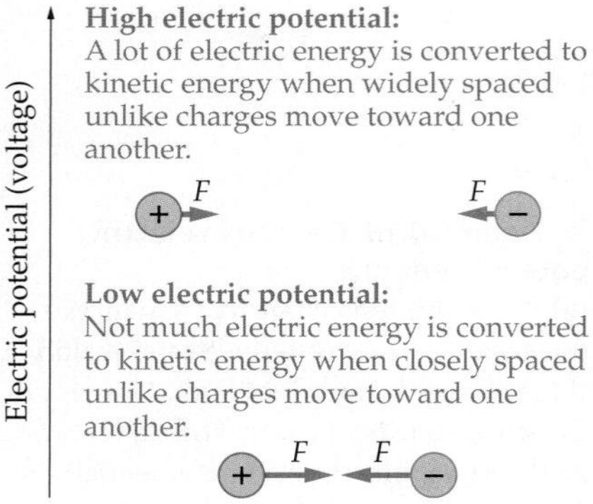

Practice Problems

18. When an ion accelerates through an electric potential difference of 2140 V, its electric potential energy decreases by 1.37×10^{-15} J. What is the charge on the ion?

19. A computer monitor accelerates electrons between two plates and sends them at high speed to form an image on the screen. If the electrons gain 4.1×10^{-15} J of kinetic energy as they go from one accelerating plate to the other, what is the voltage between the plates?

High electric potential:
A lot of electric energy is converted to kinetic energy when widely spaced unlike charges move toward one another.

Low electric potential:
Not much electric energy is converted to kinetic energy when closely spaced unlike charges move toward one another.

▲ **Figure 20.16 High and low voltage with unlike charges**
Widely separated unlike charges produce a high voltage because they store a lot of energy.

(Electric potential (voltage))

Understanding Voltage In general, a high-voltage system has a lot of electric potential energy. **Figure 20.16** shows the situation for charges of opposite sign. When the charges are widely separated, the voltage is high. If these charges are released, a lot of electrical energy is converted to kinetic energy as the charges race toward each other. Not as much energy is converted when the charges start out close together, and so the corresponding voltage is low. For the case of charges with the same sign, like those in **Figure 20.17**, the situation is reversed. Charges close together correspond to high voltage because they fly apart at high speed when released.

Electric potential is related to the electric field

There is a straightforward and useful connection between electric field and electric potential. To obtain this relationship, let's apply the definition $\Delta V = \Delta PE/q$ to the case of a charge that moves through a distance d in the direction of the electric field, as shown in **Figure 20.18**. The work done by the electric field in this case is simply the magnitude of the electric force, $F = qE$, times the distance, d:

$$W = qEd$$

Therefore, the change in electric potential is

$$\Delta V = \frac{\Delta PE}{q} = \frac{-W}{q} = \frac{-(qEd)}{q} = -Ed$$

Solving for the electric field, we find the following:

Connection between the Electric Field and the Electric Potential

$$\text{electric field} = -\frac{\text{change in electric potential}}{\text{distance}}$$

$$E = -\frac{\Delta V}{d}$$

SI units: volts/meter (V/m)

To summarize, the electric field depends on the rate of change of the electric potential with position. In terms of our gravitational analogy, you can think of the electric potential, V, as the height of a hill and the electric field, E, as the slope of the hill. This analogy is illustrated in **Figure 20.19**.

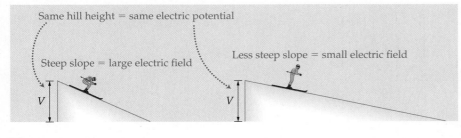

(a)

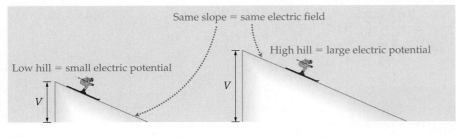

(b)

High electric potential:
A lot of energy is released when closely spaced like charges fly apart.

Low electric potential:
Less energy is released when widely spaced like charges fly apart.

▲ **Figure 20.17 High and low voltage with like charges**
Like charges close together produce a high voltage because they store a lot of energy.

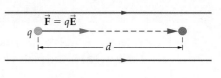

▲ **Figure 20.18 Electric field and electric potential**
As a charge q moves in the direction of the electric field, $\vec{E}$, the electric potential, V, decreases. In particular, if the charge moves a distance d, the electric potential decreases by the amount $\Delta V = -Ed$.

◄ **Figure 20.19 Gravitational analogy for electric potential and electric field**
(a) Two skiers glide down hills of the same height, though one hill is steeper than the other. These skiers are similar to two electrical systems with the same voltage (electric potential) but different electric fields. A steep slope corresponds to a large electric field. **(b)** Two skiers glide down hills with the same slope, though one hill is higher than the other. These skiers are similar to two electrical systems with the same electric field but different electric potentials (voltage). A high hill corresponds to a large voltage.

► **Figure 20.20 The electric potential for a constant electric field**

As a general rule, the electric potential, V, decreases as one moves in the direction of the electric field. In the case shown here, the electric field is constant; as a result, the electric potential decreases uniformly with distance.

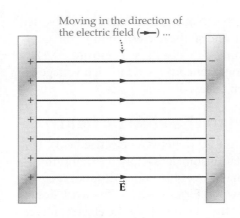

Moving in the direction of the electric field ($\longrightarrow$) ...

$\vec{E}$

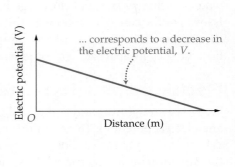

Electric potential (V)

... corresponds to a decrease in the electric potential, V.

O

Distance (m)

🔑 *Relative to the electric field, in what direction does the electric potential decrease?*

Figure 20.20 illustrates an important property of electric potentials and electric fields. 🔑 **The electric potential *decreases* in the direction of the electric field. In addition, the electric potential doesn't change at all in the direction perpendicular to the electric field.** For example, notice that as you move from left to right in Figure 20.20 (in the direction of $\vec{E}$), the electric potential decreases. This concept is applied in the next Guided Example.

GUIDED Example 20.7 | Plates at Different Potentials   **Electric Fields and Potentials**

A uniform electric field is established by connecting the plates of a parallel-plate capacitor to a 12-V battery. (**a**) If the plates have a separation of 0.75 cm, what is the magnitude of the electric field in the capacitor? (**b**) A charge of $+6.24 \times 10^{-6}$ C moves from the positive plate to the negative plate. Find the change in electric potential energy for this charge.

Picture the Problem

Our sketch shows the parallel-plate capacitor connected to a 12-V battery. The battery guarantees that the difference in electric potential between the plates is 12 V, with the positive plate at the higher potential. The separation of the plates is $d = 0.75$ cm $= 0.0075$ m, and the charge that moves from the positive to the negative plate is $q = +6.24 \times 10^{-6}$ C.

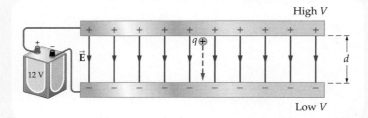

High V

q ⊕

$\vec{E}$

d

12 V

Low V

Strategy

(**a**) The electric field can be calculated using $E = -\Delta V/d$. Notice that if the charge moves in the direction of the field from the positive plate to the negative plate, the electric potential decreases by 12 V; that is, $\Delta V = -12$ V. (**b**) The change in electric potential energy is $\Delta PE = q\Delta V$.

Known

$\Delta V = -12$ V

$d = 0.0075$ m

$q = +6.24 \times 10^{-6}$ C

Unknown

(**a**) $E = ?$

(**b**) $\Delta PE = ?$

Solution

1 (**a**) Substitute $d = 0.0075$ m and $\Delta V = -12$ V in $E = -\Delta V/d$:

$$E = -\frac{\Delta V}{d} = -\frac{(-12 \text{ V})}{0.0075 \text{ m}} = \boxed{1600 \text{ V/m}}$$

2 (b) Substitute the known values into
$\Delta PE = q\Delta V$:

$$\Delta PE = q\Delta V$$
$$= (6.24 \times 10^{-6}\text{ C})(-12\text{ V})$$
$$= \boxed{-7.5 \times 10^{-5}\text{ J}}$$

Insight

The electric potential energy of the system decreases as the positive charge moves in the direction of the electric field. This is just like the gravitational potential energy of a ball decreasing as it falls to a lower height. The positive charge will speed up as it moves toward the negative plate. It will convert the decrease in electric potential energy $\Delta PE = -7.5 \times 10^{-5}\text{ J}$ into a kinetic energy of $KE = 7.5 \times 10^{5}\text{ J}$.

> **Math HELP**
> Scientific Notation
> See Math Review, Section I

Practice Problems

20. [Follow-up] Find the separation of the plates that results in an electric field of 2.0×10^3 V/m.

21. A parallel-plate capacitor has plates separated by 0.75 mm. If the electric field between the plates has a magnitude of $E = 1.2 \times 10^5$ V/m, what is the potential difference (difference in electric potential) between the plates?

Energy is conserved in electrical systems

When a ball is dropped in a gravitational field, its gravitational potential energy decreases as it falls. At the same time, its kinetic energy increases. If nonconservative forces such as air resistance can be ignored, we know that the decrease in gravitational potential energy is equal to the increase in kinetic energy. In other words, the total energy of the ball is conserved.

Energy conservation also applies to a charged object in an electric field. As a result, the sum of the object's kinetic and electric potential energies must be the same at any two points, say, A and B:

$$KE_A + PE_A = KE_B + PE_B$$

General Equation for Energy Conservation
By recalling that the object's kinetic energy is given by $\frac{1}{2}mv^2$, we can write the following general expression for energy conservation between points A and B:

$$\tfrac{1}{2}mv_A^2 + PE_A = \tfrac{1}{2}mv_B^2 + PE_B$$

This equation applies to any conservative force. Notice, however, that the PE term in the equation depends on the type of conservative force involved.

- For a uniform gravitational field the potential energy is $PE = mgy$.
- For an ideal spring the potential energy is $PE = \frac{1}{2}kx^2$.
- For an electrical system the potential energy is $PE = qV$.

Applying Energy Conservation
As a specific example, suppose a particle with mass $m = 1.75 \times 10^{-5}$ kg and charge $q = 5.20 \times 10^{-5}$ C is released from rest at a point A. The particle moves to point B, where the electric potential is lower by 60.0 V. That is,

$$V_A - V_B = 60.0\text{ V}$$

> **CONNECTING IDEAS**
>
> Chapter 6 introduced the law of conservation of energy: Energy can never be created or destroyed—it can only be changed from one form to another.
> - Here we see that energy conservation applies to charged objects in electric fields.

The particle's speed at point B can be found using energy conservation. To do so, we first solve the energy conservation equation for the kinetic energy at point B:

$$\tfrac{1}{2}mv_B^2 = \tfrac{1}{2}mv_A^2 + PE_A - PE_B$$
$$= \tfrac{1}{2}mv_A^2 + q(V_A - V_B)$$

Next we set $v_A = 0$, since the particle starts at rest, and solve for v_B:

$$\tfrac{1}{2}mv_B^2 = q(V_A - V_B)$$
$$v_B = \sqrt{\frac{2q(V_A - V_B)}{m}}$$
$$= \sqrt{\frac{2(5.20 \times 10^{-5}\ \text{C})(60.0\ \text{V})}{1.75 \times 10^{-5}\ \text{kg}}}$$
$$= \sqrt{\frac{6.24 \times 10^{-3}\ \text{J}}{1.75 \times 10^{-5}\ \text{kg}}} = 18.9\ \text{m/s}$$

Thus, the decrease in electric potential energy appears as an increase in kinetic energy—and a corresponding increase in speed. (Recall from Chapter 6 that energy is measured in joules: $1\ \text{J} = 1\ \text{kg} \cdot \text{m}^2/\text{s}^2$.)

Direction of Acceleration in an Electric Field In the preceding example a positive charge moved to a region where the electric potential was less, and its speed increased. As you might expect, the situation is just the opposite for a negative charge. In particular, the speed of a negative charge increases when it moves to a region with a higher electric potential. To summarize:

- *Positive* charges accelerate in the direction of *decreasing* electric potential.
- *Negative* charges accelerate in the direction of *increasing* electric potential.

In both cases the charge moves to a region of lower electric potential energy—like a ball rolling downhill to a position where it has lower gravitational potential energy.

CONCEPTUAL **Example 20.8 How Do the Speeds Compare?**

An electron, with a charge of -1.60×10^{-19} C, accelerates from rest through a potential difference V. A proton, with a charge of $+1.60 \times 10^{-19}$ C, accelerates from rest through a potential difference $-V$. Is the final speed of the electron greater than, less than, or the same as the final speed of the proton?

Reasoning and Discussion
The electron and proton have charges of equal magnitude, and therefore they have equal changes in electric potential energy. As a result, their final kinetic energies are equal. Since the electron has less mass than the proton, however, its speed must be greater.

Answer
The electron's final speed is greater than that of the proton.

22. An object with a charge $q = 9.2 \times 10^{-5}$ C accelerates from rest through a region where the electric potential decreases by 45 V. If the final speed of the object is 21 m/s, what is its mass?

23. An electron accelerates from rest through a region where the electric potential increases by the amount V. If the final speed of the electron is 4.8×10^4 m/s, what is the potential difference?

24. An object of mass $m = 4.2 \times 10^{-15}$ kg has an initial speed of 2.6 m/s. The charge on the object is 8.0×10^{-19} C. If the object accelerates through an electric potential difference of 7500 V, what is its final speed?

Electric potential decreases with distance

Consider a point charge $+q$ at the origin of a coordinate system. In Lesson 20.1 we learned that the electric field a distance r from the origin is given by

$$E = k\frac{q}{r^2}$$

Similarly, it can be shown that the electric potential a distance r from this charge is the following:

> **Electric Potential for a Point Charge**
>
> $$\text{electric potential} = k\frac{\text{charge}}{\text{distance}}$$
>
> $$V = k\frac{q}{r}$$
>
> SI unit: volt (V)

Notice that the electric potential is zero at an infinite distance, $r = \infty$. Also, the potential is positive for a positive charge and negative for a negative charge. The electric potential is a number (a scalar), and therefore it has no associated direction.

If a charge q_0 is in a location where the electric potential is V, the corresponding electric potential energy is

$$PE = q_0 V$$

For the special case where the electric potential is due to a point charge q, the electric potential energy is as follows:

> **Electric Potential Energy for Point Charges q and q_0**
>
> $$PE = q_0 V = k\frac{q_0 q}{r}$$
>
> SI unit: joule (J)

The electric potential energy of two charges separated by an infinite distance is zero.

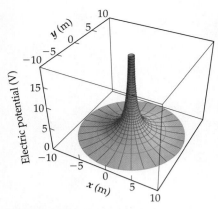

(a) Electric potential near a positive charge

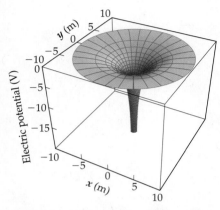

(b) Electric potential near a negative charge

▲ **Figure 20.21 The electric potential of a point charge**
Electric potential near **(a)** a positive and **(b)** a negative charge at the origin. In the case of the positive charge, the electric potential forms a potential hill near the charge. The negative charge produces a potential well.

🔑 *How is the total electric potential for two or more charges determined?*

QUICK Example 20.9 What's the Electric Potential?

Find the electric potential produced by a point charge of 6.80×10^{-7} C at a distance of 2.60 m.

Solution
Substituting $q = 6.80 \times 10^{-7}$ C and $r = 2.60$ m into the equation for the electric potential of a point charge, $V = kq/r$, and solving yields

$$V = k\frac{q}{r}$$

$$= (8.99 \times 10^9 \text{ N} \cdot \text{m}^2/\text{C}^2)\frac{(6.80 \times 10^{-7} \text{ C})}{2.60 \text{ m}}$$

$$= \boxed{2350 \text{ V}}$$

Thus, the electric potential at $r = 2.60$ m due to this point charge is 2350 V greater than the electric potential at infinity.

Practice Problems

25. The electric potential 1.1 m from a point charge q is 2.8×10^4 V. What is the value of the charge q?

26. At what distance from a charge of 2.9 μC is the electric potential due to the charge equal to 1.5×10^4 V?

27. A charge of 6.8 μC is separated from a charge of 4.4 μC by a distance of 0.13 m. What is the electric potential energy of this system?

As mentioned earlier, the sign of the electric potential V depends on the sign of the charge in question. This relationship is illustrated in **Figure 20.21**, which shows the electric potential for a positive and a negative charge at the origin. The potential for the positive charge increases to positive infinity near the origin and decreases to zero far away, forming a *potential hill*. On the other hand, the potential for the negative charge approaches negative infinity near the origin, forming a *potential well*.

Electric potentials combine by simple addition

Like many physical quantities, the electric potential obeys a simple superposition principle. 🔑 **The total electric potential due to two or more charges is equal to the algebraic sum of the potentials due to the individual charges.** By *algebraic sum* we mean that the potential of a given charge may be positive or negative. Thus, the *algebraic sign* of each potential must be taken into account when calculating the total potential. In particular, positive and negative contributions may cancel to give zero potential at a given location.

CONCEPTUAL Example 20.10 A Peak or a Valley?

Two point charges $+q$ are placed on the x axis at $x = -1$ m and $x = +1$ m. Does a graph of electric potential along the x axis look like a peak or a valley near the origin?

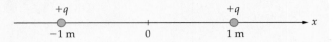

Reasoning and Discussion

We know that the electric potential is large and positive near each of the charges. As you move away from each charge, the potential decreases. In particular, at very large positive or negative values of x, the potential approaches zero.

Between $x = -1$ m and $x = +1$ m, the potential has its lowest value when you are as far away from the two charges as possible. This occurs at the origin. Moving slightly to the left or the right brings you closer to one of the two charges, resulting in an increase in the potential. Therefore, the potential has a minimum (bottom of a valley) at the origin. A graph of the potential for this case is shown to the right.

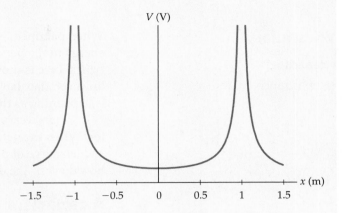

Answer

Near the origin the graph of the electric potential looks like a valley.

20.2 LessonCheck ⓂⓅ

Checking Concepts

28. 💬 **Explain** How is mechanical energy stored when work is done on an electrical system?

29. 💬 **Relate** How is the change in electric potential related to the direction of the electric field?

30. 💬 **Describe** How is the total electric potential for a system with two or more charges calculated?

31. Apply An electron is released from rest in a region of space with a nonzero electric field. As the electron moves, does it experience an increasing or decreasing electric potential? Explain.

32. Triple Choice A positive charge produces an electric potential in its vicinity. Is the electric potential at a distance of 1 m from the charge greater than, less than, or equal to the electric potential at a distance of 2 m? Explain.

Solving Problems

33. Calculate A point charge of 7.2 µC is at the origin. What is the electric potential at the following (x, y) locations? **(a)** (3.0 m, 0); **(b)** (3.0 m, −3.0 m).

34. Calculate A uniform electric field of magnitude 4.1×10^5 N/C points in the positive x direction. Find the change in electric potential energy of a 4.5-µC charge as it moves 6.0 m in the positive x direction.

35. Calculate Earth has a vertical electric field with a magnitude of approximately 100 V/m near its surface. What is the magnitude of the potential difference between a point on the ground and a point at the top of the Washington Monument, at a height of 169 m (555 ft)?

36. Think & Calculate Consider two point charges, $q_1 = +7.22$ µC and $q_2 = -26.1$ µC.
(a) How far must the charges be separated for the electric potential energy of the system to be −126 J?
(b) If the separation between the charges is increased, is the electric potential energy of the system greater than, less than, or equal to −126 J? Explain.

20.3 Capacitance and Energy Storage

Vocabulary

- capacitor
- capacitance

How is capacitance related to the charge and the voltage of a capacitor?

When paramedics revive a heart-attack victim, they apply a jolt of electrical energy to the person's heart with a device known as a *defibrillator*. A typical case is shown in **Figure 20.22**. Before they can use the defibrillator, however, they have to wait a few seconds for it to "charge up." What exactly is happening as the defibrillator charges?

The answer is that electrical energy is building up inside the defibrillator. When the defibrillator is activated, the energy stored in it is released in a sudden surge of power that can save a person's life. In this lesson you'll learn how devices like defibrillators store electrical energy.

A capacitor can store charge and energy

A common way for electrical systems to store energy is in a device known as a *capacitor*. A **capacitor** gets its name from the fact that it has a *capacity* to store both electric charge and electrical energy. Capacitors are an important element in modern electronic devices. They can provide large bursts of energy to a circuit or protect delicate circuitry from excess charge originating elsewhere. No cell phone or computer could work without capacitors.

In general, a capacitor is nothing more than two conductors, referred to as *plates*, separated by a finite distance. When the plates of a capacitor are connected to the terminals of a battery, they become charged. That is, one plate acquires a positive charge, $+Q$, and the other plate acquires an equal and opposite negative charge, $-Q$.

To be specific, suppose a certain battery produces a potential difference (or voltage) V between its terminals. When this battery is connected to a capacitor, a charge of magnitude Q appears on each plate. The ratio of the charge stored to the applied voltage—that is, the ratio Q/V—is called the **capacitance**, C. **The greater the charge Q for a given voltage V, the greater the capacitance of the capacitor.**

> **Definition of Capacitance, C**
>
> $$\text{capacitance} = \frac{\text{charge}}{\text{electric potential difference}}$$
>
> $$C = \frac{Q}{V}$$
>
> SI unit: coulomb/volt = farad (F)

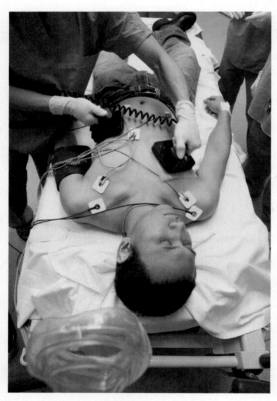

▲ **Figure 20.22 A defibrillator in action**
Paramedics revive a heart-attack victim with a jolt of electrical energy stored in a defibrillator.

In this equation Q is the magnitude of the charge on either plate and V is the magnitude of the voltage difference between the plates. By definition, then, the capacitance is always a positive quantity.

Units of Capacitance

You can see from the relation $C = Q/V$ that the units of capacitance are coulombs per volt. In the SI system this combination of units is referred to as the *farad* (F), in honor of the English physicist Michael Faraday (1791–1867), a pioneering researcher into the properties of electricity and magnetism. In particular,

$$1 \text{ F} = 1 \text{ C/V}$$

Just as the coulomb is a rather large unit of charge, so too is the farad a rather large unit of capacitance. Typical values for capacitance are in the picofarad (1 pF $=$ 10^{-12} F) to microfarad (1 μF $=$ 10^{-6} F) range.

QUICK Example 20.11 What's the Charge?

A 0.75-μF capacitor is charged to a voltage of 16 V. What is the magnitude of the charge on each plate of the capacitor?

Solution

Using $Q = CV$, we find

$$
\begin{aligned}
Q &= CV \\
&= (0.75 \times 10^{-6} \text{ F})(16 \text{ V}) \\
&= \boxed{1.2 \times 10^{-5} \text{ C}}
\end{aligned}
$$

Be careful to distinguish between the symbol used for capacitance and the symbol for the unit of charge, the coulomb. The same capital letter is used for both, but C for capacitance is italicized, while C for coulomb is not.

Practice Problems

37. A capacitor stores 2.7×10^{-5} C of charge when it is connected to a 9.0-V battery. What is the capacitance of the capacitor?

38. Suppose you want to store 5.8 μC of charge on each plate of a 3.2-μF capacitor. What potential difference is required between the plates?

A capacitor is like a bucket of water

A bucket of water provides a useful analogy when thinking about capacitors, as shown in **Figure 20.23**. For this analogy we make the following identifications:

- The cross-sectional area of the bucket is the capacitance, C.
- The amount of water in the bucket is the charge, Q.
- The depth of the water is the potential difference, V, between the plates.

Reading Support ✓
Vocabulary Builder
fibrillation
[fib ruh LAY shun]
(noun) a localized and uncontrolled twitching of muscle fibers, often those in the heart
Related terms:
Atrial fibrillation, known as *a-fib*, produces a rapid and irregular heartbeat.
Ventricular fibrillation, known as *v-fib*, is an extremely serious heart condition in which the ventricles fail to pump blood.
The doctor used a defibrillator to shock the patient out of v-fib and restore a regular heart rhythm.

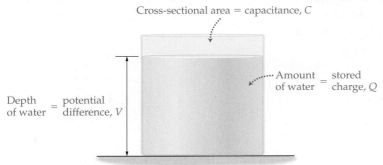

Cross-sectional area = capacitance, C

Amount of water = stored charge, Q

Depth of water = potential difference, V

◀ **Figure 20.23 Capacitor analogy**
A bucket of water has many similarities to an electrical capacitor. Filling the bucket with water to a certain level is similar to filling a capacitor with electric charge to a certain voltage.

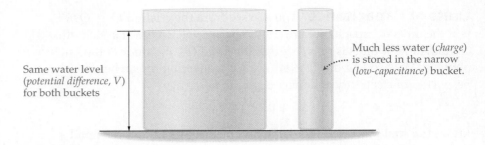

▲ **Figure 20.24 Different capacities**
A narrow bucket has a lower capacity for water than a wide bucket. Similarly, a low-capacitance capacitor has a lower capacity for electric charge than a high-capacitance capacitor.

Same water level
(*potential difference, V*)
for both buckets

Much less water (*charge*)
is stored in the narrow
(*low-capacitance*) bucket.

In terms of this analogy charging a capacitor is like pouring water into a bucket. If the capacitance is large, it's like having a wide bucket. In this case the bucket holds a lot of water when it has a given level of water. A narrow bucket with the same water level holds much less water. You can see this in **Figure 20.24**. Similarly, a capacitor with a large capacitance holds a lot of charge (water) for a given applied voltage (water level.)

GUIDED Example 20.12 | All Charged Up   **Capacitors**

A parallel-plate capacitor has a capacitance of 451 pF, where 1 pF $= 10^{-12}$ F. The separation between the plates of the capacitor is $d = 0.550$ mm. (**a**) Find the voltage difference between the two plates of the capacitor when the charge on the plates has a magnitude of 9.06×10^{-9} C. (**b**) What is the magnitude of the electric field between the plates of the capacitor?

Picture the Problem

The capacitor is shown in our sketch. Notice that the charge on each plate has the magnitude Q. The electric field is uniform and points from the positive plate to the negative plate.

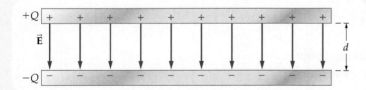

Strategy

(**a**) Capacitance is defined as $C = Q/V$. We can solve this expression for the voltage, which yields $V = Q/C$. The charge and the capacitance are given in the problem statement.
(**b**) In Lesson 20.2 we saw that the electric field is $E = -\Delta V/d$. It follows that the magnitude of the electric field is $|E| = |\Delta V|/d$. The voltage difference, ΔV, is determined in part (a). The distance d is the distance between the plates. Therefore, $d = 0.550$ mm.

Known	Unknown
$C = 451$ pF	(**a**) $V = ?$
$d = 0.550$ mm	(**b**) $E = ?$
$Q = 9.06 \times 10^{-9}$ C	

Solution

1 **(a)** Use $V = Q/C$ to find the voltage difference between the plates:

$$V = \frac{Q}{C} = \frac{9.06 \times 10^{-9}\,\text{C}}{451 \times 10^{-12}\,\text{F}} = \boxed{20.0\,\text{V}}$$

2 **(b)** The magnitude of the electric field between the plates is $|E| = |\Delta V|/d$, with $d = 0.550\,\text{mm} = 0.550 \times 10^{-3}\,\text{m}$:

$$|E| = \frac{|\Delta V|}{d} = \frac{20.0\,\text{V}}{0.550 \times 10^{-3}\,\text{m}} = \boxed{36{,}400\,\text{V/m}}$$

Insight

The total amount of charge in the capacitor is zero: $+Q + (-Q) = 0$. The fact that this charge is separated—with positive charge on one plate and negative charge on the other—means that work had to be done to create the separation. This work is stored in the capacitor as electrical energy.

Practice Problems

39. Follow-up What capacitance is necessary to store the same amount of charge as in Guided Example 20.12 if the potential difference between the plates is 35.0 V?

40. A 150-pF capacitor is connected to a 12-V battery. What is the magnitude of the charge on the plates of the capacitor?

CONCEPTUAL Example 20.13 How Does the Field Change?

A battery maintains a constant potential difference between the plates of a parallel-plate capacitor. If the plates are pulled away from one another, increasing their separation, does the magnitude of the electric field between the plates increase, decrease, or remain the same?

Reasoning and Discussion

The magnitude of the electric field is given by $|E| = \Delta V/d$. It follows that increasing the separation (d) while maintaining a constant potential difference (ΔV) results in a decrease in the electric field. That is, a smaller electric field acting over a larger distance results in the same potential difference.

Answer

The magnitude of the electric field between the plates decreases.

Size and shape affect a capacitor's capacitance

The two main factors that determine the capacitance of a capacitor are plate area and plate separation.

Plate Area If the area of the plates is increased, the capacitance goes up. This is as you would expect. After all, larger plates have more area, and this gives the plates more room to store charge. It's just like the analogy with a bucket of water—a capacitor with a large plate area is like a bucket with a large cross-sectional area.

A *theremin* is a musical instrument you play without ever touching it! This instrument has two antennas that are used to control the sound it makes. One antenna adjusts the volume; the other adjusts the pitch. When a person places a hand near one of the antennas, the situation is similar to a parallel-plate capacitor. The hand plays the role of one plate, and the antenna plays the role of the other plate. Changing the separation between hand and antenna changes the capacitance. Electronic circuitry converts the change in capacitance into a corresponding change of volume or pitch. Theremins are best known for the ethereal music sometimes used in science fiction films.

🔑 *How does increasing the charge or the voltage of a capacitor affect the amount of energy it can store?*

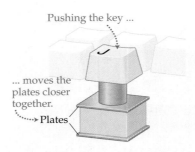

▲ **Figure 20.25 Capacitance and the computer keyboard**
The keys on many computer keyboards form part of a parallel-plate capacitor. Pushing on a key changes the plate separation. The corresponding change in capacitance can be detected by the computer's circuitry.

Plate Separation The dependence of capacitance on plate separation is just the opposite of its dependence on area. If the plate separation is decreased, the capacitance increases. The reason is that a smaller separation between the plates reduces the potential difference between them. This means that less voltage is required to store a given amount of charge—which is another way of saying that the capacitance is larger.

Physics & You: Technology The dependence of capacitance on plate separation is useful in a number of interesting applications. In fact, you rely on this dependence every time you type on a computer keyboard. Each key is connected to the upper plate of a parallel-plate capacitor, as illustrated in **Figure 20.25**. When you press on a key, the separation between the plates of that capacitor decreases. As a result, the capacitance of the key increases. The circuitry of the computer detects this change in capacitance and determines which key you have pressed.

Capacitors can store large amounts of energy

As mentioned before, capacitors store more than just charge—they also store energy. It takes energy to separate the charges on the two plates of a capacitor, and that energy is stored in the capacitor. It can be shown that the total energy, PE, stored in a capacitor with charge Q and potential difference V is

$$PE = \tfrac{1}{2}QV$$

🔑 **Increasing a capacitor's charge or voltage increases its stored energy.** The greater the charge, the greater the stored energy; the greater the voltage, the greater the stored energy.

Physics & You: Technology The energy stored in a capacitor can be put to a number of practical uses. Any time you take a flash photograph, for example, you are triggering the rapid release of energy from a capacitor. The flash unit typically contains a capacitor with a capacitance of about 400 μF. When fully charged to a voltage of 300 V, this capacitor contains roughly 15 J of energy. Activating the flash causes the stored energy, which took several seconds to accumulate, to be released in less than a millisecond. Because of the rapid release of energy, the power output of a flash unit is impressively large—about 10–20 kW. This is far in excess of the power provided by the battery that operates the unit. This release of energy by a capacitor is also used in defibrillators to help heart-attack victims.

When a person's heart undergoes ventricular fibrillation—a rapid, uncoordinated twitching of the heart muscles—it often takes a strong jolt of electrical energy to restore the heart's regular beating and save the person's life. The device that delivers this jolt of energy is a defibrillator, and it uses a capacitor to store the necessary energy. In a typical defibrillator a 175-μF capacitor is charged until the potential difference between the plates is 2240 V. (**a**) What is the magnitude of the charge on each plate of the fully charged capacitor? (**b**) How much energy is stored in the charged-up defibrillator?

Picture the Problem

Our sketch shows a simplified representation of a capacitor. The values of the capacitance and the potential difference are indicated.

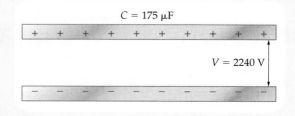

Strategy

(**a**) We can find the charge stored on the capacitor plates by rearranging the definition of capacitance, $C = Q/V$, to solve for the charge. This yields $Q = CV$. (**b**) The energy stored in the capacitor can be determined using $PE = \frac{1}{2}QV$.

Known	Unknown
$C = 175$ μF	(**a**) $Q = ?$
$V = 2240$ V	(**b**) $PE = ?$

Solution

1 (**a**) Use $Q = CV$ to find the charge on the plates:

$$Q = CV$$
$$= (175 \times 10^{-6} \text{ F})(2240 \text{ V})$$
$$= \boxed{0.392 \text{ C}}$$

2 (**b**) Find the stored energy with $PE = \frac{1}{2}QV$:

$$PE = \frac{1}{2}QV$$
$$= \frac{1}{2}(0.392 \text{ C})(2240 \text{ V})$$
$$= \boxed{439 \text{ J}}$$

Insight

Of the 439 J stored in the defibrillator's capacitor, about 200 J actually passes through the person's body in a pulse lasting about 2 ms. The power delivered by the pulse is approximately $P = PE/t = (200 \text{ J})/(0.002 \text{ s}) = 100$ kW. This is significantly larger than the power delivered by the battery, which can take up to 30 s to fully charge the capacitor.

Practice Problems

41. Follow-up Suppose the defibrillator is "fired" when the voltage and the charge are only half of the typical values. How much energy is stored in this case?

42. Calculate the work done by a 3.0-V battery as it charges a 7.8-μF capacitor in the flash unit of a camera.

43. A 0.31-μF capacitor is charged by a 12-V battery. If this capacitor is connected to an electric motor in a 16-kg go-kart, what speed can the go-kart acquire if it accelerates from rest? Assume 100% efficiency for the motor and for the energy transfer from the capacitor to the motor.

Capacitors can be dangerous

A defibrillator uses a capacitor to deliver a shock to a person's heart, restoring it to normal function. Capacitors can have the opposite effect as well. It is for this reason that they can be quite dangerous, even in electrical devices that are turned off and unplugged from the wall.

For example, a typical TV set or computer monitor contains a number of capacitors. Some of these capacitors store significant amounts of charge and energy. When a TV set is unplugged, the capacitors in it retain their charge for long periods of time—the charge just sits on their plates. If you reached into the back of an unplugged television set, there is a danger that you might come into contact with the terminals of a capacitor. This would discharge the capacitor's stored energy through your body. The resulting shock could be painful and even harmful. Knowing the physics of capacitors can save you from an unpleasant shock!

20.3 LessonCheck (MP)

Checking Concepts

44. **Relate** How are charge and voltage related to the capacitance of a capacitor?

45. **Relate** How is the charge on a capacitor related to the amount of energy it can store?

46. **Big Idea** What effect does an electric field have on an electric charge? What happens to the mechanical work that is done to separate two unlike charges?

47. Apply The plates of a particular parallel-plate capacitor are uncharged. Is the capacitance of this capacitor zero? Explain.

48. Analyze A particular parallel-plate capacitor has the charge $+Q$ on one plate and the charge $-Q$ on the other plate. If the plate separation is decreased, does the capacitance of the capacitor increase, decrease, or stay the same? Explain.

Solving Problems

49. Calculate A capacitor with a capacitance of 750 μF has a charge of 56 μC on one plate and a charge of -56 μC on the other plate. What is the potential difference between the plates?

50. Calculate A flash lamp requires a charge of magnitude 32 μC on each plate of its capacitor to operate. What capacitance is needed to store this much charge if the potential difference between the capacitor's plates is 9.0 V?

51. Calculate An automatic external defibrillator (AED) delivers 125 J of energy at a voltage of 1050 V. What is the capacitance of this device?

52. Calculate An electronic flash unit for a camera contains a capacitor with a capacitance of 890 μF. When the unit is fully charged and ready for operation, the potential difference between the capacitor plates is 330 V.
(a) What is the magnitude of the charge on each plate of the fully charged capacitor?
(b) How much energy is stored in the charged-up flash unit?

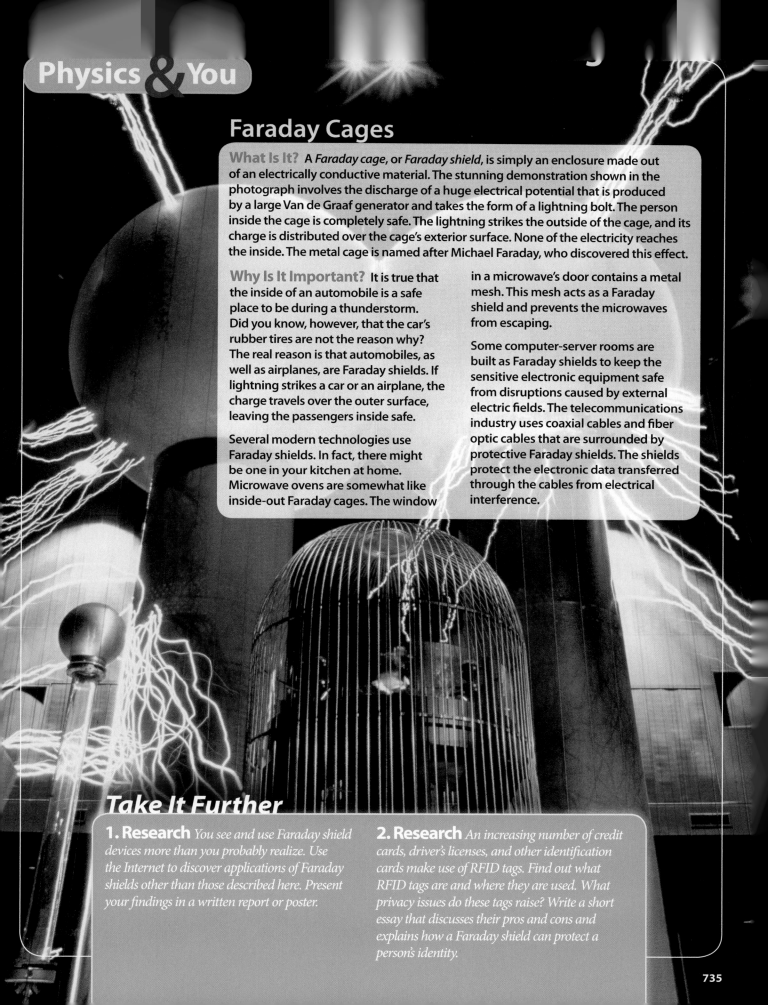

Physics & You

Faraday Cages

What Is It? A *Faraday cage*, or *Faraday shield*, is simply an enclosure made out of an electrically conductive material. The stunning demonstration shown in the photograph involves the discharge of a huge electrical potential that is produced by a large Van de Graaf generator and takes the form of a lightning bolt. The person inside the cage is completely safe. The lightning strikes the outside of the cage, and its charge is distributed over the cage's exterior surface. None of the electricity reaches the inside. The metal cage is named after Michael Faraday, who discovered this effect.

Why Is It Important? It is true that the inside of an automobile is a safe place to be during a thunderstorm. Did you know, however, that the car's rubber tires are not the reason why? The real reason is that automobiles, as well as airplanes, are Faraday shields. If lightning strikes a car or an airplane, the charge travels over the outer surface, leaving the passengers inside safe.

Several modern technologies use Faraday shields. In fact, there might be one in your kitchen at home. Microwave ovens are somewhat like inside-out Faraday cages. The window

in a microwave's door contains a metal mesh. This mesh acts as a Faraday shield and prevents the microwaves from escaping.

Some computer-server rooms are built as Faraday shields to keep the sensitive electronic equipment safe from disruptions caused by external electric fields. The telecommunications industry uses coaxial cables and fiber optic cables that are surrounded by protective Faraday shields. The shields protect the electronic data transferred through the cables from electrical interference.

Take It Further

1. Research *You see and use Faraday shield devices more than you probably realize. Use the Internet to discover applications of Faraday shields other than those described here. Present your findings in a written report or poster.*

2. Research *An increasing number of credit cards, driver's licenses, and other identification cards make use of RFID tags. Find out what RFID tags are and where they are used. What privacy issues do these tags raise? Write a short essay that discusses their pros and cons and explains how a Faraday shield can protect a person's identity.*

Physics Lab Mapping an Electric Field

This lab explores the electric field around conductors. You will map equipotential lines and electric field lines around two charged conductors.

Materials
- two nails
- two sheets of graph paper
- digital multimeter
- four electrical leads
- clear plastic tray
- modeling clay
- 9-V battery

Procedure

1. Prepare two identical sheets of graph paper by labeling the grid markings with numbers and letters as shown. Position the tray over one of the sheets of graph paper.

2. Place the electrodes (each consisting of a nail held vertically in a ball of clay) along the central horizontal grid line about 10–12 cm apart. Use the second sheet of graph paper as a data table and mark the location of each electrode.

3. Connect the electrical leads to the electrodes, the 9-V battery, and the multimeter, as shown. Record the polarity of each electrode.

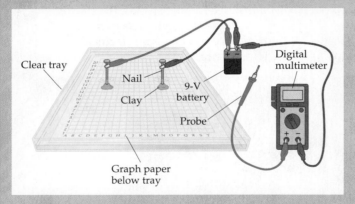

Clear tray
Nail
Clay
9-V battery
Probe
Digital multimeter
Graph paper below tray

4. Set the multimeter to an appropriate voltage scale. Add water to the tray until a small portion of each nail is submerged.

5. Use the unconnected electrical lead as a probe. Place the probe exactly halfway between the two electrodes and observe the potential. Record the reading and its location.

6. Move the probe a few centimeters at a time, finding points that have the same potential as in Step 5. Record the location of each point on the data table. Explore points across the tray until the shape of the *equipotential line* becomes clear.

7. Draw a best-fit curve through the equipotential points plotted in Steps 5 and 6 and label the potential of the line.

8. Repeat Steps 5–7 at least six more times, starting at a different point between the two electrodes each time. Each of these equipotential lines will have a different electric potential. Conduct your trials so that the equipotential lines are fairly evenly spaced throughout the region between the electrodes.

Analysis

1. Draw the lines of electric force that correspond to the plotted equipotential lines. Each line should begin at the positive electrode, end at the negative electrode, and always be perpendicular to the equipotential lines.

Conclusions

1. Explain why the plotted equipotential lines do not cross each other.

2. Describe the electric field across the tray. Where was the electric field strongest? Where was it weakest? Were there any regions with an electric field of zero?

3. Explain why the electric force lines must be at right angles to the equipotential lines.

4. Describe three properties of electric force lines.

20 Study Guide

Big Idea

Electric charges produce fields that exert forces and store energy.

All electric charges produce electric fields. An electric field exerts a force on other charges and also stores energy in the form of electric potential energy. Electric energy can be converted to other forms of energy, like kinetic energy, but the total energy is always conserved.

20.1 The Electric Field

🔑 The direction of an electric field is away from a positive charge and toward a negative charge.

🔑 When a system consists of several charges, the total electric field is found by *superposition*—that is, by calculating the vector sum of the electric fields due to the individual charges.

🔑 The distribution of charge on the surface of a conductor guarantees that the electric field within the conductor is zero.

• The force field produced by a charged object is referred to as the *electric field*.

• The electric field is the force per charge at a given location in space.

• The electric field can be visualized by drawing lines according to a set of simple rules.

• A system of two equal and opposite charges separated by a finite distance is known as an *electric dipole*.

• Any excess charge placed on a conductor moves to its exterior surface.

• A conductor can be charged without having direct physical contact with another charged object. This is charging by induction.

Key Equations

The electric field a distance r from a point charge of magnitude q has a magnitude given by

$$E = k\frac{|q|}{r^2}$$

20.2 Electric Potential Energy and Electric Potential

🔑 Separating two charges that attract each other requires mechanical work. This work is stored in the electric field as electric potential energy.

🔑 Moving two charges that repel each other closer together requires mechanical work. This work is stored in the electric field as electric potential energy.

🔑 The electric potential *decreases* in the direction of the electric field. In addition, the electric potential doesn't change at all in the direction perpendicular to the electric field.

🔑 The total electric potential due to two or more charges is equal to the algebraic sum of the potentials due to the individual charges.

Key Equations

The electric potential is defined as follows:

$$\Delta V = \frac{\Delta PE}{q}$$

The electric field is related to the electric potential as follows:

$$E = -\frac{\Delta V}{d}$$

The electric potential at a distance r from a point charge q is

$$V = k\frac{q}{r}$$

The electric potential energy of two charges separated by a distance r is

$$PE = k\frac{q_0 q}{r}$$

20.3 Capacitance and Energy Storage

🔑 The greater the charge Q for a given voltage V, the greater the capacitance of the capacitor.

🔑 Increasing a capacitor's charge or voltage increases its stored energy.

• A capacitor has a *capacity* to store both electric charge and electrical energy.

Key Equations

Capacitance is defined as the amount of charge, Q, stored in a capacitor per volt of potential difference, V, between the plates of the capacitor:

$$C = \frac{Q}{V}$$

The electrical energy stored in a capacitor is

$$PE = \frac{1}{2}QV$$

20 Assessment

For instructor-assigned homework, go to www.masteringphysics.com

ANSWERS TO SELECTED ODD-NUMBERED PROBLEMS APPEAR IN APPENDIX A.

Lesson by Lesson

20.1 The Electric Field

Conceptual Questions

53. A proton moves in a region of constant electric field. Does it follow that the proton's velocity is parallel to the electric field? Explain.

54. A proton moves in a region of constant electric field. Does it follow that the proton's acceleration is parallel to the electric field? Explain.

55. Describe the difference between charging by induction and charging by contact.

56. A system consists of two charges of equal magnitude and opposite sign separated by a distance d. Since the total electric charge of this system is zero, can we conclude that the electric field produced by the system is also zero? Does your answer depend on the separation d? Explain.

57. The force experienced by charge 1 at point A is different in direction and magnitude from the force experienced by charge 2 at point B. Can we conclude that the electric fields at points A and B are different? Explain.

58. Can an electric field exist in a vacuum? Explain.

59. Explain why electric field lines never cross.

Problem Solving

60. A $+3.6$-μC charge experiences a force of 0.80 N due to an electric field. What is the magnitude of the electric field?

61. An electric field of 4.7×10^4 N/C exerts a force of 0.61 N on a charged object. The force is opposite in direction to the electric field. What are the sign and the magnitude of the charge on the object?

62. An object with a mass of 0.017 kg and a charge of 5.6 μC experiences an acceleration of 3.3 m/s^2 due to an electric field. Assuming that all other forces acting on the object can be ignored, what is the magnitude of the electric field?

63. **Think & Calculate** A charge of 5.7 μC produces an electric field. **(a)** Does the electric field point toward or away from the charge? Explain. **(b)** What is the magnitude of the electric field 0.45 m from the charge?

64. **Think & Calculate** The electric field lines surrounding three charges are shown in **Figure 20.26**. The center charge is $q_2 = -10.0$ μC. **(a)** What are the signs of q_1 and q_3? **(b)** Find q_1. **(c)** Find q_3.

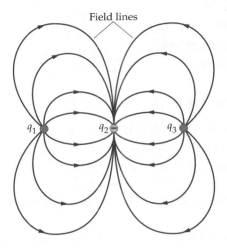

Field lines

Figure 20.26

65. Make a qualitative sketch of the electric field lines produced by two charges, $+q$ and $-q$, separated by a distance d.

66. Make a qualitative sketch of the electric field lines produced by two charges, $+q$ and $+2q$, separated by a distance d.

67. Make a qualitative sketch of the electric field lines produced by three charges, $+q$, $+q$, and $-q$, positioned at the corners of an equilateral triangle with sides of length d.

68. Two point charges lie on the x axis. A charge of $+6.2$ μC is at the origin, and a charge of -9.5 μC is at $x = 12$ cm. What are the magnitude and direction of the total electric field at $x = 4.0$ cm?

69. **Think & Calculate** Four point charges, each of magnitude q, are located at the corners of a square with sides of length d. Two of the charges are $+q$, and two are $-q$. The charges are arranged in one of the following two ways: (1) The charges alternate in sign ($+q$, $-q$, $+q$, $-q$) around the square; (2) the top two corners of the square have positive charges ($+q$, $+q$), and the bottom two corners have negative charges ($-q$, $-q$). **(a)** In which case will the electric field at the center of the square have the greater magnitude? Explain. **(b)** Calculate the electric field at the center of the square for each of these two cases. (Give your result as a multiple of kq/d^2.)

70. The electric field at $x = 5.00$ cm and $y = 0$ points in the positive x direction with a magnitude of 10.0 N/C. At $x = 10.0$ cm and $y = 0$ the electric field points in the positive x direction with a magnitude of 15.0 N/C. Assuming that this electric field is produced by a single point charge, find **(a)** its location and **(b)** the sign and the magnitude of its charge.

71. **Think & Calculate** A point charge, $q = +4.7$ μC, is placed at each corner of an equilateral triangle with sides 0.21 m in length. **(a)** What is the magnitude of the electric field at the midpoint of any of the three sides of the triangle? **(b)** Is the magnitude of the electric field at the center of the triangle greater than, less than, or the same as the magnitude at the midpoint of a side? Explain.

20.2 Electric Potential Energy and Electric Potential

Conceptual Questions

72. In one region of space the electric potential has a positive constant value. In another region of space the potential has a negative constant value. What can be said about the electric field within each of these two regions of space?

73. If the electric field is zero in some region of space, is the electric potential zero there as well? Explain.

74. **Predict & Explain** An electron is released from rest in a region of space with an electric field. **(a)** As the electron moves, does the electric potential energy of the system increase, decrease, or stay the same? **(b)** Choose the *best* explanation from among the following:

A. Because the electron has a negative charge, its electric potential energy doesn't decrease, as one might expect, but increases instead.

B. As the electron begins to move, its kinetic energy increases. The increase in kinetic energy is equal to the decrease in the electric potential energy of the system.

C. The electron will move perpendicular to the electric field, and hence its electric potential energy will remain the same.

75. **Triple Choice** In **Figure 20.27**, suppose the charge q_2 has the value necessary to produce zero electric potential at point A. Is the electric potential at point B positive, negative, or zero? Explain.

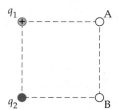

Figure 20.27

76. **Triple Choice** In Figure 20.27, suppose the charge q_2 has the value necessary to produce zero electric potential at point B. Is the electric potential at point A positive, negative, or zero? Explain.

77. **Rank** Four different arrangements of point charges are shown in **Figure 20.28**. In each case the charges are the same distance from the origin. Rank the four arrangements in order of increasing electric potential at the origin. Indicate ties where appropriate.

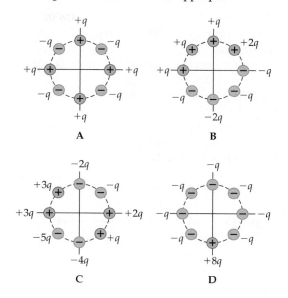

Figure 20.28

Problem Solving

78. **Cell Membranes** The electric potential inside a typical living cell is 0.070 V lower than the electric potential outside the cell. The thickness of the cell membrane is 0.10 μm. What are the magnitude and the direction of the electric field within the cell membrane?

79. Calculate the speed of a proton after it accelerates from rest through a potential difference of 275 V.

80. An object with a charge of 3.1 μC accelerates from rest through a region with a changing electric potential. If the object gains 0.0014 J of kinetic energy, what is the corresponding potential difference?

81. The electrons in a TV picture tube are accelerated from rest through a potential difference of 25 kV. What is the speed of the electrons after they have been accelerated by this potential difference?

82. **The Bohr Atom** The hydrogen atom consists of one electron and one proton. In the Bohr model of the hydrogen atom, the electron orbits the proton in a circular orbit of radius 0.529×10^{-10} m. What is the electric potential due to the proton at the electron's orbit?

83. **Think & Calculate** A spark plug in a car has electrodes separated by a gap of 0.0635 cm. To create a spark and ignite the air-fuel mixture in the engine, an electric field of 3.0×10^6 V/m must be present in the gap. **(a)** What potential difference must be applied to the spark plug to initiate a spark? **(b)** If the separation between electrodes is increased, does the required potential difference increase, decrease, or stay the same? Explain.

84. A Charged Battery A typical 12-V car battery can deliver 7.5×10^5 C of charge. If the energy supplied by the battery could be converted entirely to kinetic energy, what speed would it give to a 1400-kg car?

85. A proton has an initial speed of 4.0×10^5 m/s. What potential difference is required to bring the proton to rest?

86. Think & Calculate A uniform electric field with a magnitude of 1200 N/C points in the positive x direction, as shown in **Figure 20.29**. **(a)** What is the difference in electric potential, $\Delta V = V_B - V_A$, between the points A and B? **(b)** Is the electric potential at point B greater than or less than the electric potential at point C? Explain. **(c)** Calculate the difference in electric potential, $\Delta V = V_B - V_C$, between the points B and C.

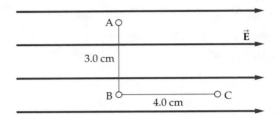

Figure 20.29

87. Find the electric potential at point P in **Figure 20.30**.

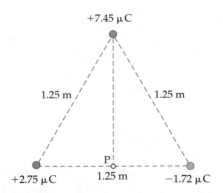

Figure 20.30

88. Figure 20.31 shows three charges at the corners of a rectangle. What is the electric potential at the empty corner of the rectangle?

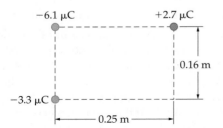

Figure 20.31

89. Think & Calculate The electric potential of a system as a function of position along the x axis is given in **Figure 20.32**. **(a)** In which of the regions, 1, 2, 3, or 4, does the electric field have its greatest magnitude? Explain. **(b)** Calculate the magnitude of the electric field in each of the regions 1, 2, 3, and 4.

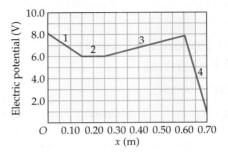

Figure 20.32

20.3 Capacitance and Energy Storage

Conceptual Questions

90. Triple Choice Two capacitors have the same charge on their plates. Capacitor 1 has a greater voltage between its plates than capacitor 2. Is the capacitance of capacitor 1 greater than, less than, or equal to the capacitance of capacitor 2? Explain.

91. In the water-bucket analogy to capacitors, what corresponds to the amount of charge stored in the capacitor?

92. In the water-bucket analogy to capacitors, what corresponds to the voltage between the plates of the capacitor?

93. What does the cross-sectional area of the water bucket correspond to in the water-bucket analogy to capacitors?

94. Triple Choice If the area of a capacitor's plates is increased, does its capacitance increase, decrease, or stay the same? Explain.

95. Triple Choice If the separation between a capacitor's plates is increased, does its capacitance increase, decrease, or stay the same? Explain.

Problem Solving

96. A capacitor has a capacitance of 4.1 μF. Find the magnitude of the charge on each of the capacitor's plates when it is connected to a 12-V battery.

97. What voltage is required to give the plates of a 320-pF capacitor a charge of 8.2×10^{-9} C?

98. A capacitor stores 6.9×10^{-9} C of charge on each of its plates when attached to a 1.5-V battery. What is its capacitance?

99. A capacitor stores 1.1×10^{-8} C of charge on each of its plates when attached to a 9.0-V battery. **(a)** What is its capacitance? **(b)** If the plate separation is 0.45 mm, what is the electric field between the plates?

100. The charge on each plate of a capacitor has a magnitude of 6.7×10^{-9} C when the electric field between the plates is 29,000 V/m. If the plate separation is 0.35 mm, what is the capacitance of the capacitor?

101. A 0.22-μF capacitor is charged by a 1.5-V battery. After being charged, the capacitor is connected to a small electric motor. Assuming 100% efficiency, **(a)** to what height can the motor lift a 5.0-g mass? **(b)** What initial voltage must the capacitor have if it is to lift a 5.0-g mass through a height of 1.0 cm?

102. A capacitor with a capacitance of 430 pF is connected to a battery with a voltage of 550 V. **(a)** What is the magnitude of the charge on each plate of the capacitor? **(b)** How much energy is stored in the capacitor? **(c)** What is the electric field between the plates if their separation is 0.89 mm?

Mixed Review

103. Triple Choice A parallel-plate capacitor is connected to a battery that maintains a constant potential difference between the plates. If the spacing between the plates is doubled, does the magnitude of the charge on the plates increase, decrease, or stay the same?

104. Predict & Explain A proton is released from rest in a region of space with a nonzero electric field. **(a)** As the proton moves, does the electric potential energy of the system increase, decrease, or stay the same? **(b)** Choose the *best* explanation from among the following:

A. As the proton begins to move, its kinetic energy increases. The increase in kinetic energy is equal to the decrease in the electric potential energy of the system.

B. Because the proton has a positive charge, its electric potential energy will always increase.

C. The proton will move perpendicular to the electric field, and hence its electric potential energy will remain the same.

105. Triple Choice The plates of a parallel-plate capacitor have constant charges of $+Q$ and $-Q$. Do the following quantities increase, decrease, or remain the same as the separation of the plates is increased? **(a)** the electric field between the plates; **(b)** the potential difference between the plates; **(c)** the capacitance; **(d)** the energy stored in the capacitor.

106. Triple Choice A parallel-plate capacitor is connected to a battery that maintains a constant potential difference V between the plates. If the plates of the capacitor are pulled farther apart, do the following quantities increase, decrease, or remain the same? **(a)** the electric field between the plates; **(b)** the charge on the plates; **(c)** the capacitance; **(d)** the energy stored in the capacitor.

107. Electric Catfish The electric catfish (*Malapterurus electricus*) shown in **Figure 20.33** is an aggressive fish, 1.0 m in length, found in tropical Africa (and depicted in Egyptian hieroglyphics). The catfish is capable of generating jolts of electricity of up to 350 V by producing a positively charged region of muscle near the head and

a negatively charged region near the tail. What is the direction of the electric field produced by the catfish?

Figure 20.33 A stunning creature from Africa—the electric catfish.

108. Computer Keyboards Many computer keyboards operate using capacitance. As shown in Figure 20.25, each key is part of a small parallel-plate capacitor whose separation is reduced when the key is depressed. Does depressing a key increase or decrease its capacitance? Explain.

109. A proton is released from rest in a uniform electric field of magnitude 1.08×10^5 N/C. Find the speed of the proton after it has traveled 0.50 m.

110. An electric field does 0.052 J of work as a $+5.7$-μC charge moves from point A to point B. Find the difference in electric potential, $\Delta V = V_B - V_A$, between the points A and B.

111. A small object with mass 0.0150 kg and charge 3.1 μC hangs from the ceiling by a thread. A second small object, with a charge of 4.2 μC, is placed 1.2 m vertically below the first charge. Find **(a)** the electric field at the position of the upper charge due to the lower charge and **(b)** the tension in the thread.

112. Think & Calculate A $+1.2$-μC charge and a -1.2-μC charge are placed at (0.50 m, 0) and (-0.50 m, 0), respectively, as shown in **Figure 20.34**. **(a)** At which of the points, A, B, C, or D, is the electric potential smallest in value? **(b)** At which of these points does the electric potential have its greatest value? Explain. **(c)** Calculate the electric potential at each of the points A, B, C, and D.

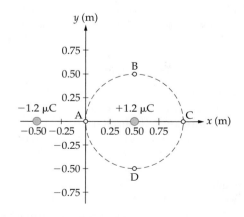

Figure 20.34

113. Repeat Problem 112 for the case where both charges are +1.2 μC.

114. Long, long ago, on a planet far, far away, a physics experiment was carried out. First, a 0.250-kg ball with zero net charge was dropped from rest from a height of 1.00 m. The ball landed 0.552 s later. Next, the ball was given a net charge of 7.75 μC and dropped from rest from the same height. This time the ball fell for 0.680 s before landing. What is the electric potential at a height of 1.00 m above the ground on this planet, given that the electric potential at ground level is zero? (Air resistance can be ignored.)

115. Figure 20.35 shows an electron entering a parallel-plate capacitor with a speed of 5.45×10^6 m/s. The electric field of the capacitor has deflected the electron downward by a distance of 0.618 cm at the point where the electron exits the capacitor. Find the magnitude of the electric field in the capacitor.

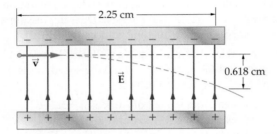

Figure 20.35

116. A charge of 24.5 μC is located at (4.40 m, 6.22 m), and a charge of −11.2 μC is located at (−4.50 m, 6.75 m). What charge must be located at (2.23 m, −3.31 m) for the electric potential to be zero at the origin?

Writing about Science

117. Write a report on automatic external defibrillators (AED). Address the following questions: How much energy is stored in an AED when ready for use? What is its voltage? What is its capacitance? Does your local mall have AEDs ready to be used in an emergency?

118. Connect to the Big **Idea** Describe an experiment that would allow you to measure the energy stored in a capacitor. Would increasing the charge on the capacitor's plates increase or decrease the energy it stores? Explain.

Read, Reason, and Respond

The Electric Eel Of the many unique and unusual animals that inhabit the rainforests of South America, including howler monkeys, freshwater dolphins, and deadly piranhas, one stands out because of its mastery of electricity. The electric eel (*Electrophorus electricus*), shown in **Figure 20.36**, one of the few creatures on Earth able to generate, store, and discharge electricity, can deliver a powerful series of high-voltage discharges reaching 650 V. These jolts of electricity are so strong, in

fact, that electric eels have been known to topple a horse crossing a stream 20 feet away, and to cause respiratory paralysis, cardiac arrhythmia, and even death in humans.

The organs that produce the eel's electricity take up most of its body and consist of thousands of modified muscle cells—called *electroplaques*—stacked together like the cells in a battery. Each electroplaque is capable of generating a voltage of 0.15 V, and together they produce a positive charge near the head of the eel and a negative charge near its tail.

Figure 20.36 An electric eel can incapacitate its prey with a powerful shock.

119. Electric eels produce an electric field within their body. In which direction does the electric field point?
A. toward the head C. upward
B. toward the tail D. downward

120. As a rough approximation, consider an electric eel to be a parallel-plate capacitor with a capacitance of 7.6 pF. How much charge does an electric eel generate at each end of its body when it produces a voltage of 650 V?
A. 1.2×10^{-14} C C. 4.9×10^{-9} C
B 5.2×10^{-11} C D. 6.1×10^{-5} C

121. How much energy is stored by an electric eel when it has the potential difference of 650 V? Use the parallel-plate model discussed in Problem 120.
A. 1.8×10^{-17} J C. 1.6×10^{-6} J
B. 1.7×10^{-8} J D. 2.0×10^{-2} J

Standardized Test Prep

Multiple Choice

Use the diagram below to answer Questions 1 and 2. The diagram shows two stationary point charges, a $+2\text{-}\mu C$ charge at $x = 1$ m and a $-4\text{-}\mu C$ charge at $x = 6$ m.

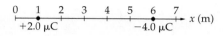

1. What is the net electric field at $x = 3$ m?
   - **(A)** 8500 N/C toward the right
   - **(B)** 4500 N/C toward the right
   - **(C)** 4000 N/C toward the right
   - **(D)** 500 N/C toward the left

2. At what point could the net electric field due to the two charges be zero?
   - **(A)** at some point between the two charges
   - **(B)** at some point to the left of $x = 1$ m
   - **(C)** at some point to the right of $x = 6$ m
   - **(D)** There is no such point.

3. An isolated point charge Q produces an electric field. The electric field has the magnitude E at a distance d from the charge. If the charge is doubled, what is the magnitude of the field at $2d$?
   - **(A)** $E/4$　　**(C)** $2E$
   - **(B)** $E/2$　　**(D)** $4E$

Use the diagram below to answer Questions 4 and 5. A 2.0-m-diameter hollow metal sphere is located with its center at $x = 2$ m. The sphere has a total charge of 10.0 μC on its surface.

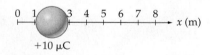

4. What is the electric field due to the sphere at $x = 8$ m?
   - **(A)** 2500 N/C toward the right
   - **(B)** 1840 N/C toward the right
   - **(C)** 3600 N/C toward the right
   - **(D)** zero

5. What is the potential energy of the system when a proton is placed at $x = 8$ m?
   - **(A)** 4.0×10^{-16} J
   - **(B)** 2.4×10^{-15} J
   - **(C)** 4.8×10^{-16} J
   - **(D)** zero

6. When a capacitor is charged to a potential difference of V, it stores charge Q and electrical energy PE. How will the stored charge and energy change if the capacitor is instead charged to a potential difference of $2V$?
   - **(A)** The stored charge will be $2Q$; the stored energy will be $2PE$.
   - **(B)** The stored charge will be $Q/2$; the stored energy will be $2PE$.
   - **(C)** The stored charge will be $2Q$; the stored energy will be $4PE$.
   - **(D)** The stored charge will be $4Q$; the stored energy will be $4PE$.

Extended Response

7. A stream of electrons enters a uniform electric field between two charged parallel plates. The electrons do not strike either plate before exiting. Describe the path the electrons take while moving through the field and after exiting the field.

> **Test-Taking Hint**
>
> Electric force and electric field are in the same direction for a positive charge and in opposite directions for a negative charge.

If You Had Difficulty With . . .

Question	1	2	3	4	5	6	7						
See Lesson	20.1	20.1	20.1	20.1	20.2	20.3	20.1						

21 Electric Current and Electric Circuits

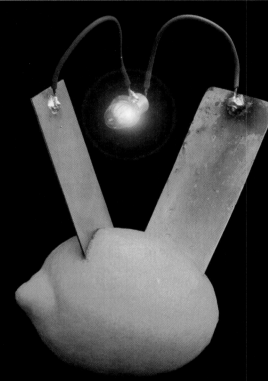

Did you know that a lemon can act as a battery? Here we see that two electrodes made of different metals experience different electric potentials when inserted in the acidic juice of a lemon. The difference in electric potential creates an electric current that powers a light.

Big Idea

Electrons flow through electric circuits in response to differences in electric potential.

As you read this, your heart is pumping blood through your arteries and veins. In many ways your circulatory system is like an electric circuit. The blood flowing in your veins is like electric charge flowing through a wire. And just as your heart must do work to pump blood through your body, a source of energy is needed to operate an electronic device.

The flow of blood is important to life, and the flow of electric charge is just as important to modern technology. In this chapter you'll learn about some of the basic properties of moving electric charges and common electric circuits.

21.1 Electric Current, Resistance, and Semiconductors

All electric circuits have one thing in common—they depend on the flow of electric charge. In general, electric charge is carried through a circuit by electrons. This lesson explores the properties of flowing electric charge.

Electric Current

When water flows from one place to another, we say that it forms a *current*. The more water that flows, and the faster it flows, the stronger the current. Much the same is true for electric charge. When electric charge flows from one place to another, we say it forms an **electric current**. The more charge that flows, and the faster it flows, the greater the electric current.

Electric current is the amount of charge that flows in a given period of time

Suppose an amount of charge ΔQ flows past a given point in a wire in the time Δt. The electric current, I, in the wire is simply defined as the amount of charge divided by the amount of time.

Vocabulary

- electric current
- electric circuit
- battery
- electromotive force
- resistance
- Ohm's law
- ohmmeter
- resistor
- diode

The following equation is used to determine the current flowing in a wire.

Definition of Electric Current, *I*

$$\text{electric current} = \frac{\text{amount of charge}}{\text{amount of time}}$$

$$I = \frac{\Delta Q}{\Delta t}$$

SI unit: coulombs per second (C/s) = ampere (A)

Measuring Current The unit of current is the *ampere* (A), or *amp* for short. It is named for the French physicist André-Marie Ampère (1775–1836). A current of 1 amp is defined as the flow of 1 coulomb of charge in 1 second:

$$1\,A = 1\,C/s$$

A 1-amp electric current is fairly strong. Many electronic devices, like cell phones and digital music players, operate on currents that are a fraction of an amp. Examples of electric currents in nature include lightning bolts, **Figure 21.1 (a)**, and the shocks delivered by electric rays, **Figure 21.1 (b)**.

CONCEPTUAL Example 21.1 Comparing Currents

Which of the following situations has the greatest electric current? Which has the smallest electric current?
(a) the flow of 6 C in 3 s
(b) the flow of 3 C in 1 s
(c) a person walking 8 m in 4 s

Reasoning and Discussion
The electric current is charge per time. Therefore, the current in situation (a) is $I = \frac{6\,C}{3\,s} = 2A$, and the current in situation (b) is $I = \frac{3\,C}{1\,s} = 3\,A$. The current in situation (c) is zero because a person's total charge is essentially zero. As a result, when a person moves from one place to another, there is practically no *net* flow of electric charge.

Answer
Situation (b) has the greatest current, 3 A; situation (c) has the smallest current, zero.

▶ **Figure 21.1 Electric currents in nature**
(a) A lightning bolt is an example of a large electric current. It flows when the difference in electric potential between a cloud and the ground becomes so great that electrons are pulled from atoms in the air. An enormous quantity of charge then flows through the bolt in a fraction of a second. **(b)** Some organisms, such as this electric torpedo ray, have internal "batteries" that produce significant electrical potentials. The resulting current is used to stun their prey.

(a)

(b)

The following Guided Example shows that the number of electrons flowing in a typical electric circuit is extremely large. The situation is similar to the large number of water molecules flowing through a garden hose.

The disk drive in a portable CD player is connected to a battery that supplies it with a current of 0.22 A. How many electrons pass through the disk drive in 4.5 s? (Recall that the charge of an electron has the magnitude $e = 1.60 \times 10^{-19}$ C.)

Picture the Problem

Our sketch shows the CD player with a current $I = 0.22$ A flowing through it. Also indicated is the time, $\Delta t = 4.5$ s, during which the current flows.

Strategy

Since we know both the current, I, and the amount of time, Δt, we can use the definition of current, $I = \Delta Q / \Delta t$, to find the amount of charge, ΔQ, that flows through the player. Once we know the amount of charge, the number of electrons, N, is that amount divided by the magnitude of the electron's charge. That is, $N = \Delta Q / e$, where $e = 1.60 \times 10^{-19}$ C.

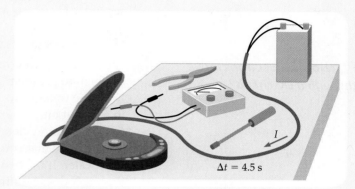

$\Delta t = 4.5$ s

Known	Unknown
$I = 0.22$ A	$\Delta Q = ?$
$\Delta t = 4.5$ s	$N = ?$
$e = 1.60 \times 10^{-19}$ C	

Solution

 Rearrange $I = \Delta Q / \Delta t$ to solve for the amount of charge, ΔQ, that flows through the drive:

$$\Delta Q = I \Delta t$$
$$= (0.22 \text{ A})(4.5 \text{ s})$$
$$= 0.99 \text{ C}$$

Math HELP
Significant Figures
See Lesson 1.4

2 Divide the amount of charge, ΔQ, by the magnitude of the electron's charge, e, to find the number of electrons, N:

$$N = \frac{\Delta Q}{e}$$
$$= \frac{0.99 \ \cancel{C}}{1.60 \times 10^{-19} \ \cancel{C}/\text{electron}}$$
$$= \boxed{6.2 \times 10^{18} \text{ electrons}}$$

Insight

Thus, even a modest current flowing for a brief time corresponds to the transport of an extremely large number of electrons.

Practice Problems

1. **Follow-up** For what period of time must the current flow if 7.5×10^{18} electrons are to pass through the disk drive?

2. What is the current if 3600 C of charge passes a point in 17 min?

3. A flashlight bulb carries a current of 0.18 A. How much time is required for 14 C of charge to pass through the bulb?

AC and DC Circuits

If you travel from home to another location, then to a different location, and then back home again, you have traveled a circuit. Similarly, when charge flows through a closed path and returns to its starting point, we say that the closed path is an **electric circuit**. In this chapter we concentrate on *direct-current circuits*, also known as *DC circuits*. In a DC circuit the current always flows in the *same* direction. Circuits that run on batteries are typically DC circuits.

Circuits with currents that periodically reverse their direction are referred to as *alternating-current circuits*, or *AC circuits*. The electricity provided by a wall plug in your house is AC, with 60 cycles of oscillation per second. AC circuits are considered in detail in Chapter 23.

Batteries produce direct currents

Although electrons move fairly freely in metal wires, something has to push on them to get them going and keep them going. It's like water in a garden hose; the water flows only when a force pushes on it. Similarly, electrons flow in a circuit only when an electrical force pushes on them. Let's look at this analogy in more detail.

Imagine that you and a friend each hold one end of a garden hose filled with water. If the two ends are held at the same level, as in **Figure 21.2 (a)**, the water does not flow. Now suppose one end is lowered below the other, as in **Figure 21.2 (b)**. In this case water flows from the high end of the hose to the low end. The difference in gravitational potential energy between the two ends of the hose results in a force on the water—which in turn produces a flow. A *battery* performs a similar function in an electric circuit.

Battery-Powered Circuits

Simply put, a **battery** uses chemical reactions to produce a difference in electric potential between its two ends, which are referred to as the *terminals*. The symbol for a battery is ⊣⊢. The positive terminal has a high electric potential and is denoted with a plus (+) sign; the negative terminal has a low electric potential and is denoted with a minus (−) sign. **When a battery is connected to a circuit, electrons move in a closed path from one terminal of the battery through the circuit and back to the other terminal of the battery. The electrons leave from the negative terminal of the battery and return to the positive terminal.** Electric charge is conserved as electrons flow from one terminal of a battery to the other.

The situation is similar to the flow of blood in your body. As you know, blood won't flow through your arteries and veins unless your heart beats and pushes it along. Your heart acts like a battery in an electrical circuit. A battery causes electric charge to flow through a closed circuit of wires, and your heart causes blood to flow through a closed circuit of vessels in your body.

🔑 *How do electrons move through a DC circuit?*

▶ **Figure 21.2 Water flow as an analogy for electric current**
Water can flow freely through a garden hose, but if both ends are at the same level **(a)**, there is no flow. If the ends are held at different levels **(b)**, water flows from the end where the gravitational potential energy is high to the end where it is low.

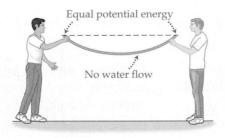

(a) Equal potential energy → no flow

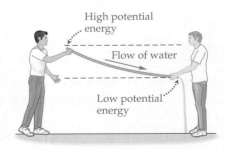

(b) Water flows from high potential energy to low.

A Simple DC Circuit

A simple electrical system consisting of a battery, a switch, and a lightbulb connected together in a flashlight is shown in **Figure 21.3 (a)**. The circuit diagram in **Figure 21.3 (b)** shows that the switch is open—creating an *open circuit*. In an open circuit there is no closed path through which the electrons can flow. As a result, the light is off. When the switch is closed—which closes the circuit—charge flows around the circuit, causing the light to glow.

Figure 21.4 shows a mechanical equivalent of the flashlight circuit. The person raising the water from a low level to a high level is like the battery. The paddle wheel is like the lightbulb, and the water is like the electric charge. The person does work in raising the water. As the water falls to its original level, it does work turning the paddle wheel.

A battery produces a voltage difference

The difference in electric potential between the terminals of a battery is the **electromotive force**, or *emf*. Symbolically, the electromotive force is represented by the symbol ε (the Greek letter epsilon). The unit of emf is the same as that of electric potential, namely, the volt. Clearly, then, the electromotive force is not really a force at all. Instead, the emf determines the amount of work a battery does to move a certain amount of charge around a circuit.

To be specific, the magnitude of the work done by a battery with the emf ε as charge ΔQ moves from one terminal to the other is

> **Work Done by a Battery**
>
> work done by a battery $=$ amount of charge $\times$ emf of battery
> $$W = (\Delta Q)\varepsilon$$
> SI unit: J

(a) A simple flashlight

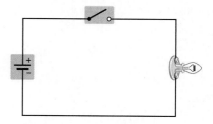

(b) Circuit diagram for flashlight

▲ **Figure 21.3 The flashlight: A simple electric circuit**
(a) A simple flashlight, consisting of a battery, a switch, and a lightbulb. (b) When the switch is in the open position, the circuit is open, and no charge can flow. When the switch is closed, electrons flow through the circuit and the light glows.

ACTIVE Example 21.3 Determine the Charge and the Work

A battery with an emf of 1.5 V delivers a current of 0.44 A to a flashlight bulb for 64 s (see Figure 21.3). Find **(a)** the charge that passes through the circuit and **(b)** the work done by the battery.

Solution (*Perform the calculations indicated in each step.*)

Part (a)

1. Rearrange the definition of current, $I = \Delta Q/\Delta t$, $\Delta Q = 28$ C
to solve for the charge that flows through the circuit:

Part (b)

2. Substitute the value for ΔQ determined in part (a) $W = 42$ J
in $W = \Delta Q \varepsilon$ to find the work done by the battery:

Insight
Notice that the more charge a battery moves through a circuit, the more work it does. Similarly, the greater the emf, the greater the work. We can see, then, that a car battery that operates at 12 V and delivers several amps of current does much more work than a flashlight battery operating at 1.5 V and delivering 0.44 A—as expected.

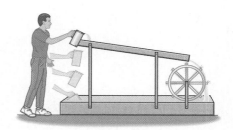

▲ **Figure 21.4 A mechanical equivalent of a flashlight circuit**
The person lifting the water corresponds to the battery in Figure 21.3, and the paddle wheel corresponds to the lightbulb.

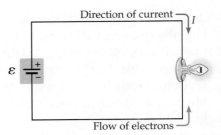

Direction of current — I

ε

Flow of electrons

▶ **Figure 21.5 Directions of current and electron flow**
In the flashlight circuit electrons flow from the battery's negative terminal to its positive terminal. The direction of the current, I, is just the opposite: from the positive terminal to the negative terminal.

Electric current flow is *opposite* that of the electrons

When drawing an electric circuit, it's helpful to include an arrow to indicate the flow of current. By convention, the direction of the current arrow is given in terms of a *positive* test charge:

● The direction of the current in an electric circuit is the direction in which a *positive* test charge would move.

In typical circuits the charges that flow are actually *negatively* charged electrons. As a result, the flow of electrons and the current arrow point in opposite directions, as indicated in **Figure 21.5**. Notice that a positive test charge would flow from a region of high electric potential (near the positive terminal of the battery) to a region of low electric potential (near the negative terminal). The negatively charged electrons do just the opposite.

Electrons move very slowly through wires

As surprising as it may seem, electrons move rather slowly through a wire. Their path is tortuous and roundabout because they are involved in numerous collisions with the atoms in the wire, as indicated in **Figure 21.6**. Like a car contending with a series of speed bumps, an electron's average speed, or *drift speed* as it is called, is limited by these repeated collisions. Typical drift speeds are about 10^{-4} m/s—that's only about a hundredth of a centimeter per second!

To put this in context, suppose you switch on the headlights of a car. It takes an electron (moving at 10^{-4} m/s) about 3 hours to go from the battery to the lightbulb! How can this be possible? We know that the lights come on as soon as the switch is turned, not hours later.

The answer is that as an electron begins to move away from the battery, it exerts a force on its neighbors. This causes them to move in the same general direction and, in turn, to exert a force on their neighbors, and so on. This influence moves through the wire at nearly the speed of light. The phenomenon is similar to a bowling ball hitting one end of a line of other bowling balls—the effect of the colliding ball travels through the line at roughly the speed of sound, even though the individual balls have very little displacement. Similarly, the electrons in a wire move with a rather small average velocity, but the *influence* they have on one another races ahead—and causes the lights to shine.

▲ **Figure 21.6 Path of an electron in a wire**
Typical path of an electron as it bounces off atoms in a metal wire. Because of the tortuous path the electron follows, its average velocity is rather small.

Resistance

Electrons flow through metal wires with relative ease. In the ideal case the electrons move with complete freedom. Real wires, however, always affect the electrons to some extent. Collisions between electrons and atoms in a wire cause a **resistance** to the electrons' motion. This effect is similar to friction resisting the motion of a box sliding across a floor, or a dense crowd reducing your walking speed.

Potential difference drives electrons through a wire

To move electrons against the resistance of a wire it is necessary to apply a potential difference between its ends. **Ohm's law** relates the applied potential difference to the current produced and the wire's resistance. 🔌 **The applied voltage equals the product of the current and the wire's resistance.**

🔌 *How is the voltage across a resistor related to the current through the resistor?*

> ### Ohm's Law
>
> applied voltage = current × resistance
> $$V = IR$$
>
> SI unit: volt (V)

Ohm's law is named for the German physicist Georg Simon Ohm (1789–1854). Rearranging Ohm's law to solve for the resistance, we find

$$R = \frac{V}{I}$$

From this expression it is clear that resistance has units of volts per amp. A resistance of 1 volt per amp defines a new unit—the *ohm*. The Greek letter omega (Ω) is used to designate the ohm. Thus,

$$1\ \Omega = 1\ \text{V/A}$$

A device for measuring resistance is called an **ohmmeter**.

A **resistor** is a small device used in electric circuits to provide a particular resistance to current. The resistance of a resistor is given in ohms, as shown in the following Quick Example.

QUICK Example 21.4 What's the Current?

A potential difference of 24 V is applied to a 150-Ω resistor. How much current flows through the resistor?

Solution
Solving Ohm's law for the current, we find

$$I = \frac{V}{R} = \frac{24\ \text{V}}{150\ \Omega} = \frac{24\ \cancel{V}}{150\ \cancel{V}/A} = \boxed{0.16\ \text{A}}$$

Practice Problems

4. What voltage is required to produce a current of 0.62 A in a 250-Ω resistor? (*Hint*: Solve Ohm's law for voltage, *V*.)

5. When a potential difference of 18 V is applied to a given wire, the wire conducts 0.35 A of current. What is the resistance of the wire? (*Hint*: Solve Ohm's law for resistance, *R*.)

In an electric circuit a resistor is signified by a zigzag line, ⌇⌇⌇, as a reminder of the zigzag path of the electrons in the resistor. The straight lines in a circuit indicate ideal wires of zero resistance. To indicate the resistance of a real wire or device, we simply include a resistor of the appropriate value in the circuit. All circuit elements used in this chapter are summarized in **Table 21.1** on the next page.

Table 21.1 Elements of Electric Circuits

Circuit Element	Symbol	Physical Characteristics
Resistor	‑\/\/\‑	Resists the flow of electric current. Converts electric energy into thermal energy.
Ideal wire	———	An ideal wire has zero resistance. It is used to connect various elements in a circuit.
Battery	⊣⊢	A device that produces a constant difference in electrical potential between its two terminals.
Switches (open and closed)	⁄○ •—○	Devices used to control whether electric current is allowed to flow through a circuit or a portion of a circuit.
Diode	▶⊦	A device that allows electric current to flow in one direction only.
Incandescent lightbulb	🔆	A device containing a resistor that gets hot enough to give off visible light.

A wire's resistance is affected by several factors

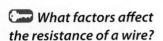 **What factors affect the resistance of a wire?**

Wires come in a variety of shapes, sizes, and lengths. In addition, they can be made from different metals and operate over a wide range of temperatures. Not surprisingly, the resistance of a wire is affected by these factors.

Type of Material 🔌 **The resistance of a wire depends on the material from which it is made.** If a wire is made of copper, for instance, its resistance is less than if it is made from iron. This is just like a concrete road providing a smoother ride—with less "resistance"—than a dirt road. The resistance of a given material is described by its *resistivity, ρ*.

Physical Dimensions 🔌 **A wire's resistance also depends on its length, *L*, and its cross-sectional area, *A*.** To understand these factors, let's consider water flowing through a hose. If the hose is very long, its resistance to the water is correspondingly large. On the other hand, a wide hose, with a greater cross-sectional area, offers less resistance to the water. After all, water flows more easily through a short fire hose than through a long soda straw. It follows that the resistance of a hose—and similarly of a wire—is proportional to its length, *L*, and inversely proportional to its cross-sectional area, *A*.

Combining our observations, we can write the following relationship:

> **Definition of Resistivity, ρ**
>
> $$\text{resistance} = \text{resistivity} \times \frac{\text{length of wire}}{\text{cross-sectional area}}$$
>
> $$R = \rho\left(\frac{L}{A}\right)$$
>
> SI units: ohm-meters ($\Omega \cdot m$)

The units of resistivity are ohm-meters ($\Omega \cdot m$), and its magnitude varies greatly with the type of material. Insulators have large resistivities, typically in the range of 10^{10} $\Omega \cdot m$, and conductors have low resistivities, in the range of 10^{-8} $\Omega \cdot m$. Thus, insulators are characterized by large resistance, and conductors have low resistance.

Table 21.2 Factors Affecting the Resistance of a Wire

Factor	Example of Lower Resistance	Example of Higher Resistance	Reason for Difference
Material	Copper	Lead	Resistance depends on the material. The lead wire has a greater resistance than the copper wire.
Length			Resistance increases with length. A long wire has a greater resistance than a short wire.
Area			Resistance increases as cross-sectional area decreases. A small-diameter wire has a greater resistance than a large-diameter wire.
Temperature			Resistance increases as temperature increases. A heated wire has a greater resistance than a cooled wire.

Temperature

You know from everyday experience that a wire carrying an electric current can become warm. It can even be quite hot, as in the case of a heating element on an electric stove or the filament in an incandescent lightbulb. Recall that electrons collide with the atoms in a wire as they flow through an electric circuit. These collisions cause the atoms to jiggle and acquire greater kinetic energy. As a result, the temperature of the wire increases.

For example, the wire filament in an incandescent lightbulb heats up to a temperature of roughly 2800 °C. For comparison, the surface of the Sun has a temperature of about 5500 °C. The heating coil on a stove has a temperature of about 750 °C. Even computers heat up as they operate. In fact, heating is a serious problem in the design of computers, and it gets worse as a computer is made smaller. Computers could be made considerably smaller than they are today, but their own heat would melt them.

As a wire is heated, its resistivity tends to increase. This effect occurs because atoms that are jiggling more rapidly are more likely to collide with electrons and slow their progress through the wire. In fact, many metals show an approximately linear increase of resistivity with temperature. Once the dependence of resistivity on temperature is known for a given material, the change in resistivity can be used as a means of measuring temperature.

Table 21.2 summarizes the four major factors that affect the resistance of a wire.

Physics & You: Technology The first practical application to make use of the temperature dependence of resistance was a device known as the *bolometer*. Invented in 1880, the bolometer is an extremely sensitive thermometer that uses the temperature-related variation in the resistivity of platinum, nickel, or bismuth as a means of detecting temperature changes as small as 0.0001 °C. Soon after its invention, a bolometer was used to detect infrared radiation from the stars. An example of a modern-day bolometer is shown in **Figure 21.7**.

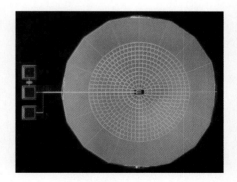

▲ **Figure 21.7 Close-up view of a modern bolometer**
A bolometer uses the variation in electric resistance with temperature to give extremely sensitive measurements of temperature.

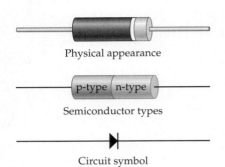

Physical appearance

p-type | n-type

Semiconductor types

Circuit symbol

▲ **Figure 21.8 A diode**
A diode is a device that allows electric current to flow freely in one direction, but not in the reverse direction. Diodes are formed by joining p-type and n-type semiconductors. The circuit symbol that indicates a diode contains an arrow pointing in the direction in which current can flow.

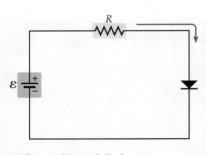

(a) Forward-biased diode

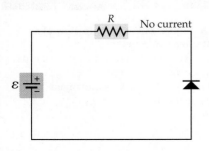

(b) Reverse-biased diode

Semiconductors

Though Ohm's law is an excellent approximation for metal wires and the resistors used in electric circuits, it does not apply to all materials. Materials knows as *semiconductors* are an important exception to Ohm's law.

Semiconductors have many interesting properties

Elements such as germanium and silicon (for which Silicon Valley in California is named) are insulators in their pure form. However, when impurities are added—which is referred to as *doping*—these substances can conduct electricity. Doping produces two types of semiconductors.

Semiconductor Types
If a small amount of arsenic is added to silicon—say, one arsenic atom per million silicon atoms—the silicon becomes a conductor. The arsenic-doped silicon conducts electricity because electrons break free from the arsenic atoms and move freely through the material. Silicon doped in this way is referred to as an *n-type semiconductor* because current is carried by the negative (n) electrons.

Silicon also becomes a conductor when it is doped with gallium instead of arsenic. In this case, however, the gallium atoms *take electrons from* the silicon atoms, forming positively charged "holes" that can carry current. Because positive (p) holes carry the current, this type of material is referred to as a *p-type semiconductor*.

Semiconductor Behavior
Unlike a typical resistor, a semiconductor has a lower resistance when its temperature increases. This is because the resistance of a semiconductor is strongly dependent on the number of electrons given up by arsenic atoms (which are free to move about) or the number of electrons that move from silicon atoms to gallium atoms (to create holes). In either case an increase in temperature makes it easier for electrons to move, and this produces more current. The result is a decrease in resistance.

Physics & You: Technology Electronic devices incorporating temperature-dependent semiconductors are known as *thermistors*. You might have used a thermistor yourself. The digital thermometers so common in today's hospitals and homes use thermistors to measure body temperature.

Diodes are the simplest semiconductor devices

Semiconductors can be used to make a variety of electronic devices. The simplest semiconducting device, the **diode**, consists of a p-type semiconductor joined to an n-type semiconductor. A diode is shown in **Figure 21.8**. The basic property of a diode is that it allows current to flow in one direction, but not in the other. For example, when the positive terminal of a battery is attached to the p-type semiconductor in an ideal diode, as in **Figure 21.9 (a)**, current flows with zero resistance. In this case we say that the diode is *forward biased*. On the other hand, if the positive terminal of a battery is connected to the n-type semiconductor of an ideal diode, as in **Figure 21.9 (b)**, no current flows at all. This case is referred to as a *reverse-biased* diode. Clearly, the behavior of diodes is *not* described by Ohm's law.

◄ **Figure 21.9 Simple diode circuits**
(a) A forward-biased diode is connected to a battery in such a way that current flows in the direction of the arrow in the diode symbol. Since the forward-biased diode allows current to flow with no resistance, the current in this circuit is $I = \varepsilon/R$. **(b)** No current flows through a reverse-biased diode.

Because of the one-way nature of diodes, they find many uses in electric circuits. One common application is the conversion of AC current (which alternates in direction) to DC current (which flows in one direction only). Another application makes use of the fact that light is emitted when electrons and holes come together in a diode. This is the basic process behind the operation of an LED, or *light-emitting diode*. LEDs are used in many practical applications, from tail lights on cars, to traffic signals and flashlights. Examples of LEDs are shown in **Figure 21.10**.

Transistors consist of three semiconductor layers

Another useful semiconductor device is produced by making a "sandwich" of three layers of semiconductor. The most common type of *transistor* has an n-type semiconductor on either side of the sandwich, and a thin p-type semiconductor in the middle, as shown in **Figure 21.11**. This is known as an *npn transistor*. Transistors can also be made with the opposite sequence of semiconductors, resulting in a *pnp transistor*.

The basic function of a transistor is to act as an electronic switch that controls the flow of current in a circuit. For example, consider the schematic view of an npn transistor shown in **Figure 21.12**. The three electrodes of the transistor are the collector, the base, and the emitter. Of these three electrodes it is the base that switches on or off the flow of current through the other two electrodes.

You might find it helpful to think of the control of the current by the base electrode as similar to turning a valve in a large-diameter water pipe. Though it doesn't take much force to turn the valve, once the valve is opened, a large volume of water flows through the pipe. Similarly, a small base current "opens the valve" that allows a large amount of current to flow from the collector to the emitter. In a typical transistor, a current *I* in the base can control the flow of a current of up to 300*I* through the other two electrodes. The transistor reacts quickly to changes in base current, and

▲ **Figure 21.10 Light from diodes: LEDs**
LEDs were introduced in the 1960s, and have become increasingly common in everyday life ever since. Part of their appeal is that they are more efficient at producing light than incandescent lightbulbs, and have a greater lifetime as well. The first LEDs were red, but they now come in a wide range of colors. LEDs also produce the infrared light used in remote controls.

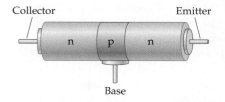

▲ **Figure 21.11 An npn transistor**
A transistor is produced by making a "sandwich" of different types of semiconductor. The transistor shown here is a sandwich of two n-type semiconductors with a p-type semiconductor in the middle. Such transistors are npn transistors. A sandwich of two p-type semiconductors with an n-type semiconductor in the middle is a pnp transistor.

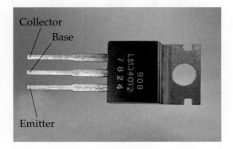

(a)

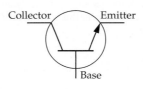

(b)

◄ **Figure 21.12 Real and symbolic transistors**
(a) A real transistor is typically a small black plastic container with three metal electrodes protruding from one side. These electrodes are the collector, the base, and the emitter. **(b)** In an electric circuit diagram a transistor is represented by the symbol shown here.

► **Figure 21.13 A transistor analogy**
The operation of a transistor is similar to the action of a valve in a large-diameter water pipe. Slight changes in the orientation of the valve can turn on or off the flow of a large volume of water. Similarly, a small current in the base of a transistor can turn on or off the flow of a large current through the collector and emitter.

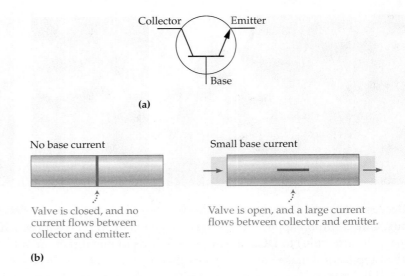

(a)

No base current

Small base current

Valve is closed, and no current flows between collector and emitter.

Valve is open, and a large current flows between collector and emitter.

(b)

therefore any signal with a changing current that comes into the base electrode is reflected accurately in a corresponding change in current flowing from the collector to the emitter—but amplified 300 times. The water valve analogy for a transistor is shown in **Figure 21.13**.

Transistors are the basis of modern computers

One of the great advantages of transistors is that a small base current can turn a transistor on, by allowing current to flow through it, or off, by preventing the flow of current. A device that can switch rapidly between on and off states is just what's needed in modern digital computers, whose language is based on the binary digit (bit), which takes on the value 1 or 0. Computers represent these two states by a transistor that is either on or off.

Many transistors are required in a computer or smart device. Most electronic devices today rely on silicon wafers, called *microchips*, that contain thousands of transistors, diodes, and resistors connected in elaborate circuits. These *integrated circuits* (ICs), as they're called, are built up layer by layer on a silicon wafer by depositing specific patterns of silicon, gallium, arsenic, and so on, to produce the desired arrangement of n-type and p-type semiconductors. The result, as shown in **Figure 21.14**, is a circuit so intricate it almost looks like an aerial photograph of a large city.

► **Figure 21.14 Integrated circuit**
An integrated circuit can contain thousands of transistors connected in intricate patterns on a chip of silicon. These devices are at the heart of modern electronic devices.

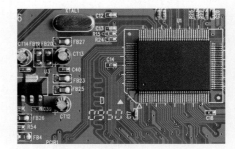

21.1 LessonCheck ⓜⓟ

Checking Concepts

6. 🔌 **Sketch** Draw a simple circuit consisting of a battery and a wire. Indicate the direction in which the electrons move through the wire.

7. 🔌 **Describe** How does the current through a wire change if the wire's resistance is increased?

8. 🔌 **Describe** What happens to the resistance of a wire if its temperature is increased?

9. Big **Idea** Explain how the flow of electrons through a circuit is related to regions of low electric potential and high electric potential.

10. Identify What is the direction of the electric current produced by an electron that falls toward the ground?

11. Determine If you want to conduct current through a diode, should you connect the positive terminal of a battery to the p-type semiconductor or the n-type semiconductor?

Solving Problems

12. Identify Resistor 1 has a resistance of 10 Ω and carries a current of 3 A. Resistor 2 has a resistance of 5 Ω and carries a current of 10 A. Which resistor has the greater potential difference?

13. Calculate A potential difference of 16 V is applied to a resistor whose resistance is 220 Ω. What is the current that flows through the resistor?

14. Calculate What voltage is required to produce a current of 1.8 A through a 140-Ω resistor?

15. Calculate How long must the flashlight battery in Active Example 21.3 operate to do 150 J of work?

21.2 Electric Circuits

Electric circuits often contain a number of resistors connected in various ways. In this lesson we consider simple circuits containing resistors and batteries. For each type of circuit you'll learn how to calculate the net effect of a group of resistors.

Series Circuits

Have you ever seen a conga line? This is the dance where people hold onto the hips or shoulders of the person in front of them, forming a line. The line then snakes across the dance floor as the dancers show off their moves. What does this have to do with electric circuits? Well, the dancers in a conga line are connected to one another in *series*—one dancer connected to the next, connected to the next, and so on.

Vocabulary

- series circuit
- parallel circuit
- ammeter
- voltmeter

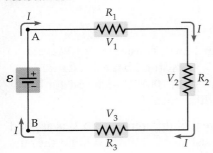

How does adding another resistor in series affect the resistance?

(a) Three resistors in series

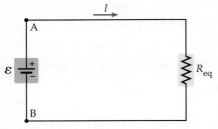

(b) Equivalent resistance has the same current.

▲ **Figure 21.15 Resistors in series**
(a) Three resistors, R_1, R_2, and R_3, connected in series. Notice that the same current, I, flows through each resistor. (b) The equivalent resistance, $R_{eq} = R_1 + R_2 + R_3$, has the same current, I, flowing through it as in the original circuit.

Series resistors are connected end to end

Resistors that are connected one after the other, end to end, are said to form a **series circuit**. **Figure 21.15 (a)** shows three resistors, R_1, R_2, and R_3, connected in series. The three resistors acting together have the same effect—that is, they draw the same current—as a single resistor, which is referred to as the *equivalent resistor*, R_{eq}. This equivalence is illustrated in **Figure 21.15 (b)**. When resistors are connected in series, the equivalent resistance is simply the sum of the individual resistances. In our case with three resistors, we have

$$R_{eq} = R_1 + R_2 + R_3$$

In general, the equivalent resistance of resistors in series is the sum of all the resistances that are connected together:

Equivalent Resistance for Resistors in Series

$\dfrac{\text{equivalent}}{\text{resistance}}$ = resistance 1 + resistance 2 + resistance 3 + $\cdots$

$$R_{eq} = R_1 + R_2 + R_3 + \cdots$$

SI unit: ohm (Ω)

The equivalent resistance is greater than the greatest resistance of any individual resistor. **In general, the more resistors connected in series, the greater the equivalent resistance.** For example, the equivalent resistance of a circuit with two identical resistors, R, connected in series is $R_{eq} = R + R = 2R$. Thus, connecting two identical resistors in series produces an equivalent resistance that is *twice* the individual resistances.

GUIDED Example 21.5 | Three Resistors in Series   **Series Circuit**

A circuit consists of three resistors connected in series to a 24.0-V battery. The current in the circuit is 0.0320 A. Given that $R_1 = 250.0\ \Omega$ and $R_2 = 150.0\ \Omega$, find (**a**) the resistance of R_3 and (**b**) the potential difference across each resistor.

Picture the Problem

The circuit is shown in our sketch. Notice that the same current, $I = 0.0320$ A, flows through each of the three resistors. This is the key characteristic of a series circuit.

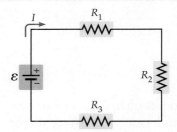

Strategy

(**a**) First, we can find the equivalent resistance of the circuit by using Ohm's law, $R_{eq} = \varepsilon/I$. Since the resistors are in series, we also know that $R_{eq} = R_1 + R_2 + R_3$. We can solve this relation for the only unknown, R_3. (**b**) We can then calculate the potential difference across each resistor using Ohm's law, $V = IR$.

Known	Unknown
$\varepsilon = 24.0$ V	(**a**) $R_3 = ?$
$I = 0.0320$ A	(**b**) $V_1 = ?$
$R_1 = 250.0\ \Omega$	$V_2 = ?$
$R_2 = 150.0\ \Omega$	$V_3 = ?$

Solution

Part (a)

> **1** Use Ohm's law to find the equivalent resistance of the circuit:

$$R_{eq} = \frac{\varepsilon}{I} = \frac{24.0 \text{ V}}{0.0320 \text{ A}}$$
$$= 750 \text{ } \Omega$$

Math HELP
Solving Simple Equations
See Math Review, Section III

> **2** Set R_{eq} equal to the sum of the individual resistances, and solve for R_3:

$$R_{eq} = R_1 + R_2 + R_3$$
$$R_3 = R_{eq} - R_1 - R_2$$
$$= 750 \text{ } \Omega - 250.0 \text{ } \Omega - 150.0 \text{ } \Omega$$
$$= \boxed{350 \text{ } \Omega}$$

Part (b)

> **3** Use Ohm's law to calculate the potential difference across R_1:

$$V_1 = IR_1$$
$$= (0.0320 \text{ A})(250.0 \text{ } \Omega)$$
$$= \boxed{8.00 \text{ V}}$$

> **4** Find the potential difference across R_2:

$$V_2 = IR_2$$
$$= (0.0320 \text{ A})(150.0 \text{ } \Omega)$$
$$= \boxed{4.80 \text{ V}}$$

> **5** Find the potential difference across R_3:

$$V_3 = IR_3$$
$$= (0.0320 \text{ A})(350 \text{ } \Omega)$$
$$= \boxed{11.2 \text{ V}}$$

Insight

Notice that the greater the resistance, the greater the potential difference. In addition, the sum of the individual potential differences is the voltage of the battery, $8.00 \text{ V} + 4.80 \text{ V} + 11.2 \text{ V} = 24.0 \text{ V}$, as expected.

Practice Problems

16. [Follow-up] Suppose we replace the resistors R_1, R_2, and R_3 with three identical resistors, R. What is the potential difference across each of these identical resistors?

17. A circuit has three resistors (12 Ω, 25 Ω, and 62 Ω) connected in series. The potential difference across the 12-Ω resistor is 4.4 V. What is the potential difference across (**a**) the 25-Ω resistor and (**b**) the 62-Ω resistor?

18. A circuit consists of a 12.0-V battery connected to three resistors (42 Ω, 17 Ω, and 110 Ω) in series. Find (**a**) the current that flows through the battery and (**b**) the potential difference across each resistor.

Parallel Circuits

Say you're riding in a car, going down a busy three-lane highway. The three lanes of the highway are independent paths for the cars to follow. Each lane has a certain "resistance" to the cars. That is, each lane has a speed limit and other cars traveling in it, so your car can only go so fast. Still, each lane carries its traffic independent of the other lanes. We say that the highway has three *parallel* lanes of traffic.

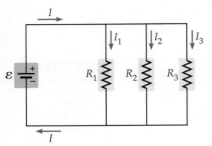

(a) Three resistors in parallel

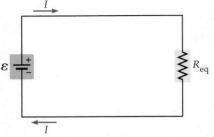

(b) Equivalent resistance has the same current.

▲ **Figure 21.16 Resistors in parallel**
(a) Three resistors, R_1, R_2, and R_3, connected in parallel. Notice that each resistor is connected across the same potential difference, ε. **(b)** The equivalent resistance, $1/R_{eq} = 1/R_1 + 1/R_2 + 1/R_3$, has the same current flowing through it as the total current, I, in the original circuit.

🔑 *How does adding another resistor in parallel affect the resistance?*

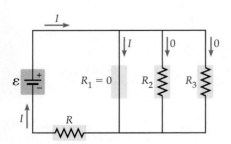

▲ **Figure 21.17 A short circuit**
If one of the resistors in parallel with others has a resistance of zero, all the current flows through that portion of the circuit, giving rise to a short circuit. In this case, resistors R_2 and R_3 are "shorted out," and the current in the circuit is $I = \varepsilon/R$.

Parallel resistors have the same potential difference

Resistors that are connected across the same potential difference are said to form a **parallel circuit**. An example of three resistors connected in parallel is shown in **Figure 21.16 (a)**. In a case like this the electrons have three parallel paths through which they can flow—like parallel lanes on the highway. The three resistors acting together draw the same current as a single equivalent resistor, R_{eq}, as indicated in **Figure 21.16 (b)**. When resistors are connected in parallel, the reciprocal of the equivalent resistance is equal to the sum of the reciprocals of the individual resistances. Thus, for our circuit of three resistors, we have

$$\frac{1}{R_{eq}} = \frac{1}{R_1} + \frac{1}{R_2} + \frac{1}{R_3}$$

In general, the inverse equivalent resistance is equal to the sum of all of the individual inverse resistances:

Equivalent Resistance for Resistors in Parallel

$$\frac{1}{\text{equivalent resistance}} = \frac{1}{\text{resistance 1}} + \frac{1}{\text{resistance 2}} + \frac{1}{\text{resistance 3}} + \cdots$$

$$\frac{1}{R_{eq}} = \frac{1}{R_1} + \frac{1}{R_2} + \frac{1}{R_3} + \cdots$$

SI unit: ohm (Ω)

Analyzing Parallel Circuits   As a simple example of parallel resistors, consider a circuit with two identical resistors, R, connected in parallel. The equivalent resistance in this case is

$$\frac{1}{R_{eq}} = \frac{1}{R} + \frac{1}{R}$$

$$\frac{1}{R_{eq}} = \frac{2}{R}$$

Solving for the equivalent resistance gives $R_{eq} = \frac{1}{2}R$. Thus, connecting two identical resistors in parallel produces an equivalent resistance that is *half* of the individual resistances. A similar calculation shows that three resistors, R, connected in parallel produces an equivalent resistance that is one-third of the original resistances, or $R_{eq} = \frac{1}{3}R$

These results show a clear trend. 🔑 **The more resistors connected in parallel, the smaller the equivalent resistance.** Each time you add a new resistor in parallel, you give the current a new path through which it can flow. It's like opening an additional lane for traffic on a busy highway. The new lane still has a "resistance" (speed limit and traffic), but it can carry additional cars. The new lane makes travel easier on that highway.

Short Circuits   In general, the equivalent resistance of a parallel circuit is less than or equal to the smallest individual resistance. So, what happens if one of the individual resistances is zero? In this case the equivalent resistance is also zero (because R_{eq} is less than or equal to the smallest individual resistance, and a resistance can't be negative). This situation, referred to as a *short circuit,* is illustrated in **Figure 21.17**. In a short circuit, all the current flows through the path of zero resistance.

Consider a circuit with three resistors, $R_1 = 250.0\ \Omega$, $R_2 = 150.0\ \Omega$, and $R_3 = 350.0\ \Omega$, connected in parallel with a 24.0-V battery. Find (**a**) the total current supplied by the battery and (**b**) the current through each resistor.

Picture the Problem

Our sketch indicates the parallel connection of the resistors with the battery. Notice that each of the resistors experiences precisely the same potential difference—namely, the 24.0 V produced by the battery. This is the feature that characterizes parallel connections.

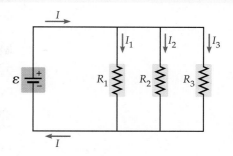

Strategy

(**a**) We can find the total current by solving Ohm's law, $V = IR$, for the current. This gives $I = \varepsilon/R_{eq}$, where $\varepsilon = V = 24.0$ V and $1/R_{eq} = 1/R_1 + 1/R_2 + 1/R_3$. (**b**) For each resistor the current is given by Ohm's law, $I = \varepsilon/R$.

Known	Unknown
$\varepsilon = 24.0$ V	(a) $I = ?$
$R_1 = 250.0\ \Omega$	(b) $I_1 = ?$
$R_2 = 150.0\ \Omega$	$I_2 = ?$
$R_3 = 350.0\ \Omega$	$I_3 = ?$

Solution

Part (a)

1 Find the equivalent resistance of the circuit:

$$\frac{1}{R_{eq}} = \frac{1}{R_1} + \frac{1}{R_2} + \frac{1}{R_3}$$

$$= \frac{1}{250.0\ \Omega} + \frac{1}{150.0\ \Omega} + \frac{1}{350.0\ \Omega}$$

$$= 0.01352\ \Omega^{-1}$$

$$R_{eq} = \frac{1}{0.01352\ \Omega^{-1}} = 73.96\ \Omega$$

2 Use Ohm's law to find the total current:

$$I = \frac{\varepsilon}{R_{eq}} = \frac{24.0\ \text{V}}{73.96\ \Omega} = \boxed{0.325\ \text{A}}$$

Part (b)

3 Calculate I_1 using $I_1 = \varepsilon/R_1$ with $\varepsilon = 24.0$ V:

$$I_1 = \frac{\varepsilon}{R_1} = \frac{24.0\ \text{V}}{250.0\ \Omega} = \boxed{0.0960\ \text{A}}$$

4 Repeat the calculation of Step 3 for resistors 2 and 3:

$$I_2 = \frac{\varepsilon}{R_2} = \frac{24.0\ \text{V}}{150.0\ \Omega} = \boxed{0.160\ \text{A}}$$

$$I_3 = \frac{\varepsilon}{R_3} = \frac{24.0\ \text{V}}{350.0\ \Omega} = \boxed{0.0686\ \text{A}}$$

Insight

As expected, the smallest resistor, R_2, carries the greatest current. The three currents combined yield the total current, as they must. That is, $I_1 + I_2 + I_3 = 0.0960\ \text{A} + 0.160\ \text{A} + 0.0686\ \text{A} = 0.325\ \text{A} = I$. Also, notice that the equivalent resistance (73.96 Ω) is less than that of the smallest resistor in the parallel circuit (150.0 Ω).

19. Follow-up (a) If R_3 in Guided Example 21.6 is increased to 525 Ω, does the current supplied by the battery increase, decrease, or stay the same? (b) Calculate the current supplied by the battery for the case where $R_3 = 525$ Ω.

20. What is the minimum number of 65-Ω resistors that must be connected in parallel to produce an equivalent resistance of 11 Ω or less?

21. A circuit consists of a battery connected to three resistors (65 Ω, 25 Ω, and 170 Ω) in parallel. The total current through the resistors is 1.8 A. Find the emf of the battery. (*Hint*: Find the equivalent resistance of the three resistors, and then use Ohm's law to calculate the potential difference across the resistors.)

Combination Circuits

The rules we've developed for series and parallel resistors can be applied to a variety of interesting circuits that aren't purely series or parallel.

Circuits can have both series and parallel resistors

The circuit shown in **Figure 21.18 (a)** contains a total of four resistors, each with resistance R, connected in a way that combines series and parallel features. Because the circuit is not strictly series or parallel, we can't directly calculate the equivalent resistance.

What we can do, however, is break the circuit into smaller subcircuits, each of which is purely series or parallel. For example, we first note that the two vertically oriented resistors on the right are connected in parallel with one another. Therefore, the equivalent resistance of this unit is given by $1/R_{eq} = 1/R + 1/R$, or $R_{eq} = R/2$.

The next step is to replace these two resistors with $R/2$. This yields the circuit shown in **Figure 21.18 (b)**. Notice that this equivalent circuit consists of three resistors in series, R, $\frac{1}{2}R$, and R. The equivalent resistance of these resistors is equal to their sum, $R_{eq} = R + \frac{1}{2}R + R = 2.5R$.

Therefore, the equivalent resistance of the original circuit is 2.5R, as indicated in **Figure 21.18 (c)**. By considering the resistors in pairs or groups that are connected in parallel or in series, you can reduce the entire circuit to one equivalent resistance. This method can be applied to many other circuits as well.

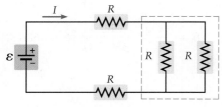

(a) Replace parallel resistors

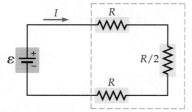

(b) Replace series resistors

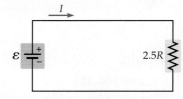

(c) Final equivalent resistance

▲ **Figure 21.18 Analyzing a complex circuit of resistors**
(a) The two vertical resistors are in parallel with one another; hence, they can be replaced with their equivalent resistance, $R/2$. **(b)** Now the circuit consists of three resistors in series. The equivalent resistance of these three resistors is 2.5R. **(c)** The original circuit reduces to a single equivalent resistance.

In the circuit shown in the diagram below, the emf of the battery is 12 V, and each resistor has a resistance of 450 Ω. Find the current supplied by the battery to this circuit.

Picture the Problem

The circuit for this problem has three resistors connected to a battery. The lower two resistors are in series with one another, and they are in parallel with the upper resistor. The battery has an emf of 12 V.

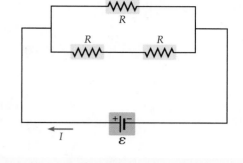

Strategy

The current supplied by the battery is given by Ohm's law, $I = \varepsilon/R_{eq}$, where R_{eq} is the equivalent resistance of the three resistors. To find R_{eq} we first note that the lower two resistors are in series, giving a net resistance of $2R$. Next, the upper resistor, R, is in parallel with the pair of resistors having resistance $2R$. Calculating the equivalent resistance of this combination yields the desired R_{eq}.

Known	Unknown
$\varepsilon = 12$ V	$I = ?$
$R = 450$ Ω	

Solution

1 Calculate the equivalent resistance of the two *lower* resistors:

$$R_{eq,lower} = R + R = 2R$$

2 Use the result of Step 1 to calculate the equivalent resistance of R in parallel with $2R$:

$$\frac{1}{R_{eq}} = \frac{1}{R} + \frac{1}{2R} = \frac{3}{2R}$$
$$R_{eq} = \tfrac{2}{3}R$$
$$= \tfrac{2}{3}(450\ \Omega)$$
$$= 300\ \Omega$$

> **Math HELP**
> Solving Simple Equations
> **See Math Review, Section III**

3 Find the current supplied by the battery, *I*:

$$I = \frac{\varepsilon}{R_{eq}}$$
$$= \frac{12\ \text{V}}{300\ \Omega}$$
$$= \boxed{0.040\ \text{A}}$$

Insight

Notice that the total resistance of the three 450-Ω resistors is less than 450 Ω—in fact, it is only 300 Ω.

Practice Problems

22. [Follow-up] Suppose the upper resistor is changed from 450 Ω to 900 Ω. The other two resistors remain the same.
(a) Will the current supplied by the battery increase, decrease, or stay the same?
(b) Find the new current.

23. Find the equivalent resistance between points A and B in **Figure 21.19**, given that $R = 25$ Ω.

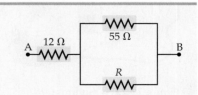

▶ **Figure 21.19**

24. [Challenge] A radio requires a 150-Ω resistor. How can a 220-Ω, a 79-Ω, and a 92-Ω resistor be connected to produce that resistance?

To measure the current flowing between points A and B in **(a)**, an ammeter is inserted into the circuit, as shown in **(b)**. An ideal ammeter would have zero resistance.

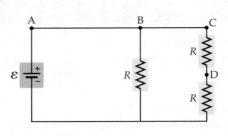

(a) Typical electric circuit

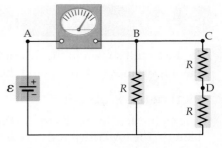

(b) Measuring the current between A and B

Measuring the voltage between C and D

▲ **Figure 21.21 Measuring the voltage in a circuit**

The potential difference, or voltage, between points C and D can be measured by connecting a voltmeter to the circuit at those points, as shown. An ideal voltmeter would have infinite resistance.

▲ **Figure 21.22 A multimeter**
A digital multimeter can measure resistance, current, or voltage. This meter is measuring the actual voltage of a 9-volt battery.

Devices for Measuring Current and Voltage

The current flowing through a circuit, or the potential difference between two points in a circuit, can be measured directly with a *meter*. In each of these cases the ideal situation is for the meter to measure the desired quantity without altering the circuit being studied.

Ammeters measure current

The device used to measure current is an **ammeter**. An ammeter is designed to measure the flow of current through a particular portion of a circuit. For example, you might want to know the current flowing between points A and B in the circuit shown in **Figure 21.20 (a)**. To measure this current, the ammeter must be added to the circuit in such a way that *all* the current flowing from A to B also flows through the meter. This is done by connecting the meter "in series" with the other circuit elements between A and B, as indicated in **Figure 21.20 (b)**.

If the ammeter has a finite resistance—which is the case for any real meter—then its presence in a circuit alters the current it is intended to measure. Thus, an *ideal* ammeter would have zero resistance (and not alter the current). Real ammeters, however, give accurate readings as long as their resistance is much less than the other resistances in the circuit.

Voltmeters measure voltage

A **voltmeter** is a device used to measure the potential difference between any two points in a circuit. Referring again to the circuit in Figure 21.20 (a), you might want to know the difference in potential between points C and D. To measure this voltage, the voltmeter is placed "in parallel" at the appropriate points, as shown in **Figure 21.21**.

Because a small current must flow through the voltmeter in order for it to work, the meter reduces the current flowing through the circuit. As a result, the measured voltage is altered from its ideal value. Thus, an *ideal* voltmeter would have infinite resistance (and draw a negligible current). Real voltmeters give accurate readings as long as their resistance is much greater than the other resistances in the circuit.

Sometimes the functions of an ammeter, a voltmeter, and an ohmmeter (a meter to measure resistance) are combined in a single device called a *multimeter*. An example of a multimeter is shown in **Figure 21.22**. Adjusting the settings on a multimeter's dial allows a variety of circuit properties to be measured.

21.2 LessonCheck (MP)

Checking Concepts

25. ⬚ **Apply** A circuit contains four resistors connected in series. What happens to the equivalent resistance when one of the resistors is replaced with an ideal wire?

26. ⬚ **Apply** Two identical resistors are connected in parallel. What happens to the equivalent resistance if a third identical resistor is added in parallel with the other two?

27. Describe Do the resistors in (**a**) a series circuit or (**b**) a parallel circuit all have the same current or the same electric potential difference?

28. Analyze Are car headlights connected in series or parallel? Give an everyday observation that supports your answer.

Solving Problems

29. Calculate An 89-Ω resistor has a current of 0.72 A and is connected in series with a 130-Ω resistor. What is the emf of the battery connected to the resistors?

30. Calculate A 210-Ω resistor has a potential difference of 7.7 V and is connected in parallel with a 130-Ω resistor. What is the current supplied by the battery to which the resistors are connected?

31. Think & Calculate A 12-V battery is connected to a 12-Ω resistor and a 36-Ω resistor. The current that flows through the battery is 0.25 A.
(**a**) What is the equivalent resistance of the resistors?
(**b**) Are the resistors connected in series or in parallel?

32. Calculate Find the equivalent resistance between points A and B for the three resistors shown in **Figure 21.23**.

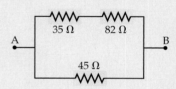

▲ **Figure 21.23**

21.3 Power and Energy in Electric Circuits

Ah, a nice relaxing evening at home. You listen to a Schubert symphony on a 200-W speaker system and read some physics by the light of a 100-W lightbulb. In this lesson you'll discover the connection between the power of an electrical device (such as 100 W or 200 W) and the current and voltage.

Electric Power

As you might expect, the power delivered by an electric circuit increases with both the current and the voltage. Increase either and the power increases. The precise connection between power, current, and voltage is very simple—as we now show.

Current and voltage determine the electric power

When a ball falls in a gravitational field, there is a change in gravitational potential energy. Similarly, when an amount of charge, ΔQ, moves across a potential difference, V, there is a change in electrical potential energy, ΔPE, given by

$$\Delta PE = (\Delta Q)V$$

Recalling that power is the rate at which energy changes, $P = \dfrac{\Delta PE}{\Delta t}$, we can express the electric power as follows:

$$P = \frac{\Delta PE}{\Delta t} = \frac{(\Delta Q)V}{\Delta t}$$

Knowing that electric current is given by $I = \dfrac{\Delta Q}{\Delta t}$ allows us to write an expression for electric power in terms of current and voltage.

The electric power used by a device is equal to the current times the voltage. It's as simple as that. For example, a current of 1 amp flowing across a potential difference of 1 V produces a power of 1 W.

QUICK Example 21.8 What's the Current?

A handheld electric fan operates on a 3.0-V battery. If the power generated by the fan is 2.2 W, what is the current supplied by the battery?

Solution
Solving $P = IV$ for the current, we find

$$I = \frac{P}{V}$$

$$= \frac{2.2 \text{ W}}{3.0 \text{ V}}$$

$$= \boxed{0.73 \text{ A}}$$

Practice Problems

33. A flashlight operates with a current of 3.0 A and a power of 4.5 W. What is the voltage of the flashlight's battery?

34. A 12-V battery delivers a current of 0.75 A to start a small gasoline engine. What is the power produced by the battery?

CONNECTINGIDEAS

Power, which is the amount of work done in a given time, was introduced in Chapter 6.
- Here you learn that the power of an electrical device is related to voltage and current.

Ohm's law yields other equations for electric power

The equation $P = IV$ applies to any electrical system. In the special case of a resistor, the electric power is dissipated in the form of heat and light, as shown in **Figure 21.24**. Applying Ohm's law, $V = IR$, which deals with resistors, we can express the power dissipated in a resistor as follows:

$$P = IV$$
$$= I(IR)$$
$$= I^2R$$

Similarly, solving Ohm's law for the current, $I = V/R$, and substituting that result gives an alternative expression for the power dissipated in a resistor:

$$P = IV$$
$$= \left(\frac{V}{R}\right)V$$
$$= \frac{V^2}{R}$$

All three equations for the power are valid. The first one ($P = IV$) applies to all electrical systems. The other two ($P = I^2R$ and $P = V^2/R$) are specific to resistors, which is why the resistance, R, appears in those equations. Which equation to use depends on the information you have about a given system. If you know current and voltage, then $P = IV$ is the equation for you. If you know current and resistance, then use $P = I^2R$, and if you know voltage and resistance, use $P = V^2/R$.

▲ **Figure 21.24 Electric power dissipated as heat**
The heating element of an electric space heater is nothing more than a length of resistive wire coiled up for compactness. As electric current flows though the wire, the power it dissipates is converted to heat and light. The coils near the center are the hottest, and hence they glow with a higher-frequency, yellowish light.

CONCEPTUAL Example 21.9 Comparing Currents and Resistances

Battery 1 produces a potential difference V and is connected to a 5-W lightbulb. Battery 2 produces a potential difference V and is connected to a 10-W lightbulb. **(a)** Which battery supplies more current? **(b)** Which lightbulb has the greater resistance?

Reasoning and Discussion
(a) Because the voltage is the same for both batteries, we can compare the currents using $P = IV$. Solving for the current yields $I = P/V$. When the voltage, V, is the same, it follows that the greater the power, the greater the current. Therefore, the current in the 10-W bulb is twice the current in the 5-W bulb. **(b)** In this case we use the equation $P = V^2/R$, which gives power in terms of voltage and resistance. Solving for the resistance gives $R = V^2/P$. With V the same in the two cases, it follows that the smaller the power, the greater the resistance. Thus, the resistance of the 5-W bulb is twice that of the 10-W bulb.

Answers
(a) Battery 2 delivers twice as much current as battery 1. **(b)** The 5-W bulb has twice as much resistance as the 10-W bulb.

Resistors get hot and dissipate power

The power dissipated by a resistor is the result of collisions between electrons moving through the circuit (conduction electrons) and atoms making up the resistor. Specifically, the potential difference produced by the battery causes the conduction electrons to accelerate until they bounce off an atom of the resistor. These collisions transfer kinetic energy from the electrons to the atoms, causing the atoms to jiggle more rapidly.

The increased kinetic energy of the atoms is reflected as an increased temperature of the resistor. After each collision the potential difference accelerates the electrons again, and the process repeats—like a car bouncing through a series of speed bumps. The result is a continuous transfer of energy from the conducting electrons to the atoms. This is why the filament in an incandescent lightbulb gets so hot. The energy given off as heat and light by a lightbulb corresponds to an equivalent decrease in energy of the battery connected to the light. As is always the case, energy is conserved.

GUIDED Example 21.10 | Heated Resistance   **Power Dissipation**

A battery with an emf of 12 V is connected to a 570-Ω resistor. How much energy is dissipated in the resistor in 65 s?

Picture the Problem

The circuit, consisting of a battery and a resistor, is shown in our sketch. Current flows from the positive terminal of the 12-V battery, through the 570-Ω resistor, to the negative terminal of the battery.

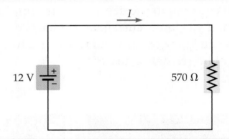

Strategy

The resistor dissipates power (energy per time). The energy dissipated in a given time is the power multiplied by the time: $\Delta PE = P\Delta t$. The time is given ($\Delta t = 65$ s), and the power can be found using $P = IV$, $P = I^2R$, or $P = V^2/R$. We'll use the last equation because the voltage and resistance are given. To summarize, we first calculate the power, then multiply by time to find the energy.

Known

$\varepsilon = 12$ V
$R = 570\ \Omega$
$\Delta t = 65$ s

Unknown

$\Delta PE = ?$

Solution

1 Calculate the power dissipated in the resistor:

$$P = \frac{V^2}{R}$$
$$= \frac{(12\ \text{V})^2}{570\ \Omega}$$
$$= 0.25\ \text{W}$$

2 Multiply the power by the time to find the energy dissipated:

$$\Delta PE = P\Delta t$$
$$= (0.25\ \text{W})(65\ \text{s})$$
$$= \boxed{16\ \text{J}}$$

Insight

The current in this circuit is $I = V/R = 0.021$ A. Using this value we find the same power with either of the other two expressions for power. In particular, $P = I^2R = IV = 0.25$ W, as expected.

35. [Follow-up] How much energy is dissipated in the resistor if the emf of the battery is increased to 15 V?

36. A portable CD player operates with a current of 22 mA at a potential difference of 4.1 V. What is the power usage of the player?

37. The power dissipated in an electric heater is 120 W. The heater is connected to a 120-V outlet. What is the resistance of the heater?

The brightness of a lightbulb is related to its power

The filament of an incandescent lightbulb is basically a resistor inside a sealed, evacuated glass bulb. The filament gets so hot that it glows, just like the heating coil on a stove or the heater coils in Figure 21.24. The power dissipated in the filament determines the brightness of the lightbulb. The higher the power, the brighter the light. This basic concept is applied in the next Conceptual Example.

CONCEPTUAL Example 21.11 Brightness of the Lights

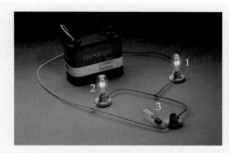

The circuit shown to the right has two identical lightbulbs (1 and 2) connected in series with a battery. The bulbs are equally bright, as expected. A third identical lightbulb (3) is disconnected from its socket and is not part of the circuit.

Suppose you put lightbulb 3 into the empty socket, so all three bulbs are connected to the battery. Does the brightness of bulb 1 increase, decrease, or stay the same? (Notice that bulbs 2 and 3 are now in parallel with one another, and their combination is in series with bulb 1.)

Reasoning and Discussion
Connecting bulb 3 adds one more path through which current can flow. Therefore, the current in the circuit increases. All of the current in the circuit flows through bulb 1, and hence its brightness increases, as shown on the right.

The increased current is split equally between bulbs 2 and 3, which are in parallel with one another. It can be shown that the current increases by a factor of $\frac{4}{3}$, so splitting the increased current in half for bulbs 2 and 3 means that each of these bulbs has $\frac{2}{3}$ of the original current. As a result, the brightness of bulb 2 decreases, as shown.

Answer
The brightness of bulb 1 increases.

Electrical Energy

The local electric company bills your family for the electricity it uses each month. How does it determine the amount of the bill? Does it charge for the amount of current it supplies, for the number of electrons, or for the power provided by the electricity? None of the above. It charges for the amount of electrical energy it supplies. It is energy that is costly.

► **Figure 21.25 The cost of electric energy**
(a) The consumption of electric energy by a household is measured with a meter.
(b) The electric bill charges for the number of kilowatt-hours of energy that have been consumed.

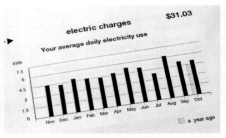

(a)

(b)

Electrical energy is measured in kilowatt-hours

🔑 *What quantity does the electric company use for billing?*

A convenient unit for measuring electric energy is the kilowatt-hour (kWh). Recall that a kilowatt is 1000 W, or equivalently 1000 J/s. Similarly, an hour is 3600 s. Combining these results, we see that a kilowatt-hour is equal to 3.6 million joules of energy:

$$1 \text{ kWh} = \left(1000 \, \frac{J}{\cancel{s}}\right)(3600 \, \cancel{s})$$
$$= 3.6 \times 10^6 \text{ J}$$

🔑 **The electric company bills your family for the number of kilowatt-hours (kWh) of energy it uses.** Take a look at the electric bill the next time it arrives and check out the number of kilowatt-hours used and the rate per kilowatt-hour. **Figure 21.25** shows the type of meter used to measure the electrical energy consumption of a household, as well as a typical electric bill.

The following Guided Example looks at the energy used and the cost of cooking a meal in an electric oven.

GUIDED Example 21.12 | Your Goose Is Cooked   **Energy Cost**

A holiday goose is cooked in an electric oven for 4.00 h. Assume that the stove draws a current of 20.0 A, operates at a voltage of 220.0 V, and uses electric energy that costs $0.085 per kilowatt-hour. How much does it cost to cook your goose?

Picture the Problem

We show a schematic representation of the stove cooking the goose in our sketch. The current in the circuit is 20.0 A, and the voltage difference across the heating coils is 220 V.

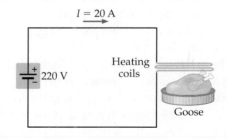

Strategy

The cost equals the amount of energy used (in kilowatt-hours) times the rate ($0.085 per kilowatt-hour). To find the energy used, we multiply power by time. The time is given, and the power is calculated using $P = IV$. To summarize, we find the power, multiply by the time, and then multiply by $0.085/\text{kWh}$ to find the cost.

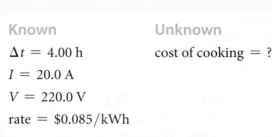

Known	Unknown
$\Delta t = 4.00$ h	cost of cooking $= ?$
$I = 20.0$ A	
$V = 220.0$ V	
rate $= \$0.085/\text{kWh}$	

Solution

1 Calculate the power delivered to the stove:

$$P = IV$$
$$= (20.0 \text{ A})(220.0 \text{ V})$$
$$= 4.40 \text{ kW}$$

2 Multiply the power by the time to determine the total energy supplied to the stove during the 4.00 h of cooking:

$$\Delta PE = P\Delta t$$
$$= (4.40 \text{ kW})(4.00 \text{ h})$$
$$= 17.6 \text{ kWh}$$

3 Multiply by the cost per kilowatt-hour to find the total cost of cooking:

$$\text{cost} = \text{energy used} \times \text{rate}$$
$$= (17.6 \text{ kWh})(\$0.085/\text{kWh})$$
$$= \boxed{\$1.50}$$

> **Math HELP**
> Dimensional Analysis
> See Lesson 1.3

Insight

Thus, your goose can be cooked for about a dollar and a half.

Practice Problems

38. Follow-up If the voltage and the current are each reduced by a factor of 2, how long must your goose be cooked to use the same amount of energy?

39. The current in a 120-V reading lamp is 2.6 A. If the cost of electrical energy is $0.075 per kilowatt-hour, how much does it cost to operate the lamp for an hour?

40. It costs 2.6¢ to charge a car battery at a voltage of 12 V and a current of 15 A for 120 min. What is the cost of electrical energy per kilowatt-hour at this location?

21.3 LessonCheck MP

Checking Concepts

41. Identify What equations can be used to calculate electric power?

42. Identify What physical quantity does the kilowatt-hour measure?

43. Explain What physical quantity changes when a charge moves across a potential difference?

44. Triple Choice Light A has four times the power rating of light B when operating at the same voltage. Is the resistance of light A greater than, less than, or equal to the resistance of light B? Explain.

Solving Problems

45. Calculate Find the power dissipated in a 25-Ω electric heater connected to a 120-V outlet.

46. Calculate A current of 2.1 A flows through an 85-Ω resistor. How much power is dissipated?

47. Calculate A circuit contains only a 12-V battery and a resistor. If a 1.4-A current flows through the resistor, how much power does the resistor dissipate?

48. Calculate A 75-W lightbulb operates at a potential difference of 95 V. Find (**a**) the current in the bulb and (**b**) the resistance of the bulb.

Semiconductor Industry

A silicon wafer.

The semiconductor industry is built on a foundation of silicon, the second most abundant element on Earth, after oxygen. A large number of companies are involved in transforming silicon-bearing ore into electronic components such as computer microprocessors. These companies provide careers at every level in physics, metallurgy, chemistry, photographic science, electronics, and computer science.

It all begins with natural silicon dioxide (SiO_2). The silicon dioxide, in the form of quartz lumps or quartzite gravel, is heated with carbon in electric arc furnaces, producing carbon dioxide (CO_2) and metallurgical-grade silicon that is about 99.5% pure. Raw material companies use chemical processes to further refine the silicon to a purity of 99.9999%. This near-pure electronic-grade silicon is called *poly* because its structure is *polycrystalline* (consisting of many crystals).

Poly, in turn, goes to companies that melt it and produce large single-crystal ingots called *boules*. The boules, which can measure up to 30 cm in diameter, are then sliced into thin wafers. Integrated circuit companies (called *foundries* if they do fabrication but not design) use ultraviolet photolithography, etching, and vacuum deposition techniques to create hundreds of integrated circuits (ICs) on the surface of a wafer. The circuits are made up mostly of transistors and their interconnections. The final steps are to dice the wafers into chips, mount the individual chips in plastic or ceramic packages, and connect fine wire leads from the chips to pins on the packages. The packaged chips are then shipped to electronics companies that manufacture consumer products such as computers, cell phones, and video game consoles, to name just a few.

Remarkable progress in computer and communication technology continues to drive growth in the semiconductor industry.

Take It Further

1. Research *Do research on the use of horizontal integration (where separate companies perform the various steps of the chip-making process) and vertical integration (where each company turns raw quartz into finished chips) in the semiconductor industry. Describe each of these business models, and list several pros and cons of each.*

A worker inspects semiconductor components in a clean-room facility.

Physics Lab Ohm's Law

This lab explores Ohm's law. You will use measurements of the potential difference across resistors and the current through resistors to demonstrate the law.

Materials
- C-size battery holder
- three C-size batteries
- three different resistors
- digital multimeter
- two connecting wires with alligator clip ends

Procedure

1. Create a circuit using one battery in the battery holder, a resistor, and the two wires.

2. Use the multimeter to measure the potential difference across the resistor.

- Make sure the multimeter is measuring volts.

- Connect the multimeter in parallel with the resistor. Do this by placing the ends of the probes on opposite ends of the resistor.

- If the multimeter gives a negative value, reverse the placement of the probes on the resistor.

- Record the potential difference in Data Table 1.

3. Use the multimeter to measure the current through the resistor.

- Make sure the multimeter is measuring amps. This might require moving one of the probes to a different spot on the multimeter.

- Connect the multimeter in series with the resistor. To do this, first detach the alligator clip that is attached

to the resistor. Then, attach one of the multimeter probes to the alligator clip and the other multimeter probe to the free end of the resistor.

- If the multimeter gives negative values, reverse the placement of the probes.

- Record the current in Data Table 1.

4. Repeat Steps 2 and 3 using two batteries in the battery holder.

5. Repeat Steps 2 and 3 using three batteries in the battery holder.

6. Remove the resistor and repeat Steps 1 through 5 using the second resistor. Record your data in Data Table 2 (which is not shown but is set up just like Data Table 1).

7. Remove the resistor and repeat Steps 1 through 5 using the third resistor. Record your data in Data Table 3 (which is not shown but is set up just like Data Tables 1 and 2).

Data Table 1: Resistor 1

Number of Batteries	Potential Difference (V)	Current (A)
1		
2		
3		

Analysis

1. On a sheet of graph paper, plot potential difference versus current for each resistor. Draw a best-fit line for each set of data points.

2. Determine the equation for each best-fit line on your graph. Label each line with its equation.

Conclusions

1. What does the slope of the line represent? Explain.

2. Are voltage and current directly or inversely related? Explain.

3. Ask your teacher for the accepted values for the three resistors. Calculate the percent error for each resistor.

21 Study Guide

Big Idea

Electrons flow through electric circuits in response to differences in electric potential.

Batteries produce a difference in potential between their terminals. This potential difference causes electrons in a wire to flow through a circuit from one terminal to the other. As the electrons flow, they convert electrical energy from the battery to mechanical work and thermal energy.

21.1 Electric Current, Resistance, and Semiconductors

🔑 When a battery is connected to a circuit, electrons move in a closed path from one terminal of the battery through the circuit and back to the other terminal of the battery. The electrons leave from the negative terminal of the battery and return to the positive terminal.

🔑 The applied voltage equals the product of the current and the wire's resistance.

🔑 The resistance of a wire depends on the material from which it is made. A wire's resistance also depends on its length, L, and its cross-sectional area, A.

🔑 As a wire is heated, its resistivity tends to increase.

• When electric charge flows from one place to another, we say that it forms an electric current.

• A battery uses chemical reactions to produce a difference in electric potential between its two terminals.

• The potential difference V necessary to produce a current I in a wire of resistance R is given by Ohm's law: $V = IR$.

• The unit of current is the ampere, or amp for short. By definition, 1 amp is 1 coulomb per second: $1 \text{ A} = 1 \text{ C/s}$.

• The direction of the current in a circuit is the direction in which a *positive* test charge would move. The actual charge carriers, however, are generally negatively charged electrons; thus, they move in the direction opposite to the current.

• The resistivity, ρ, of a material determines how much resistance it offers to the flow of electric current.

Key Equations

If the amount of charge ΔQ passes a given point in the time Δt, the corresponding electric current is

$$I = \frac{\Delta Q}{\Delta t}$$

To produce a current I through a wire with resistance R, the following potential difference, V, is required:

$$V = IR$$

The work done by a battery whose emf is ε as the amount of charge ΔQ passes through it is

$$W = (\Delta Q)\varepsilon$$

21.2 Electric Circuits

🔑 In general, the more resistors connected in series, the greater the equivalent resistance.

🔑 The more resistors connected in parallel, the smaller the equivalent resistance.

• Resistors that are connected one after the other, end to end, are said to be in series.

• Resistors that are connected across the same potential difference are said to be in parallel.

• Ammeters and voltmeters are devices for measuring current and voltage, respectively, in electric circuits.

Key Equations

The equivalent resistance, R_{eq}, of resistors connected in series is equal to the sum of the individual resistances:

$$R_{eq} = R_1 + R_2 + R_3 + \cdots = \sum R$$

The equivalent resistance, R_{eq}, of resistors connected in parallel is given by the following:

$$\frac{1}{R_{eq}} = \frac{1}{R_1} + \frac{1}{R_2} + \frac{1}{R_3} + \cdots = \sum \frac{1}{R}$$

21.3 Power and Energy in Electric Circuits

🔑 The electric power used by a device is equal to the current times the voltage.

🔑 The electric company bills your family for the number of kilowatt-hours (kWh) of energy it uses.

• When the amount of charge ΔQ moves across a potential difference V, there is a change in electric potential energy, ΔPE, given by $\Delta PE = (\Delta Q)V$.

Key Equations

If a current I flows across a potential difference V, the corresponding electric power is

$$P = IV$$

If a potential difference V produces a current I in a resistor R, the electric power dissipated in the resistor is

$$P = I^2R \quad \text{or} \quad P = V^2/R$$

The energy equivalent of 1 kilowatt-hour (kWh) is

$$1 \text{ kWh} = 3.6 \times 10^6 \text{ J}$$

21 Assessment

For instructor-assigned homework, go to www.masteringphysics.com

ANSWERS TO SELECTED ODD-NUMBERED PROBLEMS APPEAR IN APPENDIX A.

Lesson by Lesson

21.1 Electric Current, Resistance, and Semiconductors

Conceptual Questions

49. Suppose you charge a comb by rubbing it through your hair. Do you produce a current when you walk across the room carrying the charged comb?

50. Suppose you charge a comb by rubbing it through the fur on your dog's back. Do you produce a current when you walk across the room carrying the charged comb?

51. When a battery with a voltage V_0 is attached to a resistor with a resistance R_0, the resulting current is 1 A. What is the current in each of the following systems?

System	Voltage	Resistance
A	$2V_0$	$4R_0$
B	$4V_0$	$4R_0$
C	$4V_0$	$2R_0$

52. Explain how electrical devices can begin operating almost immediately after you flip a switch, even though individual electrons in the wire may take hours to reach the device.

53. Is the direction of an electric current from the positive terminal of a battery to the negative terminal or from the negative terminal to the positive terminal? Explain.

Problem Solving

54. An ammeter can measure currents as small as 12 pA. How many electrons per second flow through a wire with a 12-pA current? (Recall that $1 \text{ pA} = 10^{-12} \text{ A.}$)

55. A charge of 12 C flows past a given point in the time Δt. If the current produced by the charge is 0.15 A, what is Δt?

56. How much time is required for 0.25 C of charge to flow through a wire carrying a current of 1.1 A?

57. A battery does 59 J of work as 6.6 C of charge passes through it. What is the voltage of the battery?

58. How much charge passes through a 12-V battery that does 41 J of work?

59. A current of 2.2 A flows through a 140-Ω resistor. What is the potential difference applied to the resistor?

60. What current flows through a 210-Ω resistor when a potential difference of 15 V is applied to it?

61. A current of 3.2 A flows through a resistor when a potential difference of 95 V is applied to it. What is the resistance of the resistor?

62. Pacemaker Batteries Pacemakers designed for long-term use commonly employ a lithium–iodine battery capable of supplying 1500 C of charge. If the average current produced by the pacemaker is 5.6 μA, what is the expected lifetime of the device?

63. A 1.5-V battery delivers 9.6 C of charge to a small lightbulb in 45 s. **(a)** What is the current passing through the lightbulb? **(b)** How much work has been done by the battery?

64. Think & Calculate A car battery does 260 J of work on the charge passing through it as it starts an engine. **(a)** If the emf of the battery is 12 V, how much charge passes through the battery during the start? **(b)** If the emf is doubled to 24 V, does the amount of charge passing through the battery for the same amount of work increase or decrease? By what factor?

65. Finger Resistance and Current The interior of the human body has an electrical resistivity of 0.15 Ω·m. **(a)** Estimate the resistance for current flowing the length of your index finger. (For this calculation, ignore the much higher resistance of your skin.) **(b)** Your muscles contract when they carry a current greater than 15 mA. What voltage is required to produce this current in your finger?

21.2 Electric Circuits

Conceptual Questions

66. Is it possible to connect four resistors with resistance R in such a way that the equivalent resistance is less than R? If so, give a specific example.

67. What physical quantity do resistors connected in series have in common?

68. What physical quantity do resistors connected in parallel have in common?

69. Predict & Explain A dozen identical lightbulbs are connected to a given battery. **(a)** Will the lights be brighter if they are connected in series or in parallel? **(b)** Choose the *best* explanation from among the following:

A. When connected in parallel, each bulb experiences the maximum emf and dissipates the maximum power.

B. Resistors in series have a larger equivalent resistance and dissipate more power.

C. Resistors in parallel have a smaller equivalent resistance and dissipate less power.

70. Predict & Explain A *fuse* is a device to protect a circuit from the effects of a large current. The fuse is a small strip of metal that burns through when the current in it exceeds a certain value, thus producing an open circuit. **(a)** Should a fuse be connected in series or in parallel with the circuit it is intended to protect? **(b)** Choose the *best* explanation from among the following:

A. Either type of connection is acceptable; the main thing is to have a fuse in the circuit.

B. The fuse should be connected in parallel; otherwise, it will interrupt the current in the circuit.

C. With a fuse connected in series, the current in the circuit drops to zero as soon as the fuse burns through.

71. Rank A circuit consists of three resistors, $R_1 < R_2 < R_3$, connected in series to a battery. Rank these resistors in order of **(a)** increasing current through them and **(b)** increasing potential difference across them. Indicate ties where appropriate.

72. Predict & Explain Two resistors are connected in parallel. **(a)** If a third resistor is connected in parallel with the other two, does the equivalent resistance of the circuit increase, decrease, or remain the same? **(b)** Choose the *best* explanation from among the following:

A. Adding a resistor tends to increase the resistance, but putting it in parallel tends to decrease the resistance; therefore, the effects offset and the resistance stays the same.

B. Adding more resistance to the circuit will increase the equivalent resistance.

C. The third resistor gives yet another path for current to flow in the circuit, which means that the equivalent resistance is less.

73. Give an example of how five resistors of resistance R can be combined to produce an equivalent resistance of $2R$.

74. Four lightbulbs (A, B, C, and D) are connected in a circuit of unknown arrangement. When a single bulb is removed from the circuit, the following behavior is observed:

	A	B	C	D
A removed	—	on	on	on
B removed	on	—	on	off
C removed	off	off	—	off
D removed	on	off	on	—

Draw a circuit diagram for these bulbs.

Problem Solving

75. What is the equivalent resistance of the following systems: System A, $R = 25 \ \Omega$ and $R = 15 \ \Omega$ in series; System B, $R = 220 \ \Omega$ and $R = 430 \ \Omega$ in parallel?

76. Three resistors, 11 Ω, 53 Ω, and R, are connected in series with a 24.0-V battery. The total current flowing through the battery is 0.16 A. Find the resistance of R.

77. Three resistors, 22 Ω, 67 Ω, and R, are connected in parallel with a 12.0-V battery. The total current flowing through the battery is 0.88 A. Find the resistance of R.

78. (a) How much current flows through three resistors (170 Ω, 240 Ω, 65 Ω) connected in series with a 12-V battery? **(b)** What is the voltage across each resistor?

79. (a) How much current is supplied by a 12-V battery connected to three resistors (88 Ω, 130 Ω, 270 Ω) in parallel? **(b)** What is the current flowing through each resistor?

80. Find the equivalent resistance between points A and B for the three resistors shown in **Figure 21.26**.

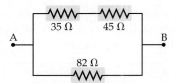

Figure 21.26

81. A 9.0-V battery is connected to terminals A and B in Figure 21.26. What is the current in each resistor?

82. Find the equivalent resistance between points A and B in **Figure 21.27**, given that $R = 62 \ \Omega$.

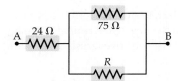

Figure 21.27

83. Think & Calculate A 12-V battery is connected to terminals A and B in Figure 21.27. **(a)** Given that $R = 62 \ \Omega$, find the current in each resistor. **(b)** Suppose the value of R is increased. For each resistor in turn, state whether the current flowing through it increases or decreases. Explain.

84. If the equivalent resistance between points A and B of the resistors shown in Figure 21.27 is 46 Ω, what is the resistance of R?

85. How many 65-W lightbulbs can be connected in parallel across a potential difference of 85 V before the total current in the circuit exceeds 2.1 A?

86. Find the equivalent resistance between points A and B in **Figure 21.28**.

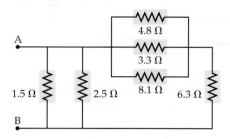

Figure 21.28

87. | **Think & Calculate** | The terminals A and B in Figure 21.28 are connected to a 9.0-V battery. **(a)** Find the current flowing through the 1.5-Ω resistor. **(b)** Is the potential difference across the 6.3-Ω resistor greater than, less than, or the same as the potential difference across the 1.5-Ω resistor? Explain.

21.3 Power and Energy in Electric Circuits

Conceptual Questions

88. Light A has five times the power rating of light B when operated at the same voltage. What is the ratio of the resistance of light A to the resistance of light B?

89. Two lightbulbs operate at the same potential difference. Bulb A has three times the power output of bulb B. What is the ratio of the current in bulb A to the current in bulb B?

90. Two lightbulbs operate on the same current. Bulb A has two times the power output of bulb B. What is the ratio of the potential difference across bulb A to that across bulb B?

91. Two identical resistors are connected to a battery. Is the power dissipated when the resistors are connected in parallel greater than, less than, or equal to the power dissipated when the resistors are connected in series? Explain.

92. When a current I_0 flows through a resistor with a resistance R_0, the power dissipated is 1 W. What is the power dissipated in each of the following systems?

System	Current	Resistance
A	$2I_0$	R_0
B	I_0	$2R_0$
C	$4I_0$	$2R_0$

93. When a voltage V_0 is applied to a resistor with a resistance R_0, the power dissipated is 1 W. What is the power dissipated in each of the following systems?

System	Voltage	Resistance
A	$2V_0$	R_0
B	V_0	$2R_0$
C	$4V_0$	$2R_0$

94. Two resistors, $R_1 = R$ and $R_2 = 2R$, are connected in series to a battery. Which resistor dissipates more power? Explain.

95. Two resistors, $R_1 = R$ and $R_2 = 2R$, are connected in parallel with a battery. Which resistor dissipates more power? Explain.

Problem Solving

96. A 75-V battery supplies 3.8 kW of power. How much current does the battery produce?

97. A battery connected to a 140-Ω resistor dissipates 2.5 W of power. If the circuit contains only the battery and the resistor, what is the voltage of the battery?

98. A resistor carries a current of 2.8 A and dissipates 17 W of power. What is the resistance of the resistor?

99. Suppose that points A and B in Figure 21.28 are connected to a 12-V battery. Find the power dissipated by the circuit.

100. How much does it cost to operate a 120-W lightbulb for 25 min if the cost of electricity is $0.086 per kilowatt-hour?

101. A portable CD player uses a current of 7.5 mA at a potential difference of 3.5 V. How much energy does the player use in 35 s?

102. A 94-Ω resistor is connected to a 15-V battery. How much energy is dissipated by the resistor in 6 min?

103. A small heater operates at a voltage of 120 V with a current of 3.3 A. If it costs 3.8¢ to operate the heater for an hour, what is the cost of electricity at that location? Give your answer in dollars per kilowatt-hour.

104. A 45-Ω resistor is connected to a 12-V battery, and a 65-Ω resistor has a current of 0.25 A. **(a)** Which resistor dissipates more energy per time? **(b)** Calculate the energy dissipated in 150 s by each resistor.

105. | **Think & Calculate** | A 65-W lightbulb operates at a potential difference of 95 V. Find **(a)** the current in the bulb and **(b)** the resistance of the bulb. **(c)** If this bulb is replaced with one whose resistance is half the value found in part (b), is the replacement bulb's power rating greater than or less than 65 W? By what factor?

106. Rating Car Batteries Car batteries are rated by the following two numbers: (1) the number of cranking amps indicates the current the battery can produce for 30.0 s while maintaining a terminal voltage of at least 7.2 V, and (2) the reserve capacity is the number of minutes for which the battery can produce a 25-A current while maintaining a terminal voltage of at least 10.5 V. One particular battery is advertised as having 905 cranking amps and a 155-min reserve capacity. Which of these two ratings represents the greater amount of energy delivered by the battery?

Mixed Review

107. Consider the circuit shown in **Figure 21.29**, in which three lights, each with a resistance R, are connected in series. The circuit also contains an open switch. **(a)** When the switch is closed, does the intensity of light 2 increase, decrease, or stay the same? Explain. **(b)** Do the intensities of lights 1 and 3 increase, decrease, or stay the same when the switch is closed? Explain.

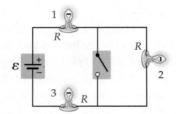

Figure 21.29

108. Predict & Explain **(a)** Referring to the circuit in Figure 21.29, does the current supplied by the battery increase, decrease, or remain the same when the switch is closed? **(b)** Choose the *best* explanation from among the following:

A. The current decreases because only two of the lights can draw current from the battery when the switch is closed.

B. Closing the switch makes no difference to the current since light 2 is still connected to the battery.

C. Closing the switch shorts out light 2, decreases the total resistance of the circuit, and increases the current.

109. Consider the circuit shown in **Figure 21.30**, in which three lights, each with a resistance R, are connected in parallel. The circuit also contains an open switch. **(a)** When the switch is closed, does the intensity of light 3 increase, decrease, or stay the same? Explain. **(b)** Do the intensities of lights 1 and 2 increase, decrease, or stay the same when the switch is closed? Explain.

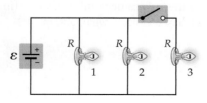

Figure 21.30

110. Predict & Explain **(a)** When the switch is closed in the circuit in Figure 21.30, does the current supplied by the battery increase, decrease, or stay the same? **(b)** Choose the *best* explanation from among the following:

A. The current increases because three lights are drawing current from the battery when the switch is closed, rather than just two.

B. Closing the switch makes no difference to the current because the voltage is the same as before.

C. Closing the switch decreases the current because an additional light is added to the circuit.

111. Suppose your car has a total charge of 85 μC. What current does it produce as it travels from Dallas to Fort Worth in 0.75 h?

112. Triple Choice **Figure 21.31** shows two circuits with identical batteries and resistors. Is the current in circuit A greater than, less than, or equal to the current in circuit B? Explain.

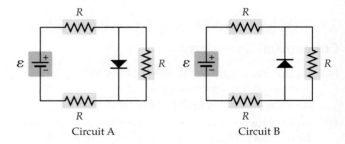

Circuit A   Circuit B

Figure 21.31

113. Triple Choice **Figure 21.32** shows two circuits with identical batteries and resistors. Is the current through the battery in circuit A greater than, less than, or equal to the current through the battery in circuit B? Explain.

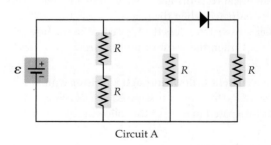

Circuit A

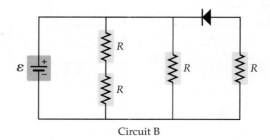

Circuit B

Figure 21.32

114. Predict & Explain The resistivity of tungsten increases with temperature. **(a)** When a light containing a tungsten filament heats up, does its power consumption increase, decrease, or stay the same? **(b)** Choose the *best* explanation from among the following:

A. The voltage is unchanged, and therefore an increase in resistance implies a reduced power, as we can see from $P = V^2/R$.

B. Increasing the resistance increases the power, as is clear from $P = I^2R$.

C. The power consumption is independent of the resistance, as we can see from $P = IV$.

115. Three resistors $(R, \frac{1}{2}R, 2R)$ are connected to a battery. If the resistors are connected in series, which one dissipates the most power?

116. Three resistors $(R, \frac{1}{2}R, 2R)$ are connected to a battery. If the resistors are connected in parallel, which one dissipates the most power?

117. Predict & Explain An electric space heater has a power rating of 500 W at a given voltage, V. **(a)** If two of these heaters are connected in series to the same voltage source, is the power consumed by the two heaters greater than, less than, or equal to 1000 W? **(b)** Choose the *best* explanation from among the following:

A. Each heater consumes 500 W; therefore, two of them will consume 500 W + 500 W = 1000 W.

B. The voltage is the same, but the resistance is doubled by connecting the heaters in series. Therefore, the power consumed $(P = V^2/R)$ is less than 1000 W.

C. Connecting two heaters in series doubles the resistance. Since power depends on resistance squared, it follows that the power consumed is greater than 1000 W.

118. You are given resistors with resistances 413 Ω, 521 Ω, and 146 Ω. Describe how these resistors can be connected to produce an equivalent resistance of 255 Ω.

119. Referring to Figure 21.31 (diode circuit with three resistors), find the current in circuit A given that $\varepsilon = 12$ V and $R = 55$ Ω.

120. Referring to Figure 21.32 (diode circuit with four resistors), find the current in circuit A given that $\varepsilon = 24$ V and $R = 62$ Ω.

121. Three-Way Lightbulb A three-way lightbulb has two filaments with resistances R_1 and R_2 connected in series. The resistors are connected to three terminals, as indicated in **Figure 21.33**, and the light switch determines which two of the three terminals are connected to a source of a potential difference of 120 V at any given time. When terminals A and B are connected to the 120-V source, the bulb uses 75.0 W of power. When terminals A and C are connected to the 120-V source, the bulb uses 50.0 W of power. **(a)** What is the resistance R_1? **(b)** What is the resistance R_2? **(c)** How much power does the bulb use when the 120-V source is connected to terminals B and C?

Figure 21.33

122. Two resistors are connected in series to a battery with an emf of 12.0 V. The voltage across the first resistor is 2.7 V, and the current through the second resistor is 0.15 A. Find the resistance of each resistor.

123. An electric heating coil is immersed in 4.6 kg of water at 22 °C. The coil, which has a resistance of 250 Ω, warms the water to 32 °C in 3.8 min. What is the potential difference at which the coil operates?

124. When two resistors, R_1 and R_2, are connected in series with a 6.0-V battery, the potential difference across R_1 is 4.0 V. When R_1 and R_2 are connected in parallel to the same battery, the current through R_2 is 0.45 A. Find the resistances of R_1 and R_2.

Writing about Science

125. Write a report about superconductors and semiconductors. How are these materials produced? How are they used in practical applications? Give specific examples. Where is the nearest semiconductor to you at this moment? The nearest superconductor?

126. Connect to the Big Idea Energy that is dissipated in a resistor appears as thermal energy. This energy can be used to heat a cup of water or some other object. Describe how measurements of temperature versus time could allow you to verify the power equations given in Lesson 21.3.

Read, Reason, and Respond

Footwear and Safety The American National Standards Institute (ANSI) specifies safety standards covering a number of potential workplace hazards. For example, ANSI requires that footwear provide protection against the effects of compression from a static weight, impact from a dropped object, puncture from a sharp tool, and cuts from saws. In addition, to protect against the potentially lethal effects of electrical shock, ANSI provides standards for the electrical resistance that a person and the person's footwear must offer to the flow of electric current.

Specifically, regulation ANSI Z41-1999 states that the resistance of a person and the person's footwear must be tested with the circuit shown in **Figure 21.34**. In this circuit the voltage supplied by the battery is $\varepsilon = 50.0$ V, and the resistance in the circuit is $R = 1.00$ MΩ. Initially the circuit is open and no current flows. When a person touches the metal sphere attached to the battery, however, the circuit is closed and a small current flows through the person and the person's shoes and back to the battery. The amount of current flowing through the person can be determined by using a voltmeter to measure the voltage drop, V, across the resistor R. To be safe, the current should not exceed 150 μA.

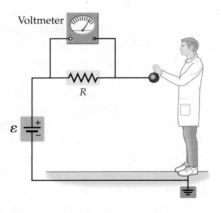

Figure 21.34

Notice that the setup in Figure 21.34 is a DC circuit with two resistors in series—R and R_{pf} (where the subscript stands for "person and footwear"). It follows that the current in the circuit is $I = \varepsilon/(R + R_{pf})$. We also know that the current is $I = V/R$, where V is the reading of the voltmeter. These relations can be combined to relate the voltage, V, to the resistance, R_{pf}, with the result shown in **Figure 21.35**. According to ANSI regulations, Type II footwear must produce a resistance R_{pf} in the range from 0.1×10^7 Ω to 100×10^7 Ω.

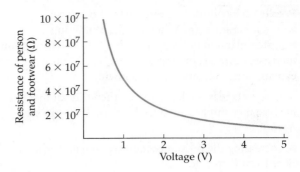

Figure 21.35

127. Suppose the voltmeter measures a potential difference of 3.70 V across the resistor. What is the current that flows through the person's body?

　A. 3.70×10^{-6} A　　C. 0.0740 A

　B. 5.00×10^{-5} A　　D. 3.70 A

128. What is the resistance of the person and footwear when the voltmeter reads 3.70 V?

　A. 1.25×10^7 Ω　　C. 4.63×10^7 Ω

　B. 1.35×10^7 Ω　　D. 1.71×10^8 Ω

129. The resistance of a given person and her footwear is 4.00×10^7 Ω. What is the reading on the voltmeter when this person is tested?

　A. 0.976 V　　　　C. 1.25 V

　B. 1.22 V　　　　D. 50.0 V

130. Suppose that during one test a person's shoes become wet when water spills onto the floor. When this happens, do you expect the reading on the voltmeter to increase, decrease, or stay the same?

Standardized Test Prep

Multiple Choice

Use the circuit diagram below to answer Questions 1–4. The circuit consists of a 6-V battery, a switch, and three resistors.

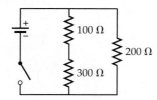

1. What is the equivalent resistance of the three resistors connected as shown?
   (A) 600 Ω
   (B) 520 Ω
   (C) 133 Ω
   (D) 55 Ω

2. Which of the following correctly ranks the three resistors in terms of the amount of *current* flowing through each?
   (A) 100 Ω > 200 Ω > 300 Ω
   (B) 300 Ω > 200 Ω > 100 Ω
   (C) 100 Ω = 300 Ω > 200 Ω
   (D) 100 Ω = 300 Ω < 200 Ω

3. What is the total power used by the circuit?
   (A) 27 W
   (B) 2.7 W
   (C) 0.27 W
   (D) 0.027 W

4. What will be the result of replacing the 200-Ω resistor by a 400-Ω resistor?
   (A) The total resistance of the circuit decreases.
   (B) The current through the 100-Ω resistor remains the same.
   (C) The potential difference across the 300-Ω resistor decreases.
   (D) The current through the 400-Ω resistor is twice as much as the current through the 200-Ω resistor.

5. A 0.5-mm-diameter graphite pencil "lead" is used as a resistor. It has resistance R. What would be the resistance of a second piece of "lead" that has a diameter of 1.0 mm and is half as long as the first piece?
   (A) R/8   (C) R
   (B) R/2   (D) 2R

Use the circuit diagram below to answer Questions 6 and 7. The circuit consists of a 12-V battery and three resistors.

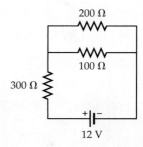

6. Which *best* describes the magnitude and direction of the current through the 300-Ω resistor?
   (A) 33 mA, upward
   (B) 16 mA, downward
   (C) 40 mA, upward
   (D) 10 mA, downward

7. Which of the following correctly ranks the three resistors according to the amount of heat dissipated by each resistor, from greatest to least?
   (A) 300 Ω > 100 Ω > 200 Ω
   (B) 200 Ω > 300 Ω > 100 Ω
   (C) 300 Ω > 200 Ω > 100 Ω
   (D) 200 Ω > 100 Ω > 300 Ω

Extended Response

8. Draw a circuit diagram showing how to connect an ammeter and a voltmeter so that they measure the current in a 100-Ω resistor and the potential difference across a 200-Ω resistor.

> **Test-Taking Hint**
>
> When calculating the equivalent resistance of resistors connected in parallel, be sure to take the inverse *after* you've added together the inverses of the individual resistances.

If You Had Difficulty With . . .

Question	1	2	3	4	5	6	7	8				
See Lesson	21.2	21.2	21.3	21.2	21.1	21.2	21.3	21.1				

22

Magnetism and Magnetic Fields

Inside

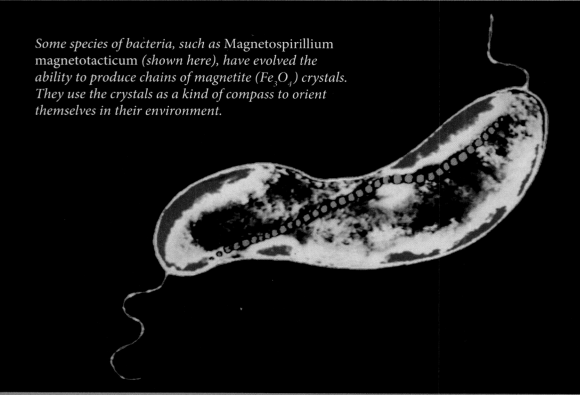

Some species of bacteria, such as Magnetospirillium magnetotacticum *(shown here), have evolved the ability to produce chains of magnetite (Fe_3O_4) crystals. They use the crystals as a kind of compass to orient themselves in their environment.*

Big Idea

Moving charges produce magnetic fields, and magnetic fields exert forces on moving charges.

The effects of magnetism have been known since antiquity. For example, a piece of the naturally occurring iron-oxide mineral known as *lodestone* can behave just like a manufactured magnet. This chapter explores the properties of magnets and the magnetic fields they produce. It also reveals the fascinating, and unexpected, connection between magnetic fields and electric currents.

Explore

1. Cut a 5-mm-wide strip from the long edge of a flexible-sheet refrigerator magnet. This strip is your probe.
2. Hold one end of the probe *very gently* between two fingers and drag it slowly from end to end along the dark unprinted side of the magnet.
3. Repeat Step 2, this time going from side to side. Be sure to explore the entire surface of the magnet.
4. Use your probe to examine the magnetic fields of the refrigerator magnets being used by other students.

Think

1. **Describe** What did you feel as you dragged the probe along the magnet from end to end and then from side to side? Describe the probe's response when used on your classmates' refrigerator magnets.
2. **Identify** On a simple sketch of your magnet, identify the regions that responded in the same manner to the probe.
3. **Predict** What do you think you would find if you were to drag one complete refrigerator magnet over another? Do you think the orientation of one magnet relative to the other would affect your results? Try it and see.

22.1 Magnets and Magnetic Fields

Have you ever heard of animal magnetism? Well, some animals *do* use magnetism. Examples include bacteria, pigeons, and honeybees. We use magnetism too, in the form of bar magnets, horseshoe magnets, and compass needles. In this lesson you'll learn about the basic properties of magnetic fields and about the important role magnets play in everyday life.

Vocabulary

- magnetic field
- magnetic domain

Bar magnets provide a good example of magnetism

Your first direct experience with magnetism was probably a playful exploration of bar magnets and their properties. From such experiences, you know that the two ends of a magnet are different.

🔑 *How are the north and south poles of a magnet defined?*

Magnetic Poles A bar magnet attracts or repels another bar magnet depending on which ends of the magnets are brought together. One end of a magnet is referred to as its *north pole* and is labeled N. The other end of a magnet is its *south pole*, which is labeled S. More specifically, the poles of a bar magnet are defined by suspending it from a string so that it is free to rotate like a compass needle.

- 🔑 **The end of a freely rotating bar magnet that points toward the north geographic pole of Earth is the north-seeking pole, or simply the north pole.**

- 🔑 **The opposite end of the magnet is the south-seeking pole, or simply the south pole.**

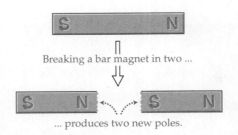

Breaking a bar magnet in two ...

... produces two new poles.

▲ **Figure 22.1 Magnets always have two poles**
When a bar magnet is broken in half, two new poles appear. Each half has both a north pole and a south pole, just like the original magnet.

An interesting aspect of magnets is that they *always* have two poles. Always! You might think that if you broke a bar magnet in two, each of the halves would have just one pole. That's not what happens. Instead, breaking a magnet in half produces two new poles on either side of the break, as illustrated in **Figure 22.1**.

This behavior is fundamentally different from that of electricity, in that the two types of electric charge (positive and negative) can exist separately. Physicists continue to look for a single magnetic pole, the elusive *magnetic monopole*, but none has been found so far. To the best of our knowledge, separate magnetic poles simply don't exist in nature.

Attraction and Repulsion The rule for whether two bar magnets attract or repel each other is learned at an early age: opposites attract and likes repel. Thus, if two magnets are brought together in such a way that their opposite poles approach each other, as in **Figure 22.2 (a)**, the force each experiences is attractive. Like poles brought close together, as in **Figure 22.2 (b)**, experience a repulsive force.

Magnets produce magnetic fields

Just as an electric charge creates an electric field, so too does a magnet create a magnetic field. A **magnetic field** is a vector force field that surrounds any magnetic material. In addition to exerting force, a magnetic field also contains energy, just like an electric field. As you would expect, the greater the energy, the more intense the field.

A magnetic field, which is represented with the symbol $\vec{\mathbf{B}}$, can be visualized using small iron filings sprinkled onto a smooth surface. In **Figure 22.3 (a)**, for example, a sheet of glass is placed on top of a bar magnet. When iron filings are sprinkled onto the glass sheet, they align with the magnetic field in their vicinity. The pattern they form gives a good idea of the overall field produced by the magnet. Similar effects are seen in

CONNECTINGIDEAS

Newton's third law, which states that an action force is always accompanied by a reaction force, was introduced in Chapter 5.
• Here Newton's third law is applied to the magnetic forces exerted by bar magnets. Notice that the action and reaction forces act on different magnets.

🔑 *How is the direction of a magnetic field determined?*

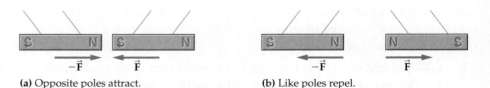

(a) Opposite poles attract. **(b)** Like poles repel.

▲ **Figure 22.2 The force between two bar magnets**
(a) Opposite poles of magnets attract one another. **(b)** When like poles of magnets are brought together, they repel each other.

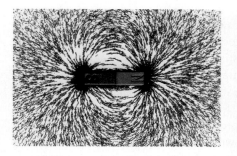

(a)

(b)

◀ **Figure 22.3 Magnetic field lines**
The field of **(a)** a bar magnet or
(b) a horseshoe magnet can be visualized
using iron filings on a sheet of glass or
paper. The filings orient themselves with
the field lines, creating a "snapshot" of the
magnetic field.

Figure 22.3 (b), with a magnet that has been bent to bring its poles close together. Because of its shape, this type of magnet is referred to as a *horseshoe magnet.*

Field Strength and Direction

Notice that the filings are bunched together near the poles of the magnets in Figure 22.3. This is where the magnetic field is most intense. We illustrate this by drawing field lines that are close to one another near the poles, as in **Figure 22.4**. As you move away from a magnet in any direction, the field weakens. This weakening is indicated by a wider separation between field lines. It's important to notice that the magnetic field lines continue even within the body of a magnet. In fact, magnetic field lines do *not* start or stop anywhere.

How do we determine the direction of the magnetic field? 🔑 **The direction of a magnetic field at a given location is defined as the direction a compass needle would point if placed at that location.** This definition is applied to a bar magnet in Figure 22.4. Imagine, for example, placing a compass near the south pole of the magnet (the left end in Figure 22.4). Because opposites attract, the north pole of the compass needle—the end with the arrowhead—points toward the south pole of the magnet. Thus, according to our definition, the direction of the magnetic field at that location is toward the magnet's south pole. Similarly, you can see that the magnetic field must point away from the north pole of the bar magnet. In general, magnetic field lines exit from the north pole of a magnet and enter at the south pole.

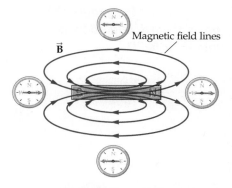

▲ **Figure 22.4 Magnetic field lines for a bar magnet**
Magnetic field lines are closely spaced near the poles, where the magnetic field, $\vec{B}$, is most intense. In addition, the lines form closed loops that leave the magnet at the north pole and enter it at the south pole.

CONCEPTUAL Example 22.1 Can They Cross?

Can magnetic field lines cross one another?

Reasoning and Discussion
Recall that the direction in which a compass points at any given location is the direction of the magnetic field at that point. Since a compass can point in only one direction, there must be only one direction for the magnetic field, $\vec{B}$. If field lines were to cross, however, there would be two directions for $\vec{B}$ at the crossing point. This is not allowed.

Answer
No. Magnetic field lines never cross.

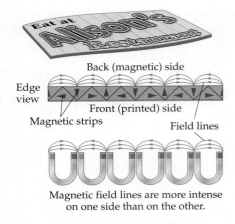

Back (magnetic) side

Edge view

Magnetic strips

Front (printed) side

Field lines

Magnetic field lines are more intense on one side than on the other.

▲ **Figure 22.5 Refrigerator magnet**
A flexible refrigerator magnet is made from a large number of narrow magnetic strips. The resulting field is similar to that of parallel horseshoe magnets placed side by side.

🔑 *How are the geographic poles of Earth related to its magnetic field?*

COOL PHYSICS
When North Becomes South

Earth's magnetic field is similar to that of a bar magnet, but far more complex in both its shape and behavior. For example, scientists know that Earth's field has *reversed* direction many times over the ages. The last reversal occurred about 780,000 years ago. From 980,000 years ago to 780,000 years ago, a compass would have pointed in the direction opposite to where it points today! Earth may be preparing for another such reversal. "Bubbles" of reversed field have appeared in Africa. These may be early signs of a coming global reversal. A compass would be of little use near one of these reversed bubbles.

An interesting example of a magnetic field is provided by the common refrigerator magnet. The flexible variety of these popular magnets have an unusual property that you can verify at home. Try sticking one side of such a magnet to the refrigerator, and then the other side. You'll find that one side sticks, but the other side (the printed side) doesn't.

Clearly, the magnetic field produced by such a magnet is not like that of a bar magnet. If it were, both sides would stick equally well. Instead, the flexible refrigerator magnet is composed of multiple magnetic strips of opposite polarity, as indicated in **Figure 22.5**. The net effect is a magnetic field similar to the field that would be produced by a large number of tiny horseshoe magnets placed side by side. Thus, the field is intense on the side containing the poles of the tiny magnets, as indicated in Figure 22.3 (b). The field is weak on the other side, where the message or image is printed. Who would have guessed that a refrigerator magnet was so sophisticated?

Earth produces its own magnetic field

Earth, like many planets, produces its own magnetic field. In many respects, Earth's magnetic field is like that of a giant bar magnet, as illustrated in **Figure 22.6**. Notice that there is a magnetic pole near each geographic pole of Earth. In addition, the field lines are essentially horizontal (parallel to Earth's surface) near the equator but enter or leave Earth vertically near the poles. If you were to stand near the north geographic pole, commonly referred to as the North Pole, your compass would try to point straight down.

Because the north pole of a compass needle points toward the north geographic pole of Earth, and because opposites attract, we conclude the following: 🔑 **The *north* geographic pole of Earth is actually near the *south* pole of Earth's magnetic field.** This is shown in Figure 22.6.

Location of Earth's Poles The axis of the magnetic poles is not perfectly aligned with the rotational axis of Earth. Instead, it is inclined at an angle that varies slowly with time. Presently, the magnetic axis is tilted away from the rotational axis by an angle of about 11.5°. The current location of

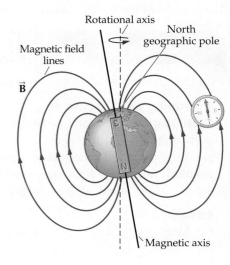

▶ **Figure 22.6 Magnetic field of Earth**
Earth's magnetic field is similar to that of a giant bar magnet tilted slightly from the rotational axis. Notice that the north geographic pole is near the south pole of Earth's magnetic field.

Table 22.1 Typical Magnetic Fields

Physical System	Magnetic Field (G)
Magnetar (a magnetic neutron star formed in a supernova explosion)	10^{15}
Strongest magnetic field produced in a lab	6×10^5
High-field MRI	15,000
Low-field MRI	2000
Sunspots	1000
Bar magnet	100
Earth	0.50

Reading Support ✓

Word Origin of

paleomagnetism [pay lee oh MAG nih tiz uhm]

(noun) a branch of geophysics; the study of the record of Earth's magnetic field in rocks

Latin *palaio-* (combining form of *palaios*) meaning "old" + Latin (from Greek) *magnēs*, shortened from *ho Magnēs lithos*, meaning "the Magnesian stone," a reference to a mineral-rich region in Greece known as Magnesia where deposits of magnetite, a magnetic iron ore, are found

the magnetic pole that is near the north geographic pole is just west of Ellef Ringnes Island, one of the Queen Elizabeth Islands in extreme northern Canada.

Strength of Earth's Magnetic Field

As magnetic fields go, Earth's field is relatively weak. To make this quantitative, we note that magnetic field strength is measured in terms of a unit called the *tesla* (T). The tesla is named in recognition of the pioneering electrical and magnetic studies of the Croatian-born American engineer Nikola Tesla (1856–1943).

A magnetic field of 1 T is rather large. In comparison, the magnetic field at the surface of Earth is roughly 5.0×10^{-5} T. Thus, another commonly used unit for magnetism is the *gauss* (G), defined as follows:

$$1 \text{ gauss} = 1 \text{ G} = 10^{-4} \text{ T}$$

In terms of this unit, Earth's magnetic field at its surface is approximately 0.5 G. The gauss is not an SI unit. Even so, it finds wide usage because of its convenient magnitude. The magnitudes of some typical magnetic fields are given in **Table 22.1**.

Earth's magnetic field verifies seafloor spreading

Earth's magnetic field reverses direction over geological time periods. Ancient magnetic field reversals have left a permanent record in the rocks of the ocean floors. To see how, look at **Figure 22.7**. Here we see that molten rock is being extruded from a mid-ocean ridge. The extruded rock has no magnetization, because of its high temperature. When the rock cools, however, it becomes magnetized in the direction of Earth's magnetic field. In effect, the direction of Earth's magnetic field becomes "frozen" in the solidified rock.

As the seafloor spreads, and more rock is formed along the mid-ocean ridge, a continuous record of Earth's magnetic field is formed. If Earth's field reverses at some point in time, the field in the solidified rocks will record that fact. Figure 22.7 shows a record of the magnetization of rock that has been formed by seafloor spreading, showing the regions of oppositely magnetized rocks that indicate the field reversals. Evidence like this shows not only that the magnetic field has reversed multiple times in the past, but also that the seafloor is spreading. Such evidence provides strong support for the theories of continental drift and plate tectonics. In fact, a new branch of geology, referred to as **paleomagnetism**, developed from these discoveries.

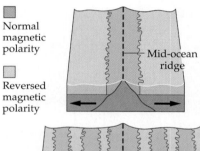

Normal magnetic polarity

Reversed magnetic polarity

Mid-ocean ridge

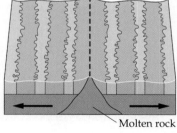

Molten rock

▲ **Figure 22.7 Mid-ocean ridge** Molten rock extruded at a mid-ocean ridge magnetizes in the direction of Earth's magnetic field when it cools to temperatures below about 770 °C. The direction of the field remains "frozen" in the cooled rock. As this rock moves away from the ridge, due to seafloor spreading, newly extruded molten rock near the ridge undergoes the same process. As a result, the magnetization of the rocks on either side of a mid-ocean ridge produces a geological record of the polarity of Earth's magnetic field over time, as well as convincing confirmation of seafloor spreading.

(a) Magnetic domains in zero external field

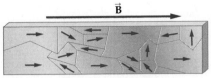

$\vec{B}$

(b) Domains in direction of external magnetic field grow in size.

▲ **Figure 22.8 Magnetic domains**
(a) A magnetic material tends to have numerous domains with magnetization in different directions. **(b)** When an external magnetic field is applied, the domains pointing in the direction of the field often grow in size at the expense of the domains pointing in other directions.

Magnets are made up of many magnetic domains

How does a magnetic material produce a magnetic field? Well, at the microscopic level, the magnetic field of a magnet is due to the magnetic fields produced by electrons in its atoms. Each electron acts like a small bar magnet. In some materials the magnetic fields of the electrons cancel, leaving zero net magnetic field. In other materials—like iron, nickel, and cobalt—the magnetic fields of the electrons don't cancel, and the electrons in neighboring atoms tend to align with one another, producing a strong magnetic field.

The magnetic field of any magnetic material is broken up into regions in which the field points in different directions, as indicated in **Figure 22.8 (a)**. A region within a magnetic material where the electrons are aligned in the same direction is referred to as a **magnetic domain**. Each domain has a strong magnetic field in a given direction. Different domains are oriented differently, however, so that the net effect may be small. The typical size of these domains is on the order of 10^{-4} cm to 10^{-1} cm.

When an external magnetic field is applied to such a material, the magnetic domains that are pointing in the direction of the applied field tend to grow in size at the expense of the domains with different orientations. This is illustrated in **Figure 22.8 (b)**. The result is a net magnetization of the material—it becomes a permanent magnet.

Magnetic effects are important to many animals

Many living organisms are known to have small magnetic crystals (magnetite) in their bodies. For example, some species of bacteria use magnetite crystals to help orient themselves with respect to Earth's magnetic field. Magnetite has also been found in the brains of bees and pigeons, where it is thought to play a role in navigation. It is even found in human brains, though its possible function there is unclear.

Whatever the role of magnetite in the human brain, observations show that a magnetic field can affect the way the brain works. Recent experiments tested volunteers who tried to interpret an optical illusion. It was found that if their brains were exposed to a strong magnetic field of 1 T, their ability to understand what they were seeing was diminished. Fortunately, fields strong enough to have such effects are not encountered in everyday life.

22.1 LessonCheck (MP)

Checking Concepts

1. 🔒 **Identify** In what direction does the north pole of a bar magnet point if it is suspended by a string?

2. 🔒 **Describe** How can you use a compass needle to determine the direction of a magnetic field at a given location?

3. 🔒 **Explain** What pole of Earth's magnetic field would you be closer to if you visited Alaska?

4. Explain Can a single isolated magnetic pole be created by breaking a magnet in two?

5. Triple Choice Is the magnitude of Earth's magnetic field at the equator greater than, less than, or equal to its magnitude at the North Pole? Explain.

6. State In what direction does Earth's magnetic field point in Antarctica?

7. Compare and Contrast How are magnetic poles similar to electric charges? How are they different?

8. Explain Why do some pieces of iron act like magnets, while others do not?

22.2 Magnetism and Electric Currents

The connection between electricity and magnetism was discovered accidentally by the Danish scientist Hans Christian Oersted (1777–1851) in 1820. Professor Oersted was giving a science lecture when, at one point, he closed a switch and allowed a current to flow through a wire. He noticed that a nearby compass needle rotated rapidly when the switch was closed. With that simple observation, Oersted discovered that electric currents can create magnetic fields. In this lesson you'll learn the details about this connection.

A current-carrying wire produces a magnetic field

Fill a sink with water. Put your finger in the water and hold it still. No waves are created. As soon as you move your finger, however, waves appear. Similarly, a stationary electric charge produces an electric field but not a magnetic field. As soon as you move the charge, however, a magnetic field is generated. Moving charges create magnetic fields, much like moving objects create water waves. As a result, a wire carrying an electric current produces a magnetic field.

Let's start with the simplest possible case—a long, straight wire that carries a current, I. To visualize the magnetic field such a wire produces, shake iron filings onto a sheet of paper that is pierced by the wire, as shown in **Figure 22.9 (a)**. The result is that the filings form circular patterns centered on the wire. Clearly, the magnetic field "circulates" around the wire.

Determining Magnetic Field Direction We can gain additional information about the magnetic field by placing a group of small compasses about the wire, as in **Figure 22.9 (b)**. In addition to confirming the circular shape of the field lines, the compass needles show the field's direction. To understand this direction, we use the magnetic field right-hand rule (RHR):

> **Magnetic Field Right-Hand Rule**
>
> 🔑 To find the direction of the magnetic field due to a current-carrying wire, point the thumb of your right hand along the wire in the direction of the current, I. Your fingers will then curl around the wire in the direction of the magnetic field.

This rule is illustrated in **Figure 22.10** on the next page. Notice that it predicts the same direction as that indicated by the compass needles in Figure 22.9 (b).

Representing Magnetic Field Direction

In some cases a magnetic field we are considering will point into or out of the page. This can be difficult to draw. Therefore, we establish the convention that the symbol ⊗ indicates that the magnetic field points into the page. The way to remember this is to think of a magnetic field vector as an arrow. At the tail end of an arrow are crossed feathers. Therefore, if you view a vector from

Vocabulary

- solenoid

🔑 *How is the direction of the magnetic field around a current-carrying wire determined?*

(a)

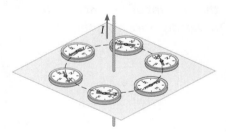

(b)

▲ Figure 22.9 **The magnetic field of a current-carrying wire**
(a) An electric current flowing through a wire produces a magnetic field. In the case of a long, straight wire, the field "circulates" around the wire. **(b)** Compass needles point along the circumference of a circle centered on the wire. This result shows that the magnetic field around the wire is circular.

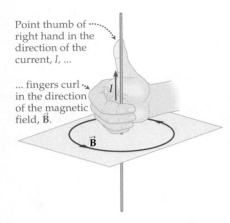

Point thumb of right hand in the direction of the current, I, ...

... fingers curl in the direction of the magnetic field, $\vec{B}$.

$\vec{B}$

▲ **Figure 22.10 The magnetic field right-hand rule**
With the thumb of the right hand pointing in the direction of the current, the fingers curl in the direction of the field.

🔑 *How is the strength of the magnetic field around a current-carrying wire related to the amount of current and the radial distance from the wire?*

behind, it looks like an X. Similarly, if an arrow points out of the page, all you see is the point at its tip. Thus, we represent a magnetic field vector pointing out of the page with the symbol ⊙, where the dot represents the tip of the arrow. We apply these conventions in the next Conceptual Example.

CONCEPTUAL Example 22.2 Which Direction?

The magnetic field shown below is due to the horizontal, current-carrying wire. Does the current in the wire flow toward the left or the right?

⊗ ⊗ ⊗ ⊗ ⊗ ⊗ ⊗
$\vec{B}$
⊗ ⊗ ⊗ ⊗ ⊗ ⊗ ⊗
⊗ ⊗ ⊗ ⊗ ⊗ ⊗ ⊗
————————————————
⊙ ⊙ ⊙ ⊙ ⊙ ⊙ ⊙
⊙ ⊙ ⊙ ⊙ ⊙ ⊙ ⊙
$\vec{B}$
⊙ ⊙ ⊙ ⊙ ⊙ ⊙ ⊙

Reasoning and Discussion
If you point the thumb of your right hand along the wire toward the left, your fingers curl into the page above the wire and out of the page below the wire, as shown in the figure. Thus, the current flows toward the left.

Answer
The current in the wire flows toward the left.

The magnetic field is proportional to the current

Experiments show that the magnetic field produced by a current-carrying wire doubles if the current, I, is doubled. In addition, the field doubles if the radial distance from the wire, r, is halved. These observations are easily summarized in one statement: 🔑 **The magnetic field produced by a current in a wire is proportional to the current and inversely proportional to the radial distance from the wire.**

The magnetic field for a long, straight wire is given by the following equation:

Magnetic Field for a Long, Straight Wire
$\text{magnetic field} = \dfrac{\text{permeability of free space} \times \text{current}}{2\pi \times \text{radial distance from wire}}$
$B = \dfrac{\mu_0 I}{2\pi r}$
SI unit: tesla (T)

In this equation, μ_0 is a constant called the *permeability of free space*. Its value is

$$\mu_0 = 4\pi \times 10^{-7}\ \text{T} \cdot \text{m/A}$$

The following Quick Example shows how to use the magnetic field equation.

Find the magnetic field at a distance of 1 m from a long, straight wire carrying a current of 1 A.

Solution

Straightforward substitution of the given values into $B = \mu_0 I / 2\pi r$ yields

$$B = \frac{\mu_0 I}{2\pi r}$$

$$= \frac{(4\pi \times 10^{-7} \text{ T} \cdot \text{m}/\text{A})(1 \text{ A})}{2\pi(1 \text{ m})}$$

$$= \boxed{2 \times 10^{-7} \text{ T}}$$

This is a rather weak magnetic field. It is less than one hundredth as strong as Earth's magnetic field.

Practice Problems

9. A long, straight wire carries a current of 7.2 A. At what distance from this wire is its magnetic field equal in strength to Earth's magnetic field, which is approximately 5.0×10^{-5} T?

10. What is the current in a long, straight wire that produces a magnetic field of 7.1×10^{-6} T at a distance of 12.5 cm?

ACTIVE Example 22.4 Determine the Magnetic Field

Two horizontal wires 22 cm apart carry currents $I_1 = 1.5$ A and $I_2 = 4.5$ A, flowing in the same direction. Find the magnetic field halfway between the wires. (Assume that magnetic fields pointing out of the page are positive and magnetic fields pointing into the page are negative.)

1. Find the magnetic field produced by wire 1. This field points into the page:

$B_1 = -2.7 \times 10^{-6}$ T (into page)

2. Find the magnetic field produced by wire 2. This field points out of the page:

$B_2 = 8.2 \times 10^{-6}$ T (out of page)

3. Calculate the total magnetic field. The sign of the total field gives its direction:

$B_{\text{net}} = B_2 + B_1 = 5.5 \times 10^{-6}$ T (out of page)

Insight

Since the field produced by I_2 has the greater magnitude, the total field points out of the page. If the currents were equal, the total field midway between the wires would be zero.

(a) The magnetic field produced by a current-carrying loop is relatively intense within the loop and falls off rapidly outside the loop. (b) A bar magnet produces a field that is very similar to the field of the loop.

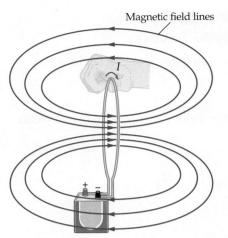
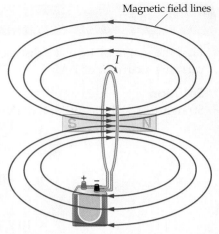

(a) Magnetic field of a current-carrying loop

(b) Magnetic field of a bar magnet is similar to that of the loop.

🔑 *What determines whether the force between current-carrying wires is attractive or repulsive?*

Wire loops and bar magnets produce similar magnetic fields

You've seen that a long, straight wire carrying an electric current produces a magnetic field. This is what Oersted noticed when he turned on a current and saw a compass needle rotate. The field around the wire consists of concentric circles centered on the wire.

What happens if a straight wire is wrapped into a circular loop instead? In this case the magnetic field looks a lot like the field produced by a bar magnet. Let's see how this comes about.

Magnetic Field of a Single Wire Loop **Figure 22.11 (a)** shows a wire loop connected to a battery producing a current in the direction indicated. Using the magnetic field RHR, as shown in the figure, we see that the magnetic field points from left to right as it passes through the loop. Notice also that the field lines are bunched together within the loop, indicating that the field is intense there. The field lines are more widely spaced outside the loop, where the field is weaker.

The most interesting aspect of the field produced by the current-carrying loop is its close resemblance to the field of a bar magnet. This similarity is illustrated in **Figure 22.11 (b)**. This figure shows a ghosted (pale) bar magnet in the middle of the loop. Notice that one side of the loop behaves like a north magnetic pole (with field lines exiting) and the other side like a south magnetic pole (with field lines entering).

The Force between Wire Loops Now, imagine placing two loops with identical currents next to one another, as in **Figure 22.12 (a)**. The force between the loops will be similar to the force between two bar magnets pointing in the same direction, as indicated in the figure. As you can see, the ghosted bar magnets would attract one another, since their opposite poles are near one another. Therefore, the current-carrying loops will attract one another as well. 🔑 **Wires with currents in the same direction experience an attractive force.**

Suppose the loops have oppositely directed currents, as in **Figure 22.12 (b)**. 🔑 **Wires with currents in opposite directions experience a repulsive force.** This is just like the force between the corresponding ghosted bar magnets.

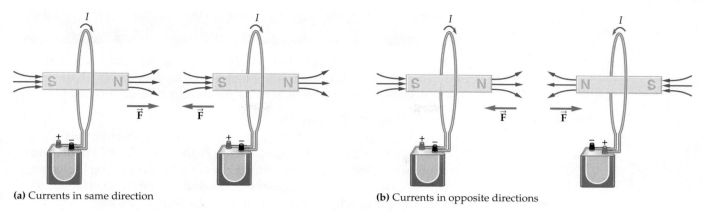

(a) Currents in same direction   **(b)** Currents in opposite directions

▲ **Figure 22.12 Magnetic forces between current-carrying loops**
To decide whether current-carrying loops experience an attractive or repulsive force, it is useful to think in terms of the corresponding bar magnets. **(a)** Wire loops with currents in the same direction are like two bar magnets lined up in the same direction; they attract each other. **(b)** Wire loops with opposite currents act like bar magnets with opposite orientations; they repel each other.

Closely spaced loops of wire form a solenoid

A **solenoid** is an electrical device in which a long wire is wound into a succession of closely spaced loops—forming a cylindrical coil of wire. A solenoid carrying an electric current produces an intense, nearly uniform, magnetic field inside the loops, as indicated in **Figure 22.13**. For this reason, solenoids are commonly referred to as *electromagnets*. Notice that each loop of a solenoid carries current in the same direction. It follows that the magnetic force between loops is attractive and serves to hold them tightly together.

Magnetic Field of a Solenoid The magnetic field lines in Figure 22.13 are tightly packed inside the solenoid but are widely spaced outside. In the ideal case of a very long, tightly packed solenoid, the magnetic field is intense and uniform inside the solenoid. If a solenoid has N loops and a length L, the magnetic field inside the solenoid is given by the following equation:

Magnetic Field of a Solenoid

$$\text{magnetic field} = \frac{\text{permeability}}{\text{of free space}} \times \left(\frac{\text{number of loops}}{\text{length of solenoid}} \right) \times \text{current}$$

$$B = \mu_0 \left(\frac{N}{L} \right) I$$

SI unit: tesla (T)

This result is independent of the cross-sectional area of the solenoid. Notice that the field depends directly on the number of loops per length and on the current.

When used as an electromagnet, a solenoid has many useful properties. First and foremost, it produces a strong magnetic field that can be turned on or off at the flip of a switch—unlike the field of a permanent magnet. In addition, filling the core of the solenoid with an iron bar intensifies the magnetic field. In such a case the magnetic field of the solenoid magnetizes the iron bar, and its field adds to that of the solenoid. These properties and others make solenoids useful devices in a variety of electrical circuits.

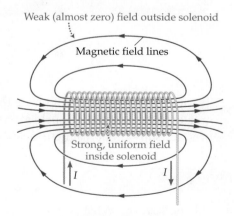

Weak (almost zero) field outside solenoid

Magnetic field lines

Strong, uniform field inside solenoid

▲ **Figure 22.13 The solenoid**
The magnetic field inside the solenoid is relatively strong and uniform. Outside the solenoid the field is weak. In the ideal case, the magnetic field is uniform inside a solenoid and zero outside.

If you want to increase the strength of the magnetic field inside a solenoid, is it better to double the number of loops while keeping the length the same or to double the length while keeping the number of loops the same?

Reasoning and Discussion

Referring to the equation $B = \mu_0(N/L)I$, we see that doubling the number of loops ($N \rightarrow 2N$) while keeping the length the same ($L \rightarrow L$) results in a doubled magnetic field ($B \rightarrow 2B$). On the other hand, doubling the length ($L \rightarrow 2L$) while keeping the number of loops the same ($N \rightarrow N$) reduces the magnetic field by a factor of 2 ($B \rightarrow B/2$). Thus, to increase the field you should pack more loops into the same length.

Answer

You should double the number of loops and keep the same length.

GUIDED Example 22.6 | The Core of a Solenoid — Solenoids

A solenoid is 20.0 cm long, has 200 loops, and carries a current of 3.25 A. Find the strength of the magnetic field inside the solenoid.

Picture the Problem

Our sketch shows the solenoid with the uniform magnetic field it produces parallel to its axis.

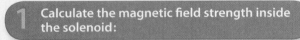

Strategy

The magnetic field produced by the solenoid is given by $B = \mu_0(N/L)I$. We can calculate the field strength (magnitude) using the data given in the problem statement.

Known	Unknown
$L = 20.0$ cm	$B = ?$
$N = 200$	
$I = 3.25$ A	

Solution

1 Calculate the magnetic field strength inside the solenoid:

$$B = \mu_0\left(\frac{N}{L}\right)I$$

$$= (4\pi \times 10^{-7}\ \mathrm{T \cdot m/A})\left(\frac{200}{0.200\ \mathrm{m}}\right)(3.25\ \mathrm{A})$$

$$= \boxed{4.08 \times 10^{-3}\ \mathrm{T}}$$

Insight

The magnetic field strength inside this fairly small solenoid is approximately 100 times greater than the magnetic field at the surface of Earth. Clearly, it is relatively easy to produce substantial magnetic fields with solenoids.

Practice Problems

11. Follow-up What current would be required to double the magnetic field strength inside the solenoid?

12. You want a solenoid that is 38 cm long with 430 loops to produce a magnetic field within its core equal to Earth's magnetic field (5.0×10^{-5} T). What current is required?

13. A solenoid that is 75 cm long produces a magnetic field of 1.3 T within its core when it carries a current of 8.4 A. How many loops of wire are contained in this solenoid?

Physics & You: Technology Magnetic resonance imaging (MRI) instruments utilize solenoids large enough to accommodate a person within their coils. Not only are these solenoids large, they are also capable of producing extremely powerful magnetic fields. Typical MRI solenoids produce magnetic fields of about 1 or 2 T (see Table 22.1). These fields are so powerful, in fact, that they can pull metallic objects such as mop buckets and stretchers across the room. A metal oxygen bottle in the same room as an MRI machine can be turned into a dangerous, high-speed projectile as it flies toward the core of the solenoid. Working with MRI solenoids is serious business.

In some cases, artificial pacemakers have been affected by MRI machines. Many types of pacemakers have what are known as *magnetic reed switches*. Such a switch allows a physician to change the operating mode of a pacemaker, without surgery, by simply placing a magnet at the appropriate location on a patient's chest. If a person with one of these pacemakers comes near an operating MRI machine, the results can be serious.

22.2 LessonCheck (MP)

Checking Concepts

14. Describe How can the thumb and fingers of your right hand be used to identify the direction of a magnetic field?

15. Explain What happens to the magnetic field from a current-carrying wire if you (**a**) double the current or (**b**) double the distance from the wire?

16. Relate Pair up the following phrases so that they correctly relate the type of magnetic force to the type of current: attractive force, repulsive force, same-direction currents, opposite-direction currents.

17. Determine If you halve the number of loops in a solenoid while doubling its length, does the magnetic field increase, decrease, or stay the same? Explain.

18. Analyze A vertical wire and a horizontal wire carry currents as shown in **Figure 22.14**.
(**a**) Does the magnetic field on the right side of the vertical wire point into the page, out of the page, upward, downward, toward the left, or toward the right? Explain.
(**b**) Does the magnetic field above the horizontal wire point into the page, out of the page, upward, downward, toward the left, or toward the right? Explain.

Solving Problems

19. Calculate Find the magnetic field 6.25 cm from a long, straight wire that carries a current of 5.81 A.

20. Calculate A wire carries a current of 4.2 A. At what distance from the wire does the magnetic field have a magnitude of 1.3×10^{-5} T?

21. Calculate The magnetic field in a solenoid that has 250 loops and a length of 12 cm is 9.4×10^{-5} T. What is the current in the solenoid?

22. Calculate The current in a superconducting solenoid is 3.75 kA. If the number of loops per meter in this solenoid is 3250, what is the magnitude of the magnetic field it produces?

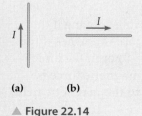

(a) (b)

▲ Figure 22.14

22.3 The Magnetic Force

Vocabulary

- plasma
- mass spectrometer
- galvanometer

🔑 **When is the force exerted by a magnetic field maximized?**

We've discussed the familiar forces that act between one bar magnet and another. We've also talked about the forces between current-carrying wires. In this lesson you'll learn about the force a magnetic field exerts on a moving electric charge. As you'll see, both the magnitude and the direction of this force have some rather interesting characteristics.

A magnetic field exerts a force on moving charges

Consider a magnetic field, $\vec{B}$, that points from left to right, as indicated in **Figure 22.15**. Suppose an object with charge q moves through this region with velocity $\vec{v}$, and the angle between $\vec{v}$ and $\vec{B}$ is θ. Experiment shows that the magnitude of the force $\vec{F}$ experienced by this object is given by the following equation:

> **Magnitude of the Magnetic Force, F**
>
> force = |charge| × velocity × magnetic field × sin θ
>
> $$F = |q|vB \sin \theta$$
>
> SI unit: newton (N)

Notice that this equation uses the *magnitude* of the charge, $|q|$, because we are calculating the *magnitude* of the force.

The equation also shows that the magnetic force depends on several different factors. Two of these factors are the same as for the electric force:

- The magnetic force depends on the charge of the object, q.
- The magnetic force depends on the magnitude of the field, in this case, of the magnetic field, B.

However, the magnetic force also depends on two factors that do not affect the strength of the electric force:

- The magnetic force depends on the speed of the object, v. An object at rest experiences no force.
- The magnetic force depends on the angle θ between the velocity vector and the magnetic field vector.

It follows that the behavior of objects in magnetic fields is significantly different from their behavior in electric fields.

In particular, an object must have a charge and must be moving if the magnetic field is to exert a force on it. Even then the force vanishes if the object moves in the direction of the field (that is, if $\theta = 0$) or in the direction opposite to the field ($\theta = 180°$). 🔑 **Maximum magnetic force is exerted when a charged object moves at right angles to the magnetic field, so $\theta = 90°$ and $\sin \theta = 1$.**

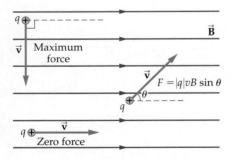

▲ **Figure 22.15 The magnetic force on a moving charged object**
An object with charge q moves through a region with magnetic field $\vec{B}$ with velocity $\vec{v}$. The magnitude of the force exerted on the charge is $F = |q|vB \sin \theta$. Notice that the force is maximum when the velocity is perpendicular to the field and is zero when the velocity is parallel to the field.

Object 1, with charge $q_1 = 3.60 \ \mu C$ and speed $v_1 = 862$ m/s, travels at right angles to a uniform magnetic field. The magnetic force it experiences is 4.25×10^{-3} N. Object 2, with charge $q_2 = 53.0 \ \mu C$ and speed $v_2 = 1.30 \times 10^3$ m/s, moves at an angle of 55.0° relative to the same magnetic field. Find (**a**) the strength of the magnetic field and (**b**) the magnitude of the magnetic force exerted on object 2.

Picture the Problem

Our sketch shows the charged objects and the magnetic field lines. Notice that $q_1 = 3.60 \ \mu C$ moves at right angles to $\vec{B}$. The charge $q_2 = 53.0 \ \mu C$ moves at an angle of 55.0° with respect to $\vec{B}$. The magnetic force also depends on the speeds of the objects: $v_1 = 862$ m/s and $v_2 = 1.30 \times 10^3$ m/s.

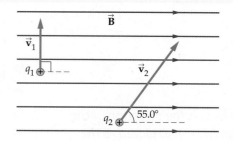

Strategy

(**a**) Find the strength of the magnetic field using the information given for object 1. Because the object moves at right angles to the magnetic field, the angle is $\theta_1 = 90°$. Therefore, the magnetic force object 1 experiences is $F_1 = |q_1| v_1 B \sin \theta_1 = q_1 v_1 B$. Since we know F_1, q_1, and v_1, we can solve for B. (**b**) Once we have determined B in part (a), we can calculate the magnetic force on object 2 using $F_2 = |q_2| v_2 B \sin \theta_2$.

Known

$q_1 = 3.60 \ \mu C$	$q_2 = 53.0 \ \mu C$
$v_1 = 862$ m/s	$v_2 = 1.30 \times 10^3$ m/s
$\theta_1 = 90°$	$\theta_2 = 55.0°$
$F_1 = 4.25 \times 10^{-3}$ N	

Unknown

(**a**) $B = ?$ (**b**) $F_2 = ?$

Solution

Part (a)

1 Set $\theta_1 = 90°$ in $F_1 = |q_1| v_1 B \sin \theta_1$ and solve for the magnetic field, B:

$$F_1 = |q_1| v_1 B \sin 90°$$
$$= |q_1| v_1 B$$
$$B = \frac{F_1}{|q_1| v_1}$$

> **Math HELP**
> Solving Simple Equations
> **See Math Review, Section III**

2 Substitute the numerical values and calculate B:

$$B = \frac{4.25 \times 10^{-3} \text{ N}}{(3.60 \times 10^{-6} \text{ C})(862 \text{ m/s})}$$
$$= \boxed{1.37 \text{ T}}$$

Part (b)

3 Use the value of B from part (a) to find the magnetic force exerted on object 2:

$$F_2 = |q_2| v_2 B \sin \theta_2$$
$$= (53.0 \times 10^{-6} \text{ C})(1.30 \times 10^3 \text{ m/s})(1.37 \text{ T}) \sin 55.0°$$
$$= \boxed{0.0773 \text{ N}}$$

Insight

The charge and the speed of an object are not enough to determine the magnetic force. The direction of the object's motion relative to the magnetic field is needed as well.

23. [Follow-up] In Guided Example 22.7, at what angle relative to the magnetic field must object 2 move for the magnetic force it experiences to be 0.0500 N?

24. What is the acceleration of a proton moving with a speed of 9.5 m/s at right angles to a magnetic field of 1.6 T?

25. An electron moves at right angles to a magnetic field of 0.12 T. What is its speed if the force exerted on it is 8.9×10^{-15} N?

A right-hand rule gives the direction of the magnetic force

🔑 *How is the direction of the magnetic force related to the direction of the magnetic field vector, $\vec{B}$, and the velocity vector, $\vec{v}$?*

We now consider the direction of the magnetic force, which is rather interesting and unexpected. The force *does not* point in the direction of the magnetic field, $\vec{B}$, or the velocity, $\vec{v}$. The magnetic force, $\vec{F}$, points in a direction that is perpendicular to both $\vec{B}$ and $\vec{v}$.

As an example, consider the vectors $\vec{B}$ and $\vec{v}$ in **Figure 22.16 (a)**. The force on a positive charge, $\vec{F}$, is perpendicular to the plane containing $\vec{B}$ and $\vec{v}$. Thus, the force is perpendicular to both $\vec{B}$ and $\vec{v}$.

The way we determine the precise direction of $\vec{F}$ is with another right-hand rule (RHR). To be specific, the direction of $\vec{F}$ is found using the magnetic force right-hand rule:

Magnetic Force Right-Hand Rule

🔑 **To find the direction of the magnetic force on a moving positive charge, start by pointing the fingers of your right hand in the direction of the velocity, $\vec{v}$. Now curl your fingers in the direction of $\vec{B}$. Your thumb points in the direction of $\vec{F}$. If the charge is negative, the force points opposite to the direction of your thumb.**

This rule is applied in **Figure 22.16 (b)** and **(c)**. Notice that $\vec{F}$ does indeed point upward for a positive charge, as indicated.

As an additional example, **Figure 22.17** shows a uniform magnetic field, $\vec{B}$, that points into the page. An object with a positive charge moves to the right. Using the magnetic force RHR—extending our fingers to the right and then curling them into the page—we see that the force exerted on this object is upward, as indicated. If the charge is negative, the direction of $\vec{F}$ is reversed, as shown.

▼ **Figure 22.16 The magnetic force right-hand rule**
(a) The magnetic force, $\vec{F}$, is perpendicular to both the velocity, $\vec{v}$, and the magnetic field, $\vec{B}$. (The magnetic force vectors shown in this figure are for a positive charge. The force on a negative charge would be in the opposite direction.) **(b)** As the fingers of the right hand are curled from $\vec{v}$ to $\vec{B}$, the thumb points in the direction of $\vec{F}$. **(c)** An overhead view, looking down on the plane defined by the vectors $\vec{v}$ and $\vec{B}$. In this two-dimensional representation, the force vector points out of the page and is indicated by a circle with a dot inside. If the charge were negative, the force would point into the page, and the symbol indicating $\vec{F}$ would be a circle with an X inside.

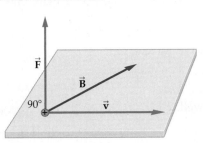

(a) $\vec{F}$ is perpendicular to both $\vec{v}$ and $\vec{B}$.

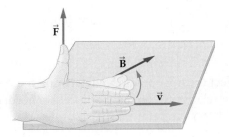

(b) Curl fingers from $\vec{v}$ to $\vec{B}$; thumb points in direction of $\vec{F}$.

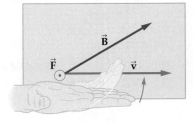

(c) Top view, looking down on $\vec{F}$

Three particles travel through a region of space where the magnetic field points out of the page, as shown below. For each of the three particles, state whether the particle's charge is positive, negative, or zero.

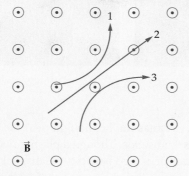

Reasoning and Discussion

The diagram below indicates the general directions of the forces required to cause the motions of particles 1 and 3. No force is required for particle 2's motion.

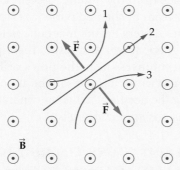

The force indicated for particle 3 is in the direction given by the magnetic force RHR. Therefore, particle 3 has a positive charge. The force acting on particle 1 is in the opposite direction from that given by the RHR, and hence that particle is negatively charged. Finally, particle 2 is undeflected. Its charge and the force acting on it are both zero.

Answer

Particle 1 has a negative charge. Particle 2 has zero charge. Particle 3 has a positive charge.

Magnetic fields deflect moving charges

The deflection of moving charges by a magnetic field is illustrated in **Figure 22.18**. Here we see a small horseshoe magnet held close to a cathode-ray tube (CRT) screen, used in many older television sets and computer monitors. The image on a CRT is produced by a beam of electrons that lights up the pixels on the screen. The magnetic field of the horseshoe magnet is deflecting the beam of electrons, resulting in a scrambled picture. A nice way to visualize how a magnetic field affects moving charges.

On a larger scale, the northern lights—or *aurora borealis*—are produced in a similar way. The story of the northern lights begins on the surface of the Sun, where the temperature is so high (about 5800 K) that atoms are ripped apart to form positively and negatively charged ions. A gas-like collection of ions is

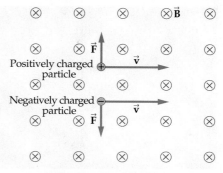

▲ Figure 22.17 **The magnetic force for positive and negative charges**
The force exerted on a negatively charged object is opposite in direction to the force exerted on a positively charged object.

▲ Figure 22.18 **Deflecting an electron beam with a magnet**
The image on this TV screen is produced by a beam of electrons that "paints" the picture on the screen by illuminating the appropriate pixels. When a magnet is held near the screen, the electrons in the beam are deflected by the magnetic force, resulting in a scrambled picture.

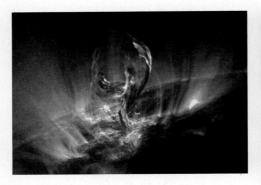

(a)

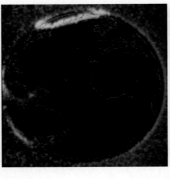

(b)

(c)

▲ **Figure 22.19 The northern lights**
(a) Plasma streams shooting outward from the Sun's surface. (b) A glowing aurora borealis (northern lights) surrounds Earth's north geographic pole in this photograph from space. (c) An auroral display seen from Earth. The characteristic red and green colors are produced by ionized nitrogen and oxygen atoms, respectively.

referred to as a **plasma**. Plasmas can be thought of as the fourth state of matter, along with gases, liquids, and solids. Though a plasma is similar to a gas, the fact that it consists of electrically charged particles means that electric and magnetic fields have a great influence on its behavior. For example, **Figure 22.19 (a)** shows streams of plasma shooting up from a storm on the surface of the Sun. The plasma follows arcing paths that trace out the magnetic field lines of the Sun.

The next step in producing the northern lights occurs when loops of plasma are shot into space from the Sun in an event known as a *coronal mass ejection*. The resulting stream of charged particles is known as the *solar wind*. When the solar wind encounters Earth's magnetic field, the charged particles are deflected by the magnetic force. As a result, these particles concentrate where the field is most intense—near the poles of Earth, as shown in **Figure 22.19 (b)**. The particles excite atoms in the atmosphere, causing them to glow and thus producing the northern lights, shown in **Figure 22.19 (c)**, as well as their southern cousins, the *aurora australis*.

Magnetic fields can cause circular motion

If the velocity of a charged object is perpendicular to a magnetic field, the result is circular motion of the object. Consider, for example, the situation shown in **Figure 22.20**. Here an object of mass m, charge $+q$, and speed v moves in a region with a constant magnetic field, $\vec{\mathbf{B}}$, pointing out of the page. Since $\vec{\mathbf{v}}$ is at right angles to $\vec{\mathbf{B}}$, the magnitude of the magnetic force is $F = |q|vB \sin 90° = |q|vB$. The force has the same magnitude at each of the points 1, 2, 3, and 4 in Figure 22.20.

Figure 22.20 also shows that the magnetic force is at right angles to the velocity (and thus always points toward the center of the circle) at every point on the object's path. This is exactly the condition required for circular motion. Recall from Chapter 9 that circular motion requires a centripetal force and that acceleration is toward the center of the circle. In this case the centripetal force is supplied by the magnetic force—in the same way that a string exerts a centripetal force on a ball being whirled in a circle.

The centripetal acceleration of an object moving with a speed v in a circle of radius r is

$$a_{cp} = \frac{v^2}{r}$$

Therefore, setting ma_{cp} equal to the magnitude of the magnetic force, $|q|vB$, yields the following condition:

$$m\frac{v^2}{r} = |q|vB$$

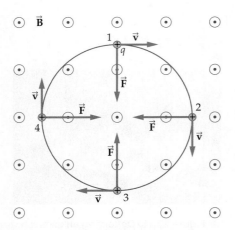

▲ **Figure 22.20 Circular motion in a magnetic field**
A charged object moves in a direction perpendicular to the magnetic field. At each point on the object's path, the magnetic force is at right angles to the velocity and hence points toward the center of a circle. This results in circular motion.

Simplifying yields

$$m\frac{v}{r} = |q|B$$

Finally, a straightforward rearrangement gives the radius of the circular path:

$$r = \frac{mv}{|q|B}$$

We can see, therefore, that the faster and the more massive the object, the larger the circle. Conversely, the stronger the magnetic field and the greater the charge, the smaller the circle.

QUICK Example 22.9 What's the Speed?

An electron moving perpendicular to a magnetic field of 4.60×10^{-3} T follows a circular path of radius 2.80 mm. What is the electron's speed?

Solution
Solving $r = mv/|q|B$ for v, we find

$$v = \frac{r|q|B}{m}$$

$$= \frac{(2.80 \times 10^{-3}\ \text{m})(1.60 \times 10^{-19}\ \text{C})(4.60 \times 10^{-3}\ \text{T})}{9.11 \times 10^{-31}\ \text{kg}}$$

$$= \boxed{2.26 \times 10^{6}\ \text{m/s}}$$

Thus, the speed of this electron is about 1% of the speed of light.

Math HELP
Scientific
Notation
See Math Review, Section I

Practice Problems

26. Find the radius of an electron's circular path when it moves perpendicular to a magnetic field of 0.45 T with a speed of 6.27×10^{5} m/s.

27. What is the magnitude of the magnetic field that is required to give an electron a circular path of radius 8.5 mm when its speed is 5.1×10^{6} m/s?

Magnetic fields can be used to separate isotopes

Physics & You: Technology A **mass spectrometer** is a device that makes use of circular motion in a magnetic field to separate *isotopes* (atoms of the same element that have different masses) and to measure atomic masses. It has many uses in medicine (anesthesiologists use it to measure respiratory gases), biology (to determine reaction mechanisms in photosynthesis), geology (to date fossils), space science (to determine the atmospheric composition of Mars), and a variety of other fields.

The basic principles of a mass spectrometer are illustrated in **Figure 22.21**. Here we see a beam of ions of mass m and charge $+q$ entering a region of constant magnetic field with a speed v. The field causes the ions to move along circular paths. Because the radius of the path depends on the mass and the charge of the ion, as described by $r = mv/|q|B$, different isotopes can be separated and identified.

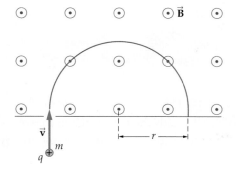

▲ **Figure 22.21 The operating principle of a mass spectrometer**
In a mass spectrometer, a beam of charged particles enters a region with a magnetic field perpendicular to the velocity. The particles then follow a circular orbit of radius $r = mv|q|B$. Particles of different mass follow different paths, allowing them to be separated from one another.

Two isotopes of uranium, U-235 and U-238, are sent into a mass spectrometer with a speed of 1.05×10^5 m/s. Given the mass of each isotope ($m_{\text{U-235}} = 3.90 \times 10^{-25}$ kg and $m_{\text{U-238}} = 3.95 \times 10^{-25}$ kg), the strength of the magnetic field ($B = 0.750$ T), and the charge of each isotope ($|q| = 1.60 \times 10^{-19}$ C), find the distance, d, between the two isotopes after they complete half of a circular path.

Picture the Problem

Our sketch shows that both isotopes enter at the same location with the same speed. Because they have different masses, however, they follow circular paths of different radii. This difference results in the separation d after half of each path.

Strategy

We begin by calculating the radius of each isotope's circular path using $r = mv/|q|B$. The masses, speeds, and magnetic field are given. The charge of the isotopes is $|q| = 1.60 \times 10^{-19}$ C. Once we know the radii, the separation between the isotopes is given by $d = 2r_{\text{U-238}} - 2r_{\text{U-235}}$, as shown.

Known	Unknown
$m_{\text{U-235}} = 3.90 \times 10^{-25}$ kg	$d = ?$
$m_{\text{U-238}} = 3.95 \times 10^{-25}$ kg	
$v = 1.05 \times 10^5$ m/s	
$|q| = 1.60 \times 10^{-19}$ C	
$B = 0.750$ T	

Solution

1 Determine the radius of the circular path of the U-235 isotope:

$$r_{\text{U-235}} = \frac{(m_{\text{U-235}})(v)}{|q|B}$$

$$= \frac{(3.90 \times 10^{-25} \text{ kg})(1.05 \times 10^5 \text{ m/s})}{(1.60 \times 10^{-19} \text{ C})(0.750 \text{ T})}$$

$$= 0.341 \text{ m}$$

Math HELP
Rules for Exponents
See Math Review, Section I

2 Determine the radius of the circular path of the U-238 isotope:

$$r_{\text{U-238}} = \frac{(m_{\text{U-238}})(v)}{|q|B}$$

$$= \frac{(3.95 \times 10^{-25} \text{ kg})(1.05 \times 10^5 \text{ m/s})}{(1.60 \times 10^{-19} \text{ C})(0.750 \text{ T})}$$

$$= 0.346 \text{ m}$$

3 Calculate the separation between the isotopes:

$$d = 2r_{\text{U-238}} - 2r_{\text{U-235}}$$

$$= 2(0.346 \text{ m} - 0.341 \text{ m})$$

$$= \boxed{0.01 \text{ m}}$$

Insight

Although the difference in masses is very small, the mass spectrometer converts this small difference into an easily measurable separation distance of 0.01 m, or 1 cm.

Practice Problems

28. Follow-up Does the separation, d, increase or decrease if the magnetic field is increased?

29. Find the radius of a proton's circular path when it moves perpendicular to a magnetic field of 0.45 T with a speed of 6.27×10^5 m/s.

30. An electron moves on a circular path with a radius of 2.1 cm and perpendicular to a magnetic field of 0.0033 T. What is the electron's speed?

Magnetic fields exert force on current-carrying wires

A charged object experiences a force when it moves across magnetic field lines. This is true whether it travels in a vacuum or inside a current-carrying wire. **Thus, a wire carrying a current in a magnetic field experiences a magnetic force that is simply the sum of all the magnetic forces experienced by the individual charges moving within it.**

Magnitude of Force on a Current-Carrying Wire For example, consider a straight wire segment of length L with a current I flowing from left to right, as shown in **Figure 22.22 (a)**. Also present is a magnetic field, $\vec{B}$, at an angle θ to the wire segment. The conducting charges move through the wire with an average speed given by

$$v = \frac{L}{\Delta t}$$

Here, Δt is the time required for the charges to move from one end of the wire to the other. The amount of charge that flows through the wire in this time is

$$q = I\Delta t$$

Therefore, the force exerted on the wire is

$$F = qvB \sin \theta$$

$$= (I\Delta t)\left(\frac{L}{\Delta t}\right)B \sin \theta$$

Canceling the time, Δt, we find that the force on a wire segment of length L with a current I at an angle θ to a magnetic field $\vec{B}$ is given by the following equation:

> **Magnetic Force on a Current-Carrying Wire**
>
> force = current × length × magnetic field × sin θ
>
> $F = ILB \sin \theta$
>
> SI unit: newton (N)

As with a single charge, the force on a wire is maximum when the current is perpendicular to the magnetic field ($\theta = 90°$) and is zero when the current is in the same direction as the magnetic field ($\theta = 0$).

Direction of Force on a Current-Carrying Wire The direction of the magnetic force on a wire is given by the same right-hand rule used earlier for single charges. Thus, to find the direction of the force in **Figure 22.22 (b)**, start by pointing the fingers of your right hand in the direction of the current, I. Now, curl your fingers in the direction of $\vec{B}$. Your thumb, which points out of the page, indicates the direction of the magnetic force, $\vec{F}$.

Of course, the current is actually caused by negatively charged electrons flowing in the direction opposite to the current. The magnetic force on these negatively charged particles moving in the opposite direction is the same as the force on positively charged particles moving in the direction of I. Thus, in all cases pertaining to current-carrying wires, you can simply think of the current as the direction in which positively charged particles move.

The magnetic force exerted on a current-carrying wire can be quite substantial. In the next Guided Example we consider the current necessary to levitate a metal rod.

How is the magnetic force on a current-carrying wire related to the forces on the individual moving charges?

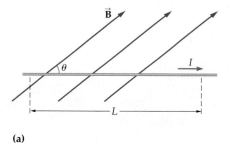

(a)

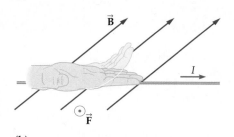

(b)

▲ Figure 22.22 **The magnetic force on a current-carrying wire**
A current-carrying wire in a magnetic field experiences a force, unless the current is parallel to the field. **(a)** For a wire segment of length L, the magnitude of the force is $F = ILB \sin \theta$. **(b)** The direction of the force is given by the magnetic force RHR; the only difference is that you start by pointing the fingers of your right hand in the direction of the current, I. In this case the force points out of the page.

A metal rod 0.150 m long and with a mass of 0.0500 kg is suspended from two thin, flexible wires, as shown in the sketch. At right angles to the rod is a uniform magnetic field of 0.550 T pointing into the page. Find (**a**) the direction and (**b**) the magnitude of the electric current needed to levitate the rod.

Picture the Problem

Our sketch shows the physical situation and indicates the direction of the current. With the current in this direction, the RHR gives an upward magnetic force, as required for the rod to be levitated.

Strategy

To find the magnitude of the current, we set the magnetic force equal in magnitude to the force of gravity. For the magnetic force we have $F = ILB \sin \theta$. In this case the field is at right angles to the current. Hence, $\theta = 90°$ and the force simplifies to $F = ILB$. Thus, setting ILB equal to mg determines the current.

Known	Unknown
$L = 0.150$ m	(**a**) direction of electric current $= ?$
$m = 0.0500$ kg	
$B = 0.550$ T	(**b**) $I = ?$

Solution

Part (a)

1 Determine the direction of *I*:

> The current I flows from left to right. To verify, point the fingers of your right hand to the right, curl them into the page, and your thumb will point upward, as desired.

Part (b)

2 Set the magnitude of the magnetic force equal to the magnitude of the force of gravity:

$$ILB = mg$$

> **Math HELP**
> Solving Simple Equations
> **See Math Review, Section III**

3 Solve for the current, *I*:

$$I = \frac{mg}{LB}$$

$$= \frac{(0.0500 \text{ kg})(9.81 \text{ m/s}^2)}{(0.150 \text{ m})(0.550 \text{ T})}$$

$$= \boxed{5.95 \text{ A}}$$

Insight

Magnetic forces, like electric forces, can easily exceed the force of gravity. In fact, gravity plays no role in the behavior of an atom—only electric and magnetic forces are important on the atomic level.

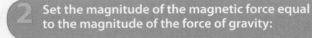

Practice Problems

31. **Follow-up** Suppose the rod is doubled in length, which also doubles its mass. Does the current needed to levitate the rod increase, decrease, or stay the same?

32. The magnetic force exerted on a 2.15-m length of wire that carries a current of 0.695 A is 0.25 N. Find the magnitude of the magnetic field, given that the wire is perpendicular to the magnetic field.

33. A wire with a current of 2.8 A is at an angle of 36.0° relative to a magnetic field of 0.88 T. Find the force exerted on a 2.25-m length of the wire.

Magnetic fields can exert a torque on current-carrying wires

The fact that a current-carrying wire experiences a force when placed in a magnetic field is one of the fundamental discoveries that makes modern applications of electric power possible. In most of these applications, including electric motors and generators, the wire is shaped into a current-carrying loop. Let's examine what happens when a simple current-carrying loop is placed in a magnetic field.

Consider a rectangular loop of height h and width w carrying a current I, as shown in **Figure 22.23**. The loop is placed in a region of space with a uniform magnetic field, $\vec{B}$, that is parallel to the plane of the loop. It is clear that the horizontal segments of the loop experience zero force because they are parallel to the field. The vertical segments, on the other hand, are perpendicular to the field. They experience forces of magnitude $F = IhB$. One of these forces is directed into the page (left side); the other points out of the page (right side). These forces tend to rotate the loop—that is, they cause a torque.

Galvanometers The torque exerted by a magnetic field finds a number of useful applications. For example, if a needle is attached to a coil, as in **Figure 22.24**, it can be used as part of a meter known as **galvanometer**, which is a device used to measure the current in a circuit. When a galvanometer is connected to an electrical circuit, its coil experiences a torque. This torque rotates the coil and deflects the needle. The deflection of the needle is directly proportional to the current in the coil—and the current in the circuit.

Electric Motors Of even greater practical importance is the fact that magnetic torque can be used to power a motor. For example, an electric current passing through the coils of a motor causes a torque that rotates the axle of the motor. As the coils rotate, a device known as the *commutator* reverses the direction of the current as the orientation of the coil reverses. This ensures that the torque is always in the same direction and that the coils continue to turn in the same direction. Electric motors, which will be discussed in greater detail in Chapter 23, are used in everything from electric razors to hybrid cars.

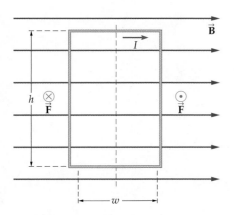

▲ Figure 22.23 **Magnetic forces on a rectangular current-carrying loop**
Only the vertical segments of the loop experience forces, and those forces tend to rotate the loop about a vertical axis.

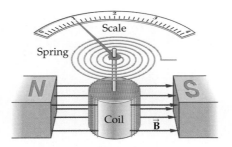

▲ Figure 22.24 **Basic elements of a galvanometer**
As current passes through the coil, a torque acts on it, causing it to rotate. The spring ensures that the angle of rotation is proportional to the current in the coil.

22.3 LessonCheck ⓂⓅ

Checking Concepts

34. 💬 **Explain** How is the force exerted on a charged object moving through a magnetic field affected by the angle of the object's velocity relative to the field?

35. 💬 **Identify** What two vectors is the magnetic force perpendicular to?

36. 💬 **Explain** Why does a current-carrying wire experience a magnetic force?

37. Big **Idea** What is the connection between moving charges and magnetic fields? What is the connection between magnetic fields and moving charges?

38. Analyze An electron moves with constant velocity through a region of space that is free of electric fields. Can one conclude that the magnetic field is zero in this region? Explain.

39. Apply An electron moves due north in a region where the magnetic field is vertically upward. Does the force on the electron point up, down, east, west, north, or south? Explain.

Solving Problems

40. Calculate An electron moves at right angles to a magnetic field of 0.12 T. What is the force exerted on the electron if its speed is 2.8×10^4 m/s?

41. Calculate A wire carrying a current of 5.3 A is at an angle of 45° relative to a magnetic field of 0.68 T. What is the force exerted on a 1.5-m length of the wire?

42. Calculate A proton with a kinetic energy of 4.9×10^{-16} J moves perpendicular to a magnetic field of 0.26 T. What is the radius of its circular path?

43. Calculate A 0.45-m copper rod with a mass of 0.17 kg carries a current of 11 A in the positive x direction. What are the magnitude and the direction of the minimum magnetic field needed to levitate the rod?

Physics & You

The world's largest superconducting solenoid magnet is used in the Large Hadron Collider.

Particle Accelerators

What Is It? A *particle accelerator* is a device that uses strong electromagnetic fields to accelerate charged particles to very high speeds while containing them in a beam. Accelerators can be linear or circular in shape. The Stanford Linear Accelerator, the world's largest linear device, is about 3 km long.

Who Invented It? Rolf Wideröe (1902–1996), a Norwegian physicist, originated many of the concepts behind particle acceleration. America physicist Ernest Lawrence (1901–1958) constructed the earliest circular accelerator, called the *cyclotron*, at the University of California, Berkeley, in the 1930s. The Large Hadron Collider (LHC) built by the European Organization for Nuclear Research (CERN) is the most current advancement of this technology (see pages 12–13 for an aerial view of the LHC). There are currently more than 25,000 particle accelerators throughout the world.

Why Is It Important? The particle accelerator is an essential tool for particle physicists. By colliding high-energy particles and observing the results, researchers are able to gain a better understanding of subatomic structures and the laws governing their behavior. These researchers hope to answer fundamental questions such as why particles have mass, what the early universe was like, and what dark matter is. Particle accelerators also have important industrial and medical uses.

How Does It Work? A particle accelerator uses thousands of superconducting magnets to produce an electromagnetic field that accelerates and directs a beam of charged particles. When the particles reach the correct velocity, they are guided to collide with a fixed target particle, or with another particle beam. The end products of the collision are recorded by sophisticated detectors and then analyzed with computers.

Take It Further

1. Research *Medical isotopes, like technetium-99m and molybdenum-99, are traditionally produced in nuclear reactors. There is currently a shortage of such isotopes due to the shutting down of many nuclear reactors. Examine the benefit to society of using particle accelerators to create isotopes.*

2. Research *The operation of the LHC sparked end-of-the-world scenarios. Research the possibility that the LHC could create a black hole. Write a short essay on your findings and on the importance of balancing scientific research and public safety.*

Physics Lab Mapping Magnetic Fields

In this lab you will use a magnetic compass to map the magnetic lines of force of bar magnets. You will map these lines for a single bar magnet and for two bar magnets with like and unlike poles facing each other.

Materials
- two bar magnets
- magnetic compass
- twelve sheets of paper
- tape

Procedure

1. Tape four sheets of paper together in a two-by-two grid, and then tape the paper grid to the table.

2. Use the compass to determine the north-south orientation of your lab station. Make sure that no bar magnets or other pieces of iron are anywhere near the compass. Draw a straight line and label the north-south direction of Earth's magnetic field.

3. Place a single bar magnet on the sheet of paper with the magnet's north pole directed exactly to the south.

4. Draw an outline of the magnet on the paper, labeling the magnet's north and south poles.

5. Place the compass near the north pole of the magnet. Mark dots on the paper (as close as possible to each end of the needle) that indicate the north-south orientation of the compass needle.

6. Carefully slide the compass in a straight line in the direction of its north pole. Stop sliding when the compass's south pole is directly above the dot previously marked on the paper for the north pole. Mark a dot at the north pole of the compass (as close as possible to the end of the needle) in its new position.

7. Repeat Step 6 until the compass reaches either the south pole of the magnet or the edge of the paper.

8. Draw a smooth curve through the dots. Add arrowheads at several points along this curve to indicate the direction of the magnetic force.

9. Repeat Steps 5–8 until you have drawn five lines of force on each side of the magnet. Note that you need to use a different starting point for each trial.

10. Repeat the mapping procedure described in Steps 3–9 with two bar magnets on the north-south line. Make sure that the magnets' ends are separated by 10 cm and that one magnet's north pole faces the other magnet's south pole.

11. Repeat the mapping procedure described in Steps 3–9 with two bar magnets on the north-south line. Make sure that the magnets' ends are separated by 10 cm and that their north poles face each other.

Analysis

1. Examine your drawings. Where are the magnetic field lines the densest?

2. Describe the magnetic lines of force observed in Steps 10 and 11.

Conclusions

1. What does the spacing of the lines of force reveal about the field?

2. How would the pattern of field lines look if two bar magnets were placed parallel to each other with their unlike poles directly across from each other?

22 Study Guide

Big Idea

Moving charges produce magnetic fields, and magnetic fields exert forces on moving charges.

As a result, electric currents in wires produce magnetic fields, and current-carrying wires experience forces in magnetic fields. Magnetic fields are similar to electric fields in that they contain energy and exert forces, but are different in that magnetic field lines form closed loops that do not start or stop anywhere.

22.1 Magnets and Magnetic Fields

The end of a freely rotating bar magnet that points toward Earth's north geographic pole is the north-seeking pole, or simply the north pole. The opposite end of the magnet is the south-seeking pole, or simply the south pole.

The direction of a magnetic field at a given location is defined as the direction a compass needle would point if placed at that location.

The *north* geographic pole of the Earth is actually near the *south* pole of Earth's magnetic field.

• Breaking a magnet in half produces two new poles on either side of the break.

• Magnetic fields can be represented with lines in much the same way as electric fields. In particular, the more closely spaced the lines, the more intense the field. Magnetic field lines point away from north poles and toward south poles, and they always form closed loops.

• Earth produces its own magnetic field. The magnetic axis is inclined at an angle of about 11.5° with Earth's rotational axis.

22.2 Magnetism and Electric Currents

To find the direction of the magnetic field due to a current-carrying wire, point the thumb of your right hand along the wire in the direction of the current, I. Your fingers will then curl around the wire in the direction of the magnetic field.

The magnetic field produced by a current in a wire is proportional to the current and inversely proportional to the distance from the wire.

Wires with currents in the same direction experience an attractive force; wires with currents in opposite directions experience a repulsive force.

• The magnetic field inside a solenoid is intense and nearly uniform.

Key Equations

The magnitude of the magnetic field at a distance r from a long, straight wire carrying a current I is

$$B = \frac{\mu_0 I}{2\pi r}$$

The magnetic field inside a solenoid with N loops and a length L that is carrying a current I is given by

$$B = \mu_0 \left(\frac{N}{L}\right) I$$

22.3 The Magnetic Force

Maximum magnetic force is exerted when a charged object moves at right angles to the magnetic field, so $\theta = 90°$ and $\sin \theta = 1$.

To find the direction of the magnetic force on a moving positive charge, start by pointing the fingers of your right hand in the direction of the velocity, $\vec{v}$. Now curl your fingers in the direction of $\vec{B}$. Your thumb points in the direction of $\vec{F}$. If the charge is negative, the force points opposite to the direction of your thumb.

A wire carrying a current in a magnetic field experiences a magnetic force that is simply the sum of all the magnetic forces experienced by the individual charges moving within it.

• In order for a magnetic field to exert a force on an object, the object must have charge and must be moving.

• If a charged particle moves perpendicular to a magnetic field, it travels with constant speed in a circle.

• A current-carrying loop placed in a magnetic field experiences a torque.

Key Equations

The magnitude of the magnetic force is

$$F = |q|vB \sin \theta$$

The radius of the circular path for a charged object moving perpendicular to a magnetic field is

$$r = \frac{mv}{|q|B}$$

The magnetic force exerted on a wire of length L carrying a current I at an angle θ to a magnetic field of magnitude B is

$$F = ILB \sin \theta$$

22 Assessment

For instructor-assigned homework, go to www.masteringphysics.com

ANSWERS TO SELECTED ODD-NUMBERED PROBLEMS APPEAR IN APPENDIX A.

Lesson by Lesson

22.1 Magnets and Magnetic Fields
Conceptual Questions

44. In which direction does Earth's magnetic field point near the north geographic pole?

45. Why is it not possible for magnetic field lines to cross?

46. Do magnetic field lines start and stop at magnetic poles? Explain.

47. A piece of iron is placed in a magnetic field. Which magnetic domains in the iron grow in size as a result?

48. How does magnetism provide evidence in favor of continental drift and seafloor spreading?

49. Why does only one side of a flexible refrigerator magnet stick to the refrigerator?

50. If you suspend a bar magnet from a string, you will find that its north pole points toward the geographic north pole of Earth. How can this be, given that like poles repel one another?

51. Region 1 has magnetic field lines that are closely spaced; region 2 has magnetic field lines that are widely spaced. In which region is the magnitude of the magnetic field larger? Explain.

22.2 Magnetism and Electric Currents
Conceptual Questions

52. A loop of wire is connected to the terminals of a battery, as indicated in **Figure 22.25**. For the loop to attract the bar magnet, which of the terminals, A or B, should be the positive terminal of the battery? Explain.

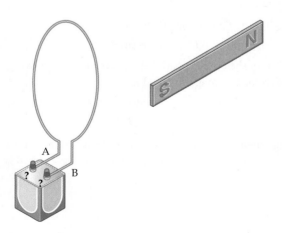

Figure 22.25

53. **Predict & Explain** The number of loops in a solenoid is doubled, and at the same time its length is doubled. **(a)** Does the magnetic field within the solenoid increase, decrease, or stay the same? **(b)** Choose the *best* explanation from among the following:

A. Doubling the number of loops in a solenoid doubles its magnetic field, and hence the field increases.

B. Making a solenoid longer decreases its magnetic field, and therefore the field decreases.

C. The magnetic field remains the same because the number of loops per length is unchanged.

54. The circular wire loop shown in **Figure 22.26** produces a magnetic field at its center that points out of the page. Is the current in the wire flowing clockwise or counterclockwise? Explain.

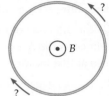

Figure 22.26

55. The current in a long, straight wire flows directly toward you. Does the magnetic field produced by the current circulate about the wire in the clockwise or counterclockwise direction?

56. A long, straight wire carries a current I_0. At a distance of r_0 from the wire, the magnetic field produced by the wire is 1 T. What is the magnetic field produced by the following wires?

Wire	Current	Distance
1	$2I_0$	r_0
2	I_0	$2r_0$
3	$4I_0$	$4r_0$

57. The four wires shown in **Figure 22.27** are long and straight. Each wire carries a current of the same magnitude, I. The currents in wires 1, 2, and 3 are flowing out of the page; the current in wire 4 is flowing into the page. What is the direction of the total magnetic field produced by the four wires at the center of the square?

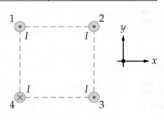

Figure 22.27

Problem Solving

58. What is the magnitude of the magnetic field 0.22 m from a wire carrying a current of 5.6 A?

59. A straight wire produces a magnetic field of 6.1×10^{-6} T at a distance of 26 cm. What is the current in the wire?

60. The current in a straight wire is 8.6 A. At what distance is the magnetic field produced by the wire equal to 3.3×10^{-5} T?

810 Chapter 22 · Assessment

61. You would like a solenoid that is 32 cm long to produce a magnetic field of 0.78 T when it carries a current of 7.7 A. How many loops should this solenoid have?

62. A solenoid has 1200 loops per meter and carries a current of 6.8 A. What is the magnetic field inside the solenoid?

63. In Oersted's experiment, suppose the compass was 0.25 m from the current-carrying wire. If a magnetic field half as strong as Earth's magnetic field of 5.0×10^{-5} T was required to give a noticeable deflection of the compass needle, what current must the wire have carried?

64. A wire carries a current of 3.6 A. At what distance is the magnetic field from this wire equal to 3.5×10^{-5} T?

65. **Pacemaker Switches** Some pacemakers employ magnetic reed switches to enable doctors to change their mode of operation without surgery. A typical reed switch can be switched from one position to another with a magnetic field of 5.0×10^{-4} T. What current must a wire carry if it is to produce a field of this magnitude at a distance of 0.50 m?

66. ⎡Think & Calculate⎤ Consider the long, straight, current-carrying wires shown in **Figure 22.28**. One wire carries a current of 6.2 A in the positive y direction; the other wire carries a current of 4.5 A in the positive x direction. **(a)** At which of the two points, A or B, do you expect the magnitude of the total magnetic field to be greater? Explain. **(b)** Calculate the magnitude of the total magnetic field at points A and B.

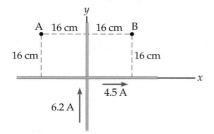

Figure 22.28

67. Two current-carrying wires are parallel to one another and have a separation of 25 cm. If the wires carry opposite currents of 11 A each, what is the magnitude of the total magnetic field midway between them?

68. Two long, straight wires are separated by a distance of 12.2 cm and carry parallel currents. One wire carries a current of 2.75 A; the other carries a current of 4.33 A. What is the magnitude of the total magnetic field midway between the two wires?

69. To construct a solenoid, you wrap a wire uniformly around a plastic tube 12 cm in diameter and 55 cm in length. You would like a 2.0-A current to produce a 0.25-T magnetic field inside your solenoid. What is the total length of wire you will need to meet these specifications?

70. ⎡Challenge⎤ Two long, straight wires are oriented perpendicular to the page, as shown in **Figure 22.29**. The current in one wire is $I_1 = 3.0$ A, flowing into the page, and the current in the other wire is $I_2 = 4.0$ A,

flowing out of the page. Find the magnitude and direction of the total magnetic field at point P.

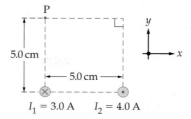

Figure 22.29

22.3 The Magnetic Force
Conceptual Questions

71. Two charged particles move at right angles to a magnetic field and deflect in opposite directions. Can one conclude that the particles have opposite charges? Explain.

72. A current-carrying wire is placed in a region with a uniform magnetic field. The wire experiences zero magnetic force. Explain how this can happen.

73. ⎡Predict & Explain⎤ Proton 1 moves with a speed v from the east coast toward the west coast in the continental United States; proton 2 moves with the same speed from the southern United States toward Canada. **(a)** Is the magnitude of the magnetic force due to Earth's magnetic field experienced by proton 2 greater than, less than, or equal to the force experienced by proton 1? **(b)** Choose the *best* explanation from among the following:

A. The protons experience the same force because the magnetic field is the same and their speeds are the same.

B. Proton 1 experiences the greater force because it moves at right angles to the magnetic field.

C. Proton 2 experiences the greater force because it moves in the same direction as the magnetic field.

74. An electron moves west to east across the continental United States. Does the magnetic force experienced by the electron point in a direction that is generally north, south, east, west, upward, or downward? Explain.

75. An electron moving in the positive x direction, at right angles to a magnetic field, experiences a magnetic force in the positive y direction. What is the direction of the magnetic field?

76. ⎡Rank⎤ Suppose particles A, B, and C in **Figure 22.30** have identical masses and charges of the same magnitude. Rank the particles in order of increasing speed. Indicate ties where appropriate.

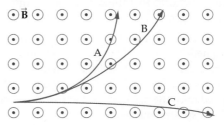

Figure 22.30

77. Referring to Figure 22.30, what is the sign of the charge for each of the three particles? Explain.

78. When the switch is closed in the circuit shown in **Figure 22.31**, the wire between the poles of the horseshoe magnet deflects downward. Is the left end of the horseshoe magnet a north magnetic pole or a south magnetic pole? Explain.

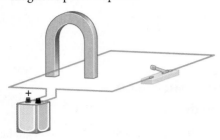

Figure 22.31

Problem Solving

79. The magnetic force exerted on a 1.2-m segment of straight wire is 1.6 N. The wire carries a current of 3.0 A in a region with a constant magnetic field of 0.50 T. What is the angle between the wire and the magnetic field?

80. A proton high above the equator approaches Earth moving straight downward with a speed of 350 m/s. Find the magnitude of the magnetic force exerted on the proton, given that the magnetic field at its location is horizontal and has a magnitude of 4.1×10^{-5} T.

81. A 0.32-μC object moves with a speed of 16 m/s through a region where a magnetic field has a strength of 0.95 T. At what angle to the field is the object moving if the magnetic force exerted on it is 4.8×10^{-6} N?

82. A 2.7-m-long wire with a mass of 0.75 kg is in a region with a horizontal magnetic field of 0.84 T. What is the minimum current needed to levitate the wire?

83. A 12.5-μC object with a mass of 2.80×10^{-5} kg moves perpendicular to a 1.01-T magnetic field in a circular path of radius 26.8 m. How fast is the object moving?

84. An object with a charge of 14 μC experiences a force of 2.2×10^{-4} N when it moves at right angles to a magnetic field with a speed of 27 m/s. What force does this object experience when it moves with a speed of 6.3 m/s at an angle of 25° relative to the magnetic field?

85. A horizontal power line 270 m in length carries a current of 110 A. A second power line produces a magnetic field of magnitude 8.8×10^{-5} T pointing upward at the location of the first power line. What is the magnitude of the force that the second power line exerts on the first?

86. Charged particles pass through a *velocity selector* with electric and magnetic fields at right angles to each other, as shown in **Figure 22.32**. If the electric field has a magnitude of 450 N/C and the magnetic field has a magnitude of 0.18 T, what speed must the particles have to pass through the selector while experiencing zero total force?

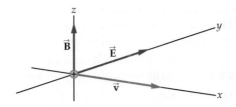

Figure 22.32

87. An electron moving with a speed of 9.1×10^5 m/s in the positive x direction experiences zero magnetic force. When it moves in the positive y direction, it experiences a force of 2.0×10^{-13} N that points in the negative z direction. What are the direction and the magnitude of the magnetic field?

88. **Think & Calculate** When a charged particle enters a region of uniform magnetic field, it follows a circular path, as indicated in **Figure 22.33**. **(a)** Is this particle positively or negatively charged? Explain. **(b)** Suppose the magnetic field has a magnitude of 0.180 T, the particle's speed is 6.0×10^6 m/s, and the radius of its path is 52.0 cm. Find the mass of the particle, given that its charge has a magnitude of 1.60×10^{-19} C. Give your result in atomic mass units (u), where $1\,u = 1.67 \times 10^{-27}$ kg.

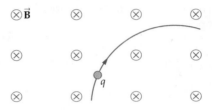

Figure 22.33

Mixed Review

89. At a point near the equator, Earth's magnetic field is horizontal and points to the north. If an electron is moving vertically upward at this point, does the magnetic force acting on it point north, south, east, west, upward, or downward? Explain.

90. **Figure 22.34** shows an electron beam whose initial direction of motion is horizontal, from right to left. A magnetic field deflects the beam downward. What is the direction of the magnetic field?

Figure 22.34

91. Brain Function Experiments have shown that thought processes in the brain can be affected if the parietal lobe is exposed to a magnetic field with a strength of 1.0 T. How much current must a long, straight wire carry if it is to produce a 1.0-T magnetic field at a distance of 0.50 m? (For comparison, a typical lightning bolt carries a current of about 20,000 A, which would melt most wires.)

92. Superconducting Solenoid A company advertises a high-field, superconducting solenoid that produces a magnetic field of 17 T with a current of 105 A. What is the number of loops per meter in this solenoid?

93. Rank A proton moves with constant speed along the path shown in **Figure 22.35**. As it moves, it passes through three regions with different uniform magnetic fields, B_1, B_2, and B_3. In each region the proton completes a half-circle, and the magnetic field is perpendicular to the page. **(a)** Rank the three magnetic fields in order of increasing magnitude. Indicate ties where appropriate. **(b)** Give the direction (into or out of the page) for each of the magnetic fields.

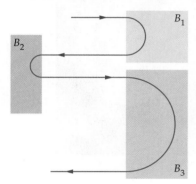

Figure 22.35

94. Predict & Explain Suppose the initial speed of the proton in Figure 22.35 is increased. **(a)** Does the radius of each half-circular path segment increase, decrease, or stay the same? **(b)** Choose the *best* explanation from among the following:

A. The radius of a circular path in a magnetic field is proportional to the speed of the particle; therefore, the radius of the proton's half-circular path segments will increase.

B. A greater speed means that the proton will experience more force from the magnetic field, resulting in a decrease in the radius.

C. The increase in speed offsets the increase in magnetic force, resulting in no change of the radius.

95. Credit-Card Data Experiments carried out on a TV show determined that a magnetic field of 1000 G is needed to corrupt the information on a credit card's magnetic strip. (The show also showed that a credit card cannot be demagnetized by an electric eel or an eel skin

wallet.) Suppose a long, straight wire carries a current of 2.5 A. How close can a credit card be held to this wire without damaging its magnetic strip?

96. A current-carrying circular loop of radius R is placed next to a long, straight wire, as shown in **Figure 22.36**. The current in the wire flows toward the right and is of magnitude I. In which direction (clockwise or counterclockwise) must current flow in the loop to have the magnetic field be zero at its center? Explain.

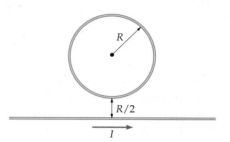

Figure 22.36

97. High above the surface of Earth, charged particles (such as electrons and protons) can become trapped in Earth's magnetic field in regions known as *Van Allen belts*. A typical electron in a Van Allen belt has a speed of 1.1×10^8 m/s and travels in a roughly circular orbit with an average radius of 220 m. What is the magnitude of Earth's magnetic field where such an electron orbits, assuming that the field is perpendicular to the plane of the orbit?

98. Lightning Bolts A powerful bolt of lightning can carry a current of 225 kA. Treating such a lightning bolt as a long, straight wire, calculate the magnitude of the magnetic field produced by the bolt at a distance of 35 m.

99. Rank A positively charged particle moves through a region with a uniform electric field pointing toward the top of the page and a uniform magnetic field pointing into the page. The particle can have one of the four velocities shown in **Figure 22.37**. **(a)** Rank the four possibilities in order of increasing magnitude of the total force the particle experiences. Indicate ties where appropriate. **(b)** Which of the four velocities could potentially result in zero total force?

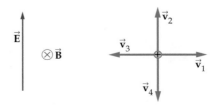

Figure 22.37

Writing about Science

100. Write a report on Earth's magnetic field. How is it produced? What is its strength at various locations on the planet? How often does it reverse direction? Where can you find a region with a reversed magnetic field today? What might be the consequences if Earth's magnetic field dropped to zero for a short time as it reversed direction?

101. Connect to the Big Idea Discuss similarities and differences between the force exerted by an electric field and that exerted by a magnetic field. Also, discuss similarities and differences in how an electric charge produces an electric field and a magnetic field.

Read, Reason, and Respond

Magnetoencephalography To read and understand this sentence, your brain must process visual input from your eyes and translate it into words and thoughts. As you do so, minute electric currents flow through the neurons in your visual cortex. These currents, like any electric current, produce magnetic fields. In fact, even your innermost thoughts and dreams produce magnetic fields that can be detected outside your head.

Magnetoencephalography (MEG) is the study of magnetic fields produced by electrical activity in the brain. Though completely noninvasive, as shown in **Figure 22.38 (a)**, MEG can provide detailed information about spontaneous brain function—like alpha waves and pathological epileptic spikes—as well as brain activity that is evoked by visual, auditory, and tactile stimuli.

The magnetic fields produced by brain activity are incredibly weak—roughly 100 million times smaller than Earth's magnetic field. Even so, sensitive detectors called *SQUIDs* (superconducting quantum interference devices), which were invented by physicists for their research, can detect fields as small as 1.0×10^{-15} T. Coupled with sophisticated electronics and software, and operating at a liquid-helium temperature ($-269\ °C$), a SQUID can localize the source of brain activity to within millimeters. When the results from MEG are overlaid with the anatomical data from an MRI scan, a richly detailed "map" of the electrical activity within the brain is produced, as shown in **Figure 22.38 (b)**.

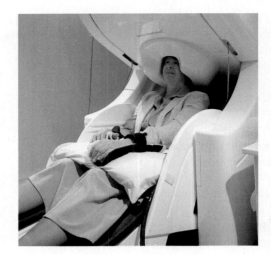

(a)

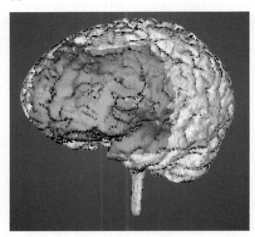

(b)

Figure 22.38 (a) A magnetoencephalograph (MEG) is made by measuring the magnetic fields produced by the brain. It is a completely noninvasive process. **(b)** The result is a map of the electrical activity within the brain.

102. Approximating a neuron by a straight wire, what electric current is needed to produce a magnetic field of 1.0×10^{-15} T at a distance of 5.0 cm?

A. 4.0×10^{-22} A   C. 2.5×10^{-10} A

B. 7.9×10^{-11} A   D. 1.0×10^{-7} A

103. Suppose a neuron in the brain carries a current of 5.0×10^{-8} A. Treating the neuron as a straight wire, what is the magnetic field it produces at a distance of 7.5 cm?

A. 1.3×10^{-13} T   C. 1.1×10^{-10} T

B. 4.2×10^{-12} T   D. 3.3×10^{-7} T

104. A given neuron in the brain carries a current of 3.1×10^{-8} A. If the SQUID detects a magnetic field of 2.8×10^{-14} T, how far away is the neuron? Treat the neuron as a straight wire.

A. 22 cm   C. 140 cm

B. 70 cm   D. 176 cm

Standardized Test Prep

Multiple Choice

Use the diagram below to answer Questions 1 and 2.
The diagram shows an electron moving with a velocity of 2.0×10^5 m/s in a region with a uniform magnetic field of magnitude $B = 200$ T.

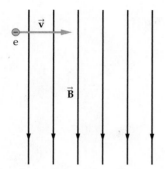

1. What are the magnitude and the direction of the magnetic force on the electron as it first enters the field?
   - **(A)** 450 N, out of the page
   - **(B)** 6.4×10^{-12} N, out of the page
   - **(C)** 450 N, into the page
   - **(D)** 6.4×10^{-12} N, into the page

2. Suppose the electron moves through the magnetic field region at an angle (toward the upper right) instead. Which of the following *best* describes the magnetic force on the electron in this case?
   - **(A)** The magnitude of the magnetic force is the same as before.
   - **(B)** The direction of the magnetic force is the same as before.
   - **(C)** The direction of the magnetic force is opposite to what it was before.
   - **(D)** There is no magnetic force on the electron.

3. Which of the following correctly describes the magnetic force exerted on a proton moving in the same direction as a magnetic field line?
   - **(A)** The force is at a maximum.
   - **(B)** The force is perpendicular to the proton's motion.
   - **(C)** The force is in the same direction as the proton's motion.
   - **(D)** The force is zero.

4. In which case is the force exerted on the charged particle the greatest?
   - **(A)** A proton moves at 60,000 m/s through an electric field with a magnitude of 400 N/m.
   - **(B)** An electron is stationary in an electric field with a magnitude of 800 N/m.
   - **(C)** A proton moves at 60,000 m/s perpendicular to a magnetic field with a magnitude of 800 T.
   - **(D)** A proton is stationary in a magnetic field with a magnitude of 800 T.

Use the diagram below to answer Questions 5–7. The diagram shows two long, parallel wires each carrying a current of 0.5 A flowing toward the right. Point P is located 0.2 cm above the top wire and 0.4 cm above the bottom wire.

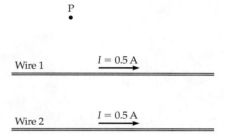

5. What is the direction of the magnetic field at point P due to the top wire?
   - **(A)** into the page
   - **(B)** out of the page
   - **(C)** toward the right
   - **(D)** toward the left

6. What are the magnitude (per meter of length) and the direction of the force that the top wire exerts on the bottom wire?
   - **(A)** 1.2×10^{-5} N, upward
   - **(B)** 1.2×10^{-5} N, downward
   - **(C)** 2.5×10^{-5} N, upward
   - **(D)** 2.5×10^{-5} N, downward

Extended Response

7. **(a)** Determine the total magnetic field at point P due to the two wires. **(b)** How does the total magnetic field change if the current in the top wire is reversed?

> **Test-Taking Hint**
>
> A charged particle must be moving in order for a magnetic field to exert a force on it.

If You Had Difficulty With . . .

Question	1	2	3	4	5	6	7						
See Lesson	22.3	22.3	22.3	22.2	22.3	22.2	22.3						

23 Electromagnetic Induction

Inside

When the steel strings of a bass guitar vibrate, they create a changing magnetic field. This produces an oscillating electric current that goes to the speakers. None of this would be possible without the phenomenon of electromagnetic induction.

Big Idea

Changing magnetic fields produce electric fields, and the electric fields can be used to generate electric currents.

Mention an electric guitar, and everyone knows what you mean. Talk about a "magnetic guitar," however, and you'll get some puzzled looks. Yet it could be argued that the second term is just as appropriate as the first. Electric guitars use devices known as *pickups* to convert vibrations into electrical signals that can be amplified. If the guitar's vibrating steel strings did not change the magnetic fields of the pickups, there would be nothing to be "picked up"—no signal to play through the speakers.

Inquiry Lab

What are the components of a simple electric motor?

Explore

1. Obtain a battery, a small nail, a strong disc magnet, and a length of wire.
2. Magnetically attach the head of the nail to the magnet.
3. Hold the battery in one hand with its positive terminal pointing downward. Now bring the tip of the nail in contact with the positive terminal. The nail and magnet should hang from the positive terminal of the battery.
4. Hold one end of the wire to the negative terminal of the battery while touching the other end of the wire to the side of the magnet. Observe what happens.

Think

1. **Observe** Describe the motion of the magnet when the wire makes contact with the battery and the magnet.
2. **Identify** What forces act on the magnet and in which directions do they act? Draw a force diagram.
3. **Predict** How do you think the results would change if two magnets were used instead of one? Explain your reasoning. If time permits, try it and see!

23.1 Electricity from Magnetism

When Hans Oersted observed that an electric current produces a magnetic field, it was pure serendipity. In contrast, Michael Faraday (1791–1867), an English chemist and physicist, was aware of Oersted's results, and purposefully set out to see if a magnetic field could produce an electric field. His ingenious experiments showed that such a connection does exist.

Vocabulary

- electromagnetic induction
- magnetic flux

A changing magnetic field induces an electric current

A boat has to change its location to produce a wave. Even a large boat produces no waves as long as it remains at rest. Something similar happens with magnetic fields. A *changing* magnetic field produces an electric current, but a magnetic field that doesn't change has no such effect. Faraday set out to study this type of behavior.

Faraday's Experiment Figure 23.1 shows a simplified version of Faraday's experiment. Two electric circuits are involved. The first, called the *primary circuit*, consists of a battery, a switch, a resistor, and a wire coil wrapped around an iron bar. When the switch is closed on the primary circuit, a current flows through the coil, producing a strong magnetic field in the iron bar.

▶ **Figure 23.1 Magnetic induction**
Basic setup of Faraday's experiment on magnetic induction. When the position of the switch on the primary circuit is changed from open to closed or from closed to open, an electromotive force (emf) is induced in the secondary circuit. The induced emf causes a current in the secondary circuit, which is detected by the ammeter. There is no induced current in the secondary circuit if the current in the primary circuit doesn't change.

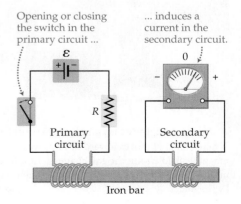

Opening or closing the switch in the primary circuit ...

... induces a current in the secondary circuit.

Primary circuit

Secondary circuit

Iron bar

The *secondary circuit* also has a wire coil wrapped around the same iron bar, and this coil is connected to an ammeter that detects any current in the circuit. There is no battery in the secondary circuit, and no direct physical contact between the two circuits. What does link the circuits, instead, is the magnetic field in the iron bar. The bar ensures that the field experienced by the secondary coil is approximately the same as the field produced by the primary coil.

Now, let's take a look at the experimental results. When the switch is closed on the primary circuit, the magnetic field in the iron bar rises from zero to some finite amount, and the ammeter in the secondary coil deflects to one side briefly, and then returns to zero. As long as the current in the primary circuit is maintained at a constant value, the ammeter in the secondary circuit gives a zero reading. If the switch on the primary circuit is then opened, so the magnetic field drops again to zero, the ammeter in the secondary circuit deflects briefly in the *opposite* direction, and then returns to zero. These observations can be summarized as follows:

- The current in the secondary circuit is zero as long as the magnetic field in the iron bar is constant. It does not matter whether the constant value of the magnetic field is zero or nonzero.

- When the magnetic field in the secondary coil increases, a current is observed to flow in one direction in the secondary circuit. When the magnetic field in the secondary coil decreases, a current is observed to flow in the opposite direction.

Induced emf It is important to note that the current in the secondary circuit appears without any physical contact between the primary and secondary circuits. For this reason, the current in the secondary circuit is referred to as an *induced current*. The process of inducing an electric current in a circuit by using a changing magnetic field is known as **electromagnetic induction**.

Because an induced current behaves the same as a current produced by an electromotive force (emf) supplied by a battery, we say that the changing magnetic field creates an *induced emf* in the secondary circuit. As far as the circuit is concerned, the changing magnetic field has the same effect as a battery with a certain emf (voltage). **Faraday observed that the magnitude of the induced emf is proportional to the *rate of change* of the magnetic field—the more rapidly the magnetic field changes, the greater the induced emf.**

In Faraday's experiment the changing magnetic field is caused by a changing current in the primary circuit. Any means of changing the magnetic field is just as effective, however. For example, **Figure 23.2** shows a common classroom demonstration of induced emf. In this case there is no primary circuit; instead, the magnetic field is changed by simply moving a bar magnet toward or away from a coil connected to an ammeter. When the magnet is moved toward the coil, the meter deflects in one direction; when it is pulled away from the coil, the meter deflects in the opposite direction. There is no induced emf (and no induced current) when the magnet is held still—just as a boat produces no waves when it is at rest.

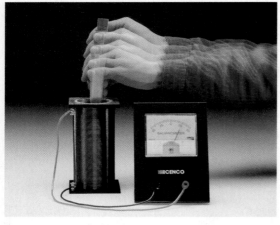

▲ **Figure 23.2 Inducing an electric current**
When a magnet moves toward or away from a coil, the magnetic field within the coil changes. This, in turn, induces an electric current in the coil, which is detected by the meter.

Induced emf is related to magnetic flux

Understanding electromagnetic induction requires a new concept—*magnetic flux*. We care about magnetic flux because a changing magnetic flux is what induces an electric current.

Defining Magnetic Flux The word *flux* basically means "flow." For example, the flux, or flow, of air through a window is related to the direction of the wind and the cross-sectional area of the window. If wind blows straight through an open window, the flux is high. The larger the window, the greater the flux. If wind blows parallel to a window, no air passes through it at all. In this case the flux is zero, no matter how large the window.

Similarly, **magnetic flux** is a measure of the number of magnetic field lines that pass through a given area. A magnetic field perpendicular to a surface gives a high flux, and the larger the surface area, the greater the flux. A magnetic field parallel to a surface gives zero flux.

Calculating Magnetic Flux Suppose a magnetic field, $\vec{B}$, crosses a surface area, A, at right angles, as in **Figure 23.3 (a)**. The magnetic flux, Φ, in this case is simply the magnitude of the magnetic field times the area:

$$\Phi = BA$$

If, on the other hand, the magnetic field is parallel to the surface—like wind blowing parallel to an open window—then *no* field lines cross through the surface. As **Figure 23.3 (b)** shows, the magnetic flux in this case is zero:

$$\Phi = 0$$

In general, only the component of $\vec{B}$ that is *perpendicular* to a surface contributes to the magnetic flux. The magnetic field in **Figure 23.3 (c)**, for example, crosses the surface at an angle θ relative to the normal, and hence its perpendicular component is $B \cos \theta$. The magnetic flux, then, is simply $B \cos \theta$ times the area, A:

> **Definition of Magnetic Flux, Φ**
>
> magnetic flux = magnitude of magnetic field $\times$ area $\times \cos \theta$
>
> $$\Phi = BA \cos \theta$$
>
> SI unit: $T \cdot m^2$ = weber (Wb)

The SI unit of magnetic flux is the *weber* (Wb), named after the German physicist Wilhelm Weber (1804–1891). It is defined as follows:

$$1 \text{ Wb} = 1 \text{ T} \cdot m^2$$

We usually use the explicit units on the right in the above equality in order to make the connection with base units clearer.

🔑 **Magnetic flux depends on the magnitude of the magnetic field, B, its orientation with respect to a surface, θ, and the area of the surface, A.** A change in any of these variables results in a change in the flux. In the case of a bar magnet moved toward or away from a coil, it is the change in the *magnitude* of the field that results in a change in the flux. In the following Guided Example we consider the effect of changing the *orientation* of a wire loop in a region of constant magnetic field.

🔑 *What factors affect magnetic flux?*

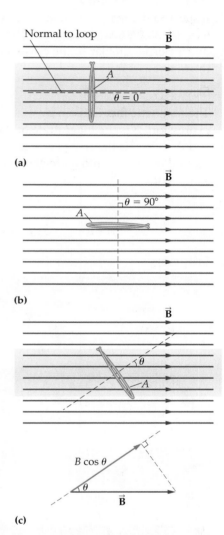

▲ Figure 23.3 **The magnetic flux through a loop**
The magnetic flux through a loop of area A is $\Phi = BA \cos \theta$, where θ is the angle between the normal to the loop and the magnetic field. **(a)** The loop is perpendicular to the field; hence, $\theta = 0$ and $\Phi = BA$. **(b)** The loop is parallel to the field; therefore, $\theta = 90°$ and $\Phi = 0$. **(c)** For a general angle θ, the component of the field that is perpendicular to the loop is $B \cos \theta$; hence, the flux is $\Phi = BA \cos \theta$.

A circular wire loop with a 2.50-cm radius is in a constant magnetic field $\vec{B}$, whose magnitude is 0.625 T. Find the magnetic flux through the loop when its normal makes an angle of 0°, 30.0°, 60.0°, and 90.0° with the direction of the magnetic field.

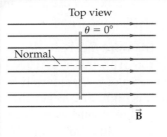
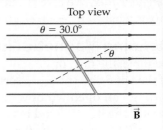
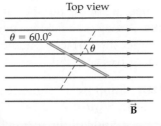
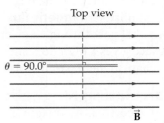

Picture the Problem

Our four sketches look down on the top edge of the circular loop. The orientation of the loop with respect to the magnetic field is shown for each case: $\theta = 0°, 30.0°, 60.0°,$ and $90.0°$. The dashed normal line is perpendicular to the cross-sectional area of the loop.

Strategy

The magnetic flux is given by $\Phi = BA\cos\theta$. In this expression, B is the magnitude of the magnetic field and $A = \pi r^2$ is the area of a circular loop of radius r. The values of B, r, and θ are given in the problem statement.

Known

$r = 2.50$ cm
$A = \pi r^2$
$B = 0.625$ T
$\theta = 0°, 30.0°, 60.0°, 90.0°$

Unknown

$\Phi = ?$

> **Math HELP**
> Scientific Notation
> See Math Review, Section I

Solution

1 Substitute $\theta = 0°$ in $\Phi = BA\cos\theta$:

$\Phi = BA\cos\theta$
$= (0.625 \text{ T})\pi(0.0250 \text{ m})^2\cos 0°$
$= \boxed{1.23 \times 10^{-3} \text{ T}\cdot\text{m}^2}$

2 Substitute $\theta = 30.0°$ in $\Phi = BA\cos\theta$:

$\Phi = BA\cos\theta$
$= (0.625 \text{ T})\pi(0.0250 \text{ m})^2\cos 30.0°$
$= \boxed{1.06 \times 10^{-3} \text{ T}\cdot\text{m}^2}$

3 Substitute $\theta = 60.0°$ in $\Phi = BA\cos\theta$:

$\Phi = BA\cos\theta$
$= (0.625 \text{ T})\pi(0.0250 \text{ m})^2\cos 60.0°$
$= \boxed{0.614 \times 10^{-3} \text{ T}\cdot\text{m}^2}$

4 Substitute $\theta = 90.0°$ in $\Phi = BA\cos\theta$:

$\Phi = BA\cos\theta$
$= (0.625 \text{ T})\pi(0.0250 \text{ m})^2\cos 90.0°$
$= \boxed{0}$

Insight

Thus, even if a magnetic field is uniform in space and constant in time, the magnetic flux through a given area changes if the orientation of the area changes. This is particularly relevant to electric motors and generators, as we will see later in this chapter.

1. Follow-up At what angle is the flux equal to 1.00×10^{-4} T·m²?

2. The magnetic flux through a circular loop is 4.6×10^{-4} T·m². If the normal to the loop makes an angle of 45° with a magnetic field of 0.50 T, what is the area of the loop?

3. A rectangular loop 3.2 cm wide and 5.1 cm long is placed in a magnetic field. The angle between the normal to the loop and the magnetic field is 32°, and the magnetic flux through the loop is 2.2×10^{-3} T·m². What is the magnitude of the magnetic field?

CONCEPTUAL Example 23.2 Does the Magnetic Flux Change?

Three loops of wire are all in a region of space with a uniform, constant magnetic field. Loop 1 swings back and forth slightly, like the bob on a pendulum; loop 2 rotates about a vertical axis; and loop 3 oscillates vertically on the end of a spring. For which loop(s) does the magnetic flux change with time?

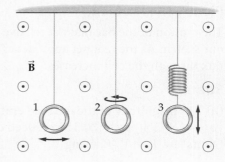

Reasoning and Discussion
Loop 1 moves back and forth, and loop 3 moves up and down, but since the magnetic field is uniform, the flux doesn't depend on the loop's position. Loop 2, on the other hand, changes its orientation relative to the field as it rotates; hence, its flux does change with time.

Answer
The magnetic flux changes with time only for loop 2.

Faraday's law relates magnetic flux and emf

Now that the magnetic flux is defined, we can be more precise about the experimental observations described earlier. In particular, Faraday found that the secondary coil experiences an induced emf only when the magnetic flux through it changes with time. In general, we define the *rate* at which the magnetic flux changes with time as follows:

$$\text{rate of change of magnetic flux} = \frac{\text{change in magnetic flux}}{\text{change in time}} = \frac{\Delta\Phi}{\Delta t}$$

If there are N loops in a coil, the induced emf is given by *Faraday's law of induction*:

Faraday's Law of Induction

$$\text{induced emf} = -\left(\begin{array}{c}\text{number} \\ \text{of loops}\end{array}\right) \times \left(\begin{array}{c}\text{rate of change} \\ \text{of magnetic flux}\end{array}\right)$$

$$\varepsilon = -N\frac{\Delta\Phi}{\Delta t}$$

The negative sign in Faraday's law indicates that the induced emf *opposes* the change in magnetic flux.

If you are only concerned about the magnitude of the emf, which will often be the case, then you can use the following equation:

$$|\varepsilon| = N\left|\frac{\Delta\Phi}{\Delta t}\right|$$

Notice that Faraday's law gives the *emf* that is induced in a coil or a loop of wire. The current that is induced as a result of the emf depends on the resistance in the circuit—just as in the case of a battery connected to a resistor. This is shown in the following Guided Example.

GUIDED Example 23.3 | Bar Magnet Induction   Induced emf and Current

A bar magnet is moved rapidly toward a 45-loop coil of wire. As the magnet moves, the magnetic flux through the coil increases from 1.3×10^{-5} T·m^2 to 3.7×10^{-3} T·m^2 in 0.25 s. **(a)** What is the magnitude of the induced emf? **(b)** If the resistance of the wire in the coil is 3.6 Ω, what is the induced current?

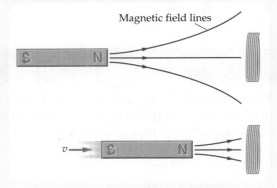

Magnetic field lines

Picture the Problem

The motion of the bar magnet relative to the coil is shown in our sketch. As the magnet approaches the coil, the magnetic flux through the coil increases.

Strategy

(a) The magnitude of the induced emf is given by Faraday's law, $|\varepsilon| = N|\Delta\Phi/\Delta t|$. **(b)** The induced current can be found with Ohm's law, $I = V/R$, using $V = |\varepsilon|$ from part (a).

Known

$\Phi_{initial} = 1.3 \times 10^{-5}$ T·m^2

$\Phi_{final} = 3.7 \times 10^{-3}$ T·m^2

$\Delta t = 0.25$ s $N = 45$ $R = 3.6$ Ω

Unknown

(a) $|\varepsilon| = ?$ **(b)** $I = ?$

Solution

1 **(a) Use Faraday's law to find the magnitude of the induced emf:**

$$|\varepsilon| = N\left|\frac{\Phi_{final} - \Phi_{initial}}{\Delta t}\right|$$

$$= (45)\left|\frac{3.7 \times 10^{-3}\,\text{T·m}^2 - 1.3 \times 10^{-5}\,\text{T·m}^2}{0.250\,\text{s}}\right|$$

$$= \boxed{0.66\,\text{V}}$$

2 **(b) Use Ohm's law to calculate the induced current:**

$$I = \frac{V}{R} = \frac{0.66\,\text{V}}{3.6\,\Omega} = \boxed{0.18\,\text{A}}$$

Math HELP
Significant Figures
See
Lesson 1.4

Insight

If the magnet is pulled back to its original position in the same amount of time, the induced emf and current will have the same magnitudes, but their directions will be reversed.

Practice Problems

4. [Follow-up] How many loops of wire are needed in the coil to give an induced emf of 1.5 V?

5. A 15-loop coil experiences an induced emf of 0.78 V when the magnetic flux is changed in 0.45 s. What is the change in magnetic flux?

6. The magnetic flux through a 25-loop wire coil changes by 2.6×10^{-4} T·m^2 and produces an induced emf of 1.7 V. How much time is required for this change in magnetic flux?

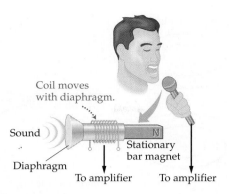

Faraday's law has many everyday applications

Physics & You: Technology A familiar example of Faraday's law in action is the *dynamic microphone*. This type of microphone uses a stationary magnet and a wire coil attached to a movable diaphragm, as illustrated in **Figure 23.4**. When a sound wave strikes the microphone, it vibrates the diaphragm back and forth. This movement changes the magnetic flux through the coil and generates an induced emf. Connecting the coil to an amplifier increases the induced emf enough that it can power a set of speakers. The same principle is used in the *seismograph* shown in **Figure 23.5**, except in this case the oscillations that produce the induced emf are generated by earthquakes.

Lenz's law describes an induced current's direction

Nature often reacts in a way that opposes change. For example, if you compress a gas, the pressure of the gas increases—and opposes the compression. Similarly, if a volcano erupts on a flat landscape, the winds that pass through the area will now be deflected up the slopes of the mountain. This wind pattern will produce rain that erodes the volcano—opposing the change.

A similar principle applies to induced electric currents. It is known as *Lenz's law,* and was first stated by the Estonian physicist Heinrich Lenz (1804–1865). **Lenz's law states that an induced current always flows in a direction that *opposes* the change that caused it.** Lenz's law is the reason for the negative sign in Faraday's law; the negative sign simply indicates that the induced current opposes the change in magnetic flux.

What determines the direction of an induced current?

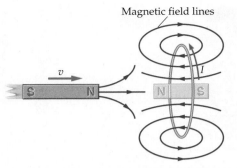 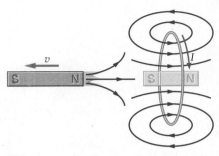

▶ Figure 23.6 Applying Lenz's law to a magnet moving toward and away from a wire loop
(a) If the north pole of a magnet is moved toward a conducting loop, the induced current produces a north pole pointing toward the magnet's north pole. This creates a repulsive force opposing the change that caused the current. **(b)** If the north pole of a magnet is pulled away from a conducting loop, the induced current produces a south magnetic pole near the magnet's north pole. The result is an attractive force opposing the motion of the magnet.

(a) Moving the magnet toward the loop induces a field that repels the magnet.

(b) Moving the magnet away from the loop induces a field that attracts the magnet.

Let's see how Lenz's law works. Consider a bar magnet that is moved toward a conducting loop, as in **Figure 23.6 (a)**. If the north pole of the magnet approaches the loop, a current is induced that tends to oppose the motion of the magnet. To be specific, the current in the loop creates a magnetic field that has a north pole (that is, diverging field lines) facing the north pole of the magnet, as indicated in the figure. This produces a repulsive force acting on the magnet, opposing its motion.

On the other hand, suppose the magnet is pulled away from the loop, as in **Figure 23.6 (b)**. The induced current is in the opposite direction in this case, and it creates a south pole facing the north pole of the magnet. The resulting attractive force tries to keep the magnet from moving away—again opposing the motion.

CONCEPTUAL Example 23.4 **Falling Magnets**

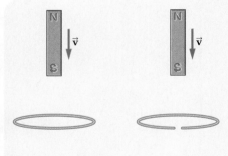

Two magnets fall through the middle of conducting rings. The ring on the right has a small break in it, but the ring on the left forms a closed loop. As the magnets drop toward the rings, does the magnet on the left have an acceleration that is greater than, less than, or equal to that of the magnet on the right?

Reasoning and Discussion
As the magnet on the left approaches the ring, it induces a circulating current. According to Lenz's law, this current produces a magnetic field that exerts a repulsive force on the magnet—to oppose its motion. In contrast, the ring on the right has a break, so it cannot have a circulating current. As a result, it exerts no force on its magnet. Therefore, the magnet on the right falls with the acceleration of gravity; the magnet on the left falls with a smaller acceleration.

Answer
The acceleration of the left magnet is less.

In the examples given so far, the change in magnetic flux was due to the motion of a bar magnet, but Lenz's law applies no matter how the magnetic flux is changed. Suppose, for example, that a magnetic field decreases with time, as in **Figure 23.7**. In this case the change is a decrease of magnetic flux through the ring. The induced current produces a field within the ring that is in the same direction as the decreasing magnetic field, as we see in Figure 23.7 (b). This, again, opposes the change.

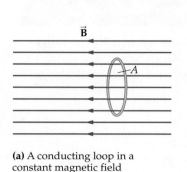

If the magnetic field decreases in magnitude ...

... the induced current produces a field that opposes the decrease.

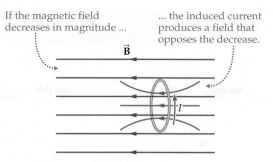

(a) A conducting loop in a constant magnetic field

(b) A change in the field induces a current.

◀ **Figure 23.7 Lenz's law applied to a decreasing magnetic field**
As the magnetic field is decreased, the induced current produces a magnetic field that acts to maintain the original field strength in the ring.

Motion in a magnetic field can induce a current

An object moving through a magnetic field may experience an induced current. As an example, consider the situation shown in **Figure 23.8 (a)**. Here we see a metal rod of length L moving in the vertical direction with a speed v through a region with a constant magnetic field B. The rod is in frictionless contact with two vertical wires, which allows a current to flow in a loop through the rod, the wires, and the lightbulb.

The magnetic field is constant in this system, but the magnetic flux through the loop still changes with time. The reason is that as the rod moves downward, the area enclosed by the loop decreases. This causes the magnetic flux to decrease as well. The motion of the rod produces an emf, called a *motional emf*. The magnitude of the motional emf is

$$\varepsilon = vBL$$

Notice that the emf depends directly on the speed of the rod, its length, and the strength of the magnetic field through which it moves.

According to Lenz's law, the direction of the motional emf—and thus the direction of the induced current—must oppose the changes caused by the motion of the rod. To see how this works, consider **Figure 23.8 (b)**. We see that a counterclockwise induced current produces two effects that oppose the changes. First, the induced current in the rod produces an upward force on the rod—opposite to the rod's motion. Second, the induced current produces a magnetic field within the loop that points out of the page. This strengthens the field inside the loop, which opposes the decrease in magnetic flux. Thus, the motional emf in this case is counterclockwise.

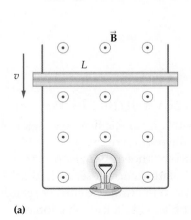

(a)

The direction of the induced current ...

... produces a force that opposes the motion of the rod.

Magnetic field due to I

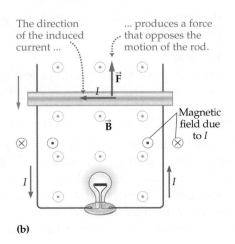

(b)

◀ **Figure 23.8 Motional emf and the direction of an induced current**
(a) Motional emf is created in this system as the rod moves downward. The result is an induced current that causes the light to shine. **(b)** The direction of the current induced by the rod's downward motion is counterclockwise, because this direction produces an upward force on the rod, opposing its downward motion.

COOL PHYSICS
Electric/Magnetic Guitar

The American pop-jazz guitarist Les Paul (1915–2009) applied the basic physics of Faraday's law to musical instruments when he made the first solid-body electric guitar in 1941. The pickup in an electric guitar is simply a small permanent magnet with a coil wrapped around it, as shown. The magnet produces a magnetization in the steel guitar string, which is the moving part in the system. When the string is plucked, it oscillates and changes the magnetic flux in the coil, inducing an emf that can be amplified. A typical electric guitar has two or three sets of pickups, each positioned to amplify a different harmonic of the vibrating strings. Rock on!

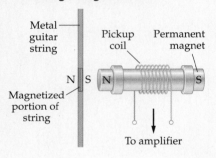

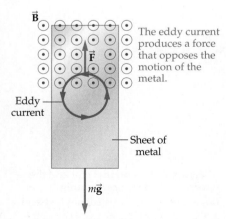

▲ **Figure 23.9 Eddy currents**
A circulating current is induced in a sheet of metal near where it is just leaving a region with a magnetic field. This eddy current produces a force that opposes the motion of the sheet.

CONCEPTUAL Example 23.5 The Direction of Induced Current

Consider a system in which a metal ring is falling out of a region with a magnetic field and into a field-free region, as shown below. According to Lenz's law, is the induced current in the ring clockwise or counterclockwise?

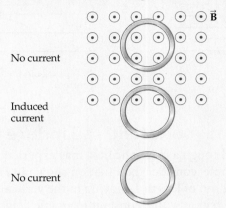

Reasoning and Discussion

The induced current must flow in a direction that opposes the change in the system. In this case, the change is that fewer magnetic field lines are piercing the area of the loop and pointing out of the page. The induced current can oppose this change by generating more field lines out of the page *within the loop*. As shown on the left in the diagram below, the induced current must be counterclockwise to accomplish this.

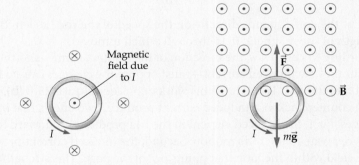

In addition, the diagram on the right above shows that the induced current generates an upward magnetic force at the top of the ring, but no magnetic force on the bottom where the magnetic field is zero. Thus, the downward motion of the ring is opposed as it drops out of the field, as you might expect.

Answer
The induced current is counterclockwise.

Motional emf has many practical applications

Suppose a sheet of metal falls from a region with a magnetic field to a region with no field, as illustrated in **Figure 23.9**. In the portion of the sheet that is just leaving the field, a localized circulating current known as an *eddy current* is induced in the metal. This is just like the circulating current set up in the ring in Conceptual Example 23.5. As in the case of the ring, the current retards the motion of the metal sheet, having an effect much like a frictional force.

Physics & You: Technology The friction-like effect of eddy currents is the basis for *magnetic braking*. An important advantage of this type of braking is that no direct physical contact occurs, thus eliminating frictional wear. In addition, the force of magnetic braking increases with the speed of the metal through the magnetic field. In contrast, kinetic friction is independent of the relative speed of the surfaces. Magnetic braking is used in everything from fishing reels to exercise bicycles to roller coasters.

Eddy currents are also used in the kitchen. In an *induction stove* a metal coil is placed just beneath the cooking surface. When an alternating current is sent through the coil, it sets up an alternating magnetic field that induces eddy currents in nearby metal objects, like a cooking pan. The pan heats up as a result of the induced current, while nonconducting objects—like the surface of the stove and glassware—remain cool to the touch.

23.1 LessonCheck (MP)

Checking Concepts

7. Explain What produces a larger induced emf, a magnetic field that changes quickly or slowly?

8. Assess How does the magnetic flux change when the area of a wire loop is doubled and the magnetic field is halved?

9. Explain How is Lenz's law related to the direction of an induced current?

10. Big Idea Do constant magnetic fields produce electric fields? Do changing magnetic fields produce electric fields?

11. Triple Choice Two identical wire loops have magnetic fields of equal magnitude but different directions. The magnetic field for loop 1 has an angle of 10° with the normal to that loop, and the magnetic field for loop 2 has an angle of 20° with the loop's normal. Is the magnetic flux for loop 1 greater than, less than, or equal to the magnetic flux for loop 2? Explain.

12. Triple Choice A region of space has a magnetic field that points out of the page. The field is uniform in the y direction, but increases in strength in the positive x direction, as indicated in **Figure 23.10**. Four wire loops move through the region in different directions. For each of the loops, state whether its induced current is clockwise, counterclockwise, or zero.

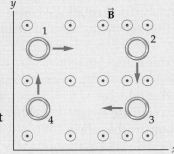

▲ Figure 23.10

Solving Problems

13. Calculate A rectangular wire loop 32 cm long and 16 cm wide is placed in a magnetic field with a magnitude of 0.77 T. The magnetic field is either **(a)** perpendicular to the plane of the loop or **(b)** parallel to the plane of the loop. Calculate the magnetic flux for each of these cases.

14. Calculate The induced emf in a single loop of wire has a magnitude of 1.48 V when the magnetic flux is changed from 0.850 T $\cdot$ m^2 to 0.110 T $\cdot$ m^2. How much time is required for this change in flux?

15. Calculate The magnetic flux on a 25-loop wire coil is changed in 0.35 s. The result is an induced emf of 2.6 V. What was the magnitude of the change in magnetic flux?

16. Calculate What is the angle between the normal to a wire loop and the magnetic field, given that the magnitude of the field is 0.45 T, the area of the loop is 0.085 m^2, and the magnetic flux is 7.1 $\times$ 10^{-4} T $\cdot$ m^2?

Electromagnetic Induction **827**

Vocabulary

- electric generator
- electric motor

Have you ever turned a crank to make a light shine or to power a radio? If so, you supplied energy to an electric generator and experienced the conversion of mechanical work (turning a crank) into electric energy. Similarly, if you've ever ridden in a hybrid car or a golf cart, you've experienced the conversion of electric energy (stored in the batteries) to mechanical energy. In this lesson you'll learn about the operation of both electric generators and electric motors.

Electric Generators

As you know, energy can take many forms and can be converted from one form to another. Sliding a box across a rough floor, for example, converts kinetic energy to thermal energy and sound. Energy can also be converted from one form to another using a mechanical device. An **electric generator** is a device designed to convert mechanical energy to electrical energy.

🔑 *How does an electric generator produce an emf?*

▼ **Figure 23.11 An electric generator** The basic operating elements of an electric generator are shown in a schematic representation. As the coil is rotated by an external source of mechanical work, it produces an emf that can be used to power an electrical circuit.

Mechanical work powers electric generators

The mechanical energy used to drive a generator can come from many different sources. Examples include falling water in a hydroelectric dam, expanding steam in a coal-fired power plant, and a gasoline-powered motor in a portable generator. 🔑 **All generators use the same basic operating principle—mechanical energy moves a conductor through a magnetic field to produce a motional emf.**

Operation of an Electric Generator The linear motion of a metal rod through a magnetic field, as in Figure 23.8 (b), results in a motional emf. To continue producing an electric current in this way, however, the rod would have to move through ever greater distances. A practical way to employ the same effect is to use a wire loop or coil that can be *rotated* in a magnetic field. Rotating the loop or coil to change the magnetic flux allows the electromagnetic induction process to continue indefinitely. Thus, rotating a coil of wire through a magnetic field is a way to transfer energy from mechanical motion to an electric emf and current.

To see how this works, imagine a wire coil of area A located in the magnetic field between the poles of a magnet, as illustrated in **Figure 23.11** (where the coil is represented as a single loop). Metal rings (called *slip rings*) are attached to either end of the wire that makes up the coil. Carbon brushes are in contact with the rings to allow the induced emf to be delivered to the outside world.

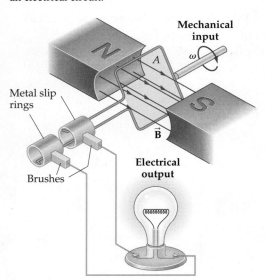

Mechanical input

Metal slip rings

Brushes

Electrical output

As mechanical work rotates the coil with an angular speed ω, the emf produced in it is given by Faraday's law. In the case of a rotating coil, it can be shown that Faraday's law gives the following result:

$$\varepsilon = NBA\omega \sin \omega t$$

This result is plotted in **Figure 23.12**. Notice that the induced emf in the coil alternates in sign, which means that the current in the coil alternates in direction. For this reason, this type of generator is referred to as an *alternating current generator* or, simply, an *AC generator*. The maximum emf occurs when $\sin \omega t = 1$. Thus,

$$\varepsilon_{max} = NBA\omega$$

We apply this result in the next Guided Example.

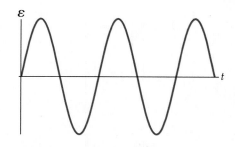

▲ **Figure 23.12 Induced emf of a rotating coil**
The emf alternates in sign, so the current changes sign as well. We call this type of current an *alternating current*.

GUIDED Example 23.6 | Generator Next   **Electric Generator**

The coil in an electric generator has 100 loops and each loop has an area of 2.5×10^{-3} m^2. It is desired that the maximum emf of the coil be 120 V when it rotates with an angular speed of 377 rad/s (60.0 cycles per second). Find the strength of the magnetic field, B, required for this generator to operate at the desired voltage.

Picture the Problem

Our sketch shows a square conducting coil (for simplicity the coil is represented by a single loop) of area A and angular speed ω rotating between the poles of a magnet. The magnetic field has strength B. As the coil rotates, the magnetic flux through it changes continuously. The result is an induced emf that varies with time.

Strategy

The induced emf of the generator is given by $\varepsilon = NBA\omega \sin \omega t$. Therefore, the maximum emf occurs when $\sin \omega t$ has its maximum value of 1. It follows that $\varepsilon_{max} = NBA\omega$. This relation can be solved for the magnitude of the magnetic field, B.

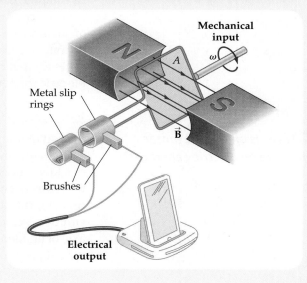

Known   **Unknown**

$N = 100 \quad A = 2.5 \times 10^{-3}$ m^2 $B = ?$

$\varepsilon_{max} = 120$ V $\quad \omega = 377$ rad/s

Solution

1 Solve the equation $\varepsilon_{max} = NBA\omega$ for the magnitude of the magnetic field B:

$$\varepsilon_{max} = NBA\omega$$

$$B = \frac{\varepsilon_{max}}{NA\omega}$$

> **Math HELP**
> Solving Simple Equations
> See Math Review, Section III

2 Substitute the known quantities and calculate B:

$$B = \frac{120 \text{ V}}{(100)(2.5 \times 10^{-3} \text{ m}^2)(377 \text{ rad/s})}$$

$$= \boxed{1.3 \text{ T}}$$

Insight

Notice that the emf of a generator coil depends on its cross-sectional area, A, but not on its shape. In addition, each loop of the coil generates the same induced emf; therefore, the total emf of the coil is directly proportional to the number of loops. Finally, the more rapidly the coil rotates, the more rapidly the magnetic flux through it changes. As a result, the induced emf is also proportional to the angular speed of the generator.

17. Follow-up What is the maximum emf of the generator in Guided Example 23.6 if its rotational frequency is reduced to 50.0 Hz? (Recall that angular speed, ω, is related to rotational frequency, f, by $\omega = 2\pi f$.)

18. What is the minimum number of loops required for a coil of area 5.1×10^{-3} m² and angular speed 350 rad/s to produce a maximum emf of 95 V in a magnetic field of strength 1.5 T?

19. A 45-loop coil produces a maximum emf of 65 V as it rotates with an angular speed of 250 rad/s in a magnetic field with a strength of 2.0 T. What is the area of the coil?

Electric Motors

In Chapter 22 you learned that a current-carrying loop in a magnetic field experiences a torque that tends to make it rotate. If such a loop is mounted on an axle, as shown in **Figure 23.13**, the magnetic torque can be used to operate machinery. This device converts electric energy to mechanical work. A device that converts electric energy into mechanical energy is called an **electric motor**.

Electric motors are powered by electric energy

How is an electric motor related to an electric generator?

Instead of doing work to turn a coil and produce an electric current, as in a generator, an electric motor uses an electric current to produce rotation of a loop or coil, which then does work. **Thus, an electric motor transforms energy from electric emf and current into mechanical motion. It follows that an electric motor is basically an electric generator run in reverse.** In fact, it's possible to have one electric motor turn the axle of an identical motor and thereby produce electricity from the "motor"-turned generator.

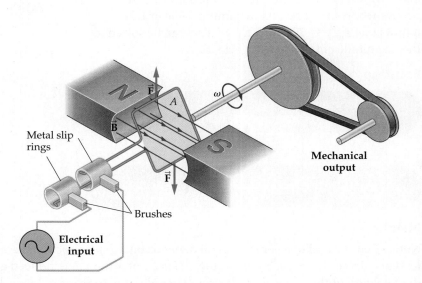

▲ Figure 23.13 **A simple electric motor**
An electric current causes the loop in this motor to rotate and deliver mechanical work to the outside world. In a practical electric motor, a coil consisting of many loops is used.

Operation of an Electric Motor To understand how an electric motor works, notice that the torque exerted on the loop at the moment shown in Figure 23.13 causes it to rotate clockwise *toward* the vertical position. As soon as it reaches this orientation and continues past it as a result of its angular momentum, the alternating current from the electrical input reverses direction. This reverses the torque on the loop and causes the loop to rotate *away* from the vertical—which means it is still rotating in the clockwise sense. The next time the loop becomes vertical, the current again reverses, causing the loop to continue rotating clockwise. The result is an axle continually turning in the same direction.

If a car is powered by an electric motor, its motor can double as a generator during braking. This results in more efficient transportation because kinetic energy that is normally converted to thermal energy in the brakes is instead used to recharge the batteries and extend the range of the car. All hybrid cars make use of this energy-recovery technology.

23.2 LessonCheck (MP)

Checking Concepts

20. ⬤ **Identify** What is the power source for an electric generator?

21. ⬤ **Compare and Contrast** In what ways are electric motors and electric generators different? In what ways are they similar?

22. Assess If the angular speed of a generator is increased, does the maximum emf produced by the generator increase, decrease, or stay the same? Explain.

23. Assess If the number of loops of a generator coil is increased, does the maximum emf produced by the generator increase, decrease, or stay the same? Explain.

Solving Problems

24. Calculate A 95-loop generator coil produces a maximum emf of 75 V when it rotates with an angular speed of 220 rad/s. If the area of the coil's loops is 0.0044 m², what is the magnitude of the magnetic field?

25. Calculate A generator has a 55-loop coil with an area of 0.0085 m². If the coil rotates with an angular speed of 310 rad/s in a magnetic field with a magnitude of 0.95 T, what is the maximum emf?

26. Calculate What area must a 27-loop coil have if it is to produce a maximum emf of 22 V when rotating in a magnetic field of 0.82 T with an angular speed of 290 rad/s?

23.3 AC Circuits and Transformers

Vocabulary

- transformer

Electricity comes in two main types—direct current and alternating current. Each has its benefits and drawbacks. Alternating current is particularly useful in the home, in part because it works so well with devices called *transformers* that change the voltage. This lesson presents the basics of alternating currents, and then explores the operation of transformers.

Alternating Current Circuits

When you switch on a lamp plugged into a wall socket, the voltage supplied to the lightbulb changes direction 60 times a second. Similarly, the current changes direction at the same rate. Because the current *alternates* in direction, we say that the wall socket provides an alternating current and that the lamp is part of an alternating current (AC) circuit.

🔑 **How do voltage and current vary in an AC circuit?**

Voltage and current vary constantly in AC circuits

A simplified AC circuit diagram for a lamp is shown in **Figure 23.14**. The bulb is represented by a resistor with equivalent resistance R and the wall socket is shown as an AC generator, represented by a circle enclosing one cycle of a sine wave.

The voltage delivered by an AC generator is plotted in **Figure 23.15 (a)**. Notice that the graph has the shape of a sine curve. In fact, the mathematical equation for the voltage is

$$V = V_{max} \sin \omega t$$

The maximum voltage, V_{max}, is the largest value of the voltage during a cycle. In household circuits the angular frequency is $\omega = 2\pi f$, with $f = 60$ Hz.
🔑 **Because the voltage in an AC circuit depends on the sine function, we say that it has a *sinusoidal dependence*.**

The current in a resistor in an AC circuit is

$$I = I_{max} \sin \omega t$$

The value of the maximum current is given by Ohm's law:

$$I_{max} = \frac{V_{max}}{R}$$

🔑 **Thus, the current in an AC circuit also has a sinusoidal dependence.** This result is plotted in **Figure 23.15 (b)**.

The voltage and current for a resistor reach their maximum values at the same times. This means that the voltage and current are *in phase* with one another. Other circuit elements, like capacitors and inductors, have different phase relationships between the current and voltage. For these elements the current and voltage reach maximum values at different times.

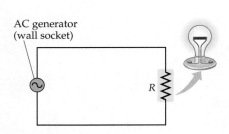

▲ **Figure 23.14 An AC generator connected to a lamp**
Simplified AC circuit diagram for a lamp plugged into a wall socket. The lightbulb is represented in the circuit as its equivalent resistance, R.

Root mean square values characterize AC circuits

Notice that both the voltage and the current in Figure 23.15 have average values that are zero. Thus, average values of AC quantities give very little information. A more useful type of average, or mean, is the *root mean square,* or *rms* for short.

rms Current To see the significance of a root mean square, start by taking the square of the AC current:

$$I^2 = I_{max}^2 \sin^2 \omega t$$

This result is plotted in **Figure 23.16**. Notice that the current squared is always positive. In fact, it varies *symmetrically* between 0 and I_{max}^2. This means that it spends equal amounts of time above and below the value $\frac{1}{2}I_{max}^2$. It follows that the average value of the current squared is

$$(I^2)_{av} = \frac{1}{2}I_{max}^2$$

Now, we take the *square root* of this average so that the final result is a current rather than a current squared. This yields the rms value of the current:

$$I_{rms} = \sqrt{(I^2)_{av}} = \frac{1}{\sqrt{2}}I_{max}$$

rms Voltage The same reasoning applies to the rms value of the voltage in an AC circuit. Therefore,

$$V_{rms} = \frac{1}{\sqrt{2}}V_{max}$$

We apply these results in the following Quick Example.

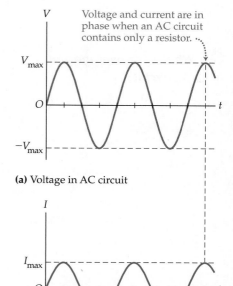

(a) Voltage in AC circuit

(b) Current in AC circuit with resistance only

▲ **Figure 23.15 AC voltage and current for a resistor circuit**
(a) An AC voltage described by $V = V_{max} \sin \omega t$. **(b)** The alternating current corresponding to the AC voltage in part (a). The voltage of the generator and the current in the resistor are in phase with one another; that is, their maximum values and minimum values occur at precisely the same times.

QUICK Example 23.7 What's the Maximum Voltage?

Typical household circuits operate with an rms voltage of 120 V. What is the maximum, or peak, value of the voltage in these circuits?

Solution
Solving $V_{rms} = (1/\sqrt{2})V_{max}$ for the maximum voltage, V_{max}, we find

$$V_{max} = \sqrt{2}V_{rms}$$
$$= \sqrt{2}(120 \text{ V})$$
$$= \boxed{170 \text{ V}}$$

Practice Problems

27. An AC circuit has a current whose maximum value is 2.7 A. What is the rms current in this circuit?

28. What is the maximum current in an AC circuit that has an rms current of 1.6 A?

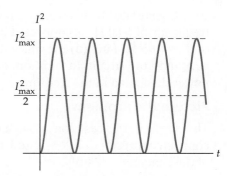

▲ **Figure 23.16 The square of a sinusoidally varying current**
Notice that I^2 varies symmetrically about the value $\frac{1}{2}I_{max}^2$. The average of I^2 over time, then, is $\frac{1}{2}I_{max}^2$.

The average power in an AC circuit depends on root mean square values

🔑 What is the average power in an AC circuit?

Let's see how rms voltage and rms current are related to the average power consumed by a circuit. As an example, recall that the power dissipated in a resistor is

$$P = I^2 R$$

The current in an AC circuit is changing constantly with time, however, and therefore the power is also changing with time. So what is the average power in the circuit?

To find the average power dissipated in a resistor, we recall that the average of the current squared is $\frac{1}{2} I_{max}^2$. Using this result, we find

$$P_{av} = (I^2)_{av} R = \tfrac{1}{2} I_{max}^2 R$$

We've also seen that the maximum current is related to the rms current by the relation $I_{max} = \sqrt{2} I_{rms}$. Therefore, the average power can be written as follows:

$$P_{av} = I_{rms}^2 R$$

Similar conclusions apply to the other power formulas as well. For example, recall that the power dissipated in a DC circuit can be written as $P = V^2/R$ or as $P = IV$. To find the average power in an AC circuit, we take those equations and simply change the current to the rms current and the voltage to the rms voltage:

$$P_{av} = \frac{V_{rms}^2}{R}$$

$$P_{av} = I_{rms} V_{rms}$$

🔑 Thus, the average power in an AC circuit is given by the same equations used in DC circuits—we just replace the DC current and voltage with the corresponding rms values. The close similarity between DC expressions and the corresponding AC expressions with rms values is one of the advantages of working with rms values.

GUIDED Example 23.8 | A Resistor Circuit **Average AC Power**

An AC generator with a maximum voltage of 24.0 V and a frequency of 60.0 Hz is connected to a resistor with a resistance of $R = 265\ \Omega$. Find (**a**) the rms voltage and (**b**) the rms current in the circuit. In addition, determine (**c**) the average power dissipated in the resistor.

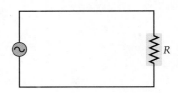

Picture the Problem

The circuit in this case consists of a 60.0-Hz AC generator connected directly to a 265-Ω resistor. The maximum voltage of the generator is 24.0 V.

Known	Unknown
$V_{max} = 24.0$ V	(**a**) $V_{rms} = ?$
$f = 60.0$ Hz	(**b**) $I_{rms} = ?$
$R = 265\ \Omega$	(**c**) $P_{av} = ?$

Strategy

(**a**) The rms voltage is $V_{rms} = V_{max}/\sqrt{2}$. (**b**) Ohm's law gives the rms current: $I_{rms} = V_{rms}/R$. (**c**) The average power is given by $P_{av} = I_{rms}^2 R$.

Solution

1 (a) Use $V_{rms} = V_{max}/\sqrt{2}$ to find the rms voltage:

$$V_{rms} = \frac{V_{max}}{\sqrt{2}}$$

$$= \frac{24.0 \text{ V}}{\sqrt{2}}$$

$$= \boxed{17.0 \text{ V}}$$

2 (b) Divide the rms voltage by the resistance, R, to find the rms current:

$$I_{rms} = \frac{V_{rms}}{R}$$

$$= \frac{17.0 \text{ V}}{265 \ \Omega}$$

$$= \boxed{0.0642 \text{ A}}$$

3 (c) Use $I_{rms}^2 R$ to find the average power:

$$P_{av} = I_{rms}^2 R$$

$$= (0.0642 \text{ A})^2 (265 \ \Omega)$$

$$= \boxed{1.09 \text{ W}}$$

Insight

The average power in Step 3 can be obtained just as well with $P_{av} = V_{rms}^2/R$ or $P_{av} = I_{rms}V_{rms}$. All three equations are equally valid.

Practice Problems

29. Follow-up Suppose we would like the average power dissipated in the resistor to be 5.00 W. Should the resistance be increased or decreased, assuming that the same AC generator is used? Find the required value of R.

30. An AC generator with a frequency of 60.0 Hz is connected to a 375-Ω resistor. If the average power dissipated in the resistor is 2.25 W, what is the maximum voltage of the generator?

31. What is the maximum current in an AC circuit that contains a 232-Ω resistor and dissipates an average power of 4.45 W?

Many devices are used to ensure electrical safety

Physics & You: Technology It is easy to forget that household electrical circuits pose potential dangers to homes and their occupants. For example, if several electrical devices are plugged into a single outlet, the current in the wires connected to that outlet may become quite large. The corresponding power dissipation in the wires ($P = I^2R$) can turn them red hot and lead to a fire. To protect against this type of danger, household circuits use fuses and circuit breakers.

Fuses In the case of a fuse, the current in a circuit must flow through a thin metal strip enclosed within the fuse. If the current exceeds a predetermined amount (typically 15 A) the metal strip becomes so hot that it melts and breaks the circuit. Thus, when a fuse "burns out," it is an indication that too many devices are operating on that circuit.

Circuit Beakers Circuit breakers like the one in **Figure 23.17 (a)** provide protection in a way similar to a fuse by means of a switch that incorporates a bimetallic strip. When the bimetallic strip is cool, it closes the switch, allowing current to flow. When the strip is heated by a large current, however, it bends enough to open the switch and stop the current. Unlike a fuse, which cannot be used after it burns out, a circuit breaker can be reset when the bimetallic strip cools and returns to its original shape.

Polarized Plugs Household circuits can also pose a threat to the occupants of a home, and several strategies are employed to reduce these dangers. The first line of defense against accidental shock is the *polarized plug* (see **Figure 23.17 (b)**), on which one prong is wider than the other prong. The corresponding wall socket will accept the plug in only one orientation, with the wide prong in the wide receptacle. The narrow receptacle of the outlet is wired to the high-potential side of the circuit; the wide receptacle is connected to the low-potential side, which is essentially at ground potential. A polarized plug provides protection by ensuring that the case of an electrical appliance, which is connected to the wide prong, is at low potential. Furthermore, when an electrical device with a polarized plug is turned off, the high potential extends only from the wall outlet to the switch, leaving the rest of the device at zero potential.

▼ **Figure 23.17 Electrical safety devices**
(a) Most homes today are protected by circuit breaker panels. If a circuit draws too much current the heat produced trips the circuit breaker by bending a bimetallic strip. The danger of electric shock from appliances can be reduced by the use of (b) polarized plugs or (c) three-prong grounded plugs. (d) Even more protection is afforded by a ground fault circuit interrupter, or GFCI. This device, which is much faster and more sensitive than an ordinary circuit breaker, applies Faraday's law and interrupts the current in a circuit that has developed a short.

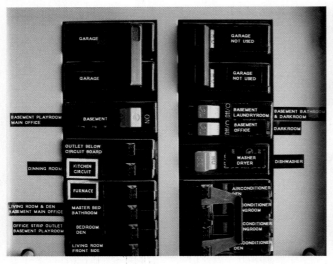

(a)

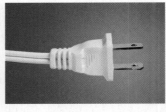

(b)

(c)

(d)

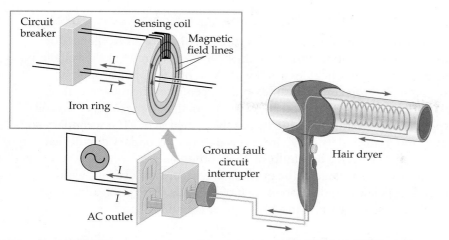

◀ **Figure 23.18 The ground fault circuit interrupter**

A short circuit in an appliance usually trips the circuit breaker. A circuit breaker is not very good at preventing electric shock, however. This is because it is activated by the heat produced when the current becomes abnormally high, and heat takes some time to build up. A ground fault circuit interrupter, in contrast, can cut off the current in a short circuit in less than a millisecond, before any harm can be done.

Grounded Plugs   The next line of defense against accidental shock is the *three-prong grounded plug*, shown in **Figure 23.17 (c)**. In this plug, the rounded third prong is connected directly to ground when plugged into a three-prong receptacle. In addition, the third prong is wired to the case of an electrical appliance. If something goes wrong within the appliance, and a high-potential wire comes into contact with the case, the resulting current flows through the third prong, rather than through the body of a person who happens to touch the case.

GFCI Devices   An even greater level of protection is provided by a device known as a *ground fault circuit interrupter* (GFCI), shown in **Figure 23.17 (d)**. The basic operating principle of an interrupter is illustrated in **Figure 23.18**. Notice that the wires carrying an AC current to the protected appliance pass through a small iron ring. When the appliance operates normally, the two wires carry equal currents in opposite directions—in one wire the current goes to the appliance, and in the other the current returns from the appliance. Each of these wires produces a magnetic field, but because their currents are in opposite directions, the magnetic fields are in opposite directions as well. As a result, the magnetic fields of the two wires cancel.

   If a malfunction occurs in the appliance—say, a wire frays and contacts the case—current that would ordinarily return through the power cord may pass through the user's body instead and into the ground. In such a situation, the wire carrying current to the appliance immediately produces a net magnetic field within the iron ring that varies with the frequency of the AC generator. The changing magnetic field in the ring induces a current in the sensing coil wrapped around the ring, and the induced current triggers a circuit breaker in the interrupter. This cuts the flow of current to the appliance within a millisecond, protecting the user. In newer homes, interrupters are built directly into the wall sockets. The same protection can be obtained, however, by plugging a ground fault interrupter into an unprotected wall socket and then plugging an appliance into the interrupter.

Electrical Transformers

It is often useful to be able to change the voltage from one value to another in an electrical system. For example, high-voltage power lines may operate at voltages as high as 750,000 V, but before the electric power can be used in homes it must be *stepped down* (lowered) to 120 V. Similarly, the 120 V from a wall socket may be stepped down again to 9 V or 12 V to power a portable CD player, or stepped up to give the 15,000 V needed in a TV tube. The electrical device that changes the voltage in an AC circuit is called a **transformer**.

How does the number of loops on each coil in a transformer affect how it changes the voltage?

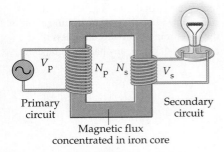

▲ Figure 23.19 **The basic elements of a transformer**
An alternating current in the primary circuit creates an alternating magnetic flux, and hence an alternating induced emf in the secondary circuit. The ratio of emfs in the two circuits, V_p/V_s, is equal to the ratio of the numbers of loops in the two coils, N_p/N_s.

How are voltage and current related in a transformer?

Transformers use two coils to change the voltage

A simple transformer is shown in **Figure 23.19**. Here an AC generator produces an alternating current in the primary (p) circuit at the voltage V_p. The primary circuit includes a coil with N_p loops wrapped around an iron core. The iron core intensifies and concentrates the magnetic flux and ensures, at least to a good approximation, that the secondary (s) coil experiences the same magnetic flux as the primary coil. The secondary coil has N_s loops around its iron core and is part of a secondary circuit that may operate a CD player, a lightbulb, or some other device.

To relate the voltage of the primary and secondary circuits, we apply Faraday's law of induction to each of the coils. The result, after some straightforward algebra, is the transformer equation:

Transformer Equation

$$\frac{\text{voltage in primary coil}}{\text{voltage in secondary coil}} = \frac{\text{turns in primary coil}}{\text{turns in secondary coil}}$$

$$\frac{V_p}{V_s} = \frac{N_p}{N_s}$$

This equation relates the voltages and the numbers of loops in the two circuits.

Let's solve the transformer equation for the voltage in the secondary circuit. The result is

$$V_s = V_p\left(\frac{N_s}{N_p}\right)$$

The transformer equation shows that if the number of loops in the secondary coil is less than the number of loops in the primary coil, the voltage is stepped down to a lower value. Similarly, if the number of loops in the secondary coil is higher, the voltage is stepped up to a higher value.

Increasing voltage with a transformer has a cost

Now, before you begin to think that transformers give you something for nothing, we note that there is more to the story. There is always a tradeoff between voltage and current in a transformer. To see why, let's look at the power in each coil.

Because energy must always be conserved, the average power in the primary circuit must be the same as the average power in the secondary circuit. Since power can be written as $P = IV$, it follows that

$$I_p V_p = I_s V_s$$

Isolating the currents, I_s and I_p, on one side of the equation and the voltages, V_s and V_p, on the other side yields the following equation:

Transformer Equation (with current and voltage)

$$\frac{\text{current in secondary coil}}{\text{current in primary coil}} = \frac{\text{voltage in primary coil}}{\text{voltage in secondary coil}}$$

$$\frac{I_s}{I_p} = \frac{V_p}{V_s}$$

This version of the transformer equation shows an important relationship. **🔑 If a transformer increases the voltage by a given factor, it decreases the current by the same factor. Similarly, if it decreases the voltage, it increases the current.** In other words, if the voltage is stepped up, the current is stepped down, and vice versa.

For example, suppose the number of loops in the secondary coil of a transformer is twice the number of loops in the primary coil. This transformer doubles the voltage in the secondary circuit, $V_s = 2V_p$, and at the same time halves the current, $I_s = I_p/2$. This behavior is similar to that of a lever, which involves a tradeoff between the force that can be exerted and the distance through which it is exerted. The tradeoff is due to energy conservation: The work done on one end of the lever, $F_1 d_1$, must be equal to the work done on the other end of the lever, $F_2 d_2$. Thus, a transformer is like an electrical version of a lever.

ACTIVE Example 23.9   Determine the Number of Loops

A common summertime sound in many backyards is the *zap* heard when an unfortunate insect flies into a high-voltage "bug zapper." Typically, such devices operate at a voltage of about 4000 V, obtained from a transformer plugged into a standard 120-V outlet. How many loops are on the secondary coil of such a transformer if the primary coil has 27 loops?

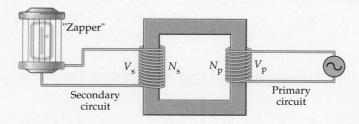

Solution *(Perform the calculations indicated in each step.)*

1. Solve the transformer equation for the number of loops in the secondary coil:

$$N_s = N_p(V_s/V_p)$$

2. Substitute the numerical values:

$$N_s = 900$$

Insight
Of course, any transformer with a loops ratio of $N_s/N_p = 900/27$ will produce the same secondary voltage.

Electrical power transmission relies on transformers

A transformer depends on a *changing* magnetic flux to create an induced emf in the secondary coil. If the current is constant—as in a DC circuit—there is simply no induced emf, and the transformer ceases to function. This is an important advantage that AC circuits have over DC circuits and is one reason why most electrical power systems today operate with alternating currents.

Transformers also play an important role in the *transmission* of electrical energy from the power plants that produce it to the communities and businesses where it is used. When electrical energy is transmitted over large distances, the resistivity of the wires that carry the current becomes significant. If a wire carries a current I and has a resistance R, the power dissipated as waste heat is $P = I^2 R$.

(a)

(b)

(c)

▲ Figure 23.20 The distribution of electric energy

The transmission of electric power over long distances would not be feasible without transformers. (a) A step-up transformer near the power plant boosts the voltage from 12,000 V to the 240,000 V carried by (b) high-voltage lines. A series of step-down transformers then reduce the voltage, first to 2400 V at local substations for distribution to neighborhoods, and next to the 240 V and 120 V supplied to most houses. (c) The gray cylinders commonly seen on utility poles are the transformers responsible for this last voltage reduction.

One way to reduce this energy loss is to reduce the current. A transformer like that in **Figure 23.20 (a)** that steps up the voltage of a power plant by a factor of 20 will at the same time reduce the current by a factor of 20, which reduces the power dissipation by a factor of $20^2 = 400$. The electric energy is then transmitted over large distances at a very high voltage (**Figure 23.20 (b)**). When the electricity reaches the location where it is to be used, step-down transformers like those in **Figure 23.20 (c)** lower the voltage to a level such as 120 V or 240 V, as typically used in homes or workplaces. This is the main reason why AC electric power dominates over DC electric power in the power grid.

23.3 LessonCheck (MP)

Checking Concepts

32. Identify What trigonometric function describes how voltage and current vary with time in an AC circuit?

33. Describe What change must be made to the DC power formula, $P = I^2R$, for it to apply to the average power of an AC circuit?

34. Apply A transformer has more loops on its secondary coil than it has on its primary coil. What effect does the transformer have on voltage?

35. Explain If a transformer increases the voltage in a circuit, what happens to the current?

36. State What is the advantage of an rms value in analyzing an AC circuit?

37. Apply A transformer has twice the number of loops on its secondary coil as on its primary coil.
(a) What is the ratio of the secondary voltage to the primary voltage?
(b) What is the ratio of the secondary current to the primary current?

Solving Problems

38. Calculate The rms current in a circuit is 2.5 A. What is the maximum current in the circuit?

39. Calculate An AC circuit has a maximum voltage of 5.0 V. What is the rms voltage of this circuit?

40. Calculate An AC circuit has an rms current of 3.2 A. What is the average power dissipated in a 180-Ω resistor?

41. Think & Calculate A transformer has 50 loops in the primary coil and 125 loops in the secondary coil. The voltage in the primary circuit is 25 V.
(a) Is the voltage in the secondary circuit greater than, less than, or equal to 25 V?
(b) What is the voltage in the secondary circuit?

42. Think & Calculate A transformer has 150 loops in the primary coil and 35 loops in the secondary coil. The current in the primary circuit is 2.1 A.
(a) Is the current in the secondary circuit greater than, less than, or equal to 2.1 A?
(b) What is the current in the secondary circuit?

Physics & You

The Induction Motor

What Is It? Perhaps the most common electric motor, the *induction motor*, is used to power fans, air conditioners, refrigerators, and washing machines, as well as almost all industrial machinery.

Who Invented It? Nikola Tesla patented the induction motor in 1888.

Why Is It Important? The rugged and inexpensive AC motor has a rotor with no permanent magnets, brushes, or slip rings. Induction motors range in size from 5 W to more than 15,000 kW (about 20,000 hp).

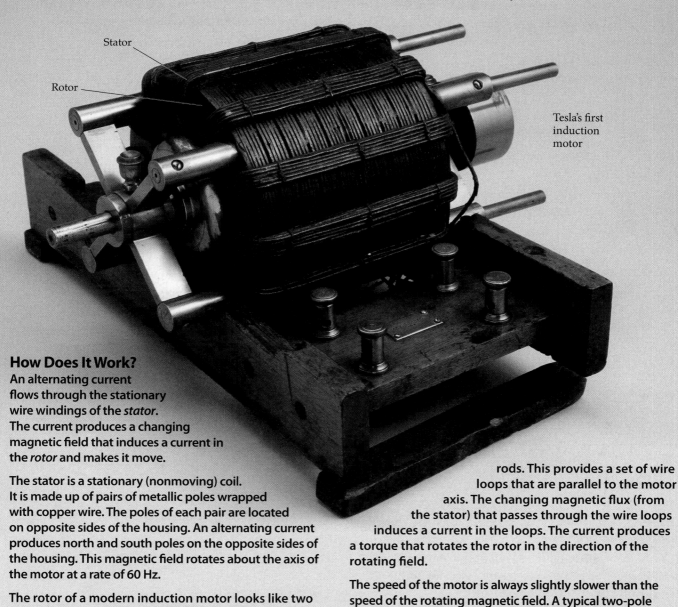

Stator

Rotor

Tesla's first induction motor

How Does It Work?

An alternating current flows through the stationary wire windings of the *stator*. The current produces a changing magnetic field that induces a current in the *rotor* and makes it move.

The stator is a stationary (nonmoving) coil. It is made up of pairs of metallic poles wrapped with copper wire. The poles of each pair are located on opposite sides of the housing. An alternating current produces north and south poles on the opposite sides of the housing. This magnetic field rotates about the axis of the motor at a rate of 60 Hz.

The rotor of a modern induction motor looks like two circular rings joined by a small number of thin metal rods. This provides a set of wire loops that are parallel to the motor axis. The changing magnetic flux (from the stator) that passes through the wire loops induces a current in the loops. The current produces a torque that rotates the rotor in the direction of the rotating field.

The speed of the motor is always slightly slower than the speed of the rotating magnetic field. A typical two-pole induction motor rotates a little slower than 3600 rpm.

Take It Further

1. Explain *Why is an induction motor safer to operate in an explosive atmosphere than a motor that uses brushes and slip rings?*

2. Predict *If you turn the shaft of an induction motor, will it work as a generator?*

Physics Lab Electromagnetic Induction

In this lab you will investigate Faraday's law of electromagnetic induction by observing the ways in which a changing magnetic flux can induce an electric current.

Materials
- galvanometer
- three bar magnets
- nesting primary and secondary coils
- 1.5-V battery
- switch

Procedure

Observe the direction and magnitude of the deflection of the galvanometer's needle in each step. Create three data tables like the one shown below to record your observations.

Part I: Single Bar Magnet

1. Connect the leads of the galvanometer to the primary coil.

2. Observe what happens when you rapidly insert the north pole of a bar magnet into the coil.

3. Observe what happens when the magnet is at rest in the coil.

4. Observe what happens when you rapidly pull the north pole of the magnet out of the coil.

5. Repeat Steps 2–4, this time inserting and removing the magnet slowly.

6. Perform Steps 2–5 using the south pole of the magnet.

7. Observe what happens when you hold magnet still and move the coil instead.

Part II: Multiple Bar Magnets

8. Observe what happens as you rapidly insert the north pole of a bar magnet into the primary coil.

9. Stack two bar magnets together with the north poles aligned. Repeat Step 8 using the stacked magnets.

10. Repeat Step 9 using three bar magnets.

Part III: Nested, End-to-End, and Perpendicular Coils

11. Connect the primary coil in series with the battery and the switch. Connect the galvanometer to the secondary coil.

12. Nest the primary coil inside the secondary coil. Briefly close the switch to send current into the primary coil, and observe what happens.

13. Line the coils up end to end. Close the battery switch and move the secondary coil quickly away from the primary coil. Observe what happens. Now move the secondary coil quickly back toward the primary coil and observe the result.

14. Repeat Step 13, this time moving the primary coil instead of the secondary coil.

15. Position the secondary coil so that it is at a right angle to and in contact with the primary coil. Briefly close the switch to send current into the primary coil, and observe what happens.

Data Table: Part I

Step	Description of Action	Observations
2		
3		
4		

Analysis

1. How does changing the speed of the magnet's motion into and out of the coil affect the magnitude and the direction of the induced current?

2. How does changing the polarity of the magnet affect the magnitude and the direction of the induced current in a coil of wire?

3. How does increasing the number of magnets thrust into a coil affect the magnitude of the induced current?

Conclusions

1. Describe a situation that would create the greatest deflection of the galvanometer needle using a magnet.

2. List the ways in which a current can be induced in a coil of wire that is not connected to a battery.

23 Study Guide

Big Idea

Changing magnetic fields produce electric fields, and the electric fields can be used to generate electric currents.

Specifically, a changing magnetic flux produces an induced emf, and the emf in turn produces an electric current. The magnetic flux can change as a result of several other changes, including a change in the magnetic field strength or a change in the orientation of a wire loop in the magnetic field.

23.1 Electricity from Magnetism

Faraday observed that the magnitude of the induced emf is proportional to the *rate of change* of the magnetic field—the more rapidly the magnetic field changes, the greater the induced emf.

Magnetic flux depends on the magnitude of the magnetic field, B, its orientation with respect to a surface, θ, and the area of the surface, A.

Lenz's law states that an induced current always flows in a direction that *opposes* the change that caused it.

• Magnetic flux is a measure of the number of magnetic field lines that cross a given area.

Key Equations

If a magnetic field of strength B crosses a surface of area A at an angle θ relative to the normal to the surface, the magnetic flux is

$$\Phi = BA \cos \theta$$

The unit of magnetic flux is the weber (Wb): $1 \text{ Wb} = 1 \text{ T} \cdot \text{m}^2$.

If the magnetic flux in a coil of N loops changes by the amount $\Delta\Phi$ in the time Δt, the induced emf is

$$\varepsilon = -N\frac{\Delta\Phi}{\Delta t}$$

The magnitude of motional emf is

$$\varepsilon = vBL$$

23.2 Electric Generators and Motors

All generators use the same basic operating principle—mechanical energy moves a conductor through a magnetic field to produce a motional emf.

An electric motor transforms energy from electric emf and current into mechanical motion. It follows that an electric motor is basically an electric generator run in reverse.

Key Equations

The emf produced by a generator is

$$\varepsilon = NBA\omega \sin \omega t$$

The maximum emf produced by a generator is

$$\varepsilon_{\text{max}} = NBA\omega$$

23.3 AC Circuits and Transformers

Because the voltage in an AC circuit depends on the sine function, we say that it has a sinusoidal dependence. The current in an AC circuit also has a sinusoidal dependence.

The average power in an AC circuit is given by the same equations used in DC circuits—we just replace the DC current and voltage with the corresponding rms values.

The transformer equation shows that if the number of loops in the secondary coil is less than the number of loops in the primary coil, the voltage is stepped down to a lower value. Similarly, if the number of loops in the secondary coil is higher, the voltage is stepped up to a higher value.

If a transformer increases the voltage by a given factor, it decreases the current by the same factor. Similarly, if it decreases the voltage, it increases the current.

• A transformer is an electrical device that changes the voltage in an AC circuit.

Key Equations

An AC generator produces a voltage that varies with time according to the following:

$$V = V_{\text{max}} \sin \omega t$$

The rms, or root mean square, current in an AC circuit is

$$I_{\text{rms}} = \frac{1}{\sqrt{2}} I_{\text{max}}$$

The rms voltage in an AC circuit is

$$V_{\text{rms}} = \frac{1}{\sqrt{2}} V_{\text{max}}$$

The equation relating voltage, V, current, I, and number of loops, N, in the primary (p) and secondary (s) coils of a transformer is

$$\frac{I_s}{I_p} = \frac{V_p}{V_s} = \frac{N_p}{N_s}$$

ANSWERS TO SELECTED ODD-NUMBERED PROBLEMS APPEAR IN APPENDIX A.

Lesson by Lesson

23.1 Electricity from Magnetism

Conceptual Questions

43. Explain the difference between a magnetic field and a magnetic flux.

44. A metal ring with a break in it is dropped from a field-free region of space into a region with a magnetic field. What effect does the magnetic field have on the ring?

45. In a common classroom demonstration, a magnet is dropped down a long, vertical copper tube. The magnet moves very slowly as it falls through the tube, taking several seconds to reach the bottom. Explain this behavior.

46. Many equal-arm balances have a small metal plate attached to one of the two arms. The plate passes between the poles of a magnet mounted in the base of the balance. Explain the purpose of this arrangement.

47. **Figure 23.21** shows a vertical iron rod with a wire coil of many loops wrapped around its base. A metal ring slides over the rod and rests on the wire coil. Initially, the switch connecting the coil to a battery is open, but when it is closed, the ring flies into the air. Explain why this happens.

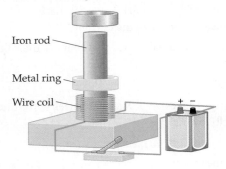

Iron rod

Metal ring

Wire coil

Figure 23.21

48. **Rank** A wire loop is placed in a magnetic field that is perpendicular to the loop's plane. The field varies with time as shown in **Figure 23.22**. Rank the six periods of time in order of increasing magnitude of the induced emf. Indicate ties where appropriate.

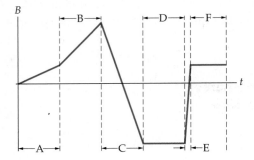

Figure 23.22

49. **Predict & Explain** A metal ring is dropped into a localized region of constant magnetic field, as indicated in **Figure 23.23**. The magnetic field is zero above and below the region. **(a)** At each of the three indicated locations (1, 2, and 3), is the induced current clockwise, counterclockwise, or zero? **(b)** Choose the *best* explanation from among the following:

A. The induced current is clockwise at 1 to oppose the field, zero at 2 because the field is uniform, and counterclockwise at 3 to maintain the field.

B. The induced current is counterclockwise at 1 to oppose the field, zero at 2 because the field is uniform, and clockwise at 3 to maintain the field.

C. The induced current is clockwise at 1 to oppose the field, clockwise at 2 to maintain the field, and clockwise at 3 to oppose the field.

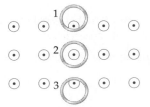

Figure 23.23

Problem Solving

50. A 0.055-T magnetic field passes through a circular ring of radius 3.1 cm at an angle of 16° to the normal. Find the magnitude of the magnetic flux through the ring.

51. A uniform magnetic field of magnitude 0.0250 T points vertically upward. Find the magnitude of the magnetic flux through each of the five sides of the open-topped rectangular box shown in **Figure 23.24**, given that the dimensions of the box are $L = 32.5$ cm, $W = 12.0$ cm, and $H = 10.0$ cm.

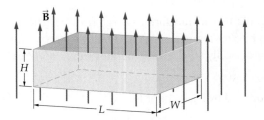

Figure 23.24

52. A magnetic field is oriented at an angle of 47° to the normal of a rectangular area 5.1 cm by 6.8 cm. If the magnetic flux through this area has a magnitude of 4.8×10^{-5} T·m², what is the strength of the magnetic field?

53. Find the magnitude of the magnetic flux through the floor of a house that measures 22 m by 18 m. Assume that the Earth's magnetic field at the location of the house has a north-pointing horizontal component of 2.6×10^{-5} T and a downward vertical component of 4.2×10^{-5} T.

54. **MRI Solenoid** The magnetic field produced by an MRI solenoid 2.5 m long and 1.2 m in diameter is 1.7 T. Find the magnitude of the magnetic flux inside the core of this solenoid.

55. A 0.45-T magnetic field is perpendicular to a circular coil of wire with 53 loops and a radius of 15 cm. If the magnetic field is reduced to zero in 0.12 s, what is the magnitude of the induced emf in the coil?

56. **Figure 23.25** shows the magnetic flux through a wire loop as a function of time. What is the induced emf in the loop at **(a)** $t = 0.050$ s, **(b)** $t = 0.15$ s, and **(c)** $t = 0.50$ s?

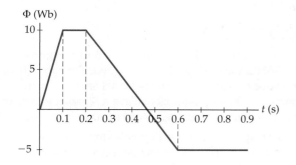

Figure 23.25

57. | Think & Calculate | The magnetic flux through a wire loop is plotted in Figure 23.25. **(a)** Is the magnitude of the induced emf at $t = 0.25$ s greater than, less than, or the same as the magnitude of the induced emf at $t = 0.55$ s? Explain. **(b)** Calculate the magnitude of the induced emf at $t = 0.25$ s and $t = 0.55$ s.

58. A conducting loop has an area of 7.4×10^{-2} m² and a resistance of 110 Ω. Perpendicular to the plane of the loop is a magnetic field of strength 0.48 T. At what rate (in T/s) must this field change if the induced current in the loop is to be 0.32 A?

59. A metal rod 0.76 m long moves with a speed of 2.0 m/s perpendicular to a magnetic field. If the induced emf between the ends of the rod is 0.45 V, what is the strength of the magnetic field?

60. **Airplane emf** A Boeing KC-135A airplane has a wingspan of 39.9 m and flies at a constant altitude near the North Pole with a speed of 850 km/h. If Earth's magnetic field is 5.0×10^{-6} T at that location, what is the induced emf between the wing tips of the airplane?

23.2 Electric Generators and Motors
Conceptual Questions

61. A metal rod with resistance R can slide without friction on two zero-resistance rails, as shown in **Figure 23.26**. The rod and the rails are within a region of constant magnetic field pointing out of the page. Describe the motion of the rod when the switch is closed and be sure to include the effects of motional emf.

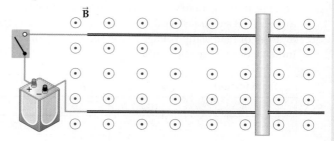

Figure 23.26

62. A conducting rod slides on two wires in a region with a magnetic field. The two wires are not connected. Is a force required to keep the rod moving with constant speed? Explain.

63. The maximum emf of a certain generator is 100 V. Suppose the magnetic field in the generator is doubled and the angular speed is tripled. What is the new maximum emf?

64. The maximum emf of a certain generator is 100 V. Suppose the magnetic field in the generator is halved and the area of the loops is tripled. What is the new maximum emf?

Problem Solving

65. The maximum induced emf in a generator rotating at 22 rad/s is 45 V. How fast must the rotor of the generator rotate if it is to generate a maximum induced emf of 55 V?

66. A rectangular coil 25 cm by 35 cm has 120 loops. This coil produces a maximum emf of 65 V when it rotates with an angular speed of 190 rad/s in a magnetic field of strength B. Find the value of B.

67. A 1.6-m wire is wound into a coil with a radius of 3.2 cm. If this coil is rotated at 85 rpm in a 0.075-T magnetic field, what is its maximum emf?

68. | Think & Calculate | A circular coil with a diameter of 22.0 cm and 155 loops rotates about a vertical axis with an angular speed of 1250 rpm. The only magnetic field experienced by the system is that of the Earth. At the location of the coil, the horizontal component of this magnetic field is 3.80×10^{-5} T, and the vertical component is 2.85×10^{-5} T. **(a)** Which component of the magnetic field is important when calculating the induced emf in this coil? Explain. **(b)** Find the maximum emf induced in the coil.

23.3 AC Circuits and Transformers

Conceptual Questions

69. How can the rms voltage of an AC circuit be nonzero when its average value is zero? Explain.

70. What changes must be made to the DC power formula, $P = IV$, for it to give the average power in an AC circuit?

71. Rank Consider four transformers (A, B, C, and D) for which the voltage in the primary coil is V_p, the number of loops in the primary coil is N_p, and the number of loops in the secondary coil is N_s. Rank the transformers in order of increasing voltage in the secondary coil. Indicate ties where appropriate.

Transformer	V_p (V)	N_p	N_s
A	100	20	100
B	100	100	20
C	20	50	50
D	50	400	800

72. Triple Choice Suppose the number of loops in the secondary coil of a transformer is decreased. **(a)** Does the voltage in the secondary circuit increase, decrease, or stay the same? Explain. **(b)** Does the current in the secondary circuit increase, decrease, or stay the same? Explain.

73. Transformer 1 has a primary voltage V_p and a secondary voltage V_s. Transformer 2 has twice the number of loops in both its primary and secondary coils as transformer 1 does. If the primary voltage of transformer 2 is $2V_p$, what is its secondary voltage? Explain.

74. Transformer 1 has a primary current I_p and a secondary current I_s. Transformer 2 has twice as many loops on its primary coil as transformer 1 does, and both transformers have the same number of loops on the secondary coil. If the primary current of transformer 2 is $3I_p$, what is its secondary current? Explain.

Problem Solving

75. An AC generator produces a peak voltage of 55 V. What is the rms voltage of this generator?

76. **European Electricity** In many European homes the rms voltage from wall sockets is 240 V. What is the maximum voltage in this case?

77. An rms voltage of 120 V produces a maximum current of 2.1 A in a certain resistor. Find the resistance of this resistor.

78. The rms current in an AC circuit with a resistance of 150 Ω is 0.85 A. What are **(a)** the average power and **(b)** the maximum power consumed by this circuit?

79. A 3.33-kΩ resistor is connected to an AC generator with a maximum voltage of 141 V. Find **(a)** the average power and **(b)** the maximum power delivered to this circuit.

80. The electric motor in a toy train requires a voltage of 3.0 V. Find the ratio of loops on the primary coil to loops on the secondary coil in a transformer that will step down the 110-V household voltage to 3.0 V.

81. Think & Calculate A disk drive plugged into a 120-V outlet operates on a voltage of 9.0 V. The transformer that powers the disk drive has 147 loops on its primary coil. **(a)** Should the number of loops on the secondary coil be greater than or less than 147? Explain. **(b)** Find the number of loops on the secondary coil.

82. A transformer with a loops ratio (secondary/primary) of 1/18 is used to step down the voltage from a 110-V wall socket to power a battery recharging unit. What is the voltage supplied to the recharger?

83. A neon sign that requires a voltage of 11,000 V is plugged into a 120-V wall outlet. What loops ratio (secondary/primary) must a transformer have to power the sign?

84. A lightbulb uses an average power of 75 W when connected to a source that supplies an rms voltage of 120 V. **(a)** What is the resistance of the lightbulb? **(b)** What is the maximum current in the bulb? **(c)** What is the maximum power used by the bulb at any given instant of time?

Mixed Review

85. You hold a circular loop of wire at the north magnetic pole of the Earth. Consider the magnetic flux through this loop due to Earth's magnetic field. Is the flux when the normal to the loop is horizontal greater than, less than, or equal to the flux when the normal is vertical? Explain.

86. A bar magnet with its north pole facing downward is falling toward the center of a horizontal conducting ring. As viewed from above, is the direction of the induced current in the ring clockwise or counterclockwise? Explain.

87. **Figure 23.27** shows a zero-resistance rod sliding toward the right on two zero-resistance rails that are separated by the distance $L = 0.450$ m. The rails are connected by a 12.5-Ω resistor, and the entire system is in a uniform magnetic field with a magnitude of 0.750 T. Find the speed at which the bar must move to produce a current of 0.125 A in the resistor.

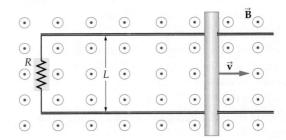

Figure 23.27

88. Figure 23.28 shows a current-carrying wire and a circuit containing a resistor R. **(a)** If the current in the wire is constant, is the induced current in the circuit clockwise, counterclockwise, or zero? Explain. **(b)** If the current in the wire increases in magnitude, is the induced current in the circuit clockwise, counterclockwise, or zero? Explain.

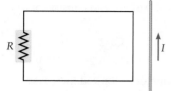

Figure 23.28

89. A Wire Loop and a Magnet A loop of wire is dropped and allowed to fall between the poles of a horseshoe magnet, as shown in **Figure 23.29**. State whether the induced current in the loop is clockwise or counterclockwise when **(a)** the loop is above the magnet and **(b)** the loop is below the magnet.

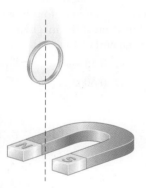

Figure 23.29

90. Interstellar Magnetic Field The *Voyager I* spacecraft moves through interstellar space with a speed of 8.0×10^3 m/s. The magnetic field in this region of space has a magnitude of 2.0×10^{-10} T. Assuming that the 5.0-m-long antenna on the spacecraft is at right angles to the magnetic field, find the motional emf between its ends.

91. Consider a 5.8 cm by 8.2 cm rectangular loop of wire in a uniform magnetic field of magnitude 1.3 T. The loop is rotated from a position where the magnetic flux is zero to a position of maximum flux in 21 ms. What is the average induced emf in the loop?

92. Predict & Explain **Figure 23.30** shows two metal disks of the same size and material oscillating in and out of a region with a magnetic field. One disk is solid; the other has a series of slots. **(a)** Is the effect of eddy currents

on the solid disk greater than, less than, or equal to their effect on the slotted disk? **(b)** Choose the *best* explanation from among the following:

A. The solid disk experiences a greater force because eddy currents in it flow freely and are not interrupted by the slots.

B. The slotted disk experiences the greater force because the slots allow more magnetic field to penetrate the disk.

C. The disks are the same size and made of the same material; therefore, they experience the same force.

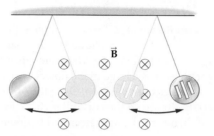

Figure 23.30

93. Transcranial Magnetic Stimulation *Transcranial magnetic stimulation* (TMS) is a noninvasive method for studying brain function. In TMS a conducting loop is held near a person's head, as shown in **Figure 23.31**. When the current in the loop is changed rapidly, the magnetic field it creates can change at the rate of 3.00×10^4 T/s. This rapidly changing magnetic field induces an electric current in a restricted region of the brain that can cause a finger to twitch, bright spots to appear in the visual field (called *magnetophosphenes*), or an overwhelming feeling of complete happiness. If the magnetic field changes at the previously mentioned rate over an area of 1.13×10^{-2} m², what is the induced emf?

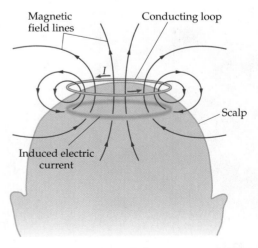

Figure 23.31 Transcranial magnetic stimulation.

Writing about Science

94. Write a report on electric guitars. When was the first one made, and who invented it? When were electric guitars first used in professional performances? How many pickups do various electric guitars have? How are the different pickups used? How does a whammy bar work?

95. Connect to the Big Idea Present several examples of everyday devices that apply Faraday's law of induction. How is the change in magnetic flux produced? How is the induced emf or the induced current used?

Read, Reason, and Respond

Loop Detectors on Roadways "Smart" traffic lights are controlled by loops of wire embedded in the road, as shown in **Figure 23.32**. These *loop detectors* sense the change in magnetic field as a large metal object—such as a car or a truck—moves over the loop. Once the object is detected, electric circuits in the controller check for cross traffic, and then turn the light from red to green.

A typical loop detector consists of three or four loops of 14-gauge wire buried about 7 cm below the pavement. You can see the marks on the road where the pavement has been cut to allow for installation of the wires. There may be more than one loop detector at a given intersection; this allows the system to recognize that a vehicle is moving as it activates first one detector and then another over a short period of time. If the system determines that a car has entered the intersection while the light is red, it can activate one camera to take a picture of the car from the front—to show the driver's face—and then a second camera to take a picture of the car and its license plate from behind.

Motorcycles are small enough that they often fail to activate the detectors, leaving cyclists waiting and waiting for a green light. Powerful neodymium magnets can be mounted to the underside of motorcycles to ensure that they are "seen" by the detectors.

96. Suppose the downward-pointing vertical component of the magnetic field increases as a car drives over a loop detector. As viewed from above, is the induced current in the loop clockwise, counterclockwise, or zero?

97. A car drives onto a loop detector and increases the downward-pointing component of the magnetic field within the loop from 1.2×10^{-5} T to 2.6×10^{-5} T in 0.38 s. What is the induced emf in the detector if it is circular, has a radius of 0.67 m, and consists of four loops of wire?

  A. 0.66×10^{-4} V   C. 2.1×10^{-4} V

  B. 1.5×10^{-4} V   D. 6.2×10^{-4} V

98. A truck drives onto a loop detector and increases the downward-pointing component of the magnetic field within the loop from 1.2×10^{-5} T to a larger value, B, in 0.38 s. The detector is circular, has a radius of 0.67 m, and consists of three loops of wire. What is B, given that the induced emf is 8.1×10^{-4} V?

  A. 3.6×10^{-5} T   C. 8.5×10^{-5} T

  B. 7.3×10^{-5} T   D. 24×10^{-5} T

99. A motorcycle can increase the downward-pointing component of the magnetic field within a loop only from 1.2×10^{-5} T to 1.9×10^{-5} T. Suppose the detector is square, 0.75 m on a side, and has four loops of wire. Over what period of time must the magnetic field increase if it is to induce an emf of 1.4×10^{-4} V?

  A. 0.028 s   C. 0.35 s

  B. 0.11 s   D. 0.60 s

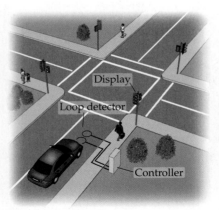

Figure 23.32

Standardized Test Prep

Multiple Choice

Use the diagram below to answer Questions 1–3. A rectangular wire loop of length L and width W is moving at speed v into and then out of a uniform magnetic field, B, which is directed into the page.

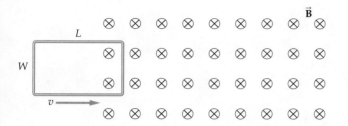

1. When is the current in the loop clockwise and at a maximum?
   (A) when the loop just enters the field
   (B) when the loop is halfway into the field
   (C) when the loop is completely in the field
   (D) when the loop has moved halfway out of the field.

2. What is the magnetic flux through the loop when it is completely within the field?
   (A) vLW
   (B) BLv
   (C) BLW
   (D) zero

3. If $B = 200$ T, $L = 0.2$ m, $W = 0.1$ m, $v = 0.5$ m/s, and the resistance of the loop is 400 Ω, find the current in the loop when two-thirds of it has moved out of the field.
   (A) 0.05 A
   (B) 0.025 A
   (C) 0.017 A
   (D) zero

4. A transformer is labeled as follows:
   (input) 120 V, 0.3 A
   (output) 9V
   What are the output current and power?
   (A) 4 A, 36 W
   (B) 4 A, 360 W
   (C) 40 A, 11 W
   (D) 40 A, 36 W

Use the diagram below to answer Questions 5 and 6. The diagram shows a circular wire loop in a magnetic field that points out of the page.

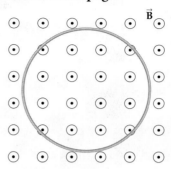

5. What is the direction of the current induced in the loop if the magnetic field is increasing?
   (A) clockwise
   (B) counterclockwise
   (C) The current alternates in direction.
   (D) No current is induced in this situation.

6. Which action will induce the largest current in the loop if the magnetic field is constant?
   (A) spinning the loop clockwise around an axis through its center that is perpendicular to the page
   (B) spinning the loop counterclockwise around an axis through its center that is perpendicular to the page
   (C) rotating the loop around an axis through its center that runs left to right across the page
   (D) moving the loop back and forth in the plane of the page while keeping it within the field

Extended Response

7. A coil of wire consists of 10 loops, each with a diameter of 0.2 m. The total resistance of the wire is 10 kΩ. A magnetic field inside the coil increases from 100 T to 200 T each second. Calculate the emf and the current induced in the coil. Show your work.

> ### Test-Taking Hint
> Induction occurs only if there is a *change* in magnetic flux—that is, a change in either magnetic field or area or both.

If You Had Difficulty With . . .

Question	1	2	3	4	5	6	7						
See Lesson	23.1	23.1	23.1, 23.2	23.3	23.1	23.1	23.1						

24

Quantum Physics

Inside

Photographs are usually taken with light waves, but they can be taken with matter waves as well. This electron micrograph was produced using a beam of electrons, whose wave properties are similar to those of light.

Big Idea

At the atomic level, energy is quantized and particles have wavelike properties.

Understanding the behavior of nature at the atomic level requires a number of new and surprising physics concepts. For example, many physical quantities—like energy—have been found to come in fixed amounts rather than having any possible value, which was the view of classical physics. In this chapter we consider the basic ideas of modern quantum physics and see how they lead to a deeper understanding of the world around us.

24.1 Quantized Energy and Photons

Suppose you get a summer job at a gold mine. If the mining company pays you in gold dust, then your weekly pay can have a wide range of values—perhaps $139.56 one week, and $121.97 the next. On the other hand, if the company pays you with $20 gold coins, your weekly pay might be $100, $120, or $140, but nothing in between those values. In this case your payment is limited to a whole-number multiple of $20 coins. A physicist would say your payment is *quantized*. Something is **quantized** if it comes in *discrete* units that cannot be broken into smaller units. Certain quantities in physics work in much the same way, as you will see.

Vocabulary

- quantized
- blackbody
- photon
- photoelectric effect
- work function

Quantized Energy

Have you ever looked through a small opening into a hot furnace? If so, you've seen the glow of light associated with its high temperature. As unlikely as it may seem, this light played a central role in the revolution of physics that occurred in the early twentieth century. It was through the study of such light that the idea of *energy quantization*—energy taking on only discrete values—was first introduced to physics.

Blackbodies emit energy over a range of frequencies

Physicists in the late nineteenth century were actively studying the physics of a system known as a **blackbody**, an ideal object that absorbs all of the light (and other electromagnetic radiation) that strikes it. An example of a

Figure 24.1 An ideal blackbody

In an ideal blackbody incident light is completely absorbed. In the case shown here, the absorption occurs as the result of multiple reflections within a cavity. The blackbody is in thermal equilibrium with the electromagnetic radiation it contains, at a temperature T.

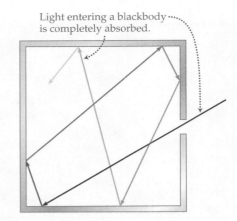

Light entering a blackbody is completely absorbed.

What determines the distribution of energies emitted by a blackbody?

blackbody is shown in **Figure 24.1**. This blackbody has a cavity with a small opening to the outside world—much like a furnace. Light that enters the cavity is reflected multiple times until it is completely absorbed. It is for this reason that the system is referred to as "black."

Though the blackbody absorbs light that falls on it, it also emits light of its own that varies with its temperature—just like the light coming out of a furnace. Physicists studied this emitted light. A basic experiment was conducted as follows:

- Heat a blackbody to a specific temperature.
- Measure the amount of electromagnetic radiation (light, radio waves, infrared, etc.) emitted by the blackbody at a given frequency.
- Repeat the measurement over a range of frequencies.
- Plot the intensity of radiation versus frequency.

The result of such an experiment is shown in **Figure 24.2** for a variety of temperatures. Notice that there is little radiation at low frequencies or high frequencies, and a peak in the radiation at intermediate frequencies. From tests on a variety of materials, a very surprising result emerged. **The distribution of energy in blackbody radiation is *independent* of the material from which the blackbody is constructed—it depends only on the temperature.**

This result means that a blackbody made of steel and one made of wood emit energy in precisely the same manner when held at the same temperature. When physicists observe a phenomenon that is independent of the details of the system, it's a clear signal that they're observing something of fundamental significance. This was certainly the case with blackbody radiation.

Figure 24.2 Blackbody radiation

Blackbody radiation as a function of frequency for various temperatures: **(a)** 3000 K and 6000 K; **(b)** 6000 K and 12,000 K. Notice that as the temperature is increased, the peak in the radiation shifts toward higher frequency.

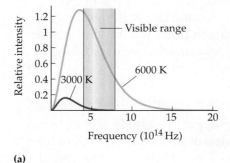

(a)

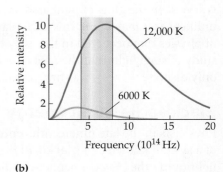

(b)

Peak blackbody radiation depends on temperature

All objects emit electromagnetic radiation over a range of frequencies. Two aspects of this radiation are of particular importance. First, as the temperature is increased, the areas under the curves in Figure 24.2 increase. The total area under each curve is a measure of the total energy emitted by the blackbody. Therefore, an object radiates more energy as it becomes hotter, as you would expect.

Second, the peak in a blackbody radiation curve moves to a higher frequency as the temperature is increased. This movement, or *displacement*, of the peak obeys what is known as *Wien's displacement law*. According to this law, the frequency of the peak is given by the following equation:

Wien's Displacement Law

peak frequency of emitted radiation = constant × temperature

$$f_{\text{peak}} = (5.88 \times 10^{10}\,\text{Hz/K})\,T$$

SI unit: $\text{Hz} = \text{s}^{-1}$

The temperature, T, in Wien's law is the absolute, or Kelvin, temperature. As the absolute temperature is increased, the frequency of the most intense radiation increases as well.

The color of a hot object indicates its temperature

An interesting result of blackbody radiation is that it provides a simple way to determine the temperature of a hot object—by its color. Let's take a closer look at the blackbody radiation curves in Figure 24.2.

Red-Hot Objects   We start with the lowest temperature shown on the graph, 3000 K. The radiation at this temperature is more intense at the low-frequency (red) end of the visible spectrum than at the high-frequency (blue) end. Therefore, an object at this temperature—like the heating coil on a stove or the bolt in **Figure 24.3**—appears red-hot to the eye. Even so, most of the energy radiated at this temperature is in the infrared portion of the spectrum, and thus is not visible.

Yellow-Hot Objects   Next, consider an object at 6000 K. A familiar example of such an object is the surface of the Sun. Objects at this temperature give out strong radiation throughout the visible spectrum, though there is more radiation at the low-frequency (red) end than at the high-frequency (blue) end. As a result, the light of the Sun appears somewhat yellowish.

Blue-Hot Objects   Finally, an object at 12,000 K appears bluish white. The object is blue-hot, which is considerably hotter than red-hot. Most of the radiation at this temperature is ultraviolet and cannot be seen. Some stars in the night sky are bluish, indicating that they have a temperature of 12,000 K or more.

▶ **Figure 24.3 A red-hot object**
The red-hot, glowing bolt in this picture radiates primarily in the infrared part of the spectrum, but it is hot enough (a few thousand kelvins) that a significant portion of its radiation falls within the red end of the visible spectrum. The other bolts are too cool to radiate any detectable amount of visible light.

Use Wien's displacement law to find the surface temperature of the star Rigel, given that its blackbody radiation peak occurs at a frequency of 1.17×10^{15} Hz.

Solution

Solve $f_{\text{peak}} = (5.88 \times 10^{10}$ Hz/K$)T$ for the temperature, T, and substitute the frequency given in the problem statement:

$$T = \frac{f_{\text{peak}}}{5.88 \times 10^{10} \text{ Hz/K}}$$

$$= \frac{1.17 \times 10^{15} \text{ Hz}}{5.88 \times 10^{10} \text{ Hz/K}}$$

$$= \boxed{19{,}900 \text{ K}}$$

This is a little more than three times the surface temperature of the Sun. Thus, blackbody radiation curves allow us to determine the temperature of a distant star that we may never visit.

Math HELP
Scientific Notation
See Math Review, Section I

Practice Problems

1. What is the peak frequency of blackbody radiation for an object with a temperature of 2500 K?

2. Concept Check Betelgeuse is a red-giant star in the constellation Orion; Rigel is a bluish star in the same constellation. Which star has the greater surface temperature? Explain.

Quantized energy explains blackbody radiation

Although experimental understanding of blackbody radiation was quite extensive in the late nineteenth century, there was a problem. Attempts to explain the blackbody radiation curves of Figure 24.2 with the physics known at that time (often referred to as *classical physics*) failed—and failed miserably.

To see the problem, consider the curves shown in **Figure 24.4**. The green curve is the experimental result for blackbody radiation. In contrast, the blue curve shows the prediction of classical physics. Clearly, the classical result cannot be valid—its curve goes to infinity at high frequency, which implies an infinite amount of energy. This disagreement—which is most obvious at high frequencies—is referred to as the *ultraviolet catastrophe*.

🔑 *What does it mean to say that the energy of radiation is quantized?*

▶ **Figure 24.4 The ultraviolet catastrophe**
Classical physics predicts a blackbody radiation curve that rises without limit as the frequency increases. This outcome is referred to as the *ultraviolet catastrophe*. By assuming energy quantization, Planck was able to derive a curve in agreement with experimental results.

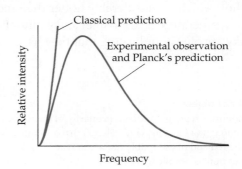

German physicist Max Planck (1858–1947) worked tirelessly on this problem. Eventually, he came up with a mathematical equation that agreed with experimental results for all frequencies. His next problem was to *derive* (show the basis for) the equation, in order to understand its physical meaning. The only way he could do this was to make an unprecedented assumption. Planck proposed that the energy of radiation in a blackbody is *quantized*. ⬤ **The quantized energy of a blackbody *only* comes in amounts equal to a constant (h) times the frequency of the radiation—it cannot come in any other amount.** Thus, Planck found that the energy in a blackbody comes in fixed units, just like gold coins. It doesn't come in arbitrary amounts, like gold dust.

Planck's Constant Planck found that the energy in a blackbody can only have the following energies for any given frequency:

Quantized Energy

energy = quantum number $\times$ Planck's constant $\times$ frequency

$$E = nhf \qquad n = 0, 1, 2, 3, \ldots$$

In this expression, n is an integer known as the *quantum number* and h is *Planck's constant*. The value of Planck's constant is as follows:

Planck's Constant, h

$$h = 6.63 \times 10^{-34}\,\text{J} \cdot \text{s}$$

SI units: $\text{J} \cdot \text{s}$

Planck's constant is one of the fundamental numbers of nature. It is on an equal footing with other fundamental constants like the speed of light and the mass of an electron.

Quantum of Energy The assumption of energy quantization is quite a departure from classical physics, in which energy can have any value. In Planck's calculation, the energy can have only the discrete values hf, $2hf$, $3hf$, and so on. Because of this quantization, the new physics developed by Planck and others is referred to as *quantum physics*. In quantum physics, the energy of a blackbody can change only in *quantum jumps* of hf or larger as the system goes from one quantum state to another. Thus, the basic unit, or *quantum*, of energy is hf.

Basic Unit (Quantum) of Energy

quantum of energy $= hf$

This amount of energy is incredibly small, as can be seen from the small magnitude of Planck's constant.

Notice that the quantum of energy depends on frequency. In fact, the smaller the frequency, the smaller the quantum of energy, and the larger the frequency, the larger the quantum of energy. In terms of the gold coin analogy used earlier, a low frequency has a quantum of energy that is like a \$5 coin, an intermediate frequency corresponds to a \$10 coin, and a high frequency is like a \$20 coin. Each frequency comes in its own "denomination."

Energy Changes Quantum numbers in typical systems are incredibly large. As a result, changing the quantum number by 1 is completely insignificant and undetectable—it's just like changing the number of water molecules in a swimming pool by one molecule. Similarly, the change in energy from one quantum state to the next is so small that it cannot be measured in a typical experiment. For all practical purposes, then, the energy of an everyday system seems to change continuously, even though it actually changes by small quantum jumps. The next Guided Example explores the size of the energy quantum and the value of the quantum number, n, for a typical everyday system.

GUIDED Example 24.2 | Quantum Numbers   Quantum Energy

A mass on a spring oscillates with a frequency of 0.86 Hz and an energy of 0.54 J. (**a**) What is the quantum of energy, hf, for this system? (**b**) Assuming that the energy of this system satisfies $E = nhf$, find the quantum number, n.

Picture the Problem

Our sketch shows a mass oscillating on a spring. The frequency of oscillation is $f = 0.86$ Hz, and the energy is $E = 0.54$ J.

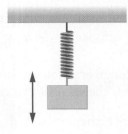

Strategy

(**a**) The energy of one quantum is hf, where f is the frequency of the mass on the spring. (**b**) The quantum number is obtained by solving $E = nhf$ for n.

Known	Unknown
$f = 0.86$ Hz	(**a**) $hf = ?$
$E = 0.54$ J	(**b**) $n = ?$

Solution

1 (a) The energy of one quantum is hf, where $f = 0.86$ Hz $= 0.86$ s^{-1}:

$$hf = (6.63 \times 10^{-34} \text{ J} \cdot \text{s})(0.86 \text{ s}^{-1})$$
$$= \boxed{5.7 \times 10^{-34} \text{ J}}$$

2 (b) Solve $E = nhf$ for the quantum number, n, and substitute the numerical values:

$$E = nhf$$
$$n = \frac{E}{hf}$$
$$= \frac{0.54 \text{ J}}{5.7 \times 10^{-34} \text{ J}}$$
$$= \boxed{9.5 \times 10^{32}}$$

> **Math HELP**
> Solving Simple Equations
> **See Math Review,
> Section III**

Insight

The quantum of energy for this system is on the order of 10^{-34} J. In comparison, the energy required to break a bond in a DNA molecule is on the order of 10^{-20} J. Thus, the quantum energy for a macroscopic (everyday) system is about 10^{14} times smaller than the energy needed to affect a molecule. Similarly, the number of quanta in this system, roughly 10^{33}, is extremely large. It is comparable to the number of atoms in four Olympic-size swimming pools.

3. Follow-up If the quantum of energy for a mass on a spring is 0.80×10^{-33} J, what is the frequency of oscillation?

4. An oscillator has a quantum number of 8.6×10^{32} and a frequency of 1.5 Hz. What is the energy of this oscillator?

5. A pendulum oscillates with a frequency of 0.50 Hz and an energy of 0.75 J. What is the quantum number for this pendulum?

Quantized Light

Planck's theory of energy quantization led to a good description of blackbody radiation. Even so, Planck was troubled by the theory, as were many other physicists. Although the idea of energy quantization worked, at least for this one case, it seemed more like a mathematical trick than a true representation of nature.

The well-founded misgivings about quantum theory began to fade away as a result of the work of Einstein. Einstein introduced the idea that light energy is also quantized, in the form of photons. **Photons** are discrete bundles of light energy that obey Planck's hypothesis of energy quantization.

A light beam is like a beam of particles

Einstein stated that light with a frequency f consists of photons with an energy hf:

> **Energy of a Photon of Frequency f**
>
> energy of a photon = Planck's constant × frequency
> $$E = hf$$
>
> SI unit: J

Thus, the energy in a beam of light of frequency f can have only the values hf, $2hf$, $3hf$, and so on. Planck's initial reaction to Einstein's suggestion was that he had gone too far with the idea of quantization. As it turns out, nothing could have been further from the truth.

🔑 In Einstein's photon model, a beam of light can be thought of as a beam of particles, each carrying the energy hf. This model is illustrated in **Figure 24.5**. If the beam of light is made more intense while the frequency remains the same, the result is that the photons in the beam are more tightly packed. In this way, more photons shine on a given surface in a given time, increasing the energy delivered to the surface per time. Even so, each photon in the more intense beam has exactly the same amount of energy as each photon in the less intense beam. The energy of a typical photon of visible light is calculated in the next Quick Example.

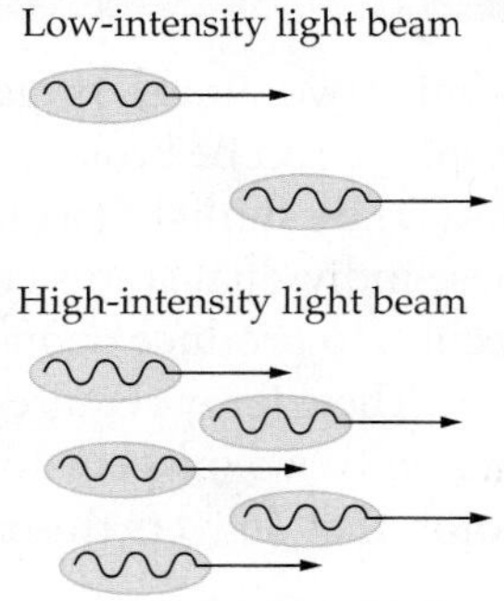

🔑 *How are photons related to energy quantization?*

Low-intensity light beam

High-intensity light beam

▲ **Figure 24.5 The photon model of light**
In the photon model of light, a beam of light consists of many photons, each with an energy hf. The more intense the beam, the more tightly packed the photons.

Calculate the energy of a photon of yellow light with a frequency of 5.25×10^{14} Hz. Give the energy in both joules (J) and electron volts (eV), where $1\ eV = 1.60 \times 10^{-19}$ J.

Solution
Applying $E = hf$, we find

$$E = hf$$

$$= (6.63 \times 10^{-34}\ J \cdot s)(5.25 \times 10^{14}\ s^{-1})$$

$$= \boxed{3.48 \times 10^{-19}\ J}$$

Converting to electron volts yields

Math HELP
Unit
Conversion
See Lesson 1.3

$$3.48 \times 10^{-19}\ J\left(\frac{1\ eV}{1.60 \times 10^{-19}\ J}\right) = \boxed{2.18\ eV}$$

Practice Problem

6. What is the frequency of light whose photons each have an energy of 2.5×10^{-19} J?

Notice that the energy of a single photon of yellow light is a very small number of joules. It's in the range of a couple of electron volts, however, which is a typical energy in atomic systems. This is why photons of visible light can be absorbed or emitted by atoms.

GUIDED Example 24.4 | When Oxygens Split **Photons**

Diatomic (two-atom) oxygen molecules (O_2) in Earth's atmosphere can be broken apart into two separate oxygen atoms (O) by sunlight (which consists of photons). Some of these individual atoms can combine with other oxygen molecules to produce ozone (O_3), an air pollutant and a lung irritant. The energy required to dissociate one molecule of oxygen into two oxygen atoms is 5.13 eV. Find the frequency of a photon that delivers this amount of energy.

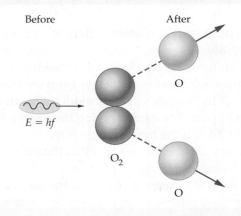

Picture the Problem

Our sketch shows a single photon breaking an O_2 molecule into two O atoms. The energy of the photon is $E = hf$.

Strategy

Since Planck's constant is given in units of J · s, we first convert the dissociation energy from electron volts to joules using $1\ eV = 1.60 \times 10^{-19}$ J. Next, we find the frequency of the photon using $E = hf$.

Known
$E = 5.13\ eV$

Unknown
$f = ?$

Solution

1 Convert the dissociation energy to joules:

$$(5.13 \text{ eV})\left(\frac{1.60 \times 10^{-19} \text{ J}}{1 \text{ eV}}\right) = 8.21 \times 10^{-19} \text{ J}$$

2 Solve $E = hf$ for the frequency, and then substitute the known values to calculate f:

$$E = hf$$

$$f = \frac{E}{h}$$

$$= \frac{8.21 \times 10^{-19} \text{ J}}{6.63 \times 10^{-34} \text{ J} \cdot \text{s}}$$

$$= \boxed{1.24 \times 10^{15} \text{ Hz}}$$

> **Math HELP**
> Scientific Notation
> **See Math Review, Section I**
> Unit Conversion
> **See Lesson 1.3**

Insight

The calculated frequency is in the ultraviolet region of the electromagnetic spectrum. In fact, ultraviolet rays in Earth's upper atmosphere cause O_2 molecules to dissociate, freeing up atomic oxygen, which can then combine with other O_2 molecules to form ozone, O_3. Similarly, ozone molecules can also absorb ultraviolet light, providing protection to organisms living on Earth's surface.

Practice Problems

7. An infrared photon has a frequency of 1.00×10^{13} Hz. How much energy is carried by 1 mole of these photons?

8. The frequency of a photon is 8.2×10^{14} Hz. What is the energy of this photon, in electron volts?

The photon model explains the photoelectric effect

Einstein used his photon model of light to explain the photoelectric effect. The **photoelectric effect** is a phenomenon in which a beam of light (*photo-*) hits the surface of a metal and ejects an electron (*-electric*). The effect can be measured using a device like that pictured in **Figure 24.6**. Notice that incoming light ejects an electron from a metal plate called the *emitter* (E). The ejected electron is known as a *photoelectron*. The photoelectron is then attracted to a *collector plate* (C), which is at a positive potential relative to the emitter. The result is an electric current that can be measured with an ammeter.

Problems with Classical Physics Just like blackbody radiation, the photoelectric effect could not be explained with the classical physics known at the time. Two of the main areas of disagreement were as follows:

- Classical physics predicts that *any* color of light (that is, any frequency) will eject electrons, as long as the beam is intense enough. This, however, does not happen. Instead, only light with a frequency *greater* than a certain minimum value, referred to as the *cutoff frequency*, f_0, causes electrons to be ejected. Light with a frequency less than the cutoff frequency does not eject electrons, no matter how intense the beam.

▼ **Figure 24.6 The photoelectric effect**
The photoelectric effect can be studied with a device similar to the one shown here. Light shines on a metal plate, ejecting electrons, which are then attracted to a positively charged collector plate. The result is an electric current that can be measured with an ammeter.

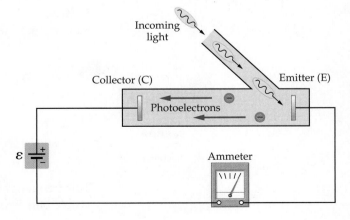

COOLPHYSICS

Amateur Astronomers

Enormous numbers of photons are involved in most everyday situations. For example, a typical lightbulb gives off about 10^{18} photons per second. This is a lot, considering that humans need only about 1000 photons per second to see an object clearly. Amateur astronomers often push the limits of human vision as they try to see incredibly dim objects in the night sky. Because photons move at the speed of light, an astronomer seeing 1000 photons per second through a telescope is actually seeing photons that are separated from one another by about 160 km. It follows that only one photon at a time is passing through the astronomer's telescope. To capture more photons, astronomers use large telescopes that they affectionately refer to as "light buckets." Every photon counts in astronomy!

- Classical physics also predicts that the maximum kinetic energy of an ejected electron should increase as the intensity of the light beam is increased. Again, this does not agree with experiments. Instead, the effect of increasing the intensity is to increase the *number* of electrons that are emitted per second—not their energy. The maximum kinetic energy of the electrons does not increase with the intensity of the light.

As you will see, the photoelectric effect is explained readily with the photon model of light.

Work Function The minimum amount of energy necessary to eject an electron from a particular metal is referred to as the **work function**, W_0, for that metal. Work functions vary from metal to metal but are typically on the order of a few electron volts. If an electron is given an energy E by a beam of light that has an energy greater than W_0, the excess energy goes into kinetic energy of the ejected electron. Thus, the maximum kinetic energy (KE) a photoelectron can have is

$$KE_{max} = E - W_0$$

The Photon Model According to Einstein's photon model, each photon has an energy determined *solely* by its frequency. Therefore, making a beam of a given frequency more intense simply means increasing the number of photons hitting the metal in a given time—not increasing the energy carried by a photon. An electron, then, is ejected only if an incoming photon has an energy that is at least equal to the work function: $E = hf_0 = W_0$. The *cutoff frequency* is thus defined as follows:

Cutoff Frequency, f_0

$$f_0 = \frac{W_0}{h}$$

SI unit: Hz $= s^{-1}$

If the frequency of the light is greater than f_0, the electron can leave the metal with a finite kinetic energy; if the frequency is less than f_0, no electrons are ejected, no matter how intense the beam.

QUICK Example 24.5 What's the Cutoff Frequency?

The work function for gold is 4.58 eV. Find the cutoff frequency, f_0, for a gold surface.

Solution
Substituting the given values into $f_0 = W_0/h$ yields

$$f_0 = \frac{W_0}{h}$$

$$= \frac{(4.58\ \cancel{eV})\left(\dfrac{1.60 \times 10^{-19}\ \cancel{J}}{1\ \cancel{eV}}\right)}{6.63 \times 10^{-34}\ \cancel{J} \cdot s}$$

$$= \boxed{1.11 \times 10^{15}\ Hz}$$

This frequency is in the ultraviolet, just slightly higher in frequency than visible blue light.

Math HELP
Scientific
Notation
**See Math
Review,
Section I**

9. What is the work function of a material whose cutoff frequency is 4.2×10^{15} Hz?

10. For the following photons, state whether they can or cannot eject electrons from gold, whose work function is 4.58 eV: (**a**) a photon with a frequency of 1.0×10^{15} Hz; (**b**) a photon with a frequency of 2.0×10^{15} Hz.

Einstein's photon model of light also shows that a more intense beam of monochromatic light simply delivers more photons per time to the metal. This, in turn, causes more electrons per time to be ejected. Since each electron receives precisely the same amount of energy, however, the maximum kinetic energy of any electron is the same, regardless of the intensity of the light.

If we return to $KE_{max} = E - W_0$ and replace the energy, E, with the energy of a photon, hf, we obtain

$$KE_{max} = hf - W_0$$

This result shows that KE_{max} depends linearly on the frequency but is independent of the intensity. A graph of KE_{max} for sodium (Na) and gold (Au) is given in **Figure 24.7**. Notice that both lines have the same slope, h, as expected from $KE_{max} = hf - W_0$, but have different cutoff frequencies. Therefore, Einstein was able to show that Planck's constant plays a clear role in the photoelectric effect and is not limited to blackbodies.

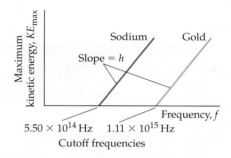

▲ **Figure 24.7 The kinetic energy of photoelectrons**
The maximum kinetic energy of photoelectrons as a function of the frequency of light. Notice that sodium and gold have different cutoff frequencies, as one might expect for different materials. On the other hand, the slope of the two lines is the same, h, as predicted by Einstein's photon model of light.

GUIDED **Example 24.6 | White Light on Sodium**   **Photoelectric Effect**

A beam of white light containing photons with frequencies between 4.00×10^{14} Hz and 7.90×10^{14} Hz is incident on a sodium surface, which has a work function of 2.28 eV. (**a**) What is the range of frequencies in this beam of light for which electrons are ejected from the sodium surface? (**b**) Find the maximum kinetic energy of the photoelectrons that are ejected from this surface.

Picture the Problem

Our sketch shows a beam of white light, represented by photons with different frequencies (colors), incident on a sodium surface. Photoelectrons ejected from this surface have kinetic energies that depend on the frequency of the photons they absorb.

Strategy

(**a**) We can find the cutoff frequency, f_0, for sodium using $f_0 = W_0/h$, with $W_0 = 2.28$ eV. Frequencies between the cutoff frequency and the maximum frequency in the beam of light, 7.90×10^{14} Hz, will eject electrons. (**b**) The maximum kinetic energy is given by $KE_{max} = hf - W_0$. Clearly, the higher the frequency, the greater the maximum kinetic energy. It follows that the greatest maximum kinetic energy corresponds to the highest frequency in the beam, 7.90×10^{14} Hz.

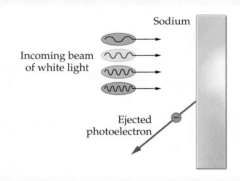

Known

frequencies of incident light:
4.00×10^{14} Hz to 7.90×10^{14} Hz

$W_0 = 2.28$ eV

Unknown

(**a**) frequency range that ejects electrons = ?

(**b**) KE_{max} = ?

Solution

Part (a)

1 Use $f_0 = W_0/h$ to calculate the cutoff frequency for sodium:

$$f_0 = \frac{W_0}{h}$$

Math HELP
Unit Conversion
See Lesson 1.3

$$= \frac{(2.28\ \text{eV})\left(\dfrac{1.60 \times 10^{-19}\ \text{J}}{1\ \text{eV}}\right)}{6.63 \times 10^{-34}\ \text{J} \cdot \text{s}}$$

$$= 5.50 \times 10^{14}\ \text{Hz}$$

2 The frequencies that eject electrons are those between the cutoff frequency and the highest frequency in the beam:

$$\boxed{5.50 \times 10^{14}\ \text{Hz to } 7.90 \times 10^{14}\ \text{Hz}}$$

Part (b)

3 Calculate KE_{max} by substituting the maximum frequency of the light beam, $f = 7.90 \times 10^{14}$ Hz, into $KE_{max} = hf - W_0$. Notice also that the value of the work function, W_0, must be converted from electron volts to joules:

$$KE_{max} = hf - W_0$$

$$= (6.63 \times 10^{-34}\ \text{J} \cdot \text{s})(7.90 \times 10^{14}\ \text{Hz})$$

$$\qquad - (2.28\ \text{eV})\left(\frac{1.60 \times 10^{-19}\ \text{J}}{1\ \text{eV}}\right)$$

$$= \boxed{1.59 \times 10^{-19}\ \text{J}}$$

Insight

Most of the photons in a beam of white light eject electrons from sodium, and the maximum kinetic energy of one of these photoelectrons is about 1 eV.

Practice Problems

11. Follow-up What frequency of light is needed to give a maximum kinetic energy of 2.00 eV to the photoelectrons ejected from the sodium surface in Guided Example 24.6?

12. Follow-up What frequency of light gives a maximum kinetic energy of 1.00 eV to the photoelectrons ejected from a gold surface?

13. Photons are incident on the surface of a metal with a work function of 2.5 eV. If the maximum kinetic energy of the ejected electrons is 1.8×10^{-19} J, what is the frequency of the photons?

The photoelectric effect has practical applications

Physics & You: Technology Applications of the photoelectric effect are in common use all around us. For example, if you've ever dashed into an elevator as its doors were closing, you were probably saved from being crushed by the photoelectric effect. Many elevators and garage-door systems use a beam of light and a photoelectric device known as a *photocell* as a safety feature. As long as the beam of light strikes the photocell, the photoelectric effect generates enough ejected electrons to produce a detectable electric current. When the light beam is blocked—by a late arrival at the elevator, for example—the electric current produced by the photocell is interrupted and the doors are signaled to open. Similar photocells automatically turn on streetlights at dusk and measure the amount of light entering a camera.

▲ **Figure 24.8 Solar panels**
The photoelectric effect is the basic mechanism used by photocells, which power calculators (left), telephones (middle), and the solar panels that supply electricity to the Hubble Space Telescope (right).

Photocells are also the basic unit in the *solar panels* that convert some of the energy in sunlight into electrical energy. Examples of such applications of photocells are shown in **Figure 24.8**. A small version of a solar panel can be found on many pocket calculators. These panels are efficient enough to operate their calculators with nothing more than dim indoor lighting. Larger outdoor panels can operate billboards, safety lights, and emergency telephones in remote areas far from commercial power lines. Truly large solar panels power the Hubble Space Telescope and the International Space Station (those on the space station are 240 feet long and make it visible to the naked eye from Earth's surface). These applications may only hint at the potential for solar energy use. Incredibly, sunlight delivers about 200,000 times more energy to Earth each day than all the world's electrical energy production combined. Now that's a natural resource!

24.1 LessonCheck (MP)

Checking Concepts

14. ⚬ **Compare** How would the energy radiated from a blackbody made of steel compare to that of a blackbody made of plywood at the same temperature?

15. ⚬ **Differentiate** What is the difference between quantized energy and continuous energy?

16. ⚬ **Explain** What determines the energy of a photon?

17. Summarize Give a brief description of the ultraviolet catastrophe.

18. Apply Can a photon of red light have more energy than a photon of blue light? Can a beam of red light deliver more energy than a beam of blue light? Explain.

19. Explain Why does the existence of a cutoff frequency for the photoelectric effect argue in favor of the photon model of light?

Solving Problems

20. Calculate The surface of the Sun has a temperature of 5800 K. What is the peak frequency of its radiation?

21. Calculate A mass on a spring oscillates with an energy of 1.2 J and a frequency of 0.82 Hz. What is its quantum number?

22. Calculate What is the frequency of (**a**) a photon with an energy of 4.2×10^{-19} J and (**b**) a photon with an energy of 1.9 eV?

23. Calculate Photons with a frequency of 9.3×10^{14} Hz strike the surface of a metal with a work function of 3.6 eV. What is the maximum kinetic energy of the ejected electrons?

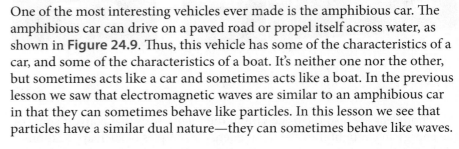

Vocabulary

- de Broglie wavelength
- wave-particle duality

🔑 *How is the de Broglie wavelength related to a particle's momentum?*

▲ **Figure 24.9 An amphibious car**
An amphibious car has some of the characteristics of a normal car and some of the characteristics of a boat.

One of the most interesting vehicles ever made is the amphibious car. The amphibious car can drive on a paved road or propel itself across water, as shown in **Figure 24.9**. Thus, this vehicle has some of the characteristics of a car, and some of the characteristics of a boat. It's neither one nor the other, but sometimes acts like a car and sometimes acts like a boat. In the previous lesson we saw that electromagnetic waves are similar to an amphibious car in that they can sometimes behave like particles. In this lesson we see that particles have a similar dual nature—they can sometimes behave like waves.

Particles have wavelengths

A significant advance in quantum physics occurred when a French graduate student, Louis de Broglie (1892–1987), put forward a most remarkable hypothesis that would later win him the Nobel Prize in physics. His suggestion was basically the following:

> Since light, which we usually think of as a wave, can exhibit particle-like behavior, perhaps a particle of matter, like an electron, can exhibit wavelike behavior.

In particular, de Broglie proposed that a particle with a momentum p has a wavelength λ given by the following equation:

de Broglie Wavelength

$$\text{wavelength} = \frac{\text{Planck's constant}}{\text{momentum}}$$

$$\lambda = \frac{h}{p}$$

SI unit: m

The wavelength in this equation—that is, the wavelength associated with a particle—is known as the **de Broglie wavelength**. 🔑 **The de Broglie wavelength equation shows that the greater a particle's momentum, the smaller its de Broglie wavelength.**

How can the idea of a wavelength for matter make sense? After all, we know that objects like baseballs and airplanes behave like particles, not like waves. To see how this is possible, let's calculate the de Broglie wavelength of a 0.13-kg apple thrown through the air at 5.0 m/s. Recalling that $p = mv$ and substituting known values into $\lambda = h/p$, we get a wavelength of $\lambda = 1.0 \times 10^{-33}$ m. Wow—that's small! This wavelength is much too small to be observed in any realistic experiment—after all, it's smaller than the diameter of an atom by a factor of 10^{23}. Thus, an apple can have the wavelength given by the de Broglie wavelength equation, and we could never detect it.

In contrast, consider an electron with a kinetic energy of 10.0 eV, a typical atomic energy. The de Broglie wavelength in this case is $\lambda = 3.88 \times 10^{-10}$ m. This wavelength, which is about the size of an atom or molecule, can clearly be significant in atomic systems. Therefore, the de Broglie wavelength is unobservable in macroscopic systems, but all important in atomic systems.

ACTIVE Example 24.7 Determine the Speed and the Wavelength

How fast is an electron moving if its de Broglie wavelength is 3.50×10^{-7} m?

Solution *(Perform the calculations indicated in each step.)*

1. Substitute $p = mv$ in the de Broglie wavelength equation, $\lambda = h/p$, to write it in terms of the electron's mass and speed:

$$\lambda = \frac{h}{mv}$$

2. Rearrange to solve for the electron's speed, v: $v = h/m\lambda$

3. Substitute the numerical values: $v = 2080$ m/s

Insight
This is a speed that is easily attained in instruments like electron microscopes.

Practice Problems

24. A proton moves with a speed of 1500 m/s. What is its de Broglie wavelength?

25. Find the de Broglie wavelength of an electron whose kinetic energy is 5.5 eV.

Particles can produce diffraction patterns

The idea of a de Broglie wavelength was a creative and inspired suggestion. Before it could be taken seriously by physicists, however, it had to be observed experimentally. A convincing way to do this was with diffraction patterns.

For example, if particles like electrons and neutrons have wavelengths comparable to atomic distances, it should be possible to produce diffraction patterns by scattering them from a crystal, just as with X-rays. The basic idea is illustrated in **Figure 24.10**. The first to demonstrate diffraction with particles were the American physicists C. J. Davisson and L. H. Germer. They

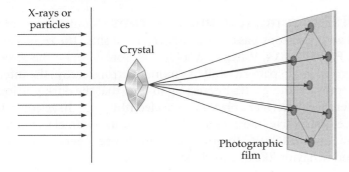

CONNECTINGIDEAS

This lesson introduces you to de Broglie wavelengths.
• De Broglie wavelengths are important in Chapter 25, where we will study the Bohr model of the atom. As you will learn, the allowed orbits for electrons in that model correspond to standing waves that have de Broglie wavelengths.

◀ **Figure 24.10 Diffraction patterns** Diffraction patterns can be observed by passing a beam of X-rays or particles through a crystal. The beams emerge from the crystal at specific angles, due to constructive interference, and can be recorded on photographic film.

▶ **Figure 24.11 Neutron diffraction**
The neutron diffraction pattern of the
protein lysozyme, a digestive enzyme.

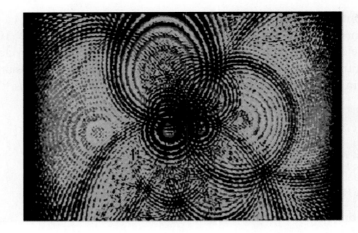

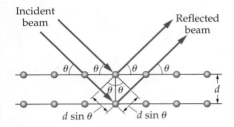

▲ **Figure 24.12 Scattering from a crystal**
Waves (X-rays or particles) reflecting from
a lower plane of atoms in a crystal have a
longer path length than waves reflecting
from an upper plane. When the path-
length difference is whole-number multiple
of a wavelength, the result is a bright spot
in the diffraction pattern.

🔑 *How does the behavior
of subatomic particles compare
to that of macroscopic objects?*

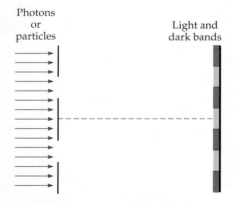

▲ **Figure 24.13 The two-slit experiment**
When photons or particles are passed
through a screen with two slits, the result
is an interference pattern of light and dark
bands.

produced diffraction patterns by scattering electrons from crystals of nickel.
The spacing between spots in the electron diffraction pattern allowed them
to determine the electron's wavelength, and their results verified the de Broglie
relation, $\lambda = h/p$. Similar diffraction patterns have since been observed
with neutrons, hydrogen atoms, and helium atoms. An example of a neutron
diffraction pattern is shown in **Figure 24.11**.

To see how this works, imagine directing a beam of particles at a crystalline
substance composed of layers of regularly spaced atoms, as in **Figure 24.12**.
Notice that the reflected beam is made up of particles that have followed
different paths and hence have different path lengths. This path-length difference results in interference. Destructive interference occurs if the path
lengths differ by half of a de Broglie wavelength, three halves of a wavelength, and so on. Similarly, constructive interference results if the path
lengths differ by one wavelength, two wavelengths, and so on.

Waves and particles have dual natures

We have now come full circle in our study of waves and particles. We've seen
that light is a wave, but it also comes in discrete units called *photons*. We've also
seen that electrons and other particles have well-defined masses and charges, but
they have wave properties as well. This deep connection, whereby waves are like
particles and particles are like waves, is known as the **wave-particle duality**.

Two-Slit Experiment Using Light   To understand the wave-
particle duality, consider a two-slit experiment, as shown in **Figure 24.13**.
When light is passed through these slits, it forms interference patterns of
dark and light fringes. If the intensity of the light is reduced to a very low
level, it is possible to have only a single photon at a time passing through the
apparatus. This photon lands on the distant screen. Eventually, as more and
more photons land on the screen, the interference pattern emerges.

Two-Slit Experiment Using Electrons   Similar behavior is
observed with electrons passing through a pair of slits. The results are
shown in **Figure 24.14**. When only a few electrons have hit the screen,
there seems to be no pattern at all. As more electrons arrive, the interference
pattern begins to emerge. It's interesting to note that the dark portions of
the interference pattern are places where electron waves have combined
to produce destructive interference. At these locations an electron has
essentially "canceled itself" to produce a dark fringe. There is no way to
understand behavior like this in classical physics.

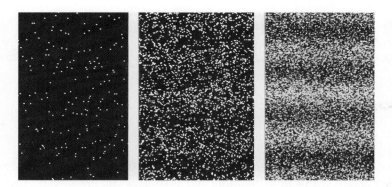

▲ **Figure 24.14 Creation of an interference pattern by electrons passing through two slits** At the beginning, electrons seem to arrive at the screen in random locations (left). As the number of electrons increases, however, the interference pattern begins to show itself (center). With more and more electrons arriving at the screen, the pattern becomes increasingly distinct (right).

Clearly, then, subatomic particles are quite different from objects that we observe on a macroscopic level. In fact, one of the most profound and unexpected insights of quantum physics is that even though baseballs, apples, and people are composed of electrons, protons, and other particles, the behavior of these subatomic particles is nothing like the behavior of baseballs, apples, and people. In short, an electron is not like a billiard ball that is simply reduced in size. No billiard ball can cancel itself out. ⌬ **Subatomic particles like electrons and protons have properties that are different from those of any macroscopic object you might encounter.** To try to force light and electrons into categories like waves and particles is to miss the essence of their existence—they are neither one nor the other, though they have characteristics of both. Again, one is reminded of the saying "The universe is not only stranger than we imagine, it is stranger than we can imagine."

C⊙LPHYSICS
Electron Microscopes

An *electron microscope* uses a beam of electrons to form an image of an object in much the same way that a normal microscope uses a beam of light. A key difference, however, is that the wavelength of an electron is about 1000 times smaller than the shortest wavelength of visible (blue) light. Since the ability to resolve small objects depends on using a wavelength that is smaller than the object to be imaged, an electron microscope can see much finer detail than a light microscope. The image below shows bacteria on the surface of a person's tooth.

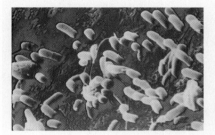

24.2 LessonCheck ⓂⓅ

Checking Concepts

26. ⌬ **Determine** If a particle's momentum is increased, does its de Broglie wavelength increase, decrease, or stay the same?

27. ⌬ **Critique** A classmate states that all forms of matter are made up of atoms, and that all atoms are made of subatomic particles. He then says, however, that a macroscopic object like a baseball does not behave like the subatomic particles it is made of. Is he correct? Explain.

28. Explain What does *wave-particle duality* mean?

29. Big Idea Why don't we notice energy quantization or particle wavelengths in everyday objects?

30. ⎡**Triple Choice**⎤ An electron and a neutron have the same kinetic energy. Is the de Broglie wavelength of the electron greater than, less than, or the same as the de Broglie wavelength of the neutron? Explain.

Solving Problems

31. Calculate A proton has a de Broglie wavelength of 6.4×10^{-7} m. What is its speed?

32. Calculate What is the de Broglie wavelength of a helium atom that has a mass of 6.6×10^{-27} kg and is moving with a speed of 4.1×10^4 m/s?

33. Calculate What is the kinetic energy of a neutron whose de Broglie wavelength is 3.1×10^{-11} m? Give your answer in electron volts (eV).

34. ⎡**Think & Calculate**⎤ An electron and a proton both have a momentum equal to 5.7×10^{-26} kg · m/s. **(a)** Is the de Broglie wavelength of the electron greater than, less than, or equal to the de Broglie wavelength of the proton? Explain. **(b)** Calculate the de Broglie wavelength of the electron.

24.3 The Heisenberg Uncertainty Principle

Vocabulary

- Heisenberg uncertainty principle
- quantum tunneling

Let's say you go to a local pond to take pictures of the ducks. A beautiful female mallard swims near the shore, but as soon as you point your camera at her, she reacts and veers off in a different direction. If you hadn't paid any attention to the duck, it would have continued on its original path. The fact that you observed it (by pointing your camera at it) caused the duck to change its behavior. Something similar happens in physics, where observing a particle changes its behavior and limits our ability to make precise measurements.

The fate of individual particles is uncertain

One of the interesting aspects about the two-slit experiment with electrons (Figure 24.14) is that the electrons appear on the screen in random order. Each time the experiment is run, the electrons appear in different locations and in different sequence—only the final pattern that emerges after a large number of electrons are observed remains the same. The point is that as any given electron passes through the two-slit apparatus, *it is not possible to predict exactly where that one electron will land on the screen.* We can give the probability that it lands at different locations, but the fate of each individual electron is uncertain. Now, you might say that we could follow each electron by shining a light on it and watching its motion. If we do that, however, the photons of light will collide with the electron and send it off on a different path—like the duck reacting to your camera at the pond.

This kind of uncertainty is a fundamental feature of quantum physics, and it is due to the fact that matter has wavelike properties. As a simple example, consider a beam of electrons moving in the x direction and passing through a single slit, as shown in **Figure 24.15**. The electrons diffract and form a pattern similar to that produced by light. In particular, if the beam passes through a slit of width W, it produces a large central maximum—where the probability of detecting an electron is high—with dark fringes on either side located at an angle θ:

$$\sin \theta = \frac{\lambda}{W}$$

▶ **Figure 24.15 Diffraction pattern of electrons**
Central region of the diffraction pattern formed on a screen by electrons passing through a single slit. The curve to the right is a measure of the number of electrons detected at any given location. Notice that no electrons are detected at the dark fringes. This diffraction pattern is identical in form with that produced by light passing through a single slit.

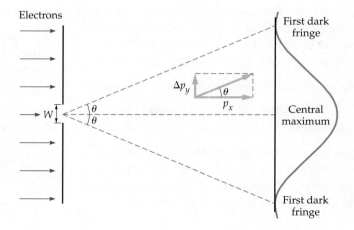

The wavelength here is the de Broglie wavelength, given by $\lambda = h/p_x$, where p_x is the momentum of the electron in the x direction.

We interpret this experiment as follows: When the beam of electrons enters the slit, their location in the y direction has an uncertainty (Δy) roughly equal to the width of the slit, W. In other words, $\Delta y \approx W$. After passing through the slit, the beam spreads out to form a diffraction pattern—with some electrons acquiring a y component of momentum. Thus, there is now an uncertainty in the y component of momentum, Δp_y. (Notice that the delta symbol, Δ, has a different meaning in quantum physics. In particular, Δy means the uncertainty in y—or the spread in the values of y—and not the final value of y minus the initial value of y, as in previous chapters. Similar comments apply to Δp_y.)

The uncertainties in a particle's position and its momentum are related

There is a reciprocal relationship between the uncertainty in position, Δy, and the uncertainty in momentum, Δp_y. This is shown in **Figure 24.16**. Notice that as the width W of the slit is made smaller (to decrease Δy) the angle increases, and the diffraction pattern spreads out (increasing Δp_y). The opposite is also true—increasing the width of the slit makes the diffraction pattern narrower. To summarize:

How are the uncertainties in a particle's position and momentum related?

- Knowing the position of a particle with greater precision makes its momentum more uncertain. Knowing the momentum with greater precision makes its position more uncertain.

In 1927, German physicist Werner Heisenberg (1901–1976) derived the specific relationship between Δp_y and Δy. This relationship is known as the **Heisenberg uncertainty principle**, and it states that it is impossible to know the precise position and precise momentum of a particle at the same time. The uncertainty principle is expressed as follows:

The Heisenberg Uncertainty Principle (for Momentum and Position)

$$\left(\begin{array}{c}\text{uncertainty in}\\ y \text{ momentum}\end{array}\right) \times \left(\begin{array}{c}\text{uncertainty in}\\ y \text{ position}\end{array}\right) \geq \frac{\text{Planck's constant}}{2\pi}$$

$$(\Delta p_y)(\Delta y) \geq \frac{h}{2\pi}$$

Heisenberg showed that this relationship is a general principle and not restricted in any way to the single-slit system considered here. There is simply an intrinsic (built-in) uncertainty in nature that is the result of the wave behavior of matter. Since the wavelength of matter, $\lambda = h/p$, depends directly on the magnitude of Planck's constant, so too does the uncertainty.

According to the uncertainty principle, if the position is known precisely, so Δy approaches zero, then the uncertainty in the momentum, Δp_y, must approach infinity. This implies that the y component of momentum is completely uncertain. Likewise, complete knowledge of the momentum, $(\Delta p_y \rightarrow 0)$, implies that the position is completely uncertain. As one might expect, the uncertainty principle has negligible impact on macroscopic (everyday) systems, but is all important in atomic and nuclear systems.

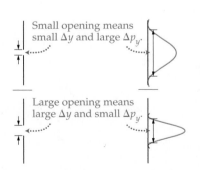

▲ **Figure 24.16 Uncertainty in position and momentum**
Reciprocal relationship between the uncertainty in position (Δy) and the uncertainty in momentum (Δp_y).

Uncertainty also applies to energy and time

The uncertainty relationship for position and momentum, $(\Delta p_y)(\Delta y) \geq \dfrac{h}{2\pi}$, is just one of many forms of the uncertainty principle. There are a number of other quantities besides y and p_y that have the same kind of uncertainty relationship. Perhaps the most important of these other forms of the uncertainty principle is the one relating the uncertainty in energy to the uncertainty in time:

> **The Heisenberg Uncertainty Principle**
> **(for Energy and Time)**
>
> $$\begin{pmatrix} \text{uncertainty} \\ \text{in energy} \end{pmatrix} \times \begin{pmatrix} \text{uncertainty} \\ \text{in time} \end{pmatrix} \geq \frac{\text{Planck's constant}}{2\pi}$$
>
> $$(\Delta E)(\Delta t) \geq \frac{h}{2\pi}$$

The meaning of this equation is quite simple—the shorter the time of a measurement, the greater the uncertainty in the energy.

CONCEPTUAL Example 24.8 More Certain or Less?

If Planck's constant were magically reduced to zero, would the uncertainties in position and momentum and in energy and time increase, decrease, or stay the same?

Reasoning and Discussion
If h were zero, the products of the uncertainties, $\Delta y \Delta p_y$ and $\Delta E \Delta t$, would be reduced to zero. The result would be that position and momentum could be determined simultaneously with zero uncertainty, and similarly for energy and time. With h equal to zero, particles would behave as predicted in classical physics, with well-defined positions and momenta at all times. All the strange behavior of quantum physics is due to the fact that h is nonzero.

Answer
The uncertainties would decrease; in fact, they would decrease to zero.

Nature is profoundly affected by Planck's constant

Planck's constant is extremely small. It's for this reason that it took so long for scientists to appreciate the quantum aspects of nature.

If h were relatively large, the wavelike properties of matter would be apparent even on the macroscopic level. For example, consider pitching a baseball toward a catcher's glove in a universe where h is much larger. If the ball is thrown with a well-defined momentum toward the glove, its position will be uncertain, and the ball could end up anywhere. Similarly, if the catcher gives the glove a relatively small uncertainty in position, its uncertainty in momentum is large, meaning that the glove is moving rapidly about its average position. So, even if the pitcher could aim the ball so that it went to the desired location—which is not possible—there is no way to know where to aim it. If h were significantly larger than it actually is, our experience of the natural world would be very different indeed.

On the other hand, if Planck's constant were zero, quantum behavior would cease to exist, as indicated in Conceptual Example 24.8. In that case energy could take on any value, particles would not behave like waves, and position and momentum could be measured simultaneously with arbitrary precision. It's remarkable how profoundly nature is affected by the value of a single number.

A particle can "tunnel" across a gap

Because particles have wavelike properties, any behavior seen in waves can be found in particles as well. An example is the phenomenon known as **quantum tunneling**, which occurs when a particle moves, or *tunnels*, through a region of space that is forbidden to it in classical physics. For example, if you throw a ball against a wall, classical physics says that it will bounce back to you every time. In quantum physics there is a very small— but finite—probability that the ball will pass through the wall to the other side. A ball tunneling through a wall is extremely unlikely, but an electron tunneling from one atom to another, or from one material to another, is much more likely. Thus, tunneling is important on the atomic level.

Figure 24.17 shows a case of tunneling by light. In Figure 24.17 (a) you can see a beam of light traveling within glass and undergoing total internal reflection when it encounters a glass–air interface. If light were composed of classical particles, this would be the end of the story—the particles of light would simply change direction at the interface and continue traveling within the glass.

What actually happens, however, is much more interesting. Figure 24.17 (b) shows a second piece of glass close to the first piece, with a small vacuum gap between them. If the gap is small, a weak beam of light is observed in the second piece of glass, traveling in the same direction as the original beam. We say that the light waves have *tunneled* across the gap.

Because of their wave properties, electrons and other particles can also tunnel across gaps that would be forbidden to them if they were classical particles. An electron, for instance, is not simply a point of mass; instead, it has wave properties that extend outward from it like ripples on a pond. These waves can "feel" the surroundings of an electron, allowing it to tunnel through a barrier to a region on the other side.

Physics & You: Technology   An example of electrons tunneling across a vacuum gap is shown in **Figure 24.18**, which illustrates the operation of a *scanning tunneling microscope* (STM). The lower part of the figure shows the specimen to be investigated with the microscope. The upper portion shows the key element of the microscope—a small, pointed tip of metal that can be moved up and down with piezoelectric supports. This tip is brought very close to the specimen being observed, leaving a small vacuum gap between it and the specimen. Classically, electrons in the specimen would not be able to move across the gap to the tip. In reality, electrons can tunnel to the tip and create a small electric current. The number of electrons that tunnel, and thus the magnitude of the tunneling current, depends on the width of the gap—the wider the gap, the smaller the current.

In one version of the STM, the tunneling current between the specimen and the tip is held constant. Imagine, for example, that the tip in Figure 24.18 is moved horizontally from left to right. Since the tunneling current is very

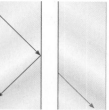

(a) Total internal reflection   **(b)** Light tunnels across the gap.

▲ **Figure 24.17 Optical tunneling**
(a) An incident beam of light undergoes total internal reflection from the glass–air interface. **(b)** If a second piece of glass is brought near the first piece, a weak beam of light is observed continuing in the same direction as the incident beam. We say that the light has *tunneled* across the gap.

Tip of microscope

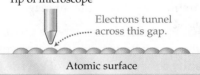

Electrons tunnel across this gap.

Atomic surface

▲ **Figure 24.18 Operation of a scanning tunneling microscope**
Schematic diagram showing the operation of a scanning tunneling microscope. Electrons tunnel across the gap between the atomic surface and the tip of the microscope. Moving the tip up and down, following the contours of the surface, keeps the tunneling current constant and allows the microscope to map out the surface.

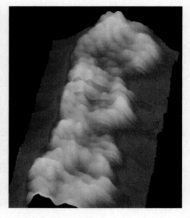

▲ **Figure 24.19 Scanning tunneling microscopy (STM)**
STM images are particularly good at recording the "terrain" of surfaces at the atomic level. The image on the left shows clusters of antimony atoms on a silicon surface. On the right a DNA molecule sits on a substrate of graphite. Three turns of the double helix are visible, magnified about 2,000,000 times.

sensitive to the size of the gap between the tip and the specimen, the tip must be moved up and down with the contours of the specimen in order to maintain a gap of constant width. The tip is moved by sending electrical voltage to the piezoelectric supports. Thus, the voltage going to the supports is a measure of the height of the surface being scanned, and its variations can be converted to a visual image of the surface, as in **Figure 24.19**. The resolution of these microscopes is on the atomic level, showing the hills and valleys created by atoms on the surface of the material being examined.

24.3 LessonCheck (MP)

Checking Concepts

35. 🔑 **Explain** If the uncertainty in a particle's position is decreased, does the uncertainty in its momentum increase, decrease, or stay the same?

36. Explain If you increase the time interval during which an energy measurement is made, does the uncertainty about the amount of energy increase, decrease, or stay the same?

37. Identify What do quantum physicists call the movement of a particle through a region that is forbidden to it in classical physics?

38. Reason Suppose the width of the vacuum gap in a scanning tunneling microscope is decreased. As a result, do you expect the tunneling current to increase or decrease? Explain.

Physics & You

Solar Installation

The Sun showers us with photons every day, and these tiny packets of light provide an enormous source of clean energy. But that energy isn't directly available for powering our homes and other buildings. That's where solar panels come in. These technological marvels convert light energy into energy that we can use. Converting photons to electricity is generating many new careers, such as solar shingling.

Solar shingles are small solar panels roughly the size of a typical roofing shingle. The technology has generated a career opportunity that is a combination of roofer and solar electrician. These solar shingles are laid down just like roofing shingles, which requires the skills of an experienced roofer. But, installing the shingles also requires knowledge of electronics. Each shingle is an individual electronic unit and must be wired to all of the others to form a single functioning system. Finding people with a combination of roofing and electronic skills isn't easy. Companies have begun hiring roofers and training them as solar electricians, or hiring solar electricians and training them to install roofs.

Being a solar shingle installer is just one of many solar-related careers. Many technical and community colleges offer courses that provide valuable training in construction and electronics.

Solar shingle installation is largely outdoor work that is conducted in all types of weather. Anyone considering this career needs to be safety conscious and should not be afraid of heights. The work requires attention to detail and good mathematical skills.

▲ Traditional solar panels are installed on top of an existing roof.

▲ A solar shingle is a roof shingle and a solar panel in one.

Take It Further

1. Research *Use the Internet and other resources to research other solar-related careers, such as solar panel manufacturing or solar design engineering. Write a short report on your findings. Be sure to include information on working conditions, salary, and educational requirements.*

2. Identify *Though solar power is clean and renewable, using it is not without environmental impact. Describe some of the environmental costs associated with manufacturing solar panels.*

Physics Lab Investigating Quanta

In this lab you will determine a "quantum" mass using methods similar to those used by Planck and Einstein.

Materials
• 20 labeled envelopes
• electronic scale or triple beam balance

Procedure

1. Obtain a set of labeled envelopes from your teacher. Do not open any of the envelopes.

2. Carefully measure the mass of each envelope to the nearest gram. Record your results in the data table and on the board.

3. Create additional data tables as needed in order to copy the data obtained by the rest of the class. Each team should have all of the data collected by the whole class.

Data Table: Mass Data

Envelope Identification	Mass (g)	Number of Cards

Analysis

1. Rearrange your data in order of increasing mass. Write out the sequence of increasing masses.

2. Create a bar graph of mass versus envelope identification, using the data as arranged Step 1. Label each bar with the identification code marked on one of the envelopes.

3. Observe the plotted data and look for a pattern in the results. Describe the pattern.

4. If each envelope contains a whole number of index cards, what is the "quantum" mass of a single index card? Explain the method you used to determine this.

5. Determine the number of index cards each envelope contains and record it in the data table.

6. Write a set of data pairs that give the mass of each envelope and the number of cards in it, for example: (13 g, 1 card), (22 g, 2 cards), and so on. Delete repeated values from this list.

7. Create a graph of mass versus number of cards based on your data pairs from Step 6.

8. What does the slope of this graph tell you? What does the *y* intercept tell you?

9. Measure the mass of an empty envelope and record it. How is this mass related to the graph of mass versus number of cards?

10. Measure the mass of a single index card and record it. How is this mass related to the graph of mass versus number of cards?

11. How does your answer in Step 10 affect your conclusions about the numbers of index cards in the envelopes? What was the quantum mass in this lab? Explain.

Conclusions

1. Did you make any assumptions that led you to incorrect results? Explain.

2. Explain why it was important to use whole index cards for this lab.

3. Relate what you did in this experiment to the discoveries made by Planck and Einstein. Use your textbook to assist you in answering this question.

24 Study Guide

Big Idea

At the atomic level, energy is quantized and particles have wavelike properties.

As a result, particles can form diffraction and interference patterns, just like light. In addition, the location and the momentum of particles are subject to uncertainty, which leads to particles' ability to tunnel through regions that are forbidden by classical physics.

24.1 Quantized Energy and Photons

🔑 The distribution of energy in blackbody radiation is *independent* of the material from which the blackbody is constructed—it depends only on the temperature.

🔑 The quantized energy of a blackbody *only* comes in amounts equal to a constant (h) times the frequency of the radiation—it cannot come in any other amount.

🔑 In Einstein's photon model, a beam of light can be thought of as a beam of particles, each carrying the energy hf.

• Einstein proposed that light comes in bundles of energy, called *photons*, that obey Planck's hypothesis of energy quantization.

• The photoelectric effect occurs when photons of light eject electrons from the surface of a metal.

• The minimum energy required to eject an electron from a particular metal is the work function, W_0.

• The constant h is known as *Planck's constant*.

Key Equations

The frequency at which the radiation from a blackbody is maximum is given by Wien's displacement law:

$$f_{\text{peak}} = (5.88 \times 10^{10} \text{ Hz/K})T$$

The energy in a blackbody at a frequency f must be an integer multiple of the constant $h = 6.63 \times 10^{-34} \text{ J} \cdot \text{s}$:

$$E = nhf \qquad n = 0, 1, 2, 3, \ldots$$

The energy of a photon depends on its frequency:

$$E = hf$$

The cutoff frequency, f_0, below which incident light does not produce the photoelectric effect is

$$f_0 = \frac{W_0}{h}$$

The maximum kinetic energy of a photoelectron is

$$KE_{\text{max}} = hf - W_0$$

24.2 Wave-Particle Duality

🔑 The de Broglie wavelength equation shows that the greater a particle's momentum, the smaller its de Broglie wavelength.

🔑 Subatomic particles like electrons and protons have properties that are different from those of any macroscopic object you might encounter.

• The fact that waves are like particles and particles are like waves is known as *wave-particle duality*.

Key Equation

The de Broglie wavelength of a particle of momentum p is

$$\lambda = \frac{h}{p}$$

24.3 The Heisenberg Uncertainty Principle

🔑 There is a reciprocal relationship between the uncertainty in position, Δy, and the uncertainty in momentum, Δp_y.

• Knowing the position of a particle with greater precision makes its momentum more uncertain. Conversely, knowing the momentum with greater precision makes the position more uncertain.

• In quantum tunneling a particle moves through a region of space that is forbidden to it in classical physics.

• Because particles have wavelengths and can behave like waves, their position and momentum cannot be determined simultaneously with exact precision.

Key Equations

The Heisenberg uncertainty principle for momentum and position is

$$(\Delta p_y)(\Delta y) \geq \frac{h}{2\pi}$$

The Heisenberg uncertainty principle for energy and time is

$$(\Delta E)(\Delta t) \geq \frac{h}{2\pi}$$

Lesson by Lesson

24.1 Quantized Energy and Photons

Conceptual Questions

39. **Predict & Explain** The radiation emitted by blackbody A peaks at a longer wavelength than that of blackbody B. **(a)** Is the temperature of blackbody A greater than or less than the temperature of blackbody B? **(b)** Choose the *best* explanation from among the following:

A. Blackbody A has the higher temperature because the higher the temperature, the greater the wavelength.

B. Blackbody B has the higher temperature because an increase in temperature means an increase in frequency, which corresponds to a decrease in wavelength.

40. How can an understanding of blackbody radiation allow us to determine the temperatures of distant stars?

41. **Differential Fading** Many vehicles in the United States have a small American flag decal in one of their windows, as shown in **Figure 24.20**. If the decal has been in place for a long time, the colors show some fading from exposure to the Sun. In fact, the red stripes are generally more faded than the blue background for the stars. Photographs and posters react in the same way, with red colors showing the most fading. Explain this effect in terms of the photon model of light.

Figure 24.20

42. A source of light is monochromatic. What can you say about the photons emitted by this source?

43. **Rank** A source of red light, a source of green light, and a source of blue light produce beams of light that have the same power. Rank these sources in order of increasing **(a)** wavelength of light, **(b)** frequency of light, and **(c)** number of photons emitted per second. Indicate ties where appropriate.

44. **Predict & Explain** A source of red light has a higher power than a source of green light. **(a)** Is the energy of the photons emitted by the red source greater than, less than, or equal to the energy of the photons emitted by the green source? **(b)** Choose the *best* explanation from among the following:

A. The photons emitted by the red source have greater energy because that source has the greater power.

B. The photons from the red source have less energy than the photons from the green source because they have a lower frequency. The power of the source doesn't matter.

C. Photons from the red source have a lower frequency, but that source also has the greater power. The two effects cancel, so the photons have equal energy.

45. **Predict & Explain** A source of yellow light has a higher power than a source of blue light. **(a)** Is the number of photons emitted per second by the yellow source greater than, less than, or equal to the number of photons emitted per second by the blue source? **(b)** Choose the *best* explanation from among the following:

A. The yellow source emits more photons per second, because (1) it emits more energy per second than the blue source and (2) its photons have less energy than those from the blue source.

B. The yellow source has the higher power, which means its photons have higher energy than the photons from the blue source. Therefore, the yellow source emits fewer photons per second.

C. The two sources emit the same number of photons per second because the higher power of the yellow source compensates for the higher energy of the blue photons.

Problem Solving

46. **Surface Temperature** Betelgeuse, a red-giant star in the constellation Orion, has a peak in its radiation at a frequency of 1.82×10^{14} Hz. What is the surface temperature of Betelgeuse?

47. What is the frequency of the most intense radiation emitted by your body? Assume a skin temperature of 95 °F. What is the wavelength of this radiation?

48. **Cosmic Background Radiation** Outer space is filled with a sea of photons, created in the early moments of the universe. The frequency distribution of this *cosmic background radiation* matches that of a blackbody at a temperature of 2.7 K. **(a)** What is the peak frequency of this radiation? **(b)** What is the wavelength that corresponds to the peak frequency?

49. When people use a tanning salon, they absorb photons of ultraviolet (UV) light to get the desired tan. What are the frequency and the wavelength of a UV photon whose energy is 6.5×10^{-19} J?

50. A flashlight emits 2.5 W of light energy. Assuming that the light has a frequency of 5.2×10^{14} Hz, determine the number of photons given off by the flashlight per second.

51. Light of frequency 9.95×10^{14} Hz ejects electrons from the surface of silver. If the maximum kinetic energy of the ejected electrons is 0.180×10^{-19} J, what is the work function of silver?

52. (a) How many 350-nm (UV) photons are needed to provide a total energy of 2.5 J? **(b)** How many 750-nm (red) photons are needed to provide the same energy?

53. The maximum wavelength an electromagnetic wave can have and still eject an electron from a copper surface is 264 nm. What is the work function of copper?

54. ⟨Think & Calculate⟩ Albireo in the constellation Cygnus, which appears as a single star to the naked eye, is actually a beautiful double-star system. One of the two stars is referred to as A and has a surface temperature of $T_A = 4700$ K; its companion is B, with a surface temperature of $T_B = 13,000$ K. **(a)** When viewed through a telescope, one star is a brilliant blue color, and the other has a warm golden color, as shown in **Figure 24.21**. Is star A or star B the one that is blue? Explain. **(b)** What is the ratio of the peak frequencies emitted by the two stars (f_A/f_B)?

Figure 24.21 The double star Albireo in the constellation Cygnus.

55. ⟨Think & Calculate⟩ Modern halogen lightbulbs have filaments that operate at a higher temperature than the filaments in standard incandescent bulbs. For comparison, the filament in a standard lightbulb operates at about 2900 K, whereas the filament in a halogen bulb operates at 3400 K. **(a)** Which bulb has the higher peak frequency? **(b)** The human eye is most sensitive to a frequency around 5.5×10^{14} Hz. Which bulb produces a peak frequency closer to this value?

56. ⟨Think & Calculate⟩ A typical lightbulb contains a tungsten filament that reaches a temperature of about 2850 K, roughly half the surface temperature of the Sun. **(a)** Treating the filament as a blackbody, determine the frequency for which its radiation is a maximum. **(b)** Do

you expect the lightbulb to radiate more energy in the visible or in the infrared part of the spectrum? Explain.

57. Exciting an Oxygen Molecule An oxygen molecule (O_2) vibrates with an energy identical to that of a single particle of mass $m = 1.340 \times 10^{-26}$ kg attached to a spring with a spring constant $k = 1215$ N/m. The energy levels of the system are uniformly spaced, as indicated in **Figure 24.22**, with a separation given by hf. **(a)** What is the vibration frequency of this molecule? **(b)** How much energy must be added to the molecule to excite it from one energy level to the next higher level?

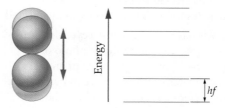

Figure 24.22

58. ⟨Think & Calculate⟩ You have two different lightbulbs, as shown in **Figure 24.23**. One is a 150-W red bulb, and the other is a 25-W blue bulb. **(a)** Which bulb emits more photons per second? **(b)** Which bulb emits photons of higher energy? **(c)** Calculate the number of photons emitted per second by each bulb. Take $\lambda_{red} = 650$ nm and $\lambda_{blue} = 460$ nm. (Most of the electromagnetic radiation given off by incandescent lightbulbs is in the infrared portion of the spectrum. For the purposes of this problem, however, assume that all of the radiated power is at the wavelengths indicated.)

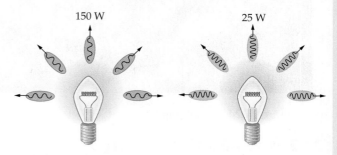

Figure 24.23

59. ⟨Think & Calculate⟩ Aluminum and calcium have the work functions $W_{Al} = 4.28$ eV and $W_{Ca} = 2.87$ eV, respectively. **(a)** Which metal requires higher-frequency light to produce photoelectrons? Explain. **(b)** Calculate the minimum frequency that will produce photoelectrons from each metal.

60. ⟨Think & Calculate⟩ Two beams of light, A and B, with different wavelengths ($\lambda_A > \lambda_B$) are used to produce photoelectrons from a given metal surface. **(a)** Which beam produces photoelectrons with greater kinetic energy? Explain. **(b)** Find KE_{max} for cesium ($W_0 = 1.9$ eV) if $\lambda_A = 620$ nm and $\lambda_B = 410$ nm.

61. ⃞Think & Calculate⃞ Zinc and cadmium have the work functions $W_{Zn} = 4.33$ eV and $W_{Cd} = 4.22$ eV, respectively. **(a)** If both metals are illuminated by UV radiation of the same wavelength, which one gives off photoelectrons with the greater maximum kinetic energy? Explain. **(b)** Calculate the maximum kinetic energy of photoelectrons from each metal if $\lambda = 275$ nm.

62. White light, with frequencies ranging from 4.00×10^{14} Hz to 7.90×10^{14} Hz, is incident on a potassium surface. Given that the work function of potassium is 2.24 eV, find **(a)** the maximum kinetic energy of electrons ejected from this surface and **(b)** the range of frequencies for which no electrons are ejected.

63. **Owl Vision** Owls have large, sensitive eyes that provide good night vision, as is apparent in the photo of an owl in **Figure 24.24**. Typically, the pupil of an owl's eye can have a diameter of 8.5 mm (as compared with a maximum diameter of about 7.0 mm for a human eye). In addition, an owl's eye is about 100 times more sensitive to light of low intensity than a human eye, allowing owls to detect light with an intensity as low as 5.0×10^{-13} W/m². Find the minimum number of photons per second an owl can detect, assuming a frequency of 7.0×10^{14} Hz for the light.

Figure 24.24 An apt pupil.

24.2 Wave-Particle Duality
Conceptual Questions

64. ⃞Predict & Explain⃞ **(a)** As you accelerate your car away from a stoplight, does the de Broglie wavelength of the car increase, decrease, or stay the same? **(b)** Choose the *best* explanation from among the following:

A. The de Broglie wavelength will increase because the momentum of the car increases.

B. The momentum of the car increases. It follows that the de Broglie wavelength will decrease, because it is inversely proportional to the wavelength.

C. The de Broglie wavelength of the car depends only on its mass, which doesn't change by pulling away from the stoplight. Therefore, the de Broglie wavelength stays the same.

65. By what factor does the de Broglie wavelength of a particle change if **(a)** its momentum is doubled or **(b)** its kinetic energy is doubled?

66. Why can an electron microscope resolve smaller objects than a light microscope?

67. A hydrogen atom and a helium atom have the same speed. Is the de Broglie wavelength of the hydrogen atom greater than, less than, or the same as the de Broglie wavelength of the helium atom? Explain.

68. A proton is about 2000 times more massive than an electron. Is it possible for an electron to have the same de Broglie wavelength as a proton? Explain.

Problem Solving

69. A particle with a mass of 6.69×10^{-27} kg has a de Broglie wavelength of 7.22 pm. What is the particle's speed? (Recall that 1 pm $= 10^{-12}$ m.)

70. What speed must a neutron have if its de Broglie wavelength is to be equal to 0.282 nm, the spacing between planes of atoms in crystals of table salt?

71. A 79-kg jogger runs with a speed of 4.2 m/s. What is the jogger's de Broglie wavelength?

72. Find the kinetic energy of an electron whose de Broglie wavelength is 0.15 nm.

73. A beam of particles consists of neutrons with a de Broglie wavelength of 0.250 nm. What is the speed of the neutrons?

74. ⃞Think & Calculate⃞ An electron and a proton have the same speed. **(a)** Which has the longer de Broglie wavelength? Explain. **(b)** Calculate the ratio of the wavelengths ($\lambda_{electron}/\lambda_{proton}$).

75. ⃞Think & Calculate⃞ An electron and a proton have the same de Broglie wavelength. **(a)** Which has the greater kinetic energy? Explain. **(b)** Calculate the ratio of the electron's kinetic energy to the proton's kinetic energy.

76. Diffraction effects become significant when the width of an aperture is comparable to the wavelength of the waves being diffracted. **(a)** At what speed will the de Broglie wavelength of a 65-kg student be equal to the 0.76-m width of a doorway? **(b)** At this speed, how long will it take the student to travel a distance of 1.0 mm? (For comparison, the age of the universe is approximately 4×10^{17} s.)

77. A particle has a mass m and a charge q. The particle is accelerated from rest through a potential difference V. What is the particle's de Broglie wavelength, expressed in terms of m, q, and V?

24.3 The Heisenberg Uncertainty Principle
Conceptual Questions

78. An electron and a proton have the same uncertainty in speed. Is the electron's uncertainty in momentum greater than, less than, or equal to the proton's uncertainty in momentum? Explain.

79. An electron and a proton are confined to boxes of the same size. Is the uncertainty in momentum of the electron greater than, less than, or equal to the uncertainty in momentum of the proton? Explain.

80. An electron and a helium atom are confined in separate boxes. The uncertainty in the electron's kinetic energy is the same as the uncertainty in the helium atom's kinetic energy. All other things being equal, is the box containing the electron larger, smaller, or the same size as the box containing the helium atom? Explain.

Mixed Review

81. Suppose you perform an experiment on the photoelectric effect using light with a frequency high enough to eject electrons. If the intensity of the light is increased while the frequency is held constant, predict whether the following quantities will increase, decrease, or stay the same: **(a)** the maximum kinetic energy of an ejected electron; **(b)** the minimum de Broglie wavelength of an electron; **(c)** the number of electrons ejected per second; **(d)** the electric current in the phototube.

82. Suppose you perform an experiment on the photoelectric effect using light with a frequency high enough to eject electrons. If the frequency of the light is increased while the intensity is held constant, predict whether the following quantities will increase, decrease, or stay the same: **(a)** the maximum kinetic energy of an ejected electron; **(b)** the minimum de Broglie wavelength of an electron; **(c)** the number of electrons ejected per second; **(d)** the electric current in the phototube.

83. **Predict & Explain** Light of a particular wavelength does not eject electrons from the surface of a given metal. **(a)** Should the wavelength of the light be increased or decreased in order to cause electrons to be ejected? **(b)** Choose the *best* explanation from the following:

A. The photons have too little energy to eject electrons. To increase their energy, their wavelength should be increased.

B. The energy of a photon is proportional to its frequency, that is, inversely proportional to its wavelength. To increase the energy of the photons so that they can eject electrons, their wavelength should be decreased.

84. Two 57.5-kW radio stations broadcast at different frequencies. Station A broadcasts at a frequency of 892 kHz, and station B broadcasts at a frequency of 1410 kHz. **(a)** Which station emits more photons per second? Explain. **(b)** Which station emits photons of higher energy?

85. You want to construct a photocell that works with visible light. Three materials are readily available: aluminum ($W_0 = 4.28$ eV), lead ($W_0 = 4.25$ eV), and cesium ($W_0 = 2.14$ eV). Which material(s) would be suitable?

86. **Human Vision** Studies have shown that some people can detect 545-nm light with as few as 100 photons entering the eye per second. What is the power delivered by such a beam of light?

87. To allow a family to listen to a radio station, a certain home receiver must pick up a signal of at least 1.0×10^{-10} W. If the radio waves have a frequency of 96 MHz, how many photons must the receiver absorb per second to pick up the signal?

88. The latent heat for converting ice at 0 °C to water at 0 °C is 33.5×10^4 J/kg. How many photons of frequency 6.0×10^{14} Hz must be absorbed by a 1.0-kg block of ice at 0 °C to melt it to water at 0 °C?

89. How many 550-nm photons would have to be absorbed to raise the temperature of 1 g of water by 1.0 °C?

90. Light with a frequency of 2.11×10^{15} Hz ejects electrons from a surface of lead, which has a work function of 4.25 eV. What is the minimum de Broglie wavelength of the ejected electrons?

91. An electron moving with a speed of 2.7×10^6 m/s has the same momentum as a proton. Find **(a)** the de Broglie wavelength of the electron, **(b)** the de Broglie wavelength of the proton, and **(c)** the speed of the proton.

92. **Firefly Light** Fireflies, like the one shown in **Figure 24.25**, are often said to give off "cold" light. Given that the peak in a firefly's radiation occurs at about 5.4×10^{14} Hz, determine the temperature of a blackbody that would have the same peak frequency. From your result, would you say that firefly radiation is well approximated by blackbody radiation? Explain.

Figure 24.25 How cool is that?

93. A pendulum consisting of a 0.15-kg mass attached to a 0.78-m string undergoes simple harmonic motion. **(a)** What is the frequency of oscillation for this pendulum? **(b)** Assuming that the energy of this system satisfies $E = nhf$, find the maximum speed of the 0.15-kg mass when the quantum number, n, is 1.0×10^{33}.

94. When light with a wavelength of 545 nm shines on a metal surface, electrons are ejected with speeds of 3.10×10^5 m/s or less. Determine the work function and cutoff frequency for this metal.

95. **Think & Calculate** A hydrogen atom absorbs a 486.2-nm photon. A short time later, the same atom emits a photon with a wavelength of 97.23 nm. **(a)** Has the net energy of the atom increased or decreased? Explain. **(b)** Calculate the change in energy of the hydrogen atom.

96. **Think & Calculate** **(a)** Does the de Broglie wavelength of a particle increase or decrease as its kinetic energy increases? Explain. **(b)** Show that the de Broglie wavelength of an electron, in nanometers, is given by $\lambda = (1.23 \text{ nm})/\sqrt{K}$, where K is the kinetic energy of the electron in electron volts.

97. **Think & Calculate** Light of frequency 8.22×10^{14} Hz ejects electrons from surface A that have a maximum kinetic energy that is 2.00×10^{-19} J greater than the maximum kinetic energy of electrons ejected from surface B. **(a)** If the frequency of the light is increased, does the difference in the maximum kinetic energies of electrons from the two surfaces increase, decrease, or stay the same? Explain. **(b)** Calculate the difference in work function for the two surfaces.

Writing about Science

98. Write a report on solar energy panels. When were they first developed? What is their efficiency in converting the energy of sunlight to electrical energy? How large a panel is needed to power a typical house?

99. **Connect to the** Big **Idea** Discuss the operation of an electron microscope. Why does it have a higher magnification than a light microscope? How is this related to wave-particle duality?

Read, Reason, and Respond

Millikan and the Photoelectric Effect Robert A. Millikan (1868–1953), best known for his oil-drop experiment, in which he measured the charge of an electron, also performed pioneering research on the photoelectric effect. In fact, the 1923 Nobel Prize in physics was awarded to Millikan "for his work on the elementary charge of electricity and on the photoelectric effect." Initially convinced that Einstein's theory of the photoelectric effect was wrong—because of overwhelming evidence for the wave nature of light—Millikan undertook a decade-long experimental program to study the effect. In the end, his experiments confirmed Einstein's theory in every detail and ushered in the modern view of light as having wave-particle duality.

Millikan carried out an exhaustive set of experiments on a variety of materials. In experiments on lithium, for example, he observed a maximum kinetic energy of 0.550 eV when electrons were ejected by 433.9-nm light. When light of 253.5 nm was used, he observed a maximum kinetic energy of 2.57 eV. Using results like this, Millikan was able to measure the value of Planck's constant and to show that the value obtained from the experiments on the photoelectric effect is in complete agreement with the value obtained from studying blackbody radiation.

100. What is the work function, W_0, for lithium, as determined from Millikan's results?

    A. 0.0112 eV   C. 1.63 eV

    B. 0.951 eV   D. 2.29 eV

101. What value did Millikan obtain for Planck's constant, based on the lithium measurements? (His value is close to, but not the same as, the currently accepted value.)

    A. 1.12×10^{-34} J · s

    B. 3.84×10^{-34} J · s

    C. 6.14×10^{-34} J · s

    D. 6.57×10^{-34} J · s

102. What maximum kinetic energy do you predict Millikan found when he used light with a wavelength of 365.0 nm?

    A. 0.805 eV   C. 2.29 eV

    B. 1.08 eV   D. 2.82 eV

Standardized Test Prep

Multiple Choice

1. Which of the following statements about the Heisenberg uncertainty principle is *false*?
   - (A) Measurements of position become more uncertain as measurements of momentum are more exact.
   - (B) Measurements of momentum become more uncertain as measurements of time are more exact.
   - (C) Measurements of momentum become more uncertain as measurements of position are more exact.
   - (D) Measurements of energy become more uncertain as measurements of time are more exact.

2. Which of the following energies is closest to that of a photon with a frequency of 1×10^{15} Hz?
   - (A) 1.2×10^{-27} J
   - (B) 2.0×10^{-21} J
   - (C) 6.6×10^{-19} J
   - (D) 4.2×10^{-12} J

3. What is the momentum of a proton with a wavelength of 200 nm?
   - (A) 3.3×10^{2} kg · m/s
   - (B) 3.3×10^{-10} kg · m/s
   - (C) 3.3×10^{-15} kg · m/s
   - (D) 3.3×10^{-27} kg · m/s

4. Planck's constant was initially introduced to explain
   - (A) blackbody radiation curves.
   - (B) the photoelectric effect.
   - (C) photon scattering.
   - (D) X-ray diffraction.

5. According to the concept of wave-particle duality, the de Broglie wavelength
   - (A) of an electron is too small to be significant.
   - (B) increases as momentum increases.
   - (C) is too large to be significant for macroscopic objects.
   - (D) is inversely proportional to momentum.

6. A certain metal has a cutoff frequency of 1.5×10^{15} Hz. What is the work function for this metal?
   - (A) 9.9×10^{-19} J
   - (B) 1.9×10^{-16} J
   - (C) 9.9×10^{-15} J
   - (D) 1.9×10^{-12} J

Use the graph below to answer Questions 7–9. The graph shows the results of a photoelectric experiment in which light of varying frequency strikes a zinc metal strip.

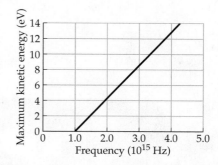

7. Which is the *best* estimate of the cutoff frequency for the ejection of electrons from the zinc surface?
   - (A) 5.0×10^{14} Hz
   - (B) 1.0×10^{15} Hz
   - (C) 2.0×10^{15} Hz
   - (D) 2.5×10^{15} Hz

8. What does the slope of the line on the graph represent?
   - (A) the energy of the incident photons
   - (B) the work function for zinc
   - (C) Planck's constant
   - (D) the *KE* of the emitted electrons

Extended Response

9. Discuss the meaning of the graph's *y* intercept and relate it to the equation for the photoelectric effect. Estimate the value of the *y* intercept from the graph—be sure to include units.

> **Test-Taking Hint**
>
> **Think about conservation of energy when solving problems in quantum physics.**

If You Had Difficulty With . . .

Question	1	2	3	4	5	6	7	8	9				
See Lesson	24.3	24.1	24.2	24.1	24.2	24.1	24.1	24.1	24.1				

25

Atomic Physics

Inside

This colorful lighted sign has great eye appeal. It's even more interesting when you realize that the light is produced by electrons in various atoms jumping from one energy level to another.

Big Idea

The wave properties of matter mean that the atomic-level world must be described in terms of probability.

In today's world we take for granted that everything around us is made of atoms. Although it may seem surprising at first, this belief in atoms has not always been universal. As recently as the first part of the twentieth century, there was still serious debate about the microscopic nature of matter. With the advent of quantum physics, however, that debate quickly faded away as atomic structure came to be understood in ever greater detail.

Inquiry Lab   How did Rutherford discover the nucleus?

Explore

1. Obtain a model of the Rutherford experiment from your teacher. Each model has a board that conceals an unknown "atomic structure" below and a set of metal marbles to use as bombardment particles.
2. Roll the marbles under the board one at a time and carefully note how they behave. Where do they enter? Where do they exit? Are they deflected as they pass through? What happens when you change the incident angle of a marble?
3. Repeat Step 2 several times, taking detailed notes and adjusting the paths of your bombardment particles in order to develop an accurate picture of the unseen "atom" below the board.

Think

1. **Describe** Review your results from Steps 2 and 3. Based on your data, draw a picture of the unknown "atomic structure" below the board. Explain how your data support the picture.
2. **Predict** How would the results differ if you used a tennis ball instead of a marble as the bombardment particle?
3. **Extend** Remove the board and see what is under it. Evaluate the accuracy of your experimental results. Explain one change you could make in this lab to improve your results.

25.1   Early Models of the Atom

Speculations about the microscopic structure of matter have intrigued people for thousands of years. The ancient Greek philosophers Leucippus and Democritus considered what would happen if one were to repeatedly cut a block of copper in half. They reasoned that eventually the block would be reduced to a single speck of copper that could not be divided further. This smallest piece of an element was called the *atom* ($a + tom$), which means, literally, "without division."

It took until the late nineteenth century, however, before scientists started unlocking the mystery of the atom. We start this chapter with a look at some of the earliest scientific models of atoms.

Vocabulary

- nucleus
- line spectrum

Thomson compared the atom to a plum pudding

In 1897 the English physicist J. J. Thomson (1856–1940) discovered the *electron*, a particle smaller in size and thousands of times less massive than the smallest atom. Unlike atoms, which are electrically neutral, electrons were found to have a negative charge. Thomson proposed, therefore, that atoms have an internal structure that includes both electrons and a quantity of positively charged matter. The positive matter would account for most of the atom's mass, and its charge would be equal to but opposite that of the electrons.

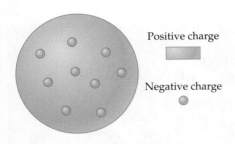

Positive charge

Negative charge

▲ **Figure 25.1 The plum-pudding model of an atom**
The model of an atom proposed by J. J. Thomson consists of matter having a uniform positive charge, which accounts for most of the mass of the atom, with small negatively charged electrons scattered throughout it, like raisins in a pudding.

Thomson's concept of the atom was known as the *plum-pudding model*. He envisioned negative electrons embedded in a more or less uniform distribution of positive charge—like raisins spread throughout a pudding. **Figure 25.1** shows the basic idea. Although the plum-pudding model agreed with everything known about atoms at the time, experimental results quickly discredited it. The plum-pudding model was soon replaced.

The Nuclear Model

Thomson's findings and speculations inspired other physicists to investigate atomic structure. In 1909 Ernest Rutherford (1871–1937) and his coworkers Hans Geiger (1882–1945) and Ernest Marsden (1889–1970) tested Thomson's model by shooting a beam of positively charged *alpha particles* (the nuclei of helium atoms) at a thin strip of gold foil.

Rutherford found the atom to be mostly empty space

Rutherford, Geiger, and Marsden expected the *positively* charged alpha particles to be deflected as they passed through the *positively* charged "pudding" in the gold foil. After all, like charges repel. They had other expectations as well:

- The deflections should be relatively small, since the alpha particles have a substantial mass and the positive charge in the atom is spread out.
- The alpha particles should deflect in roughly the same way, since the positive pudding fills virtually all of an atom.

The results of the gold-foil experiment did not support these predictions. In fact, most of the alpha particles passed right through the foil—as if it were not there. Could it be that the atoms in the foil were mostly empty space? Because the results were rather surprising, Rutherford suggested repeating the experiment to look for alpha particles that might be deflected through large angles.

This suggestion turned out to be an inspired hunch. Not only were large-angle deflections observed, some of the alpha particles were found to have *reversed* their direction of motion! Rutherford was stunned. In his own words, "It was almost as incredible as if you fired a 15-inch [artillery] shell at a piece of tissue paper and it came back and hit you."

How does the nuclear model depict atomic structure?

Rutherford proposed a nuclear model for the atom

Rutherford proposed a new atomic model to account for the experimental results. The model was similar to the Solar System, as illustrated in **Figure 25.2.** **Rutherford's model depicted an atom in which lightweight, negatively charged electrons orbit an extremely small but massive nucleus.** The nucleus is the region of space at the center of the atom that contains all of the atom's positive charge and almost all of its mass.

According to Rutherford's *nuclear model*, most of an atom is empty space. This explained why the majority of alpha particles passed right through the gold foil. Furthermore, the atom's positive charge is highly concentrated in a small nucleus rather than spread throughout the atom. As a result, an alpha particle that makes a head-on collision with the nucleus can actually be turned around, just as observed in the experiments.

Rutherford estimated the diameter of a nucleus to be about 10,000 times smaller than the diameter of the atom. To put this into perspective, imagine enlarging an atom until its nucleus is the size of the Sun. An electron in this case would be the same distance from the nucleus as Pluto is from the Sun. Inside the electron's orbit are empty space and the nucleus. This means that an atom has an even larger fraction of empty space than the Solar System!

Rutherford's atomic model contained serious flaws

Though the nuclear model seems reasonable, it contains fatal flaws. One of these is that an orbiting electron experiences a centripetal acceleration toward the nucleus. As you may recall, any accelerating electric charge gives off energy in the form of electromagnetic radiation. This means that an orbiting electron would constantly radiate away energy as it orbits. The situation is similar to a satellite losing energy to air resistance when it orbits too close to the Earth's atmosphere. Just like such a satellite, an electron radiating energy would spiral inward and eventually plunge into the nucleus. Because this process of collapse would happen very quickly, the atom of Rutherford's model would simply not be stable. This did not agree with the observed stability of atoms in nature.

A second flaw has to do with the radiation emitted by the orbiting electron. The frequency of the radiation should be the same as the frequency of the orbit. If electrons spiraled inward, the frequency would increase continuously, and the light emitted by the atom would span a continuous range of frequencies. This prediction does not agree with experiments, which show that atoms emit only certain discrete frequencies of light.

Atomic Line Spectra

A device known as a *gas discharge tube* is used to study the light given off by atoms. A discharge tube is a sealed glass tube containing a gas at low pressure. When a large voltage is applied to the ends of the tube, the atoms in the gas become excited and emit electromagnetic radiation. Passing the radiation through a diffraction grating separates it into its various wavelengths, as shown in **Figure 25.3**. The result is a series of brightly colored lines reminiscent of the bar codes used on products sold in supermarkets.

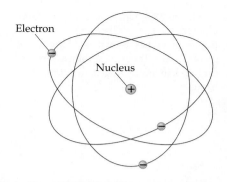

▲ **Figure 25.2 The solar system model of an atom**
Ernest Rutherford proposed that an atom is like a miniature Solar System, with a massive positively charged nucleus orbited by lightweight negatively charged electrons.

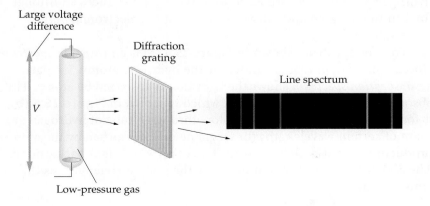

◀ **Figure 25.3 The line spectrum of an atom**
The light given off by individual atoms, such as those of a gas at low pressure, consists of a series of discrete wavelengths corresponding to different colors.

Each element has its own distinctive set of spectral lines. The spectra shown here are for the elements hydrogen, mercury, and neon.

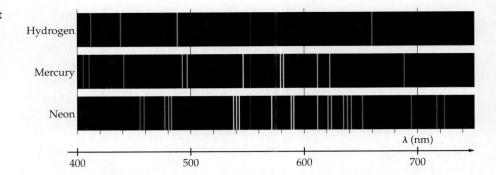

This type of spectrum emitted by atoms, with bright lines at specific frequencies (colors), is referred to as a **line spectrum.** The precise colors of the lines provide a sort of "fingerprint" that identifies a particular type of atom, just as each product in a supermarket has its own unique bar code. Examples of line spectra for several different elements are shown in **Figure 25.4**.

Line spectra differ from blackbody radiation curves

How does an atomic line spectrum differ from a blackbody radiation curve?

Let's contrast the line spectra of atoms with the blackbody radiation curves discussed in Chapter 24. Blackbody radiation is electromagnetic radiation is given off over a continuous range of frequencies. The spectrum, or distribution, of this radiation depends only on temperature, and it is the same for all materials. **Blackbody radiation is produced by the blackbody as a whole, and not its individual atoms. A line spectrum, on the other hand, is produced by individual atoms giving off electromagnetic radiation at precise frequencies.**

Physicists study atomic spectra in gas discharge tubes. The atoms in the gas inside the discharge tube are well separated from one another, and the light they give off is the result of their acting separately. As we will see later in this chapter, atoms have energies that are similar to standing waves on a string. Because only certain energies produce a standing wave, it follows that only certain energies (frequencies) are emitted in the radiation given off by an atom. The lines in an atomic spectrum are like the harmonics of a vibrating string, and they show the unique character of the atom's radiation.

Atoms have characteristic spectra

Scientists can learn a lot about atoms from the light they give off. As an example, consider the visible part of the line spectrum of atomic hydrogen shown in **Figure 25.5 (a)**. (Hydrogen produces additional lines in the infrared and ultraviolet parts of the electromagnetic spectrum.)

The line spectrum shown in Figure 25.5 (a) is an *emission spectrum*, since it shows light that is emitted by the hydrogen atoms. Similarly, an *absorption spectrum* shows the light that is absorbed by atoms. The absorption spectrum of hydrogen, which is shown in **Figure 25.5 (b)**, is formed by passing light of all colors through a tube of hydrogen gas. Light of certain wavelengths (or frequencies) is absorbed by the atoms, producing a set of dark lines against an otherwise bright background. The absorption lines occur at precisely the same wavelengths as the emission lines.

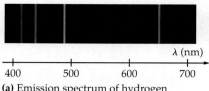

(a) Emission spectrum of hydrogen

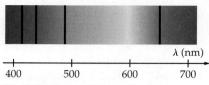

(b) Absorption spectrum of hydrogen

◀ **Figure 25.5 The line spectrum of hydrogen**
The **(a)** emission and **(b)** absorption spectra of hydrogen. Notice that the wavelengths absorbed by hydrogen (dark lines) are the same as those emitted by hydrogen (colored lines).

The fact that atoms have unique line spectra has interesting implications. For example, did you know that helium was first found on the Sun, rather than on Earth? How could that be? Well, astronomers observing a total eclipse of the Sun in 1868 noticed a spectral line that didn't correspond to any known element on Earth. They proposed that this line was due to a new element, which they named *helium*—for the Greek god of the Sun, Helios. It wasn't until 1903 that helium was found on Earth. Similarly, measuring the spectra of distant stars tells us about their composition.

Physics & You: Technology Here on Earth, *spectroscopy*—the study of atomic line spectra—has many practical applications. For example, a technique known as *laser induced breakdown spectroscopy* (LIBS) is used to study the chemical composition of materials. In LIBS a short laser pulse vaporizes a small sample of a material, creating a plasma of excited atoms. The atoms give off light, with a unique spectral signature for each element. Analyzing the light gives the chemical composition of the sample. Applications of LIBS range from testing human hair and teeth for metal poisoning, cancer, or bacterial infection, to testing for the presence of gunpowder residue at a crime scene. LIBS is also used by geologists to analyze soil and minerals, and by architects to determine the quality of materials used in the construction of buildings.

So why do atoms give off just certain frequencies of radiation and not others? Quantum physics gives the answers, as we will see in the next two lessons.

25.1 LessonCheck (MP)

Checking Concepts

1. 🔲 **Explain** How is Rutherford's model of the atom similar to the Solar System?

2. 🔲 **Explain** Why is sunlight made up of a continuous range of colors, whereas the light emitted from a hydrogen atom is not?

3. Identify What aspects of Rutherford's atomic model are not supported by experimental results?

4. Describe How do the wavelengths of lines on an atom's emission spectrum compare with those on its absorption spectrum?

25.2 Bohr's Model of the Hydrogen Atom

Vocabulary

- Bohr radius
- energy-level diagram
- ground state
- excited state

🔑 *When does an atom radiate energy in the Bohr model?*

CONNECTINGIDEAS

In Chapter 9 you learned how to calculate the orbits of planets subject to the attractive force of gravity.

- The same basic calculation is used here to calculate the allowed Bohr orbits for electrons. The difference is that the force acting on the electron is due to electric charge (and determined with Coulomb's law), not gravity.

Our understanding of the hydrogen atom took a giant leap forward in 1913. In that year Niels Bohr (1885–1962), a Danish physicist who had just earned his doctorate in 1911, introduced a model that allowed him to understand the line spectrum of hydrogen. Bohr, seen riding a motorcycle in **Figure 25.6**, combined elements of classical physics with the ideas of quantum physics introduced by Planck and Einstein. Thus, his model is a hybrid that spanned the gap between the classical physics of Newton and the newly emerging quantum physics.

Bohr's model is based on a few simple assumptions

Bohr's model of the hydrogen atom is based on four assumptions. The first two are specific to his model. They are as follows:

- The electron in a hydrogen atom moves in a circular orbit about the nucleus.
- Only certain circular orbits are allowed. In these orbits the angular momentum of the electron is quantized, just like energy is quantized in Einstein's photon model of light.

The next two assumptions are more general:

- Electrons do not give off electromagnetic radiation when they are in an allowed orbit. Thus, these orbits are stable.
- Electromagnetic radiation is given off or absorbed only when an electron changes from one allowed orbit to another. If the energy difference between two allowed orbits is ΔE, then the frequency, f, of the photon that is emitted or absorbed is given by $|\Delta E| = hf$.

To summarize, Bohr's model retains the classical picture of an electron orbiting a nucleus, as in Rutherford's model. It also adds the quantum requirements, however, that only certain orbits are allowed and that no radiation is given off from these orbits. 🔑 **Radiation is given off *only* when an electron shifts from one orbit to another, and then the radiation is in the form of a photon that obeys Einstein's quantum relation $E = hf$.**

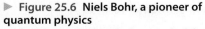

▶ **Figure 25.6 Niels Bohr, a pioneer of quantum physics**
Niels Bohr was known for his zest for life, and for his leading role in the development of quantum physics. Here he's seen applying the principles of classical physics with some of his family members.

Bohr's electron orbits have specific radii

All Bohr orbits are quantized—they have specific radii *and* specific energies. To find the allowed orbits in the Bohr model, we apply two conditions. First, the electron moves in a circular orbit of radius r with a speed v, as depicted in **Figure 25.7**. As a result, the electron experiences a centripetal acceleration toward the nucleus of magnitude $a = v^2/r$. From Newton's second law of motion, we know that a force, $F = ma$, is required to produce the acceleration. In this case, the force is the electrostatic attraction between the electron and the nucleus (a single proton). The magnitude of this force is $F = ke^2/r^2$, where e is the magnitude of the charge on both the electron and the proton. Combining these results yields the following relationship:

$$F = ma$$

$$k\frac{e^2}{r^2} = m\frac{v^2}{r}$$

Next, Bohr assumed that the angular momentum in an allowed orbit must be an integer n (the quantum number) times $h/2\pi$, where h is Planck's constant. Since the electron moves with a speed v in a circular path of radius r, its angular momentum is $L = rmv$. Thus, this condition is

$$L_n = r_n mv_n = \frac{nh}{2\pi}$$

Combining the force and angular momentum equations allows us to solve for the radii of the allowed orbits. The result is

$$r_n = \left(\frac{h^2}{4\pi^2 mke^2}\right)n^2 \qquad n = 1, 2, 3, \ldots$$

Substituting the known values for h, π, m, k, and e yields the following simple result:

> **Orbital Radius of an Electron in a Hydrogen Atom**
>
> $$\text{orbital radius of electron} = (5.29 \times 10^{-11}\ \text{m}) \times (\text{quantum number})^2$$
>
> $$r_n = (5.29 \times 10^{-11}\ \text{m})n^2 \qquad n = 1, 2, 3, \ldots$$

The radius of the first Bohr orbit, which corresponds to $n = 1$, is

$$r_1 = 5.29 \times 10^{-11}\ \text{m}$$

This is known as the **Bohr radius.** It sets the typical size of a hydrogen atom (which has a diameter twice its value). The higher orbits in the Bohr model increase in radius as n^2, as indicated in **Figure 25.8**. There is no upper limit to the value of the quantum number, n, and hence there is no upper limit to the size of the orbital radius. In principle, a hydrogen atom can have any size from the Bohr radius to infinity.

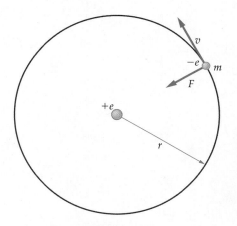

▲ **Figure 25.7 A Bohr orbit**
In the Bohr model of hydrogen, electrons orbit the nucleus in circular orbits. The centripetal acceleration of the electron is produced by the electrostatic force of attraction between it and the nucleus.

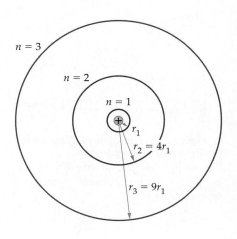

▶ **Figure 25.8 The first three Bohr orbits**
The first Bohr orbit has a radius $r_1 = 5.29 \times 10^{-11}$ m. The second and third Bohr orbits have radii $r_2 = 2^2 r_1 = 4r_1$ and $r_3 = 3^2 r_1 = 9r_1$, respectively. (*Note:* For clarity, the nucleus is drawn larger than its true scale relative to the size of the atom.)

What is the radius of the $n = 4$ Bohr orbit?

Solution

Substitute $n = 4$ in the equation $r_n = (5.29 \times 10^{-11}\text{ m})n^2$. This yields

$$r_4 = (5.29 \times 10^{-11}\text{ m})(4^2)$$

$$= \boxed{8.46 \times 10^{-10}\text{ m}}$$

Practice Problems

5. Which Bohr orbit has a radius of 1.90×10^{-9} m?

6. What is the Bohr orbital radius that lies in the range from 9.5×10^{-10} m to 1.7×10^{-9} m?

Bohr's electron orbits have specific energies

The total energy of the hydrogen atom is also quantized. In fact, a straightforward calculation, combining both the kinetic energy $(\frac{1}{2}mv^2)$ and the potential energy $(-ke^2/r)$, shows that the total energy of the nth Bohr orbit is

$$E_n = -\left(\frac{2\pi^2 mk^2 e^4}{h^2}\right)\frac{1}{n^2} \qquad n = 1, 2, 3, \ldots$$

This simplifies to the following:

CONNECTINGIDEAS

Electric potential energy was introduced in Chapter 20.

• Here it plays a role in determining the allowed energies for orbiting electrons in the Bohr model.

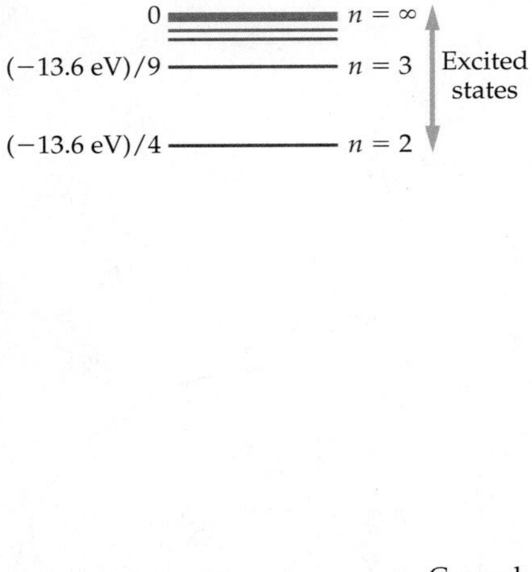

Energy of a Hydrogen Atom
$\text{total energy of a hydrogen atom} = -(13.6\text{ eV}) \times \dfrac{1}{(\text{quantum number})^2}$
$E_n = -(13.6\text{ eV})\dfrac{1}{n^2} \qquad n = 1, 2, 3, \ldots$

These energies are plotted in **Figure 25.9** for various values of n. This type of graph is referred to as an **energy-level diagram**. The **ground state** $(n = 1)$ of the atom corresponds to its lowest possible energy. Energy levels above the ground state are referred to as **excited states**. In general, an allowed Bohr orbit is known as a *state* of hydrogen—for example, the $n = 2$ Bohr orbit is referred to as the $n = 2$ state of hydrogen.

Figure 25.9 also shows that the energy of the excited states approaches zero as the quantum number n becomes infinitely large. Zero is the energy an electron and proton have when at rest and separated by an infinite distance. Thus, to *ionize* a hydrogen atom—that is, to remove the electron from the atom—requires a minimum energy of 13.6 eV. This value, which is a specific prediction of the Bohr model, is in complete agreement with experimental results.

▲ **Figure 25.9 Energy-level diagram for the Bohr model of hydrogen**
The energy of the ground state of hydrogen is −13.6 eV. Excited states of hydrogen approach zero energy.

What is the energy of the $n = 4$ Bohr orbit?

Solution

Substitute $n = 4$ in the equation $E_n = -(13.6 \text{ eV})\dfrac{1}{n^2}$. This yields

$$E_4 = -(13.6 \text{ eV})\left(\dfrac{1}{4^2}\right)$$

$$= \boxed{-0.85 \text{ eV}}$$

Practice Problems

7. Which Bohr orbit has an energy of -0.278 eV?

8. What is the quantum number of the Bohr orbit whose energy lies in the range from -0.72 eV to -0.46 eV?

Bohr's model predicts the spectrum of hydrogen

The energy values given by the Bohr model agree with the experimentally observed line spectrum of hydrogen. Let's see how this occurs.

The basic idea is that when an electron "jumps" from an excited state of hydrogen to a lower energy level, the change in energy is given off as a photon. If the electron does a quantum jump from the $n = 3$ state to the $n = 2$ state, for example, the photon that is given off corresponds to the red line shown in **Figure 25.10**. If the quantum jump is from the $n = 4$ state to the $n = 2$ state, the photon corresponds to the teal green line in Figure 25.10. **Each possible jump of an electron from one energy state to another results in a different line of the hydrogen spectrum.** This is why hydrogen gives off a line spectrum, and not a continuous spectrum.

Now, according to Einstein's photon model, the energy of a photon is given by hf. Thus, if the energy of an electron in its initial state is E_i, and the energy in its final state is E_f, its change in energy is

$$\Delta E = E_f - E_i$$

The initial and final energies are calculated with $E_n = -(13.6 \text{ eV})/n^2$ using different values of n. Setting the magnitude of the energy change equal to hf yields

$$hf = |\Delta E|$$

How are the spectral lines of hydrogen related to energy level changes?

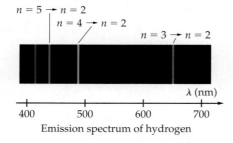

$n = 5 \rightarrow n = 2$
$n = 4 \rightarrow n = 2$
$n = 3 \rightarrow n = 2$

λ (nm)

400   500   600   700

Emission spectrum of hydrogen

◀ **Figure 25.10 Spectral lines correspond to quantum jumps**
The red spectral line of hydrogen corresponds to an electron making a quantum jump from the $n = 3$ state to the $n = 2$ state. The teal green line corresponds to a quantum jump from the $n = 4$ state to the $n = 2$ state. In general, each line in the spectrum corresponds to a different quantum jump.

It is convenient to calculate the energy of a Bohr orbit in terms of electron volts (eV), since this is a typical unit of atomic energy. To use the expression $hf = |\Delta E|$, however, the change in energy must be converted to joules (J), using 1.60×10^{-19} J = 1 eV.

Recall that photons travel at the speed of light, c. Therefore, a photon's frequency and wavelength are related by $c = \lambda f$, or equivalently, $f = c/\lambda$. Making this substitution allows us to solve for the wavelength of a photon:

$$h\frac{c}{\lambda} = |\Delta E| \quad \text{or} \quad \lambda = \frac{hc}{|\Delta E|}$$

As before, the energy change in this expression must be given in joules. The wavelengths calculated using this equation agree with experimental results, and with the spectral lines in Figure 25.10.

GUIDED Example 25.3 | The Hydrogen Lineup   Hydrogen Spectrum

The blue spectral line of hydrogen corresponds to a quantum jump from the $n = 5$ state to the $n = 2$ state. Calculate the wavelength of this line.

Picture the Problem

Our sketch is based on Figure 25.10. It shows the first several lines in the visible part of the hydrogen spectrum, along with the corresponding colors. The line for the jump from $n = 5$ to $n = 2$ is blue.

Strategy

Calculate the energy of both the initial state ($n = 5$) and the final state ($n = 2$) of the electron, using the energy equation for hydrogen, $E_n = -(13.6 \text{ eV})/n^2$. Once the energies are known, calculate the difference in energy, and convert it to joules (J). The wavelength can then be found using the equation $\lambda = hc/|\Delta E|$.

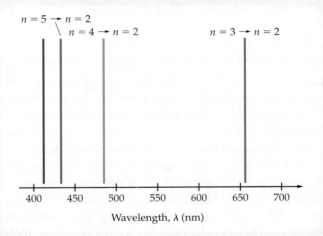

Wavelength, λ (nm)

Known	Unknown
initial state: $n = 5$	$\lambda = ?$
final state: $n = 2$	

Solution

1 Calculate the energy of the $n = 5$ state of hydrogen. This is the initial energy:

$$E_5 = -(13.6 \text{ eV})\frac{1}{5^2} = -0.544 \text{ eV}$$

2 Calculate the energy of the $n = 2$ state of hydrogen. This is the final energy:

$$E_2 = -(13.6 \text{ eV})\frac{1}{2^2} = -3.40 \text{ eV}$$

3 Use the energies from Steps 1 and 2 to calculate the magnitude of the change in energy. Convert the result to joules:

$$|\Delta E| = |E_2 - E_5|$$
$$= |-3.40 \text{ eV} - (-0.544 \text{ eV})|$$
$$= 2.86 \text{ eV}$$
$$= (2.86 \text{ eV})\left(\frac{1.60 \times 10^{-19} \text{ J}}{1 \text{ eV}}\right)$$
$$= 4.58 \times 10^{-19} \text{ J}$$

Math HELP
Unit Conversion
See
Lesson 1.3

4 Use the equation $\lambda = hc/|\Delta E|$ to find the wavelength for this spectral line:

$$\lambda = \frac{hc}{|\Delta E|}$$

$$= \frac{(6.63 \times 10^{-34}\ \text{J} \cdot \text{s})(3.00 \times 10^8\ \text{m/s})}{4.58 \times 10^{-19}\ \text{J}}$$

$$= \boxed{4.34 \times 10^{-7}\ \text{m, or 434 nm}}$$

Insight

The same procedure can be used to find the wavelength of any spectral line in hydrogen.

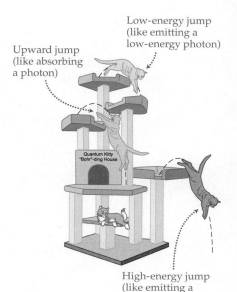

▲ Figure 25.11 An analogy for quantum jumps

Practice Problems

9. Follow-up What is the wavelength of the red spectral line of hydrogen? This line corresponds to the jump from $n = 3$ to $n = 2$.

10. What is the frequency of light corresponding to the quantum jump from $n = 5$ to $n = 4$ by an electron in a hydrogen atom?

11. Concept Check Is the wavelength corresponding to the quantum jump from $n = 7$ to $n = 2$ greater than or less than the wavelength corresponding to the quantum jump from $n = 4$ to $n = 2$? Explain.

Spectral lines correspond to energy states

Have you ever watched a cat playing on an indoor cat tree, like the one shown in **Figure 25.11**? The cat can perch on platforms at different levels and jump from one platform to another. The platforms of the cat tree are like the energy levels of the hydrogen atom, and the jumping cat is like an electron making a quantum jump. Just like the cat, an electron can land at different levels, releasing different amounts of energy. **Figure 25.12** illustrates the situation for hydrogen.

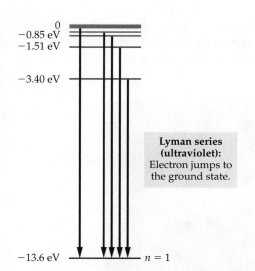
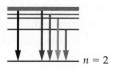
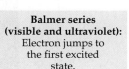

Lyman series (ultraviolet):	Balmer series (visible and ultraviolet):	Paschen series (infrared):
Electron jumps to the ground state.	Electron jumps to the first excited state.	Electron jumps to the second excited state.

▲ Figure 25.12 **The origin of spectral series for hydrogen**
Each series of spectral lines for hydrogen is the result of electrons jumping from an excited state to a lower-energy state. Note that the electrons jump to a specific lower-energy state for each series.

The first three series of spectral lines in the spectrum of hydrogen. The shortest wavelengths appear in the Lyman series. There is no upper limit to the number of series for hydrogen or to the wavelengths that can be emitted.

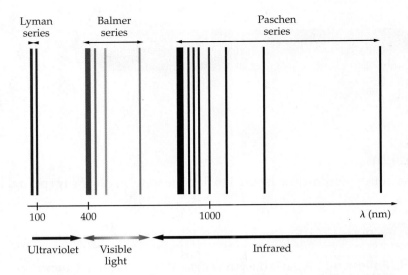

Electron jumps that land at the lowest energy level ($n = 1$) release the most energy—like a cat jumping to the ground. The light that is emitted in these jumps is ultraviolet. All possible electron jumps that end at $n = 1$ produce the ultraviolet *Lyman series* of spectral lines. Electrons that jump down to the $n = 2$ level give off less energy, in the form of visible light. All possible electrons jumps ending at $n = 2$ produce the visible *Balmer series*, and all possible jumps ending at $n = 3$ produce the infrared *Paschen series*, and so on. The wavelengths of these first three spectral series are shown in **Figure 25.13**.

The Balmer spectral line corresponding to the electron jump from $n = 3$ to $n = 2$ is seen in many astronomical objects. An example is the Lagoon Nebula shown in **Figure 25.14**. The red light in this photo has a wavelength of 656 nm.

Electrons jump to higher energy levels when light is absorbed

When hydrogen absorbs light, the process is basically the reverse of what happens when it emits light. It's like Figure 25.12, only with all the arrows pointing upward instead of downward. Specifically, the energy of an absorbed photon causes an electron to jump to a *higher* energy level—like a cat jumping to a higher platform in Figure 25.11. Photons are absorbed only if they have just the right amount of energy to raise an electron from one level to another—otherwise, the photon just keeps on going. This concept is applied in the following Active Example.

▶ **Figure 25.14 Observing hydrogen emissions in space**
Emission nebulas, like the Lagoon Nebula in Sagittarius shown here, are masses of glowing interstellar gas. The gas is excited by high-energy radiation from nearby stars and emits light at wavelengths characteristic of the atoms present, chiefly hydrogen. Much of the visible light from such nebulas corresponds to the red Balmer line of hydrogen (produced by electrons jumping from $n = 3$ to $n = 2$) with a wavelength of 656 nm.

Find the frequency a photon must have if it is to raise an electron in a hydrogen atom from the $n = 3$ state to the $n = 5$ state.

Solution *(Perform the calculations indicated in each step.)*

1. Calculate the energy of the $n = 5$ state: $\qquad$ -0.544 eV

2. Calculate the energy of the $n = 3$ state: $\qquad$ -1.51 eV

3. Calculate the magnitude of the energy change between these two states, and convert the result to joules: $\qquad$ 1.55×10^{-19} J

4. Set the energy of a photon equal to the magnitude of the energy change: $\qquad$ $hf = 1.55 \times 10^{-19}$ J

5. Solve for the frequency of the photon: $\qquad$ $f = 2.34 \times 10^{14}$ Hz

Insight
This frequency corresponds to one of the lines in the absorption spectrum of hydrogen. In this case the line is in the infrared.

De Broglie matter waves explain the Bohr orbits

In 1923 de Broglie used his idea of *matter waves* (the notion that particles such as electrons have wave properties) to show that one of Bohr's assumptions can be thought of as a condition for standing waves. As we saw earlier in this lesson, Bohr assumed that the angular momentum of an electron in an allowed orbit must be an integer times $h/2\pi$. Specifically,

$$rmv = n\frac{h}{2\pi}$$

In Bohr's model there is no particular reason for this condition other than it produces results in agreement with experiment.

However, de Broglie imagined his matter waves as being analogous to a wave on a string—except that in this case the "string" is not tied down at both ends. Instead, it forms a circle of radius r representing an electron's orbit about the nucleus, as illustrated in **Figure 25.15**.

The relation $E = hf$ was used in Chapter 24 to explain blackbody radiation and the photoelectric effect.
• The same relationship applies here to the photons emitted by an electron in a hydrogen atom.

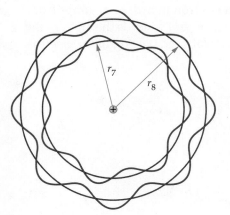

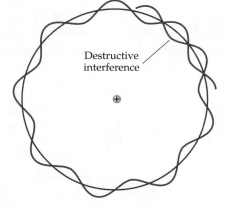

(a) $\qquad\qquad\qquad\qquad$ (b)

◀ **Figure 25.15 De Broglie wavelengths and Bohr orbits**
Bohr's condition that the angular momentum of an allowed orbit must be an integer, n, times $h/2\pi$ is equivalent to the condition that n de Broglie wavelengths must fit into the circumference of an orbit. **(a)** De Broglie waves for the $n = 7$ and $n = 8$ orbits. **(b)** If an integral number of wavelengths do not fit the circumference of an orbit, the result is destructive interference.

▲ Figure 25.16 Mechanical analogy for de Broglie waves in hydrogen
A loop of wire is oscillated vertically about the support point at the bottom of the loop. The oscillations set up standing waves on the circumference of the loop—like de Broglie waves on a Bohr orbit in hydrogen. Different frequencies of oscillation produce different standing wave patterns.

The condition for a standing wave in this case is that an integer number of wavelengths will fit into the circumference of the orbit. Stated mathematically, the condition is

integer number of wavelengths = circumference of circular orbit

$$n\lambda = 2\pi r$$

Figure 25.15 (a) shows the cases corresponding to seven and eight wavelengths fitting around the circumference. Other intermediate wavelengths result in destructive interference, as indicated in Figure 25.15 (b). The standing waves on a vibrating metal ring in **Figure 25.16** further illustrate de Broglie's condition.

Next, de Broglie combined his expression for the wavelength of matter, $\lambda = h/p$, with the condition for standing waves. Since the linear momentum is $p = mv$, the standing wave condition can be written as follows:

$$n\lambda = 2\pi r$$

$$n\frac{h}{mv} = 2\pi r$$

A simple rearrangement yields $rmv = nh/2\pi$, which is precisely Bohr's orbital condition. Thus, matter waves are seen to have a direct connection to physics on the atomic level.

25.2 LessonCheck (MP)

Checking Concepts

12. **Describe** According to the Bohr model, under what conditions does an electron give off radiation? How is the frequency of that radiation determined?

13. **Explain** Why do the quantized energies of electrons in hydrogen atoms produce emission spectra made up of individual lines at specific frequencies?

14. Triple Choice If the state of a hydrogen atom changes to one with a larger value of the quantum number, n, does the radius of the electron's orbit increase, decrease, or stay the same? Explain.

15. Triple Choice If the state of a hydrogen atom changes to one with a larger value of the quantum number, n, does the energy of the atom increase, decrease, or stay the same? Explain.

Solving Problems

16. **Calculate** An electron is in the $n = 6$ state of the Bohr model.
(a) What is the radius of its orbit?
(b) What is the energy of this state?

17. **Calculate** Which Bohr electron orbit has a radius of 2.56×10^{-8} m?

18. **Calculate** Which Bohr electron orbit has an energy of -0.54 eV?

19. **Calculate** What is the wavelength of light that is emitted when an electron in hydrogen jumps from the $n = 2$ state to the $n = 1$ state? Is this light in the ultraviolet, visible, or infrared portion of the electromagnetic spectrum?

25.3 The Quantum Physics of Atoms

While the Bohr model works well in describing hydrogen, attempts to extend it to other atoms failed. Clearly, the Bohr model isn't the full story. For example, how can an electron accelerating in a circular orbit give off no radiation? The Bohr model gives no explanation. The Bohr model also assumes that the electron has a well-defined momentum, which means—according to the Heisenberg uncertainty principle—that its position should be completely uncertain. How can this be if the electron moves in a well-defined circular orbit? All of these questions and inconsistencies were resolved by Erwin Schrödinger (1887–1961), who developed a whole new approach to quantum physics.

Schrödinger's Equation

Schrödinger took de Broglie's idea of matter waves seriously, going so far as to develop an equation that describes their behavior. The *Schrödinger equation* applies to all atoms—not just hydrogen—and is completely consistent with the uncertainty principle. In fact, Schrödinger's equation plays the same role in quantum physics as Newton's laws in classical mechanics and Maxwell's equations in electromagnetism.

Schrödinger's equation describes probabilities

The interpretation of Schrödinger's equation is quite a departure from classical physics. When you solve the Schrödinger equation, the result is a **wave function** that gives the probability of finding the electron in a given location. This is an important point: The wave function doesn't tell you where the electron is—as in the case of classical physics—it tells you the *probability* that the electron is here, or there, or someplace else.

To visualize a wave function, think of it as a *probability cloud*. An example is shown in **Figure 25.17**, which shows the probability cloud for the ground state of hydrogen. The probability cloud in this case is spherical, somewhat like an orange. 🔑 **Interpreting a probability cloud is simple— the electron is most likely to be found where the cloud is most dense.** For the cloud in Figure 25.17, the highest probability is in the dark band, which is actually a spherical shell. Notice that the probability of finding the electron near the nucleus (at the center of the cloud) is small, as is the probability of finding the electron far from the nucleus.

Probability clouds for excited states of hydrogen have the same interpretation as that for the ground state, but their shapes become more interesting. For example, a probability cloud for the first excited state of hydrogen is shown in **Figure 25.18** on the next page. Notice that there are two distinct clouds in this case, separated by a *node*—a plane of zero probability. Just as standing waves on a string have more and more nodes as the frequency of the waves increases, probability clouds have more and more nodes as the energy goes up. A node on a string is a location of zero displacement; a node on a wave function is a location of zero probability.

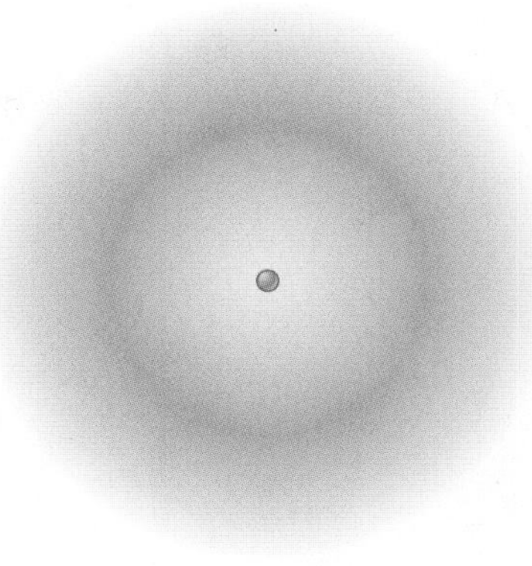

Vocabulary

- wave function
- quantum mechanics
- laser
- hologram
- fluorescence

🔑 *How does a probability cloud describe an electron in a hydrogen atom?*

▲ **Figure 25.17 Probability cloud for the ground state of hydrogen**
In Schrödinger's model of hydrogen, the electron can be found at any distance from the nucleus. The probability of finding the electron at a given location is proportional to the density of the probability cloud.

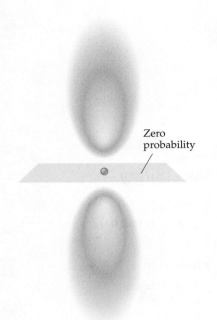

Figure 25.18 Probability cloud for an excited state of hydrogen
The probability cloud for an electron in hydrogen increases in complexity as the quantum number increases. In the case shown here, the nucleus lies on the node of the probability cloud (the plane of zero probability).

Zero probability

CONCEPTUAL Example 25.5 Finding the Electron

An electron in the state shown in Figure 25.18 can never be found halfway between the two probability clouds. How is it possible, then, that the electron is equally likely to be found both above and below the zero-probability plane?

Reasoning and Discussion
The probability cloud of an electron is the result of a standing wave pattern, similar to the standing waves found on a string tied down at both ends. Both the node on a string and the location of zero probability of an electron are the result of destructive interference. For example, waves moving back and forth on a string can cancel in the middle. Similarly, an electron's matter wave can move back and forth and cancel in the middle. The main thing is that an electron isn't just a small ball of charge moving around from place to place; it's a matter wave that can form a standing wave pattern with nodes in certain locations.

Quantum mechanics applies to all of matter

The matter-wave physics described by Schrödinger's equation is known as **quantum mechanics** to distinguish it from the *classical mechanics* of Newton's laws of motion. Quantum mechanics is the most thoroughly tested theory in the history of science. It applies to everything in the physical world, from atoms to molecules, solids to liquids, metals to semiconductors. The technological advances of modern society are primarily a result of our ability to understand and apply quantum mechanics.

One of the most important applications of quantum mechanics is the development of a new way to produce light.

Lasers

The production of light by humans advanced significantly when flames were replaced by the lightbulb. An even greater advance occurred in 1960, when the first laser was developed. Lasers produce light that is intense, travels in one direction, and has a single, pure color. Because of these properties, lasers are used in a multitude of technological applications, ranging from supermarket scanners to CD players, from laser pointers to eye surgery. In fact, lasers are now a common source of light in everyday life.

Lasers are a means of light amplification

To understand just what a laser is and what makes it so special, let's start with its name. The word **laser** is an acronym for *light amplification by the stimulated emission of radiation*. As we will see, the properties of stimulated emission lead directly to the amplification of light.

Spontaneous Emission To understand how a laser works, consider two energy levels in an atom. Suppose an electron occupies the higher of the two levels. If the electron is left alone, it will "eventually" drop to the lower level in a time that is typically about 10^{-8} s, giving off a photon in the process, which is referred to as *spontaneous emission*. The photon given off in this process can travel away from the atom in any direction.

Stimulated Emission In a laser the excited-state electron is not left alone. Instead, a photon with an energy equal to the energy difference between the two energy levels passes near the electron. This photon *increases* the probability that the electron will drop to the lower level. That is, the incident photon can *stimulate* the emission of a photon by the electron. The photon given off in this process of *stimulated emission* has the same energy and the same phase and travels in the same direction as the incident photon, as illustrated in **Figure 25.19**. This matchup accounts for the fact that laser light is highly focused and of a single color.

Light Amplification A single photon encountering an excited atom can cause two identical photons to exit the atom. If each of these two photons encounters another excited atom and undergoes the same process, the number of photons increases to four. Continuing in this manner, the photons undergo a sort of "chain reaction" that doubles their number with each generation. It is this property of stimulated emission that results in *light amplification.*

In order for the light amplification process to work, photons must continue to encounter atoms with electrons in excited states. Under ordinary conditions this is not the case, since most electrons are in the lowest possible energy levels. For laser action to occur, atoms must first be raised to an excited state. Then, before the electrons have a chance to drop to a lower level by way of spontaneous emission, the process of stimulated emission can proceed.

Helium-neon lasers produce red light

A specific example of a laser is the *helium-neon laser,* shown schematically in **Figure 25.20**. Basically, this laser consists of a glass tube containing helium and neon atoms. The tube has a reflecting mirror at one end and a semitransparent mirror at the other end. Light bounces back and forth between the mirrors, increasing in intensity as it stimulates more and more atoms to emit light. Eventually the amplified light exits through the semitransparent mirror.

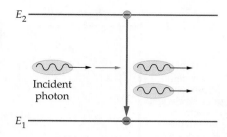

▲ **Figure 25.19 Stimulated emission** A photon with an energy equal to the difference $E_2 - E_1$ can enhance the probability that an electron in the state with E_2 will drop to the state with E_1 and emit a photon.

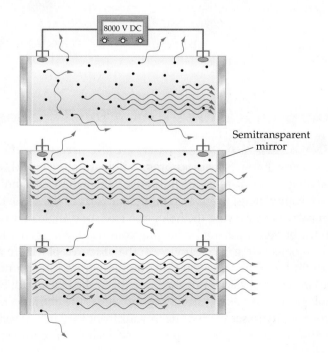

◄ **Figure 25.20 The helium-neon laser** A schematic representation of the basic features of a helium-neon laser. The process begins with the application of a high voltage to a tube containing helium and neon atoms (dots in this figure). The applied voltage causes electrons and helium atoms to move through the tube, colliding with neon atoms and exciting them until they emit light. The light is reflected back and forth within the tube—raising its intensity to a high level and resulting in the laser light that exits the device.

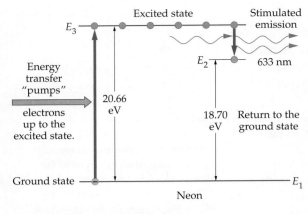

E_3 Excited state — Stimulated emission

Energy transfer "pumps" electrons up to the excited state.

20.66 eV

E_2 — 633 nm

18.70 eV — Return to the ground state

Ground state —— E_1

Neon

▲ **Figure 25.21 "Pumping" electrons and stimulated emission**
Collisions with rapidly moving electrons and helium atoms in a helium-neon laser "pump" the electrons in the neon atoms up to an excited state with energy E_3. These excited electrons participate in stimulated emission, which results in laser light of wavelength 633 nm.

The neon atoms in a helium-neon laser are the ones that produce the laser light. **Figure 25.21** shows the energy-level diagram for the neon atoms involved in the lasing process. Electrons are excited, or "pumped up," to the E_3 energy level by an electrical power supply connected to the tube. This results in a large number of electrons that are available to engage in stimulated emission by dropping from the E_3 level to the E_2 level. (The electrons subsequently proceed through a variety of intermediate steps to the ground state, but only the emission of light between the levels E_3 and E_2 results in laser light.) The difference in energy between the E_3 and E_2 levels is 1.96 eV, and hence the light coming out of the laser is red, with a wavelength of 633 nm.

Lasers are used in many medical procedures

Physics & You: Technology One medical application of lasers is in *laser eye surgery*. In such surgery a laser that emits high-energy photons in the ultraviolet range is used to reshape the cornea and correct nearsightedness. For example, in LASIK eye surgery the procedure begins with a small mechanical shaver known as a *microkeratome* cutting a flap in the cornea, leaving a portion of the cornea uncut to serve as a hinge. After the mechanical cut is made, the corneal flap is folded back, exposing the middle portion of the cornea, as shown in **Figure 25.22**. Next, the laser sends pulses of UV light onto the cornea, each pulse vaporizing a small layer of corneal material (0.1–0.5 μm in thickness) with no heating. This process continues until the cornea is flattened just enough to correct the nearsightedness, after which the corneal flap is put back into place.

▶ **Figure 25.22 Laser vision correction**
In LASIK eye surgery a flap of the cornea is cut and folded back. Next, a UV laser (known as an *excimer laser*) is used to vaporize some of the underlying corneal material. When the flap is replaced, the cornea is flatter than it was originally, correcting the patient's nearsightedness.

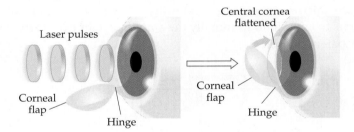

Laser pulses

Corneal flap

Hinge

Central cornea flattened

Corneal flap

Hinge

Lasers can produce three-dimensional images

Physics & You: Technology Lasers are also used to produce three-dimensional images known as **holograms.** A hologram is a true three-dimensional image, as **Figure 25.23** indicates. When you view a hologram, you can move your vantage point to see different parts of the scene that is recorded. In particular, viewing the hologram from one angle may obscure an object in the background, but by moving your head, you can look around the foreground objects to get a clear view of the obscured object. In addition, you have to adjust your focus as you shift your gaze from foreground to background objects, just as you do when looking at the real world. Finally, if you cut a hologram into pieces, each piece still shows the entire scene! This is similar to your ability to see everything in your front yard just as well through a small window as through a large window—both show the entire scene.

▲ **Figure 25.23 Viewing a hologram**
A hologram is a three-dimensional image produced in empty space.

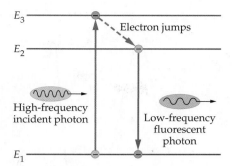

Fluorescence

Have you ever wondered how a fluorescent lightbulb works? As its name implies, a fluorescent lightbulb makes use of a phenomenon known as *fluorescence*. **Fluorescence** is the emission of light of a lower frequency by a material that has been illuminated with light of a higher frequency. In essence, fluorescence can be thought of as a conversion process in which photons of high frequency are converted to photons of lower frequency. Let's take a closer look at the process involved.

Energy level transitions can produce fluorescence

Consider the energy levels shown in **Figure 25.24**. If an atom with these energy levels absorbs a photon that has an energy equal to the difference $E_3 - E_1$, the photon can excite an electron from the E_1 level to the E_3 level. In some atoms, the most likely way for the electron to return to the ground state is by first jumping to level E_2 and then jumping to level E_1. Each of the photons emitted in these jumps has *less* energy than the photon that caused the excitation in the first place. ⬤▬ **This is the process of fluorescence; an atom is illuminated with a photon of one frequency, and it subsequently emits photons of lower energy and lower frequency.**

⬤▬ *What happens during the process of fluorescence?*

Physics & You: Technology A fluorescent lightbulb uses fluorescence to convert high-frequency UV light to lower-frequency visible light. The bulb basically consists of a sealed tube filled with mercury vapor. When electricity is applied to the ends of the tube, a filament is heated, producing electrons. These electrons are then accelerated through the tube by an applied voltage. When the electrons strike mercury atoms in the tube, the atoms become excited and then give off UV light as they decay to their ground state. This process is not particularly useful in itself, since the UV light is not visible to us. However, the inside of the tube is coated with a material known as a *phosphor*. The phosphor coating absorbs the UV light and then emits lower-frequency visible light. Thus, several different physical processes must take place before a fluorescent lightbulb produces the light that we see.

Fluorescence finds many less familiar applications as well. In forensics, the analysis of a crime scene is enhanced by the fact that human bones and teeth are fluorescent. Thus, illuminating a crime scene with ultraviolet light can make items of interest stand out for easy identification. In addition, the use of a fluorescent dye can make fingerprints clearly visible, as shown in **Figure 25.25**.

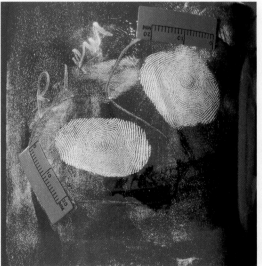

▲ Figure 25.25 **Fluorescence at a crime scene**
Fingerprints become clearly visible when treated with fluorescent dye and illuminated by UV light.

▲ **Figure 25.26 Fluorescence in nature**
A variety of creatures, including scorpions, show their natural fluorescence when illuminated with UV light.

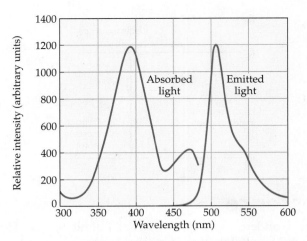

▲ **Figure 25.27 The fluorescence spectrum of GFP**
The green fluorescent protein (GFP) strongly absorbs light with a wavelength of about 400 nm (violet in color). It reemits green light with a wavelength of 509 nm.

Fluorescence often occurs in nature

Many creatures produce fluorescence in their bodies. For example, several types of coral glow brightly when illuminated with UV light. Scorpions are also strongly fluorescent, giving off a distinctive green light, as shown in **Figure 25.26**. In fact, it is often possible to discern a greenish cast when viewing a scorpion in sunlight. At night in the desert, scorpions stand out with a bright green glow when someone illuminates the area with UV light. This aids researchers who want to find certain scorpions for study, as well as campers who are just as interested in avoiding scorpions altogether.

The green fluorescence of the jellyfish *Aequorea victoria* finds many uses in biological experiments. The gene that produces the green fluorescent protein (GFP) can serve as a marker to identify whether an organism has incorporated a new segment of DNA into its genome. For example, bacterial colonies that incorporate the GFP gene can be screened by eye simply by viewing them under a UV light. Recently, GFP has been inserted into the genome of a white rabbit, giving rise to the *GFP bunny*. The bunny appears normal in white light, but when viewed under light with a wavelength of 392 nm, it glows with a bright green light at 509 nm. The fluorescence spectrum of GFP is shown in **Figure 25.27**.

25.3 LessonCheck ⓂⓅ

Checking Concepts

20. ▭ **Explain** What is a probability cloud?

21. ▭ **Describe** How is the energy of the absorbed photons related to the energies of the emitted photons during fluorescence?

22. Describe What is the physical meaning of a node in a probability cloud?

23. Identify What are some of the ways in which laser light differs from the light given off by the Sun?

24. Big Idea Identify some ways in which the quantum physics description of an atom differs from Rutherford's solar system model.

Technology and Society

itm

Hydrogen as Fuel

The Technology A *fuel cell* is a device that uses a chemical reaction to convert a fuel directly into electric current. Hydrogen is currently used in fuel cells that power cars and other vehicles.

Why Hydrogen? Hydrogen is the simplest atom, which is one reason why early physicists studied it. Today, hydrogen is in the news as one of the fuels of the future. The fundamental reason is hydrogen's reactivity with oxygen. When two hydrogen atoms bond to an oxygen atom, the chemical reaction releases a great deal of energy. Hydrogen-powered fuel cells convert this energy into electrical energy. Hydrogen is also appealing as a fuel because the only product of the fuel-cell reaction is H_2O, water!

Concerns Hydrogen fuel cells do not produce carbon dioxide as the burning of fossil fuels does. Even so, there are environmental concerns about the use of hydrogen as a fuel. Currently, fossil fuels are the main source of hydrogen. Extracting hydrogen from fossil fuels requires energy and produces carbon dioxide, a gas implicated in climate change. Evidence suggests that significant amounts of hydrogen, released into the atmosphere during fueling, for example, could damage Earth's protective ozone layer. Another concern is that water vapor is a potentially problematic heat-trapping gas, much like carbon dioxide. Some argue that the wide use of hydrogen as a fuel could lead to environmental problems.

The Future The most abundant source of hydrogen on Earth is water. But the hydrogen atoms in water are strongly bonded to oxygen. It takes a lot of energy to tear them away. If that energy comes mostly from fossil fuels, as it does now, then producing hydrogen for use as a fuel would indirectly add to carbon dioxide emissions. But there are promising alternative methods. For example, scientists have found that some types of bacteria can work to greatly reduce the amount of electricity needed to extract hydrogen from water. That technology could allow us to produce hydrogen fuel in a more environmentally friendly way.

Take It Further

1. Propose *It takes energy to produce hydrogen for use as a fuel. Suggest several energy sources other than fossil fuels that could be used to extract hydrogen from water.*

2. Infer *How do you think the fuel-tank receptacles and fueling nozzles for a hydrogen-fueled car would be different from those at a gas station?*

Physics Lab 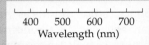 Spectra of Common Light Sources

In this lab you will use a spectroscope to observe the spectra of several common light sources. You will also observe the spectrum of a light source as its brightness slowly increases.

Materials
• various light sources
• quantitative spectroscope

Procedure

Caution: *Never look directly at the Sun when using the spectroscope!*

Part I

1. In your notebook or on a clean sheet of paper, make up to eight copies of the scale shown below (the number of light sources your teacher provides determines how many scales you will need).

> 400   500   600   700
> Wavelength (nm)

2. View an incandescent lightbulb through the spectroscope. Be careful to aim the aperture slit on the front of the spectroscope directly at the bulb. Observe the spectrum displayed on the spectroscope's scale. Notice that each color has an associated wavelength in nanometers (nm).

3. Use crayons or colored pencils to sketch the observed colors on one of the scales you created in Step 1. Identify the sketch as "Incandescent Bulb."

4. Repeat Steps 2 and 3 for up to seven other light sources.

Part II

5. Using your unaided eyes, observe changes in the light produced by a clear showcase bulb when a light dimmer is used to gradually increase the bulb's brightness. Record your observations.

6. Repeat Step 5, using a spectroscope to view the showcase bulb's filament as the brightness of the bulb is slowly increased. Carefully note changes in the observed spectrum as the light from the bulb becomes brighter. Record your observations.

Analysis

1. Describe the similarities and differences of the light sources you have observed. Consider their spectra, temperatures, and light-emitting materials in your response.

2. Give an example of a light source that produces **(a)** a continuous spectrum, **(b)** a line spectrum, and **(c)** both continuous and line spectra.

Conclusions

1. Based on your observations, what type of materials produce continuous spectra? Line spectra?

2. In Part II which colors were visible through the spectroscope when the bulb was glowing dull red? Yellow? White-hot? How do you explain these observations?

3. Is it possible to determine the spectral composition of a light source by looking at the source with the unaided eye? Explain your answer.

25 Study Guide

Big Idea

The wave properties of matter mean that the atomic-level world must be described in terms of probability.

The probabilities, in turn, are determined by wave functions, which are solutions to Schrödinger's equation. Wave functions can be visualized as probability clouds, whose density is proportional to the probability of the electron's being found in a location.

25.1 Early Models of the Atom

🔑 Rutherford's model depicted an atom in which light-weight, negatively charged electrons orbit an extremely small but massive nucleus.

🔑 Blackbody radiation is produced by the blackbody as a whole, and not its individual atoms. A line spectrum, on the other hand, is produced by individual atoms giving off electromagnetic radiation at precise frequencies.

• Atoms are the smallest unit of a given element.

• The spectrum produced by the atoms of a substance, with its bright lines of different colors, is referred to as a *line spectrum*.

• Rutherford discovered that an atom is somewhat like the Solar System: mostly empty space, with most of its mass concentrated in the nucleus.

25.2 Bohr's Model of the Hydrogen Atom

🔑 Radiation is given off *only* when an electron shifts from one orbit to another, and then the radiation is in the form of a photon that obeys Einstein's quantum relation $E = hf$.

🔑 Each possible jump from one energy state to another results in a different line of the hydrogen spectrum.

• All Bohr orbits have specific radii *and* specific energies.

• De Broglie showed that the allowed orbits of the Bohr model correspond to standing matter waves of the electrons. In particular, an allowed orbit in Bohr's model has a circumference equal to an integer times the wavelength of the electron in that orbit.

Key Equations

The radii of allowed orbits in the Bohr model are given by

$$r_n = (5.29 \times 10^{-11}\,\text{m})n^2 \qquad n = 1, 2, 3, \ldots$$

The energy of an allowed Bohr orbit is

$$E_n = -(13.6\,\text{eV})\frac{1}{n^2} \qquad n = 1, 2, 3, \ldots$$

The wavelength of a photon emitted by a hydrogen atom is

$$\lambda = \frac{hc}{|\Delta E|}$$

25.3 The Quantum Physics of Atoms

🔑 Interpreting a probability cloud is simple—the electron is most likely to be found where the cloud is most dense.

🔑 In the process of fluorescence, an atom is illuminated with a photon of one frequency, and it subsequently emits photons of lower energy and lower frequency.

• The correct description of the hydrogen atom is derived from Schrödinger's equation.

• A solution to the Schrödinger equation is called a *wave function*; the wave function gives the probability of finding an electron in a given location.

• A laser is a device that produces light amplification by the stimulated emission of radiation.

ANSWERS TO SELECTED ODD-NUMBERED PROBLEMS APPEAR IN APPENDIX A.

Lesson by Lesson

25.1 Early Models of the Atom

Conceptual Questions

25. Give a reason why the Thomson plum-pudding model does not agree with experimental results.

26. What observation led Rutherford to propose that an atom has a small nucleus containing most of the atom's mass?

27. How does the line spectrum of an atom differ from the blackbody radiation curve of a hot object?

28. How can elements be identified just by looking at the light they emit?

Problem Solving

29. The electron in a hydrogen atom is typically found at a distance of about 5.3×10^{-11} m from the nucleus, which has a diameter of about 1.0×10^{-15} m. If you assume that the hydrogen atom is a sphere of radius 5.3×10^{-11} m, what fraction of its volume is occupied by the nucleus?

30. Referring to Problem 29, suppose the nucleus of the hydrogen atom were enlarged to the size of a baseball (diameter $= 7.3$ cm). At what typical distance from the center of the baseball would you expect to find the electron?

31. Suppose an electron orbits a stationary proton at a radius of 5.3×10^{-11} m. What speed must the electron have if its mass times its centripetal acceleration is to be equal to the electrostatic force of attraction between the electron and the proton?

32. In Rutherford's scattering experiments, alpha particles (charge $= +2e$) were fired at a piece of gold foil. Consider an alpha particle with an initial kinetic energy, *KE*, heading directly for the nucleus of a gold atom (charge $= +79e$). The alpha particle will come to rest when all of its initial kinetic energy has been converted to electrical potential energy. Find the distance of closest approach between the alpha particle and the gold nucleus if the initial kinetic energy of the alpha particle is $KE = 3.0$ MeV.

25.2 Bohr's Model of the Hydrogen Atom

Conceptual Questions

33. Is there an upper limit to the radius of an allowed Bohr orbit? Explain.

34. (a) Is there a lower limit to the wavelengths of lines in the spectrum of hydrogen? Explain. (b) Is there an upper limit? Explain.

35. In principle, how many spectral lines are there in any given series of hydrogen? Explain.

36. Is the wavelength corresponding to the quantum jump from $n = 6$ to $n = 5$ greater than or less than the wavelength corresponding to the quantum jump from $n = 5$ to $n = 4$? Explain.

37. What is the value of the quantum number n for the Bohr orbit shown in **Figure 25.28**?

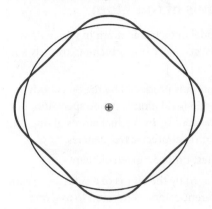

Figure 25.28

38. An electron in the $n = 1$ Bohr orbit has the de Broglie wavelength λ_1. In terms of λ_1, what is the de Broglie wavelength of an electron in the $n = 3$ Bohr orbit?

Problem Solving

39. How much energy is required to ionize a hydrogen atom (that is, to separate its electron from the nucleus) that is initially in the $n = 4$ state?

40. Find the energy of the photon required to excite the electron in a hydrogen atom from the $n = 2$ state to the $n = 5$ state.

41. Find the de Broglie wavelength of an electron in the ground state of the hydrogen atom.

42. What is the radius of the Bohr orbit of a hydrogen atom shown in **Figure 25.29**?

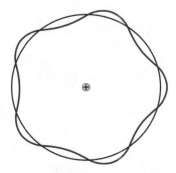

Figure 25.29

43. Find the wavelength of the light emitted when an electron in hydrogen jumps from the $n = 6$ state to the $n = 5$ state.

44. What is the orbital radius of the $n = 8$ Bohr state? What is its energy?

45. What is the orbital radius of the Bohr state whose energy is -1.51 eV?

46. What is the energy of the Bohr state whose orbital radius is 1.32×10^{-9} m?

47. Initially, an electron is in the $n = 3$ state of hydrogen. If this electron acquires an additional 1.23 eV of energy, what is the value of n for the final state of the electron?

48. Identify the initial and final states if an electron in hydrogen emits a photon with a wavelength of 656 nm.

49. **Think & Calculate** An electron in hydrogen absorbs a photon and jumps to a higher state. **(a)** Find the energy the photon must have if the initial state is $n = 3$ and the final state is $n = 5$. **(b)** If the initial state was $n = 5$ and the final state $n = 7$, would the energy of the photon be greater than, less than, or the same as that found in part (a)? Explain. **(c)** Calculate the photon energy for part (b).

50. Consider the following four quantum jumps in a hydrogen atom:

A. $n_i = 2, n_f = 6$ **C.** $n_i = 7, n_f = 8$
B. $n_i = 2, n_f = 8$ **D.** $n_i = 6, n_f = 2$

Find **(a)** the longest-wavelength and **(b)** the shortest-wavelength photon that can be emitted or absorbed by these transitions. Give the value of the wavelength in each case.

51. **Think & Calculate** The potential energy of a hydrogen atom with its electron in a particular Bohr orbit is -1.20×10^{-19} J. **(a)** Which Bohr orbit does the electron occupy in this atom? **(b)** Suppose the electron moves away from the nucleus to the next higher Bohr orbit. Does the potential energy of the atom increase, decrease, or stay the same? Explain.

25.3 The Quantum Physics of Atoms
Conceptual Questions

52. **Triple Choice** The probability of finding an electron at point A is greater than the probability of finding it at point B. Is the density of the probability cloud at point A greater than, less than, or equal to the density of the probability cloud at point B?

53. What is the probability of finding an electron at a node of a probability cloud?

54. What are some of the properties of laser light that make it different from other kinds of light?

55. **Triple Choice** Is the wavelength of the radiation that excites a fluorescent material greater than, less than, or equal to the wavelength of the radiation that the material emits? Explain.

Problem Solving

56. **Photorefractive Keratectomy** A person's vision may be improved significantly by having the cornea reshaped with a laser beam, in a procedure known as *photorefractive keratectomy*. The excimer laser used in this procedure produces UV light with a wavelength of 193 nm. **(a)** What is the difference in energy between the two levels that participate in stimulated emission in the excimer laser? **(b)** How many photons from this laser are required to deliver an energy of 1.58×10^{-13} J to the cornea?

Mixed Review

57. **Rank** Consider the following three quantum jumps in a hydrogen atom:

A. $n_i = 5, n_f = 2$
B. $n_i = 7, n_f = 2$
C. $n_i = 7, n_f = 6$

Rank the jumps in order of increasing **(a)** wavelength and **(b)** frequency of the emitted photon. Indicate ties where appropriate.

58. Suppose a hydrogen atom is in the ground state. **(a)** What is the highest-energy photon this system can absorb without separating the electron from the proton? Explain. **(b)** What is the lowest-energy photon this system can absorb? Explain.

59. Find the smallest frequency a photon can have if it is to ionize a hydrogen atom in the ground state.

60. **Rydberg Atoms** There is no limit to the size a hydrogen atom can attain, provided it is free from disruptive outside influences. In fact, radio astronomers have detected radiation from large, so-called *Rydberg atoms* in the diffuse hydrogen gas of interstellar space. Find the smallest value of n such that the Bohr radius of a single hydrogen atom is greater than 8.0 microns, the size of a typical single-celled organism.

61. The electron in a hydrogen atom makes a quantum jump from the $n = 4$ state to the $n = 2$ state, as shown in **Figure 25.30**. **(a)** Given that the momentum of a photon is $p = h/\lambda$, find the momentum of the photon emitted in this quantum jump. **(b)** Find the recoil speed of the hydrogen atom, assuming that it was at rest before the photon was emitted.

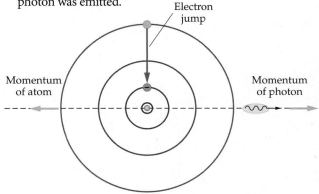

Figure 25.30

62. Consider a particle with mass m, charge q, and constant speed v moving perpendicular to a uniform magnetic field of magnitude B, as shown in **Figure 25.31**. The particle follows a circular path. Suppose the angular momentum of the particle about the center of its circular path is quantized in the following way:

$$mvr = n\hbar \qquad n = 1, 2, 3, \ldots \text{ and } \hbar = h/2\pi$$

(a) Show that the radii of its allowed orbits have the following values:

$$r_n = \sqrt{\frac{n\hbar}{qB}}$$

(b) Find the speed of the particle in each allowed orbit.

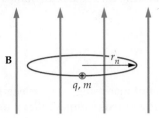

Figure 25.31

Writing about Science

63. Write a report on lasers and their development. When was the laser invented? Who invented it? What different types of lasers are in use today? What are the most powerful lasers? What are some interesting real-world uses of lasers?

64. Connect to the Big Idea What are some of the concepts of quantum physics that improved our understanding of atoms? How does quantization play a role? What is the role of wave-particle duality? What is the interpretation of probability clouds in an atom?

Read, Reason, and Respond

Welding a Detached Retina As a person ages, a normal part of the process is a shrinkage of the vitreous gel—the gelatinous substance that fills the interior of the eye. When this happens, the usual result is that the gel pulls away cleanly from the retina, with little or no adverse effect on the person's vision. This is referred to as *posterior vitreous detachment*. In some cases, however, the vitreous membrane that surrounds the vitreous gel pulls on the retina as the gel contracts, eventually creating a hole or a tear in the retina itself. Fluid can then seep through the hole in the retina and separate it from the underlying supporting cells—the retinal pigment epithelium. This process, known as *rhegmatogenous retinal detachment*, causes a blind spot in the person's vision. If not treated immediately, a retinal detachment can lead to permanent vision loss.

One way to treat a detached retina is to "weld" it back in place using a laser beam. This procedure is performed with an argon laser because the blue-green light it produces passes through the vitreous gel without being absorbed or causing damage, but is strongly absorbed by the red pigments in the retina and the retinal epithelium. An argon laser produces light consisting primarily of two wavelengths, 488.0 nm (blue-green) and 514.5 nm (green), and it has a power output ranging between 1 W and 20 W.

65. Suppose an argon laser emits 1.49×10^{19} photons per second, half with a wavelength of 488.0 nm and half with a wavelength of 514.5 nm. What is the power output of this laser in watts?

    A. 1.49 W   C. 5.92 W

    B. 5.76 W   D. 6.07 W

66. A different type of laser also emits 1.49×10^{19} photons per second. If all of its photons have a wavelength of 414.0 nm, is its power output greater than, less than, or equal to the power output of the argon laser in Problem 65?

67. What is the power output of the laser in Problem 66?

    A. 1.27 W   C. 4.80 W

    B. 2.39 W   D. 7.16 W

68. What is the energy difference (in electron volts) between the states of an argon atom that are involved in the emission of a photon with a wavelength of 514.5 nm?

    A. 2.13 eV   C. 3.87 eV

    B. 2.42 eV   D. 6.40 eV

Standardized Test Prep

Multiple Choice

1. Which of the following is *not* an assumption of the Bohr model of the atom?
   (A) Electrons move in circular orbits about the nucleus.
   (B) Electrons in allowed orbits have quantized angular momentums.
   (C) Electrons in allowed orbits give off electromagnetic radiation.
   (D) Radiation is emitted only when electrons jump from one allowed orbit to another.

2. Whose early model described the atom as a positively charged pudding with electrons embedded throughout it?
   (A) Thomson
   (B) Rutherford
   (C) Bohr
   (D) Millikan

3. If R is the orbital radius for an electron in the first orbit in the Bohr model, what is the radius for an electron in the third orbit?
   (A) $2R$
   (B) $3R$
   (C) $6R$
   (D) $9R$

4. If E is energy of an electron in the first orbit of the Bohr model, what is the energy of an electron in an orbit with quantum number n?
   (A) nE
   (B) E/n
   (C) n^2E
   (D) E/n^2

Test-Taking Hint

Because energy is conserved, the energy absorbed or emitted by an electron during a jump from one energy level to another is equal to the energy of the photon absorbed or emitted.

Use the energy-level diagram below to answer Questions 5 and 6. The diagram shows the first four energy levels for an electron in a hydrogen atom.

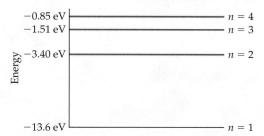

5. What is the frequency of the light emitted when an electron jumps from the $n = 4$ level to the $n = 3$ level?
   (A) 1.6×10^{15} Hz
   (B) 3.2×10^{15} Hz
   (C) 3.2×10^{14} Hz
   (D) 1.6×10^{14} Hz

6. Which of the following electron jumps gives off visible light? (Visible light has wavelengths in the range 400–700 nm.)
   (A) $n = 4$ to $n = 2$
   (B) $n = 4$ to $n = 1$
   (C) $n = 3$ to $n = 1$
   (D) $n = 2$ to $n = 3$

7. How much energy would an electron in the ground state of a hydrogen atom need to absorb in order to ionize the atom?
   (A) 0.85 eV
   (B) 12.1 eV
   (C) 12.8 eV
   (D) 13.6 eV

Extended Response

8. Use the concept of de Broglie wavelengths for electrons to explain the allowed orbits in the Bohr model of the atom.

If You Had Difficulty With . . .														
Question	1	2	3	4	5	6	7	8						
See Lesson(s)	25.3	25.1	25.3	25.3	25.3	25.2, 25.3	25.3	25.3						

26 Nuclear Physics

Inside

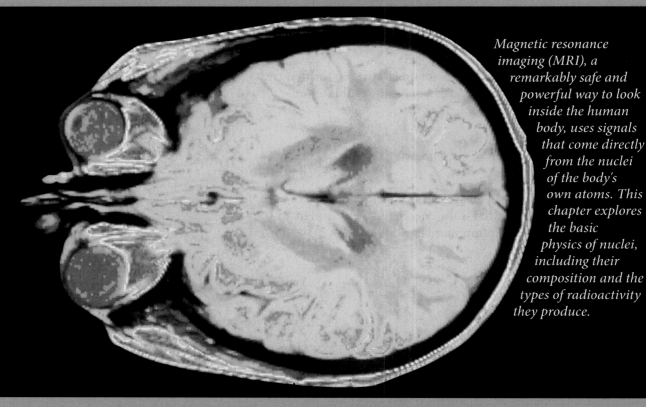

Magnetic resonance imaging (MRI), a remarkably safe and powerful way to look inside the human body, uses signals that come directly from the nuclei of the body's own atoms. This chapter explores the basic physics of nuclei, including their composition and the types of radioactivity they produce.

Big Idea

The nuclei of atoms can release tremendous amounts of energy when part of their mass is converted to energy.

The preceding chapter looked at electrons and their orbits. The nucleus played little role in that discussion—it was treated as a point at the center of the atom. The nucleus is much more than a point, however. In fact, most nuclei contain tightly packed particles that interact strongly with one another. In this chapter you'll learn about the physics of the nucleus, and you'll discover how this knowledge is put to use in the real world.

26.1 The Nucleus

Even though the nucleus is small compared to the atom, it plays an important role in everyday life. For example, warmth and light from the Sun are the result of reactions that take place within the nuclei of atoms. Let's take a close look at the inner workings of the nucleus.

Nuclei consist of protons and neutrons

The simplest nucleus is that of the hydrogen atom. This nucleus consists of a single proton, with an electric charge of $+e$. All other nuclei contain neutrons in addition to protons. Recall that the neutron is an electrically neutral particle (its electric charge is zero) with a mass just slightly greater than that of the proton.

    Collectively, protons and neutrons are known as **nucleons**. Nuclear physics is the study of how nucleons interact with one another in a nucleus. **Figure 26.1** shows how nucleons are represented in this text.

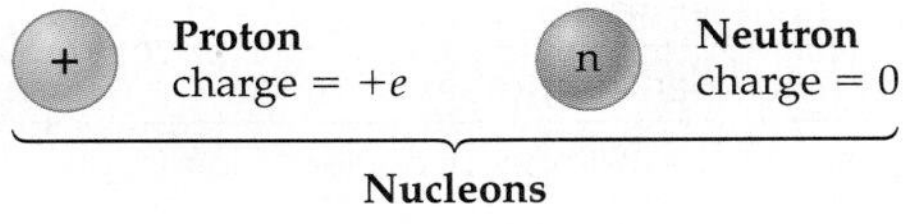

Vocabulary

- nucleon
- atomic number
- neutron number
- mass number
- isotopes
- atomic mass unit
- strong nuclear force

◀ **Figure 26.1 Nucleons**
The particles that make up a nucleus are called *nucleons*. Nucleons come in two types—protons with charge $+e$, and neutrons with zero charge.

Proton charge = $+e$ **Neutron** charge = 0

Nucleons

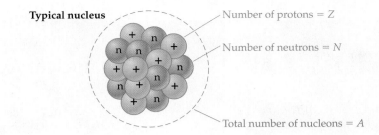

► Figure 26.2 A typical nucleus
A typical nucleus contains several protons and several neutrons. The number of protons is the atomic number, Z. The number of neutrons is N. The total number of nucleons (protons + neutrons) is the mass number, A. In this case, $Z = 8$, $N = 7$, and $A = 15$.

How are mass number, neutron number, and atomic number related?

A good way to think of a nucleus is like a bag of marbles, with the closely packed marbles similar to the nucleons. Since a nucleus contains nearly all of an atom's mass packed into an incredibly small volume, it is not surprising that its density is enormous. In fact, all nuclei have nearly the same density, which is approximately 2.3×10^{17} kg/m^3. With this density, a single teaspoon of nuclear matter would weigh about a trillion tons!

Describing Nuclei A typical nucleus, like the one illustrated in **Figure 26.2**, is described in terms of the numbers and types of nucleons it contains. The **atomic number**, Z, is defined as the number of protons in a nucleus. (Because atoms are electrically neutral, the number of electrons must also equal Z.) The number of neutrons in a nucleus is designated by the **neutron number**, N. Finally, the total number of nucleons in a nucleus is the **mass number**, A. Thus, the mass number is the *sum* of the atomic number and the neutron number:

$$A = Z + N$$

The definitions of atomic number, neutron number, and mass number are summarized in **Table 26.1**.

Representing Nuclei The composition of a nucleus is expressed with special notation. In general, the nucleus of an element, X, with atomic number Z and mass number A is written as follows:

$$^{A}_{Z}\text{X}$$

For example, the nucleus of carbon-14 is written as follows:

$$^{14}_{6}\text{C}$$

Notice that the letter C is the chemical symbol of carbon. The atomic number, $Z = 6$, is written as a subscript in front of the chemical symbol. Likewise, the mass number, $A = 14$, is written as a superscript. Thus, carbon-14 has 14 nucleons in its nucleus—6 of them are protons, and the remaining 8 are neutrons.

A similar type of notation is used for subatomic particles—such as the nucleons. Neutrons and protons are represented as follows:

$$^{1}_{0}\text{n} = \text{neutron (mass number 1, charge 0)}$$

$$^{1}_{1}\text{p} = \text{proton (mass number 1, charge 1)}$$

Table 26.1 Numbers That Characterize a Nucleus

Z	Atomic number = number of protons in nucleus
N	Neutron number = number of neutrons in nucleus
A	Mass number = number of nucleons in nucleus

Simplified representation of nucleus

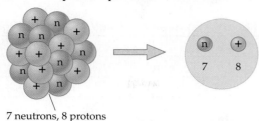

7 neutrons, 8 protons

◀ **Figure 26.3 A simple representation of a nucleus**
We use a shorthand method for representing the composition of a nucleus. Instead of drawing each proton and neutron in a nucleus (left), we use a simple circle for the nucleus and show the number of protons and neutrons it contains (right).

When nuclei contain large numbers of neutrons and protons, a drawing like that in Figure 26.2 becomes cumbersome. To simplify things, we draw a circle for the nucleus. Inside the circle we write the symbols for neutrons and protons, along with the number of each type of particle. An example is shown in **Figure 26.3**.

QUICK Example 26.1 What's the Symbol?

Write the symbol for a nucleus of aluminum that contains 14 neutrons.

Solution
Looking up aluminum in the periodic table in Appendix C, we find that $Z = 13$. In addition, we are given that the number of neutrons is $N = 14$. Therefore, the mass number is

$$A = Z + N$$
$$= 13 + 14$$
$$= 27$$

It follows that the symbol for this nucleus is

$$^{27}_{13}\text{Al}$$

Practice Problems

1. Tritium is a type of "heavy" hydrogen. The nucleus of tritium can be represented as $^{3}_{1}\text{H}$. What are the numbers of protons and neutrons in a tritium nucleus?

2. Write the symbol for a nucleus with 7 protons and 6 neutrons.

Isotopes contain different numbers of neutrons

All nuclei of a given element have the same number of protons. For example, carbon nuclei *always* have 6 protons, and oxygen nuclei *always* have 8 protons. Atoms of a given element can have different numbers of neutrons in their nuclei, however.

Nuclei with the same number of protons (the same value of Z) but different numbers of neutrons (different values of N) are referred to as **isotopes**. For example, $^{12}_{6}\text{C}$ and $^{13}_{6}\text{C}$ are two isotopes of carbon, with $^{12}_{6}\text{C}$ being the most common one, constituting about 98.89% of naturally occurring carbon. About 1.11% of natural carbon is $^{13}_{6}\text{C}$. Values for the percent abundances of various isotopes can be found in Appendix C.

Table 26.2 Mass and Charge of Particles in the Atom

Particle	Mass (kg)	Mass (u)	Energy Equivalence (MeV)	Charge (C)
Proton	1.672623×10^{-27}	1.007276	938.28	$+1.6022 \times 10^{-19}$
Neutron	1.674929×10^{-27}	1.008665	939.57	0
Electron	9.109390×10^{-31}	0.0005485799	0.511	-1.6022×10^{-19}

Atomic Mass Unit   Because different isotopes have different numbers of neutrons, they also have different masses. Appendix C gives the atomic masses of many common isotopes. Notice that these masses are *not* given in grams or kilograms. Instead, the *atomic mass unit* is used.

- The **atomic mass unit** (u) is a unit of mass exactly equal to $\frac{1}{12}$ of the mass of a carbon-12 atom.

In other words, the mass of one atom of $^{12}_{6}C$ is exactly 12 u. The value of the atomic mass unit in kilograms is as follows:

> **Definition of the Atomic Mass Unit (u)**
>
> $1 \text{ u} = 1.660539 \times 10^{-27} \text{ kg}$
>
> SI unit: kg

Table 26.2 puts the atomic mass unit into perspective. Notice, for example, that protons have a mass just a little greater than 1 u, and that the neutron is slightly more massive than the proton. The mass of an electron is much less than 1 u, as expected.

Nuclei can be described in terms of mass or energy

How did Einstein relate mass to energy?

One of Albert Einstein's most famous results is the relationship between mass and energy. **According to Einstein, mass and energy are equivalent, with energy equal to mass times the speed of light squared.**

> **Mass-Energy Equivalence**
>
> energy $=$ mass $\times$ (speed of light)^2
>
> $\qquad E = mc^2$
>
> SI unit: J

As an example, let's apply the mass-energy equivalence to a mass equal to the atomic mass unit, 1 u. Substituting 1.660539×10^{-27} kg (the mass of 1 u in kilograms) for m yields

$$E = mc^2$$
$$= (1.660539 \times 10^{-27} \text{ kg})(2.998 \times 10^8 \text{ m/s})^2$$
$$= 1.492 \times 10^{-10} \text{ J}$$

Converting to electron volts gives

$$E = (1.492 \times 10^{-10} \text{ J})\left(\frac{1 \text{ eV}}{1.6022 \times 10^{-19} \text{ J}}\right)$$
$$= 9.315 \times 10^8 \text{ eV}$$

This is a significant amount of energy—more than 900 million electron volts, in fact. For comparison, recall that the energy required to ionize hydrogen is only 13.6 electron volts. As a general rule, atomic energies are in the range of electron volts (eV), whereas nuclear energies are in the range of millions of electron volts (MeV), where $1 \text{ MeV} = 10^6 \text{ eV}$.

We conclude, then, that 1 atomic mass unit, or 1 u, is equivalent to an amount of energy, E_u, where

$$E_u = 931.5 \text{ MeV}$$

This relationship is quite useful, because mass and energy are often converted from one form to the other within a nucleus. For example, if the mass of a nucleus decreases by 0.102 u, the energy released is

$$(0.102)E_u = (0.102)(931.5 \text{ MeV})$$
$$= 95.0 \text{ MeV}$$

We'll use this conversion between mass and energy later in the chapter when we study nuclear reactions. For convenience, the energy equivalents of the proton, neutron, and electron are given in millions of electron volts (MeV) in Table 26.2.

On a fundamental level, all the energy in the universe is the result of mass being converted to energy. We've seen how this concept applies to nuclei, but it applies everywhere. For example, if an atom emits a photon—which has energy but no mass—the mass of the atom decreases by the energy of the photon divided by the speed of light squared. Similarly, if you compress a spring, increasing its potential energy, it will have more mass. The change in mass in cases like these is generally too small to be noticed, but the basic concept of mass-energy equivalence still applies.

Nuclei are held together by the strong nuclear force

We know that like charges repel one another, and that this repulsive force increases rapidly as the charges are brought closer together. It follows that protons in a tightly packed nucleus exert relatively large forces on one another. Coulomb's law shows that the electrostatic force acting on two protons separated by 10^{-15} m (a typical nuclear distance) is

$$F = \frac{ke^2}{r^2} = 230 \text{ N}$$

If this force alone acted on a proton, it would cause the following acceleration:

$$a = \frac{F}{m} = \frac{230 \text{ N}}{1.67 \times 10^{-27} \text{ kg}} = 1.4 \times 10^{29} \text{ m/s}^2$$

This is about 10^{28} times greater than the acceleration due to gravity! Thus, if protons in a nucleus experienced only the electrostatic force, the nucleus would fly apart in an instant. Because this does not happen, it follows that large attractive forces also act within the nucleus. The attractive force that holds a nucleus together is called the **strong nuclear force**. The properties of the strong nuclear force are as follows:

- 🔌 The strong nuclear force acts over a very short range ($\sim 10^{-15}$ m).

- 🔌 The strong nuclear force is always attractive and acts with nearly equal strength on protons and neutrons.

- 🔌 The strong nuclear force does not act on electrons; therefore, it has no effect on the chemical properties of an atom.

🔌 **What are the properties of the strong nuclear force?**

Not all nuclei are stable

The competition between the repulsive electrostatic forces and the attractive strong nuclear force determines whether a given nucleus is stable. **Figure 26.4** shows the neutron number, N, and the atomic number, Z, for stable nuclei.

Stability of Small Nuclei

Small nuclei—those with relatively small atomic numbers—are most stable when they have nearly equal numbers of neutrons (N) and protons (Z). For example, $^{12}_{6}\text{C}$ and $^{13}_{6}\text{C}$ are both stable. The condition $N = Z$ is indicated by the straight line in Figure 26.4.

Stability of Larger Nuclei

As the atomic number increases, stable nuclei deviate from the line $N = Z$. In fact, large stable nuclei tend to contain significantly more neutrons than protons. An example is $^{185}_{75}\text{Re}$, which has 110 neutrons but only 75 protons. Since all nucleons experience the strong nuclear force, but only protons experience the electrostatic force, the "extra" neutrons effectively "dilute" the nuclear charge density. This reduces the repulsive forces that would otherwise cause the nucleus to disintegrate.

As the number of protons in a nucleus increases, however, a point is reached at which the strong nuclear force is no longer able to compensate for the repulsive forces between protons. In fact, the largest number of protons in a stable nucleus is $Z = 83$, corresponding to the element bismuth. Nuclei with more than 83 protons are simply not stable. In the next lesson we take a look at what happens when an unstable nucleus breaks apart.

CONNECTINGIDEAS

Previously you learned about energy conservation (Chapter 6), momentum conservation (Chapter 7), and electrostatic repulsion (Chapter 19).
• All of these concepts are important in understanding nuclear stability and radioactive decay.

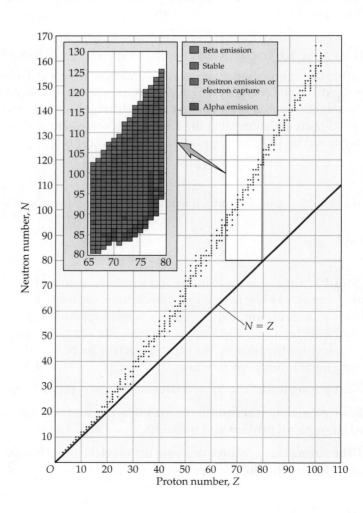

► **Figure 26.4** *N* and *Z* for stable and unstable nuclei

Stable nuclei with proton numbers less than 104 are indicated by small dots. Notice that large nuclei have significantly more neutrons, *N*, than protons, *Z*. The inset shows unstable nuclei and their decay modes for proton numbers between 65 and 80.

26.1 LessonCheck (MP)

Checking Concepts

3. **Explain** How do the atomic number, neutron number, and mass number characterize the nucleus of an atom?

4. **Describe** How is energy related to mass?

5. **Describe** What effect does the strong nuclear force have on protons, neutrons, and electrons?

6. **Triple Choice** If a nucleus gains a neutron, does each of the following quantities increase, decrease, or stay the same?
(**a**) atomic number
(**b**) neutron number
(**c**) mass number

Solving Problems

7. **Calculate** A nucleus with mass number $A = 182$ and atomic number $Z = 74$ loses two neutrons. What are the resulting (**a**) mass number, (**b**) atomic number, and (**c**) neutron number?

8. **Calculate** A nucleus has 25 protons and 30 neutrons. What is the symbol for this nucleus?

9. **Calculate** Identify the mystery nucleus (represented by the question mark) in the following reaction:

$$\ _{0}^{1}\text{n} + \ _{92}^{235}\text{U} \longrightarrow \ _{56}^{144}\text{Ba} + ? + 2\ _{0}^{1}\text{n}$$

(*Hint*: Find the mystery nucleus by requiring the atomic numbers and the mass numbers to add up to the same amount on both sides of the reaction.)

26.2 Radioactivity

If you've ever built a house of cards, you know it's very unstable. The slightest touch can send it tumbling down, with each falling card knocking down another in a sort of chain reaction. Large nuclei are similar; they are very unstable and can decay in a number of ways. This lesson explores the main types of nuclear decay.

Unstable nuclei are radioactive

An unstable nucleus isn't like a diamond—it doesn't last forever. When an unstable nucleus decays, it emits particles or high-energy photons. The particles and photons emitted when a nucleus decays are known as **radioactivity**. In analogy with the term *X-ray*, the radioactive products of decay are generally referred to as *rays*. Thus, you can think of an unstable nucleus as emitting "ray"-dioactivity.

Vocabulary

- radioactivity
- alpha particle
- beta particle
- positron
- gamma ray
- nuclear binding energy

Types of Radioactive Decay Four types of radioactive decay are most common. Three involve the emission of particles, and one involves the emission of an energetic photon.

- **Alpha particles** Alpha particles (denoted by the Greek letter α) consist of two protons and two neutrons. They are actually the nuclei of helium atoms, ^4_2He. When a nucleus decays by giving off alpha particles, we say that it emits α rays.

- **Beta particles** Beta particles (denoted by the Greek letter β) are electrons that have been given off during radioactive decay. When a nucleus gives off an electron, we say that it emits a β^- ray. (The negative sign is a reminder that the charge of an electron is $-e$.) In a nuclear reaction equation, we write an electron as e^-.

- **Positrons** A positron, short for "positive electron," is the *antiparticle* to an electron. Positrons have the same mass as an electron, but the opposite charge $(+e)$. If a nucleus gives off positrons when it decays, we say that it emits β^+ rays. In a nuclear reaction equation, we write a positron as e^+.

- **Gamma rays** A nucleus in an excited state can drop to a lower-energy state and emit a high-energy photon known as a *gamma ray*. We denote gamma rays with the Greek letter γ.

The positron is your first encounter with an antiparticle, or *antimatter*. All particles have antiparticle counterparts that have the opposite charge. Particles and antiparticles annihilate one another in a burst of energy when they come into contact.

The following Conceptual Example examines the behavior of radioactivity in a magnetic field.

CONCEPTUAL Example 26.2 Identify the Radiation

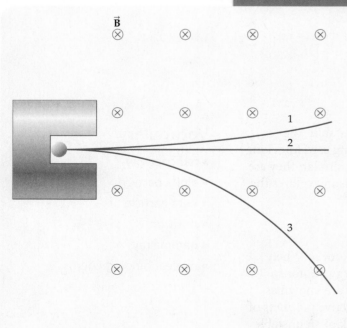

A radioactive material is placed at the bottom of a small hole drilled into a piece of lead. The material emits α rays, β^- rays, and γ rays into a region of constant magnetic field. It is observed that the radiation follows three distinct paths, 1, 2, and 3, as shown. Match each path to the corresponding type of radiation.

Reasoning and Discussion
First, because γ rays are uncharged, they are not deflected by the magnetic field. It follows that path 2 corresponds to the γ rays. Next, the right-hand rule for the magnetic force indicates that positively charged particles are deflected upward and negatively charged particles are deflected downward. As a result, path 1 corresponds to the α rays, and path 3 corresponds to the β^- rays.

Answer
Path 1, α rays; path 2, γ rays; path 3, β^- rays.

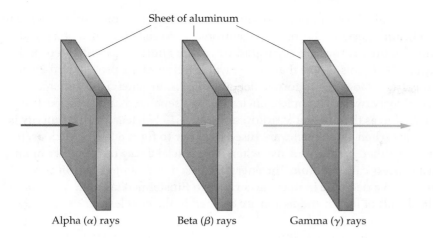

Sheet of aluminum

Alpha (α) rays   Beta (β) rays   Gamma (γ) rays

◀ **Figure 26.5 Penetrating power of radiation**
Alpha rays are stopped as soon as they encounter a sheet of aluminum. Beta rays (electrons) can penetrate a few millimeters into the aluminum before stopping. Gamma rays (high-energy photons) can pass right through a thin aluminum sheet.

Penetrating Power

Radioactivity was discovered by the French physicist Antoine Henri Becquerel (1852–1908) in 1896. He observed that uranium was able to expose photographic film, even when the film was covered. This was proof of the penetrating ability of radioactivity.

The types of radioactivity were initially named according to their ability to penetrate materials. The least penetrating rays were called alpha rays (because alpha is the first letter of the Greek alphabet). Beta and gamma rays were increasingly penetrating. Typical penetrating abilities for these three types of rays are as follows:

- Alpha (α) rays can barely penetrate a sheet of paper, and they are stopped by a sheet of aluminum.
- Beta rays (both β^- and β^+) can penetrate a few millimeters of aluminum.
- Gamma (γ) rays pass right through a thin aluminum sheet and can even penetrate several centimeters of lead.

The penetrating power of radiation is illustrated in **Figure 26.5**.

Nuclear Decay and Mass-Energy Equivalence

One of the most interesting aspects of radioactivity is completely contrary to everyday experience. Suppose a bunch of grapes rests on a kitchen counter. The bunch of grapes has a certain mass. Now, suppose you split the grapes into two smaller bunches and put them on opposite sides of the counter. Is the total mass of the grapes the same as before? Of course it is. You would think that this observation would apply to the particles in a nucleus. It doesn't!

When a large nucleus (larger than iron) undergoes radioactive decay, the mass of the system decreases. **Figure 26.6** shows that the mass of a large nucleus before decay is *greater than* the mass of the resulting nucleus plus the mass of the particles it emits. It's as if the grapes in the two small bunches had slightly less mass than the grapes in the original bunch.

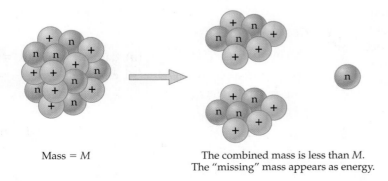

Mass = M

The combined mass is less than M. The "missing" mass appears as energy.

◀ **Figure 26.6 Radioactive decay: the conversion of mass to energy in large nuclei**
The pieces that are left after a large nucleus decays have less total mass than the original nucleus. The "missing" mass is converted to energy.

What happens to the mass of a nucleus when it undergoes a nuclear reaction?

A similar change in mass occurs with small nuclei. For example, the mass of a helium nucleus (2 protons, 2 neutrons) *is less* than the mass of 2 protons and 2 neutrons that are separated from one another, as indicated in **Figure 26.7**. This means that energy will be released if 2 protons and 2 neutrons are put together to form a helium nucleus. In general, the difference in energy between a complete nucleus and its separated individual parts is referred to as the **nuclear binding energy.** Nuclear binding energy is released when small nuclei are fused together to form a larger nucleus, in a process called *fusion*, and also when large nuclei decay into smaller nuclei, in a process called *fission*. The energy released in fusion or fission corresponds to a decrease in mass, according to Einstein's relation $E = |\Delta m|c^2$. The details of fusion and fission are covered in the next lesson.

Alpha decay emits neutrons and protons

When a nucleus decays by giving off an α particle ($^4_2$He), it loses two protons and two neutrons. As a result, its atomic number, Z, decreases by 2, and its mass number, A, decreases by 4. Symbolically, this process can be written as follows:

$$^A_Z X \longrightarrow ^{A-4}_{Z-2}Y + ^4_2He + energy$$

In this decay, X is referred to as the *parent nucleus*, and Y is the *daughter nucleus*. The sum of the atomic numbers on the right side of the reaction is equal to the atomic number on the left side. The same is true of the mass numbers. The energy released is given by the decrease in mass during the reaction times the speed of light squared. The next Guided Example considers the alpha decay of uranium-238.

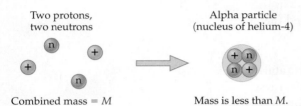

Two protons, two neutrons

Combined mass = M

Alpha particle (nucleus of helium-4)

Mass is less than M. The "missing" mass appears as energy.

▲ **Figure 26.7 Radioactive decay: the conversion of mass to energy in small nuclei**
Combining 2 protons and 2 neutrons into a helium-4 nucleus (alpha particle) results in a decrease in mass, accompanied by a release of energy.

GUIDED Example 26.3 | Alpha Decay of Uranium-238   **Alpha Decay**

Determine (**a**) the daughter nucleus and (**b**) the energy released in MeV when $^{238}_{92}$U undergoes alpha decay. The mass of the system decreases by 0.004587 u as a result of the decay. (Recall that the energy released by a mass decrease of 1 u is $E_u = 931.5$ MeV.)

Picture the Problem

Our sketch shows the decay of $^{238}_{92}$U into a daughter nucleus plus an α particle. Notice that the numbers of neutrons and protons are indicated for $^{238}_{92}$U. The α particle consists of two neutrons and two protons.

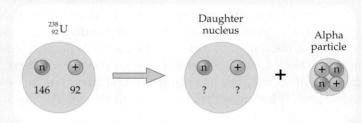

Strategy

(**a**) We can identify the daughter nucleus by requiring that the total numbers of neutrons and protons be the same before and after the decay. (**b**) The decrease in mass is 0.004587 u, and therefore the energy released is $E = (0.004587)E_u$.

Known

parent nucleus: $^{238}_{92}$U
decay type: alpha decay
$\Delta m = -0.004587$ u

Unknown

(**a**) daughter nucleus = ?
(**b**) energy released: $E = ?$

Solution

Part (a)

1 Determine the numbers of neutrons and protons in the daughter nucleus. Add these numbers together to obtain the mass number of the daughter nucleus:

$$N = 146 - 2$$
$$= 144$$
$$Z = 92 - 2$$
$$= 90$$
$$A = N + Z$$
$$= 144 + 90$$
$$= 234$$

2 Referring to Appendix C, identify the daughter nucleus as thorium-234:

$$\boxed{^{234}_{90}\text{Th}}$$

Part (b)

3 Calculate the energy released using the relation $E = (0.004587)E_u$:

$$E = (0.004587)E_u$$
$$= (0.004587)(931.5 \text{ MeV})$$
$$= \boxed{4.273 \text{ MeV}}$$

Insight

This result confirms that nuclear reactions involve energies on the order of millions of electron volts (MeV).

Practice Problems

10. $^{226}_{88}\text{Ra}$ undergoes alpha decay. What is the daughter nucleus?

11. **Think & Calculate** The alpha decay of $^{226}_{88}\text{Ra}$ releases 4.871 MeV. **(a)** Does the mass of the system increase or decrease as a result of the decay? Explain. **(b)** Determine the change in mass that occurs during this process.

12. Identify the daughter nucleus (represented by the question mark) in the following alpha decay:

$$^{242}_{96}\text{Cm} \longrightarrow \text{?} + {}^{4}_{2}\text{He}$$

Beta decay is the emission of an electron

The basic process that occurs in beta decay is the conversion of a neutron to a proton, an electron (e^-), and an *antineutrino*:

$$\text{neutron} \longrightarrow \text{proton} + \text{electron} + \text{antineutrino}$$
$$^{1}_{0}\text{n} \longrightarrow {}^{1}_{1}\text{p} + e^- + \bar{\nu}_e$$

The symbol $\bar{\nu}_e$ in this reaction represents the antineutrino, the antiparticle to the *neutrino*, ν_e (literally, "little neutral one"). Neutrinos have very little mass (about a hundred-thousandth of the mass of an electron) and little interaction with matter. Only 1 in every 200 million neutrinos that pass through the Earth interacts with it in any way. In fact, billions of neutrinos pass through your body each second without effect.

COOL PHYSICS
Smoke Detectors

The operation of many smoke detectors depends on a radioactive isotope of americium, $^{241}_{95}\text{Am}$. This isotope decays primarily by the release of α particles. A minute quantity of $^{241}_{95}\text{Am}$ is placed between two metal plates that are connected to a battery. The emitted α particles ionize the air, allowing a small electric current to flow between the plates. As long as the current flows, the smoke detector remains silent. During a fire, however, the ionized air molecules bind to smoke particles and become neutralized. This reduces the current and triggers the alarm. Take a look at the smoke detectors in your home and see if any of them are ionization detectors. If so, don't be concerned about radioactivity. The levels are much too low to have any harmful effect on living creatures.

The force responsible for beta decay is known as the *weak nuclear force*. This force is short range and is important only within the nucleus of an atom. As you will see in Lesson 26.4, the weak nuclear force is intermediate in strength between the gravitational and electromagnetic forces.

When a nucleus decays by giving off an electron, its mass number is unchanged (since protons and neutrons count equally in determining A), but its atomic number increases by 1. This process can be represented symbolically as follows:

$$^A_Z X \longrightarrow\ ^A_{Z+1} Y + e^- + \bar{\nu}_e + \text{energy}$$

In some cases a nucleus gives off a positron (e^+) rather than an electron. This process can be written

$$^A_Z X \longrightarrow\ ^A_{Z-1} Y + e^+ + \nu_e + \text{energy}$$

In the next Guided Example we determine the energy that is released when carbon-14 undergoes beta decay. (Nuclear decay results in energy being given off in the form of the kinetic energy of the decay products. Because this energy release is understood, we often omit the term "+ energy" in an equation for a nuclear reaction.)

GUIDED Example 26.4 | Beta Decay of Carbon-14   **Beta Decay**

Find **(a)** the daughter nucleus and **(b)** the energy released (in MeV) when $^{14}_6 C$ undergoes beta decay. The mass decreases by 0.000168 u as a result of the decay. (Recall that the energy released by a mass decrease of 1 u is $E_u = 931.5$ MeV.)

Picture the Problem

Our sketch shows $^{14}_6 C$ giving off an electron and converting to a daughter nucleus. The numbers of neutrons and protons in the $^{14}_6 C$ nucleus are indicated.

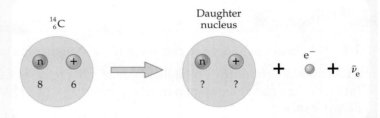

Strategy

(a) We can identify the daughter nucleus by requiring that the total numbers of nucleons be the same before and after the decay. The number of neutrons will be decreased by 1, and the number of protons will be increased by 1. **(b)** The decrease in mass is 0.000168 u, and therefore the energy released is $E = (0.000168)E_u$.

Known

parent nucleus: $^{14}_6 C$
decay type: beta decay
$\Delta m = -0.000168$ u

Unknown

(a) daughter nucleus = ?
(b) energy released: E = ?

Solution

Part (a)

> 1   Determine the numbers of neutrons and protons in the daughter nucleus. Add these numbers together to obtain the mass number of the daughter nucleus:

$$N = 8 - 1 = 7$$
$$Z = 6 + 1 = 7$$
$$A = N + Z$$
$$\quad = 7 + 7$$
$$\quad = 14$$

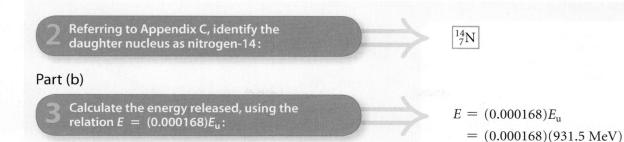

2 Referring to Appendix C, identify the daughter nucleus as nitrogen-14: $\qquad\Longrightarrow\qquad$ $^{14}_{7}\text{N}$

Part (b)

3 Calculate the energy released, using the relation $E = (0.000168)E_{\text{u}}$: $\qquad\Longrightarrow\qquad$

$$E = (0.000168)E_{\text{u}}$$
$$= (0.000168)(931.5 \text{ MeV})$$
$$= \boxed{0.156 \text{ MeV}}$$

Insight

Because the emitted electron is *not* a nucleon, the total number of nucleons in the nucleus is unchanged. The energy released in beta decay is typically less than the energy released in alpha decay.

Practice Problems

13. Identify the daughter nucleus produced when $^{234}_{90}\text{Th}$ undergoes beta decay.

14. Think & Calculate The beta decay of $^{234}_{90}\text{Th}$ releases 0.274 MeV.
(a) Does the mass of the system increase or decrease as a result of the decay? Explain.
(b) Determine the change in mass that occurs during this process.

15. Identify the daughter nucleus in the following beta decay:

$$^{238}_{94}\text{Pu} \longrightarrow ? + e^- + \bar{\nu}_e$$

Gamma decay is the emission of a photon

An atom in an excited state can emit a photon when one of its electrons drops to a lower energy level. Similarly, a nucleus in an excited state can emit a photon as it decays to a state of lower energy. Since nuclear energies are so much greater than typical atomic energies, the photons given off by decaying nuclei are highly energetic. In fact, these photons have energies that place them well above X-rays in the electromagnetic spectrum. High-energy photons emitted by nuclei are known as *gamma* (γ) *rays*.

Consider the following decay process:

$$^{14}_{6}\text{C} \longrightarrow ^{14}_{7}\text{N}^* + e^- + \bar{\nu}_e$$

The asterisk on the nitrogen symbol indicates that the nitrogen nucleus has been left in an excited state as a result of this beta decay. Subsequently, the nitrogen nucleus may decay to its ground state, with the emission of a γ ray:

$$^{14}_{7}\text{N}^* \longrightarrow ^{14}_{7}\text{N} + \gamma$$

Notice that neither the atomic number nor the mass number is changed by the emission of a γ ray.

Radioactive decays can occur in series

Consider an unstable nucleus that decays and produces a daughter nucleus. If the daughter nucleus is also unstable, it eventually decays and produces its own daughter nucleus, which may in turn be unstable. In such cases, an

When $_{92}^{235}$U decays, it changes to a number of intermediate nuclei before becoming the stable nucleus at the end of the series, $_{82}^{207}$Pb. Some intermediary nuclei can decay in only one way, whereas others have two decay possibilities.

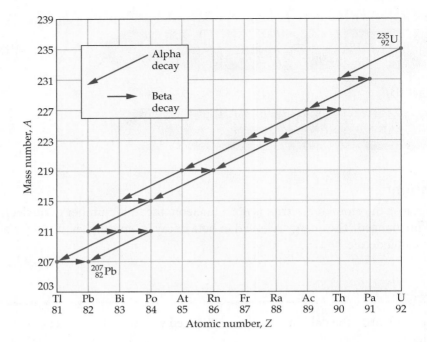

original parent nucleus can produce a series of related nuclei in a process referred to as a *radioactive decay series*. An example of a radioactive decay series is shown in **Figure 26.8**. In this series the parent nucleus is $_{92}^{235}$U, and the nucleus at the end of the series is $_{82}^{207}$Pb, which is stable.

Notice that several of the intermediate nuclei in this series can decay in two different ways—either by alpha decay or by beta decay. Thus, there are a variety of paths a $_{92}^{235}$U nucleus can follow as it transforms into a $_{82}^{207}$Pb nucleus. In addition, the intermediate nuclei in this series decay fairly rapidly, at least on a geological time scale. For example, any actinium-227 that was present when the Earth formed decayed away long ago. The fact that actinium-227 is still found on Earth today in natural uranium deposits is due to its continual production in this and other decay series.

The rate of radioactive decay is the activity

The rate at which nuclear decay occurs—that is, the number of decays per second—is referred to as the *activity*. A highly active material has many nuclear decays occurring every second. For example, a typical sample of radium (usually a fraction of a gram) might have 10^5 to 10^{10} radioactive decays per second.

The unit we use to measure activity is the *curie*, named in honor of Pierre Curie (1859–1906) and Marie Curie (1867–1934), pioneers in the study of radioactivity. The curie (Ci) is defined as follows:

$$1 \text{ curie} = 1 \text{ Ci} = 3.7 \times 10^{10} \text{ decays/s}$$

The reason for this choice is that 1 Ci is roughly the activity of 1 g of radium. In SI units, activity is measured in terms of the *becquerel* (Bq):

$$1 \text{ becquerel} = 1 \text{ Bq} = 1 \text{ decay/s}$$

As you can see, an activity of 1 curie is much larger (more than ten trillion times larger!) than an activity of 1 becquerel.

26.2 LessonCheck (MP)

Checking Concepts

16. ⬤ **Compare** How does the mass of a large nucleus before it decays compare with the mass of the system after the decay?

17. Explain How is a radioactive decay series produced?

18. Describe What happens when a particle of matter encounters its antimatter counterpart?

19. Summarize Create a table to summarize how the atomic number, neutron number, and mass number are affected by (**a**) alpha decay, (**b**) beta decay, and (**c**) gamma decay.

Solving Problems

20. Calculate A nucleus emits an α particle. (**a**) If the mass number of this parent nucleus was A, what is the mass number of the daughter nucleus? Repeat part (a) for (**b**) the atomic number, Z, and (**c**) the neutron number, N.

21. Calculate A nucleus emits a β^- particle. (**a**) If the mass number of this parent nucleus was A, what is the mass number of the daughter nucleus? Repeat part (a) for (**b**) the atomic number, Z, and (**c**) the neutron number, N.

22. Identify Determine the daughter nucleus produced when $^{227}_{89}\text{Ac}$ undergoes alpha decay.

23. Identify Determine the daughter nucleus produced when $^{215}_{83}\text{Bi}$ undergoes beta decay.

24. Identify Determine the parent nucleus when $^{219}_{86}\text{Rn}$ is produced by (**a**) alpha decay and (**b**) beta decay.

26.3 Applications of Nuclear Physics

Chemical reactions involve changes to the electron clouds that surround a nucleus; nuclear reactions involve changes to the neutrons and protons *within* a nucleus. Nuclear reactions generally have no effect on chemical reactions, and chemical reactions don't alter the behavior of a nucleus. These two types of reactions have very different energies, and they are almost completely independent of one another.

Just as chemical reactions are an important part of our everyday lives, so too are nuclear reactions. Radioactivity makes significant contributions to everything from electric power generation to archeology to medicine. We explore these applications in this lesson.

Vocabulary

- nuclear fission
- nuclear fusion
- half-life

Nuclear Fission

A new type of natural phenomenon was discovered in 1939 when Otto Hahn (1879–1968) and Fritz Strassmann (1902–1980) discovered that a uranium nucleus can break into two pieces. In general, the process in which a large nucleus splits into two smaller nuclei is called **nuclear fission**. The mass of the system decreases during the fission process, and energy is released.

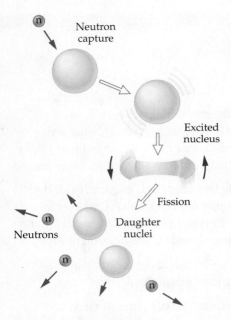

▲ Figure 26.9 Nuclear fission
When a large nucleus captures a neutron, it can become excited and ultimately split into two smaller nuclei. This is the process of fission.

Nuclear fission releases far more energy than any chemical reaction. This clearly follows if you recall the earlier discussion of mass-energy equivalence, which pointed out that nuclear energies are in the range of millions—or even hundreds of millions—of electron volts, compared with atomic (chemical) energies, which are in the range from 1 to 10 electron volts. Thus, a typical nuclear reaction can give off about a hundred million times more energy than a typical chemical reaction! This is why nuclear fission is used in many of today's electric power plants.

Large unstable nuclei can decay by splitting in two

The first step in a typical fission reaction occurs when a slow neutron is absorbed, or *captured*, by a uranium-235 nucleus. This step increases the mass number of the nucleus by 1 and leaves it in an excited state:

$$\,^1_0n + \,^{235}_{92}U \longrightarrow \,^{236}_{92}U^*$$

The excited nucleus (indicated by the asterisk) oscillates wildly and becomes highly distorted, as depicted in **Figure 26.9**. In many respects, the nucleus behaves like a spinning drop of water. Like a drop of water, the nucleus can distort only so much before it breaks apart (fissions) into smaller pieces.

There are about 90 different ways in which the uranium-235 nucleus can undergo fission. Typically, 2 or 3 neutrons (2.47 on average) are released during the fission process, and two smaller nuclei are formed. One of the possible fission reactions for $\,^{235}_{92}U$ is considered in the following Guided Example.

GUIDED Example 26.5 | A Fission Reaction of Uranium-235   **Fission**

When uranium-235 captures a neutron, one of the fission reactions it can undergo is the following:

$$\,^1_0n + \,^{235}_{92}U \longrightarrow \,^{236}_{92}U^* \longrightarrow \,^{141}_{56}Ba + ? + 3\,^1_0n$$

(**a**) Identify the mystery nucleus (represented by the question mark). (**b**) If the mass of the system decreases by 0.186 u during this process, how much energy (in MeV) is released? (Recall that the energy released by a mass decrease of 1 u is $E_u = 931.5$ MeV.)

Picture the Problem

The specified fission reaction is shown in our sketch, starting with the $\,^{236}_{92}U^*$ nucleus. The known numbers of protons and neutrons are indicated for both $\,^{236}_{92}U^*$ and $\,^{141}_{56}Ba$. The corresponding numbers for the unidentified nucleus are to be determined.

Strategy

(**a**) The total number of protons must be the same before and after the reaction, as must the number of neutrons. By conserving protons and neutrons, we can determine the missing numbers in the unidentified nucleus. (**b**) The decrease in mass is 0.186 u, and therefore the energy released is $E = (0.186)E_u$.

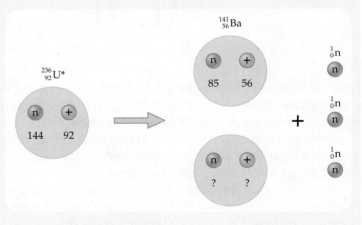

Known

(**a**) $\,^{236}_{92}U^*$ decays to $\,^{141}_{56}Ba$ plus another nucleus and 3 neutrons
(**b**) $|\Delta m| = 0.186$ u

Unknown

(**a**) unidentified nucleus $= ?$
(**b**) energy released: $E = ?$

Solution

Part (a)

1 Determine the numbers of protons, Z, neutrons, N, and nucleons, A, in the unidentified nucleus. When calculating the number of neutrons, be sure to include the three neutrons released in the reaction:

$$Z = 92 - 56$$
$$= 36$$
$$N = 144 - 85 - 3$$
$$= 56$$
$$A = Z + N$$
$$= 36 + 56$$
$$= 92$$

2 Use Z and A to specify the unidentified nucleus:

$$\boxed{{}^{92}_{36}\text{Kr} \ (\text{krypton-92})}$$

Part (b)

3 Calculate the energy released, using the relation $E = (0.186)E_u$:

$$E = (0.186)E_u$$
$$= (0.186)(931.5 \text{ MeV})$$
$$= \boxed{173 \text{ MeV}}$$

Insight

The energy given off in a typical fission reaction is on the order of 200 MeV.

Practice Problems

25. Identify the mystery nucleus in the following reaction:

$$^{1}_{0}\text{n} + {}^{235}_{92}\text{U} \longrightarrow {}^{236}_{92}\text{U}^* \longrightarrow {}^{140}_{54}\text{Xe} + \ ? + 2{}^{1}_{0}\text{n}$$

26. **Think & Calculate** The reaction in Problem 25 releases 184.7 MeV.
(a) Does the mass of the system increase or decrease as a result of the reaction? Explain.
(b) Determine the change in mass that occurs during this reaction.

27. Identify the mystery nucleus in the following reaction:

$$^{1}_{0}\text{n} + {}^{235}_{92}\text{U} \longrightarrow {}^{236}_{92}\text{U}^* \longrightarrow {}^{148}_{57}\text{La} + \ ? + 3{}^{1}_{0}\text{n}$$

Nuclear fissions can produce a chain reaction

Have you ever set up a large number of dominos in a line? As you know, knocking down one domino can set off a *chain reaction*, where each falling domino hits the next one and causes it to fall also. Sometimes one domino knocks down two dominos, and each of them knocks down two more, and so on. The same kind of thing can happen with nuclei. When one nucleus decays, it releases one or more neutrons, each of which can trigger another nucleus to decay and give off more neutrons.

Suppose, for example, that a fission reaction gives off two neutrons and that both neutrons cause additional fissions in other nuclei. These nuclei in turn give off two neutrons. Starting with one nucleus to begin the chain reaction, we have two nuclei in the second generation of the chain, four

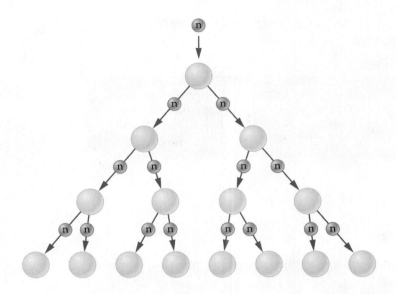

► **Figure 26.10 A chain reaction**
A chain reaction occurs when a neutron emitted in one fission reaction induces a fission reaction involving another nucleus. In the case shown here, two neutrons from one reaction induce two additional reactions.

nuclei in the third generation, and so on, as indicated in **Figure 26.10**. After only 100 generations, the number of nuclei undergoing fission is 1.3×10^{30}. If each of these reactions gives off a typical 200 MeV of energy, the total energy released will be 4.1×10^{19} J. Put into everyday terms, this is enough energy to supply the needs of the entire United States for half a year! Clearly, a rapidly developing *runaway chain reaction*, like the one just described, will produce the explosive release of an enormous amount of energy.

A nuclear reactor controls a fission chain reaction

Physics & You: Technology A great deal of effort has gone into controlling chain reactions that begin with uranium-235. If the release of energy can be kept to a manageable and usable level—avoiding explosions or meltdowns—a powerful source of energy is at our disposal.

The first controlled nuclear chain reaction was achieved by Enrico Fermi (1901–1954) in 1942, using a racquetball court at the University of Chicago for his improvised laboratory. His reactor consisted of blocks of uranium (the fuel) stacked together with blocks of graphite (the moderator) to form a large pile. In such a nuclear reactor, the *moderator* slows the neutrons given off during fission, making it more likely that they will be captured by other uranium nuclei and cause additional fissions. The reaction rate is adjusted with *control rods* made of a material (such as cadmium) that is very efficient at absorbing neutrons. With the control rods fully inserted into the pile, any reaction that begins quickly dies out because neutrons are absorbed rather than allowed to cause additional fissions. As the control rods are pulled partway out of the pile, more neutrons become available to induce reactions.

🔑 **When, on average, one neutron given off by a fission reaction produces an additional reaction, the reactor is said to be *critical*.** If the control rods are pulled out of the pile even farther, the number of neutrons causing fission reactions in the next generation becomes greater than one, and a runaway reaction begins to occur. Nuclear reactors that produce power for practical applications are kept near their critical condition by continuous adjustment of the placement of the control rods.

🔑 *What occurs when a nuclear fission reactor is critical?*

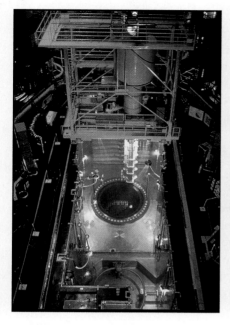

(a)

(b)

(c)

Fermi's original reactor was designed to test the basic scientific principles of fission reactions. It produced a power of only about 0.5 W when operating near its critical condition. This is the power necessary to operate a single flashlight bulb. In comparison, today's nuclear reactors typically produce 1000 MW of power, enough to power an entire city. A modern nuclear reactor is shown in **Figure 26.11**.

Safety is an issue with all forms of energy

All forms of energy come with risks: A campfire can get out of control and burn down a forest; a coal plant can release potentially harmful materials into the atmosphere; a dam might flood a valley after being destroyed in an earthquake. Nuclear power is no exception. An accident at a nuclear power plant can have immediate consequences, and it can also release into the environment radioactive materials that might be dangerous for generations. It's important for people to know some of the basic facts about different types of power generation, including nuclear power, so they can make intelligent decisions when weighing the potential benefits and risks.

Nuclear Fusion

Not only can nuclei split apart, they can also be made to join, or fuse, together. When two light nuclei combine to form a more massive nucleus, the reaction is referred to as **nuclear fusion**. The mass of the system decreases when light nuclei fuse, and energy is released.

Energy is released when small nuclei combine

Fusion reactions can release large amounts of energy, but they aren't easy to initiate. Combining two small nuclei into a single nucleus requires tremendous kinetic energy—enough to overcome the electrostatic repulsions acting on the protons. The temperature required for combining nuclei is about 10^7 K! When the temperature is high enough to initiate fusion, the process is called a *thermonuclear fusion reaction*.

▲ **Figure 26.11 Using nucleus fission to generate power**
(a) The core of a nuclear reactor at a power plant. The core sits in a pool of water, which provides cooling while absorbing stray radiation. The blue glow suffusing the pool is radiation from electrons traveling through the water at extremely high velocities. Above the core is a crane used to replace the fuel rods when their radioactive material becomes depleted.
(b) A closer view of a reactor core. The structures projecting above the core house the mechanisms that raise and lower the control rods. The placement of the rods regulates the rate of fission in the reactor. The tubes that are nearly flush with the top of the core house the fuel rods.
(c) A technician loads pellets of fissionable material into fuel rods.

▲ Figure 26.12 Solar energy
The Sun is powered by nuclear reactions that fuse hydrogen into helium.

Thermonuclear fusion occurs in the cores of stars. In fact, all stars generate energy through thermonuclear reactions. Most stars (including the Sun) fuse hydrogen to produce helium. At this very moment, the Sun is converting roughly 600,000,000 tons of hydrogen into helium every second. Some stars also fuse helium or other heavier elements. The Sun, shown in **Figure 26.12**, will be able to burn its hydrogen for about 10 billion years, producing a remarkably stable output of energy that is vital to life on Earth.

Nuclear fusion might be a future source of energy

Scientists are working to control nuclear fusion reactions and use them to generate electric power. In fact, fusion has many potential advantages over fission. For example, fusion yields more energy per mass of fuel than fission. In addition, fusion reactors could be fueled by hydrogen, which is readily obtained from seawater. To date, however, sustained laboratory fusion reactions require more energy than they produce. Still, researchers are close to the break-even point, and many are confident that success is possible.

Two methods are commonly used to produce controlled nuclear fusion. Both methods use extremely high temperatures to completely ionize atoms into a gaslike collection of electrons and nuclei known as a *plasma*. In *magnetic confinement* powerful magnetic fields trap the plasma and keep it away from the walls of the container, preventing the walls from melting. In *inertial confinement* a solid pellet of fuel is dropped into a vacuum chamber and vaporized by high-powered laser beams that hit it from all sides. An example of a plasma fusion reactor is shown in **Figure 26.13**.

Radioactive Dating

In addition to being a source of energy, radioactivity is also useful in the fields of archeology and paleontology. For example, have you ever been to a natural history museum and seen the skeletal remains of a now extinct animal? If so, you might have wondered how the age of the skeleton was determined. For instance, how is a museum sure that its skeleton of a saber-toothed tiger is 10,000 years old and not 100,000 years old? The answer likely involves an application of radioactive decay known as *radioactive dating*.

Radioactive decay is similar to flipping a coin

A useful way to see the connection between radioactivity and dating is to compare radioactive decay to flipping a coin. Imagine that a coin represents a nucleus, and the side that comes up when the coin is flipped determines whether the nucleus decays. For example, suppose you flip 64 coins, and then remove all of those that come up tails. On average, you would expect

▶ Figure 26.13 A plasma fusion reactor
Nuclear fusion, which powers the stars, may eventually turn out to be the clean, inexpensive, and renewable energy source that our society needs. But fusion requires enormous temperatures and pressures, and so far the problems involved in creating practical fusion technology have not been overcome. Several different approaches have been explored in the effort to produce sustained nuclear fusion. One of them, embodied by the Z machine at Sandia National Laboratories in New Mexico, shown here, involves the compression of a plasma by an intense burst of X-rays. So far, temperatures of nearly 2 million degrees have been attained, but only for very brief periods of time.

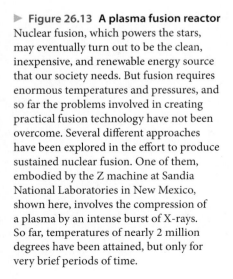

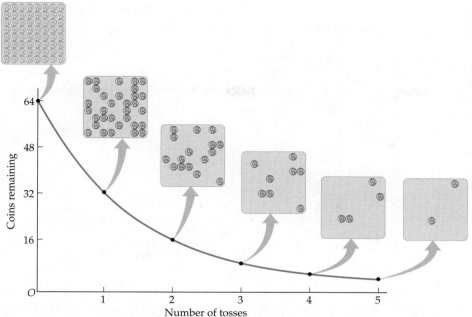

to remove 32 coins. The particular coins that are removed, and the exact number of coins removed, cannot be known. The flip of a coin—like the decay of a nucleus—is a random process.

When you flip the remaining 32 coins, you expect an average of 16 more to be removed, and so on. Each time you flip a set of coins, the number that remain is on average one half of the original number. The points in **Figure 26.14** show the results after a few tosses.

Half-Life Nuclear decay is similar to the coin flips described above, each round of which cuts the number of coins in half. In fact, we characterize the nuclear decay of a substance by its *half-life*. ⚷ **The time required for half of the nuclei in a sample of a radioactive material to decay is known as the half-life, $T_{1/2}$.** For example, if a radioactive isotope has a half-life of 5 days, then half of the nuclei decay in 5 days, half of the remaining nuclei decay in the next 5 days, and so on.

⚷ *What is the half-life of a radioactive material?*

Decay Rate as a Clock The number of nuclei decaying at any moment in time is proportional to the number of nuclei present. In other words, as the nuclei decrease in number, so too does the number of them that are decaying. This fact turns out to be very useful. Because the decay rate depends on time in a straightforward way, it can be used as a sort of clock. In fact, radioactivity can be used to date items from the past.

To be specific, if you know both the initial activity rate of a sample, $R_{initial}$, and its current activity rate, $R_{current}$, you can calculate the amount of time that has passed with the following equation:

> **Elapsed Time Based on Activity**
>
> $$\text{time} = \left(\frac{\text{half-life}}{0.693}\right) \times \ln\left(\frac{\text{initial activity rate}}{\text{current activity rate}}\right)$$
>
> $$t = \left(\frac{T_{1/2}}{0.693}\right) \ln\left(\frac{R_{initial}}{R_{current}}\right)$$

Notice that the elapsed time is directly proportional to the half-life of the radioactive material in the sample, as you might expect.

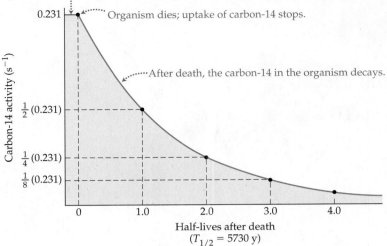

Figure 26.15 Activity of carbon-14
While an organism is living and taking in carbon from the surroundings, its carbon-14 activity remains constant. When the organism dies, the carbon-14 activity decays exponentially with a half-life of 5730 years.

Activity is constant during life because organism takes up carbon-14 continuously.

Organism dies; uptake of carbon-14 stops.

After death, the carbon-14 in the organism decays.

Carbon-14 is commonly used for radioactive dating

Physics & You: Technology Understanding how carbon-14 dating works requires some background knowledge. The carbon in Earth's biosphere consists of both carbon-12 and carbon-14. Carbon-14 is unstable with a half-life of 5730 years. Carbon-12 is stable and does not decay.

Even though carbon-14 decays, the ratio of carbon-14 to carbon-12 in Earth's atmosphere remains constant over time at about one carbon-14 atom for every trillion carbon-12 atoms. This constant ratio occurs because high-energy cosmic rays from outer space continuously enter Earth's upper atmosphere and cause nuclear reactions involving nitrogen-14 (a stable isotope). These reactions result in a steady production of carbon-14. Thus, the steady level of carbon-14 in the atmosphere is a result of the balance between the *production rate* due to cosmic rays and the *decay rate* due to the properties of the carbon-14 nucleus.

Living organisms absorb carbon-14

Living organisms have the same ratio of carbon-14 to carbon-12 as the atmosphere, since they continuously take in carbon from their surroundings. When an organism dies, however, the absorption of carbon ceases and the carbon-14 in the organism (in its wood, bone, shell, etc.) begins to decay. **Figure 26.15** illustrates this process. Notice that the carbon-14 activity of an organism is constant until it dies, at which point the activity decreases exponentially with a half-life of 5730 years.

The only other piece of information needed to implement carbon-14 dating is knowledge of the initial activity. Careful measurements show that the initial activity of a 1-gram sample of carbon is 0.231 decays of carbon-14 per second. This initial activity is shown in Figure 26.15. It follows that 5730 years after an organism dies, its carbon-14 activity per gram of carbon will have decreased by half ($\frac{1}{2} \times 0.231$ decays/s = 0.116 decays/s). The passing of another 5730 years will reduce the activity by an additional factor of two, and so on. **Measuring the current carbon-14 activity in a sample from a once-living organism gives a direct indication of the time that has elapsed since the death of the organism.**

The next Guided Example applies this basic idea to a real-world case of some interest—the Iceman of the Alps, shown in **Figure 26.16**.

How does carbon-14 dating work?

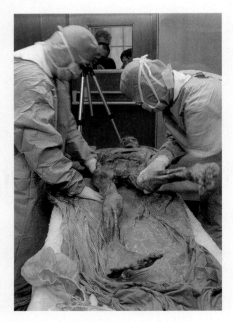

▲ **Figure 26.16 Examining Ötzi the Iceman**
The Iceman of the Italian Alps (known as Ötzi) was found in 1991. His age was established by means of carbon-14 dating.

Early in the afternoon of September 19, 1991, a German couple hiking in the Italian Alps noticed something brown sticking out of the ice ahead of them. At first they thought the object might be a doll or some rubbish. As they got closer, however, it became clear that it was the body of a person trapped in the ice. The hikers had discovered the remarkably well-preserved body of a Stone Age man who had died in the mountains and become entombed in the ice. The carbon-14 dating method was later applied to the remains of the Iceman. The current carbon-14 activity was found to be about 0.121 Bq per gram of carbon. Using this information, date the remains of the Iceman. (Recall that 1 Bq = 1 decay/s.)

Picture the Problem

Our sketch shows the decay of carbon-14 as a function of time. The initial activity of a 1-g carbon sample is 0.231 Bq. After one half-life the activity will decrease to 0.116 Bq.

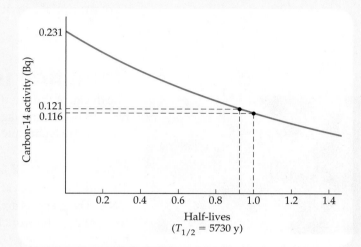

Strategy

The age of the remains is given by

$$t = \left(\frac{T_{1/2}}{0.693}\right)\ln\left(\frac{R_{\text{initial}}}{R_{\text{current}}}\right)$$

In this case the initial activity rate is $R_{\text{initial}} = 0.231$ Bq, and the current activity rate is $R_{\text{current}} = 0.121$ Bq. The half-life of carbon-14 is $T_{1/2} = 5730$ years (y).

Known

$R_{\text{initial}} = 0.231$ Bq

$R_{\text{current}} = 0.121$ Bq

$T_{1/2} = 5730$ y

Unknown

$t = ?$

Solution

1 Substitute the known activity rates and half-life into $t = \left(\dfrac{T_{1/2}}{0.693}\right)\ln\left(\dfrac{R_{\text{initial}}}{R_{\text{current}}}\right)$ and calculate the time, t:

$$t = \left(\frac{T_{1/2}}{0.693}\right)\ln\left(\frac{R_{\text{initial}}}{R_{\text{current}}}\right)$$

$$= \left(\frac{5730 \text{ y}}{0.693}\right)\ln\left(\frac{0.231 \text{ Bq}}{0.121 \text{ Bq}}\right)$$

$$= \boxed{5340 \text{ y}}$$

> **Math HELP**
> Natural
> Logarithms
> **See Math
> Review,
> Section II**

Insight

Notice that the observed activity of 0.121 Bq is slightly *greater* than $\frac{1}{2}(0.231$ Bq$) = 0.116$ Bq; hence, we expect the age of the remains to be slightly *less* than carbon-14's half-life of 5730 years—as our calculation shows. It follows that Ötzi the Iceman died in the mountains during the Stone Age, some 5340 years ago. Detailed examination of Ötzi's body and possessions indicated that he was probably a sheepherder or hunter.

28. **Follow-up** If the remains of another iceman of the same age as Ötzi (see Guided Example 26.6) are found in the year 2991, what will be the carbon-14 activity of those remains per gram of carbon?

29. Scientists can determine the age of an ancient fire by measuring the activity of the charcoal that is left behind. If the charcoal has an activity of 0.185 Bq per gram of carbon, how long ago did the fire burn?

30. The radioactive isotope $^{241}_{95}$Am, with a half-life of 432 y, is essential to the operation of many smoke detectors. Suppose such a detector will no longer function if the activity of the $^{241}_{95}$Am it contains drops below $\frac{1}{525}$ of the initial activity. How long will this kind of smoke detector work?

Radiation in Medicine

The decay of radioactive isotopes has many applications in medicine. It can be used both as a diagnostic tool, to look for indications of disease, and as a therapeutic technique to treat certain ailments.

Radioactivity can be harmful or beneficial

The radiation given off by a radioactive nucleus can kill a cell or damage its DNA. These effects are of concern if the cells that are affected are healthy. That's why people who work around radioactive materials must take precautions to limit their exposure. One way to determine the amount of exposure a person has experienced is with a film badge, like the one shown in **Figure 26.17 (a)**.

On the other hand, if radiation is used to kill cancerous cells, the result can be beneficial. An example of this medical use of radioactivity is shown in **Figure 26.17 (b)**, where a malignant tumor is being treated with radioactive cobalt. As with many technologies, the effects of radioactivity can be good or bad depending on the circumstances.

Physics & You: Technology Perhaps the most amazing use of radioactive decay in everyday life is the diagnostic technique known as *positron-emission tomography*, or PET. A PET scan is produced with radiation that *emerges from within* the body, as opposed to radiation that is generated externally and is then passed through the body. Remarkably, the radiation that creates a PET scan is produced by the *annihilation of matter and antimatter* within the patient's body! It sounds like science fiction, but this technique is, in fact, a valuable medical tool.

To produce a PET scan, a patient is given a *radioactive tracer*, which is a substance containing a radioactive isotope. For example, the tracer fluoro-deoxyglucose (FDG) contains a fluorine-18 atom attached to a molecule of glucose, the basic fuel of cells. The unstable fluorine-18 atom, which has a half-life of 110 minutes, emits a positron as it decays.

$$^{18}_{9}\text{F} \longrightarrow {}^{18}_{8}\text{O} + e^+ + \nu_e$$

The emitted positron almost immediately encounters an electron, and the two particles annihilate each other and produce a burst of energy. The annihilation generates two powerful γ rays moving in opposite directions (which conserves momentum). As these γ rays emerge from the body, specialized

(a)

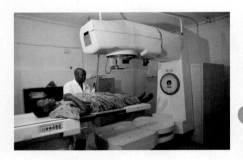

(b)

▲ **Figure 26.17 Radioactivity and the human body**
(a) Because radiation can kill or damage living cells, people who work with radioactive materials or other sources of ionizing radiation (such as X-ray tubes) must be careful to monitor their exposure. One way to do this is with a film badge. When the photographic film (contained inside the rectangular holder) is developed, the degree of darkening records the amount of radiation it has received.
(b) The fact that radiation can be lethal to cells also makes it a valuable therapeutic tool in the fight against cancer. This patient is receiving a highly targeted dose of radiation to treat a malignant tumor.

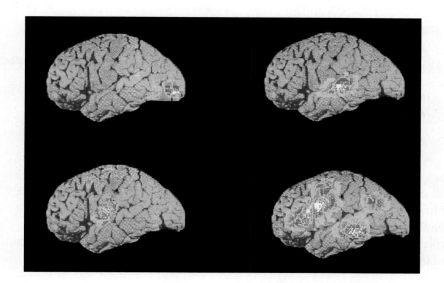

◀ **Figure 26.18 PET scans**
These PET scans show brain activity when a person sees words (upper left), hears words (upper right), says words (lower left), and thinks about words while speaking them (lower right). (The image of the brain is superimposed for reference and is not part of the PET scan.)

detectors observe them and determine their point of origin. The resulting computerized image shows the areas in the body where glucose metabolism is most intense.

PET scans are particularly useful in examining the brain, as illustrated in **Figure 26.18**. For example, a PET scan can show which regions of the brain are most active when a person is performing a specific mental task, like counting, speaking, or translating a foreign language. The scans can also show abnormalities in brain function, as when one side of a brain becomes more active than the other during an epileptic seizure. In addition, PET scans can be used to locate tumors in the brain, or other parts of the body, and to monitor the progress of their treatment.

26.3 LessonCheck (MP)

Checking Concepts

31. ⬛ **Explain** Does a runaway chain reaction occur when a nuclear reactor is critical?

32. ⬛ **Apply** How many half-lives are required to reduce the number of radioactive nuclei in a sample to 1/8 of their original number?

33. ⬛ **Explain** A sample from a fossil has an activity of 0.190 Bq per gram of carbon. Without doing a calculation, explain how you know the fossil is less than 5730 years old.

34. Big **Idea** What is the physical mechanism that allows the small nuclei of atoms to release such large amounts of energy?

35. Apply A sample contains N radioactive nuclei. How many nuclei decay in the time interval between one half-life and two half-lives?

Solving Problems

36. Calculate One of the decay series that occurs when uranium-235 captures a neutron is the following:

$$^{1}_{0}n + ^{235}_{92}U \longrightarrow ^{236}_{92}U^* \longrightarrow ^{137}_{55}Cs + ? + 4^{1}_{0}n$$

(a) Identify the mystery nucleus.
(b) How much energy is released in this reaction if the mass of the system decreases by 0.205 u?

37. Calculate A sample of radon contains N atoms. How long does it take for the number of radon atoms to decrease to $0.25N$? (The half-life of radon is 3.8 d.)

38. | **Think & Calculate** | An ancient ax handle is found to have an activity of 0.105 Bq per gram of carbon.
(a) Is the age of the wood in the handle greater than, less than, or equal to 5730 years? Explain.
(b) Find the age of the wood.

26.4 Fundamental Forces and Elementary Particles

Vocabulary

- lepton
- hadron
- quark

Scientists have long sought to identify the fundamental building blocks of all matter. These basic units are referred to as *elementary particles*. Similarly, physicists would like to identify and combine the *fundamental forces* of nature with a single unifying theory. In this lesson you'll learn about the roles fundamental forces and elementary particles play in our understanding of the universe.

Fundamental Forces

Although nature presents us with a myriad of different physical phenomena—from tornadoes and volcanoes to sunspots and comets to galaxies and black holes—all are the result of just a few fundamental forces. This is one example of the simplicity that physicists see in nature.

There are four fundamental forces in nature

The fundamental forces are the strong nuclear force, the electromagnetic force, the weak nuclear force, and the force of gravity. **Table 26.3** gives the relative strengths and ranges of these forces. All objects with mass experience the gravitational force. This is one reason why gravity is such an important force in the universe, even though it is spectacularly weak. Similarly, all objects with electric charge experience electromagnetic forces. As for the weak and strong nuclear forces, some particles experience only the weak force, but others experience both the weak and the strong forces.

The four fundamental forces were once unified

The universe today has four fundamental forces. This has not always been the case, however. Just after the Big Bang the four forces were merged into a single force known as the *unified force*. This situation lasted for an incredibly brief interval of time. As the early universe expanded and cooled, it underwent a type of "phase transition" in which the gravitational force took on a separate identity. This transition occurred at approximately 10^{-43} s after the Big Bang, when the temperature of the universe was about 10^{32} K.

This phase transition was the first of three such transitions to occur in the early universe, as we see in **Figure 26.19**. At 10^{-35} s, when the temperature was 10^{28} K, the strong nuclear force became a separate force. Similarly, the

Table 26.3 The Fundamental Forces

Force	Relative Strength	Range
Strong nuclear	1	$\sim 10^{-15}$ m
Electromagnetic	10^{-2}	Infinite (inverse square, or $1/r^2$)
Weak nuclear	10^{-6}	$\sim 10^{-3}$ fm
Gravitational	10^{-43}	Infinite (inverse square, or $1/r^2$)

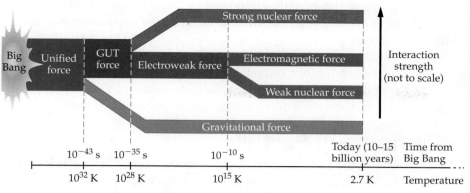

◀ **Figure 26.19 The evolution of the four fundamental forces**
The four forces we observe in today's universe began as a single unified force just after the Big Bang. As the universe cooled, the unified force went through a series of "phase transitions" in which the various forces took on different characteristics.

weak nuclear force became a separate force at 10^{-10} s, when the temperature was 10^{15} K. From 10^{-10} s until the present, the situation has remained the same, even as the temperature of the universe has dropped to a chilly 2.7 K.

Unified field theories account for the four fundamental forces

Let's take a closer look at the fundamental forces and the transitions between them. As a first example, the electromagnetic force combines the forces associated with both electricity and magnetism. Although electricity and magnetism were originally thought to be separate forces, the work of James Clerk Maxwell (1831–1879) and others showed that these forces are simply different aspects of the same underlying force. For example, changing electric fields generate magnetic fields, and changing magnetic fields generate electric fields. In fact, the theory of electromagnetism can be thought of as the first *unified field theory*, in which seemingly different forces are combined through one all-encompassing theory.

Electroweak Theory At times earlier than 10^{-10} s after the Big Bang, the weak nuclear force was indistinguishable from the electromagnetic force. Thus, even though these forces seem very different today, we can recognize them as different aspects of the same underlying force—much like the two faces of a coin look different but are part of the same object. The theory that joins the weak nuclear force and the electromagnetic force is called the *electroweak theory*. It was developed by Sheldon Glashow, Abdus Salam, and Steven Weinberg.

Grand Unified Theory Going farther back in time, the strong nuclear force was indistinguishable from the electroweak force before the universe was 10^{-35} s old. Although no one has yet succeeded in producing a theory combining the electroweak force and the strong nuclear force, most physicists feel confident that such a theory exists. This hypothetical theory is referred to as the *grand unified theory*, or GUT.

Theory of Everything Finally, a theory that encompasses gravity and the other three forces of nature is one of the ultimate goals of physics. Many physicists, including Einstein in his later years, have worked long and hard toward such an end, but so far with little success.

As things stand today, the electroweak theory is probably the best tested of all the descriptions of the forces of nature. Still, it includes only electromagnetism and the weak nuclear force. The *Standard Model*, an attempt to include the strong nuclear force in the electroweak theory, is considered by many to be the best overall theory of the electromagnetic and nuclear forces (hence its name). Other grand unified theories that combine the electromagnetic and nuclear forces are under study, but none has won popular support, and none includes gravity. *String theory* is an approach that does incorporate gravity, but many of its predictions (like extra dimensions) are still controversial. The best we can say at this point is that unifying the forces of nature is an area of active research that may require a new generation of physicists for its resolution.

Elementary Particles

We turn now to the basic building blocks of matter, elementary particles. At one point it was thought that atoms were elementary particles—one type for each element. As you've learned, however, this idea was put to rest when atoms were discovered to be made up of electrons, protons, and neutrons. Of these three particles, only the electron is currently considered to be elementary—protons and neutrons are now known to be composed of still smaller particles.

Leptons experience the weak nuclear force

Why are leptons considered elementary particles?

Particles that are acted on by the weak nuclear force but not by the strong nuclear force are referred to as **leptons**. Only six leptons are known to exist, and all of these are listed in **Table 26.4**. The most familiar leptons are the electron and its corresponding neutrino—both of which are stable. **No internal structure has ever been detected in any of the leptons. As a result, all six leptons have the status of true elementary particles.**

The weak nuclear force is responsible for most radioactive decay processes, such as beta decay. It is also a force of extremely short range. In fact, the weak force has practically no effect at all on particles that are separated by more than roughly one-thousandth of the diameter of a nucleus.

Hadrons are made up of smaller particles

What types of particles combine to make hadrons?

Hadrons are particles that are acted on by both the weak and the strong nuclear forces. They are also acted on by gravity, since all hadrons have mass. The two most familiar hadrons are the proton and the neutron. A partial list of the hundreds of hadrons known to exist is given in **Table 26.5**. Notice that the proton is the only stable hadron (though some theories suggest that even it may decay with an incredibly long half-life of 10^{35} years).

The strong nuclear force is the only force powerful enough to hold a nucleus together. It is a short-range force, extending only to distances comparable to the diameter of a nucleus. Within that range, however, it is strong enough to counteract the intense electrostatic repulsions between positively charged protons. Outside the nucleus the strong nuclear force is of negligible strength.

Table 26.4 Leptons

Particle	Particle Symbol	Antiparticle Symbol	Rest Energy (MeV)	Lifetime (s)
Electron	e^- or β^-	e^+ or β^+	0.511	Stable
Muon	μ^-	μ^+	105.7	2.2×10^{-6}
Tau	τ^-	τ^+	1784	10^{-13}
Electron neutrino	v_e	$\bar{v}_e$	~ 0	Stable
Muon neutrino	v_μ	$\bar{v}_\mu$	~ 0	Stable
Tau neutrino	v_τ	$\bar{v}_\tau$	~ 0	Stable

Table 26.5 Hadrons

	Particle	Particle Symbol	Antiparticle Symbol	Rest Energy (MeV)	Lifetime (s)
MESONS	Pion	π^+	π^-	139.6	2.6×10^{-8}
		π^0	π^0	135.0	0.8×10^{-16}
	Kaon	K^+	K^-	493.7	1.2×10^{-8}
		K^0_S	$\overline{K}^0_S$	497.7	0.9×10^{-10}
		K^0_L	$\overline{K}^0_L$	497.7	5.2×10^{-8}
	Eta	η^0	η^0	548.8	$< 10^{-18}$
BARYONS	Proton	p	$\overline{p}$	938.3	Stable
	Neutron	n	$\overline{n}$	939.6	900
	Sigma	Σ^+	$\overline{\Sigma}^-$	1189	0.8×10^{-10}
		Σ^0	$\overline{\Sigma}^0$	1192	6×10^{-20}
		Σ^-	$\overline{\Sigma}^+$	1197	1.6×10^{-10}
	Omega	Ω^-	Ω^+	1672	0.8×10^{-10}

None of the hadrons are elementary particles. 🔑 **All hadrons are composed of either two or three smaller particles called quarks.** Hadrons formed from two quarks are referred to as *mesons*; hadrons formed from three quarks are *baryons*. The properties of quarks are considered next.

Quarks have fractional charge

To account for the internal structure observed in hadrons, Murray Gell-Mann (b. 1929) and George Zweig (b. 1937) independently proposed in 1964 that all hadrons are composed of a number of truly elementary particles that Gell-Mann dubbed *quarks*.

Originally, it was proposed that there are three types of quarks, arbitrarily named *up* (u), *down* (d), and *strange* (s). Discoveries of more massive hadrons, such as the J/ψ particle in 1974, necessitated the addition of three more quarks. The equally whimsical names for these new quarks are *charmed* (c), *top* or *truth* (t), and *bottom* or *beauty* (b). **Table 26.6** lists the six quarks, some of their properties, and their antiparticles.

Quarks are unique among the elementary particles in a number of ways. For one, they all have charges that are *fractions* of the charge of the electron—their charge is quantized, it's just quantized to a fractional value. As can be seen in Table 26.6, some quarks have a charge of $+\frac{2}{3}e$ or $-\frac{2}{3}e$, and others have a charge of $+\frac{1}{3}e$ or $-\frac{1}{3}e$. No other particles are known to have charges that differ from integer multiples of the electron's charge.

Table 26.6 Quarks and Antiquarks

Name	Rest Energy (MeV)	Quarks Symbol	Quarks Charge	Antiquarks Symbol	Antiquarks Charge
Up	360	u	$+\frac{2}{3}e$	$\overline{u}$	$-\frac{2}{3}e$
Down	360	d	$-\frac{1}{3}e$	$\overline{d}$	$+\frac{1}{3}e$
Strange	540	s	$-\frac{1}{3}e$	$\overline{s}$	$+\frac{1}{3}e$
Charmed	1500	c	$+\frac{2}{3}e$	$\overline{c}$	$-\frac{2}{3}e$
Top	173,000	t	$+\frac{2}{3}e$	$\overline{t}$	$-\frac{2}{3}e$
Bottom	5000	b	$-\frac{1}{3}e$	$\overline{b}$	$+\frac{1}{3}e$

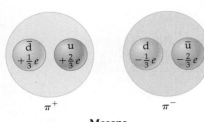

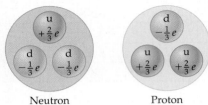

Mesons

Baryons

Neutron   Proton

🔑 *How do the partial charges of quarks determine the charges of larger particles?*

Table 26.7 Quark Composition of Some Hadrons

Particle	Quark Composition
MESONS	
π^+	$u\bar{d}$
π^-	$\bar{u}d$
K^+	$u\bar{s}$
K^-	$\bar{u}s$
K^0	$d\bar{s}$
BARYONS	
p	uud
n	udd
Σ^+	uus
Σ^0	uds
Σ^-	dds
Ξ^0	uss
Ξ^-	dss
Ω^-	sss

◀ **Figure 26.20 The quark composition of mesons and baryons**
Mesons and baryons are composed of various quark combinations. Mesons always have a quark and an antiquark; baryons always have three quarks. Notice that even though quarks have fractional charges (in units of the electron's charge, *e*), the resulting mesons and baryons always have integer charges.

Quarks are confined within hadrons

It might seem that the fractional charge of a quark would make it easy to identify experimentally. In fact, a number of experiments have searched for quarks in just that way, by looking for particles with fractional charge. No such particle has ever been observed, however. It is now believed that a free, independent quark cannot exist; quarks must always be bound with other quarks inside a hadron. This concept is referred to as *quark confinement*. The physical reason behind confinement is that the force between two quarks increases with separation—like two particles connected by a spring. Thus, an infinite amount of energy is required to increase the separation between two quarks to infinity.

The smallest system of bound quarks that can be observed as an independent particle is a pair of quarks. In fact, mesons consist of bound pairs of quarks and antiquarks, as illustrated schematically in **Figure 26.20**. For example, the π^+ meson is composed of a $u\bar{d}$ pair of quarks. This combination of quarks gives the π^+ meson a net charge of $+e$. The π^- meson, the antiparticle to the π^+ meson, consists of a $\bar{u}d$ pair with a charge of $-e$. 🔑 **Quarks are always bound in combinations that result in a net charge that is an integer multiple of *e*.**

Baryons are bound systems consisting of three quarks, as is also shown in Figure 26.20. The proton, for example, has the composition uud, with a net charge of $+e$. The neutron, on the other hand, is formed from the combination udd, with a net charge of 0. The quark compositions of a variety of hadrons are given in **Table 26.7**.

Quarks come in different varieties

Not long after the quark model of elementary particles was introduced, it was found that some quark compositions implied a violation of basic principles of quantum mechanics. To resolve these discrepancies, it was suggested that quarks must come in three different varieties, which were given the completely arbitrary but colorful names *red*, *green*, and *blue*. Though these quark "colors" have nothing to do with visible colors in the electromagnetic spectrum, they explain other experimental observations that were difficult to understand before the introduction of this new property. The theory of how quarks of different "colors" interact with one another is called *quantum chromodynamics*, or QCD. This theory is in close analogy with the theory describing interactions between charged particles, which is known as *quantum electrodynamics*, or QED.

26.4 LessonCheck 🄼🄿

Checking Concepts

39. 🔑 **Identify** Which type of particle is considered elementary, leptons or hadrons?

40. 🔑 **Identify** What particles are made up of combinations of quarks?

41. Identify Give an example of a specific particle that is formed from three quarks. Give an example of one that is formed from two quarks.

42. Identify Which fundamental force is strongest, the force of gravity or the weak nuclear force?

Archeologist

Archeology is a branch of anthropology that focuses on the study of human society—people and their culture. The job often involves fieldwork at the site of an archeological dig. Archeologists typically survey a site and develop a systematic plan before excavating begins. The collected artifacts and remains are then studied to gain insight into how the people of the ancient civilization lived.

The analysis portion of the job often makes use of radioactive dating methods. Determining the age of artifacts helps archeologists put things in a chronological context. Scientists have used this technique on archeological sites around the world, from the fascinating structures at Stonehenge in England to the ancient temple at Angkor Wat in Cambodia, shown here.

Archeologists often work for universities, museums, and state and federal governments. For many of these professionals, the job involves teaching, conducting field investigations, analyzing collected artifacts, and ultimately publishing their findings. To work as a field archeologist requires a bachelor's degree with a major in anthropology or archeology and some field experience. Higher-level positions involving supervisory roles generally require a master's degree or doctorate.

The temple at Angkor Wat

Fieldwork at an archeological dig.

Take It Further

1. Research *An archeologist might tell you that it is important to study the past in order to prepare for the future. Research the ancient inhabitants of Rapa Nui, and explain some of the ideas archeologists have about the various factors that negatively impacted the well-being of their society. How might their experiences apply to modern society?*

Physics Lab Modeling Radioactive Decay

In this lab you will model radioactive decay using pennies and dice and determine the half-lives of these two simulated elements.

Materials
- 100 pennies
- 100 dice
- shoe box with lid
- graph paper

Procedure

Part I: Radioactive Pennies

1. Place the pennies in the box with their tail side up. Each tail-side-up penny represents an active nucleus.

2. Put the lid on the box and shake well.

3. Remove the lid and take out all the pennies with their head side up. These represent nuclei that have decayed.

4. Count the number of pennies left in the box. This represents the number of active nuclei remaining. Record this number in Data Table 1 for shake 1.

5. Repeat Steps 2–4 until four or fewer pennies remain in the box. Remove the pennies from the box at that point.

Data Table 1: Radioactive Pennies

Number of Shakes	Number of Active Nuclei
0	100
1	
2	

Part II: Radioactive Dice

6. Put the dice in the box, making sure that none of them have the six side up. Each of the dice represents an active nucleus.

7. Put the lid on the box and shake well.

8. Remove the lid and take out all the dice with their six side up. These represent nuclei that have decayed.

9. Count the number of dice left in the box. This represents the number of active nuclei remaining. Record this number in Data Table 2 for shake 1.

10. Repeat Steps 7–9 until four or fewer dice remain in the box. Remove these dice from the box.

Data Table 2: Radioactive Dice

Number of Shakes	Number of Active Nuclei
0	100
1	
2	

Analysis

1. Plot a graph of number of active nuclei versus number of shakes for the radioactive pennies and the radioactive dice on a single sheet of graph paper. Use a different color to represent each type of "nucleus."

2. Draw a best-fit curve through each set of data points.

2. How does the probability of a penny "nucleus" or a die "nucleus" decaying change with time?

3. Which simulated element, radioactive pennies or radioactive dice, decays faster? Explain why this is so.

4. What are the half-lives of the two simulated elements in terms of the number of shakes?

5. What effect would increasing the numbers of pennies and dice used have on the outcome of the experiment?

Conclusions

1. During each shake, what was the probability that any individual penny would "decay"? During each shake, what was the probability that any individual die would "decay"?

26 Study Guide

Big Idea

The nuclei of atoms can release tremendous amounts of energy when part of their mass is converted to energy. The energy released when a nucleus decays is generally in the form of the kinetic energy of the decay products. In some cases energy is released in the form of high-energy photons, referred to as gamma rays. For all nuclear decay processess the mass of the final system is less than the mass of the initial system.

26.1 The Nucleus

🔑 The mass number is the *sum* of the atomic number and the neutron number: $A = Z + N$.

🔑 According to Einstein, mass and energy are equivalent, with energy equal to mass times the speed of light squared.

🔑 The strong nuclear force acts over a very short range ($\sim 10^{-15}$ m). It is always attractive and acts with nearly equal strength on protons and neutrons. It does not act on electrons; therefore, it has no effect on the chemical properties of an atom.

• Nuclei are composed of protons and neutrons.

• The atomic number, Z, is equal to the number of protons in a nucleus. The neutron number, N, is the number of neutrons in a nucleus. The mass number of a nucleus, A, is the total number of nucleons it contains.

Key Equations

The mass number is the *sum* of the atomic number and the neutron number:

$$A = Z + N$$

A nucleus X with atomic number Z and mass number A is represented as follows:

$$^{A}_{Z}X$$

The atomic mass unit (u) is defined as follows:

$$1\ u = 1.660539 \times 10^{-27}\ kg$$

The energy equivalent of 1 u is 931.5 MeV.

26.2 Radioactivity

🔑 Nuclear binding energy is released when small nuclei are fused together to form a larger nucleus, in a process called *fusion*, and also when large nuclei decay into smaller nuclei, in a process called *fission*. The energy released in fusion or fission corresponds to a decrease in mass, according to Einstein's relation $E = |\Delta m|c^2$.

• Alpha particles are the nuclei of helium atoms. Beta decay refers to the emission of an electron. Gamma rays are high-energy photons.

Key Equations

A nucleus that emits an α particle decreases its mass number by 4 and its atomic number by 2:

$$^{A}_{Z}X \longrightarrow ^{A-4}_{Z-2}Y + ^{4}_{2}He$$

During beta decay a neutron is converted into a proton, an electron, and an antineutrino:

$$^{1}_{0}n \longrightarrow ^{1}_{1}p + e^{-} + \bar{\nu}_e$$

26.3 Applications of Nuclear Physics

🔑 When, on average, one neutron given off by a fission reaction produces an additional reaction, the reactor is said to be *critical*.

🔑 The time required for half of the nuclei in a sample of radioactive material to decay is known as the *half-life*, $T_{1/2}$.

🔑 Measuring the current carbon-14 activity in a sample from a once-living organism gives a direct indication of the time that has elapsed since the death of the organism.

Key Equation

$$t = \left(\frac{T_{1/2}}{0.693}\right)\ln\left(\frac{R_{\text{initial}}}{R_{\text{current}}}\right)$$

26.4 Fundamental Forces and Elementary Particles

🔑 No internal structure has ever been detected in any of the leptons. As a result, all six leptons have the status of true elementary particles.

🔑 All hadrons are composed of either two or three smaller particles called *quarks*.

🔑 Quarks are always bound in combinations that result in a net charge that is an integer multiple of e.

• There are four fundamental forces in nature. In order of *decreasing* strength, they are the strong nuclear force, the electromagnetic force, the weak nuclear force, and the gravitational force.

26 Assessment

ANSWERS TO SELECTED ODD-NUMBERED PROBLEMS APPEAR IN APPENDIX A.

Lesson by Lesson

26.1 The Nucleus
Conceptual Questions

43. How many neutrons are in the nucleus of carbon-13?

44. What is the symbol for a nucleus with 37 protons and 59 neutrons?

45. **Rank** Consider the following four isotopes: $^{202}_{80}\text{Hg}$, $^{93}_{41}\text{Nb}$, $^{220}_{86}\text{Rn}$, and $^{98}_{41}\text{Nb}$. Rank the isotopes in order of increasing atomic number, Z. Indicate ties where appropriate.

46. **Rank** Referring to Problem 45, rank the isotopes in order of increasing neutron number, N. Indicate ties where appropriate.

47. **Rank** Referring to Problem 45, rank the isotopes in order of increasing mass number, A. Indicate ties where appropriate.

Problem Solving

48. How many neutrons are released in the following reaction?
$$^{1}_{0}\text{n} + ^{235}_{92}\text{U} \longrightarrow ^{132}_{50}\text{Sn} + ^{101}_{42}\text{Mo} + \underline{} \text{ neutrons}$$

49. Determine the mystery nucleus in the following reaction:
$$^{2}_{1}\text{H} + ^{3}_{1}\text{H} \longrightarrow ? + ^{1}_{0}\text{n}$$

50. A nucleus with mass number $A = 224$ and atomic number $Z = 89$ emits two neutrons and two protons. What are the resulting **(a)** mass number, **(b)** atomic number, and **(c)** neutron number?

51. **The Evaporating Sun** The Sun radiates energy at the prodigious rate of 3.90×10^{26} W. **(a)** At what rate, in kilograms per second, does the Sun convert mass into energy? **(b)** Assuming that the Sun has radiated at this same rate for its entire lifetime of 4.50×10^{9} y and that its current mass is 2.00×10^{30} kg, what percentage of its original mass has been converted to energy?

26.2 Radioactivity
Conceptual Questions

52. One of the three decay processes (alpha, beta, or gamma) does not result in a new element. Identify the process.

53. Complete the following nuclear reaction:
$$^{7}_{3}\text{Li} + ^{1}_{1}\text{H} \longrightarrow ^{4}_{2}\text{He} + ?$$

54. Complete the following nuclear reaction:
$$^{234}_{90}\text{Th} \longrightarrow ^{230}_{88}\text{Ra} + ?$$

55. Complete the following nuclear reaction:
$$? \longrightarrow ^{14}_{7}\text{N} + \text{e}^{-} + \bar{\nu}$$

56. One radioactive decay series for the isotope $^{238}_{92}\text{U}$ includes the following four decays:
$$^{238}_{92}\text{U} \longrightarrow ^{234}_{90}\text{Th} \longrightarrow ^{234}_{91}\text{Pa} \longrightarrow ^{234}_{92}\text{U} \longrightarrow ^{230}_{90}\text{Th}$$
Identify, in the order shown, the type of each decay.

Problem Solving

57. The following nuclei are observed to decay by emitting an α particle: **(a)** $^{212}_{84}\text{Po}$; **(b)** $^{239}_{94}\text{Pu}$. Write the decay reaction for each of these nuclei.

58. The following nuclei are observed to decay by emitting a β^{-} particle: **(a)** $^{35}_{16}\text{S}$; **(b)** $^{212}_{82}\text{Pb}$. Write the decay reaction for each of these nuclei.

59. The following nuclei are observed to decay by emitting a β^{+} particle: **(a)** $^{18}_{9}\text{F}$; **(b)** $^{22}_{11}\text{Na}$. Write the decay reaction for each of these nuclei.

60. The alpha decay of $^{218}_{84}\text{Po}$ releases 6.115 MeV. **(a)** Identify the daughter nucleus that results from this decay. **(b)** Determine the change in mass that occurs during this process.

61. The beta decay of $^{206}_{80}\text{Hg}$ releases 1.308 MeV. **(a)** Identify the daughter nucleus that results from this decay. **(b)** Determine the change in mass that occurs during this process.

62. It is observed that $^{66}_{28}\text{Ni}$, with an atomic mass of 65.9291 u, decays by β^{-} emission. **(a)** Identify the daughter nucleus that results from this decay. **(b)** If the mass of the system decreases by 0.0002705 u as a result of the decay, find the energy released (in MeV).

26.3 Applications of Nuclear Physics
Conceptual Questions

63. Complete the following fission reaction:
$$^{1}_{0}\text{n} + ^{235}_{92}\text{U} \longrightarrow ^{133}_{51}\text{Sb} + ? + 5^{1}_{0}\text{n}$$

64. Complete the following fission reaction:
$$^{1}_{0}\text{n} + ^{235}_{92}\text{U} \longrightarrow ^{88}_{38}\text{Sr} + ^{136}_{54}\text{Xe} + \underline{} ^{1}_{0}\text{n}$$

65. The half-life of carbon-14 is 5730 y. **(a)** Is it possible for a particular nucleus in a sample of carbon-14 to decay after only 1 s has passed? Explain. **(b)** Is it possible for a particular nucleus to remain stable for 10,000 y before decaying? Explain.

66. Suppose we were to discover that the ratio of carbon-14 to carbon-12 in the atmosphere was significantly smaller 10,000 years ago than it is today. How would this affect the ages we have assigned to objects on the basis of carbon-14 dating? In particular, would the true age of an object be greater than or less than the age we had previously assigned to it? Explain.

Problem Solving

67. A radioactive sample is placed in a closed container. Two days later only one-quarter of the sample is still radioactive. What is the half-life of this material?

68. The half-life of $^{15}_{8}O$ is 122 s. How long does it take for the number of $^{15}_{8}O$ nuclei in a given sample to decrease to $\frac{1}{128}$ of its original value?

69. Charcoal from an ancient fire pit is found to have a carbon-14 activity of 0.121 Bq per gram of carbon. What is the age of the fire pit?

70. An archeologist on a dig finds a fragment of an ancient basket woven from grass. Later, it is determined that the carbon-14 activity of the grass in the basket is 9.25% of the activity of a sample of present-day grass with an equal carbon content. What is the age of the basket?

71. The bones of a saber-toothed tiger are found to have an activity per gram of carbon that is 15.0% of that found in a similar live animal. How old are these bones?

72. **Radioactivity in the Bones** Because of its chemical similarity to calcium, $^{90}_{38}Sr$ can collect in a person's bones and present a health risk. What approximate percentage of $^{90}_{38}Sr$ present initially still exists after a period of **(a)** 30 y, **(b)** 60 y, and **(c)** 90 y? The half-life of $^{90}_{38}Sr$ is approximately 30 y.

26.4 Fundamental Forces and Elementary Particles

Conceptual Questions

73. What is the difference between a hadron and a lepton?

74. For each of the following particles, state whether it is a hadron or a lepton: electron, proton, muon, neutron, positron.

75. Which of these particles—electron, proton, muon, neutron, or positron—is considered to be truly elementary?

76. Which of the fundamental forces are long-range forces? Which are short-range forces?

77. What is unusual about the electric charges of quarks?

Mixed Review

78. Identify Z, N, and A for the following isotopes: **(a)** $^{238}_{92}U$, **(b)** $^{239}_{94}Pu$, **(c)** $^{144}_{60}Nd$.

79. Complete the following nuclear reaction:
$$^{3}_{1}H \longrightarrow\ ^{3}_{2}He\ +\ ?\ +\ ?$$

80. An α particle (charge $+2e$) and a β particle (charge $-e$) deflect in opposite directions when they pass through a magnetic field. Which particle deflects by a greater amount, given that both particles have the same speed? Explain.

81. [Triple Choice] To produce a given amount of electrical energy, is the amount of coal burned in a coal-burning power plant greater than, less than, or the same as the amount of $^{235}_{92}U$ consumed in a nuclear power plant? Explain.

82. The two radioactive decay series that begin with $^{232}_{90}Th$ and end with $^{208}_{82}Pb$ are shown in **Figure 26.21**. Identify the 10 intermediary nuclei that appear in these series.

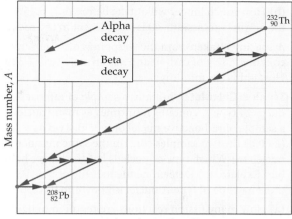

Figure 26.21

83. Energy is released when three α particles fuse to form carbon-12. Is the mass of carbon-12 greater than, less than, or the same as the mass of three α particles? Explain.

84. Determine the number of neutrons and protons in **(a)** $^{232}_{90}Th$, **(b)** $^{211}_{82}Pb$, and **(c)** $^{60}_{27}Co$.

85. Identify the daughter nucleus that results when **(a)** $^{210}_{82}Pb$ undergoes alpha decay, **(b)** $^{239}_{92}U$ undergoes β^{-} decay, and **(c)** $^{11}_{6}C$ undergoes β^{+} decay. (β^{-} indicates an electron, and β^{+} indicates a positron.)

86. Identify the nucleus whose β^{-} decay (β^{-} indicates an electron) produces the same nucleus as that produced by the alpha decay of $^{214}_{84}Po$.

87. **Mantles in Gas Lanterns** Gas lanterns used on camping trips have mantles (small mesh bags that give off light) made from rayon that is impregnated with thorium and other materials. An example is shown in **Figure 26.22**. Thorium is used, even though it is radioactive, because it forms an oxide that can withstand being incandescent for long periods of time. Almost all natural thorium is $^{232}_{90}Th$, which has a half-life of 1.405×10^{10} y. A typical mantle contains 328 mg of thorium. How much time is required for this amount of thorium to decrease to 82 mg?

Figure 26.22 The glowing mantle of a gas lantern.

88. Moon Rocks In one of the rocks brought back from the Moon, 75% of the initial potassium-40 was found to have decayed to argon-40. **(a)** If the half-life for potassium-40 is 1.20×10^9 years, how old is the rock? **(b)** How much longer will it take before only 6.25% of the original potassium-40 is still present in the rock?

89. A specimen taken from the wrappings of a mummy contains 7.82 g of carbon and has an activity of 1.38 Bq. How old is the mummy?

90. Think & Calculate Initially, a sample of radioactive nuclei of type A contains four times as many nuclei as a sample of radioactive nuclei of type B. Two days later (2.00 d) the two samples contain the same number of nuclei. **(a)** Which type of nucleus has the longer half-life? Explain. **(b)** Determine the half-life of type B nuclei if the half-life of type A nuclei is known to be 0.500 d.

91. (a) How many fission reactions are required to light a 120-W lightbulb for 2.5 d? Assume an energy release of 212 MeV per fission reaction and a 32% conversion efficiency. **(b)** What mass of $^{235}_{92}U$ corresponds to the number of fission reactions found in part (a)?

Writing about Science

92. Write a report on ionization smoke detectors, which use a radioactive material to detect smoke. What is the radioactive material, and how is it obtained? How common are these smoke detectors, and how do the detect smoke? What are the advantages or disadvantages of these detectors compared with those that use photocells?

93. Connect to the Big Idea How does Einstein's famous equation $E = mc^2$ relate to the energy given off in nuclear reactions? How much mass is lost by a typical nuclear reactor in a day's operation?

Read, Reason, and Respond

Treating a Hyperactive Thyroid Of the many endocrine glands in the body, the thyroid is one of the most important. Weighing about an ounce and situated just below the Adam's apple, the thyroid produces hormones that regulate the metabolic rate of every cell in the body. To produce these hormones the thyroid uses iodine from the food we eat—in fact, the thyroid specializes in absorbing iodine.

The central role played by the thyroid is indicated by the symptoms produced when it ceases to function properly. For example, a person experiencing *hyperthyroidism* (an overactive thyroid) can experience weight loss, ravenous appetite, anxiety, fatigue, hyperactivity, apathy, palpitations, arrhythmias, and nausea, just to mention a few of the symptoms.

The most common treatment for hyperthyroidism is to destroy the overactive thyroid tissues with radioactive iodine-131. This treatment takes advantage of the fact that only thyroid cells absorb and concentrate iodine. To begin the treatment a patient swallows a single, small capsule containing iodine-131. The radioactive isotope quickly enters the bloodstream and is taken up by the overactive thyroid cells, which are destroyed as the iodine-131 decays with a half-life of 8.03 d. Other cells in the body experience very little radiation damage, which minimizes side effects. In one or two months the thyroid activity is reduced to an acceptable level. Sometimes too much—or even all—of the thyroid is killed, which can result in *hypothyroidism* (an underactive thyroid). This is easily treated, however, with dietary supplements to replace the missing thyroid hormones.

94. How is the isotope iodine-131 represented?

A. $^{131}_{52}I$ C. $^{131}_{78}I$

B. $^{131}_{53}I$ D. $^{78}_{53}I$

95. Iodine-131 undergoes beta decay. What is the daughter nucleus that results?

A. $^{130}_{53}Xe$ C. $^{131}_{54}Xe$

B. $^{131}_{52}Xe$ D. $^{77}_{54}Xe$

96. When iodine-131 decays, the energy released is 0.971 MeV. What is the corresponding decrease in mass?

A. 0.00104 u   C. 0.904 u

B. 0.0959 u   D. 931.5 u

97. A sample of iodine-131 contains 4.00×10^{16} nuclei. How much time is required for the number of nuclei to decrease to 2.50×10^{15}?

A. 8.03 d   C. 24.1 d

B. 16.1 d   D. 32.1 d

Standardized Test Prep

Multiple Choice

1. Which of the following lists the fundamental forces of nature in order of increasing strength?
   - (A) strong nuclear force, electromagnetic force, weak nuclear force, gravitational force
   - (B) electromagnetic force, weak nuclear force, gravitational force, strong nuclear force
   - (C) gravitational force, weak nuclear force, electromagnetic force, strong nuclear force
   - (D) gravitational force, strong nuclear force, electromagnetic force, weak nuclear force

2. Which statement about the atomic nucleus is incorrect?
   - (A) All atomic nuclei have about the same density, regardless of mass number.
   - (B) A nucleus is held together by the strong nuclear force.
   - (C) Protons and neutrons in the nucleus are collectively called *nucleons*.
   - (D) All stable nuclei contain equal numbers of protons and neutrons.

3. Which type of particle is formed from a quark-antiquark pair?
   - (A) hadron
   - (B) fermion
   - (C) meson
   - (D) boson

4. Which statement is true for an excited nucleus that drops to a lower-energy state and emits a photon?
   - (A) Neither the mass number nor the atomic number is changed.
   - (B) The mass number decreases by 1.
   - (C) The atomic number decreases by 1.
   - (D) The atomic number increases by 1.

5. What is 931 MeV/c^2 a measurement of?
   - (A) momentum
   - (B) mass
   - (C) energy
   - (D) velocity

6. Identify the particle that completes the following nuclear reaction:

$$^{3}_{1}\text{H} \longrightarrow \, ^{3}_{2}\text{He} + \, ?$$

   - (A) proton ($^{1}_{1}\text{p}$)
   - (B) β^{-} particle (e^{-})
   - (C) α particle ($^{4}_{2}\text{He}$)
   - (D) positron (e^{+})

Use the graph below to answer Questions 7 and 8. The graph shows the decay curve for a 100-g sample of the radioactive isotope polonium-210.

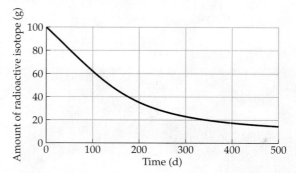

7. The half-life of Po-210 is approximately
   - (A) 500 d.
   - (B) 250 d.
   - (C) 180 d.
   - (D) 140 d.

Extended Response

8. When an atom of Po-210 ($^{210}_{84}\text{Po}$) decays, it emits an alpha particle ($^{4}_{2}\text{He}$) and releases some energy as it is converted to a daughter nucleus. **(a)** Identify the daughter nucleus, and write the complete nuclear reaction. **(b)** Is the sum of the masses of the decay products in part (a) greater than, less than, or equal to the mass of a Po-210 atom? Explain.

> **Test-Taking Hint**
>
> When balancing nuclear equations, use conservation of charge to balance the bottom numbers and conservation of mass to balance the top numbers in each equation.

If You Had Difficulty With . . .

Question	1	2	3	4	5	6	7	8					
See Lesson	26.4	26.1	26.4	26.2	26.1	26.2	26.3	26.3					

27 Relativity

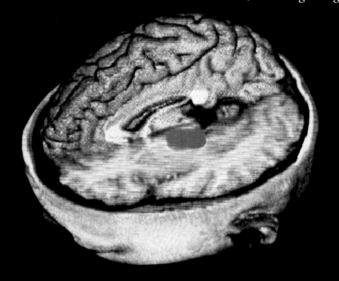
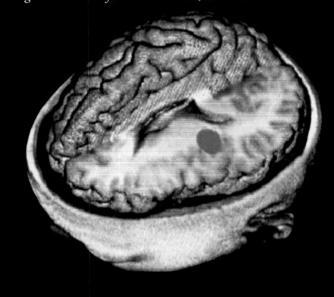

These images are positron-emission tomography (PET) scans of a brain—they show the brain performing different mental tasks. The images are produced when matter and antimatter annihilate, releasing energy according to Einstein's famous relation, $E = mc^2$.

Big Idea

Nature behaves differently near the speed of light.

Modern physics, which started around the beginning of the twentieth century, introduced two fundamentally new ways of looking at nature. Quantum physics deals with atomic systems, and Albert Einstein's theory of relativity deals with the surprising behavior that occurs when an object's speed approaches the speed of light. In this chapter you'll discover that clocks run slow and metersticks become shorter as their speed increases. You'll also learn about the key role that relativity plays in modern society, in everything from GPS systems to medical research.

Inquiry Lab   Is velocity always relative?

Read the activity, obtain the required materials, and create a data table. This lab explores an important concept, the observation of velocity from the point of view of different observers— that is, from different frames of reference. Keep the intuitive results of this lab in mind as you learn about objects with velocities near the speed of light.

Explore

1. Observe the velocity of a constant-velocity car as it moves across a table.
2. Place the car on a sheet of cardboard, and observe its velocity while pulling the cardboard in the direction of the car's motion.
3. Repeat Step 2 while increasing the speed of the cardboard.
4. Repeat Step 2 while pulling the cardboard in the direction opposite the car.
5. Repeat Step 4 while attempting to make the car appear to stand still.

Think

1. **Explain** From your point of observation, did the velocity of the car change as you changed the velocity of the cardboard? Explain your observations in Steps 2–5.
2. **Infer** What would an ant sitting on the cardboard have observed in Steps 2–5?
3. **Apply** Are you at rest when sitting on a moving train? Explain your answer.

27.1 The Postulates of Relativity

You've probably heard the expression "It's all relative." In physics this refers to the fact that different observers can see the same object or event in different ways. For example, you might observe a passenger on a train zipping by as you stand motionless next to the tracks. The passenger, however, sees herself as motionless and you zipping by in the opposite direction. Different observers can even disagree about the rate at which a clock ticks or the length of a meterstick. The physics theory that explains how different observers moving with speeds near the speed of light can see things differently is called the **theory of relativity**.

Relativity theory is based on two postulates

By the early twentieth century many scientists thought physics had completely described nature, with only minor details to be straightened out. This view was changed forever with the introduction of the theory of relativity in 1905. Published by Albert Einstein (1879–1955), a 26-year-old patent clerk in Berne, Switzerland, this theory fundamentally altered our understanding of such basic physical concepts as time, length, mass, and energy. **Figure 27.1** shows Albert Einstein at the time he developed the theory of relativity.

Vocabulary

● theory of relativity

▲ Figure 27.1 **Albert Einstein in his twenties**

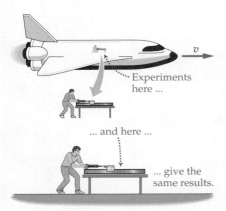

Figure 27.2 Inertial frames of reference

The two observers shown are in different inertial frames of reference. According to the first postulate of relativity, physics experiments give identical results, and the laws of nature are the same, in the two frames.

What basic principles is relativity theory based on?

Einstein's theory of relativity is based on just two very simple postulates.

- **First Postulate: Equivalence of Physical Laws**
  The laws of physics are the same in all *inertial frames of reference.* (Your classroom is a good approximation to an inertial frame, and any observer moving with constant velocity relative to you is also in an inertial frame.)
- **Second Postulate: Constancy of the Speed of Light**
  The speed of light in a vacuum, $c = 3.00 \times 10^8$ m/s, is the same in all inertial frames of reference, *regardless* of the motion of the source or the observer of the light.

All of the consequences of relativity follow directly from these two postulates.

The first postulate relates to the laws of nature

An inertial frame of reference is one in which Newton's laws of motion are obeyed. For example, an object with no force acting on it has zero acceleration in *all* inertial frames. Einstein's first postulate simply says that the concept of an inertial frame of reference applies not only to Newton's laws, but to *all* the known laws of physics, including those dealing with thermodynamics, electricity, magnetism, and electromagnetic waves.

For example, a physics experiment performed on the surface of Earth (which approximates an inertial frame of reference) gives the same results as when it is carried out in an airplane moving with constant velocity. In addition, the behavior of heat, magnets, and electrical circuits is the same in the airplane as on the ground, as indicated in **Figure 27.2**.

Special Theory versus General Theory Einstein's original theory of relativity is often referred to as the *special* theory of relativity. The word "special" means that it applies to the special case of frames of reference with no acceleration—like all inertial frames of reference. The more general case, in which accelerated motion is considered, is the subject of the *general* theory of relativity, which we will discuss later in this chapter. Einstein published the theory of general relativity in 1916.

The second postulate relates to the speed of light

Einstein's second postulate states that light travels with the constant speed c regardless of whether the source or the observer is in motion. This fact has particularly interesting implications. To see why, let's first consider waves on water. **Figure 27.3 (a)** shows two moving sources (boats) generating waves and an observer in another boat at rest on the water. The waves produced by each boat travel at the characteristic speed of water waves, v_w, once they are generated. Thus, the observer sees a wave speed that is independent of the speed of the source. The same is true for light, as indicated in **Figure 27.3 (b)**.

On the other hand, suppose the observer is in motion with a speed v relative to the water. If the observer is moving to the right and the water waves are moving to the left with a speed v_w, as in **Figure 27.4 (a)**, the waves move past the observer with a high relative speed, $v + v_w$. Similarly, if the water waves are moving to the right, as in **Figure 27.4 (b)**, the observer sees them as having a low relative speed, $v - v_w$. Clearly, the fact that the observer is in motion relative to the medium through which the waves are traveling (water in this case) means that the speed of the waves depends on the speed of the observer.

Reading Support
Vocabulary Builder
hypothetical
[hahy puh THET i kuhl]

(adjective) having the nature of a hypothesis; supposed or thought to exist; existing only in concept

The professor often discussed hypothetical, or "what-if," situations.

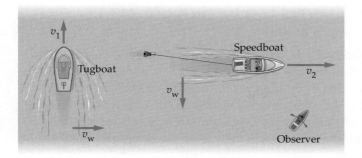

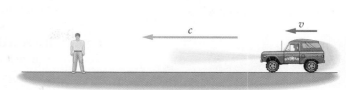

(a) Speed of water waves independent of speed of source

(b) Speed of light waves independent of speed of source

▲ **Figure 27.3 Wave speed versus source speed**
The speed of waves is independent of the speed of the source that generates the waves. **(a)** Water waves produced by a slow-moving tugboat have the same speed as those produced by a high-powered speedboat. **(b)** The speed of a beam of light is independent of the speed of its source.

Experimental results support the second postulate

Before Einstein's theory of relativity, it was generally thought that light behaves just like water waves. In addition, it was thought that light travels through a **hypothetical** medium (called *ether*) that permeates all space. Because Earth rotates about its axis and orbits the Sun, it follows that Earth must move relative to the ether. Thus, it should be possible to detect Earth's motion by measuring differences in the speed of light traveling in different directions—just as in the case of water waves produced by a boat.

From 1883 to 1887, American physicists A. A. Michelson (1852–1931) and E. W. Morley (1838–1923) conducted extremely precise experiments designed to measure the speed of light in different directions. They were unable to detect *any* difference in the speed of light. More recent and more accurate experiments have come to precisely the same conclusion. Thus, experimental results support the behavior of light as described by the second postulate, and they invalidate the idea that light travels through an ether.

The second postulate can seem counterintuitive

To see how counterintuitive the second postulate can be, consider the situation illustrated in **Figure 27.5**. In this case a ray of light is moving to the right with a speed *c* relative to observer 1. A second observer is moving to the right as well, with a speed of 0.9*c* relative to observer 1. Now, it seems natural that observer 2 would see the ray of light passing with a speed of only 0.1*c*. This is *not* the case, however. Observer 2, like all observers in an inertial frame of reference, sees the ray go by with the speed of light, *c*. It seems strange, but nature does work this way.

For the observations illustrated in Figure 27.5 to be valid—that is, for both observers to measure the same speed of light—the behavior of space and time must differ from our everyday experience when objects move at speeds that approach the speed of light. This is indeed the case. In everyday circumstances, however, the physics described by Newton's laws are perfectly adequate. In fact, Newton's laws are valid at low speeds (relative to *c*), whereas Einstein's theory of relativity gives correct results for all speeds from zero up to the speed of light.

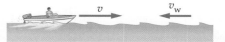

(a) High relative speed

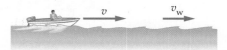

(b) Low relative speed

▲ **Figure 27.4 Wave speed versus observer speed**
The speed of water waves depends on the speed of the observer relative to the water. **(a)** Water waves move relative to the observer with a high relative speed, $v + v_w$. **(b)** Water waves move relative to the observer with a low relative speed, $v - v_w$.

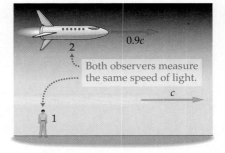

▲ **Figure 27.5 The speed of light for different observers**
A beam of light is moving to the right with a speed *c* relative to observer 1. Observer 2 is moving to the right with a speed of 0.9*c*. Even so, observer 2 *also* sees the beam of light moving to the right with a speed of *c*, in agreement with the second postulate of relativity.

All inertial observers can be thought of as being at rest

What is the highest possible speed in the universe?

Since all inertial observers measure the same speed for light, they are all equally correct in claiming that *they* are at rest. For example, observer 1 in Figure 27.5 may say that he is at rest and that observer 2 is moving to the right with a speed of 0.9*c*. Observer 2, however, is equally justified in saying that she is at rest and that observer 1 is in motion with a speed of 0.9*c* to the left. From the point of view of relativity, both observers are equally correct. There is no absolute rest or absolute motion, only motion relative to something else—"It's all relative."

Finally, notice that it doesn't make sense for observer 2 in Figure 27.5 to have a speed greater than the speed of light. If this were the case, it would not be possible for the light ray to pass her, much less to pass her with the speed *c*. This leads to an important conclusion. **The ultimate speed in the universe is the speed of light in a vacuum.** Light travels slower in substances like air, water, and glass, but its greatest possible speed is in a vacuum, and nothing can travel faster.

27.1 LessonCheck (MP)

Checking Concepts

1. Recall (**a**) What is an inertial reference frame? (**b**) What does the second postulate of relativity say about the speed of light in an inertial reference frame?

2. Explain Is it possible for an object to move with a speed greater than the speed of light? Explain why or why not.

3. Identify An experiment performed in a car moving in a straight line at 100 km/h gives the same result as when it is performed in a lab. What physics concept guarantees this?

4. Infer Does an experiment performed in an accelerating car give the same result as in a lab? Explain.

5. Triple Choice You are traveling in a spaceship in deep space, moving along a straight line with a speed of 100 m/s. As you approach a stationary observer, you turn on a flashlight and shine it at her. Does the observer see the light approaching her with a speed that is greater than, less than, or equal to the speed of light? Explain.

952 Chapter 27 • Lesson 27.1

27.2 The Relativity of Time and Length

Certain things in life seem obvious. For instance, we all "know" that time moves forward at a constant rate and that the length of an object, such as a car, is the same whether it is moving or not. Neither of these statements is true, however, as you'll discover in this lesson.

Time Dilation

The passing of time seems unwavering and steady. If you could move at nearly the speed of light, however, you would discover that time is not as constant as it seems. It turns out that nature is much more interesting than you might imagine.

A moving clock runs slow

Suppose you're standing at rest holding a clock. As you do so, a spaceship carrying an identical clock moves past you with a speed of $0.5c$. How do you think the readings on the two clocks would compare? If you could do this experiment, you would observe that the clock on the spaceship was running slow compared to your clock—even if the clocks were identical in all other respects.

A Stationary Light Clock To calculate the difference between the rate of a moving clock and that of one at rest, consider the light clock shown in **Figure 27.6**. For this clock, a cycle begins when a burst of light is emitted from the light source, S. The light travels a distance d to a mirror, where it is reflected. It then travels back a distance d to the detector, D, and triggers the next burst of light. Each round-trip of light can be thought of as one "tick" of this clock.

Let's begin by calculating the time interval between the ticks of this clock when it is at rest. Since the light covers a total distance $2d$ with a speed c, the time between ticks is

$$\Delta t_0 = \frac{2d}{c}$$

The subscript zero on Δt indicates the clock is at rest ($v = 0$) when the measurement is made. In general, we refer to the time interval of a clock that is at rest relative to an observer as the **proper time**.

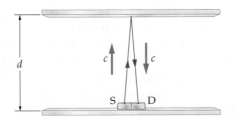

◀ **Figure 27.6 A stationary light clock**
Light emitted by the source, S, travels to a mirror a distance d away and is reflected back into the detector, D. The time between emission and detection is one cycle, or one "tick," of the clock.

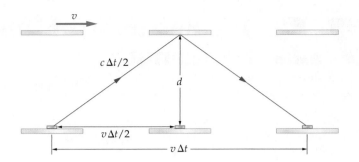

A moving light clock requires a time Δt to complete one cycle. Notice that the light follows a zigzag path that is longer than $2d$; hence, the time between ticks is greater for the moving clock than it is for the clock that is at rest.

🔑 *What happens to the time between ticks of a moving clock?*

Reading Support ✓
Vocabulary Builder
dilate [dahy LAYT]

(verb) to spread out or expand

from the Latin *dilatare*, meaning "spread out," from *di-* ("apart") + *latus* ("wide")

A Moving Light Clock

Now consider the same light clock moving with speed v, as shown in **Figure 27.7**. Notice that the light follows a zigzag path to complete a tick of the clock. Because this path is longer than it was for the stationary clock and because the speed of the light is the same (according to the second postulate of relativity), the time between ticks *must be* greater than Δt_0. 🔑 **With more time elapsing between ticks, the moving clock runs slower than the clock at rest.**

The conclusion that a moving clock runs slow applies to any type of clock, not just the light clock. If this were not the case—if different clocks ran at different rates when moving at the same speed—the first postulate of relativity would be violated.

The slowing of a moving clock is known as **time dilation**. The reason for this terminology is that time intervals for a moving clock are expanded—or *dilated*—just like the pupils of your eyes are said to be **dilated** when they expand.

Time dilation is easily calculated

Before calculating the time dilation of a moving clock, let's carefully define a few quantities. We define the speed of the moving clock to be v. The dilated time interval associated with the moving clock is Δt, and the corresponding time interval for the clock at rest is the proper time, Δt_0. The speed of light, of course, is c. With these definitions, we write the time dilation equation as follows:

Time Dilation

$$\text{dilated time interval} = \frac{\text{proper time}}{\sqrt{1 - \dfrac{(\text{speed of clock})^2}{(\text{speed of light})^2}}}$$

$$\Delta t = \frac{\Delta t_0}{\sqrt{1 - \dfrac{v^2}{c^2}}}$$

SI unit: s

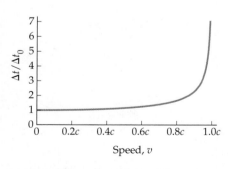

▲ **Figure 27.8 Time dilation as a function of speed**
As the speed of a clock increases, the time required for one tick increases slowly at first, and then rapidly to infinity at the speed of light.

As expected, the time intervals Δt_0 and Δt are the same for a clock at rest; that is, $\Delta t = \Delta t_0$ at $v = 0$. As the speed v increases, so does the dilated time interval, Δt. This is illustrated in **Figure 27.8**, where the ratio $\Delta t/\Delta t_0$ is shown as a function of the speed v. Notice that $\Delta t/\Delta t_0$ goes to infinity as the speed approaches the speed of light. This means that it would take an infinite amount of time for one tick of the clock—a clock would basically slow to a stop if it moved at the speed of light.

A spaceship carrying a light clock moves with a speed of $0.500c$ relative to Earth. According to an observer on Earth, how long does it take for the spaceship's clock to advance 1.00 s?

Solution

To an observer on the spaceship, the light clock is at rest, and its time interval is the proper time, $\Delta t_0 = 1.00$ s. The observer on Earth sees the clock moving with the speed $v = 0.500c$. As a result, the amount of time, Δt, that passes as the observer on Earth watches the spaceship clock advance 1.00 s is given by the time dilation equation:

$$\Delta t = \frac{\Delta t_0}{\sqrt{1 - \dfrac{v^2}{c^2}}}$$

$$= \frac{1.00 \text{ s}}{\sqrt{1 - \dfrac{(0.500c)^2}{(c)^2}}}$$

$$= \frac{1.00 \text{ s}}{\sqrt{1 - 0.25}}$$

$$= \boxed{1.15 \text{ s}}$$

Even at this high speed, the relativistic effect is fairly small—only about 15%. Still, the observer on Earth sees the clock as running slow, taking 1.15 s for every tick.

Math HELP
Delta, Δ
See Math Review, Section I

Practice Problems

6. Follow-up What is the dilated time interval if the spaceship moves at $v = 0.750c$ instead?

7. A clock makes one tick every 1.00 s when at rest relative to observer 1. What is the speed of this clock relative to observer 2, who says that it takes 1.25 s for one tick of the clock? Give your answer as a multiple of the speed of light.

8. Triple Choice An observer notices that a clock moving with a speed of $0.650c$ advances 2.00 s.
(a) Is the corresponding elapsed time on an identical clock at rest relative to the observer greater than, less than, or equal to 2.00 s?
(b) How much time elapses on the identical clock at rest?

Time dilation is insignificant in everyday situations

You might be wondering why time dilation was not observed long before Einstein put forward the theory of relativity. The reason is that everyday speeds are very small compared with the speed of light. As a result, the corresponding time dilation is insignificant in most situations and cannot be detected.

C☉L PHYSICS
Time Dilation and GPS

It isn't easy to measure time to an accuracy of 1 second per century. Modern atomic clocks can do the job, however. The physicists J. C. Hafele and R. E. Keating conducted a direct test of relativity in 1971 by placing one atomic clock on board a jet airplane and leaving an identical clock at rest in a laboratory. After many hours of flight, they found that the moving clock had run more slowly than the clock left in the lab. The discrepancy in times–though very small–agreed precisely with the predictions of the theory of relativity. Today, relativistic time dilation is taken into account for all atomic clocks that are in motion. This adjustment is particularly important for the global positioning system (GPS). The system consists of 24 satellites, each carrying its own atomic clock. Since the GPS system relies on accurate time measurements, the system would be useless if relativistic time dilation weren't taken into account.

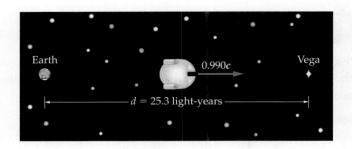

▶ Figure 27.9 **A relativistic trip to Vega**
An astronaut travels the 25.3 light-years from Earth to Vega at a speed of 0.990c.

For example, the greatest speed a human might reasonably attain today is the speed of an astronaut in orbit. This speed is typically about 7700 m/s (~17,000 mi/h). Although this is a rather large speed, it is still only 1/39,000 of the speed of light. An orbiting clock is therefore slowed by a factor of 1.00000000033. At this rate it would take almost 100 years for an orbiting clock to lose 1.0 s compared with a clock on Earth.

Time dilation also applies to aging

What types of processes does time dilation apply to?

To this point, our discussion of time dilation has been applied solely to clocks. Clocks are not the only objects that show time dilation, however. In fact, time dilation applies to people and other living things as well. Relativistic time dilation applies to *all physical processes*, including chemical reactions and biological functions like aging.

Thus, an astronaut in a moving spaceship *ages more slowly* than one who remains on Earth. Even so, the moving astronaut thinks that time in the spaceship is progressing at its normal pace. This is yet more evidence of everything being relative.

As a specific example, suppose astronaut Jenny travels to Vega, the fifth brightest star in the night sky, leaving her 35-year-old twin brother Benny behind on Earth. Jenny travels with a speed of 0.990c, and Vega is 25.3 light-years from Earth, as indicated in **Figure 27.9**. (A *light-year* is the distance light travels in 1 year.) When Jenny arrives at Vega, she is only 38.6 years old, whereas Benny, who stayed on Earth, is now 60.6 years old. Relativistic speeds make quite a difference.

Length Contraction

Just as time is altered for an observer moving with a speed close to the speed of light, so too is distance. As speeds approach *c*, the effect becomes more pronounced.

A moving object becomes shorter

What happens to the length of a moving object as its speed approaches c?

Let's begin by looking at a meterstick moving with a speed of 0.5c. Is this meterstick 1 meter in length? Far from it—the moving meterstick is only 0.866 meter in length. The shrinking in length of a moving object is referred to as **length contraction**. As the speed of an object approaches *c*, its length approaches zero.

Calculating length contraction is similar to calculating time dilation, and the variables are defined in a similar way as well. The length of an object at rest, with zero speed, is its **proper length**, L_0. The contracted length of the object at the speed *v* is *L*. The connection between the contracted and proper lengths is easily calculated using the following equation.

Length Contraction

$$\text{contracted length} = \text{proper length} \times \sqrt{1 - \frac{(\text{speed of object})^2}{(\text{speed of light})^2}}$$

$$L = L_0\sqrt{1 - \frac{v^2}{c^2}}$$

SI unit: m

As expected, the lengths L and L_0 are equal when $v = 0$. As v approaches the speed of light, however, the contracted length, L, approaches zero. **Figure 27.10** shows L as a function of speed v for a meterstick. Again, we see that a speed greater than the speed of light isn't possible, since the length of a meterstick can't be less than zero.

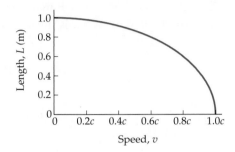

◀ **Figure 27.10 Length contraction**
The graph shows the length of a meterstick as a function of its speed. The length shrinks and approaches zero as the speed approaches the speed of light.

QUICK Example 27.2 What's the Length?

A spaceship moves with a speed of $0.750c$ relative to Earth. According to an observer on Earth, what is the length of a meterstick on the spaceship?

Solution
Substituting $L_0 = 1.00$ m and $v = 0.750c$ in the equation for length contraction yields

$$L = L_0\sqrt{1 - \frac{v^2}{c^2}}$$

$$= 1.00 \text{ m}\sqrt{1 - \frac{(0.750c)^2}{(c)^2}}$$

$$= \boxed{0.661 \text{ m}}$$

Practice Problems

9. **Follow-up** What is the length of the meterstick if the spaceship travels at nearly the speed of light, $0.999c$?

10. How long is a plank of wood at rest if its length when moving at $0.995c$ is 2.00 m?

11. What speed must a meterstick have for an observer at rest to see its length as 0.500 m? Give your answer as a multiple of the speed of light.

27.2 LessonCheck (MP)

Checking Concepts

12. 🔊 **Explain** Is the time between ticks of a moving clock greater than, less than, or the same as the time between ticks when the same clock at rest?

13. 🔊 **Critique** A classmate states that time dilation affects clocks but does not apply to the beating of a human heart. Is he correct? Explain.

14. 🔊 **Explain** Is the length of a moving object greater than, less than, or the same as the length of the same object at rest?

15. Big Idea How do the concepts of time and length differ for objects with speeds close to the speed of light?

Solving Problems

16. Calculate An unstable isotope decays in 25 s when at rest. How long does it take to decay when it moves with a speed of $0.750c$?

17. Calculate What is the length of a meterstick moving with a speed of $0.910c$?

18. Calculate The clock on a spaceship moving with a speed of $0.885c$ advances by 13.2 s. How much time elapses for an observer at rest?

19. Calculate A meterstick in a passing spaceship has a length of 0.950 m. What is the speed of the spaceship?

27.3 $E = mc^2$

Vocabulary

- rest mass
- rest energy

🔊 *What does it mean to say that mass and energy are equivalent?*

Perhaps the most famous equation in all of science is $E = mc^2$. Though most people have heard of this equation, few truly understand its significance. In this lesson you will explore the meaning of $E = mc^2$.

Mass is a form of energy

The equation $E = mc^2$ shows that mass and energy are equivalent and are related by the speed of light squared. 🔊 **The equivalence of mass and energy means that mass and energy can be thought of as being the same, and that one can be converted into the other.** This result, like time dilation and length contraction, was completely unanticipated before the introduction of the theory of relativity.

The mass of an object at rest is referred to as its **rest mass**, represented as m_0. Einstein showed that when the object moves with a speed v, its relativistic energy, E, is given by a fairly simple equation.

Relativistic Energy

$$\text{relativistic energy} = \frac{\text{rest mass} \times (\text{speed of light})^2}{\sqrt{1 - \dfrac{(\text{speed of object})^2}{(\text{speed of light})^2}}}$$

$$E = \frac{m_0 c^2}{\sqrt{1 - \dfrac{v^2}{c^2}}}$$

SI unit: J

This is Einstein's most famous result from his theory of relativity. Notice that the relativistic energy, E, does not vanish when the object's speed goes to zero. The energy of an object at rest is known as its **rest energy**, E_0. The rest energy of an object is given by the following:

Rest Energy

$$\text{rest energy} = \text{rest mass} \times (\text{speed of light})^2$$

$$E_0 = m_0 c^2$$

SI unit: J

When people write $E = mc^2$, they are just dropping the zero subscripts for simplicity.

Because the speed of light is so large, it follows that the mass of an object times the speed of light squared is a truly enormous amount of energy. This is verified in the following Quick Example.

QUICK Example 27.3 What's the Rest Energy?

Find the rest energy of a 0.12-kg apple.

Solution
Substituting $m_0 = 0.12$ kg and $c = 3.00 \times 10^8$ m/s into $E_0 = m_0 c^2$ gives

$$E_0 = m_0 c^2$$
$$= (0.12 \text{ kg})(3.00 \times 10^8 \text{ m/s})^2$$
$$= \boxed{1.1 \times 10^{16} \text{ J}}$$

Math HELP
Scientific
Notation
See Math
Review,
Section I

To put this result in perspective, if the rest energy of an apple could be converted entirely to usable forms of energy, it would supply the energy needs of the entire United States for about an hour.

Practice Problem

20. A typical household in the United States uses 18,000 kilowatt-hours (kWh) of energy per year. How much mass is required to have a rest energy of 18,000 kWh? (Recall that 1 watt = 1 J/s.)

Nuclear reactions convert mass to energy

Quick Example 27.3 shows that even a modest mass can be converted to a large amount of energy. This is the basic principle behind the operation of nuclear power plants, in which a small decrease in mass (due to various nuclear reactions) is converted into electrical energy. For example, the nucleus of a uranium-235 atom can fission into two smaller nuclei and a number of neutrons. Since the mass of the uranium nucleus is greater than the sum of the masses of the products of this fission reaction, the reaction releases energy. In fact, fission reactions in 0.45 kg (1 pound) of uranium can produce about 3 million times more energy than the combustion of 0.45 kg of coal.

The Sun is also powered by the conversion of mass to energy. In this case the energy is released by fusion reactions, in which two small nuclei combine to form a larger nucleus. In fact, the Sun radiates about 3.9×10^{26} J of energy per second. This corresponds to a decrease in mass of 4.4×10^9 kg—roughly the equivalent of 2000 space shuttles. The mass of the Sun is so large, however, that even at this rate it loses only 0.01% of its mass every 1.5 billion years. Clearly, the Sun will not "evaporate" into space anytime soon.

CONCEPTUAL Example 27.4 **Compare the Mass**

When you compress a spring between your fingers, does its mass increase, decrease, or stay the same?

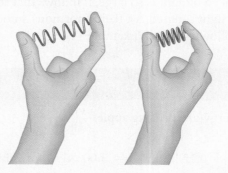

Reasoning and Discussion
When the spring is compressed by a distance x, its energy is increased by the amount $\Delta E = \frac{1}{2}kx^2$, as you learned in Chapter 6. Since the energy of the spring has increased, its mass increases as well, by the amount $\Delta m = \Delta E/c^2$. In everyday systems like this one, the change in mass is far too small to be observed.

Answer
The mass of the spring increases.

Matter and antimatter can annihilate each other

What happens when matter and antimatter collide?

A particularly interesting aspect of mass-energy equivalence is the existence of antimatter. Recall that every elementary particle has a corresponding antimatter particle that has precisely the same mass but the opposite charge.

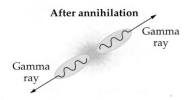

Before annihilation

Electron ●━━▶ ◀━━● Positron

After annihilation

Gamma ray

Gamma ray

▲ **Figure 27.11 Electron-positron annihilation**
An electron and a positron annihilate one another when they come into contact. The result is the emission of two energetic gamma rays with no mass. The mass of the original particles has been converted into the energy of the gamma rays.

Antimatter is frequently created in accelerators, where particles collide at speeds approaching the speed of light. In fact, it is even possible to create antiatoms made entirely of antimatter in labs. An intriguing possibility is that the universe may actually contain entire antigalaxies of antimatter.

If this is indeed the case, one would have to be a bit careful about visiting such a galaxy, because particles of matter and antimatter have a rather interesting behavior when they meet—they *annihilate* one another. The annihilation of matter and antimatter is illustrated in **Figure 27.11**, which shows an electron and a positron (the antimatter particle of an electron) coming into contact. The result is that the particles cease to exist, which satisfies charge conservation, since the net charge of the system is zero before and after the annihilation. As for energy conservation, the mass of the two particles is converted into two gamma rays, which are similar to X-rays only more energetic. **Thus, in matter-antimatter annihilation the particles vanish, producing a burst of radiation.**

27.3 LessonCheck (MP)

Checking Concepts

21. Identify What is the equation that describes the connection between energy and mass?

22. Describe What happens during matter-antimatter annihilation?

23. Explain Why does even a small amount of mass correspond to an enormous amount of energy?

Solving Problems

24. Calculate What is the rest energy of a 1.0-kg mass?

25. Calculate What is the relativistic energy of a 1.0-kg mass moving with a speed of 0.50c?

26. Think & Calculate Heating a pan of soup on the stove increases its energy by 120 J.
(**a**) As a result of the heating, does the mass of the pan of soup increase, decrease, or stay the same? Explain.
(**b**) Calculate the change in mass (if any) of the pan of soup as a result of its being heated.

27.4 General Relativity

Vocabulary

- gravitational lensing
- black hole
- gravity waves

Relativity has revolutionized our understanding of the universe. If you think back over the results presented in this chapter—time dilation, length contraction, mass-energy equivalence—it's clear that relativity reveals a universe far richer and more varied than was ever imagined before. And adding even more to this richness is *general relativity*, which applies to accelerating frames of reference and gravity. We examine the main results of Einstein's general theory of relativity in this lesson.

🔑 *What is the connection between an accelerated frame of reference and gravity?*

General relativity applies to gravity

Imagine you're riding in an elevator when the cable breaks. Remember, we're just imagining this! What happens? Well, as you and the elevator fall, you feel weightless—just as if you were an astronaut in orbit. Now imagine riding in an elevator in outer space that accelerates upward. To you, the upward acceleration feels like gravity. Simple thought experiments like these led Einstein to his general theory of relativity. To explore general relativity further, we'll consider observers inside three different elevators.

Comparing Observers Figure 27.12 (a) shows an observer in an elevator that is at rest on the surface of Earth. If this observer drops or throws a ball (or other object), it falls toward the floor of the elevator with an acceleration equal to the acceleration due to gravity.

Now consider an identical elevator located in deep space. If this elevator is at rest, or moving with a constant velocity, an observer within the elevator experiences weightlessness, as shown in **Figure 27.12 (b)**. If a ball is released by this observer, it remains in place.

(a) A frame of reference in a gravitational field

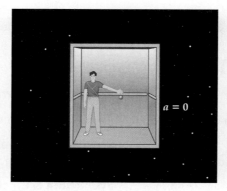

(b) An inertial frame of reference with no gravitational field

(c) An accelerated frame of reference

▲ Figure 27.12 **Comparing an accelerated frame of reference with gravity**
(a) An observer in an elevator at rest on Earth's surface. If the observer drops or throws a ball, it falls with a downward acceleration of *g*.
(b) An observer in an elevator in deep space experiences weightlessness—as long as the elevator is at rest or moves with constant velocity. If the observer releases a ball, it remains at rest. (c) If an elevator in deep space is given an upward acceleration of magnitude *g*, the observer in the elevator observes that dropped objects fall toward the floor of the elevator with an acceleration *g*, just as on Earth. The acceleration of the elevator causes an "artificial" gravity.

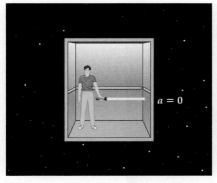

$a = 0$

g

(a) Nonaccelerating elevator   **(b)** Accelerating elevator

▲ **Figure 27.13 A light experiment in two different frames of reference**
(a) In a nonaccelerating elevator, a beam of light travels on a straight line as it crosses the elevator. **(b)** An accelerating elevator moves upward as the light crosses the elevator; hence, the light strikes the opposite wall at a lower level. The path of the light in this case appears parabolic to the observer riding in the elevator.

What if the elevator in deep space is given an upward acceleration equal to the acceleration due to gravity, g, as in **Figure 27.12 (c)**? In this case a ball that is released remains at rest relative to the background stars while the floor of the elevator accelerates upward toward it with the acceleration g. Similarly, if this observer throws the ball horizontally, it follows a parabolic path to the floor, just as for the observer in Figure 27.12 (a). In addition, the floor of the elevator exerts a force mg on the feet of this observer (whose mass is m) to give him an upward acceleration g.

Principle of Equivalence Based on the above observations, we conclude that when the observer in Figure 27.12 (c) conducts an experiment in his accelerating elevator, the results are the same as those obtained by the observer in Figure 27.12 (a) in her elevator at rest on Earth. Einstein extended these kinds of observations into the *principle of equivalence*.
🔑 The principle of equivalence states that any physical experiment conducted in a uniform gravitational field and in an accelerated frame of reference gives identical results. Thus, the two observers cannot tell, without looking outside the elevator, whether they are at rest in a gravitational field or in deep space in an accelerating elevator.

The principle of equivalence implies that gravity bends light

Let's apply the principle of equivalence to a simple experiment involving light. If the observer in **Figure 27.13 (a)** shines a flashlight toward the opposite wall of the elevator (which has zero acceleration), the light strikes the wall at its initial height. If the same experiment is conducted in the elevator in **Figure 27.13 (b)**, the elevator is accelerating upward during the time the light travels across the elevator. Thus, when the light reaches the far wall, it strikes the wall at a lower level. In fact, the light has followed a parabolic path, just as one would expect for a ball that was projected horizontally.

Applying the principle of equivalence, Einstein concluded that a beam of light in a gravitational field must also bend downward, just as it does in an accelerating elevator; that is, gravity bends light. This phenomenon is illustrated in **Figure 27.14**, where the amount of bending has been exaggerated for clarity.

▲ **Figure 27.14 The principle of equivalence**
By the principle of equivalence, a beam of light in a gravitational field should follow a parabolic path, just as in an accelerating elevator. The bending of the light's path has been exaggerated here for clarity.

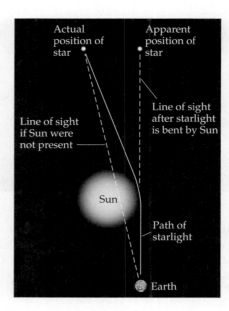

▶ **Figure 27.15 Gravitational bending of light**
As light from a distant star passes close to the Sun, it is bent. The result is that an observer on Earth sees the star along a line of sight that is displaced away from the center of the Sun.

The Sun bends the light from distant stars

In order to put Einstein's prediction to the test, it's necessary to increase the amount of bending the light undergoes as much as possible to make it large enough to be measured. Thus, we need to use the strongest gravitational field available. In our Solar System, the strongest gravitational field is provided by the Sun; hence, an experiment to detect the bending of light produced by the Sun was needed.

To see the effect of the Sun's gravitational field, consider a ray of light from a distant star passing near the Sun, as shown in **Figure 27.15**. As the light passes the Sun, it is bent—that is, its direction changes. It follows that an observer on Earth must look in a direction that is farther from the Sun than the actual direction of the distant star—the Sun's gravitational field displaces the apparent positions of the distant stars farther from the Sun. If we imagine the Sun moving in front of a background field of stars, as in **Figure 27.16**, the stars near the Sun are displaced outward. It is almost as if the Sun were a lens, slightly distorting everything behind it.

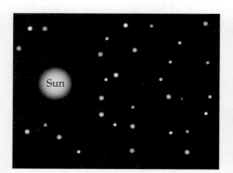  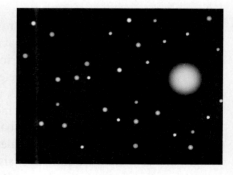

▲ **Figure 27.16 Bending of light near the Sun**
As the Sun moves across a starry background, the stars near it appear to be displaced outward, away from the center of the Sun. This is the effect that was used in the first experimental confirmation of general relativity.

Light bending by the Sun can be observed

Because the Sun is so bright, an experiment like that shown in Figures 27.15 and 27.16 can only be carried out during a total eclipse of the Sun, when the Sun's light is blocked by the Moon. During the eclipse, photographs can be taken to show the positions of the background stars. Later, these photographs can be compared with photographs of the same star field taken 6 months later, when the Sun is on the other side of Earth. Comparing the photographs allows one to measure the displacement of the stars. This experiment was carried out during an expedition to Africa in 1919 by Sir Arthur Eddington. His results confirmed the predictions of the general theory of relativity and made Einstein a household name.

Since gravity can bend light, the more powerful the gravitational force, the more the bending, and the more dramatic the results. In **Figure 27.17** we see what can happen when a large galaxy or a cluster of galaxies, with its immense gravitational field, lies between us and more distant galaxies. The intermediate galaxy or cluster can produce significant bending of light, resulting in multiple images of a more distant galaxy that can take the form of arcs or crosses. Some examples of such gravitational lensing are shown in **Figure 27.18**.

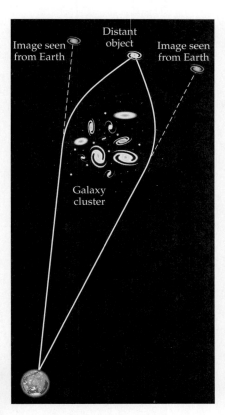

▲ **Figure 27.17 Bending of light by a galaxy**
When light passes by a galaxy or a cluster of galaxies it can be bent by a large amount. When the light reaches Earth, it seems to be coming from a direction that is different from the direction to its source.

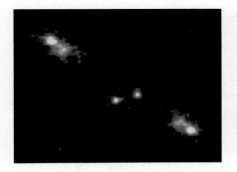 

▲ **Figure 27.18 Gravitational lensing**
The images created by gravitational lensing take a number of forms. Sometimes the light from a distant object is stretched out into an arc or even a complete ring. In other cases, such as that of the distant quasar at left, we may see a pair of images. (The quasar appears at the upper left and lower right. The lensing galaxy is not visible in the photo—the small dots at the center are unrelated objects.) In still other instances, four images may be produced, as in the famous *Einstein cross* shown at right. The lensing galaxy at the center is some 400 million light-years from us, while the quasar whose multiple images surround it is about 20 times farther away.

Black holes can bend light by an extreme amount

An intense gravitational field can also be produced when a star burns up its nuclear fuel and collapses to a very small size. In such a case the gravitational field can become strong enough to actually trap light—that is, to bend it so much that it cannot escape the star. This extreme bending of light is illustrated in **Figure 27.19**. **A star that bends light so much that the light cannot escape is referred to as a black hole.**

Black holes, by definition, cannot be directly observed; however, their presence can be inferred by their gravitational effects on other bodies. It is also possible to detect the intense radiation emitted by ionized matter as it falls into a black hole. By these and other means, the existence of black holes in the centers of many galaxies has been firmly established. In fact, it is now thought that black holes may be relatively common in the universe.

What distinguishes a black hole from any other star?

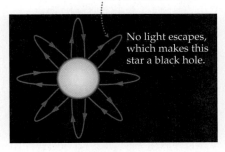

▲ **Figure 27.19 A black hole traps light**
Light emitted by a collapsed star can be bent so much that it cannot escape from the star. In this case, the star no longer emits light and is referred to as a *black hole*.

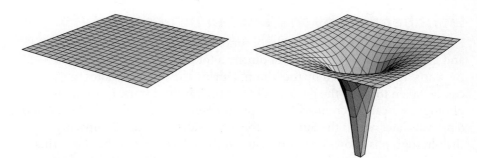

(a) Flat space, away from massive objects   **(b)** Warped space, near a massive object

▲ **Figure 27.20 Warped space and black holes**
(a) Regions of space that are far from any large masses can be thought of as flat. In these regions light propagates in straight lines. **(b)** Near a large concentrated mass, such as a black hole, space can be thought of as warped. In these regions the paths of light rays are bent.

Moving masses cause gravity waves

A convenient way to visualize the effects of intense gravitational fields is to think of space as a sheet of rubber with a square array of grid lines, as shown in **Figure 27.20**. In the absence of mass, the sheet is flat, and the grid lines are straight, as in Figure 27.20 (a). In this case, a beam of light follows a straight-line path, parallel to a grid line. If a large mass is present, however, the sheet is deformed, as in Figure 27.20 (b), and light rays follow the curved paths of the grid lines. In cases where one large mass orbits another, as in **Figure 27.21**, the result is a series of ripples moving outward through the rubber sheet. These ripples represent **gravity waves**, one of the many intriguing phenomena predicted by the general theory of relativity.

Experiments are attempting to detect gravity waves

When a gravity wave passes through a given region of space, it causes a local deformation of space, as Figure 27.21 suggests. Early attempts to detect gravity waves were based on measuring the distortion such a wave would produce in a large metal bar. Unfortunately, the sensitivity of these devices was too low to detect the weak waves that are thought to pass through

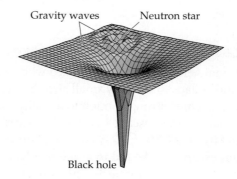

Gravity waves   Neutron star

Black hole

▲ **Figure 27.21 Gravity waves**
Gravity waves can be thought of as ripples in the warped space around a large mass. In the case illustrated here, a neutron star orbits a black hole. As a result of its acceleration, the neutron star emits gravity waves. (A neutron star is an extremely small, dense star produced by the collapse of a massive star. During the collapse, electrons and protons are combined to form neutrons. Thus, a neutron star is basically a city-sized ball of neutrons.)

Earth all the time. Ironically, these detectors *were* sensitive enough to detect the relatively strong gravity waves that must have accompanied the 1987 supernova explosion in the nearby Large Magellanic Cloud (a small satellite galaxy of our Milky Way). As luck would have it, however, none of the detectors were operating at that time.

The next generation of gravity wave detector is starting to become operational. This type of detector, which goes by the name *Laser Interferometer Gravitational Wave Observatory*, or LIGO for short, should be sensitive enough to detect several gravitational wave events per year. One type of event that LIGO will be looking for is the final death spiral of a neutron star as it plunges into a black hole. The neutron star might orbit the black hole for 150,000 years, but only in the last 15 minutes of its life does it find fame, because during those few minutes its acceleration is great enough to produce gravity waves detectable by LIGO.

There is still much to discover in physics

We have come far in our study of physics. We started with the behavior of falling objects and simple forces and have reached the physics of relativity and black holes. Still, there are many unresolved questions in physics, like how to combine general relativity and quantum mechanics. The next generation of physicists will discover aspects of nature that no one can predict at this moment. In so doing, they will experience the pure joy and wonder of discovery.

27.4 LessonCheck (MP)

Checking Concepts

27. Explain What is the connection between an experiment performed in an accelerating frame of reference and the same experiment performed in a uniform gravitational field?

28. Describe How much does a star have to bend light to become a black hole?

29. Summarize Explain how a thought experiment involving elevators can be used to predict that gravity bends light.

Physics & You

Miniature Nuclear Reactors

What Is It? As the name implies, miniature nuclear reactors are much smaller than the reactors currently used in nuclear power plants in the United States. These portable nuclear reactors use fission reactions to produce electrical power. However, they are self-contained and designed to run from 10 to 30 years without needing maintenance or any other sort of outside intervention. The reactors are said to be portable because they can be shipped to a site completely assembled and simply installed.

How Does It Work? Miniature nuclear reactors generate energy from fission reactions. In order to generate power for up to 30 years, however, such a reactor has to operate as a *breeder reactor*. Breeder reactors use neutrons from fission reactions to convert nonfissionable U-238 into fissionable Pt-239. A moving reflector moves along a stationary fuel core, reflecting neutrons back into the fissionable fuel and sustaining the fission reactions.

Why Is It Important? The demand for energy is increasing. One way to meet this demand is through the use of nuclear power. However, the risks of radiation exposure and of fissionable fuel being used for weapons are serious issues that must considered. Miniature nuclear reactors attempt to address these problems. The basic concept is simple: Deliver a sealed, tamper-proof reactor to a site; allow it to generate power unattended for up to 30 years; retrieve the reactor when the fuel is used up; and recycle the spent fuel at a secure facility.

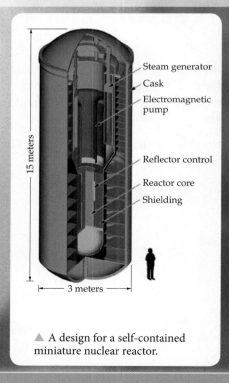

▲ A design for a self-contained miniature nuclear reactor.

Take It Further

1. Research *Write a report or create a poster on how a breeder reactor works. Research the* *pros and cons of this type of nuclear reactor and summarize your findings.*

Physics Lab Time Dilation

In this lab you will derive the equation for time dilation.

Materials
- large sheet of paper (about 1 m wide by at least 2 m long)
- piece of cardboard (about 1 m wide by 30 cm long)
- constant-velocity car
- meterstick
- stopwatch
- tape

Procedure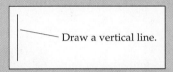

Before you begin, read the entire procedure and create a data table to record your results. Answer the questions in each section as you work through the lab.

Part I: Measuring Independent Velocities

1. Tape the ends of the large sheet of paper to a table. Draw a vertical line as shown and measure its length to the nearest 0.1 cm. Record this length as the distance $d_{vertical}$.

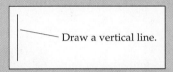
Draw a vertical line.

2. Place the constant-velocity car at the beginning of the line and measure the time, $t_{vertical}$, required for the car to reach the end of the line. Record $t_{vertical}$ to the nearest 0.01 s. Use the value to calculate the velocity of the car (in m/s): $v_{car} = d_{vertical}/t_{vertical}$.

3. Draw a long horizontal line that begins at the vertical line drawn in Step 1. Place a tick mark at the 1.0-m point, as shown.

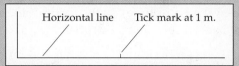
Horizontal line   Tick mark at 1 m.

4. Draw a vertical line down the center of the cardboard. Align the line on the cardboard with the vertical line on the paper. Pull the cardboard to the right at a constant velocity and measure the time, $t_{horizonal}$, required for the vertical line on the cardboard to reach the 1.0-m mark along the horizontal line. Record $t_{horizonal}$ to the nearest 0.01 s. : Use the value to calculate the velocity of the cardboard (in m/s): $v_{cardboard} = 1.0 \text{ m}/t_{horizonal}$.

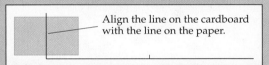
Align the line on the cardboard with the line on the paper.

Part II: Galilean Relativity: Newtonian Frames of Reference

5. Repeat Step 4 with the car placed on the vertical line on the cardboard. Position the car so it is at the start of the vertical line on the paper beneath the cardboard. Release the car at the same instant you begin pulling the cardboard. (Try to pull the cardboard at the same constant speed used in Step 4.) Measure the time t' required for the car to reach the 1.0-m mark on the horizontal line. Record t' to the nearest 0.1 s.

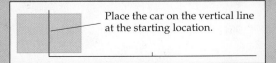
Place the car on the vertical line at the starting location.

(a) In your data table, draw the path you *see* the car take. (*Hint:* The path should be the hypotenuse of a right triangle.)

(b) What determines the length of each side of the triangle? Draw a picture and explain.

(c) Which quantity (distance, velocity, or time) does each side of the triangle have in common?

(d) Calculate the velocity you observed for the car, v'. (*Hint:* The velocity v' is the car's velocity along the hypotenuse of the triangle.)

(e) What happened to the velocity of the car? Why?

(f) Show how the Pythagorean formula can be used to calculate the velocity in part (d) using the horizontal and vertical velocities from Steps 2 and 4.

Part III: Einstein's Relativity

6. Now imagine that the car is actually a photon traveling at the speed of light, c. Recall that Einstein's second postulate requires that the speed of light be constant for all observers.

(a) What would you measure the velocity of the photon to be when it travels along the vertical line on the paper?

(b) What would you measure the velocity of the photon to be when it travels along the hypotenuse?

(c) How do these results differ from the results in Part II?

(d) Draw the triangle for a photon. Which quantity (distance, velocity, or time) do the vertical side and the diagonal have in common?

(e) Redraw your triangle. Notice that the time along the vertical and horizontal paths is the same, $t_{vertical}$; call the time along the diagonal t'. This time, label the lengths in terms of c, $v_{cardboard}$, $t_{vertical}$, and t'.

(f) You are at rest in the Earth's frame of reference. Which path, or leg of the triangle, did you see the photon travel? Which time do you measure, t' or $t_{vertical}$?

(g) Does an ant sitting on the moving cardboard measure time t' or $t_{vertical}$?

(h) Use the Pythagorean theorem to derive the time dilation equation.

27 Study Guide

Big Idea

Nature behaves differently near the speed of light.

For example, clocks run slow and meter-sticks shrink when moving at very high speeds. In addition, the relativistic energy approaches infinity as the speed of light is approached. This means that no mass can ever be given enough energy to move as fast as the speed of light. Nothing can move faster than the speed of light.

27.1 The Postulates of Relativity

Einstein's theory of relativity is based on two postulates:

First Postulate: Equivalence of Physical Laws

The laws of physics are the same in all inertial frames of reference.

Second Postulate: Constancy of the Speed of Light

The speed of light in a vacuum, $c = 3.00 \times 10^8$ m/s, is the same in all inertial frames of reference, regardless of the motion of the source or the observer of the light.

The ultimate speed in the universe is the speed of light in a vacuum.

27.2 The Relativity of Time and Length

With more time elapsing between ticks, a moving clock runs slower than a clock at rest.

Relativistic time dilation applies to *all physical processes*, including chemical reactions and biological functions such as aging.

As the speed of an object approaches c, its length approaches zero.

• The slowing of a moving clock is referred to as *time dilation*.

• The decrease in length of a moving object is referred to as *length contraction*.

Key Equations

The time dilation equation, relating the elapsed time at rest, Δt_0, and the elapsed time, Δt, at a speed v is

$$\Delta t = \frac{\Delta t_0}{\sqrt{1 - \dfrac{v^2}{c^2}}}$$

The length contraction equation, relating the length of an object at rest, L_0, and the length L at a speed v is

$$L = L_0\sqrt{1 - \frac{v^2}{c^2}}$$

27.3 $E = mc^2$

The equivalence of mass and energy means that mass and energy can be thought of as being the same, and that one can be converted into the other.

In matter-antimatter annihilation the particles vanish, producing a burst of radiation.

• One of the most important results of relativity is that mass is another form of energy.

• The rest mass of an object, m_0, is the mass it has when its speed is zero.

Key Equations

The relativistic energy, E, of an object with rest mass m_0 and speed v is

$$E = \frac{m_0c^2}{\sqrt{1 - \dfrac{v^2}{c^2}}}$$

When an object is at rest, its energy, E_0, is

$$E_0 = m_0c^2$$

27.4 General Relativity

The principle of equivalence states that any physical experiment conducted in a uniform gravitational field and in an accelerated frame of reference gives identical results.

A star that bends light so much that the light cannot escape is referred to as a *black hole*.

27 Assessment

ANSWERS TO SELECTED ODD-NUMBERED PROBLEMS APPEAR IN APPENDIX A.

Lesson by Lesson

27.1 The Postulates of Relativity

Conceptual Questions

30. Some distant galaxies are moving away from us with speeds of 0.5c. What is the speed of the light that reaches Earth from these galaxies? Explain.

31. What do we call a frame of reference in which physics experiments give the same results as they would if carried out in your classroom?

32. The speed of light in glass is less than c. Why is this not a violation of the second postulate of relativity?

33. ⬚Predict & Explain⬚ You are in a spaceship, traveling directly away from the Moon with a speed of 0.9c. A light signal is sent in your direction from the surface of the Moon. **(a)** As the signal passes your ship, do you measure its speed to be greater than, less than, or equal to 0.1c? **(b)** Choose the *best* explanation from among the following:

A. The speed you measure will be greater than 0.1c; in fact, it will be c, since all observers in inertial frames measure the same speed of light.

B. You will measure a speed less than 0.1c because light travels slower in the vacuum of space.

C. You will measure a speed of 0.1c, which is the difference between c and 0.9c.

27.2 The Relativity of Time and Length

Conceptual Questions

34. Describe some of the everyday consequences that would follow if the speed of light were 20 km/h.

35. What is the difference between the proper time interval and the dilated time interval?

36. What is the difference between the proper length and the contracted length of an object?

37. What happens to the length of an object as its speed approaches the speed of light? Why does it not make sense for the object's speed to be greater than the speed of light?

38. ⬚Predict & Explain⬚ Suppose you are a traveling salesman for SSC, the Spacely Sprockets Company. You travel on a spaceship that reaches speeds near the speed of light, and you are paid by the hour. **(a)** When you return to Earth after a sales trip, would you prefer to be paid according to the clock at Spacely Sprockets headquarters on Earth, or according to the clock on the spaceship in which you travel, or would your pay be the same in either case? **(b)** Choose the *best* explanation from among the following:

A. You want to be paid according to the clock on Earth, because the clock on the spaceship runs slow when it approaches the speed of light.

B. You want to be paid according to the clock on the spaceship, because from your viewpoint the clock on Earth has run slow.

C. Your pay would be the same in either case because motion is relative, and all inertial observers will agree on the amount of time that has elapsed.

39. If the speed of light were greater than 3.00×10^8 m/s, would the effects of time dilation and length contraction be greater or less than they are now? Explain.

Problem Solving

40. A neon sign in front of a café flashes on and off once every 4.1 s, as measured by the head cook. How much time elapses between flashes of the sign as measured by an astronaut in a spaceship moving toward Earth at a speed of 0.84c?

41. A lighthouse sweeps its beam of light around in a circle once every 7.5 s. To an observer in a spaceship moving away from Earth, the beam of light completes one full circle every 15 s. What is the speed of the spaceship relative to Earth?

42. Usain Bolt set a world record for the 100-m dash on August 16, 2009. Suppose observers on a spaceship moving with a speed of 0.7705c relative to Earth saw Bolt's run and measured his time to be 15.03 s. Find the time that was recorded on Earth.

43. How fast does a 250-m-long spaceship move relative to an observer who measures the ship's length to be 150 m?

44. Suppose the speed of light in a vacuum were only 25.0 km/h. Find the length of a bicycle being ridden at a speed of 20.0 km/h as measured by an observer sitting on a park bench, given that the bicycle's rest length is 1.89 m.

45. A rectangular painting is 124 cm wide and 80.5 cm high, as indicated in **Figure 27.22**. At what speed, v, must the painting move parallel to its width if it is to appear to be square? (Only lengths parallel to the direction of motion are contracted.)

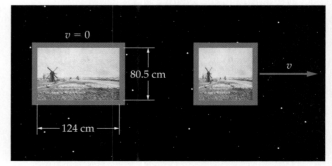

Figure 27.22

46. The radar antenna on a navy ship rotates with an angular speed of 0.29 rad/s. What is the angular speed of the antenna as measured by an observer moving away from the antenna with a speed of 0.82c?

47. Think & Calculate (a) Is it possible for you to travel far enough and fast enough that when you return from the trip, you are younger than your stay-at-home sister, who was born 5.0 y after you? (b) Suppose you travel on a rocket ship at the speed $v = 0.99c$ for 1 y, according to the ship's clocks and calendars. How much time elapses on Earth during your 1-y trip? (c) If you were 22 when you left home and your sister was 17, what are your ages when you return?

27.3 $E = mc^2$

Conceptual Questions

48. An electron (mass m) and a positron (mass m) at rest annihilate each other, producing two high-energy gamma ray photons. Each gamma ray has an energy equal to mc^2 and carries a certain amount of momentum. Is it possible for the electron and the positron to annihilate each other and send out a single gamma ray with an energy of $2mc^2$? Would this conserve energy? Would it conserve momentum?

49. What happens to the relativistic energy of an object as its speed approaches the speed of light? Why would it not make sense for the object's speed to exceed the speed of light?

50. Rank Particles A through D have the rest energies and relativistic energies shown below. Rank these particles in order of (a) increasing rest mass and (b) increasing speed. Indicate ties where appropriate.

Particle	Rest Energy	Relativistic Energy
A	6E	6E
B	2E	4E
C	4E	6E
D	3E	4E

Problem Solving

51. If a neutron moves with a speed of 0.99c, what are its (a) relativistic energy and (b) rest energy?

52. A spring with a force constant of 574 N/m is compressed a distance of 39 cm. Find the resulting increase in the spring's mass.

53. When a certain capacitor is charged, its mass increases by 8.3×10^{-16} kg. How much energy is stored in the capacitor?

54. When a proton encounters an antiproton, the two particles annihilate each other, producing two gamma rays. Assuming that the particles were at rest when they annihilated one another, find the magnitude of the energy of each of the two gamma rays produced. (*Note*: The rest energies of an antiproton and a proton are identical.)

55. A rocket with a rest mass of 2.7×10^6 kg has a relativistic energy of 2.7×10^{23} J. How fast is the rocket moving?

56. An object has a relativistic energy that is 5.5 times its rest energy. What is its speed?

57. A nuclear power plant produces an average of 1.0×10^3 MW of power during a year of operation. Find the corresponding change in mass of the reactor's fuel, assuming that all of the energy released by the fuel can be converted directly to electrical energy. (In a real-world reactor, only a relatively small fraction of the released energy can be converted to electricity.)

27.4 General Relativity

Conceptual Questions

58. An elevator moves in deep space with a constant upward velocity. What happens if you drop a ball in this elevator?

59. An elevator moves in deep space with a constant upward acceleration. What happens if you drop a ball in this elevator?

60. The light given off by a black hole is trapped and cannot escape. How then can we observe a black hole?

61. During a total eclipse of the Sun, an astronomer observes a distant star near the edge of the Sun. Does the bending of light by the Sun make the distant star appear to be closer to or farther from the edge of the Sun than it actually is? Explain.

Mixed Review

62. Can the proper time interval ever be larger than the dilated time interval? Can the proper time interval ever be smaller than the dilated time interval? Can these two time intervals ever be the same? Explain.

63. Two observers are moving relative to one another. Which of the following quantities will they always measure as having the same value: (a) their relative speed, (b) the time between two events, (c) the length of an object, (d) the speed of light in a vacuum?

64. Which clock runs slower relative to a clock on the North Pole: clock 1, which is on an airplane flying from New York to Los Angeles, or clock 2, which is on an airplane flying from Los Angeles to New York? Explain, assuming that each plane has the same speed relative to Earth's surface.

65. Predict & Explain Consider two apple pies that are identical in every respect, except that pie 1 is piping hot from the oven and pie 2 is at room temperature. (a) If identical forces are applied to the two pies, is the acceleration of pie 1 greater than, less than, or equal to the acceleration of pie 2? (b) Choose the *best* explanation from among the following:

A. The acceleration of pie 1 is greater than that of pie 2 because it is hot and thus has the greater energy.

B. The fact that pie 1 is hot means that it has more mass than pie 2, and therefore it has a smaller acceleration.

C. The pies have the same acceleration regardless of their temperatures because they have identical rest masses.

66. A girl swings back and forth on a homemade swing hung from a tree branch, as shown in **Figure 27.23**. The length of the swing is $L = 2.4$ m, and it can be treated as a simple pendulum. What is period of the girl's oscillations as observed by a boy in a spaceship moving past the Earth with a speed of $0.84c$?

Figure 27.23

67. **Think & Calculate** An astronaut moving with a speed of $0.65c$ relative to Earth measures her heart rate to be 72 beats per minute. **(a)** When an Earth-based observer measures the astronaut's heart rate, is the result greater than, less than, or equal to 72 beats per minute? Explain. **(b)** Calculate the astronaut's heart rate as measured on Earth.

Writing about Science

68. Write a report on relativity in everyday life. How does relativistic time dilation affect the GPS system? How is $E = mc^2$ related to PET scans? How is it related to nuclear power plants?

69. **Connect to the Big Idea** Describe ways in which nature is different at speeds near the speed of light than it is at everyday speeds. Give some examples of how our everyday experience of the world would be different if the speed of light were only 40 km/h.

Read, Reason, and Respond

Relativity in a TV Set The first televisions used cathode-ray tubes, or CRTs, to display the picture. A CRT uses a beam of electrons to "paint" a picture on a fluorescent screen (see **Figure 27.24**). A heated coil at the negative terminal of the tube (the cathode) produces electrons, which are accelerated toward the positive terminal of the tube (the anode). This beam of electrons is the cathode ray. A series of horizontal and vertical deflecting plates direct the beam to any spot on the fluorescent screen, where it produces a glowing dot. Moving the glowing dot rapidly around the screen and varying its intensity produces the image.

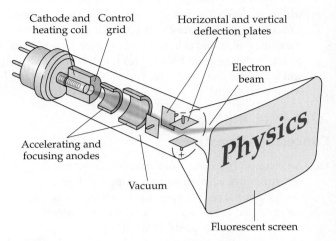

Figure 27.24 A cathode-ray tube.

The interior of a CRT must be a very good vacuum to ensure that the electrons on their way to the screen aren't scattered by air molecules. The electrons are accelerated through a potential difference of 25.0 kV, which is sufficient to give them speeds comparable to the speed of light. As a result, relativity must be used to accurately determine their behavior.

70. Ignoring relativistic effects, what is the speed of an electron, as a fraction of the speed of light, that is accelerated through a potential difference of 25.0 kV? (Speeds greater than 0.1c are generally regarded as relativistic.)

   A. $0.221c$ C. $0.312c$

   B. $0281c$ D. $0.781c$

71. When relativistic effects are included, do you expect the speed of the electron to be greater than, less than, or the same as the result in Problem 70?

72. What is the speed of an electron accelerated through a potential difference of 25.0 kV, if relativistic effects are considered? (The relativistic kinetic energy of an object with rest mass m_0 is $KE = E - m_0c^2$, where E is the total relativistic energy.)

   A. $0.301c$ C. $0.412c$

   B. $0.312c$ D. $0.953c$

73. Suppose the accelerating potential difference in a CRT were increased by a factor of 10. What would the relativistic speed of an electron be in this case? (Use the equation for KE given in Problem 72.)

   A. $0.205c$ C. $0.740c$

   B. $0.672c$ D. $0.862c$

Standardized Test Prep

Multiple Choice

1. Which of the following are addressed by the general theory of relativity?
   - (A) proper time and time dilation
   - (B) accelerated frames of reference and proper time
   - (C) accelerated frames of reference and gravity
   - (D) relativity of length and velocity

2. Which of the following statements *does not* describe length, according to the theory of relativity?
   - (A) Proper length must be measured by an observer at rest relative to the moving object.
   - (B) The length of an object depends on its speed relative to the observer.
   - (C) Length contraction occurs only in the direction of an object's motion.
   - (D) An object's contracted length is independent of its proper length.

3. Which of the following is the *best* description of a moving clock, according to the theory of relativity?
   - (A) To a stationary observer it appears to keep the same time as a stationary clock.
   - (B) It runs faster than a clock that is stationary relative to the same observer.
   - (C) It runs slower than a clock that is stationary relative to the same observer.
   - (D) It keeps proper time relative to a moving observer.

4. The period of a tick of a stationary clock is exactly 1.00 s. What is the period of a tick of the clock when it is moving at $0.800c$ relative to an observer?
   - (A) 0.811 s
   - (B) 0.958 s
   - (C) 1.15 s
   - (D) 1.67 s

5. Which curve *best* describes the relativistic momentum of an object as its speed approaches the speed of light?

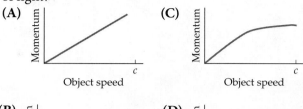

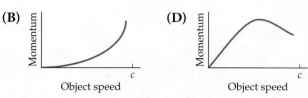

6. Which statement about relativistic time is true?
   - (A) An astronaut in a moving spaceship ages more rapidly than a person on Earth.
   - (B) Time dilation applies to all physical processes except chemical reactions.
   - (C) Observers in relative motion measure different times between the same events.
   - (D) Time is measured as being the same, regardless of relative speed.

7. A pencil has a proper length of 10 cm. How fast must the pencil move in order to have a relativistic length of 5 cm?
   - (A) c
   - (B) $0.87c$
   - (C) $0.5c$
   - (D) $0.25c$

Extended Response

8. An object has a mass of 2.5 kg. (a) Calculate the relativistic energy of the object when it moves at 100 m/s and at $0.50c$. (b) Calculate the classical kinetic energy ($KE = \frac{1}{2} mv^2$) of the object in part (a) at the speeds 100 m/s and $0.50c$. (c) Compare the results of parts (a) and (b) and comment on their significance.

> **Test-Taking Hint**
>
> In relativistic terms, length contracts and time dilates by a factor of $\sqrt{1 - v^2/c^2}$.

If You Had Difficulty With . . .

Question	1	2	3	4	5	6	7	8				
See Lesson	27.4	27.2	27.1	27.2	27.2	27.1	27.2	27.3				

Math Review

Inside

Mathematics can be used to describe many of the patterns we see in nature. Here, grains of salt collect on nodes on a vibrating metal plate.

This text is designed for students with a working knowledge of basic arithmetic, algebra, and trigonometry. Even so, it's useful to review some of the mathematical tools that are of particular importance in the study of physics. This math review covers a number of topics related to mathematical notation, numbers, equations, and vectors. Use this material as a quick math refresher course before beginning the text, or as needed when working through a particular chapter—it will help you "do the math."

I. Notation, Numbers, and Arithmetic

Mathematical notation (like $+$, $-$, $\times$, and $\div$) is used to indicate when numbers are added or subtracted, multiplied or divided, and so on. We also use specific notations, or symbols, to show that a particular value is less than or greater than another value, if it is infinite or an average value, and so on. Let's take a quick look at the mathematical notation that is used in this text.

Addition and subtraction

The first mathematical operations you are likely to encounter are addition and subtraction. The symbol for **addition**, or *summation*, is $+$. We also refer to the addition symbol as the *plus* sign. For example,

$$5 + 3 = 8$$

The line above, which represents the sum of 5 and 3, is read aloud as "five plus three equals eight."

The symbol for **subtraction** is $-$, which is also referred to as the *minus* sign. For example,

$$5 - 3 = 2$$

This line is read aloud as "five minus three equals two."

Positive and negative

What happens if you subtract 5 from 3? As you know, the result is negative 2; that is, $3 - 5 = -2$. Notice that a minus sign is written in front of a number to indicate that it is negative, as we did with -2. When a number is positive, we sometimes write it with a plus sign, as in $+2$, in order to be clear about its sign. Usually, though, we omit the plus sign when no confusion will occur.

A few additional features of positive and negative numbers are worth pointing out. **Adding a negative number to a positive number is the same as subtraction.** For example,

$$75 + (-32) = 75 - 32$$
$$= 43$$

Subtracting a negative number is the same as adding it. In other words, the negative of a negative number is positive.

$$75 - (-32) = 75 + 32$$
$$= 107$$

Double Negatives To understand why the negative of a negative is positive, consider the situation shown in **Figure 1** on the next page. Here we see a marching band musician facing in the positive direction. The musician is instructed to "about face," which results in her facing in the opposite

How are negative numbers added and subtracted?

▶ **Figure 1 The negative of a negative is positive**
A band member faces in the positive direction initially. After an "about face" the musician faces in the negative direction. A second "about face" results in the musician again facing in the positive direction.

Positive direction

direction—the negative direction. When a second command of "about face" is given, the musician turns around again and points in the positive direction. Two negatives (two "about faces") result in a positive.

A similar example is the double negative in language. Suppose a friend tells you that he is a nonswimmer—he doesn't swim. If you say that you are not a nonswimmer, it's the same as saying you *are* a swimmer. The *double negatives* (*not* and *non-*) combine to make a positive statement.

Multiplication and division

In mathematics, **multiplication** is indicated in several different ways. The multiplication symbol, $\times$, which is also called the *times* sign, is most commonly used. For example,

$$4 \times 6 = 24$$

Alternatively, a dot or parentheses can be used. For example,

$$4 \cdot 6 = 24$$
$$(4)(6) = 24$$

Thus, $4 \times 6 = 4 \cdot 6 = (4)(6) = 24$. Each of these forms is read aloud as "four times six equals twenty-four." The result of multiplying one number by another is the *product*. The product in our example is 24. Calculators use the symbol $\times$ to indicate multiplication.

To indicate **division**, the / symbol and the $\div$ symbol are commonly used. For example,

$$8 \div 2 = 4$$
$$8/2 = 4$$
$$\frac{8}{2} = 4$$

Each line above is read aloud as "eight divided by two equals four." The result obtained by dividing one number by another is the *quotient*. The quotient in the previous example is 4. Some calculators use the symbol $\div$ to indicate division, and others use /.

As an example of different ways to indicate multiplication and division, the units for force are shown in Chapter 5 to be $kg \cdot m/s^2$, where kg = kilogram, m = meter, and s = second. The meaning of this combination of units is the following:

$$kg \cdot m/s^2 = \frac{kg \times m}{s \times s}$$

Clearly, the version on the left side is much more compact and easier to write.

Other mathematical symbols

Table 1 presents other common mathematical symbols, along with their meanings in English. Though most of these symbols are no doubt familiar, it's worthwhile to make sure their interpretation is completely clear.

Two of the symbols in Table 1 might be less familiar than the others. Let's take a closer look at them.

Delta (Δ) 🔑 The symbol **Δ means "change in."** Thus, the expression Δx means the *change in x*. Pronounced "delta x," it is defined as the final value of x (which we write as x_f) minus the initial value of x (which is x_i):

$$\Delta x = x_f - x_i$$

Thus, Δx is *not* the Greek letter Δ times x. Instead Δx is a shorthand way of writing $x_f - x_i$, just like BTW is a shorthand way of writing "by the way." (BTW, river deltas get their name from the resemblance of their shape to the Greek letter Δ, as shown in **Figure 2**.)

The delta notation can be applied to any quantity—it does not have to be x. In general, we can say that

$$\Delta(\text{anything}) = (\text{anything})_f - (\text{anything})_i$$

For example, $\Delta t = t_f - t_i$ is the change in time, and $\Delta m = m_f - m_i$ is the change in mass.

Sigma (Σ) The Greek letter Σ (capital sigma) is encountered frequently in physics. 🔑 **In general, Σ is shorthand for "sum."** You can remember this by noting that both *sigma* and *sum* start with the letter s.

As an example, suppose a system has several different objects in it, each with its own mass, m. The total mass M of the system is the sum of all the individual masses, which we write as

$$M = \sum m$$

Thus, if a system has three masses, with the values $m_1 = 1$ kg (kg is short for kilogram, the scientific unit of mass), $m_2 = 4$ kg, and $m_3 = 5$ kg, the total mass of the system is

$$M = \sum m = m_1 + m_2 + m_3 = 1 \text{ kg} + 4 \text{ kg} + 5 \text{ kg} = 10 \text{ kg}$$

Notice that we indicated the three masses with subscripts 1, 2, and 3, just like we indicated initial and final values with the subscripts i and f. Using subscripts in this way is quite common.

Exponents

An **exponent** is the power to which a *base number* is raised. For example, in the expression 10^3, we say that 3 is the exponent and 10 is the base number. To evaluate 10^3 we simply multiply 10 by itself three times:

$$10^3 = 10 \times 10 \times 10 = 1000$$

Similarly, a negative exponent implies an *inverse*, as in the relation $10^{-1} = 1/10$. Thus, to evaluate the number 10^{-4}, for example, we multiply $1/10$ by itself four times:

$$10^{-4} = \frac{1}{10} \times \frac{1}{10} \times \frac{1}{10} \times \frac{1}{10} = \frac{1}{10,000} = 0.0001$$

🔑 *What do the Greek letters Δ and Σ represent?*

Table 1 Mathematical Symbols and Their Meanings

Symbol	Meaning
$=$	equal to
$\neq$	not equal to
$\sim$	approximately equal to
$\propto$	proportional to
$>$	greater than
$\geq$	greater than or equal to
$<$	less than
$\leq$	less than or equal to
$\pm$	plus or minus
x_{av} or $\bar{x}$	average value of x
Δx	change in x (pronounced "delta x")
$\lvert x \rvert$	absolute value of x
Σ	sum of (pronounced "sigma")
∞	infinity

▲ **Figure 2 Aerial view of a river delta**
A river delta, where the river enters a large body of water, generally has a triangular shape, like the Greek letter Δ. In fact, that's how it got its name.

The relations just given apply not just to powers of 10, of course, but to powers of any number at all. For example, the number N raised to the fourth power is

$$N^4 = N \times N \times N \times N$$

Similarly, N^{-3} is

$$N^{-3} = \frac{1}{N} \times \frac{1}{N} \times \frac{1}{N} = \frac{1}{N^3}$$

Combining Exponents

Exponents add when two or more numbers are multiplied together. For example,

$$N^2 N^3 = (N \times N)(N \times N \times N)$$
$$= N \times N \times N \times N \times N$$
$$= N^{2+3} = N^5$$

Exponents subtract when numbers are divided. For example,

$$\frac{N^2}{N^3} = \frac{N \times N}{N \times N \times N}$$
$$= N^{2-3} = N^{-1}$$

On the other hand, exponents multiply when a number is raised to a power:

$$(N^2)^3 = (N \times N) \times (N \times N) \times (N \times N)$$
$$= N \times N \times N \times N \times N \times N$$
$$= N^{2 \times 3} = N^6$$

The general rules obeyed by exponents are summarized in **Table 2**.

Fractional Exponents

Fractional exponents, such as $1/n$, indicate the nth root of a number. Specifically, the **square root** of x is written as

$$\sqrt{x} = x^{1/2}$$

For n greater than 2, we write the nth root in the following form:

$$\sqrt[n]{x} = x^{1/n}$$

The **cube root** of x corresponds to $n = 3$, and it is usually written as $x^{1/3}$. In general, the nth root of x is the value that gives x when multiplied by itself n times:

$$(x^{1/n})^n = x^{n/n} = x^1 = x$$

Examples of the various mathematical operations discussed so far are given in the clock shown in **Figure 3**.

Special Cases

An interesting special case occurs when the exponent is zero. To see how to evaluate x^0, consider x^n divided by x^n. Since any number divided by itself is 1, we have

$$\frac{x^n}{x^n} = 1$$

Using the rules for division with exponents, we can also say that

$$\frac{x^n}{x^n} = x^{n-n} = x^0$$

Thus, any number raised to the zero power is equal to 1: $x^0 = 1$.

Table 2 Rules for Exponents

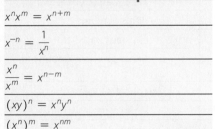

$$x^n x^m = x^{n+m}$$

$$x^{-n} = \frac{1}{x^n}$$

$$\frac{x^n}{x^m} = x^{n-m}$$

$$(xy)^n = x^n y^n$$

$$(x^n)^m = x^{nm}$$

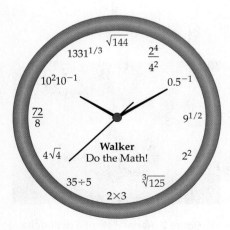

▲ **Figure 3 Time to do the math**
The numbers on this clock are examples of common mathematical operations. To tell the time, you have to do the math!

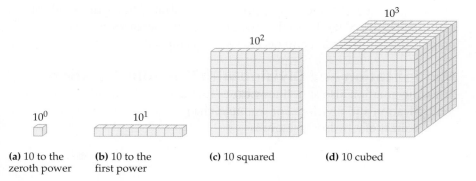

▲ **Figure 4 Various powers of a number**
(a) Raising any number to the zeroth power results in 1. (b) The first power of a number is simply the number itself. (c) When a number is squared, the result corresponds to a square arrangement of basic elements. In this case, 10 squared is represented by a 10 × 10 square grid consisting of $10^2 = 100$ small cubes. (d) When we cube 10, the result is a 10 × 10 × 10 cube consisting of $10^3 = 1000$ small cubes.

(a) 10 to the zeroth power   **(b)** 10 to the first power   **(c)** 10 squared   **(d)** 10 cubed

Let's consider some larger exponents. For example, notice that x raised to the first power, x^1, is simply x. Thus, when the exponent is 1, it is generally not written, with the understanding that $x = x^1$. When a number is raised to the second power, we say that it is **squared**; thus, x^2 is read aloud as "x squared." When the exponent is 3, we say that the number is **cubed**. For example, 6^3 is read aloud as "six cubed." Examples of 10 raised to various powers are given in **Figure 4**.

Recalling our earlier discussion of double negatives, it follows that raising -1 to the power n results in $+1$ when n is even and -1 when n is odd. For example, $(-1)^2 = (-1)(-1) = +1$, and $(-1)^3 = (-1)(-1)(-1) = -1$. To summarize,

$$(-1)^n = +1 \qquad n = 0, 2, 4, \ldots$$
$$(-1)^n = -1 \qquad n = 1, 3, 5, \ldots$$

Scientific notation

In physics the numerical value of a physical quantity can cover an enormous range, from the astronomically large to the microscopically small. For example, the mass of Earth is roughly

$$M_{\text{Earth}} = 5970000000000000000000000 \text{ kg}$$

In contrast, the mass of a hydrogen atom is approximately

$$M_{\text{hydrogen}} = 0.00000000000000000000000000167 \text{ kg}$$

Clearly, representing such large and small numbers with a long string of zeros is clumsy and can result in errors.

The preferred method for handling such numbers is to replace the zeros with the appropriate power of 10. For example, the mass of Earth can be written as follows:

$$M_{\text{Earth}} = 5.97 \times 10^{24} \text{ kg}$$

The factor 10^{24} simply means that the decimal point for the mass of Earth is 24 places to the right of its location in 5.97. Similarly, the mass of a hydrogen atom is

$$M_{\text{hydrogen}} = 1.67 \times 10^{-27} \text{ kg}$$

In this case, the correct location of the decimal point is 27 places to the left of its location in 1.67. This type of representation, using a coefficient (1.67 in this case) multiplied by a power of 10, is referred to as **scientific notation**.

🔑 *How are numbers written in scientific notation?*

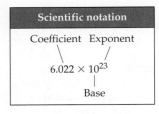

Scientific notation

Coefficient Exponent

6.022×10^{23}

Base

(a)

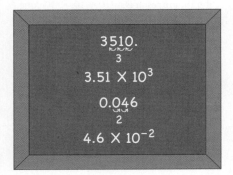

$3\underset{\underset{3}{\curvearrowleft}}{510.}$

3.51×10^3

$0.0\underset{\underset{2}{\curvearrowright}}{46}$

4.6×10^{-2}

(b)

▲ **Figure 5 Scientific notation**
(a) Avogadro's number (Chapter 12) represented in scientific notation. The basic elements of scientific notation include the coefficient, the base, and the exponent.
(b) Moving the decimal point of a number to the left increases the power of 10 (the exponent); moving the decimal point to the right decreases the power of 10.

The basic elements of scientific notation are illustrated in **Figure 5 (a)**. In addition, **Figure 5 (b)** shows that the power of 10 increases as the decimal point is moved to the left and decreases as it is moved to the right.

Multiplication and Division with Scientific Notation

Scientific notation also simplifies various mathematical operations, such as multiplication and division. For example, the product of the mass of Earth and the mass of a hydrogen atom is

$$M_{Earth} M_{hydrogen} = (5.97 \times 10^{24} \text{ kg})(1.67 \times 10^{-27} \text{ kg})$$
$$= (5.97 \times 1.67)(10^{24} \times 10^{-27}) \text{ kg}^2$$
$$= (9.97) \times (10^{24-27}) \text{ kg}^2$$
$$= 9.97 \times 10^{-3} \text{ kg}^2$$

Similarly, the mass of a hydrogen atom divided by the mass of Earth is

$$\frac{M_{hydrogen}}{M_{Earth}} = \frac{1.67 \times 10^{-27} \text{ kg}}{5.97 \times 10^{24} \text{ kg}}$$
$$= \left(\frac{1.67}{5.97}\right) \times \left(\frac{10^{-27}}{10^{24}}\right)$$
$$= 0.280 \times 10^{-27-24}$$
$$= 0.280 \times 10^{-51}$$
$$= 2.80 \times 10^{-52}$$

Notice the change in location of the decimal point in the last two expressions, and the corresponding change in the power of 10. **Typically, numbers written in scientific notation have one nonzero digit to the left of the decimal point followed by the appropriate power of 10.** Many calculators have keys that automatically display numbers in scientific notation.

Practice Problems

1. Evaluate the following expressions:

(a) $\dfrac{5^6}{5^4}$

(b) $(3^4)^2$

(c) $8^{12}8^{-10}$

(d) $59^{-1/4}$

2. Express **(a)** 1210000 and **(b)** 0.00076 in scientific notation.

3. Perform the following operations and express the answers in scientific notation:
(a) $(6.8 \times 10^{-13})(3.1 \times 10^8)$
(b) $(5.9 \times 10^7)/(8.3 \times 10^{-4})$

Answers: **1.** (a) $5^2 = 25$; (b) 6,561; (c) 64; (d) 0.36
2. (a) 1.21×10^6; (b) 7.6×10^{-4} **3.** (a) 2.1×10^{-4}; (b) 7.1×10^{10}

II. Logarithmic and Exponential Functions

Logarithmic functions are useful for determining the power to which a number is raised. We use logarithmic functions when we consider the loudness of sound in Chapter 14. Exponential functions are useful in describing the decay of radioactive elements, which we do in Chapter 26.

Logarithms

A general method for calculating the exponent of a number is provided by the **logarithm**. For example, suppose x is equal to 10 raised to the power n:

$$x = 10^n$$

The exponent, n, is equal to the logarithm (log) of x:

$$n = \log x$$

Common Logarithms   The notation *log* is known as the *common logarithm*, and it refers specifically to base 10. As an example, suppose that $x = 1205 = 10^n$. To find the exponent for this value of x, we use the log button on a calculator. The result is

$$n = \log 1205 = 3.081$$

Thus, 10 raised to the power 3.081 gives 1205.

In Section I we saw that any number raised to the zeroth power is equal to 1. It follows that the log of 1 is zero:

$$\log 1 = 0$$

🔑 **The log of a number between 0 and 1 is negative, and the log of a number greater than 1 is positive.**

Natural Logarithms   Another common base for calculating exponents is $e = 2.718\ldots$. To represent $x = 1205$ with this base, we write

$$x = 1205 = e^n$$

The logarithm to the base e is known as the *natural logarithm*, and it is represented by the notation *ln*. Using the ln button on a calculator, you can verify that

$$n = \ln 1205 = 7.094$$

Thus, e raised to the power 7.094 gives 1205. The connection between the common and natural logarithms is as follows:

$$\ln x = 2.3026 \log x$$

In the example just given, we have

$$\ln 1205 = 7.094 = 2.3026 \log 1205 = 2.3026 \times 3.081$$

🔑 *When is the log of a number positive, and when it is negative?*

▶ Figure 6 The exponential function
The exponential function, $y = e^x$, becomes large with increasing x. The inverse of the exponential function, $y = 1/e^x = e^{-x}$, approaches zero as x increases.

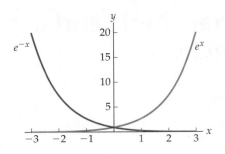

Table 3   Rules for Logarithms
$\ln(xy) = \ln x + \ln y$
$\ln\left(\dfrac{x}{y}\right) = \ln x - \ln y$
$\ln x^n = n \ln x$

Combining Logarithms The basic rules obeyed by logarithms follow directly from the rules given for combining exponents. **Table 3** summarizes several important rules. Though these rules are stated in terms of natural logarithms, they are satisfied by logarithms with any base.

As an example of the above results, consider $\log(1/2)$. Using a calculator, we find $\log(1/2) = -0.301$. The result is negative, as expected for a number between 0 and 1. We can also write this result in the following way:

$$\log\left(\frac{1}{2}\right) = \log 1 - \log 2$$
$$= 0 - \log 2$$
$$= -\log 2$$

Thus, not only is $\log(1/2)$ negative, it is the negative of $\log 2$. Noting that $\log 2 = 0.301$, we see that the results are consistent.

The exponential function

 How does the exponential function, e^x, change in value as x increases?

Raising e to the power x results in $y = e^x$, which is referred to as the **exponential function**. This function is equal to 1 when $x = 0$. For $x < 0$ the exponential function is less than 1, and for $x > 0$ it is greater than 1. A plot of $y = e^x$ is shown in **Figure 6**. **Notice that the exponential function e^x increases rapidly with increasing x.** Calculators usually have a button labeled e^x to represent e raised to the power x.

The inverse of the exponential function is $y = 1/e^x = e^{-x}$. This function is greater than 1 for $x < 0$, equal to 1 for $x = 0$, and less than 1 for $x > 0$. Figure 6 also shows a plot of e^{-x}.

The exponential function plays an important role in many real-world situations. Examples are given in **Figure 7**.

▼ Figure 7 The exponential function in nature
The exponential function has many applications in the natural world, in everything from the growth of the human population to the shape of Romanesco broccoli and seashells.

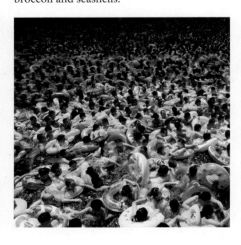

Combining Exponential Functions The rules for combining exponential functions are similar to those for combining exponents. **Table 4** summarizes several important rules.

Table 4 Rules for Exponential Functions

$e^{nx}e^{mx} = e^{(n+m)x}$
$e^{-nx} = \dfrac{1}{e^{nx}}$
$\dfrac{e^{nx}}{e^{mx}} = e^{(n-m)x}$
$(e^{nx})^m = e^{nmx}$

Practice Problems

4. Find the exponent n for the following expressions:

(a) $10^n = 12$ **(b)** $e^n = 12$ **(c)** $e^{-n} = 0.032$ **(d)** $10^n = 0.25$

5. Simplify the following expressions:

(a) $\log(x^2 y^{-3})$ **(b)** $\ln\left(\dfrac{\sqrt{x}}{y^2}\right)$ **(c)** $e^{3x}e^{-x}$ **(d)** $\dfrac{e^{2x}}{(e^{3x})^4}$

Answers: **4.** (a) 1.08; (b) 2.48; (c) 3.44; (d) -0.602
5. (a) $2\log x - 3\log y$; (b) $\frac{1}{2}\ln x - 2\ln y$; (c) e^{2x}; (d) e^{-10x}

III. Algebra and Equations

Solving simple equations

The **equations** used in this textbook are a shorthand way to show that two different expressions are equal. For example, you might want to show that a number y is equal to three times the number x plus seven. Rather than writing this out in English, we simplify things with an equation that expresses the same relationship:

$$y = 3x + 7$$

Now, suppose you would like to solve this equation for x; that is, you want to write the equation $x = $ (something). What is the "something"? 🔑 The basic idea in solving any equation is very simple: **Any operation that is done to one side of an equation must be done to the other side of the equation as well.** That's it. It's a simple matter of being fair to the two sides of the equation.

For the equation $y = 3x + 7$, we want to isolate x by itself on one side. Let's start by subtracting 7 from both sides:

$$y - 7 = 3x + 7 - 7$$
$$y - 7 = 3x$$

🔑 *What basic approach is used to solve any equation?*

Next, we divide both sides by 3:

$$\frac{y - 7}{3} = \frac{3x}{3}$$

$$\frac{y - 7}{3} = x$$

So here's our answer: $x = (y - 7)/3$. For example, if $y = 1$ it follows that $x = -2$, and if $y = 10$ then $x = 1$. First-order equations (those with no x^2 or higher-power terms) can always be solved with a simple rearrangement of terms.

The quadratic equation

> 🔑 What condition does a solution to a quadratic equation satisfy?

Some physics equations have an unknown raised to the second power. The result is the well-known **quadratic equation.** The general form of the quadratic equation is

$$ax^2 + bx + c = 0$$

In this equation, a, b, and c are constants and x is a variable. 🔑 **When we refer to the solutions of the quadratic equation, we mean the values of x that result in $(ax^2 + bx + c)$ being equal to zero.** These values are given by the following expression:

Solutions to the Quadratic Equation

$$x = \frac{-b \pm \sqrt{b^2 - 4ac}}{2a}$$

Multiple Solutions In general, there are two solutions to a quadratic equation. One solution corresponds to using the plus sign in front of the square root, the other to using the minus sign in front of the square root. In the special case where the quantity under the square root vanishes, there will be only a single solution. If the quantity under the square root is negative, the result for x is not physical, because numbers that represent physical quantities can't be negative when squared. A negative value under the square root usually means a mistake has been made in the calculation.

Applying the Equation To illustrate the use of the quadratic equation and its solutions, consider a standard one-dimensional kinematics problem, such as you might encounter in Chapter 3:

> A ball is thrown straight upward with an initial speed of 11 m/s. How much time does it take for the ball to first reach a height of 4.5 m above its launch point?

The first step in solving this problem is to write the equation giving the height of the ball, y, as a function of time. Referring to Chapter 3, we have

$$y = y_i + v_i t - \frac{1}{2}gt^2$$

To make this look more like a quadratic equation, let's move all the terms to the left-hand side, which yields

$$\frac{1}{2}gt^2 - v_i t + y - y_i = 0$$

Though it may not look like it at first, the equation above has the same general form as the quadratic equation introduced earlier: $ax^2 + bx + c = 0$. Seeing the connection simply requires identifying the values for a, b, c, and x. In this case these values are as follows: $a = \frac{1}{2}g$, $b = -v_i$, $c = y - y_i$, and $x = t$. The desired result comes from substituting a, b, c, and x into the solutions equation:

$$x = \frac{-b \pm \sqrt{b^2 - 4ac}}{2a}$$

$$t = \frac{v_i \pm \sqrt{v_i^2 - 2g(y - y_i)}}{g}$$

The final step is to insert the appropriate numerical values: $g = 9.81 \text{ m/s}^2$, $v_i = 11 \text{ m/s}$, and $y - y_i = 4.5 \text{ m}$. Straightforward calculation then gives $t = 0.54 \text{ s}$ and $t = 1.7 \text{ s}$. Therefore, the time it takes the ball to first reach a height of 4.5 m is 0.54 s; the second solution (1.7 s) is the time when the ball is again at a height of 4.5 m, only this time on its way down.

Two equations and two unknowns

In some problems, two unknown quantities are determined by two interlinked equations. In such cases it often seems at first that you have not been given enough information to obtain a solution. By patiently writing out what is known, however, you can generally use straightforward algebra to solve the problem. As an example, consider the following problem:

> A father and daughter share the same birthday. On one birthday the father announces to his daughter, "Today I am four times older than you, but in five years I will be only three times older." How old are the father and daughter now?

You might be able to solve this problem by guessing, but there's an easier way. Let's approach the problem systematically. First, we write what is given in the form of equations. Letting F be the father's age in years and D the daughter's age in years, we know that on this birthday

$$F = 4D$$

We also know that in five years the father's age will be $F + 5$, the daughter's age will be $D + 5$, and the father will be three times as old as the daughter. We write this relationship as

$$F + 5 = 3(D + 5)$$

Multiplying through with the factor 3 on the right-hand side gives

$$F + 5 = 3D + 15$$

After subtracting 5 from each side of the equation, we have

$$F = 3D + 10$$

These two equations, $F = 4D$ and $F = 3D + 10$, express everything we know from the problem statement.

Eliminating a Variable **The simplest way to solve a situation described by two equations and two unknowns is to substitute from one equation into the other.** Here's how it works. The first equation is $F = 4D$. Substitute this result for F into the second equation. This eliminates F and gives an equation with only one unknown, D.

$$4D = 3D + 10$$

Solving for D, we find

$$4D - 3D = 10$$
$$D = 10$$

How can a system of two equations with two unknowns be solved?

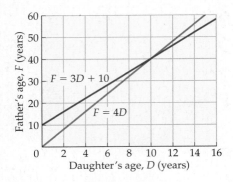

▲ Figure 8 Solving two equations with two unknowns
The solution of the two equations $F = 4D$ (red line) and $F = 3D + 10$ (blue line) is given by the point where the two straight lines intersect.

Now, we substitute this value of D back into the first equation to find F:

$$F = 4D$$

$$= 4(10) = 40$$

Currently, the father is 40 years old and the daughter is 10 years old. In five years, the father will be 45 and the daughter will be 15. As desired, the ages are related by a factor of 4 now and by a factor of 3 in five years.

Graphical Solution

Another way to solve a problem involving two equations is to plot each equation and look for their intersection point. The intersection corresponds to the solution.

As an example, in the father-daughter age problem, we want to find the values of F and D that satisfy the equations $F = 4D$ and $F = 3D + 10$. In a graphical solution, we plot both equations on the same graph. This is shown in **Figure 8**. Each of the straight lines in Figure 8 represents one of the equations for this problem. It follows that the intersection point satisfies *both* equations at the same time. As you can see, the intersection point is indeed at $F = 40$ and $D = 10$, as expected.

Practice Problems

6. Solve the following equations for x:
(a) $y = 6x - 13$ (c) $x^2 + x - 1 = 0$
(b) $y = x/3 + 32$ (d) $-5x^2 + 2x + 3 = 0$

7. A ball is thrown straight upward with an initial speed of 7.4 m/s at the time $t = 0$.
(a) At what time does the ball first reach a height of 1.5 m above its launch point?
(b) At what time is the ball 1.5 m above the release point on the way down?

8. In 1990, Jackson was four times older than Abby. In 2010, Jackson was twice as old as Abby. In what years were Jackson and Abby born?

9. It takes you 1.5 h to drive with a speed v from home to a nearby town, a distance d away. Later, on the way back, the traffic is lighter, and you are able to increase your speed by 28 km/h. With this higher speed, you get home in just 1.0 h. Find your initial speed, v, and the distance to the town, d.

Answers: **6.** (a) $x = (y + 13)/6$; (b) $x = 3(y - 32)$; (c) $-1.62, 0.62$; (d) $-0.6, 1.0$ **7.** (a) $t = 0.241$ s; (b) $t = 1.27$ s **8.** Jackson in 1950, Abby in 1980 **9.** $v = 56$ km/h, $d = 84$ km

IV. Mathematical Relationships and Graphs

Physical quantities are related to one another in a number of different ways. In this section we see how some of the more common relationships can be represented graphically.

Linear relationship

The most straightforward relationship between two physical quantities is *linear* dependence. A simple example is the connection between position and time for an object moving with constant velocity. The relationship in this case is given by the following equation (Chapter 2):

$$x = vt + x_i$$

In this equation, x is position, v is velocity, t is time, and x_i is the initial position. A graph of this equation is a straight line, as indicated in **Figure 9**, with a slope equal to v and an intercept equal to x_i. Notice that the slope of the straight line can be positive (upward sloping), negative (downward sloping), or zero (horizontal). In general, a straight-line graph indicates a **linear relationship** between the plotted quantities.

Calculating Slope 🔑 A straight line has a constant **slope**, which is calculated by dividing the *rise* by the *run*. Recall that the *rise* is the change in the vertical direction, and the *run* is the change in the horizontal direction.

🔑 *What are the key characteristics of a graph that displays a linear relationship?*

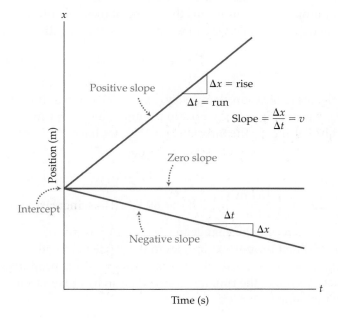

◄ **Figure 9 Linear relationship**
When two physical quantities have a linear relationship, plotting their values results in a straight line.

(a) A warning sign indicating a steep uphill
road ahead with a 6% grade. (b) A 6% grade
means a rise that is 6 percent of the run.

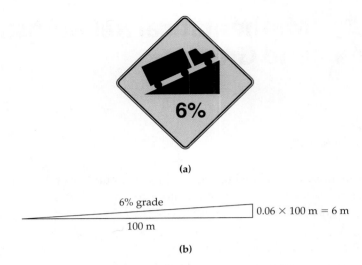

In the case shown in Figure 9, the rise is Δx and the run is Δt. Thus, the slope is

$$\text{slope} = \frac{\text{rise}}{\text{run}} = \frac{\Delta x}{\Delta t} = v$$

The slope is equal to the velocity v because $\Delta x / \Delta t$ is distance per time, like kilometers per hour. For example, a car that travels 80 km in an hour has a velocity of 80 km/h.

An everyday example of rise over run is the steepness, or grade, of a road. You might have seen signs like the one shown in **Figure 10 (a)**, which indicates a 6% grade. What does this mean, exactly? Well, a 6% grade has a rise that is 6 percent, or 0.06, of the run. Thus, on a 6% grade a run of 100 m corresponds to a rise of 0.06 × 100 m = 6 m, as indicated in **Figure 10 (b)**.

Direct Proportion 🔑 **In a linear relationship, the change in one quantity is in direct proportion to the change in the other quantity.** For example, the change in position, Δx, and the change in time, Δt, in Figure 9 are in direct proportion. This means that if one of these quantities doubles, so does the other. Symbolically, we can write this direct proportion as follows:

$$\frac{\Delta x_1}{\Delta t_1} = \frac{\Delta x_2}{\Delta t_2}$$

For example, if the position changes by $\Delta x_1 = 40$ km in a time of $\Delta t_1 = 1$ h, how much does the position change, Δx_2, in a time of $\Delta t_2 = 1.5$ h? Setting up the direct proportion, we have

$$\frac{40 \text{ km}}{1 \text{ h}} = \frac{\Delta x_2}{1.5 \text{ h}}$$

This yields $\Delta x_2 = \left(\frac{40 \text{ km}}{1 \text{ h}} \right)(1.5 \text{ h}) = 60$ km. Thus, increasing the time by a factor of 1.5 increases the change in position by a factor of 1.5.

A similar direct proportion applies to the grade of a road. For example, on a 6% grade, as in Figure 10, the rise is always in direct proportion to the run—in fact, it is 6% of the run. Increasing the run by a given factor increases the rise by the same factor.

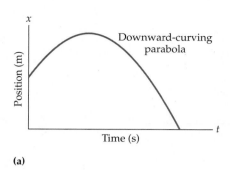

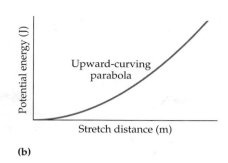

◀ **Figure 11 Quadratic relationship**
(a) A graph of position versus time for an object in free fall is a downward-curving parabola. (b) A graph of a spring's potential energy versus stretch distance is an upward-curving parabola.

(a)   (b)

Quadratic relationship

A **quadratic relationship** occurs when one quantity is related to the square of another quantity. A good example is the position of an object in free fall moving with constant acceleration, which can be described by the following equation (Chapter 3):

$$x = x_i + vt - \frac{1}{2}gt^2$$

This equation shows how position, x, is related to initial position, x_i, initial velocity, v_i, time, t, and the acceleration due to gravity, g. Notice that position depends on time squared, t^2, because of the term $-\frac{1}{2}gt^2$. This term produces a parabolic shape with a downward curvature (due to the minus sign), as can be seen in **Figure 11 (a)**.

Another example of a quadratic relationship in physics is the potential energy of a spring (Chapter 6), which is given by

$$PE = \frac{1}{2}kx^2$$

In this expression, k is the spring constant and x is the stretch distance. A graph of PE versus x is shown in **Figure 11 (b)**, and you can see that it has an upward-curving parabolic shape. **A quadratic relationship always results in a parabolic shape with an upward or downward curvature, depending on the sign of the squared term.**

What are the key characteristics of a graph that displays a quadratic relationship?

Inverse and inverse-square relationships

Sometimes an increase in one physical quantity results in a decrease of another quantity. This is referred to as an **inverse relationship**. For example, if the pressure, P, of a gas at constant temperature is increased, its volume, V, decreases (Chapter 12). This can be written mathematically as follows:

$$P = \frac{constant}{V}$$

A graph of pressure versus volume is given in **Figure 12**. **Plotting an inverse relationship produces a curve that is a hyperbola, and it has the property that as one quantity gets large, the other quantity gets small.** For example, if the volume is doubled, the pressure is halved.

What are the key characteristics of a graph that displays an inverse relationship?

▶ **Figure 12 Inverse dependence**
In this graph, pressure is inversely proportional to volume. The result is a curve with the shape of a hyperbola.

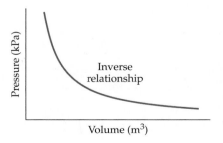

An inverse-square relationship is similar to an inverse relationship, but the graph in this case is more sharply curved.

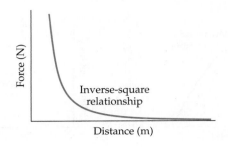

In some cases, one physical quantity depends inversely on the *square* of the other quantity. This results in an **inverse-square relationship**. An example is the dependence of the force of gravity, F, on distance, r (Chapter 9). The equation describing this relationship is

$$F = \frac{\text{constant}}{r^2}$$

Notice that the force decreases rapidly with an increase in distance. In particular, if the distance is doubled, the force decreases by a factor of $2^2 = 4$. **Figure 13** shows a graph of F versus r.

Practice Problems

10. The position of a train as a function of time is given by

$$x = 45 \text{ km} + (56 \text{ km/h})t$$

What are (**a**) the velocity and (**b**) the initial position of the train?

11. A car moving with constant speed covers 2 km in 180 s. What distance does the car cover in 450 s?

12. What is the shape of each of the following functions when plotted on a y-versus-x graph?
(**a**) $y = -2x^2$ (**c**) $y = 17x$
(**b**) $y = 5/x^2$ (**d**) $y = 23/x$

13. Identify the type of dependence for each of the following cases where a physical quantity y depends on another physical quantity x:
(**a**) Doubling x reduces y by a factor of 4.
(**b**) Tripling x increases y by a factor of 9.
(**c**) Tripling x triples y.
(**d**) Doubling x halves y.

Answers: **10.** (a) 56 km/h; (b) 45 km **11.** 5 km **12.** (a) parabola, downward curving; (b) inverse square; (c) linear; (d) hyperbola **13.** (a) inverse square; (b) quadratic; (c) linear; (d) inverse

V. Lengths, Areas, Volumes, and Geometry

Lengths, areas, and volumes

Physics problems often involve calculating the length, area, or volume of an object. **Table 5** presents results for common shapes.

Geometry

Basic geometry is helpful in any problem involving angles. Some of the most useful results of geometry are shown in **Figure 14** on the next page. Notice that the sum of angles created when one or more lines touch a long straight line is 180°, as indicated in Figure 14 (c) and (d).

Table 5 Finding Lengths, Areas, and Volumes of Common Shapes

LENGTHS

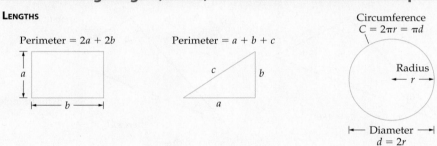

AREAS

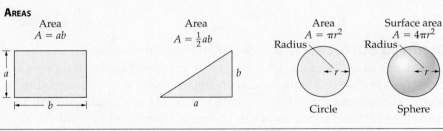

VOLUMES

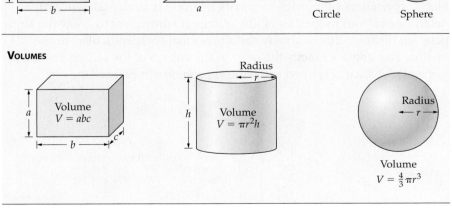

Figure 14 Straight lines and angles
(a) When straight lines cross, the angles on either side of the crossing point are equal. (b) A line crossing a pair of parallel lines makes the same angle with respect to each line. (c) The sum of two angles created by a line touching a long straight line is 180°. (d) The sum of three angles created by two lines touching a long straight line is 180°.

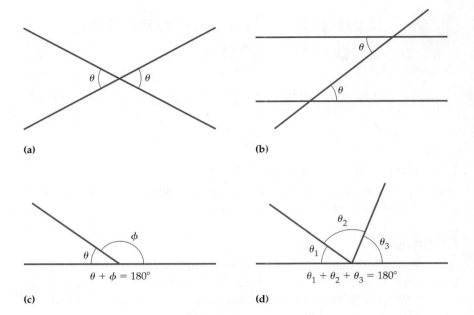

(a)

(b)

$\theta + \phi = 180°$

(c)

$\theta_1 + \theta_2 + \theta_3 = 180°$

(d)

Figure 15 (a) illustrates the most useful geometrical property of triangles—namely, that the sum of angles inside a triangle is always 180°. This property is verified in **Figure 15 (b)**.

Figure 15 Angles in a triangle
(a) The sum of the angles inside a triangle is always equal to 180°. (b) To verify the statement in part (a), add all the changes in direction as you circumnavigate a triangle. The sum of these changes in direction must equal 360°, since you return to your original direction after completing the round-trip. Setting the sum of the changes of direction equal to 360° yields $\theta_1 + \theta_2 + \theta_3 = 180°$.

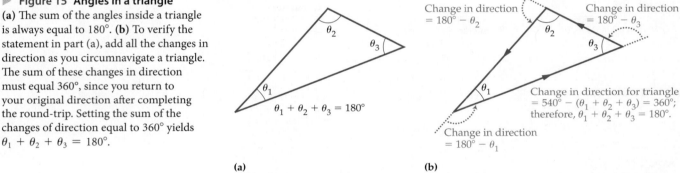

(a)

(b)

Inclined Surfaces Many physics problems involve an object on an angled, or *inclined*, surface (Chapter 5). For example, **Figure 16** shows a block on a surface that is inclined at an angle θ above the horizontal. The block experiences a force due to gravity that points straight downward. To determine various properties of this system, it's important to know the angle between the force due to gravity and the normal (perpendicular) to the incline. This angle is exactly the same as the angle θ of the incline above the horizontal, which is verified with simple geometry in Figure 16.

Figure 16 Angles on an inclined surface
The angle between the force due to gravity and the normal to an inclined surface is the same as the angle of the incline above the horizontal.

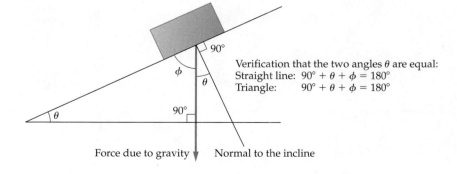

Verification that the two angles θ are equal:
Straight line: $90° + \theta + \phi = 180°$
Triangle: $90° + \theta + \phi = 180°$

Force due to gravity   Normal to the incline

14. Find (**a**) the circumference and (**b**) the area of a circle of radius $r = 1.1$ m.

15. Find (**a**) the surface area and (**b**) the volume of a sphere of radius $r = 2.50$ m.

16. Two angles of a triangle are 90° and 55°. What is the third angle?

Answers: **14.** (a) 6.9 m; (b) 3.8 m²   **15.** (a) 78.5 m²; (b) 65.4 m³   **16.** 35°

VI. Trigonometry and Vectors

Degrees and radians

We all know the definition of a degree; there are 360 degrees in a circle. The definition of a **radian** is somewhat less well known; there are 2π radians in a circle. An equivalent definition of the radian is the following:

🔑 **A radian is the angle for which the arc length is equal to the radius.**

Consider a pie with a piece cut out, as shown in **Figure 17 (a)**. Notice that a piece of pie has three sides—two radial lines from the center and an arc of crust. If a piece of pie is cut with an angle of 1 radian, all three sides are equal in length, as shown in **Figure 17 (b)**. Since a radian is about 57.3°, this amounts to a fairly good-sized piece of pie. Thus, if you want a hearty helping of pie, just tell the server, "One radian, please."

🔑 *What is a radian?*

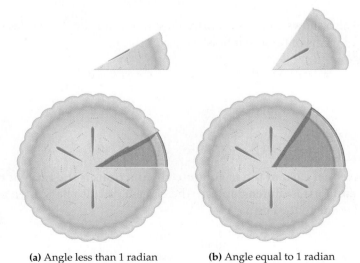

(a) Angle less than 1 radian   **(b)** Angle equal to 1 radian

◀ Figure 17 **The definition of a radian**
(a) This piece of pie is cut with an angle less than a radian. Thus, the two radial sides (coming out from the center) are longer than the arc of crust. (b) The angle for this piece of pie is equal to 1 radian (about 57.3°). Thus, all three sides of the piece are of equal length.

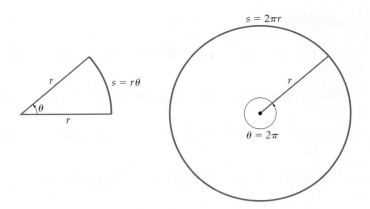

▶ **Figure 18 The length of an arc**
The arc created by rotating a radius r through an angle θ has a length $s = r\theta$. If the radius is rotated through a full circle, the angle is $\theta = 2\pi$, and the arc length is the circumference of a circle, $s = 2\pi r$.

🔑 **How are radians used to determine arc lengths?**

Radians are particularly convenient when we are interested in the length of an arc. **Figure 18** shows a circular arc corresponding to a radius r and an angle θ. 🔑 **If an angle θ is measured in radians, then the length of the arc, s, is the product of the radius, r, and the angle θ.** In equation form:

$$s = r\theta$$

This simple relation is *not valid* when θ is measured in degrees. For a full circle, for which $\theta = 2\pi$, the length of the arc (which is the circumference of the circle) is $2\pi r$, as expected.

Trig functions and the Pythagorean theorem

🔑 **What dimensions are associated with each of the trigonometric functions?**

Next, we consider some of the more important and frequently used results from trigonometry. We start with a right triangle, as shown in **Figure 19**. The basic **trigonometric functions**, $\sin \theta$ (read as "sine theta"), $\cos \theta$ ("cosine theta"), and $\tan \theta$ ("tangent theta") are defined as follows:

$$\cos \theta = \frac{\text{adjacent side}}{\text{hypotenuse}}$$

$$\sin \theta = \frac{\text{opposite side}}{\text{hypotenuse}}$$

$$\tan \theta = \frac{\text{opposite side}}{\text{adjacent side}}$$

Figure 19 also shows common shorthand symbols for the three sides of a triangle. In terms of these symbols, we have

$$\cos \theta = \frac{x}{r} \quad \text{or} \quad x = r \cos \theta$$

$$\sin \theta = \frac{y}{r} \quad \text{or} \quad y = r \sin \theta$$

$$\tan \theta = \frac{y}{x} \quad \text{or} \quad y = x \tan \theta$$

🔑 **Notice that each of the trigonometric functions is the ratio of two lengths, and thus is dimensionless.** The tangent can also be written as the ratio of the sine to the cosine:

$$\tan \theta = \frac{\sin \theta}{\cos \theta}$$

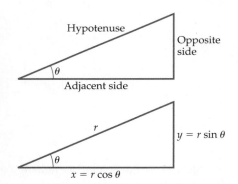

▲ **Figure 19 Basic trigonometric functions**
The trigonometric functions $\sin \theta$ and $\cos \theta$ are useful in relating the lengths of the short sides of a right triangle to the length of its hypotenuse.

The trigonometric functions cot θ ("cotangent theta"), sec θ ("secant theta"), and csc θ ("cosecant theta") are not used in this textbook. It's a good idea to know how to use them, however, since you may encounter them in other contexts.

$$\cot \theta = \frac{1}{\tan \theta}$$

$$\sec \theta = \frac{1}{\cos \theta}$$

$$\csc \theta = \frac{1}{\sin \theta}$$

According to the **Pythagorean theorem**, the lengths of the sides of a right triangle (see Figure 19) are related. The relationship is as follows:

$$x^2 + y^2 = r^2$$

Dividing by r^2 yields

$$\frac{x^2}{r^2} + \frac{y^2}{r^2} = 1$$

This can be rewritten in terms of the sine and the cosine to give

$$\sin^2 \theta + \cos^2 \theta = 1$$

Figure 19 also shows how the sine and the cosine are used in a typical calculation. In many cases, the hypotenuse of a triangle, r, and an angle, θ, are given. The short sides of the triangle are then calculated with $x = r \cos \theta$ for the adjacent side and $y = r \sin \theta$ for the opposite side. The following Guided Example applies this type of calculation to the case of an inclined roadway.

GUIDED Example 1 | Highway to Heaven   Using Trig Functions

You are driving on a long straight road that slopes uphill at an angle of 6.4° above the horizontal. At one point you notice a sign that reads "Elevation 450 m." What is your elevation after you have driven another 2.0 km?

Picture the Problem

From our sketch we see that the car is moving along the hypotenuse of a right triangle. The length of the hypotenuse is 2.0 km.

Strategy

The elevation gain is the vertical side of the triangle, y. We find y by multiplying the hypotenuse r by the sine of θ; that is, $y = r \sin \theta$.

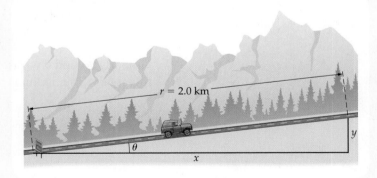

Known	Unknown
$r = 2.0$ km	$y = ?$
$\theta = 6.4°$	new elevation $=$ old elevation $+ y$

Solution

1 Calculate the elevation gain, y:

$$y = r \sin \theta$$
$$= (2.0 \text{ km}) \sin 6.4°$$
$$= (2.0 \text{ km})(0.11)$$
$$= 0.22 \text{ km}$$
$$= (0.22 \text{ km})\left(\frac{1000 \text{ m}}{1 \text{ km}}\right) = 220 \text{ m}$$

2 Add the elevation gain to the original elevation to obtain the new elevation:

$$\text{new elevation} = 450 \text{ m} + 220 \text{ m}$$
$$= \boxed{670 \text{ m}}$$

Insight

As surprising as it may seem, the horizontal distance covered by the car is $x = r \cos \theta = 1990$ m, only 10 m less than the total distance covered by the car. At the same time the car rises a distance of 220 m.

Practice Problem

17. **Follow-up** How far along the road from the first sign in Guided Example 1 should the road crew put another sign reading "Elevation 750 m"?

Answer: **17.** 2200 m

Finding the Hypotenuse and the Angle

In some problems the sides of a triangle (x and y) are given and it is desired to find the corresponding hypotenuse, r, and angle, θ. For example, suppose that $x = 5.0$ m and $y = 2.0$ m. Using the Pythagorean theorem, the hypotenuse is given by the following:

$$r = \sqrt{x^2 + y^2} = \sqrt{(5.0 \text{ m})^2 + (2.0 \text{ m})^2} = 5.4 \text{ m}$$

Similarly, to find θ we use the definition of tangent: $\tan \theta = y/x$. The inverse of this relation is

$$\theta = \tan^{-1}\left(\frac{y}{x}\right)$$
$$= \tan^{-1}\left(\frac{2.0 \text{ m}}{5.0 \text{ m}}\right)$$
$$= \tan^{-1}(0.40)$$
$$= 22°$$

Notice that the expression $\tan^{-1}$ is the *inverse tangent function*—it does not mean 1 divided by the tangent. Expressed in words, $\tan^{-1} 0.40$ means "the angle whose tangent is equal to 0.40." Your calculator should have a button on it labeled $\tan^{-1}$. If you enter 0.40 and then press $\tan^{-1}$, you will get 22° (rounded off to two significant figures). Inverse sine and cosine functions work in the same way.

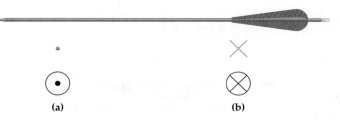

(a)   (b)

▲ **Figure 20 Vectors pointing out of and into the page**
(a) A vector pointing out of the page is represented by a dot in a circle. The dot stands for the tip of the vector's arrow. (b) A vector pointing into the page is represented by an X in a circle. The X looks like the tail feathers of the vector's arrow.

Vector notation

When we draw a **vector** to represent a physical quantity, we typically use an arrow whose length is proportional to the magnitude of the quantity and whose direction is the direction of the quantity. (This and other aspects of vector notation are discussed in Chapter 4.) A slight problem arises, however, when a physical quantity points into or out of the page. In such a case we use the conventions illustrated in **Figure 20**.

Figure 20 (a) shows a vector pointing out of the page. Notice that we see only the tip. We also show the corresponding convention, which is a dot inside a circle. The dot represents the point of the vector's arrow coming out of the page toward you.

A similar convention is employed in Figure 20 (b) for a vector pointing into the page. In this case the arrow moves directly away from you, giving a view of its tail feathers. The feathers are arranged in an X-shaped pattern if you look at the arrow from that end, so we represent the vector as an X inside a circle.

These conventions are used in Chapter 22 to represent the magnetic field vector.

Practice Problems

18. Convert the following angles to radians:
(**a**) 180° (**b**) 90° (**c**) 270° (**d**) 40.0°

19. A right triangle like the one in Figure 19 has a hypotenuse $r = 3.7$ m and an angle $\theta = 28°$. What are the lengths of (**a**) the x side and (**b**) the y side of the triangle?

20. A right triangle like the one in Figure 19 has an x side of length 1.50 m and a y side of length 0.660 m. What are (**a**) the angle θ and (**b**) the hypotenuse of the triangle?

Answers: **18.** (a) π; (b) $\pi/2$; (c) $3\pi/2$; (d) 0.698 **19.** (a) 3.3 m; (b) 1.7 m
20. (a) 23.7°; (b) 1.64 m

Math Review Assessment

ANSWERS TO SELECTED ODD-NUMBERED PROBLEMS APPEAR IN APPENDIX A.

Section by Section

I. Notation, Numbers, and Arithmetic

21. **Rank** Evaluate the following expressions and rank them in order of increasing numerical value.
    (a) $520 - (-23)$
    (b) $450 + (-55)$
    (c) $7^6 5^{-2}$
    (d) $(3^4)^{1/5}$

22. Which of the following numbers are positive?
    (a) $(-1)^4$
    (b) $(-6)^3$
    (c) $(-25)(-4)^3$
    (d) $(-12)/(-1)$

23. **Rank** Four numerical values are expressed in (a) through (d). Rank the expressions in order of increasing numerical value.
    (a) 7300000
    (b) 7.3×10^5
    (c) 0.73×10^8
    (d) 73×10^3

24. Perform the following operations and express the answers in scientific notation.
    (a) $(-5.2 \times 10^{-11})(343)(9.3 \times 10^4)$
    (b) $(-11.2)(-4.7 \times 10^5)/(6.8 \times 10^{-3})$

II. Logarithmic and Exponential Functions

25. **Triple Choice** (a) Is $\log 0.2$ greater than, less than, or equal to $-\log 5$? (b) Is $-\log 0.1$ greater than, less than, or equal to $\log 5$?

26. Write the following expressions in terms of $\log x$ and $\log y$.
    (a) $\log\left(\dfrac{x^5 y^2}{\sqrt{y}}\right)$ (b) $\log(x^3 y^{3.1}) + \log(x^{-2}/y^{5.4})$

27. **Triple Choice** (a) Is the exponent in the expression $10^n = 9$ greater than, less than, or equal to the exponent in $e^n = 9$? (b) Is the exponent in the expression $10^n = 0.3$ greater than, less than, or equal to the exponent in $e^n = 0.3$?

28. Find the exponent n for each of the following expressions.
    (a) $e^{-n} = 0.12$ (c) $e^n = 76$
    (b) $10^n = 25$ (d) $10^{-n} = 0.45$

III. Algebra and Equations

29. Solve the following equations for x.
    (a) $y = 4.5x + 12$
    (b) $y = \dfrac{3x}{8} - 26$
    (c) $2x^2 + 3x - 4 = 0$
    (d) $-x^2 + 5x + 7 = 6$

30. A golfer scores 68, 72, 71, and 65 on four rounds at a local golf course. If the golfer's total score for the four rounds is 4 below par, what is par for the course?

31. A ball thrown upward has the position x at the time t given by the following expression:
    $$x = -\tfrac{1}{2}(9.81 \text{ m/s}^2)t^2 + (6.00 \text{ m/s})t + (4.00 \text{ m})$$
    At what times is the position of the ball given by $x = 5.00$ m?

32. In 1995, Allan was three times older than Maria. In 2005, Allan was only twice as old as Maria. In what years were Allan and Maria born?

IV. Mathematical Relationships

33. A car moves with constant velocity. Its position as a function of time is given by
    $$x = 13 \text{ km} + (42 \text{ km/h})t$$
    What are (a) the velocity and (b) the initial position of the car? (c) Is the relationship between x and t linear, parabolic, or inverse? Explain.

34. A boat moves with constant velocity. Its position as a function of time is given by
    $$x = 68 \text{ km} + (-13 \text{ km/h})t$$
    What are (a) the velocity and (b) the initial position of the boat? (c) Is the relationship between x and t linear, parabolic, or inverse? Explain.

35. An airplane covers a distance of 840 m in 3.0 s. How much time is required for the airplane to cover a distance of 1500 m?

36. What is the shape of the following functions when plotted on a y-versus-x graph?
    (a) $y = 21/x^2$
    (b) $y = 52x^2$
    (c) $y = -15x + 3$
    (d) $y = 7/x$

V. Lengths, Areas, Volumes, and Geometry

37. Find **(a)** the radius and **(b)** the area of a circle whose circumference is 1.8 m.

38. The surface area of a sphere is 2.5 m². What is the volume of the sphere?

39. Find **(a)** the surface area and **(b)** the volume of a cylinder whose radius is 0.72 m and whose height is 0.95 m. **(c)** If the radius of the cylinder is doubled, but the height remains the same, by what factor does the volume increase? **(d)** If the height of the cylinder is doubled, but the radius remains the same, by what factor does the volume increase?

40. One of the angles in a right triangle is 32°. What are the other two angles?

VI. Trigonometry and Vectors

41. Convert the following angle measures to degrees.

    (a) 2.5 rad

    (b) π rad

    (c) $3\pi/2$ rad

    (d) 5.1 rad

42. A right triangle like the one in Figure 19 has a hypotenuse $r = 4.5$ m and a side length $x = 3.8$ m. What are **(a)** the side length y and **(b)** the angle θ?

43. A right triangle like the one in Figure 19 has a hypotenuse $r = 6.2$ m and a side length $y = 4.1$ m. What are **(a)** the side length x and **(b)** the angle θ?

44. A right triangle like the one in Figure 19 has an x side of length 2.35 m and a y side of length 1.12 m. What are **(a)** the angle θ and **(b)** the hypotenuse of the triangle?

Appendix A

Selected Answers

Chapter 1

11. (a) 33.2 kilodollars (b) 0.0332 megadollars

13. 620 m^3

15. 0.23 lb

23. 6.8 s

25. 13 m^2

27. 86.6 cm

29. 1.414

49. (a) 0.114 gigadollars (b) 1.14×10^{-4} teradollars

51. 1.4×10^{13} cm^3

53. 3.3×10^4 mm

55. 1.368×10^8 calculations

57. 2.0×10^6 m^2

59. $p = 1$

63. (a) 3.14 (b) 3.1416

65. 10^5 seats

67. (a) 10^3 km/h (b) 10^4 km (c) 10^4 km

69. B, C, and D

71. 2.7×10^{-3} km/h

73. 3.05 m and 5.14 m/s

75. (a) 2.3 g/day (b) 3.3 days

77. $p = -1$

Chapter 2

1. 12.8 km

11. 1.2 km

13. 6.5 m/s

35. (a) 8.3 m (b) 2.2 m/s

39. (a) $x_{f,car} = (26 \text{ m/s})t$;
 $x_{f,truck} = (420 \text{ m}) + (-31 \text{ m/s})t$ (c) 7.4 s

51. (a) 1.95 km (b) 0.75 km

53. 10.7 m

55. 30 m

57. (a) 5 m; 5 m (b) 2 m; -2 m

59. (a) 130 m; 100 m (b) 260 m; 0 m

67. 0.099 m/s

69. 0.30 s

71. 11 m

73. (a) less than 25.0 m/s (b) 24.0 m/s

81. (a) 2.0 m/s (b) 0 m/s (c) 1.0 m/s (d) -1.5 m/s

85. (a) 35 m (b) 10 m

87. 0.50 m/s

95. $x_f = (10 \text{ m}) + (4.3 \text{ m/s})t$

97. (a) 15.0 m (b) 4.0 s

99. (a) -3.8 m (b) 2.9 s

101. (a) soccer player 2 (b) 3.0 s (c) 4.4 m

103. yes

105. 10^{-3} m/s

107. object 1

109. (a) -0.92 m/s (b) 0.92 m/s

111. (b) 3.2 s

113. (b) 7.4 m/s

115. (a) 14 m/s (b) 22 m

Chapter 3

3. 4.0 m/s

5. -1.9 m/s^2

9. 3.6 s

21. 7.6 m/s

23. (a) 13 m/s (b) 19 m

25. 1.2 s

27. 19 m

29. 18.8 m

31. 3.8×10^4 m/s^2

33. -2.1 m/s^2

45. (a) $x_{f,1} = (20.0 \text{ m/s})t + (1.25 \text{ m/s}^2)t^2$,
 $x_{f,2} = 1000 \text{ m} - (30.0 \text{ m/s})t + (1.6 \text{ m/s}^2)t^2$ (b) 24 s

71. 3.8 m/s

73. -2.7 m/s^2

75. (a) 1.0 m/s^2 (b) -1.0 m/s^2 (c) zero

77. (a) factor of 2 (b) 3.8 s (c) 7.6 s

85. 9.1 m/s^2

87. 32 m/s^2

89. (a) 10 m (b) 20 m (c) 40 m

91. (a) 21 m (b) greater than 6.0 m/s; 8.49 m/s

99. -8.2 m

101. 79.8 m

103. (a) $x_f = (1.2 \text{ m/s}^2)t^2$ (b) 1.0 m (c) 4.8 m

111. 59 m/s

113. (a) 2.8 m (b) 7.4 m/s

115. 0.10 s

117. (a) more than 2.0 m (b) 8 m

119. (a) 2.8 s (b) -30 m/s

121. 23.3 m/s

123. (a) 0.61 s (b) 1.22 s

127. (a) 5.0 s (b) 8.0 m/s

129. 2.6 m/s

131. (a) 3.8 m/s^2 (b) 15 m/s

Chapter 4

3. (a) increase (b) 180 m

5. (a) $r_x = 61.4$ m; $r_y = 43.0$ m (b) $r_x = 31.7$ m; $r_y = 68.0$ m

7. (a) decrease (b) 8.13°

9. 17°

29. 79°

37. 0.33 s

39. 39.3° below horizontal

41. $x = 5.1$ m, $y = 2.0$ m

55. (a) 38.7 m; 45° (b) 27.4 m; 90° (c) zero

57. (a) 760 m (b) about 30° north of east (c) 27° north of east

59. (a) 29.7 m; 47.7° west of north (b) 24.2 m; 65.6° west of north (c) 29.7 m; 47.7° east of north (d) 24.2 m; 65.6° east of north

65. (b) about 120 m at 120°

67. (b) 320 km at 170°

73. 25 s

75. 3.29 m/s at 43° west of north

77. 11 m/s

83. 0.786 m

85. (a) 9.9 m (b) 0.99 m

87. 2.6 m/s

89. (a)

t (s)	x (m)	y (m)
0.25	0.45	2.7
0.50	0.90	1.8
2.0	1.35	0.24

91. (a) 121 m (b) 24.3 m/s

93. (a) no (b) yes

95. less than

99. (a) 64.9° (b) 21 m/s (c) 1.8 s

Chapter 5

3. 481 kg

5. 41 kg

19. -3.11 m/s^2

21. 19 N/m

31. 1.18 m/s^2

33. (a) less than (b) 9.7 N

53. 1.4 kN

55. 2.6 m/s^2

57. 2.1 kN

59. 0.072 m/s^2

61. (a) 5.1 kN, opposite the motion (b) 15.3 m

63. $F_1 = \frac{1}{2}m(a_1 + a_2)$ and $F_2 = \frac{1}{2}m(a_1 - a_2)$

77. (b) The free-body diagram does not change.

79. 6.6 N

81. 51 N/m

83. 3.3×10^5 N/m

85. 1.5 m/s^2

87. (a) 100 N at 35° above the positive x axis (b) 830 kg

89. (a) $\frac{1}{2}mg$ (b) mg (c) $2mg$

97. (a) stay the same (b) -4.5 m/s^2

99. 1.2

101. (a) 2.4 m/s^2 (b) 6.4 s

103. $\mu_s = 0.109$; $\mu_k = 0.0547$

109. (a) northward (b) northward (c) zero (d) zero (e) downward

111. 1.7 kN/m

113. (a) 415 N (b) greater than

115. (a) 0.0119 N (b) the same as

117. (a) 4 geckos (b) 1.8 N/cm²

119. 74 kg

Chapter 6

1. remain the same

3. 25 J

5. 0.17 kJ

7. 0.0062 J

17. 1200 kg

19. equal to; 19.9 J

21. 0.54 kN

23. 3.2 J

25. 7.2 cm

37. the same as

39. 5.2 J; 9.7 m/s

49. 3.86 s

51. 386 W

67. 0.81 J

69. 0.37 m

71. (a) 60 J (b) zero (c) −60 J

73. (a) negative (b) −7.8 kJ

75. 1.6 kJ

85. 12.7 MJ

87. 15 m/s

89. 36 mJ

91. 0.16 kN/m

93. (a) −7.2 kJ (b) 2.1 kN

95. (a) 2.3 kg (b) 25 J (c) 11 J

101. 0.942 m

103. (a) 15 m/s (b) 43 m

105. 1.4 m/s

107. (a) 0.95 J (b) 0.95 J (c) 41°

113. 3.6 MJ

115. 0.82 m/s

117. (a) 2.3 MJ (b) 2.0 candy bars

121. (a) graph B (b) graph C (c) graph A

123. 386 W

125. 21.6 mJ

127. (a) 12 J (b) decrease

129. (a) 4.8 W (b) 0.43 kJ

133. (a) 0.62 m/s (b) 1.6 s

135. (a) 10 kJ (b) −10 kJ

137. (a) 46 J (b) 610 W (c) more than

139. (a) less than (b) 7.65 m/s

141. 1.04 m

143. 1.1 m

Chapter 7

1. 6.40 kg·m/s

5. 1.40×10^4 kg·m/s

15. 92 N

17. 16.7 kg·m/s

19. 12.1 m/s

21. (a) 2.7 N (b) 3.8 kg·m/s

31. canoe 1: −73 kg·m/s; canoe 2: 73 kg·m/s

33. 14 m

43. 4.79 m/s

45. 5.0 m/s

47. red car: 22 m/s; blue minivan: 14 m/s

65. 5.2 m/s

67. 12 kg·m/s

69. −1.3 m/s

71. (a) −0.70 kg·m/s (b) 0.56 kg·m/s (c) 1.26 kg·m/s

81. 7.44 kg·m/s

83. 2.6 m/s

85. 0.33 kg

93. 31 kg

95. (a) 1.9 m/s (b) less than (c) 1.5 m/s

97. (a) 0.60 m/s (b) less than (c) 0.51 m/s

105. (a) 1.75 m/s (b) 14,100 J

107. (b) 690 kg·m/s (c) −690 kg·m/s (d) They are equal and opposite. (e) 6.16 m/s

109. (a) 1.9 m/s (b) 0.23 m/s

111. (a) 1.0×10^2 m/s (b) less than (c) $KE_i = 850$ J, $KE_f = 47$ J

113. (a) $v_{4,f} = \frac{1}{3}v$; $v_{2,f} = \frac{4}{9}v$; $v_{1,f} = \frac{16}{9}v$

117. 29 s

119. (a) 2.7 N (b) 3.8 kg $\cdot$ m/s

121. (a) 8.3 m/s (b) 8.7 N

123. 2.6×10^3 kg

125. 2.20 m

Chapter 8

1. $\frac{\pi}{6}$ rad, $\frac{\pi}{4}$ rad, $\frac{\pi}{2}$ rad, π rad

3. 210 rpm

5. 1300 rad, $3.7 \times 10^{4\circ}$

7. 2.84 cm

9. 2.1 rad/s^2

21. 0.50 m

33. 1.4 N $\cdot$ m

35. 55 N

37. 7.1 N

49. 1.05 kg

51. 0.080 kg

63. 60°; 72°; 306°; 1080°

65. (a) 0.38 m/s (b) 0.24 m/s

67. (a) 0.70 m/s (b) The linear speed would double.

71. 0.609 m/s

73. 0.0049 kg $\cdot$ m^2/s

75. 37.5 rad/s

81. 25 N $\cdot$ m

83. (a) -2.14 N $\cdot$ m (b) clockwise

87. (a) 42 kg (b) 0.74 m

89. 2.3 m

91. (a) greater than (b) 0.44 kg

93. 370 rad/s; 3500 rpm

95. half a revolution

Chapter 9

1. 0.94 m

3. $F_1 = F_2 = 93.4$ N

17. 2.64×10^6 m

19. (a) 2.9×10^7 m (b) 0.48 m/s^2

29. 0.33

55. (a) 2.7×10^{-11} N (b) 6.8×10^{-12} N

57. (a) 2.4×10^{-12} N (b) 2.4×10^{-12} N

59. 2.40×10^{20} N toward the Sun

61. 4.7×10^{-8} N at 16° to the left of downward

67. (a) 3.70 m/s^2 (b) 8.87 m/s^2

69. 0.00270 m/s^2

71. 1.8×10^7 m

73. $0.11 M_{\text{Earth}}$

79. 2.9 m/s

81. 7.2 kN

83. 19 m/s

85. (b) 0.56 kN

93. 12 h

95. 7.64 h

97. (a) 7.31 h (b) The period is independent of mass.

99. 3.07 km/s

101. (b) 8.9×10^{16} kg

103. (a) satellite 2 (b) 5.59 km/s

107. 6.3 km/s^2

109. increase

111. 0.21

113. $v = \sqrt{Mrg/m}$

Chapter 10

3. 43 °C

5. 15 °C

7. (a) 5.7×10^3 °C (b) 1.0×10^4 °F

19. 42 K

31. 3.47 Cal

33. 23.3 °C

35. 22.9 °C

65. -272.2 °C

67. (a) 1473 K (b) 2192 °F

69. 0.23 K/s

77. 1.183 cm

79. 1.9×10^{-5} K^{-1}

81. 0.0037 cm

87. 25.5 °C

89. 15 kJ

91. 0.12 kW; 0.16 hp

93. 23.9 °C

95. 385 J/(kg · °C); copper

103. about 13 °C

105. 22 kJ

107. 30 °C; no ice remains

109. 32.4 s

111. 0 °C; 44.2 g

113. too long

115. (a) no (b) yes (c) no

121. 43 J/s

123. 13.2 min

125. (a) decreased (b) −83 °C

127. 25.5 °C

Chapter 11

1. -3.7×10^5 J

3. (a) 0 J (b) 0 J (c) −100 J

5. 0.28

17. 60 Pa

19. 4.1×10^4 J

21. 470 J

31. 0.224 kg

49. (a) 100 J (b) −100 J (c) 200 J

51. 23%

53. 93 J

55. (a) 119 J (b) 35 J (c) 0 J

63. 92 kJ

65. (a) decrease (b) -5.4×10^{-4} m^3

67. (a) on the system (b) −890 J

75. (a) 670 J (b) 14%

77. (a) 1.2 GW (b) 1.7 GW

79. (a) 15 kJ (b) 12 kJ (c) decrease

81. (a) 382 K (b) decreased (c) 327 K

83. 12 W/K

85. second law of thermodynamics

87. second law of thermodynamics

89. 300 kJ

Chapter 12

1. 2.7 cm

3. 144 cm^2

5. decrease

7. 0.010 m^3

19. 1.4 kg/m^3; oxygen

23. 2.17×10^6 Pa

25. 4.6 cm

37. 2.4 cm

47. 19 N/m

63. 4.3

65. 0.303 mol

67. 99.2 cm^2

75. 2.5 kN

77. 1.9 m

79. (a) 995 m (b) greater than

81. 210 kg

89. 0.66 cm

91. (a) 6.4 m/s (b) less than

97. 5.9 N

99. 47 N/m

101. 1.6 kN/m

103. 2.9×10^5 N/m

107. 9.56×10^4 Pa

109. 4.9×10^2 N; 1.1×10^2 lb

Chapter 13

1. 5 Hz; 0.2 s

5. 0.10 m

7. 2.0 m

11. 0.50 kg

13. 7.68 cm

25. (a) 9.83 m/s^2 (b) 1.42 s

37. (a) 4.0 Hz (b) 0.25 s (c) 12 m/s

47. 8.46 Hz

61. 0.38 s

63. 6.7 s

65. (a) 20 N/m (b) 0.071 s

67. (a) 87.0 beats/min (b) increase (c) 93.0 beats/min

75. 4.3 m

77. 9.6 m/s^2

79. (a) 2.000 s (b) 9.803 m/s^2

81. 8.95 m

87. 2 Hz; 0.50 s; 0.18 m; 0.36 m/s

89. 7.9 m/s

91. 1.2 m

97. 43 m/s

101. (a) third (b) 44 cm

103. 48 kg

105. 0.0029 Hz

107. 4

109. (a) $t = 1.0$ s, 3.0 s, 5.0 s (b) $t = 0$, 2.0 s, 6.0 s
 (c) 4.7 N

111. (a) less than (b) $T = \pi\left(\sqrt{\dfrac{\ell}{g}} + \sqrt{\dfrac{L}{g}}\right)$ (c) 1.5 s

Chapter 14

1. 0.50 s

3. less than twice

7. initial 444 Hz; final 446 Hz

9. 445 Hz

19. 390 Hz

21. 0.22 m

23. 0.26 m

27. 2.5 m

37. 630 Hz

39. 8.58 m/s

49. 1.95 m

65. (a) 0.098 s (b) 28 mm

67. 258 Hz or 264 Hz

69. (a) less than (b) 10.6 m

75. 25 Hz

77. 57 Hz; 172 Hz; 286 Hz

79. (a) 50 Hz (b) 150 Hz

81. (a) 93.5 Hz (b) 31.2 Hz

87. 150 Hz

89. (a) 35.3 kHz (b) higher (c) 35.4 kHz

91. (a) 903 Hz (b) 2.83 kHz

97. (a) 0.014 W/m^2 (b) 0.0045 W/m^2 (c) 1.4×10^6 m

99. (a) 60 dB (b) equal to

103. 0.43 mm

105. 108 machines

107. (a) 355 Hz (b) greater than (c) 380 Hz

Chapter 15

1. 2.2×10^{11} m

3. 7.0×10^{-5} s

5. 1.122

17. 1.0×10^{18} Hz

27. 1.7 W/m^2

31. 45°

33. 0.40 W/m^2

49. 1.5×10^4 s

51. 0.878

53. 2×10^8 m

55. (a) 2.2×10^7 m/s (b) toward

57. 4.945×10^{14} Hz

59. -2.38 kHz

71. 6.5×10^{14} Hz

73. 2.4 m

75. 30.0 km

77. (a) violet (b) blue, 6.4×10^{14} Hz; red, 4.4×10^{14} Hz

79. 2.27 m

81. 670 kHz

89. 0.213

91. 78.5°

93. (a) d-glutamic acid (b) l-leucine, 0.576 mW/m^2; d-glutamic acid, 0.732 mW/m^2

95. (a) $I_f = 0.375\, I_i \cos^2(\theta - 30.0°)$ (c) 30.0° or 210.0°

103. 2.00×10^{-11} m

105. (a) green (b) red (c) blue

107. 0.40 km

109. 9.0 mm

111. (a) increase (b) 915 rev/s

113. 2.10 kHz

115. 40 m²

Chapter 16

7. increase; $3h/4$

9. equal to

17. decrease

19. (a) 11 cm (b) −3.33 cm

21. 10 cm

23. 8.3 cm

25. −3.2 cm; 0.021

43. 46°

45. (a) 5 times (b) 4 times

47. 0.55 m

53. 5.0 m

55. 61 m

67. −7.88 cm

69. −14 cm

71. −0.50; inverted

73. (a) −40 cm; 8.4 cm tall (b) upright

75. 9.6 cm

77. 21 cm

79. (a) −4.4 cm (b) 2.0 cm

81. (a) 109 m (b) real (c) −5.45

83. 11 m

85. zero

87. increases

89. arrow 2

91. (a) real (b) 2f (c) inverted (d) −1.00

93. 8.6 cm; −2.5 cm high

95. 28.5 cm, real, inverted; 9.5 cm, virtual, upright

97. (a) 1.7 m (b) 80.3 m

99. (a) −8.0 cm (b) −11 cm

Chapter 17

3. 21.0°

5. 42°

7. 1.5

17. greater than; 1.56

19. 1.38

31. closer to; object at 6.7 cm; image at 20 cm

33. upright and virtual

59. 1.85×10^8 m/s

61. 1.25

63. 2.1

65. 1.8

67. (a) less than (b) 39°

69. 3.6 cm

77. 1.3

79. 0.24°

81. (a) less than (b) 1.0

83. 1.7

91. (b) upright (c) −7 cm, 3 cm tall

93. (a) The image distance is 2f. (b) inverted (c) real

95. −13 cm; 0.57

97. 45.9 mm

99. (a) concave (b) −42 cm

107. 14 cm

109. −4.7 cm

111. (a) right lens (b) 2.6

113. (a) The image is just to the left of lens 2. (b) inverted (c) virtual

117. (a) nearsighted (b) diverging

119. nearsighted

121. (b) upright (c) about 13 cm to the left of the lens and about 10 cm tall

123. 1.33 cm

125. (a) The image is to the right of lens 2, just beyond F_2. (b) inverted (c) real

Chapter 18

1. greater than; 2.5 km

3. 134 m

5. 1.2 μm

7. 623 nm

17. 216 fringes

19. dark fringe

21. 260 nm

31. 1.8 μm

33. 1.98 μm

47. 3.0°

63. 0.44 mm

65. 300 m

67. 34 s

69. 1.9 m

71. (a) 415 nm (b) decrease

73. (a) 660 nm (b) farther apart

81. 1.31

83. 630 nm

85. 533 nm

87. 439 nm and 549 nm

89. 493 nm and 658 nm

91. (a) 103 nm (b) No

99. 0.085°

101. 1.5 μm

103. 0.27°

105. 3.20 cm

107. 7.4 mm

109. (a) 0.073° (b) 120 m

117. 4.0×10^{-6} m

119. 4.1°

121. 920 nm

123. (a) 490 nm (b) greater than

125. 3.94 m

127. decrease

129. farther away

133. (a) 490 nm (b) greater than

Chapter 19

1. -6.2×10^{-12} C

13. 1.29 m

25. -1.2 N

27. (a) 0.20 kN in the positive x direction (b) The magnitude of the force would be cut to a ninth of its value, and the direction would be unchanged.

41. 1×10^{6} C

43. -2.6×10^{11} C

45. 6.70×10^{-25} kg

47. 21 cm

49. 2700 protons

57. 1.7×10^{-5} C

59. (a) 1.93 N (b) the same as

61. 2.1×10^{-8} N

63. $y = 2.9 \times 10^{5}$ m

65. $q_1 = 12.2$ μC and $q_2 = 49.8$ μC

71. 0.38 N

73. 32 N in the positive x direction

75. 9.8×10^{-8} C/m^2

77. (a) 1.7 N (b) 0.43 N

79. 32 N

81. negative

85. at the center of the square

87. (a) 1.58×10^{4} electrons (b) 2.73×10^{4} protons

93. 5.71×10^{13} C

95. $2\dfrac{kq^2}{d^2} \cos 30°$

97. (a) along the positive x axis (b) kq^2/R^2

99. (a) 1.2 N in the negative y direction (b) greater than (c) 9.7 N

101. (a) 5.0×10^{-4} N (b) 1.5×10^{-3} N (c) 4.8×10^{-9} C

Chapter 20

1. 8.9×10^{4} N/C

3. $+7.4$ μC

5. 1.9 m

7. 1.8×10^{5} N/C

19. 2.6×10^{4} V

21. 90 V

23. 0.0066 V

25. 3.4 μC

27. 2.1 J

37. 3.0×10^{-6} F

39. 2.6×10^{-10} F

41. 110 J

43. 0.0017 m/s

61. -13 μC

63. (a) away from (b) 2.5×10^5 N/C

69. (a) configuration (2) (b) $\vec{E}_{1,\text{net}} = 0$;
$$\vec{E}_{2,\text{net}} = \frac{4\sqrt{2}\,kq}{d^2}$$

71. (a) 1.3×10^6 N/C (b) less than

79. 2.29×10^5 m/s

81. 9.4×10^7 m/s

83. (a) 1.9 kV (b) increase

85. 0.84 kV

87. 76.7 kV

89. (a) region 4 (b) $E_1 = 13$ V/m; $E_2 = 0$;
$E_3 = 5.1$ V/m; $E_4 = 68$ V/m

97. 26 V

99. (a) 1.2×10^{-9} F (b) 2.0×10^4 V/m

101. (a) 5.0 μm (b) 67 V

107. away from the head and toward the tail

109. 3.21×10^6 m/s

111. (a) 2.6×10^4 N/C (b) 0.066 N

113. (a) C (b) A (c) $V_A = 43$ kV; $V_B = 31$ kV;
$V_C = 29$ kV; $V_D = 31$ kV

115. 4.13×10^3 N/C

Chapter 21

1. 5.5 s

3. 78 s

5. 51 Ω

17. (a) 9.2 V (b) 23 V

19. (a) decrease (b) 0.302 A

21. 29 V

23. 29 Ω

33. 1.5 V

35. 26 J

37. 0.12 kΩ

39. $0.023

55. 1.3 min

57. 8.9 V

59. 0.31 kV

61. 30 Ω

63. (a) 0.21 A (b) 14 J

65. (a) 50 Ω (b) 1 V

75. system A, 40 Ω; system B, 150 Ω

77. 77 Ω

79. (a) 0.27 A (b) $I_{88\,\Omega} = 0.14$ A; $I_{130\,\Omega} = 0.092$ A;
$I_{270\,\Omega} = 0.044$ A

81. $I_{35\,\Omega} = I_{45\,\Omega} = 0.11$ A; $I_{82\,\Omega} = 0.11$ A

83. (a) $I_{24\,\Omega} = 0.21$ A; $I_{75\,\Omega} = 0.093$ A; $I_R = 0.11$ A
(b) The currents through the 24-Ω resistor and R
decrease, and the current through the 75-Ω resistor
increases.

85. two

87. (a) 6.0 A (b) less than

97. 19 V

99. 170 W

101. 0.91 J

103. $0.095/kWh

105. (a) 0.68 A (b) 0.14 kΩ (c) greater than

111. 3.1×10^{-8} A

115. the $2R$ resistor

119. 0.11 A

121. (a) 192 Ω (b) 96 Ω (c) 150 W

123. 0.46 kV

Chapter 22

9. 2.9 cm

11. 6.50 A

13. 9.2×10^4 loops

23. 32.0°

25. 4.6×10^5 m/s

27. 3.4×10^{-3} T

29. 1.5 cm

31. stay the same

33. 3.3 N

59. 7.9 A

61. 2.6×10^4 loops

63. 31 A

65. 1.3 kA

67. 3.5×10^{-5} T

69. 21 km

79. 63°

81. 81°

83. 12.1 m/s

85. 2.6 N

87. 1.4 T in the negative x direction

89. east

91. 2.5×10^6 A

95. 5.0×10^{-6} m

97. 2.8×10^{-6} T

Chapter 23

1. 85.3°

3. 1.6 T

5. 0.023 Wb

17. 100 V

19. 0.0029 m^2

27. 1.9 A

29. decreased; 57.6 Ω

31. 0.196 A

51. 9.75×10^{-4} Wb

53. 1.7×10^{-2} Wb

55. 14 V

57. (a) the same as (b) 0.04 kV

59. 0.30 T

65. 27 rad/s

67. 0.017 V

75. 39 V

77. 81 Ω

79. (a) 2.99 W (b) 5.97 W

81. (a) less than (b) 9.4 loops

83. 92

87. counterclockwise

89. (a) zero (b) clockwise

93. 0.29 V

Chapter 24

1. 1.5×10^{14} Hz

3. 1.2 Hz

5. 2.3×10^{33}

7. 3.99 kJ

9. 17 eV

11. 1.03×10^{15} Hz

13. 8.7×10^{14} Hz

25. 5.2×10^{-10} m

47. 1.81×10^{13} Hz; 1.66×10^{-5} m

49. 9.8×10^{14} Hz; 310 nm

51. 4.01 eV

53. 4.71 eV

55. (a) halogen (b) halogen

57. (a) 4.792×10^{13} Hz (b) 0.199 eV

59. (a) aluminum (b) aluminum, 1.03×10^{14} Hz; calcium, 6.93×10^{14} Hz

61. (a) cadmium (b) zinc, 0.19 eV; cadmium, 0.30 eV

63. 61 photons/s

69. 13.7 km/s

71. 2.0×10^{-36} m

73. 1.58 km/s

75. (a) the electron (b) 1836

77. $\lambda = \dfrac{h}{\sqrt{2mqV}}$

81. (a) stay the same (b) stay the same (c) increase (d) increase

85. cesium

87. 1.6×10^{15} photons/s

89. 1.2×10^{19} photons

91. (a) 0.27 nm (b) 0.27 nm (c) 1.5×10^3 m/s

93. (a) 0.56 Hz (b) 2.2 m/s

95. (a) decreased (b) -10.2 eV

97. (a) stay the same (b) 1.25 eV

Chapter 25

5. $n = 6$

7. $n = 7$

9. 659 nm

29. 8.4×10^{-16}

31. 2.2×10^6 m/s

39. 0.850 eV

41. 0.332 nm

43. 7.48×10^{-6} m

45. 4.76×10^{-10} m

47. $n = 7$

49. (a) 0.967 eV (b) less than (c) 0.266 eV

51. (a) $n = 6$ (b) increase

59. 3.28×10^{15} Hz

61. (a) 1.36×10^{-27} kg·m/s (b) 0.814 m/s

Chapter 26

1. $Z = 1; N = 2$

11. (a) decrease (b) -0.005229 u

13. $^{234}_{91}$Pa

15. $^{238}_{95}$Am

25. $^{94}_{38}$Sr

27. $^{85}_{35}$Br

29. 1840 y

49. $^4_2$He

51. (a) 4.33×10^9 kg/s (b) 0.0307%

57. (a) $^{212}_{84}$Po $\rightarrow$ $^{208}_{82}$Pb $+$ $^4_2$He (b) $^{239}_{94}$Pu $\rightarrow$ $^{235}_{92}$U $+$ $^4_2$He

59. (a) $^{18}_9$F $\rightarrow$ $^{18}_8$O $+$ e$^+$ $+$ ν_e
(b) $^{22}_{11}$Na $\rightarrow$ $^{22}_{10}$Ne $+$ e$^+$ $+$ ν_e

61. (a) $^{218}_{81}$Tl (b) -0.001404 u

67. 1.00 d

69. 5350 y

71. 15,700 y

79. $^3_1$H $\rightarrow$ $^3_2$He $+$ e$^-$ $+$ $\bar{\nu}_e$

83. less than

85. (a) $^{206}_{80}$Hg (b) $^{239}_{93}$Np (c) $^{11}_5$B

87. 2.81×10^{10} y

89. 2250 y

91. (a) 2.4×10^{18} reactions (b) 0.93 mg

Chapter 27

7. $0.600c$

9. 0.0447 m

11. $0.866c$

41. $0.87c$

43. $80c$

45. $0.761c$

47. (a) yes (b) 7.1 y (c) Your age is 23 y; your sister's age is 24 y.

51. (a) 1.07×10^{-9} J (b) 1.51×10^{-10} J

53. 75 J

55. $0.32c$

57. 0.35 kg

67. (a) less than (b) 55 beats/min

Appendix B Additional Problems

Chapter 1

85. Use dimensional analysis to determine whether the surface area of a sphere is given by $2\pi r$ or $4\pi r^2$. Explain.

86. For each of the following expressions, state whether its dimensions are those of distance, speed, or acceleration. Note that x is distance, v is speed, a is acceleration, and t is time. Refer to Table 1.5 for the corresponding dimensions.
(a) x/t^2 (b) vt (c) $(2ax)^{1/2}$ (d) $\frac{1}{2}at^2$

87. What is the volume of a sphere whose radius is 0.735 m?

88. An object accelerates from rest with a constant acceleration a. The distance x covered by the object in the time t is $\frac{1}{2}at^p$. Find the power p that makes this equation dimensionally consistent.

89. Give an order-of-magnitude estimate for the rate at which a human fingernail grows. Express your answer in meters per second.

90. Helium-neon lasers produce a red beam of light with a wavelength of 632.8 nm. Express this wavelength in kilometers.

91. According to NASA, the top speed of the Mars rover *Curiosity* is 1.5 inches per second. What is this speed in (a) meters per second and (b) kilometers per hour?

92. Give an order-of-magnitude estimate of the number of baseballs it would take to fill your physics classroom.

Chapter 2

123. A car moves in a straight line through a displacement of 22 m. If its initial position is $x_i = 13$ m, what is its final position?

124. As you walk along a straight sidewalk, your displacement is −7.6 m. If your final position is 2.8 m, what was your initial position?

125. A dog runs to greet you with a speed of 2.9 m/s. If it takes 3.1 s for the dog to get to you, how much distance does it cover?

126. You walk with a speed of 1.1 m/s for 6.0 s and then run in the same direction with a speed of 3.3 m/s for 6.0 s. What was your average speed for these 12 s?

127. You walk with a speed of 1.1 m/s for 7.5 m and then run in the same direction with a speed of 3.3 m/s for 7.5 m. What was your average speed over the 15 m?

128. An object is at $x_i = 1.5$ m at the time $t = 0$ and moves with a constant velocity of 2.5 m/s. Plot the motion of the object on a position-time graph from $t = 0$ to $t = 5.0$ s.

129. The equation of motion for a baseball player running in a straight line is $x = 2.5$ m $+ (1.3$ m/s$)t$. (a) Where is the player at $t = 2.2$ s? (b) At what time is the player at 7.1 m?

130. In a soccer match, two players run in a straight line directly toward one another. Their equations of motion are as follows:
$$x_1 = 5.8 \text{ m} + (-2.7 \text{ m/s})t$$
$$x_2 = -2.4 \text{ m} + (3.6 \text{ m/s})t$$
(a) Which player is running with the greater speed?
(b) At what time do the players collide?

Chapter 3

137. A skateboarder starts from rest and accelerates down a ramp with a constant acceleration of 2.1 m/s². What is the speed of the skateboarder 2.5 s after she starts?

138. At the start of a race, a sprinter accelerates out of the starting blocks and attains a speed of 3.9 m/s in 1.7 s. What was the sprinter's average acceleration?

139. A horse is running with a speed of 4.6 m/s. If the horse comes to rest in 15 m, what is the magnitude of its acceleration?

140. A car is traveling with a speed of 18 m/s. If it slows with an acceleration of magnitude 3.3 m/s², what distance is required for it to come to rest?

141. The initial position of a motorcycle is 24 m, and its initial velocity is 19 m/s. If the motorcycle has a constant acceleration of 1.4 m/s², what is its position at the time $t = 3.5$ s?

142. As a train pulls into a station, its initial position is 31 m and its initial velocity is 12 m/s. If the train has a constant acceleration of −0.87 m/s², what is its position at the time $t = 4.2$ s?

143. A coconut free-falls from a palm tree to the ground. If the speed of the coconut just before it lands is 11 m/s, from what height did it fall?

144. After winning a game, you toss your baseball glove straight up into the air from a height of 1.2 m with an initial speed of 3.7 m/s. How much time does it take for the glove to hit the ground, assuming that it moves in free fall?

Chapter 4

105. In its daily prowl of the neighborhood, a cat makes a displacement of 120 m due north, followed by a 72-m displacement due west. Find the magnitude and the direction of the displacement required for the cat to return home.

106. A ball thrown straight upward returns to its original level in 2.75 s. A second ball is thrown at an angle of 40.0° above the horizontal. What is the initial speed of the second ball if it also returns to its original level in 2.75 s?

107. An off-roader explores the open desert. First she drives 25° west of north with a speed of 6.5 km/h for 15 min, then due east with a speed of 12 km/h for 7.5 min. She completes the final leg of her trip in 22 min. What are her direction and speed of travel on the final leg? (Assume that her speed is constant on each leg and that she returns to her starting point at the end of the final leg.)

108. Two airplanes taxi as they approach the terminal. Plane 1 taxis with a speed of 12 m/s due north. Plane 2 taxis with a speed of 7.5 m/s in a direction 20° north of west. **(a)** What are the direction and magnitude of the velocity of plane 1 relative to plane 2? **(b)** What are the direction and magnitude of the velocity of plane 2 relative to plane 1?

109. A shopper at the supermarket follows the path indicated by vectors $\vec{A}$, $\vec{B}$, $\vec{C}$, and $\vec{D}$ in **Figure 4.45**. Given that the vectors have the magnitudes $A = 15.5$ m, $B = 13.7$ m, $C = 10.6$ m, and $D = 3.96$ m, find the total displacement of the shopper using **(a)** the graphical method and **(b)** the component method of vector addition. Give the direction of the displacement relative to the direction of vector $\vec{A}$.

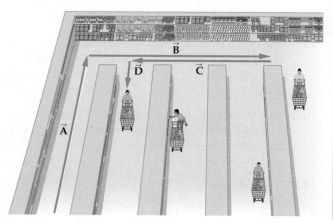

Figure 4.45

110. A Big Clock The clock that rings the bell known as Big Ben has an hour hand that is 2.74 m (9.0 feet) long and a minute hand that is 4.27 m (14 feet) long, where the distances are measured from the center of the clock to the tip of each hand. What is the tip-to-tip distance between these two hands when the clock reads 12 minutes after 4 o'clock?

111. Two canoeists start paddling at the same time and head toward a small island in a lake, as shown in **Figure 4.46**. Canoeist 1 paddles with a speed of 1.35 m/s at an angle of 45° north of east. Canoeist 2 starts on the opposite shore of the lake, a distance of 1.5 km due east of canoeist 1. **(a)** In what direction relative to north must canoeist 2 paddle to reach the island? **(b)** What speed must canoeist 2 have if the two canoes are to arrive at the island at the same time?

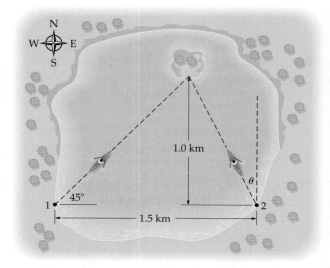

Figure 4.46

112. A soccer ball is kicked from ground level. How much time does it take for the ball to return to the ground if it is kicked with an initial speed of 14 m/s at an angle of 53° above the horizontal?

Chapter 5

127. What net force must act on an airplane with a mass of 12,000 kg if it is to accelerate at 0.33 m/s²?

128. It takes 3.1 N of force to compress a given spring by 6.9 cm. How much force is required to stretch the spring by 7.3 cm?

129. A newborn baby's brain grows rapidly. In fact, it has been found to increase in mass by about 1.6 mg per minute. **(a)** How much does the brain's weight increase in 1 day? **(b)** How much time does it take for the brain's weight to increase by 0.15 N?

130. Your groceries are in a bag with paper handles. The handles will tear off if a force greater than 51.5 N is applied to them. What is the greatest mass of groceries that can be lifted safely with this bag if it is raised (a) with constant speed or (b) with an upward acceleration of 1.25 m/s²?

131. A friend tells you that since his car is at rest, there are no forces acting on it. How would you reply?

132. The coefficient of static friction between a rope and the table on which it rests is μ_s. Find the fraction of the rope that can hang over the edge of the table before it begins to slip.

133. Two springs are attached end-to-end to form one large compound spring. The spring constant of one of the original springs is 49 N/m, and the spring constant of the other original spring is 63 N/m. (a) What is the increase in length of the compound spring when a 56-N force pulls on it? (b) What is the spring constant of the compound spring?

134. When you lift a bowling ball with a force of 82 N, the ball accelerates upward with an acceleration a. If you lift with a force of 92 N, the ball's acceleration is $2a$. Find (a) the weight of the bowling ball and (b) the acceleration a.

Chapter 6

149. [Rank] As the three small sailboats shown in **Figure 6.22** drift next to a dock, because of wind and water currents, three students pull on the lines attached to the bows and exert forces of equal magnitude, F. Each boat drifts through the same distance, d. Rank the three boats (A, B, and C) in order of increasing work done on the boat by the force F. Indicate ties where appropriate.

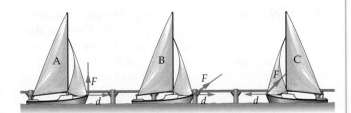

Figure 6.22

150. A youngster rides on a skateboard with a speed of 2 m/s. After a force acts on the youngster, her speed is 3 m/s. Was the work done by the force positive, negative, or zero? Explain.

151. Car 1 has four times the mass of car 2, but they both have the same kinetic energy. If the speed of car 2 is v, is the speed of car 1 equal to $v/4$, $v/2$, $2v$, or $4v$? Explain.

152. If the force on an object is zero, does that mean the potential energy of the system is zero? If the potential energy of a system is zero, is the force zero? Explain.

153. You push a cardboard box across the floor with a force of 32 N at an angle of 15° below the horizontal. If the work you did on the box is 47 J, through what distance did you move it?

154. What is the mass of a rock that has 26 J of kinetic energy when it moves with a speed of 2.7 m/s?

155. A force of 23 N increases the potential energy of a spring by 1.6 J. What is the spring constant of this spring?

156. You hit a golf ball from ground level with your favorite club and give it an initial speed of 16 m/s. (a) How fast is the ball moving when it is 4.5 m above the ground? Assume that air resistance can be ignored. (b) How does your answer to part (a) depend on the takeoff angle of the ball? Explain.

Chapter 7

132. What is the speed of a small airplane with a mass of 1100 kg and a momentum of 69,000 kg·m/s?

133. Two cars move directly toward one another in a parking lot. Car 1 has a mass of 1600 kg and a speed of 1.5 m/s in the positive direction. Car 2 has a mass of 2100 kg and a speed of 1.4 m/s in the negative direction. (a) What is the total momentum of the two cars? (b) What speed should car 2 have for the total momentum of the two cars to be zero?

134. When a 0.15-kg baseball is struck by a bat, its velocity changes from −12 m/s to 27 m/s in a certain direction. (a) What is the impulse delivered to the ball? (b) If the bat is in contact with the ball for 3.4×10^{-3} s, what is the force it exerts on the ball?

135. As a 0.43-kg soccer ball rolls on a grassy field, its speed decreases from 4.9 m/s to zero in 3.1 s. What is the average force exerted by the grass on the ball?

136. A 61-kg astronaut pushes on a satellite, giving it a speed of 0.095 m/s in one direction and giving her a speed of 1.1 m/s in the opposite direction. What is the mass of the satellite?

137. Two carts collide and stick together on a frictionless air track. Cart 1 has a mass of 0.15 kg and an initial speed of 0.43 m/s. Cart 2 has a mass of 0.25 kg and is initially at rest. What is the speed of the carts after the collision?

138. Two carts collide elastically on a frictionless air track. Cart 1 has a mass of 0.19 kg and is initially at rest. Cart 2 has a mass of 0.22 kg and an initial speed of 0.57 m/s. What is the final speed (a) of cart 1 and (b) of cart 2?

139. A 36,000-kg train car is moving in the positive direction with a speed of 2.8 m/s when it collides and sticks to an identical car moving in the negative direction with a speed of 2.4 m/s. **(a)** What is the initial kinetic energy of the system? **(b)** What is the final kinetic energy of the system?

Chapter 8

102. Referring to Problem 86, suppose the surface holding the trophies is slowly but steadily tilted, as shown in **Figure 8.38**. Does the first-place trophy tip over first, does the second-place trophy tip over first, or do both trophies tip over at the same time? Explain.

Figure 8.38

103. Find the angular speed of **(a)** the minute hand and **(b)** the hour hand of the famous clock in London, England, that rings the bell known as Big Ben.

104. Find the angular speed of the Earth as it spins about its axis. Give your result in rad/s.

105. The tip of the minute hand on a clock moves with a tangential speed of 2.1×10^{-4} m/s. What is the length of the minute hand?

106. Two children ride on the merry-go-round shown in Conceptual Example 8.4. Child 1 is 2.0 m from the axis of rotation, and child 2 is 1.5 m from the axis. If the merry-go-round completes one revolution every 4.5 s, find **(a)** the angular speed and **(b)** the linear speed of each child.

107. The outer edge of a rotating flying disk with a diameter of 29 cm has a linear speed of 3.7 m/s. What is the angular speed of the disk?

108. Rank Rank the following three examples in order of increasing angular speed:

A. an automobile tire rotating at 2.00×10^3 deg/s

B. an electric drill rotating at 400.0 rev/min

C. an airplane propeller rotating at 40.0 rad/s

109. Predict & Explain The Taipei 101 tower in Taiwan rises to a height of 508 m (1667 ft). **(a)** If you stand on the top floor of the building, is your angular speed due to the Earth's rotation greater than, less than, or equal to your angular speed when you stand on the ground floor? **(b)** Choose the *best* explanation from among the following:

A. The angular speed is the same at all distances from the axis of rotation.

B. At the top of the building you are farther from the axis of rotation and hence you have a greater angular speed.

C. You a spinning faster when you are closer to the axis of rotation.

110. A spot of paint on a bicycle tire moves in a circular path of radius 0.33 m. When the spot has traveled a linear distance of 1.95 m, through what angle has the tire rotated? Give your answer in radians.

111. Two spheres have identical radii and masses. How might you tell which of these spheres is hollow and which is solid?

112. Rank The L-shaped object in **Figure 8.39** can be rotated in one of the following three ways: case 1, rotation about the x axis; case 2, rotation about the y axis; and case 3, rotation about the z axis (which passes through the origin perpendicular to the plane of the figure). Rank these three cases in order of increasing moment of inertia. Indicate ties where appropriate.

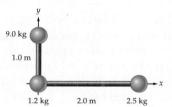

Figure 8.39

113. Triple Choice Two children with equal masses ride on a rotating merry-go-round. Child 1 rides at the outer rim of the merry-go-round, while child 2 rides halfway between the center and the outer rim. Is the angular momentum of child 1 greater than, less than, or equal to the angular momentum of child 2? Explain.

114. What is the angular momentum of the electric fan in Problem 72?

115. What is the rotational kinetic energy of the disk in Problem 73?

116. Rank Suppose a race like the one in Conceptual Example 8.7 is run with three different objects: a hoop, a hollow sphere, and a solid sphere. All three objects have the same mass and radius. Rank the objects in the order in which they finish the race. Indicate ties where appropriate.

117. The moment of inertia of a 0.98-kg bicycle wheel rotating about its center is 0.13 kg·m². What is the radius of this wheel, assuming that the weight of the spokes can be ignored?

118. The tires on a car have a radius of 31 cm. What is the angular speed of these tires when the car is driven at 15 m/s?

119. A tightrope walker uses a long pole to aid in balancing. Why?

120. A 1.6 kg bowling trophy is held at arm's length, a distance of 0.62 m from the shoulder joint. What torque does the trophy exert about the shoulder if the arm is at an angle of 22° below the horizontal?

121. **Triple Choice** Suppose you rotate on a nearly frictionless piano stool with your arms held against your sides. If you then extend your arms straight out, does your angular speed increase, decrease, or stay the same? Explain.

122. Give an example of a system in which the net torque is zero but the net force is nonzero.

123. Give an example of a system in which the net force is zero but the net torque is nonzero.

124. The L-shaped object in **Figure 8.40** consists of three masses connected by light rods. What torque must be applied to this object to give it an angular acceleration of 1.20 rad/s² if it rotates about (a) the x axis, (b) the y axis, or (c) the z axis (which is through the origin and perpendicular to the page)?

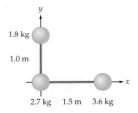

Figure 8.40

125. A front-wheel drive car in static equilibrium presses against the ground with more force on the front tires than on the rear tires. Explain why this is so.

126. Why is an object not in static equilibrium if the total torque acting on it is nonzero?

127. Referring to Problem 85, which combination of mass and distance results in the teeter-totter rotating (a) clockwise and (b) counterclockwise?

128. A hand-held shopping basket 62.0 cm long has a 1.81-kg carton of milk at one end and a 0.722-kg box of cereal at the other end. Where should a 1.80-kg container of orange juice be placed so that the basket balances at its center?

129. **Think & Calculate** An 11-kg block is placed on the x axis at $x = 1.0$ m. A second block with a mass of 21 kg is placed on the x axis at $x = 2.0$ m. (a) Is the center of mass of these two blocks at a location that is greater than, less than, or equal to $x = 1.5$ m? (b) Where is the center of mass of these two blocks?

130. **Maximum Overhang** Three identical, uniform books of width L are stacked one on top of another, as shown in **Figure 8.41**. Find the maximum overhang distance, d, such that the books do not fall off the table. (The books will be stable as long as their combined center of mass is not beyond the edge of the table.)

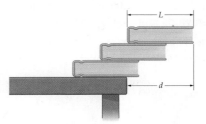

Figure 8.41

131. To loosen the lid on a jar of jam 8.9 cm in diameter, a torque of 8.5 N·m must be applied to the circumference of the lid. If a jar wrench whose handle extends 15 cm from the center of the jar is attached to the lid, what is the minimum force required to open the jar?

132. A uniform disk stands upright on its edge and rests on a sheet of paper placed on a tabletop. (a) If the paper is pulled horizontally to the right, as indicated in **Figure 8.42**, does the disk rotate clockwise or counterclockwise about its center? Explain. (b) Does the center of the disk move to the right, to the left, or does it stay in the same location?

Figure 8.42

133. A diver completes $2\frac{1}{2}$ somersaults (revolutions) during a 2.3-s dive. What was the diver's average angular speed during the dive?

134. Object 1 has a moment of inertial I and an angular speed ω. Object 2 has a moment of inertia equal to $2I$. What angular speed must object 2 have if its angular momentum is to be the same as the angular momentum of object 1?

135. A beetle sits near the rim of a turntable that rotates without friction about a vertical axis. The beetle then begins to walk toward the center of the turntable. As a result, does the angular speed of the turntable increase, decrease, or stay the same?

136. | Predict & Explain | A disk and a hoop (bicycle wheel) of equal radius and mass each have a string wrapped around the circumference. Hanging from the strings, halfway between the disk and the hoop, is a block of mass m, as shown in **Figure 8.43**. The disk and the hoop are free to rotate about their centers. **(a)** When the block is allowed to fall, does it stay on the center line, move toward the right, or move toward the left? **(b)** Choose the *best* explanation from among the following:

A. The disk is harder to rotate, and hence its angular acceleration is less than that of the wheel.

B. The wheel has the greater moment of inertia, and its string unwinds more slowly than the disk's.

C. The system is symmetrical, with equal mass and radius on either side.

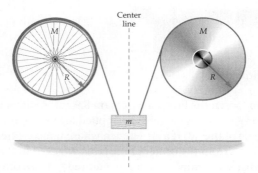

Figure 8.43

137. A motorcycle accelerates from rest, and both the front and rear tires rotate without slipping. **(a)** Is the force exerted by the ground on the rear tire in the forward or the backward direction? **(b)** Is the force exerted by the ground on the front tire in the forward or the backward direction?

138. **Balancing a *T. rex*** Paleontologists believe that *Tyrannosaurous rex* stood and walked with its spine almost horizontal, as indicated in **Figure 8.44**, and that its tail was held off the ground to balance its upper torso about the hip joint. Given that the total mass of *T. rex* was 5400 kg and that the placement of the center of mass of the tail and the upper torso was as shown in Figure 8.44, find the mass of the tail required for balance—that is, to have the center of mass of the entire *T. rex* be at its hip.

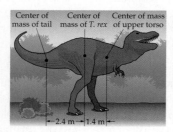

Figure 8.44

139. **Roller Pigeons** Pigeons are bred to display a number of interesting characteristics. One breed of pigeon, the "roller," is remarkable for the fact that it does a number of backward somersaults as it drops straight down toward the ground. Suppose a roller pigeon drops from rest and free falls downward for a distance of 14 m. If the pigeon somersaults at the rate of 12 rad/s, how many revolutions has it completed by the end of its fall?

140. A city park features a large fountain surrounded by a circular pool. Two high school students start at the northernmost point of the pool and walk slowly around it in opposite directions. **(a)** If the angular speed of the student walking in the clockwise direction (as viewed from above) is 0.045 rad/s and the angular speed of the other student is 0.023 rad/s, how long does it take before they meet? **(b)** At what angle, measured clockwise from due north, do the students meet?

141. **Pulling a Weed** The gardening tool shown in **Figure 8.45** is used to pull weeds. If a 1.23-N · m torque is required to pull a given weed, what force does the weed exert on the tool?

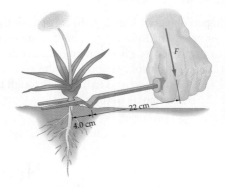

Figure 8.45

Chapter 9

119. | Rank | Humans lose consciousness if exposed to prolonged accelerations of more than about 7g. This is of concern to jet fighter pilots, who may experience centripetal accelerations of this magnitude when making high-speed turns. Suppose we would like to decrease the centripetal acceleration of a jet. Rank the following changes in flight path in order of how effective they would be in decreasing the centripetal acceleration, starting with the least effective:

A. decrease the turning radius by a factor of 2

B. decrease the speed by a factor of 3

C. increase the turning radius by a factor of 4

120. **Gravitropism** As plants grow, they tend to align their stems and roots along the direction of the gravitational field. This tendency, which is related to differential concentrations of plant hormones known as *auxins*, is referred to as *gravitropism*. Interestingly, experiments show that seedlings placed in pots on the rim of a rotating turntable do not grow in the vertical direction. Do you expect their stems to tilt inward—toward the axis of rotation—or outward—away from the axis of rotation?

121. **Predict & Explain** The orbital speed of Earth is greatest around January 4 and least around July 4. **(a)** Is the distance from Earth to the Sun on January 4 greater than, less than, or equal to that distance on July 4? **(b)** Choose the *best* explanation from among the following:

A. Earth's orbit is circular, with an equal distance to the Sun at all times.

B. Earth sweeps out equal areas in equal times; thus, it must be closer to the Sun when it is moving faster.

C. The greater the speed of Earth, the greater its distance from the Sun.

122. **Path of the Moon** Earth and Moon exert gravitational forces on one another as they orbit the Sun. As a result, the path they follow is not the simple circular orbit you would expect if either one orbited the Sun alone. Occasionally, you will see a suggestion that the Moon follows a path like a sine wave centered on a circular path, as in the upper part of **Figure 9.37**. Though few realize it, this is *incorrect*. The Moon's path is qualitatively like that shown in the lower part of Figure 9.37. Explain.

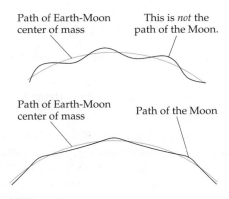

Figure 9.37

123. Two 6.7-kg bowling balls, each with a radius of 0.11 m, are in contact with one another. What is the gravitational attraction felt by each ball?

124. A satellite is placed in Earth orbit 2000 km higher than the altitude of a geosynchronous satellite. **(a)** Is the period of this satellite greater than or less than 24 hr? **(b)** As viewed from the surface of Earth, does this satellite move eastward or westward? Explain.

125. **Think & Calculate** Suppose a planet is discovered that has the same total mass as Earth, but half its radius.

(a) Is the acceleration due to gravity on this planet more than, less than, or the same as the acceleration due to gravity on Earth? Explain. **(b)** Calculate the acceleration due to gravity on this planet.

126. Show that the speed of a satellite in a circular orbit at a height h above the surface of Earth is given by

$$v = \sqrt{\frac{GM_E}{R_E + h}}$$

127. You swing a 3.25-kg bucket of water in a vertical circle of radius 0.950 m. At the top of the circle the speed of the bucket is 3.23 m/s; at the bottom of the circle its speed is 6.91 m/s. Find the tension in the rope tied to the bucket at **(a)** the top and **(b)** the bottom of the circle.

128. Referring to Figure 9.31 and Problem 85, suppose the Ferris wheel rotates fast enough to make you feel "weightless" at the top. **(a)** How many seconds does it take to complete one revolution in this case? **(b)** How does your answer to part (a) depend on your mass? Explain.

129. **Think & Calculate** Suppose a planet is discovered that has the same amount of mass in a given volume as Earth, but has half Earth's radius. **(a)** Is the acceleration due to gravity on this planet more than, less than, or the same as the acceleration due to gravity on Earth? Explain. **(b)** Calculate the acceleration due to gravity on this planet. (*Hint:* Use $V = \frac{4}{3}(\pi)r^3$ to calculate the volume of a sphere.)

130. **Think & Calculate** On a typical mission a space shuttle ($m = 2.00 \times 10^6$ kg) orbits at an altitude of 250 km above Earth's surface. **(a)** Does the orbital speed of the shuttle depend on its mass? Explain. **(b)** Find the speed of the shuttle in its orbit. **(c)** How long does it take for the shuttle to complete one orbit of Earth?

131. An equilateral triangle 10.0 m on a side has a 1.00-kg mass at one corner, a 2.00-kg mass at another corner, and a 3.00-kg mass at the third corner, as shown in **Figure 9.38**. Find the magnitude and direction of the net force acting on the 1.00-kg mass.

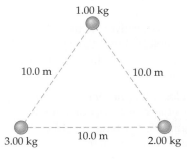

Figure 9.38

132. **A Conical Pendulum** A 0.075-kg toy airplane is tied to the ceiling with a string. When the airplane's motor is started, it moves with a constant speed of 1.21 m/s in a horizontal circle of radius 0.44 m, as illustrated in **Figure 9.39**. Find **(a)** the angle the string makes with the vertical and **(b)** the tension in the string.

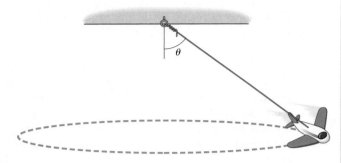

Figure 9.39

Chapter 10

132. We know that $-40 \,°F$ is the same as $-40 \,°C$. Is there a temperature for which the Kelvin and Celsius scales agree? Explain.

133. What Fahrenheit temperature is three times greater than the corresponding Celsius temperature?

134. Triple Choice A hollow pipe is cooled. Does the cylindrical volume within the pipe increase, decrease, or stay the same?

135. When you touch a piece of metal and a piece of wood that are both at room temperature, the metal feels cooler. Why?

136. Predict & Explain Two objects are made of the same material but have different temperatures. Object 1 has mass m, and object 2 has mass $2m$. **(a)** If the objects are brought into contact so that they can exchange thermal energy, is the temperature change of object 1 greater than, less than, or equal to the temperature change of object 2? **(b)** Choose the *best* explanation from among the following:

A. The larger object (object 2) gives up more thermal energy, and therefore its temperature change is greater.

B. The thermal energy given up by one object is taken up by the other object. Since the objects have the same specific heat capacity, the temperature changes are also the same.

C. One object loses thermal energy of magnitude Q; the other gains thermal energy of magnitude Q. With the same magnitude of energy involved, the smaller object (object 1) has the greater temperature change.

137. Consider the apparatus that Joule used in his experiments on the mechanical equivalent of heat, shown in Figure 10.13. Suppose both blocks have a mass of 0.95 kg and that they fall through a distance of 0.48 m.

Find the expected rise in temperature of the water, given that 6200 J are needed for every 1.0 °C increase. Give your answer in Celsius degrees.

138. An 97.6-g lead ball is dropped from rest from a height of 5.43 m. The collision between the ball and the ground is completely inelastic. Assuming that all the ball's kinetic energy goes into heating the ball, find the change in its temperature.

139. A drop of water on a kitchen counter evaporates in a matter of minutes. Explain why the drop evaporates even though its temperature is well below the boiling point of water.

140. An 825-g iron block is heated to 352 °C and placed in a container that absorbs a negligible amount of thermal energy and holds 40.0 g of water at 20.0 °C. What is the equilibrium temperature of this system? If your answer is 100 °C, determine the amount of water that has vaporized. The average specific heat capacity of iron over this temperature range is 560 J/(kg·C).

141. A 35-g ice cube at 0.0 °C is added to 110 g of water in a 62-g aluminum cup. The cup and the water have an initial temperature of 23 °C. **(a)** Find the equilibrium temperature of the cup and its contents. **(b)** Suppose the aluminum cup is replaced with one of equal mass made from silver. Is the equilibrium temperature with the silver cup greater than, less than, or the same as with the aluminum cup? Explain.

142. Celsius and Fahrenheit temperatures are the same at -40 degrees. Which temperature scale has the larger temperatures when the temperature is above -40?

143. How does a lava lamp illustrate convection?

144. How are vaporization and sublimation similar? How are they different?

145. What temperature is the same on both the Kelvin and Fahrenheit scales?

146. Suppose you could convert the 525 Calories in a cheeseburger into mechanical energy with 100% efficiency. **(a)** How high could you throw a 0.145-kg baseball with the energy contained in the cheeseburger? **(b)** How fast would the ball be moving at the moment of release?

147. Suppose the 0.550-kg of ice in Figure 10.24 has just begun to melt. How much ice will be left in the system after **(a)** 5.00×10^4 J, **(b)** 1.00×10^5 J, and **(c)** 1.50×10^5 J of thermal energy have been added to the system?

148. To make steam you add 5.60×10^5 J of thermal energy to 0.220 kg of water at an initial temperature of 50.0 °C. Find the final temperature of the steam.

149. You turn a crank on a device similar to that shown in Figure 10.13 and produce a power of 0.18 hp. If the paddles are immersed in 0.65 kg of water, for what length of time must you turn the crank to increase the temperature of the water by 5.0 °C?

150. A 0.22-kg steel pot sitting on a stove contains 2.1 L of water at 22 °C. When the burner is turned on, the water begins to boil after 8.5 min. At what rate is thermal energy transferred from the burner to the pot of water?

Chapter 11

95. Explain why pushing down rapidly on the plunger in Figure 11.11 causes the bits of paper at the bottom of the cylinder to ignite. Why must the plunger be pushed down rapidly? Which thermal process is involved?

96. Predict & Explain (a) A gas is expanded isothermally. Does the entropy of the universe increase, decrease, or stay the same? (b) Choose the *best* explanation from among the following:

A. Heat will be drawn from the surroundings to keep the gas from cooling as it expands. This flow of heat will decrease the entropy.

B. Heat must be added to the gas to maintain its constant temperature. This flow of heat will increase the entropy.

C. This is an isothermal process, which means that the entropy of the universe remains the same.

97. The heat that flows into a particular heat engine from the hot reservoir is 4.0 times greater than the work it performs. What is the engine's efficiency?

98. The heat that flows into a particular heat engine is 5.0 times greater than the work it performs. (a) If the hot reservoir has a temperature of 490 K, what is the temperature of the cold reservoir? (b) What fraction of the heat that goes into the engine is expelled to the cold reservoir?

99. Think & Calculate Thermal energy is added to a 0.25-kg block of ice at 0 °C, causing 0.13 kg of the ice to melt. (a) Does the entropy of the ice increase, decrease, or stay the same as a result of this process? Explain. (b) What is the change in the entropy of the ice?

100. An inventor claims that his new heat engine uses organic grape juice as its working material. According to the claim, the engine takes 1250 J of heat from a reservoir of juice at 1010 K and performs 1120 J of work. The waste heat is exhausted to the atmosphere at a temperature of 302 K. (a) What is the efficiency that is implied by this claim? (b) What is the maximum efficiency of a heat engine operating between the given high and low temperatures? (Should you invest in this invention?)

101. Think & Calculate A small dish containing 530 g of water is placed outside for the birds. During the night the outside temperature drops to -5.0 °C and stays at that point for several hours. (a) When the water in the dish freezes, does its entropy increase, decrease, or stay the same? Explain. (b) Calculate the change in entropy that occurs as the water freezes. (c) When the water freezes, is there an entropy change anywhere else in the universe? If so, specify where the increase occurs.

102. Which would result in a greater increase in the efficiency of a heat engine: (a) raising the temperature of the high-temperature reservoir by 50 K or (b) lowering the temperature of the low-temperature reservoir by 50 K? Justify your answer by calculating the change in efficiency for each of these cases. Choose any values you like for the initial temperatures, T_h and T_c.

Chapter 12

116. Triple Choice Is the number of molecules in 1 mole of N_2 greater than, less than, or equal to the number of molecules in 1 mole of O_2?

117. Standard temperature and pressure (STP) is defined as a temperature of 0 °C and a pressure of 101.3 kPa. What is the volume occupied by 1 mole of an ideal gas at STP?

118. A balloon contains 3.7 L of nitrogen gas at a temperature of 87 K and a pressure of 101 kPa. If the temperature of the gas is allowed to increase to 24 °C while the pressure remains constant, what volume will the gas occupy?

119. Consider the system shown in Figure 12.30. The container holds an ideal gas at a constant temperature of 313 K. Initially, the pressure applied by the piston and the mass is 137 kPa, and the height of the piston above the base of the flask is 23.4 cm. When additional mass is added to the piston, the height of the piston decreases to 20.0 cm. Find the new pressure applied by the piston.

120. Think & Calculate A spherical balloon is filled with helium at a pressure of 2.4×10^5 Pa. The balloon is at a temperature of 18 °C and has a radius of 0.25 m. (a) How many helium atoms are contained in the balloon? (b) Suppose we double the number of helium atoms in the balloon, keeping the pressure and the temperature fixed. By what factor does the volume of the balloon change? Explain.

121. The air inside a hot-air balloon has a temperature of 79.2 °C. The outside air has a temperature of 20.3 °C. The pressure is the same inside and outside the balloon. What is the ratio of the density of the air in the balloon to the density of the air in the surrounding atmosphere? (The lower density of the air inside the balloon allows it to float.)

122. One day, while snorkeling near the surface of a crystal-clear ocean, it occurs to you that you could go considerably deeper by simply lengthening the snorkel tube. Unfortunately, this does not work well at all. Why?

123. Since metal is more dense than water, how is it possible for a metal boat to float?

124. Two drinking glasses, 1 and 2, are filled with water to the same depth. Glass 1 has twice the diameter of glass 2. **(a)** Is the weight of the water in glass 1 greater than, less than, or equal to the weight of the water in glass 2? **(b)** Is the pressure at the bottom of glass 1 greater than, less than, or equal to the pressure at the bottom of glass 2?

125. Think & Calculate As a stunt you want to sip some water through a very long, vertical straw. **(a)** First, explain why the liquid moves upward, against gravity, into your mouth when you sip. **(b)** What is the longest vertical straw that you could, in principle, drink water with?

126. A hot-air balloon has a mass of 1890 kg (including its cargo) and a volume of 11,430 m^3. The balloon is floating at a constant height of 6.25 m above the ground. What is the density of the hot air in the balloon?

127. A solid block is attached to a spring scale. When the block is suspended in air, the scale reads 20.0 N; when it is completely immersed in water, the scale reads 17.7 N. What are **(a)** the volume and **(b)** the density of the block?

128. Think & Calculate A log floats in a river with one-fourth of its volume above the water. **(a)** What is the density of the log? **(b)** If the river carries the log into the ocean, does the portion of the log above the water increase, decrease, or stay the same? Explain.

129. Water flows through horizontal tube 1, which has a diameter of 2.8 cm and is joined to horizontal tube 2, whose diameter is 1.6 cm. The pressure difference between the tubes is 7.5 kPa. **(a)** Which tube has the higher water pressure? **(b)** In which tube does the water have the higher flow speed?

130. Two springs are attached end to end to form one compound spring. The spring constant of one of the original springs is 53 N/m, and the spring constant of the other is 71 N/m. **(a)** By how much does the compound spring increase in length when a 49-N force pulls on it? **(b)** What is the spring constant of the compound spring?

131. A solid block is attached to a spring scale. When the block is suspended in air, the scale reads 21.5 N.

When the block is completely immersed in water, the scale reads 15.3 N. What is the volume of the block?

132. Plastic bubble wrap is used as a protective packing material. Is the bubble wrap more effective on a cold day or on a warm day? Explain.

133. Two adjacent rooms in a hotel are equal in size and connected by an open door. Room 1 is warmer than room 2. Which room contains more air molecules? Explain.

134. Water flows through a pipe with a speed of 2.1 m/s. Find the flow rate in kg/s if the diameter of the pipe is 3.8 cm.

135. At what depth below the ocean surface is the pressure equal to 2 atm?

136. Predict & Explain A person floats in a boat in a small backyard swimming pool. Inside the boat with the person are some bricks. **(a)** If the person drops the bricks overboard and they sink to the bottom of the pool, does the water level in the pool increase, decrease, or stay the same? **(b)** Choose the *best* explanation from among the following:

A. When the bricks sink, they displace less water than when they were floating in the boat; hence, the water level decreases.

B. The same total mass (boat plus bricks plus person) is in the pool in either case, and therefore the water level remains the same.

C. The bricks displace more water when they sink to the bottom than they did when they were above the water in the boat; therefore, the water level increases.

137. Predict & Explain On Wednesday, August 15, 1934, William Beebe and Otis Barton made history by descending in the *Bathysphere*—basically a steel sphere 4.75 ft in diameter—3028 ft below the surface of the ocean, deeper than anyone had been before. **(a)** As the *Bathysphere* was lowered, was the buoyant force exerted on it at a depth of 10 m greater than, less than, or equal to the buoyant force exerted on it at a depth of 50 m? **(b)** Choose the *best* explanation from among the following:

A. The buoyant force depends on the density of the water, which is essentially the same at 10 m and at 50 m.

B. The pressure increases with depth, and the higher pressure increases the buoyant force.

C. The buoyant force decreases as an object sinks below the surface of the water.

138. A solid block is suspended from a spring scale. When the block is in air, the scale reads 35.0 N; when it is immersed in water, the scale reads 31.1 N; and when it is immersed in oil, the scale reads 31.8 N. **(a)** What is the density of the block? **(b)** What is the density of the oil?

139. **Think & Calculate** A backyard swimming pool is cylindrical in shape and contains water to a depth of 48 cm. It is 2.1 m in diameter and is not completely filled. **(a)** What is the pressure at the bottom of the pool? **(b)** If a person gets in the pool and floats motionlessly, does the pressure at the bottom of the pool increase, decrease, or stay the same? **(c)** Calculate the pressure at the bottom of the pool if the floating person has a mass of 72 kg.

140. A person weighs 685 N in air but only 497 N when standing in freshwater up to the hips. Find **(a)** the volume of each of the person's legs and **(b)** the mass of each leg, assuming that they have a density that is 1.05 times the density of water.

141. **Hydrostatic Paradox, Part I** Consider the lightweight containers shown in **Figure 12.42**. Both containers have bases of area $A_{base} = 24$ cm^2 and contain water to a depth of 18 cm. As a result, the downward force on the base of container 1 is equal to the downward force on the base of container 2, even though the containers clearly hold different weights of water. This is referred to as the *hydrostatic paradox*. **(a)** Given that container 2 has an annular (ring-shaped) region of area $A_{ring} = 72$ cm^2, determine the *net* downward force exerted on the container by the water. **(b)** Show that your result from part (a) is equal to the weight of the water in container 2.

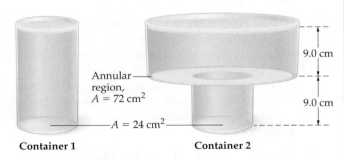

Figure 12.42

142. **Hydrostatic Paradox, Part II** Consider the two lightweight containers shown in **Figure 12.43**. As in Problem 141, these containers have equal forces on their bases but contain different weights of water. This is another version of the hydrostatic paradox. **(a)** Determine the *net* downward force exerted by the water on the base of container 2. Note that the bases of the containers have the area $A_{base} = 24$ cm^2, the annular region has the area $A_{ring} = 18$ cm^2, and the depth of the water is 18 cm. **(b)** Show that your result from part (a) is equal to the weight of the water in container 2.

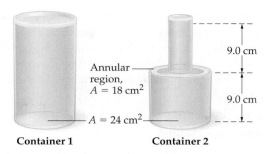

Figure 12.43

Chapter 13

116. The blades of a wind turbine rotate with a period of 4.2 s. What are **(a)** the frequency and **(b)** the angular speed (in rad/s) of the blades?

117. When a mass of 0.75 kg is attached to a spring, it oscillates with a period of 0.66 s. How far does this spring stretch when a 1.1-kg mass is suspended from it?

118. **(a)** What is the period of a pendulum with a length of 1.26 m in a location where the acceleration due to gravity is 9.79 m/s^2? **(b)** How does your answer to part (a) depend on the mass of the bob attached to the pendulum?

119. A pendulum oscillates with a frequency of 0.454 Hz in a location where the acceleration due to gravity is 9.78 m/s^2. What is the length of the pendulum?

120. A wave on a string has a wavelength of 33 cm and an amplitude of 7.2 cm. **(a)** What is the vertical distance between a crest and a trough on this wave? **(b)** What is the horizontal distance from a crest to the nearest trough?

121. You dip your finger in a pool of water three times a second, producing waves that travel with a speed of 1.8 m/s. What are the frequency, period, and wavelength of these waves?

122. Both ends of a 1.1-m-long string are tied down. If waves on this string travel with a speed of 38 m/s, what are **(a)** the frequency and **(b)** the wavelength of the first harmonic?

123. Waves on a string tied down at both ends travel with a speed of 41 m/s. How long is this string if the frequency of the first harmonic is 25 Hz?

Chapter 14

114. Describe how the sound of a symphony played by an orchestra would be altered if the speed of sound depended on its frequency.

115. On a rainy day, while driving your car, you notice that your windshield wipers are moving in synchrony with the wipers of the car in front of you. After several cycles, however, your wipers and the wipers of the other car are moving opposite to one another. A short time later the wipers are synchronized again. What wave phenomena do the wipers illustrate? Explain.

116. What is the difference between ultrasonic and infrasonic sound waves?

117. Marching soldiers crossing a bridge are instructed to "break step"—that is, they are told to walk out of step with each other. Why is this a wise procedure?

118. A rock is thrown downward into a well in which the distance to the water is 8.85 m. If the splash is heard 1.20 s later, what was the initial speed of the rock?

119. What is the fundamental frequency? How are the frequencies of higher harmonics related to the fundamental frequency?

120. Suppose the length of the pipe in Figure 14.9 is $L = 0.5$ m. What is the wavelength of the standing wave shown in Figure 14.9 (a), (b), and (c)?

121. Suppose the length of the pipe in Figure 14.10 is $L = 0.5$ m. What is the wavelength of the standing wave shown in Figure 14.10 (a), (b), and (c)?

122. An organ pipe closed at one end is 66 cm long and contains a standing wave that has three antinodes. **(a)** Which harmonic is this? **(b)** What is the wavelength of this wave?

123. An organ pipe open at both ends has a harmonic with a frequency of 440 Hz. The next higher harmonic in the pipe has a frequency of 538 Hz. Find **(a)** the fundamental frequency and **(b)** the length of the pipe.

124. With what speed must you approach a source of sound to observe a 15% increase in frequency?

125. Hearing the siren of an approaching fire truck, you pull over to the side of the road and stop. As the truck approaches, you hear a tone of 460 Hz; as the truck recedes, you hear a tone of 410 Hz. How much time will it take for the truck to get from your position to the fire 5.0 km away, assuming that it maintains a constant speed?

126. Bullet Train The Shinkansen, the Japanese bullet train, runs at high speed from Tokyo to Nagoya. Riding on the Shinkansen, you notice that the frequency of a crossing signal changes markedly as you pass the crossing. As you approach the crossing, the frequency you hear is f. As you recede from the crossing the frequency you hear is $2f/3$. What is the speed of the train?

127. If the distance to a point source of sound is tripled, by what factor does the intensity of the sound change? Explain.

128. Thundersticks A popular noisemaking device used at many sporting events is known as a *thunderstick*. A typical thunderstick is a hollow plastic tube about 82 cm long and 8.5 cm in diameter. Sharply striking two thundersticks together produces a lot of noise. Suppose a single pair of thundersticks produces sound with an intensity level of 95 dB. What is the intensity level of 1000 pairs of thundersticks struck simultaneously?

129. A point source of sound that emits uniformly in all directions is located in the middle of a large, open field. The intensity at Brittany's location directly north of the source is twice that at Phillip's position due east of the source. What is the distance between Brittany and Phillip if Brittany is 12.5 m from the source?

130. Sitting peacefully in your living room one stormy day, you see a flash of lightning through the windows. Eight and a half seconds later thunder shakes the house. Estimate the distance from your house to the bolt of lightning.

131. Loudest Animal The loudest sound produced by a living organism on Earth is made by bowhead whales (*Balaena mysticetus*). These whales can produce a sound that, if heard in air at a distance of 3.0 m, would have an intensity of 5.0 W/m². This is roughly the equivalent of 5000 trumpeting elephants. How far away can you be from this sound and still just barely hear it? (Assume a point source, and ignore reflection and absorption. Recall that the minimum intensity of sound a human can hear is 10^{-12} W/m².)

132. Lowest Frequency in Nature Astronomers using the Chandra X-ray Observatory have observed that the Perseus Black Hole, some 250 million light years away, produces sound waves in the gaseous halo that surrounds it. The frequency of this sound is 8.09×10^{-16} Hz. How long does it take for this sound wave to complete one cycle? Give your answer in years.

133. When you blow across the top of a bottle, you hear a fundamental frequency of 206 Hz. Suppose the air in the bottle is replaced with helium. **(a)** Does the fundamental frequency increase, decrease, or stay the same? Explain. **(b)** Find the new fundamental frequency. (Assume that the speed of sound in helium is three times that in air.)

134. Two ships in a heavy fog are blowing their horns simultaneously. Both horns produce sound with a frequency of 175.0 Hz (**Figure 14.26**). One ship is at rest; the other moves on a straight line that passes through the one at rest. If people on the stationary ship hear a beat frequency of 3.5 Hz, what are the two possible speeds and directions of motion of the moving ship?

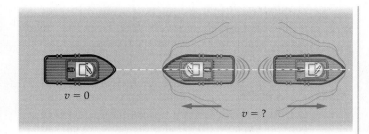

Figure 14.26

135. Hearing a Pin Drop The ability to hear a pin drop is the sign of sensitive hearing. Suppose a 0.55-g pin is dropped from a height of 28 cm, and the pin emits sound for 1.5 s when it lands. Assuming that all of the mechanical energy of the pin is converted to sound energy and that the sound radiates uniformly in all directions, find the maximum distance from which a person can hear the pin drop. (This is the ideal maximum distance, but atmospheric absorption and other factors will make the actual maximum distance considerably smaller. Recall that the minimum intensity of sound a human can hear is 10^{-12} W/m^2.)

136. Noise Standards OSHA (the Occupational Safety and Health Administration) has established standards for workplace exposure to noise. According to OSHA's Hearing Conservation Standard, the permissible noise exposure per day is 3.16×10^{-3} W/m^2 for 4 h or 3.16×10^{-2} W/m^2 for 1 h. Assuming that the eardrum is 9.5 mm in diameter, find the energy absorbed by the eardrum with exposure to **(a)** 3.16×10^{-3} W/m^2 for 4 h and **(b)** 3.16×10^{-2} W/m^2 for 1 h. **(c)** Is OSHA's safety standard simply a measure of the amount of energy absorbed by the ear? Explain.

137. A friend in another city tells you that she has two organ pipes of different lengths, one open at both ends and the other open at one end only. In addition, she has determined that the beat frequency caused by the second-lowest frequency of each pipe is equal to the beat frequency caused by the third-lowest frequency of each pipe. Her challenge to you is to calculate the length of the organ pipe that is open at both ends, given that the length of the other pipe is 1.00 m.

Chapter 15

121. The large spiral galaxy that is closest to the Milky Way is the Andromeda galaxy, which is 2.2 million light years away. Andromeda is approaching the Milky Way at the rate of 140 km/s. At this rate, how much time (in years) will elapse before the galaxies collide?

122. A race car approaches a radar unit with a speed of 56 m/s. The radar unit sends out electromagnetic waves that have a frequency of 8.5×10^9 Hz. What frequency difference does the driver of the moving race car observe if he detects the waves from the radar unit?

123. Which electromagnetic waves have the greater frequency, infrared waves or radio waves?

124. Which electromagnetic waves have the greater wavelength, blue light or microwaves?

125. What color do you see if all the pixels on your TV screen have only the green and blue color dots lit?

126. Remote controls typically emit electromagnetic waves that have a wavelength of 940 nm. **(a)** What is the frequency of these waves? **(b)** To what portion of the electromagnetic spectrum do these waves belong?

127. Vertically polarized light passes through a polarizer whose transmission axis is at an angle to the vertical. If the incident intensity of the light is 0.85 W/m^2 and the transmitted intensity is 0.22 W/m^2, what angle does the transmission axis of the polarizer make with the vertical?

128. An unpolarized beam of light with an intensity of 1.2 W/m^2 passes through two polarizers whose transmission axes form an angle of 37° with one another. What is the intensity of the light beam **(a)** after it passes through the first polarizer and **(b)** after it passes through the second polarizer?

Chapter 16

107. An object (red arrow) is placed in front of a convex mirror, as shown in **Figure 16.43**. Find the location and height of the image.

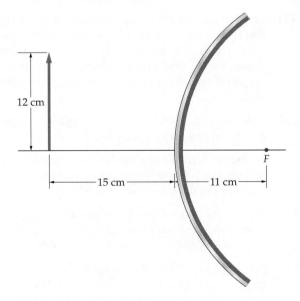

Figure 16.43

108. Get Me Out of Here! Imagine that a *T. Rex* dinosaur is chasing you as you desperately try to escape in a car. You see the *T. Rex* closing in quickly in the rear-view mirror. Near the bottom of the mirror you also see the following helpful message: "Objects in the mirror are closer than they appear." Is this mirror concave or convex? Explain.

109. Which of the four points (1, 2, 3, or 4) in **Figure 16.44** best represents the location of **(a)** the focal point, *F*, and **(b)** the center of curvature, *C*, of the convex mirror?

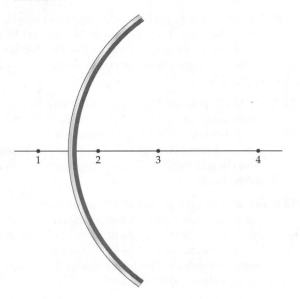

Figure 16.44

110. A beam of light strikes a plane mirror at an angle of 28° to the normal. **(a)** What is the angle between the incident and reflected beams? **(b)** What is the change in direction of the incident beam?

111. You and your brother are in front of a large mirror. You sit on a chair 1.9 m in front of the mirror, and your brother, who stands behind you, is 3.3 m from the mirror. At what distance should you focus your camera if you want to take a picture of your brother in the mirror?

112. What is the diameter of a reflecting globe in a garden, given that its focal length is 28 cm?

113. Where is the image of a small object placed 25.5 cm in front of a concave mirror, given that the radius of curvature of the mirror is 53.0 cm?

114. What is the magnification produced by a convex mirror with a radius of curvature of 1.5 m when an object is placed 62 cm in front of the mirror?

115. An object with a height of 31 cm is placed 1.8 m in front of a convex mirror that has a focal length of −0.75 m. **(a)** Using a ray diagram, determine the approximate location and size of the image produced by the mirror. **(b)** Referring to your ray diagram, is the image upright or inverted? **(c)** Does your ray diagram indicate that the image is larger or smaller than the object?

116. A makeup mirror produces an upright image that is magnified by a factor of 4.5 when your face is 28 cm in front of the mirror. **(a)** Is the mirror concave or convex? Explain. **(b)** What is the radius of curvature of the mirror?

117. Two objects are placed in a line directly in front of a convex mirror. One object is 12 cm from the mirror, and the other is 24 cm from the mirror. If the focal length of the mirror is −0.95 m, what is the distance between the images of the two objects?

Chapter 17

131. What is the smallest possible value for the index of refraction of a substance? Explain.

132. A glass slab surrounded by air causes a sideways displacement in a beam of light. If the slab is now placed in water, does the displacement it causes increase, decrease, or stay the same? Explain.

133. A submerged scuba diver looks up toward the calm surface of a freshwater lake and notes that the Sun appears to be 35° from the vertical. The diver's friend is standing on the shore of the lake. At what angle above the horizon does the friend see the Sun?

134. Light is refracted as it travels from point A in substance 1 to point B in substance 2. If the index of refraction is 1.33 for substance 1 and 1.51 for substance 2, how much time does it take light to go from A to B assuming that it travels 331 cm in substance 1 and 151 cm in substance 2?

135. Two identical containers are filled with different transparent liquids. The container with liquid A appears to have a greater depth than the container with liquid B. Which liquid has the greater index of refraction? Explain.

136. A beam of light traveling in an unknown material encounters a boundary with flint glass ($n = 1.66$). If the critical angle for total internal reflection is 32°, what is the index of refraction of the unknown material?

137. When the prism in Problem 80 is immersed in a fluid with an index of refraction of 1.21, the internal reflections shown in Figure 17.42 are still total. The reflections are no longer total, however, when the prism is immersed in a fluid with $n = 1.43$. Use this information to set upper and lower limits on the possible values of the prism's index of refraction.

138. **Think & Calculate** Suppose the glass paperweight in Figure 17.43 has the index of refraction $n = 1.48$.

(a) Find the value of θ for which the reflection at the vertical surface of the paperweight exactly satisfies the condition for total internal reflection. **(b)** If θ is increased, is the reflection at the vertical surface still total? Explain.

139. A laser beam enters one of the sloping faces of the equilateral glass prism ($n = 1.42$) in **Figure 17.56** and is refracted by the prism. Within the prism the light travels horizontally. What is the angle θ between the direction of the incident ray and the direction of the outgoing ray?

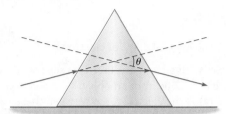

Figure 17.56

140. If a lens is cut in half, does it form only half an image?

141. An object with a height of 2.54 cm is placed 36.3 mm to the left of a lens with a focal length of 35.0 mm. **(a)** Where is the image located? **(b)** What is the height of the image?

142. A small insect is viewed through a convex lens that is 1.2 cm from the insect. The insect appears twice its actual size. What is the focal length of the lens?

143. **Think & Calculate** An object is located to the left of a concave lens whose focal length is -36 cm. The magnification produced by the lens is $m = \frac{1}{3}$. **(a)** To decrease the magnification to $\frac{1}{4}$, should the object be moved closer to the lens or farther away? **(b)** Find the object distance that gives a magnification of $\frac{1}{4}$.

144. **Predict & Explain** To focus, the eye of an octopus does not change the shape of its lens, as is the case in humans. Instead, an octopus's eye moves its rigid lens back and forth, as in a camera. This changes the distance from the lens to the retina and brings an object into focus. **(a)** If an object moves closer to an octopus, must the octopus's lens move closer to or farther from the retina to keep the object in focus? **(b)** Choose the *best* explanation from among the following:

A. The lens must move closer to the retina—that is, farther away from the object—to compensate for the object moving closer to the eye.

B. When the object moves closer to the eye, the image produced by the lens will be farther behind the lens; therefore, the lens must move farther from the retina.

145. If your near point is at a distance N, how close can you stand to a mirror and still be able to focus on your image?

146. **Think & Calculate** Ariel is nearsighted, and without her eyeglasses she can only focus on objects less than 2.2 m away. **(a)** Are Ariel's eyeglasses concave or convex? Explain. **(b)** To correct Ariel's nearsightedness, her eyeglasses must produce a virtual, upright image at a distance of 2.2 m when she is viewing an infinitely distant object. What is the focal length of the lenses in Ariel's eyeglasses?

147. A grade-school student plans to build a 35× telescope as a science fair project. She starts with a magnifying glass with a focal length of 5.0 cm as the eyepiece. What focal length should her objective lens have?

148. If a lens is immersed in water, its focal length increases, as noted in Conceptual Example 17.7. If a spherical mirror is immersed in water, does its focal length increase, decrease, or stay the same? Explain.

149. The three laser beams shown in **Figure 17.57** meet at a point at the back of a solid, transparent sphere. What is the index of refraction of the sphere?

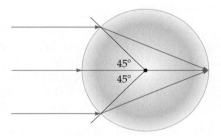

Figure 17. 57

150. A film of oil, with an index of refraction of 1.48 and a thickness of 1.50 cm, floats on a pool of water, as shown in **Figure 17.58**. A beam of light is incident on the oil at an angle of 60.0° to the vertical. Find the angle θ that the light beam makes with the vertical as it travels through the water

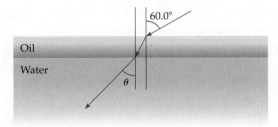

Figure 17.58

151. The sign in **Figure 17.59** has words that are either blue or red. Look near the center of the sign for several seconds. If you're like most people, you'll notice that words in one of these colors appear to be "deeper" in the sign (farther away from your eyes)

than words in the other color. Which words appear farther away to you? Use the fact that blue light refracts more than red light to explain this effect.

Figure 17.59

Chapter 18

139. A light source emits two distinct wavelengths, $\lambda_1 = 430$ nm (violet) and $\lambda_2 = 630$ nm (orange). The light strikes a diffraction grating with a slit spacing of 2.2×10^{-6} m. Identify the colors of the first three principal maxima on either side of the midpoint of the diffraction pattern.

140. When blue light with a wavelength of 465 nm shines on a diffraction grating, it produces a first-order principal maximum but no second-order principal maximum. **(a)** Explain the absence of the higher-order principal maxima. **(b)** What is the minimum slit spacing of this grating?

141. White light reflected at normal incidence from a soap bubble ($n = 1.33$) in air produces destructive interference at $\lambda = 575$ nm. What are the possible thicknesses of the soap film?

142. The headlights of a pickup truck are 1.32 m apart. What is the greatest distance at which these headlights can be resolved as separate points of light on a photograph taken with a camera whose aperture (opening) has a diameter of 12.5 mm? Let $\lambda = 555$ nm. (*Hint:* Set the angle for the first-order dark fringe in the diffraction pattern equal to the angle between the headlights.)

143. The yellow light of sodium, with wavelengths of 588.99 nm and 589.59 nm, is normally incident on a diffraction grating with 6025 lines/cm. Find the linear distance between the first-order principal maxima for these two wavelengths on a screen 3.55 m from the grating.

144. $\boxed{\text{Think \& Calculate}}$ A thin soap film ($n = 1.33$) suspended in air has a uniform thickness. When white light strikes the film, violet light ($\lambda_V = 420$ nm) experiences destructive interference. **(a)** If you wanted green light ($\lambda_G = 560$ nm) to experience destructive interference instead, should the film's thickness be increased or decreased? **(b)** Find the new thickness

of the film. (Assume that the film has the minimum thickness that can produce destructive interference.)

145. Light with a wavelength of 625 nm is used in a two-slit experiment. If the separation between the slits is 5.3×10^{-4} m, what is the angle to the first bright fringe above the central fringe?

146. The angle to the second-order dark fringe ($m = 2$) in a single-slit diffraction experiment is 26°. What is the width of the slit, given that the wavelength of the light is 575 nm?

Chapter 19

107. What is the mass of 1.5 C of electrons?

108. An object with a charge of 0.22 C is separated by a distance of 0.67 m from an object with a charge of -1.6 C. **(a)** Is the electric force between these two objects attractive or repulsive? Explain. **(b)** What is the magnitude of the electric force experienced by the object whose charge is 0.22 C? **(c)** What is the magnitude of the electric force experienced by the object whose charge is -1.6 C?

109. The electric force between two charges (4.1 µC and 7.9 µC) has a magnitude of 0.12 N. What is the separation between the charges?

110. A charge Q and a charge $2Q$ experience an electric force of magnitude 0.38 N when their separation is 0.95 m. What is the magnitude of the charge Q?

111. Three identical charges of 3.5 µC lie on the x axis. Charge 1 is at $x = 0$, charge 2 is at $x = 15$ cm, and charge 3 is at $x = 23$ cm. **(a)** Is the total electric force exerted on charge 2 in the positive or negative x direction? Explain. **(b)** What is the magnitude of the total electric force exerted on charge 2?

112. Four identical charges lie at the corners of a square with sides of length 21 cm. If the magnitude of each charge is 5.5 µC, what is the magnitude of the total electric force experienced by each of the charges?

113. The center of a sphere with a surface area of 0.025 m^2 and a surface charge density of 5.1 µC/m^2 is 0.18 m from a point charge of -3.3 µC. What is the magnitude of the electric force between the charge and the sphere?

114. A sphere with a surface charge density of 6.9 µC/m^2 exerts an electric force of 2.7×10^{-3} N on a point charge of 4.4 µC that is 0.11 m from the center of the sphere. What is the surface area of the sphere?

Chapter 20

122. An object of mass $m = 3.7$ g and charge $Q = +44$ µC is attached to a string and placed in a uniform electric field that is inclined at an angle of 30.0° with the horizontal, as shown in **Figure 20.37**. The object is in static equilibrium when the string is horizontal. Find **(a)** the magnitude of the electric field and **(b)** the tension in the string.

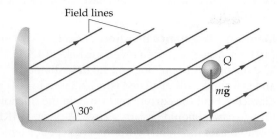

Figure 20.37

123. Think & Calculate A point charge of mass 0.081 kg and charge +6.77 μC is suspended by a thread between the vertical parallel plates of a parallel-plate capacitor, as shown in **Figure 20.38**. **(a)** If the charge deflects to the right of the vertical, as indicated in the figure, which of the two plates is at the higher electric potential? **(b)** If the angle of deflection is 22°, and the separation between the plates is 0.025 m, what is the potential difference between the plates?

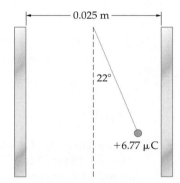

Figure 20.38

124. The electric potential a distance r from a point charge q is 2.70×10^4 V. One meter farther away from the charge the potential is 6140 V. Find the charge q and the initial distance r.

125. When the potential difference between the plates of a capacitor is increased by 3.25 V, the magnitude of the charge on each plate increases by 13.5 μC. What is the capacitance of this capacitor?

Chapter 21

131. What is the potential difference across a 250-Ω resistor if the current flowing through it is 0.75 A?

132. How much time is required for 0.85 C of charge to flow through a 220-Ω resistor when a potential difference of 12 V is applied to it?

133. Three resistors (65 Ω, 550 Ω, and 320 Ω) are connected in series. **(a)** Is the equivalent resistance of these resistors greater than, less than, or equal to 550 Ω? Explain. **(b)** What is the equivalent resistance of the three resistors?

134. Three resistors (65 Ω, 550 Ω, and 320 Ω) are connected in parallel. **(a)** Is the equivalent resistance of these resistors greater than, less than, or equal to 65 Ω? Explain. **(b)** What is the equivalent resistance of the three resistors?

135. A 9.0-V battery is connected in series with four resistors (220 Ω, 510 Ω, 330 Ω, and 160 Ω). **(a)** How much current is supplied by the battery? **(b)** What is the potential difference across the 330-Ω resistor?

136. A 9.0-V battery is connected in parallel with four resistors (220 Ω, 510 Ω, 330 Ω, and 160 Ω). **(a)** How much current is supplied by the battery? **(b)** What is the current flowing through the 330-Ω resistor?

137. When a resistor is connected to a 12-V battery, it dissipates 140 J of energy in 2.5 min. What is the resistance of the resistor?

138. How much energy does a 280-Ω resistor dissipate in 75 s when attached to a 12-V battery?

Chapter 22

105. For each of the three situations shown in **Figure 22.39**, indicate whether there will be a tendency for the square current-carrying loop to rotate clockwise, counterclockwise, or not at all, when viewed from above along the indicated axis.

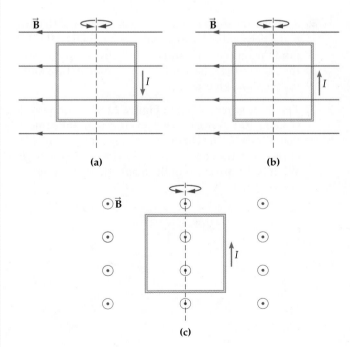

Figure 22.39

106. The velocity selector in **Figure 22.40** is designed to allow charged particles with a speed of 4.5×10^3 m/s to pass through while experiencing zero total force. Find the direction and magnitude of the required electric field, given that the magnetic field has a magnitude of 0.96 T.

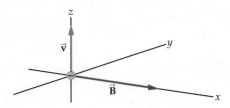

Figure 22.40

107. The long, thin wire shown in **Figure 22.41** is in a region of constant magnetic field, $\vec{B}$. The wire carries a current of 6.2 A and is oriented at an angle of 7.5° to the direction of the magnetic field. **(a)** If the magnetic force exerted on this wire per meter is 0.033 N, what is the magnitude of the magnetic field? **(b)** At what angle will the force exerted on the wire per meter be equal to 0.015 N?

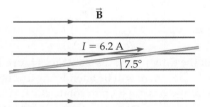

Figure 22.41

108. An electron and a proton move in circular paths in a plane perpendicular to a uniform magnetic field $\vec{B}$. Find the ratio of the radii of their circular paths when the two particles have **(a)** the same momentum and **(b)** the same kinetic energy.

109. The three wires shown in **Figure 22.42** are long and straight, and they each carry a current of the same magnitude, I. The currents in wires 1 and 3 flow out of the page; the current in wire 2 flows into the page. What is the direction of the magnetic force experienced by wire 3?

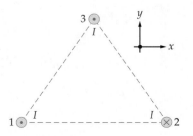

Figure 22.42

110. A mixture of two isotopes is injected into a mass spectrometer. One isotope follows a curved path of radius $R_1 = 48.9$ cm; the other follows a curved path of radius $R_2 = 51.7$ cm. Find the mass ratio, m_1/m_2, assuming that the two isotopes have the same charge and speed.

111. Magnetic Resonance Imaging The solenoid of an MRI machine produces a magnetic field of 1.5 T. The solenoid is 2.5 m long and 1.0 m in diameter, and is wound with insulated wires 2.2 mm in diameter. Find the current that flows in the solenoid. (Your answer should be rather large. A typical MRI solenoid uses niobium-titanium wire kept at a temperature where helium is liquid and where the wire is superconducting.)

112. Think & Calculate A long, straight wire carries a current of 14 A. Next to the wire is a square loop with sides 1.0 m in length, as shown in **Figure 22.43**. The loop carries a current of 2.5 A in the direction indicated. **(a)** What is the direction of the net force exerted on the loop? Explain. **(b)** Calculate the magnitude of the net force acting on the loop.

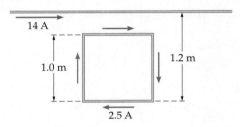

Figure 22.43

113. Two parallel wires, each carrying a current of 2.2 A in the same direction, are shown in **Figure 22.44**. Find the direction and magnitude of the net magnetic field at points A, B, and C.

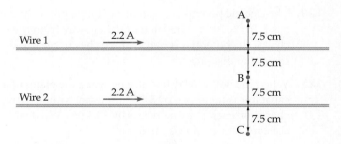

Figure 22.44

114. Think & Calculate Consider the two current-carrying wires shown in **Figure 22.45**. The current in wire 1 is 3.7 A; the current in wire 2 is adjusted to make the total magnetic field at point A equal to zero. **(a)** Is the magnitude of the current in wire 2 greater than, less than, or the same as that in wire 1?

Explain. **(b)** Find the magnitude and the direction of the current in wire 2.

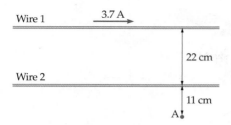

Figure 22.45

115. Solenoids produce magnetic fields that are relatively intense for the amount of current they carry. To make a direct comparison, consider a solenoid with 55 loops per centimeter, a radius of 1.05 cm, and a current of 0.124 A. **(a)** Find the magnetic field at the center of the solenoid. **(b)** What current must a long, straight wire carry to have the same magnetic field as that found in part (a)? Let the distance from the wire be the same as the radius of the solenoid, 1.05 cm.

116. An object with a charge of 34 μC moves with a speed of 62 m/s in the positive x direction. The magnetic field in this region of space has a component of 0.40 T in the positive y direction and a component of 0.85 T in the positive z direction. What are the magnitude and the direction of the magnetic force on the object?

117. Consider a system consisting of two concentric solenoids, as illustrated in **Figure 22.46**. The current in the outer solenoid is $I_1 = 1.25$ A, and the current in the inner solenoid is $I_2 = 2.17$ A. Given that the number of loops per centimeter is 105 for the outer solenoid and 125 for the inner solenoid, find the magnitude and the direction of the magnetic field **(a)** between the solenoids and **(b)** inside the inner solenoid.

Figure 22.46

Chapter 23

100. The 325-turn rectangular coil in a generator measures 11 cm by 17 cm. What is the maximum emf produced by this generator when the coil rotates with an angular speed of 525 rpm in a magnetic field of 0.45 T?

101. A long, straight wire carries a current I, as indicated in **Figure 23.33**. Three small metal rings are placed near the current-carrying wire (A and C) or directly on top of it (B). If the current in the wire is increasing with time, indicate whether the induced emf in each of the rings is clockwise, counterclockwise, or zero. Explain your answer for each ring.

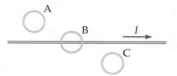

Figure 23.33

102. A car with a vertical radio antenna 85 cm long drives due east at 25 m/s. The Earth's magnetic field at this location is horizontal and has a magnitude of 5.9×10^{-5} T. Find the motional emf between the ends of the antenna.

103. | Predict & Explain | A metal ring is dropped into a localized region of constant magnetic field, as indicated in Figure 23.23. The magnetic field is zero above and below the region. **(a)** At each of the three indicated locations (1, 2, and 3), is the magnetic force exerted on the ring upward, downward, or zero? **(b)** Choose the *best* explanation from among the following:

A. The magnetic force is upward at 1 to oppose the ring entering the field, zero at 2 because the field is uniform, and downward at 3 to help the ring leave the field.

B. The magnetic force is upward at 1 to oppose the ring entering the field, upward at 2 where the field is strongest, and upward at 3 to oppose the ring leaving the field.

C. The magnetic force is upward at 1 to oppose the ring entering the field, zero at 2 because the field is uniform, and upward at 3 to oppose the ring leaving the field.

104. **Figure 23.34** shows the magnetic flux through a coil as a function of time. At what times shown in this graph does the magnetic flux have the greatest magnitude? (Recall that 1 Wb = 1 T·m².)

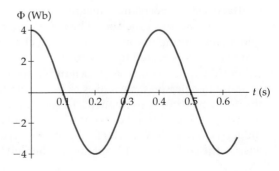

Figure 23.34

105. The area of a 120-loop coil oriented with the plane of its loops perpendicular to a 0.20-T magnetic field is 0.050 m². Find the average induced emf in this coil if the magnetic field reverses its direction in 0.34 s.

106. A magnetic field increases from zero to 0.25 T in 1.8 s. How many loops of wire are needed in a circular coil 12 cm in diameter to produce an induced emf of 6.0 V?

107. A generator is designed to produce a maximum emf of 170 V while rotating with an angular speed of 360 rad/s. Each coil of the generator has loops with an area of 0.016 m². If the magnetic field used in the generator has a magnitude of 0.050 T, how many loops of wire are needed?

108. A step-down transformer produces a voltage of 6.0 V across its secondary coil when the voltage across the primary coil is 120 V. What voltage appears across the primary coil of this transformer if 120 V is applied to the secondary coil?

109. A step-up transformer has 25 loops on the primary coil and 750 loops on the secondary coil. If this transformer is to produce a 12-mA current at a voltage of 4800 V, what input current and voltage are needed?

110. You hold a circular loop of wire at the equator. Consider the magnetic flux through this loop due to Earth's magnetic field. Is the flux when the normal to the loop is oriented in a north-south horizontal direction greater than, less than, or equal to the flux when the normal is vertical? Explain.

111. Consider the physical system shown in Figure 23.28. If the current in the wire changes direction but keeps the same magnitude, is the induced current in the circuit clockwise, counterclockwise, or zero? Explain.

112. Electrognathography Computerized jaw tracking, or *electrognathography* (EGN), is an important tool for diagnosing and treating *temporomandibular disorders* (TMDs) that affect a person's ability to bite effectively. The first step in applying EGN is to attach a small permanent magnet to the patient's gum below the lower incisors. Then, as the jaw undergoes a biting motion, the resulting change in magnetic flux is picked up by wire coils placed on either side of the mouth, as shown in **Figure 23.35**. Suppose this person's jaw moves to her right and the north pole of the permanent magnet is also toward her right. From her point of view, is the induced current flowing clockwise or counterclockwise in the coil to **(a)** her right and **(b)** her left? Explain.

Figure 23.35

113. Figure 23.36 shows four different situations (1, 2, 3, and 4) in which a metal ring moves to the right with constant speed through a region with a varying magnetic field. The intensity of the color indicates the intensity of the field, and in each case the field either increases or decreases at a uniform rate from the left edge of the colored region to the right edge. The direction of the field in each region is indicated. For each of the four cases, state whether the induced emf is clockwise, counterclockwise, or zero.

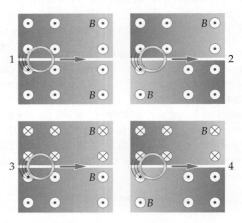

Figure 23.36

114. A rectangular loop of wire 24 cm by 72 cm is bent into an L shape, as shown in **Figure 23.37**. The magnetic field in the vicinity of the loop has a magnitude of 0.035 T and points in a direction 25° below the y axis. The magnetic field has no x component. Find the magnitude of the magnetic flux through the loop.

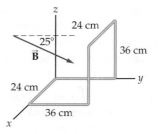

Figure 23.37

115. A magnetic field with the time dependence shown in **Figure 23.38** is at right angles to a 155-turn circular coil with a diameter of 3.75 cm. What is the induced emf in the coil at **(a)** $t = 2.50$ ms, **(b)** $t = 7.50$ ms, **(c)** $t = 15.0$ ms, and **(d)** $t = 25.0$ ms?

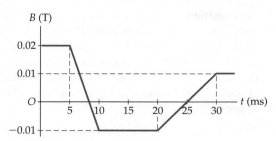

Figure 23.38

116. Think & Calculate A rectangular wire loop of width W and length L moves parallel to its length with a speed v. The loop moves from a region with a magnetic field $\vec{B}$ that is perpendicular to the plane of the loop to a region where the magnetic field is zero, as shown in **Figure 23.39**. Find the rate of change in the magnetic flux through the loop **(a)** before any of it enters the region of zero field, **(b)** just after it first enters the region of zero field, and **(c)** once it is fully within the region of zero field. **(d)** For each of the cases considered in parts (a), (b), and (c), state whether the induced current in the loop is clockwise, counterclockwise, or zero. Explain in each case.

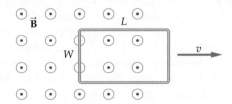

Figure 23.39

117. Think and Calculate A conducting rod of mass m is in contact with two vertical conducting rails separated by a distance L, as shown in **Figure 23.40**. The entire system is immersed in a magnetic field of magnitude B pointing out of the page. Assume that the rod slides without friction. **(a)** Describe the motion of the rod after it is released from rest. **(b)** What is the direction of the induced current (clockwise or counterclockwise) in the circuit? **(c)** Find the speed of the rod after it has fallen for a long time.

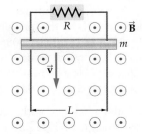

Figure 23.40

Chapter 24

103. Sirius, the brightest star in the night sky, has a surface temperature of 9940 K. What is the peak frequency of its radiation?

104. The work function of iron is 4.5 eV. What is the maximum wavelength of an electromagnetic wave that can eject an electron from iron?

105. A laser pointer emits light with a wavelength of 635 nm. If the power of the laser beam is 5.0 mW, how many photons per second are emitted by the laser?

106. What is the speed of a proton whose de Broglie wavelength is 2.0 Å?

107. What is the frequency of a photon whose energy is 3.9×10^{-19} J?

108. A blackbody gives off radiation whose peak frequency is 2.7×10^{14} Hz. What is the temperature of the blackbody?

109. What momentum must an electron have if its wavelength is to be the same as that of a photon with a frequency of 1.5×10^{14} Hz?

110. Electrons are ejected from the surface of lead by light with a frequency of 4.4×10^{14} Hz. If the work function of lead is 4.25 eV, what is the maximum kinetic energy of the ejected electrons?

Chapter 25

69. Is the frequency of a photon corresponding to a quantum jump from $n = 5$ to $n = 4$ greater than, less than, or equal to the frequency corresponding to a quantum jump from $n = 5$ to $n = 3$? Explain.

70. Is the wavelength of a photon corresponding to a quantum jump from $n = 6$ to $n = 5$ greater than, less than, or equal to the wavelength corresponding to a quantum jump from $n = 6$ to $n = 4$? Explain.

71. If the de Broglie wavelength of an electron in the $n = 1$ Bohr orbit is λ, what is its wavelength in the $n = 2$ Bohr orbit?

72. A hydrogen atom's electron is in the $n = 3$ Bohr orbit. What is the minimum energy required to ionize this hydrogen atom?

73. What is the frequency of the photon emitted when the electron in a hydrogen atom drops from the $n = 3$ Bohr orbit to the $n = 2$ Bohr orbit?

74. What are **(a)** the orbital radius and **(b)** the energy of the $n = 5$ Bohr state of hydrogen?

75. Find the orbital radius of the Bohr state with an energy of -0.85 eV.

76. An electron in a hydrogen atom jumps from the $n = 5$ state to the $n = 2$ state. Find the wavelength of the light that is emitted.

Chapter 26

98. Complete the following nuclear reaction:
$$^{6}_{3}\text{Li} + {}^{2}_{1}\text{H} \rightarrow {}^{4}_{2}\text{He} + \text{?}$$

99. Complete the following nuclear reaction:
$$^{2}_{1}\text{H} + {}^{14}_{7}\text{N} \rightarrow {}^{3}_{2}\text{He} + \text{?}$$

100. A radioactive material decays in such a way that only 1/8 of the material remains after 1 year. What is the half-life of the material?

101. The half-life of bismuth-212 ($^{212}_{83}\text{Bi}$) is approximately 1 hr. What percentage of an original amount of $^{212}_{83}\text{Bi}$ remains after **(a)** 2 hr, **(b)** 3 hr, and **(c)** 5 hr?

102. A sample of organic material from an archeological dig has a carbon-14 activity of 0.101 Bq per gram of carbon. What is the age of the material?

103. What is the fundamental difference between mesons and baryons?

104. Identify the numbers Z, N, and A for the following isotopes: **(a)** $^{10}_{5}\text{B}$, **(b)** $^{40}_{19}\text{K}$, **(c)** $^{138}_{57}\text{La}$.

105. The alpha decay of $^{238}_{92}\text{U}$ releases an energy of 4.25 MeV. **(a)** Identify the daughter nucleus that results from this decay. **(b)** Determine the change in mass that occurs as a result of this reaction.

Chapter 27

74. Albert is piloting his spaceship, heading east with a speed of 0.90c. Albert's ship sends a light beam in the forward (eastward) direction, which travels away from his ship at the speed c. Meanwhile, Isaac is piloting his ship in the westward direction, also at 0.90c, toward Albert's ship. With what speed does Isaac see Albert's light beam pass his ship?

75. Predict & Explain A clock in a moving rocket runs slow as measured by a stationary observer. **(a)** If the rocket reverses direction but moves with the same speed, does the clock run slow, fast, or at the normal rate? **(b)** Choose the *best* explanation from among the following:

A. The clock will run slow, just as before. The rate of the clock depends only on relative speed, not on direction of motion.

B. When the rocket reverses direction, the rate of the clock reverses too, and this makes it run fast.

C. Reversing the direction of the rocket undoes the time dilation effect, and so the clock will run at the normal rate.

76. An astronaut travels to Mars with a speed of 8350 m/s. After a month (30.0 d) of travel, as measured by clocks on Earth, how much difference is there between the Earth clocks and the spaceship clock? Give your answer in seconds.

77. When parked, your car is 5.0 m long. Unfortunately, your garage is only 4.0 m long. **(a)** How fast would your car have to be moving for an observer on the ground to see your car as shorter than your garage? **(b)** When you are driving at this speed, how long is your garage, as measured in the car's frame of reference?

78. Think & Calculate You and a friend travel through space in identical spaceships. Your friend informs you that he has made some length measurements and that his ship is 150 m long but that yours is only 120 m long. From your point of view, **(a)** how long is your friend's ship, **(b)** how long is your ship, and **(c)** what is the speed of your friend's ship relative to yours?

79. A spaceship moves at a constant speed $v = 0.825c$. As the ship passes a planet, the ship's captain notices that her watch and clocks on the planet both read 1:00 P.M. At 3:00 P.M., according to the captain's watch, the ship passes another planet. If the clocks on the two planets are synchronized with one another, what time do the clocks on the second planet read when the spaceship passes by?

80. The Pi Meson An elementary particle called a *pi meson* (or *pion*, for short) has an average lifetime of 2.6×10^{-8} s when at rest. If a pion moves with a speed of 0.99c relative to Earth, find **(a)** the average lifetime of the pion as measured by an observer on Earth and **(b)** the average distance traveled by the pion as measured by the same observer. **(c)** How far would the pion have traveled relative to Earth if relativistic time dilation did not occur?

81. When traveling past an observer with a relative speed v, a rocket is measured to be 9.00 m long. When the rocket moves with a relative speed 2v, its length is measured to be 5.00 m. What is the speed v?

82. Show that the units of mc^2 are joules (the units of energy), as expected.

83. Is it ever possible for the relativistic energy of an object to be less than its rest energy? Explain.

84. A helium atom has a rest mass of $m_{\text{Helium}} = 4.002603$ u. Its constituent particles (2 protons, 2 neutrons, 2 electrons), when well-separated, have the following masses: $m_{\text{proton}} = 1.007276$ u, $m_{\text{neutron}} = 1.008665$ u, and $m_{\text{electron}} = 0.000549$ u. How much work is required to completely disassemble a helium atom? (Recall that 1 u of mass has a rest energy of 931.5 MeV.)

85. Think & Calculate Consider a baseball with a rest mass of 0.145 kg. **(a)** How much work is required to increase the speed of the baseball from 25.0 m/s to

35.0 m/s? (For these low speeds, you can use the classical expression for the kinetic energy, $KE = \frac{1}{2}mv^2$.) **(b)** Is the work required to increase the speed of the baseball from 200,000,025 m/s to 200,000,035 m/s greater than, less than, or the same as the amount found in part (a)? Explain. **(c)** Calculate the work required to produce the increase in speed given in part (b). (For these high speeds, you must use the equation for relativistic energy.)

86. A lump of putty with a mass of 0.240 kg and a speed of $0.980c$ collides head-on with and sticks to an identical lump of putty moving with the same speed in the opposite direction. After the collision the system is at rest. What is the mass of the system after the collision?

87. A cluster of galaxies bends the light from a much more distant galaxy. Describe how the cluster of galaxies can produce multiple images of the distant galaxy.

88. Newly sprouted sunflowers can grow at the rate of 0.30 in. per day. One such sunflower is left on Earth, and an identical one is placed on a spacecraft that is traveling away from Earth with a speed of $0.94c$. How tall is the sunflower on the spacecraft when the sunflower on Earth is 2.0 in. high?

89. An apple falls from a tree, landing on the ground 3.7 m below. For what period of time is the apple in the air, as measured by an observer moving toward Earth with a speed of $0.89c$?

90. A *pulsar* is a collapsed, rotating star that sends out a narrow beam of radiation, like the light from a lighthouse. With each revolution we observe a brief, intense pulse of radiation from the pulsar. Suppose a pulsar is moving directly away from Earth with a speed of $0.800c$. If an observer on Earth finds that 153 pulses are emitted by the pulsar every second, at what rate are the pulses emitted from the viewpoint of an observer at rest relative to the pulsar?

Appendix C Tables

Table C.1

Color Conventions

Position vector ($\vec{r}$)	
Velocity vector ($\vec{v}$)	
Acceleration vector ($\vec{a}$)	
Force vector ($\vec{F}$)	
Momentum vector ($\vec{p}$)	
Light rays	
Object (in a ray diagram)	
Virtual image (in a ray diagram)	
Real image (in a ray diagram)	
Negative charge	⊖
Positive charge	⊕
Electric current direction in a wire	
Electric field	
Magnetic field	
Electron (subatomic particle)	⊖
Proton (subatomic particle)	⊕
Neutron (subatomic particle)	Ⓝ
Coordinate axes	y, x

Table C.2

Electric Circuit Symbols

Circuit Element	Symbol	Physical Characteristics
Conductor		Allows the flow of electric current with no resistance.
Resistor		Resists the flow of electric current. Converts electrical energy to thermal energy.
Capacitor		Stores electrical energy in the form of an electric field.
Inductor		Stores electrical energy in the form of a magnetic field.
Incandescent lightbulb		A device containing a resistor that gets hot enough to give off visible light.
Battery		A device that produces a constant difference in electrical potential between its terminals.
AC generator		A device that produces a potential difference between its terminals that oscillates with time.
Switches (open and closed)		Devices to control whether electric current is allowed to flow through a portion of a circuit.
Ground		Sets the electric potential at a point in a circuit equal to a constant value usually taken to be $V = 0$.
Transistor	Collector / Emitter / Base	A semiconductor device that allows a small current in the base to turn on or off a large current flowing through the collector and emitter, allowing the transistor to act as an amplifier.
Diode		A semiconductor device that allows current to flow only in the direction of the arrow and not in the opposite direction.

Table C.3

Multiples and Prefixes for Metric Units

Multiple	Prefix (Abbreviation)	Pronunciation
10^{24}	yotta- (Y)	yot´ta (*a* as in *a*bout)
10^{21}	zetta- (Z)	zet´ta (*a* as in *a*bout)
10^{18}	exa- (E)	ex´a (*a* as in *a*bout)
10^{15}	peta- (P)	pet´a (as in *peta*l)
10^{12}	tera- (T)	ter´a (as in *terra*ce)
10^{9}	giga- (G)	ji´ga (*ji* as in *ji*ggle, *a* as in *a*bout)
10^{6}	mega- (M)	meg´a (as in *mega*phone)

Table C.3

Multiples and Prefixes for Metric Units (cont.)		
Multiple	**Prefix (Abbreviation)**	**Pronunciation**
10^3	kilo- (k)	kil´o (as in *kilo*watt)
10^2	hecto- (h)	hek´to (*heck-toe*)
10	deka- (da)	dek´a (*deck* plus *a* as in *a*bout)
10^{-1}	deci- (d)	des´i (as in *deci*mal)
10^{-2}	centi- (c)	sen´ti (as in *senti*mental)
10^{-3}	milli- (m)	mil´li (as in *mili*tary)
10^{-6}	micro- (μ)	mi´kro (as in *micro*phone)
10^{-9}	nano- (n)	nan´oh (*an* as in *ann*ual)
10^{-12}	pico- (p)	pe´ko (*peek-oh*)
10^{-15}	femto- (f)	fem´toe (*fem* as in *fem*inine)
10^{-18}	atto- (a)	at´toe (as in an*ato*my)
10^{-21}	zepto- (z)	zep´toe (as in *zep*pelin)
10^{-24}	yocto- (y)	yock´toe (as in *sock*)

Table C.4

SI Base Units		
Physical Quantity	**Name of Unit**	**Symbol**
Length	meter	m
Mass	kilogram	kg
Time	second	s
Electric current	ampere	A
Temperature	kelvin	K
Amount of substance	mole	mol
Luminous intensity	candela	cd

Table C.5

Some SI Derived Units			
Physical Quantity	**Name of Unit**	**Symbol**	**SI Unit**
Frequency	hertz	Hz	s^{-1}
Energy	joule	J	$kg \cdot m^2/s^2$
Force	newton	N	$kg \cdot m/s^2$
Pressure	pascal	Pa	$kg/(m \cdot s^2)$
Power	watt	W	$kg \cdot m^2/s^3$
Electric charge	coulomb	C	$A \cdot s$
Electric potential	volt	V	$kg \cdot m^2/(A \cdot s^3)$
Electric resistance	ohm	Ω	$kg \cdot m^2/(A^2 \cdot s^3)$
Capacitance	farad	F	$A^2 \cdot s^4/(kg \cdot m^2)$
Inductance	henry	H	$kg \cdot m^2/(A^2 \cdot s^2)$
Magnetic field	tesla	T	$kg/(A \cdot s^2)$
Magnetic flux	weber	Wb	$kg \cdot m^2/(A \cdot s^2)$

Table C.6

SI Units of Some Other Physical Quantities	
Physical Quantity	**SI Unit**
Density (d)	kg/m^3
Speed (v)	m/s
Acceleration (a)	m/s^2
Momentum, impulse (p)	$kg \cdot m/s$
Angular speed (ω)	rad/s
Angular acceleration (α)	rad/s^2
Torque (τ)	$kg \cdot m^2/s^2$ or $N \cdot m$
Specific heat (c)	$J/(kg \cdot K)$
Thermal conductivity (k)	$W/(m \cdot K)$ or $J/(s \cdot m \cdot K)$
Entropy (S)	J/K or $kg \cdot m^2/(K \cdot s^2)$ or $N \cdot m/K$
Electric field (E)	N/C or V/m

Table C.7

Fundamental Constants		
Quantity	**Symbol**	**Approximate Value**
Speed of light	c	3.00×10^8 m/s $= 3.00 \times 10^{10}$ cm/s $= 186,000$ mi/s
Universal gravitational constant	G	6.67×10^{-11} N $\cdot$ m^2/kg^2
Stefan-Boltzmann constant	σ	5.67×10^{-8} W/(m$^2 \cdot$ K^4)
Boltzmann's constant	k	1.38×10^{-23} J/K
Avogadro's number	N_A	6.022×10^{23} mol^{-1}
Gas constant	$R = N_A k$	8.31 J/(mol $\cdot$ K) $= 1.99$ cal/(mol $\cdot$ K)
Coulomb's law constant	$k = 1/4\pi\varepsilon_0$	8.99×10^9 N $\cdot$ m^2/C^2
Electron charge	e	1.60×10^{-19} C
Permittivity of free space	ε_0	8.85×10^{-12} C^2/(N $\cdot$ m^2)
Permeability of free space	μ_0	$4\pi \times 10^{-7}$ T $\cdot$ m/A $= 1.26 \times 10^{-6}$ T $\cdot$ m/A
Planck's constant	h	6.63×10^{-34} J $\cdot$ s
Atomic mass unit	u	1.6605×10^{-27} kg $\Leftrightarrow 931.5$ MeV
Electron mass	m_e	9.10939×10^{-31} kg $\Leftrightarrow 0.511$ MeV
Neutron mass	m_n	$1.675\,00 \times 10^{-27}$ kg $\Leftrightarrow 939.57$ MeV
Proton mass	m_p	$1.672\,65 \times 10^{-27}$ kg $\Leftrightarrow 938.28$ MeV

Table C.8

Useful Physical Data	
Acceleration due to gravity (surface of Earth)	9.81 m/s^2 $= 32.2$ ft/s^2
Absolute zero	0 K $= -273.15\,°C = -459.67\,°F$
Standard temperature and pressure (STP)	$0\,°C = 273.15$ K 1 atm $= 101.325$ kPa
Density of air (at STP)	1.29 kg/m^3
Speed of sound in air (at 20 °C)	343 m/s
Density of water (at 4 °C)	1.000×10^3 kg/m^3
Latent heat of fusion of water	3.35×10^5 J/kg
Latent heat of vaporization of water	2.26×10^6 J/kg
Specific heat of water	4186 J/(kg $\cdot$ K)

Table C.9

Typical Values

Mass

Sun	2.00×10^{30} kg
Earth	5.97×10^{24} kg
Moon	7.35×10^{22} kg
747 airliner (maximum takeoff weight)	3.5×10^{5} kg
Blue whale	178,000 kg $=$ 197 tons
Elephant	5400 kg
Mountain gorilla	180 kg
Human	70 kg
Bowling ball	7 kg
Half gallon of milk	1.81 kg $=$ 4 lbs
Baseball	0.141–0.148 kg
Golf ball	0.045 kg
Female calliope hummingbird (smallest bird in North America)	3.5×10^{-3} kg $= \frac{1}{8}$ oz
Raindrop	3×10^{-5} kg
Antibody molecule	2.5×10^{-22} kg
Hydrogen atom	1.67×10^{-27} kg

Length

Orbital radius of Earth (around Sun)	1.5×10^{8} km
Orbital radius of Moon (around Earth)	3.8×10^{5} km
Altitude of geosynchronous satellite	35,800 km $=$ 22,300 mi
Radius of Earth	6370 km
Altitude of Earth's ozone layer	50 km
Height of Mt. Everest	8848 m
Height of Washington Monument	169 m $=$ 555 ft
Pitcher's mound to home plate	18.44 m
Baseball bat	1.067 m
CD (diameter)	120 mm
Aorta (diameter)	18 mm
Period in sentence (diameter)	0.5 mm
Red blood cell	7.8 mm $= \frac{1}{3300}$ in.
Typical bacterium (*E. coli*)	2 μm
Wavelength of green light	550 nm
Virus	20–300 nm
Large protein molecule	25 nm
Diameter of DNA molecule	2.0 nm
Radius of hydrogen atom	5.29×10^{-11} m

Time

Estimated age of Earth	About 4.6 billion y $\approx 10^{17}$ s
Estimated age of human species	About 150,000 y $\approx 5 \times 10^{12}$ s
Half-life of carbon-14	5730 y $= 1.81 \times 10^{11}$ s
Period of Halley's comet	76 y $= 2.40 \times 10^{9}$ s
Half-life of technetium-99	6 h $= 2.16 \times 10^{4}$ s
Time for driver of car to apply brakes	0.46 s

Table C.9

Typical Values *(cont.)*	
Time *(cont.)*	
Human reaction time	60–180 ms
Air bag deployment time	10 ms
Period of middle C sound wave	3.9 ms
Collision time for batted ball	2 ms
Decay of excited atomic state	10^{-8} s
Period of green light wave	1.8×10^{-15} s
Speed	
Light	3×10^8 m/s
Meteor	35–95 km/s
Space shuttle (orbital velocity)	8.5 km/s = 19,000 mi/h
Rifle bullet	700–750 m/s
Sound in air (STP)	340 m/s
Fastest human nerve impulses	140 m/s
747 at takeoff	80.5 m/s
Kangaroo	18.1 m/s = 40.5 mi/h
200-m dash (Olympic record)	10.1 m/s
Butterfly	1 m/s
Blood speed in aorta	0.35 m/s
Giant tortoise	0.076 m/s = 0.170 mi/h
Mer de Glace glacier (French Alps)	4×10^{-6} m/s
Acceleration	
Protons in particle accelerator	9×10^{13} m/s^2
Ultracentrifuge	3×10^6 m/s^2
Meteor impact	10^5 m/s^2
Baseball struck by bat	3×10^4 m/s^2
Loss of consciousness	$7.14g$ = 70 m/s^2
Acceleration due to gravity on Earth (g)	9.81 m/s^2
Braking auto	8 m/s^2
Acceleration due to gravity on Moon	1.62 m/s^2
Rotation of Earth at equator	3.4×10^{-2} m/s^2

Table C.10

Moments of Inertia for Common Shapes

Hoop or
cylindrical shell
$I = mr^2$

Disk or
solid cylinder
$I = \frac{1}{2}mr^2$

Disk or
solid cylinder
(axis at rim)
$I = \frac{3}{2}mr^2$

Long thin rod
(axis through midpoint)
$I = \frac{1}{12}mL^2$

Long thin rod
(axis at one end)
$I = \frac{1}{3}mL^2$

Hollow sphere
$I = \frac{2}{3}mr^2$

Solid sphere
$I = \frac{2}{5}mr^2$

Solid sphere
(axis at rim)
$I = \frac{7}{5}mr^2$

Solid plate
(axis through center,
in plane of plate)
$I = \frac{1}{12}mL^2$

Solid plate
(axis perpendicular
to plane of plate)
$I = \frac{1}{12}m(L^2 + W^2)$

Table C.11

Densities of Common Substances

Substance	Density (kg/m³)
SOLIDS	
Gold	19,300
Mercury	13,600
Lead	11,300
Silver	10,500
Iron	7860
Aluminum	2700
Ebony (wood)	1220
Ice	917
Cherry (wood)	800
Balsa (wood)	120
FLUIDS	
Ethylene glycol (antifreeze)	1114
Whole blood (at 37 °C)	1060
Seawater	1025
Freshwater	1000
Olive oil	920
Ethyl alcohol	806
GASES	
Oxygen	1.43
Air	1.29
Nitrogen	1.25
Helium	0.179
Hydrogen	0.0899

Table C.12

Melting Points and Boiling Points of Selected Substances

Substance	Melting Point (°C)	Boiling Point (°C)
Iron	1535	2750
Copper	1083	2567
Gold	1064	2808
Silver	962	2212
Germanium	937	2830
Aluminum	660	2467
Zinc	420	907
Lead	328	1740
Water	0	100
Mercury	−38.8	357
Nitrogen	−210	−196

Table C.13

Specific Heats of Selected Substances

Substance	Specific Heat, c (J/kg · K)
Water	4186
Ice	2090
Steam	2010
Beryllium	1820
Air	1004
Aluminum	900
Glass	837
Carbon	710
Silicon	703
Iron (steel)	448
Copper	387
Brass	376
Silver	234
Gold	129
Lead	128

Table C.14

Heats of Fusion and Vaporization of Selected Substances

Material	Latent Heat of Fusion, L_f (J/kg)	Latent Heat of Vaporization, L_v (J/kg)
Water	33.5×10^4	22.6×10^5
Ammonia	33.2×10^4	13.7×10^5
Iron	26.6×10^4	62.9×10^5
Copper	20.7×10^4	47.3×10^5
Benzene	12.6×10^4	3.94×10^5
Ethyl alcohol	10.8×10^4	8.55×10^5
Gold	6.28×10^4	17.2×10^5
Nitrogen	2.57×10^4	2.00×10^5
Lead	2.32×10^4	8.59×10^5
Oxygen	1.39×10^4	2.13×10^5
Mercury	1.15×10^4	2.72×10^5

Table C.15

Coefficients of Thermal Expansion

Substance	Coefficient of Thermal Expansion, α (K^{-1})
Lead	29×10^{-6}
Aluminum	24×10^{-6}
Brass	19×10^{-6}
Copper	17×10^{-6}
Iron (steel)	12×10^{-6}
Concrete	12×10^{-6}
Window glass	11×10^{-6}
Pyrex glass	3.3×10^{-6}
Quartz	0.50×10^{-6}

Table C.16

Indexes of Refraction for Common Substances	
Substance	**Index of Refraction, _n_**
SOLIDS	
Diamond	2.42
Cubic zirconia	2.20
Flint glass	1.66
Sodium chloride	1.54
Crown glass	1.52
Fused quartz (glass)	1.46
Ice	1.31
LIQUIDS	
Benzene	1.50
Glycerine	1.47
Ethyl alcohol	1.36
Water	1.33
GASES	
Carbon dioxide	1.00045
Air	1.000293

Table C.17

Wavelengths of Visible Light	
Color	**Wavelength (nm)**
Violet	380–430
Indigo	430–450
Blue	450–500
Cyan	500–520
Green	520–565
Yellow	565–590
Orange	590–625
Red	625–745

Table C.18

Solar System Data	
Equatorial radius of Earth	6.37×10^3 km = 3950 mi
Mass of Earth	5.97×10^{24} kg
Radius of Moon	1740 km = 1080 mi
Mass of Moon	7.35×10^{22} kg ≈ 1/81 mass of Earth
Average distance of Moon from Earth	3.84×10^5 km = 2.39×10^5 mi
Radius of Sun	6.95×10^5 km = 432,000 mi
Mass of Sun	2.00×10^{30} kg
Average distance of Earth from Sun	1.50×10^8 km = 93.0×10^6 mi

Table C.19

Planetary Data								
Name	**Equatorial Radius (km)**	**Mass (relative to Earth's)***	**Mean Density (kg/m³)**	**Surface Gravity (relative to Earth's)**	**Orbital Semimajor Axis (×10⁶ km)**	**Orbital Semimajor Axis (AU)**	**Escape Speed (km/s)**	**Orbital Period (y)**	**Orbital Eccentricity**
Mercury	2440	0.0553	5430	0.38	57.9	0.387	4.2	0.240	0.206
Venus	6052	0.816	5240	0.91	108.2	0.723	10.4	0.615	0.007
Earth	6370	1	5510	1	149.6	1	11.2	1.000	0.017
Mars	3394	0.108	3930	0.38	227.9	1.523	5.0	1.881	0.093
Jupiter	71,492	318	1360	2.53	778.4	5.203	60	11.86	0.048
Saturn	60,268	95.1	690	1.07	1427.0	9.539	36	29.42	0.054
Uranus	25,559	14.5	1270	0.91	2871.0	19.19	21	83.75	0.047
Neptune	24,776	17.1	1640	1.14	4497.1	30.06	24	163.7	0.009
Pluto**	1137	0.0021	2060	0.07	5906	39.84	1.2	248.0	0.249

*Mass of Earth = 5.97×10^{24} kg   **Pluto is a dwarf planet.

Table C.20

Periodic Table of the Elements

Transition elements

Key:
- 26 — Atomic number
- Fe — Symbol
- 55.85 — Atomic mass
- 3d⁶4s² — Outer electron configuration

Main table (PERIODS down the left; GROUPS across):

Period	GROUP I	GROUP II	Transition elements	GROUP III	GROUP IV	GROUP V	GROUP VI	GROUP VII	GROUP VIII
1	1 H 1.01 $1s^1$								2 He 4.00 $1s^2$
2	3 Li 6.94 $2s^1$	4 Be 9.01 $2s^2$		5 B 10.81 $2p^1$	6 C 12.01 $2p^2$	7 N 14.01 $2p^3$	8 O 16.00 $2p^4$	9 F 19.00 $2p^5$	10 Ne 20.18 $2p^6$
3	11 Na 22.99 $3s^1$	12 Mg 24.31 $3s^2$		13 Al 26.98 $3p^1$	14 Si 28.09 $3p^2$	15 P 30.97 $3p^3$	16 S 32.07 $3p^4$	17 Cl 35.45 $3p^5$	18 Ar 39.95 $3p^6$

Transition element blocks (period 4–7):

Period	Sc/Y/La/Ac	Ti/Zr/Hf/Rf	V/Nb/Ta/Db	Cr/Mo/W/Sg	Mn/Tc/Re/Bh	Fe/Ru/Os/Hs	Co/Rh/Ir/Mt	Ni/Pd/Pt/Ds	Cu/Ag/Au/Rg	Zn/Cd/Hg/Cn
4	21 Sc 44.96 $3d^14s^2$	22 Ti 47.88 $3d^24s^2$	23 V 50.94 $3d^34s^2$	24 Cr 52.00 $3d^54s^1$	25 Mn 54.94 $3d^54s^2$	26 Fe 55.85 $3d^64s^2$	27 Co 58.93 $3d^74s^2$	28 Ni 58.69 $3d^84s^2$	29 Cu 63.55 $3d^{10}4s^1$	30 Zn 65.39 $3d^{10}4s^2$
5	39 Y 88.96 $4d^15s^2$	40 Zr 91.22 $4d^25s^2$	41 Nb 92.91 $4d^45s^1$	42 Mo 95.94 $4d^55s^1$	43 Tc (98) $4d^55s^2$	44 Ru 101.07 $4d^75s^1$	45 Rh 102.91 $4d^85s^1$	46 Pd 106.42 $4d^{10}5s^6$	47 Ag 107.87 $4d^{10}5s^1$	48 Cd 112.41 $4d^{10}5s^2$
6	57 La 138.91 $5d^16s^2$ *	72 Hf 178.49 $5d^26s^2$	73 Ta 180.95 $5d^36s^2$	74 W 183.85 $5d^46s^2$	75 Re 186.21 $5d^56s^2$	76 Os 190.2 $5d^66s^2$	77 Ir 192.22 $5d^76s^2$	78 Pt 195.08 $5d^96s^1$	79 Au 196.97 $5d^{10}6s^1$	80 Hg 200.59 $5d^{10}6s^2$
7	89 Ac 227.03 $6d^17s^2$ †	104 Rf (261) $6d^27s^2$	105 Db (262) $6d^37s^2$	106 Sg (266) $6d^47s^2$	107 Bh (264) $6d^57s^2$	108 Hs (269) $6d^67s^2$	109 Mt (268) $6d^77s^2$	110 Ds (281) $6d^87s^2$	111 Rg (281) $6d^97s^2$	112 Cn (285) $6d^{10}7s^2$

p-block (period 4–7):

Period	GROUP I	GROUP II	GROUP III	GROUP IV	GROUP V	GROUP VI	GROUP VII	GROUP VIII
4	19 K 39.10 $4s^1$	20 Ca 40.08 $4s^2$	31 Ga 69.72 $4p^1$	32 Ge 72.61 $4p^2$	33 As 74.92 $4p^3$	34 Se 78.96 $4p^4$	35 Br 79.90 $4p^5$	36 Kr 83.80 $4p^6$
5	37 Rb 85.47 $5s^1$	38 Sr 87.62 $5s^2$	49 In 114.82 $5p^1$	50 Sn 118.71 $5p^2$	51 Sb 121.76 $5p^3$	52 Te 127.60 $5p^4$	53 I 126.90 $5p^5$	54 Xe 131.29 $5p^6$
6	55 Cs 132.91 $6s^1$	56 Ba 137.33 $6s^2$	81 Tl 204.36 $6p^1$	82 Pb 207.2 $6p^2$	83 Bi 208.98 $6p^3$	84 Po (209) $6p^4$	85 At (210) $6p^5$	86 Rn (222) $6p^6$
7	87 Fr (223) $7s^1$	88 Ra 226.03 $7s^2$	113 Uut (284) $7p^1$	114 Fl (289) $7p^2$	115 Uup (288) $7p^3$	116 Lv (293) $7p^4$	117 Uus (294) $7p^5$	118 Uuo (294) $7p^6$

* Lanthanides (f):

58 Ce 140.12 $5d^14f^16s^2$	59 Pr 140.91 $4f^36s^2$	60 Nd 144.24 $4f^46s^2$	61 Pm (145) $4f^56s^2$	62 Sm 150.36 $4f^66s^2$	63 Eu 151.96 $4f^76s^2$	64 Gd 157.25 $5d^14f^76s^2$	65 Tb 158.93 $4f^96s^2$	66 Dy 162.50 $4f^{10}6s^2$	67 Ho 164.93 $4f^{11}6s^2$	68 Er 167.26 $4f^{12}6s^2$	69 Tm 168.93 $4f^{13}6s^2$	70 Yb 173.04 $4f^{14}6s^2$	71 Lu 174.97 $5d^14f^{14}6s^2$

† Actinides (f):

90 Th 232.04 $6d^27s^2$	91 Pa 231.04 $5f^26d^17s^2$	92 U 238.03 $5f^36d^17s^2$	93 Np 237.05 $5f^46d^17s^2$	94 Pu (244) $5f^66d^07s^2$	95 Am (243) $5f^76d^07s^2$	96 Cm (247) $5f^76d^17s^2$	97 Bk (247) $5f^86d^17s^2$	98 Cf (251) $5f^{10}6d^07s^2$	99 Es (252) $5f^{11}6d^07s^2$	100 Fm (257) $5f^{12}6d^07s^2$	101 Md (258) $5f^{13}6d^07s^2$	102 No (259) $5f^{14}6d^07s^2$	103 Lr (262) $5f^{14}6d^17s^2$

Table C.21

Properties of Selected Isotopes

Atomic Number (Z)	Element	Symbol	Mass Number (A)	Atomic Mass	Abundance (%) or Decay Mode* (if radioactive)	Half-Life (if radioactive)
0	(Neutron)	n	1	1.008665	β^-	10.6 min
1	Hydrogen	H	1	1.007825	99.985	
	Deuterium	D	2	2.014102	0.015	
	Tritium	T	3	3.016049	β^-	12.33 y
2	Helium	He	3	3.016029	0.00014	
			4	4.002603	~100	
3	Lithium	Li	6	6.015123	7.5	
			7	7.016003	92.5	
4	Beryllium	Be	7	7.016930	EC, γ	53.3 d
			8	8.005305	2α	6.7×10^{-17} s
			9	9.012183	100	
5	Boron	B	10	10.012938	19.9	
			11	11.009305	80.1	
			12	12.014353	β^-	20.2 ms
6	Carbon	C	11	11.011433	β^+, EC	20.3 min
			12	12.000000	98.89	
			13	13.003355	1.11	
			14	14.003242	β^-	5730 y
7	Nitrogen	N	13	13.005739	β^-	9.96 min
			14	14.003074	99.63	
			15	15.000109	0.37	
8	Oxygen	O	15	15.003065	β^+, EC	122 s
			16	15.994915	99.76	
			18	17.999159	0.204	
9	Fluorine	F	19	18.998403	100	
			18	18.000938	EC	109.77 min
10	Neon	Ne	20	19.992439	90.51	
			22	21.991384	9.22	
11	Sodium	Na	22	21.994435	β^+, EC, γ	2.602 y
			23	22.989770	100	
			24	23.990964	β^-, γ	15.0 h
12	Magnesium	Mg	24	23.985045	78.99	
13	Aluminum	Al	27	26.981541	100	
14	Silicon	Si	28	27.976928	92.23	
			31	30.975364	β^-, γ	2.62 h
15	Phosphorus	P	31	30.973763	100	
			32	31.973908	β^-	14.28 d
16	Sulfur	S	32	31.972072	95.0	
			35	34.969033	β^-	87.4 d
17	Chlorine	Cl	35	34.968853	75.77	
			37	36.965903	24.23	
18	Argon	Ar	40	39.962383	99.60	
19	Potassium	K	39	38.963708	93.26	
			40	39.964000	β^-, EC, γ, β^+	1.28×10^9 y
20	Calcium	Ca	30	39.962591	96.94	

*EC stands for "electron capture."

Atomic Number (*Z*)	Element	Symbol	Mass Number (*A*)	Atomic Mass	Abundance (%) or Decay Mode* (if radioactive)	Half-Life (if radioactive)
24	Chromium	Cr	52	51.940510	83.79	
25	Manganese	Mn	55	54.938046	100	
26	Iron	Fe	56	55.934939	91.8	
27	Cobalt	Co	59	58.933198	100	
			60	59.933820	β^-, γ	5.271 y
28	Nickel	Ni	58	57.935347	68.3	
			60	59.930789	26.1	
			64	63.927968	0.91	
29	Copper	Cu	63	62.929599	69.2	
			64	63.929766	β^-, β^+	12.7 h
			65	64.927792	30.8	
30	Zinc	Zn	64	63.929145	48.6	
			66	65.926035	27.9	
33	Arsenic	As	75	74.921596	100	
35	Bromine	Br	79	78.918336	50.69	
36	Krypton	Kr	84	83.911506	57.0	
			89	88.917563	β^-	3.2 min
			92	91.926153	β^-	1.84 s
38	Strontium	Sr	86	85.909273	9.8	
			88	87.905625	82.6	
			90	89.907746	β^-	28.8 y
39	Yttrium	Y	89	89.905856	100	
41	Niobium	Nb	98	97.910331	β^-	2.86 s
43	Technetium	Tc	98	97.907210	β^-, γ	4.2×10^6 y
47	Silver	Ag	107	106.905095	51.83	
			109	108.904754	48.17	
48	Cadmium	Cd	114	113.903361	28.7	
49	Indium	In	115	114.90388	95.7; β^-	5.1×10^{14} y
50	Tin	Sn	120	119.902199	32.4	
51	Antimony	Sb	133	132.915237	β^-	2.5 min
53	Iodine	I	127	126.904477	100	
			131	130.906118	β^-, γ	8.04 d
54	Xenon	Xe	132	131.90415	26.9	
			136	135.90722	8.9	
55	Cesium	Cs	133	132.90543	100	
56	Barium	Ba	137	136.90582	11.2	
			138	137.90524	71.7	
			141	140.914406	β^-	18.27 min
			144	143.92273	β^-	11.9 s
61	Promethium	Pm	145	144.91275	EC, α, γ	17.7 y
74	Tungsten (Wolfram)	W	184	183.95095	30.7	
76	Osmium	Os	191	190.96094	β^-, γ	15.4 d
			192	191.96149	41.0	
78	Platinum	Pt	195	194.96479	33.8	
79	Gold	Au	197	196.96656	100	
81	Thallium	Tl	205	204.97441	70.5	

Atomic Number (Z)	Element	Symbol	Mass Number (A)	Atomic Mass	Abundance (%) or Decay Mode* (if radioactive)	Half-Life (if radioactive)
			210	209.990069	β^-	1.3 min
82	Lead	Pb	204	203.973044	1.48; β^-	1.4×10^{17} y
			206	205.97446	24.1	
			207	206.97589	22.1	
			208	207.97664	52.3	
			210	209.98418	α, β^-, γ	22.3 y
			211	210.98874	β^-, γ	36.1 min
			212	211.99188	β^-, γ	10.64 h
			214	213.99980	β^-, γ	26.8 min
83	Bismuth	Bi	209	208.98039	100	
			211	210.98726	α, β^-, γ	2.15 min
			212	211.991272	α	60.55 min
84	Polonium	Po	210	209.98286	α, γ	138.38 d
			212	211.988852	α	0.299 μs
			214	213.99519	α, γ	164 μs
86	Radon	Rn	222	222.017574	α, β	3.8235 d
87	Francium	Fr	223	223.019734	α, β^-, γ	21.8 min
88	Radium	Ra	226	226.025406	α, γ	1.60×10^3 y
			228	228.031069	β^-	5.76 y
89	Actinium	Ac	227	227.027751	α, β^-, γ	21.773 y
90	Thorium	Th	228	228.02873	α, γ	1.9131 y
			231	231.036297	α, β^-	25.52 h
			232	232.038054	100; α, γ	1.41×10^{10} y
			234	234.043596	β^-	24.10 d
91	Protactinium	Pa	234	234.043302	β^-	6.70 h
92	Uranium	U	232	232.03714	α, γ	72 y
			233	233.039629	α, γ	1.592×10^5 y
			235	235.043925	0.72; α, γ	7.038×10^8 y
			236	236.045563	α, γ	2.342×10^7 y
			238	238.050786	99.275; α, γ	4.468×10^9 y
			239	239.054291	β^-, γ	23.5 min
93	Neptunium	Np	239	239.052932	β^-, γ	2.35 d
94	Plutonium	Pu	239	239.052158	α, γ	2.41×10^4 y
95	Americium	Am	243	243.061374	α, γ	7.37×10^3 y
96	Curium	Cm	245	245.065487	α, γ	8.5×10^3 y
97	Berkelium	Bk	247	247.07003	α, γ	1.4×10^3 y
98	Californium	Cf	249	249.074849	α, γ	351 y
99	Einsteinium	Es	254	254.08802	α, γ, β^-	276 d
100	Fermium	Fm	253	253.08518	EC, α, γ	3.0 d
101	Mendelevium	Md	255	255.0911	EC, α	27 min
102	Nobelium	No	255	255.0933	EC, α	3.1 min
103	Lawrencium	Lr	257	257.0998	α	~35 s

Appendix D

Safety in the Laboratory

The experiments in this book have been carefully designed to minimize the risk of injury. However, safety is also your responsibility. The following rules are essential for keeping you safe in the laboratory.

Pre-Lab Preparation

- Read the entire procedure before you begin. When in doubt about a procedure, ask your teacher.

- Do only assigned experiments. Do not work on an experiment unless your teacher is present and has given you permission to do so.

- Get approval from your teacher for any student-designed procedure before attempting it.

- Know the location of the fire extinguisher and the fire blanket, and how to use them.

- Know the location of emergency exits and escape routes. Do not block walkways with furniture.

- Keep your work area orderly and free of personal belongings such as coats and backpacks.

Safe In-Lab Practices

General Safety

- Listen to and follow all of your teacher's instructions.

  ⚠ **Caution:** *Immediately report any accident, no matter how minor, to your teacher.*

Eye Safety

- Wear goggles at all times when working in the laboratory—goggles are designed to protect your eyes from injury.

- Never look directly at the Sun through any optical device. Do not use direct sunlight to illuminate a microscope.

- Avoid wearing contact lenses when the experiment involves chemicals. If contact lenses must be worn, then wear eye-cup safety goggles over them.

Clothing Safety

- Avoid wearing bulky or loose-fitting clothing.

- Remove dangling jewelry.

- Wear closed-toe shoes at all times in the laboratory.

Thermal Safety

- Protect your clothing and hair from sources of heat. Tie back long hair and roll up loose sleeves when working in the laboratory.

- Use extreme caution when working with any type of heating device.

- Do not handle hot glassware or equipment. You can prevent burns by being aware that hot and cold equipment can look exactly the same.

  ⚠ **Caution:** *If you are burned, immediately run cold water over the burned area for several minutes until the pain is reduced. Cooling helps the burn heal. Ask a classmate to notify your teacher.*

Chemical Safety

- Never taste any chemical used in the laboratory, including food products that are the subject of an investigation. Treat all items as though they are contaminated with unknown chemicals that may be toxic.

- Keep all food and drink that is not part of an experiment out of the laboratory. Do not eat, drink, or chew gum in the laboratory.

- Wear safety goggles, an apron, and gloves when working with corrosive chemicals.

- Report any chemical spills immediately to your teacher. Follow your teacher's instructions for cleaning up spills. Warn other students about the identity and location of spilled chemicals.

- To reduce danger, waste, and cleanup, always use the minimal amount of any chemical specified for an experiment.

  ⚠ **Caution:** *If a corrosive chemical gets on your skin or clothing, immediately wash the affected area with cold, running water for several minutes. Ask a classmate to notify your teacher.*

Sharp Object Safety

- Don't use chipped or cracked glassware. Don't handle broken glass. If glassware breaks, tell your teacher and nearby classmates. Discard broken glass as instructed by your teacher.

  ⚠ **Caution:** *If you receive a minor cut, allow it to bleed for a short time, then wash the injured area under cold,* *running water. Ask a classmate to notify your teacher. More serious cuts or puncture wounds require immediate medical attention.*

Electrical Safety

- Recognize that the danger of an electrical shock is greater in the presence of water. Keep electrical appliances away from sinks and faucets to minimize the risk of electrical shock. Be careful not to spill water or other liquids in the vicinity of an electrical appliance.

- Do not place electrical cords in such a way that they become trip hazards.

- Do not let cords hang off tables in such a way that could lead to the equipment being pulled off the table accidentally.

- Always be sure that any piece of electrical equipment is off before plugging it in and before unplugging it.

- Have your teacher inspect and approve all electrical circuits before they are used.

⚠ **Caution:** *If you spill water near an electrical appliance, stand back, notify your teacher, and warn other students in the area.*

Post-Lab Procedures

- Always follow your teacher's directions for cleanup and disposal. Dispose of used chemicals in a way that protects you, your classmates, and the environment.

- Wash your hands thoroughly with soap and water before leaving the laboratory.

One or more of the following safety symbols will appear in every experiment. Take appropriate precautions.

SAFETY SYMBOLS

 Eye Safety Wear safety goggles at all times.

 Clothing Protection Wear a lab coat or apron when using corrosive chemicals or chemicals that can stain clothing.

 Skin Protection Wear plastic gloves when using chemicals that can irritate or stain your skin.

 Broken Glass Do not use chipped or cracked glassware. Do not heat the bottom of a test tube.

 Open Flame Tie back hair and loose clothing. Never reach across a lit burner.

 Flammable Substance Do not have a flame near flammable materials.

 Corrosive Substance Wear safety goggles, an apron, and gloves when working with corrosive chemicals.

 Poison Never taste a chemical in the laboratory.

 Fume Avoid inhaling substances that can irritate your respiratory system.

 Thermal Burn Do not touch hot glassware or equipment.

 Electrical Equipment Keep electrical equipment away from water or other liquids.

 Sharp Object Avoid puncture wounds by using scissors and other sharp objects only as intended.

 Disposal Dispose of chemicals and other supplies only as directed.

 Hand Washing Wash your hands thoroughly with soap and water.

Glossary

A

absolute zero: the lowest possible temperature *(348)*
　cero absoluto: temperatura más baja posible

absorption spectrum: the line spectrum showing the wavelengths of light absorbed by an atom *(886)*
　espectro de absorción: espectro de líneas que muestra las longitudes de onda que son absorbidas por un átomo

AC circuit: a circuit in which the current periodically reverses its direction; also known as an *alternating-current circuit* *(748; 832)*
　circuito de CA: circuito en el que la corriente cambia de dirección periódicamente; también es conocido como *circuito de corriente alterna*

acceleration: the rate at which velocity changes with time *(73)*
　aceleración: magnitud que indica el cambio de la velocidad por unidad de tiempo

accuracy: a measure of how close the measured value of a quantity is to the actual value *(25)*
　precisión: medida de lo cercana que es la medición de una cantidad con respecto a su valor real

action force: one of the forces always present in any pair of action-reaction forces *(158)*
　fuerza de acción: fuerza que está siempre presente en un par de fuerzas de acción y reacción

activity: the rate at which nuclear decay occurs; the number of decays per second *(924)*
　actividad: tasa a la cual se produce la desintegración nuclear; el número de desintegraciones por segundo

additive primary colors: the three colors of light—red, green, and blue—that add together to produce white light *(541)*
　colores primarios aditivos: los tres colores de la luz (rojo, verde y azul) que, al mezclarse, producen luz blanca

adiabatic: occurring with no exchange of thermal energy; characteristic of some thermal processes *(397)*
　adiabático: que ocurre cuando no hay intercambio de energía térmica; característica de algunos procesos térmicos

alpha particle: a particle consisting of two protons and two neutrons (a helium nucleus) that is emitted in some radioactive decay processes *(918)*
　partícula alfa: partícula formada por dos protones y dos neutrones (núcleo de helio) que se emite en ciertos procesos de desintegración radiactiva

alternating-current generator: a generator that produces an alternating current *(829)*
　generador de corriente alterna: generador que produce corriente alterna

ammeter: a device designed to measure the flow of current through a portion of a circuit *(764)*
　amperímetro: aparato diseñado para medir el flujo de la corriente en una sección de un circuito

amplitude: the maximum displacement from equilibrium *(457)*
　amplitud: desplazamiento máximo con respecto a la posición de equilibrio

angle of incidence: the angle between the incident ray and the normal at the point of reflection *(566)*
　ángulo de incidencia: ángulo que se forma entre el rayo incidente y la normal en el punto de reflexión

angle of reflection: the angle between the reflected ray and the normal at the point of reflection *(566)*
　ángulo de reflexión: ángulo que se forma entre el rayo reflejado y la normal en el punto de reflexión

angular displacement: the difference between an object's final angle and its initial angle *(269)*
　desplazamiento angular: diferencia entre el ángulo final e inicial de un objeto que se desplaza circularmente

angular momentum: momentum associated with a rotating object *(230)*; the product of an object's moment of inertia and its angular velocity *(280)*
　momento angular: momento asociado con un objeto en rotación; el producto del momento de inercia de un objeto y su velocidad angular

angular position: the angle an object makes with respect to a given reference line *(267)*
　posición angular: ángulo que forma un objeto con respecto a una línea de referencia

angular speed: the magnitude of the angular velocity *(270)*
　rapidez angular: magnitud de la velocidad angular

antinode: a point of maximum displacement on a standing wave *(478)*
　antinodo: punto de desplazamiento máximo en una onda estacionaria

apparent depth: the phenomenon in which an object submerged in water appears to be closer to the water's surface than it really is *(603)*
　profundidad aparente: fenómeno en el cual un objeto sumergido en agua parece estar más cerca de la superficie de lo que realmente está

apparent weight: the sensation of having a different weight due to a force being exerted on a person *(165)*

peso aparente: sensación de tener un peso distinto debido a una fuerza ejercida sobre una persona

Archimedes' principle: the principle that states that the buoyant force on an object is equal to the weight of fluid that the object displaces *(432)*

principio de Arquímedes: principio que establece que la fuerza de flotación ejercida sobre un objeto es igual al peso del líquido que ese objeto desplaza

atomic mass: the mass of one atom of an element *(422)*

masa atómica: masa de un átomo de un elemento

atomic mass unit: a unit of mass equal to exactly $\frac{1}{12}$ of the mass of a carbon-12 atom *(914)*

unidad de masa atómica: unidad de masa que es igual a exactamente $\frac{1}{12}$ de la masa de un átomo de carbono 12

atomic number: the number of protons in a nucleus *(912)*

número atómico: número de protones de un núcleo

average acceleration: the change in velocity divided by the change in time *(75)*

aceleración media: cambio de la velocidad dividido por el intervalo de tiempo

average angular acceleration: the change in angular velocity in a given interval of time *(273)*

aceleración angular media: cambio de la velocidad angular en un intervalo de tiempo

average angular velocity: the angular displacement divided by the time during which the displacement occurs *(269)*

velocidad angular media: desplazamiento angular dividido por el tiempo durante el cual ocurre el desplazamiento

average speed: the speed of an object averaged over a given period of time *(48)*

rapidez media: promedio de la rapidez de un objeto durante un periodo de tiempo

average velocity: the displacement of an object per unit of time *(50)*

velocidad media: desplazamiento de un objeto por unidad de tiempo

Avogadro's number: the number of atoms in 1 mole of a substance *(420)*

número de Avogadro: número de átomos en 1 mol de sustancia

B

baryon: a hadron formed from three quarks *(939)*

barión: hadrón formado por tres quarks

battery: a device that uses chemical reactions to produce a difference in electric potential between its two terminals *(748)*

batería: aparato que usa reacciones químicas para producir una diferencia de potencial eléctrica entre sus dos terminales

beat: a change in loudness heard when sounds of different frequencies are produced at the same time *(498)*

batimiento: cambio en la intensidad del sonido que ocurre cuando se producen sonidos de frecuencias distintas al mismo tiempo

beat frequency: the frequency at which a beat repeats *(499)*

frecuencia de batimiento: frecuencia con la que se repite un batimiento

Bernoulli's principle: the principle stating that the pressure exerted by a fluid decreases as its flow speed increases *(437)*

principio de Bernoulli: principio que establece que la presión ejercida por un fluido disminuye al aumentar la rapidez del fluido

beta particle: an electron given off during radioactive decay of an isotope *(918)*

partícula beta: electrón cedido durante la desintegración radiactiva de un isótopo

bias: a preference for a particular point of view due to personal rather than scientific reasons *(10)*

sesgo: preferencia por un punto de vista en particular basada más en razones personales que científicas

black hole: an object such as a star with gravity so strong that light can't escape it *(319; 965)*

agujero negro: objeto, como una estrella, cuya gravedad es tan fuerte que la luz no puede escapar de él

blackbody: an ideal object that absorbs all of the light (and other electromagnetic radiation) that strikes it *(851)*

cuerpo negro: objeto ideal que absorbe toda la luz (y otras radiaciones electromagnéticas) que incide sobre él

Bohr radius: the radius of the first orbit (with quantum number $n = 1$) of the electron in the Bohr model of the hydrogen atom, which has a diameter twice as large as this radius *(889)*

radio de Bohr: radio de la primera órbita (con número cuántico $n = 1$) del electrón en el modelo de Bohr del átomo de hidrógeno, que tiene un diámetro que equivale al doble de este radio

boiling point: the temperature at which a substance boils; the temperature at which the vapor pressure of a substance equals the external pressure *(368)*

punto de ebullición: temperatura a la que hierve una sustancia; temperatura a la que la presión del vapor de una sustancia es igual a la presión externa

Boltzmann constant: a constant, denoted by k, that is used in the ideal gas law *(418)*

constante de Boltzmann: constante, denotada por k, que se usa en la ley del gas ideal

buoyant force: an upward force due to a surrounding fluid *(431)*

fuerza boyante: fuerza ascendente causada por un líquido circundante

C

Calorie: a unit of thermal energy used in nutrition science: 1 Cal = 1 kcal *(359)*

   Caloría: unidad de energía térmica usada en la ciencia nutricional: 1 Cal = 1 kcal

calorimeter: a lightweight, insulated flask used to determine the specific heat capacity of a substance *(363)*

   calorímetro: matraz ligero y aislante que se usa para determinar la capacidad de calor específica de una sustancia

capacitance: the ratio of the charge stored in a capacitor to the applied voltage *(728)*

   capacidad eléctrica: relación entre la carga almacenada en un condensador eléctrico y el voltaje aplicado

capacitor: a device that has the capacity to store both electric charge and electrical energy *(728)*

   condensador eléctrico: aparato que tiene la capacidad de almacenar carga y energía eléctricas

Celsius scale: the temperature scale on which water freezes at 0 °C and boils at 100 °C *(346)*

   escala Celsius: escala de temperatura en la que el agua se congela a 0 °C y hierve a 100 °C

center of curvature: the center of the sphere of which a curved mirror is a section *(576)*

   centro de curvatura: centro de la esfera de la cual un espejo curvo es una sección

center of mass: the point where an object can be balanced *(292)*

   centro de masa: punto donde se puede balancear un objeto

centripetal acceleration: the center-directed acceleration of an object in circular motion *(320)*

   aceleración centrípeta: aceleración dirigida hacia el centro de un objeto que se mueve de forma circular

centripetal force: the force that causes circular motion *(321)*

   fuerza centrípeta: fuerza que causa el movimiento circular

charge quantization: the fact that electric charge comes in amounts that are always integer multiples of e *(677)*

   cuantificación de la carga: el hecho de que la carga aparece en cantidades que son siempre múltiplos enteros de e

charging by induction: the charging of an object without direct contact *(716)*

   carga por inducción: carga de un objeto sin que ocurra un contacto directo

chromatic aberration: the deviation from ideal lens behavior that occurs because a lens bends light rays of different colors by different amounts *(622)*

   aberración cromática: desviación con respecto al comportamiento de una lente ideal que ocurre debido a que los rayos de luz de diferentes colores son curvados de forma distinta por una lente normal

coefficient of kinetic friction: the coefficient that relates the force of kinetic friction to the normal force *(171)*

   coeficiente de fricción cinética: coeficiente que relaciona la fuerza de fricción cinética con la fuerza normal

coefficient of static friction: the coefficient that relates the force of static friction to the normal force *(173)*

   coeficiente de fricción estática: coeficiente que relaciona la fuerza de fricción estática con la fuerza normal

coefficient of thermal expansion: the constant of proportionality relating a temperature change to an object's change in length *(351)*

   coeficiente de dilatación térmica: constante de proporcionalidad que relaciona un cambio de temperatura con el cambio en la longitud de un objeto

coherent light: light from sources that maintain a constant phase relative to one another *(637)*

   luz coherente: luz proveniente de fuentes que mantienen una relación de fase constante entre sí

collision: a situation in which two objects free from external forces strike one another *(248)*

   choque: situación en la que dos objetos libres de fuerzas externas chocan uno contra otro

completely inelastic collision: a collision in which the colliding objects stick together *(249)*

   choque perfectamente inelástico: choque después del cual los objetos permanecen unidos

compound microscope: a simple microscope consisting of two converging lenses fixed at either end of a tube *(620)*

   microscopio compuesto: microscopio simple formado por dos lentes convergentes colocadas a ambos extremos de un tubo

concave mirror: a mirror that curves inward, forming a sort of cave with its reflecting surface *(575)*

   espejo cóncavo: espejo que se curva hacia dentro y forma una especie de cavidad con su superficie reflejante

conduction: a form of thermal energy exchange that occurs through collisions between particles of matter *(354)*

   conducción: tipo de intercambio de energía térmica que ocurre mediante los choques de las partículas de la materia

conductor: a material that is good at conducting thermal energy *(355)*; a material that is good at conducting electric charge *(681)*

   conductor: material que conduce bien la energía térmica; material que conduce bien la carga eléctrica

conservation of angular momentum: the law stating that if the total torque acting on an object is zero, then the angular momentum of the object cannot change *(288)*

   conservación del momento angular: ley que establece que si el torque total que actúa sobre un objeto es igual a cero, entonces el momento angular del objeto no puede variar

conservation of momentum: the law stating that if the total force acting on an object is zero, then the momentum of the object cannot change *(242)*

conservación del momento: ley que establece que si la fuerza total que actúa sobre un objeto es igual a cero, entonces el momento del objeto no puede variar

constant acceleration: acceleration that is the same at every instant of time *(76)*

aceleración constante: aceleración que permanece igual en cada instante de tiempo

constant-pressure process: a process that occurs with no change in pressure *(393)*

proceso a presión constante: proceso que ocurre sin cambios en la presión

constant-volume process: a process in which the volume of the system remains constant *(393)*

proceso a volumen constante: proceso en el que el volumen del sistema permanece constante

constructive interference: interference that occurs when waves combine to form a larger wave *(477)*

interferencia constructiva: interferencia que ocurre cuando varias ondas se combinan para formar una onda más grande

convection: a form of thermal energy exchange that is due to the physical movement of material *(356)*

convección: una forma de intercambio de energía térmica provocado por el movimiento físico de la materia

convex mirror: a mirror that bulges outward like the surface of a ball *(575)*

espejo convexo: espejo que se curva hacia fuera como la superficie de una pelota

cooling: the process of removing thermal energy from a system *(344)*

enfriamiento: proceso de quitar energía térmica de un sistema

coordinate system: the frame of reference used to describe a system or an event *(43)*

sistema de coordenadas: sistema de referencia usado para describir un sistema o un evento

corner reflector: a mirror made up of three plane mirrors joined at right angles *(574)*

retrorreflector: espejo formado por tres espejos planos unidos en ángulo recto

coulomb: a unit of electric charge *(677)*

culombio: unidad de carga eléctrica

Coulomb's law: the law that relates the strength of the electrostatic force between point charges to the magnitude of the charges and the distance between them *(683)*

ley de Coulomb: ley que relaciona la intensidad de la fuerza electrostática entre cargas puntuales con la magnitud de las cargas y la distancia entre ellas

crest: the highest point on a wave *(473)*

cresta: punto más alto de una onda

critical: the status of a nuclear reactor when, on average, one neutron given off by a fission reaction produces an additional reaction *(928)*

reactor crítico: estado de un reactor nuclear cuando, como promedio, un neutrón cedido por una reacción de fisión produce una reacción adicional

critical angle: the angle of incidence at which a refracted beam travels parallel to the boundary between two materials *(606)*

ángulo crítico: ángulo de incidencia al que un rayo refractado viaja de forma paralela a la superficie de contacto entre dos materiales

crossed polarizers: polarizers with transmission axes at right angles to one another *(549)*

polarizadores cruzados: polarizadores con ejes de transmisión que forman ángulos rectos entre sí

cutoff frequency: the lowest frequency of electromagnetic radiation that will eject electrons from a metal surface *(859)*

frecuencia de corte: frecuencia más baja de radiación electromagnética capaz de desprender electrones de una superficie metálica

cycle: a series of events that repeat in the same order *(453)*

ciclo: serie de eventos que se repiten en un mismo orden

D

DC circuit: a circuit in which the current always flows in the same direction; also known as a *direct-current circuit* *(748)*

circuito CD: circuito en el que la corriente fluye en la misma dirección; también conocido como *circuito de corriente directa*

de Broglie wavelength: the wavelength associated with a particle that has momentum *(864)*

longitud de onda de *de Broglie*: longitud de onda asociada con una partícula que tiene momento

deceleration: the change in motion that occurs when the speed of an object decreases because the velocity and the acceleration have opposite signs *(79)*

desaceleración: cambio en el movimiento que ocurre cuando la rapidez de un objeto decrece debido a que la velocidad y la aceleración tienen signos opuestos

decibel: a unit commonly used to measure the loudness of sounds *(517)*

decibel: unidad usada comúnmente para medir la intensidad de los sonidos

density: the mass of a substance divided by its volume *(424)*

densidad: masa de una sustancia dividida por su volumen

dependent variable: the variable that is measured in an experiment to see how it depends on the independent variable *(8)*

variable dependiente: variable que se mide en un experimento para observar cómo depende de la variable independiente

destructive interference: interference that occurs when waves combine to form a smaller wave *(477)*

   interferencia destructiva: interferencia que ocurre cuando se combinan ondas para formar una onda más pequeña

diffraction: the bending of a wave around a barrier or though an opening *(654)*

   difracción: curvatura o deformación de una onda alrededor de un obstáculo o una abertura

diffraction grating: a screen with a large number of slits *(662)*

   red de difracción: red con un gran número de ranuras

diffraction pattern: the pattern of bright and dark fringes (bands) formed by the interference of light waves that have been diffracted *(654)*

   patrón de difracción: patrón de franjas (bandas) brillantes y oscuras formado por la interferencia de ondas de luz difractadas

diffuse reflection: reflection that sends light rays off in a variety of directions *(567)*

   reflexión difusa: reflexión que propaga rayos de luz en varias direcciones

dimensional analysis: a type of calculation written in terms of dimensions *(20)*

   análisis dimensional: tipo de cálculo escrito en términos dimensionales

dimensionally consistent: a characteristic of an equation in which each term has the same dimensions *(19)*

   dimensionalmente consistente: característica de una ecuación en la que cada término tiene las mismas dimensiones

diode: a simple semiconducting device consisting of a p-type semiconductor joined to an n-type semiconductor *(754)*

   diodo: dispositivo semiconductor simple que consiste en un semiconductor tipo *p* unido a un semiconductor tipo *n*

dispersion: the spreading out of a refracted light beam according to color *(609)*

   dispersión: propagación de la luz refractada de acuerdo con su color

displacement: the change in the position of an object *(45)*

   desplazamiento: cambio de posición de un objeto

distance: the total length of the path taken on a trip *(44)*

   distancia: largo total del camino recorrido en un viaje

Doppler effect: the change in pitch due to the relative motion between a source of sound and the person hearing the sound *(507)*

   efecto Doppler: cambio en el tono debido al movimiento relativo entre una fuente de sonido y la persona que escucha el sonido

Doppler shift: the change in frequency due to the Doppler effect *(534)*

   corrimiento Doppler: cambio en la frecuencia debido al efecto Doppler

driven oscillations: oscillations caused by an applied force *(468)*

   oscilaciones forzadas: oscilaciones causadas por una fuerza aplicada

E

eddy current: a localized circulating current induced in a portion of a metal sheet that is just leaving a magnetic field *(826)*

   corriente de Foucault: corriente inducida localizada en una porción de una lámina de metal cuando sale de un campo magnético

efficiency: the fraction of the heat supplied to a heat engine that is converted to work *(390)*

   eficiencia: fracción del calor suministrado a un motor térmico que es convertido en trabajo

elastic: having the ability to return to the original size and shape after being distorted *(205; 441)*

   elástico(a): que tiene la habilidad de retornar al tamaño y forma originales después de haber sido deformado(a)

elastic collision: a collision in which the kinetic energy is conserved *(248)*

   choque elástico: choque en el que la energía cinética se conserva

elastic limit: the point beyond which an object cannot be stretched without experiencing permanent deformation *(441)*

   límite elástico: punto límite a partir del cual un objeto no puede alargarse más sin experimentar una deformación permanente

elastic potential energy: the energy stored in a distorted elastic material *(205)*

   energía potencial elástica: energía almacenada en un material elástico deformado

electric circuit: a closed path through which electric charge flows and returns to its starting point *(748)*

   circuito eléctrico: ruta o camino cerrado por el que fluyen las cargas eléctricas y retornan a su punto de partida

electric current: the flow of electric charge from one place to another *(745)*

   corriente eléctrica: flujo de cargas eléctricas de un lugar a otro

electric dipole: a system consisting of a positive charge and a negative charge *(715)*

   dipolo eléctrico: sistema que consiste en una carga positiva y una carga negativa

electric field: the force field that exists around an electrically charged object *(705)*

   campo eléctrico: campo de fuerzas que existe alrededor de un objeto cargado eléctricamente

electric generator: a device that converts mechanical energy to electrical energy *(828)*

   generador eléctrico: aparato que convierte la energía mecánica en energía eléctrica

electric motor: a device that converts electric energy to mechanical energy *(830)*

   motor eléctrico: aparato que convierte la energía eléctrica en energía mecánica

electric potential: the change in electric potential energy for a given amount of charge *(719)*

   potencial eléctrico: cambio en la energía potencial eléctrica para una determinada cantidad de carga

electric potential energy: mechanical work that is stored as electrical energy *(718)*

   energía potencial eléctrica: trabajo mecánico que se almacena como energía eléctrica

electromagnet: a solenoid carrying an electric current that produces an intense, nearly uniform, magnetic field inside its loops *(793)*

   electroimán: solenoide por donde pasa una corriente que produce un campo magnético intenso y casi uniforme en su interior

electromagnetic induction: the process of inducing an electric current in a circuit using a changing magnetic field *(818)*

   inducción electromagnética: proceso de inducir una corriente eléctrica en un circuito por medio de un campo magnético variable

electromagnetic spectrum: the full range of frequencies of electromagnetic waves *(537)*

   espectro electromagnético: rango completo de las frecuencias de las ondas electromagnéticas

electromagnetic wave: any wave produced by oscillating electric and magnetic fields *(537)*

   onda electromagnética: toda onda producida por campos magnéticos y eléctricos oscilantes

electromotive force: the difference in electric potential between the terminals of a battery *(749)*

   fuerza electromotriz: diferencia de potencial eléctrico entre los terminales de una batería

electron: a negatively charged particle located in a "cloud" that surrounds the nucleus of an atom *(676)*

   electrón: partícula cargada negativamente que se localiza en una "nube" que rodea al núcleo de un átomo

electroweak theory: the theory that encompasses the weak nuclear force and the electromagnetic force *(937)*

   modelo electrodébil: teoría que unifica la fuerza nuclear débil y la fuerza electromagnética

ellipse: an oval shape defined by a point moving in a plane in such a way that the sum of its distances from two other points, called *foci*, is constant *(328)*

   elipse: figura en forma de óvalo definida por un punto que se mueve en un plano de tal forma que la suma de las distancias desde ese punto hasta otros dos puntos, llamados *focos*, es constante

emission spectrum: the line spectrum showing the wavelengths of light emitted by an atom *(886)*

   espectro de emisión: espectro de línea que muestra las longitudes de onda de la luz emitida por un átomo

energy-level diagram: a graph that plots the total energy of an atom for various values of the quantum number n *(890)*

   diagrama de niveles energéticos: gráfica que representa la energía total de un átomo para varios valores del número cuántico n

energy quantization: the idea that energy can have only discrete values *(851)*

   cuantización de la energía: idea de que la energía solo puede tener valores discretos

entropy: a measure of the amount of disorder in a system *(403)*

   entropía: medida de la magnitud del desorden de un sistema

equation: a mathematical expression that relates physical quantities *(23)*

   ecuación: expresión matemática que relaciona cantidades físicas

equation of continuity: an equation that applies conservation of mass to fluid flow *(435)*

   ecuación de continuidad: ecuación que aplica la conservación de la masa a un fluido en movimiento

equilibrium: a state in which an object has zero acceleration and is subject to zero net force *(167)*

   equilibrio: estado en el que un objeto tiene aceleración igual a cero y la fuerza total ejercida sobre este es cero

equilibrium vapor pressure: the pressure of a gas in equilibrium with a liquid *(368)*

   presión de vapor: presión de un gas en equilibrio con un líquido

equivalent resistor: a single resistor that has the same resistance as several resistors in a circuit *(758)*

   resistencia equivalente: resistencia única que tiene la misma resistencia que varias resistencias en un circuito

evaporation: the process in which particles leave the liquid phase and go into the gas phase *(369)*

   evaporación: proceso en el cual las partículas pasan del estado líquido al estado gaseoso

excited state: an energy level in an atom that is higher than the ground state *(890)*

   estado excitado: nivel de energía en un átomo que es mayor que el del estado básico

external force: a force exerted on a system by something outside the system *(242)*

   fuerza externa: fuerza ejercida en un sistema desde fuera de él

F

Fahrenheit scale: the temperature scale on which water freezes at 32 °F and boils at 212 °F *(346)*

   escala Fahrenheit: escala de temperatura en la que el agua se congela a 32 °F y hierve a 212 °F

farsighted: the condition in which a person can see clearly beyond a certain distance but cannot focus on closer objects *(624)*

> **hipermetropía:** condición en la cual una persona puede ver objetos lejanos con claridad, pero no puede enfocar bien los objetos cercanos

first harmonic: the fundamental mode *(478)*

> **primer armónico:** tono fundamental

first law of thermodynamics: the law stating that the change in a system's thermal energy is equal to the heat added to the system minus the work done by the system *(387)*

> **primera ley de la termodinámica:** ley que establece que la variación de la energía térmica de un sistema es igual al calor suministrado al sistema menos el trabajo ejercido por el sistema

fluid: a substance that can flow from one location to another and has no definite shape *(415)*

> **fluido:** sustancia que puede fluir de un lugar a otro y que no tiene forma definida

fluorescence: the emission of light of a lower frequency after illumination with light of a higher frequency *(901)*

> **fluorescencia:** emisión de luz de frecuencia baja después de una iluminación con una luz de una frecuencia más alta

focal length: the distance from the surface of a curved mirror to the focal point *(576)*

> **distancia focal:** distancia desde la superficie de un espejo curvo hasta el foco

focal point: the point at which rays reflecting from the surface of a mirror converge *(576)*

> **foco:** punto donde convergen los rayos reflejados por la superficie de un espejo curvo

foci: the two points used to define an ellipse *(328)*

> **focos:** dos puntos usados para definir una elipse

force: a push or a pull *(151)*

> **fuerza:** empuje o tirón (halón)

free-body diagram: a sketch that shows all the forces acting on an object *(161)*

> **diagrama de cuerpo libre:** dibujo que muestra todas las fuerzas que actúan sobre un objeto

free fall: motion that is determined by gravity alone *(97)*

> **caída libre:** movimiento afectado únicamente por la gravedad

frequency: the number of oscillations per unit of time *(454)*

> **frecuencia:** número de oscilaciones por unidad de tiempo

friction: the force that opposes the motion of one surface past another *(170)*

> **fricción:** fuerza que se opone al movimiento de una superficie sobre otra

fundamental mode: a standing wave with a node at each end and an antinode in the middle *(478)*

> **frecuencia fundamental:** onda estacionaria con un nodo en cada extremo y un antinodo en el medio

G

galvanometer: a device used to measure the current in a circuit *(805)*

> **galvanómetro:** dispositivo usado para medir la corriente en un circuito

gamma rays: electromagnetic waves with frequencies above 10^{20} Hz *(541)*; high-energy photons emitted in some radioactive decay processes *(918)*

> **rayos gamma:** ondas electromagnéticas con frecuencias mayores de 10^{20} Hz; fotones de alta energía emitidos en ciertos procesos de desintegración radioactiva

gas discharge tube: a sealed glass tube containing a gas at low pressure that is used to study the light emitted by atoms *(885)*

> **tubo de descarga:** tubo sellado relleno de gas a baja presión que se usa para estudiar la luz emitida por los átomos

gauge pressure: the pressure of a gas as measured on a gauge, which equals the actual pressure minus the atmospheric pressure *(416)*

> **presión manométrica:** presión de un gas medida en un manómetro, igual a la presión real menos la presión atmosférica

geosynchronous satellite: a satellite that orbits above Earth's equator with a period equal to 1 day *(331)*

> **satélite geosíncrono:** satélite que orbita sobre el Ecuador terrestre con un periodo de rotación de 1 día

gravitational lensing: an optical effect created when a massive object in space bends light rays *(319; 965)*

> **lente gravitacional:** efecto óptico creado cuando un objeto masivo en el espacio curva los rayos de luz

gravitational potential energy: the energy stored due to an object's position relative to the ground *(203)*

> **energía potencial gravitatoria:** energía almacenada debido a la posición de un objeto con respecto a la superficie

gravity: the force of nature that attracts one mass to another mass *(307)*

> **gravedad:** fuerza natural que atrae una masa hacia otra masa

gravity waves: ripples in the warped space around a large mass, predicted by the general theory of relativity *(966)*

> **ondas gravitacionales:** ondulaciones en la curvatura del espacio-tiempo cercano a un objeto masivo, predichas por la teoría general de la relatividad

ground state: the lowest possible energy state of an atom *(890)*

> **estado fundamental:** estado de energía más bajo que puede tener un átomo

grounded plug: a three-pronged electrical plug whose rounded prong is connected directly to ground when plugged in *(837)*

> **enchufe con conexión a tierra:** enchufe eléctrico de tres varillas en el que la varilla de forma cilíndrica está conectada directamente a tierra cuando se conecta al tomacorriente

H

hadron: a particle that is acted on by both the weak and the strong nuclear force *(938)*

hadrón: partícula sobre la que actúan tanto la fuerza nuclear débil como la fuerte

half-life: the time required for half of the nuclei in a radioactive sample to decay *(931)*

periodo de semidesintegración: tiempo requerido para que se desintegren la mitad de los núcleos de una muestra radiactiva

harmonic: a standing wave having more nodes and antinodes than the fundamental mode *(479)*

armónico: onda estacionaria con más nodos y antinodos que la onda de frecuencia fundamental

heat: energy transferred between objects because of a temperature difference *(345)*

calor: energía transferida entre objetos debido a una diferencia en su temperatura

heat engine: a device that converts heat into work *(389)*

motor térmico: aparato que convierte el calor en trabajo

heating: the process of adding thermal energy to a system *(344)*

calentar: proceso de suministrar energía térmica a un sistema

heating curve: a graph showing the temperature of a substance in response to heating *(372)*

curva de calentamiento: gráfica que muestra la temperatura de una sustancia en respuesta a su calentamiento

Heisenberg uncertainty principle: the principle stating that it is impossible to know the precise position and the precise momentum of a particle at the same time *(869)*

principio de incertidumbre de Heisenberg: principio que establece que es imposible saber, al mismo tiempo, la posición y el momento exactos de una partícula

hertz: a unit of frequency *(454)*

hercio: unidad de frecuencia

hologram: a three-dimensional image made with a laser *(900)*

holograma: imagen tridimensional hecha con un láser

Hooke's law: the law that states that the force exerted by an ideal spring is proportional to the distance by which it is stretched or compressed *(166)* or, equivalently, that the change in length of a solid is proportional to the applied force *(440)*

ley de Hooke: ley que establece que la fuerza ejercida por un resorte ideal es proporcional a la distancia a la que este se alarga o comprime o, de forma equivalente, que el cambio en las dimensiones de un cuerpo sólido es proporcional a la fuerza que se le aplica

horsepower: a unit of power equal to 746 watts: 1 hp = 746 W *(212)*

caballo de fuerza: unidad de potencia igual a 746 vatios: 1 hp = 746 W

Huygens's principle: the principle stating that every point on a wave front acts like a point source for new waves *(641)*

principio de Huygens: principio que establece que todo punto de un frente de onda puede considerarse una fuente de ondas nuevas

hypothesis: a detailed scientific explanation for a set of observations that can be verified or rejected by careful experiments *(7)*

hipótesis: explicación científica detallada para un conjunto de observaciones que puede verificarse o rechazarse mediante experimentos bien planeados

I

ideal ammeter: an ammeter with zero resistance that would not alter the current in a circuit *(764)*

amperímetro ideal: amperímetro con cero resistencia que no alteraría la corriente en un circuito

ideal gas: a gas whose particles have no effect on one another *(418)*

gas ideal: gas cuyas partículas no interactúan

ideal gas law: the law relating the pressure exerted by an ideal gas to the number of gas particles, the temperature of the gas, the volume of the gas, and the Boltzmann constant *(418)*

ley del gas ideal: ley que relaciona la presión ejercida por un gas ideal con su número de partículas, su temperatura, su volumen y la constante de Boltzmann

ideal projectile: any projectile that follows a path determined solely by the influence of gravity *(132)*

proyectil ideal: proyectil que sigue una trayectoria solamente determinada por la fuerza de gravedad

ideal voltmeter: a voltmeter with infinite resistance that would draw a negligible current *(764)*

voltímetro ideal: voltímetro de resistencia infinita que variaría la corriente del circuito de manera insignificante

impulse: the product of the force that acts on an object and the time over which it acts *(234)*

impulso: producto de la fuerza que actúa sobre un objeto y el tiempo durante el que esta actúa

in phase: a characteristic of interacting waves whose crests are aligned *(478)*

en fase: característica de ondas que interactúan cuyas crestas están alineadas

incoherent light: light from sources for which the phase difference varies randomly with time *(637)*

luz incoherente: luz proveniente de fuentes en las cuales la diferencia de fase varía de forma aleatoria en el tiempo

independent variable: the variable that is changed in an experiment *(8)*

variable independiente: variable que se manipula en un experimento

index of refraction: the factor by which a material reduces the speed of light *(598)*

   índice de refracción: factor por el cual el material reduce la velocidad de la luz

induced emf: the electromotive force induced in a secondary circuit by a changing magnetic field *(818)*

   fuerza electromotriz inducida: fuerza electromotriz inducida en un circuito secundario por un campo magnético cambiante

inelastic collision: a collision in which the kinetic energy changes and is not conserved *(249)*

   choque inelástico: choque en el que la energía cinética cambia y no se conserva

inertia: the tendency of an object to resist any change in its motion *(153)*

   inercia: tendencia de un objeto a resistir cualquier cambio en su movimiento

inference: a logical interpretation based on observations *(7)*

   inferencia: interpretación lógica basada en observaciones

infrared (IR) rays: electromagnetic waves with frequencies just below that of red light, from roughly 10^{12} Hz to 4.3×10^{14} Hz *(540)*

   rayos infrarrojos (RI): ondas electromagnéticas con frecuencias apenas por debajo de la luz roja, desde aproximadamente 10^{12} Hz hasta 4.3×10^{14} Hz

infrasonic: having a frequency less than 20 Hz *(497)*

   infrasónico: que tiene un frecuencia menor de 20 Hz

instantaneous acceleration: the acceleration of an object at a given instant of time *(76)*

   aceleración instantánea: aceleración de un objeto en un instante de tiempo determinado

instantaneous velocity: the velocity of an object at a given instant of time *(56)*

   velocidad instantánea: velocidad de un objeto en un instante de tiempo determinado

insulator: a material that is poor at conducting thermal energy *(355)*; a material that is poor at conducting electric charge *(681)*

   aislante: material que no conduce bien la energía térmica; material que no conduce bien las cargas eléctricas

intensity: the amount of sound energy passing through a given area in a given time; also known as *loudness* *(513)*

   intensidad de sonido: cantidad de energía sonora que pasa por un área determinada en un tiempo determinado; también conocida como *volumen*

intensity level: the loudness of a sound in decibels *(517)*

   nivel de intensidad de sonido: volumen de un sonido en decibeles

interference pattern: a pattern made up of alternating regions of constructive and destructive interference of light waves *(641)*

   patrón de interferencia: patrón formado por regiones alternadas de interferencias constructivas y destructivas de ondas de luz

internal energy: thermal energy *(344)*

   energía interna: energía térmica

internal force: a force acting between objects within a system *(242)*

   fuerza interna: fuerza que actúa entre los objetos dentro de un sistema

inverse relationship: a relationship between two quantities in which one gets larger as the other gets smaller *(32)*

   proporcionalidad inversa: relación entre dos cantidades en la que cuando una se hace mayor la otra se hace menor

ion: an atom that has gained or lost one or more electrons *(679)*

   ión: átomo que ha ganado o perdido uno o más electrones

iridescence: coloration that is produced by the interference of light waves *(664)*

   iridiscencia: coloración producida por la interferencia de ondas de luz

isothermal: occurring at a constant temperature *(396)*

   isotérmico(a): que ocurre a una temperatura constante

isotopes: nuclei with the same number of protons but different numbers of neutrons *(913)*

   isótopos: núcleos con el mismo número de protones pero con números diferentes de neutrones

J

joule: the unit of work that is a combination of the newton and the meter: $1\,\text{J} = 1\,\text{N} \cdot \text{m}$ *(190)*

   julio: unidad de trabajo que es una combinación del Newton y el metro: $1\,\text{J} = 1\,\text{N} \cdot \text{m}$

K

Kelvin scale: the temperature scale based on absolute zero *(348)*

   escala Kelvin: escala de temperatura basada en el cero absoluto

kilocalorie: a unit of thermal energy; one kilocalorie (kcal) is the amount of heat needed to raise the temperature of 1 kg of water from 14.5 °C to 15.5 °C *(359)*

   kilocaloría: unidad de energía térmica; una kilocaloría (kcal) es la cantidad de calor necesaria para elevar la temperatura de 1 kg de agua desde 14.5 °C hasta 15.5 °C

kilogram: the base unit of mass, abbreviated as kg *(15)*

   kilogramo: unidad básica de masa; su abreviatura es kg

kinetic energy: the energy of an object due to its motion *(198)*

   energía cinética: energía de un cuerpo debida a su movimiento

kinetic friction: the friction that arises when surfaces slide against one another *(170)*

   fricción dinámica: fricción que surge cuando dos superficies se deslizan una contra otra

kinetic molecular theory: theory that describes phase transitions in terms of the kinetic behavior of molecules *(422)*

   teoría cinético-molecular: teoría que describe los cambios de estado en términos del movimiento de las moléculas

kinetic theory of gases: the theory that describes a gas as consisting of innumerable particles flying about randomly at high speeds *(422)*

   teoría cinética de los gases: teoría que describe un gas como innumerables partículas que vuelan de forma aleatoria a altas velocidades

L

laser: a device that produces light that is intense, travels in one direction, and is a single color; the word originated as an acronym for "**l**ight **a**mplification by the **s**timulated **e**mission of **r**adiation" *(898)*

   láser: aparato que produce luz intensa de un solo color que viaja en una solo dirección; la palabra se originó como un acrónimo de "**l**ight **a**mplification by the **s**timulated **e**mission of **r**adiation" (amplificación de luz mediante emisión estimulada de radiación)

latent heat: the heat required to change 1 kilogram of a substance from one phase to another *(371)*

   calor latente: calor necesario para cambiar 1 kilogramo de sustancia de un estado a otro

length contraction: the shrinking in length of a moving object *(956)*

   contracción de Lorentz: contracción de la longitud de un objeto en movimiento

lens: a device that uses refraction to focus light *(612)*

   lente: dispositivo que usa la refracción para enfocar la luz

lepton: a particle that is acted on by the weak nuclear force but not by the strong nuclear force *(938)*

   leptón: partícula sobre la que actúan las fuerzas nucleares débiles, pero no las fuerzas nucleares fuertes

lift: an upward force due to a pressure difference *(437)*

   sustentación: fuerza ascendente producida por una diferencia de presión

line spectrum: the spectrum emitted by atoms, consisting of bright lines at specific frequencies, or colors *(886)*

   espectro de líneas: espectro emitido por los átomos, que consiste en líneas brillantes para frecuencias específicas, o colores

linear momentum: the momentum associated with straight-line motion *(230)*

   momento lineal: momento asociado con el movimiento rectilíneo uniforme

linear relationship: the relationship that exists between two quantities that form a straight line on a graph *(31; 100)*

   relación lineal: relación entre dos cantidades que forma una línea recta en una gráfica

longitudinal wave: a wave in which the particles oscillate parallel to the direction in which the wave travels *(471)*

onda longitudinal: onda en la que las partículas oscilan de forma paralela a la dirección en la que viaja la onda

M

magnetic braking: the friction-like effect of eddy currents *(827)*

   freno electromagnético: efecto parecido a la fricción que es provocado por las corrientes de Foucault

magnetic domain: a region within a magnetic material where the electrons are aligned in the same direction *(788)*

   dominio magnético: región dentro de un campo magnético donde los electrones están alineados en la misma dirección

magnetic field: a vector force field that surrounds any magnetic material *(784)*

   campo magnético: campo de fuerza vectorial que rodea todo material magnético

magnetic flux: a measure of the number of magnetic field lines that pass through a given area *(819)*

   flujo magnético: medida del número de líneas del campo magnético que pasan a través de un área dada

magnetic reed switch: a magnetic switch in a pacemaker that allows a physician to change the pacemaker's operating mode by simply placing a magnet at the appropriate location on the patient's chest *(795)*

   interruptor magnético de lengüeta: interruptor magnético en un marcapasos que permite a un médico cambiar el modo de funcionamiento del marcapasos con solo colocar un imán en el lugar apropiado de la región torácica del paciente

magnification: the ratio of the height of the image to the height of the object *(583)*

   magnificación: relación entre la altura de la imagen y la altura del objeto

magnitude: a numerical value with a unit but no direction *(32)*; the length of a vector *(114)*

   magnitud: valor numérico que denota cantidad pero no dirección; longitud de un vector

mass: a measure of the amount of matter in an object *(16)*

   masa: medida de la cantidad de materia de un objeto

mass number: the total number of nucleons in a nucleus *(912)*

   número másico: número total de nucleones en un núcleo

mass spectrometer: a device that makes use of circular motion in a magnetic field to separate isotopes (atoms of the same element that have different masses) and to measure atomic masses *(801)*

   espectrómetro de masas: instrumento que usa el movimiento circular dentro de un campo magnético para separar isótopos (átomos del mismo elemento que tienen masas distintas) y para medir masas atómicas

mechanical energy: the sum of the potential and kinetic energies of an object *(207)*

   energía mecánica: suma de las energías potencial y cinética de un objeto

mechanical wave: a wave that requires a medium to travel through *(474)*

   onda mecánica: onda que requiere de un medio para desplazarse

medium: the material a wave travels through *(474)*

   medio: material a través del cual viaja una onda

meson: a hadron formed from two quarks *(939)*

   mesón: hadrón formado por dos quarks

meter: the base unit of length, abbreviated as m *(15)*

   metro: unidad básica de distancia; su abreviatura es m

microchip: a silicon wafer containing thousands of transistors, diodes, and resistors connected in complex circuits *(756)*

   circuito integrado (microchip): pastilla de silicio que contiene miles de transistores, diodos y resistencias conectados en circuitos complejos

microwaves: electromagnetic waves with frequencies from roughly 10^9 Hz to about 10^{12} Hz *(539)*

   microondas: ondas electromagnéticas con frecuencias desde aproximadamente 10^9 Hz hasta 10^{12} Hz

mirage: an optical illusion caused by the refraction of light *(601)*

   espejismo: ilusión óptica causada por la refracción de la luz

mirror: an object that is particularly good at reflecting light waves *(565)*

   espejo: objeto que refleja específicamente bien las ondas de luz

mirror equation: the precise mathematical relationship between the object distance, the image distance, and the focal length for a given mirror *(581)*

   ecuación de espejos: relación matemática precisa entre la distancia al objeto, la distancia a la imagen y la distancia focal para un espejo determinado

molar mass: the mass in grams of 1 mole of a substance *(422)*

   masa molar: masa en gramos de 1 mol de sustancia

mole: the amount of a substance that contains as many particles as there are atoms in 12 grams of carbon-12 *(420)*

   mol: cantidad de sustancia que contiene tantas partículas como átomos hay en 12 gramos de carbono 12

moment arm: the perpendicular distance from the axis of rotation to the line extending through the force vector that is creating the torque *(283)*

   brazo del momento: distancia perpendicular desde el eje de rotación hasta la línea extendida del vector fuerza que está creando el torque

moment of inertia: a quantity that determines how easy or hard it is to change an object's rotation *(277)*

   momento de inercia: cantidad que determina qué tan fácil o difícil es cambiar la rotación de un objeto

momentum: a quantity defined as mass times velocity *(230)*

   momento: cantidad definida como masa por velocidad

monochromatic light: light of a single color, or frequency *(637)*

   luz monocromática: luz de un solo color o frecuencia

motional emf: the electromotive force produced by motion *(825)*

   fuerza electromotriz de movimiento: fuerza electromotriz producida por el movimiento

multimeter: a device that can function as an ammeter, a voltmeter, and an ohmmeter *(764)*

   multímetro: dispositivo que puede funcionar como amperímetro, voltímetro y ohmímetro

N

natural frequency: the frequency at which an object oscillates by itself *(468)*

   frecuencia natural: frecuencia a la que un objeto oscila por sí mismo

neap tide: a smaller tide produced when the Sun and the Moon are at right angles relative to the Earth *(326)*

   marea muerta: marea pequeña que se produce cuando el Sol y la Luna se encuentran y forman un ángulo recto con respecto a la Tierra

nearsighted: the condition in which a person's relaxed eyes do not focus at infinity as they should *(624)*

   miopía: condición en la que los ojos relajados de una persona no enfocan en el infinito, como debieran

negative charge: one of two types of electrical charge; the charge produced in amber when it is rubbed on a glass rod *(676)*

   carga negativa: uno de los dos tipos de carga eléctrica; la carga que se produce en el ámbar cuando se frota con una barra de vidrio

negative ion: an atom that has gained one or more electrons *(679)*

   ión negativo: átomo que ha ganado uno o más electrones

net force: the sum of the individual forces acting on an object *(152)*

   fuerza total: suma de todas las fuerzas individuales que actúan sobre un objeto

neutral: having zero total charge *(676)*

   neutro(a): que tiene una carga total igual a cero

neutron: a neutral particle located in the nucleus of an atom *(677)*

   neutrón: partícula neutra localizada en el núcleo de un átomo

neutron number: the number of neutrons in a nucleus *(912)*

   número neutrónico: número de neutrones de un núcleo

newton: a unit of force, abbreviated as N *(155)*

   newton: unidad de fuerza; su abreviatura es N

node: a point on a standing wave that does not move *(478)*

   nodo: punto de una onda estacionaria que no se mueve

normal: a line drawn perpendicular to a surface (566)

normal: línea trazada de forma perpendicular a una superficie

normal force: a force exerted perpendicular to the surface of contact (163)

fuerza normal: fuerza ejercida de forma perpendicular a la superficie de contacto

nuclear binding energy: the difference in energy between a complete nucleus and its separated particles (920)

energía de enlace nuclear: la diferencia entre la energía de las partículas en un núcleo y la de cada una de sus partículas por separado

nuclear fission: the process in which a large nucleus splits into two smaller nuclei (925)

fisión nuclear: proceso en el cual un núcleo grande se descompone en dos núcleos más pequeños

nuclear fusion: when two light nuclei combine to form a more massive nucleus (929)

fusión nuclear: ocurre cuando dos núcleos ligeros se combinan para formar un núcleo más pesado

nucleon: particles in the nucleus; protons and neutrons (911)

nucleón: partículas del núcleo; protones y neutrones

nucleus: the region of space at the center of the atom that contains all of the atom's positive charge and almost all of its mass (884)

núcleo: región en el centro del átomo que contiene todas las cargas positivas del átomo y casi toda su masa

O

observation: a logical and orderly description of an event (7)

observación: descripción lógica y ordenada de un evento

Ohm's law: the law relating the potential difference applied to a wire to the current produced and the wire's resistance (751)

ley de Ohm: ley que relaciona la diferencia de potencial aplicada a un cable con la corriente producida y la resistencia del cable

ohmmeter: a device used to measure resistance to the flow of current (751)

ohmímetro: dispositivo usado para medir la resistencia al flujo de la corriente

open circuit: a circuit in which there is no closed path through which electric charge can flow (749)

circuito abierto: circuito en el que la corriente eléctrica no puede fluir por no tener un camino cerrado

optical fibers: thin fibers that use total internal reflection to transmit light and are generally composed of a glass or plastic core with a high index of refraction surrounded by an outer coating, or cladding, with a low index of refraction (608)

fibras ópticas: fibras finas que usan la reflexión interna total para transmitir luz y que, generalmente, están formadas por un núcleo de plástico o vidrio con un índice de refracción alto, rodeado por una cobertura o revestimiento externo con un índice de refracción bajo

orbital period: the time it takes for a planet to complete one orbit (330)

periodo orbital: tiempo que se demora un planeta en completar una órbita

order-of-magnitude calculation: a rough, or ballpark, estimate designed to be accurate to within the nearest power of 10 (30)

cálculo de orden de magnitud: estimación diseñada para ser preciso dentro de la potencia más cercana de 10

origin: the location along a coordinate axis that corresponds to a value of zero (43)

origen: lugar de un eje de coordenadas que se corresponde con el valor cero

oscillation: a back-and-forth motion (452)

oscilación: movimiento de vaivén

out of phase: a characteristic of interacting waves for which the crest of one wave aligns with the trough of the other (478)

fuera de fase: característica de ondas que interactúan, en la que la cresta de una onda se alinea con el valle de la otra

P

parabola: a curved shape on a graph indicating that one quantity depends on the square of another (31)

parábola: forma curvada de una gráfica que indica que una cantidad depende del cuadrado de la otra

parabolic relationship: a relationship between two quantities in which one depends on the square of the other (31; 100)

relación parabólica: relación entre dos cantidades en la que una depende del cuadrado de la otra

parallel circuit: a circuit in which two or more resistors are connected across the same potential difference (760)

circuito en paralelo: circuito en el que dos o más resistencias están conectadas a la misma diferencia de potencial

parallel-plate capacitor: a device consisting of two parallel plates (separated by a finite distance) that have opposite charge (715)

condensador de placas paralelas: dispositivo que consiste en dos placas paralelas (separadas por una distancia finita) que poseen cargas opuestas

Pascal's principle: the principle stating that an external pressure applied to an enclosed fluid is transmitted unchanged to every point within the fluid (430)

ley de Pascal: principio que establece que una presión externa aplicada sobre un líquido encerrado es transmitida con igual intensidad a cada punto del fluido

peer review: the process in which a report is sent to several experts in the field to look for errors, biases, and oversights (11)

revisión por pares: proceso en el cual se envía un informe a varios expertos en ese campo para identificar errores, sesgos y omisiones

period: the time required to complete one full cycle of periodic motion *(454)*

    periodo: tiempo requerido para completar un ciclo completo de movimiento periódico

periodic motion: any motion that repeats itself over and over *(453)*

    movimiento periódico: cualquier movimiento que se repite una y otra vez

phase: a physical state of matter, such as solid, liquid, or gas *(366)*

    estado: estado físico de la materia, como sólido, líquido o gaseoso

phosphor: a material that absorbs ultraviolet light and then emits lower-frequency visible light *(901)*

    sustancia fosforescente: material que absorbe luz ultravioleta y emite luz visible de frecuencia más baja

photoconductive: able to conduct electricity when exposed to light but acting as an insulator when in the dark *(682)*

    fotoconductible: capaz de conducir electricidad cuando se expone a la luz, pero que actúa como aislante en la oscuridad

photoelastic stress analysis: a stress analysis technique that makes use of crossed polarizers *(550)*

    análisis de fotoelástico: técnica de análisis para la medición de esfuerzos que usa polarizadores cruzados

photoelectric effect: a phenomenon in which a beam of light hits the surface of a metal and ejects electrons *(859)*

    efecto fotoeléctrico: fenómeno en el cual un rayo de luz desprende electrones de una superficie de metal al incidir sobre ella

photon: a "particle" of light that carries energy but has no mass *(534)*; a discrete bundle of light energy that obeys Planck's hypothesis of energy quantization *(857)*

    fotón: "partícula" de luz que contiene energía pero no tiene masa; paquete de energía lumínica discreta que cumplen con la hipótesis de la cuantización de la energía de Planck

physical quantity: a property of a physical system that can be measured *(23)*

    cantidad física: propiedad de un sistema físico que puede ser medida

physics: the study of the fundamental laws of nature *(3)*

    física: estudio de las leyes fundamentales de la naturaleza

pitch: the perceived highness or lowness of a sound, directly related to the frequency of the sound wave *(497)*

    tono: percepción de lo alto o bajo que es un sonido, directamente relacionada con la frecuencia de la onda de sonido

plane mirror: a flat mirror *(570)*

    espejo plano: espejo aplanado, sin curvatura

plane waves: waves having flat wave fronts and parallel rays all pointing in the same direction *(566)*

    ondas planas: ondas con frentes de ondas planos y rayos paralelos que apuntan a la misma dirección

plasma: a phase of matter; a gaslike collection of charged ions *(800)*

    plasma: estado de la materia; masa de iones cargados en un estado parecido al gaseoso

polarization: the direction of the electric field in an electromagnetic wave *(545)*

    polarización: dirección del campo eléctrico en una onda electromagnética

polarized plug: a two-pronged electrical plug in which one prong is wider than the other *(836)*

    enchufe polarizado: enchufe eléctrico de dos varillas en el que una varilla es más ancha que la otra

polarizer: a filter that transmits light waves with only one direction of polarization *(546)*

    polarizador: filtro que transmite ondas de luz en una sola dirección de polarización

Porro prism: a type of prism found in many binoculars that uses total internal reflection to "fold" a relatively long light path into a short length *(608)*

    prisma de Porro: tipo de prima en muchos binoculares que usa la reflexión interna total para "reducir" una aparente trayectoria larga de la luz hasta formar una más corta

position-time graph: a graph in which position is plotted on the vertical axis, or y axis, and time is plotted on the horizontal axis, or x axis *(54)*

    gráfica de posición contra tiempo: gráfica en que la posición se indica en el eje vertical, o eje y, y el tiempo en el eje horizontal, o eje x

position vector: a vector drawn from the origin to an object's location *(44)*

    vector de posición: vector dibujado desde el origen hasta el lugar donde está un objeto

positive charge: one of two types of electrical charge; the charge produced in a glass rod when it is rubbed with amber *(676)*

    carga positiva: uno de los dos tipos de carga; la carga que se produce en una varilla de vidrio cuando se frota con ámbar

positive ion: an atom that has lost one or more electrons *(679)*

    ión positivo: átomo que ha perdido uno o más electrones

positron: the antiparticle to an electron; the name is short for "positive electron" *(918)*

    positrón: antipartícula de un electrón; el nombre tiene su origen en "electrón positivo"

potential energy: energy that is stored for later use *(203)*

    energía potencial: energía almacenada para uso posterior

power: the amount of work done in a given amount of time *(211)*

    potencia: cantidad de trabajo hecho en una cantidad de tiempo

precision: a measure of how close together the values of a series of measurements are to one another *(25)*

    precisión: medida de qué tan cerca están entre sí los valores de una serie de medidas

pressure: the amount of force exerted on a given area (367)

    presión: cantidad de fuerza ejercida en un área

primary colors: the three main colors human eyes can perceive: red, green, and blue (541)

    colores primarios: rojo, verde y azul: los tres colores básicos que el ojo humano puede percibir

principal axis: the straight line drawn through the center of curvature and the midpoint of a curved mirror (576)

    eje óptico: línea recta que se traza por el centro de curvatura y el punto medio de un espejo curvo

principal maxima: the sharp, widely spaced bright fringes in an interference pattern formed by a diffraction grating (662)

    máximo principal: bandas nítidas, brillantes y bien separadas en un patrón de interferencia formado por una red de difracción

principal rays: the rays used in ray tracing (577)

    rayos principales: rayos usados para trazar o dar seguimiento a los demás rayos

principle of equivalence: the principle stating that all physical experiments conducted in a uniform gravitational field and in an accelerated frame of reference give identical results (963)

    principio de equivalencia: principio que establece que todos los experimentos físicos conducidos en un campo gravitacional uniforme y en un sistema de referencia acelerado, dan resultados idénticos

principle of superposition: the principle stating that a resultant wave is simple the sum of the individual waves that make it up (476)

    principio de superposición: principio que plantea que una onda resultante es simplemente la suma de las ondas individuales que la forman

projectile: an object that is thrown, kicked, batted, or otherwise launched into motion and then allowed to follow a path determined solely by the influence of gravity (131)

    proyectil: objeto lanzado, pateado, bateado o puesto en movimiento y luego liberado para seguir una trayectoria que solo depende de la influencia de la fuerza de gravedad

proper length: the length of an object at rest, whose speed is zero (956)

    largo propio: longitud de un objeto en reposo, cuya velocidad es igual a cero

proper time: the time interval of a clock that is at rest relative to an observer (953)

    tiempo propio: intervalo de tiempo en un reloj que está en reposo con respecto a un observador

proton: a positively charged particle located in the nucleus of an atom (677)

    protón: partícula cargada positivamente localizada en el núcleo de un átomo

Q

quantize: to exist in discrete units that cannot be broken into smaller units (851)

    cuantizar: existir en unidades discretas que no pueden ser separadas en unidades más pequeñas

quantum chromodynamics: the theory of how quarks of different "colors" interact with one another (940)

    cromodinámica cuántica: teoría sobre cómo interactúan los quarks de diferentes "colores"

quantum electrodynamics: the theory describing interactions between charged particles (940)

    electrodinámica cuántica: teoría que describe las interacciones entre partículas cargadas

quantum mechanics: wave-matter physics described by Schrödinger's equation (898)

    mecánica cuántica: materia de la física de las ondas descrita por la ecuación de Schrödinger

quantum tunneling: the process in which a particle moves, or tunnels, through a region of space that is forbidden to it in classical physics (871)

    efecto túnel: proceso en el que una partícula se mueve, o pasa como por un túnel, a través de una región del espacio que le está prohibida en la física clásica

quark confinement: the concept that quarks must always be bound to other quarks inside a hadron (940)

    confinamiento de color: concepto que indica que los quarks siempre están unidos a otros quarks dentro de un hadrón

quarks: the particles that make up hadrons (939)

    quarks: partículas que forman el hadrón

R

radian: the angle for which the length of the corresponding circular arc is equal to the radius of the circle (268)

    radián: ángulo para el cual la longitud del arco circular correspondiente es igual al radio del círculo

radiation: the transfer of thermal energy in the form of electromagnetic waves (356)

    radiación: transferencia de energía térmica en forma de ondas electromagnéticas

radio waves: the lowest-frequency electromagnetic waves of practical importance, having a frequency range from roughly 10^6 Hz to 10^9 Hz (539)

    ondas de radio: ondas electromagnéticas de baja frecuencia con importancia práctica que tienen un rango de frecuencia entre aproximadamente 10^6 Hz y 10^9 Hz

radioactive decay series: a series of related decay reactions of nuclei (924)

    series (cadenas) de desintegración radioactiva: series de reacciones de desintegración de núcleos que están relacionadas

radioactivity: the particles and photons emitted when a nucleus decays *(917)*

  radioactividad: partículas y fotones que se emiten cuando un núcleo se desintegra

range: the horizontal distance traveled by a projectile before it lands *(138)*

  rango: distancia horizontal que recorre un proyectil antes de tocar tierra

ray: an arrow that points in the direction in which light is traveling *(566)*

  rayo: flecha que apunta a la dirección en la que que viaja la luz

ray tracing: a method that traces the paths of light rays as they reflect from a mirror or pass through a lens in order to find the location of the image *(577; 613)*

  trazado de rayos: método que traza la trayectoria de rayos de luz al reflejarse en un espejo o cuando pasan a través de una lente, con el fin de encontrar la posición de la imagen

reaction force: one of the forces always present in any pair of action-reaction forces *(158)*

  fuerza de reacción: fuerza que está siempre presente en un par de fuerzas de acción y reacción

real image: the type of image formed by converging light rays *(578)*

  imagen real: tipo de imagen formada por rayos de luz convergentes

recoil: the "backward" motion that results from conservation of momentum *(246)*

  retroceso: movimiento "hacia atrás" que resulta de la conservación del momento

reflection: the process in which a wave hits a surface or boundary and bounces off in a different direction *(565)*

  reflexión: proceso en el que una onda incide en una superficie o frontera y rebota en otra dirección

refraction: a change in direction due to a change in the speed of a wave *(599)*

  refracción: cambio en la dirección debido al cambio de velocidad de una onda

relative motion: the motion of one object relative to another object *(127)*

  movimiento relativo: movimiento de un objeto con respecto a otro

repeat time: the period of a wave; the time required for one wavelength to pass a given point *(473)*

  tiempo de repetición: periodo de una onda; tiempo requerido por una longitud de onda para pasar un punto determinado

resistance: the opposition to current flow in a wire due to collisions between electrons and atoms *(750)*

  resistencia: la oposición al flujo de la corriente debido a choques entre los electrones y los átomos

resistivity: a quantity that describes the resistance of a particular material *(752)*

resistividad: cantidad que describe la resistencia de un material en particular

resistor: a small device used in electrical circuits to provide a particular resistance to current flow *(751)*

  resistencia: dispositivo pequeño usado en los circuitos eléctricos para suministrar una resistencia determinada al flujo de la corriente

resolution: the sharpness of vision; in particular, the ability to visually separate closely spaced objects *(658)*

  resolución: nitidez de la visión; específicamente, la habilidad de separar visualmente objetos alejados por espacios pequeños

resonance: the state that exists when a system is subject to driven oscillations at its natural frequency *(468)*

  resonancia: estado que aparece cuando el sistema es sujeto a oscilaciones forzadas en su frecuencia natural

rest energy: the energy of an object at rest *(959)*

  energía en reposo: energía de un objeto en reposo

rest mass: the mass of an object at rest *(958)*

  masa en reposo: masa de un objeto en reposo

restoring force: a force that acts to bring an object back to equilibrium *(456)*

  fuerza de restauración: fuerza que actúa para llevar un objeto de nuevo a su punto de equilibrio

resultant: the result of adding two or more vectors *(121)*

  resultante: resultado de sumar dos o más vectores

resultant wave: the combined result of adding the amplitudes of overlapping waves *(476)*

  onda resultante: resultado combinado al sumar las amplitudes de ondas superpuestas

rms current: the root mean square of the current *(833)*

  corriente efectiva (rms): raíz media cuadrática de la corriente

rms voltage: the root mean square of the voltage *(833)*

  voltaje efectivo (rms): raíz media cuadrática del voltaje

rolling motion: the combination of rotational motion and linear motion as a wheel rolls without slipping *(276)*

  movimiento de rodadura: combinación del movimiento de rotación y del movimiento rectilíneo, como cuando un aro rueda sin deslizarse

rotational kinetic energy: the energy of motion due to rotation, which is equal to half the product of an object's moment of inertia and the square of its angular speed *(279)*

  energía cinética rotacional: energía del movimiento debido a la rotación, igual al producto del momento de inercia de un objeto por el cuadrado de su rapidez angular, dividido entre dos

rotational motion: motion along a circular path about a central axis of rotation *(267)*

  movimiento de rotación: movimiento de trayectoria circular alrededor de un eje de rotación central

round-off error: error caused when numerical results are rounded off at different times during a calculation *(29)*

   error de redondeo: error causado cuando un resultado numérico se redondea varias veces durante un cálculo

S

scalar: a quantity represented by a numerical value and a unit *(32)*

   escalar: cantidad representada por un valor numérico y una unidad

science: an organized way of thinking about nature and understanding how it works *(6)*

   ciencia: modo organizado de pensar sobre la naturaleza y de entender cómo funciona

scientific method: the systematic approach scientists use to learn about the laws of nature *(6)*

   método científico: acercamiento sistemático que usan los científicos para aprender sobre las leyes de la naturaleza

scientific notation: a method used to express numerical values as a number between 1 and 10 times an appropriate power of 10 *(28)*

   notación científica: método usado para expresar valores numéricos con potencias de base 10

second: the base unit of time, abbreviated as s *(15)*

   segundo: la unidad básica de tiempo; su abreviatura es s

second law of thermodynamics: the law stating that of all the processes that conserve energy (and satisfy the first law) only those that proceed in a certain direction will occur *(401)*

   segunda ley de la termodinámica: ley que establece que de todos los procesos que conservan energía (y que satisfacen la primera ley) solo ocurrirán aquellos que van en una dirección determinada

secondary maxima: the less bright fringes in the regions between principal maxima in an interference pattern formed by a diffraction grating *(662)*

   máximo secundario: las bandas menos brillantes que se encuentran en las regiones entre el máximo principal en un patrón de interferencia formado por una red de difracción

semiconductor: a material with properties intermediate between those of a good conductor and a good insulator *(682)*

   semiconductor: material con propiedades intermedias entre las de un buen conductor y un buen aislante

series circuit: a circuit in which two or more resistors are connected one after the other, end to end *(758)*

   circuito en serie: circuito en el que dos o más resistencias están conectadas una a continuación de la otra, extremo con extremo

shielding: the effect that is observed when the distribution of charge on the surface of a conductor guarantees that the electric field within the conductor is zero *(716)*

   blindaje electromagnético: efecto que se observa cuando la distribución de cargas en la superficie de un conductor garantiza que el campo eléctrico dentro de este sea igual a cero

SI units: the unit system also known as the Système International d'Unités *(15)*

   unidades del SI: unidades del Sistema Internacional de Unidades

significant figure: the digits actually measured plus one estimated digit in a properly expressed measurement *(25)*

   cifra significativa: dígitos de la medición más un dígito estimado en una medición expresada apropiadamente

simple harmonic motion: motion that occurs when the force pushing or pulling an object toward the equilibrium position is proportional to the object's displacement from that position *(456)*

   movimiento armónico simple: movimiento que ocurre cuando la fuerza que empuja o tira de un objeto hacia la posición de equilibrio es proporcional al desplazamiento del objeto desde esa posición

simple pendulum: a device consisting of a mass suspended by a string or a rod *(462)*

   péndulo simple: aparato que consiste en una masa suspendida de un hilo o varilla

slope: the rise over the run for a line on a graph *(55)*

   pendiente: inclinación de una línea en una gráfica

solenoid: an electrical device in which a long wire is wound into a succession of closely spaced loops, forming a cylindrical coil *(793)*

   solenoide: dispositivo eléctrico en el que un alambre largo se enrolla sucesivamente en espiras muy juntas y forma un rollo cilíndrico

specific heat capacity: the thermal energy required to change the temperature of 1 kilogram of a substance by 1 °C *(361)*

   capacidad calorífica específica: energía térmica necesaria para cambiar la temperatura de 1 kilogramo de sustancia en 1 °C

spectroscopy: the study of atomic line spectra *(887)*

   espectroscopía: estudio del espectro atómico de líneas

specular reflection: reflection from a smooth surface, with all the reflected light rays moving in a single direction *(567)*

   imagen especular: reflexión de una superficie lisa, con todos los rayos reflejados moviéndose en una sola dirección

speed: a scalar quantity that describes the rate of motion *(32)*

   rapidez: cantidad escalar que describe la tasa de movimiento

spherical aberration: the deviation from ideal lens behavior that occurs when a lens has a spherical surface *(621)*

   aberración esférica: desviación del comportamiento de la lente ideal que ocurre en una lente esférica

spherical waves: wave fronts having rays that point radially outward in all directions *(566)*

    ondas esféricas: frentes de onda con rayos que apuntan radialmente en todas las direcciones

spring constant: the constant k in Hooke's law, whose units are newtons per meter, or N/m *(166)*

    constante elástica del muelle: constante k en la ley de Hooke, cuyas unidades son los newtons por metro, o N/m

spring tide: a large tide produced when the Sun aligns with the Moon relative to Earth *(326)*

    marea viva: marea alta que se produce cuando el Sol está alineado con la Luna y la Tierra

standing wave: a wave that oscillates in a fixed location *(478)*

    onda estacionaria: onda que oscila en un lugar fijo

static electric field: the electric field produced by charges that remain in fixed positions in a system *(710)*

    campo electrostático: campo eléctrico producido por cargas que no se mueven de sus posiciones en un sistema

static equilibrium: a state in which a nonmoving object has zero acceleration and is subject to zero net force *(290)*

    equilibrio estático: estado en el que un objeto inmóvil tiene una aceleración igual a cero y está sometido a una fuerza total igual a cero

static friction: the force that opposes the sliding of one nonmoving surface past another *(173)*

    fricción estática: fuerza que se opone al deslizamiento de una superficie en reposo contra otra

strong nuclear force: the attractive force that holds a nucleus together *(915)*

    fuerza nuclear fuerte: fuerza de atracción que mantiene unido un núcleo

sublimation: the change in phase from a solid directly to a gas *(371)*

    sublimación: cambio de estado en el que un sólido pasa directamente a gas

subtractive primary colors: the three pigment colors—cyan, magenta, and yellow—that can combine to produce any desired color by subtracting light *(543)*

    colores primarios substractivos: son los tres colores (cian, magenta y amarillo) que pueden combinarse para producir cualquier color deseado mediante la sustracción de luz

superconductor: a conductor whose resistance drops to zero at extremely low temperatures *(756)*

    superconductor: conductor cuya resistencia baja hasta cero a temperaturas extremadamente bajas

superposition: the addition of displacements, forces, or other quantities by treating them as vectors *(310)*

    superposición: suma de desplazamientos, fuerzas u otras cantidades al tratarlas como vectores

supersonic: moving faster than the speed of sound *(497)*

    supersónico: movimiento más rápido que la velocidad del sonido

surface charge density: the charge per area on a surface *(693)*

    densidad de carga superficial: carga por unidad de área en una superficie

surface tension: the force that tries to minimize the surface area of a fluid *(439)*

    tensión superficial: fuerza que trata de minimizar la superficie por unidad de área de un fluido

T

tangent line: a line that touches a curve on a graph at a single point and has a slope equal to the slope of the curve at that point *(56)*

    línea tangente: línea de una gráfica que toca una curva en un punto determinado y que tiene una pendiente que es igual a la pendiente de la curva en ese punto

tangential acceleration: acceleration in a direction along a tangent to an object's circular path *(275)*

    aceleración tangencial: aceleración en dirección de la tangente a la trayectoria circular de un objeto

tangential speed: speed in a direction along a tangent to an object's circular path *(271)*

    velocidad tangencial: velocidad en dirección de la tangente a la trayectoria circular de un objeto

temperature: a measured quantity that is proportional to the average kinetic energy of the particles in a substance *(343)*

    temperatura: medida cuantitativa proporcional a la energía cinética promedio de las partículas de una sustancia

tension: the force exerted by a string, rope, or wire that is pulled tight *(167)*

    tensión: fuerza ejercida por un resorte, soga o alambre que está siendo halado

theory: a detailed explanation of some aspect of nature that accounts for a set of well-tested hypotheses *(9)*

    teoría: explicación detallada de algún aspecto de la naturaleza que resulta en un conjunto de hipótesis bien comprobadas

theory of relativity: Einstein's theory that explains how different observers moving with speeds near the speed of light can see things differently *(949)*

    teoría de la relatividad: teoría de Einstein que explica cómo observadores distintos que se mueven a velocidades cercanas a la velocidad de la luz pueden ver las cosas de manera distinta

thermal energy: the sum of all the kinetic and potential energy of an object; also known as *internal energy* *(344)*

    energía térmica: suma de toda la energía cinética y potencial de un objeto; también conocida como *energía interna*

thermal equilibrium: a state in which a system has a constant temperature and experiences no net transfer of energy *(345)*

    equilibrio térmico: estado en el que un sistema tiene una temperatura constante y no experimenta transferencia de energía

thermal expansion: the increase in length of an object due to an increase in its temperature *(351)*

    expansión térmica: aumento en la longitud de un objeto debido a un aumento de su temperatura

thermal process: a process that changes a system's thermal energy *(393)*

    proceso térmico: proceso que cambia la energía térmica de un sistema

thermal reservoir: an object, like a large body of water, that supplies or receives thermal energy with essentially no change in temperature *(390)*

    reserva térmica: un objeto, como una gran cantidad de agua, que suministra o recibe energía térmica sin variar su temperatura

thermistor: an electronic device incorporating temperature-dependent semiconductors *(754)*

    termistor: dispositivo electrónico que incorpora semi-conductores que dependen de la temperatura

thermonuclear fusion reaction: a process in which the temperature is high enough to initiate fusion of small nuclei *(929)*

    reacción de fusión termonuclear: proceso en el que la temperatura es lo suficientemente alta como para iniciar la fusión de núcleos pequeños

thin-lens equation: a precise mathematical relationship between the object distance, the image distance, and the focal length for a given lens *(616)*

    ecuación de la lente delgada: relación matemática precisa entre la distancia del objeto, la distancia imagen y la distancia focal, para una lente determinada

third law of thermodynamics: the law stating that there is no temperature lower than absolute zero, which is unattainable *(405)*

    tercera ley de la termodinámica: ley que establece que no existe una temperatura más baja que el cero absoluto, al cual no es posible llegar

time dilation: the slowing of a moving clock *(954)*

    dilatación del tiempo: fenómeno que hace más lento el paso del tiempo en un reloj en movimiento

torque: a physical quantity that is the product of force and distance and causes rotation *(281)*

    torque: cantidad física que es el producto de la fuerza por la distancia y que causa una rotación

total internal reflection: the complete reflection of light back into the material in which it is traveling *(606)*

    reflexión interna total: reflexión total de la luz hacia dentro del material por donde está viajando

total momentum: the vector sum of the momentums of all the individual objects that make up a system *(232)*

    momento total: vector suma de los momentos de todos los objetos individuales que forman un sistema

transformer: an electrical device that changes the voltage in an AC circuit *(837)*

    transformador: aparato eléctrico que cambia el voltaje en un circuito de CA

transmission axis: the direction of the polarized light that is transmitted by a polarizer *(546)*

    eje de transmisión: dirección de la luz polarizada que es transmitida por un polarizador

transverse wave: a wave in which the particles oscillate at right angles to the direction in which the wave travels *(471)*

    onda transversal: onda en la que las partículas oscilan de forma perpendicular a la dirección de propagación de la onda

triboelectric charging: the transfer of charge by rubbing objects together *(679)*

    efecto triboeléctrico: transferencia de carga al frotar dos objetos

trough: the lowest point on a wave *(473)*

    valle: punto más bajo de una onda

U

ultrasonic: having a frequency greater than 20,000 Hz *(497)*

    ultrasónico: que tiene una frecuencia mayor de 20.000 Hz

ultraviolet (UV) rays: electromagnetic waves with frequencies just above that of violet light, from roughly 7.5×10^{14} Hz to 10^{17} Hz *(540)*

    rayos ultravioleta (UV): ondas electromagnéticas con frecuencias mayores que la luz violeta, desde aproximadamente 7.5×10^{14} Hz hasta 10^{17} Hz

unified field theory: an all-encompassing theory that combines seemingly different forces into a single force *(937)*

    teoría del campo unificado: teoría unificadora que combina fuerzas aparentemente diferentes en una fuerza única

unified force: the single force that combined the four fundamental forces in a brief period just after the Big Bang *(936)*

    fuerza unificada: la fuerza única que combinaba las cuatro fuerzas fundamentales en un breve periodo de tiempo inmediatamente después del Big Bang

universal gas constant: a constant, denoted by R, used in the ideal gas law when the amount of gas is measured in moles *(421)*

    constante universal del gas ideal: constante denotada por R, usada en la ley del gas ideal cuando la cantidad de gas se mide en moles

universal gravitation constant: a constant, denoted by G, that has a value of 6.67×10^{-11} N $\cdot$ m^2/kg^2 *(308)*

    constante de gravitación universal: constante, denotada por G, cuyo valor es de 6.67×10^{-11} N $\cdot$ m^2/kg^2

universal law of gravitation: the law stating that the force of gravity between any two objects with masses m_1 and m_2 separated by a distance r is attractive and has a magnitude F given by $F = Gm_1m_2/r^2$ *(308)*

    ley de gravitación universal: ley que establece que la fuerza de gravedad entre dos objetos con masas m_1 y m_2 separados por una distancia r es de atracción y con una magnitud F establecida por $F = Gm_1m_2/r^2$

unpolarized: having random polarization directions *(546)*
  despolarizado: que tiene direcciones de polarización aleatorias

V

vector: a quantity consisting of both a numerical value with its unit and a direction *(32)*
  vector: cantidad que consiste en un valor numérico con unidad y dirección

vector components: the lengths of a vector along specified directions *(115)*
  componentes vectoriales: largo de un vector en una determinada dirección

velocity: a vector quantity that describes the rate of motion and its direction *(32)*
  velocidad: vector de cantidad que describe la tasa de movimiento y su dirección

velocity-time graph: a graph in which velocity is plotted on the vertical axis, or y axis, and time is plotted on the horizontal axis, or x axis *(76)*
  gráfica de velocidad contra tiempo: gráfica en la que la velocidad se indica en el eje vertical, o eje y, y el tiempo en el eje horizontal, o eje x

virtual image: an image formed behind a mirror *(571)*
  imagen virtual: imagen que se forma detrás de un espejo

viscosity: the tendency of a fluid to resist flow *(438)*
  viscosidad: tendencia de un fluido a resistirse al movimiento

visible light: any electromagnetic wave that human eyes can detect *(537)*
  luz visible: ondas electromagnéticas que pueden ser detectadas por el ojo humano

volt: the unit used to measure the amount of electric potential energy for a given amount of charge *(720)*
  voltio: unidad usada para medir la cantidad de energía potencial eléctrica de una cantidad de carga determinada

voltage: the electric potential *(720)*
  voltaje: potencial eléctrico

voltmeter: a device used to measure the potential difference between any two points in a circuit *(764)*
  voltímetro: dispositivo usado para medir la diferencia de potencial entre dos puntos de un circuito

W

watt: a unit of power equal to 1 joule per second: $1 \text{ W} = 1 \text{ J/s}$ *(212)*
  vatio: unidad de potencia igual a 1 julio por segundo: $1 \text{ W} = 1 \text{ J/s}$

wave: a disturbance that travels from one place to another carrying energy along with it *(470)*
  onda: perturbación que viaja de un lugar a otro llevando consigo energía

wave front: the shape produced by a wave crest moving away from a source *(566)*
  frente de onda: forma producida por la cresta de una onda cuando se aleja de una fuente

wave function: a solution to the Schrödinger equation giving the probability of finding the electron in a given location *(897)*
  función de onda: solución de la ecuación de Schrödinger que da la probabilidad de encontrar al electrón en un lugar determinado

wave-particle duality: the connection between waves and particles in which waves are like particles and particles are like waves *(866)*
  dualidad onda-partícula: conexión entre las ondas y las partículas en la que las ondas se comportan como partículas y las partículas se comportan como ondas

wavelength: the distance from one crest to the next or from one trough to the next; the repeat length of a wave *(473)*
  longitud de onda: distancia desde una cresta a la siguiente, o desde un valle al siguiente en una onda; la longitud repetida de una onda

weak nuclear force: the force responsible for beta decay *(922)*
  fuerza nuclear débil: fuerza responsable de la desintegración beta

weight: a measure of the gravitational force acting on an object *(16)*
  peso: medida de la fuerza gravitatoria ejercida sobre un objeto

Wien's displacement law: the law that states that the peak in a blackbody radiation curve moves to a higher frequency as the temperature increases *(853)*
  ley de desplazamiento de Wien: ley que establece que el máximo de la curva de radiación de un cuerpo negro se mueve a frecuencias más altas si aumenta la temperatura

work: a quantity defined as the product of force exerted on an object and the distance the object moves *(189)*
  trabajo: cantidad definida como el producto de la fuerza ejercida sobre un objeto y la distancia que se mueve el objeto

work function: the minimum amount of energy necessary to eject an electron from a particular metal *(860)*
  función de trabajo: la menor cantidad de energía necesaria para desprender un electrón de un metal determinado

X

X-rays: electromagnetic waves having a frequency range from roughly 10^{17} Hz to 10^{20} Hz *(541)*
  rayos X: ondas electromagnéticas en el rango de frecuencias entre aproximadamente 10^{17} Hz y 10^{20} Hz

Credits

Index

The page on which a term is defined is in **boldface** type.